Houshang H. Sohrab

Basic Real Analysis

Second Edition

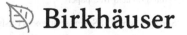 Birkhäuser

Houshang H. Sohrab
Mathematics
Towson University
Towson, MD, USA

ISBN 978-1-4939-3714-1   ISBN 978-1-4939-1841-6 (eBook)
DOI 10.1007/978-1-4939-1841-6
Springer New York Heidelberg Dordrecht London

Mathematics Subject Classification (2010): 26-01, 26Axx, 28-01, 40-01, 46-01, 54-01, 60-01

To Shohreh, Mahsa, and Zubin

Preface

Ah, Love! could thou and I with Fate conspire
To grasp this sorry Scheme of Things entire,
Would not we shatter it to bits—and then
Re-mould it nearer to the Heart's Desire!

Omar Khayyam, Rubaiyat

More than 10 years have passed since the publication of the first edition of this textbook. During these years, a large number of monographs dealing with the same topics have appeared. Some of them have been included in the new bibliography. In addition, a wealth of material is now freely available online, some of it posted by the very best (cf., e.g., [Tao11]). So one may question the wisdom of offering a new edition of the old *Basic Real Analysis*, henceforth abbreviated *BRA*. And yet, as is always the case, different people look at the same material in different ways depending on their tastes. What should or should not be included and to what extent may vary considerably, and all choices have their legitimate and logical justification.

Despite the fact that I have looked at a large number of real analysis textbooks and have benefited from all of them, I still prefer not to modify the organization of the material in *BRA*. The initial idea of a new edition came from Tom Grasso of Birkhäuser, and I want to use this opportunity to thank him for suggesting it. He pointed out that for the project to be justified, a reasonable number of changes must be made. The most substantial change in the new edition is that I rewrote Chaps. 10 and 11 on Lebesgue measure and integral entirely. In doing so, I decided to abandon F. Riesz's method used in the first edition in favor of the more traditional approach of treating Lebesgue measure *before* introducing the integral. I have come to believe that measure theory is a fundamental part of analysis and the sooner one learns it, the better.

Lebesgue measure and integral *on the real line* are now covered in Chap. 10. Chapter 11 contains additional topics, including a quick look at improper Riemann integrals, integrals depending on a parameter, the classical L^p-spaces, other modes of convergence, and a final section on the *differentiation* problem. This last section contains Lebesgue's theorem on the differentiability of monotone functions (with F. Riesz's *Rising Sun Lemma* used in the proof) and his versions of the *Fundamental Theorem(s) of Calculus*. Abstract measure and integration are treated in Chap. 12, where I have included the Radon–Nikodym theorem which is used in the last section on probability.

Although the newly written chapters on Lebesgue's theory constitute the major change in this edition, all other chapters have been affected to various degrees. For example, the treatment of convex functions has been modified and (hopefully) improved. I have added a number of exercises in the text and many new problems at the end of all chapters. A large number of typographical as well as more serious errors have been corrected. I am particularly indebted to Professor Giorgio Giorgi of Università Degli Studi Di Pavia for pointing out a serious one. Of course, as always, other undetected errors may still be there and I take full responsibility for them. Needless to say, I would be grateful to careful readers for pointing them out to me (hsohrab@towson.edu).

Ideally, a book at this level should include some spectral theory, say, at least the spectral properties of compact, self-adjoint operators. Unfortunately, this would increase the size of the book beyond what I consider to be reasonable. I have decided to include some of this and similar interesting material in the end-of-chapter problems, and the interested readers may try as many of them as they want. A complete solution manual is available from the publisher for the benefit of the potential instructors. I have decided to use sequential numbering of all the items throughout. I believe that this simplifies the navigation considerably even though it may have its problems.

It is a great pleasure to thank Mitch Moulton, Birkhäuser's assistant editor, for his help and patience during the preparation of the manuscript. I am also grateful for the technical assistance I received from Birkhäuser. One of the people I completely forgot to thank in the first edition of *BRA* (shame on me!) is Loren Spice. He was 16 when he started enrolling in mathematics courses at Towson University, right when I was preparing the first draft of the textbook. He read the first five chapters very thoroughly and made a large number of suggestions and corrections. I am truly indebted to him for his valuable comments which resulted in many improvements. Also, I owe so much to the brilliant, anonymous reviewer of the first edition of *BRA* whose excellent critical comments had a great influence, even though I couldn't possibly live up to his high expectations. I hope he finds this new edition to be closer to his taste. In addition, the anonymous reviewers of this new edition have made a number of excellent comments for which I am truly grateful.

Finally, I would like to thank my wife, Shohreh, and my children, Mahsa and Zubin, whose love and support are the greatest driving force in my life.

Towson, MD, USA   Houshang Sohrab

4.7   Approximation by Step, Piecewise Linear,
        and Polynomial Functions 167
4.8   Problems 174

5   Metric Spaces 181
5.1   Metrics and Metric Spaces 181
5.2   Topology of a Metric Space 186
5.3   Limits, Cauchy Sequences, and Completeness 191
5.4   Continuity 198
5.5   Uniform Continuity and Continuous Extensions 207
5.6   Compact Metric Spaces 216
5.7   Connected Metric Spaces 226
5.8   Problems 232

6   The Derivative 241
6.1   Differentiability 242
6.2   Derivatives of Elementary Functions 247
6.3   The Differential Calculus 250
6.4   Mean Value Theorems 255
6.5   L'Hôpital's Rule 262
6.6   Higher Derivatives and Taylor's Formula 267
6.7   Convex Functions 278
6.8   Problems 283

7   The Riemann Integral 291
7.1   Tagged Partitions and Riemann Sums 291
7.2   Some Classes of Integrable Functions 300
7.3   Sets of Measure Zero and Lebesgue's Integrability Criterion 306
7.4   Properties of the Riemann Integral 315
7.5   Fundamental Theorem of Calculus 324
7.6   Functions of Bounded Variation 330
7.7   Problems 336

8   Sequences and Series of Functions 345
8.1   Complex Numbers 345
8.2   Pointwise and Uniform Convergence 349
8.3   Uniform Convergence and Limit Theorems 356
8.4   Power Series 361
8.5   Elementary Transcendental Functions 373
8.6   Fourier Series 381
8.7   Problems 400

9   Normed and Function Spaces 411
9.1   Norms and Normed Spaces 411
9.2   Banach Spaces 420
9.3   Hilbert Spaces 432

Contents

1   Set Theory 1
   1.1   Rings and Algebras of Sets 1
   1.2   Relations and Functions.. 6
   1.3   Basic Algebra, Counting, and Arithmetic 17
   1.4   Infinite Direct Products, Axiom of Choice, and Cardinal
         Numbers 29
   1.5   Problems 34

2   Sequences and Series of Real Numbers 39
   2.1   Real Numbers 39
   2.2   Sequences in $\mathbb{R}$ 50
   2.3   Infinite Series 61
   2.4   Unordered Series and Summability 78
   2.5   Problems 86

3   Limits of Functions 97
   3.1   Bounded and Monotone Functions 97
   3.2   Limits of Functions 10
   3.3   Properties of Limits 10
   3.4   One-Sided Limits and Limits Involving Infinity 10
   3.5   Indeterminate Forms, Equivalence, and Landau's Little
         "oh" and Big "Oh" 11
   3.6   Problems 12

4   Topology of $\mathbb{R}$ and Continuity 1
   4.1   Compact and Connected Subsets of $\mathbb{R}$ 1
   4.2   The Cantor Set... 1
   4.3   Continuous Functions 1
   4.4   One-Sided Continuity, Discontinuity, and Monotonicity 1
   4.5   Extreme Value and Intermediate Value Theorems 1
   4.6   Uniform Continuity 1

9.4   Function Spaces 445
9.5   Problems 454

10 Lebesgue Measure and Integral in ℝ 465
10.1 Outer Measure 467
10.2 (Lebesgue) Measurable Functions 478
10.3 The Lebesgue Integral 487
10.4 Convergence Theorems 502
10.5 Littlewood's Other Principles and Modes of Convergence 511
10.6 Problems 521

11 More on Lebesgue Integral and Measure 527
11.1 Lebesgue vs. Riemann 527
11.2 Dependence on a Parameter 536
11.3 L^p-Spaces 540
11.4 More on Modes of Convergence 548
11.5 Differentiation 553
11.6 Problems 567

12 General Measure and Probability 575
12.1 Measures and Measure Spaces 575
12.2 Measurable Functions 592
12.3 Integration 596
12.4 Product Measures 610
12.5 Probability 616
12.6 Problems 637

A   Construction of Real Numbers 655

Bibliography 663

Index 667

Chapter 1
Set Theory

Set theory is an important part of the foundation of mathematics and its rigorous treatment is beyond the scope of this book. The material in this chapter is included mainly to make the book self-contained as far as elementary set theory is concerned. *Consequently, the readers need not go through the entire chapter to proceed further. In fact, they can skip most sections and return to the corresponding topics later if a reference makes it necessary.* The first section introduces rings and algebras of sets and the standard set theory notation to be used throughout the book. Next, relations and functions are briefly covered, including equivalence and order relations. The third section is a quick review of algebraic structures (groups, rings, etc.) and some elementary number theory and combinatorial questions. Finally, the last section covers infinite direct products and cardinal numbers. Most readers of this book may have already seen the material covered in this chapter before, possibly in slightly different form. It should be pointed out that the topics in this chapter will be referred to on several occasions in subsequent chapters and most of the results will be needed later.

1.1 Rings and Algebras of Sets

A *set*, S, will be defined as a "collection" (or "family") of "objects" called *elements*. The statement "s is an element of S" will be denoted $s \in S$, and its negation will be denoted $s \notin S$. The set with no elements will be called the *empty set* and denoted $\emptyset$. Given a pair of sets, S and T, we say that S is a *subset* of T, and write $S \subset T$, if each element of S is an element of T. Again the negation of the statement will be denoted $S \not\subset T$. One obviously has $\emptyset \subset S$ for any set S. We write $S = T$ if both $S \subset T$ and $T \subset S$. S is called a *proper subset* of T if $S \subset T$, but $S \neq T$. In this case we also say that the inclusion $S \subset T$ is a *proper inclusion* and use the notation $S \subsetneq T$. We shall constantly use the notation $S = \{t \in T : P(t)\}$ to denote the set of all elements in T for which the property P holds. In most problems, all

© Springer Science+Business Media New York 2014
H.H. Sohrab, *Basic Real Analysis*, DOI 10.1007/978-1-4939-1841-6_1

the sets we consider are subsets of a fixed (large) set, called the *universal set* or
the *universe of discourse*, which we denote by U. We will usually assume that such
a universe has been chosen, especially when complements of sets (to be defined
below) are involved in the discussion. Before defining the basic operations on sets,
let us introduce a notation which will be used throughout the book.

Notation 1.1.1. Given a pair of mathematical *expressions* P and Q, involving one
or more variables, we write $P := Q$ if the right side (i.e., the expression Q) is the
definition of the left side (i.e., the expression P).

Definition 1.1.2 (Union, Intersection, etc.). Given two sets S and T, both subsets
of a universal set U, we define their *union* to be the set

$$S \cup T := \{x \in U : x \in S \text{ or } x \in T\}.$$

Their *intersection* is defined by

$$S \cap T := \{x \in U : x \in S \text{ and } x \in T\}.$$

We define the *complement* of S (with respect to the fixed set U) to be the set

$$S^c := \{x \in U : x \notin S\}.$$

We also define the *difference set* $S \setminus T$ by

$$S \setminus T := \{x \in S : x \notin T\} = S \cap T^c,$$

and the *symmetric difference* of S and T by

$$S \triangle T := (S \setminus T) \cup (T \setminus S) = (S \cup T) \setminus (S \cap T).$$

Finally, we define the *power set* of S, denoted $\mathcal{P}(S)$, to be the set of all subsets of S:

$$\mathcal{P}(S) := \{A \subset U : A \subset S\}.$$

Two sets S and T are called *disjoint* if $S \cap T = \emptyset$. Given any set S, we obviously
have $S \cap S = S \cup S = S$, $S \cap S^c = \emptyset$, $S \cap U = S$, and $S \cup U = U$. The following
properties are also immediate consequences of the definitions. The reader is invited
to prove them as an exercise. Recall that $S = T$ if and only if $S \subset T \subset S$. Thus,
one can prove $S = T$ by the "elementwise method", i.e., by showing that *every*
element of S belongs to T and *vice versa*:

$$A \cup B = B \cup A, \quad \text{and} \quad A \cap B = B \cap A,$$

$$A \subset A \cup B, \quad \text{and} \quad A = A \cup B \iff B \subset A,$$

$$A \cap B \subset A, \quad \text{and} \quad A = A \cap B \iff A \subset B,$$

$$A \cup (B \cup C) = (A \cup B) \cup C = A \cup B \cup C,$$
$$A \cap (B \cap C) = (A \cap B) \cap C = A \cap B \cap C,$$
$$A \cup (B \cap C) = (A \cup B) \cap (A \cup C),$$
$$A \cap (B \cup C) = (A \cap B) \cup (A \cap C),$$
$$\emptyset^c = U, \quad \text{and} \quad U^c = \emptyset,$$
$$(A^c)^c = A, \quad A \cup A^c = U, \quad A \cap A^c = \emptyset,$$
$$A \subset B \iff B^c \subset A^c.$$

We will frequently use the following laws, called *De Morgan's laws*, that relate the complements to unions and intersections:

$$(A \cup B)^c = A^c \cap B^c,$$
$$(A \cap B)^c = A^c \cup B^c.$$

Unions and intersections may also be defined for families of sets: Suppose that we are given an *index set*, Λ, and that, for each $\lambda \in \Lambda$, we are given a set $A_\lambda \subset U$. We then define the union and intersection of the A_λ by

$$\bigcup_{\lambda \in \Lambda} A_\lambda := \{x \in U : (\exists \lambda \in \Lambda)(x \in A_\lambda)\},$$

$$\bigcap_{\lambda \in \Lambda} A_\lambda := \{x \in U : (\forall \lambda \in \Lambda)(x \in A_\lambda)\},$$

where the *universal quantifier* "$\forall$" means "for all", and the *existential quantifier* "$\exists$" means "there exists", or "for some".

If the index set is the set of natural numbers $\mathbb{N} := \{1, 2, 3, \ldots\}$, then we have a *sequence* of sets A_n, and their union and intersection are given by

$$\bigcup_{n \in \mathbb{N}} A_n = \{x \in U : (\exists n \in \mathbb{N})(x \in A_n)\},$$

$$\bigcap_{n \in \mathbb{N}} A_n = \{x \in U : (\forall n \in \mathbb{N})(x \in A_n)\}.$$

De Morgan's laws are also valid for families of sets:

$$\left(\bigcup A_\lambda\right)^c = \bigcap A_\lambda^c,$$
$$\left(\bigcap A_\lambda\right)^c = \bigcup A_\lambda^c,$$

where the unions and intersections are obviously over all $\lambda \in \Lambda$.

To demonstrate the *elementwise method*, let us prove the first De Morgan law: $(\bigcup_{\lambda \in \Lambda} A_\lambda)^c = \bigcap_{\lambda \in \Lambda} A_\lambda^c$. Now, $x \in (\bigcup_{\lambda \in \Lambda} A_\lambda)^c$ if and only if $x \notin \bigcup_{\lambda \in \Lambda} A_\lambda$, which (by the definition of $\bigcup_{\lambda \in \Lambda} A_\lambda$) happens if and only if $x \notin A_\lambda$ *for all* $\lambda \in \Lambda$. But this is equivalent to $x \in A_\lambda^c$ *for all* $\lambda \in \Lambda$, which (by the definition of $\bigcap_{\lambda \in \Lambda} A_\lambda^c$) means $x \in \bigcap_{\lambda \in \Lambda} A_\lambda^c$.

Exercise 1.1.3. Prove the following properties of the symmetric difference for arbitrary sets A, B, C, and D:

(a) $\emptyset \triangle A = A$,
(b) $A \triangle A = \emptyset$,
(c) $A \triangle B = B \triangle A$,
(d) $A \triangle B = A^c \triangle B^c$,
(e) $A \triangle (B \triangle C) = (A \triangle B) \triangle C$,
(f) $A \triangle B \subset (A \triangle C) \cup (C \triangle B)$,
(g) $(A \cup B) \triangle (C \cup D) \subset (A \triangle C) \cup (B \triangle D)$,
(h) $(A \cap B) \triangle (C \cap D) \subset (A \triangle C) \cup (B \triangle D)$,
(i) $(A \setminus B) \triangle (C \setminus D) \subset (A \triangle C) \cup (B \triangle D)$,
(j) $A \triangle B = C \Longleftrightarrow B = A \triangle C$.

Hints: (a), (b), (c), and (d) follow at once from the definition. To prove the associative property (e), first use the definition to show that the left side is equal to

$$[A \setminus (B \cup C)] \cup [B \setminus (C \cup A)] \cup [C \setminus (A \cup B)] \cup (A \cap B \cap C).$$

Now show that this is also the right side using the commutativity

$$(A \triangle B) \triangle C = C \triangle (A \triangle B)$$

and simple substitutions. For (f), use the inclusions

$$A \setminus B \subset (A \setminus C) \cup (C \setminus B), \qquad B \setminus A \subset (C \setminus A) \cup (B \setminus C).$$

To show (g), use the inclusion

$$(A \cup B) \setminus (C \cup D) \subset (A \setminus C) \cup (B \setminus D).$$

Next, note that (h) follows from (g), (d), and De Morgan's laws and (i) follows from (h) and the fact that $A \setminus B = A \cap B^c$. Finally, to prove (j), show that $A \triangle (A \triangle B) = B$.

Notation 1.1.4 (Standard Set Notation). Throughout the book we shall use the following standard notation for some frequently used sets: The set of *natural numbers* will be denoted by $\mathbb{N} := \{1, 2, 3, \ldots\}$ and the set of *nonnegative integers* by $\mathbb{N}_0 := \{0, 1, 2, \ldots\}$; the set of all *integers* by $\mathbb{Z} := \{0, \pm 1, \pm 2, \pm 3, \ldots\}$, and the set of *rational numbers* by $\mathbb{Q} := \{m/n : m, n \in \mathbb{Z}, n \neq 0\}$. We shall use $\mathbb{R}$ for the set of all *real* numbers and $\mathbb{C} := \mathbb{R} + i\mathbb{R}$ (with $i = \sqrt{-1}$) for the set of all *complex* numbers. The sets of all *positive* elements of $\mathbb{Z}$, $\mathbb{Q}$, and $\mathbb{R}$ will be denoted by $\mathbb{Z}^+$, $\mathbb{Q}^+$, and $\mathbb{R}^+$, respectively. Note that $\mathbb{Z}^+ = \mathbb{N}$.

Definition 1.1.5 (Rings and Algebras of Sets). A nonempty set $\mathcal{R} \subset \mathcal{P}(U)$ is called a *ring* (of subsets of U) if $A \cup B$ and $A \setminus B$ are in $\mathcal{R}$ whenever A and B are. A ring $\mathcal{A}$ is called an *algebra* if $U \in \mathcal{A}$. The ring $\mathcal{R}$ (resp., the algebra $\mathcal{A}$) is called a *σ-ring* (resp., a *σ-algebra*) if it is closed under *countable* unions:

$$A_n \in \mathcal{R}, \quad n = 1, 2, 3, \ldots \quad \Longrightarrow \quad \bigcup_{n \in \mathbb{N}} A_n \in \mathcal{R},$$

and similarly for $\mathcal{A}$.

Remark 1.1.6. The set U is called the *unit*. It is easily seen that $\{\emptyset\}$ and the set of all *finite* subsets of U are rings, and the latter is an algebra if and only if U itself is finite. Also, the power set $\mathcal{P}(U)$ is an algebra of sets, and so is $\{\emptyset, U\}$. It should be noted that, by Exercise 1.1.3, a subset $\mathcal{R} \subset \mathcal{P}(U)$ is a ring if and only if $A \triangle B$ and $A \cap B$ are in $\mathcal{R}$ whenever A and B are.

Exercise 1.1.7.

(a) Show that a ring $\mathcal{R}$ is *closed under* symmetric differences as well as finite unions and intersections. In other words,

$$A \in \mathcal{R}, \ B \in \mathcal{R} \Longrightarrow A \triangle B \in \mathcal{R},$$

$$A_1, \ldots, A_n \in \mathcal{R} \Longrightarrow \bigcup_{k=1}^{n} A_k \in \mathcal{R},$$

$$A_1, \ldots, A_n \in \mathcal{R} \Longrightarrow \bigcap_{k=1}^{n} A_k \in \mathcal{R}.$$

(b) Show that a ring $\mathcal{R}$ is an algebra if and only if, for each $A \in \mathcal{R}$, we have $A^c \in \mathcal{R}$.

Remark 1.1.8. One can show, as an exercise, that a ring $\mathcal{R}$ (resp., an algebra $\mathcal{A}$) in the above sense is, indeed, a ring (resp., a ring with unit element) in the *algebraic sense* (cf. Definition 1.3.7 and Remark 1.3.8(c) in this chapter) under the operations of addition and multiplication defined by

$$A + B := A \triangle B, \quad \text{and} \quad AB := A \cap B.$$

The following proposition shows that *any* collection of subsets of a nonempty set U can *generate* an algebra (or σ-algebra):

Proposition 1.1.9. *Let U be a nonempty set and let $C \subset \mathcal{P}(U)$ be any collection of subsets of U. Then there is a smallest algebra (resp., σ-algebra) $\mathcal{A}_C$ such that $C \subset \mathcal{A}_C$. In other words, $\mathcal{A}_C$ is an algebra (resp., σ-algebra) containing C, and if $\mathcal{B}$ is any algebra (resp., σ-algebra) with $C \subset \mathcal{B}$, then $\mathcal{A}_C \subset \mathcal{B}$.*

Proof. We prove the existence of a smallest algebra. The case of a σ-algebra is obtained by minor modifications. Let $\mathcal{F}$ be the family of all algebras (of subsets of U) which contain C, and note that $\mathcal{P}(U) \in \mathcal{F}$. Define $\mathcal{A}_C := \bigcap_{\mathcal{B} \in \mathcal{F}} \mathcal{B}$. Then we have $C \subset \mathcal{B}$, for each $\mathcal{B} \in \mathcal{F}$, so that $C \subset \mathcal{A}_C$. Next, if A, $B \in \mathcal{A}_C$, then A, $B \in \mathcal{B}$ for each $\mathcal{B} \in \mathcal{F}$. Since $\mathcal{B}$ is an algebra, $A \cup B \in \mathcal{B}$ and $A \setminus B \subset \mathcal{B}$ for each $\mathcal{B} \in \mathcal{F}$. Therefore, $A \cup B \in \mathcal{A}_C$ and $A \setminus B \in \mathcal{A}_C$. $\square$

1.2   Relations and Functions

To define relations we need the concept of Cartesian product of sets, which we now define.

Definition 1.2.1 (Cartesian Product). Given two elements a, $b \in U$, the set $\{a, b\}$ will be called an *unordered pair*. We also define the *ordered pair* $(a, b) := \{\{a\}, \{a, b\}\}$, in which a is the *first* element and b is the *second* element. Thus $(a, b) = (c, d)$ if and only if $a = c$ and $b = d$. Now let A and B be two sets. We define their *Cartesian (or direct) product* by

$$A \times B := \{(a, b) : a \in A, \, b \in B\}.$$

In a similar way, we define *(ordered) triples, quadruples,..., n-tuples*, which we denote by (a, b, c), $(a, b, c, d), \dots,$ $(a_1, a_2, \dots, a_n)$. The Cartesian product of the sets A_1, $A_2, \dots,$ A_n is defined to be

$$\prod_{k=1}^{n} A_k = A_1 \times \cdots \times A_n := \{(a_1, \dots, a_n) : a_1 \in A_1, \dots, a_n \in A_n\},$$

and $(a_1, a_2, \dots, a_n) = (b_1, b_2, \dots, b_n)$ if and only if $a_j = b_j$ for $1 \le j \le n$.

Exercise 1.2.2. For arbitrary sets A, B, C, and D, show that

(a) $A \times B = \emptyset \Leftrightarrow A = \emptyset$,   or $B = \emptyset$,
(b) $A \subset B$ and $C \subset D \Rightarrow A \times C \subset B \times D$,
(c) $(A \cup B) \times C = (A \times C) \cup (B \times C)$,
(d) $(A \cap B) \times (C \cap D) = (A \times C) \cap (B \times D)$.

Definition 1.2.3 (Relation, Inverse, Composition). Given any sets X and Y, a *relation* from (a subset of) X to Y is a subset $R \subset X \times Y$. We say that x is R-*related* to y if $(x, y) \in R$, which we also write xRy. If $f \subset X \times Y$ is a relation, we define its *domain* and *range* by

$$\mathrm{dom}(f) := \{x \in X : (\exists y \in y)((x, y) \in f)\},$$
$$\mathrm{ran}(f) := \{y \in Y : (\exists x \in X)((x, y) \in f)\}.$$

The *inverse* of a relation $f \subset X \times Y$ is the relation

$$f^{-1} := \{(y, x) : (x, y) \in f\} \subset Y \times X.$$

Given two relations $f \subset X \times Y$ and $g \subset Y \times Z$, their *composition* (or *composite*) is the relation from (a subset of) X to Z defined by

$$g \circ f := \{(x, z) \in X \times Z : (\exists y \in Y)((x, y) \in f, (y, z) \in g)\} \subset X \times Z.$$

Note that we have

$$g \circ f \neq \emptyset \iff \mathrm{ran}(f) \cap \mathrm{dom}(g) \neq \emptyset.$$

Definition 1.2.4 (Restriction, Extension). Let $f, g \subset X \times Y$ be two relations. If $f \subset g$, we say that f is a *restriction* of g or that g is an *extension* of f. If $\mathrm{dom}(f) = D$, then $f \subset g$ is also denoted by $f = g|D$.

Definition 1.2.5 (Equivalence Relation). A relation $R \subset X \times X$ is called an *equivalence relation* on X if it is

reflexive:	$(x, x) \in R \quad \forall x \in X,$
symmetric:	$(x, y) \in R \Rightarrow (y, x) \in R,$ and
transitive:	$(x, y) \in R$ and $(y, z) \in R \Rightarrow (x, z) \in R.$

Example 1.2.6.

(a) The simplest example of an equivalence relation on a set X is *equality*; i.e., "x is related to y" simply means $x = y$. The corresponding subset of $X \times X$ is then the *diagonal* $\Delta_X := \{(x, x) : x \in X\}$.

(b) A more interesting and widely used example is the relation of *congruence modulo n* (where $n \in \mathbb{N}$ is a fixed positive integer) on the set $\mathbb{Z}$ of all integers: For two integers $a, b \in \mathbb{Z}$, it is defined by

$$a \equiv b \pmod{n} \iff b - a = kn \text{ for some } k \in \mathbb{Z}.$$

Notation 1.2.7. Notice the use of xRy to indicate $(x, y) \in R$ in the above examples. It is also a common practice to use notation such as $\sim$ or $\approx$ (rather than R) to denote an *equivalence relation*. Hence we write, for instance, $\sim \subset X \times X$ and $x \sim y$ will then mean that x and y are *equivalent*.

Definition 1.2.8 (Equivalence Class, Quotient Set). Let $R \subset X \times X$ be an *equivalence relation* on X. For each element $x \in X$, the set

$$[x] := \{y \in X : yRx\}$$

is said to be the *equivalence class* of x and the element x is called a *representative* of the class $[x]$. The set of *all* equivalence classes is denoted by X/R and is called the *quotient set of X by R*:

$$X/R = \{[x] : x \in X\}.$$

Definition 1.2.9 (Partition). A *partition* of a nonempty set X is a collection of sets $\{X_\lambda\}_{\lambda \in \Lambda}$ such that $\emptyset \neq X_\lambda \subset X$ for all $\lambda \in \Lambda$, $X_\lambda \cap X_\mu = \emptyset$ for all λ, $\mu \in \Lambda$ with $\lambda \neq \mu$, and $X = \bigcup_{\lambda \in \Lambda} X_\lambda$. In other words, a partition divides the set X into a collection of *pairwise disjoint* and *nonempty* subsets whose *union* is X.

The following theorem which shows that, for a given (nonempty) set X, the sets of "equivalence relations" on X and "partitions" of X are in *one-to-one correspondence* has many applications including some interesting ones in combinatorial questions.

Theorem 1.2.10 (Equivalence Relations and Partitions). *Let X be a nonempty set and let $R \subset X \times X$ be an equivalence relation on X. Then the equivalence classes of the elements of X form a partition of X. Conversely, given any partition $\{X_\lambda\}_{\lambda \in \Lambda}$ of X, the relation*

$$R := \{(x, y) \in X \times X : x, \ y \in X_\lambda \ \text{for some} \ \lambda \in \Lambda\} \tag{†}$$

is an equivalence relation on X whose equivalence classes are precisely the sets X_λ.

Proof. Let R be an equivalence relation on X. Then for each $x \in X$, we have xRx and hence $x \in [x]$, so that $[x] \neq \emptyset$. Next, we show that for any x, $y \in X$, *either* $[x] = [y]$ *or* $[x] \cap [y] = \emptyset$. Indeed, if xRy, then $z \in [y]$ implies zRy (or, by *symmetry*, yRz) and hence (by *transitivity*) xRz so that $z \in [x]$. Thus, we have the inclusion $[y] \subset [x]$. A similar argument shows that $[x] \subset [y]$ and hence $[x] = [y]$. On the other hand, if $x\cancel{R}y$ (i.e., if $(x, y) \notin R$), then we must have $[x] \cap [y] = \emptyset$ since otherwise $z \in [x] \cap [y]$ implies $[x] = [z] = [y]$, by what we just proved, and we get xRy. Finally, since $x \in [x]$ $\forall x \in X$, we have $X = \bigcup'[x]$, where $\bigcup'$ is the union of *pairwise disjoint* classes. Conversely, let $\{X_\lambda\}_{\lambda \in \Lambda}$ be a *partition* of X and let R be the relation defined by (†). Then R is immediately seen to be reflexive, symmetric, and transitive. (Why?)   □

Example 1.2.11.

(a) As we saw above, *equality*, which corresponds to the *diagonal* $\Delta_X := \{(x, x) : x \in X\}$, is a trivial equivalence relation on an arbitrary set $X \neq \emptyset$. For each $x \in X$, we have $[x] = \{x\}$ and hence

$$X/\Delta_X = \{\{x\} : x \in X\},$$

which is why it is customary to *identify* the two sets and write $X/\Delta_X = X$ even though this is *not* really an *equality*.

(b) On the set $\mathbb{Z}$ of all integers, *congruence modulo n*, defined above, is an equivalence relation. For each $a \in \mathbb{Z}$, its equivalence class is $[a] = a + n\mathbb{Z}$, where $n\mathbb{Z} := \{nk : k \in \mathbb{Z}\}$, and the set of all equivalence classes is denoted by $\mathbb{Z}_n$. Since the possible remainders upon division by n are $0, 1, 2, \ldots, n-1$, we have

$$\mathbb{Z}_n = \{0, 1, 2, \ldots, n-1\},$$

where, for simplicity, we write $[k] = k$.

Definition 1.2.12 (Partial Ordering). Given a set X, a relation $R \subset X \times X$ is called a *partial ordering* on X if it is

reflexive: $\quad (x, x) \in R \quad \forall x \in X,$

antisymmetric: $\quad (x, y) \in R \quad \text{and} \quad (y, x) \in R \implies x = y, \quad \text{and}$

transitive: $\quad (x, y) \in R \quad \text{and} \quad (y, z) \in R \implies (x, z) \in R.$

If R is a partial ordering on a set X, then we say that X is a *partially ordered* set.

Notation 1.2.13. If $R \subset X \times X$ is an arbitrary partial ordering on a set X, then xRy will be denoted by $x \preceq y$. We also use $x \prec y$ to mean $x \preceq y$ and $x \neq y$.

Note that the usual ordering "$\leq$" on the sets $\mathbb{N}, \mathbb{Z}, \mathbb{Q}$, and $\mathbb{R}$ is obviously a partial ordering (in fact a *total ordering*, as defined below) on those sets. Also, the *inclusion* "$\subset$" is a partial ordering on $\mathcal{P}(U)$ which is *not* total if U contains *more than* one element.

Definition 1.2.14 (Linear (or Total) Ordering, Chain). Let X be a partially ordered set with partial ordering "$\preceq$." Two elements x and y are called *comparable* if $x \preceq y$ or $y \preceq x$. The set X is called *linearly ordered*, or *totally ordered*, if for any $x, y \in X$, x and y are comparable. A linearly ordered set is also called a *chain*.

Definition 1.2.15 (Maximal and Minimal Elements). Let X be a partially ordered set. An element $u \in X$ is called *maximal* if, for any $v \in X$, $u \preceq v$ implies $u = v$. Similarly, an element $t \in X$ is called *minimal* if $s \in X$ and $s \preceq t$ imply $s = t$.

Definition 1.2.16 (Upper and Lower Bounds, Sup, Inf, etc.). Let S be a subset of a partially ordered set X. Then an element $u \in X$ (resp., $t \in X$) is called an *upper bound* (resp., a *lower bound*) of S if $s \preceq u$ (resp., $t \preceq s$) for all $s \in S$. u is called the *least upper bound* or *supremum* of S and denoted by $u = \sup(S)$ (or $\sup S$) if u is an upper bound of S and if, for any upper bound v of S, we have $u \preceq v$. The *greatest lower bound* or *infimum* of S, denoted by $t = \inf(S)$ (or $\inf S$), is defined similarly. If $u = \sup(S) \in S$, then we write $u = \sup(S) = \max(S)$ (or $\max S$). u is then called the *greatest element* or *maximum* of S. Similarly, if $t = \inf(S) \in S$, then we write $t = \inf(S) = \min(S)$ (or min S). The element t is then the *least element* or *minimum* of S.

Definition 1.2.17 (Bounded Set). Let X be a partially ordered set, and let $S \subset X$. We say that S is *bounded* if it is bounded *both* above and below; in other words, if there are elements t, $u \in X$ such that $t \preceq s \preceq u$ *for all* $s \in S$.

The most important fact about partially ordered sets, chains, upper bounds, and maximal elements is the following lemma which is equivalent to the Axiom of Choice:

Lemma 1.2.18 (Zorn's Lemma). *If X is a partially ordered set in which every chain has an upper bound, then X contains a maximal element.*

Example 1.2.19.

(a) Consider the power set $\mathcal{P}(U)$ with the partial ordering $\subset$ and suppose that, for each element of an index set Λ, we are given a set $A_\lambda \subset U$. Then we have

$$\sup\{A_\lambda : \lambda \in \Lambda\} = \bigcup_{\lambda \in \Lambda} A_\lambda,$$

$$\inf\{A_\lambda : \lambda \in \Lambda\} = \bigcap_{\lambda \in \Lambda} A_\lambda.$$

In particular, $\sup \mathcal{P}(U) = \max \mathcal{P}(U) = U$, and $\inf \mathcal{P}(U) = \min \mathcal{P}(U) = \emptyset$.

(b) Consider the set $\mathbb{Z}$ of integers, ordered by the partial ordering: "$m \preceq n$, if and only if $m|n$, i.e., if and only if m *divides* n." Then we have

$$\sup\{m, n\} = \text{lcm}(m, n) = \text{least common multiple of } m \text{ and } n,$$

$$\inf\{m, n\} = \gcd(m, n) = \text{greatest common divisor of } m \text{ and } n.$$

Warning. There is an important distinction between *minimal* and *least* elements. For example, for the collection $\mathcal{P}(U) \setminus \{\emptyset\}$ of all nonempty subsets of U (partially ordered by "$\subset$"), each singleton $\{x\}$, $x \in U$, is a minimal element, but, unless U itself is a singleton, there is no least element. Similarly one should distinguish between *maximal* and *greatest* elements.

Definition 1.2.20 (Well Ordering). A partial ordering "$\preceq$" on a set X is called a *well ordering*, and the set X is called *well ordered*, if for any subset $S \subset X$, $S \neq \emptyset$, we have $\inf(S) \in S$; in other words, if every nonempty subset of X has a *least* element.

Remark 1.2.21. Note that a well-ordered set X is automatically a chain (i.e., linearly ordered). Indeed, for any pair of elements a, $b \in X$, the nonempty subset $\{a, b\}$ must have a least element, so that $a \preceq b$ or $b \preceq a$.

Example 1.2.22.

(a) (Well-Ordering Axiom). Under the usual ordering "$\leq$," the set $\mathbb{N}$ of all natural numbers is well ordered. More generally, for any $n \in \mathbb{Z}$, the set $T_n = \{k \in \mathbb{Z} : k \geq n\}$ is well ordered.
(b) The sets $\mathbb{Z}$ and $\{1/n : n \in \mathbb{N}\}$ are not well ordered.

The following consequence of the Well-Ordering Axiom of the set $\mathbb{N}$ of natural numbers is a powerful tool for many proofs:

Proposition 1.2.23 (Principle of Mathematical Induction). *If a subset $S \subset \mathbb{N}$ satisfies the following two conditions:*

(i) $1 \in S$,
(ii) $n \in S \Rightarrow n + 1 \in S$,

then we have $S = \mathbb{N}$.

Proof. If $S \neq \mathbb{N}$, let $m = \min(\mathbb{N} \setminus S)$. Then $m - 1 \notin \mathbb{N} \setminus S$, which means that $m - 1 \in S$. But then, by (ii), $(m - 1) + 1 = m \in S$, which is absurd. $\qquad\square$

Another way of stating the principle is this: If, for each $n \in \mathbb{N}$, $P(n)$ is a statement about n, and if we are given that $P(1)$ is true and that, for each natural number k, $P(k + 1)$ is true whenever $P(k)$ is, then $P(n)$ is true for all $n \in \mathbb{N}$. Indeed, we then simply define $S := \{n \in \mathbb{N} : P(n)\}$.

Definition 1.2.24 (Initial Segment). Let X be a partially ordered set. For each $x \in X$, the *initial segment* determined by x, denoted by $s(x)$, is the subset $s(x) := \{y \in X : y \prec x\}$.

Proposition 1.2.25 (Principle of Transfinite Induction). *Let X be a well-ordered set and let $S \subset X$ satisfy the following condition:*

$$\forall x \in X, \quad s(x) \subset S \Rightarrow x \in S.$$

Then $S = X$.

Proof. First, note that $S \neq \emptyset$. Indeed, if $x_0 := \min(X) \in X$, then $s(x_0) = \emptyset \subset S$, which implies $x_0 \in S$. Next, if $X \setminus S \neq \emptyset$, let $\xi := \min(X \setminus S) \in X \setminus S$. Then $s(\xi) \subset S$. But this implies $\xi \in S$, which is absurd. $\qquad\square$

Corollary 1.2.26 (Principle of Strong Induction). *If $S \subset \mathbb{N}$ satisfies the condition $(\forall n \in \mathbb{N})(\{k \in \mathbb{N} : k < n\} \subset S \Rightarrow n \in S)$, then $S = \mathbb{N}$. Equivalently, let $P(n)$ be a statement about n for each $n \in \mathbb{N}$. If (i) $P(1)$ is true and (ii) $P(n)$ is true whenever $P(k)$ is true for all $1 \leq k < n$, then $P(n)$ is true for all $n \in \mathbb{N}$.*

Remark 1.2.27. The Principle of Strong Induction is in fact equivalent to the Principle of Mathematical Induction (Proposition 1.2.23). The reader is invited to supply the proof (cf. Exercise 1.2.28 below). There are many situations where this "strong" version is the appropriate one to use. An important example is the proof of the *Prime Factorization Theorem* (cf. Corollary 1.3.45 of Proposition 1.3.39 below).

Exercise 1.2.28. Show that the Principle of Strong Induction is in fact (logically) *equivalent* to the Principle of Mathematical Induction (i.e., Proposition 1.2.23).

The most important fact about well ordering is the following theorem which is equivalent to the Axiom of Choice. The proof, which we omit, may be found, e.g., in Suppes's *Axiomatic Set Theory* [Sup60].

Theorem 1.2.29 (Well-Ordering Theorem of Zermelo). *Every set X can be well ordered. In other words, there exists an order relation "$\preceq$" on X which is a well ordering.*

Definition 1.2.30 (Directed Set, Lattice). Let "$\preceq$" be a partial ordering on a set X. We say that X is *directed* if every pair of elements a, $b \in X$ has an upper bound in X. We say that X is a *lattice* if, for every pair of elements a, $b \in X$, we have $\sup\{a,b\}$, $\inf\{a,b\} \in X$; we then write $a \vee b := \sup\{a,b\}$ and $a \wedge b := \inf\{a,b\}$. It is obvious that if a and b are *comparable*, then $a \vee b = \max\{a,b\}$ is the greater of a and b, and $a \wedge b = \min\{a,b\}$ is the lesser of a and b.

Using the above definitions and notation, one can prove the following identities, which are quite obvious for the usual ordering on the set of real numbers.

Proposition 1.2.31 (Lattice Identities). *Let X be a partially ordered set and let $\{x, y, z\}$ be a chain in X. Then the following identities are satisfied:*

1. $x \wedge x = x$, $\quad x \vee x = x$ *(idempotent);*
2. $x \wedge y = y \wedge x$, $\quad x \vee y = y \vee x$ *(commutative);*
3. $x \wedge (y \wedge z) = (x \wedge y) \wedge z$, $\quad x \vee (y \vee z) = (x \vee y) \vee z$ *(associative);*
4. $x \wedge (x \vee y) = x \vee (x \wedge y) = x$ *(absorption);*
5. $x \preceq y \iff x \wedge y = x \iff x \vee y = y$ *(consistency);*
6. $x \wedge (y \vee z) = (x \wedge y) \vee (x \wedge z)$, $\quad x \vee (y \wedge z) = (x \vee y) \wedge (x \vee z)$ *(Distributive).*

Proof. Exercise!   □

Remark 1.2.32.

(a) A lattice X in which the identity (6) holds for *all* elements x, y, $z \in X$ is called a *distributive lattice*. Note that, in the distributive property (6) above, the two distributive relations are equivalent; i.e., each is a consequence of the other.

(b) Using the associative property, one can inductively define the $\wedge$ and $\vee$ operations for any finite chain $\{x_1, x_2, \ldots, x_n\} \subset X$. In this case we write

$$\bigwedge_{i=1}^{n} x_i := x_1 \wedge \cdots \wedge x_n, \quad \text{and} \quad \bigvee_{i=1}^{n} x_i := x_1 \vee \cdots \vee x_n.$$

Exercise 1.2.33 (Lexicographic Ordering on $\mathbb{N} \times \mathbb{N}$). On the set $\mathbb{N} \times \mathbb{N}$, where $\mathbb{N}$ is the set of natural numbers, define the partial ordering

$$(a,b) \preceq (c,d) \iff a < c, \quad \text{or} \quad a = c \quad \text{and} \quad b \leq d,$$

where "$<$" and "$\leq$" have their usual meanings. Show that "$\preceq$" is a *well ordering* on $\mathbb{N} \times \mathbb{N}$. This *dictionary ordering* is one of the many possible ways of ordering a (Cartesian) product.

Definition 1.2.34 (Function). Given two sets X and Y, a relation $f \subset X \times Y$ is called a *function* (also called a *mapping*, or simply a *map*) if it is *single-valued*; in other words, if we have

$$(x, y_1), \ (x, y_2) \in f \Rightarrow y_1 = y_2.$$

If $(x, y) \in f$, the (unique) element $y \in Y$ is denoted by $y = f(x)$. The set $\mathrm{dom}(f) := \{x \in X : (\exists y \in Y)((x, y) \in f)\} \subset X$ is called the *domain* of f, and the set $\mathrm{ran}(f) := \{y \in Y : (\exists x \in X)((x, y) \in f)\} = \{f(x) : x \in \mathrm{dom}(f)\} \subset Y$ is called the *range* of f. If $\mathrm{dom}(f) = X$, then we say that f is a function *from X to Y* and write $f : X \rightarrow Y$. In this case we also define f, informally, as a *rule* which assigns to each $x \in X$ a *unique* $y = f(x) \in Y$. We may also use the notation $x \mapsto y = f(x)$.

Notation 1.2.35. The set of all functions from a set X to a set Y will be denoted by Y^X. Occasionally, we may also use $\mathcal{F}(X, Y)$.

Definition 1.2.36 (Sequence). Let X be a set. A *sequence* in X is a function $x : \mathbb{N} \rightarrow X$. We write $x(n) = x_n$, and $x = (x_n) = (x_n)_{n \in \mathbb{N}} = (x_1, x_2, \ldots, x_n, \ldots)$. The element $x_n := x(n)$ is called the *nth term* of the sequence x. Using the above notation, the set of all sequences in X will be denoted by $X^{\mathbb{N}}$.

Definition 1.2.37 (Direct and Inverse Images). Given a function $f : X \rightarrow Y$ and subsets $A \subset X$, $B \subset Y$, we define the *(direct) image of A under f* by

$$f(A) := \{f(a) : a \in A\}$$

and the *inverse image of B under f* by

$$f^{-1}(B) := \{x \in X : f(x) \in B\}.$$

Note, in particular, that $\mathrm{ran}(f) = f(X)$. Also, if $B = \{y\} \subset Y$, then we write

$$f^{-1}(y) := f^{-1}(\{y\}).$$

Example 1.2.38. Let $f : \mathbb{R} \rightarrow \mathbb{R}$ be the function $f(x) = x^2$. We then have $f(\mathbb{R}) = [0, \infty)$ and $f^{-1}([0, 1]) = [-1, 1]$, while $f^{-1}([-1, 0)) = \emptyset$.

Exercise 1.2.39. Let $f : X \rightarrow Y$ be a function. Suppose that for an index set I we have for each $i \in I$ a subset $A_i \subset X$ and a subset $B_i \subset Y$. Show that

(a) $f(\bigcup_{i \in I} A_i) = \bigcup_{i \in I} f(A_i)$,
(b) $f(\bigcap_{i \in I} A_i) \subset \bigcap_{i \in I} f(A_i)$,
(c) $f^{-1}(\bigcup_{i \in I} B_i) = \bigcup_{i \in I} f^{-1}(B_i)$,

(d) $f^{-1}(\bigcap_{i \in I} B_i) = \bigcap_{i \in I} f^{-1}(B_i)$,

(e) $f^{-1}(Y \setminus B) = X \setminus f^{-1}(B)$ for any subset $B \subset Y$,

(f) $f^{-1}(B_1 \setminus B_2) = f^{-1}(B_1) \setminus f^{-1}(B_2)$ for any subsets $B_1, B_2 \subset Y$, and

(g) $f^{-1}(B_1 \triangle B_2) = f^{-1}(B_1) \triangle f^{-1}(B_2)$ for any $B_1, B_2 \subset Y$.

Definition 1.2.40 (One-to-One, Onto, etc.). Let $f : X \to Y$ be a function. We say that f is *one-to-one*, or *injective*, if for all $x_1, x_2 \in X$ with $x_1 \neq x_2$, we have $f(x_1) \neq f(x_2)$. We say that f is *onto*, or *surjective*, if $\mathrm{ran}(f) = f(X) = Y$. Finally we say that f is a *one-to-one correspondence*, or is *bijective*, if it is both one-to-one *and* onto.

Example 1.2.41.

(a) (Canonical Projection) Let $R \subset X \times X$ be an *equivalence relation* on a set X and let X/R be the corresponding *quotient set* as in Definition 1.2.8 (i.e., the set of all equivalence classes). The *canonical projection* is then the function $\pi : X \to X/R$ defined by

$$\pi(x) := [x] \qquad \forall x \in X.$$

It is obviously a *surjective* (i.e., *onto*) map.

(b) Let $A = \{a_1, a_2, \ldots, a_m\}$ and $B = \{b_1, b_2, \ldots, b_n\}$ be two finite sets. Define $f : A \times B \to \{1, 2, \ldots, mn\}$ by $f(a_i, b_j) := (i - 1)n + j$, $1 \leq i \leq m$, $1 \leq j \leq n$. Then f is *bijective* (why?).

(c) Let $A_1, A_2, \ldots, A_k$ be sets and define $g : A_1 \times A_2 \times \cdots \times A_k \to (A_1 \times A_2 \times \cdots \times A_{k-1}) \times A_k$ by $g(a_1, a_2, \ldots, a_{k-1}, a_k) := \big((a_1, a_2, \ldots, a_{k-1}), a_k\big)$. Then g is clearly a *bijective* map.

Definition 1.2.42 (Permutation). Given a set X, a *one-to-one correspondence* $f : X \to X$ is called a *permutation* of X.

Example 1.2.43. Let f, g, and h be functions from $\mathbb{R}$ to $\mathbb{R}$ defined by $f(x) := x^2$, $g(x) := x^3 + x^2$, and $h(x) := x^3$ $\forall x \in \mathbb{R}$. Then f is neither one-to-one nor onto, g is onto but not one-to-one, and h is a one-to-one correspondence, i.e., a *permutation* of $\mathbb{R}$.

Definition 1.2.44 (Composite Functions). Given the sets X, Y, Z and functions $f : X \to Y$, $g : Y \to Z$, the *composite* function $g \circ f : X \to Z$ is defined by

$$(g \circ f)(x) := g(f(x)) \qquad \forall x \in X.$$

Definition 1.2.45 (Inverse Function). Given a *one-to-one* function $f : X \to Y$, the inverse relation $f^{-1} = \{(y, x) : y = f(x)\} \subset Y \times X$ is a *function* called the *inverse of* f. If the function f is a *one-to-one correspondence*, then the inverse function f^{-1} has domain Y, i.e., $f^{-1} : Y \to X$, and we have $f \circ f^{-1} = \mathrm{id}_Y$, while $f^{-1} \circ f = \mathrm{id}_X$, where, for any set S, the *identity function* id_S is defined by $\mathrm{id}_S(s) := s$ $\forall s \in S$.

Remark 1.2.46.

(a) (Associativity of Composition). The composition of functions is *associative* in the sense that, for any functions $f : S \to T$, $g : T \to U$, and $h : U \to V$, we have

$$h \circ (g \circ f) = (h \circ g) \circ f.$$

This follows at once from the definition of composition: $\forall s \in S$,

$$(h \circ (g \circ f))(s) = h((g \circ f)(s)) = h(g(f(s))) = (h \circ g)(f(s)) = ((h \circ g) \circ f)(s).$$

(b) If $f : X \to Y$ is *not* injective, the inverse *relation* $f^{-1} := \{(y, x) : (x, y) \in f\}$ is *not* a function since one can then find x_1, $x_2 \in X$ and $y \in Y$ with $x_1 \neq x_2$ but $f(x_1) = f(x_2) = y$, so that f^{-1} contains the pairs (y, x_1), (y, x_2) and hence is *not* single-valued.

Exercise 1.2.47. Let X and Y be sets, $A \subset X$, $B \subset Y$, $f : X \to Y$, and $g : Y \to X$.

1. Show that, if $X = Y$ and if f and g are *permutations* of X, then so are f^{-1} and $g \circ f$.
2. Show that, if $g \circ f = \mathrm{id}_X$, then f is *one-to-one* and g is *onto*.
3. Show that $f(f^{-1}(B)) \subset B$ and that equality holds if f is *onto*. Show by example that the inclusion is *proper* in general.
4. Show that $A \subset f^{-1}(f(A))$ and that equality holds if f is *one-to-one*. Show by example that the inclusion is *proper* in general.

Exercise 1.2.48. Let $f : X \to Y$ be a *one-to-one* function. Show that for any subsets A, $B \subset X$, we have

(a) $f(A \cap B) = f(A) \cap f(B)$,
(b) $f(X \setminus A) = f(X) \setminus f(A)$.
(c) Show by examples that the statements in (a) and (b) are false if f is *not* one-to-one.

The following exercise provides a large class of *equivalence relations* that includes most commonly encountered cases.

Exercise 1.2.49. Let $X \neq \emptyset$ and S be sets and let $f : X \to S$ be an arbitrary map. Define the relation

$$R_f := \{(x, y) \in X \times X : f(x) = f(y)\}.$$

Show that R_f is an *equivalence relation*. Also, show that the *equivalence classes* are the sets

$$f^{-1}(s) := \{x \in X : f(x) = s\} \quad \forall s \in f(X).$$

Recall that the equivalence class of an element $x \in X$ is the set

$$[x] := \{y \in X : (x, y) \in R_f\}.$$

Definition 1.2.50 (Finite and Infinite Sets). A set S is called *finite* if either $S = \emptyset$ or there is a *one-to-one correspondence* between S and the set $\{1, 2, 3, \ldots, n\}$, for some $n \in \mathbb{N}$. We then say that S has n *elements* and write $|S| = n$. A set that is *not* finite is called *infinite*.

Example 1.2.51. It follows from Example 1.2.41(b) that if A and B are *finite* sets with m and n elements, respectively, then their Cartesian product $A \times B$ is a finite set with mn elements. Inductively, using Example 1.2.41(c), if A_j is a finite set with n_j elements, $1 \le j \le k$, then the Cartesian product $\Pi_{j=1}^{k} A_j$ is a finite set with $n_1 n_2 \cdots n_k$ elements. It is obvious that the standard sets $\mathbb{N}$, $\mathbb{N}_0$, $\mathbb{Z}$, $\mathbb{Q}$, $\mathbb{R}$, and $\mathbb{C}$ are all *infinite*.

Definition 1.2.52 (Characteristic Function). Let X be a set and $A \subset X$. The *characteristic function* (or *indicator function*) of the set A, denoted by χ_A (or $\mathbf{1}_A$), is defined by

$$\chi_A(x) = \mathbf{1}_A(x) := \begin{cases} 1 & \text{if } x \in A, \\ 0 & \text{otherwise.} \end{cases}$$

Example 1.2.53. Given a universal set U, we obviously have $\chi_U = 1$ and $\chi_\emptyset = 0$, where by 1 and 0 we mean the constant functions identically equal to 1 and 0, respectively. Let $S \subset U$ be any set, and let $\Delta = \Delta_S = \{(s, s) : s \in S\}$ be the *diagonal* in $S \times S$. Then *Kronecker's delta*, $\delta := \chi_\Delta$, is the characteristic function of Δ:

$$\delta_{xy} := \delta(x, y) := \begin{cases} 1 & \text{if } x = y, \\ 0 & \text{if } x \ne y. \end{cases}$$

Exercise 1.2.54. Let X be a set and $\mathcal{P}(X)$ its power set. Consider the set $\{0, 1\}^X$ of all functions from X to $\{0, 1\}$, which we also denote by 2^X, and define the function $\chi : \mathcal{P}(X) \to 2^X$ by $\chi(A) = \chi_A$. Show that χ is a *one-to-one correspondence*.

Exercise 1.2.55. Prove the following properties of the characteristic function. Here, A, B, $A_1, \ldots, A_k$ are arbitrary subsets of a universal set U:

(a) $A \subset B \Leftrightarrow \chi_A \le \chi_B$,
(b) $\chi_{A \cap B} = \chi_A \chi_B = \min\{\chi_A, \chi_B\}$,
(c) $\chi_{A^c} = 1 - \chi_A$,
(d) $\chi_{A \cup B} = 1 - (1 - \chi_A)(1 - \chi_B) = \max\{\chi_A, \chi_B\} = \chi_A + \chi_B - \chi_A \chi_B$,
(e) $\chi_{A \triangle B} = |\chi_A - \chi_B|$,
(f) $\chi_{A_1 \cap A_2 \cap \cdots \cap A_k} = \prod_{j=1}^{k} \chi_{A_j}$, and $\chi_{A_1 \cup A_2 \cup \cdots \cup A_k} = 1 - \prod_{j=1}^{k}(1 - \chi_{A_j})$,

(g) if A is *finite*, then $|A| = \Sigma_{x \in U} \chi_A(x)$, and
(h) (Inclusion–Exclusion Principle) if A and B are finite, then $|A \cup B| = |A| + |B| - |A \cap B|$.

Definition 1.2.56 (Bounded Function). Let X and Y be sets, and assume that Y is *partially ordered*. A function $f : X \to Y$ is called *bounded above* (resp., *below*) if its range $f(X) \subset Y$ is bounded above (resp., below) in Y; i.e., if there exists $z \in Y$ (resp., $y \in Y$) such that $f(x) \preceq z$ (resp., $y \preceq f(x)$) for all $x \in X$. The function f is called *bounded* if it is bounded *both* above and below; otherwise, we call it *unbounded*.

Example 1.2.57. Let f, g, h be the functions from $\mathbb{R}$ to $\mathbb{R}$ defined by $f(x) := \sin x$, $g(x) := x^2$, and $h(x) := x^3$ $\forall x \in \mathbb{R}$. Then f is bounded $(-1 \leq \sin x \leq 1 \ \forall x)$, g is bounded below $(0 \leq x^2 \ \forall x)$ but *not* above, and h is neither bounded above nor bounded below.

1.3 Basic Algebra, Counting, and Arithmetic

Our goal in this section is to give a brief summary of the most basic definitions and terminology in *algebra*, *counting*, and *arithmetic*. These will be needed on various occasions in the upcoming chapters of the text. Most readers have already encountered in other courses more detailed accounts of the topics we briefly cover here. We start with the definitions of commonly used *algebraic structures*, namely, *groups, rings, fields, vector spaces*, and *algebras*. We then introduce the *Basic Counting Principle* as well as the *Inclusion–Exclusion Principle* and end the section with some elementary facts from *arithmetic*. The reader is referred to the excellent textbooks *Topics in Algebra* by Herstein [Her75] and *A Survey of Modern Algebra* by Birkhoff and MacLane [BM77] for details. For vector spaces, we also recommend Halmos's beautifully written *Finite-Dimensional Vector Spaces* [Hal58].

Definition 1.3.1 (Group). A *group* is a *set* G together with a *binary operation* denoted by $\cdot$ (i.e., a map $\cdot : G \times G \to G$) satisfying the following axioms:

1. $a \cdot (b \cdot c) = (a \cdot b) \cdot c$ $\quad \forall a, b, c \in G$ (associativity).
2. There exists an element $e \in G$ (called the *identity element*) such that $a \cdot e = e \cdot a = a$ $\quad \forall a \in G$ (existence of an identity element).
3. For every $a \in G$ there exists $a^{-1} \in G$ (called the *inverse* of a) such that $a \cdot a^{-1} = a^{-1} \cdot a = e$ (existence of inverses).

Remark 1.3.2.

(a) The existence of (the) identity element [in axiom (2)] implies that a group is *never empty*.

(b) The binary operation $\cdot : G \times G \to G$ is usually called the *product* even though it may have nothing in common with the ordinary product of two numbers. Also, one usually omits the "$\cdot$" and writes ab instead of $a \cdot b$.

(c) A group is a *structured set*; i.e., it is a set *together with* a binary operation. Thus, to be precise, one should write "a group $(G, \cdot)$" rather than "a group G." Nevertheless, the latter is often used if there is no confusion over the binary operation.

Definition 1.3.3 (Subgroup, Abelian Group). Let G be a group. A subset $H \subset G$ is said to be a *subgroup* of G if, with the product $\cdot|H$ (i.e., with the product of G restricted to H), the set H itself is a group. A group G is said to be *Abelian* (or *commutative*) if

$$ab = ba \quad \forall a, b \in G.$$

Example 1.3.4.

(a) The set $\mathbb{Z}$ of all integers with the operation of *addition* (i.e., $(a, b) \mapsto a + b$) is an Abelian group. The identity element is 0 and, for each integer $a \in \mathbb{Z}$, its inverse is the *opposite* number $-a$. The subset

$$2\mathbb{Z} := \{2n : n \in \mathbb{Z}\}$$

of all *even* integers is a *subgroup*, as the reader can check at once. The subset of all *odd integers*, however, is *not* a subgroup, nor is the subset $\mathbb{N}_0 := \mathbb{N} \cup \{0\}$ of all *nonnegative* integers. (Why?)

(b) The set $\mathbb{Q}^* := \mathbb{Q} \setminus \{0\}$ of all *nonzero rational* numbers is a group with the operation of *multiplication*, i.e., $(r, s) \mapsto rs$. The subset $\mathbb{Q}^+$ of all *positive* rationals is a subgroup and so is the subset $\{-1, 1\}$.

(c) Let X be an arbitrary *nonempty* set and let $\mathfrak{S}_X$ denote the set of all *permutations* of X, i.e., the set of all maps $f : X \to X$ that are *bijective* (i.e., one-to-one and onto). Then $(\mathfrak{S}_X, \circ)$ is a group (called the *symmetric group* of X) where $\circ$ denotes the *composition* of maps: $(f, g) \mapsto f \circ g$. This follows from Exercise 1.2.47(1). Here, the identity element is the *identity map* $\mathrm{id}_X : X \to X$ defined by $\mathrm{id}_X(x) := x \quad \forall x \in X$.

Proposition 1.3.5 (Uniqueness of Identity and Inverse). *Let $(G, \cdot)$ be a group. Then the identity element $e \in G$ is unique. Also, for each $a \in G$, the inverse a^{-1} is unique.*

Proof. If e and e' are both identity elements, then we have $ee' = e$ since e' is an identity element, and $ee' = e'$ since e is an identity element. Thus

$$e = ee' = e'.$$

Next, if b, $c \in G$ are both inverses of $a \in G$, then, by associativity, we have

$$b = be = b(ac) = (ba)c = ec = c$$

and the proof is complete.   □

Exercise 1.3.6. Let G be a group and let $\emptyset \neq H \subset G$. Show that H is a subgroup of G if and only if the following is true:

$$ab^{-1} \in H \quad \forall a,\, b \in H.$$

Definition 1.3.7 (Ring). A *ring* is a set R together with *two* binary operations $+ :$ $R \times R \to R$ and $\cdot : R \times R \to R$, called *addition* and *multiplication*, such that, for *arbitrary* elements x, y, $z \in R$, the following axioms are satisfied:

1. $x + y = y + x$,
2. $(x + y) + z = x + (y + z)$,
3. $\exists 0 \in R$ such that $x + 0 = x$,
4. $\exists -x \in R$ such that $x + (-x) = 0$,
5. $x \cdot (y \cdot z) = (x \cdot y) \cdot z$,
6. $x \cdot (y + z) = x \cdot y + x \cdot z$,   and $(y + z) \cdot x = y \cdot x + z \cdot x$.

Remark 1.3.8.

(a) Axioms (1)–(4) simply indicate that $(R, +)$ is a *commutative group*.
(b) (Commutative Ring). If $x \cdot y = y \cdot x \quad \forall x,\, y \in R$, then the ring R is said to be *commutative*.
(c) (Ring with Unit Element). If there exists an element $1 \in R$ such that $1 \cdot x = x \cdot 1 = x \quad \forall x \in R$, then the element $1 \in R$ is called a *unit element* and the ring R is said to be a *ring with unit element*.
(d) (Division Ring). A ring R with unit element is said to be a *division ring* if $R \setminus \{0\}$ is a *group* under multiplication, i.e., if each $x \in R \setminus \{0\}$ has an *inverse* $x^{-1} \in R$ (so that $xx^{-1} = x^{-1}x = 1$, where $1 \in R$ is the *unit element*).

Example 1.3.9.

(a) With the usual addition and multiplication, the set $\mathbb{Z}$ of all integers is a commutative ring with unit element 1. The set $2\mathbb{Z} := \{2n : n \in \mathbb{Z}\}$ of all *even* integers is a commutative ring but has *no* unit element.
(b) With the usual addition and multiplication, the set $\mathbb{Q}$ of rational numbers is a commutative (division) ring with unit element 1.
(c) Let $X \neq \emptyset$ be an arbitrary set and let $\mathcal{F}(X, \mathbb{Q}) := \mathbb{Q}^X$ be the set of all functions from X to $\mathbb{Q}$. Then, as the reader can easily check, $\mathcal{F}(X, \mathbb{Q})$ is a commutative ring with addition and multiplication defined, for arbitrary f, $g \in \mathcal{F}(X, \mathbb{Q})$ and $x \in X$, by

$$(f + g)(x) := f(x) + g(x) \quad \text{and} \quad (f \cdot g)(x) := f(x)g(x).$$

Definition 1.3.10 (Subring, Ideal). Let R be a ring and let $S \subset R$. We say that S is a *subring* of R, if with the addition and multiplication (of R) *restricted to* S, the set S is itself a ring. In particular, $(S, +)$ is a *subgroup* of $(R, +)$. We say that S is a *(two-sided) ideal* (or simply an *ideal*) if S is a *subgroup* of $(R, +)$ and if, in addition, we have

$$(\forall x \in S)(\forall y \in R)(x \cdot y, \ y \cdot x \in S).$$

Thus, every ideal is obviously a subring. The converse is, however, *false* in general.

Example 1.3.11.

(a) The set $\mathbb{Z}$ is a subring of $\mathbb{Q}$ but *not* an ideal. The set $2\mathbb{Z}$ of *even* integers is an ideal of $\mathbb{Z}$.
(b) Let $\mathcal{F}(X, \mathbb{Z}) = \mathbb{Z}^X$ be the set of all *integer-valued* elements of $\mathcal{F}(X, \mathbb{Q})$. Then $\mathcal{F}(X, \mathbb{Z})$ is a subring of $\mathcal{F}(X, \mathbb{Q})$ but *not* an ideal. On the other hand, if $x_0 \in X$ is a *fixed* element, then the subset $\mathcal{F}_{x_0} := \{f \in \mathcal{F}(X, \mathbb{Q}) : f(x_0) = 0\}$ is an *ideal* of $\mathcal{F}(X, \mathbb{Q})$. (Why?)

Definition 1.3.12 (Field, Subfield). A *field* F is a commutative ring with unit element $1 \neq 0$ such that $(F \setminus \{0\}, \cdot)$ is a *group*; i.e., each $x \in F \setminus \{0\}$ has a *multiplicative inverse* x^{-1} (so that $xx^{-1} = x^{-1}x = 1$). In other words, a field is simply a *commutative division ring*. A subset $K \subset F$ is said to be a *subfield* of F if K is a subring of F and $K \setminus \{0\}$ is a subgroup of $(F \setminus \{0\}, \cdot)$.

Example 1.3.13. The set $\mathbb{Q}$ of *rational* numbers is a field and so is the set $\mathbb{R}$ (resp., $\mathbb{C}$) of all *real* (resp., *complex*) numbers (to be defined later). In fact, $\mathbb{Q}$ is a *subfield* of $\mathbb{R}$ which is itself a subfield of $\mathbb{C}$.

Definition 1.3.14 (Vector Space, Subspace). Let F be a field. A nonempty set V, whose elements will be called *vectors*, is said to be a *vector space (over F)* if V is an *Abelian group* under an operation $+ : V \times V \to V$ called *(vector) addition* and if there is a map $\cdot : F \times V \to V$, called *scalar multiplication* and written as $\cdot(a, v) = av$ for all $a \in F$, $v \in V$, such that the axioms

1. $a(u + v) = au + av$,
2. $(a + b)u = au + bu$,
3. $a(bu) = (ab)u$, and
4. $1u = u$

are satisfied for arbitrary elements a, $b \in F$ and vectors u, $v \in V$. A subset $U \subset V$ is said to be a (vector) *subspace* of V if, with the addition and scalar multiplication *restricted* to U, the set U is itself a vector space.

Example 1.3.15.

(a) Any *field* F is a vector space *over itself* and, of course, over each of its *subfields*. Thus, the field $\mathbb{R}$ is a vector space over itself and over $\mathbb{Q}$. Also, the field $\mathbb{C}$ of complex numbers is a vector space over $\mathbb{C}$, $\mathbb{R}$, and $\mathbb{Q}$.

(b) Let $X \neq \emptyset$ be an arbitrary set and let F be any field. Then the set

$$\mathcal{F}(X, F) := F^X$$

of all functions from X to F is a vector space over F.

(c) [A special case of (b)] Consider the set

$$\mathbb{Q}^n := \mathbb{Q}^{\{1,2,\ldots,n\}} = \{(x_1, x_2, \ldots, x_n) : x_k \in \mathbb{Q}, \ 1 \leq k \leq n\}.$$

For any $x = (x_1, \ldots, x_n)$, $y = (y_1, \ldots, y_n) \in \mathbb{Q}^n$ and any $r \in \mathbb{Q}$, define the vector addition and scalar multiplication *componentwise*, i.e.,

$$x + y := (x_1 + y_1, \ldots, x_n + y_n), \qquad rx = (rx_1, \ldots, rx_n).$$

Then $\mathbb{Q}^n$ is a vector space over $\mathbb{Q}$. Similarly, $\mathbb{R}^n$ is a vector space over $\mathbb{R}$ (and $\mathbb{Q}$) and $\mathbb{C}^n$ is a vector space over $\mathbb{C}$, $\mathbb{R}$, and $\mathbb{Q}$.

Remark 1.3.16 (Module). If in the above definition of vector space the field F is replaced by a *ring* R, then the resulting (structured set) V is said to be an *R-module* (or a *module over R*). Since every field is a ring, it is obvious that every vector space is a module. Note that, if the ring R has no unit element, then the axiom (4) (i.e., $1u = u \ \forall u \in V$) must be omitted. On the other hand, if R has a unit 1 and $1u = u \ \forall u \in V$ is satisfied, then V is called a *unital* R-module. Given an arbitrary set $X \neq \emptyset$ and an arbitrary ring R, the set R^X of all R-valued functions on X is an R-module.

Definition 1.3.17 (Direct Sum, Complement). Let V_1 and V_2 be two subspaces of a vector space V. We say that V is the *direct sum* of V_1 and V_2, and we write $V = V_1 \oplus V_2$, if every $v \in V$ can be written *uniquely* as $v = v_1 + v_2$ with $v_1 \in V_1$ and $v_2 \in V_2$. The subspace V_1 (resp., V_2) is then said to be a *complement* of V_2 (resp., V_1).

Definition 1.3.18 (Span, Finite-Dimensional). Let V be a vector space (over a field F) and $S \subset V$. The *span of S* is the subspace of all *finite linear combinations of vectors in S*; i.e.,

$$\mathrm{Span}(S) := \{a_1 v_1 + \cdots + a_n v_n : a_1, \ldots, a_n \in F, v_1, \ldots, v_n \in S\}.$$

For $S = \emptyset$, we define $\mathrm{Span}(\emptyset) := \{0\}$. The space V is said to be *finite-dimensional* if $V = \mathrm{Span}(S)$, for a *finite* set $S \subset V$. If V is *not* finite-dimensional, we call it *infinite-dimensional*.

Definition 1.3.19 (Linear Independence, Basis). Given a vector space V over a field F, a set $S \subset V$ is said to be *linearly independent* if, for any finite subset $\{v_1, \ldots, v_n\} \subset S$, we have

$$\sum_{k=1}^{n} c_n v_k = 0 \implies c_1 = c_2 = \cdots = c_n = 0,$$

where $c_k \in F$, for $1 \le k \le n$. A set $B \subset V$ is said to be a *basis* for V if B is linearly independent and $\text{Span}(B) = V$.

We state the following well-known fact without proof:

Theorem 1.3.20 (Dimension). *Any two bases of a finite-dimensional vector space V have the same number of elements; this number, denoted* $\dim(V)$, *is called the dimension of V.*

Definition 1.3.21 (Algebra, Subalgebra). Let F be a *field*. A *ring* A is called an *algebra* (*over* F) if A is a *vector space* over F such that for any x, $y \in A$ and any $a \in F$ we have $a(xy) = (ax)y = x(ay)$. A subset $B \subset A$ is said to be a *subalgebra* of A if, with the operations of A *restricted* to B, the set B is itself an algebra over F.

Definition 1.3.22 (Commutative Algebra, Division Algebra). Let A be an algebra over a field F. We say that A is *commutative* if $xy = yx$ $\forall x$, $y \in A$. We say that A is a *division algebra* if A has a *unit element* 1 and if each $x \in A \setminus \{0\}$ has an *inverse* x^{-1}; i.e., $xx^{-1} = x^{-1}x = 1$.

Example 1.3.23.

(a) Every *field* is a commutative division algebra over *itself* and, of course, over any of its *subfields*. Thus, the field $\mathbb{C}$ is a commutative division algebra over $\mathbb{R}$ and also over $\mathbb{Q}$.
(b) Given any field F and any set $X \ne \emptyset$, the set $\mathcal{F}(X, F) := F^X$ is a commutative algebra over F and for each subfield $K \subset F$, the set K^X is a *subalgebra* of F^X. Note, however, that these algebras are *not* division algebras. (Why?)

The following exercise will give a rather involved but important example of a *noncommutative* division algebra.

Exercise 1.3.24 (Real Quaternions). Let Q be the set

$$Q := \{\mathbf{a} = a_0 + a_1 i + a_2 j + a_3 k : a_0, a_1, a_2, a_3 \in \mathbb{R}\},$$

where i, j, and k are *symbols* having the following *multiplication table*.

$\cdot$	i	j	k
i	-1	k	$-j$
j	$-k$	-1	i
k	j	$-i$	-1

The elements of Q will be called *(real) quaternions*. Given two quaternions $\mathbf{a} = a_0 + a_1 i + a_2 j + a_3 k$ and $\mathbf{b} = b_0 + b_1 i + b_2 j + b_3 k$, we write $\mathbf{a} = \mathbf{b}$ if and only if $a_t = b_t$ for $t = 0, 1, 2, 3$. On the set Q one defines three operations as follows. For arbitrary quaternions $\mathbf{a}$, $\mathbf{b} \in Q$ and real number $c \in \mathbb{R}$, the operations of *addition* and *scalar multiplication* are defined componentwise:

$$\mathbf{a} + \mathbf{b} = (a_0 + a_1 i + a_2 j + a_3 k) + (b_0 + b_1 i + b_2 j + b_3 k)$$
$$:= (a_0 + b_0) + (a_1 + b_1)i + (a_2 + b_2)j + (a_3 + b_3)k,$$
$$c\mathbf{a} = c(a_0 + a_1 i + a_2 j + a_3 k) := ca_0 + (ca_1)i + (ca_2)j + (ca_3)k,$$

while *multiplication* is defined by

$$\mathbf{a} \cdot \mathbf{b} = (a_0 + a_1 i + a_2 j + a_3 k)(b_0 + b_1 i + b_2 j + b_3 k)$$
$$:= c_0 + c_1 i + c_2 j + c_3 k,$$

where the real numbers c_0, c_1, c_2, and c_3 are defined to be

$$c_0 := a_0 b_0 - a_1 b_1 - a_2 b_2 - a_3 b_3,$$
$$c_1 := a_0 b_1 + a_1 b_0 + a_2 b_3 - a_3 b_2,$$
$$c_2 := a_0 b_2 + a_2 b_0 + a_3 b_1 - a_1 b_3, \quad \text{and}$$
$$c_3 := a_0 b_3 + a_3 b_0 + a_1 b_2 - a_2 b_1.$$

The above definition of the *product* of two quaternions is indeed complicated, to say the least, but it can be obtained by formally expanding $(a_0 + a_1 i + a_2 j + a_3 k)(b_0 + b_1 i + b_2 j + b_3 k)$, collecting the terms, and simplifying them using the above multiplication table for i, j, k.

(a) Show that the eight elements $\{\pm 1, \pm i, \pm j, \pm k\}$ form a *non-Abelian* group with the product defined by the above multiplication table.

(b) Show that Q is a *noncommutative division ring* with *zero element* $\mathbf{0} := 0 + 0i + 0j + 0k = 0 \in \mathbb{R}$ and *unit element* $\mathbf{1} := 1 + 0i + 0j + 0k = 1 \in \mathbb{R}$. *Hint:* For each $\mathbf{a} = a_0 + a_1 i + a_2 j + a_3 k \in Q$, define its *absolute value* $|\mathbf{a}|$ to be

$$|\mathbf{a}| := \sqrt{a_0^2 + a_1^2 + a_2^2 + a_3^2}$$

and note that $\mathbf{a} = \mathbf{0}$ if and only if $|\mathbf{a}| = 0$. Now show that, given any $\mathbf{a} \neq \mathbf{0}$, its inverse is the quaternion

$$\mathbf{a}^{-1} = \frac{a_0}{|\mathbf{a}|} - \frac{a_1}{|\mathbf{a}|}i - \frac{a_2}{|\mathbf{a}|}j - \frac{a_3}{|\mathbf{a}|}k.$$

(c) Deduce that Q is a *(noncommutative) division algebra* (over $\mathbb{R}$).

Definition 1.3.25 (Homomorphism, Isomorphism).

(a) Given two *groups* G and G', a map $\phi : G \to G'$ is called a (group) *homomorphism* if we have $\phi(ab) = \phi(a)\phi(b) \; \forall a, b \in G$.
(b) Given two *rings* R and R', a map $\phi : R \to R'$ is a (ring) *homomorphism* if $\phi(x + y) = \phi(x) + \phi(y)$ and $\phi(xy) = \phi(x)\phi(y) \; \forall x, y \in R$.
(c) Given two *fields* F and F', a map $\phi : F \to F'$ is called a (field) *homomorphism* if it is a *ring homomorphism*, i.e., if $\phi(x + y) = \phi(x) + \phi(y)$ and $\phi(xy) = \phi(x)\phi(y) \; \forall x, y \in F$.
(d) Given two *vector spaces* V and V' over the *same* field F, a map $\phi : V \to V'$ is called a (vector space) *homomorphism* (or a *linear map*) if $\phi(u + v) = \phi(u) + \phi(v)$ and $\phi(au) = a\phi(u)$ for all $a \in F$ and $u, v \in V$.
(e) Given two *algebras* A and A' over the *same* field F, a map $\phi : A \to A'$ is called an (algebra) *homomorphism* if it is a vector space homomorphism (i.e., $\phi(x + y) = \phi(x) + \phi(y)$ and $\phi(ax) = a\phi(x)$ for all $a \in F$ and $x, y \in A$) and if, in addition, we have $\phi(xy) = \phi(x)\phi(y) \quad \forall x, y \in A$.

In each of the above cases, the map ϕ is said to be an *isomorphism* if (in addition to the conditions above) it is *bijective* (i.e., one-to-one and onto). The corresponding groups (rings, fields, etc.) are then said to be *isomorphic*.

We now look at some basic methods of counting the number of elements of a *finite* set. Counting plays a fundamental role in many parts of mathematics and numerous proofs are heavily dependent on the ability to count the number of elements of various sets. Here are a few questions we can answer as soon as we are familiar with the *Basic Counting Principle* and the (general) *Inclusion–Exclusion Principle* (to be defined below): Given two finite sets X and Y, the set Y^X of all functions from X to Y is obviously also finite. How many elements does this set have? How many of these elements are one-to-one maps? How many are onto? For more difficult questions and methods to answer them, the reader can consult, e.g., Pólya–Szegö [PSz72]. The following proposition is an immediate consequence of Example 1.2.51.

Proposition 1.3.26 (Basic Counting Principle). *If a task can be performed in k steps, and if for each $j = 1, 2, \ldots, k$, the jth step can be performed in n_j ways, regardless of the choices made for the preceding steps, then the total number of ways in which the entire task can be performed is $n_1 n_2 \cdots n_k$.*

Definition 1.3.27 (Permutation, Combination). A *permutation* of n objects taken k at a time is any *ordered arrangement* of k of the n objects. A *combination* of n

objects taken k at a time is simply a selection of k objects from the n objects, with order disregarded. The number of all permutations of n objects taken k at a time is denoted by $P(n, k)$, and the number of all combinations of the n objects taken k at a time is denoted by $C(n, k) = \binom{n}{k}$ [read n *choose* k] and is also called a *binomial coefficient*.

Remark 1.3.28. The number of permutations or combinations of n objects taken k at a time can be easily computed using the Basic Counting Principle: For the number of *permutations*, note that to form a permutation, we have n choices for the first object, $n - 1$ choices for the second object, ..., $n - k + 1$ choices for the kth object. Thus $P(n, k) = n(n - 1)(n - 2) \cdots (n - k + 1) = n!/(n - k)!$, where "$n$ factorial" is defined by $n! := 1 \cdot 2 \cdot 3 \cdots n$ for each $n \in \mathbb{N}$, and $0! := 1$. In particular, $P(n) := P(n, n) = n!$. Note that in this particular case, a permutation is indeed a one-to-one correspondence of the set of n objects with itself. For the *combinations*, note that each *subset* of k objects produces $k!$ permutations, so that $P(n, k) = k! C(n, k)$, from which we get $\binom{n}{k} = C(n, k) = P(n, k)/k! = n!/k!(n - k)!$.

Exercise 1.3.29. For $n \in \mathbb{N}$ and $k = 0, 1, 2, \ldots, n$, prove the identities

$$\binom{n}{k} = \binom{n}{n - k} \quad \text{and} \quad \binom{n + 1}{k} = \binom{n}{k} + \binom{n}{k - 1},$$

where $\binom{n}{0} = 1 = \binom{n}{n}$ follows from $0! := 1$.

The numbers $C(n, k) = \binom{n}{k}$ are called *binomial coefficients* for the following reason:

Proposition 1.3.30 (Binomial and Multinomial Formulas). *For any integer* $n \in \mathbb{N}$ *and any real numbers* $x, y \in \mathbb{R}$, *we have*

$$(x + y)^n = \sum_{k=0}^{n} \binom{n}{k} x^{n-k} y^k = x^n + n x^{n-1} y + \frac{n(n - 1)}{2} x^{n-2} y^2 + \cdots + y^n. \quad (*)$$

More generally, for any real numbers $x_1, x_2, \ldots, x_k \in \mathbb{R}$, *we have*

$$(x_1 + x_2 + \cdots + x_k)^n = \sum \binom{n}{n_1, n_2, \cdots, n_k} x_1^{n_1} x_2^{n_2} \cdots x_k^{n_k}, \quad (**)$$

where the sum on the right is over all $n_1, n_2, \ldots, n_k \in \mathbb{N}_0$ *with* $n_1 + n_2 + \cdots + n_k = n$, *and where the multinomial coefficients are defined by*

$$\binom{n}{n_1, n_2, \cdots, n_k} := \frac{n!}{n_1! n_2! \cdots n_k!}.$$

Proof. Note that the expansion of the product $(x_1 + y_1)(x_2 + y_2) \cdots (x_n + y_n)$ consists of 2^n terms, each containing n factors that are either x_j or y_j, for some $j = 1, 2, \ldots, n$. Now the number of terms containing as factors k of the y_j's is $\binom{n}{k}$. (Why?) Setting $x_j = x$, $y_j = y$, $j = 1, 2, \ldots, n$, the *binomial formula* (*) follows at once. As for the *multinomial formula* (**), it can be deduced from the binomial formula (*) using induction and the identity

$$\binom{n}{n_1} \frac{(n - n_1)!}{n_2! n_3! \cdots n_k!} = \frac{n!}{n_1! n_2! \cdots n_k!}.$$

We leave the details as an exercise for the reader.   □

Remark 1.3.31. The binomial and multinomial formulas are valid in *any field.* Thus the numbers x and y in (*) and the x_i, $1 \le i \le k$ in (**) may be *complex.*

The following proposition has numerous applications in counting problems and will be needed in Exercises 1.3.35 and 1.3.36 below. For the following and other proofs and applications we refer to [PSz72].

Proposition 1.3.32 (Inclusion–Exclusion Principle). *Suppose we have a set of N objects. Let N_j be the number of those objects that have the property P_j, $1 \le j \le n$, N_{jk} the number of those having simultaneously the properties P_j and P_k, $1 \le j, k \le n, \ldots$, and $N_{123\cdots n}$ the number of objects having simultaneously all the properties $P_1, P_2, \ldots, P_n$. Then the number N_0 of those objects having none of the properties $P_1, P_2, \ldots, P_n$ is given by*

$$N_0 = N - \sum_i N_i + \sum_{i<j} N_{ij} - \sum_{i<j<k} N_{ijk} + \cdots + (-1)^n N_{123\cdots n}.$$

Proof. Let U be the set of N objects and let A_j be the subset of those objects that satisfy the property P_j, $j = 1, 2, \ldots, n$. Similarly, let $A_{ij} := A_i \cap A_j$, $A_{ijk} := A_i \cap A_j \cap A_k, \ldots$, be the subsets of all objects that simultaneously satisfy the two properties P_i and P_j, the three properties P_i, P_j, and P_k, etc. Now the set A_0 of the objects that satisfy *none* of the properties $P_1, P_2, \ldots, P_n$ can be written as $A_0 = A_1^c \cap A_2^c \cap \cdots \cap A_n^c$, and by Exercise 1.2.55, we have

$$\chi_{A_0} = (1 - \chi_{A_1})(1 - \chi_{A_2}) \cdots (1 - \chi_{A_n})$$

$$= 1 - \sum_i \chi_{A_i} + \sum_{i<j} \chi_{A_i} \chi_{A_j} - \cdots + (-1)^n \chi_{A_1} \chi_{A_2} \cdots \chi_{A_n}$$

$$= 1 - \sum_i \chi_{A_i} + \sum_{i<j} \chi_{A_{ij}} - \sum_{i<j<k} \chi_{A_{ijk}} + \cdots + (-1)^n \chi_{A_{12\cdots n}}.$$

Summing the (values of the) two sides over all $x \in U$ and noting that $|A| = \sum_{x \in U} \chi_A(x)$ [cf. Exercise 1.2.55(g)], the proposition follows.   □

Exercise 1.3.33. Extend part (h) of Exercise 1.2.55 to the case of n finite sets, for any $n \geq 2$. In other words, if $A_1, A_2, \ldots, A_n$ are finite sets, find $|A_1 \cup A_2 \cup \cdots \cup A_n|$ in terms of the number of elements in various intersections of the A_i, $1 \leq i \leq n$.

Notation 1.3.34 (P_n^m, S_n^m). Let X and Y be two finite sets with m and n elements $(1 \leq n \leq m)$, respectively. The number of all *partitions* of X into n (pairwise disjoint nonempty) subsets (with union X) will be denoted by P_n^m. The number of all *surjective* (i.e., onto) functions from X onto Y will be denoted by S_n^m. Note that we obviously have $P_1^m = 1 = P_m^m$ and $S_1^m = 1$. Also, $S_m^m = P(m) = m!$. (Why?)

Exercise 1.3.35. Let X and Y be finite sets with m and n elements $(1 \leq n \leq m)$, respectively, and let P_n^m and S_n^m be defined as above. Show that

(a) $P_n^{m+1} = nP_n^m + P_{n-1}^m$,
(b) $S_n^{m+1} = nS_n^m + nS_{n-1}^m$,
(c) $S_n^m = n!P_n^m$.

Hints: How are the numbers P_n^m and S_n^m affected if we adjoin an element to X? Also, for (c), note that for any surjection $f : X \to Y$, $\{f^{-1}(y) : y \in Y\}$ is a partition of X.

Exercise 1.3.36. Let X and Y be finite sets with $|X| = m$, $|Y| = n$.

(a) Using the Basic Counting Principle, find $|Y^X|$. Use the special case $|2^X|$ to find $|\mathcal{P}(X)|$.
(b) Find the number of *one-to-one* maps from X to Y.
(c) Show that the number of *surjective* (i.e., onto) maps from X onto Y is given by

$$S_n^m = n^m - \binom{n}{1}(n-1)^m + \binom{n}{2}(n-2)^m - \cdots + (-1)^n 0^m.$$

(d) Find a formula for P_n^m.

Hint for (c): Let $Y = \{y_1, y_2, \ldots, y_n\}$, and for each $j = 1, 2, \ldots, n$, define, for a function $f \in Y^X$, the property P_j by "$y_j \notin f(X)$." Now use Proposition 1.3.32.

For our next application of Proposition 1.3.32, we need a few elementary facts about integers. The proofs can be found in any textbook on abstract algebra (e.g., Herstein's *Topics in Algebra* [Her75]).

Proposition 1.3.37 (Division Algorithm). *Let a and b be integers and $a \neq 0$. Then there are unique integers q and r, with $0 \leq r < |a|$, such that $b = aq + r$. We call q the quotient and r the remainder of the division of b by a. If $r = 0$, then we say that a is a divisor of b, or that a divides b, and write $a|b$.*

Definition 1.3.38 (Greatest Common Divisor, Relatively Prime). The *greatest common divisor* of two integers a and b, not both zero, is the *largest positive* integer that divides *both* a and b. It is denoted by $\gcd(a, b)$. In details, $d = \gcd(a, b)$ if the following two conditions are satisfied:

(i) $d > 0$, $d \mid a$, and $d \mid b$,

(ii) $c > 0$, $c \mid a$, and $c \mid b$, which imply $c \mid d$.

We say that two integers a and b are *relatively prime* if $\gcd(a, b) = 1$.

Proposition 1.3.39. *Let a and b be two integers, not both zero. One can then find integers m and n such that $\gcd(a, b) = ma + nb$.*

Corollary 1.3.40. *If a and b are relatively prime, one can find integers m and n such that $ma + nb = 1$.*

Corollary 1.3.41. *Let a, b, and c be nonzero integers such that $\gcd(a, b) = 1$. Then $a \mid bc$ implies $a \mid c$.*

Corollary 1.3.42. *Let a and b be nonzero integers with $\gcd(a, b) = 1$. Then*

$$a \mid c \quad \text{and} \quad b \mid c \quad \Longleftrightarrow \quad ab \mid c.$$

Corollary 1.3.43. *Let $a_1, a_2, \ldots, a_k$ be pairwise relatively prime, nonzero integers. Then we have*

$$a_i \mid b, \quad i = 1, 2, \ldots, n \quad \Longleftrightarrow \quad a_1 a_2 \cdots a_k \mid b.$$

Corollary 1.3.44. *Let p be a prime number (i.e., an integer $p \geq 2$ whose only positive divisors are 1 and p). Then, for any nonzero integers a and b,*

$$p \mid ab \implies p \mid a \quad \text{or} \quad p \mid b.$$

Corollary 1.3.45 (Prime Factorization). *Let $n > 1$ be an integer. Then there are unique primes $p_1, p_2, \ldots, p_m$ and unique positive integers $r_1, r_2, \ldots, r_m$, such that $p_1 > p_2 > \cdots > p_m$ and*

$$n = p_1^{r_1} p_2^{r_2} \cdots p_m^{r_m}.$$

Some Hints for the Proofs. For the proof of the division algorithm, let $r :=$ $\min\{m \geq 0 : (\exists k \in \mathbb{Z}) \, (b = ak + m)\}$, and define q accordingly. For Proposition 1.3.39, show that $\gcd(a, b) = \min\{d > 0 : (\exists m, n \in \mathbb{Z})(d = ma + nb)\}$. For Corollaries 1.3.40 and 1.3.41, pick $m, n \in \mathbb{Z}$, with $ma + nb = 1$, and use $mac + nbc = c$. For Corollary 1.3.43, use induction. For Corollary 1.3.44, note that, if $p \nmid a$, then $\gcd(a, p) = 1$. Finally, for Corollary 1.3.45, use (strong) induction.

Definition 1.3.46 (Euler's Phi-Function). For each $n \in \mathbb{N}$, we define $\phi(n)$ to be the number of positive integers less than or equal to n that are *relatively prime* to n.

Exercise 1.3.47. Let $a_1, a_2, \ldots, a_m$ be pairwise relatively prime positive integers and let $n > 1$ be any integer.

(a) Show that the number of integers in $\{1, 2, \ldots, n\}$ that are divisible by a_i, $1 \leq i \leq m$, is $\left[\frac{n}{a_i}\right]$, where $[r]$ is the greatest integer less than or equal to r.
(b) More generally, show that the number of integers in $\{1, 2, \ldots, n\}$ that are divisible by $a_1, a_2, \ldots, a_k$, $1 \leq k \leq m$, is $\left[\frac{n}{a_1 a_2 \cdots a_k}\right]$.
(c) Show that the number n_0 of integers in $\{1, 2, \ldots, n\}$ that are *not* divisible by any of the integers $a_1, a_2, \ldots, a_m$ is

$$n_0 = n - \sum_i \left[\frac{n}{a_i}\right] + \sum_{i<j} \left[\frac{n}{a_i a_j}\right] + \cdots + (-1)^m \left[\frac{n}{a_1 a_2 \cdots a_m}\right].$$

(d) Using (c), show that if $n = p_1^{r_1} p_2^{r_2} \cdots p_m^{r_m}$ is the prime factorization of n, as in Corollary 1.3.45, then we have

$$\phi(n) = n\left(1 - \frac{1}{p_1}\right)\left(1 - \frac{1}{p_2}\right) \cdots \left(1 - \frac{1}{p_m}\right).$$

1.4   Infinite Direct Products, Axiom of Choice, and Cardinal Numbers

In this section we shall introduce *infinite direct* (Cartesian) products and cardinal numbers. Unlike finite sets, infinite sets have properties that are quite surprising and are fundamentally different. Our definition of *cardinal number*, which extends the idea of *number of elements* to arbitrary sets, will be vague, although more precise definitions can be given. For instance, one way to define cardinal numbers is to use the concept of *ordinal* numbers. However, the introduction of ordinal numbers requires additional sections and we prefer, instead, to send the reader to the books mentioned in the introduction, particularly Halmos's *Naive Set Theory* [Hal60], where details on these and other topics can be found.

Definition 1.4.1 (Direct Product, Choice Function). Let Λ be a (nonempty) *index set*, and let X_λ be a set for each $\lambda \in \Lambda$. The *direct product* (also called *Cartesian product*) of the sets X_λ, denoted by $\prod_{\lambda \in \Lambda} X_\lambda$, is the set of all functions $x : \Lambda \to \bigcup_{\lambda \in \Lambda} X_\lambda$ such that $x(\lambda) \in X_\lambda$ for each $\lambda \in \Lambda$. Each such function x is called a *choice function* for the family $\{X_\lambda\}_{\lambda \in \Lambda}$ and the element $x_\lambda := x(\lambda) \in X_\lambda$ is called the λth *coordinate* of x.

Remarks and Notation 1.4.2. It follows from the definition that we have $\prod_{\lambda \in \Lambda} X_\lambda = \emptyset$ if $X_\lambda = \emptyset$ for some $\lambda \in \Lambda$. If $X_\lambda = X$ for all $\lambda \in \Lambda$, then we write $\prod_{\lambda \in \Lambda} X_\lambda = X^\Lambda$. In other words, we obtain the set of all *functions* from the index set Λ to the set X. If $\Lambda = \mathbb{N}$, then $X^\Lambda = X^\mathbb{N}$ is the set of all *sequences* in X. Finally, if $\Lambda = \{1, 2, \ldots, n\}$ for some $n \in \mathbb{N}$, then we write

$\prod_{\lambda \in \Lambda} X_\lambda = \prod_{k=1}^{n} X_k = X_1 \times X_2 \times \cdots \times X_n$; i.e., we obtain the *finite* Cartesian product defined earlier. In the latter case, if $X_k = X$ for $1 \leq k \leq n$, then we use the notation $X^n := X^\Lambda = X^{\{1, 2, \ldots, n\}}$.

Example 1.4.3. The set $\mathbb{R}^n = \{(x_1, x_2, \ldots, x_n) : x_k \in \mathbb{R}\}$ is called the *Euclidean n-space*, and the set $\mathbb{C}^n = \{(z_1, z_2, \ldots, z_n) : z_k \in \mathbb{C}\}$ is called the *unitary n-space*.

Axiom of Choice. Let $\{X_\lambda\}_{\lambda \in \Lambda}$ *be a family of sets indexed by a nonempty set* Λ. *If* $X_\lambda \neq \emptyset$, *for all* $\lambda \in \Lambda$, *then the direct product* $\prod_{\lambda \in \Lambda} X_\lambda$ *is nonempty. In other words, the family has at least one choice function.*

Remark 1.4.4.

(a) As a special case, the *Axiom of Choice* may be used for the family $\mathcal{P}(U) \setminus \{\emptyset\}$ of all nonempty subsets of a nonempty set U. In this case, a choice function *chooses* an element from each nonempty subset of U, guaranteeing the possibility of simultaneously choosing elements from a (*possibly infinite*) collection of sets.

(b) The *Axiom of Choice* is logically equivalent to the *Well-Ordering Theorem* which, in turn, is logically equivalent to *Zorn's Lemma*.

Definition 1.4.5 (Equivalent Sets). Two sets S and T are called *equivalent* (also called *equipotent* or *equipollent*) if there is a *one-to-one correspondence* between them. This equivalence will be denoted by $S \sim T$.

Exercise 1.4.6.

(a) Show that the set equivalence $S \sim T$ defined above is indeed an equivalence relation on $\mathcal{P}(U)$ for each fixed set U.

(b) Show that $\mathbb{N} \sim 2\mathbb{N}$ and that $\mathbb{N} \sim 2\mathbb{N} - 1$, where $2\mathbb{N} := \{2k : k \in \mathbb{N}\}$ and $2\mathbb{N} - 1 := \{2k - 1 : k \in \mathbb{N}\}$ are the subsets of *even* and *odd* positive integers, respectively.

(c) Let A and B be sets, $a \in A$ and $b \in B$. Show that $A \sim B$ if and only if $A \setminus \{a\} \sim B \setminus \{b\}$.

Definition 1.4.7 (Cardinal Number). To each set, X, we associate a symbol, $|X|$, called the *cardinal number* (also called *cardinality*) of X in such a way that $|X| = |Y|$ if and only if $X \sim Y$.

Notation 1.4.8 ($\aleph_0$, c). We define $|\emptyset| := 0$ and $|\{1, 2, \ldots, n\}| := n$, for each $n \in \mathbb{N}$. Thus, if $X \sim \{1, 2, \ldots, n\}$, then $|X| := n$; i.e., $|X|$ is the number of elements in X. The cardinality of the set $\mathbb{N}$ of all natural numbers is denoted by $|\mathbb{N}| = \aleph_0$ [read "aleph naught"]. Finally, we write $|\mathbb{R}| = \mathfrak{c}$ [read "continuum"] for the cardinality of the set of real numbers.

Definition 1.4.9 (Countable and Uncountable Sets). We say that a set X is *countable* if either X is finite or $X \sim \mathbb{N}$; i.e., either $|X| = n$ for some $n \in \mathbb{N}_0$ or $|X| = \aleph_0$. In the latter case X is also called *countably infinite* (or *denumerable*) and any one-to-one correspondence $x : \mathbb{N} \to X$ is called an *enumeration* of X. A set that is not countable is called *uncountable*.

Exercise 1.4.10.

(i) Show that $\mathbb{N}^2 = \mathbb{N} \times \mathbb{N}$ is countably infinite by showing that the map f : $\mathbb{N} \times \mathbb{N} \to \mathbb{N}$ defined by $f(m,n) := 2^{m-1}(2n-1)$ is a one-to-one correspondence. Deduce that if D is countable, then so is $D \times D$ (and, inductively, D^k, $k \in \mathbb{N}$).

(ii) Show that $\mathbb{Z} \sim \mathbb{N}$ by defining (explicitly) a bijection $g : \mathbb{Z} \to \mathbb{N}$.

Our first application of the Axiom of Choice is in the proof of the following proposition, even though our informal proof does not mention the axiom explicitly. The proof can, of course, be made precise by defining the appropriate choice function.

Proposition 1.4.11. *Every infinite set contains a countably infinite subset.*

Proof. Let X be an infinite set. Then, in particular, $X \neq \emptyset$, and we can pick an element $x_1 \in X$. Since X is infinite, $X \setminus \{x_1\} \neq \emptyset$, and we can pick an element $x_2 \in X \setminus \{x_1\}$. Using once again the fact that X is infinite, we have that $X \setminus \{x_1, x_2\} \neq \emptyset$ and hence contains an element x_3. Continuing this process indefinitely (this is where the Axiom of Choice is used), we obtain the countably infinite subset $\{x_1, x_2, x_3, \ldots\} \subset X$. $\square$

Proposition 1.4.12. *Let D be an infinite subset of $\mathbb{N}$. Then D is countably infinite. In fact, there is a unique enumeration $d : \mathbb{N} \to D$ of D such that*

$$d_1 < d_2 < \cdots < d_n < d_{n+1} < \cdots,$$

where $d_n := d(n)\ \forall n \in \mathbb{N}$.

Proof. Let $d_1 = \min(D)$ and define, inductively, $d_{n+1} := \min(D \setminus \{d_1, \ldots, d_n\})$. The map $n \mapsto d_n$ is the desired enumeration of D. $\square$

Corollary 1.4.13. *If Y is a countable set and if $X \subset Y$, then X is countable.*

Exercise 1.4.14. Prove the corollary.

Here is a couple of other characterizations of countable sets:

Proposition 1.4.15. *A set X is countable if and only if there is an injective map $g : X \to \mathbb{N}$ if and only if there is a surjective map $f : \mathbb{N} \to X$.*

Proof. If X is countable, the very definition of *countability* implies that there exists a *one-to-one* map $g : X \to \mathbb{N}$. Conversely, if g exists, then $g(X) \subset \mathbb{N}$ is countable (by Corollary 1.4.13) and hence so is $X \sim g(X)$. Next, if $g : X \to \mathbb{N}$ is *injective*, then we pick a fixed $x_0 \in X$ and define a surjective map $f : \mathbb{N} \to X$ by setting $f(n) = x$ if $n = g(x)$ and $f(n) = x_0$ if $n \notin g(X)$. Conversely, if $f : \mathbb{N} \to X$ is *surjective*, then for each $x \in X$ we can pick an integer $n_x \in f^{-1}(x)$, getting $f(n_x) = x$. We now define the map $g : X \to \mathbb{N}$ by $g(x) := n_x$. Since g is one-to-one (why?), the proof is complete. $\square$

Exercise 1.4.16. Let $A_k = 2^{k-1} \cdot \{1, 3, 5, \ldots\} := \{2^{k-1} \cdot 1, 2^{k-1} \cdot 3, 2^{k-1} \cdot 5, \ldots\}$, $k \in \mathbb{N}$. Show that $\{A_k\}$ is a *partition* of $\mathbb{N}$, i.e., that $A_j \cap A_k = \emptyset$ if $j \neq k$ and that $\mathbb{N} = \bigcup_{k=1}^{\infty} A_k$. Also show that $A_k \sim \mathbb{N}$ for all $k \in \mathbb{N}$. *Hint:* See Exercise 1.4.10.

Proposition 1.4.17. *A countable union of countable sets is countable.*

Proof. Let X_k be a countable set for each $k \in \mathbb{N}$, and let the A_k be as in Exercise 1.4.16. Now, by Proposition 1.4.15, we can find *surjective* (i.e., onto) maps $f_k : A_k \to X_k$ $\forall k \in \mathbb{N}$. Define the map $f : \mathbb{N} \to \bigcup_{k=1}^{\infty} X_k$ by $f(n) := f_k(n)$ if $n \in A_k$. It follows easily that f is *onto*, and another application of Proposition 1.4.15 completes the proof. $\square$

Corollary 1.4.18. *The set $\mathbb{Z}$ of all integers and the set $\mathbb{Q}$ of all rational numbers are countably infinite.*

Proof. For the set $\mathbb{Z}$, note that $\mathbb{Z} = \mathbb{N} \cup \{0\} \cup (-\mathbb{N})$, where $-\mathbb{N} := \{-1, -2, -3, \ldots\}$. For $\mathbb{Q} = \{m/n : m \in \mathbb{Z}, n \in \mathbb{N}\}$, define, for each $k \in \mathbb{Z}$, the set $\mathbb{Q}_k := \{k/n : n \in \mathbb{N}\} \subset \mathbb{Q}$. Then each $\mathbb{Q}_k$ is countably infinite, and $\mathbb{Q} = \bigcup_{k \in \mathbb{Z}} \mathbb{Q}_k$. $\square$

Definition 1.4.19 (Domination). Given two sets X and Y, we say that Y *dominates* X, and write $X \preceq Y$, if there is a *one-to-one* map from X into Y. If $X \preceq Y$, then we write $|X| \leq |Y|$. We also write $|X| < |Y|$ if $X \prec Y$, i.e., if $X \preceq Y$, but $X \not\sim Y$. In the latter case we say that Y *strictly dominates* X.

Remark 1.4.20.

(a) If $f : X \to Y$ is one-to-one, then $X \sim f(X)$. Therefore, an equivalent definition of domination is the following:

$$X \preceq Y \quad \Longleftrightarrow \quad X \sim Y_1 \quad \text{for some} \quad Y_1 \subset Y.$$

(b) We can define the *countability* of sets in terms of *set domination* as follows:

$$X \quad \text{is countable} \quad \Longleftrightarrow \quad X \preceq \mathbb{N} \quad \Longleftrightarrow \quad |X| \leq \aleph_0.$$

The relation "$\leq$" between the cardinal numbers of two subsets of a universal set U (or, equivalently, the domination relation between the subsets themselves) is easily seen to be *reflexive* and *transitive* (check it!). Therefore, to prove that it is in fact a *partial ordering* on the cardinalities of all subsets of a fixed given set, all we need is to show that it is *antisymmetric*. That this is indeed the case is a consequence of the following important theorem.

Theorem 1.4.21 (Schröder–Bernstein). *Let X and Y be two sets, and suppose that there are one-to-one functions $f : X \to Y$ and $g : Y \to X$. Then $X \sim Y$. In other words, $|X| \leq |Y|$ and $|Y| \leq |X|$ imply $|X| = |Y|$.*

Proof. Define the function $\phi : \mathcal{P}(X) \to \mathcal{P}(X)$ by

$$\phi(S) := X \setminus g(Y \setminus f(S)) \qquad (\forall S \in \mathcal{P}(X)), \qquad (*)$$

and note that we have

$$S \subset T \subset X \implies \phi(S) \subset \phi(T). \qquad (**)$$

Indeed, $S \subset T \Rightarrow f(S) \subset f(T) \Rightarrow Y \setminus f(T) \subset Y \setminus f(S)$. Thus $g(Y \setminus f(T)) \subset g(Y \setminus f(S))$, from which $(**)$ follows at once. Now let $\boldsymbol{S} := \{S \in \mathcal{P}(X) : S \subset \phi(S)\}$ and note that $\emptyset \in \boldsymbol{S}$. If $Z := \bigcup_{S \in \boldsymbol{S}} S$, then for each $S \in \boldsymbol{S}$ we have $S \subset Z$ and $(**)$ implies $S \subset \phi(S) \subset \phi(Z)$. Thus $Z \subset \phi(Z)$ and another application of $(**)$ gives $\phi(Z) \subset \phi(\phi(Z))$, which implies $\phi(Z) \in \boldsymbol{S}$. But then, $\phi(Z) \subset Z$ and we get $\phi(Z) = Z$. Therefore, by $(*)$, we have $Z = X \setminus g(Y \setminus f(Z))$ and hence

$$X \setminus Z = g(Y \setminus f(Z)). \qquad (\dagger)$$

Using $(\dagger)$, it is now obvious that the function

$$h(x) := \begin{cases} f(x) & \text{if } x \in Z, \\ g^{-1}(x) & \text{if } x \in X \setminus Z \end{cases}$$

is a bijection of X onto Y. □

Exercise 1.4.22.

(a) Show that any two cardinal numbers are *comparable*; i.e., for any sets X and Y, we have $|X| \leq |Y|$ or $|Y| \leq |X|$. *Hint:* Consider the set $\mathcal{F}$ of all *injective* maps f with $\mathrm{dom}(f) \subset X$ and $\mathrm{ran}(f) \subset Y$. Partially order $\mathcal{F}$ by "inclusion" and find a *maximal* element $h \in \mathcal{F}$. Show that we must have $\mathrm{dom}(h) = X$ or $\mathrm{ran}(h) = Y$.

(b) Show that, if X is a *proper* subset of a *finite* set Y, then $X \not\sim Y$ (i.e., $X \prec Y$). Deduce that $\mathbb{N}$ is *infinite*. *Hint:* Put $Y = \{1, 2, \ldots, n\}$ and proceed by induction on n. Assuming the case n, let $Y = \{1, 2, \ldots, n+1\}$ and suppose that there is an injection f from Y onto a *proper* subset $X \subset Y$. Show that both $n + 1 \notin X$ and $n + 1 \in X$ result in contradictions.

(c) Show that, if X is *infinite* and if C is *countable*, then $|X \cup C| = |X|$.

(d) Show that X is *infinite* if and only if $\aleph_0 \leq |X|$.

(e) Show that, if $|X| \leq |Y|$ and $|Y| \leq |Z|$, and if at least one of these is a *strict* inequality, then $|X| < |Z|$.

(f) Show that X is *finite* if and only if $|X| < \aleph_0$.

The reader may have observed that we still have not proved the existence of *uncountable* sets. That such sets indeed exist is a consequence of the following theorem of Georg Cantor.

Theorem 1.4.23 (Cantor). *For any set* X, *we have* $X \prec \mathcal{P}(X) \sim 2^X$. *In other words,* $|X| < |\mathcal{P}(X)| = |2^X|$.

Proof. First, note that $\mathcal{P}(X) \sim 2^X$ follows from Exercise 1.2.54. Next, note that the map $f : X \to \mathcal{P}(X)$ defined by $f(x) := \{x\} \ \forall x \in X$ is *one-to-one*, so that we have $X \preceq \mathcal{P}(X)$. To show that we have a *strict* domination, we must show that there are no maps from X *onto* $\mathcal{P}(X)$. Suppose that $g : X \to \mathcal{P}(X)$ is onto, and let S be the set $\{x \in X : x \notin g(x)\} \subset X$. Since g is *onto*, there exists $\xi \in X$ with $g(\xi) = S$. Now if $\xi \in g(\xi)$, then the definition of S implies $\xi \notin g(\xi)$, and if $\xi \notin g(\xi) = S$, then, once again, the definition of S implies that $\xi \in g(\xi) = S$. In other words, in either case we reach a contradiction, and the theorem is proved. $\square$

Corollary 1.4.24. *The set* $\mathcal{P}(\mathbb{N}) \sim 2^{\mathbb{N}}$ *is uncountable. In other words, the set of all sequences* $(x_1, x_2, x_3, \ldots)$, *where* $x_n \in \{0, 1\}$ *for each* $n \in \mathbb{N}$, *is uncountable.*

Remark 1.4.25.

1. In fact, one can prove that $|\mathcal{P}(\mathbb{N})| = |2^{\mathbb{N}}| = |\mathbb{R}| = \mathfrak{c}$. In particular, the set $\mathbb{R}$ of real numbers is uncountable. We will return to this fact (and its proof) after the axiomatic definition of real numbers.

2. One can define an *arithmetic* on cardinal numbers as follows: Given two *disjoint* sets X and Y, the *sum* of the cardinal numbers $|X|$ and $|Y|$ is defined by $|X| + |Y| := |X \cup Y|$. If X and Y are *not* disjoint, one considers the disjoint sets $X' = X \times \{0\}$ and $Y' = Y \times \{1\}$ and defines $|X| + |Y| := |X' \cup Y'|$. For any sets X and Y, not necessarily disjoint, the *product* of their cardinal numbers is defined by $|X||Y| := |X \times Y|$. One can also define *exponentiation* by $|X|^{|Y|} := |X^Y|$. It can be proved that these operations have the properties satisfied by the corresponding *numerical* operations. Checking these properties is a rewarding exercise, and we encourage the reader to try some of them or to consult the references for details. The following result will be needed later.

Exercise 1.4.26. Show that, if A is an *infinite* set, then $|A \times \mathbb{N}| = |A|$. *Hint:* Let $\mathcal{F}$ denote the set of all *bijective* maps $f : S \times \mathbb{N} \to S$, where $S \subset A$. Since $|\mathbb{N} \times \mathbb{N}| = |\mathbb{N}|$, we have $\mathcal{F} \neq \emptyset$. (Why?) Show that *Zorn's Lemma* can be applied in $\mathcal{F}$ to produce a *maximal* bijection $h : B \times \mathbb{N} \to B$, with $B \subset A$, and that we must have $B = A$, by examining the cases where $S \setminus B$ is finite or infinite.

1.5 Problems

1. Show that, for any subsets A, B of a universal set U, we have

$$A \subset B \iff A^c \supset B^c \iff A \cap B = A \iff A \cup B = B.$$

2. Show that, for any sets A, B, and C, we have

(a) $A \setminus (A \setminus B) = A \cap B$,   (b) $A \cap (B \setminus C) = (A \cap B) \setminus C$.

3. Show that, for any sets A and B, there is a set C such that $A \triangle C = B$. Is C *unique?*

4. For any sets A, B, and C, show that

$$A \cap (B \triangle C) = (A \cap B) \triangle (A \cap C).$$

5. Given two sets A and B, show that

$$A \cup B = A \triangle B \triangle (A \cap B) \quad \text{and} \quad A \setminus B = A \triangle (A \cap B).$$

Deduce that, if $A \cap B = \emptyset$, then $A \cup B = A \triangle B$.

6. Show that, given two sets A and B, we have

$$A = \emptyset \iff B = A \triangle B.$$

7. Let U be an *infinite* set and let $\mathcal{C} \subset \mathcal{P}(U)$ be the collection of all *countable* subsets of U and their *complements*. Show that $\mathcal{C}$ is a σ-algebra. Let $\mathcal{F}$ denote the set of all *finite* subsets of U and their *complements*. Is $\mathcal{F}$ a σ-algebra?

8. Let U be a nonempty set and $\mathcal{A} \subset \mathcal{P}(U)$ a σ-algebra. Show that, for any set $S \subset U$, the collection $\{A \cap S : A \in \mathcal{A}\}$ is a σ-algebra of subsets of S.

9. Let $(A_n)_{n=1}^\infty$ be a *partition* of a (nonempty) set U; i.e., the A_n are nonempty, pairwise disjoint, and $\bigcup A_n = U$. Show that the set of all unions of the A_n (including the "empty union" which we *define* to be $\emptyset$) is a σ-algebra.

10. Let U be a nonempty set. Show that, given any family $(R_j)_{j \in J}$ of *equivalence relations* on U, the intersection $R := \bigcap_{j \in J} R_j$ is also an equivalence relation on U. Give an example of two equivalence relations on a set U whose *union* is *not* an equivalence relation.

11. Let R and S be two equivalence relations on a set U. Show that $R \circ S$ is an equivalence relation on U if and only if $R \circ S = S \circ R$ and that, in this case, $R \circ S$ is the intersection of all the equivalence relations on U that contain *both* R and S.

12. Let U be a partially ordered set. Show that we can write $U = S \cup T$, with $S \cap T = \emptyset$, such that S is *well ordered* (with respect to the ordering in U) and T has *no* least element. *Hint:* Look at the union of all subsets of U that have no least element.

13. Using induction, prove the following statements for all $n \in \mathbb{N}$:

$$\sum_{k=1}^n k = \frac{n(n+1)}{2}, \qquad \sum_{k=1}^n k^2 = \frac{n(n+1)(2n+1)}{6},$$

$$\sum_{k=1}^n k^3 = \frac{n^2(n+1)^2}{4}, \qquad \prod_{k=1}^n \left(1 + \frac{1}{k}\right)^k = \frac{(n+1)^n}{n!}.$$

14. Let $f : S \to T$ be a function and $B \subset T$. Show that $f(f^{-1}(B)) = B \cap f(S)$. If $A \subset S$, show that $(f|A)^{-1}(B) = A \cap f^{-1}(B)$, where $f|A$ denotes the *restriction* of f to A.

15. Let $\preceq$ be a *well ordering* on a set X and let $f : X \to X$ be a *permutation* (i.e., a bijection of X onto itself). Show that, if f is *order preserving* (i.e., $f(x) \preceq f(y)$ whenever $x \preceq y$), then $f = \mathrm{id}_X$.

16. Let U be a nonempty set. Show that a *ring* $\mathcal{R} \subset \mathcal{P}(U)$ (resp., an *algebra* $\mathcal{A} \subset \mathcal{P}(U)$) is indeed a ring (resp., a ring with unit element) in the *algebraic sense* [cf. Definition 1.3.7 and Remark 1.3.8(c)] under the operations of addition and multiplication defined by

$$A + B := A \triangle B, \quad \text{and} \quad AB := A \cap B.$$

17. Show that, if F is a *field*, then the only ideals in F are $\{0\}$ and F.

18 (Maximal Ideal). Let R be a commutative ring with unit element $1 \neq 0$. Show that R has at least one *maximal ideal*, i.e., an ideal $M \subset R$ such that there is no ideal $N \subset R$ satisfying the *proper* inclusions $M \subset N \subset R$. *Hint:* Consider the set $\mathcal{I}$ of all ideals $I \subsetneq R$ and note that $\{0\} \in \mathcal{I}$. Partially order $\mathcal{I}$ by *inclusion* and show that, if $(I_\alpha)_{\alpha \in A}$ is a *chain* in $\mathcal{I}$, then $\bigcup_{\alpha \in A} I_\alpha \in \mathcal{I}$. Now use *Zorn's Lemma*.

19. Let R be a commutative ring with unit element $1 \neq 0$ and let $M \subset R$ be a *maximal ideal*. For any $x, y \in R$, let us write $x \sim y$ if $x - y \in M$. Show that $\sim$ is an *equivalence* relation on R and that, for each $x \in R$, its equivalence class is $[x] = x + M := \{x + m : m \in M\}$. Now define $[x] + [y] := [x + y]$ and $[x] \cdot [y] := [xy]$ for any $x, y \in R$. Show that these are *well-defined* binary operations on the quotient set $R/M := R/\sim$ (i.e., if $x \sim x'$ and $y \sim y'$, then $[x] + [y] = [x'] + [y']$ and $[x][y] = [x'][y']$). Finally, show that (with these operations) R/M is a *field*.

20. For each $n \in \mathbb{N}$, prove the identities

$$\sum_{k=0}^{n} \binom{n}{k} = 2^n \quad \text{and} \quad \sum_{k=0}^{n} (-1)^k \binom{n}{k} = 0.$$

Deduce from the latter that, for any (nonempty) *finite* set, the number of *even size* subsets equals the number of *odd size* ones.

21. For any $m, n \in \mathbb{N}$, apply the *binomial formula* to $(1 + x)^m (1 + x)^n$ to prove the identity

$$\sum_{j=0}^{k} \binom{m}{j} \binom{n}{k - j} = \binom{m + n}{k} \quad (0 \leq k \leq m + n),$$

where we define $\binom{\ell}{i} := 0$, for $i > \ell$. Deduce that

$$\sum_{j=0}^{n} \binom{n}{j}^2 = \binom{2n}{n} \quad (\forall n \in \mathbb{N}).$$

22. Given any $n \in \mathbb{N}$, show that the number of ordered pairs (i, j) of integers with $1 \leq i \leq j \leq n$ (resp., $1 \leq i < j \leq n$) is $n(n + 1)/2$ (resp., $n(n - 1)/2$).

23. For any integers $0 \leq n \leq m$, prove the identities

$$\sum_{k=0}^{n} \binom{m}{k} \binom{m - k}{n - k} = 2^n \binom{m}{n} \quad \text{and} \quad \sum_{k=0}^{n} (-1)^k \binom{m}{k} \binom{m - k}{n - k} = 0.$$

Hint: Given a set of size m, look at the subsets of size n that contain a *given* subset of size k, $0 \leq k \leq n$.

24. Let A and B be *totally ordered* sets with m and n elements, respectively. How many *strictly increasing* functions are there from A to B?

25. Show that, for any integers $1 \le n \le m$, we have

$$n^m = S_n^m + \binom{n}{1} S_{n-1}^n + \binom{n}{2} S_{n-2}^n + \cdots + \binom{n}{n-1}.$$

Hint: Let A and B have m and n elements, respectively. Look at the number of maps $f \in B^A$ whose ranges contain all but one element, two elements, etc. in B.

26. How many equivalence relations are there on a set U with n elements?

27. Given any m, $n \in \mathbb{N}$ with $\gcd(m, n) = 1$, show that $\phi(mn) = \phi(m)\phi(n)$, where ϕ is Euler's Phi-Function (cf. Definition 1.3.46 and Exercise 1.3.47).

28. Show that, if $\mathcal{A}$ is a σ-algebra containing an *infinite* number of sets, then this (cardinal) number is *uncountable*. *Hint:* Start by showing that $\mathcal{A}$ contains a sequence $(A_n)_{n=1}^\infty$ of (nonempty) *pairwise disjoint* sets and use Problem 9.

29. For each set S, let $\mathcal{F}_S$ denote the set of all *finite* subsets of S. Show that, if S is *countably infinite*, then $|\mathcal{F}_S| = |S|$. Actually, this holds for all *uncountable* sets S as well, but the proof is harder (cf. Problem 35 below).

30. Let A and B be nonempty sets. Show that if there is a *surjective* (i.e., onto) map $f : A \to B$, then $|B| \le |A|$.

31.

(a) Show that a set S is *finite* if and only if each nonempty subset of $\mathcal{P}(S)$ (partially ordered by inclusion) has a *minimal element*.
(b) Show that a set S is *infinite* if and only if S is *equivalent* to some *proper* subset of itself. *Hint:* Recall that any infinite set contains a *countably infinite* subset.

32.

(a) Show that, if A is an *infinite* set, then $|A| + |A| = |A|$. Deduce that, if $|B| \le |A|$, then $|A| + |B| = |A|$. *Hint:* Show, as in Exercise 1.4.26, that $|A \times \{1, 2\}| = |A|$.
(b) Show that, if A is an infinite set and (A_n) is a (finite or infinite) sequence of pairwise disjoint sets with $A_n \sim A$ for all n, then $\sum_n |A_n| := |\bigcup_n A_n| = |A|$. *Hint:* As in part (a), show that $|A \times \{1, \ldots, k\}| = |A|$, for all $k \in \mathbb{N}$.

33.

(a) Let J be an infinite index set and let $\{A_j : j \in J\}$ be a family of infinite sets such that $A_j \sim A$ for all $j \in J$ and a set A. Show that we have

$$|A| \le \Big| \bigcup_{j \in J} A_j \Big| \le |J \times A|.$$

Deduce, in particular, that if $A \sim \mathbb{N}$, then $|A| \le |J \times \mathbb{N}| = |J|$, and if $J \sim \mathbb{N}$, then $|\bigcup_{j \in J} A_j| = |A|$.

34. Extend Exercise 1.4.26 by showing that, for any infinite set A, we have $|A \times A| = |A|$. *Hint:* Show that the set $\mathcal{F}$ of all bijective maps $f : S \times S \to S$ (where $S \subset A$), partially ordered by inclusion, has a maximal bijection $h : B \times B \to B$. Now consider the cases $|A \setminus B| \le |B|$ and $|A \setminus B| > |B|$. In the latter case, pick $C \subset A \setminus B$ with $|C| = |B|$, and produce a bijection $g : (D \times D) \setminus (B \times B) \to C$, where $D := B \cup C$. Now extend h to $h \cup g : D \times D \to D$.

35. Extend Problem 29 by showing that, given *any* infinite set S, we have $|\mathcal{F}_S| = |S|$, where $\mathcal{F}_S$ denotes the set of all *finite* subsets of S.

Chapter 2
Sequences and Series of Real Numbers

The elementary theories of real-valued functions of a real variable and of numerical sequences and series are treated in any standard calculus text. In most cases, however, the proofs are given in appendices and omitted from the main body of the course. To give rigorous proofs of the basic theorems on convergence, continuity, and differentiability, one needs a precise definition of real numbers. One way to achieve this is to start with the *construction* of real numbers from the rational ones by means of *Dedekind Cuts*. We shall not follow this path. Instead, we will give a set of *axioms* for the real numbers from which all their properties can be deduced. These axioms will be divided into three categories: First, we introduce the *algebraic* ones. Next, we discuss the *order* axioms, and finally, we discuss the very deep and fundamental *Completeness* Axiom. After outlining the axiomatic definition of the real numbers, we will look at the *sequences* in $\mathbb{R}$ and their *limits*. Here, the most important concept is that of a *Cauchy sequence*. It will be used in Appendix A for a brief discussion of Cantor's construction of real numbers from the Cauchy sequences in the set $\mathbb{Q}$ of rational numbers. The properties of sequences will be used in a short section on infinite series of real numbers. We shall return to infinite series in another chapter to discuss series of functions, such as power series and Fourier series. Finally, the last section is a brief introduction to *unordered series* and *summability*. Throughout this chapter, our universal set will be $U = \mathbb{R}$, so that a *set* will automatically mean a subset of $\mathbb{R}$.

2.1 Real Numbers

The set $\mathbb{R}$ of *real* numbers (whose detailed construction is given in Appendix A) is an *ordered field*. That $\mathbb{R}$ is a *field* means that, on the set of real numbers, there are two *(binary) operations*, i.e., two maps from $\mathbb{R} \times \mathbb{R}$ to $\mathbb{R}$, denoted by "+" and "·" and called *addition* and *multiplication*, respectively, satisfying the following nine axioms. Here a, b, and c are *arbitrary* real numbers.

© Springer Science+Business Media New York 2014
H.H. Sohrab, *Basic Real Analysis*, DOI 10.1007/978-1-4939-1841-6_2

Algebraic Axioms:

(A_1) $a + b = b + a$ (commutativity of addition);
(A_2) $(a + b) + c = a + (b + c)$ (associativity of addition);
(A_3) $\exists\, 0 \in \mathbb{R}$ with $0 + a = a$ (existence of zero);
(A_4) $\exists\, -a \in \mathbb{R}$ with $a + (-a) = 0$ (existence of negative elements);
(M_1) $a \cdot b = b \cdot a$ (commutativity of multiplication);
(M_2) $(a \cdot b) \cdot c = a \cdot (b \cdot c)$ (associativity of multiplication);
(M_3) $\exists\, 1 \in \mathbb{R} \setminus \{0\}$ with $a \cdot 1 = a$ (existence of a unit element);
(M_4) $\forall\, a \in \mathbb{R} \setminus \{0\}\ \exists\, 1/a \in \mathbb{R}$ with $a \cdot (1/a) = 1$ (existence of reciprocals);
(D) $a \cdot (b + c) = a \cdot b + a \cdot c$ (distributivity of "$\cdot$" over "$+$").

Exercise 2.1.1. If a is any real number, show that

1. $a \cdot 0 = 0$;
2. $(-1) \cdot a = -a$;
3. $-(-a) = a$; and
4. $(-1)(-1) = 1$.

Exercise 2.1.2. Let a, b, $c \in \mathbb{R}$. Show that

(a) $a \neq 0 \implies 1/a \neq 0$ and $1/(1/a) = a$;
(b) $a \cdot b = a \cdot c$ and $a \neq 0 \implies b = c$; and
(c) $a \cdot b = 0 \implies a = 0$ or $b = 0$.

Notation 2.1.3. Henceforth, the product $a \cdot b$ will be denoted by ab, as long as there is no danger of confusion.

Definition 2.1.4 (Subtraction, Division, Integral Exponents). We define the binary operation "$-$" of *subtraction* by $a - b := a + (-b)\ \forall a,\ b \in \mathbb{R}$. *Division* is defined by $a/b = \frac{a}{b} = a \div b := a \cdot (1/b) = a(1/b)\ \forall a,\ b \in \mathbb{R},\ b \neq 0$. *Exponentiation* is defined as follows: For each real number a, we define $a^1 := a$, $a^2 := aa$, $a^3 := aaa$, and, more generally, for any positive integer n, $a^{n+1} := (a^n)a$. We next define, for each $a \in \mathbb{R} \setminus \{0\}$, $a^0 := 1$ and $a^{-1} := 1/a$. Finally, for each $a \in \mathbb{R} \setminus \{0\}$ and each $n \in \mathbb{N}$, we define $a^{-n} := 1/a^n = (1/a)^n$.

Remark 2.1.5. Using the above definition of a^n, for $n \in \mathbb{Z}$ (and $a \neq 0$, if $n \leq 0$), one can easily check the usual laws of exponents: $a^m a^n = a^{m+n}$, $a^m/a^n = a^{m-n}$, $(a^m)^n = a^{mn}$, $(ab)^n = a^n b^n$, $(a/b)^n = a^n/b^n$, etc. $\forall\, a$, b, m, n such that the symbols are defined.

Next, we look at the three axioms that define the usual ordering on the set of real numbers.

Order Axioms: There is a subset $P \subset \mathbb{R}$ satisfying the following three axioms:

(O_1) $a,\ b \in P \implies a + b \in P$;
(O_2) $a,\ b \in P \implies ab \in P$; and
(O_3) for each $a \in \mathbb{R}$, exactly one of the following holds:

$$a \in P, \qquad a = 0, \qquad -a \in P \qquad \text{(Trichotomy)}$$

Notation and Remarks 2.1.6. Given any subsets A, $B \subset \mathbb{R}$, we define $A + B :=$ $\{a + b : a \in A, \ b \in B\}$, $A \cdot B := \{ab : a \in A, \ b \in B\}$, and $-A := \{-a : a \in A\}$. With this notation, the order axioms can be written as follows:

(O_1) $P + P \subset P$;
(O_2) $P \cdot P \subset P$; and
(O_3) $\mathbb{R} = P \cup \{0\} \cup (-P)$ is a partition of $\mathbb{R}$.

Definition 2.1.7 (Positive, Negative). Let $a \in \mathbb{R}$. We say that a is *(strictly) positive* if $a \in P$ and that a is *(strictly) negative* if $-a \in P$ (equivalently if $a \in -P$). A real number a will be called *nonnegative* if $a \in P \cup \{0\}$ and *nonpositive* if $a \in (-P) \cup \{0\}$.

Definition 2.1.8 (Inequalities). Given two real numbers a and b, if $a - b \in P$, then we write $a > b$ or $b < a$ and say that a is *greater than* b or that b is *less than* a. If $a - b \in P \cup \{0\}$, then we write $a \geq b$ or $b \leq a$ and say that a is *greater than or equal* to b or that b is *less than or equal* to a.

Remarks and Notation 2.1.9. Note that, by *trichotomy* (Axiom O_3), for each $a \in \mathbb{R}$, exactly one of $a > 0$, $a = 0$, or $a < 0$ holds. For any real number a, we have $a \leq a$ because $a - a = 0$. Next, $P + P \subset P$ implies that, if $a \leq b$ and $b \leq c$, then $a \leq c$. It also follows from *trichotomy* that, for any real numbers a and b, exactly one of the following holds: $a < b$, $a = b$, $a > b$. Thus, if $a \leq b$ and $b \leq a$, then $a = b$. If $a < b$ and $b \leq c$, then we write the combined inequality in the form $a < b \leq c$. Similar notation is used for other types of inequalities.

Exercise 2.1.10. Prove each of the following:

(a) $a \in \mathbb{R} \setminus \{0\} \implies a^2 > 0$;
(b) $1 > 0$; and
(c) $n \in \mathbb{N} \implies n > 0$.

Exercise 2.1.11. Deduce the following properties from the above definitions. Here a, b, c, and d are real numbers:

1. $a < b \implies a + c < b + c \ \forall \, c \in \mathbb{R}$;
2. $a < b$ and $c < d \implies a + c < b + d$;
3. $a < b$ and $c > 0 \implies ac < bc$;
4. $a < b$ and $c < 0 \implies ac > bc$;
5. $a > 0 \implies 1/a > 0$, and $a < 0 \implies 1/a < 0$; and
6. $0 < a < b \implies 1/a > 1/b$.

Exercise 2.1.12. Prove each of the following statements. Here, a and b are real numbers:

1. $a > 0 \implies 0 < a/2 < a$.
2. $a < b \implies a < (a + b)/2 < b$.
3. If $ab > 0$, then a and b are both positive or both negative.
4. If $ab < 0$, then a and b have opposite signs.

Theorem 2.1.13. *Let a, $b \in \mathbb{R}$ be arbitrary, and assume that $a < b + \varepsilon$ for every $\varepsilon > 0$. Then $a \leq b$.*

Proof. If $a > b$, then, setting $\varepsilon = (a - b)/2 > 0$ and using Exercise 2.1.12, we have $b + \varepsilon = b + (a - b)/2 = (a + b)/2 < a$, contradicting the assumption. □

Corollary 2.1.14. *If $a \in \mathbb{R}$, and if $0 \leq a < \varepsilon$ for every $\varepsilon > 0$, then $a = 0$.*

Exercise 2.1.15. Prove the corollary.

Definition 2.1.16 (Absolute Value). For any real number $a \in \mathbb{R}$, we define its *absolute value*, denoted by $|a|$, to be

$$|a| := \begin{cases} a & \text{if } a \geq 0, \\ -a & \text{if } a < 0. \end{cases}$$

Proposition 2.1.17. *Let a, $b \in \mathbb{R}$, and let $c \geq 0$. Then we have:*

1. $|a| = 0 \iff a = 0$;
2. $|-a| = |a|$;
3. $|ab| = |a||b|$;
4. $|a/b| = |a|/|b|$, if $b \neq 0$;
5. $-|a| \leq a \leq |a|$;
6. $|a| \leq c \iff -c \leq a \leq c$; and
7. $|a| \geq c \iff a \leq -c$ or $a \geq c$.

Exercise 2.1.18. Prove the proposition.

Proposition 2.1.19 (Triangle Inequality). *Given any real numbers a, $b \in \mathbb{R}$, we have $|a + b| \leq |a| + |b|$. More generally, if $a_1, \ldots, a_n$ are real numbers, we have $|a_1 + a_2 + \cdots + a_n| \leq |a_1| + |a_2| + \cdots + |a_n|$.*

Proof. By part (5) of Proposition 2.1.17, we have $-|a| \leq a \leq |a|$ and $-|b| \leq b \leq |b|$, from which we get $-(|a| + |b|) \leq a + b \leq |a| + |b|$. Therefore, the first part of the proposition follows from part (6) of Proposition 2.1.17. The second part is proved by induction. □

Corollary 2.1.20. *Given any a, $b \in \mathbb{R}$, we have the following:*

1. $|a - b| \leq |a| + |b|$;
2. $||a| - |b|| \leq |a - b|$.

Exercise 2.1.21. Prove the corollary.

Exercise 2.1.22. Following our notation for *lattice operations*, for any numbers $a, b \in \mathbb{R}$, we define $a \vee b = \max\{a, b\}$ and $a \wedge b = \min\{a, b\}$. Show that the following are true for any numbers $a, b, c \in \mathbb{R}$:

1. $a \wedge b + a \vee b = a + b$;
2. $(-a) \wedge (-b) = -(a \vee b)$;
3. $a \vee b + c = (a + c) \vee (b + c)$;

4. $c(a \vee b) = (ca) \vee (cb)$ if and only if $c \geq 0$;
5. $|a| = a \vee (-a)$; and
6. $a \vee b = (a + b + |a - b|)/2, \quad a \wedge b = (a + b - |a - b|)/2.$

Proposition 2.1.23 (Bernoulli, Cauchy, and Triangle Inequalities). *Let $x > -1$ and $x_1, x_2, \ldots, x_n, y_1, y_2, \ldots, y_n$ be arbitrary real numbers. Then the following inequalities hold:*

Bernoulli's Inequality:

$$(1 + x)^n \geq 1 + nx, \quad \forall \quad n \in \mathbb{N}.$$

Cauchy's Inequality:

$$\left(\sum_{i=1}^{n} x_i y_i \right)^2 \leq \left(\sum_{i=1}^{n} x_i^2 \right) \left(\sum_{i=1}^{n} y_i^2 \right).$$

Triangle Inequality:

$$\left(\sum_{i=1}^{n} (x_i + y_i)^2 \right)^{1/2} \leq \left(\sum_{i=1}^{n} x_i^2 \right)^{1/2} + \left(\sum_{i=1}^{n} y_i^2 \right)^{1/2}.$$

Exercise 2.1.24. Prove the proposition and show that in Bernoulli's inequality, *equality* holds if and only if $n = 1$ or $x = 0$. Also prove the following consequence of Bernoulli's inequality: If $x > -1$, then we have

$$(1 + x)^{1/n} \leq 1 + x/n \quad \forall n \in \mathbb{N}.$$

Hints: For Bernoulli's inequality, use induction on n. For Cauchy's inequality, set $X := \sum_{i=1}^{n} x_i^2$, $Y := \sum_{i=1}^{n} y_i^2$, and $Z := \sum_{i=1}^{n} x_i y_i$. Observe that for any $t \in \mathbb{R}$, $F(t) := \sum_{i=1}^{n} (x_i - ty_i)^2 \geq 0$, and look at the discriminant $Z^2 - XY$ of $F(t)$. Finally, note that the Triangle Inequality is a consequence of Cauchy's inequality.

The next inequality is important enough to be stated separately. It is the famous *Arithmetic–Geometric Means Inequality*. We give a well-known inductive proof and ask the reader to give another one in Exercise 2.1.26 below.

Proposition 2.1.25 (Arithmetic–Geometric Means Inequality). *For each natural number $n \geq 2$, let $a_1, a_2, \ldots, a_n$ be real numbers with $a_i \geq 0$ for $i = 1, 2, \ldots, n$. If their "arithmetic mean" is defined to be $A_n := (a_1 + a_2 + \cdots + a_n)/n$ and their "geometric mean" to be $G_n := (a_1 a_2 \cdots a_n)^{1/n}$, then we have*

$$G_n \leq A_n,$$

with equality holding if and only if $a_1 = a_2 = \cdots = a_n = A_n$.

Proof. Assume first that $n = 2^m$ for some $m \in \mathbb{N}$. Now for $m = 1$, i.e., $n = 2$, the inequality $\sqrt{a_1 a_2} \leq (a_1 + a_2)/2$ is equivalent to $(\sqrt{a_1} - \sqrt{a_2})^2 \geq 0$, which is obviously true and is an equality if and only if $a_1 = a_2$. Next, assuming the inequality is true for $n = 2^m$, we must prove it for $2n = 2^{m+1}$. However, using the case $n = 2^m$ and the (already proven) inequality $G_2 \leq A_2$, we have

$$G_{2n} = \sqrt[2n]{a_1 a_2 \cdots a_{2n}} = \sqrt{\sqrt[n]{a_1 \cdots a_n} \sqrt[n]{a_{n+1} \cdots a_{2n}}}$$

$$\leq \frac{1}{2}(\sqrt[n]{a_1 \cdots a_n} + \sqrt[n]{a_{n+1} \cdots a_{2n}})$$

$$\leq \frac{1}{2}[(a_1 + \cdots + a_n)/n + (a_{n+1} + \cdots + a_{2n})/n]$$

$$= (a_1 + a_2 + \cdots + a_{2n})/(2n)$$

$$= A_{2n}.$$

To prove the inequality for arbitrary n, we pick $m \in \mathbb{N}$ such that $n < 2^m$ and set $k = 2^m - n$. Setting $a_{n+1} = a_{n+2} = \cdots = a_{2^m} = A_n$, and applying the inequality to the 2^m numbers $a_1, a_2, \ldots, a_{2^m}$, we get

$$(a_1 \cdots a_n)(A_n)^k \leq [(a_1 + \cdots + a_n + kA_n)/2^m]^{2^m}$$

$$= [(nA_n + kA_n)/2^m]^{2^m} = (A_n)^{2^m},$$

and our inequality follows if we divide the two sides by A_n^k. The last statement (about the cases where equality holds) also follows inductively.   □

Exercise 2.1.26. Prove Proposition 2.1.25 by induction along the following lines: First, the inequality is trivial if the a_i are *all equal*. If not, show that (after renaming the a_i, if necessary) we may assume that $a_1 < A_n < a_2$. Write $\bar{a}_1 = A_n$ and $\bar{a}_2 = a_1 + a_2 - A_n$, and show that $a_1 a_2 < \bar{a}_1 \bar{a}_2$, deducing that it suffices to prove the inequality $\bar{a}_1 \bar{a}_2 a_3 \cdots a_n \leq A_n^n$. Prove the latter inequality by induction, using the case $n = 1$ (or $n = 2$) as your first step and the case of $n - 1$ numbers $\bar{a}_2, a_3, \ldots, a_n$, as the inductive step.

Definition 2.1.27 (Intervals).

(A) Given any $a, b \in \mathbb{R}$ with $a < b$, we define the following sets to be the *bounded intervals* with *endpoints* a and b:

  1. $(a, b) := \{x \in \mathbb{R} : a < x < b\}$;
  2. $[a, b) := \{x \in \mathbb{R} : a \leq x < b\}$;
  3. $(a, b] := \{x \in \mathbb{R} : a < x \leq b\}$; and
  4. $[a, b] := \{x \in \mathbb{R} : a \leq x \leq b\}$.

  (a, b) is called *open* and $[a, b]$ is called *closed*; $[a, b)$ and $(a, b]$ are called *half-open* (or *half-closed*).

(B) Given any $a \in \mathbb{R}$, we define the following sets to be the *unbounded intervals* with (finite) *endpoint a*:

5. $(a, \infty) := \{x \in \mathbb{R} : x > a\}$;
6. $[a, \infty) := \{x \in \mathbb{R} : x \geq a\}$;
7. $(-\infty, a) := \{x \in \mathbb{R} : x < a\}$; and
8. $(-\infty, a] := \{x \in \mathbb{R} : x \leq a\}$.

Here, ∞ (also denoted $+\infty$) and $-\infty$ are two symbols, called *plus infinity* and *minus infinity*, respectively, that are *not* real numbers. The intervals (a, ∞) and $(-\infty, a)$ are called *open*, while $[a, \infty)$ and $(-\infty, a]$ are called *closed*. Furthermore, we consider $\mathbb{R}$ itself to be an *unbounded* interval (that is *both* open and closed):

9. $(-\infty, \infty) := \mathbb{R}$.

Remark 2.1.28 (Degenerate Interval). We may sometimes find it useful to include the empty set $\emptyset \subset \mathbb{R}$ and *singletons* (i.e., sets containing a single real number) in the set of all intervals. Thus, for any $a \in \mathbb{R}$, we have $(a, a) = (a, a] = [a, a) = \emptyset$, while $[a, a] = \{a\}$.

Finally, the last axiom of real numbers, which is fundamental in all aspects of *analysis*, is the following:

Completeness Axiom (or Supremum Property)

(C) Every nonempty subset of $\mathbb{R}$ that is bounded above has a least upper bound (in $\mathbb{R}$).

In other words, if $\emptyset \neq S \subset \mathbb{R}$, and if there exists an element $u \in \mathbb{R}$ such that $s \leq u$ for all $s \in S$, then $\sup(S) \in \mathbb{R}$.

Exercise 2.1.29. Show that the *Supremum Property* is equivalent to the *Infimum Property*: Every nonempty subset of real numbers that is bounded *below* has a greatest lower bound in $\mathbb{R}$. In other words, if $\emptyset \neq S \subset \mathbb{R}$, and if there exists a number $t \in \mathbb{R}$ such that $t \leq s$ for all $s \in S$, then $\inf(S) \in \mathbb{R}$. *Hint:* Show that $\inf(S) = -\sup(-S)$, where $-S := \{-s \in \mathbb{R} : s \in S\}$.

In most applications, the following characterization of the least upper bound of a nonempty set of real numbers is more convenient than the general definition, given in Chap. 1, for nonempty subsets of *any* partially ordered set.

Proposition 2.1.30. *Let $\emptyset \neq S \subset \mathbb{R}$, and assume that S is bounded above. Then $u = \sup(S)$ if and only if (i) $s \leq u \; \forall \; s \in S$ (i.e., u is an upper bound of S) and (ii) $\forall \; \varepsilon > 0 \; \exists \; s_\varepsilon \in S$ such that $u - \varepsilon < s_\varepsilon$.*

Proof. If $u = \sup(S)$, then (i) is obviously satisfied. Also, for each $\varepsilon > 0$, we have $u - \varepsilon < u$, so that $u - \varepsilon$ is *not* an upper bound of S. Therefore, there exists $s_\varepsilon \in S$ such that $u - \varepsilon < s_\varepsilon$. Conversely, if (i) and (ii) hold, then u is an upper bound of S by (i). If v is any other upper bound and if $v < u$, then, setting $\varepsilon = u - v > 0$

and using (ii), we can find $s_\varepsilon \in S$ such that $v = u - \varepsilon < s_\varepsilon \le u$, contradicting the assumption that v is an upper bound of S. $\square$

One of the important consequences of the Completeness Axiom is the following.

Theorem 2.1.31 (Archimedean Property of $\mathbb{R}$). *Given any $x \in \mathbb{R}$, there exists an integer $n_x \in \mathbb{N}$ such that $x < n_x$.*

Proof. Suppose not. Then there exists $x_0 \in \mathbb{R}$ such that $n \le x_0$ for all $n \in \mathbb{N}$; i.e., $\mathbb{N}$ is bounded above. Since $\emptyset \ne \mathbb{N} \subset \mathbb{R}$, the *Supremum Property* implies that $u = \sup(\mathbb{N}) \in \mathbb{R}$. Now, using Proposition 2.1.30 (with $\varepsilon = 1$), we can find $m \in \mathbb{N}$ such that $u - 1 < m \le u$. But then $u < m + 1 \in \mathbb{N}$, contradicting the fact that u is an upper bound of $\mathbb{N}$. $\square$

Corollary 2.1.32. *Let x and y be positive numbers. Then we have:*

(a) $\exists\, n \in \mathbb{N}$ such that $x < ny$;
(b) $\exists\, n \in \mathbb{N}$ such that $0 < 1/n < x$; and
(c) $\exists\, n \in \mathbb{N}$ such that $n - 1 \le x < n$.

Exercise 2.1.33. Prove the corollary. *Hint for (c):* Look at $\min(\{k \in \mathbb{N} : x < k\})$.

The next proposition which guarantees the existence of *square roots* of *positive* numbers is our first application of the Archimedean Property.

Proposition 2.1.34 (Existence of Square Roots). *Given any $a > 0$, there is a unique $x > 0$ such that $x^2 = a$. This unique x is denoted by $\sqrt{a}$.*

Proof. Since $1^2 = 1$, we assume that $a > 1$; otherwise, one can consider $1/a$. Now introduce the set $S = \{s > 0 : s^2 < a\}$. Since $1^2 = 1 < a$, we have $S \ne \emptyset$. Also, $1 < a$ implies that $s < a$ for all $s \in S$. In other words, S is *bounded above*. By the Supremum Property, $x := \sup(S) \in \mathbb{R}$. We prove $x^2 = a$ by showing that both $x^2 > a$ and $x^2 < a$ result in contradictions. Assume first that $x^2 < a$. Then the inequality $1/n^2 \le 1/n$, true for all $n \in \mathbb{N}$, implies that $(x + 1/n)^2 \le x^2 + (2x + 1)/n$. Now using $(a - x^2)/(2x + 1) > 0$ and the Archimedean Property, we can pick $n \in \mathbb{N}$ such that $1/n < (a - x^2)/(2x + 1)$, from which we get $(x + 1/n)^2 < a$. However, this gives $x + 1/n \in S$, contradicting $x = \sup(S)$. Next, assume that $x^2 > a$. Then for any $m \in \mathbb{N}$, we have $(x - 1/m)^2 > x^2 - 2x/m$. Using the inequality $(x^2 - a)/(2x) > 0$ and the Archimedean Property, we find an $m \in \mathbb{N}$ such that $1/m < (x^2 - a)/(2x)$. But this implies that $s^2 < a < (x - 1/m)^2$ for all $s \in S$. In other words, $x - 1/m$ is an upper bound of S, again contradicting $x = \sup(S)$. The uniqueness follows from the fact that $0 < x < y$ implies $y^2 - x^2 = (y - x)(y + x) > 0$; i.e., $y^2 > x^2$. $\square$

Remarks and Notation 2.1.35. A similar argument, using the binomial formula, can be used to show that any *positive* number a has a *(unique) positive nth root*, denoted by $\sqrt[n]{a}$ and such that $(\sqrt[n]{a})^n = a$, for any $n \in \mathbb{N}$. This will be obtained in Chap. 4 by a more abstract argument involving the continuity of inverse functions. For $a = 0$, we have $\sqrt[n]{0} = 0\ \forall n \in \mathbb{N}$. If $a < 0$ and n is *odd*, then we define $\sqrt[n]{a} = -\sqrt[n]{-a}$. For n *even* and $a < 0$, $\sqrt[n]{a}$ cannot be defined as a *real* number.

What is needed is the set of *complex numbers*, which will be defined later. Using nth roots, one can introduce *fractional powers* by defining $a^{1/n} := \sqrt[n]{a}$ when the right side is real and, more generally, $a^{m/n} := (a^{1/n})^m = \sqrt[n]{a^m}$ for any $n \in \mathbb{N}$, $m \in \mathbb{Z}$, when the right side is real. It is then easy to check that all the laws of exponents hold for these fractional powers as well. Finally, note that $\sqrt{a^2} = |a| \ \forall a \in \mathbb{R}$ and, more generally, $\sqrt[n]{a^n} = a^n/|a|^{n-1} \ \forall a \in \mathbb{R} \setminus \{0\}$.

The existence of square roots can be used to prove that *irrational numbers* exist; in other words, that $\mathbb{R} \setminus \mathbb{Q} \neq \emptyset$. The following theorem and its proof can be found in Euclid's *Elements*:

Theorem 2.1.36 (Irrationality of $\sqrt{2}$). $\sqrt{2} \notin \mathbb{Q}$.

Proof. Suppose that $\sqrt{2} = m/n$, where m, $n \in \mathbb{N}$ and $\gcd(m, n) = 1$. Then $m^2/n^2 = 2$, so that $m^2 = 2n^2$, and m is even, i.e., $m = 2m_0$ for some $m_0 \in \mathbb{N}$. But then $4m_0^2 = 2n^2$ implies $n^2 = 2m_0^2$ and hence n is also even, contradicting $\gcd(m, n) = 1$. □

Remark 2.1.37. More generally, one can show that $\sqrt{n} \notin \mathbb{Q}$ if $n \geq 2$ has prime factorization $n = p_1^{r_1} p_2^{r_2} \cdots p_m^{r_m}$, where at least one of the positive exponents r_j is *odd*.

Our next application of the Archimedean Property shows that the set of rational numbers is *dense* in the set of all real numbers:

Theorem 2.1.38 (Density of $\mathbb{Q}$ in $\mathbb{R}$). *If x and y are real numbers with $x < y$, then there exists a rational number $r \in \mathbb{Q}$, such that $x < r < y$.*

Proof. We may assume that $x > 0$. (Why?) By the Archimedean Property, there is $n \in \mathbb{N}$ such that $1/n < y - x$; i.e., that $ny - nx > 1$. By part (c) of Corollary 2.1.32, we can pick $m \in \mathbb{N}$ such that $m - 1 \leq nx < m$. Since $m \leq nx + 1 < ny$, we get $nx < m < ny$; in other words, with $r = m/n$, we have $x < r < y$. □

Corollary 2.1.39. *For any real numbers x and y with $x < y$, there is an irrational number t such that $x < t < y$.*

Exercise 2.1.40. Prove the corollary. *Hint:* Look at $x/\sqrt{2}$ and $y/\sqrt{2}$.

Our next application of completeness of $\mathbb{R}$ is the following.

Proposition 2.1.41 (Characterization of Intervals). *A set $I \subset \mathbb{R}$ is a nondegenerate interval if and only if for any a, $b \in I$, $a < b$, we have $[a, b] := \{x \in \mathbb{R} : a \leq x \leq b\} \subset I$.*

Proof. Clearly, any interval satisfies the condition in the proposition. To prove the converse, let $\alpha = \inf(I)$, $\beta = \sup(I)$, where we agree to write $\alpha = -\infty$ (resp., $\beta = +\infty$) if I is *not* bounded below (resp., above). Let $c \in I$. If $\alpha = -\infty$, then for any $x < c$ there is $y \in I$ such that $y < x < c$; hence $x \in I$, and we get $(-\infty, c] \subset I$. If $\alpha \in \mathbb{R}$ and $\alpha < c$, then again for each $x \in \mathbb{R}$ satisfying $\alpha < x < c$, we can find $y \in I$ with $y < x < c$, which implies that $x \in I$ and hence that $(\alpha, c] \subset I$. Similarly, we can show that $[c, \beta) \subset I$, and hence $(\alpha, \beta) \subset I$. In view of the definitions of α and β, it follows that I must be one of the four possible intervals with endpoints α and β. It is obvious that $\alpha \notin I$ (resp., $\beta \notin I$) if $\alpha = -\infty$ (resp., $\beta = +\infty$). $\square$

The following corollary may be worth stating.

Corollary 2.1.42. *Let $\{I_\lambda\}_{\lambda \in \Lambda}$ be any collection of intervals. If $\bigcap_{\lambda \in \Lambda} I_\lambda \neq \emptyset$, then the union $J = \bigcup_{\lambda \in \Lambda} I_\lambda$ is an interval.*

Proof. Let $c \in \bigcap_{\lambda \in \Lambda} I_\lambda$ be a fixed point. Given any a, $b \in J$ with $a < b$, we can pick λ_a, $\lambda_b \in \Lambda$ with $a \in I_{\lambda_a}$ and $b \in I_{\lambda_b}$ and note that if $a < c < b$, say, then (since $c \in I_{\lambda_a} \cap I_{\lambda_b}$) we have $[a,c] \subset I_{\lambda_a} \subset J$ and $[c,b] \in I_{\lambda_b} \subset J$ so that $[a,b] \subset J$, as desired. The cases $c \leq a < b$ and $a < b \leq c$ are simpler. $\square$

Here is another important application of the Completeness Axiom:

Theorem 2.1.43 (Nested Intervals Theorem). *Let $\{ I_n = [a_n, b_n] \ n \in \mathbb{N}\}$ be a sequence of closed bounded intervals in $\mathbb{R}$ that is "nested," i.e., $I_{n+1} \subset I_n \ \forall \ n \in \mathbb{N}$. Then $\bigcap_{n=1}^{\infty} I_n \neq \emptyset$. If, in addition, $\inf(\{b_n - a_n : n \in \mathbb{N}\}) = 0$, then $\bigcap_{n=1}^{\infty} I_n = \{\xi\}$, for a unique $\xi \in \mathbb{R}$.*

Proof. The set $A = \{a_n : n \in \mathbb{N}\}$ is bounded above (by b_1). Let $\xi = \sup(A)$. Since the I_n are nested, for any positive integers m and n, we have $a_m \leq a_{m+n} \leq b_{m+n} \leq b_n$, so that $\xi \leq b_n$ for each $n \in \mathbb{N}$. Since we obviously have $a_n \leq \xi$ for each n, we have $a_n \leq \xi \leq b_n$, for all n, which implies $\xi \in \bigcap_{n=1}^{\infty} I_n$. Finally, if ξ, $\eta \in \bigcap_{n=1}^{\infty} I_n$, with $\xi \leq \eta$, then we get $0 \leq \eta - \xi \leq b_n - a_n$, for all $n \in \mathbb{N}$, so that $0 \leq \eta - \xi \leq \inf(\{b_n - a_n : n \in \mathbb{N}\}) = 0$. $\square$

Exercise 2.1.44. Let $\{[a_n, b_n]\}_{n \in \mathbb{N}}$ be a sequence of nested intervals. If $\alpha = \sup\{a_n : n \in \mathbb{N}\}$ and $\beta = \inf\{b_n : n \in \mathbb{N}\}$, show that

$$\bigcap_{n=1}^{\infty} [a_n, b_n] = [\alpha, \beta].$$

As the above proof shows, the Nested Intervals Theorem is a consequence of the Supremum Property of $\mathbb{R}$. In fact, as the next theorem shows, the converse is also true, if the Archimedean Property is assumed as well.

Theorem 2.1.45. *The Supremum Property of $\mathbb{R}$ is a consequence of the Nested Intervals Theorem and the Archimedean Property. More precisely, if the Completeness Axiom is replaced by the Nested Intervals Theorem and the Archimedean Property, but all other axioms remain, then the Supremum Property holds.*

Proof. Let $\emptyset \neq S \subset \mathbb{R}$ be *bounded above*. Pick an arbitrary $s \in S$. For each $n \in \mathbb{N}$, the Archimedean Property of $\mathbb{R}$ implies that $s + m/2^n$ is an upper bound of S, for some $m \in \mathbb{N}$. Let k_n be the *smallest* such m, and set $I_n := [s + (k_n - 1)/2^n, s + k_n/2^n]$. We then have $I_n \cap S \neq \emptyset$. (Why?) Thus while (by definition) $s + k_n/2^n = s + (2k_n)/2^{n+1}$ is an upper bound of S, $s + (2k_n - 2)/2^{n+1} = s + (k_n - 1)/2^n$ is *not*. Therefore, either $k_{n+1} = 2k_n$ or $k_{n+1} = 2k_n - 1$ and $I_{n+1} \subset I_n$ follow. The Nested

Intervals Theorem now implies that $\bigcap_{n=1}^{\infty} I_n = \{u\}$ for a *unique* $u \in \mathbb{R}$. Indeed, if $u < v$ and $u, v \in \bigcap_{n=1}^{\infty} I_n$, then $v - u > 1/2^n$ for some $n \in \mathbb{N}$, which contradicts $u, v \in I_n$, since I_n has length 2^{-n}. We claim that $u = \sup(S)$. First, u is an upper bound of S for, otherwise, there is $t \in S$ with $u < t$ and hence $t - u > 1/2^n$ for some $n \in \mathbb{N}$. Since $u \in I_n$, we get $s + k_n/2^n < t$, which contradicts the definition of k_n. Next, if v is any upper bound of S and if $v < u$, then we can pick $n \in \mathbb{N}$ such that $u - v > 1/2^n$, and since $u \in I_n$, it follows that $v < s + (k_n - 1)/2^n$. In particular, $s + (k_n - 1)/2^n$ is an upper bound of S, which again contradicts the definition of k_n. □

Nested intervals can be used to obtain *decimal* (or *binary, ternary*, etc.) representations of real numbers. We will give a short account of this. The details may be supplied by the reader or found in the references.

Binary, Ternary, Decimal, etc. Expansions.

Let $x \in [0, 1)$ and let $p > 1$ be a fixed integer. Then $x \in [0, 1) = \bigcup_{j=0}^{p-1} [j/p, (j + 1)/p)$, where the union is disjoint. Therefore, there is a *unique* integer $x_1 \in \{0, 1, \ldots, p - 1\}$ such that $x \in [x_1/p, x_1/p + 1/p)$. Dividing the latter interval into p equal parts, there is a unique integer x_2 such that $0 \le x_2 < p$ and $x \in [x_1/p + x_2/p^2, x_1/p + x_2/p^2 + 1/p^2)$. Continuing this process, at the nth stage we have uniquely determined integers x_j with $0 \le x_j < p$ for $1 \le j \le n$, and

$$x \in \left[\sum_{j=1}^{n} \frac{x_j}{p^j}, \sum_{j=1}^{n} \frac{x_j}{p^j} + \frac{1}{p^n} \right).$$

Now let $I_0 := [0, 1]$ and

$$I_n := \left[\sum_{j=1}^{n} \frac{x_j}{p^j}, \sum_{j=1}^{n} \frac{x_j}{p^j} + \frac{1}{p^n} \right], \qquad n \in \mathbb{N}.$$

Then the intervals $I_0, I_1, I_2, \ldots$ are nested and I_k has length $1/p^k$ for all $k \in \mathbb{N}_0$. Since $x \in I_k$ for all k, it follows from the Nested Intervals Theorem that $\bigcap_{k=0}^{\infty} I_k = \{x\}$. The *base p expansion* of x is now denoted by $x = (0.x_1 x_2 \cdots)_p$ and it is unique by construction. If $x \in (0, 1)$ is an endpoint of some I_n, say, $x = \sum_{k=1}^{n} x_k/p^k$, with $x_n \ge 1$, then the above construction gives the unique base p expansion $x = (0.x_1 \cdots x_n 000 \cdots)_p$ with $x_k = 0$ for all $k > n$. It turns out, however, that in this case we have a second expansion $x = (0.x_1' x_2' \cdots)_p$, where $x_j' = x_j$ for $1 \le j \le n - 1$, $x_n' = x_n - 1$, and $x_k' = p - 1$ for all $k > n$. We shall come back to this in Chap. 4 when we discuss the *Cantor's ternary set*. We shall also see that $x = 1$ has the unique expansion $1 = (0.x_1 x_2 \cdots)_p$ with $x_n = p - 1$ for all n.

If in the above procedure we take $p = 2$, $p = 3$, $p = 10, \ldots$, then we obtain the *binary, ternary, decimal*, etc. expansions of the real number x. For example, the binary (i.e., *base two*) expansion of x has the form $x = (0.x_1 x_2 x_3 \cdots)_2 :=$ $x_1/2 + x_2/2^2 + x_3/2^3 + \cdots$, where each x_n is either 0 or 1. Similarly, the ternary (i.e., *base three*) expansion of x has the form $x = (0.x_1 x_2 x_3 \cdots)_3 := x_1/3 + x_2/3^2 +$ $x_3/3^3 + \cdots$, where each x_n is 0, 1, or 2. As above, the expansion is unique unless $x \in (0, 1)$ is a subdivision point at some stage, in which case two expansions exist. In the binary case, one of the two binary expansions ends with an infinite string of 0's and the other with an infinite string of 1's. If we always choose the latter, for example, then each $x \in [0, 1]$ has a *unique* binary expansion. In other words, we get a one-to-one map from $[0, 1]$ into the set $2^{\mathbb{N}}$ of all sequences $x = (x_1, x_2, \ldots)$, where each x_n is either 0 or 1. Conversely, to any such sequence we can assign the *unique* real number $(0.x_1 x_2 x_3 \cdots)_2 \in [0, 1]$. This defines a one-to-one map from $2^{\mathbb{N}}$ into $[0, 1]$. Therefore, by the Schröder–Bernstein theorem, we have

Proposition 2.1.46. $|[0, 1]| = |2^{\mathbb{N}}| = |\mathbb{R}| = \mathfrak{c}$.

Proof. We must only show that $[0, 1] \sim \mathbb{R}$. But $[0, 1] \subset \mathbb{R}$, and the map $f : \mathbb{R} \to$ $[0, 1]$ defined by $f(x) = x/(2\sqrt{1 + x^2}) + 1/2$ is one-to-one, so the equivalence follows again from the Schröder–Bernstein theorem. $\square$

For certain applications it is convenient to extend the set of real numbers by adjoining two elements called *(plus) infinity* and *minus infinity*, denoted by $\infty = +\infty$ and $-\infty$, respectively; these symbols are distinct and are *not* real numbers.

Definition 2.1.47 (Extended Real Line). The *extended real line* is the disjoint union $\overline{\mathbb{R}} := [-\infty, \infty] := \{-\infty\} \cup \mathbb{R} \cup \{\infty\}$, with the following properties:

1. $x \in \mathbb{R} \implies -\infty < x < \infty$;
2. $x \in \mathbb{R} \implies x + \infty = \infty, \quad x - \infty = -\infty, \quad x/(\pm\infty) = 0$;
3. $x > 0 \implies x \cdot \infty = \infty, \quad x \cdot (-\infty) = -\infty$;
4. $x < 0 \implies x \cdot \infty = -\infty, \quad x \cdot (-\infty) = \infty$;
5. $\infty + \infty = \infty, \quad -\infty - \infty = -\infty$; and
6. $\infty \cdot (\pm\infty) = \pm\infty, \quad -\infty \cdot (\pm\infty) = \mp\infty$.

Remark 2.1.48. The operation $\infty - \infty$ is left *undefined*. Also, by *arbitrary convention*, we define $0 \cdot \infty := 0$. To distinguish the *real* numbers from the *extended real* numbers, we call the former *finite*. Next, given any subset $S \subset \mathbb{R}$, $S \neq \emptyset$, we define $\sup(S) := \infty$ if S is *not bounded above*, and $\inf(S) := -\infty$ if S is *not bounded below*. It then follows that *every* nonempty subset of $\overline{\mathbb{R}} = [-\infty, \infty]$ has both a least upper bound and a greatest lower bound. Finally, a function with values in $\overline{\mathbb{R}} = [-\infty, \infty]$ is called an *extended real-valued* function.

2.2 Sequences in $\mathbb{R}$

In this section we summarize some of the basic facts about sequences of real numbers. The first fundamental notion here is the *convergence* of a sequence and is based on the usual concept of *distance* between two real numbers:

Definition 2.2.1 (Distance, Epsilon-Neighborhood).

1. The *distance* between any two real numbers a and b is defined to be $|b - a|$.
2. Given any $x \in \mathbb{R}$, the *ε-neighborhood* of x is the open interval $B_\varepsilon(x) = (x - \varepsilon, x + \varepsilon)$, *centered* at x. The set $\dot{B}_\varepsilon(x) := B_\varepsilon(x) \setminus \{x\}$ is called the *deleted ε-neighborhood* of x.

Definition 2.2.2 (Open and Closed Sets). A set $O \subset \mathbb{R}$ is called *open* if for each $x \in O$, there is $\varepsilon = \varepsilon(x) > 0$ such that $B_\varepsilon(x) \subset O$. A set $C \subset \mathbb{R}$ is called *closed* if its complement $C^c = \mathbb{R} \setminus C$ is open.

Example 2.2.3. The sets $\emptyset$ and $\mathbb{R}$ are both open *and* closed, and they are the *only* subsets with this property. Open intervals are open and closed intervals are closed. A half-open interval $[a, b)$ (or $(a, b]$), $a < b$, is neither open nor closed.

Exercise 2.2.4.

1. Let $\{O_\lambda\}_{\lambda \in \Lambda}$ be a family of open sets in $\mathbb{R}$ indexed by a set Λ. Show that $\bigcup_{\lambda \in \Lambda} O_\lambda$ is open. If Λ is *finite*, show that $\bigcap_{\lambda \in \Lambda} O_\lambda$ is also open. Using the infinite collection $(-1/n, 1 + 1/n)$, $n \in \mathbb{N}$, show that the latter statement is *false* if Λ is *infinite*.
2. Let $\{C_\lambda\}_{\lambda \in \Lambda}$ be a family of closed subsets of $\mathbb{R}$. Show that $\bigcap_{\lambda \in \Lambda} C_\lambda$ is closed. If Λ is *finite*, show that $\bigcup_{\lambda \in \Lambda} C_\lambda$ is also closed. Using the infinite collection $[1/n, 1 - 1/n]$, $n \in \mathbb{N}$, show that the latter statement is *false* if Λ is *infinite*.
3. Show that $\mathbb{N}$ and $\mathbb{Z}$ are closed, whereas the set $\{1/n : n \in \mathbb{N}\}$ is neither closed nor open.

Definition 2.2.5 (Convergence, Limit). Given a sequence $(x_n) \in \mathbb{R}^{\mathbb{N}}$, we say that (x_n) *converges* to a real number ξ, and we write $\lim(x_n) := \lim_{n \to \infty} x_n = \xi$, if given any $\varepsilon > 0$, there is an integer $N = N(\varepsilon)$ such that $n \geq N$ implies $|x_n - \xi| < \varepsilon$. The number ξ is called the *limit* of the sequence (x_n), and, if it exists, we say that the sequence is *convergent*. A sequence that has no limit is called *divergent*.

Example 2.2.6.

1. (Ultimately Constant Sequences). If for some $N \in \mathbb{N}$ and $c \in \mathbb{R}$ we have $x_n = c$ for all $n \geq N$, then $\lim(x_n) = c$. Indeed, given any $\varepsilon > 0$, we have $|x_n - c| = 0 < \varepsilon$ for all $n \geq N$.
2. Show that $\lim(1/\sqrt{n}) = 0$. Well, for any $\varepsilon > 0$, the inequality $|1/\sqrt{n} - 0| < \varepsilon$ gives $n > 1/\varepsilon^2$ and hence we can use any $N > 1/\varepsilon^2$.
3. The sequence $((-1)^n)$ is *divergent*. Suppose, to get a contradiction, that $\lim_{n \to \infty} (-1)^n = a \in \mathbb{R}$ and let $\varepsilon = 1$. Then for some $N \in \mathbb{N}$ we have $|(-1)^n - a| < 1$ for all $n \geq N$. Taking $n \geq N$ to be *even* or *odd*, we see that $|a - 1| < 1$ and $|a + 1| < 1$ must hold simultaneously. But then both $a > 0$ and $a < 0$ must be satisfied, which is absurd.
4. We have $\lim_{n \to \infty} \frac{5n+2}{2n+1} = \frac{5}{2}$. Here, the inequality $\left| \frac{5n+2}{2n+1} - \frac{5}{2} \right| < \varepsilon$ gives $n > 1/(4\varepsilon) - 1/2$ and hence any $N > 1/(4\varepsilon) - 1/2$ will do.

Exercise 2.2.7.

1. Show that, if $a > 0$, then $\lim(1/(1 + na)) = 0$.
2. Show that, if $0 < b < 1$, then $\lim(b^n) = 0$. Deduce that $\lim(1/b^n) = 0$ if $b > 1$.
3. Show that, if $c > 0$, then $\lim(c^{1/n}) = 1$.
4. Show that $\lim(n^{1/n}) = 1$.

Hints: For (1) and (2) use Bernoulli's inequality. For (3), set $d_n := 1 - c^{1/n}$ and use Bernoulli's inequality again. Finally, for (4) set $k_n := n^{1/n} - 1$ for all $n > 1$ and, expanding $n = (1 + k_n)^n$ (by the binomial formula), show that $k_n^2 \leq 2/n$.

Definition 2.2.8 (m-Tail, Ultimately True).

1. Given a real sequence (x_n) and any $m \in \mathbb{N}$, the m-*tail* of (x_n) is the sequence $(x_m, x_{m+1}, x_{m+2}, \ldots)$.
2. A property of sequences is said to be *ultimately true* for a sequence (x_n) (resp., sequences (x_n), (y_n), etc.), if there is an integer $m \in \mathbb{N}$ such that the property is satisfied by the m-tail of (x_n) (resp., m-tails of (x_n), (y_n), etc.).

Remark 2.2.9. Using m-tails and ε-neighborhoods, the above definition of the limit of a real sequence can be rephrased as follows: *A sequence (x_n) converges to a limit ξ if, for any $\varepsilon > 0$, there is an integer $N = N(\varepsilon) \in \mathbb{N}$ such that the ε-neighborhood $B_\varepsilon(\xi)$ contains the N-tail of (x_n).*

Exercise 2.2.10.

1. Show that, if $\lim(x_n) = \xi$, then all m-tails of (x_n) also converge to ξ and conversely: If an m-tail of (x_n) converges to $\xi \in \mathbb{R}$, then so does (x_n).
2. Show that, if two sequences (x_n) and (y_n) are *ultimately equal* (i.e., for some $m \in \mathbb{N}$, $x_n = y_n$ $\forall n \geq m$), then $\lim(x_n) = \xi \implies \lim(y_n) = \xi$ $\forall \xi \in \mathbb{R}$.
3. Let $\emptyset \neq S \subset \mathbb{R}$ be *bounded*. Show that there are sequences (s_n) and (t_n) in S such that $\lim(s_n) = \inf(S)$ and $\lim(t_n) = \sup(S)$. *Hint:* Use Proposition 2.1.30, with $\varepsilon = 1/n$, $n \in \mathbb{N}$.

For arbitrary subsets of real numbers, the following definition of *limit point* will replace the definition given above for limits of sequences:

Definition 2.2.11 (Limit Point, Isolated Point). Let $S \subset \mathbb{R}$. A number $x \in \mathbb{R}$ is said to be a *limit point* (or *accumulation point*) of S if for every $\varepsilon > 0$, $B_\varepsilon(x)$ contains an element $s \in S \setminus \{x\}$. A number $x \in S$ is called an *isolated point* of S if it is *not* a limit point of S; i.e., if for some $\varepsilon = \varepsilon(x)$, we have $B_\varepsilon(x) \cap S = \{x\}$.

Exercise 2.2.12.

1. Show that, if x is a *limit point* of a set $S \subset \mathbb{R}$, then there is a sequence (s_n) in S such that $\lim(s_n) = x$; conversely if $(s_n) \in S^\mathbb{N}$ is *not ultimately constant* and $\lim(s_n) = x$, then x is a limit point of S.
2. Show that, if x is a limit point of $S \subset \mathbb{R}$, then for each $\varepsilon > 0$ the intersection $B_\varepsilon(x) \cap S$ is *infinite*. Deduce that a *finite* set has no limit points. (*Hint:* Suppose that the intersection is finite and get a contradiction).
3. Show that a set $F \subset \mathbb{R}$ is *closed* if and only if every limit point of F belongs to F.

Definition 2.2.13 (Perfect Set). A set $S \subset \mathbb{R}$ is called *perfect* if it is closed and if *every* point of S is a *limit point* of S.

Example 2.2.14. Any *closed interval* $I \subset \mathbb{R}$ is perfect. The closed set $S = [0, 1] \cup \{2\}$ is *not* perfect, because 2 is an isolated point of S. Proving that a set is perfect is not simple in general. An important example is *Cantor's ternary set*, to be introduced in Chap. 4.

Remark 2.2.15.

1. The limit points of a set *need not* necessarily belong to the set, but an isolated point of a set always belongs to the set.
2. The limit ξ of a convergent sequence (x_n) is *not* necessarily a limit point of the set $\{x_n : n \in \mathbb{N}\}$. Indeed, a *constant* sequence is obviously convergent, but the set of its terms, being a singleton, has no limit points.
3. Note that a set may have many (possibly infinite) limit points, whereas the limit of a convergent sequence is *unique*:

Proposition 2.2.16. *The limit of a convergent sequence is unique.*

Proof. Let (x_n) be a sequence such that $\lim(x_n) = \xi$ and $\lim(x_n) = \eta$. Then, given *any* $\varepsilon > 0$, we can find positive integers N_1 and N_2 such that $n \geq N_1$ implies $|x_n - \xi| < \varepsilon/2$, and $n \geq N_2$ implies $|x_n - \eta| < \varepsilon/2$. But then, with $N = \max(N_1, N_2)$, $n \geq N$ implies $|\eta - \xi| \leq |x_n - \xi| + |x_n - \eta| < \varepsilon/2 + \varepsilon/2 = \varepsilon$, and since ε was arbitrary, we get $\xi = \eta$. $\square$

Definition 2.2.17 (Increasing, Decreasing, Monotone, Bounded). We say that a real sequence (x_n) is *increasing* (resp., *strictly increasing*) if $x_n \leq x_{n+1}$ (resp., $x_n < x_{n+1}$), for all $n \in \mathbb{N}$. We say that it is *decreasing* (resp., *strictly decreasing*) if $x_n \geq x_{n+1}$ (resp., $x_n > x_{n+1}$), for all $n \in \mathbb{N}$. The sequence is called *monotone* if it is either increasing or decreasing (strictly or not). Finally, the sequence is called *bounded* (resp., *bounded above, bounded below*) if the set $\{x_n : n \in \mathbb{N}\}$ is bounded (resp., bounded above, bounded below). A sequence that is *not* bounded is called *unbounded*.

Examples and Remarks 2.2.18.

(a) A sequence $x = (x_n)$ is decreasing (resp., strictly decreasing) if and only if the sequence $-x$ is increasing (resp., strictly increasing). (Why?)
(b) The sequences (n), (n^2), and (2^n) are strictly increasing and unbounded. The sequence $(1/n)$ is strictly decreasing and bounded, and the sequence $(1 - 1/n)$ is strictly increasing and bounded.
(c) If (x_n) is bounded, then there are a, $b \in \mathbb{R}$ such that $a \leq x_n \leq b$ for all $n \in \mathbb{N}$; or, equivalently, there is $A > 0$ such that $|x_n| \leq A$ for all $n \in \mathbb{N}$.

Proposition 2.2.19 (Convergence and Boundedness). *Every convergent sequence is bounded. In particular, an unbounded sequence is divergent.*

Proof. Suppose that $\lim(x_n) = \xi$. Then, we can find $N \in \mathbb{N}$ such that $n \geq N$ implies $|x_n - \xi| < 1$, and hence $|x_n| < |\xi| + 1$. Therefore, we have $|x_n| \leq A$ for all $n \in \mathbb{N}$, where $A := \max(\{|x_1|, |x_2|, \ldots, |x_{N-1}|, |\xi| + 1\})$.   □

Definition 2.2.20 (Subsequence, Subsequential Limit). Let $x = (x_n)$ be a real sequence, and let $\nu = (\nu_1, \nu_2, \nu_3, \ldots) : \mathbb{N} \to \mathbb{N}$ be any *strictly increasing* sequence in $\mathbb{N}$; i.e., assume $\nu_1 < \nu_2 < \nu_3 < \cdots$. Then the sequence $x \circ \nu = (x_{\nu_1}, x_{\nu_2}, x_{\nu_3}, \ldots)$ is called a *subsequence* of (x_n). If (x_{ν_k}) converges, its limit is called a *subsequential limit* of (x_n).

Example 2.2.21.

1. Given any real sequence (x_n) and any $m \in \mathbb{N}$, the m-tail $(x_{m+k-1})_{k \in \mathbb{N}}$ is a subsequence of (x_n).
2. The *even subsequence* of (x_n) is defined to be $(x_{2k})_{k \in \mathbb{N}}$, and the *odd subsequence* is defined to be $(x_{2k-1})_{k \in \mathbb{N}}$.

Exercise 2.2.22.

1. Let $\nu = (\nu_1, \nu_2, \nu_3, \ldots)$ be a strictly increasing sequence in $\mathbb{N}$. Show that $\nu_k \geq k$ for all $k \in \mathbb{N}$.
2. Let $(x_n) \in \mathbb{R}^{\mathbb{N}}$. Show that if $\lim(x_n) = \xi$, then $\lim(x_{\nu_k}) = \xi$ for *any* subsequence $(x_{\nu_k})_{k \in \mathbb{N}}$. Show by an example that the converse is *false*.
3. Show that a real sequence (x_n) converges to $\xi \in \mathbb{R}$ if and only if its even and odd subsequences (x_{2k}) and (x_{2k-1}) *both* converge to ξ. Deduce that the sequence $((-1)^n)$ is divergent.

Proposition 2.2.23 (Existence of Monotone Subsequence). *Let (x_n) be a real sequence. Then (x_n) has a monotone subsequence.*

Proof. Let us call the mth term x_m of x a *peak* if $x_m \geq x_n$, for all $n \geq m$. If the sequence has *infinitely many* peaks, then, ordering their subscripts increasingly, we get the peaks $x_{m_1}, x_{m_2}, \ldots$, with $m_1 < m_2 < \cdots$. But then, by definition, $x_{m_1} \geq x_{m_2} \geq \cdots$. If (x_n) has a *finite* number of peaks, then there is a subscript $k \geq 1$ such that x_n is *not* a peak for all $n > k$. Let $m_1 := k + 1$. Then, since x_{m_1} is not a peak, there is $m_2 > m_1$ such that $x_{m_1} < x_{m_2}$. Since x_{m_2} is not a peak, there is $m_3 > m_2$ such that $x_{m_2} < x_{m_3}$. Continuing this process, we obtain a strictly increasing subsequence: $x_{m_1} < x_{m_2} < x_{m_3} < \cdots$.   □

Our first convergence result is the following.

Theorem 2.2.24 (Monotone Convergence Theorem). *Let (x_n) be a real sequence. If (x_n) is increasing and bounded above (resp., decreasing and bounded below), then (x_n) is convergent and we have $\lim(x_n) = \sup\{x_n : n \in \mathbb{N}\}$ (resp., $\inf\{x_n : n \in \mathbb{N}\}$).*

Proof. We treat the increasing (bounded above) case and leave the decreasing one as an exercise for the reader. Thus (x_n) is increasing, and there is $u \in \mathbb{R}$ such that $x_n \leq u$ for all $n \in \mathbb{N}$. Let $\xi = \sup\{x_n : n \in \mathbb{N}\}$ and let $\varepsilon > 0$ be arbitrary. Then there is $N = N(\varepsilon)$ such that $\xi - \varepsilon < x_N$, and hence $0 \le \xi - x_N < \varepsilon$. But (x_n) is increasing, so that $x_N \le x_n$ for all $n \ge N$, and we get $0 \le \xi - x_n \le \xi - x_N < \varepsilon$ for all $n \ge N$. □

As an application, the next exercise will show that for any $a > 0$, the *positive square root* $\sqrt{a}$ can be obtained as the limit of a decreasing sequence. This method was known to the Babylonians before 1500 B.C.

Exercise 2.2.25.

1. Show that $(t + a/t)^2 \ge 4a$, for all $a \in \mathbb{R}$ and $t \ne 0$.
2. Given $a > 0$, let $x_1 > 0$ be arbitrary and define $x_{n+1} = (x_n + a/x_n)/2$ recursively for all $n \in \mathbb{N}$. Using (1) show that $x_{n+1} \le x_n \ \forall \ n \ge 2$, so that (x_n) is *ultimately decreasing* and *bounded below*.
3. Deduce that (x_n) is convergent, and show that $\lim(x_n) = \sqrt{a}$.

Next, we give a list of properties of convergent sequences. These properties are all familiar to the reader from calculus and will be used frequently in what follows.

Theorem 2.2.26 (Limit Theorems). *Let (x_n) and (y_n) be convergent sequences with $\lim(x_n) = \xi$, $\lim(y_n) = \eta$. Then the following statements are true:*

1. $\lim(x_n \pm y_n) = \lim(x_n) \pm \lim(y_n) = \xi \pm \eta$;
2. $\lim(cx_n) = c\lim(x_n) = c\xi$ $(\forall c \in \mathbb{R})$;
3. $\lim(x_n y_n) = \lim(x_n) \cdot \lim(y_n) = \xi\eta$;
4. $\lim(x_n/y_n) = \lim(x_n)/\lim(y_n) = \xi/\eta$ if $y_n \ne 0 \ \forall \ n$ and $\eta \ne 0$;
5. $\lim(1/y_n) = 1/\lim(y_n) = 1/\eta$ if $y_n \ne 0 \ \forall \ n$ and $\eta \ne 0$;
6. *if $x_n \ge 0$ is ultimately satisfied, then $\xi \ge 0$*;
7. *if $x_n \le y_n$ is ultimately satisfied, then $\xi \le \eta$; and*
8. *(Squeeze Theorem) If $\xi = \eta$ and if $x_n \le z_n \le y_n$ is ultimately satisfied, then (z_n) converges and we have $\lim(z_n) = \xi = \eta$.*

Proof. We leave (1) and (2) as exercises for the reader. Note that (5) follows from (4) if (x_n) is the constant sequence $(1, 1, 1, \ldots)$, and (4) follows from (3) and (5). Also, (7) follows from (6) applied to the sequence $(y_n - x_n)$. To prove (3), note that for each $n \in \mathbb{N}$, we have

$$|x_n y_n - \xi\eta| = |\eta(x_n - \xi) + x_n(y_n - \eta)| \le |\eta||x_n - \xi| + |x_n||y_n - \eta|. \qquad (*)$$

Since the convergent sequence (x_n) is bounded, there is $A > 0$ with $|x_n| \le A$ for all $n \in \mathbb{N}$. Now define $B = \max\{A, |\eta|\}$, and pick $N \in \mathbb{N}$ so large that $n \ge N$ implies $|x_n - \xi| < \varepsilon/(2B)$ and $|y_n - \eta| < \varepsilon/(2B)$. It then follows from $(*)$ that

$$|x_n y_n - \xi\eta| \le B(|x_n - \xi| + |y_n - \eta|) < \varepsilon.$$

To prove (5), note first that we can find $N_1 \in \mathbb{N}$ such that $n \ge N_1$ implies $|y_n - \eta| < |\eta|/2$, which in turn implies $|y_n| > |\eta|/2$. (Why?) It then follows that $|1/y_n - 1/\eta| = |y_n - \eta|/(|y_n||\eta|) < 2|y_n - \eta|/|\eta|^2$, for all $n \ge N_1$. Next, given $\varepsilon > 0$, pick $N_2 \in \mathbb{N}$ such that $n \geq N_2$ implies $|y_n - \eta| < \varepsilon |\eta|^2/2$. But then, with $N = \max\{N_1, N_2\}$, we get $|1/y_n - 1/\eta| < \varepsilon$. Looking at (6), pick $m \in \mathbb{N}$ such that $x_n \geq 0 \ \forall\, n \geq m$. If $\xi < 0$, then, with $\varepsilon = -\xi/2 > 0$, we can find $N_1 \in \mathbb{N}$, with $N_1 \geq m$, such that $n \geq N_1$ implies $|x_n - \xi| < -\xi/2$. But then, for $n \geq N := \max\{N_1, m\}$, we get $x_n < \xi - \xi/2 = \xi/2 < 0$, contradicting the fact that $x_n \geq 0$ for all $n \geq m$. Finally, to prove (8), let $m \in \mathbb{N}$ be such that $x_n \leq z_n \leq y_n \ \forall\, n \geq m$. For each $\varepsilon > 0$, we can find $N \in \mathbb{N}$, with $N \geq m$, such that $n \geq N$ implies the inequalities $|x_n - \xi| < \varepsilon/3$ and $|y_n - \eta| = |y_n - \xi| < \varepsilon/3$, from which we get $|y_n - x_n| < 2\varepsilon/3 \ \forall\, n \geq N$. But then, for $n \geq N$, we have

$$|z_n - \xi| \leq |z_n - x_n| + |x_n - \xi| \leq |y_n - x_n| + |x_n - \xi| < 2\varepsilon/3 + \varepsilon/3 = \varepsilon,$$

which completes the proof. $\square$

Remark 2.2.27. Note that, although *nonstrict* inequalities are preserved when we pass to the limit (as in parts (6) and (7) of Theorem 2.2.26), this is not necessarily true for *strict* inequalities. For example, while $1/n^2 < 1/n, \ \forall\, n \geq 2$, we have $\lim(1/n^2) = \lim(1/n) = 0$.

Example 2.2.28.

1. We have $\lim_{n \to \infty} \sin n/n = 0$. Indeed, $|\sin n/n| \leq 1/n$ gives $-1/n \leq \sin n/n \leq 1/n$ for all $n \in \mathbb{N}$. Since $\lim(-1/n) = 0 = \lim(1/n)$ (why?), the assertion follows from the *Squeeze Theorem*.
2. The sequence $(\sin n)$ is *divergent*. Suppose, to get a contradiction, that $\lim(\sin n) = b \in \mathbb{R}$. Letting $n \to \infty$ in the identity $\sin(n + 1) = \sin n \cos 1 + \cos n \sin 1$, we see that $a := \lim(\cos n)$ also exists and that $a^2 + b^2 = 1$. (Why?) But then the identities $\sin(n \pm 1) = \sin n \cos 1 \pm \cos n \sin 1$ give the equations $b = b \cos 1 \pm a \sin 1$, which in turn give $a = b = 0$, a contradiction.

Exercise 2.2.29. Let (x_n) be a real sequence with $\lim(x_n) = \xi$.

1. Show that $\lim(x_n^k) = \xi^k$ for all positive integers k. Show that the same also holds for all integers $k \leq 0$, if $x_n \neq 0$ for all n and $\xi \neq 0$.
2. Show that, if $x_n \geq 0 \ \forall\, n \in \mathbb{N}$, then $\lim(\sqrt{x_n}) = \sqrt{\xi}$. *Hint:* Consider the cases $\xi = 0$ and $\xi > 0$. In the latter case, use $x_n - \xi = (\sqrt{x_n} - \sqrt{\xi})(\sqrt{x_n} + \sqrt{\xi})$ and the fact that $\sqrt{x_n} + \sqrt{\xi} \geq \sqrt{\xi} > 0 \ \forall\, n \in \mathbb{N}$.
3. Show that $\lim |x_n| = |\xi|$.

Definition 2.2.30 (Null Sequence). A real sequence (x_n) is called a *null sequence* if $\lim(x_n) = 0$.

Exercise 2.2.31.

1. Show that $\lim(x_n) = \xi \in \mathbb{R}$ if and only if $(x_n - \xi)$ is a null sequence.
2. Show that (x_n) is a null sequence if and only if $(|x_n|)$ is.

3. Show that, if (x_n) and (y_n) are null sequences, then so are $(x_n \pm y_n)$ and (cx_n) for any constant $c \in \mathbb{R}$.
4. Show that, if (x_n) is a null sequence and (y_n) is *bounded*, then $(x_n y_n)$ is also a null sequence.

Proposition 2.2.32. *If (x_n) is a null sequence, then so is the sequence*

$$(\bar{x}_n) := \left(\frac{x_1 + x_2 + \cdots + x_n}{n} \right).$$

Proof. Given $\varepsilon > 0$, pick $m \in \mathbb{N}$ such that $n \geq m$ implies $|x_n| < \varepsilon/2$. Now for any $n \geq m$, we have

$$|\bar{x}_n| \leq \frac{|x_1 + x_2 + \cdots + x_{m-1}|}{n} + \left(\frac{n - m + 1}{n} \right) \frac{\varepsilon}{2} \leq \frac{|\sum_{k=1}^{m-1} x_k|}{n} + \frac{\varepsilon}{2}.$$

If we now pick $N \geq m$ such that $n \geq N$ implies $|x_1 + x_2 + \cdots + x_{m-1}|/n < \varepsilon/2$, it follows that $|\bar{x}_n| < \varepsilon$ for all $n \geq N$. $\qquad \square$

Corollary 2.2.33. *If $\lim(x_n) = \xi$ and if $(\bar{x}_n)$ is as in Proposition 2.2.32, then $\lim(\bar{x}_n) = \xi$.*

Proof. Simply apply Proposition 2.2.32 to the null sequence $(x_n - \xi)$. $\qquad \square$

Exercise 2.2.34. Show that $(\bar{x}_n)$ may converge for a *divergent* sequence (x_n). *Hint:* Let $x_n := 1 - (-1)^n$.

Definition 2.2.35 (Infinite Limits). Given a real sequence (x_n), we say that (x_n) *converges to ∞* and write $\lim(x_n) = \infty$, if for every $B \in \mathbb{R}$ there exists $N \in \mathbb{N}$ such that $n \geq N$ implies $x_n > B$. Similarly, we say that (x_n) *converges to $-\infty$* and write $\lim(x_n) = -\infty$, if for every $A \in \mathbb{R}$ there exists $N \in \mathbb{N}$ such that $n \geq N$ implies $x_n < A$.

Exercise 2.2.36. For real sequences (x_n) and (y_n) prove the following:

1. $\lim(x_n) = \pm\infty \implies \lim(-x_n) = \mp\infty$;
2. $\lim(x_n) = +\infty$ (resp., $-\infty$) if and only if $\exists m \in \mathbb{N}$ with $x_n > 0$ (resp., $x_n < 0$) $\forall n \geq m$ and $\lim_{k \to \infty}(1/x_{m+k}) = 0$;
3. $\lim(x_n) = \pm\infty$ and $\lim(y_n) = \pm\infty \implies \lim(x_n + y_n) = \pm\infty$;
4. $\lim(x_n) = \xi > 0$ and $\lim(y_n) = \pm\infty \implies \lim(x_n y_n) = \pm\infty$;
5. $\lim(x_n) = \xi < 0$ and $\lim(y_n) = \pm\infty \implies \lim(x_n y_n) = \mp\infty$;
6. if $x_n \leq y_n$ *ultimately* holds, then $\lim(x_n) = \infty$ implies $\lim(y_n) = \infty$ and $\lim(y_n) = -\infty$ implies $\lim(x_n) = -\infty$;
7. if $x_n > 0$, $y_n > 0 \ \forall n \in \mathbb{N}$ and $\lim(x_n/y_n) = \rho > 0$, then $\lim(x_n) = \infty$ if and only if $\lim(y_n) = \infty$;
8. if (x_n) is *ultimately increasing* and *not* bounded above (resp., *ultimately decreasing* and *not* bounded below), then $\lim(x_n) = \infty$ (resp., $\lim(x_n) = -\infty$); and
9. if (x_n) is *bounded*, $\lim(y_n) = \pm\infty$, and $y_n \neq 0 \ \forall n$, then $\lim(x_n/y_n) = 0$.

Definition 2.2.37 (Upper Limit, Lower Limit).   Let (x_n) be a real sequence.

(a) If (x_n) is *bounded above*, then we define its *upper limit* (denoted $\lim \sup(x_n)$ or $\overline{\lim}(x_n)$) to be

$$\lim \sup(x_n) = \overline{\lim}(x_n) := \lim_{n \to \infty} (\sup\{x_k : k \geq n\}).$$

If (x_n) is *not* bounded above, then we define $\lim \sup(x_n) = \overline{\lim}(x_n) := \infty$.
(a) If (x_n) is *bounded below*, then we define its *lower limit* (denoted $\lim \inf(x_n)$ or $\underline{\lim}(x_n)$) to be

$$\lim \inf(x_n) = \underline{\lim}(x_n) := \lim_{n \to \infty} (\inf\{x_k : k \geq n\}).$$

If (x_n) is *not* bounded below, then we define $\lim \inf(x_n) = \underline{\lim}(x_n) := -\infty$.

Example 2.2.38.
   Let $x_n = (-1)^n$, $y_n = n^2$, $z_n = (-1)^n n$. Then $\underline{\lim}(x_n) = -1$ and $\overline{\lim}(x_n) = 1$; $\underline{\lim}(y_n) = \overline{\lim}(y_n) = \infty$; $\underline{\lim}(z_n) = -\infty$ and $\overline{\lim}(z_n) = \infty$.

Proposition 2.2.39. *Let (x_n) be a bounded sequence and for each $n \in \mathbb{N}$ define $u_n := \inf\{x_k : k \geq n\}$ and $v_n := \sup\{x_k : k \geq n\}$.*

(a) *If the inequalities $a \leq x_n \leq b$ are ultimately satisfied (i.e., hold for all $n \geq m$ with some $m \in \mathbb{N}$), then $\underline{\lim}(x_n) \geq a$ and $\overline{\lim}(x_n) \leq b$.*
(b) *(u_n) is increasing, (v_n) is decreasing, and we have $\lim(u_n) = \underline{\lim}(x_n) \leq \overline{\lim}(x_n) = \lim(v_n)$.*
(c) *We have $\underline{\lim}(x_n) \leq \underline{\lim}(x_{v_k}) \leq \overline{\lim}(x_{v_k}) \leq \overline{\lim}(x_n)$ for every subsequence (x_{v_k}) of (x_n).*
(d) *If $\alpha < \underline{\lim}(x_n)$ [resp., $\beta > \overline{\lim}(x_n)$], then there is an $N \in \mathbb{N}$ with $x_n > \alpha$ [resp., $x_n < \beta$] for all $n \geq N$.*
(e) *The sets $\{n : x_n > \overline{\lim}(x_n) - \varepsilon\}$ and $\{n : x_n < \underline{\lim}(x_n) + \varepsilon\}$ are both infinite $\forall \varepsilon > 0$.*
(f) *$\lim(x_n) = \xi$ if and only if $\underline{\lim}(x_n) = \xi = \overline{\lim}(x_n)$.*

Proof. We leave parts (a), (b), (c), and (d) as exercises for the reader. For (e), let $v = \overline{\lim}(x_n)$ and suppose that $\{n : x_n > v - \varepsilon\}$ is *finite*. Then for some $m \in \mathbb{N}$ we have $x_n \leq v - \varepsilon$ for all $n \geq m$ and part (a) gives $v \leq v - \varepsilon$, which is absurd. A similar argument shows that $\{n : x_n < \underline{\lim}(x_n) + \varepsilon\}$ is also infinite. To prove (f), note that if $\lim(x_n) = \xi$, then for any $\varepsilon > 0$ we can find $N \in \mathbb{N}$ so that $\xi - \varepsilon < x_n < \xi + \varepsilon$ for all $n \geq N$. But then parts (a) and (b) show that $\xi - \varepsilon \leq \underline{\lim}(x_n) \leq \overline{\lim}(x_n) \leq \xi + \varepsilon$ holds for all $\varepsilon > 0$ and hence $\underline{\lim}(x_n) = \xi = \overline{\lim}(x_n)$. Conversely, if $\underline{\lim}(x_n) = \xi = \overline{\lim}(x_n)$ holds and if $\varepsilon > 0$ is arbitrary, then part (d) implies that there exists $N \in \mathbb{N}$ with $\xi - \varepsilon < x_n < \xi + \varepsilon$ for all $n \geq N$ and hence $\lim(x_n) = \xi$. $\square$

Proposition 2.2.40. *Let S be the set of all subsequential limits of a bounded real sequence (x_n); i.e., the set of all $\xi \in (-\infty, \infty)$ such that $\xi = \lim(x_{v_k})$ for some* *subsequence* (x_{ν_k}) *of* (x_n). *Then* $\inf(S)$ *and* $\sup(S)$ *are both in* S *and we have* $\liminf(x_n) = \inf(S)$ *and* $\limsup(x_n) = \sup(S)$.

Proof. Since the proofs for $\inf(S)$ and $\sup(S)$ are similar, we show that $\sup(S) = \overline{\lim}(x_n) \in S$ and leave the other case as an exercise. Now given any $\xi \in S$, there is a subsequence (x_{ν_k}) of (x_n) with $\xi = \lim(x_{\nu_k})$. By parts (c) and (f) of Proposition 2.2.39, we get

$$\xi = \lim(x_{\nu_k}) = \overline{\lim}(x_{\nu_k}) \le \overline{\lim}(x_n).$$

Therefore, $\sup(S) \le \overline{\lim}(x_n)$. To finish the proof, let $s := \overline{\lim}(x_n)$ and note that (by part (e) of Proposition 2.2.39) the set $\{n : s - 1/k < x_n < s + 1/k\}$ is *infinite* for every $k \in \mathbb{N}$. So for each $k \in \mathbb{N}$ we can pick $\nu_k \in \{n : s - 1/k < x_n < s + 1/k\}$ in such a way that $\nu_{k+1} > \nu_k$ for all k. We then have a subsequence (x_{ν_k}) such that $s - 1/k < x_{\nu_k} < s + 1/k$ for all k and hence $\lim(x_{\nu_k}) = s \in S$. $\square$

Exercise 2.2.41. Given any real sequences (x_n) and (y_n), prove the following:

1. $\liminf(x_n) + \liminf(y_n) \le \liminf(x_n + y_n) \le \limsup(x_n) + \liminf(y_n) \le \limsup(x_n + y_n) \le \limsup(x_n) + \limsup(y_n)$, if none of the sums is $\infty - \infty$ (or $-\infty + \infty$);
2. $\limsup(x_n y_n) \le (\limsup(x_n))(\limsup(y_n))$ if $x_n \ge 0$ and $y_n \ge 0$ for all n (and the right side is not of the form $0 \cdot \infty$); and
3. $\liminf(x_n) = -\limsup(-x_n)$.

The next theorem is very important and will appear in two versions. We first give the version for the sequences and then the version for bounded *infinite* subsets of $\mathbb{R}$.

Theorem 2.2.42 (Bolzano–Weierstrass Theorem for Sequences). *Every bounded sequence of real numbers has a convergent subsequence.*

Proof. Let (x_n) be a bounded sequence in $\mathbb{R}$. Then, by Proposition 2.2.23, it has a monotone subsequence $(x_{\nu_k})_{k \in \mathbb{N}}$, which is bounded because (x_n) is. The convergence of the subsequence (x_{ν_k}) now follows from the Monotone Convergence Theorem (Theorem 2.2.24). $\square$

Theorem 2.2.43 (Bolzano–Weierstrass Theorem for Infinite Sets). *Every bounded infinite subset of real numbers has a limit point in* $\mathbb{R}$.

Proof. Let X be a *bounded infinite* set of real numbers. By Proposition 1.4.11, there is a *sequence* (x_n) in X with $x_j \ne x_k$ if $j \ne k$. Now (x_n) is *bounded* because X is and hence, by Theorem 2.2.42, it has a convergent subsequence (x_{ν_k}). Let $\lim(x_{\nu_k}) = \xi$. Since $\{x_{\nu_k} : k \in \mathbb{N}\}$ is *countably infinite* (why?), it cannot be ultimately constant. Thus (cf. Exercise 2.2.12) ξ is a limit point of X. $\square$

The next definition is of fundamental importance and will allow us to *construct* the set of real numbers from the set of rational ones.

Definition 2.2.44 (Cauchy Sequence). A real sequence (x_n) is said to be a *Cauchy sequence* if for each $\varepsilon > 0$ there is an integer $N \in \mathbb{N}$ such that, if $m, n \geq N$, then $|x_m - x_n| < \varepsilon$.

The first immediate consequence of the above definition is the following

Proposition 2.2.45. *Every Cauchy sequence in $\mathbb{R}$ is bounded.*

Proof. Let (x_n) be a Cauchy sequence. Then, for $\varepsilon = 1$, we can find an integer $N > 0$ such that $|x_m - x_n| < 1$, for all $m \geq N$, $n \geq N$. In particular, $|x_n - x_N| < 1$ for all $n \geq N$, which implies $|x_n| < |x_N| + 1$ for all $n \geq N$. But then, with $B := \max\{|x_1|, \ldots, |x_{N-1}|, |x_N| + 1\}$, we get $|x_n| \leq B$ for all $n \in \mathbb{N}$. $\square$

It is intuitively obvious that a convergent sequence must be a Cauchy sequence. In fact, not only is this the case but also the converse, which is not obvious at all, is true:

Theorem 2.2.46 (Cauchy's Criterion). *A real sequence (x_n) is convergent if and only if it is a Cauchy sequence.*

Proof. If $\lim(x_n) = \xi$, then for each $\varepsilon > 0$ we can find $N \in \mathbb{N}$ such that $n \geq N$ implies $|x_n - \xi| < \varepsilon/2$. But then, if $m, n \geq N$, we have $|x_m - x_n| \leq |x_m - \xi| + |x_n - \xi| < \varepsilon/2 + \varepsilon/2 = \varepsilon$. Conversely, if (x_n) is Cauchy, then it is bounded (by Proposition 2.2.45), and hence (by Theorem 2.2.42) has a convergent subsequence (x_{n_k}). Now, given $\varepsilon > 0$, we can find $N \in \mathbb{N}$ such that $|x_m - x_n| < \varepsilon/2$ for all $m, n \geq N$. Next, if $\lim(x_{n_k}) = \xi$, then $|x_{n_K} - \xi| < \varepsilon/2$ for some $K \in \mathbb{N}$. Assuming, as we may, that $K \geq N$, it follows that, for all $n \geq N$, we have $|x_n - \xi| \leq |x_n - x_{n_K}| + |x_{n_K} - \xi| < \varepsilon/2 + \varepsilon/2 = \varepsilon$. $\square$

Remark 2.2.47.

1. The above proof of Cauchy's Criterion contains the following important fact: *If a Cauchy sequence (x_n) has a subsequence (x_{n_k}) that converges to ξ, then* $\lim(x_n) = \xi$.

2. (Supremum Property $\Longleftrightarrow$ Cauchy's Criterion) The above proof also shows that Cauchy's Criterion is a consequence of the Supremum Property (Completeness Axiom). In fact, the converse is also true and can be proved by the following *nested intervals* argument. Let S be a *nonempty* subset of $\mathbb{R}$ that is *bounded above*; i.e., there is a number $u \in \mathbb{R}$ such that $s \leq u \; \forall s \in S$. Construct a sequence of nested intervals $[a_n, b_n]$ as follows: Pick $I_1 = [a_1, b_1], a_1 < b_1$, such that I_1 meets S and b_1 is an upper bound of S. Divide I_1 in two *equal* parts. Let I_2 be the right half if it meets S; otherwise, let it be the left half. Define $I_3, I_4, \ldots$, similarly. Now, for each $n \in \mathbb{N}$ choose a point $s_n \in I_n \cap S$. The sequence (s_n) is Cauchy (why?). Let $\sigma = \lim(s_n)$. We invite the reader to show that $\sigma = \sup(S)$. Note that, by construction, each b_n is an upper bound for S.

Exercise 2.2.48 (Contractive Sequence).

(a) A sequence (x_n) is said to be *contractive* if there exists a constant $r \in (0, 1)$ such that $|x_{n+2} - x_{n+1}| \leq r|x_{n+1} - x_n|$ for all $n \in \mathbb{N}$. Show that a contractive sequence is a Cauchy sequence. If $x_1 > 0$ and $x_{n+1} := 1/(x_n + 2)$ for all $n \geq 1$, show that (x_n) is contractive and find its limit. *Hint:* Use the fact that $\sum_{k=1}^{m} r^k = (1 - r^{m+1})/(1 - r)$.

(b) Define the sequence (x_n) as follows: $x_1 := 1$, $x_2 := 2$, and $x_n := (x_{n-2} + x_{n-1})/2$, for $n > 2$. Show that (x_n) is a Cauchy sequence and find its limit. *Hint:* For the limit, look at the *odd* subsequence (x_{2n+1}).

(c) If $0 < r < 1$, and if $|x_{n+1} - x_n| < r^n$ for all $n \in \mathbb{N}$, show that (x_n) is a Cauchy sequence.

2.3   Infinite Series

Recall that for a *finite* set $\{x_1, x_2, \dots, x_n\} \subset \mathbb{R}$, we denote the sum of its elements by $\sum_{k=1}^{n} x_k$. It is tempting to extend this summation to a *countably infinite* subset of $\mathbb{R}$, but, as we shall presently see, the corresponding *infinite sums*, which we shall call *(infinite) series*, may not exist. The present section contains a brief discussion of such series and the conditions under which they are *summable*. As the reader will notice, we shall occasionally use exponentials with *real* exponents. These will be defined *rigorously* later.

Definition 2.3.1 (Infinite Series, Partial Sums). Given a sequence (x_n) of real numbers, the *formal sum*

$$\sum_{n=1}^{\infty} x_n = x_1 + x_2 + x_3 + \cdots + x_n + \cdots$$

is called an *infinite series*, or simply *series*, and, for each $n \in \mathbb{N}$, x_n is called the *nth term* of the series. Also, for integers $n \in \mathbb{N}$, the *finite* sums

$$s_n = \sum_{k=1}^{n} x_k$$

are called the *partial sums* of the series. If the *sequence* (s_n) converges to a number $s \in \mathbb{R}$, we say that the series *converges* (or is *convergent*) and write

$$\sum_{n=1}^{\infty} x_n = s.$$

The number s is then called the *sum* of the series. If (s_n) diverges, we say that the series *diverges* or that it is *divergent*. Unless the *index n* takes other values than $1, 2, 3, \dots$, we sometimes replace $\sum_{n=1}^{\infty} x_n$ by $\sum x_n$.

Remark 2.3.2. It is obvious from the above definition that, for each $n > 1$, we have $x_n = s_n - s_{n-1}$, so that, if we set $x_1 = s_1$, any statement about series can be written as a statement about sequences and vice versa. In particular, since the convergence of the series is, by definition, the convergence of the sequence of its partial sums, Cauchy's Criterion may be applied to (s_n) and implies the following theorem for the series:

Theorem 2.3.3 (Cauchy's Criterion for Series). *A series $\sum_{n=1}^{\infty} x_n$ is convergent if and only if, for each $\varepsilon > 0$, there is an integer $N \in \mathbb{N}$ such that*

$$m \geq n \geq N \implies \left| \sum_{k=n}^{m} x_k \right| < \varepsilon.$$

The next proposition gives a necessary (but not sufficient) condition for the convergence of series.

Proposition 2.3.4. *If $\sum_{n=1}^{\infty} x_n$ is convergent, then $\lim(x_n) = 0$.*

Proof. Indeed, it follows from Cauchy's Criterion (with $m = n$) that, for each $\varepsilon > 0$, there exists $N \in \mathbb{N}$ with $|x_n| < \varepsilon$ for all $n \geq N$.   □

Remark 2.3.5.

1. As pointed out above, the condition in Proposition 2.3.4 is *not sufficient*. Indeed, as we shall see below, the *harmonic series* $\sum_{n=1}^{\infty} 1/n$ is *divergent*, even though we obviously have $\lim(1/n) = 0$.
2. If $\sum_{n=1}^{\infty} x_n$ is a series of *nonnegative terms*, i.e., if $x_n \geq 0 \; \forall \; n \in \mathbb{N}$, then the sequence (s_n) of its partial sums is obviously *increasing*. The following theorem is then a consequence of the Monotone Convergence Theorem.

Theorem 2.3.6. *A series of nonnegative terms is convergent if and only if the sequence of its partial sums is bounded.*

Definition 2.3.7 (Geometric Series, p-Series).

1. Given the real numbers a and $r \neq 0$, the series

$$\sum_{n=0}^{\infty} ar^n = a + ar + ar^2 + ar^3 + \cdots$$

is called a *geometric series* with *first term a* and *ratio r*.
2. Given a real number p, the series

$$\sum_{n=1}^{\infty} 1/n^p = 1 + 1/2^p + 1/3^p + \cdots$$

is called a *p-series*. In particular, for $p = 1$, we get the *harmonic series* $\sum_{n=1}^{\infty} 1/n$.

Proposition 2.3.8. *Assuming (to avoid trivial cases) that $a \neq 0 \neq r$, the geometric series $\sum_{n=0}^{\infty} ar^n$ is convergent if and only if $|r| < 1$, in which case we have*

$$\sum_{n=0}^{\infty} ar^n = \frac{a}{1-r}. \qquad (*)$$

Proof. First, we have the identity

$$1 - r^{n+1} = (1-r)(1 + r + r^2 + \cdots + r^n),$$

which is proved by expanding and simplifying the right side. It follows that (if $r \neq 1$)

$$s_n = \sum_{k=0}^{n} ar^k = \frac{a(1 - r^{n+1})}{1-r}.$$

Now, if $|r| < 1$, then (by Exercise 2.2.7) $\lim(r^{n+1}) = 0$ and $(*)$ follows at once. If, however, $|r| \geq 1$, then, since (by assumption) $a \neq 0$, we *cannot* have $\lim(ar^n) = 0$ (why?), and hence (by Proposition 2.3.4) the series diverges. $\qquad\square$

Before we consider the convergence of p-series, we prove the following *comparison test*:

Theorem 2.3.9 (First Comparison Test).   *Let $\sum x_n$ and $\sum y_n$ be two series of nonnegative terms, and suppose that we have $x_n \leq y_n$, for all $n \in \mathbb{N}$. Then the following are true:*

(a) If $\sum y_n$ is convergent, then so is $\sum x_n$.
(b) If $\sum x_n$ is divergent, then so is $\sum y_n$.

Proof. Define $s_n := \sum_{k=1}^{n} x_n$ and $t_n := \sum_{k=1}^{n} y_n$. We then have $s_n \leq t_n$ $\forall n \in \mathbb{N}$. The theorem is therefore an immediate consequence of Theorem 2.3.6. $\qquad\square$

Exercise 2.3.10.

1. Given a real sequence (x_n) and a natural number $m \in \mathbb{N}$, show that the series $\sum_{n=1}^{\infty} x_n$ is convergent if and only if the series $\sum_{n=m}^{\infty} x_n$ is convergent.
2. Show that, in Theorem 2.3.9, the condition $x_n \leq y_n$ $\forall n \in \mathbb{N}$ can be replaced by $\exists m \in \mathbb{N}$ such that $\forall n \geq m$, we have $x_n \leq y_n$.
3. Show that, if $\sum x_n$ is a *convergent* series of nonnegative terms, and if (ξ_n) is a *bounded* sequence of nonnegative real numbers, then the series $\sum \xi_n x_n$ is also convergent.
4. Show that, if $\sum y_n$ is a *divergent* series of nonnegative terms, and if (η_n) is a sequence of positive reals that is bounded below by a *positive* number $\eta > 0$, then the series $\sum \eta_n y_n$ is also divergent.

5. Let $0 < a < b$, and let (c_n) be a real sequence satisfying $a \le c_n \le b \; \forall n \in \mathbb{N}$. Show that the series of nonnegative terms $\sum x_n$ converges if and only if $\sum c_n x_n$ converges.

Corollary 2.3.11 (Limit Comparison Test). *Let $\sum x_n$ and $\sum y_n$ be two series with positive terms such that $\ell := \lim(x_n/y_n)$ exists.*

(a) If $\ell > 0$, then $\sum x_n$ converges if and only if $\sum y_n$ converges.
(b) If $\ell = 0$ and $\sum y_n$ converges, then $\sum x_n$ converges.

Proof. Exercise!   □

Proposition 2.3.12. *The p-series $\sum 1/n^p$ is convergent for $p > 1$ and divergent for $p \le 1$. In particular, the harmonic series $\sum 1/n$ is divergent.*

Proof. Let us first show that the harmonic series $\sum 1/n$ is divergent by proving that the partial sums $s_n = \sum_{k=1}^{n} 1/k$ are *unbounded*. Given $B > 0$, pick $k \in \mathbb{N}$ with $k > 2B$, and let $n \ge 2^k$ be arbitrary. Now we have

$$s_n \ge \left(1 + \frac{1}{2}\right) + \left(\frac{1}{3} + \frac{1}{4}\right) + \cdots + \left(\frac{1}{2^{k-1}+1} + \cdots + \frac{1}{2^k}\right)$$

$$> \frac{1}{2} + 2\left(\frac{1}{4}\right) + \cdots + 2^{k-1}\left(\frac{1}{2^k}\right) = \frac{k}{2} > B.$$

Therefore, (s_n) is unbounded. Now if $p \le 1$, then $1/n^p \ge 1/n \; \forall n \in \mathbb{N}$ and, by the First Comparison Test, $\sum 1/n^p$ is divergent. Next, suppose that $p > 1$. To find an upper bound for the partial sum $\sum_{k=1}^{n} 1/k^p$, choose $k \in \mathbb{N}$ so large that $n < 2^k$, and note that we have

$$s_{2^k-1} = 1 + \left(\frac{1}{2^p} + \frac{1}{3^p}\right) + \cdots + \left(\frac{1}{(2^{k-1})^p} + \cdots + \frac{1}{(2^k-1)^p}\right)$$

$$\le 1 + \frac{2}{2^p} + \frac{4}{4^p} + \cdots + \frac{2^{k-1}}{(2^{k-1})^p}.$$

Therefore, with $r = 1/2^{p-1}$, we have

$$s_n \le s_{2^k-1} \le \sum_{j=1}^{k-1} r^j = \frac{1 - r^k}{1 - r} < \frac{1}{1 - r}.$$

Since this is true for each $n \in \mathbb{N}$, the proof is complete.   □

Exercise 2.3.13. Give another proof of the convergence of $\sum 1/n^p$ for $p \ge 2$, using the First Comparison Test and the fact that, for $k > 1$, we have $1/k^2 < 1/k(k-1) = 1/(k-1) - 1/k$.

The number e, called the *natural base*, is probably the most important number in mathematics. We first define it as the sum of an infinite series and in the next proposition show that it is also the limit of the sequence $((1 + 1/n)^n)$.

Definition 2.3.14 (The Number e). $e := \sum_{n=0}^{\infty} 1/n!$.

Proposition 2.3.15. We have $2 < e < 3$, and $e = \lim_{n \to \infty} (1 + 1/n)^n$.

Proof. It is obvious that $e > 2$. Since $1/n! \leq 1/(2 \cdot 3^{n-2})$ for all $n \geq 3$ and $\sum_{n=3}^{\infty} 1/3^{n-2} = 1/2$, we have that, for all $n \geq 3$,

$$s_n = \sum_{k=0}^{n} 1/k! < 1 + 1 + 1/2 + \sum_{k=3}^{\infty} 1/(2 \cdot 3^{k-2}) = 11/4,$$

which shows that the series converges and that $e \leq 11/4 < 3$. To prove the last statement, define $t_n := (1 + 1/n)^n$ and note that by the binomial formula,

$$t_n = 1 + 1 + \frac{1}{2!}\left(1 - \frac{1}{n}\right) + \cdots + \frac{1}{n!}\left(1 - \frac{1}{n}\right)\left(1 - \frac{2}{n}\right)\cdots\left(1 - \frac{n-1}{n}\right).$$

It follows that $t_n \leq s_n$, so that

$$\limsup(t_n) \leq e.$$

Next, for any *fixed* $m \in \mathbb{N}$ and $n \geq m$, we have

$$t_n \geq 1 + 1 + \frac{1}{2!}\left(1 - \frac{1}{n}\right) + \cdots + \frac{1}{m!}\left(1 - \frac{1}{n}\right)\cdots\left(1 - \frac{m-1}{n}\right),$$

so that letting $n \to \infty$, we get

$$s_m = \sum_{k=0}^{m} \frac{1}{k!} \leq \liminf(t_n),$$

and, since m was arbitrary, it follows that

$$e \leq \liminf(t_n).$$

The proposition now follows from $\limsup(t_n) \leq e \leq \liminf(t_n)$ and Proposition 2.2.39(f). $\qquad\square$

Exercise 2.3.16. Let $x_n := (1 + 1/n)^n$ and $y_n := (1 + 1/n)^{n+1}$, $n \in \mathbb{N}$. Show that (x_n) is *increasing*, while (y_n) is *decreasing*. Deduce that $\{[x_n, y_n] : n \in \mathbb{N}\}$ is a sequence of *nested intervals*. Next, show that $\lim(y_n - x_n) = 0$, and deduce that $\lim(x_n) = e = \lim(y_n)$. *Hint:* To prove that (x_n) is increasing, note that

$$\left(\frac{n+1}{n}\right)^n \le \left(\frac{n+2}{n+1}\right)^{n+1} \iff \frac{n}{n+1} \le \left[\frac{n(n+2)}{(n+1)^2}\right]^{n+1}$$

$$\iff 1 - \frac{1}{n+1} \le \left[1 - \frac{1}{(n+1)^2}\right]^{n+1}$$

and use Bernoulli's inequality. The proof of $y_n \ge y_{n+1}$ is similar.

Proposition 2.3.17 (The Irrationality of e). *The number $e := \sum_{n=0}^{\infty} 1/n!$ is irrational.*

Proof. If $e = p/q$ with $p, q \in \mathbb{N}$, then $q > 1$ and

$$q!e - \left(\sum_{k=0}^{q} \frac{1}{k!}\right)q! = \frac{1}{q+1} + \frac{1}{(q+1)(q+2)} + \cdots . \qquad (\dagger)$$

Now the left side of $(\dagger)$ is an integer while the right side satisfies

$$0 < \frac{1}{q+1} + \frac{1}{(q+1)(q+2)} + \cdots < \frac{1}{q+1} + \frac{1}{(q+1)^2} + \cdots = \frac{1}{q} < 1.$$

This contradiction proves that e is indeed irrational. $\qquad\qquad\qquad \square$

Theorem 2.3.18 (Second Comparison Test). *Let $\sum c_n$ and $\sum d_n$ be two series of positive terms (i.e., $c_n > 0$, $d_n > 0 \; \forall n \in \mathbb{N}$) such that $\sum c_n$ is convergent while $\sum d_n$ is divergent. Given a series $\sum x_n$ of positive terms, we have:*

1. *If the inequalities $x_{n+1}/x_n \le c_{n+1}/c_n$ are ultimately true, then $\sum x_n$ is convergent.*
2. *If the inequalities $x_{n+1}/x_n \ge d_{n+1}/d_n$ are ultimately true, then $\sum x_n$ is divergent.*

Proof. To prove (1), note that, if the inequalities hold as stated, then for some $m \in \mathbb{N}$, we have $x_{n+1}/c_{n+1} \le x_n/c_n$ for all $n \ge m$, so that the sequence $(\xi_n) := (x_n/c_n)$ is *ultimately* decreasing and bounded below (by 0); it is therefore bounded and (1) follows from Exercise 2.3.10, part (3). For (2), set $\eta_n := x_n/d_n$, and note that, if again the inequalities hold, then for some $m \in \mathbb{N}$ we have $x_{n+1}/d_{n+1} \ge x_n/d_n$ for all $n \ge m$. In other words, (η_n) is *ultimately* increasing and $\eta_n > 0 \; \forall n \in \mathbb{N}$. Therefore, there exists $\eta > 0$ with $\eta_n \ge \eta$ for all $n \in \mathbb{N}$, and (2) follows from Exercise 2.3.10, part (4). $\qquad\qquad \square$

Many series have nonnegative terms that decrease monotonically. For such series the following theorem of Cauchy is usually helpful:

Theorem 2.3.19 (Cauchy's Condensation Theorem). *Suppose that $x_1 \geq x_2 \geq x_3 \geq \cdots \geq 0$. Then the series $\sum x_n$ is convergent if and only if the series*

$$\sum_{k=0}^{\infty} 2^k x_{2^k} = x_1 + 2x_2 + 4x_4 + 8x_8 + \cdots$$

is convergent.

Proof. Let $s_n := \sum_{j=0}^{n} x_n$ and $t_k := \sum_{j=0}^{k} 2^j x_{2^j}$. Then, for $n < 2^k$, we have

$$s_n \leq x_1 + (x_2 + x_3) + \cdots + (x_{2^k} + \cdots + x_{2^{k+1}-1})$$

$$\leq x_1 + 2x_2 + \cdots + 2^k x_{2^k} = t_k.$$

Next, for $n > 2^k$, we have

$$s_n \geq x_1 + x_2 + (x_3 + x_4) + \cdots + (x_{2^{k-1}+1} + \cdots + x_{2^k})$$

$$\geq \frac{1}{2}x_1 + x_2 + 2x_4 + \cdots + 2^{k-1}x_{2^k} = \frac{1}{2}t_k.$$

It follows that the sequences (s_n) and (t_k) are either both bounded or both unbounded, so that the two series in the theorem either *both* converge or *both* diverge. □

Exercise 2.3.20. Using the above *condensation theorem*, show that the p-series $\sum 1/n^p$ converges for $p > 1$ and diverges for $p \leq 1$. Show that the same is also true for the series $\sum_{n=2}^{\infty} 1/n(\log n)^p$, where $\log n = \ln n$ is the *natural logarithm* of n (to the *base e*). The properties of logarithms are known to the reader from calculus. However, we shall see them again later in more detail.

Using the comparison tests we can prove the following *Root* and *Ratio* Tests, which can be used in most cases to decide whether a series is convergent or divergent. In order to have a test for series whose terms may also be negative, we first give the following definition.

Definition 2.3.21 (Absolute vs. Conditional Convergence). A series $\sum x_n$ of real numbers is called *absolutely convergent* if the series $\sum |x_n|$ is convergent. If $\sum x_n$ is *convergent* but $\sum |x_n|$ is *divergent*, then we say that $\sum x_n$ is *conditionally convergent*.

The following *comparison test* is an immediate consequence of Cauchy's Criterion and the Triangle Inequality:

Proposition 2.3.22. *If $\sum c_n$ is a convergent series of nonnegative terms and if, for some $N_0 \in \mathbb{N}$, we have $|x_n| \leq c_n \ \forall \ n \geq N_0$, then $\sum x_n$ is convergent. In particular, an absolutely convergent series is convergent.*

Remark 2.3.23. Note that there are *conditionally convergent* series. The standard example is the *alternating harmonic series* $\sum(-1)^{n-1}/n$. The proof of this and other similar facts will follow from the discussion of alternating series below.

Theorem 2.3.24 (Root Test). *Given a series* $\sum x_n$, *define (the extended real number)* $\xi := \overline{\lim}(\sqrt[n]{|x_n|})$. *Then the following are true:*

1. *If* $\xi < 1$, $\sum x_n$ *is convergent.*
2. *If* $\xi > 1$, $\sum x_n$ *is divergent.*
3. *If* $\xi = 1$, *the test is inconclusive.*

Proof. If $\xi < 1$, we can pick η with $\xi < \eta < 1$ and $N \in \mathbb{N}$ such that

$$\sqrt[n]{|x_n|} < \eta \quad \forall\, n \geq N.$$

(Why?) In other words,

$$|x_n| \leq \eta^n \quad \forall\, n \geq N.$$

Since $\sum \eta^n$ converges, the convergence of $\sum |x_n|$ (and hence $\sum x_n$, by Proposition 2.3.22) follows from the First Comparison Test. Next, if $\xi > 1$, then $\xi \geq \lim(|x_{n_k}|^{1/n_k}) > 1$, for some subsequence (x_{n_k}) of (x_n). But then it follows at once that $\lim(x_n) \neq 0$, and the series $\sum x_n$ is therefore divergent. Finally, to prove (3), note that $\xi = 1$ both for the *harmonic series* $\sum 1/n$ which diverges and for the convergent p-series $\sum 1/n^2$.   □

Theorem 2.3.25 (Ratio Test). *Given a real series* $\sum x_n$ *such that* $x_n \neq 0$ *is ultimately true, we have*

1. $\sum x_n$ *converges if* $\limsup(|x_{n+1}/x_n|) < 1$.
2. $\sum x_n$ *diverges if* $\exists\, m \in \mathbb{N}$ *such that* $|x_{n+1}/x_n| \geq 1 \,\forall n \geq m$.
3. *If* $\underline{\lim}(|x_{n+1}/x_n|) \leq 1 \leq \overline{\lim}(|x_{n+1}/x_n|)$, *then the test is inconclusive.*

Proof. If (1) holds, then we can find $\xi \in (0,1)$ and $N \in \mathbb{N}$ such that $|x_{n+1}/x_n| < \xi$ for all $n \geq N$. It then follows that, for $n \geq N$, we have $|x_n| \leq |x_N|\xi^{-N} \cdot \xi^n$, and the convergence of $\sum x_n$ follows from the First Comparison Test. If (2) holds, then the condition $\lim(x_n) = 0$ is *not* satisfied, and the series $\sum x_n$ diverges. Finally, if we consider once again the *divergent* series $\sum 1/n$ and the *convergent* series $\sum 1/n^2$, then in both cases we have $\lim(|x_{n+1}/x_n|) = 1$; i.e., the condition (3) holds.   □

Remark 2.3.26.

(a) The Ratio Test can also be deduced from the Second Comparison Test.
(b) In the Root Test, if $\sqrt[n]{|x_n|} \geq 1$ for *infinitely many* distinct values of n, then the series $\sum x_n$ diverges (why?).
(c) The Ratio Test is, for most series, easier to apply than the Root Test. As the following theorem shows, however, the Root Test has *wider* scope. In fact, if the Ratio Test implies convergence, so does the Root Test; also, when the Root

Test is inconclusive, so is the Ratio Test. But there are series for which the Ratio Test is inconclusive and the Root Test implies convergence.

Theorem 2.3.27. *Given a sequence (x_n) of positive numbers, we have*

$$\liminf(x_{n+1}/x_n) \leq \liminf(\sqrt[n]{x_n}) \leq \limsup(\sqrt[n]{x_n}) \leq \limsup(x_{n+1}/x_n).$$

Proof. Let $\xi = \liminf(x_{n+1}/x_n)$ and $\eta = \limsup(x_{n+1}/x_n)$. We must show that $\liminf(\sqrt[n]{x_n}) \geq \xi$ and $\limsup(\sqrt[n]{x_n}) \leq \eta$. For the latter, if $\eta = \infty$, we have nothing to prove. Otherwise, pick $\eta_1 > \eta$ and $N \in \mathbb{N}$ such that $n \geq N$ implies $x_{n+1}/x_n \leq \eta_1$. Thus, with $c := x_N \eta_1^{-N}$, we get $x_n \leq c\eta_1^n$ for all $n \geq N$. But then, taking nth roots and using the fact that $\lim(\sqrt[n]{c}) = 1$, we get $\limsup(\sqrt[n]{x_n}) \leq \eta_1$. Since this holds for *all* $\eta_1 > \eta$, we get $\limsup(\sqrt[n]{x_n}) \leq \eta$. Next, note that $\liminf(\sqrt[n]{x_n}) \geq 0$. So, assume that $\xi > 0$ and pick any $\xi_1 \in (0, \xi)$. Then there exists $N \in \mathbb{N}$ such that $n \geq N$ implies $x_{n+1}/x_n \geq \xi_1$. Let $c := x_N \xi_1^{-N}$, and note that $n \geq N$ implies $\sqrt[n]{x_n} \geq \xi_1 \sqrt[n]{c}$. Since $\lim(\sqrt[n]{c}) = 1$, we get $\liminf(\sqrt[n]{x_n}) \geq \xi_1$. This is true for *all* $\xi_1 \in (0, \xi)$; it follows that $\liminf(\sqrt[n]{x_n}) \geq \xi$. $\square$

Example 2.3.28. Let $x_n := \xi_1 \xi_2 \cdots \xi_n$, with $(\xi_n) = (3, 1/9, 3, 1/9, \ldots)$. Then $x_{n+1}/x_n = \xi_{n+1}$, so that $\liminf(x_{n+1}/x_n) = 1/9$ whereas $\limsup(x_{n+1}/x_n) = 3$. The Ratio Test is therefore inconclusive. Now $x_n = (1/3)^{n/2}$ for n *even* and $x_n = 3(1/3)^{(n-1)/2}$ for n *odd*. Thus $\lim(\sqrt[n]{x_n}) = 1/\sqrt{3}$, which implies that the series $\sum x_n$ is *convergent*.

Exercise 2.3.29. Investigate the convergence or divergence of the following series:

1. $\sum(\sqrt[n]{n} - 1)^n$;
2. $\sum_{n=2}^{\infty} 1/(\log n)^n$;
3. $\sum n!/n^n$;
4. $\sum_{n=2}^{\infty} 1/(\log n)^{\log n}$;
5. $\frac{1}{2} + \frac{1}{3} + \frac{1}{2^2} + \frac{1}{3^2} + \frac{1}{2^3} + \frac{1}{3^3} + \cdots$; and
6. $\frac{1}{2} + 1 + \frac{1}{8} + \frac{1}{4} + \frac{1}{32} + \frac{1}{16} + \cdots$.

Hint: For (5) and (6), note that the Ratio Test is inconclusive.

For many series for which the Ratio and Root Tests are both inconclusive, the following test and its corollaries may be useful.

Theorem 2.3.30 (Kummer's Test). *Let (x_n) and (d_n) be two sequences of positive numbers. Suppose that $\sum 1/d_n = +\infty$ and let $t_n := d_n - (x_{n+1}/x_n)d_{n+1}$. Then the series $\sum x_n$ converges if $t_n \geq h$ is ultimately true for some $h > 0$ (equivalently, if $\liminf t_n > 0$) and diverges if $t_n \leq 0$ is ultimately true (which is the case if, e.g., $\limsup t_n < 0$).*

Proof. If $t_n \geq h$ for some $h > 0$ and all $n \geq m_0 \in \mathbb{N}$, then

$$hx_n < x_n d_n - x_{n+1} d_{n+1} \qquad (n \geq m_0). \qquad (*)$$

Adding the inequalities $(*)$ for $n = m_0$, $m_0 + 1, \ldots, m$, we get

$$h \sum_{n=m_0}^{m} x_n \leq x_{m_0} d_{m_0} - x_{m+1} d_{m+1} < x_{m_0} d_{m_0} \qquad (\forall m \geq m_0).$$

This implies that $\sum x_n$ has *bounded partial sums* and hence is convergent. If, on the other hand, $t_n \leq 0$ for all $n \geq K \in \mathbb{N}$, then $x_{n+1} d_{n+1} \geq x_n d_n$ for all $n \geq K$, so that $x_n \geq (x_K d_K)/d_n$ for all $n \geq K$ and the divergence of $\sum x_n$ follows from that of $\sum 1/d_n$. $\qquad \square$

Corollary 2.3.31 (Raabe's Test). *Let (x_n) be a sequence of positive numbers. Then $\sum x_n$ converges if $x_{n+1}/x_n \leq 1 - r/n$ is ultimately true for some $r > 1$ (equivalently, if $\liminf(n(1 - x_{n+1}/x_n)) > 1$) and diverges if $x_{n+1}/x_n \geq 1 - 1/n$ is ultimately true (which is the case if, e.g., $\limsup(n(1 - x_{n+1}/x_n)) < 1$).*

Proof. This follows at once from Kummer's Test if we take $d_n := n$. $\qquad \square$

Before stating the next corollary, let us introduce a convenient notation. Let (c_n) be a sequence of *positive* numbers. Given a sequence (b_n), we write $b_n = O(c_n)$ if there is a constant $K > 0$ such that $|b_n| \leq K c_n$ for all *sufficiently large n*. We shall also assume the following facts:

$$\lim_{n \to \infty} \frac{\log n}{n^\alpha} = 0 \quad (\forall \alpha > 0), \qquad \lim_{n \to \infty} n \log \left(\frac{n}{n+1} \right) = -1. \qquad (\dagger)$$

These facts are immediate consequences of *l'Hôpital's Rule*, as we shall see in Chap. 6, and are true even when n is replaced by $x \in (0, \infty)$.

Corollary 2.3.32 (Gauss's Test). *Let (x_n) be a sequence of positive numbers such that, for some constants $r \in \mathbb{R}$ and $p > 1$, we have*

$$\frac{x_{n+1}}{x_n} = 1 - \frac{r}{n} + O\left(\frac{1}{n^p} \right).$$

Then $\sum x_n$ converges if $r > 1$ and diverges if $r \leq 1$.

Proof. The condition implies that $\lim_{n \to \infty} n(1 - x_{n+1}/x_n) = r$, so that the result follows from Raabe's Test if $r \neq 1$. For $r = 1$, we use Kummer's Test with $d_n = n \log n$ (cf. Exercise 2.3.20). Now, with $t_n := d_n - (x_{n+1}/x_n) d_{n+1}$ as before,

$$t_n = n \log n - \left[1 - \frac{1}{n} + O\left(\frac{1}{n^p} \right) \right](n+1) \log(n+1)$$

$$= n \log \left(\frac{n}{n+1} \right) + \frac{1}{n} \log(n+1) + O\left(\frac{1}{n^p} \right)(n+1) \log(n+1).$$

Therefore, by $(\dagger)$, we have $\lim(t_n) = -1$. $\qquad \square$

The next proposition will be needed in our treatment of *alternating series.*

Proposition 2.3.33 (Abel's Partial Summation Formula). *Given a pair of real sequences (x_n) and (y_n), set $s_n := \sum_{k=1}^{n} x_k$ and $s_0 := 0$. Then, if $1 \le m \le n$, we have*

$$\sum_{k=m}^{n} x_k y_k = \sum_{k=m}^{n-1} s_k (y_k - y_{k+1}) + s_n y_n - s_{m-1} y_m.$$

Proof. Using the fact that $x_n = s_n - s_{n-1}$, we have

$$\sum_{k=m}^{n} x_k y_k = \sum_{k=m}^{n} (s_k - s_{k-1}) y_k = \sum_{k=m}^{n} s_k y_k - \sum_{k=m-1}^{n-1} s_k y_{k+1},$$

from which the proposition follows at once. $\square$

Here is a nice application:

Theorem 2.3.34 (Kronecker's Lemma). *Let (x_n) be a real sequence with $\sum_{n=1}^{\infty} x_n = s < \infty$. If (b_n) is an increasing sequence of positive numbers with $\lim(b_n) = \infty$, then*

$$\lim_{n \to \infty} \frac{1}{b_n} \sum_{k=1}^{n} b_k x_k = 0.$$

Proof. Using the above proposition with $s_n := \sum_{k=1}^{n} x_k$, $y_n = b_n$, and $m = 1$, we have

$$\frac{1}{b_n} \sum_{k=1}^{n} b_k x_k = s_n - \frac{1}{b_n} \sum_{k=1}^{n-1} (b_{k+1} - b_k) s_k. \qquad (*)$$

Now note that for $n > N$ the right side of $(*)$ is

$$= s_n - \frac{1}{b_n} \sum_{k=1}^{N-1} (b_{k+1} - b_k) s_k - \frac{1}{b_n} \sum_{k=N}^{n-1} (b_{k+1} - b_k) s_k$$

$$= s_n - \sum_{k=1}^{N-1} \left(\frac{b_{k+1} - b_k}{b_n} \right) s_k - \sum_{k=N}^{n-1} \left(\frac{b_{k+1} - b_k}{b_n} \right) s - \sum_{k=N}^{n-1} \left(\frac{b_{k+1} - b_k}{b_n} \right) (s_k - s),$$

which simplifies to

$$\left[s_n - \left(\frac{b_n - b_N}{b_n} \right) s \right] - \frac{1}{b_n} \sum_{k=1}^{N-1} (b_{k+1} - b_k) s_k - \frac{1}{b_n} \sum_{k=N}^{n-1} (b_{k+1} - b_k)(s_k - s). \qquad (**)$$

If $n \to \infty$, then $s_n \to s$ and (as b_n increases to ∞) $(b_n - b_N)/b_n$ increases to 1. Thus the difference inside the brackets in $(**)$ goes to zero. The second term in $(**)$ also approaches zero because the sum is independent of n. So, given $\varepsilon > 0$ the first and second terms will each be $< \varepsilon/3$ if $n \geq N$ and N is large enough. Finally, since (b_n) is increasing, the absolute value of the last term is bounded by $[(b_n - b_N)/b_n]\varepsilon/3 \leq \varepsilon/3$ if N is so large that $|s_k - s| < \varepsilon/3$ for $k \geq N$. Thus $(**)$ is less than ε if $n \geq N$ and the proof is complete.   □

Here is another important application:

Theorem 2.3.35 (Dirichlet's Test). *Let $\sum x_n$ be a real series whose partial sums $s_n = \sum_{k=1}^{n} x_k$ form a bounded sequence. If (y_n) is a decreasing sequence (of nonnegative terms) with $\lim(y_n) = 0$, then the series $\sum x_n y_n$ converges.*

Proof. Pick $B > 0$ such that $|s_n| \leq B \; \forall \; n \in \mathbb{N}$. Now, given any $\varepsilon > 0$, there is an integer $N \in \mathbb{N}$ such that $y_N < \varepsilon/(2B)$. Using Proposition 2.3.33 and the fact that $y_n - y_{n+1} \geq 0$ for all n, it follows that, if $N \leq m \leq n$, then

$$\left| \sum_{k=m}^{n} x_n y_n \right| = \left| \sum_{k=m}^{n-1} s_k (y_k - y_{k+1}) + s_n y_n - s_{m-1} y_m \right|$$

$$\leq B \left[\sum_{k=m}^{n-1} (y_k - y_{k+1}) + y_n + y_m \right]$$

$$= 2B y_m \leq 2B y_N < \varepsilon.$$

By Cauchy's Criterion, the series $\sum x_n y_n$ is therefore convergent.   □

Corollary 2.3.36 (Abel's Test). *Suppose that $\sum x_n$ is convergent. Then, for any bounded monotone sequence (y_n), the series $\sum x_n y_n$ is also convergent.*

Proof. We may assume that (y_n) is *decreasing*. Let $y := \inf\{y_n : n \in \mathbb{N}\}$. Then $\lim(y_n) = y$. Put $z_n := y_n - y$. Then (z_n) is monotone, nonnegative, and $\lim(z_n) = 0$. Since the *convergent* series $\sum x_n$ has *bounded* partial sums, $\sum x_n z_n$ converges by Dirichlet's Test and the convergence of $\sum x_n y_n$ follows at once.   □

Dirichlet's Test can be used to give a convergence criterion for *alternating series* which we now define:

Definition 2.3.37 (Alternating Series). Let (a_n) be a sequence of *positive* real numbers. The series $\sum (-1)^{n+1} a_n$ (or $\sum (-1)^n a_n$) is then called an *alternating series*.

Theorem 2.3.38 (Leibniz's Test). *Let (a_n) be a sequence of positive numbers. If (a_n) is a decreasing null sequence; i.e., if $a_1 \geq a_2 \geq a_3 \geq \cdots$ and $\lim(a_n) = 0$, then the alternating series $\sum (-1)^{n+1} a_n = a_1 - a_2 + a_3 - \cdots$ (and hence also $\sum (-1)^n a_n$) is convergent.*

Proof. Simply apply Theorem 2.3.35, with $x_n = (-1)^{n+1}$ and $y_n = a_n$. $\qquad\square$

Next, we consider the arithmetic operations on convergent series. As usual, the addition and multiplication by constants are quite elementary, and we leave the proof of the next theorem for the reader.

Theorem 2.3.39. *If $\sum x_n = \xi$, and $\sum y_n = \eta$, then $\sum (x_n + y_n) = \xi + \eta$, and $\sum c x_n = c\xi$ for any constant $c \in \mathbb{R}$.*

Exercise 2.3.40. Prove Theorem 2.3.39.

The multiplication of two convergent series is more involved, especially since there are many ways to define a *product*. We therefore begin by defining the *Cauchy product*. To have a more convenient notation, we begin our summations at $n = 0$ rather than $n = 1$. Note that, in general, $\sum_{n=1}^{\infty} x_n = \sum_{n=0}^{\infty} x_{n+1}$.

Definition 2.3.41 (Cauchy Product). For two series $\sum_{n=0}^{\infty} x_n$ and $\sum_{n=0}^{\infty} y_n$, we define their *Cauchy product* (or simply *product*) to be $\sum_{n=0}^{\infty} z_n$, where

$$z_n = \sum_{k=0}^{n} x_k y_{n-k} \quad (n \in \mathbb{N}_0).$$

Exercise 2.3.42. Consider the alternating series $\sum_{n=0}^{\infty} (-1)^n / \sqrt{n+1}$, which is convergent (why?). Show that the (Cauchy) product of this series with *itself* is $\sum_{n=0}^{\infty} z_n$, where

$$z_n = (-1)^n \sum_{k=0}^{n} \frac{1}{\sqrt{(n-k+1)(k+1)}},$$

and that this series is *divergent*.
Hint: $(n - k + 1)(k + 1) = (n/2 + 1)^2 - (n/2 - k)^2$.

The series in Exercise 2.3.42 was *conditionally convergent*. The following theorem shows that if at least one of two convergent series is *absolutely convergent*, then the (Cauchy) product of the series converges to the product of their sums.

Theorem 2.3.43 (Mertens). *Suppose that $\sum_{n=0}^{\infty} |x_n|$ is convergent, $\sum_{n=0}^{\infty} x_n = \xi$, $\sum_{n=0}^{\infty} y_n = \eta$, and that $z_n = \sum_{k=0}^{n} x_k y_{n-k}$, $(n = 0, 1, 2, \ldots)$. Then $\sum_{n=0}^{\infty} z_n = \xi\eta$.*

Proof. For each $n \geq 0$, let $X_n := \sum_{k=0}^{n} x_k$, $Y_n := \sum_{k=0}^{n} y_k$, $Z_n := \sum_{k=0}^{n} z_k$, and $\rho_n := Y_n - \eta$. Then we have

$$Z_n = z_0 + z_1 + \cdots + z_n$$
$$= x_0 y_0 + (x_0 y_1 + x_1 y_0) + \cdots + (x_0 y_n + x_1 y_{n-1} + \cdots + x_n y_0)$$

$$= x_0 Y_n + x_1 Y_{n-1} + \cdots + x_n Y_0$$

$$= X_n \eta + \delta_n \qquad (n = 0, 1, 2, \ldots),$$

where we have defined (for each $n \in \mathbb{N}$)

$$\delta_n := x_0 \rho_n + x_1 \rho_{n-1} + \cdots + x_n \rho_0 \qquad (n = 0, 1, 2, \ldots).$$

Since $\lim(X_n \eta) = \xi \eta$, it suffices to show that $\lim(\delta_n) = 0$. To show this, let $\varepsilon > 0$ be given. Then, since $\lim(\rho_n) = 0$ (why?), we can pick $N \in \mathbb{N}$ such that $|\rho_n| < \varepsilon \ \forall n \geq N$. For each such n, we therefore have

$$|\delta_n| \leq |\rho_0 x_n + \cdots + \rho_N x_{n-N}| + |\rho_{N+1} x_{n-N-1} + \cdots + \rho_n x_0| \qquad (*)$$

$$< |\rho_0 x_n + \cdots + \rho_N x_{n-N}| + \varepsilon \sigma,$$

where we have defined

$$\sigma := \sum_{n=0}^{\infty} |x_n|.$$

But $\lim(x_n) = 0$ (why?), so keeping N fixed and letting $n \to \infty$ in $(*)$, we get

$$\limsup(|\delta_n|) \leq \varepsilon \sigma.$$

Since ε was arbitrary, we get $\lim(\delta_n) = 0$ and hence $\lim(Z_n) = \xi \eta$. □

Exercise 2.3.44. Show that the Cauchy product of two *absolutely* convergent series is *absolutely* convergent and that its sum is the product of the sums of the two series.

The next theorem, due to Abel, shows that what one wishes to be true is indeed true! The proof will be given later when we look at power series. Note that no absolute convergence is required.

Theorem 2.3.45 (Abel). *If* $\sum_{n=0}^{\infty} x_n = \xi$, $\sum_{n=0}^{\infty} y_n = \eta$, *and* $\sum_{n=0}^{\infty} z_n = \zeta$, *where* $z_n = \sum_{k=0}^{n} x_k y_{n-k}$ $(n = 0, 1, 2, \ldots)$, *then we have* $\zeta = \xi \eta$.

The next definition introduces the concept of *rearrangement* of an infinite series. Here the important fact is that if a series is *absolutely convergent*, then all its rearrangements converge to the same sum.

Definition 2.3.46 (Rearrangement). Let $\sum x_n$ be a real series and let $\nu = (\nu_1, \nu_2, \nu_3, \ldots)$ be a *permutation* of $\mathbb{N}$; i.e., let $\nu : \mathbb{N} \to \mathbb{N}$ be a *one-to-one correspondence*. If we set $x_n' := x_{\nu_n} \ \forall n \in \mathbb{N}$, then the series $\sum x_n'$ is called a *rearrangement* of $\sum x_n$.

Remark 2.3.47. As the next theorem (due to Riemann) shows, a rearrangement of a *conditionally convergent* series need not converge to the same limit and may even diverge. The next exercise gives an example of this phenomenon.

Exercise 2.3.48. Let $s := \sum (-1)^{n+1}/n$, and consider the rearrangement

$$\sum_{k=1}^{\infty} \left(\frac{1}{4k-3} + \frac{1}{4k-1} - \frac{1}{2k} \right) = 1 + \frac{1}{3} - \frac{1}{2} + \frac{1}{5} + \frac{1}{7} - \frac{1}{4} + \frac{1}{9} + \frac{1}{11} - \frac{1}{6} + \cdots$$

of the series, where we always have two positive terms followed by one negative term. If (s'_n) denotes the sequence of partial sums of this rearrangement, show that $s < s'_3 = 5/6$ and that $s'_3 < s'_6 < s'_9 < \cdots$. Deduce that $\limsup(s'_n) > 5/6$, which implies $\lim(s'_n) \neq s$. Show, however, that (s'_n) is *convergent*.

Theorem 2.3.49 (Riemann). *Let $\sum x_n$ be a conditionally convergent real series, and let $\xi,\ \eta \in [-\infty, +\infty]$ be given extended real numbers with $\xi \leq \eta$. Then there exists a rearrangement $\sum x'_n$ of $\sum x_n$ with partial sums s'_n such that*

$$\liminf(s'_n) = \xi, \qquad \limsup(s'_n) = \eta.$$

In particular, if $\xi = \eta$, then $\sum x'_n = \xi$.

Proof. Define the sequences (p_n) and (q_n) by

$$p_n := \frac{|x_n| + x_n}{2}, \qquad q_n := \frac{|x_n| - x_n}{2}.$$

We then have $p_n - q_n = x_n$ and $p_n + q_n = |x_n|$, $p_n \geq 0$, $q_n \geq 0$. Since $\sum x_n$ is *not* absolutely convergent, both series $\sum p_n$ and $\sum q_n$ are *divergent* (why?).

Now let (P_n) be the sequence of *nonnegative* terms of $\sum x_n$ in their proper order, and let (Q_n) be the sequence of *absolute values* of the *negative* terms of $\sum x_n$, also in their proper order. Then the series $\sum P_n$ and $\sum Q_n$ are both divergent, since they differ from the series $\sum p_n$ and $\sum q_n$ by zero terms only. We want our rearrangement $\sum x'_n$ to have the form

$$(P_1 + \cdots + P_{m_1}) - (Q_1 + \cdots + Q_{k_1})$$
$$+ (P_{m_1+1} + \cdots + P_{m_2}) - (Q_{k_1+1} + \cdots + Q_{k_2}) + \cdots, \qquad (*)$$

where the sequences (m_j) and (k_j) are to be constructed. To do so, pick sequences (ξ_n) and (η_n) such that $\lim(\xi_n) = \xi$, $\lim(\eta_n) = \eta$, $\xi_n < \eta_n\ \forall n \in \mathbb{N}$, and $\eta_1 > 0$. Now let $m_1,\ k_1$ be the *smallest* integers such that

$$\sum_{i=1}^{m_1} P_i > \eta_1 \quad \text{and} \quad \sum_{i=1}^{m_1} P_i - \sum_{j=1}^{k_1} Q_j < \xi_1.$$

Next, let m_2, k_2 be the *smallest* integers such that

$$\sum_{i=1}^{m_1} P_i - \sum_{j=1}^{k_1} Q_j + \sum_{i=m_1+1}^{m_2} P_i > \eta_2 \text{ and}$$

$$\sum_{i=1}^{m_1} P_i - \sum_{j=1}^{k_1} Q_j + \sum_{i=m_1+1}^{m_2} P_i - \sum_{j=k_1+1}^{k_2} Q_j < \xi_2,$$

and continue the process. The reason why this can be done is that both $\sum P_n$ and $\sum Q_n$ *diverge*. Now if (s_n) and (t_n) are the partial sums of the series $(*)$ whose last terms are P_{m_n} and $-Q_{k_n}$, respectively, then we have

$$|s_n - \eta_n| \le P_{m_n}, \quad |t_n - \xi_n| \le Q_{k_n},$$

and since $\lim(P_n) = \lim(Q_n) = 0$, we get $\lim(s_n) = \eta$ and $\lim(t_n) = \xi$. Finally, it is obvious that all limit points of the partial sums of the series $(*)$ must be between ξ and η. $\square$

We can now prove the theorem that characterizes the infinite series all of whose rearrangements converge.

Theorem 2.3.50. *All rearrangements of a real series $\sum x_n$ converge if and only if the series is absolutely convergent.*

Proof. Suppose that $\sum x_n$ is *absolutely convergent*. Let $v : \mathbb{N} \to \mathbb{N}$ be a *permutation* of $\mathbb{N}$, and let $\sum x_n'$ be the corresponding rearrangement, i.e., $x_n' = x_{v_n}$. Now given any $\varepsilon > 0$, there is an integer $N \in \mathbb{N}$ such that

$$m \ge n \ge N \implies \sum_{j=m}^{n} |x_j| < \varepsilon. \tag{$*$}$$

Let $k \in \mathbb{N}$ be such that $\{1, 2, \ldots, N\} \subset \{v_1, v_2, \ldots, v_k\}$, and let (s_n) and (s_n') denote the sequences of partial sums of $\sum x_n$ and $\sum x_n'$, respectively. Then, for $n > k$, the numbers $x_1, x_2, \ldots, x_N$ are all canceled in the difference $s_n - s_n'$, and $(*)$ implies $|s_n - s_n'| < \varepsilon$, hence we get $\lim(s_n) = \lim(s_n')$. Next, if $\sum x_n$ is *divergent*, then the trivial "identity" rearrangement is divergent. If $\sum x_n$ is *conditionally convergent*, then by Theorem 2.3.49 there is a rearrangement $\sum x_n'$ that is *divergent*, and the proof is complete. $\square$

An immediate consequence is the following:

Corollary 2.3.51. *If all rearrangements of a series $\sum x_n$ converge, then they all converge to the same sum.*

We end this section with an important example which uses most of the concepts introduced above. This example plays a fundamental role in the theory of *Fourier series* and will be used later.

Example 2.3.52 (The Class $\ell^2(\mathbb{N})$). The class of all *square-summable* (real) sequences is defined to be the set

$$\ell^2(\mathbb{N}) := \left\{ x = (x_n) \in \mathbb{R}^{\mathbb{N}} : \sum_{n=1}^{\infty} x_n^2 < \infty \right\},$$

and for each $x = (x_n) \in \ell^2(\mathbb{N})$ its ℓ^2-*norm* is defined to be the nonnegative number

$$\|x\|_2 := \left(\sum_{n=1}^{\infty} x_n^2 \right)^{1/2}.$$

Thus $\ell^2(\mathbb{N})$ is the set of all sequences with *finite* ℓ^2-*norm*. Then, given any sequences $x = (x_n)$, $y = (y_n) \in \ell^2(\mathbb{N})$, the series $\sum_{n=1}^{\infty} x_n y_n$ is *absolutely convergent* and we have the *Cauchy–Schwarz inequality*:

$$\left| \sum_{n=1}^{\infty} x_n y_n \right| \leq \sum_{n=1}^{\infty} |x_n y_n| \leq \sqrt{\sum x_n^2} \cdot \sqrt{\sum y_n^2} = \|x\|_2 \|y\|_2. \tag{†}$$

To see this, recall *Cauchy's inequality* (cf. Proposition 2.1.23):

$$\left(\sum_{i=1}^{n} x_i y_i \right)^2 \leq \left(\sum_{i=1}^{n} x_i^2 \right) \left(\sum_{i=1}^{n} y_i^2 \right),$$

which is a fortiori valid if its right side is replaced by $\|x\|_2^2 \|y\|_2^2$. The inequality (†) then follows if we take square roots and pass to the limit as $n \to \infty$. An important consequence of the Cauchy–Schwarz inequality is that the class $\ell^2(\mathbb{N})$ is a *vector space* over $\mathbb{R}$. Indeed, it is obvious that for each $x = (x_n) \in \ell^2(\mathbb{N})$ and each $c \in \mathbb{R}$, we have $cx := (cx_n) \in \ell^2(\mathbb{N})$ and

$$\|cx\|_2 = |c| \|x\|_2.$$

Now given any sequences $x = (x_n)$, $y = (y_n) \in \ell^2(\mathbb{N})$, we must show that $x + y := (x_n + y_n) \in \ell^2(\mathbb{N})$. Recall (cf. Proposition 2.1.23) that we have the *Triangle Inequality*:

$$\left(\sum_{i=1}^{n} (x_i + y_i)^2 \right)^{1/2} \leq \left(\sum_{i=1}^{n} x_i^2 \right)^{1/2} + \left(\sum_{i=1}^{n} y_i^2 \right)^{1/2}.$$

Replacing the right side by $\|x\|_2 + \|y\|_2$ and taking the limit as $n \to \infty$, we obtain the *Minkowski's inequality*

$$\|x + y\|_2 \leq \|x\|_2 + \|y\|_2,$$

which implies indeed that $x + y \in \ell^2(\mathbb{N})$. The following properties of the ℓ^2-norm are now easily checked. Here, $x, \ y \in \ell^2(\mathbb{N})$ and $c \in \mathbb{R}$ are arbitrary:

1. $\|x\|_2 \geq 0$;
2. $\|x\|_2 = 0 \Leftrightarrow x = 0$;
3. $\|cx\|_2 = |c|\|x\|_2$; and
4. $\|x + y\|_2 \leq \|x\|_2 + \|y\|_2$ (Minkowski's inequality).

Remark 2.3.53. In a similar fashion one defines the classes $\ell^2(\mathbb{N}_0)$ and $\ell^2(\mathbb{Z})$ with the corresponding ℓ^2-norms. Thus,

$$\ell^2(\mathbb{Z}) := \left\{ x = (x_n) \in \mathbb{R}^{\mathbb{Z}} : \sum_{n=-\infty}^{\infty} x_n^2 < \infty \right\},$$

and the partial sums are defined to be

$$s_n := \sum_{k=-n}^{n} x_k^2.$$

The above properties (1)–(4) remain valid for the new ℓ^2-norms, as may be checked at once.

2.4   Unordered Series and Summability

Recall that a *real (resp., complex) sequence* is simply a function $x : \mathbb{N} \to \mathbb{R}$ (resp., $x : \mathbb{N} \to \mathbb{C}$). The value $x_n := x(n)$ is the *nth term* and the corresponding *infinite series* is the *formal sum*

$$\sum_{n=1}^{\infty} x_n$$

of all the values of x. Its *sum*, if it exists, is the limit of *partial sums*:

$$s_n := \sum_{k=1}^{n} x_k = x_1 + x_2 + \cdots + x_n.$$

Now $\mathbb{N}$ has a natural *order*, $1 < 2 < 3 < \cdots$ and is in fact *well ordered* under this ordering. If we replace $\mathbb{N}$ by an arbitrary set $X \neq \emptyset$ with no a priori order and consider a function $u : X \to \mathbb{F}$ (where $\mathbb{F}$ is either $\mathbb{R}$ or $\mathbb{C}$), can we still define the "unordered sum"

$$\sum_{x \in X} u_x$$

of all the values $u_x := u(x)$ of u? Well, if X is *finite*, then the answer is obviously *yes:* Indeed, if $X = \{x_1, x_2, \ldots, x_n\}$, then

$$\sum_{x \in X} u_x = \sum_{k=1}^{n} u_{x_k} = u(x_1) + u(x_2) + \cdots + u(x_n),$$

where the finite sum on the right is *independent of the order of its terms*. This suggests that we look at the set $\mathcal{F}_X$ of all *finite subsets* of X. Thus,

$$\mathcal{F}_X := \{F \subset X : |F| < \infty\},$$

where $|F|$ is the *cardinality* (i.e., the number of elements) of F.

Definition 2.4.1 (Summable, Partial Sum, Sum). Let $X \neq \emptyset$ be a set and $u : X \to \mathbb{F}$, where $\mathbb{F}$ is either $\mathbb{R}$ or $\mathbb{C}$. For each finite set $F \in \mathcal{F}_X$, the sum

$$s_F = s_F(u) := \sum_{x \in F} u_x,$$

where $u_x := u(x)$, is called a *partial sum* of u. We say that u (or the corresponding *unordered series* $\sum_{x \in X} u_x$) is *summable* if there is a number $s \in \mathbb{F}$ such that

$$(\forall \varepsilon > 0)(\exists F_\varepsilon \in \mathcal{F}_X)(\forall F \in \mathcal{F}_X)(F \supset F_\varepsilon \Rightarrow |s_F - s| < \varepsilon).$$

The number s is then called the *(unordered) sum* of u (over X) (or of the unordered series $\sum_{x \in X} u_x$), and we write

$$s = \sum_{x \in X} u_x.$$

Remark 2.4.2. It can be easily checked that $u : X \to \mathbb{C}$ is summable if and only if the real-valued functions $\mathrm{Re}(u)$ and $\mathrm{Im}(u)$ are both summable. (Why?) Therefore, in what follows we shall only look at *real-valued* functions.

Example 2.4.3 (Multiple Sequences and Series).   Unlike $\mathbb{N}$, the sets $\mathbb{N}^k$, $k = 2, 3, 4 \ldots$, have no natural order. If $X := \mathbb{N}^k$ and $k \geq 2$, then a function $u : \mathbb{N}^k \to \mathbb{R}$ is called a *multiple (real) sequence* and the *unordered series*

$$\sum_{(n_1, \ldots, n_k) \in \mathbb{N}^k} u(n_1, \ldots, n_k)$$

is the corresponding *multiple series*. Thus a *double sequence* is a function $u : \mathbb{N}^2 \rightarrow$ $\mathbb{R}$ and the corresponding *double series* is the unordered series $\sum_{(m,n) \in \mathbb{N}^2} u_{mn}$, where $u_{mn} := u(m,n)$.

The following propositions follow easily from Definition 2.4.1.

Proposition 2.4.4 (Uniqueness of the Sum). *The sum of a summable function* $u :$ $X \rightarrow \mathbb{R}$ *is unique.*

Exercise 2.4.5. Prove the proposition.

Proposition 2.4.6 (Linearity of the Sum). *Let* $X \neq \emptyset$ *be a set and* $u, v : X \rightarrow \mathbb{R}$. *If* u *and* v *are summable, then, for any constants* $\alpha, \beta \in \mathbb{R}$, *the function* $\alpha u + \beta v$ *is also summable and we have*

$$\sum_{x \in X} (\alpha u + \beta v)_x = \sum_{x \in X} (\alpha u_x + \beta v_x) = \alpha \sum_{x \in X} u_x + \beta \sum_{x \in X} v_x.$$

Exercise 2.4.7. Prove the proposition.

Proposition 2.4.8. *If* $u : X \rightarrow \mathbb{R}$ *is nonnegative (i.e.,* $u_x \geq 0$ *for all* $x \in X$), *then* u *is summable if and only if its partial sums are (uniformly) bounded; i.e., there is a number* $M > 0$ *such that* $s_F \leq M$ *for all* $F \in \mathcal{F}_X$. *In this case, we have*

$$s = \sum_{x \in X} u_x = \sup\{s_F : F \in \mathcal{F}_X\}. \tag{$*$}$$

Proof. If u is summable with sum s, then for $\varepsilon = 1$ we can find $F_1 \in \mathcal{F}_X$ such that $F \supset F_1$ implies $|s - s_F| < 1$ and hence $s_F < s + 1$. If now $F \in \mathcal{F}_X$ is *any* finite set, then $F_1 \cup F \supset F_1$ and hence $s_F \leq s_{F_1 \cup F} < s + 1$. Thus, all partial sums are bounded above by $M := s + 1$. Conversely, if M is an upper bound for all partial sums and if s is defined by ($*$), then, by the very definition of "sup," for each $\varepsilon > 0$ we can find $F_\varepsilon \in \mathcal{F}_X$ such that $s - \varepsilon < s_{F_\varepsilon} \leq s$. But then, we have

$$F_\varepsilon \subset F \in \mathcal{F}_X \Rightarrow s - \varepsilon < s_F \leq s,$$

which means precisely that $\sum_{x \in X} u_x = s$. $\qquad\qquad\qquad\qquad\qquad\qquad\qquad$ $\square$

Definition 2.4.9 (Infinite Sum). We say that $u : X \rightarrow \mathbb{R}$ has sum $+\infty$ and write $\sum_{x \in X} u_x = +\infty$, if for every $M \in \mathbb{R}$ there exists $F_M \in \mathcal{F}_X$ such that $s_{F_M} = \sum_{x \in F_M} u_x > M$.

The following corollary of Proposition 2.4.8 is now obvious:

Corollary 2.4.10. *If a nonnegative* $u : X \rightarrow \mathbb{R}$ *is not summable, then we have* $\sum_{x \in X} u_x = +\infty$.

For $X = \mathbb{N}$ and a sequence $u : \mathbb{N} \rightarrow \mathbb{R}$, the definition of summability implies that, if u is *summable*, then the (ordered) series $\sum_{n=1}^{\infty} u_n$ is *convergent*. That the converse is not true, however, is a consequence of the following theorem (cf. Proposition 2.4.20 below). Before stating it, let us recall that, for any $a \in \mathbb{R}$, we define $a^+ := \max\{a, 0\}$ and $a^- := \max\{-a, 0\}$. It is then obvious that

$$|a| = a^+ + a^-, \quad \text{and} \quad a = a^+ - a^- = 2a^+ - |a|.$$

Theorem 2.4.11 (Summable $\Leftrightarrow$ Absolutely Summable). *For a function $u : X \rightarrow \mathbb{R}$, the following are equivalent:*

1. *u is summable;*
2. *u is absolutely summable (i.e., $|u|$ is summable);*
3. *There is a number $M > 0$ such that $\forall F \in \mathcal{F}_X$ we have $\sum_{x \in F} |u_x| \leq M$.*

Proof. Let us first prove the equivalence (1) $\Leftrightarrow$ (2). If $\sum_{x \in X} |u_x|$ is summable, then so is $\sum_{x \in X} u_x^+$, by Proposition 2.4.8, because $0 \leq u_x^+ \leq |u_x|$, for all $x \in X$. But then u is also summable (by Proposition 2.4.6) because $u_x = 2u_x^+ - |u_x|$, for all $x \in X$. Conversely, suppose that $\sum_{x \in X} u_x = s$. Then we can pick $F_1 \in \mathcal{F}_X$ such that $|\sum_{x \in F} u_x - s| < 1$ (and hence $|\sum_{x \in F} u_x| < 1 + |s|$) if $F_1 \subset F \in \mathcal{F}_X$. Now, given *any* $F \in \mathcal{F}_X$, we have

$$\sum_{x \in F} u_x = \sum_{x \in F \cup F_1} u_x - \sum_{x \in F_1 \setminus F} u_x \leq 1 + |s| + \sum_{x \in F_1} |u_x|. \qquad (*)$$

If we set $F^+ := \{x \in F : u_x \geq 0\}$, then $(*)$ implies

$$\sum_{x \in F} u_x^+ = \sum_{x \in F^+} u_x \leq 1 + |s| + \sum_{x \in F_1} |u_x|,$$

and hence $\sum_{x \in X} u_x^+$ is summable by Proposition 2.4.8. Since $|u_x| = 2u_x^+ - u_x$, the summability of $|u|$ now follows from Proposition 2.4.6. Finally, the equivalence (2) $\Leftrightarrow$ (3) is an immediate consequence of Proposition 2.4.8, and the proof is complete. $\square$

The following corollaries are obvious consequences.

Corollary 2.4.12. *Any summable function $u : X \rightarrow \mathbb{R}$ (nonnegative or not) has bounded partial sums; i.e., there is a number $M > 0$ such that $|s_F| \leq M$ for all $F \in \mathcal{F}_X$.*

Corollary 2.4.13. *Let X be a set and $u : X \rightarrow \mathbb{R}$ a summable function. Then, for any subset $Y \subset X$, the restriction $u|Y$ is summable (over Y).*

Definition 2.4.14 (Cauchy's Criterion). We say that $u : X \rightarrow \mathbb{R}$ satisfies *Cauchy's Criterion* if

$$(\forall \varepsilon > 0)(\exists F_\varepsilon \in \mathcal{F}_X)(\forall F \in \mathcal{F}_X)(F \cap F_\varepsilon = \emptyset \Rightarrow |s_F| < \varepsilon).$$

Exercise 2.4.15.

1. Show that, if $u : X \to \mathbb{R}$ satisfies Cauchy's Criterion, then so does the restriction $u|Y$ for any $Y \subset X$. Deduce that the functions $u^+ := \sup\{u, 0\}$ and $u^- := \sup\{-u, 0\}$ satisfy Cauchy's Criterion as well.
2. Show that, if $u : X \to \mathbb{R}$ satisfies Cauchy's Criterion, then it has *bounded partial sums*; i.e., there is a constant $M > 0$ with $|\sum_{x \in F} u(x)| \le M$ for any $F \in \mathcal{F}_X$.
3. Show that a *summable* function $u : X \to \mathbb{R}$ satisfies Cauchy's Criterion. *Hint:* Let $s = \sum_{x \in X} u(x)$ and pick $F_\varepsilon \in \mathcal{F}_X$ with $|s - s_F| < \varepsilon/2$ for all $F \supset F_\varepsilon$. If $F \in \mathcal{F}_X$ and $F \cap F_\varepsilon = \emptyset$, note that

$$|s_F| = |s_{F \cup F_\varepsilon} - s_{F_\varepsilon}| \le |s - s_{F \cup F_\varepsilon}| + |s - s_{F_\varepsilon}|.$$

The following corollary of Theorem 2.4.11 shows that the condition in part (3) of the above exercise is also *sufficient:*

Corollary 2.4.16 (Cauchy's Criterion). *A function $u : X \to \mathbb{R}$ is summable if and only if it satisfies Cauchy's Criterion.*

Proof. By Exercise 2.4.15, u^+ and u^- satisfy Cauchy's Criterion and $u = u^+ - u^-$. But then both u^+ and u^- have *bounded* partial sums and hence are summable. $\square$

Here is another interesting corollary:

Corollary 2.4.17. *If $u : X \to \mathbb{R}$ is summable, then $D := \{x \in X : u_x \ne 0\}$ is countable.*

Proof. Since u is *absolutely summable*, we have $S := \sum_{x \in X} |u_x| < \infty$. Let $D_n := \{x \in X : |u_x| > 1/n\}$. Then D_n is *finite*. Actually, $|D_n| < nS$, where $|D_n|$ is the number of elements in D_n. Since $D = \bigcup_{n=1}^{\infty} D_n$, the corollary follows. $\square$

Exercise 2.4.18. Deduce Corollary 2.4.17 from *Cauchy's Criterion*. (*Hint:* For each $n \in \mathbb{N}$, pick $F_n \in \mathcal{F}_X$ such that $x \notin F_n$ implies $|u(x)| < 1/n$ and look at $\bigcup F_n$.)

Theorem 2.4.19. *Suppose that $u : X \to \mathbb{R}$ is summable with sum s. If $(F_n)_{n=1}^{\infty}$ is a sequence of finite subsets of X such that*

$$F_1 \subset F_2 \subset F_3 \subset \cdots , \quad and \quad X = \bigcup_{n=1}^{\infty} F_n,$$

then we have

$$s = \lim_{n \to \infty} \sum_{x \in F_n} u_x.$$

In particular, if $(E_n)_{n=1}^\infty$ is another sequence of finite subsets of X satisfying the same conditions as $(F_n)_{n=1}^\infty$, then we have

$$\lim_{n\to\infty} \sum_{x\in F_n} u_x = s = \lim_{m\to\infty} \sum_{x\in E_m} u_x.$$

Proof. Given any $\varepsilon > 0$, we can pick $F_\varepsilon \in \mathcal{F}_X$ such that $F_\varepsilon \subset F \in \mathcal{F}_X$ implies $|s_F - s| < \varepsilon$. Now pick $N \in \mathbb{N}$ such that $F_\varepsilon \subset F_N$. Since $F_N \subset F_n$ for all $n \geq N$, it follows that $|\sum_{x\in F_n} u_x - s| < \varepsilon$ for all $n \geq N$ and the proof is complete. $\square$

The following proposition is an important consequence:

Proposition 2.4.20. *Let $u : \mathbb{N} \to \mathbb{R}$ be a sequence. Then the unordered series $\sum_{n\in\mathbb{N}} u_n$ is summable if and only if the (ordered) series $\sum_{n=1}^\infty |u_n|$ is convergent (i.e., the series $\sum_{n=1}^\infty u_n$ is absolutely convergent) and, in this case, we have*

$$\sum_{n\in\mathbb{N}} u_n = \sum_{n=1}^\infty u_n.$$

Proof. In view of Theorem 2.4.11, the summability of $\sum_{n\in\mathbb{N}} u_n$ and the convergence of the series $\sum_{n=1}^\infty |u_n|$ are both equivalent to the existence of a number $M > 0$ such that

$$|u_1| + |u_2| + \cdots + |u_n| \leq M \qquad (\forall n \in \mathbb{N}).$$

To prove the last statement, if u is summable, we can apply Theorem 2.4.19 with $X = \mathbb{N}$ and $F_n := \{1, 2, \ldots, n\}$ to deduce that

$$\sum_{n\in\mathbb{N}} u_n = \lim_{n\to\infty} \sum_{k\in F_n} u_k = \lim_{n\to\infty} \sum_{k=1}^n u_k = \sum_{n=1}^\infty u_n,$$

and the proof is complete. $\square$

Example 2.4.21. We have seen that the alternating series $\sum_{n=1}^\infty (-1)^{n+1}/n$ is *convergent*. We also know that the *harmonic series* $\sum_{n=1}^\infty 1/n$ is *divergent*. Therefore, the unordered series $\sum_{n\in\mathbb{N}} (-1)^{n+1}/n$ is *not* summable.

Our next application of Theorem 2.4.19 is to the convergence of *iterated sums* which arise in the study of *double series* (i.e., unordered series over the set $\mathbb{N}^2$). Actually, since we shall need the results later in our study of *power series*, it is more convenient to use the set $X := \{(m, n) : m, n = 0, 1, 2, 3, \ldots\}$. The values u_{mn} of a function $u : X \to \mathbb{R}$ can then be arranged as a rectangular *array*:

$$
\begin{array}{llll}
u_{00} & u_{01} & u_{02} & \cdots \\
u_{10} & u_{11} & u_{12} & \cdots \\
u_{20} & u_{21} & u_{22} & \cdots \\
\vdots & \vdots & \vdots & \ddots
\end{array}
$$

It is then natural to ask whether or not, leaving aside their precise meaning, the iterated sums

$$
\sum_{m=0}^{\infty} \left(\sum_{n=0}^{\infty} u_{mn} \right) \quad \text{and} \quad \sum_{n=0}^{\infty} \left(\sum_{m=0}^{\infty} u_{mn} \right)
$$

exist and (if they do) are equal. To answer such questions, we need the following.

Theorem 2.4.22 (Repeated Summation or Associativity). *Let $(X_n)_{n=1}^{\infty}$ be a partition of X; i.e., $X_n \neq \emptyset \; \forall n \in \mathbb{N}$, $X_j \cap X_k = \emptyset$ for $j \neq k$, and $X = \bigcup_{n=1}^{\infty} X_n$. If $u : X \to \mathbb{R}$ is summable, then the restriction $u | X_n$ is summable (over X_n) for each $n \in \mathbb{N}$, and the unordered series $\sum_{n \in \mathbb{N}} (\sum_{x \in X_n} u_x)$ is summable. In this case, we have*

$$
\sum_{x \in X} u_x = \sum_{n \in \mathbb{N}} \left(\sum_{x \in X_n} u_x \right) = \sum_{n=1}^{\infty} \left(\sum_{x \in X_n} u_x \right).
$$

If we assume the stronger condition that $\sum_{n \in \mathbb{N}} (\sum_{x \in X_n} |u_x|)$ is summable, then the converse is also true.

Proof. Suppose that u is summable with sum $s = \sum_{x \in X} u_x$. Then Corollary 2.4.13 implies that each restriction $u | X_n$ is also summable. Let $s_n = \sum_{x \in X_n} u_x$. We want to show that, given $\varepsilon > 0$, there exists a finite set $J_\varepsilon \subset \mathbb{N}$ such that

(i) $J_\varepsilon \subset J \in \mathcal{F}_{\mathbb{N}} \Rightarrow \left| s - \sum_{n \in J} s_n \right| < \varepsilon.$

Pick a finite set $F_\varepsilon \in \mathcal{F}_X$ such that

(ii) $F_\varepsilon \subset F \in \mathcal{F}_X \Rightarrow \left| s - \sum_{x \in F} u_x \right| < \varepsilon/2.$

Also, for each $n \in \mathbb{N}$, pick a finite set $E_{n\varepsilon} \in \mathcal{F}_{X_n}$ such that

(iii) $E_{n\varepsilon} \subset E_n \in \mathcal{F}_{X_n} \Rightarrow \left| s_n - \sum_{x \in E_n} u_x \right| < \varepsilon/2^{n+1}.$

Since F_ε is *finite*, the set $J_\varepsilon = \{ n \in \mathbb{N} : F_\varepsilon \cap X_n \neq \emptyset \}$ is finite as well. Also, we may (and will) assume that $F_\varepsilon \cap X_n \subset E_{n\varepsilon}$, for all $n \in J_\varepsilon$. It then follows from (iii) that $J_\varepsilon \subset J \in \mathcal{F}_{\mathbb{N}}$ implies

(iv) $\left| \sum_{n \in J} s_n - \sum_{x \in \bigcup_{n \in J} E_{n\varepsilon}} u_x \right| < \sum_{n \in J} \dfrac{\varepsilon}{2^{n+1}} < \sum_{n=1}^{\infty} \dfrac{\varepsilon}{2^{n+1}} = \dfrac{\varepsilon}{2}.$

Since $F_\varepsilon = \bigcup_{n \in J_\varepsilon} F_\varepsilon \cap X_n \subset \bigcup_{n \in J_\varepsilon} E_{n\varepsilon} \subset \bigcup_{n \in J} E_{n\varepsilon}$, it follows from (ii) that

(v) $\left| s - \displaystyle\sum_{x \in \bigcup_{n \in J} E_{n\varepsilon}} u_x \right| < \dfrac{\varepsilon}{2}.$

In view of (iv) and (v), we have $|s - \sum_{n \in J} s_n| < \varepsilon$ and (i) follows. Finally, if $\sum_{n \in \mathbb{N}} (\sum_{x \in X_n} |u_x|)$ is summable, i.e., if (by Proposition 2.4.20) we have

$$\sum_{n \in \mathbb{N}} \left(\sum_{x \in X_n} |u_x| \right) = \sum_{n=1}^{\infty} \sum_{x \in X_n} |u_x| < \infty,$$

and if $F \in \mathcal{F}_X$, then $F \subset \bigcup_{n=1}^{N} X_n$, for some $N \in \mathbb{N}$, so that

$$\sum_{x \in F} |u_x| = \sum_{n=1}^{N} \sum_{x \in F \cap X_n} |u_x| \leq \sum_{n=1}^{N} \sum_{x \in X_n} |u_x| \leq \sum_{n=1}^{\infty} \sum_{x \in X_n} |u_x| < \infty.$$

In other words, all partial sums of $|u|$ are (uniformly) bounded. This implies (Proposition 2.4.8) that $|u|$ is summable and hence (Theorem 2.4.11) so is u.   □

Example 2.4.23. Consider the *identity* function $u : \mathbb{Z} \to \mathbb{Z}$. Thus, $u_n = n \;\; \forall n \in \mathbb{Z}$. Then u is *not* summable because $\sum_{n=1}^{\infty} u_n = +\infty$. On the other hand, consider the partition $\mathbb{Z} = \bigcup_{k=0}^{\infty} X_k$, where $X_0 := \{0\}$ and $X_k := \{-k, k\}$, for all $k \geq 1$. We then have, for each k, that $s_k = \sum_{x \in X_k} u_x = -k + k = 0 = s_0$. Therefore $\sum_{k \in \mathbb{N}_0} s_k = 0$. This shows that the stronger condition in the last sentence of Theorem 2.4.22 is indeed necessary.

Exercise 2.4.24. Deduce Proposition 2.4.20 from Theorem 2.4.22.

Corollary 2.4.25. *If $u : X \to \mathbb{R}$ is summable and if $(X_n)_{n=1}^{\infty}$ and $(X_n')_{n=1}^{\infty}$ are two partitions of X (as in Theorem 2.4.22), then the corresponding unordered series $\sum_{n \in \mathbb{N}} (\sum_{x \in X_n} u_x)$ and $\sum_{n \in \mathbb{N}} (\sum_{x \in X_n'} u_x)$ are summable and we have*

$$\sum_{n \in \mathbb{N}} \left(\sum_{x \in X_n} u_x \right) = \sum_{x \in X} u_x = \sum_{n \in \mathbb{N}} \left(\sum_{x \in X_n'} u_x \right).$$

The following special case has many applications.

Theorem 2.4.26. *Let $u_{mn} \in \mathbb{R}$ for all integers $m \geq 0$ and $n \geq 0$. Then we have*

$$\sum_{m=0}^{\infty} \sum_{n=0}^{\infty} u_{mn} = \sum_{n=0}^{\infty} \sum_{m=0}^{\infty} u_{mn} = \sum_{k=0}^{\infty} \sum_{m+n=k} u_{mn} = \lim_{N \to \infty} \sum_{m=0}^{N} \sum_{n=0}^{N} u_{mn},$$

provided any one of the above iterated series converges when u_{mn} is replaced by $|u_{mn}|$.

Exercise 2.4.27. Show that the unordered series

$$\sum_{(m,n)\in\mathbb{N}^2} \frac{1}{(m+n)^p}$$

is *summable* if and only if $p > 2$. *Hint:* $\sum_{m=1}^{n}(m+n)^{-p} > n/(2n)^p$.

Recall that the *Cauchy product* of the (ordered) series $\sum_{n=0}^{\infty} a_n$ and $\sum_{n=0}^{\infty} b_n$ is the series $\sum_{n=0}^{\infty} c_n$, where $c_n = \sum_{k=0}^{n} a_k b_{n-k}$ for all nonnegative integers n. The following consequence of Theorem 2.4.26 is a special case of *Mertens' Theorem* (Theorem 2.3.43):

Theorem 2.4.28 (Cauchy Product).   *Suppose that the (ordered) series $\sum_{n=0}^{\infty} a_n$ and $\sum_{n=0}^{\infty} b_n$ are absolutely convergent. Then so is their Cauchy product and we have*

$$\sum_{n=0}^{\infty} c_n = \left(\sum_{n=0}^{\infty} a_n\right)\left(\sum_{n=0}^{\infty} b_n\right).$$

Proof. Let $u_{mn} := a_m b_n$. Then

$$\sum_{m=0}^{\infty}\left(\sum_{n=0}^{\infty}|u_{mn}|\right) = \sum_{m=0}^{\infty}\left(\sum_{n=0}^{\infty}|a_m||b_n|\right) = \left(\sum_{m=0}^{\infty}|a_m|\right)\left(\sum_{n=0}^{\infty}|b_n|\right) < \infty$$

and Theorem 2.4.26 implies

$$\left(\sum_{m=0}^{\infty} a_m\right)\left(\sum_{n=0}^{\infty} b_n\right) = \sum_{m=0}^{\infty}\sum_{n=0}^{\infty} u_{mn} = \sum_{k=0}^{\infty}\sum_{m+n=k} a_m b_n = \sum_{k=0}^{\infty} c_k,$$

completing the proof.   □

2.5   Problems

1. Let $s := \sum_{k=1}^{n} a_k$, where $a_k > -1$ for $1 \le k \le n$. Prove the inequality

$$\prod_{k=1}^{n}(1 + a_k) \le \left(1 + \frac{s}{n}\right)^n.$$

Hint: Use the *Arithmetic–Geometric Means* Inequality.

2.

(i) Show that, if $0 \le a \le b \le c$ and $b > 0$, then

$$2\sqrt{\frac{a}{c}} \le \frac{a}{b} + \frac{b}{c} \le 1 + \frac{a}{c}.$$

(ii) Let $m := \min\{a_1, \ldots, a_n\}$ and $M := \max\{a_1, \ldots, a_n\}$, where $(a_k)_{k=1}^n$ is a *finite* sequence of *positive* numbers. Show that we have

$$2n\sqrt{\frac{m}{M}} \le \sum_{k=1}^n \frac{a_k}{M} + \sum_{k=1}^n \frac{m}{a_k} \le n\left(1 + \frac{m}{M}\right).$$

(iii) With notation as in (ii), prove the inequalities

$$n^2 \le \left(\sum_{k=1}^n a_k\right)\left(\sum_{k=1}^n \frac{1}{a_k}\right) \le n^2 \frac{(m+M)^2}{4mM}.$$

Hint: Use *Cauchy's inequality*, the right inequality in (ii), and the trivial inequality $\alpha\beta \le (\alpha + \beta)^2/4$.

3. Show that, for all $n \in \mathbb{N}$, we have the inequalities

$$\sum_{k=1}^n \frac{1}{k} \ge \frac{2n}{n+1} \quad \text{and} \quad \sum_{k=1}^n \frac{1}{k^3} \ge \frac{4}{(n+1)^2}.$$

Hint: $\sum_{k=1}^n k = n(n+1)/2$ and $\sum_{k=1}^n k^3 = n^2(n+1)^2/4$. (Why?)

4. Show that for each $n \in \mathbb{N}$ we have

$$\binom{n}{1}\binom{n}{2}\cdots\binom{n}{n-1} \le \left(\frac{2^n - 2}{n - 1}\right)^{n-1}.$$

5 (Geometric–Harmonic Means Inequality). Show that for any *positive* numbers $a_1, a_2, \ldots, a_n$, we have

$$\frac{n}{\frac{1}{a_1} + \frac{1}{a_2} + \cdots + \frac{1}{a_n}} \le \sqrt[n]{a_1 a_2 \cdots a_n}.$$

6. For each of the following sets, find its sup, inf, max, and min, if they exist:

$$\{1/n : n \in \mathbb{N}\}, \quad \{1/n : n \in \mathbb{N}\} \cup \{0\}, \quad [0, \sqrt{2}] \cap \mathbb{Q}, \quad [0, \sqrt{2}] \cap \mathbb{Q}^c.$$

7. Let $B \subset \mathbb{R}$ be a *bounded* set. Show that, if $\max(B)$ and $\min(B)$ exist, then $\max(B) = \sup(B)$ and $\min(B) = \inf(B)$.

8. For any nonempty *bounded* subsets A and B of $\mathbb{R}$, define the sets $A + B := \{a + b : a \in A, b \in B\}$, $AB := \{ab : a \in A, b \in B\}$, and (if $0 \notin A$) $1/A := \{1/a : a \in A\}$. Prove the following:

(a) $A \subset B$ implies $\inf(B) \le \inf(A) \le \sup(A) \le \sup(B)$.
(b) $\sup(A \cup B) = \max\{\sup(A), \sup(B)\}$.
(c) If $a \le b$ for all $a \in A$ and $b \in B$, then $\sup(A) \le \inf(B)$.
(d) $\inf(A+B) = \inf(A)+\inf(B)$ and $\sup(A+B) = \sup(A)+\sup(B)$. In particular, if $A = \{a\}$, then $\sup(a + B) = a + \sup(B)$ and $\inf(a + B) = a + \inf(B)$, where $a + B := \{a\} + B$.
(e) If $A, B \subset [0, \infty)$, then $\sup(AB) = \sup(A)\sup(B)$. Furthermore, if $\inf(A) > 0$, then $\sup(1/A) = 1/\inf(A)$.

9.

(i) Let $A := \{x + 1/x : x > 0\}$. Show that $\sup(A) = \infty$ and $\inf(A) = 2$.

(ii) Let $B := \{m/n + 4n/m : m, n \in \mathbb{N}\}$. Show that $\sup(B) = \infty$ and $\inf(B) = 4$. *Hint:* Use $\sqrt{ab} \leq (a+b)/2$.

(iii) Let $C := \{m/(m+n) : m, n \in \mathbb{N}\}$. Show that $\sup(C) = 1$ and $\inf(C) = 0$. *Hint:* Fix $n = 1$ (resp., $m = 1$).

10.

(i) Show that, given any $r \in \mathbb{Q}$ and any $\varepsilon > 0$, there is an irrational $x \in \mathbb{Q}^c$ such that $|x - r| < \varepsilon$.

(ii) (Greatest Integer Function). Show that, given any $x \in \mathbb{R}$, there is a *unique* $m \in \mathbb{Z}$ such that $m \leq x < m + 1$. This m is denoted by $[x]$ and the function $x \mapsto [x]$ is called the *greatest integer function*. Note that $x = [x] + (x - [x])$, where $[x] \in \mathbb{Z}$ and $x - [x] \in [0, 1)$ are unique.

11 (Dirichlet).

(a) Given any *irrational* number $\alpha \in \mathbb{R} \setminus \mathbb{Q}$, show that the set $S := \{m + n\alpha : m, n \in \mathbb{Z}\}$ is *dense* in $\mathbb{R}$; i.e., every nonempty open interval contains a point of S. *Hint:* For each $k \in \mathbb{N} \cup \{0\}$, let $x_k := k\alpha - [k\alpha] \in [0, 1)$, where (as in Problem 10) $[k\alpha]$ is the greatest integer $\leq k\alpha$. Now for each nonempty open interval (a, b), pick $N \in \mathbb{N}$ with $b - a > 1/N$ and note that, among the $N + 1$ numbers $x_k \in [0, 1)$ with $0 \leq k \leq N$, there must be at least two, say x_{k_2} and x_{k_1}, such that $|x_{k_2} - x_{k_1}| < 1/N$. Conclude that, for some $\ell \in \mathbb{Z}$, we then have $\ell(x_{k_2} - x_{k_1}) \in (a, b) \cap S$.

(b) With notation as in part (a), show that, for each $N \in \mathbb{N}$, there are integers $p_N \in \mathbb{Z}$ and $q_N \in \mathbb{N}$ such that $q_N \to \infty$ as $N \to \infty$ and

$$\left| \alpha - \frac{p_N}{q_N} \right| < \frac{1}{Nq_N} \leq \frac{1}{q_N^2}.$$

(c) Deduce from part (a) that the set $\{n\alpha - [n\alpha] : n \in \mathbb{Z}\}$ is *dense* in $[0, 1]$. In fact, even $\{n\alpha - [n\alpha] : n \in \mathbb{N}\}$ is dense in $[0, 1]$, but the proof is harder.

12. Let $\alpha > 0$ be *irrational*. Show that for each $n \in \mathbb{Z}$ there is a *unique* integer $k_n \in \mathbb{Z}$ such that $k_n\alpha \leq n < (k_n + 1)\alpha$. Let $x_n := n - k_n\alpha$. Show that $\{x_n : n \in \mathbb{Z}\}$ is *dense* in $[0, \alpha]$. Taking $\alpha = \pi \approx 3.1415\ldots$ (which is known to be irrational), it follows that $\{n - k_n\pi : n \in \mathbb{Z}\}$ is dense in $[0, \pi]$. Now, we shall see later that the function $\cos x$ is *continuous*, maps $[0, \pi]$ onto $[-1, 1]$, and $\cos(n - k_n\pi) = \pm \cos n = \pm \cos |n|$. In particular, if $\lim_{j \to \infty}(n_j - k_{n_j}\pi) = \theta \in [0, \pi]$, then $\lim_{j \to \infty} \cos(n_j - k_{n_j}\pi)) = \cos \theta \in [-1, 1]$. Therefore, $\{\cos n : n \in \mathbb{N}\}$ is dense in $[-1, 1]$.

13. Show that, for any $n \in \mathbb{N}$, either $\sqrt{n} \in \mathbb{N}$ or $\sqrt{n} \in \mathbb{Q}^c$; i.e., $\sqrt{n} \in \mathbb{Q}$ if and only if n is a *perfect square*. *Hint:* Let $n = p_1^{k_1} \cdots p_m^{k_m}$ be the prime factorization of n. What if at least one of the k_i is *odd*?

14. For any *distinct* integers $m, n \in \mathbb{N}$, show that

$$\sqrt{m} + \sqrt{n} \in \mathbb{Q} \iff \sqrt{m} - \sqrt{n} \in \mathbb{Q} \iff \sqrt{m}, \sqrt{n} \in \mathbb{Q}.$$

Deduce that $\sqrt{n+1} \pm \sqrt{n} \in \mathbb{Q}^c$ for all $n \in \mathbb{N}$. Note that $\sqrt{n+1} - \sqrt{n} \to 0$ as $n \to \infty$. (Why?)

15. Using the Nested Intervals Theorem (Theorem 2.1.43), show that $I := [0, 1]$ is *uncountable*. *Hint:* Suppose, to get a contradiction, that I is countable: $I = \{x_1, x_2, \ldots\}$. Pick a closed interval $I_1 \subset I$ such that $x_1 \notin I_1$ and, then a closed interval $I_2 \subset I_1$ such that $x_2 \notin I_2$, etc.

16. Show that, for each $x \in \mathbb{R}$, we have $\bigcap_{\varepsilon \in \mathbb{R}^+} B_\varepsilon(x) = \{x\}$.

17. Show that, if $F \subset \mathbb{R}$ is a *closed, bounded* set, then $\inf(F)$, $\sup(F) \in F$.

18 (Derived Set). For each nonempty $A \subset \mathbb{R}$, let A' denote the set of all *limit points* of A. We call A' the *derived set* of A. Show that A' is *closed*. Show by an example that we may have $(A')' \subsetneq A'$. *Hint:* Construct a set A such that $A' = \{1/n : n \in \mathbb{N}\} \cup \{0\}$.

19. Determine all the limit points of each set.

(a) $\left\{ \dfrac{1}{m} + \dfrac{1}{n} : m, n \in \mathbb{N} \right\}$;

(b). $\left\{ (-1)^n + \dfrac{1}{m} : m, n \in \mathbb{N} \right\}$;

(c) $\{ 2^{-m} + 3^{-n} : m, n \in \mathbb{N} \}$;

(d) $\left\{ \dfrac{(-1)^n n}{n+1} : n \in \mathbb{N} \right\}$.

20. Prove the following statements:

(a) $\lim(\sqrt{n^2 + 1} - n) = 0$;

(b) $\lim(\sqrt{n^2 + n} - n) = 1/2$;

(c) $\lim \left(\dfrac{1 + \cdots + n}{n^2} \right) = \dfrac{1}{2}$;

(d) $\lim \left(\dfrac{n!}{n^n} \right) = 0$.

21. Let (x_n) be a sequence of *positive* numbers. Show that, if $\lim(x_{n+1}/x_n) = \ell < 1$, then $\lim(x_n) = 0$. Deduce that $\lim(n^k/2^n) = 0$ for any $k \in \mathbb{N}$ and that, more generally,

$$\lim_{n \to \infty} \frac{n^k}{(1+p)^n} = 0 \qquad (\forall p > 0). \tag{†}$$

Also show that $\lim_{n \to \infty} n!/2^{n^2} = 0$. Can you prove (†) using the *binomial formula?*

22 (Euler's Constant). Using the inequalities $x/(x+1) < \log(1+x) < x$, for all $x \in (0, 1)$, show that

$$\frac{1}{n+1} < \log(n+1) - \log n < \frac{1}{n}. \tag{*}$$

Deduce that the sequence (x_n), where

$$x_n := 1 + \frac{1}{2} + \cdots + \frac{1}{n} - \log n,$$

is *decreasing* and $x_n \geq 0$ for all $n \in \mathbb{N}$, and hence $\gamma := \lim(x_n)$ exists. The number $\gamma \approx 0.5772156\ldots$ is called *Euler's constant*. Using the fact that $\lim(\log n) = \infty$, conclude that $\left(\sum_{k=1}^{n} 1/k \right)$ is *divergent*.

23. Let $y_n := \sum_{k=1}^{2n} (-1)^{k-1}/k$. Show that $\lim(y_n) = \log 2$. *Hint:* $y_n = x_{2n} - x_n + \log 2$, where x_n is as in Problem 22.

24.

(i) Show that the following sequence (x_n) is *increasing* and *bounded above* and hence *convergent*:

$$x_n := \frac{1}{n+1} + \frac{1}{n+2} + \cdots + \frac{1}{2n}.$$

(ii) Let $0 < a_1 < b_1$ and define

$$a_{n+1} = \sqrt{a_n b_n}, \qquad b_{n+1} = \frac{a_n + b_n}{2} \qquad (\forall n \in \mathbb{N}).$$

Show that the sequences (a_n) and (b_n) are both *convergent* and have the *same* limit. *Hint:* Show (inductively) that the intervals $[a_n, b_n]$ are *nested*.

25. Let $x_1 = \sqrt{2}$ and $x_{n+1} := \sqrt{2 + x_n}$ for all $n \in \mathbb{N}$. Show that $\lim(x_n) = 2$. *Hint:* First show that (x_n) is *increasing* and *bounded above*.

26. Let (x_n) be a real sequence and $\xi \in \mathbb{R}$. Suppose that *every* subsequence of (x_n) has, in turn, a subsequence that converges to ξ. Show that we then have $\lim(x_n) = \xi$. *Hint:* Prove this by contradiction.

27. Using the fact that $\lim_{n \to \infty}(1 + 1/n)^n = e$, find each limit.

$$\text{(a)} \quad \lim_{n \to \infty} \left(1 - \frac{1}{n}\right)^n; \qquad\qquad \text{(b)} \quad \lim_{n \to \infty} \left(1 - \frac{1}{n^2}\right)^{n^2};$$

$$\text{(c)} \quad \lim_{n \to \infty} \left(1 - \frac{1}{n^2}\right)^n; \qquad\qquad \text{(d)} \quad \lim_{n \to \infty} \left(1 - \frac{2}{n}\right)^{n^2}.$$

Hint: For (c), note that $1 - 1/n^2 = (1 + 1/n)(1 - 1/n)$. For (d), find $\lim_{n \to \infty}(1 - 2/n)^n$ first.

28. Using the *Squeeze Theorem*, find the following limits:

$$\text{(a)} \quad \lim \left(\sum_{k=1}^{n} \frac{1}{\sqrt{n^2 + k}}\right); \qquad \text{(b)} \quad \lim(\sqrt[n]{n^2 + n}); \qquad \text{(c)} \quad \lim\left((n!)^{1/n^2}\right).$$

29. Use $\sum_{k=1}^{n} k = n(n + 1)/2$ and the *Squeeze Theorem* to show that

$$\lim_{n \to \infty} \left(\frac{1}{n^2 + 1} + \frac{2}{n^2 + 2} + \cdots + \frac{n}{n^2 + n}\right) = \frac{1}{2}.$$

30. Find the following limit in two different ways: (i) by using the identity $\sum_{k=1}^{n} k^2 = n(n + 1)(2n + 1)/6$ and (ii) by using the *Squeeze Theorem*

$$\lim_{n \to \infty} \sqrt[n]{1^2 + 2^2 + \cdots + n^2}.$$

31. Given any *nonnegative* numbers $a_1, a_2, \ldots, a_k$, find the following limit:

$$\lim_{n \to \infty} (a_1^n + a_2^n + \cdots + a_k^n)^{1/n}.$$

32. Prove the following statements:

$$\text{(a)} \quad \lim \left(\frac{1}{n} \sum_{k=1}^{n} \frac{1}{k}\right) = 0; \qquad\qquad \text{(b)} \quad \lim \left(\frac{1}{n} \sum_{k=1}^{n} \sqrt[k]{k}\right) = 1.$$

33.

(i) Let $\xi := \lim(x_n) > 0$, where $x_n > 0$ for all $n \in \mathbb{N}$, and consider the sequence $(y_n) := (\sqrt[n]{x_1 \cdots x_n})$ of *geometric means*. Show that $\lim(y_n) = \xi$. *Hint:* Look at $\log y_n$ and use Corollary 2.2.33 and the fact (to be proved later) that $\lim(\log x_n) = \log \xi$ if and only if $\lim(x_n) = \xi$.

(ii) Show that $\lim(\sqrt[n]{n}) = 1$. *Hint:* $n = 1 \cdot (2/1) \cdots [n/(n - 1)]$.

(iii) Using $\lim((1 + 1/n)^n) = e$ and $\prod_{k=1}^{n}[(k + 1)/k]^k = (n + 1)^n/n!$, show that

$$\lim_{n\to\infty} \frac{n + 1}{\sqrt[n]{n!}} = \lim_{n\to\infty} \frac{n}{\sqrt[n]{n!}} = e.$$

34. Show that, if a sequence (x_n) satisfies $|x_{n+1} - x_n| < 1/2^n$ for all $n \in \mathbb{N}$, then it is a *Cauchy* sequence and hence convergent. Does the same hold if we only insist on the weaker inequalities $|x_{n+1} - x_n| < 1/n$?

35. Let $x_1 \neq x_2$ be real numbers and let $c \in (0, 1)$. For each $n \geq 3$, define $x_n := cx_{n-1} + (1 - c)x_{n-2}$. Show that (x_n) is *contractive* (cf. Exercise 2.2.48) and find its limit.

36. It is known that $x^3 - 4x + 1 = 0$ has a (unique) *zero*, say ζ, in $(0, 1)$. Let $x_1 \in (0, 1)$ be arbitrary and define $x_{n+1} := (x_n^3 + 1)/4$, for all $n \in \mathbb{N}$. Show that (x_n) is *contractive* and $\lim(x_n) = \zeta$.

37.

(i) Let $a_1 = 4$ and $a_{n+1} = 3 - 2/a_n$, for all $n \in \mathbb{N}$. Show that $\lim(a_n) = 2$.
(ii) Let $b_1 = 1$ and $b_n = b_{n-1} + 1/b_{n-1}$, $\forall n \in \mathbb{N}$. Show that (b_n) is *divergent*.

38. Show that $\lim_{n\to\infty}(n!)^{1/n} = \infty$.

39. Show that $(\sin n)$ is *divergent* and so is $(\sin \sqrt{n})$. *Hint:* For each $k \in \mathbb{N}$ consider the intervals I_n (resp., J_n) defined by $I_k := (\pi/6 + 2(k - 1)\pi, 5\pi/6 + 2(k - 1)\pi)$ (resp., $J_k := (7\pi/6 + 2(k - 1)\pi, 11\pi/6 + 2(k - 1)\pi))$. Note that each I_k (resp., J_k) has length > 2. Let m_k (resp., n_k) be the *smallest* integer in I_k (resp., J_k), and look at the subsequence $(\sin m_k)$ (resp., $(\sin n_k)$).

40. Find $\underline{\lim}(\cos n)$ and $\overline{\lim}(\cos n)$. *Hint:* Use Problem 12.

41. Find $\underline{\lim}$ and $\overline{\lim}$ of each sequence (x_n):

(a) $x_n := (-1)^n \left(1 + \dfrac{1}{n}\right)$;

(b) $x_n := \left((-1)^n + \dfrac{1}{n}\right)$;

(c) $x_n := \sin\left(\dfrac{n\pi}{2}\right) + \cos\left(\dfrac{n\pi}{2}\right)$;

(d) $x_n := \left(2 + \dfrac{(-1)^n}{n}\right)$.

42. Show that, if $\lim(x_n) = \xi$ exists, then for any sequence (y_n),

$$\underline{\lim}(x_n + y_n) = \xi + \underline{\lim}(y_n) \quad \text{and} \quad \overline{\lim}(x_n + y_n) = \xi + \overline{\lim}(y_n).$$

Assuming that $\lim(x_n) = \xi > 0$, show that we also have

$$\underline{\lim}(x_n y_n) = \xi \underline{\lim}(y_n) \quad \text{and} \quad \overline{\lim}(x_n y_n) = \xi \overline{\lim}(y_n).$$

43. Find the sum of each series, where $k \in \mathbb{N}$ and $\sum := \sum_{n=1}^{\infty}$.

(a) $\displaystyle\sum \frac{2n + 1}{n^2(n + 1)^2}$;

(b) $\displaystyle\sum (-1)^n \frac{2n + 1}{n(n + 1)}$;

(c) $\displaystyle\sum \frac{1}{4n^2 - 1}$;

(d) $\displaystyle\sum \frac{n - \sqrt{n^2 - 1}}{\sqrt{n(n + 1)}}$;

(e) $\displaystyle\sum \frac{1}{n(n + k)}$;

(f) $\displaystyle\sum \frac{1}{n(n + 1)(n + 2)}$.

Hint: Note that $\dfrac{2n + 1}{n^2(n + 1)^2} = \dfrac{1}{n^2} - \dfrac{1}{(n + 1)^2}$, etc.

44. Show that the following series *diverge*. Here, $p > 0$ is arbitrary.

(a) $\sum \dfrac{1}{n\sqrt[n]{n}}$;   (b) $\sum \sin n$;   (c) $\sum \dfrac{1}{\sqrt{n}(\log n)^p}$;

(d) $\sum \dfrac{n}{\sqrt[n]{n!}}$;   (e) $\sum \sqrt[n]{1+\cdots+n}$;   (f) $\sum \dfrac{\sqrt{n+1}-\sqrt{n}}{\sqrt{n}}$.

45 (Abel). Show that, if (x_n) is a *decreasing* sequence of *positive* numbers such that $\sum_{n=1}^{\infty} x_n$ is convergent, then $\lim(n x_n) = 0$. *Hint:* Note that $n x_{2n} \le \sum_{k=1}^{n} x_{n+k} \to 0$, as $n \to \infty$. Similarly, $\lim(n x_{2n+1}) = 0$.

46.

(a) Show by a counterexample that in the previous problem we *cannot* remove the condition that (x_n) be *decreasing*.
(b) Show, however, that for *any* convergent series $\sum_n x_n$ of positive numbers, we have

$$\lim_{n \to \infty} n(x_1 x_2 \cdots x_n)^{1/n} = 0.$$

47. Let $\sum x_n$ be a convergent series of *positive* terms.

(i) Show that the series $\sum \sqrt{x_n x_{n+1}}$ is convergent.
(ii) Show that the series $\sum \sqrt{x_n}/n$ is convergent.

48. Let (x_n) be a sequence of *positive* numbers with $\sum x_n = \infty$. Prove the following statements:

(a) The series $\sum x_n/(1 + x_n)$ is *divergent*.
(b) The series $\sum x_n/(1 + n x_n)$ may be convergent *or* divergent.
(c) The series $\sum x_n/(1 + n^2 x_n)$ is *convergent*.
(d) The series $\sum x_n/(1 + x_n^2)$ may be convergent *or* divergent. *Hint:* For (b), use the sequence $x_n = 1$ if $n = m^2$, for some $m \in \mathbb{N}$, and $x_n = 1/n^2$ otherwise.

49. Using the *First Comparison Test*, show that each series *converges*.

(a) $\sum \dfrac{\sqrt{n+1}-\sqrt{n}}{n}$;   (b) $\sum \dfrac{1}{n^2 - \log n}$;

(c) $\sum \dfrac{1}{n} \log \left(1 + \dfrac{1}{n}\right)$;   (d) $\sum \left(1 - \cos \dfrac{1}{n}\right)$.

Hint: Use the inequalities $\log(1 + x) < x$ for all $x > -1$, $\log x < x$ for all $x > 0$, and $1 - \cos 2\theta = 2 \sin^2 \theta \le 2\theta^2$ for all $\theta \ge 0$.

50. Test each series for convergence. In (a), $k \in \mathbb{N}$ is arbitrary.

(a) $\sum \dfrac{n^k}{2^n}$;   (b) $\sum (\sqrt[n]{n} - 1)^n$;

(c) $\sum \left(\dfrac{n}{n+1}\right)^{n(n+1)}$;   (d) $\sum 3^n \left(\dfrac{n}{n+1}\right)^{n^2}$.

51. Test each series for convergence.

(a) $\sum \dfrac{(n!)^2}{(2n)!}$;

(b) $\sum \dfrac{2^n n!}{n^n}$;

(c) $\sum \dfrac{3^n n!}{n^n}$;

(d) $\sum \dfrac{2^{n^2}}{n!}$.

52. Show that the following series are *divergent*:

(a) $\sum (\sqrt[n]{n} - 1)$;

(b) $\displaystyle\sum_{n=2}^{\infty} \dfrac{1}{(\log n)^{\log(\log n)}}$.

Hint: For (a), let $x_n = \sqrt[n]{n} - 1$ so that $n = (1 + x_n)^n$ and hence $\log n = n \log(1 + x_n)$. Deduce that $\lim(n x_n) = \infty$. For (b), use $[\log(\log n)]^2 \leq \log n$ for all large n, which follows from the fact (to be proved later) that $\lim_{x \to \infty} \log x / x^{\alpha} = 0$ for any $\alpha > 0$ (cf. (†) before Gauss's Test).

53. Show that *Raabe's Test* (Corollary 2.3.31) implies the following one (which is due to *Schlömilch*):
Let $x_n > 0 \ \forall n \in \mathbb{N}$. Then $\sum x_n$ converges if $n \log(x_n/x_{n+1}) \geq r$ is ultimately true for some $r > 1$ (equivalently, if $\underline{\lim}(n \log(x_n/x_{n+1})) > 1$) and diverges if $n \log(x_n/x_{n+1}) \leq 1$ is ultimately true (which is the case if, e.g., $\overline{\lim}(n \log(x_n/x_{n+1})) < 1$). *Hint:* Use the inequalities $x/(x + 1) \leq \log(1 + x) \leq x$ for all $x > -1$.

54. Test the following series for convergence:

(a) $\sum \dfrac{1}{2^{\log n}}$;

(b) $\sum \dfrac{1}{3^{\log n}}$.

55. Determine the values of a for which the series

$$\sum_{n=1}^{\infty} \frac{n!}{(a + 1)(a + 2) \cdots (a + n)}$$

converges. Hint: Use *Raabe's Test*.

56. For which values of p does the series

$$\sum_{n=1}^{\infty} \left(\frac{1 \cdot 3 \cdot 5 \cdots (2n - 1)}{2 \cdot 4 \cdot 6 \cdots (2n)} \right)^p$$

converge? *Hint:* Assume $p \in \mathbb{N}_0 := \mathbb{N} \cup \{0\}$ (although this is not necessary) and use *Gauss's Test* (Corollary 2.3.32).

57. Show that the *Cauchy product* of $\sum (-1)^{n-1}/n$ with *itself* is the series

$$\sum_{n=1}^{\infty} (-1)^{n-1} \frac{1}{n+1} \left(1 + \frac{1}{2} + \cdots + \frac{1}{n} \right) \tag{$*$}$$

and that this series converges to $(\log 2)^2$. *Hint:* Use Problem 23.

58. Let $a_n = b_n := (-1)^n/(n+1)^p$, where $p \in [0, 1/2]$. Show that the *Cauchy product* of $\sum a_n$ and $\sum b_n$ is *divergent*.

59. Show that the *Cauchy product* of the following *divergent* series is *convergent*:

$$1 - \sum_{n=1}^{\infty} \left(\frac{3}{2}\right)^n \quad \text{and} \quad 1 + \sum_{n=1}^{\infty} \left(\frac{3}{2}\right)^{n-1} \left(2^n + \frac{1}{2^{n+1}}\right).$$

60. Test the following series for *conditional* and *absolute* convergence:

(a) $\sum \frac{(-1)^n}{\sqrt{n}}$;

(b) $\sum_{n=2}^{\infty} \frac{(-1)^n}{\log n}$;

(c) $\sum \frac{(-1)^n \log n}{n}$;

(d) $\sum (-1)^n \frac{n+1}{2n+1}$;

(e) $\sum \left(\frac{-n}{n+1}\right)^n$;

(f) $\sum (-1)^n (\sqrt[n]{n} - 1)^n$.

61. Show that, if $\sum x_n$ is *absolutely convergent*, then $\sum x_n^2 < \infty$. Give an example of a *conditionally convergent* series $\sum x_n$ such that $\sum x_n^2 = \infty$.

62. Show that the series $\sum_{n=1}^{\infty} (-1)^{n-1} (2n+1)/[n(n+1)]$ is convergent and find its sum.

63. For any $\theta \neq 2k\pi$, $k \in \mathbb{Z}$, prove the identities:

$$\sin \theta + \sin 2\theta + \cdots + \sin n\theta = \frac{\sin(n\theta/2) \sin(n+1)\theta/2}{\sin(\theta/2)}.$$

$$\cos \theta + \cos 2\theta + \cdots + \cos n\theta = \frac{\sin(n\theta/2) \cos(n+1)\theta/2}{\sin(\theta/2)}.$$

Hint: Recall that $2 \sin \alpha \sin \beta = \cos(\beta - \alpha) - \cos(\beta + \alpha)$ and $2 \sin \alpha \cos \beta = \sin(\beta + \alpha) - \sin(\beta - \alpha)$.

64. Let (b_n) be a *decreasing* sequence of *positive* numbers with $\lim(b_n) = 0$. Show that the series

$$\sum b_n \cos n\theta, \qquad \sum b_n \sin n\theta \qquad (\forall \theta \neq 2k\pi, \quad k \in \mathbb{Z})$$

are convergent. In particular, the series $\sum (\cos n\theta)/n^p$ and $\sum (\sin n\theta)/n^p$ are convergent for all $p > 0$. *Hint:* Use *Dirichlet's Test* and Problem 63.

65. Show that the series $\sum |\sin n|/n$ and $\sum |\cos n|/n$ are both *divergent*. *Hint:* For any *three* consecutive integers, at least one satisfies $|\sin n| \geq 1/2$ and similarly for $\cos n$.

66.

(a) Show that the series $\sum \sin^k n/n$ and $\sum \cos^k n/n$ are *divergent* if the (integer) exponent $k \geq 0$ is *even* and *convergent* if k is *odd*.

(b) Show, however, that the series $\sum (-1)^n \sin^k n/n$ and $\sum (-1)^n \cos^k n/n$ are *convergent* for *all* (integer) exponents $k \geq 0$.

(c) Extending the Problem 65, deduce from (a) and (b) that $\sum |\sin^k n|/n$ and $\sum |\cos^k n|/n$ are *divergent* for all $k \in \mathbb{N}_0$.

Hint: Use the trigonometric identities in Theorem 8.5.17 and Problems 64 and 65.

67. Test the following series for convergence:

(a) $\sum_{n=2}^{\infty} \frac{\sin n}{\log n}$;

(b) $\sum \frac{\sin(\log n)}{n}$.

Hint: For (b), look at blocks of consecutive n's for which $\sin n \geq 1/2$.

68. Rearrange the series $\sum(-1)^{n-1}/n$ to obtain the series

$$1 - \frac{1}{2} - \frac{1}{4} + \frac{1}{3} - \frac{1}{6} - \frac{1}{8} + \frac{1}{5} - \cdots,$$

where each positive term is followed by two negative ones. Find the sum of this series. *Hint:* Let S_n denote the nth partial sum of the series. Show that $S_{3n} = (\sum_{k=1}^{2n}(-1)^{k-1}/k)/2$ so that $\lim(S_{3n}) = (\log 2)/2$.

69. Show that the following rearrangement of $\sum(-1)^n/n$ is *divergent*:

$$\frac{1}{2} + \frac{1}{4} + \frac{1}{6} + \cdots + \frac{1}{2^8} - 1 + \frac{1}{2^8 + 2} + \cdots + \frac{1}{2^{16}} - \frac{1}{3} + \cdots.$$

Hint: Recall (as in Proposition 2.3.12) that $\sum_{k=1}^{2^n-1} 1/(2k) > n/4$.

70. Which of the following sequences are in $\ell^2(\mathbb{N})$?

(a) $\left(\dfrac{1}{n \log n}\right)$;

(b) $\left(\dfrac{1}{\sqrt{n} \log n}\right)$;

(c) $\left(\dfrac{\log n}{\sqrt{n}}\right)$;

(d) $\left(\dfrac{\sin n}{n}\right)$.

71. Show that if $(a_{k-1})_{k=1}^{\infty} \in \ell^2(\mathbb{N})$, then $\sum_{n=0}^{\infty} a_n x^n$ is (absolutely) convergent for every $x \in (-1, 1)$. Here we set $x^0 := 1$ (even for $x = 0$).

72. Let $a, b \in (0, 1)$. Show that the unordered series

$$\sum_{(m,n)\in\mathbb{N}\times\mathbb{N}} a^m b^n$$

is *summable* and find its sum.

73. Show that the double series

$$\sum_{m, n\in\mathbb{N}} \frac{1}{m^p n^q}$$

is convergent if and only if $p > 1$ and $q > 1$.

74. Show that the double series

$$\sum_{m, n\in\mathbb{N}} \frac{1}{(m^2 + n^2)^p}$$

is *convergent* if and only if $p > 1$. Deduce that

$$\sum_{(m,n)\in\mathbb{Z}\times\mathbb{Z}} \frac{1}{m^2 + n^2 + 1}$$

is *divergent*.

75. Show that, with m, $n \in \mathbb{N}_0 := \mathbb{N} \cup \{0\}$, we have

$$\sum_{(m,n) \neq (0,0)} \frac{1}{(m+n)^p} = \sum_{k=1}^{\infty} \frac{k+1}{k^p}.$$

Hint: Look at the terms with $m + n = k$.

76. Show that the following sum *converges* if and only if $p > k$:

$$\sum_{(n_1,\ldots,n_k) \in \mathbb{N}^k} \frac{1}{(n_1 + \cdots + n_k)^p}.$$

Hint: First show that $(n_1 + \cdots + n_k)^k \geq n_1 \cdots n_k$ and

$$\sum_{1 \leq n_2,\ldots,n_k \leq n_1} \frac{1}{(n_1 + \cdots + n_k)^p} \geq \frac{n_1^{k-1}}{(k n_1)^p}.$$

Chapter 3
Limits of Functions

As was pointed out in Chap. 2, the central idea in analysis is that of *limit*, which was introduced and studied for *sequences* of real numbers, i.e., for functions $x : \mathbb{N} \to \mathbb{R}$. In particular, the behavior of the term $x_n := x(n)$ was studied under the assumption that the element n in the domain of our sequence was *approaching infinity*.

Our goal in this chapter will be to define limits of functions whose domains are more general subsets of $\mathbb{R}$. We shall see that the study of this extended limit concept can, if one wishes, be reduced to the study of suitable *sequences*. The important and related concept of *continuity* of real-valued functions of a real variable and the more refined concept of *uniform continuity* will be introduced and studied in Chap. 4. Further topics on limits will appear in Chap. 8.

What we needed for the definition of limit was the *distance* (or *metric*) in the set $\mathbb{R}$. It is therefore tempting to study limits of functions whose domains and ranges are subsets of more general *metric spaces*. This will be done in Chap. 5, which can be skipped, without loss of continuity, by those who wish to study the subject later.

To avoid unnecessary repetitions the following should be pointed out: *Throughout this chapter, X,Y, Z, etc. will denote subsets of $\mathbb{R}$. Also, I and J (possibly with subscript) will always denote intervals that may be open or closed, bounded or unbounded. Finally, all functions are from subsets of $\mathbb{R}$ to $\mathbb{R}$.*

3.1 Bounded and Monotone Functions

Recall that in Chap. 2 we defined the concepts of *bounded*, *increasing*, and *decreasing* real sequences. In this introductory section we want to define the same concepts for real-valued functions defined on arbitrary subsets of $\mathbb{R}$.

Definition 3.1.1 (Bounded (Above, Below), Unbounded). A function $f : X \to \mathbb{R}$ is called *bounded above* (resp., *below*) if there exists $B \in \mathbb{R}$ (resp., $A \in \mathbb{R}$) such that, $\forall x \in X$, we have $f(x) \leq B$ (resp., $f(x) \geq A$). The function f is called *bounded* if it is bounded above *and* below, equivalently (why?), if there exists

© Springer Science+Business Media New York 2014

H.H. Sohrab, *Basic Real Analysis*, DOI 10.1007/978-1-4939-1841-6__3

$B > 0$ such that $|f(x)| \leq B$ $\forall x \in X$. Finally, f is called *unbounded* if it is *not* bounded. Given any subset $S \subset X$, we say that f is bounded above (resp., bounded below, bounded, unbounded) *on* S if the *restriction* $f|S$ is bounded above (resp., bounded below, bounded, unbounded).

Remark 3.1.2.

(a) The above definition may be restated as follows: a function $f : X \to \mathbb{R}$ is bounded above (resp., bounded below, bounded) if and only if its *range*, $f(X)$, is bounded above (resp., bounded below, bounded).
(b) It is obvious that a function $f : X \to \mathbb{R}$ is bounded above (resp., below) if and only if the function $-f$ defined by $(-f)(x) = -f(x)$ $\forall x \in X$ is bounded below (resp., above). Also, f is bounded if and only if $-f$ is.
(c) Note that the above definition makes sense even if the domain X of the function f is an *arbitrary set*.

In view of the above remarks, it is natural to introduce the concepts of *supremum, infimum, maximum,* and *minimum* for real-valued functions.

Definition 3.1.3 (Supremum, Infimum, Maximum, Minimum). Let $f : X \to \mathbb{R}$. The *supremum* (resp., *infimum*) of f is defined to be the *extended real number* $\sup(f) := \sup f(X) = \sup\{f(x) : x \in X\}$ (resp., $\inf(f) := \inf f(X) = \inf\{f(x) : x \in X\}$). It is obvious that f is bounded above (resp., bounded below) if and only if $\sup(f) < \infty$ (resp., $\inf(f) > -\infty$). We say that f *attains* its *maximum* (resp., *minimum*) if and only if , for some $x_0 \in X$, we have $f(x_0) = \sup(f)$ (resp., $f(x_0) = \inf(f)$); we then write $f(x_0) = \max(f)$ (resp., $f(x_0) = \min(f)$).

Example 3.1.4.

(a) The functions $x \mapsto \sin x$; $x \mapsto \cos x$; and $x \mapsto e^{-x^2}$ are bounded (on $\mathbb{R}$).
(b) The functions $x \mapsto |x|$; $x \mapsto e^x$; and $x \mapsto x^2 - 1$ are bounded below but *not* above.
(c) The functions $x \mapsto 1 - |x|$; $x \mapsto 4 - x^2$; and $x \mapsto -1/\sin x$ $(0 < x < \pi)$ are bounded above but *not* below.
(d) Finally, the functions $x \mapsto 1/x$; $x \mapsto \log x$ $(0 < x < \infty)$; and $x \mapsto x^3$ are neither bounded above nor bounded below.

The following definition is an extension of the corresponding one we gave for real sequences in Chap. 2.

Definition 3.1.5 (Increasing, Decreasing, Monotone, Constant). A function $f :$ $X \to \mathbb{R}$ is called *increasing* (resp., *strictly increasing*) if $x_1, x_2 \in X$ and $x_1 < x_2$ imply $f(x_1) \leq f(x_2)$ (resp., $f(x_1) < f(x_2)$). Similarly, f is called *decreasing* (resp., *strictly decreasing*) if $x_1, x_2 \in X$ and $x_1 < x_2$ imply $f(x_1) \geq f(x_2)$ (resp., $f(x_1) > f(x_2)$). The function f is called *monotone* (resp., *strictly monotone*) if it is increasing *or* decreasing (resp., *either* strictly increasing *or* strictly decreasing). Finally, f is called *constant* if $f(x_1) = f(x_2)$ $\forall x_1, x_2 \in X$. Given any subset $S \subset X$, we say that f is increasing (resp., strictly increasing, decreasing, strictly decreasing, monotone, strictly monotone, constant) *on* S if the *restriction* $f|S$ is increasing (resp., strictly increasing, decreasing, etc).

Example 3.1.6. The functions $x \mapsto e^x$ and $x \mapsto \log x$ are both strictly increasing (on their domains). Also, $x \mapsto 1/x$ and $x \mapsto e^{-x}$ are both strictly decreasing, the former on $(-\infty, 0)$ and $(0, \infty)$ and the latter on $\mathbb{R}$.

Remark 3.1.7.

(a) $f : X \to \mathbb{R}$ is increasing (resp., strictly increasing, decreasing, strictly decreasing) if and only if $-f$ is decreasing (resp., strictly decreasing, increasing, strictly increasing).

(b) $f : X \to \mathbb{R}$ is *simultaneously* increasing *and* decreasing if and only if it is *constant*. (Why?)

(c) Monotone functions send "in-between points" to "in-between points." In other words, if $f : X \to \mathbb{R}$ is monotone and if x_1, x_2, $x_3 \in X$ satisfy $x_1 < x_2 < x_3$, then $f(x_1) \leq f(x_2) \leq f(x_3)$ in the increasing case and $f(x_1) \geq f(x_2) \geq f(x_3)$ in the decreasing case. Thus, if f is *not* monotone, then there must exist x_1, x_2, $x_3 \in X$ satisfying $x_1 < x_2 < x_3$ and $f(x_2)$ *is not between* $f(x_1)$ *and* $f(x_3)$ —i.e., if we have, e.g., $f(x_1) \leq f(x_3)$, then either $f(x_2) < f(x_1)$ or $f(x_2) > f(x_3)$.

The following exercise, which is intuitively obvious and contains most of the concepts introduced above, will be useful later. Recall that, if S is a subset of the domain of a function f, then $f|S$ denotes the *restriction of f to S*.

Exercise 3.1.8. Let a, $b \in \mathbb{R}$, $a < b$, and let $f : (a, b) \to \mathbb{R}$ be increasing (resp., decreasing) and bounded. Show that, for any $c \in (a, b)$, we have $\sup(f) = \sup(f|(c, b))$ and $\inf(f) = \inf(f|(a, c))$ [resp., $\sup(f) = \sup(f|(a, c))$ and $\inf(f) = \inf(f|(c, b))$].

We end the section with another intuitively obvious fact, namely that a strictly increasing (resp., decreasing) function is necessarily one-to-one and that its inverse function has the same property.

Proposition 3.1.9. *Let $X \subset \mathbb{R}$, and let $f : X \to \mathbb{R}$ be a strictly increasing (resp., strictly decreasing) function. Then f is invertible and its inverse $f^{-1} : f(X) \to X$ is also strictly increasing (resp., strictly decreasing).*

Proof. Suppose f is strictly increasing; the decreasing case is obtained by changing f to $-f$. That f is *injective* is obvious. Indeed, if $x_1 \neq x_2$, then we have either $x_1 < x_2$ or $x_2 < x_1$. In the first case $f(x_1) < f(x_2)$ and, in the second, $f(x_2) < f(x_1)$. Next, for any y_1, $y_2 \in f(X)$ satisfying $y_1 < y_2$, there are *unique* x_1, $x_2 \in X$ with $f(x_1) = y_1$ and $f(x_2) = y_2$. If $x_1 \geq x_2$—i.e., if $f^{-1}(y_1) \geq f^{-1}(y_2)$—then $y_1 = f(x_1) \geq f(x_2) = y_2$, contradicting the assumption $y_1 < y_2$.   □

3.2 Limits of Functions

Our goal in this section is to define the limits of real-valued functions whose domains are *intervals* or unions of intervals of real numbers. The properties of such limits will be deduced, in the next section, from similar properties already proved for real sequences in Chap. 2. Before defining limits, let us define for subsets of $\mathbb{R}$ the notions of *interior point*, *interior*, *cluster point*, and *closure*. These concepts will be introduced for subsets of metric spaces in Chap. 5.

Definition 3.2.1 (Interior Point, Interior, Cluster Point, Closure). Let $X \subset \mathbb{R}$. We say that a point $x \in X$ is an *interior point* of X if, for some $\varepsilon > 0$, we have $B_\varepsilon(x) := (x - \varepsilon, x + \varepsilon) \subset X$. The set of all interior points of X is denoted by X° and is called the *interior* of X. A point $x \in \mathbb{R}$ is called a *cluster point* of X if $x \in X$ or x is a *limit point* of X (cf. Definition 2.2.11). The set of all cluster points of X is denoted by X^- and is called the *closure* of X.

Remark 3.2.2.

1. For each $X \subset \mathbb{R}$, X° is *open* and X^- is *closed*. (Why?) The *interior* I° of an interval I is obviously the open interval obtained by *removing* the endpoint(s) it contains, and its closure I^- is the closed interval obtained by *adjoining* the endpoint(s) not in it. Here by *endpoint* we obviously mean *finite* endpoints and not $\pm\infty$. Thus, if $a < b$, then all bounded intervals (a, b), $(a, b]$, $[a, b)$, and $[a, b]$ have interior (a, b) and closure $[a, b]$; the unbounded intervals (a, ∞), $[a, \infty)$ have interior (a, ∞) and closure $[a, \infty)$; the unbounded intervals $(-\infty, b)$, $(-\infty, b]$ have interior $(-\infty, b)$ and closure $(-\infty, b]$; and, finally, the interior and closure of the unbounded interval $\mathbb{R} = (-\infty, \infty)$ are both the set $\mathbb{R}$ itself.
2. The *extended* real line $[-\infty, \infty]$, introduced in Chap. 2, was denoted by $\overline{\mathbb{R}}$. This should not be confused with the *closure* of $\mathbb{R}$, which we denote by $\mathbb{R}^- = \mathbb{R} = \mathbb{R}^\circ$.
3. If $a \geq b$, then $(a, b) = \emptyset$ and $[a, b] = \{a\}$. We have $\emptyset^\circ = \emptyset^- = \emptyset$, $\{a\}^\circ = \emptyset$, and $\{a\}^- = \{a\}$, as follows easily from the definition.

Definition 3.2.3 (Limit of a Function). Let I be an interval and let $x_0 \in I^-$. Let f be a function defined on I (*except possibly* at x_0). We say that $y_0 \in \mathbb{R}$ is the *limit* of $f(x)$ at x_0 (or that $f(x)$ *converges* to y_0 as x approaches x_0) and write $\lim_{x \to x_0} f(x) = y_0$, if the following is satisfied:

$$(\forall \varepsilon > 0)(\exists \delta = \delta(\varepsilon) > 0)(\forall x \in I)(0 < |x - x_0| < \delta \Rightarrow |f(x) - y_0| < \varepsilon). \qquad (*)$$

Remark 3.2.4.

1. We shall see below that the limit y_0 is *unique* when it exists.
2. The point x_0 *need not* be in I (e.g., it may be an endpoint not in I) and, even if $x_0 \in I$, $f(x_0)$ *may* be undefined. This is the reason for the *strict* inequality $0 < |x - x_0|$ in $(*)$. Note also that, even if $f(x_0)$ is defined, we *may* have $y_0 \neq f(x_0)$.

3. If $x_0 = b$ is the *right endpoint* of I, then $0 < |x - x_0| < \delta$ must be replaced by $-\delta < x - b < 0$, and if $x_0 = a$ is the *left endpoint* of I, then $0 < |x - x_0| < \delta$ must be replaced by $0 < x - a < \delta$.

4. If the restriction $0 < |x - x_0|$ is dropped (i.e., if we allow $x = x_0$ when $x_0 \in I$), then the new definition will be quite different. For example, if one adopts it and if $x_0 \in I$, then $\lim_{x \to x_0} f(x)$ does *not* exist when the point $(x_0, f(x_0))$ is removed from the graph of f and is replaced by (x_0, y_0) for *any* $y_0 \neq f(x_0)$. In other words, with this new definition, if $x_0 \in I$ and if $\lim_{x \to x_0} f(x)$ exists, then we *must* have $\lim_{x \to x_0} f(x) = f(x_0)$. (Why?)

Proposition 3.2.5 (Limits Are Unique). *Let $x_0 \in I^-$, and let f be a function defined on I (except possibly at x_0). If $\lim_{x \to x_0} f(x)$ exists, then it is unique.*

Proof. Suppose we have *two* limits, y_0 and z_0. Then, for any $\varepsilon > 0$, it follows from $\lim_{x \to x_0} f(x) = y_0$ that there is a $\delta_1 > 0$ such that $x \in I$ and $0 < |x - x_0| < \delta_1$ imply $|f(x) - y_0| < \varepsilon/2$. Similarly, $\lim_{x \to x_0} f(x) = z_0$ implies that, for some $\delta_2 > 0$, we have $|f(x) - z_0| < \varepsilon/2$ whenever $x \in I$ and $0 < |x - x_0| < \delta_2$. If now we set $\delta := \min(\delta_1, \delta_2)$, then $x \in I$ and $0 < |x - x_0| < \delta$ imply

$$|y_0 - z_0| \leq |f(x) - y_0| + |f(x) - z_0| < \varepsilon/2 + \varepsilon/2 = \varepsilon. \qquad (*)$$

Since $(*)$ is true for *every* $\varepsilon > 0$, we get $y_0 = z_0$. $\qquad\square$

Definition 3.2.6 (True Near, Sufficiently Close). Let $x_0 \in I^-$, and let $f, g, \ldots$ be functions defined on I (except possibly at x_0). We say that a property $P(f, g, \ldots)$ involving the functions $f, g, \ldots$ is *true near* x_0 (or for all $x \neq x_0$ *sufficiently close* to x_0), and we write

$$P(f(x), g(x), \ldots) \quad (x \approx x_0),$$

if there exists a $\delta > 0$ such that the given property is true for all $x \in I$ satisfying $0 < |x - x_0| < \delta$, i.e., for all $x \in \dot{B}_\delta(x_0)$.

In terms of this definition, $\lim_{x \to x_0} f(x) = y_0$ if, given any $\varepsilon > 0$, the inequality $|f(x) - y_0| < \varepsilon$ is *satisfied near* x_0 (or satisfied for all $x \neq x_0$ *sufficiently close* to x_0).

Exercise 3.2.7.

(a) Let $x_0 \in I^-$, and let f be a function defined on I (except possibly at x_0), and assume that f is *constant near* x_0; i.e., suppose that for some constants $\delta > 0$ and $c \in \mathbb{R}$, we have $f(x) = c$ for all $x \in I$ satisfying $0 < |x - x_0| < \delta$. Show that $\lim_{x \to x_0} f(x) = c$.

(b) Let $f(x) = x/|x|$, for all $x \neq 0$. Using the definition, show that $\lim_{x \to 0} f(x)$ does *not* exist.

(c) Let $f(x) = 1 - |x|$ if $x \neq 0$ and let $f(0) = 0$. Find $y_0 = \lim_{x \to 0} f(x)$ and observe that $y_0 \neq 0 = f(0)$.

(d) Using the definition, show that, if $x_0 \in I^-$ and f is a function defined on I (except possibly at x_0), then we have

$$\lim_{x \to x_0} f(x) = 0 \iff \lim_{x \to x_0} |f(x)| = 0.$$

Example 3.2.8. Throughout, let $x_0 \in \mathbb{R}$ be arbitrary.

1. Let $f(x) = 2x + 3$. To show the intuitively obvious fact that $\lim_{x \to x_0} f(x) = 2x_0 + 3$, we must prove that

$$(\forall \varepsilon > 0)(\exists \delta > 0)(0 < |x - x_0| < \delta \Rightarrow |(2x + 3) - (2x_0 + 3)| < \varepsilon).$$

We note, however, that $|(2x + 3) - (2x_0 + 3)| = 2|x - x_0| < \varepsilon$ follows (from $|x - x_0| < \delta$) for any δ satisfying $0 < \delta \le \varepsilon/2$.

2. Let $f(x) = \frac{x^2 - x_0^2}{x - x_0}$, $x \ne x_0$. Then x_0 is a limit point of the domain $X = (-\infty, x_0) \cup (x_0, \infty)$, and $x_0 \notin X$. Let us show that $\lim_{x \to x_0} f(x) = 2x_0$. We must prove that

$$(\forall \varepsilon > 0)(\exists \delta > 0)\left(0 < |x - x_0| < \delta \Rightarrow \left|\frac{x^2 - x_0^2}{x - x_0} - 2x_0\right| < \varepsilon\right).$$

Now $\left|\frac{x^2 - x_0^2}{x - x_0} - 2x_0\right| = |(x + x_0) - 2x_0| = |x - x_0| < \varepsilon$ follows from $0 < |x - x_0| < \delta$ if δ satisfies $0 < \delta \le \varepsilon$.

3. Let $f(x) = x^2$. We want to show that $\lim_{x \to x_0} f(x) = x_0^2$; i.e., we want to prove that

$$(\forall \varepsilon > 0)(\exists \delta > 0)\left(0 < |x - x_0| < \delta \Rightarrow |x^2 - x_0^2| < \varepsilon\right).$$

Since we want x to *approach* x_0, we may restrict x to be *within one unit of* x_0; i.e., we may require that $|x - x_0| \le 1$. This implies that $|x + x_0| \le 2|x_0| + 1$ and hence that $|x^2 - x_0^2| = |x + x_0||x - x_0| \le (2|x_0| + 1)|x - x_0|$. It is now clear that $|x^2 - x_0^2| < \varepsilon$ follows (from $|x - x_0| < \delta$) for any δ satisfying $0 < \delta \le \min(1, \varepsilon/(2|x_0| + 1))$.

Remark 3.2.9. The reader may have observed that, in Example 3.2.8(1) and (2), the number δ depended *only* on ε and *not* on x_0. On the other hand, Example 3.2.8(3) shows that δ may, in general, depend on *both* ε and x_0.

3.3   Properties of Limits

As we pointed out in the introduction, the study of limits of functions may be reduced to that of sequential limits. To achieve this reduction we shall need the following.

Theorem 3.3.1 (Sequential Definition of Limit). *Let $x_0 \in I^-$, and let f be a function defined on I (except possibly at x_0). Then $\lim_{x \to x_0} f(x) = y_0$ if and only if $\lim_{n \to \infty} f(x_n) = y_0$ for all sequences $(x_n) \in I^{\mathbb{N}}$ such that $x_n \neq x_0 \, \forall n \in \mathbb{N}$ and $\lim(x_n) = x_0$.*

Proof. Suppose that $\lim_{x \to x_0} f(x) = y_0$, and pick a sequence $(x_n) \in I^{\mathbb{N}}$ satisfying $\lim(x_n) = x_0$, and $x_n \neq x_0 \, \forall n \in \mathbb{N}$. Then we have

$$(\forall \varepsilon > 0)(\exists \delta > 0)(\forall x \in I)(0 < |x - x_0| < \delta \Rightarrow |f(x) - y_0| < \varepsilon). \tag{$*$}$$

Since $\lim(x_n) = x_0$ and $x_n \neq x_0$ for all n, we also have

$$(\exists N \in \mathbb{N})(n \geq N \Rightarrow 0 < |x_n - x_0| < \delta). \tag{$**$}$$

It now follows from $(*)$ and $(**)$ that, for $n \geq N$, $|f(x_n) - y_0| < \varepsilon$ and hence $\lim_{n \to \infty} f(x_n) = y_0$. Conversely, suppose that $\lim_{n \to \infty} f(x_n) = y_0$, for all sequences (x_n) satisfying the conditions of the theorem. If $\lim_{x \to x_0} f(x) \neq y_0$, then we have

$$(\exists \varepsilon_0 > 0)(\forall \delta > 0)(\exists x_\delta \in I)(0 < |x_\delta - x_0| < \delta \quad and \quad |f(x_\delta) - y_0| \geq \varepsilon_0).$$

Choosing $\delta = 1/n, n \in \mathbb{N}$, we can then find $x_n \in I \setminus \{x_0\}$ with $|x_n - x_0| < 1/n$ and $|f(x_n) - y_0| \geq \varepsilon_0$. We have thus constructed a sequence (x_n) with $x_n \in I \setminus \{x_0\}$ $\forall n \in \mathbb{N}$, $\lim(x_n) = x_0$, and yet $\lim(f(x_n)) \neq y_0$. This contradiction completes the proof. $\qquad\square$

Exercise 3.3.2. Using Theorem 3.3.1, show that $\lim_{x \to 0} \sin(1/x)$ does *not* exist. *Hint:* You may use the sequences (x_n) and (x_n'), where $x_n := 1/(n\pi)$ and $x_n' := 1/(2n\pi + \pi/2) \, \forall n \in \mathbb{N}$.

We can now use Theorem 3.3.1 to prove the properties of limits of functions by reducing them to the corresponding properties for *sequential limits*, already proved in Chap. 2.

Theorem 3.3.3 (Limit Properties). *Let $x_0 \in I^-$ and let f and g be functions defined on I (except possibly at x_0). Suppose that $\lim_{x \to x_0} f(x) = y_0$ and $\lim_{x \to x_0} g(x) = z_0$. Then, for any constant $c \in \mathbb{R}$, we have*

1. $\lim_{x \to x_0} (f \pm g)(x) = y_0 \pm z_0$;
2. $\lim_{x \to x_0} (fg)(x) = y_0 z_0$;
3. $\lim_{x \to x_0} (cf)(x) = cy_0$;
4. $\lim_{x \to x_0} (f/g)(x) = y_0/z_0 \quad if \quad z_0 \neq 0$;
5. If $(\exists \delta > 0)(\forall x \in I)(0 < |x - x_0| < \delta \Rightarrow f(x) \geq 0)$, then $y_0 \geq 0$; and
6. If $(\exists \delta > 0)(\forall x \in I)(0 < |x - x_0| < \delta \Rightarrow f(x) \leq g(x))$, then $y_0 \leq z_0$.

Proof. This is an immediate consequence of Theorem 3.3.1, above, and Theorem 2.2.26 (Limit Theorems). $\qquad\square$

Remark 3.3.4. Regarding the properties (5) and (6) in Theorem 3.3.3, it should be pointed out that *strict* inequalities need *not* be preserved when we pass to the limits. For example, we obviously have $x^2 < |x|$ if $0 < |x| < 1$, but $\lim_{x \to 0} x^2 = 0 = \lim_{x \to 0} |x|$.

Exercise 3.3.5.

(a) Show that $\lim_{x \to 2} (x^2 - 4)/(5x - 10) = 4/5$.
(b) Let $f = p/q$, where p and q are *polynomial functions* of *degrees n* and m, respectively, i.e., $p(x) = a_0 + a_1 x + a_2 x^2 + \cdots + a_n x^n$ and $q(x) = b_0 + b_1 x + b_2 x^2 + \cdots + b_m x^m$, where the *coefficients* a_0, a_1, ..., a_n and b_0, b_1, ..., b_m are real numbers, $a_n \neq 0 \neq b_m$. Show that, if $q(x_0) \neq 0$, then $\lim_{x \to x_0} f(x) = p(x_0)/q(x_0)$.

Next we show that the *Squeeze Theorem*, which was proved for sequences, is also true for limits of functions:

Theorem 3.3.6 (Squeeze Theorem). *Let $x_0 \in I^-$, and let f and g be functions defined on I (except possibly at x_0). Suppose that $\lim_{x \to x_0} f(x) = y_0 = \lim_{x \to x_0} g(x)$. If h is a function defined on I (except possibly at x_0) such that*

$$(\exists \delta > 0)(\forall x \in I)(0 < |x - x_0| < \delta \Rightarrow f(x) \leq h(x) \leq g(x)),$$

then $\lim_{x \to x_0} h(x) = y_0$.

Exercise 3.3.7. Prove Theorem 3.3.6, using the definition of limit and an argument similar to the one used in Theorem 2.2.26 for sequences.

Example 3.3.8.

1. We have $\lim_{x \to 0} x \sin(1/x) = 0$. Indeed, if we define $h(x) := x \sin(1/x)$ for $x \neq 0$, then we have the inequalities

$$-|x| \leq h(x) \leq |x| \quad \forall x \neq 0,$$

which follow from the well-known inequality $|\sin(\theta)| \leq 1 \ \forall \theta \in \mathbb{R}$. It is also obvious (why?) that $\lim_{x \to 0}(-|x|) = 0 = \lim_{x \to 0} |x|$ and hence the Squeeze Theorem may be used.
2. We have $\lim_{x \to 0} \log(1 + x)/x = 1$, where log denotes the *natural logarithm* also sometimes written ln.
   To see this, note first that for any $x \in (-1, 1)$ we have

$$\frac{x}{1 + x} \leq \log(1 + x) \leq x. \qquad (*)$$

The second inequality, $\log(1 + x) \leq x$, is in fact true for all $x > -1$ (simply look at the graphs of $y = \log(1 + x)$ and $y = x$) and will be proved rigorously later. The first inequality in $(*)$ follows from the second: Indeed, replacing x by

$-x$, we first get $\log(1 - x) \le -x$ for all $x < 1$. In this inequality we replace x by $x/(1 + x)$, noting that $x/(1 + x) < 1$ for $x > -1$. It follows that

$$-\log(1 + x) = \log\left(\frac{1}{1 + x}\right) = \log\left(1 - \frac{x}{1 + x}\right) \le -\frac{x}{1 + x},$$

from which $x/(1 + x) \le \log(1 + x)$ follows. Now $(*)$ implies that

$$\frac{1}{1 + x} \le \frac{\log(1 + x)}{x} \le 1 \quad \forall x \in (0, 1) \tag{$**$}$$

and

$$1 \le \frac{\log(1 + x)}{x} \le \frac{1}{1 + x} \quad \forall x \in (-1, 0). \tag{$***$}$$

If we let $f(x) := 1/(1 + x)$ for $x \in (0, 1)$ and $f(x) := 1$ for $x \in (-1, 0)$ and let $g(x) := 1$ for $x \in (0, 1)$ and $g(x) := 1/(1 + x)$ for $x \in (-1, 0)$, it is then obvious that $\lim_{x \to 0} f(x) = 1 = \lim_{x \to 0} g(x)$ and [by $(**)$ and $(***)$] that

$$f(x) \le \log(1 + x)/x \le g(x) \quad \forall x \in (-1, 1) \setminus \{0\}.$$

Thus, once again, the Squeeze Theorem may be applied.

In Example 3.3.8(1) we deduced $\lim_{x \to 0} x \sin(1/x) = 0$ from the Squeeze Theorem. Now $x \sin(1/x) = f(x)g(x)$, where $f(x) := x$ and $g(x) := \sin(1/x)$ satisfy $\lim_{x \to 0} f(x) = 0$ and $|g(x)| \le 1 \ \forall x \ne 0$. In other words, as $x \to 0$, $f(x)$ *converges to zero* while $g(x)$ is *bounded*. The following theorem, which is also an immediate consequence of the Squeeze Theorem (cf. Exercise 3.3.10 below), shows that this example is a special case of a general result.

Theorem 3.3.9. *Let $x_0 \in I^-$, and let f and g be functions defined on I (except possibly at x_0). Suppose that g is bounded near x_0, and that $\lim_{x \to x_0} f(x) = 0$. Then we have*

$$\lim_{x \to x_0} f(x)g(x) = 0.$$

Proof. Since g is *bounded near* x_0, there exist constants $\delta_1 > 0$ and $B > 0$ such that $|g(x)| \le B$ whenever $x \in I$ and $0 < |x - x_0| < \delta_1$. Now $\lim_{x \to x_0} f(x) = 0$ implies that, for any $\varepsilon > 0$, we can find a $\delta_2 > 0$ such that $x \in I$ and $0 < |x - x_0| < \delta_2$ imply $|f(x)| < \varepsilon/B$. Therefore, with $\delta := \min(\delta_1, \delta_2)$, we have

$$(x \in I \text{ and } 0 < |x - x_0| < \delta) \Rightarrow |f(x)g(x)| \le B|f(x)| < B(\varepsilon/B) = \varepsilon.$$

Since $\varepsilon > 0$ was arbitrary, the theorem follows. $\qquad \square$

Exercise 3.3.10. Show that Theorem 3.3.9 also follows from the Squeeze Theorem.

Exercise 3.3.11. Let $x_0 \in I^-$, and let f be a function defined on I (except possibly at x_0). Assume that $\lim_{x \to x_0} f(x) = y_0 \in \mathbb{R}$.

(a) Show that f is *bounded near* x_0, i.e., that we can find constants $\delta > 0$ and $B > 0$ such that $x \in I$ and $0 < |x - x_0| < \delta$ imply $|f(x)| \le B$.
Hint: Take $\varepsilon = 1$ in the definition of $\lim_{x \to x_0} f(x) = y_0$, and observe that one may set $B := |y_0| + 1$ if $x_0 \notin I$ and $B := \max(|f(x_0)|, |y_0| + 1)$ if $x_0 \in I$.
(b) Let $|f|$ be the function defined by $|f|(x) := |f(x)| \; \forall x \in \mathrm{dom}(f)$. Show that $\lim_{x \to x_0} |f|(x) = |y_0|$. Give an example where $\lim_{x \to x_0} |f|(x)$ exists but $\lim_{x \to x_0} f(x)$ does *not* exist.
(c) Assume that f is *nonnegative near* x_0, i.e., $\exists \delta > 0$ such that $x \in I$ and $0 < |x - x_0| < \delta$ imply $f(x) \ge 0$. Show that $\lim_{x \to x_0} \sqrt{f(x)} = \sqrt{y_0}$.

Exercise 3.3.12. Let $x_0 \in I^-$, and let g be a function defined on I (except possibly at x_0). Suppose that $\lim_{x \to x_0} g(x) = z_0 \ne 0$. Show that there exists $\delta > 0$ such that $0 < |x - x_0| < \delta$ implies $g(x) \ne 0$. This result, which is needed in the proof of part (4) of Theorem 3.3.3, can also be stated as follows: If $g(x)$ has a *nonzero* limit as $x \to x_0$, then $g(x)$ is *nonzero near* x_0. Show that the statement remains true if *nonzero* is replaced throughout by *positive* or *negative*.

Cauchy's Criterion for sequences and series (cf. Theorems 2.2.46 and 2.3.3) has the following analog for functions:

Exercise 3.3.13 (Cauchy's Criterion). Let $x_0 \in I^-$ and let f be a function defined on I (except possibly at x_0). Show that $\lim_{x \to x_0} f(x)$ *exists* if and only if for any given $\varepsilon > 0$ there exists a $\delta > 0$ such that, if $x_1, x_2 \in I \setminus \{x_0\}$, then the inequalities $|x_1 - x_0| < \delta$ and $|x_2 - x_0| < \delta$ imply $|f(x_1) - f(x_2)| < \varepsilon$.

We have seen that limits behave *nicely* with respect to the algebraic operations and order relations on functions. There is one operation, however, that we have not yet considered; it is the fundamental operation of *composition* (i.e., "chaining together") of functions. As the following theorem shows, limits also behave nicely with respect to this operation.

Theorem 3.3.14. *Let I, J be intervals, $x_0 \in I^-$, and $y_0 \in J^-$. Let f be a function defined on I (except possibly at x_0) satisfying $f(I \setminus \{x_0\}) \subset J$ and $\lim_{x \to x_0} f(x) = y_0$ such that for some $\delta_0 > 0$, we have $f(x) \ne y_0$ for all $0 < |x - x_0| < \delta_0$. Finally, let g be a function defined on J (except possibly at y_0). If $\lim_{y \to y_0} g(y) = z_0$, then*

$$\lim_{x \to x_0} g(f(x)) = z_0.$$

Proof. Given $\varepsilon > 0$, pick $\delta_1 > 0$ such that $y \in J$ and $0 < |y - y_0| < \delta_1$ imply $|g(y) - z_0| < \varepsilon$. Now, given this $\delta_1 > 0$, there exists a $\delta > 0$ with $0 < \delta < \delta_0$ such that $x \in I$ and $0 < |x - x_0| < \delta$ imply $0 < |f(x) - y_0| < \delta_1$. But then, with $y = f(x)$, we have $0 < |y - y_0| < \delta_1$ whenever $x \in I$ and $0 < |x - x_0| < \delta$; i.e.,

$$(x \in I \text{ and } 0 < |x - x_0| < \delta) \Longrightarrow 0 < |f(x) - y_0| < \delta_1 \Longrightarrow |g(f(x)) - z_0| < \varepsilon,$$

and the proof is complete. $\qquad\square$

It is a fact (cf. Exercise 3.3.11(a) above) that, if $\lim_{x \to x_0} f(x) = y_0$, then f is bounded near x_0. The converse is *not* true in general. Indeed, the function $\sin(1/x)$ is bounded on $\mathbb{R} \setminus \{0\}$, but $\lim_{x \to 0} \sin(1/x)$ does not exist. Observe, however, that the graph of $\sin(1/x)$ *oscillates* rapidly as $x \to 0$. The following theorem shows that, for *monotone* functions, boundedness *does* imply the existence of limit.

Theorem 3.3.15 (Monotone Limit Theorem). *Let $a < b$ and let $f : (a, b) \to \mathbb{R}$. Suppose that f is increasing (resp., decreasing) and bounded. Then $\lim_{x \to b} f(x) = \sup(f)$ and $\lim_{x \to a} f(x) = \inf(f)$ (resp., $\lim_{x \to b} f(x) = \inf(f)$ and $\lim_{x \to a} f(x) = \sup(f)$).*

Proof. Since f is decreasing and bounded below if and only if $-f$ is increasing and bounded above, and one then has $\inf(f) = -\sup(-f)$ (why?), it suffices to look at the *increasing* case. If $u = \sup(f) := \sup\{f(x) : a < x < b\}$ and if $\varepsilon > 0$, then we can pick $\xi \in (a, b)$ such that $u - \varepsilon < f(\xi) \le u$. Set $\delta := b - \xi$. We then have $\delta > 0$ and, since f is increasing,

$$-\delta < x - b < 0 \Rightarrow \xi < x < b \Rightarrow u - \varepsilon < f(\xi) \le f(x) \le u \Rightarrow |f(x) - u| < \varepsilon.$$

Therefore, $\lim_{x \to b} f(x) = u = \sup(f)$. The proof that $\lim_{x \to a} f(x) = \inf(f)$ is similar. $\square$

3.4   One-Sided Limits and Limits Involving Infinity

In our definition of $\lim_{x \to x_0} f(x) = y_0$, the restriction $0 < |x - x_0| < \delta$ allows x to be on *either* side of x_0. There are situations, however, where we want x to approach x_0 from *one side only*. This is formalized in the following:

Definition 3.4.1 (Left and Right Limits). Let $x_0 \in I^-$, and let f be a function defined on I (except possibly at x_0).

1. Assume that x_0 is *not the left endpoint* of I. We say that $y_0 \in \mathbb{R}$ is the *left limit* of f *at x_0*, and we write $y_0 = f(x_0-) = f(x_0 - 0) = \lim_{x \to x_0-} f(x)$, if the following is satisfied:

$$(\forall \varepsilon > 0)(\exists \delta > 0)(\forall x \in I)(-\delta < x - x_0 < 0 \Rightarrow |f(x) - y_0| < \varepsilon). \qquad (*-)$$

2. Similarly, assume that x_0 is *not the right endpoint* of I. We say that $y_0 \in \mathbb{R}$ is the *right limit* of f at x_0, and we write $y_0 = f(x_0+) = f(x_0 + 0) = \lim_{x \to x_0+} f(x)$, if the following is satisfied:

$$(\forall \varepsilon > 0)(\exists \delta > 0)(\forall x \in I)(0 < x - x_0 < \delta \Rightarrow |f(x) - y_0| < \varepsilon). \qquad (*+)$$

Remark 3.4.2.

1. Note that $0 < |x - x_0| < \delta$ is equivalent to the pair of inequalities: $-\delta < x - x_0 < 0$ *or* $0 < x - x_0 < \delta$. We have used the first (which is equivalent to $x_0 - \delta < x < x_0$) in (*−) and the second (which is equivalent to $x_0 < x < x_0 + \delta$) in (*+).
2. The notation $x \to x_0-$ (resp., $x \to x_0+$) is sometimes written as $x \downarrow x_0$ (resp., $x \uparrow x_0$).
3. As the reader can check easily, we have $f(x_0 \pm 0) = \lim_{\delta \to 0+} f(x_0 \pm \delta)$. This fact is the reason behind the notation $f(x_0 \pm 0)$.

Example 3.4.3.

(a) Let $f : \mathbb{R} \setminus \{0\} \to \mathbb{R}$ be defined by $f(x) := x/|x|$. Then, as the reader can check at once, we have $f(0-) = \lim_{x \to 0-} f(x) = -1$ and $f(0+) = \lim_{x \to 0+} f(x) = 1$.
(b) Let $g : [-2, 2] \to \mathbb{R}$ be defined by $g(x) := \sqrt{4 - x^2}$. Then we have $g(2 - 0) = 0 = g(-2 + 0)$. Notice that, in this example, the right endpoint 2 *must* be approached from the *left* and that the left endpoint -2 *must* be approached from the *right*.

To show that all the properties we proved (in Sect. 3.3) for the limits of functions remain true (with minor modifications) for one-sided limits, we need the following analog of Theorem 3.3.1. The proof is a copy of the one given for Theorem 3.3.1 (with obvious modifications) and is left as an exercise for the reader.

Theorem 3.4.4. *Let $x_0 \in I^-$, and let f be a function defined on I (except possibly at x_0). Assume that x_0 is not the left (resp., right) endpoint of I. Then $f(x_0-) = y_0$ (resp., $f(x_0+) = y_0$), if and only if $\lim_{n \to \infty} f(x_n) = y_0$ for all sequences $(x_n) \in I^{\mathbb{N}}$ such that $x_n < x_0 \ \forall n \in \mathbb{N}$ (resp., $x_n > x_0 \ \forall n \in \mathbb{N}$) and $\lim(x_n) = x_0$.*

Exercise 3.4.5. Using Theorem 3.4.4, show that all the limit properties proved in Sect. 3.3 remain valid (with properly modified statements) for one-sided limits.

The next theorem gives the relation between the limit and the one-sided limits of a function.

Theorem 3.4.6. *Let $x_0 \in I^-$, and let f be a function defined on I (except possibly at x_0). If x_0 is not an endpoint of I, then $y_0 = \lim_{x \to x_0} f(x)$ if and only if $f(x_0-) = y_0 = f(x_0+)$. If $x_0 = b$ is the right endpoint of I and if $y_0 = \lim_{x \to x_0} f(x)$, then $y_0 = f(x_0-)$. Finally, if $x_0 = a$ is the left endpoint of I and if $y_0 = \lim_{x \to x_0} f(x)$, then $y_0 = f(x_0+)$.*

Exercise 3.4.7. Prove Theorem 3.4.6.

Remark 3.4.8.

(a) Theorem 3.4.6 shows that, if x_0 is the right endpoint of I, then the concepts of *limit* and *left limit* are the same. Similarly, if x_0 is the left endpoint of I, then the concepts of *limit* and *right limit* are the same.

(b) We saw that $\lim_{x \to 0-} x/|x| = -1$ and $\lim_{x \to 0+} x/|x| = 1$. Since $-1 \neq 1$, Theorem 3.4.6 shows that $\lim_{x \to 0} x/|x|$ does *not* exist.

In all the limits we have considered so far—i.e., $y_0 = \lim_{x \to x_0} f(x)$ and $y_0 = \lim_{x \to x_0 \pm} f(x)$—we have assumed x_0, $y_0 \in \mathbb{R}$. We now want to extend the concept of limit (one-sided or not) to the cases where x_0, $y_0 \in \overline{\mathbb{R}} := [-\infty, \infty]$. Therefore, we want to allow x_0 or y_0 (or both) to be $\pm\infty$. We first begin by looking at the case where $x_0 \in \mathbb{R}$ but $y_0 = \pm\infty$:

Definition 3.4.9 (Infinite Limits). Let $x_0 \in I^-$, and let f be a function defined on I (except possibly at x_0).

1. We say that $f(x)$ *converges to* $+\infty$ *at* x_0, and write $\lim_{x \to x_0} f(x) = +\infty$, if the following is satisfied:

$$(\forall B \in \mathbb{R})(\exists \delta = \delta(B) > 0)(\forall x \in I)(0 < |x - x_0| < \delta \Rightarrow f(x) > B).$$

2. Similarly, we say that $f(x)$ *converges to* $-\infty$ *at* x_0, and write $\lim_{x \to x_0} f(x) = -\infty$, if the following is satisfied:

$$(\forall A \in \mathbb{R})(\exists \delta = \delta(A) > 0)(\forall x \in I)(0 < |x - x_0| < \delta \Rightarrow f(x) < A).$$

With minor modifications, we can also give the definition of one-sided infinite limits:

Definition 3.4.10 (One-Sided Infinite Limits). Let $x_0 \in I^-$, and let f be a function defined on I (except possibly at x_0).

1. Assume that x_0 is *not* the *left* (resp., *right*) endpoint of I. We say that $f(x)$ *converges to* $+\infty$ *as* x *approaches* x_0 *from the left* (resp., *right*), and write $\lim_{x \to x_0-} f(x) = +\infty$ (resp., $\lim_{x \to x_0+} f(x) = +\infty$), if for every $B \in \mathbb{R}$ there exists $\delta = \delta(B) > 0$ such that $x \in I$ and $-\delta < x - x_0 < 0$ (resp., $x \in I$ and $0 < x - x_0 < \delta$) imply $f(x) > B$.
2. Similarly, assume that x_0 is *not* the *left* (resp., *right*) endpoint of I. We say that $f(x)$ *converges to* $-\infty$ *as* x *approaches* x_0 *from the left* (resp., *right*), and write $\lim_{x \to x_0-} f(x) = -\infty$ (resp., $\lim_{x \to x_0+} f(x) = -\infty$), if for every $A \in \mathbb{R}$ there exists $\delta = \delta(A) > 0$ such that $x \in I$ and $-\delta < x - x_0 < 0$ (resp., $x \in I$ and $0 < x - x_0 < \delta$) imply $f(x) < A$.

Example 3.4.11.

(a) $\lim_{x \to 0} 1/x^2 = +\infty$. Indeed, given any $B > 0$, the inequality $1/x^2 > B$ follows from $0 < |x| < \delta$ for any δ satisfying $0 < \delta \leq 1/\sqrt{B}$.
(b) We have $\lim_{x \to 0-} 1/(1 - e^x) = +\infty$ and $\lim_{x \to 0+} 1/(1 - e^x) = -\infty$. Let us prove the first assertion. Given any $B > 1$, the inequality $1/(1 - e^x) > B$ is equivalent to $e^x > 1 - 1/B$, which is equivalent to $x > \log(1 - 1/B)$. Therefore, if δ satisfies $0 < \delta \leq -\log(1 - 1/B)$, the inequality $-\delta < x < 0$ implies $1/(1 - e^x) > B$.

Exercise 3.4.12. Let $x_0 \in I^-$, and let f be a function defined on I (except possibly at x_0). Show that $\lim_{x \to x_0} f(x) = \pm\infty$ if and only if $\lim_{x \to x_0}(-f)(x) = \mp\infty$. Show that the same holds also for one-sided infinite limits.

Definition 3.4.13 (Vertical Asymptote). Let $x_0 \in I^-$, and let f be a function defined on I (except possibly at x_0). The line $\{(x, y) \in \mathbb{R}^2 : x = x_0\}$ is called a *vertical asymptote* of f (or of the *graph* of f) if f has an *infinite limit* as x approaches x_0 *(possibly from one side only)*, i.e., if any one of the limits $\lim_{x \to x_0} f(x)$, $\lim_{x \to x_0-} f(x)$, or $\lim_{x \to x_0+} f(x)$ is $+\infty$ or $-\infty$.

The following theorem, which is similar to Theorems 3.3.1 and 3.4.4, gives a sequential characterization of (possibly one-sided) infinite limits:

Theorem 3.4.14. *Let $x_0 \in I^-$, and let f be a function defined on I (except possibly at x_0).*

1. *We have $\lim_{x \to x_0} f(x) = \pm\infty$ if and only if $\lim(f(x_n)) = \pm\infty$ for all sequences $(x_n) \in I^{\mathbb{N}}$ such that $x_n \neq x_0$ $\forall n \in \mathbb{N}$ and $\lim(x_n) = x_0$.*
2. *Assume that x_0 is not the left (resp., right) endpoint of I. Then $f(x_0-) = \pm\infty$ (resp., $f(x_0+) = \pm\infty$) if and only if $\lim(f(x_n)) = \pm\infty$ for all sequences $(x_n) \in I^{\mathbb{N}}$ such that $x_n < x_0$ $\forall n \in \mathbb{N}$ (resp., $x_n > x_0$ $\forall n \in \mathbb{N}$) and $\lim(x_n) = x_0$.*

Exercise 3.4.15. Prove Theorem 3.4.14.

Finally, we are going to look at the cases where $x_0 = \pm\infty$ and where y_0 is finite or $\pm\infty$. First, we look at the case $y_0 \in \mathbb{R}$:

Definition 3.4.16 (Finite Limits at Infinity). Let $f : X \to \mathbb{R}$, and let $y_0 \in \mathbb{R}$.

1. Suppose that $(a, \infty) \subset X$ for some $a \in \mathbb{R}$. We say that y_0 is the *limit* of $f(x)$ (or that $f(x)$ *converges to* y_0) as x approaches $+\infty$, and we write $\lim_{x \to +\infty} f(x) = y_0$, if the following is true:

$$(\forall \varepsilon > 0)(\exists B = B(\varepsilon) > a)(x > B \Rightarrow |f(x) - y_0| < \varepsilon).$$

2. Suppose that $(-\infty, b) \subset X$ for some $b \in \mathbb{R}$. We say that y_0 is the *limit* of $f(x)$ (or that $f(x)$ *converges to* y_0) as x approaches $-\infty$, and we write $\lim_{x \to -\infty} f(x) = y_0$, if the following is true:

$$(\forall \varepsilon > 0)(\exists A = A(\varepsilon) < b)(x < A \Rightarrow |f(x) - y_0| < \varepsilon).$$

Notation 3.4.17.

(a) We will often write $x \to \infty$ instead of $x \to +\infty$.
(b) If $\lim_{x \to +\infty} f(x) = y_0 = \lim_{x \to -\infty} f(x)$, then we write $\lim_{x \to \pm\infty} f(x) = y_0$.

The following exercise involving $\lim_{x\to\infty} f(x)$ is formulated for functions defined on $[a, \infty)$ for some $a \in \mathbb{R}$, but it can obviously be stated and proved for functions defined on $(-\infty, b]$ and $\lim_{x\to-\infty} f(x)$ as well.

Exercise 3.4.18 (Cauchy's Criterion). Let $f : [a, \infty) \to \mathbb{R}$. Show that $\lim_{x\to\infty} f(x)$ exists (in $\mathbb{R}$) if and only if

$$(\forall \varepsilon > 0)(\exists B > a)(\forall x, x' \in \mathbb{R})(x, x' > B \Rightarrow |f(x) - f(x')| < \varepsilon).$$

Hint: The necessity of the condition is easy to see. For the sufficiency, pick $(x_n) \in \mathbb{R}^{\mathbb{N}}$ with $\lim(x_n) = +\infty$. Show that $(f(x_n))$ is a Cauchy sequence in $\mathbb{R}$. Deduce that $y_0 := \lim(f(x_n))$ exists and that $\lim_{x\to\infty} f(x) = y_0$.

Recall that we defined a property to be *true near* x_0 if it is true for all $x \in \dot{B}_\delta(x_0)$ for some $\delta > 0$. Now that the concept of "limit as $x \to x_0$" has been extended to one-sided limits ("$x \to x_0\pm$") and limits at infinity ("$x \to \pm\infty$"), we must extend the definition of "true near x_0" accordingly:

Definition 3.4.19 (True Near, Sufficiently Close, Sufficiently Large). Let I be an interval, and let $x_0 \in I^-$ or $x_0 = \pm\infty$ (if I is unbounded). Suppose that $f, g, \ldots$ are functions defined on I (except possibly at x_0), and let $P(f, g, \ldots)$ be a *property* (proposition, statement, etc.) involving $f, g, \ldots$.
(a) If $x_0 \in I$, we say that $P(f(x), g(x), \ldots)$ is *true near* x_0, and we write

$$P(f(x), g(x), \ldots) \quad (x \approx x_0),$$

if $\exists \delta > 0$ such that $P(f(x), g(x), \ldots)$ is true for all $x \in I$ with $0 < |x - x_0| < \delta$.
(b$\gtrless$) If x_0 is *not* the *left* (resp., *right*) endpoint of I, we say that $P(f(x), g(x), \ldots)$ is *true near* x_0- (resp., *true near* x_0+), and we write

$$P(f(x), g(x), \ldots) \quad (x \approx x_0-) \quad [\text{resp.,} \quad (x \approx x_0+)],$$

if $\exists \delta > 0$ such that $P(f(x), g(x), \ldots)$ is true for all $x \in I$ satisfying $x_0 - \delta < x < x_0$ (resp., $x_0 < x < x_0 + \delta$).
(c$\pm\infty$) If $x_0 = +\infty$ (resp., $x_0 = -\infty$) and $(a, +\infty) \subset I$ for some $a \in \mathbb{R}$ (resp., $(-\infty, b) \subset I$ for some $b \in \mathbb{R}$), we say that $P(f(x), g(x), \ldots)$ is *true near* $+\infty$ (resp., *true near* $-\infty$), and we write

$$P(f(x), g(x), \ldots) \quad (x \approx +\infty) \quad [\text{resp.,} \quad (x \approx -\infty)],$$

if $\exists A > a$ (resp., $B < b$) such that $P(f(x), g(x), \ldots)$ is true for all $x > A$ (resp., $x < B$).
In case (a) [resp., (b$\leftarrow$), (b$\rightarrow$)], we also say that $P(f(x), g(x), \ldots)$ is true for all $x \neq x_0$ (resp., $x < x_0$, $x > x_0$) *sufficiently close* to x_0. Finally, in case (c$+\infty$) (resp., (c$-\infty$)), we also say that $P(f(x), g(x), \ldots)$ is true for all $x > a$ (resp., $x < b$) *sufficiently large*.

Remark 3.4.20. As we saw in Chap. 2, the limit of a sequence depends only on the behavior of its *tails*. Similarly, the limit of a function, say as $x \to x_0$, depends only on the behavior of the function *near* x_0. Thus, if $f(x) = g(x)$ $(x \approx x_0)$ and $\lim_{x \to x_0} f(x) = y_0$, then we also have $\lim_{x \to x_0} g(x) = y_0$. In fact, the same is true under weaker conditions, as we shall see in the next section (cf. the definition of *equivalent functions* in Sect. 3.5).

Definition 3.4.21 (Horizontal Asymptote). Let $f : X \to \mathbb{R}$, where X contains $(a, +\infty)$ or $(-\infty, b)$ or both, for some a, $b \in \mathbb{R}$, and let $y_0 \in \mathbb{R}$. The horizontal line $\{(x, y) \in \mathbb{R}^2 : y = y_0\}$ is called a *horizontal asymptote* of f (or of the graph of f) if $f(x)$ *converges to* y_0 as $x \to +\infty$ or $x \to -\infty$ or $x \to \pm\infty$.

Exercise 3.4.22. Show that, if $\lim_{x \to \infty} f(x)$ (resp., $\lim_{x \to -\infty} f(x)$) exists (in $\mathbb{R}$), then it is *unique*. Under the same assumption, show that f is *bounded near* $+\infty$ (resp., $-\infty$).

Example 3.4.23.

(a) We have $\lim_{x \to \pm\infty} 1/(ax + b) = 0$ for any $a \neq 0$ and b in $\mathbb{R}$. Well, it is easy to see that $|1/(ax + b)| = 1/|ax + b| < \varepsilon$ follows from $|x| > (|b| + 1/\varepsilon)/|a|$, and hence any $B = B(\varepsilon) \geq (|b| + 1/\varepsilon)/|a|$ will do.
(b) We have $\lim_{x \to \pm\infty} \sin(1/x) = 0$. Indeed, as we shall see later, $|\sin \theta| \leq |\theta|$ $\forall \theta \in \mathbb{R}$. Therefore, $|\sin(1/x)| \leq 1/|x|$ $\forall x \neq 0$, and we can use part (a).

The following proposition shows that *(finite) limits at infinity* may be reduced to (possibly one-sided) ordinary limits.

Proposition 3.4.24. *The following statements are true.*

1. $\lim_{x \to +\infty} f(x) = y_0 \iff \lim_{x \to 0+} f(1/x) = y_0$
2. $\lim_{x \to -\infty} f(x) = y_0 \iff \lim_{x \to 0-} f(1/x) = y_0$
3. $\lim_{x \to \pm\infty} f(x) = y_0 \iff \lim_{x \to 0} f(1/x) = y_0$

Exercise 3.4.25. Prove Proposition 3.4.24 and, combining it with Exercise 3.4.5, deduce that all the limit properties proved in Sect. 3.3 remain valid, with suitably modified statements, for *finite* limits *at infinity*.

Example 3.4.26. We have $\lim_{x \to -\infty} x/(1 + |x|) = -1$ and $\lim_{x \to +\infty} x/(1 + |x|) = 1$. Indeed (Proposition 3.4.24), the first claim follows from the fact that $\lim_{x \to 0-}(1/x)/(1 + 1/|x|) = \lim_{x \to 0-} 1/(x + x/|x|) = -1$, and the second is treated similarly.

Exercise 3.4.27. Let $f = p/q$, where p and q are *polynomial functions* of *degrees* n and m, respectively, i.e., $p(x) = a_0 + a_1 x + a_2 x^2 + \cdots + a_n x^n$ and $q(x) = b_0 + b_1 x + b_2 x^2 + \cdots + b_m x^m$, where the *coefficients* a_0, a_1, ..., a_n and b_0, b_1, ..., b_m are real, $a_n \neq 0$, and $b_m \neq 0$.

1. Show that, if $m = n$, then $\lim_{x \to \pm\infty} f(x) = a_n/b_n$.
2. Show that, if $m > n$, then $\lim_{x \to \pm\infty} f(x) = 0$.

Hint: Use Proposition 3.4.24.

Once again, we can give a sequential version of the above definition of "limits at infinity":

Theorem 3.4.28. *Suppose that, for some $a \in \mathbb{R}$ (resp., $b \in \mathbb{R}$), the domain of f contains the interval (a, ∞) (resp., $(-\infty, b)$). Then we have $\lim_{x \to +\infty} f(x) = y_0$ (resp., $\lim_{x \to -\infty} f(x) = y_0$) if and only if $\lim(f(x_n)) = y_0$ for all sequences (x_n) satisfying $x_n > a \;\forall n \in \mathbb{N}$ and $\lim(x_n) = +\infty$ (resp., $x_n < b \;\forall n \in \mathbb{N}$ and $\lim(x_n) = -\infty$).*

Proof. Exercise! $\qquad\square$

Example 3.4.29. The limits $\lim_{x \to +\infty} \sin x$ and $\lim_{x \to +\infty} \cos x$ do *not* exist, and the same is true if $+\infty$ is replaced by $-\infty$. We prove the claim for $\lim_{x \to +\infty} \sin x$ and leave the others for the reader. Now, if we define the sequences (x_n) and (x_n') by $x_n := n\pi$, and $x_n' := 2n\pi + \pi/2$, then we obviously have $\lim(x_n) = \lim(x_n') = +\infty$, but $\lim(\sin(n\pi)) = 0 \neq 1 = \lim(\sin(2n\pi + \pi/2))$, and the claim is proved. Actually, we can also show directly that $\lim_{n \to \infty} \sin n$ and $\lim_{n \to \infty} \cos n$, where $n \in \mathbb{N}$, do *not* exist [cf. Example 2.2.28(2)]. Indeed, the existence of $\alpha := \lim_{n \to \infty} \sin n \in [-1, 1]$ and the identities

$$\sin(n \pm 1) = \sin n \cos 1 \pm \cos n \sin 1 \qquad (*)$$

would imply the existence of $\beta := \lim_{n \to \infty} \cos n \in [-1, 1]$. But then $(*)$ would give the system of equations

$$\alpha \cos 1 + \beta \sin 1 = \alpha$$
$$\alpha \cos 1 - \beta \sin 1 = \alpha,$$

from which we get $\alpha = \beta = 0$, contradicting $\alpha^2 + \beta^2 = 1$.

Finally, the next definition covers all the remaining cases, namely, those in which *both* x_0 and y_0 are *infinite:*

Definition 3.4.30 (Infinite Limits at Infinity). Let $f : X \to \mathbb{R}$.

1. Suppose that $(a, \infty) \subset X$ for some $a \in \mathbb{R}$. We say that $f(x)$ *converges to* $+\infty$ (resp., $-\infty$) as x *approaches* $+\infty$ if the following is true:

$$(\forall B \in \mathbb{R})(\exists x_B > a)(x > x_B \Rightarrow f(x) > B) \quad [\text{resp., } (x > x_B \Rightarrow f(x) < B)].$$

2. Suppose that $(-\infty, b) \subset X$, for some $b \in \mathbb{R}$. We say that $f(x)$ *converges to* $+\infty$ (resp., $-\infty$) as x *approaches* $-\infty$ if the following is true:

$$(\forall A \in \mathbb{R})(\exists x_A < b)(x < x_A \Rightarrow f(x) > A) \quad [\text{resp., } (x < x_A \Rightarrow f(x) < A)].$$

As before, the following theorem shows that this definition can be replaced by an equivalent "sequential" version, which is more convenient to use in many cases.

Theorem 3.4.31. *Let* $f : X \to \mathbb{R}$, *and suppose that* $(a, \infty) \subset X$ *for some* $a \in \mathbb{R}$ *(resp.,* $(-\infty, b) \subset X$ *for some* $b \in \mathbb{R}$*). Then* $\lim_{x \to +\infty} f(x) = \pm\infty$ *(resp.,* $\lim_{x \to -\infty} f(x) = \pm\infty$*) if and only if* $\lim(f(x_n)) = \pm\infty$ *for all sequences* (x_n) *satisfying* $x_n > a$ $\forall n \in \mathbb{N}$ *and* $\lim(x_n) = +\infty$ *(resp.,* $x_n < b$ $\forall n \in \mathbb{N}$ *and* $\lim(x_n) = -\infty$*).*

Proof. Exercise! □

We have seen that all the limit properties are valid for *finite limits*, i.e., the cases where the limit y_0 in $\lim_{x \to x_0} f(x) = y_0$ is a *real number*. For infinite limits (i.e., $y_0 = \pm\infty$), however, one has to be more careful in handling the operations on limits. Indeed, when we defined the algebraic operations on *extended* real numbers, $\overline{\mathbb{R}} = [-\infty, \infty]$, expressions such as $+\infty + (-\infty)$, $-\infty + \infty$, and $\pm\infty / \pm\infty$ were left *undefined*. Limits leading to such expressions are among the so-called *indeterminate forms* $(0/0, \infty/\infty, 0 \cdot \infty, \infty - \infty, 0^0, 1^\infty, \infty^0)$, which will be discussed briefly below and more extensively later when we introduce *l'Hôpital's Rule*.

Theorem 3.4.32. *Let* f, g, $h : I \to \mathbb{R}$, *and let* $x_0 \in I^-$ *or* $x_0 = \pm\infty$ *(if* I *is unbounded). Assume that* $\lim_{x \to x_0} f(x) = y_0$ *and* $\lim_{x \to x_0} g(x) = z_0$ *for some* y_0, $z_0 \in \overline{\mathbb{R}}$. *Then the following statements are true.*

1. *If* $y_0 = \pm\infty = z_0$, *then* $\lim_{x \to x_0} (f + g)(x) = y_0 + z_0 = (\pm\infty) + (\pm\infty) = \pm\infty$ *and* $\lim_{x \to x_0} (fg)(x) = y_0 z_0 = (\pm\infty)(\pm\infty) = +\infty$.
2. *If* $y_0 = \pm\infty$ *and* $z_0 = \mp\infty$, *i.e.,* $z_0 = -y_0$, *then* $\lim_{x \to x_0} (fg)(x) = y_0 z_0 = (\pm\infty)(\mp\infty) = -\infty$.
3. *If* $0 < y_0 < \infty$ *and* $z_0 = \pm\infty$, *then* $\lim_{x \to x_0} (fg)(x) = y_0 z_0 = y_0 \cdot (\pm\infty) = \pm\infty$ *and* $\lim_{x \to x_0} (g/f)(x) = z_0/y_0 = (\pm\infty)/y_0 = \pm\infty$.
4. *If* $-\infty < y_0 < 0$ *and* $z_0 = \pm\infty$, *then* $\lim_{x \to x_0} (fg)(x) = y_0 z_0 = y_0(\pm\infty) = \mp\infty$ *and* $\lim_{x \to x_0} (g/f)(x) = z_0/y_0 = \pm\infty/y_0 = \mp\infty$.
5. *If* $y_0 \in \mathbb{R}$ *and* $z_0 = \pm\infty$, *then* $\lim_{x \to x_0} (f + g)(x) = y_0 + z_0 = y_0 + (\pm\infty) = \pm\infty$ *and* $\lim_{x \to x_0} (f/g)(x) = y_0/(\pm\infty) = 0$.
6. *Suppose that, for some constant* $c > 0$, *we have*

$$f(x) \le cg(x) \quad (x \approx x_0).$$

   Then $y_0 = +\infty$ *implies* $z_0 = +\infty$, *and* $z_0 = -\infty$ *implies* $y_0 = -\infty$.
7. *Suppose that* $\lim_{x \to x_0} f(x)/g(x) = \lambda \in \mathbb{R}$. *Then* $y_0 = \pm\infty \Leftrightarrow z_0 = \pm\infty$ *if* $\lambda > 0$, *and* $y_0 = \pm\infty \Leftrightarrow z_0 = \mp\infty$ *if* $\lambda < 0$.
8. *(Squeeze Theorem) Suppose that* $y_0 = z_0$ *and that we have*

$$f(x) \le h(x) \le g(x) \quad (x \approx x_0).$$

   Then we also have $\lim_{x \to x_0} h(x) = y_0 = z_0$.

Proof. We shall prove the case $\lambda > 0$ (and $z_0 = +\infty$) of property (7) and leave the proofs of the remaining properties to the reader as an exercise! If $x_0 \in \mathbb{R}$, then we can find $\delta > 0$ such that $x \in I$ and $0 < |x - x_0| < \delta$ imply

$$0 < \lambda/2 < f(x)/g(x) < 3\lambda/2. \qquad (*)$$

If $x_0 = +\infty$ (resp., $-\infty$), then there exists $B \in \mathbb{R}$ (resp., $A \in \mathbb{R}$) such that $x \in I$ and $x > B$ (resp., $x \in I$ and $x < A$) imply $(*)$. Now, $(*)$ is equivalent to $\lambda g(x)/2 < f(x) < 3\lambda g(x)/2$ (if $g(x) > 0$), and we can apply property (6). □

Remark 3.4.33. Although, for simplicity, Theorem 3.4.32 is stated for infinite limits "as $x \to x_0$," it is obviously satisfied for *one-sided* infinite limits as well.

Example 3.4.34.

(a) Show that $\lim_{x\to+\infty}(\sin x)/x = 0$. Well, we simply note that

$$(\forall \varepsilon > 0)(x > 1/\varepsilon \Rightarrow |\sin x/x| \le 1/|x| < \varepsilon).$$

(b) Find $\lim_{x\to+\infty}(\sqrt{x^2 + 2x} - \sqrt{x^2 + x})$. First note that this limit has the *indeterminate form* $\infty - \infty$. Now, for all $x > 0$,

$$\sqrt{x^2 + 2x} - \sqrt{x^2 + x} = \frac{(x^2 + 2x) - (x^2 + x)}{\sqrt{x^2 + 2x} + \sqrt{x^2 + x}}$$

$$= \frac{x}{x\sqrt{1 + 2/x} + x\sqrt{1 + 1/x}}$$

$$= \frac{1}{\sqrt{1 + 2/x} + \sqrt{1 + 1/x}}.$$

Thus,

$$\lim_{x\to\infty} (\sqrt{x^2 + 2x} - \sqrt{x^2 + x}) = \lim_{x\to\infty} 1/(\sqrt{1 + 2/x} + \sqrt{1 + 1/x}) = 1/2.$$

(c) Find $\lim_{x\to\infty} x \log(1 + 1/x)$. Note that $\lim_{x\to\infty} x = \infty$ and $\lim_{x\to\infty} \log(1 + 1/x) = \log 1 = 0$. Thus, the desired limit has the *indeterminate form* $\infty \cdot 0$ (or $0 \cdot \infty$). To find the limit, recall that we have, for any $x \in (-1, 1)$, the inequalities $x/(1 + x) \le \log(1 + x) \le x$ [cf. Example 3.3.8(2)]. Replacing x by $1/x$ and simplifying, we get the inequalities

$$\frac{1}{1 + 1/x} \le x \log(1 + 1/x) \le 1, \qquad \text{if} \quad x > 1.$$

It now follows from property (8) of Theorem 3.4.32 (Squeeze Theorem) that

$$\lim_{x \to +\infty} x \log(1 + 1/x) = 1.$$

Remark 3.4.35. In all the limits considered above, the requirement $x \to x_0$ (resp., $x \to x_0\pm$) can always be reduced to $x \to 0$ (resp., $x \to 0\pm$). Indeed, if $x_0 \in \mathbb{R}$, then we can make the change of variable $x' = x - x_0$, and if $x_0 = \pm\infty$, we can make the change of variable $x' = 1/x$.

We end this section with an extension of Theorem 3.3.15 (*Monotone Limit Theorem*) which includes one-sided limits, limits at infinity, and infinite limits. For simplicity, we state the theorem for *increasing* functions, but a similar result (with obvious modifications) holds for *decreasing* functions as well.

Theorem 3.4.36 (Monotone Limit Theorem).   *Let $I \subset \mathbb{R}$ be an interval with endpoints a, $b \in \overline{\mathbb{R}} = [-\infty, +\infty]$, $a < b$, and let $f : I \to \mathbb{R}$ be an increasing function. Then, for each interior point $x_0 \in I^\circ$, the one-sided limits $f(x_0 \pm 0)$ both exist (i.e., are finite), and we have*

$$f(x_0 - 0) = \sup\{f(x) : a < x < x_0\}$$

$$\leq f(x_0) \leq \inf\{f(x) : x_0 < x < b\} = f(x_0 + 0). \qquad (*)$$

Moreover, if $a < x < y < b$, then

$$f(x + 0) \leq f(y - 0). \qquad (\dagger)$$

Finally, $\lim_{x \to a} f(x) = \inf(f) := \inf\{f(x) : x \in I\}$ and $\lim_{x \to b} f(x) = \sup(f) := \sup\{f(x) : x \in I\}$ also exist (possibly as extended real numbers).

Proof. First, since f is increasing on (a, b), it must be *bounded* on any subinterval $(c, d) \subset (a, b)$ with $a < c < d < b$. (Why?) Thus, if $x_0 \in I^\circ$, then f is bounded *near* x_0 and the inequalities $(*)$ are an immediate consequence of Theorem 3.3.15. Next, the inequality $(\dagger)$ follows from $f(x + 0) = \inf\{f(t) : x < t < y\}$ and $f(y - 0) = \sup\{f(t) : x < t < y\}$, which follow from $(*)$ applied on the interval (x, y) (Exercise 3.1.8). To prove the last statement of the theorem, note that if I and f are both *bounded*, then Theorem 3.3.15 may be applied again. Thus, we may assume that I is unbounded or f is unbounded (or both). Let us first assume that $b = +\infty$ and that f is *bounded above*, i.e., $u := \sup(f) \in \mathbb{R}$. Then, given any $\varepsilon > 0$, we can find $A \in I$ such that $u - \varepsilon < f(A) \leq u$. But then, f being increasing, we get

$$x > A \Rightarrow u - \varepsilon < f(A) \leq f(x) \leq u \Longrightarrow |f(x) - u| < \varepsilon,$$

which shows that $\lim_{x \to +\infty} f(x) = u = \sup(f)$, as claimed. Suppose next that $b = +\infty$ and $\sup(f) = +\infty$. The latter implies that $\forall B > 0$ we can find $A \in I$

such that $f(A) > B$. Again, using the fact that f is increasing, $x > A$ implies $f(x) \geq f(A) > B$. In other words, we have

$$(\forall B > 0)(\exists A \in I)(x > A \Rightarrow f(x) > B),$$

which shows, indeed, that $\lim_{x \to +\infty} f(x) = +\infty = \sup(f)$. The remaining cases are similar and will be left to the reader.   □

3.5   Indeterminate Forms, Equivalence, and Landau's Little "oh" and Big "Oh"

The *indeterminate forms* have already been mentioned a few times (cf. Example 3.4.34(b) and (c) above). As we pointed out before stating Theorem 3.4.32, some limit properties do not extend to *infinite limits*. For example, if $f(x) = cx$, where $c \neq 0$ is *arbitrary*, and if $g(x) = x$, then $\lim_{x \to 0} f(x)/g(x) = \lim_{x \to \pm\infty} f(x)/g(x) = c$, whereas $\lim_{x \to 0} f(x)/\lim_{x \to 0} g(x) = 0/0$ and $\lim_{x \to \pm\infty} f(x)/\lim_{x \to \pm\infty} g(x) = \pm\infty/\pm\infty$ (or $\mp\infty/\pm\infty$, if $c < 0$) are *both meaningless*. Since c was arbitrary, expressions such as $0/0$ or ∞/∞ are called *indeterminate forms*. We now give a formal definition of all indeterminate forms to be studied later (when l'Hôpital's Rule is introduced).

Definition 3.5.1 (Indeterminate Forms).   Let I be an interval, and let $x_0 \in I^-$ or $x_0 = \pm\infty$ (if I is unbounded). Suppose that f, g are two functions defined on I (except possibly at x_0). Then, with lim denoting $\lim_{x \to x_0}$, we define the *indeterminate forms* $0/0$, ∞/∞, $0 \cdot \infty$, $\infty - \infty$, 0^0, 1^∞, and ∞^0 as follows:

1. $\lim f(x)/g(x)$ is said to have the indeterminate form $0/0$ if $\lim f(x) = \lim g(x) = 0$.
2. $\lim f(x)/g(x)$ is said to have the indeterminate form ∞/∞ if $|\lim f(x)| = \infty = |\lim g(x)|$.
3. $\lim f(x) \cdot g(x)$ is said to have the indeterminate form $0 \cdot \infty$ if $\lim f(x) = 0$ and $\lim g(x) = \pm\infty$ (or if $\lim g(x) = 0$ and $\lim f(x) = \pm\infty$).
4. $\lim(f(x) - g(x))$ is said to have the indeterminate form $\infty - \infty$ if $\lim f(x) = \pm\infty = \lim g(x)$.
5. $\lim f(x)^{g(x)}$ is said to have the indeterminate form 0^0 if f is *ultimately positive* and $\lim f(x) = 0 = \lim g(x)$.
6. $\lim f(x)^{g(x)}$ is said to have the indeterminate form 1^∞ if $\lim f(x) = 1$ and $\lim g(x) = \pm\infty$.
7. $\lim f(x)^{g(x)}$ is said to have the indeterminate form ∞^0 if $\lim f(x) = +\infty$ and $\lim g(x) = 0$.

If $x_0 \in \mathbb{R}$, then the above indeterminate forms can also be defined, in a similar fashion, for *one-sided limits* $x \to x_0-$ and $x \to x_0+$.

Remark 3.5.2.

(a) The exponential functions $f^g := e^{g \log f}$, where f is assumed to be *positive* on
    its domain, will be defined later when we give the precise definitions of the exp
    and log functions. We are, however, assuming that the reader is already familiar
    with such functions from calculus.
(b) Recall that in Chap. 2, we defined $0 \cdot \pm\infty = \pm\infty \cdot 0 := 0$ by an *arbitrary*
    convention. This should not be confused with the *indeterminate form* $0 \cdot \infty$,
    which is a *limit* of the form $\lim_{x \to x_0} f(x)g(x)$, where $\lim_{x \to x_0} f(x) = 0$
    and $\lim_{x \to x_0} g(x) = \pm\infty$. For instance, $\lim_{x \to +\infty} x \log(1 + 1/x)$ has the
    indeterminate form $0 \cdot \infty$ (or, more accurately, $\infty \cdot 0$), but $\lim_{x \to +\infty} x \log(1 + 1/x) = 1 \neq 0$, as was proved in Example 3.4.34(c).
(c) All the above indeterminate forms can be transformed to the form $0/0$, but, as
    we shall see, this is not necessarily a good practice. If desired, the transformation
    to $0/0$ is carried out as follows, where lim denotes $\lim_{x \to x_0}$:

   (i) The identity $f(x)/g(x) = (1/g(x))/(1/f(x))$ transforms ∞/∞ into $0/0$.
   (ii) The identity $f(x)g(x) = f(x)/(1/g(x))$ transforms $0 \cdot \infty$ into $0/0$.
   (iii) The identity

   $$f(x) - g(x) = \frac{1}{1/f(x)} - \frac{1}{1/g(x)} = \frac{1/g(x) - 1/f(x)}{1/(f(x) \cdot g(x))}$$

   transforms $\infty - \infty$ to $0/0$.
   (iv) If $\lim f(x)^{g(x)}$ has the indeterminate form 0^0, we first write $f(x)^{g(x)} := e^{g(x) \log f(x)}$ and note that $\lim \log f(x) = -\infty$. Thus, $\lim g(x) \log f(x)$
    has the indeterminate form $0 \cdot \infty$ and is transformed to $0/0$ as above.
   (v) If $\lim f(x)^{g(x)}$ has the indeterminate form 1^∞, we first write $f(x)^{g(x)} := e^{g(x) \log f(x)}$ and note that $\lim \log f(x) = 0$. Hence $\lim g(x) \log f(x)$ has
    the indeterminate form $0 \cdot \infty$ (or, more accurately, $\infty \cdot 0$), which has been
    discussed.
   (vi) Finally, if $\lim f(x)^{g(x)}$ has the indeterminate form ∞^0, write $f(x)^{g(x)} := e^{g(x) \log f(x)}$ and note that $\lim \log f(x) = +\infty$. Thus, $\lim g(x) \log f(x)$
    has the indeterminate form $0 \cdot \infty$ and can be transformed to $0/0$ as before.

Exercise 3.5.3.

(a) Using the fact that $\lim_{\theta \to 0}(\sin \theta)/\theta = 1$ (to be proved rigorously later), show
    that (i) $\lim_{x \to \pm\infty} x \sin(1/x) = 1$; (ii) $\lim_{x \to 0}(1 - \cos x)/x^2 = 1/2$. *Hints:* For
    (i), use Proposition 3.4.24. For (ii), use the identity $1 - \cos x = 2 \sin^2(x/2)$.
(b) Show that $\lim_{x \to +\infty}(\log x)/x = 0$. *Hint:* Using the inequality $\log(1 + x) \leq x$,
    valid for all $x > -1$, show that $\log x = 2 \log \sqrt{x} < 2\sqrt{x} \; \forall x > 0$ and apply
    the Squeeze Theorem.

Definition 3.5.4 (Equivalent Functions). Let I be an interval, and let $x_0 \in I^-$ or
$x_0 = \pm\infty$ (if I is unbounded). Suppose that $f, \ g$ are two functions defined on I

(except possibly at x_0). We say that f and g are *equivalent* as $x \to x_0$, and we write $f \sim g$ $(x \to x_0)$, if there exists a function u *defined near* x_0 such that

$$f(x) = g(x)u(x) \quad (x \approx x_0) \quad \text{and} \quad \lim_{x \to x_0} u(x) = 1.$$

When $x_0 \in I^\circ$, we define the equivalences $f \sim g$ $(x \to x_0-)$ and $f \sim g$ $(x \to x_0+)$ in a similar way.

Remark 3.5.5.

1. In what follows, we state all the properties of the equivalence (defined above) for the case $x \to x_0$ (where $x_0 = \pm\infty$ is allowed). It is obvious that the same properties also hold for the one-sided limits $x \to x_0-$ and $x \to x_0+$ when $x_0 \in \mathbb{R}$.
2. If $f \sim g$ $(x \to x_0)$ and if $g = 0$ (*the zero function*) near x_0, then the definition implies that $f = 0$ near x_0. On the other hand, if $g = c$ is constant near x_0 with $c \neq 0$, then it does *not* follow that $f = c$ near x_0.
3. If $g(x) \neq 0$ $(x \approx x_0)$, then

$$f \sim g \quad (x \to x_0) \iff \lim_{x \to x_0} f(x)/g(x) = 1. \quad \text{(Why?)}$$

Example 3.5.6.

(a) $\sin x \sim x$ $(x \to 0)$. Indeed, $\lim_{x \to 0}(\sin x)/x = 1$, as we shall see later.
(b) Let $p(x) = a_0 + a_1 x + \cdots + a_n x^n$, where $a_0, a_1, \ldots, a_n \in \mathbb{R}$ and $a_n \neq 0$, be a polynomial of degree n. Then we have $p(x) \sim a_n x^n$ $(x \to \pm\infty)$. Indeed, we have $p(x) = a_n x^n u(x)$, with

$$u(x) = 1 + \left(\frac{a_{n-1}}{a_n}\right)\frac{1}{x} + \left(\frac{a_{n-2}}{a_n}\right)\frac{1}{x^2} + \cdots + \left(\frac{a_0}{a_n}\right)\frac{1}{x^n},$$

and we obviously have $\lim_{x \to \pm\infty} u(x) = 1$.
(c) We have $\log(1 + 1/x) \sim 1/x$ $(x \to +\infty)$. This, of course, is an immediate consequence of Example 3.4.34(c).

The following proposition shows that the use of the word *equivalence* for the relation $f \sim g$ $(x \to x_0)$ is justified.

Proposition 3.5.7. *Let $x_0 \in I^-$ or $x_0 = \pm\infty$ (for I unbounded), and let $\mathcal{F}_{x_0}$ denote the set of all functions defined on I (except possibly at x_0). Then the equivalence $f \sim g$ $(x \to x_0)$, defined above, is indeed an equivalence relation on $\mathcal{F}_{x_0}$.*

Proof. First, $\forall f \in \mathcal{F}_{x_0}$, we obviously have $f \sim f$ $(x \to x_0)$. Next, if $f \sim g$ $(x \to x_0)$, then $f(x) = g(x)u(x)$ $(x \approx x_0)$ and $\lim_{x \to x_0} u(x) = 1$. In particular, $u(x) \neq 0$ $(x \approx x_0)$ and we can write $g(x) = f(x)(1/u(x))$ $(x \approx x_0)$. Since $\lim_{x \to x_0} 1/u(x) = 1$, it follows that $g \sim f$ $(x \to x_0)$. Finally, if

$f \sim g$ $(x \to x_0)$ and $g \sim h$ $(x \to x_0)$, then we can find functions u and v, *defined near* x_0, such that $f(x) = g(x)u(x)$, $g(x) = v(x)h(x)$, and $\lim_{x \to x_0} u(x) = \lim_{x \to x_0} v(x) = 1$. But then, $f(x) = h(x)u(x)v(x)$ $(x \approx x_0)$, uv is defined near x_0, and $\lim_{x \to x_0}(uv)(x) = 1$. This shows that $f \sim h$ $(x \to x_0)$ and the proof is complete.   □

The next theorem shows that, if two functions have the *same (finite) nonzero* limit as $x \to x_0$, then they are *equivalent* as $x \to x_0$.

Theorem 3.5.8. *Let* I *be an interval, and let* $x_0 \in I^-$ *or* $x_0 = \pm\infty$ *(if* I *is unbounded). Suppose that* f, g *are two functions defined on* I *(except possibly at* x_0*). If* $\lim_{x \to x_0} f(x) = \lim_{x \to x_0} g(x) = y_0 \in \mathbb{R} \setminus \{0\}$, *then* $f \sim g$ $(x \to x_0)$.

Proof. Since $y_0 \neq 0$, the function $u(x) := f(x)/g(x)$ is *defined near* x_0, and $\lim_{x \to x_0} u(x) = y_0/y_0 = 1$. Also, $f(x) = g(x)u(x)$ $(x \approx x_0)$, and the theorem follows.   □

Remark 3.5.9. Note that the condition $y_0 \in \mathbb{R} \setminus \{0\}$ is *necessary*. Indeed, if $f(x) = |x|$ and $g(x) = x^2$, then $\lim_{x \to 0} f(x) = 0 = \lim_{x \to 0} g(x)$, and yet $\lim_{x \to 0} f(x)/g(x) = +\infty$. Also, if $f(x) = 1/|x|$ and $g(x) = 1/x^2$, then we have $\lim_{x \to 0} f(x) = \lim_{x \to 0} g(x) = +\infty$, and yet $\lim_{x \to 0} f(x)/g(x) = 0$.

In view of Theorem 3.5.8, we may wonder whether two *equivalent* functions must have the *same limit*. Well, the following theorem shows that this is indeed the case if (at least) one of the functions *does* have a limit (even 0 or $\pm\infty$).

Theorem 3.5.10. *Let* I *be an interval, and let* $x_0 \in I^-$ *or* $x_0 = \pm\infty$ *(if* I *is unbounded). Suppose that* f, g *are two functions defined on* I *(except possibly at* x_0*). If* $f \sim g$ $(x \to x_0)$ *and if* $\lim_{x \to x_0} f(x) = y_0$, *then* $\lim_{x \to x_0} g(x) = y_0$, *even if* $y_0 = 0$ *or* $y_0 = \pm\infty$.

Proof. Indeed, there exists a function u, *defined near* x_0, such that $g(x) = f(x)u(x)$ $(x \approx x_0)$ and $\lim_{x \to x_0} u(x) = 1$. Thus,

$$\lim_{x \to x_0} g(x) = \lim_{x \to x_0} f(x)u(x) = y_0 \cdot 1 = y_0,$$

and this is true even if $y_0 = 0$ or $y_0 = \pm\infty$.   □

Theorem 3.5.10 shows that when looking for the limit of a function, one may replace it with *any* equivalent one. Together with the following theorem, this provides a powerful tool for finding and manipulating limits in a simpler way.

Theorem 3.5.11. *Let* I *be an interval, and let* $x_0 \in I^-$ *or* $x_0 = \pm\infty$ *(if* I *is unbounded). Suppose that* f_1, f_2, g_1, g_2 *are functions defined on* I *(except possibly at* x_0*) such that* $f_1 \sim g_1$ $(x \to x_0)$ *and* $f_2 \sim g_2$ $(x \to x_0)$. *Then* $f_1 f_2 \sim g_1 g_2$ $(x \to x_0)$ *and (if* f_2 *and* g_2 *are nonzero near* x_0*)* $f_1/f_2 \sim g_1/g_2$ $(x \to x_0)$. *More generally, if* $f_1, f_2, \ldots, f_n, g_1, g_2, \ldots, g_n$ *are defined on* I *(except possibly at* x_0*) and if* $f_j \sim g_j$ $(x \to x_0)$ *for* $j = 1, 2, \ldots, n$, *then* $f_1 f_2 \cdots f_n \sim g_1 g_2 \cdots g_n$ $(x \to x_0)$.

Proof. Indeed, there are functions u_1, u_2, defined near x_0, such that $f_1(x) = g_1(x)u_1(x)$, $f_2(x) = g_2(x)u_2(x)$ $(x \approx x_0)$, and $\lim u_1(x) = 1 = \lim u_2(x)$ as $x \to x_0$. It then follows that $f_1 f_2 = g_1 g_2 u_1 u_2$ and (if f_2 and g_2 are nonzero near x_0) $f_1/f_2 = (g_1/g_2)(u_1/u_2)$ are satisfied near x_0, and we have $\lim_{x \to x_0}(u_1 u_2)(x) = 1 = \lim_{x \to x_0}(u_1/u_2)(x)$. The last statement follows by induction. $\qquad\square$

Example 3.5.12.

(a) We have $\tan x \sim x$ $(x \to 0)$. Indeed, $\sin x \sim x$ $(x \to 0)$ and $\cos x = 1 - 2\sin^2(x/2) \sim 1$ $(x \to 0)$ so that, by Theorem 3.5.11, $\tan x = \sin x / \cos x \sim x/1$ $(x \to 0)$.

(b) Let m, $n \in \mathbb{N}$, and let $p(x) = a_0 + a_1 x + \cdots + a_n x^n$, $q(x) = b_0 + b_1 x + \cdots + b_m x^m$ be polynomials of degrees n and m, respectively. Thus $a_0, a_1, \ldots, a_n, b_0, b_1, \ldots, b_m$ are real and $a_n \neq 0 \neq b_m$. We have seen above that $p(x) \sim a_n x^n$ $(x \to \pm\infty)$ and $q(x) \sim b_m x^m$ $(x \to \pm\infty)$. It follows from Theorem 3.5.11 that $p(x)/q(x) \sim a_n x^n / b_m x^m$ $(x \to \pm\infty)$.

Warning!

(a) If $f_1 \sim g_1$ $(x \to x_0)$, $f_2 \sim g_2$ $(x \to x_0)$, it *does not* follow (in general) that $f_1 \pm f_2 \sim g_1 \pm g_2$ $(x \to x_0)$. For example, as $x \to 0$, we have $x + x^2 \sim x + x^3$ and $x \sim x$, but $(x + x^2) - x = x^2 \not\sim x^3 = (x + x^3) - x$. Also, we have $\cos x \sim 1$ $(x \to 0)$ and (obviously) $1 \sim 1$ $(x \to 0)$, but $1 - \cos x \not\sim 0$. In fact, $1 - \cos x \neq 0$ for all x sufficiently close to 0.

(b) If g is *nonzero near x_0* and $f \sim g$ $(x \to x_0)$, then $f(x)$ and $g(x)$ are *approximately equal* (i.e., $|f(x) - g(x)|$ is *very small*) when $x \approx x_0$. Note that we *need not* have $f(x) = g(x)$ $(x \approx x_0)$. If, in addition, g is *bounded* near x_0, then $f(x) - g(x) = g(x)(f(x)/g(x) - 1) \to 0$ as $x \to x_0$. However, if $\lim_{x \to x_0} f(x) = \pm\infty = \lim_{x \to x_0} g(x)$, we *do not* (in general) have $\lim_{x \to x_0}(f(x) - g(x)) = 0$. For example, $1/|x| + 1/x^2 \sim 1/x^2$ $(x \to 0)$, because $\lim_{x \to 0}(1/|x| + 1/x^2)/(1/x^2) = \lim_{x \to 0}(1 + |x|) = 1$, but $\lim_{x \to 0}((1/|x| + 1/x^2) - 1/x^2) = \lim_{x \to 0} 1/|x| = +\infty$.

Definition 3.5.13 (Landau's Little "oh" and Big "Oh").   Let I be an interval, and let $x_0 \in I^-$ or $x_0 = \pm\infty$ (if I is unbounded). Suppose that f, g are two functions defined on I (except possibly at x_0).

(o)   We say that f is *negligible compared to g* as $x \to x_0$ and write $f = o(g)$ $(x \to x_0)$, if there exists a function ζ, defined near x_0, such that

$$f(x) = g(x)\zeta(x) \quad (x \approx x_0) \quad \text{and} \quad \lim_{x \to x_0} \zeta(x) = 0.$$

(O)   We say that f has the *same order as g* as $x \to x_0$ and write $f = O(g)$ $(x \to x_0)$, if there exists a function β, defined near x_0, such that

$$f(x) = g(x)\beta(x) \quad (x \approx x_0) \quad \text{and} \quad \beta \text{ is } bounded.$$

Remark 3.5.14. Note that $f = o(g)$ $(x \to x_0)$ obviously implies $f = O(g)(x \to x_0)$ but the converse is *false* in general.

Proposition 3.5.15. *Let I be an interval, and let $x_0 \in I^-$ or $x_0 = \pm\infty$ (if I is unbounded). Suppose that f, g are two functions defined on I (except possibly at x_0). If g is nonzero near x_0, then*

$$f = o(g) \quad (x \to x_0) \iff \lim_{x \to x_0} f(x)/g(x) = 0,$$

and

$$f = O(g) \quad (x \to x_0) \iff f/g \quad is\ bounded\ near \quad x_0.$$

In particular, $f = o(1)$ $(x \to x_0)$ if and only if $\lim_{x \to x_0} f(x) = 0$, and $f = O(1)$ $(x \to x_0)$ if and only if f is bounded near x_0.

Proof. Exercise!   □

Example 3.5.16.

(a) Let m, $n \in \mathbb{N}$, and let $p(x) = a_0 + a_1 x + \cdots + a_n x^n$, $q(x) = b_0 + b_1 x + \cdots + b_m x^m$ be polynomial functions of degrees n and m, respectively. Here, $a_0, a_1, \ldots, a_n, b_0, b_1, \ldots, b_m$ are real, $a_n \neq 0 \neq b_m$. Then we have $p = o(q)$ $(x \to \pm\infty)$ if and only if $m > n$, and $p = O(q)$ $(x \to \pm\infty)$ if and only if $m \geq n$.
(b) We have $\log x = o(x)$ $(x \to +\infty)$. This follows from Exercise 3.5.3(b).
(c) Since $\lim_{x \to 0} \sin x = 0 = \lim_{x \to 0} \tan x$, we have $\sin x = o(1)$ and $\tan x = o(1)$ as $x \to 0$.

Definition 3.5.17 (Infinitesimal). Let I be an interval, and let $x_0 \in I^-$ or $x_0 = \pm\infty$ (if I is unbounded). Suppose that f is a function defined on I (except possibly at x_0). We say that f is an *infinitesimal* (or *infinitely small*) *at x_0* (or, as $x \to x_0$) if $f = o(1)$ $(x \to x_0)$, i.e., if $\lim_{x \to x_0} f(x) = 0$.

Example 3.5.18. As we saw above, $\sin x$ and $\tan x$ are infinitesimals at $x = 0$, and $\log x/x = o(1)$ $(x \to +\infty)$. If p and q are polynomial functions of degrees n and m, respectively, and if $m > n$, then p/q is an infinitesimal at $x = \pm\infty$.

Remark 3.5.19.

1. Using infinitesimals, we can rephrase many statements. Thus $f = o(g)$ $(x \to x_0)$ can also be written as $f = g \cdot o(1)$ $(x \to x_0)$, which means (if g is nonzero near x_0) that f/g is an infinitesimal at $x = x_0$.
2. As $x \to 0$, we have an important sequence of infinitesimals, namely, the sequence of *monomials* x, x^2, $x^3, \ldots, x^n, \ldots$. It is obvious that the larger the exponent n, the faster x^n converges to 0. We shall see that many infinitesimals at 0 are *equivalent* to an infinitesimal of the form ax^n, where $n \in \mathbb{N}$ and a is a *nonzero constant*.

Our next theorem will summarize the behavior of "o" and "O" under algebraic operations, as well as composition. Before stating it, however, we need the following:

Definition 3.5.20 (Bounded Away From Zero). Let I be an interval, and let $x_0 \in I^-$ or $x_0 = \pm\infty$ (if I is unbounded). Suppose that f is a function defined on I (except possibly at x_0). We say that f is *bounded away from zero* as $x \to x_0$ if there exists $\varepsilon > 0$ such that

$$|f(x)| \geq \varepsilon \quad (x \approx x_0).$$

Exercise 3.5.21. Let f be as in the above definition. Show that f is bounded away from zero as $x \to x_0$ if and only if $1/f$ is *bounded near* x_0, i.e., if and only if $1/f = O(1)$ $(x \to x_0)$.

Theorem 3.5.22. *Let I be an interval, and let $x_0 \in I^-$ or $x_0 = \pm\infty$ (if I is unbounded). Suppose that f, g, h are functions defined on I (except possibly at x_0). Then, as $x \to x_0$, the following are true.*

1. *If $f = o(h)$ and $g = o(h)$, then $f \pm g = o(h)$ and $fg = o(h^2)$.*
2. *If $f = o(h)$ and $c \in \mathbb{R}$, then $cf = o(h)$ and (if $c \neq 0$) $f = o(ch)$.*
3. *If $f = o(h)$ and g is bounded away from zero $(x \to x_0)$, then $f/g = o(h)$.*
4. *If $f = o(g)$ and if $g = O(h)$ (e.g., if $g = o(h)$), then $f = o(h)$.*
5. *If $f = o(1)$, then $1/(1 + f) = 1 - f + o(f)$.*
6. *If $f = O(h)$ and $g = O(h)$, then $f \pm g = O(h)$ and $fg = O(h^2)$.*
7. *If $f = O(h)$ and $c \in \mathbb{R}$, then $cf = O(h)$ and (if $c \neq 0$) $f = O(ch)$.*
8. *If $f = O(h)$ and g is bounded away from zero $(x \to x_0)$, then $f/g = O(h)$.*
9. *If $f = O(g)$ and $g = O(h)$, then $f = O(h)$.*
10. *If $f = O(1)$ and $g = O(h)$, then $fg = O(h)$.*
11. *If $f = o(1)$ and $g = O(h)$, then $fg = o(h)$.*

Proof. We shall only prove the properties (5) and (11) and leave the rest as exercises for the reader (cf. Exercise 3.5.23 below). To prove (5), note that

$$\frac{1}{1 + f} - (1 - f) = \frac{f^2}{1 + f} = f \left(\frac{f}{1 + f} \right)$$

and that $\lim_{x \to x_0} f(x)/(1 + f(x)) = 0$, i.e., $f/(1 + f) = o(1)$. Thus, $1/(1 + f) - (1 - f) = f \cdot o(1) = o(f)$. For property (11), note that $\lim_{x \to x_0} f(x) = 0$ and we have $g(x) = h(x)\beta(x)$, where β is *bounded near* x_0. Thus, $f(x)g(x) = (f(x)\beta(x))h(x)$, where, by Theorem 3.3.9 (which is also valid if $x \to x_0\pm$ or $x \to \pm\infty$), $\lim_{x \to x_0} f(x)\beta(x) = 0$. This shows, indeed, that $fg = o(h)$. $\square$

Exercise 3.5.23. Prove the remaining properties in Theorem 3.5.22, i.e., all properties except (5) and (11).

Remark 3.5.24. Notice that, in property (1) of Theorem 3.5.22, $f = o(h)$ and $g = o(h)$ imply $fg = o(h^2)$ and *not* $fg = o(h)$, which is *false* in general. For instance, we have $e^{2x/3} = o(e^x)$ $(x \to +\infty)$, but $e^{2x/3} \cdot e^{2x/3} = e^{4x/3} \neq o(e^x)$. The same remark can be made for the property (6).

Despite the above remark, the behavior of "o" and "O" under the algebraic operations is much nicer for the special case where h is a nonzero constant function, in which case we may (without loss of generality) assume $h = 1$. The following exercise is basically an immediate consequence of Theorem 3.5.22.

Exercise 3.5.25. Let $x_0 \in I^-$ or $x_0 = \pm\infty$ (for unbounded I), and let $\mathcal{F}_{x_0}(I)$ denote the set of all functions defined on I (except possibly at x_0). Let $\mathbf{O} := \{f \in \mathcal{F}_{x_0}(I) : f = O(1) \ (x \to x_0)\}$ and $\mathbf{o} := \{f \in \mathcal{F}_{x_0}(I) : f = o(1) \ (x \to x_0)\}$. Show that $\mathbf{O}$ is a (commutative) *ring* with identity and that $\mathbf{o} \subset \mathbf{O}$ is an *ideal*. This means that, for *any* f, g, $h \in \mathbf{O}$, the following properties are satisfied:

1. $f + g$, $fg \in \mathbf{O}$;
2. $f + g = g + f$, $fg = gf$, $(f + g) + h = f + (g + h)$, and $(fg)h = f(gh)$;
3. $\exists\, 0 \in \mathbf{o}$, $1 \in \mathbf{O}$ such that $0 + f = f$, $1 \cdot f = f$;
4. $\exists -f \in \mathbf{O}$ such that $f + (-f) = 0$;
5. $f(g + h) = fg + fh$;
6. if f, $g \in \mathbf{o}$, then $f - g \in \mathbf{o}$; and
7. if $f \in \mathbf{o}$, then $fg \in \mathbf{o}$ $\ \forall g \in \mathbf{O}$.

As was pointed out before, we can always reduce $x \to x_0$ to $x \to 0$ by introducing the new variable $x' = x - x_0$ (if $x_0 \in \mathbb{R}$) or $x' = 1/x$ (if $x_0 = \pm\infty$). We also introduced the distinguished sequence x, x^2, $x^3, \dots$ of infinitesimals at $x = 0$ and mentioned that many infinitesimals are equivalent to constant multiples of these powers of x as $x \to 0$. Before giving the formal definition, we prove a *uniqueness* result:

Proposition 3.5.26. *If* $f(x) \sim ax^n$ $(x \to 0)$ *for some* $n \in \mathbb{N}$ *and* $a \in \mathbb{R} \setminus \{0\}$ *(which implies that* f *is an infinitesimal at* 0*), then the constants* a *and* n *are uniquely determined by* f.

Proof. Suppose that we also have $f(x) \sim bx^m$ $(x \to 0)$ with $m \in \mathbb{N}$, $b \in \mathbb{R} \setminus \{0\}$. Then $ax^n \sim bx^m$ $(x \to 0)$. This means that $\lim_{x \to 0} ax^n/bx^m = 1$. Now, if $m > n$, then $\lim_{x \to 0} |ax^n/bx^m| = +\infty$, and if $m < n$, then $\lim_{x \to 0} ax^n/bx^m = 0$. Thus, we must have $m = n$, which also forces $a = b$. $\square$

Definition 3.5.27 (Principal Part and Order of Infinitesimals). If $f(x) \sim ax^n$ $(x \to 0)$ for some constants $n \in \mathbb{N}$ and $a \in \mathbb{R} \setminus \{0\}$, then we say that ax^n is the *principal part* of f and that f is an infinitesimal of *order* n at $x = 0$.

Example 3.5.28. We have seen that $\lim_{x \to 0}(\sin x)/x = \lim_{x \to 0}(\tan x)/x = \lim_{x \to 0}(1 - \cos x)/(x^2/2) = \lim_{x \to 0}(\log(1 + x))/x = 1$. It follows that $\sin x$, $\tan x$, and $\log(1 + x)$ are infinitesimals at 0 of order 1 and principal part x, while $1 - \cos x$ is an infinitesimal (at 0) of order 2 and principal part $x^2/2$.

3.6 Problems

Throughout this section, A, B, C,..., X, Y, Z will denote nonempty subsets of $\mathbb{R}$.

1. Let $f : X \to \mathbb{R}$ be a *bounded* function. Show that, for any $c \in \mathbb{R}$,

$$\sup(c + f) = c + \sup(f) \quad \text{and} \quad \inf(c + f) = c + \inf(f).$$

2. Let $f, g : X \to \mathbb{R}$ be *bounded*. Show that

$$\sup(f) + \inf(g) \le \sup(f + g) \le \sup(f) + \sup(g).$$

3. Let $f : X \to \mathbb{R}$ and $g : Y \to \mathbb{R}$ be *bounded* and assume that $f(x) \le g(y)$ for all $x \in X$ and $y \in Y$. Show that $\sup(f) \le \inf(g)$.

4. Consider the function $f(x) := (ax + b)/(cx + d)$, where $ad - bc \ne 0$. Show *directly* that f is *strictly increasing* (resp., *strictly decreasing*) if $ad - bc > 0$ (resp., $ad - bc < 0$) and find $\sup(f)$ and $\inf(f)$ in each case. *Hint:* First look at the case where $c = 0$. Next, assume that $c \ne 0$ and reduce to the case where $f(x) := \alpha + \beta/(cx + d)$, for some α and β.

5. Give an example of a *one-to-one* function $f : (0, \infty) \to \mathbb{R}$ that is *not* monotone.

6. Let us say that a function $f \in \mathbb{R}^{\mathbb{R}}$ is *increasing at a point* x_0 if there is a $\delta > 0$ such that $f(x) \le f(x_0) \le f(y)$ for all $x \in (x_0 - \delta, x_0)$ and $y \in (x_0, x_0 + \delta)$. Show that f is *increasing* (on $\mathbb{R}$) if and only if it is increasing at *every* $x_0 \in \mathbb{R}$.

7. Let $A, B \subset \mathbb{R}$. Show that $(A \cap B)^{\circ} = A^{\circ} \cap B^{\circ}$ and $(A \cup B)^{-} = A^{-} \cup B^{-}$. On the other hand, show that $(A \cup B)^{\circ} \supset A^{\circ} \cup B^{\circ}$ and $(A \cap B)^{-} \subset A^{-} \cap B^{-}$, and give examples to show that both inclusions may be *proper*.

8. Using the *definition*, prove each statement.

(a) $\displaystyle \lim_{x \to -1/2} \frac{2x + 1}{1 - 4x^2} = \frac{1}{2}$; (b) $\displaystyle \lim_{x \to 2} \frac{x + 1}{2x - 3} = 3$; (c) $\displaystyle \lim_{x \to -2} (x^2 + x) = 2$;

(d) $\displaystyle \lim_{x \to 0} \frac{\sqrt{x + 4} - 2}{x} = \frac{1}{4}$; (e) $\displaystyle \lim_{x \to 0} \frac{x}{1 + 1/x} = 0$; (f) $\displaystyle \lim_{x \to 2} \frac{1}{x^2} = \frac{1}{4}$.

9. Show that

$$\lim_{x \to x_0} f(x) = y_0 \iff \lim_{x \to x_0} [f(x) - y_0] = 0 \iff \lim_{h \to 0} f(x_0 + h) = y_0.$$

10. Show that the following limits do *not* exist. Recall that $[x]$ denotes the *greatest integer* $\le x$ and that, for each $A \subset \mathbb{R}$, $\chi_A(x) = 1$ if $x \in A$ and $\chi_A(x) = 0$ if $x \notin A$.

(a) $\displaystyle \lim_{x \to 1} \frac{|x - 1|}{x - 1}$; (b) $\displaystyle \lim_{x \to n} [x] \quad (\forall n \in \mathbb{Z})$;

(c) $\displaystyle \lim_{x \to x_0} \chi_{\mathbb{Q}}(x) \quad (\forall x_0 \in \mathbb{R})$; (d) $\displaystyle \lim_{x \to 0} \cos(1/x^2)$.

11. Using $\lim_{x \to 0} (\sin x)/x = 1$ and the properties of limits (including Theorem 3.3.14), find each limit.

(a) $\displaystyle \lim_{x \to 0} \frac{\sin 3x}{\sin 2x}$; (b) $\displaystyle \lim_{x \to 0} \frac{x \sin x}{\cos x}$; (c) $\displaystyle \lim_{x \to 0} \frac{\tan^2 x}{2x^2}$;

(d) $\lim\limits_{x\to 0}\dfrac{1-\cos x}{x^2}$; (e) $\lim\limits_{x\to 1}\dfrac{\tan(x^2-1)}{1-x}$; (f) $\lim\limits_{x\to 0}\dfrac{x\sin x}{1-\cos x}$.

12. Using $\lim\limits_{x\to 0}(\sin x)/x = 1$, show that $\lim\limits_{x\to 0}(\sin^{[n]} x)/x = 0$, where we have used the "nth iterate", $\sin^{[n]}$, defined by $\sin^{[n]} := \sin\circ\sin\circ\cdots\circ\sin$, with n iterations.

13. Show that, if $\lim\limits_{x\to 0} f(x)/x = \ell \in \mathbb{R}$ and $a \neq 0$, then $\lim\limits_{x\to 0} f(ax)/x = a\ell$. What if $a = 0$?

14. Let $f \in \mathbb{R}^{\mathbb{R}}$ satisfy $f(x+y) = f(x)+f(y)$, for all x, $y \in \mathbb{R}$. Show that $\lim\limits_{x\to 0} f(x) = 0$, if the limit exists, and that, in this case, $\lim\limits_{x\to x_0} f(x) = f(x_0)$, $\forall\ x_0 \in \mathbb{R}$. *Hint:* $f(2x) = 2f(x)$.

15. Consider the functions

(a) $f(x) := \begin{cases} x & \text{if } x \in \mathbb{Q}, \\ 0 & \text{if } x \notin \mathbb{Q}. \end{cases}$ (b) $g(x) := \begin{cases} x & \text{if } x \in \mathbb{Q}, \\ -x & \text{if } x \notin \mathbb{Q}. \end{cases}$

Show that $\lim\limits_{x\to 0} f(x) = \lim\limits_{x\to 0} g(x) = 0$ but that $\lim\limits_{x\to x_0} f(x)$ and $\lim\limits_{x\to x_0} g(x)$ *do not exist for any* $x_0 \neq 0$.

16. Find each limit *if it exists*; if it doesn't, explain why.

(a) $\lim\limits_{x\to 1}\dfrac{\sqrt{x}-x^2}{1-\sqrt{x}}$; (b) $\lim\limits_{x\to 0}\left(\dfrac{1}{x\sqrt{x+1}}-\dfrac{1}{x}\right)$; (c) $\lim\limits_{x\to 2}\dfrac{\sqrt{6-x}-2}{\sqrt{3-x}-1}$;

(d) $\lim\limits_{x\to 1}\dfrac{|x-1|^{3/2}}{x-1}$; (e) $\lim\limits_{x\to 0}\dfrac{|x-1|-|x+1|}{x}$; (f) $\lim\limits_{x\to 1}\dfrac{x\sqrt{x}-1}{x^2-1}$.

17. Prove or disprove each statement.

(a) If $\lim\limits_{x\to x_0} f(x)$ and $\lim\limits_{x\to x_0} [f(x)+g(x)]$ exist, then $\lim\limits_{x\to x_0} g(x)$ exists.
(b) If $\lim\limits_{x\to x_0} f(x)$ and $\lim\limits_{x\to x_0} [f(x)\cdot g(x)]$ exist, then $\lim\limits_{x\to x_0} g(x)$ exists.

18. Let f, $g : \mathbb{R} \to \mathbb{R}$ be such that $\lim\limits_{x\to x_0} f(x) = y_0$ and $\lim\limits_{y\to y_0} g(y) = z_0$. Does it necessarily follow that $\lim\limits_{x\to x_0} (g \circ f)(x) = z_0$? Why or why not?

19. For each function $f(x)$, find $\lim\limits_{x\to 0+}$, $\lim\limits_{x\to 0-}$, and $\lim\limits_{x\to 0}$, if they exist.

(a) $f(x) := x[x]$; (b) $f(x) := x - [x]$; (c) $f(x) := [1-x^2]$;

(d) $f(x) := [x^2-1]$; (e) $f(x) := \dfrac{x^2}{|x|}$; (f) $f(x) := \dfrac{\sin x}{|x|}$.

20. Show that, if $\lim\limits_{x\to a-} f(x) < \lim\limits_{x\to a+} f(x)$, then there is a $\delta > 0$ such that $|x-a| < \delta$, $|y-a| < \delta$, and $x < a < y$, imply $f(x) < f(y)$. Is the *converse* also true?

21. Assuming that all the limits involved exist, prove each statement.

(a) $\lim\limits_{x\to 0+} f(x) = \lim\limits_{x\to 0-} f(-x)$; (b) $\lim\limits_{x\to 0} f(|x|) = \lim\limits_{x\to 0+} f(x) = \lim\limits_{x\to 0} f(x^2)$.

22. Find each limit if it exists.

(a) $\lim\limits_{x\to -\infty}\dfrac{4x^2+1}{x}$; (b) $\lim\limits_{x\to\infty}\dfrac{3x+2}{\sqrt{x+1}}$;

(c) $\displaystyle\lim_{x\to-\infty} (\sqrt{x^2 + x} + x)$;

(d) $\displaystyle\lim_{x\to\infty} x \cos \frac{1}{x}$.

23. Find each limit if it exists. Recall that $\lim_{x\to 0}(\sin x)/x = 1$.

(a) $\displaystyle\lim_{x\to\infty} \frac{\sin x}{\sqrt{x}}$;

(b) $\displaystyle\lim_{x\to\infty} x \sin \frac{1}{x}$;

(c) $\displaystyle\lim_{x\to\infty} \frac{x \sin^2 x}{x + 1}$;

(d) $\displaystyle\lim_{x\to\infty} \frac{x^2(1 + \sin^2 x)}{(x + \sin x)^2}$.

24. Let $f : (a, \infty) \to \mathbb{R}$ and assume that $\lim_{x\to\infty} xf(x) = \ell \in \mathbb{R}$. Show that we then have $\lim_{x\to\infty} f(x) = 0$.

25. For each $x \in \mathbb{R}$, let $\langle x \rangle$ denote the *distance from x to the integer nearest x* and, recall that $[x]$ denotes the *greatest* integer $\leq x$. Find each limit if it exists. If it doesn't, explain why!

(a) $\displaystyle\lim_{x\to\infty} [x]$;

(b) $\displaystyle\lim_{x\to\infty} (x - [x])$;

(c) $\displaystyle\lim_{x\to\infty} \frac{[x]}{x}$;

(d) $\displaystyle\lim_{x\to\infty} \langle x \rangle$;

(e) $\displaystyle\lim_{x\to\infty} (x - \langle x \rangle)$;

(f) $\displaystyle\lim_{x\to\infty} \frac{\langle x \rangle}{x}$.

26. Find each limit if it exists.

(a) $\displaystyle\lim_{x\to 0+} \left(\frac{1}{x} - \frac{1}{|x|} \right)$;

(b) $\displaystyle\lim_{x\to 0-} \left(\frac{1}{x} - \frac{1}{|x|} \right)$;

(c) $\displaystyle\lim_{x\to\infty} x^2 \sin \frac{1}{x}$;

(d) $\displaystyle\lim_{x\to 0+} \sqrt{x} \sin \frac{1}{x}$.

27. Find each limit if it exists.

(a) $\displaystyle\lim_{x\to\infty} \frac{(\log x)^\beta}{x^\alpha}$ $(\alpha, \beta > 0)$;

(b) $\displaystyle\lim_{x\to 0+} x^\alpha(|\log x|)^\beta$ $(\alpha, \beta > 0)$;

(c) $\displaystyle\lim_{x\to\infty} x^{1/x}$;

(d) $\displaystyle\lim_{x\to\infty} (1 + 1/x)^x$.

28. Find each limit if it exists.

(a) $\displaystyle\lim_{x\to\infty} x^2 \left(1 - \cos \frac{1}{x} \right)$;

(b) $\displaystyle\lim_{x\to 0} (1 - \sin x)^{1/x}$;

(c) $\displaystyle\lim_{x\to 0+} x^x$;

(d) $\displaystyle\lim_{x\to 0+} (\tan x)^{1/x}$.

29. Prove each statement. Here, $\alpha > 0$ and $\beta > 0$ are arbitrary.

(a) $(\log x)^\alpha = o(x^\beta)$ $(x \to \infty)$;

(b) $\sec 2x = 1 + 2\sin^2 x + o(1)$ $(x \to 0)$;

(c) $x^\alpha |\log x|^\beta = o(1)$ $(x \to 0+)$;

(d) $\left(1 - \sin \frac{1}{x} \right)^x = O(1)$ $(x \to \infty)$.

30. Find the *order* and the *principal part* of each infinitesimal (at 0).

(a) $\tan x - \sin x$;

(b) $\sqrt{1 - \cos x}$;

(c) $\sin x \tan x$;

(d) $\log \sqrt{1 + x^2}$.

Chapter 4
Topology of $\mathbb{R}$ and Continuity

Roughly speaking, a quantity y is said to depend *continuously* on a quantity x if "small" changes in x result in small changes in y. Our goal in this chapter is to make this statement mathematically precise. Now most interesting sets in mathematics have *structures* (algebraic, geometric, topological, ...). For example, the set $\mathbb{R}$ of real numbers is, algebraically, a *field*; i.e., it satisfies the nine axioms $A_1 - A_4$, $M_1 - M_4$, and D listed at the beginning of Chap. 2. Given this field structure, the most *(algebraically) desirable* functions $\phi : \mathbb{R} \to \mathbb{R}$ are those that are *faithful* to the field properties, i.e., *preserve* them. Such maps are called the *morphisms* of the field $\mathbb{R}$.

For instance, *addition* is a map (binary operation) $+ : \mathbb{R} \times \mathbb{R} \to \mathbb{R}$, given by $+(x, y) = x + y \; \forall x, \, y \in \mathbb{R}$. A function $\phi : \mathbb{R} \to \mathbb{R}$ is *additive* (or *faithful to the addition*) if it satisfies *Cauchy's functional equation:*

$$\phi(x + y) = \phi(x) + \phi(y) \quad \forall x, \, y \in \mathbb{R}.$$

This can be written in a more suggestive way, using the *composition* of maps:

$$(\phi \circ +)(x, y) = (+ \circ \tilde{\phi})(x, y)) \quad \forall x, \, y \in \mathbb{R},$$

where $\tilde{\phi} := (\phi, \phi)$ is defined by $\tilde{\phi}(x, y) = (\phi, \phi)(x, y) = (\phi(x), \phi(y))$. In other words,

$$\phi \circ + = + \circ \tilde{\phi}. \tag{$*$}$$

If $(*)$ is satisfied, we say (by abuse of language) that ϕ *commutes* with the addition. Thus, a map ϕ is "faithful" to $+$ if it *commutes* with it.

Now, we have repeatedly mentioned that the fundamental notion in analysis is that of *limit*. Therefore, in the study of limits of functions (defined "near" a point x_0) as $x \to x_0$, the most *desirable* functions are those that *commute with* $\lim_{x \to x_0}$, i.e.,

© Springer Science+Business Media New York 2014
H.H. Sohrab, *Basic Real Analysis*, DOI 10.1007/978-1-4939-1841-6_4

$$f\left(\lim_{x \to x_0} x\right) = \lim_{x \to x_0} f(x),$$

and such functions are precisely what we call *continuous functions* (at x_0). Before giving the formal definition of continuity, we shall introduce some important facts dealing with the *topology* of the real line, i.e., with its *open sets*. These facts will be crucial when we introduce *uniform continuity*, a refinement of the concept of continuity, and will play a fundamental role in our study of (Riemann) integration, sequences and series of functions, and approximation.

4.1 Compact and Connected Subsets of ℝ

Recall (cf. Definition 2.2.2) that a subset $O \subset \mathbb{R}$ is called *open* if, $\forall x \in O$, there exists $\varepsilon = \varepsilon(x) > 0$ such that $B_{\varepsilon(x)}(x) := (x - \varepsilon, x + \varepsilon) \subset O$. It is therefore obvious that $O = \bigcup_{x \in O} B_{\varepsilon(x)}(x)$ is a union of open intervals and that for $O \neq \emptyset$ this union is *uncountable* (why?) and is *not disjoint*. Theorem 4.1.2 below will show that we can do much better than this. But let us first show that any union of open subsets of R can be written as a *countable* union:

Proposition 4.1.1 (Lindelöf). *Let $\{O_\lambda\}_{\lambda \in \Lambda}$ be a collection of open subsets of $\mathbb{R}$. Then there is a countable subset $\{\lambda_1, \lambda_2, \ldots\} \subset \Lambda$ such that*

$$\bigcup_{\lambda \in \Lambda} O_\lambda = \bigcup_{k=1}^{\infty} O_{\lambda_k}.$$

Proof. Let $\mathbf{O} := \bigcup_{\lambda \in \Lambda} O_\lambda$. Then, $\forall x \in \mathbf{O}$, we have $x \in O_{\lambda_x}$ for some $\lambda_x \in \Lambda$ and, since O_{λ_x} is open, we can find $\varepsilon_x > 0$ with $x \in B_{\varepsilon_x}(x) \subset O_{\lambda_x}$. Using the fact that the set $\mathbb{Q}$ of rational numbers is *dense* in $\mathbb{R}$, we can find a *rational* number $\rho_x > 0$ such that $x \in B_{\rho_x}(x) \subset B_{\varepsilon_x}(x)$. Now the set $\{\rho_x : x \in \mathbf{O}\} \subset \mathbb{Q}$ is *countable* and hence can be written as $\{\rho_x : x \in \mathbf{O}\} = \{\rho_1, \rho_2, \ldots\}$, where $\rho_k = \rho_{x_k}$ for some $x_k \in \mathbf{O}$. If for each $k \in \mathbb{N}$ we pick $\lambda_k \in \Lambda$ such that $B_{\rho_k}(x_k) \subset O_{\lambda_k}$, then we have a countable subcollection $\{O_{\lambda_k}\}_{k \in \mathbb{N}} \subset \{O_\lambda\}_{\lambda \in \Lambda}$ which satisfies $\mathbf{O} = \bigcup_{k=1}^{\infty} O_{\lambda_k}$. $\square$

Proposition 4.1.1 actually shows that $\mathbf{O} = \bigcup_{k=1}^{\infty} B_{\rho_k}$ is in fact a countable union of open *intervals*. These intervals need not, however, be *disjoint*. The next theorem shows that this extra requirement can also be met. Before stating the theorem, let us recall that a *partition* of a set S is a collection of nonempty subsets of S that are pairwise disjoint and whose union is the set S. We saw in Chap. 1 that each partition of S corresponds an *equivalence relation* on S and vice versa. Thus, it should not be surprising to encounter an equivalence relation in the proof of the theorem.

Theorem 4.1.2. *A set $O \subset \mathbb{R}$ is open if and only if it is a countable union of pairwise disjoint open intervals.*

Proof. If $O \subset \mathbb{R}$ is a disjoint union of open intervals, then it is obviously open. To prove the converse, define an equivalence relation on O as follows. For any $a, b \in O$, let us say that a is *equivalent* to b, and write $a \sim b$, if the (possibly empty) open interval with *endpoints* a and b is contained in O. Now, this is obviously *reflexive* and *symmetric*. (Why?) To prove the *transitivity* property, let a, b, c be the three (distinct) elements of O such that $a \sim b$ and $b \sim c$. Then, assuming (without loss of generality) that $a < b$, we have the three possible cases $a < c < b$, $c < a < b$, and $a < b < c$, and it follows at once from $a \sim b$ and $b \sim c$ that we have $(a, c) \subset O$ in all these cases. Now, for each $x \in O$, let $[x]$ denote its *equivalence class*. Since $x \in [x]$ and since two equivalence classes are either identical or disjoint, $\{[x] : x \in O\}$ is a partition of O, so we need only show that each $[x]$ is an *open interval*; because then, by the density of $\mathbb{Q}$, each $[x]$ contains a (necessarily different) rational number and hence $\{[x] : x \in O\}$ must be *countable*. First, to prove that $[x]$ is an interval, let y, $z \in [x]$ be any pair of distinct elements with, say, $y < z$. Then, if $y < u < z$, we have $y \sim u \sim z$ (why?) and hence $(y, z) \subset [x]$. Finally, to show that $[x]$ is *open*, note that, if $y \in [x]$, then (since O is open) there exists $\varepsilon > 0$ such that $(y - \varepsilon, y + \varepsilon) \subset O$. But then $(y - \varepsilon, y + \varepsilon) \subset [y] = [x]$ and the proof is complete.   □

In Theorem 4.1.2, the set O was *covered* by a collection of open intervals. Such a collection is said to be an *open cover* of the set O. This suggests the following:

Definition 4.1.3 (Open Cover, Subcover, Finite Subcover). Let $S \subset \mathbb{R}$. A collection $\mathcal{O} = \{O_\lambda\}_{\lambda \in \Lambda}$ of open subsets of $\mathbb{R}$ is called an *open cover* of S if $S \subset \bigcup_{\lambda \in \Lambda} O_\lambda$. If, for some $\Lambda' \subset \Lambda$, we also have $S \subset \bigcup_{\lambda \in \Lambda'} O_\lambda$, then the *subcollection* $\mathcal{O}' = \{O_\lambda\}_{\lambda \in \Lambda'}$ is called a *subcover* (of $\mathcal{O}$). If, in addition, Λ' is *finite*, then the subcollection $\mathcal{O}'$ is called a *finite subcover* (of $\mathcal{O}$).

Example 4.1.4. The collection $\{(n, 2n + 1)\}_{n=1}^{\infty}$ is an open cover of $(2, 1024)$; the subcollection $\{(2k, 4k + 1)\}_{k=1}^{\infty}$ is a subcover, and the subcollection $\{(2k, 4k + 1)\}_{k=1}^{256}$ is a *finite* subcover.

We are now ready to define the important notion of *compactness* in $\mathbb{R}$:

Definition 4.1.5 (Compact Set). A set $K \subset \mathbb{R}$ is called *compact* if *every* open cover $\mathcal{O} = \{O_\lambda\}_{\lambda \in \Lambda}$ of K has a *finite* subcover.

Exercise 4.1.6.

(a) Let $\{K_\lambda\}_{\lambda \in \Lambda}$ be a collection of *compact* subsets of $\mathbb{R}$. Show that $\bigcap_{\lambda \in \Lambda} K_\lambda$ is compact and that, if Λ is *finite*, then $\bigcup_{\lambda \in \Lambda} K_\lambda$ is also compact. Deduce, for instance, that any *finite* set is compact.

(b) Show that the set $\{0, 1, 1/2, 1/3, \ldots\}$ is compact but the subset $\{1, 1/2, 1/3, \ldots\}$ is not.

Can we replace, in the definition of compactness, the open *sets* O_λ of our covers by open *intervals*? The answer is *yes*.

Proposition 4.1.7. *A set* $K \subset \mathbb{R}$ *is compact if and only if every open cover* $\mathcal{I} = \{I_\lambda\}_{\lambda \in \Lambda}$ *of* K *by (open) intervals* I_λ *has a finite subcover.*

Proof. Well, if K is compact, then any cover of K by open intervals is an open cover and hence has a *finite* subcover. Assume, conversely, that all covers of K by open intervals have finite subcovers and let $\mathcal{O} = \{O_\lambda\}_{\lambda \in \Lambda}$ be an open cover of K. Then, for each $\lambda \in \Lambda$, the open set O_λ is the union of a (countable) collection $\mathcal{I}_\lambda$ of (pairwise disjoint) open intervals. Thus, the collection $\bigcup_{\lambda \in \Lambda} \mathcal{I}_\lambda$ is a cover of K by open intervals, and hence there exist intervals $I_1, I_2, \ldots, I_n$ in this collection such that $K \subset I_1 \cup I_2 \cup \cdots \cup I_n$. If we now pick $\lambda_1, \lambda_2, \ldots, \lambda_n \in \Lambda$ such that $I_j \subset O_{\lambda_j}$, then $K \subset \bigcup_{j=1}^{n} O_{\lambda_j}$. $\qquad\square$

The (closed) unbounded set $[1, \infty)$ is *not* compact because the open cover $\{(0, n)\}_{n \in \mathbb{N}}$ has no finite subcover. Similarly, the (bounded) open set $(0, 1)$ is *not* compact because the open cover $\{(1/n, 1)\}_{n \in \mathbb{N}}$ has no finite subcover. These examples suggest that compact sets *cannot* be unbounded or open. It is in fact easy to show that they have to be *bounded and closed:*

Proposition 4.1.8. *Compact subsets of* $\mathbb{R}$ *are necessarily closed and bounded.*

Proof. Let $K \subset \mathbb{R}$ be *compact*. Then the open cover $\{(-n, n)\}_{n \in \mathbb{N}}$ has a finite subcover, say $\{(-n_1, n_1), \ldots, (-n_k, n_k)\}$. If $N = \max\{n_1, \ldots, n_k\}$, then $K \subset [-N, N]$ and hence is bounded. Next, if K has no limit points, then it is closed. If, to get a contradiction, we assume that $\xi \notin K$ is a limit point of K, then the open cover $\{(-\infty, \xi - 1/n) \cup (\xi + 1/n, \infty)\}_{n \in \mathbb{N}}$ has no finite subcover. (Why?) $\qquad\square$

The deeper fact, however, is that the converse of Proposition 4.1.8 is also true. Before presenting it in full generality, let us first prove it for the special case of closed and bounded *intervals:*

Proposition 4.1.9. *Given any* a, $b \in \mathbb{R}$ *such that* $a \leq b$, *the interval* $[a, b]$ *is compact.*

Proof. Let $\mathcal{O} = \{O_\lambda\}_{\lambda \in \Lambda}$ be an open cover of $[a, b]$, and define $S \subset [a, b]$ to be the set of all $x \in [a, b]$ such that $[a, x]$ can be covered by a *finite* subcover (of $\mathcal{O}$). Then $a \in S$, b is an upper bound of S, and $c := \sup(S) \in [a, b]$. There is $\lambda \in \Lambda$ such that $c \in O_\lambda$ and, since O_λ is open, we have $(c - \varepsilon, c + \varepsilon) \subset O_\lambda$ for some $\varepsilon > 0$. Now, if $c < b$, then we can pick $d \in (c, c + \varepsilon)$ such that $c < d < b$, and it follows that $[a, d]$ can also be covered by a finite subcover, i.e., $d \in S$. This, however, contradicts $c = \sup(S)$. Thus $c = b$, and the proof is complete. $\qquad\square$

We can now prove the following important theorem that completely characterizes the *compact* subsets of $\mathbb{R}$:

Theorem 4.1.10 (Heine–Borel). *A set* $K \subset \mathbb{R}$ *is compact if and only if it is closed and bounded.*

Proof. In view of Proposition 4.1.8, we need only show that any closed and bounded subset of R is compact. So let $K \subset \mathbb{R}$ be *closed and bounded*, and let $\mathcal{O} = \{O_\lambda\}_{\lambda \in \Lambda}$ be an open cover of K. Since K is bounded, we can pick a, $b \in \mathbb{R}$ such that

$a < b$ and $K = K^- \subset [a, b]$. Now, K^c is open and if we let $O' := K^c$, then the collection $\mathcal{O}' := \mathcal{O} \cup \{O'\}$ is an open cover of $[a, b]$. Since $[a, b]$ is compact (by Proposition 4.1.9), we can find a *finite* subcover $\mathcal{O}'' \subset \mathcal{O}'$. If $O' := K^c \notin \mathcal{O}''$, then $\mathcal{O}''$ is the desired finite subcover of K. If, however, $O' \in \mathcal{O}''$, then $\mathcal{O}'' \setminus \{O'\}$ is the finite subcover we are looking for. $\qquad\square$

Remark 4.1.11. Notice that, using Theorem 4.1.10, Exercise 4.1.6(a) becomes trivial. (Why?) Also, it follows from Theorem 4.1.10 that *closed subsets of compact sets are compact.* (Why?)

There are other useful characterizations of compact sets in terms of sequences and limit points, and the next theorem shows that they are equivalent to the one given in Theorem 4.1.10.

Theorem 4.1.12. *For a set $K \subset \mathbb{R}$ the following statements are equivalent:*

(a) K is compact.
(b) K is closed and bounded.
(c) Every infinite subset of K has a limit point in K.
(d) Every sequence in K has a subsequence that converges (to an element of K).

Proof. The equivalence (a) $\Leftrightarrow$ (b) is, of course, the Heine–Borel Theorem. Let us then prove the implications (b) $\Rightarrow$ (c) $\Rightarrow$ (d) $\Rightarrow$ (b). Suppose that K is closed and bounded, and let $S \subset K$ be an *infinite* subset. Then S is *bounded* (because K is) and hence, by (the Bolzano–Weierstrass) Theorem 2.2.43, it has a limit point, say ξ. Since K is *closed* we must have $\xi \in K$ and (c) follows. Next, suppose that (c) is satisfied and let $(x_n) \in K^{\mathbb{N}}$. If $\{x_n : n \in \mathbb{N}\}$ is *finite*, then we can find integers n_0 and n_k, $k \in \mathbb{N}$, such that $n_1 < n_2 < n_3 < \cdots$ and $x_{n_k} = x_{n_0} \, \forall k \in \mathbb{N}$. (Why?) Thus $\lim(x_{n_k}) = x_{n_0} \in K$ as desired. If, on the other hand, $\{x_n : n \in \mathbb{N}\}$ is *infinite*, then [by (c)] it has a limit point $\xi \in K$. By the very definition of *limit point*, for each $k \in \mathbb{N}$, we can find increasing $n_k \in \mathbb{N}$ such that $|x_{n_k} - \xi| < 1/k$. It is then obvious that $\lim(x_{n_k}) = \xi \in K$ and (d) is satisfied. Finally, suppose that (d) is satisfied, and let ξ be a limit point of K. We can find (using the definition of limit point) a sequence $(x_n) \in K^{\mathbb{N}}$ such that $\lim(x_n) = \xi$. This implies that *all* subsequences of (x_n) also converge to ξ, and hence [by (d)] we must have $\xi \in K$. Thus, K contains all its limit points and is therefore *closed.* If K is *unbounded*, then we can find a sequence $(x_n) \in K^{\mathbb{N}}$ such that $|x_n| > n \, \forall n \in \mathbb{N}$. But then *no* subsequence of (x_n) converges, contradicting (d). $\qquad\square$

We now introduce the concept of *connectedness* for subsets of $\mathbb{R}$. Intuitively, a subset of $\mathbb{R}$ should be *connected* if it is in "one piece". For example, we "expect" any *interval* to be connected but the set $(0, 1) \cup (2, 3)$, for instance, should not be connected. The precise definition is as follows:

Definition 4.1.13 (Connected, Disconnected, Totally Disconnected). A set $S \subset \mathbb{R}$ is said to be *disconnected* if there are open sets U, $V \subset \mathbb{R}$ such that $\{S \cap U, S \cap V\}$ is a *partition* of S, i.e., $S \cap U \neq \emptyset$, $S \cap V \neq \emptyset$, $S \cap U \cap V = \emptyset$, and $S = (S \cap U) \cup (S \cap V)$. A set S that is *not* disconnected is said to be *connected*.

Finally, a set $S \subset \mathbb{R}$ is called *totally disconnected* if for any x, $y \in S$ such that $x < y$, there exists $z \in (x, y)$ such that $z \notin S$.

Exercise 4.1.14. Show that if $\{S_\lambda\}_{\lambda \in \Lambda}$ is a family of *connected* subsets of $\mathbb{R}$ and if $\bigcap_{\lambda \in \Lambda} S_\lambda \neq \emptyset$, then $\bigcup_{\lambda \in \Lambda} S_\lambda$ is also connected.

The following theorem shows that what is intuitively obvious is indeed the case, namely, that the connected subsets of $\mathbb{R}$ are precisely the *intervals*.

Theorem 4.1.15. *A subset $S \in \mathbb{R}$ is connected if and only if it is an interval.*

Proof. Recall that, by the *characterization of intervals* (Proposition 2.1.41), a subset $S \subset \mathbb{R}$ is an interval if and only if, for any a, $b \in S$ such that $a < b$, we have $(a, b) \subset S$. If we suppose that S is *connected* and a, $b \in S$ but $x \notin S$ for some $x \in (a, b)$, then $U := (-\infty, x)$ and $V := (x, \infty)$ are *disjoint open* sets and $\{U \cap S, V \cap S\}$ is obviously a partition of S, contradicting the fact that S is connected. Conversely, suppose that S is an *interval* with endpoints α, $\beta \in \overline{\mathbb{R}} = [-\infty, \infty]$, $\alpha < \beta$. Also, to get a contradiction, suppose U, $V \subset \mathbb{R}$ are open sets such that $\{U \cap S, V \cap S\}$ is a partition of S. Pick $a \in U \cap S$, $b \in V \cap S$, and assume, for instance, that $a < b$. Let $x = \sup([a, b] \cap (U \cap S)) = \sup([a, b] \cap U)$. If $x \in U \cap S$, then $x < b$ and we have $[x, x + \delta) \subset [a, b] \cap (U \cap S) = [a, b] \cap U$, for some $\delta > 0$, contradicting the definition of x. If on the other hand $x \in V \cap S$, then $x > a$ and we have $(x - \delta, x] \subset [a, b] \cap (V \cap S) = [a, b] \cap V$, for some $\delta > 0$, which again contradicts the definition of x. Thus we must have $x \notin U \cap S$ and $x \notin V \cap S$. This, however, is absurd since $[a, b] \subset S$.   □

4.2 The Cantor Set

In this section we end our *topological* preliminaries with a discussion of Cantor's *ternary set*. This is a remarkable subset of the unit interval $[0, 1]$ which has many applications and can be used to define numerous interesting examples and counterexamples in analysis. Recall that a *closed* set of real numbers is called *perfect* if every one of its elements is a *limit point*. For example, a closed interval is a perfect set. Let us introduce one more "topological" definition before we look at the Cantor set:

Definition 4.2.1 (Nowhere Dense). A set $S \subset \mathbb{R}$ is called *nowhere dense* if $S^{-\circ} = \emptyset$, i.e., if the closure of S contains *no* open intervals. Equivalently, S is nowhere dense if its *exterior* (i.e., the interior of its complement) is *dense:* $((S^c)^\circ)^- = ((S^-)^c)^- = \mathbb{R}$.

Example. The sets $\mathbb{N}$ and $\mathbb{Z}$ are nowhere dense, because all their elements are *isolated points*. It *does not* follow, however, that sets with limit points cannot be nowhere dense. For example, the set $\{1, 1/2, 1/3, \ldots\}$ which has 0 as its *unique* limit point is obviously nowhere dense. In fact, as the example of Cantor's ternary set will show, even *perfect* sets can be nowhere dense! Let us finally point out that the sets $\mathbb{Q}$ and $\mathbb{Q}^c = \mathbb{R} \setminus \mathbb{Q}$ are at the other extreme: they are *dense*, i.e., $\mathbb{Q}^- = (\mathbb{Q}^c)^- = \mathbb{R}$.

We now indicate how Cantor's ternary set is constructed. We begin with the (closed) *unit interval* $C_0 = [0, 1]$. From C_0 we remove its *open middle third* to obtain C_1; thus,

$$C_1 = C_0 \setminus (1/3, 2/3) = [0, 1/3] \cup [2/3, 1].$$

Next, to obtain C_2, we remove the open middle thirds of the two subintervals $[0, 1/3]$ and $[2/3, 1]$ of C_1:

$$C_2 = [0, 1/9] \cup [2/9, 3/9] \cup [6/9, 7/9] \cup [8/9, 1].$$

Continuing this construction, we obtain C_{n+1} by removing the open middle thirds of all 2^n subintervals of C_n. Observe that the C_n are *nested*: $C_1 \supset C_2 \supset C_3 \supset \cdots$. The *Cantor set* C is now defined to be the intersection of the C_n:

Definition 4.2.2 (Cantor Set). Let $\{C_n\}_{n \in \mathbb{N}}$ be the collection of sets constructed above. We define the *Cantor set* (or *Cantor's ternary set*) C to be

$$C := \bigcap_{n=1}^{\infty} C_n.$$

Remark 4.2.3.

1. That $C \neq \emptyset$ follows, for instance, from the Nested Intervals Theorem. In fact, since at each stage of the construction we remove the (open) middle thirds of the remaining (closed) subintervals, the Cantor set contains all the endpoints of the subintervals in C_n $\forall n \in \mathbb{N}$. We shall see, however, that C is actually *uncountable*!
2. The sets C_n can be defined (more explicitly) by the following *recursive* formula:

$$C_{n+1} = C_n \setminus \bigcup_{k=0}^{\infty} \left(\frac{3k+1}{3^{n+1}}, \frac{3k+2}{3^{n+1}} \right) \qquad (n = 0, 1, 2 \ldots).$$

The concept of *length* (or *measure*) will be discussed in detail when we study the Lebesgue integral. For our next exercise, however, we shall need the following temporary:

Definition 4.2.4 (Length).

1. For a *bounded* (possibly degenerate) interval I with endpoints a, $b \in \mathbb{R}$, $a \leq b$, we define the *length* (or *measure*) of I to be $\lambda(I) := b - a$. Thus,

$$\lambda((a, b)) = \lambda([a, b)) = \lambda((a, b]) = \lambda([a, b]) = b - a.$$

If I is an *unbounded* interval, we define its length to be $\lambda(I) := +\infty$.

2. For a *countably infinite* collection $\{I_n\}_{n \in \mathbb{N}}$ of *pairwise disjoint* intervals, we define the length of $S = \bigcup_{k=1}^{\infty} I_k$ to be

$$\lambda(S) = \lambda\left(\bigcup_{k=1}^{\infty} I_k\right) := \sum_{n=1}^{\infty} \lambda(I_n), \qquad (*)$$

where, once again, the length is defined to be $+\infty$ if the above series is *divergent*.
3. If A, B are countable unions of disjoint intervals, $A \subset B$, and $\lambda(A) < \infty$, then we define the length of $B \setminus A$ to be $\lambda(B \setminus A) := \lambda(B) - \lambda(A)$.

Exercise 4.2.5.

(a) Show that $\lambda(\emptyset) = 0$. More generally, show that for any *countable* set $S \subset \mathbb{R}$, we have $\lambda(S) = 0$.
(b) Prove the relation $(*)$ above for a *finite* collection $\{I_k\}_{k=1}^{n}$ of pairwise disjoint intervals. (*Hint:* Let $I_k := \emptyset$ for all $k > n$). Deduce that

$$\lambda\left(\bigcup_{k=1}^{\infty} I_k\right) = \lim_{n \to \infty} \lambda\left(\bigcup_{k=1}^{n} I_k\right).$$

(c) Show that, for every *open* set $O \subset \mathbb{R}$, $\lambda(O)$ is a well-defined *extended* real number in $[0, +\infty]$.
(d) Show that if $A \subset B \subset \mathbb{R}$ are as in (3) above, then $\lambda(A) \leq \lambda(B)$.
(e) Show that, for any intervals I and J, we have $\lambda(I \cup J) = \lambda(I) + \lambda(J) - \lambda(I \cap J)$. More generally, show that, if $\{I_k\}_{k=1}^{n}$ is any finite collection of intervals, then

$$\lambda\left(\bigcup_k I_k\right) = \sum_i \lambda(I_i) - \sum_{i<j} \lambda(I_i \cap I_j) + \sum_{i<j<k} \lambda(I_i \cap I_j \cap I_k) - \cdots + (-1)^{n-1} \lambda\left(\bigcap_k I_k\right).$$

Hint: Use the *Inclusion–Exclusion Principle* (Proposition 1.3.32).
(f) Let C_n be as in the construction of the Cantor set. Find $\lambda(C_n)$ for each $n \in \mathbb{N}$ and show that $\lim_{n \to \infty} \lambda(C_n) = \lim_{n \to \infty} \lambda\left(\bigcap_{k=1}^{n} C_k\right) = 0$.
(g) Show that $\lambda(C) = 0$, where C is the Cantor set. (*Hint:* Find $\lambda(D)$, where D is the union of all the open middle thirds *deleted* in the construction of the C_n, and note that $C = [0, 1] \setminus D$).

Actually, the length of an interval and hence that of a *finite, disjoint* union of intervals can be defined as the limit of a sequence of *normalized cardinalities*:

Exercise 4.2.6.

(a) Let $x \in \mathbb{R}$ and consider the sequence $([nx])_{n=1}^{\infty}$, where, as usual, $[t]$ denotes the greatest integer $\leq t$. Show that we have

$$\lim_{n \to \infty} \frac{[nx]}{n} = x.$$

Deduce, in particular, that $\mathbb{Q}$ is *dense* in $\mathbb{R}$ (cf. Theorem 2.1.38).

(b) Let I be an interval with endpoints $a \leq b$. Show that its length $\lambda(I) := b - a$ is given by

$$\lambda(I) = \lim_{n \to \infty} \frac{1}{n} |I \cap (\mathbb{Z}/n)|,$$

where $|S|$ denotes the *cardinality* of S and $\mathbb{Z}/n := \{m/n : m \in \mathbb{Z}\}$. *Hint:* For each $n \in \mathbb{N}$ count the number of integers $m \in \mathbb{Z}$ such that $na < m \leq nb$ and use part (a).

To study the properties of the Cantor set C, we need a simple characterization of its elements. Since C was obtained by deleting open middle thirds, it is not surprising that the desired characterization of its elements is provided by their *ternary expansions*.

Recall (from Chap. 2) that if $p > 1$ is a fixed integer, then any $x \in [0, 1)$ has a base p expansion $x = (0.x_1 x_2 \cdots)_p$, where we have

$$x \in \left[\sum_{j=1}^n \frac{x_j}{p^j}, \sum_{j=1}^n \frac{x_j}{p^j} + \frac{1}{p^n} \right), \quad \forall n \in \mathbb{N}.$$

In particular, $|x - \sum_{k=1}^n x_k/p^k| < 1/p^n$ for all $n \in \mathbb{N}$ and hence (why?)

$$x = (0.x_1 x_2 \cdots)_p = \sum_{n=1}^{\infty} \frac{x_n}{p^n}.$$

As was pointed out in Sect. 2.1, if $x = \sum_{k=1}^n x_k/p^k$, then besides the expansion $x = (0.x_1 x_2 \cdots x_n 000 \cdots)_p$, we also have the second expansion $x = (0.x_1' x_2' \cdots)_p$, where $x_j' = x_j$ for $1 \leq j \leq n-1$, $x_n' = x_n - 1$, and $x_k' = p - 1$ for all $k > n$. Indeed,

$$\sum_{n=1}^{\infty} \frac{x_n'}{p^n} = \sum_{k=1}^n x_k/p^k - \frac{1}{p^n} + \sum_{k=n+1}^{\infty} \frac{p-1}{p^k} = \sum_{k=1}^n x_k/p^k$$

because $\sum_{k=n+1}^{\infty} \frac{p-1}{p^k} = 1/p^n$. Also, note that $1 = (0.x_1 x_2 \cdots)_p$ with $x_n = p - 1$ for all n because

$$\sum_{n=1}^{\infty} \frac{p-1}{p^n} = \frac{(p-1)/p}{1 - 1/p} = 1.$$

In particular, if $p = 3$, then each $x \in [0, 1]$ has a (not necessarily unique) ternary expansion

$$x = (0.x_1 x_2 x_3 \ldots)_3 := \frac{x_1}{3} + \frac{x_2}{3^2} + \frac{x_3}{3^3} + \cdots,$$

where each x_k is either 0, 1, or 2. Note that, if $x = (0.x_1 x_2 x_3 \ldots)_3$ is an arbitrary ternary expansion, then, since $x_k \leq 2$ for all $k \in \mathbb{N}$, we indeed have

$$x = \sum_{k=1}^{\infty} \frac{x_k}{3^k} \leq \sum_{k=1}^{\infty} \frac{2}{3^k} = \frac{2/3}{1 - 1/3} = 1.$$

Example.

(a) We have $1/2 = (0.111\ldots)_3$, since

$$(0.111\ldots)_3 = \sum_{k=1}^{\infty} \frac{1}{3^k} = \frac{1/3}{1 - 1/3} = \frac{1}{2}.$$

(b) We have $11/26 = (0.102102102\ldots)_3$, since

$$(0.102102102\ldots)_3 = \frac{1}{3} + \frac{0}{3^2} + \frac{2}{3^3} + \frac{1}{3^4} + \frac{0}{3^5} + \frac{2}{3^6} + \cdots$$

$$= \left(\frac{1}{3} + \frac{1}{3^4} + \frac{1}{3^7} + \cdots \right) + \left(\frac{2}{3^3} + \frac{2}{3^6} + \frac{2}{3^9} + \cdots \right)$$

$$= \frac{1/3}{1 - 1/27} + \frac{2/27}{1 - 1/27}$$

$$= \frac{11}{26}.$$

As was pointed out, the *ternary* (or *base 3*) expansion of a number $x \in [0, 1]$ is *unique* except when $x = m/3^n$ for some positive integers m, n, where we may assume that 3 does *not* divide m. In these exceptional cases, we have *two* ternary expansions, one ending with a string of 0's and, the other with a string of 2s:

$$x = (0.x_1 x_2 \ldots x_n 1000\ldots)_3 = (0.x_1 x_2 \ldots x_n 0222\ldots)_3,$$

for some n. For example, $1/3 = (0.1000\ldots)_3 = (0.0222\ldots)_3$, since

$$\sum_{k=2}^{\infty} \frac{2}{3^k} = \frac{2}{9} \sum_{k=0}^{\infty} \left(\frac{1}{3} \right)^k = \frac{1}{3}.$$

It is therefore possible to write all the exceptional x's—i.e., those with a *terminating* ternary expansion—so that their expansions end with strings of 2s and, as we shall see presently, have no 1's.

First, we observe that if in the expansion $x = (0.x_1 x_2 x_3 \ldots)_3$ we have $x_k = 1$ for some $k \in \mathbb{N}$, then x must belong to (the closure of) one of the middle thirds deleted in the construction of the Cantor set. This can be seen inductively. For example, if

$x_1 = 1$, then $x \in [1/3, 2/3]$. Indeed, we have $x = x_1/3 + t$, where the "tail" t satisfies

$$t = \sum_{n=2}^{\infty} \frac{x_n}{3^n} \leq \sum_{n=2}^{\infty} \frac{2}{3^n} = \frac{1}{3}.$$

Also, $x_2 = 1$ implies that $x \in [1/9, 2/9]$, if $x_1 = 0$; $x \in [4/9, 5/9]$, if $x_1 = 1$; and $x \in [7/9, 8/9]$, if $x_1 = 2$. Thus, deleting the middle third $(1/3, 2/3)$ removes all numbers x whose *unique* ternary expansions satisfy $x_1 = 1$. Similarly, deleting the middle third $(1/9, 2/9)$ of $[0, 1/3]$ removes all x's whose *unique* expansions satisfy $x_1 = 0$ and $x_2 = 1$, while deleting the middle third $(7/9, 9/9)$ of $[2/3, 1]$ removes the x's whose *unique* expansions satisfy $x_1 = 2$ and $x_2 = 1$. If all middle thirds are deleted up to the nth stage, then deleting the middle thirds of the remaining subintervals will remove all numbers x whose *unique* expansions satisfy $x_j = 0$ or 2 for $1 \leq j \leq n$ and $x_{n+1} = 1$. Also, if x is the left *endpoint* of one of the deleted middle thirds, then its expansion has the form $x = (0.x_1 x_2 \cdots x_n 1000 \cdots)_3$, with $x_k = 0$ or 2 for $1 \leq k \leq n$. But then we also have $x = (0.x_1 x_2 \cdots x_n 0222 \cdots)_3$. Summarizing these observations, we have

Proposition 4.2.7. *The Cantor set C is the set of all $x \in [0, 1]$ such that $x = (0.x_1 x_2 x_3 \ldots)_3$, where each x_k is either 0 or 2.*

Example 4.2.8. We have $1/3 = (0.0222\ldots)_3 \in C$, $2/3 = (0.2000\ldots)_3 \in C$, and, in general, C contains all the endpoints of the subintervals in all the C_n. But we also have $1/4 = (0.020202\ldots)_3 \in C$ and $3/4 = (0.202020\ldots)_3 \in C$. The next theorem shows that, in fact, C and $[0, 1]$ have the *same* cardinality!

Theorem 4.2.9. *The Cantor set C is compact, nowhere dense, totally disconnected, and perfect. Moreover, there is a surjective map from C onto $[0, 1]$ and hence C has the cardinality of the continuum:* $|C| = |[0, 1]| = |\mathbb{R}| = \mathfrak{c}$.

Proof. First, C is compact because it is a *closed* subset of the compact set $[0, 1]$ (cf. Remark 4.1.11). Next, to show that C is nowhere dense—i.e., that $C^{-\circ} = C^{\circ} = \emptyset$—note that if $(\alpha, \beta) \subset C$ for some $\alpha < \beta$ in $\mathbb{R}$, then $(\alpha, \beta) \subset C_n \; \forall n \in \mathbb{N}$. Since C_n is a union of 2^n pairwise disjoint intervals of equal length $1/3^n$, we must have $\beta - \alpha \leq (2/3)^n \; \forall n \in \mathbb{N}$, which is absurd. This also shows that C is totally disconnected (why?). To prove that C is perfect, let $x \in C$ be arbitrary. Given any $\varepsilon > 0$, we can pick $n \in \mathbb{N}$ such that $1/3^n < \varepsilon$. Let I_n denote the subinterval in C_n containing x. If $x_n \neq x$ is an *endpoint* of I_n, then we have ($x_n \in C$ and) $0 < |x - x_n| < \varepsilon$, which proves that x is a *limit point* of C. Finally, if we define the map $\phi : C \to [0, 1]$ by

$$\phi\left(\sum_{n=1}^{\infty} \frac{x_n}{3^n}\right) = \sum_{n=1}^{\infty} \frac{x_n/2}{2^n},$$

then $\phi((0.x_1x_2x_3 \ldots)_3) = (0.y_1y_2y_3 \ldots)_2$, where $y_n = x_n/2$. Since each $y \in [0, 1]$ has a *binary* expansion, the map ϕ is onto and we get $|C| \geq |[0, 1]|$. But $C \subset [0, 1]$ implies that we also have $|[0, 1]| \leq |C|$ and the proof is complete. □

Remark 4.2.10. We shall see later that the map $\phi : C \rightarrow [0, 1]$ defined above is actually *continuous* and can be extended to a continuous, monotone map $\kappa :$ $[0, 1] \rightarrow [0, 1]$, called the *Cantor (ternary) function.*

4.3 Continuous Functions

We are now ready to define the notion of *continuity* for real-valued functions of a real variable. We will assume that *all* sets in our discussions are subsets of ℝ and that, as before, I and J (possibly with subscripts or superscripts) will always denote (not necessarily bounded) *intervals.* Since intervals are the most basic subsets of ℝ, we shall mainly consider functions defined on intervals. Most concepts discussed below can, however, be defined for functions whose domains are arbitrary subsets of ℝ.

Definition 4.3.1 (Continuous, Discontinuous). Let $f : I \rightarrow \mathbb{R}$ and let $x_0 \in I$. We say that f is *continuous at* x_0 if $\lim_{x \to x_0} f(x) = f(x_0)$, i.e., if f *commutes with* "$\lim_{x \to x_0}$":

$$\lim_{x \to x_0} f(x) = f\left(\lim_{x \to x_0} x \right).$$

Using the (ε, δ)-definition of the limit, this means the following:

$$(\forall \varepsilon > 0)(\exists \delta = \delta(\varepsilon, x_0) > 0)(\forall x \in I)(|x - x_0| < \delta \Rightarrow |f(x) - f(x_0)| < \varepsilon).$$
$$(*)$$

If f is continuous at x *for all* $x \in I$, then we say that f is *continuous on* I. Finally, if f is *not* continuous at $x_0 \in I$, then we say that f is *discontinuous* at x_0.

Notation 4.3.2. The set of all functions $f \in \mathbb{R}^I$ that are *continuous on* I will be denoted by $C(I)$. In particular, $C(\mathbb{R})$ denotes the set of all $f \in \mathbb{R}^{\mathbb{R}}$ such that f is continuous at *every* $x \in \mathbb{R}$. To simplify the notation, we shall often write $C(a, b)$ and $C[a, b]$ instead of $C((a, b))$ and $C([a, b])$, respectively.

Remark 4.3.3.

(a) If $S \subset \mathbb{R}$ is any subset and $x_0 \in S$, then a function $f : S \rightarrow \mathbb{R}$ is continuous at x_0 if $(*)$ (with $(\forall x \in I)$ replaced by $(\forall x \in S)$) is satisfied. Also, note that continuity is defined "locally," i.e., *pointwise.* Thus, a function f is said to be continuous on a subset S of its domain if it is continuous at *every point* of S. In this case, we write $f \in C(S)$.

(b) In terms of ε-neighborhoods, "$f : S \to \mathbb{R}$ is continuous at $x_0 \in S$" means

$$(\forall\, \varepsilon > 0)(\exists \delta = \delta(\varepsilon, x_0) > 0)\,(f(S \cap B_\delta(x_0)) \subset B_\varepsilon(f(x_0)))\,.$$

(c) The reader may have noticed that, in the above (ε, δ)-definition, we have used the restriction $|x - x_0| < \delta$ rather than $0 < |x - x_0| < \delta$. This is justified because f is *defined* at x_0 and, for $x = x_0$, we certainly have $|f(x) - f(x_0)| = 0 < \varepsilon$.

As the Remark (b) suggests, we should be able to define continuity by means of *open* sets. The next theorem confirms this:

Theorem 4.3.4. *We have $f \in C(S)$, where $S \subset \mathbb{R}$ may or may not be an interval, if and only if given any open set $O' \subset \mathbb{R}$ there is an open set $O \subset \mathbb{R}$ such that $f^{-1}(O') = S \cap O$. In particular, if S is open, then f is continuous (on S) if and only if the inverse image (under f) of every open set is open.*

Proof. Well, suppose first that $f \in C(S)$. Let O' be any open set, and let $x_0 \in f^{-1}(O')$, i.e., $f(x_0) \in O'$. Since O' is open, we have $B_\varepsilon(f(x_0)) \subset O'$ for some $\varepsilon > 0$. Now, by the continuity of f at x_0, we can find $\delta > 0$ such that $f(S \cap B_\delta(x_0)) \subset B_\varepsilon(f(x_0)) \subset O'$. Therefore, $f^{-1}(O') = S \cap O$ for some open set O. (Why?) Conversely, if this condition is satisfied for any open set O' and if $x_0 \in S$, then, given any $\varepsilon > 0$, $B_\varepsilon(f(x_0))$ is open and hence $f^{-1}(B_\varepsilon(f(x_0))) = S \cap O_\varepsilon$ for some open set O_ε and we obviously have $x_0 \in O_\varepsilon$. Pick $\delta > 0$ such that $B_\delta(x_0) \subset O_\varepsilon$. It is then clear that $f(S \cap B_\delta(x_0)) \subset B_\varepsilon(f(x_0))$ so that f is continuous at x_0. Since $x_0 \in S$ was arbitrary, the proof is complete. $\square$

Remark 4.3.5. Since each open set in $\mathbb{R}$ is a (countable) union of (pairwise disjoint) open *intervals*, in Theorem 4.3.4 we can replace "any open set O'" by "any open interval I'."

Corollary 4.3.6. *We have $f \in C(S)$ if and only if given any closed set $F' \subset \mathbb{R}$ there is a closed set $F \subset \mathbb{R}$ such that $f^{-1}(F') = S \cap F$. In particular, if S is closed, then f is continuous (on S) if and only if the inverse image (under f) of every closed set is closed.*

Proof. Exercise! $\square$

Definition 4.3.7. Let $f : I \to \mathbb{R}$. Let $x_0 \in I$ and recall that, for each $\delta > 0$, $\dot{B}_\delta(x_0) := (x_0 - \delta, x_0 + \delta) \setminus \{x_0\}$ is a *deleted* neighborhood of x_0.

(Upper and Lower Limits). The *upper* and *lower* limits of f at x_0 are (the extended real numbers) defined, respectively, by

$$\limsup_{x \to x_0} f(x) := \inf \left\{ \sup\{f(x) : x \in \dot{B}_\delta(x_0) \cap I\} : \delta > 0 \right\} \quad \text{and}$$

$$\liminf_{x \to x_0} f(x) := \sup \left\{ \inf\{f(x) : x \in \dot{B}_\delta(x_0) \cap I\} : \delta > 0 \right\}.$$

One also uses $\overline{\lim}_{x \to x_0} f(x)$ and $\underline{\lim}_{x \to x_0} f(x)$, respectively, to denote the upper and lower limits of f at x_0.

(Upper and Lower Envelopes).   The *upper* and *lower* envelopes of f are the (extended real-valued) functions $\overline{f}$ and $\underline{f}$ defined, respectively, by

$$\overline{f}(x_0) := \inf\left\{\sup\{f(x) : x \in B_\delta(x_0) \cap I\} : \delta > 0\right\}.$$
$$\underline{f}(x_0) := \sup\left\{\inf\{f(x) : x \in B_\delta(x_0) \cap I\} : \delta > 0\right\}.$$

(Oscillation).   The *oscillation* of f *at* x_0 is defined by

$$\omega_f(x_0) := \overline{f}(x_0) - \underline{f}(x_0).$$

Exercise 4.3.8.

(a) Let $f \in C(a,b)$. Show that, for any $y_0 \in \mathbb{R}$, the set $\{x \in (a,b) : f(x) \neq y_0\}$ is *open*. In particular, the set $\{x \in (a,b) : f(x) \neq 0\}$ of all points at which f *does not vanish* is open.

(b) Let $f \in C[a,b]$. Show that, for any $y_0 \in \mathbb{R}$, the set $f^{-1}(y_0) := \{x \in [a,b] : f(x) = y_0\}$ is *closed*. In particular, the set $Z_f = \{z \in [a,b] : f(z) = 0\}$ of all *zeros* of f is closed.

(c) Let $f : I \to \mathbb{R}$. Show that $\liminf_{x \to x_0} f(x) \leq \limsup_{x \to x_0} f(x)$, for all $x_0 \in I$, with equality (for $\limsup_{x \to x_0} f(x) \neq \pm\infty$) if and only if $\lim_{x \to x_0} f(x)$ exists and, in this case, $\liminf_{x \to x_0} f(x) = \lim_{x \to x_0} f(x) = \limsup_{x \to x_0} f(x)$.

(d) Let $f : I \to \mathbb{R}$. Show that $\underline{f}(x) \leq f(x) \leq \overline{f}(x)$ for all $x \in I$ and that f is *continuous* at $x_0 \in I$ if and only if $\omega_f(x_0) = 0$.

Finally, in view of Theorem 3.3.1, our definition of continuity is also equivalent to the following *sequential* version:

Theorem 4.3.9. *Let* $f : I \to \mathbb{R}$ *and let* $x_0 \in I$. *Then* f *is continuous at* x_0 *if and only if* $\lim_{n \to \infty}(f(x_n)) = f(x_0)$ *for all sequences* $(x_n) \in I^{\mathbb{N}}$ *satisfying* $\lim(x_n) = x_0$.

Definition 4.3.10 (Cauchy's Functional Equation). A function $f : \mathbb{R} \to \mathbb{R}$ is said to be *additive*, or to satisfy *Cauchy's functional equation*, if

$$f(s+t) = f(s) + f(t) \qquad (\forall s,\, t \in \mathbb{R}). \tag{†}$$

Theorem 4.3.11. *If* f *satisfies Cauchy's functional equation and is continuous at a point* $x_0 \in \mathbb{R}$, *then it is continuous on* $\mathbb{R}$ *and is linear; i.e., we have* $f(x) = ax$, *for all* $x \in \mathbb{R}$ *and a constant* $a \in \mathbb{R}$.

Proof. First note that $f(0) = f(0+0) = f(0) + f(0)$ and hence $f(0) = 0$. Now, for any $x \in \mathbb{R}$, we have $0 = f(x - x) = f(x) + f(-x)$, so that $f(-x) = -f(x)$. From the continuity of f at x_0, it follows that

$$\lim_{x \to 0} f(x) = \lim_{x \to x_0} f(x - x_0) = \lim_{x \to x_0} [f(x) - f(x_0)] = 0.$$

Therefore, f is continuous at $x = 0$. But then, given *any* $x \in \mathbb{R}$, we have

$$\lim_{h \to 0} f(x + h) = f(x) + \lim_{h \to 0} f(h) = f(x),$$

and f is indeed continuous on $\mathbb{R}$. On the other hand, it follows from (†) and a simple inductive argument that, $\forall x_1, \ldots, x_n \in \mathbb{R}$,

$$f(x_1 + x_2 + \cdots + x_n) = f(x_1) + f(x_2) + \cdots + f(x_n). \tag{‡}$$

Thus, taking $x_1 = \cdots = x_n = x$ in (‡), we get $f(nx) = nf(x)$, for all $x \in \mathbb{R}$ and all $n \in \mathbb{N}$. In particular, we have $f(n) = nf(1)$, for all $n \in \mathbb{N}$. Since $f(-n) = -f(n)$, we actually have $f(n) = nf(1)$, for all $n \in \mathbb{Z}$. If $m, \ n \in \mathbb{Z}$ and $n \neq 0$, then $mf(1) = nf(m/n)$ implies that $f(m/n) = (m/n)f(1)$. Therefore, we have $f(r) = rf(1)$, for all $r \in \mathbb{Q}$. Summing up, we have proved that

$$f(r) = ar \qquad (\forall r \in \mathbb{Q}), \tag{$*$}$$

where $a := f(1)$. Finally, if $x \notin \mathbb{Q}$, pick a sequence $(r_n) \in \mathbb{Q}^{\mathbb{N}}$ with $\lim_{n \to \infty} r_n = x$. The continuity of f at x and ($*$) now imply that

$$f(x) = \lim_{n \to \infty} f(r_n) = \lim_{n \to \infty} ar_n = ax,$$

and the proof is complete. □

Exercise 4.3.12 (Dirichlet Function). Show that the *Dirichlet function* f, defined by

$$f(x) = \begin{cases} 1 & \text{if } x \in \mathbb{Q}, \\ 0 & \text{if } x \notin \mathbb{Q}, \end{cases}$$

is *discontinuous at every* $x \in \mathbb{R}$. *Hint:* Use the sequential definition of continuity and the fact that both $\mathbb{Q}$ and $\mathbb{Q}^c$ are *dense* in $\mathbb{R}$.

Example 4.3.13 (Cantor's Ternary Function). Given a point $x \in [0, 1]$ with $x = (0.x_1 x_2 x_3 \ldots)_3$, define $N := +\infty$ if $x_n \neq 1 \ \forall n \in \mathbb{N}$, and $N := \min\{n : x_n = 1\}$ otherwise. Let $y_n = x_n/2$ for $n < N$ and let $y_N = 1$. Now define *Cantor's ternary function* $\kappa : [0, 1] \to [0, 1]$ by setting

$$\kappa(x) := \sum_{n=1}^{N} \frac{y_n}{2^n}.$$

Claim 1: κ is *well defined*; i.e., $\sum_{n=1}^{N} y_n/2^n$ is *independent* of the ternary expansion of x if x has *two* such expansions.

Claim 2: κ is a *monotone, continuous* function from $[0, 1]$ *onto* $[0, 1]$. In particular, the function $\phi : C \to [0, 1]$ defined in the proof of Theorem 4.2.9 is continuous.

Claim 3: κ is *constant* on each interval contained in the *complement* of the Cantor set C.

To prove the first claim, note that x has two ternary expansions if and only if it is an endpoint of a removed middle third. In this case, we either have

$$x = (0.x_1 x_2 \cdots x_{N-1} 1 \bar{0})_3 = (0.x_1 x_2 \cdots x_{N-1} 0 \bar{2})_3,$$

where $x_j = 0$ or 2 for $1 \leq j \leq N - 1$ and $\bar{0}$ and $\bar{2}$ indicate (infinite) strings of 0's and 2's, respectively, in which case we obtain the same value

$$\kappa(x) = \left(0.\frac{x_1}{2}\frac{x_2}{2} \cdots \frac{x_{N-1}}{2} 1 \bar{0}\right)_2 = \left(0.\frac{x_1}{2}\frac{x_2}{2} \cdots \frac{x_{N-1}}{2} 0 \bar{1}\right)_2,$$

or we have

$$x = (0.x_1 x_2 \cdots x_{N-1} 2 \bar{0})_3 = (0.x_1 x_2 \cdots x_{N-1} 1 \bar{2})_3,$$

in which case we have the obviously unique value

$$\kappa(x) = \left(0.\frac{x_1}{2}\frac{x_2}{2} \cdots \frac{x_{N-1}}{2} 1 \bar{0}\right)_2 = \left(0.\frac{x_1}{2}\frac{x_2}{2} \cdots \frac{x_{N-1}}{2} 1 \bar{0}\right)_2.$$

For the second claim, let $x = (0.x_1 x_2 \cdots)_3$ and $x' = (0.x'_1 x'_2 \cdots)_3$ be two points in $[0, 1]$ with $x < x'$ and let $N_x \leq \infty$ and $N_{x'} \leq \infty$ be as above. If m is the *smallest* index with $x_m \neq x'_m$ then $x_m < x'_m$ and hence $y_m \leq y'_m$ so that $\kappa(x) = \sum_{k=1}^{N_x} y_k / 2^k \leq \sum_{k=1}^{N_{x'}} y'_k / 2^k = \kappa(x')$. Also, by Theorem 4.2.9, ϕ is onto $[0, 1]$ and hence so is κ. To prove the continuity, let $\varepsilon > 0$ be given and pick ℓ so that $1/2^\ell \leq \varepsilon$. If x, $x' \in [0, 1]$ satisfy $|x - x'| < 1/3^\ell$, then we can pick ternary expansions $x = (0.x_1 x_2 \cdots)_3$ and $x' = (0.x'_1 x'_2 \cdots)_3$ with $x_k = x'_k$ for $1 \leq k \leq \ell$. It follows that the first ℓ digits of the binary expansions of $\kappa(x)$ and $\kappa(x')$ are equal and hence $|\kappa(x) - \kappa(x')| \leq 1/2^{\ell+1} < \varepsilon$. In fact, as we shall see (cf. Theorem 4.4.7), since κ is *increasing*, it can only have *jump* discontinuities. However, being onto, κ satisfies the *Intermediate Value Property* and hence cannot have jump discontinuities (cf. Theorem 4.5.16) and must indeed be *continuous*.

Finally, to prove the third claim, note that if $x = (0.x_1 x_2 \cdots)_3$ and $x' = (0.x'_1 x'_2 \cdots)_3$ belong to the same middle third in the complement $[0, 1] \setminus C$, then the smallest index m with $x_m = 1$ is also the smallest with $x'_m = 1$ and hence $\kappa(x) = \kappa(x')$; i.e., κ is *constant* on all such middle thirds.

Definition 4.3.14 (Period, Periodic Function). Let $f : \mathbb{R} \to \mathbb{R}$. A number $p \in \mathbb{R}$ is said to be a *period* of f if $f(x + p) = f(x)$ for all $x \in \mathbb{R}$. Let P denote the set of all periods of f. Since $0 \in P$, we have $P \neq \emptyset$. We say that f is *periodic* if $P \neq \{0\}$.

Proposition 4.3.15. *The set P of all periods of a function f : $\mathbb{R} \to \mathbb{R}$ is a subgroup of $\mathbb{R}$, i.e., (i) $0 \in P$, (ii) $-p \in P$ whenever $p \in P$, and (iii) $p + q \in P$ whenever $p, q \in P$. In particular, for each $p \in P$, we have*

$$p\mathbb{Z} := \{kp : k \in \mathbb{Z}\} \subset P.$$

Proof. Exercise! □

Theorem 4.3.16 (Continuity and Period). *Let $f \in C(\mathbb{R})$. Then the set P of all periods of f is a closed subgroup of $\mathbb{R}$. In fact, either $P = \mathbb{R}$ or $P = \{0\}$ or $P = a\mathbb{Z}$ for some $a > 0$, where, as in Proposition 4.3.15, $a\mathbb{Z} := \{ka : k \in \mathbb{Z}\}$.*

Proof. By Proposition 4.3.15, P is a subgroup of $\mathbb{R}$. For each $x \in \mathbb{R}$, set

$$F_x := \{t \in \mathbb{R} : f(x + t) = f(x)\}.$$

Since the function $f_x : t \mapsto f(x + t)$ is *continuous* on $\mathbb{R}$ (why?) and $\{f(x)\}$ is *closed* in $\mathbb{R}$, the set $F_x = f_x^{-1}(\{f(x)\})$ is *closed* as well (Corollary 4.3.6). Therefore

$$P = \bigcap_{x \in \mathbb{R}} F_x$$

is also closed as claimed. Now, if 0 is a *limit point* of P, then for each $\varepsilon > 0$ there exists $p = p_\varepsilon \in P$ with $0 < |p| < \varepsilon$. This implies that, for any (nonempty) open interval $I \subset \mathbb{R}$ of length $> \varepsilon$, we have $kp \in I$ for some $k \in \mathbb{Z}$. (Why?) It follows that P is *dense* in $\mathbb{R}$ and, since it is *closed*, we must have $P = \mathbb{R}$. If 0 is an *isolated point* of P, i.e., $P \cap (-\varepsilon, \varepsilon) = \{0\}$ for some $\varepsilon > 0$, then $P \cap (p - \varepsilon, p + \varepsilon) = \{p\}$ for every $p \in P$ (why?) and hence P is *discrete;* i.e., every point in P is isolated. But then, since P is closed, $P \cap [-c, c]$ is *compact and discrete* and hence *finite*, for any $c > 0$. (Why?) Therefore, either $P = \{0\}$ or the set P^+ of all *positive* elements of P has a *smallest* element, say a. By Proposition 4.3.15, we have $a\mathbb{Z} \subset P$. If $P \setminus a\mathbb{Z} \neq \emptyset$, then we can find $p \in P$ and $n \in \mathbb{Z}$ such that $a(n - 1) < p < an$. But then $an - p \in P$ and we have $0 < an - p < a$, contradicting the choice of a. This shows that $P = a\mathbb{Z}$ and completes the proof. □

Corollary 4.3.17 (Continuous Periodic Function). *If $f \in C(\mathbb{R})$ is periodic, then either f is constant or it has a smallest positive period, p. This (unique) period $p > 0$ is then said to be "the" period of f.*

Proof. Since f is *periodic*, we have $P \neq \{0\}$. Therefore, the corollary follows at once from Theorem 4.3.16 if we note that $P = \mathbb{R}$ implies that f is *constant*. □

The next theorem, which shows that sums, differences, constant multiples, products, and ratios of continuous functions are continuous, is an immediate consequence of Theorem 3.3.3.

Theorem 4.3.18. *If f, $g : I \to \mathbb{R}$ are continuous at $x_0 \in I$, then the functions $f \pm g$, cf (where $c \in \mathbb{R}$ is any constant), and fg are continuous at x_0. If, in addition, $g(x_0) \neq 0$ (resp., $g(x_0) > 0$, $g(x_0) < 0$), then $g(x) \neq 0$ (resp., $g(x) > 0$, $g(x) < 0$) is true near x_0 and the function f/g is also continuous at x_0.*

Also, the continuity of composites of continuous functions follows at once from Theorem 3.3.14:

Theorem 4.3.19. *Let $f : I \to J$ and $g : J \to \mathbb{R}$. Assume that f is continuous at $x_0 \in I$ and that g is continuous at $y_0 = f(x_0) \in J$. Then $g \circ f : I \to \mathbb{R}$ is continuous at x_0.*

Example 4.3.20.

(a) Let $c \in \mathbb{R}$ be arbitrary. Then the *constant function* $x \mapsto c$ is continuous on $\mathbb{R}$, as follows at once from the definition.
(b) Since $\lim_{x \to x_0} x = x_0$, the *identity function* $x \mapsto x$ is continuous on $\mathbb{R}$.
(c) Examples (a) and (b) and Theorem 4.3.18 imply that any polynomial function $x \mapsto p(x) = a_0 + a_1 x + \cdots + a_n x^n$ is continuous on $\mathbb{R}$. Also, if $x \mapsto q(x) = b_0 + b_1 x + \cdots + b_m x^m$ is any other (*nonzero*) polynomial function, then the *rational function* $x \mapsto p(x)/q(x)$ is continuous at all x_0 such that $q(x_0) \neq 0$.
(d) The functions sin, cos, and $x \mapsto e^x$ are continuous on $\mathbb{R}$. Also, $x \mapsto \log x$ is continuous on $(0, \infty)$. These facts will be proved later when the precise definitions of these functions are given (in terms of power series). Note, however, that using the inequalities $|\sin \theta| \leq |\theta|$ and $|\cos \theta| \leq 1 \; \forall \theta \in \mathbb{R}$ and the identity

$$\sin x - \sin x_0 = 2 \cos \frac{x + x_0}{2} \cdot \sin \frac{x - x_0}{2},$$

we get, for any $\varepsilon > 0$, the implication

$$|x - x_0| < \varepsilon \implies |\sin x - \sin x_0| \leq |x - x_0| < \varepsilon,$$

which proves that sin is continuous on $\mathbb{R}$. A similar argument (or the identity $\cos x = \sin(\pi/2 - x)$, $\forall x \in \mathbb{R}$ which shows that $\cos x$ is the composite of the continuous functions sin and $x \mapsto \pi/2 - x$) implies that cos is also continuous on $\mathbb{R}$.
(e) Let us define the function $f : \mathbb{R} \to \mathbb{R}$ by

$$f(x) = \begin{cases} x \sin(1/x) & \text{if } x \neq 0, \\ 0 & \text{if } x = 0. \end{cases}$$

Then f is continuous on $\mathbb{R}$. Indeed, $x \mapsto \sin x$ is continuous on $\mathbb{R}$ and $x \mapsto 1/x$ is continuous on $\mathbb{R} \setminus \{0\}$. Thus, f is certainly continuous on $\mathbb{R} \setminus \{0\}$. On the other hand, we have already seen (by *squeezing*) that $\lim_{x \to 0} x \sin(1/x) = 0 = f(0)$. This proves the continuity at the remaining point $x = 0$.

Exercise 4.3.21. Let f, $g : I \to \mathbb{R}$ be continuous at $x_0 \in I$.

(a) Show that $|f|$ is continuous at x_0.
(b) Show that, if $f(x) \geq 0 \; \forall x \in I$, then $\sqrt{f}$ is continuous at x_0.
(c) Show that the functions $f \vee g$ and $f \wedge g$ defined by $(f \vee g)(x) = f(x) \vee g(x) :=$ $\max\{f(x), g(x)\}$ and $(f \wedge g)(x) = f(x) \wedge g(x) := \min\{f(x), g(x)\}$ are continuous at x_0.

Hints: For parts (a) and (b), first prove the continuity of $x \mapsto |x|$ and $x \mapsto \sqrt{x}$, and then use Theorem 4.3.19. For (c), note that $\forall a, \; b \in \mathbb{R}$, we have $a \vee b = (a + b + |a - b|)/2$ and $a \wedge b = (a + b - |a - b|)/2$, and use part (a).

4.4   One-Sided Continuity, Discontinuity, and Monotonicity

By definition, a function f (defined at x_0) is continuous at x_0 if it *commutes* with "$\lim_{x \to x_0}$." If this limit is replaced by a *one-sided* limit, then we obtain the following definition of *one-sided* continuity which is a convenient tool for the study of *discontinuities*.

Definition 4.4.1 (Left Continuous, Right Continuous). Let $f : I \to \mathbb{R}$, and let $x_0 \in I$. If x_0 is not the *left* (resp., *right*) endpoint of I, we say that f is *left continuous* (resp., *right continuous*) at x_0 if $f(x_0 - 0) := \lim_{x \to x_0-} f(x) = f(x_0)$ (resp., $f(x_0 + 0) := \lim_{x \to x_0+} f(x) = f(x_0)$).

Remark 4.4.2.

(a) It is obvious that, if x_0 is the right (resp., left) endpoint of I, then "f is *left* (resp., *right*) *continuous* at x_0" simply means that f is *continuous* at x_0.
(b) We can also give an (ε, δ)-definition of left (resp., right) continuity at x_0. For example, if x_0 is *not* the left endpoint of I, then "f is left continuous at x_0" means that

$$(\forall \varepsilon > 0)(\exists \delta = \delta(\varepsilon, x_0) > 0)(\forall x \in I)(x_0 - \delta < x < x_0 \Rightarrow |f(x) - f(x_0)| < \varepsilon).$$

A similar definition can be given (hopefully by the reader!) for the right continuity at x_0 (if x_0 is *not* the right endpoint of I).
(c) The reader can also provide the corresponding (equivalent) *sequential* definitions of the above one-sided continuities.

The following theorem is an immediate consequence of Theorem 3.4.6.

Theorem 4.4.3. *Let $f : I \to \mathbb{R}$, and let $x_0 \in I^\circ$ be an interior point of I. Then f is continuous at x_0 if and only if it is both left and right continuous at x_0, i.e., if and only if*

$$f(x_0 - 0) = f(x_0) = f(x_0 + 0).$$

If a function f is *undefined* at x_0, then it obviously cannot be continuous at x_0. Since, however, one can always *extend* the domain of f so that it includes x_0 [simply by assigning *any* value to $f(x_0)$], we shall only consider the discontinuities of a function *on its domain*. Using Theorem 4.4.3, we can classify the most common *discontinuities* of a function $f : I \to \mathbb{R}$ as follows:

Definition 4.4.4 (Removable, Jump, and Infinite Discontinuities). Let $f : I \to \mathbb{R}$, and let $x_0 \in I$.

1. If $\lim_{x \to x_0} f(x) = y_0 \in \mathbb{R}$ and $y_0 \neq f(x_0)$, then we say that f has a *removable discontinuity* at x_0
2. If x_0 is an *interior* point of I and $f(x_0 - 0)$ and $f(x_0 + 0)$ both exist (i.e., are *finite*) but $f(x_0 - 0) \neq f(x_0 + 0)$, then we say that f has a *jump discontinuity* at x_0. In this case, the difference

$$f(x_0 + 0) - f(x_0 - 0)$$

is called the *jump* of f at x_0.
3. If $x = x_0$ is a *vertical asymptote* of f (and note that $f(x_0)$ is still assumed to be *defined*), then we say that f has an *infinite discontinuity* at x_0.

Remark 4.4.5.

(a) Let $f : I \to \mathbb{R}$, and let $x_0 \in I$. Suppose that f has a *removable* discontinuity at x_0; i.e., assume that $\lim_{x \to x_0} f(x) = y_0 \in \mathbb{R}$, but $y_0 \neq f(x_0)$. Define the function $\tilde{f} : I \to \mathbb{R}$ by

$$\tilde{f}(x) = \begin{cases} f(x) & \text{if } x \in I \setminus \{x_0\}, \\ y_0 & \text{if } x = x_0. \end{cases}$$

Then $\tilde{f}$ is *continuous at x_0*. In fact, even if the original f is *undefined* at x_0, the new function $\tilde{f}$ is an *extension* of f that is continuous at x_0. We can, therefore, *remove* the discontinuity of f at x_0.
(b) Jump discontinuities are also called *discontinuities of the first kind*.
(c) If at least one of the one-sided limits $\lim_{x \to x_0\pm} f(x)$ *does not exist* as a real number, then the discontinuity of f at x_0 is called a *discontinuity of the second kind*. Thus, infinite discontinuities are of the second kind. Note, however, that $x \mapsto \sin(1/x)$, which is *bounded*, has a discontinuity of the second kind at 0. Indeed, *neither* of the one-sided limits $\lim_{x \to 0\pm} \sin(1/x)$ exists, as we saw in Chap. 3. Finally, note that the *Dirichlet function* (cf. Exercise 4.3.12) has a discontinuity of the second kind at *every* $x \in \mathbb{R}$.

Example 4.4.6 (Greatest Integer Function). For each $x \in \mathbb{R}$ let $[x] = \max\{n \in \mathbb{Z} : n \leq x\}$ be the *greatest integer $\leq x$*. The function $x \mapsto [x]$ is called the *greatest integer function*. This function is continuous on $\mathbb{R} \setminus \mathbb{Z}$. It is, however, *right continuous* at each *integer* $n \in \mathbb{Z}$ but *not left continuous*. Indeed, $[x] = n \ \forall x \in$

$[n, n+1)$, so that $\lim_{x \to n+}[x] = n = [n]$, whereas $\lim_{x \to n-}[x] = n - 1$. It follows that $[x]$ has a jump of *one unit* at each integer: $[n + 0] - [n - 0] = n - (n - 1) = 1$ $\forall n \in \mathbb{Z}$. The greatest integer function is a *step function with an infinite number of steps*, i.e., a function whose domain is a union of (an infinite number of) intervals on each of which the function assumes a *constant* value.

The class of *monotone functions* plays an important role in analysis and will be examined on several occasions. One remarkable feature of monotone functions is that they have no discontinuities of the "second kind" and, in fact, are continuous at "almost all" points of their domains. The precise statements of these facts are summarized in the following.

Theorem 4.4.7. *A monotone function* $f : I \to \mathbb{R}$ *can only have jump discontinuities (i.e., discontinuities of the first kind). Moreover, the set* D *of all points* $x_0 \in I$ *at which* f *is discontinuous is countable.*

Proof. Recall that f has a *discontinuity of the second kind* at an interior point $x_0 \in I$ if at least one of the one-sided limits $f(x_0 \pm 0)$ does *not* exist. Now, by Theorem 3.4.36 (*Monotone Limit Theorem*), this does not happen for monotone functions. The same theorem also shows that, if at an interior point $x_0 \in I^\circ$ we have $f(x_0 - 0) = f(x_0 + 0)$, then this common value is necessarily $f(x_0)$ (why?) and f is therefore continuous at x_0. To prove the last statement, let us assume that f is *increasing*; the decreasing case may be treated by a change of f to $-f$. Now, for each $x \in D$, pick a rational number $r(x) \in \mathbb{Q}$ such that

$$f(x - 0) < r(x) < f(x + 0).$$

Since (by Theorem 3.4.36) $x_1 < x_2$ implies $f(x_1 + 0) \le f(x_2 - 0)$, we have $r(x) \ne r(y)$, $\forall x \ne y$, and the map $r : x \to r(x)$ is a *one-to-one* function from D to $\mathbb{Q}$. The countability of $\mathbb{Q}$ now implies that D is countable. $\square$

Exercise 4.4.8. Let $f : \mathbb{R} \to \mathbb{R}$ be a *monotone* function satisfying Cauchy's functional equation:

$$f(s + t) = f(s) + f(t) \qquad (\forall s, t \in \mathbb{R}).$$

Show that f is *linear*, i.e., $f(x) = ax$ for all $x \in \mathbb{R}$ and some constant $a \in \mathbb{R}$. *Hint:* cf. Theorem 4.3.11.

Remark 4.4.9.

(a) Since D is a *countable* subset of I, we can write $D = \{x_1, x_2, \ldots\}$. Now, assuming f is increasing, to each $x_n \in D$ we associate the interval $J_{x_n}(f) := (f(x_n - 0), f(x_n + 0))$ and note that, with λ denoting the *length*, $\lambda(J_{x_n}(f)) = f(x_n + 0) - f(x_n - 0)$ is precisely the *jump* of f at x_n. If $x_m < x_n$, then it follows from $f(x_m + 0) \le f(x_n - 0)$ that $J_{x_m}(f) \cap J_{x_n}(f) = \emptyset$. Therefore, $\{J_{x_n}(f)\}_{x_n \in D}$ is a countable collection of *pairwise disjoint* open subintervals of the range $f(I)$.

(b) Note that, in Theorem 4.4.7, the points of D are *not* necessarily *isolated*. In fact, D may even be *dense*, as the following construction of *jump functions* will show.

Definition 4.4.10 (Jump Function). Let I be any interval, $(x_n) \in I^{\mathbb{N}}$ any sequence, and $(h_n) \in (0, \infty)^{\mathbb{N}}$ any sequence of *positive* numbers such that

$$\sum_{n=1}^{\infty} h_n < \infty.$$

The function $f : I \to (0, \infty)$ defined by

$$f(x) := \sum_{x_n < x} h_n \quad \forall x \in I, \tag{†}$$

where the sum is over all n for which $x_n < x$, is called a *jump function*. Note that the sum in (†) is defined to be 0 if $x_n \geq x$ for all $n \in \mathbb{N}$.

Remark 4.4.11.

(a) Since the series $\sum_{n=1}^{\infty} h_n$ converges *absolutely*, its sum is unaffected by any rearrangement of the terms.
(b) If, in the above definition, we assume that $x_1 < x_2 < x_3 < \cdots$, then our jump function is a "step function" with infinite number of steps.

The following theorem shows that the jump function (†) is *left continuous* on its domain and has a jump equal to h_n at $x_n \; \forall n \in \mathbb{N}$. It also shows that a left-continuous *increasing* function is "continuous modulo a jump function":

Theorem 4.4.12.

(a) *Given any sequence $(x_n) \in I^{\mathbb{N}}$ in I and any sequence $(h_n) \in (0, \infty)^{\mathbb{N}}$ of positive numbers satisfying $\sum_{n=1}^{\infty} h_n < \infty$, let $f : I \to (0, \infty)$ be the corresponding jump function (†). Then f is increasing, continuous on the set $I \setminus \{x_1, x_2, \ldots\}$, and left continuous on I. Furthermore, f has jump discontinuities at $x_n \; \forall n \in \mathbb{N}$ with the jump at x_n equal to h_n.*
(b) *Let $f : I \to \mathbb{R}$ be any bounded, left continuous, increasing function. Then $f = \phi + \psi$, where $\phi : I \to \mathbb{R}$ is a continuous, increasing function and $\psi : I \to \mathbb{R}$ is a jump function.*

Proof. That f is *increasing* is clear from its definition. Let $X := \{x_1, x_2, \ldots\}$ and suppose that $x \in I \setminus X$. If x is *not* a limit point of X, then $B_\delta(x) \cap X = \emptyset$ for some $\delta > 0$. Thus, f is *constant* on $I \cap B_\delta(x)$ and hence continuous at x. If, however, x is a limit point of X, then for any $\varepsilon > 0$ let $N \in \mathbb{N}$ be such that $\sum_{n>N} h_n < \varepsilon$. Pick $\delta > 0$ such that

$$B_\delta(x) \cap \{x_1, x_2, \ldots, x_N\} = \emptyset.$$

It is then obvious that for any $y \in I \cap B_\delta(x)$, we have

$$|f(y) - f(x)| \le \sum_{n > N} h_n < \varepsilon.$$

Therefore, f is *continuous* at every $x \in I \setminus X$. Exactly the same argument with $B_\delta(x)$ replaced by $(x - \delta, x)$ shows that, if x is *not* the left endpoint of I, then f is *left continuous* at x, i.e., $f(x - 0) = f(x)$ (even if $x \in X$). On the other hand, f is *not right continuous* at any $x_n \in X$. Indeed, if x_n is *not* the right endpoint of I, we clearly have the inequalities

$$f(x_n + \delta) - f(x_n) \ge h_n \qquad\qquad (*)$$

for all $n \in \mathbb{N}$ and all *sufficiently small* $\delta > 0$. Finally, we show that

$$f(x_n + 0) - f(x_n - 0) = f(x_n + 0) - f(x_n) = h_n \quad (\forall n \in \mathbb{N}), \qquad (**)$$

assuming that the x_n are all *interior points* of I. The case where some x_n is an endpoint is treated similarly and will be left to the reader. Now, if for some $\delta > 0$ we have $(x_n, x_n + \delta) \cap X = \emptyset$, then $(**)$ follows at once. (Why?) Assume, then, that $(x_n, x_n + \delta) \cap X$ is *infinite* for every $\delta > 0$. For each $\varepsilon > 0$ pick $N > n$ such that $\sum_{k > N} h_k < \varepsilon$. If $\delta > 0$ is chosen so that

$$(x_n, x_n + \delta) \cap \{x_1, x_2, \ldots, x_N\} = \emptyset,$$

then $(*)$ implies that, for every $x \in (x_n, x_n + \delta)$, we have

$$h_n \le f(x) - f(x_n) < h_n + \varepsilon.$$

Since ε was arbitrary, $(**)$ follows in this case as well.

To prove (b), note that f is *increasing* and hence has a countable number of jump discontinuities, say at $x_1, x_2, \ldots$, with *(positive) jumps*

$$h_n := f(x_n + 0) - f(x_n - 0), \quad (n = 1, 2, \ldots).$$

Also, the *boundedness* of f implies that

$$\sum_{n=1}^{\infty} h_n \le \sup(f) - \inf(f) < \infty.$$

Now, for each $x \in I$, we define $\psi(x)$ and $\phi(x)$ to be

$$\psi(x) := \sum_{x_n < x} h_n, \qquad \phi(x) := f(x) - \psi(x).$$

For any x, $y \in I$, $x < y$, we have

$$\phi(y) - \phi(x) = (f(y) - f(x)) - (\psi(y) - \psi(x)) \geq 0,$$

for the *net change* $f(y) - f(x)$ of the increasing function f on $[x, y]$ is at least as large as the *sum of its jumps* in the same interval. Thus, ϕ is increasing. To prove the continuity of ϕ, let $x \in I^\circ$. If $x \notin X := \{x_1, x_2, \ldots\}$, then f and ψ are both *continuous* at x and

$$\phi(x+0) - \phi(x-0) = (f(x+0) - f(x-0)) - (\psi(x+0) - \psi(x-0)) = 0 - 0 = 0,$$

while if $x = x_n$ for some n, then f and ψ have the same jump h_n at x_n and we still have

$$\phi(x+0) - \phi(x-0) = (f(x+0) - f(x-0)) - (\psi(x+0) - \psi(x-0)) = h_n - h_n = 0.$$

Thus, ϕ is continuous on I°. The continuity at the endpoints is proved similarly. $\square$

We end this section by showing that, for a *continuous* function to be monotone (say, *increasing*) on an interval, the condition "$x < y \Rightarrow f(x) \leq f(y)$" can be replaced by a much weaker one:

Proposition 4.4.13. *Let D be a countable subset of an interval $I \in \mathbb{R}$ and let $f : I \to \mathbb{R}$ be continuous. If for every interior point $x \in I^\circ$ with $x \notin D$ and every $\delta > 0$ there exists $y \in (x, x + \delta)$ such that $f(x) \leq f(y)$, then f is increasing on I.*

Proof. Let s, $t \in I$ with $s < t$. We claim that the following holds.

$$\eta < f(s) \quad \text{and} \quad \eta \notin f(D) \quad \Longrightarrow \quad \eta \leq f(t). \tag{†}$$

Define $S := \{x \in [s, t] : \eta \leq f(x)\}$ and note that $s \in S$ so that $S \neq \emptyset$. Also, since f is *continuous*, $S = [s, t] \cap f^{-1}([\eta, \infty))$ is *closed* (cf. Corollary 4.3.6). Therefore, $\xi := \sup(S) \in S$. We claim that $\xi = t$. Suppose that $\xi < t$ and note that we cannot have $\eta < f(\xi)$ because, f being continuous, this would imply that ξ is an *interior* point of S, which is impossible as $\xi = \sup(S)$. (Why?) Therefore, $\eta = f(\xi)$. But then $\xi \notin D$ because $\eta = f(\xi) \notin f(D)$. By assumption, we can then find a point $\zeta \in (\xi, t)$ with $\eta = f(\xi) \leq f(\zeta)$. This, however, implies that $\zeta \in S$, contradicting $\xi = \sup(S)$. Therefore, $\xi = t$ and hence $\eta \leq f(t)$, establishing (†). Since $f(D)$ is *countable*, there are numbers $\eta \notin f(D)$ arbitrarily close to $f(s)$ and hence we have $f(s) \leq f(t)$. $\square$

4.5 Extreme Value and Intermediate Value Theorems

We shall now look at the (topological) properties of continuous functions. The two most important properties are that continuous functions map compact sets onto compact sets and connected sets onto connected ones (i.e., intervals onto intervals).

Theorem 4.5.1 (Continuity and Compactness). *Let $f \in C(I)$. If K is compact and $K \subset I$, then the (direct) image $f(K)$ is also compact.*

Proof. Suppose K is a compact subset of I, and let $\{O'_\lambda\}_{\lambda \in \Lambda}$ be an open cover of $f(K)$. Then the continuity of f implies that for each $\lambda \in \Lambda$, we have $f^{-1}(O'_\lambda) = I \cap O_\lambda$ for some open set O_λ. Thus, the collection $\{O_\lambda\}_{\lambda \in \Lambda}$ is an open cover of K (why?). Since K is compact, it can be covered by a *finite* subcollection:

$$K \subset O_{\lambda_1} \cup O_{\lambda_2} \cup \cdots \cup O_{\lambda_n}.$$

It then follows that

$$f(K) \subset O'_{\lambda_1} \cup O'_{\lambda_2} \cup \cdots \cup O'_{\lambda_n}.$$

Since $\{O'_\lambda\}_{\lambda \in \Lambda}$ was an *arbitrary* open cover of $f(K)$, the proof is complete. $\qquad\square$

The following theorem is an immediate corollary of Theorem 4.5.1. It states the important fact that a continuous function on a compact set K assumes its *(absolute) maximum* and *minimum* values in K.

Theorem 4.5.2 (Weierstrass's Extreme Value Theorem). *Let $f \in C(I)$ and let $K \subset I$ be a compact set. Then f is bounded on K and there are $\alpha, \beta \in K$ such that $f(\alpha) = \inf\{f(x) : x \in K\}$ and $f(\beta) = \sup\{f(x) : x \in K\}$. In particular, if $f \in C[a,b]$, then, for some $\alpha, \beta \in [a,b]$, we have $f(\alpha) = \min(f) = \inf(f)$ and $f(\beta) = \max(f) = \sup(f)$.*

Proof. Indeed, by Theorem 4.5.1, $f(K)$ is compact—hence *closed and bounded*—and, as such, contains both $\sup\{f(x) : x \in K\}$ and $\inf\{f(x) : x \in K\}$. $\qquad\square$

Remark 4.5.3. Note that the *compactness* of K and the *continuity* of the function f are both *necessary* for the conclusion of Theorem 4.5.2, as the following simple examples demonstrate:

(a) The function $f : [0, \infty) \to \mathbb{R}$ defined by $f(x) := x^2$ is certainly continuous but it is *not* bounded above. This is because the domain $[0, \infty)$ is unbounded and hence *not* compact.

(b) Define $f : [-1, 1] \to \mathbb{R}$ by

$$f(x) := \begin{cases} 1/x & \text{if } x \neq 0, \\ 0 & \text{if } x = 0. \end{cases}$$

Here, the domain $[-1, 1]$ is compact but f is *unbounded*. This is because f is *discontinuous* at 0.

(c) Consider the function $f : [0, 1) \to \mathbb{R}$ defined by $f(x) := x^3$. Here, f is continuous and bounded on the domain $[0, 1)$, but $\sup(f) = 1$ is *not attained* in the domain. The reason is that the interval $[0, 1)$ (which is bounded but *not* closed) is *not* compact.

The next fundamental result is the fact that continuous functions *preserve connectedness*. This is classically known as *Bolzano's Intermediate Value Theorem*. Since the connected subsets of $\mathbb{R}$ are the intervals and the latter have been characterized (cf. Proposition 2.1.41), the theorem states that once a continuous function assumes a pair of (distinct) values, it also assumes all the values *in between*. We first give the abstract version of the result and, although Bolzano's Theorem is an immediate corollary, we still include (for the sake of concreteness) a direct proof.

Theorem 4.5.4 (Continuity and Connectedness). *Let I be an open interval and let $f \in C(I)$. Then, for any connected set S with $S \subset I$, the (direct) image $f(S)$ is connected.*

Proof. If $f(S)$ is *disconnected*, then we can find open sets U, $V \subset \mathbb{R}$ such that $\{U \cap f(S), V \cap f(S)\}$ is a *partition* of $f(S)$. Since f is continuous on the *open* interval I, the inverse images $f^{-1}(U)$, $f^{-1}(V)$ are both *open*. Also, from $U \cap f(S) \neq \emptyset$, $V \cap f(S) \neq \emptyset$ we deduce that the sets $f^{-1}(U) \cap S$ and $f^{-1}(V) \cap S$ are *nonempty* and *disjoint*, the latter following from $(U \cap f(S)) \cap (V \cap f(S)) = \emptyset$. Finally, we have $S \subset f^{-1}(U) \cup f^{-1}(V)$. But this means that S is *disconnected*, which is absurd.   □

Definition 4.5.5 (Intermediate Value Property). Let $I \subset \mathbb{R}$ be any interval and let $J \subset I$ be any subinterval of I. A function $f : I \to \mathbb{R}$ is said to have the *Intermediate Value Property on J*, if for any a, $b \in J$ the function f takes on all the values *between* $f(a)$ and $f(b)$, i.e., in view of Proposition 2.1.41 (*Characterization of Intervals*), if $f(J)$ is an *interval*.

Theorem 4.5.6 (Bolzano's Intermediate Value Theorem). *Let $f \in C(I)$. Then f has the Intermediate Value Property on I. In other words, given any a, $b \in I$, $a < b$, and any number y_0 between $f(a)$ and $f(b)$, there is a number $x_0 \in [a, b]$ such that $f(x_0) = y_0$. In particular, the range $f(I)$ is an interval.*

Proof. As mentioned above, this is an immediate consequence of Theorem 4.5.4, but here is a direct proof: Suppose, for instance, that $f(a) \leq f(b)$, and let $S := \{x \in [a, b] : f(x) \leq y_0\}$. Since $a \in S$, we have $S \neq \emptyset$. Also, S is certainly bounded above by b. Let $x_0 := \sup(S)$. Then we have $a \leq x_0 \leq b$. (Why?) Now, if $f(x_0) < y_0$, then $f(a) \leq y_0 \leq f(b)$ implies that $x_0 < b$. Since f is *continuous* at x_0, we can find $x_1 \in (x_0, b)$ such that $f(x_1) < y_0$. Thus $x_1 \in S$. But then $x_1 > x_0$ contradicts the fact that $x_0 = \sup(S)$. Next, assume that $f(x_0) > y_0$. Then, using the continuity of f at x_0, we can find $x_2 \in (a, x_0)$ such that $f(x) > y_0$ for all $x \in [x_2, x_0]$. But this contradicts the fact that x_0 is the *least* upper bound of S. Thus, we must have $f(x_0) = y_0$. The last statement follows from Proposition 2.1.41 (*Characterization of Intervals*).   □

Combining Theorems 4.5.2 and 4.5.4, we conclude that a continuous function maps closed bounded intervals onto closed bounded intervals:

Theorem 4.5.7. *If* $f \in C[a,b]$, *then its range* $f([a,b])$ *is a closed, bounded interval.*

Proof. By Theorem 4.5.2, we can find α, $\beta \in [a,b]$ such that $f(\alpha) = m := \inf(f)$ and $f(\beta) = M := \sup(f)$. Now Theorem 4.5.4 implies that $[m,M] \subset f([a,b])$. Since $f([a,b]) \subset [m,M]$ obviously holds, we get $f([a,b]) = [m,M]$.   □

Definition 4.5.8 (Zero, Consecutive zeros). Given a function $f : I \to \mathbb{R}$, a number $z \in I$ is called a *zero* of f if $f(z) = 0$. Two zeros z_1, $z_2 \in I$, $z_1 < z_2$ of f are called *consecutive zeros* of f if $f(x) \neq 0$ for all $x \in (z_1, z_2)$.

The Intermediate Value Theorem is in fact equivalent to the following special case, which can be used to investigate the *zeros* of continuous functions:

Theorem 4.5.9 (Location of Zeros Theorem). *Let* $a < b$ *and* $f \in C[a,b]$. *If* $f(a)$ *and* $f(b)$ *have opposite signs (i.e., if* $f(a)f(b) < 0$), *then* $f(c) = 0$ *for at least one* $c \in (a,b)$. *More generally, let* f *and* g *be continuous real-valued functions on a closed bounded interval* $[a,b]$, $a < b$. *If* $f(a)-g(a)$ *and* $f(b)-g(b)$ *have opposite signs, then* $f(c) = g(c)$ *for at least one* $c \in (a,b)$.

Proof. If $f(a)f(b) < 0$, then either $f(a) < 0 < f(b)$ or $f(b) < 0 < f(a)$, and we can apply the Intermediate Value Theorem. The second part can be reduced to the first by considering the continuous function $f - g$.   □

Corollary 4.5.10 (Fixed Point Theorem). *If* $f : [a,b] \to [a,b]$ *is continuous, then there is a point* $p \in [a,b]$ *such that* $f(p) = p$.

Proof. If $f(a) = a$ or $f(b) = b$, then there is nothing to prove. Hence we may assume that $f(a) > a$ and $f(b) < b$. Now consider the function

$$g(x) := x - f(x) \qquad (\forall x \in [a,b]).$$

Then g is continuous on $[a,b]$ and we have $g(a) < 0$ while $g(b) > 0$. The corollary now follows from the theorem.   □

Corollary 4.5.11. *Let* $f \in C(I)$ *and let* z_1, $z_2 \in I$, $z_1 < z_2$, *be consecutive zeros of* f. *Then either* $f(x) > 0$ $\forall x \in (z_1, z_2)$ *or* $f(x) < 0$ $\forall x \in (z_1, z_2)$. *In other words, f does not change sign on* (z_1, z_2).

Remark 4.5.12.

(a) As pointed out above, Theorem 4.5.9 implies the Intermediate Value Theorem. Indeed, if in Theorem 4.5.6 we replace f by the continuous function $\tilde{f} := f - y_0$, where y_0 denotes the *constant function* $x \mapsto y_0$, then $\tilde{f}(a)$ and $\tilde{f}(b)$ have opposite signs and Theorem 4.5.9 may be applied.

(b) The converse of the Intermediate Value Theorem is *false*. Indeed, consider the
    function $f : \mathbb{R} \to \mathbb{R}$ defined by

$$f(x) := \begin{cases} \sin(1/x) & \text{if } x \neq 0, \\ 0 & \text{if } x = 0. \end{cases}$$

Then, as was seen before, neither of the one-sided limits $\lim_{x \to 0\pm} \sin(1/x)$
exists and hence f is *discontinuous* at 0, although the range of f is clearly
the closed interval $[-1, 1]$ and hence f has the *Intermediate Value Property*.
We shall see below, however, that the converse is *true* for *piecewise monotone
functions*. In fact, it is also true for *derivatives* of differentiable functions (cf.
Darboux's Theorem in Chap. 6).

Exercise 4.5.13 (Zeros of Odd-Degree Polynomials). Let $p(x) = a_0 + a_1 x + \cdots + a_n x^n$ be a polynomial of degree n, and assume that n is *odd*. Show that p
has at least one real zero, i.e., $p(z) = 0$ for at least one $z \in \mathbb{R}$. *Hint:* Note that
$p(x) \sim a_n x^n$ $(x \to \pm\infty)$, and look at the sign of $a_n x^n$ for $|x|$ *sufficiently large.*

Exercise 4.5.14 (Bisection Method). Let $I_0 = [a_0, b_0]$, $a_0 < b_0$, and let
$f \in C[a_0, b_0]$ be such that $f(a_0)$ and $f(b_0)$ have *opposite signs*; e.g., assume
that $f(a_0) < 0 < f(b_0)$. Define (inductively) a sequence $\{I_n, n = 0, 1, 2, \ldots\}$ of
closed intervals as follows. Let $z_0 = (a_0 + b_0)/2$ be the midpoint of I_0. If $f(z_0) \neq 0$,
let $I_1 = [a_1, b_1] := [a_0, z_0]$ if $f(z_0) > 0$ and $I_1 = [a_1, b_1] := [z_0, b_0]$ if $f(z_0) < 0$.
Suppose $I_n = [a_n, b_n]$ is already defined. Let $z_n := (a_n + b_n)/2$ be its midpoint and,
if $f(z_n) \neq 0$, let $I_{n+1} = [a_{n+1}, b_{n+1}]$ be the half of I_n on which f *changes sign.*

(a) Show that, if $f(z_n) \neq 0$ $\forall n \in \mathbb{N}$, then

$$\bigcap_{n=0}^{\infty} [a_n, b_n] = \{z\},$$

   where $z = \lim(z_n) \in (a_0, b_0)$ and $f(z) = 0$. [*Hint:* Use the Nested Intervals
   Theorem and the fact that $|z_n - z| \leq (b_0 - a_0)/2^{n+1}$(why?).]
(b) Apply the Bisection Method to the function $f : [1, 2] \to \mathbb{R}$ defined by $f(x) := x^2 - 2$. Here we obviously have $z = \sqrt{2}$, and the approximation $z_n \approx \sqrt{2}$ can be
   made accurate to any prescribed number of decimal places by a suitable choice
   of n. Find n such that this approximation is accurate to ten decimal places.

Definition 4.5.15 (Piecewise Monotone, Piecewise Continuous). A function $f :
[a, b] \to \mathbb{R}$ with $a < b$ is called *piecewise monotone* (resp., *piecewise continuous*)
if, for some $n > 1$, there exist $n - 1$ numbers $x_1, x_2, \ldots, x_{n-1} \in (a, b)$ satisfying

$$x_0 := a < x_1 < x_2 < \cdots < x_{n-1} < x_n := b$$

such that f is *monotone* on (resp., has a *continuous extension* to) each $[x_{k-1}, x_k]$, $1 \le k \le n$. Note that the continuous extension of f to $[x_{k-1}, x_k]$ is guaranteed if and only if $f(x_{k-1} + 0)$ and $f(x_k - 0)$ are both *finite*.

We are now ready to prove the converse of the Intermediate Value Theorem for functions that are piecewise monotone on any finite subinterval of their domain.

Theorem 4.5.16. *Suppose* $f : I \to \mathbb{R}$ *is piecewise monotone on any subinterval* $[a, b] \subset I$ *with* $a < b$ *and satisfies the Intermediate Value Property (on* I*). Then* $f \in C(I)$.

Proof. Since f is monotone on closed subintervals of the form $[x_{k-1}, x_k]$, it is sufficient to show that a *monotone* function f on a closed interval $[a, b]$ with $a < b$ is continuous on $[a, b]$ if it satisfies the *Intermediate Value Property* (on $[a, b]$). Changing f to $-f$, if necessary, we may assume that f is *increasing*. If f has a *discontinuity* at point $x_0 \in [a, b]$, then (by Theorem 4.4.7) it must be a *jump discontinuity*; i.e., we must have $f(x_0 - 0) < f(x_0)$ or $f(x_0) < f(x_0 + 0)$ (or both). Assume, for instance, that $f(x_0) < f(x_0 + 0)$. Then the Intermediate Value Property *fails* on $[x_0, x_0 + \delta]$ for any $\delta > 0$. (Why?) Thus, we must have $f(x_0 + 0) = f(x_0)$. Similarly, we have $f(x_0 - 0) = f(x_0)$. $\qquad\square$

The following proposition shows that if, in Theorem 4.5.16, we assume f to be *injective* rather than *piecewise monotone*, then f will automatically be (strictly) monotone and hence continuous.

Proposition 4.5.17. *Suppose* $f : I \to \mathbb{R}$ *is injective and satisfies the Intermediate Value Property. Then* f *is strictly monotone and continuous.*

Proof. Let a, $b \in I$, $a < b$. Since (f being one-to-one) $f(a) \ne f(b)$, let us assume $f(a) < f(b)$. We want to show that, for *any* x, $y \in I$, $x < y$ implies $f(x) < f(y)$. Now, if $a < c < b$, then $f(a) < f(c) < f(b)$. Indeed, if $f(c) > f(b)$, then $f(a) < f(b) < f(c)$, and by the Intermediate Value Property there exists $x \in (a, c)$ with $f(x) = f(b)$, contradicting the injectivity of f. Likewise, we cannot have $f(c) < f(a)$. Similar arguments show that, if we start with $b < c$ (resp., $c < a$), then $f(b) < f(c)$ (resp., $f(c) < f(a)$). In general, given any numbers x, $y \in I$ such that $x < y$, and any $a \in I$, $x \ne a \ne y$, we may apply the preceding arguments to the triple a, x, y, to deduce $f(x) < f(y)$. The continuity of f follows from Theorem 4.5.16. $\qquad\square$

Our goal for the rest of this section is to investigate the continuity of the *inverse* of an injective, continuous function. We begin by making the following.

Remark 4.5.18. If f is *strictly monotone*, say strictly increasing, then the intervals $J_{x_n} := \big(f(x_n - 0), f(x_n + 0)\big)$ in Remark 4.4.9(a) correspond to the "vertical gaps" in the graph of f. We may "fill" these gaps by vertical line segments. The resulting graph is *not* necessarily the graph of a *function*, but its "inverse graph", i.e., its reflection in the bisector $y = x$, is (draw a picture), and it actually provides a *continuous (left) inverse* for f:

Proposition 4.5.19. *If* $f : I \to \mathbb{R}$ *is strictly monotone, then it has a continuous (left) inverse. In other words, there is a continuous function* $g : J \to \mathbb{R}$, *where* J *is the interval whose endpoints are (the extended real numbers)* $\inf\{f(x) : x \in I\}$ *and* $\sup\{f(x) : x \in I\}$, *such that* $g(f(x)) = x$ *for all* $x \in I$.

Proof. We assume that f is *strictly increasing*. The decreasing case is similar, or we can use the "flipped" function $\check{f}(x) := f(-x)$. As the above remark suggests, we define

$$g(y) := \sup\{t \in I : f(t) \le y\}, \quad \forall y \in J.$$

Then, since f is strictly increasing, g is well defined and increasing (although *not* necessarily strictly) and we have $g(f(x)) = x$ for all $x \in I$. Also, for any $a < b$ in I, we have $g([f(a), f(b)]) = [a, b]$ and hence g satisfies the Intermediate Value Property on J. The continuity of g now follows from Theorem 4.5.16. □

If a continuous, injective function has a *continuous inverse function*, then we call it a *homeomorphism:*

Definition 4.5.20 (Homeomorphism). Let I, $J \subset \mathbb{R}$ be intervals. A function $f : I \to J$ is called a *homeomorphism* of I *onto* J if it is a continuous, one-to-one correspondence whose inverse f^{-1} is also continuous. If such a function f exists, the intervals I and J are then called *homeomorphic*.

Example 4.5.21. Let $f : \mathbb{R} \to (-1, 1)$ be the function $f(x) := x/\sqrt{1 + x^2}$. Then f is a homeomorphism of $\mathbb{R}$ onto $(-1, 1)$. Indeed, the inverse is $f^{-1}(x) = x/\sqrt{1 - x^2}$, and both f and f^{-1} are continuous. Moreover, the restriction $f|(0, \infty)$ is a homeomorphism of $(0, \infty)$ onto $(0, 1)$, and the restriction $f|[0, \infty)$ is a homeomorphism of $[0, \infty)$ onto $[0, 1)$.

Exercise 4.5.22. Let I, $J \subset \mathbb{R}$ be *nontrivial* intervals (i.e., each containing more than one point).

(a) Show that, if I and J are both *open*, then they are homeomorphic. (*Hint:* In the bounded case $I = (a, b)$, $J = (c, d)$, show that there is an *affine* function $f(x) := \alpha x + \beta$ (with suitable constants $\alpha \ne 0$ and β) mapping (a, b) homeomorphically onto (c, d). If at least one of I, J is unbounded, use the homeomorphism $f(x) := x/\sqrt{1 + x^2}$ given in Example 4.5.21.)

(b) Show that, if I, J are both *closed and bounded*, then they are homeomorphic. Also show that all closed unbounded intervals $(-\infty, b]$ and $[a, +\infty)$, a, $b \in \mathbb{R}$ are homeomorphic to each other as well as to any *bounded half-open* intervals $(c, d]$ and $[c, d)$, $-\infty < c < d < +\infty$.

(c) Show that, if I is open and $J \ne \mathbb{R}$ is closed or half-open, then I and J are *not* homeomorphic. Also, if I is closed and bounded and J is half-open, show that I and J are *not* homeomorphic.

The following theorem shows that the homeomorphisms between intervals of $\mathbb{R}$ are precisely the *strictly monotone, continuous* functions.

Theorem 4.5.23 (Homeomorphism Theorem). *For a function $f : I \to \mathbb{R}$ the following statements are equivalent and each implies that $J := f(I)$ is an interval.*

(a) f *is a homeomorphism onto* J.
(b) f *is injective and continuous.*
(c) f *is injective and satisfies the Intermediate Value Property.*
(d) f *is strictly monotone and satisfies the Intermediate Value Property.*
(e) f *is strictly monotone and continuous.*

Proof. First note that, if f satisfies any one of the above statements, then it satisfies the Intermediate Value Property and hence the range $J := f(I)$ is an *interval*. Let us now prove the implications (a) $\Rightarrow$ (b) $\Rightarrow$ (c) $\Rightarrow$ (d) $\Rightarrow$ (e) $\Rightarrow$ (a). Observe that (a) $\Rightarrow$ (b) follows from the definition of "homeomorphism," (b) $\Rightarrow$ (c) follows from the Intermediate Value Theorem, (c) $\Rightarrow$ (d) follows from Proposition 4.5.17, and (d) $\Rightarrow$ (e) follows from Theorem 4.5.16. To prove (e) $\Rightarrow$ (a), note that, if f is strictly monotone, then it is obviously one-to-one. Thus, to finish the proof, we must only show that $f^{-1} : J \to I$ is *continuous*. Since f^{-1} is also strictly monotone (cf. Proposition 3.1.9), it suffices (by Theorem 4.5.16) to show that it satisfies the Intermediate Value Property. However, this is clearly the case because the *range* of f^{-1} is the *interval* I.   □

Example 4.5.24. For any integer $n \in \mathbb{N}$, consider the *monomial* $p(x) = x^n$. If n is *odd*, then p is continuous and strictly increasing on $\mathbb{R}$ and, if n is *even*, it is continuous and strictly increasing on $[0, \infty)$. By Theorem 4.5.23, p maps $\mathbb{R}$ *onto* itself if n is odd and maps $[0, \infty)$ *onto* itself if n is even. In other words, *every* real number has a unique nth root if n is odd, and every *nonnegative* real number has a unique (nonnegative) nth root if n is even. Moreover, the *inverse function* $x \mapsto x^{1/n}$ is *continuous* (on $\mathbb{R}$ for odd n's and on $[0, \infty)$ for even n's).

4.6   Uniform Continuity

Recall that a function $f : I \to \mathbb{R}$ is *continuous* at $x_0 \in I$ if

$$(\forall \varepsilon > 0)(\exists \delta = \delta(\varepsilon, x_0) > 0)(\forall x \in I)(|x - x_0| < \delta \Rightarrow |f(x) - f(x_0)| < \varepsilon).$$

Note that the dependence of δ on *both* ε *and* x_0 appears explicitly: $\delta = \delta(\varepsilon, x_0)$. To remove the dependence of δ on x_0, we need a refinement of the concept of continuity which we now define.

Definition 4.6.1 (Uniformly Continuous). A function $f : I \to \mathbb{R}$ is said to be *uniformly continuous* (on I) if

$$(\forall \varepsilon > 0)(\exists \delta = \delta(\varepsilon) > 0)(\forall x, x' \in I)(|x - x'| < \delta \Rightarrow |f(x) - f(x')| < \varepsilon).$$

Remark 4.6.2.

(a) If, for each (fixed) $x_0 \in I$, we set $x' = x_0$ in the above definition, it follows that a uniformly continuous function (on I) is *continuous* (on I). Unfortunately, the converse is *false* in general, as the following example shows: Consider the function $f(x) := x^2$ defined on $\mathbb{R}$. Suppose that f is uniformly continuous on $\mathbb{R}$. Then, setting $\varepsilon = 1$, we can find $\delta > 0$ such that

$$(\forall x, \ x' \in \mathbb{R})(|x - x'| < \delta \Rightarrow |x^2 - x'^2| < 1).$$

Now, let $x = 2/\delta$ and $x' = \delta/2 + 2/\delta$. Then $|x - x'| = \delta/2 < \delta$, but

$$|x^2 - x'^2| = \left| \frac{4}{\delta^2} - \left(\frac{\delta^2}{4} + 2 + \frac{4}{\delta^2} \right) \right| = 2 + \frac{\delta^2}{4} > 1.$$

Observe that $2/\delta \to +\infty$ as $\delta \to 0+$; hence $\lim_{\delta \to 0+} x = \lim_{\delta \to 0+} x' = +\infty$. But even if the domain of our function is *bounded*, uniform continuity does not follow in general. As we shall see below, *compactness* of the domain is a *sufficient* (but not necessary) condition for a continuous function to be uniformly continuous.

(b) Uniform continuity may be defined on sets other than intervals. For any set $S \subset \mathbb{R}$, we say that $f : S \to \mathbb{R}$ is *uniformly continuous on S* if

$$(\forall \varepsilon > 0)(\exists \delta = \delta(\varepsilon) > 0)(\forall x, \ x' \in S)(|x - x'| < \delta \Rightarrow |f(x) - f(x')| < \varepsilon).$$

It is obvious that, if f is uniformly continuous on S, then it is also uniformly continuous on any *subset* of S. On the other hand, a function not uniformly continuous on a subset of S cannot be uniformly continuous on S.

(c) For a function $f : S \to \mathbb{R}$ to be *nonuniformly continuous on S*, the following must be satisfied:

$$(\exists \varepsilon_0 > 0)(\forall \delta > 0)(\exists x_\delta, \ x'_\delta \in S)(|x_\delta - x'_\delta| < \delta \quad and \quad |f(x) - f(x')| \geq \varepsilon_0).$$

Using $\delta = 1/n, n \in \mathbb{N}$, we get the following criterion for *nonuniform continuity* of a function $f : S \to \mathbb{R}$: $\exists \ \varepsilon_0 > 0$ *and two sequences* $(x_n), (x'_n) \in S^{\mathbb{N}}$ *such that* $\lim(x_n - x'_n) = 0$ *and* $|f(x_n) - f(x'_n)| \geq \varepsilon_0 \ \forall n \in \mathbb{N}$.

Example 4.6.3.

(a) Let $f : \mathbb{R} \to \mathbb{R}$ be an *affine* function, i.e., $f(x) = \alpha x + \beta$, for all $x \in \mathbb{R}$ and some $\alpha, \ \beta \in \mathbb{R}$. Then f is uniformly continuous on $\mathbb{R}$. Indeed, if $\alpha = 0$, then f is *constant* and $|f(x) - f(x')| = |\beta - \beta| = 0 < \varepsilon$ for any $x, \ x' \in \mathbb{R}$. If $\alpha \neq 0$, then any $\delta \in (0, \varepsilon/|\alpha|]$ will do, since $|x - x'| < \delta \leq \varepsilon/|\alpha|$ implies

$$|(\alpha x + \beta) - (\alpha x' + \beta)| = |\alpha||x - x'| < \varepsilon.$$

(b) Consider the function $x \mapsto \sin x$ on $\mathbb{R}$. The inequalities $|\sin \theta| \leq |\theta|$ and $|\cos \theta| \leq 1$, valid for every $\theta \in \mathbb{R}$, and the identity

$$\sin x - \sin x' = 2 \cos \frac{x + x'}{2} \sin \frac{x - x'}{2}$$

imply that for any $x, \ x' \in \mathbb{R}$ we have

$$|\sin x - \sin x'| \leq |x - x'|.$$

Therefore, sin is uniformly continuous on $\mathbb{R}$. The same is of course true for cos, since $\cos x = \sin(\pi/2 - x) \ \forall x \in \mathbb{R}$.

We now prove that a continuous function is uniformly continuous on *compact* subsets of its domain. Due to the importance of this result, we give *two* proofs for it!

Theorem 4.6.4 (Uniform Continuity and Compactness). *Let $f \in C(I)$ and let $K \subset I$ be compact. Then f is uniformly continuous on K. In particular, any $f \in C[a, b]$ is uniformly continuous.*

First Proof. If f is *not* uniformly continuous on K, then (by Remark 4.6.2(c) above) $\exists \, \varepsilon_0 > 0$ and two sequences $(x_n), \ (x'_n) \in K^{\mathbb{N}}$ such that $\lim(x_n - x'_n) = 0$ and $|f(x_n) - f(x'_n)| \geq \varepsilon_0 \ \forall n \in \mathbb{N}$. Since K is *compact*, there is a subsequence (x_{n_k}) such that $\lim(x_{n_k}) = x_0$ for some $x_0 \in K$. But then $\lim(x_{n_k} - x'_{n_k}) = 0$ implies $\lim(x'_{n_k}) = x_0$ as well. The *continuity* of f at x_0 now implies that

$$\lim_{k \to \infty} f(x_{n_k}) = f(x_0) = \lim_{k \to \infty} f(x'_{n_k}).$$

This, however, is impossible since

$$|f(x_{n_k}) - f(x'_{n_k})| \geq \varepsilon_0 \quad \forall k \in \mathbb{N}. \square$$

Second Proof. Let $\varepsilon > 0$ be given. For each $x \in K$ the continuity of f at x implies that

$$(\exists \delta_x > 0)(\forall x' \in K)(|x - x'| < \delta_x \Rightarrow |f(x) - f(x')| < \varepsilon/2).$$

The open intervals $I_x := (x - \delta_x/2, x + \delta_x/2), \ x \in K$ form an open cover of the *compact* set K and hence we can find finitely many points $x_1, \ x_2, \ldots, \ x_n \in K$ such that

$$K \subset \bigcup_{k=1}^{n} I_{x_k}.$$

Now let $\delta := \min\{\delta_{x_1}/2, \ \delta_{x_2}/2, \ \ldots, \ \delta_{x_n}/2\}$. For any x, $x' \in K$ satisfying $|x - x'| < \delta$, we then have x, $x' \in (x_k - \delta_{x_k}, x_k + \delta_{x_k})$ for some k. Indeed, if $x \in I_{x_k}$, then we have

$$|x' - x_k| \le |x' - x| + |x - x_k| < \delta + \delta_{x_k}/2 \le \delta_{x_k}/2 + \delta_{x_k}/2 = \delta_{x_k}.$$

Thus, $|x - x'| < \delta$ implies

$$|f(x) - f(x')| \le |f(x) - f(x_k)| + |f(x') - f(x_k)| < \varepsilon/2 + \varepsilon/2 = \varepsilon,$$

and the proof is complete.   □

We shall now look at some of the advantages of having uniform continuity. As we saw in Chap. 3, the study of limits of functions can be reduced to the study of suitably related *sequences*. The following proposition shows that uniformly continuous functions map Cauchy sequences to Cauchy sequences.

Proposition 4.6.5 (Cauchy Sequences and Uniform Continuity). *If $f : S \to \mathbb{R}$ is a uniformly continuous function on a set $S \subset \mathbb{R}$ and if $(x_n) \in S^{\mathbb{N}}$ is a Cauchy sequence in S, then $(f(x_n)) \in \mathbb{R}^{\mathbb{N}}$ is a Cauchy sequence (in $\mathbb{R}$).*

Proof. Let $\varepsilon > 0$ be given. Pick $\delta > 0$ such that

$$(\forall x, \ x' \in S)(|x - x'| < \delta \Rightarrow |f(x) - f(x')| < \varepsilon). \qquad (*)$$

Since (x_n) is Cauchy, we can find $N \in \mathbb{N}$ such that $|x_m - x_n| < \delta$ for all m, $n \ge N$. But then $(*)$ implies that $|f(x_m) - f(x_n)| < \varepsilon$ holds for all m, $n \ge N$.   □

We are now ready to prove that a function $f : (a, b) \to \mathbb{R}$ is uniformly continuous if it has a *continuous extension* to the closed interval $[a, b]$.

Theorem 4.6.6 (Continuous Extension Theorem). *A function $f : (a, b) \to \mathbb{R}$ is uniformly continuous on (a, b) if and only if it can be extended to a continuous function on $[a, b]$.*

Proof. If f has a continuous extension $\tilde{f} : [a, b] \to \mathbb{R}$, then ($[a, b]$ being compact) $\tilde{f}$ is *uniformly continuous* on $[a, b]$ and hence also on (a, b), where $\tilde{f} = f$. Conversely, suppose f is uniformly continuous on (a, b). If $(x_n) \in (a, b)^{\mathbb{N}}$ and $\lim(x_n) = a$, then $(f(x_n))$ is Cauchy and (since $\mathbb{R}$ is *complete*) we have $\lim(f(x_n)) = \alpha$ for some $\alpha \in \mathbb{R}$. Moreover, if $(x'_n) \in (a, b)^{\mathbb{N}}$ is any other sequence with $\lim(x_n) = a$, then $\lim(x_n - x'_n) = 0$ and the uniform continuity of f implies

$$\lim(f(x_n)) = \alpha = \lim(f(x'_n)).$$

A similar argument shows that there is a unique $\beta \in \mathbb{R}$ such that $\lim(f(x_n)) = \beta$ for *every* sequence $(x_n) \in (a, b)^{\mathbb{N}}$ satisfying $\lim(x_n) = b$. If we now define

$$\tilde{f}(x) := \begin{cases} \alpha & \text{if } x = a, \\ f(x) & \text{if } a < x < b, \\ \beta & \text{if } x = b, \end{cases}$$

then $\tilde{f} = f$ on (a, b) and hence is continuous there. Also, by the sequential definition of continuity, $\tilde{f}$ is continuous at the endpoints a and b. □

Exercise 4.6.7. Show that the function $x \mapsto \sin(1/x)$ is *not* uniformly continuous on $(0, 1]$.

Remark 4.6.8.

1. As we have seen above, a continuous function with *compact* domain is automatically uniformly continuous. But compactness is *not* a necessary condition. For example, any *affine* function $x \mapsto \alpha x + \beta$ is uniformly continuous on $\mathbb{R}$ and so are sin and cos. Given a uniformly continuous function on a *noncompact* domain, it is not easy (in general) to check its uniform continuity. In the next definition we introduce a condition that guarantees uniform continuity and holds for many functions.

2. In view of Theorem 4.6.6, the definition of *piecewise continuous* functions (cf. Definition 4.5.15) is equivalent to the following, where we use the notation of Definition 4.5.15: $f : [a, b] \to \mathbb{R}$ is piecewise continuous if it is *uniformly continuous* on each (x_{k-1}, x_k), $1 \le k \le n$.

Definition 4.6.9 (Lipschitz Function, Contraction). Let $f : I \to \mathbb{R}$.

(a) We say that f is *Lipschitz* (or *satisfies a Lipschitz condition*) and write $f \in \boldsymbol{Lip}(I)$, if there is a constant $A > 0$ such that

$$|f(x) - f(x')| \le A|x - x'| \quad \forall x, x' \in I.$$

The constant A is then called a *Lipschitz constant* for f.

(b) We say that f is *locally Lipschitz* and write $f \in \boldsymbol{Lip}_{loc}(I)$ if, for each $x \in I$, there exists $\varepsilon = \varepsilon(x) > 0$ such that f is Lipschitz on $I \cap B_\varepsilon(x) := I \cap (x - \varepsilon, x + \varepsilon)$. The Lipschitz constant will (in general) vary with x.

(c) We say that f is a *contraction* (or a *contraction mapping*) if it is Lipschitz with a Lipschitz constant $A < 1$.

(d) We say that f is *Lipschitz of order* α, $0 < \alpha \le 1$ and write $f \in \boldsymbol{Lip}^\alpha(I)$, if there is a constant $A > 0$ (still called a *Lipschitz constant*) such that

$$|f(x) - f(x')| \le A|x - x'|^\alpha \quad \forall x, x' \in I.$$

(e) We say that f is *locally Lipschitz of order* α, $0 < \alpha \le 1$ and write $f \in \boldsymbol{Lip}^\alpha_{loc}(I)$ if, for each $x \in I$, there exists $\varepsilon = \varepsilon(x) > 0$ such that $f \in \boldsymbol{Lip}^\alpha(I \cap B_\varepsilon(x))$.

Remark 4.6.10.

(a) Geometrically, if $f : I \to \mathbb{R}$ satisfies the *Lipschitz condition*

$$|f(x) - f(x')| \le A|x - x'| \quad \forall x, \, x' \in I,$$

then for any $x, \, x' \in I, \, x \ne x'$, the inequality

$$\left| \frac{f(x) - f(x')}{x - x'} \right| \le A$$

indicates that the *slope* of the *chord* joining the points $(x, f(x))$ and $(x', f(x'))$ on the graph of f is bounded by A.

(b) A Lipschitz function of order $\alpha = 1$ is simply a Lipschitz function, i.e., $Lip^1(I) = Lip(I)$.

Exercise 4.6.11. Let $f : I \to \mathbb{R}$ be *locally Lipschitz* of order α, $0 < \alpha \le 1$. Show that if K is *compact* and $K \subset I$, then f is *Lipschitz* of order α (on K).

Example 4.6.12.

(a) We have seen before that

$$|\sin x - \sin x'| \le |x - x'| \quad \forall x, \, x' \in \mathbb{R}.$$

Thus, sin is Lipschitz on $\mathbb{R}$ with Lipschitz constant $A = 1$. The same is of course true for $x \mapsto \cos x = \sin(\pi/2 - x)$.

(b) Any *affine* function $f(x) := ax + b$ is Lipschitz on $\mathbb{R}$. Indeed,

$$|f(x) - f(x')| = |(ax + b) - (ax' + b)| = |a||x - x'|$$

holds for all $x, x' \in \mathbb{R}$, and the *smallest* Lipschitz constant is of course $A = |a|$.

(c) The function $x \mapsto x^2$ defined on $\mathbb{R}$ is *locally* Lipschitz. In fact, it is Lipschitz on any *bounded* subset of $\mathbb{R}$. Indeed, if $S \subset \mathbb{R}$ and if $|x| \le M$ for all $x \in S$ and some $M > 0$, then

$$|x^2 - x'^2| = |x + x'||x - x'| \le 2M|x - x'| \quad \forall x, \, x' \in S.$$

Note, however, that this function is *not Lipschitz on* $\mathbb{R}$, as follows, for instance, from the next theorem.

Exercise 4.6.13. Show that the function $x \mapsto x \sin x$ is *locally* Lipschitz on $\mathbb{R}$ but it is *not* Lipschitz on $\mathbb{R}$.

Theorem 4.6.14.

(a) If $f : I \to \mathbb{R}$ is Lipschitz of order α, $0 < \alpha \le 1$, then it is uniformly continuous.

(b) If $f,\ g : I \to \mathbb{R}$ are both Lipschitz functions of (the same) order α, then so are $f \pm g$ and cf for any $c \in \mathbb{R}$. If (in addition) the functions $f,\ g$ are bounded, then the product fg is also Lipschitz of order α.

(c) If $f : I \to \mathbb{R}$ and $g : J \to \mathbb{R}$ are Lipschitz functions of orders α and β, respectively, and if $f(I) \subset J$, then the composite function $g \circ f$ is Lipschitz of order $\alpha\beta$ (on I). In particular, if f and g are Lipschitz, then so is $g \circ f$.

Proof. To prove (a), let $\varepsilon > 0$ be arbitrary and let $A > 0$ be a Lipschitz constant for f. Then, for any $0 < \delta \le (\varepsilon/A)^{1/\alpha}$, we have

$$(x,\ x' \in I,\ |x - x'| < \delta) \Rightarrow |f(x) - f(x')| \le A|x - x'|^{\alpha} < A((\varepsilon/A)^{1/\alpha})^{\alpha} = \varepsilon.$$

For (b), assume that f and g have Lipschitz constants A and B, respectively. Then, for any $x,\ x' \in I$, we have the inequalities

$$|(f \pm g)(x) - (f \pm g)(x')| \le |f(x) - f(x')| + |g(x) - g(x')| \le (A + B)|x - x'|^{\alpha},$$

$$|(cf)(x) - (cf)(x')| = |c||f(x) - f(x')| \le |c|A|x - x'|^{\alpha}.$$

If $f,\ g$ are bounded, i.e., $|f(x)| \le M$ and $|g(x)| \le M'$ for all $x \in I$ and some constants $M > 0,\ M' > 0$, then, for any $x,\ x' \in I$, we have

$$|(fg)(x) - (fg)(x')| \le |g(x)||f(x) - f(x')| + |f(x')||g(x) - g(x')|$$

$$\le (M'A + MB)|x - x'|^{\alpha}.$$

Finally, for (c), given any $x,\ x' \in I$, we have $f(x),\ f(x') \in J$. So, if A and B are Lipschitz constants for f and g, respectively, then for all $x,\ x' \in I$,

$$|g(f(x)) - g(f(x'))| \le B|f(x) - f(x')|^{\beta} \le BA^{\beta}|x - x'|^{\alpha\beta}.$$

$\square$

Remark 4.6.15.

(a) As the reader may have noticed, in the above proof we have used the fact that, if $0 < x < y$ and $r > 0$, then $x^r < y^r$. This is obvious if $r \in \mathbb{Q}$. (Why?) For *irrational* values of r, we have not yet defined x^r for a positive base x. This will be defined later, and we shall see that $x \mapsto x^r$ is *increasing* on $[0, \infty)$ for any fixed exponent $r > 0$.

(b) By part (b) of Theorem 4.6.14, the product of two *bounded* Lipschitz functions is Lipschitz. On the other hand, the function $x \mapsto x \sin x$ on $\mathbb{R}$, which is *not* Lipschitz on $\mathbb{R}$ (cf. Exercise 4.6.13 above), is the product of the *unbounded Lipschitz* function $x \mapsto x$ and the *bounded Lipschitz* function $x \mapsto \sin x$.

(c) If the interval I is *bounded*, then any $f \in \boldsymbol{Lip}^{\alpha}(I)$ is automatically bounded. (Why?) Therefore, on a bounded interval, sums, differences, constant multiples, and products of Lipschitz functions (of order α) are all Lipschitz (of order α).

Note, however, that the *boundedness* of f, g in part (b) of Theorem 4.6.14 is *not* a necessary condition for fg to be Lipschitz. Indeed, the function $x \mapsto \sqrt{x}$ ($x \geq 1$) is obviously unbounded on $[1, \infty)$ and is Lipschitz, since

$$|\sqrt{x} - \sqrt{x'}| = \frac{|x - x'|}{\sqrt{x} + \sqrt{x'}} \leq \frac{1}{2}|x - x'| \quad \forall\, x,\, x' \in [1, \infty),$$

and the function $(f^2)(x) = (\sqrt{x})^2 = x$ is also Lipschitz on $[1, \infty)$.

(d) Although any Lipschitz function is automatically uniformly continuous, the converse is *not* true. For example, consider the function $x \mapsto \sqrt{x}$ on $[0, \infty)$. As pointed out in part (c) above, the restriction of this function to the interval $[1, \infty)$ is Lipschitz, hence uniformly continuous on $[1, \infty)$. On the other hand, $x \mapsto \sqrt{x}$ is *continuous* on the *compact* set $[0, 2]$ and hence is uniformly continuous there. It follows that f is uniformly continuous on $[0, \infty) = [0, 2] \cup [1, \infty)$. (Why?) The reader should check, however, that $f : x \to \sqrt{x}$ is *not* Lipschitz on $[0, 2]$ (why not?) and hence not Lipschitz on $[0, \infty)$.

(e) Any affine map $f : x \to ax + b$ with $a \neq 0$ is a *homeomorphism* of $\mathbb{R}$ onto itself. The inverse function $f^{-1} : x \mapsto x/a - b/a$ is also affine. In particular, both f and f^{-1} are Lipschitz (hence uniformly continuous) on $\mathbb{R}$. In general, however, the inverse of a (one-to-one) uniformly continuous function is *not* uniformly continuous. Thus, the inverse of a Lipschitz function need *not* be Lipschitz:

Exercise 4.6.16. Let $f : \mathbb{R} \to (-1, 1)$ be the function $x \mapsto x/(1 + |x|)$.

(a) Show that f is a *homeomorphism* [onto $(-1, 1)$].
(b) Show that f is Lipschitz (hence *uniformly continuous*). Show, however, that the inverse function f^{-1} is *not* uniformly continuous (hence not Lipschitz).

We are now going to prove that a *contraction*, i.e., a Lipschitz function with a Lipschitz constant $A < 1$, has a (*unique*) *fixed point*. This "elementary" result turns out to be an extremely powerful tool in proving the existence of solutions to differential and integral equations.

Theorem 4.6.17 (Fixed Point Theorem). *Let I be a closed interval, and let $f :$ $I \to \mathbb{R}$ be a contraction, i.e., $f \in \mathbf{Lip}(I)$ with a Lipschitz constant $A < 1$. Then f has a unique fixed point; i.e., there is a unique point $x \in I$ such that $f(x) = x$. If $x_0 \in I$ is an arbitrary point and if we define the sequence $(x_0, x_1, x_2, \ldots)$ by $x_{n+1} := f(x_n)$ for all $n \geq 0$, then $x = \lim(x_n)$.*

Proof. Let $x_0 \in I$ be *arbitrary* and let $x_{n+1} := f(x_n) \; \forall n \in \{0, 1, 2, \ldots\}$. Then, by induction, we have

$$|x_{n+1} - x_n| = |f(x_n) - f(x_{n-1})| \leq A^n |x_1 - x_0| \quad \forall n \in \mathbb{N}. \tag{$*$}$$

Indeed, for $n = 1$, we have

$$|x_2 - x_1| = |f(x_1) - f(x_0)| \le A|x_1 - x_0|,$$

and for any $k \ge 2$, $|x_k - x_{k-1}| \le A^{k-1}|x_1 - x_0|$ implies

$$|x_{k+1} - x_k| = |f(x_k) - f(x_{k-1})| \le A|x_k - x_{k-1}|$$
$$\le A \cdot A^{k-1}|x_1 - x_0| = A^k|x_1 - x_0|.$$

Now, if $m \ge n$, then, using $(*)$ and the Triangle Inequality repeatedly,

$$|x_m - x_n| \le |x_m - x_{m-1}| + |x_{m-1} - x_{m-2}| + \cdots + |x_{n+1} - x_n|$$
$$\le (A^{m-1} + A^{m-2} + \cdots + A^n)|x_1 - x_0|.$$

Therefore, for $m \ge n$, we have

$$|x_m - x_n| \le \frac{A^n}{1 - A}|x_1 - x_0|, \tag{†}$$

where, we recall, $1 - A > 0$ by assumption. Since $\lim(A^n) = 0$, it follows from (†) that (x_n) is a Cauchy sequence and hence converges: $\lim(x_n) := x \in \mathbb{R}$. But then, since I is assumed to be *closed*, we have $x \in I$. To show that x is a *fixed point*, we note that f, being Lipschitz, is uniformly continuous and hence continuous on I. Thus,

$$f(x) = f(\lim(x_n)) = \lim(f(x_n)) = \lim(x_{n+1}) = x.$$

Finally, to prove the *uniqueness* of the fixed point x, suppose that $x' \in I$ is another one, i.e., $f(x') = x'$. Then the inequalities

$$|x - x'| = |f(x) - f(x')| \le A|x - x'|$$

and $A < 1$ imply that we must have $x = x'$.   □

4.7   Approximation by Step, Piecewise Linear, and Polynomial Functions

The basic goal of *analysis* is to break up the objects of its study into "simpler" *pieces*, study these new objects, and then use a "synthesis" to get information on the original objects. For example, to study continuous functions, one may need to "approximate" them by other, more "elementary" functions. The following three theorems show how uniform continuity can be used to approximate continuous functions by *step functions, piecewise linear functions*, and *polynomial functions*.

Definition 4.7.1 (Step and Piecewise Linear Functions). Let I be an interval with endpoints $a < b$.

(a) A function $\phi : I \to \mathbb{R}$ is called a *step function* if there are finite sequences $(x_k)_{k=0}^n$ and $(c_j)_{j=1}^n$, with

$$a = x_0 < x_1 < x_2 < \cdots < x_{n-1} < x_n = b, \qquad (*)$$

such that $\phi(x) = c_j \ \forall x \in (x_{j-1}, x_j)$; i.e., the restriction of ϕ to $I_j :=$ (x_{j-1}, x_j) is the constant function c_j, $j = 1, 2, \ldots, n$.
(b) A function $\psi : I \to \mathbb{R}$ is called *piecewise linear* (or, more accurately, *piecewise affine*) if there is a finite sequence $(x_k)_{k=0}^n$, satisfying $(*)$ above, and such that ψ is *affine* on each $I_j := (x_{j-1}, x_j)$, $1 \le j \le n$, i.e., $\psi(x) = \alpha_j x + \beta_j$ $\forall x \in I_j$ and some constants $\alpha_j, \beta_j \in \mathbb{R}$, $1 \le j \le n$.

Theorem 4.7.2 (Step Function Approximation). *Given any $f \in C[a, b]$ and any $\varepsilon > 0$, there exists a step function $\phi_\varepsilon : [a, b] \to \mathbb{R}$ such that*

$$|f(x) - \phi_\varepsilon(x)| < \varepsilon \quad \forall x \in [a, b]. \qquad (**)$$

Proof. Since f is uniformly continuous on (the *compact* set) $[a, b]$, given any $\varepsilon > 0$, we can find $\delta = \delta(\varepsilon) > 0$ such that, for any $x, \ x' \in [a, b]$ with $|x - x'| < \delta$, we have $|f(x) - f(x')| < \varepsilon$. Now pick $n \in \mathbb{N}$ such that $h := (b - a)/n < \delta$, and let $x_k := a + kh, k = 0, 1, \ldots, n$. Define ϕ_ε as follows: $\phi(a) := f(a)$ and

$$\phi_\varepsilon(x) := f(x_j) \quad \forall x \in (x_{j-1}, x_j], \quad 1 \le j \le n,$$

so that, on each $I_j := (x_{j-1}, x_j)$, the function ϕ_ε is constantly equal to the value of f at the *right endpoint* of I_j. Now, for each $x \in (a, b]$, we have $x \in (x_{j-1}, x_j]$ for some (unique) j, and hence

$$|f(x) - \phi_\varepsilon(x)| = |f(x) - f(x_j)| < \varepsilon,$$

since $|x - x_j| < \delta$. Therefore, $(**)$ is satisfied on $[a, b]$.   □

Remark 4.7.3.

(a) Recall that the *characteristic function* of set $S \subset \mathbb{R}$ is defined by

$$\chi_S(x) := \begin{cases} 1 & \text{if } x \in S, \\ 0 & \text{if } x \notin S. \end{cases}$$

In terms of characteristic functions, the step function ϕ_ε defined in Theorem 4.7.2 can also be written as

$$\phi_\varepsilon = f(a)\chi_{\{a\}} + \sum_{j=1}^{n} f(x_j)\chi_{(x_{j-1}, x_j]}.$$

(b) Note that the construction of the step function ϕ_ε in Theorem 4.7.2 actually shows that given any continuous function f on $[a, b]$ and any $\varepsilon > 0$, we can approximate f by a step function that takes constant values on n subintervals having the *same* length $h := (b - a)/n$ for a suitable integer $n \in \mathbb{N}$.

Since (nontrivial) step functions have jump discontinuities, it is natural to ask whether we can approximate a continuous function f by *elementary* but *continuous* functions. One such approximation is by means of *piecewise linear* functions:

Theorem 4.7.4 (Piecewise Linear Approximation). *Given any $f \in C[a, b]$ and any $\varepsilon > 0$, there exists a piecewise linear function $\psi_\varepsilon : [a, b] \to \mathbb{R}$ such that*

$$|f(x) - \psi_\varepsilon(x)| < \varepsilon \quad \forall x \in [a, b]. \tag{†}$$

Proof. Since f is uniformly continuous on $[a, b]$, for every $\varepsilon > 0$, we can pick $\delta = \delta(\varepsilon) > 0$ such that, for any x, $x' \in [a, b]$ with $|x - x'| < \delta$, we have $|f(x) - f(x')| < \varepsilon$. Let $n \in \mathbb{N}$ be such that $h := (b - a)/n < \delta$ and let $x_k := a + kh$, $k = 0, 1, \ldots, n$. Now on each $[x_j - 1, x_j]$ define ψ_ε to be the *affine* function whose graph is the line segment joining the two points

$$(x_{j-1}, f(x_{j-1})) \quad \text{and} \quad (x_j, f(x_j)).$$

Explicitly, each $x \in [x_{j-1}, x_j]$ has the form $x = x_{j-1} + t(x_j - x_{j-1}) = x_{j-1} + th$ for some $t \in [0, 1]$, and we have

$$\psi_\varepsilon(x) = f(x_{j-1}) + t(f(x_j) - f(x_{j-1})).$$

But $f(x_{j-1}) + t(f(x_j) - f(x_{j-1}))$ belongs to the (closed) interval with endpoints $f(x_{j-1})$ and $f(x_j)$, so the Intermediate Value Theorem implies that $f(x_{j-1}) + t(f(x_j) - f(x_{j-1})) = f(\xi_j)$ for some $\xi_j \in [x_{j-1}, x_j]$. Since each $x \in [a, b]$ belongs to some $[x_{j-1}, x_j]$, the choice of δ implies that

$$|f(x) - \psi_\varepsilon(x)| = |f(x) - f(\xi_j)| < \varepsilon,$$

and hence (†) is satisfied on $[a, b]$. $\qquad\qquad\qquad\qquad\qquad\qquad\qquad\qquad\square$

There is a much deeper approximation theorem, due to Weierstrass, in which continuous functions are approximated by *polynomials*. There are several proofs of this important theorem, but the most "elementary" treatment is by means of

Bernstein polynomials to be defined below. These polynomials involve, in an essential way, the *binomial formula:*

$$(a + b)^n = \sum_{k=0}^{n} \binom{n}{k} a^k b^{n-k},$$

valid for any a, $b \in \mathbb{R}$ and $n \in \mathbb{N}$. Recall that the *binomial coefficients* are defined to be

$$\binom{n}{k} := \frac{n!}{k!(n-k)!} \qquad (k = 0, 1, \ldots, n).$$

Proposition 4.7.5. *For each* $n \in \mathbb{N}$ *and* $k \in \{0, 1, \ldots, n\}$, *let us define the polynomial function* p_k^n *(or, for simplicity,* p_k*) by*

$$p_k(x) := \binom{n}{k} x^k (1-x)^{n-k} \qquad (0 \le x \le 1).$$

The following identities are then valid $\forall x \in [0, 1]$:

$$\sum_{k=0}^{n} p_k(x) = 1,$$

$$\sum_{k=0}^{n} k p_k(x) = nx, \quad and$$

$$\sum_{k=0}^{n} k^2 p_k(x) = n^2 x^2 - nx^2 + nx.$$

Proof. The first identity follows directly from the binomial formula:

$$1 = 1^n = (x + (1-x))^n = \sum_{k=0}^{n} \binom{n}{k} x^k (1-x)^{n-k} = \sum_{k=0}^{n} p_k(x).$$

To prove the second identity, note that

$$\sum_{k=0}^{n} k p_k(x) = \sum_{k=1}^{n} k \frac{n!}{k!(n-k)!} x^k (1-x)^{n-k}$$

$$= \sum_{k=1}^{n} \frac{n!}{(k-1)!(n-k)!} x^k (1-x)^{n-k}$$

$$= nx \sum_{k=1}^{n} \frac{(n-1)!}{(k-1)!(n-k)!} x^{k-1} (1-x)^{n-k}$$

$$= nx \sum_{k=1}^{n} \binom{n-1}{k-1} x^{k-1} (1-x)^{n-k}$$

$$= nx,$$

where the last equality follows from

$$\sum_{k=1}^{n} \binom{n-1}{k-1} x^{k-1} (1-x)^{n-k} = \sum_{j=0}^{n-1} \binom{n-1}{j} x^{j} (1-x)^{(n-1)-j}$$

$$= (x + (1-x))^{n-1} = 1.$$

A slightly more involved but similar argument is needed for the third identity, whose proof we leave as an exercise for the reader (cf. Exercise 4.7.7 below).   □

Corollary 4.7.6. *With notation as in Proposition 4.7.5, we have*

$$\sum_{k=0}^{n} (x - k/n)^2 p_k(x) = \frac{x(1-x)}{n}. \qquad (\ddagger)$$

Proof. Let S denote the sum on the left side of ($\ddagger$). Using the identities in Proposition 4.7.5, we have

$$S = x^2 \sum_{k=0}^{n} p_k(x) - \frac{2x}{n} \sum_{k=0}^{n} k p_k(x) + \frac{1}{n^2} \sum_{k=0}^{n} k^2 p_k(x)$$

$$= x^2 - \frac{2x}{n}(nx) + \frac{1}{n^2}(n^2 x^2 - nx^2 + nx)$$

$$= \frac{x(1-x)}{n} \qquad \forall x \in I. \square$$

Exercise 4.7.7. Prove the third identity in Proposition 4.7.5, i.e., show that

$$\sum_{k=0}^{n} k^2 p_k(x) = \sum_{k=0}^{n} k^2 \binom{n}{k} x^k (1-x)^{n-k} = n^2 x^2 - nx^2 + nx \qquad (\forall x \in I).$$

Hint: Note that for each $x \in I$, we have

$$\sum_{k=0}^{n} k^2 p_k(x) = \sum_{k=2}^{n} k(k-1) p_k(x) + \sum_{k=1}^{n} k p_k(x);$$

rewrite the first sum as

$$\sum_{k=2}^{n} k(k-1)p_k(x) = n(n-1)x^2 \sum_{k=2}^{n} \frac{(n-2)!}{(k-2)!(n-k)!} x^{k-2}(1-x)^{n-k},$$

and observe that the sum on the right equals 1. (Why?)

We are now ready to define the *Bernstein polynomials*, which are going to be used in the polynomial approximation of a continuous function $f : [0, 1] \to \mathbb{R}$. The readers who have studied elementary probability theory will certainly suspect a connection to *binomial random variables*. Indeed, as we shall see later, Bernstein's discovery of his remarkable polynomials had its origin in probability theory.

Definition 4.7.8 (Bernstein Polynomials). For each $n \in \mathbb{N}$, the *nth Bernstein polynomial* of a function $f : [0, 1] \to \mathbb{R}$ is defined to be

$$B_n(x) = B_n(x, f) := \sum_{k=0}^{n} f\left(\frac{k}{n}\right)\binom{n}{k} x^k (1-x)^{n-k} = \sum_{k=0}^{n} f\left(\frac{k}{n}\right) p_k(x),$$

where $p_k(x)$ is as in Proposition 4.7.5.

Theorem 4.7.9 (Bernstein Approximation Theorem). *If $f \in C[0, 1]$ and $\varepsilon > 0$, then there exists $N = N(\varepsilon) \in \mathbb{N}$ (independent of $x \in [0, 1]$) such that*

$$(\forall n > N)(\forall x \in [0, 1])(|f(x) - B_n(x)| < \varepsilon).$$

Proof. From the identity

$$f(x) = \sum_{k=0}^{n} f(x) p_k(x), \quad \forall x \in I,$$

which is a consequence of the first identity in Proposition 4.7.5, we get

$$f(x) - B_n(x) = \sum_{k=0}^{n}(f(x) - f(k/n)) p_k(x), \quad \forall x \in I,$$

which in turn implies

$$|f(x) - B_n(x)| \le \sum_{k=0}^{n} |f(x) - f(k/n)| p_k(x), \quad \forall x \in I. \qquad (*)$$

Now f (which is continuous on the compact set $[0, 1]$) is *uniformly continuous* and *bounded*, so $|f(x)| \le M$ for all $x \in [0, 1]$ and some $M > 0$. To make the right side of $(*)$ small, we observe that since f is *continuous* at x, the difference

$|f(x) - f(k/n)|$ can be made arbitrarily small if k/n is sufficiently close to x. Otherwise, we can only assert that $|f(x) - f(k/n)| \leq 2M$. So, we shall split the sum on the right side of $(*)$ in two parts: the first containing the terms for which k/n is *near* x and the second consisting of the remaining terms. Given $\varepsilon > 0$, the *uniform continuity* of f implies that we can find $\delta = \delta(\varepsilon) > 0$ such that $|x - x'| < \delta$ implies $|f(x) - f(x')| < \varepsilon/2$. Now pick $N \in \mathbb{N}$ such that $N \geq \max\{1/\delta^4, M^2/\varepsilon^2\}$, and assume $n > N$. Write, for $x \in I$ arbitrary,

$$\sum_{k=0}^{n} |f(x) - f(k/n)| p_k(x) = \sum{}' + \sum{}'' ,$$

where $\sum'$ is the sum over the values of k for which $|x - k/n| < 1/\sqrt[4]{n}$ and $\sum''$ is the sum over the remaining values of k. If $|x - k/n| < 1/\sqrt[4]{n} \leq \delta$, then $|f(x) - f(k/n)| < \varepsilon/2$ and hence

$$\sum{}' < \sum_{k=0}^{n} \frac{\varepsilon}{2} p_k(x) = \frac{\varepsilon}{2} \sum_{k=0}^{n} p_k(x) = \frac{\varepsilon}{2}.$$

To estimate $\sum''$, note that, for the k's in this sum, we have $(x - k/n)^2 \geq 1/\sqrt{n}$. Thus, using the identity $(\ddagger)$ of Corollary 4.7.6 and the fact that $|f(x) - f(k/n)| \leq 2M$, we have

$$\sum{}'' \leq 2M \sum{}'' p_k(x) = 2M \sum{}'' \frac{(x - k/n)^2}{(x - k/n)^2} p_k(x)$$

$$\leq 2M \sqrt{n} \sum_{k=0}^{n} (x - k/n)^2 p_k(x)$$

$$\leq 2M \sqrt{n} \, \frac{x(1 - x)}{n}$$

$$\leq \frac{M}{2\sqrt{n}} < \frac{\varepsilon}{2},$$

where the "$\leq$" at the beginning of the last line follows from the fact that $x(1 - x) \leq 1/4$ on $[0, 1]$. (Why?) Therefore, for $n > N$, the inequality

$$\sum_{k=0}^{n} |f(x) - f(k/n)| p_k(x) = \sum{}' + \sum{}'' < \frac{\varepsilon}{2} + \frac{\varepsilon}{2} = \varepsilon$$

holds for all $x \in [0, 1]$ and the proof is complete. $\square$

As an immediate corollary, we obtain the celebrated Weierstrass Approximation Theorem:

Corollary 4.7.10 (Weierstrass Approximation Theorem). *Given any* $f \in C[a, b]$ *and any* $\varepsilon > 0$, *there exists a polynomial function* p_ε *such that*

$$|f(x) - p_\varepsilon(x)| < \varepsilon \qquad (\forall x \in [a, b]).$$

Proof. The *affine* function $\psi : x \to (b-a)x + a$ is a *homeomorphism* of $[0, 1]$ onto $[a, b]$, where we are obviously assuming $a < b$. The composite function $g = f \circ \psi$ is therefore a continuous function on $[0, 1]$, and (by Theorem 4.7.9) we can choose $n_\varepsilon \in \mathbb{N}$ so that $|g(x) - B_{n_\varepsilon}(x)| < \varepsilon \ \forall x \in [0, 1]$. If we now let $p_\varepsilon(x) = B_{n_\varepsilon}((x - a)/(b - a))$, then $|f(x) - p_\varepsilon(x)| < \varepsilon \ \forall x \in [a, b]$ as desired.   □

4.8 Problems

1. Let $(U_\alpha)_{\alpha \in A}$ be an open cover of a *compact* set $K \subset \mathbb{R}$. Show that there is an $\varepsilon > 0$ such that, $\forall x \in K$, we have $B_\varepsilon(x) \subset U_\alpha$ for some $\alpha \in A$.

2. Show that a set $K \subset \mathbb{R}$ is *compact* if and only if every *countable* open cover $(U_n)_{n \in \mathbb{N}}$ of K has a *finite* subcover.

3. For any sets $A, B \subset \mathbb{R}$, define $A + B := \{a + b : a \in A, b \in B\}$. (i) Show that, if A and B are *compact*, then so is $A + B$. (ii) Give an example to show that if A and B are *closed*, then $A + B$ need *not* be closed. (iii) Show, however, that if A is *compact* and B is *closed*, then $A + B$ is closed. *Hints: Use sequences.* For (ii), try the sets $A = \mathbb{N}$ and $B = \{n^{-1} - n : n = 2, 3, \ldots\}$.

4. Show that any *compact* set $K \subset \mathbb{R}$ is *complete*; i.e., every *Cauchy* sequence $(x_n) \in K^{\mathbb{N}}$ converges to a limit in K.

5 (Cantor). Let $(K_n)_{n \in \mathbb{N}}$ be a family of *nonempty, compact* subsets of $\mathbb{R}$. Show that, if $K_{n+1} \subset K_n$ for all $n \in \mathbb{N}$, then $\bigcap_{n \in \mathbb{N}} K_n \neq \emptyset$. *Hint:* $K_1 \not\subset (\bigcap_{n \geq 2} K_n)^c$.

6 (Finite Intersection Property). Prove the following extension of the previous problem. A set $K \subset \mathbb{R}$ is *compact* if and only if, given any collection $(F_\alpha)_{\alpha \in A}$ of *nonempty, closed* sets such that $K \cap (\bigcap_{\alpha \in A'} F_\alpha) \neq \emptyset$ for every *finite* subset $A' \subset A$, we have $K \cap (\bigcap_{\alpha \in A} F_\alpha) \neq \emptyset$.

7. Prove the following characterizations of continuity. Recall that A° and A^- denote the *interior* and *closure* of $A \subset \mathbb{R}$, respectively.

$$f \in C(\mathbb{R}) \iff f(A^-) \subset (f(A))^- \iff f^{-1}(A^\circ) \subset (f^{-1}(A))^\circ \qquad (\forall A \subset \mathbb{R}).$$

8. Consider the function

$$f(x) := \begin{cases} x & \text{if } x \in [0, 1] \cap \mathbb{Q}, \\ 1 - x & \text{if } x \in [0, 1] \cap \mathbb{Q}^c. \end{cases}$$

(a) Show that $f(f(x)) = x$ and $f(x) + f(1 - x) = 1$ for all $x \in [0, 1]$.
(b) Show that f is *onto* $[0, 1]$ and is continuous *only* at $x = 1/2$.
(c) Show that $f(x + y) - f(x) - f(y) \in \mathbb{Q}$ for all $x, y \in [0, 1]$.

9. Let I be a nonempty interval and $f : I \to \mathbb{R}$. Show that $f \in C(I)$ if and only if f is continuous on every *compact* set $K \subset I$. *Hint:* If $\lim(x_n) = \xi$, then $\{\xi, x_1, x_2, \ldots\}$ is compact.

10. Let $f : \mathbb{R} \to \mathbb{R}$ be defined as follows. For $x \in (0, 1]$, we set

$$f(x) := \begin{cases} \dfrac{1}{q} & \text{if } x = \dfrac{p}{q} \in (0, 1] \quad (p, q \in \mathbb{N}, \ \gcd(p, q) = 1), \\ 0 & \text{if } x \in \mathbb{Q}^c \cap (0, 1), \end{cases}$$

and for $x \notin (0, 1]$ we define $f(x)$ by *periodicity*, i.e. $f(x) := f(x - n)$ where $n \in \mathbb{Z}$ is the *unique* integer with $x \in (n, n + 1]$. Show that f is *continuous* at each $x \in \mathbb{Q}^c$ and *discontinuous* at each $x \in \mathbb{Q}$. *Hint:* If (p_n/q_n) is a sequence of (reduced) rationals in $(0, 1)$ with $\lim(p_n/q_n) = x \in \mathbb{Q}^c \cap (0, 1)$, show that $\lim(q_n) = \infty$.

11 (Volterra's Theorem). Let f, $g : (a, b) \to \mathbb{R}$ and let C_f and C_g denote the *continuity sets* of f and g, respectively. Thus $C_f := \{x \in (a, b) : f \text{ is continuous at } x\}$ and C_g defined similarly. Show that if C_f and C_g are both *dense*, then so is $C_f \cap C_g$. Deduce that there is no function that is *continuous* on $\mathbb{Q}$ and *discontinuous* on $\mathbb{Q}^c$.

12. Show that there can be no *continuous* function $g : \mathbb{R} \to \mathbb{R}$ that maps rational numbers to irrational ones and vice versa.

13. Show that there can be no *continuous* function $f : \mathbb{R} \to \mathbb{R}$ that satisfies

$$f(x) \in \mathbb{Q} \iff f(x + 1) \in \mathbb{Q}^c.$$

14.

(a) Give an example of a function $f \in C(\mathbb{R})$ and a *compact* set $K \subset \mathbb{R}$ such that $f^{-1}(K)$ is *not* compact.

(b) Give an example of a function $f \in C(\mathbb{R})$ and a *connected* set $C \subset \mathbb{R}$ such that $f^{-1}(C)$ is *not* connected.

15. Let f, $g \in C[a, b]$ satisfy $f(x) < g(x)$ for all $x \in [a, b]$. Show that there is a constant $c < 1$ such that $f(x) \leq cg(x)$ for all $x \in [a, b]$. *Hint:* Pick $N \geq 0$ such that $f + N > 0$ on $[a, b]$ and look at $(f + N)/(g + N)$.

16. Let $f \in C(\mathbb{R})$ satisfy $\lim_{x \to \pm\infty} f(x) = \infty$. Show that f attains its *(absolute) minimum;* i.e., there is a point $x_0 \in \mathbb{R}$ such that $f(x) \geq f(x_0)$ for all $x \in \mathbb{R}$.

17. Show that a function $f \in \mathbb{R}^{\mathbb{R}}$ is *continuous* if and only if (i) $f(I)$ is an *interval* for each interval $I \subset \mathbb{R}$ and (ii) $f^{-1}(y)$ is *closed* for each $y \in \mathbb{R}$. In fact, show that (ii) may be replaced by: $f^{-1}(y)$ is *closed* for each $y \in \mathbb{Q}$. *Hint:* Show that $f^{-1}(y_0)$ is *not* closed if $\underline{f}(x_0) < y_0 < \overline{f}(x_0)$ and $y_0 \neq f(x_0)$, by constructing a sequence (x_n) with $\lim(x_n) = x_0$ and $f(x_n) = y_0$ for all $n \in \mathbb{N}$. Here, $\underline{f}$ and $\overline{f}$ are the lower and upper envelopes of f, respectively (cf. Definition 4.3.7).

18. Show that a function $f \in C[a, b]$ is *bounded*, using the following *Bisection Method:* Assume that f is *unbounded* on $[a, b]$. Then it must be unbounded on either $[a, (a+b)/2]$ or $[(a+b)/2, b]$; let I_1 be the one on which f is unbounded. Bisect I_1 and let I_2 be the (closed) half on which f is unbounded, etc. Now use the *Nested Intervals Theorem* to get a contradiction.

19. Show that, if $f : [a, b] \to [a, b]$ is a *homeomorphism*, then either a and b are *fixed points* or $f(a) = b$ and $f(b) = a$.

20. Show that, if $f : (a, b) \to \mathbb{R}$ satisfies the *Intermediate Value Property* and $|f|$ is *constant*, then so is f. Deduce that if f is continuous and if $f^n = C$ is constant for an *even* integer n, then either $f = \sqrt[n]{C}$ or $f = -\sqrt[n]{C}$ on (a, b).

21. Show that, if $f \in \mathbb{R}^{\mathbb{R}}$ satisfies the *Intermediate Value Property* and $f(x \pm 0)$ both exist at every $x \in \mathbb{R}$, then $f \in C(\mathbb{R})$.

22. Locate and classify the discontinuities of the following functions, where each function satisfies $f(0) := 0$ and is defined for $x \neq 0$ by

(a) $f(x) := (\sin x)/|x|$,
(b) $f(x) := e^{1/x}$,
(c) $f(x) := e^{1/x} + \sin(1/x)$,
(d) $f(x) := 1/(1 - e^{1/x})$.

23.

(a) Find a function $f \in \mathbb{R}^{\mathbb{R}}$ such that f is continuous *nowhere* but $|f|$ is continuous *everywhere*.
(b) Given any $a \in \mathbb{R}$, find a function $f \in \mathbb{R}^{\mathbb{R}}$ such that f is continuous at a but *discontinuous* at every other point.

24 (Characterization of Monotone Functions).

(a) Show that $f \in \mathbb{R}^{\mathbb{R}}$ is *monotone* if and only if $f^{-1}(J)$ is an interval for every interval $J \subset \mathbb{R}$.
(b) Give an example of a function $f \in C(\mathbb{R})$ and an *interval* $J \subset \mathbb{R}$ such that $f^{-1}(J)$ is *not* an interval.

25. Let $\mathbb{Q} = \{q_1, q_2, \ldots\} \subset \mathbb{R}$ be an *enumeration* of the rationals and consider the *jump function*

$$f(x) := \sum_{\{n : q_n \leq x\}} 2^{-n} \qquad (\forall x \in \mathbb{R}).$$

Show that f is *right* (but *not* left) *continuous* at each $x \in \mathbb{Q}$ and *continuous* at each $x \notin \mathbb{Q}$.

26. Let $\kappa : [0, 1] \to [0, 1]$ be the *Cantor ternary function* (cf. Example 4.3.13) and define the set

$$D := \left\{ \frac{\kappa(x + h) - \kappa(x)}{h} : x \in [0, 1], \, h \neq 0 \right\}.$$

Show that $\inf(D) = 0$ and $\sup(D) = \infty$.

27. Let $f \in C[a, b]$ satisfy $f(a) < f(b)$. Show that there are $c, d \in [a, b]$ such that $a \leq c < d \leq b$ and $f(a) = f(c) < f(x) < f(d) = f(b)$ for all $x \in (c, d)$.

28 (Local Maxima and Minima). Let $\emptyset \neq I \subset \mathbb{R}$ be an interval and $f \in \mathbb{R}^{I}$. We say that f has a *local maximum* (resp., *strict local maximum*) at $x_0 \in I$ if there is an $\varepsilon > 0$ such that $f(x) \leq f(x_0)$ (resp., $f(x) < f(x_0)$) for all $x \in I \cap \dot{B}_{\varepsilon}(x_0)$. A similar definition can be given for a *local minimum* (resp., *strict local minimum*) at x_0.

(a) Show that, if f has a local maximum at *every* point $x \in I$, then the range of f is *countable*. *Hint:* For each $y \in f(I)$, pick an interval J_y with rational endpoints such that $y = \max\{f(x) : x \in I \cap J_y\}$. Assuming in addition that $f \in C(I)$, show that f must be *constant*.
(b) Show that $M := \{x \in I : f \text{ has a strict local maximum at } x\}$ is *countable*.

29. Let $f \in C[a, b]$. Show that, if f has a local maximum (see Problem 28) at x_1 and a local maximum at x_2, for distinct points $x_1, x_2 \in (a, b)$, then there is a point ξ *between* x_1 and x_2 where f has a local *minimum*.

30. Show that, if $f \in C([a, b])$ does *not* have a local maximum or minimum at *any* point in (a, b), then it must be *monotone*.

31. Show that, if $f \in C(\mathbb{R})$ is *strictly decreasing,* then $f(x) = x$ for a *unique* $x \in \mathbb{R}$; i.e., f has a unique *fixed point.*

32. If $f \in C[0, 1]$ satisfies $f([0, 1]) \subset \mathbb{Q}$, what can you say about f?

33. Show that *no* function $f \in C(\mathbb{R})$ assumes each of its values *exactly twice.*

34. Let I and J be (nonempty) intervals.

(a) Show that, if f, $g : I \to \mathbb{R}$ are *uniformly* continuous (on I), then so is $f + g$. Show that the same is true for fg provided f and g are *both bounded* (on I). Give an example to show that, if one of f, g is *unbounded,* then fg need *not* be uniformly continuous.
(b) Show that, if $f : I \to \mathbb{R}$ and $g : J \to \mathbb{R}$ are *uniformly* continuous and $f(I) \subset J$, then $g \circ f$ is also uniformly continuous.

35. Show by a Bisection Method similar to the one used in Problem 18 that *any* function $f \in C[a, b]$ is *uniformly continuous.*

36. Show that, if $f \in C[a, \infty)$ for some $a \in \mathbb{R}$ [resp., $f \in C(\mathbb{R})$] and if $\lim_{x\to\infty} f(x) = \ell \in \mathbb{R}$ (resp., $\lim_{x\to-\infty} f(x) = k \in \mathbb{R}$ and $\lim_{x\to\infty} f(x) = \ell \in \mathbb{R}$), then f is *bounded* and *uniformly* continuous. Give an example of a bounded uniformly continuous $f \in C(\mathbb{R})$ such that $\lim_{x\to\pm\infty} f(x)$ do *not* exist.

37. Show that, if $S \subset \mathbb{R}$ is *bounded,* then any *uniformly* continuous $f : S \to \mathbb{R}$ is *bounded.* Show (by example) that f need not be bounded if S is *unbounded.*

38. Show that, if $f : [1, \infty) \to \mathbb{R}$ is *uniformly continuous,* then $f(x) = O(x) \quad (x \to \infty)$.

39. Let $I \subset \mathbb{R}$ be an interval and $f : I \to \mathbb{R}$. Show that, if for each $\varepsilon > 0$ there is a *uniformly* continuous function $g : I \to \mathbb{R}$ such that $|f(x) - g(x)| < \varepsilon \;\; \forall \, x \in I$, then f is also *uniformly* continuous.

40. Show that, if $f \in C(\mathbb{R})$ is *periodic,* then it is *uniformly* continuous.

41. Give an example of a *bounded continuous* function on a *bounded* interval that is *not* uniformly continuous.

42. Let $I \subset \mathbb{R}$ be an interval. Show that $f : I \to \mathbb{R}$ is *uniformly* continuous if and only if

$$(\forall \varepsilon > 0)(\exists M > 0)\left(\left| \frac{f(x) - f(y)}{x - y} \right| > M \implies |f(x) - f(y)| < \varepsilon \right). \qquad (*)$$

Hint: If $(*)$ holds and $\varepsilon > 0$ is given, let $\delta := \varepsilon/M$ and show that $|f(x) - f(y)| \geq \varepsilon$ implies $|x - y| \geq \delta$. If f *is* uniformly continuous and $\varepsilon > 0$ is given, pick $\delta > 0$ so that $|f(x) - f(y)| \geq \varepsilon$ implies $|x - y| \geq \delta$ and set $M := 2\varepsilon/\delta$. Assume (without loss of generality) that $f(x) < f(y)$ and pick $n \in \mathbb{N}$ such that $\eta := (f(y) - f(x))/n \in [\varepsilon, 2\varepsilon]$. Divide $[f(x), f(y)]$ into n equal parts using the partition points $f(x) + k\eta, 0 \leq k \leq n$. Using the *Intermediate Value Theorem,* pick $x_0 := x, x_1, \ldots, x_n := y$ such that $f(x_k) = f(x) + k\eta$. Deduce that $|x - y| \geq n\delta$ and hence $|f(x) - f(y)|/|x - y| \leq M$.

43. Let $I \subset \mathbb{R}$ be an interval. Show that, if $f \in \boldsymbol{Lip}^\alpha(I)$ with $\alpha > 1$, then f is *constant* (on I). *Hint:* Fix $x_0 \in I$. Now, for each $x \in I$, estimate $|f(x) - f(x_0)| = |\sum_{k=1}^{n} f(x_0 + kh) - f(x_0 + (k - 1)h)|$, where $h := (x - x_0)/n$ and let $n \to \infty$.

44. Let $\emptyset \neq S \subset \mathbb{R}$. Show that the function

$$d_S(x) := \inf\{|x - s| : s \in S\} \qquad (\forall x \in \mathbb{R})$$

is *Lipschitz:* $|d_S(x) - d_S(y)| \leq |x - y| \;\; \forall x, y \in \mathbb{R}$.

45.

(a) Let $F \subset \mathbb{R}$ be a *closed* set. Show that there is a function $f \in C(\mathbb{R})$ such that $F = \{x : f(x) = 0\}$.

(b) Let E, $F \subset \mathbb{R}$ be *closed* sets with $E \cap F = \emptyset$. Show that there is a function $f \in C(\mathbb{R})$ such that $E = \{x : f(x) = 1\}$ and $F = \{x : f(x) = 0\}$. *Hint:* Use the functions d_E and d_F introduced in Problem 44.

46. Define $f : (-1, 1) \to \mathbb{R}$ by

$$f(x) := \begin{cases} 0 & \text{if } x = 0, \\ 1/\log|x| & \text{if } 0 < |x| < 1. \end{cases}$$

Show that $f \in \boldsymbol{Lip}_{loc}((-1, 1) \setminus \{0\})$ but that $f \notin \boldsymbol{Lip}^{\alpha}(I)$ for *any* $\alpha \in (0, 1]$ and *any* open interval I with $0 \in I \subset (-1, 1)$.

47 (Kepler's Equation). Show that, for any constants $a \in (0, 1)$ and $b \in \mathbb{R}$, the equation $x = a \sin x + b$ has a *unique* solution.

48. Let $a < b$ and suppose that $f : [a, b] \to [a, b]$ satisfies $|f(x) - f(y)| \leq |x - y|$ for all x, $y \in [a, b]$. Show that the sequence (x_n), defined recursively by $x_{n+1} = (x_n + f(x_n))/2$ and an *arbitrary* $x_1 \in [a, b]$, converges to a *fixed point* of f. *Hint:* Show that (x_n) is *monotone* by using the identity

$$x_{n+2} - x_{n+1} = \frac{1}{2}[f(x_{n+1}) - f(x_n) + x_{n+1} - x_n] \qquad (\forall n \in \mathbb{N}).$$

49. Let $a < b$ and let $f : [a, b] \to [a, b]$ be *Lipschitz* with constant 1, i.e., $|f(x) - f(y)| \leq |x-y|$ for all x, $y \in [a, b]$. Show that the set of all *fixed points* of f is a subinterval $[\alpha, \beta] \subset [a, b]$, possibly reduced to a *single point*.

50 (Contractive Map). Let $\emptyset \neq X \subset \mathbb{R}$. A map $f : X \to X$ is said to be *contractive* if

$$|f(x) - f(y)| < |x - y| \qquad (\forall x, y \in X, x \neq y).$$

(a) Show that a contractive map has *at most* one fixed point.

(b) Show that the function $f(x) := x + 1/x$ is contractive on $[1, \infty)$ and does *not* have fixed points.

(c) Show that $g(x) := (x + \sin x)/2$ is contractive on $\mathbb{R}$. Is there a fixed point?

(d) Show that, if f is contractive on a (nonempty) *compact* set $K \subset \mathbb{R}$, then it has a *unique* fixed point. *Hint:* Look at $\inf\{|f(x) - x| : x \in K\}$.

51 (Expansive Map). Let $\emptyset \neq X \subset \mathbb{R}$. A map $f : X \to X$ is said to be *expansive* if

$$|f(x) - f(y)| \geq |x - y| \qquad (\forall x, y \in X).$$

Show that if $f : \mathbb{R} \to \mathbb{R}$ is both *continuous* and *expansive*, then it is a *homeomorphism* with *Lipschitz* inverse $f^{-1} : \mathbb{R} \to \mathbb{R}$.

52. Let $f : X \to X$, where $\emptyset \neq X \subset \mathbb{R}$. For each $n \in \mathbb{N}$, let $f^{[n]}$ denote the *nth iterate* of f, i.e., $f^{[n]} := f \circ f \circ \cdots \circ f$, with n copies of f. Show that, if $f^{[n]}$ has a *unique* fixed point x_0 for some $n \in \mathbb{N}$, then x_0 is a fixed point of f, i.e., $f(x_0) = x_0$.

53 (Absolute Continuity). A function $F : [a, b] \to \mathbb{R}$ is said to be *absolutely continuous* (on $[a, b]$) if for each $\varepsilon > 0$ there is a $\delta > 0$ such that, given any collection $\{(a_k, b_k) : 1 \le k \le n\}$ of *pairwise disjoint* open subintervals of $[a, b]$, we have

$$\sum_{k=1}^{n} (b_k - a_k) < \delta \implies \sum_{k=1}^{n} |F(b_k) - F(a_k)| < \varepsilon.$$

(a) Show that an absolutely continuous function on $[a, b]$ is *uniformly continuous* there. To show that the converse is *false*, consider the function $f(x) := x \sin(\pi/x)$ for $x \ne 0$ and $f(0) := 0$. Show that f is *uniformly continuous* but *not* absolutely continuous on $[0, 1]$. *Hint:* Let $\varepsilon = 1$ and, for each $\delta > 0$, pick $M, N \in \mathbb{N}$ with $1/\delta < M < N$ such that $\sum_{k=M}^{N} a_k > 1$, where $a_k := 2/(4k + 1)$. Now let $b_k := 2/(4k)$ and consider the disjoint intervals (a_k, b_k), with $M \le k \le N$.

(b) Show that, if $F \in \boldsymbol{Lip}([a, b])$, then it is absolutely continuous. The converse is *false* again. Indeed, as we know, the function $f(x) := \sqrt{x}$ is *not* Lipschitz on $[0, 1]$. Show, however, that it *is* absolutely continuous on $[0, 1]$ as follows. Given $\varepsilon > 0$, let $\delta = \varepsilon^2/2$ and let $(a_j, b_j) \subset [0, 1]$ be pairwise disjoint with $\sum_{j=1}^{n} (b_j - a_j) < \delta$. If $\delta/2 \in (a_j, b_j)$, for some j, then insert it as an endpoint, getting two subintervals $(a_j, \delta/2)$ and $(\delta/2, b_j)$. Now write $\sum_{j=1}^{n} (\sqrt{b_j} - \sqrt{a_j}) = \sum_1 + \sum_2$, where $\sum_1$ is over all j with $b_j \le \delta/2$ and $\sum_2$ is over the other j's. Finally, observe that $\sum_1 \le \varepsilon/2$ and

$$\sum_2 \le \frac{1}{\varepsilon} \sum (b_j - a_j) < \frac{\varepsilon}{2}.$$

54.

(a) Let $f(x) := x^2$. Show that $B_n(x, f) = (n - 1)x^2/n + x/n$ and hence $B_n(x, f) \to x^2$, as $n \to \infty$.

(b) Let $f(x) := x^3$ on $[0, 1]$. Find the *Bernstein polynomials* $B_n(x) = B_n(x, f)$ and prove that $B_n \to f$.

55. Let $f \in C[a, b]$, where $\emptyset \ne [a, b] \subset (0, 1)$. For each $n \in \mathbb{N}$, define the polynomials

$$\tilde{B}_n(x) := \sum_{k=0}^{n} \left[\binom{n}{k} f(k/n) \right] x^k (1 - x)^{n-k},$$

where $[t]$ denotes the greatest integer $\le t$. Show that, with Bernstein polynomials $B_n(x)$, we have

$$\sup\{|B_n(x) - \tilde{B}_n(x)| : x \in [a, b]\} \to 0, \quad \text{as} \quad n \to \infty.$$

Deduce that $\sup\{|f(x) - \tilde{B}_n(x)| : x \in [a, b]\} \to 0$, as $n \to \infty$, i.e., f can be *uniformly* approximated (on $[a, b]$) by polynomials with *integer coefficients*.

Chapter 5
Metric Spaces

Our goal in this chapter is to show that most of the concepts introduced in the previous chapters for the set $\mathbb{R}$ of real numbers can be extended to any abstract *metric space*, i.e., a set on which the concept of *metric* (or *distance*) can be defined. Indeed, as we have already seen, the basic concept of *limit* which we studied in Chaps. 2 and 3, and used to define (in Chap. 4) the related concept of continuity, is defined in terms of *distance*. Let us recall that the distance between two real numbers x and y is defined to be $d(x, y) := |x - y|$ and satisfies three simple properties: For any numbers x, y, $z \in \mathbb{R}$ we have: (i) $|x - y| \geq 0$, and equality holds if and only if $x = y$; (ii) $|x - y| = |y - x|$; and (iii) $|x - y| \leq |x - z| + |y - z|$. Property (iii) is called the *Triangle Inequality* for obvious geometrical reasons. Using this distance, we defined, in Chap. 2, the concepts of ε-neighborhood, open set, closed set, limit point, isolated point, convergent sequence, and Cauchy sequence. We then defined the concept of limit for general real-valued functions of a real variable and proved that such limits can also be defined in terms of limits of sequences. Also, before introducing the related notion of *continuity*, we introduced (in Chap. 4) the *topological* concepts of *compactness* and *connectedness*. All these notions can be defined, in essentially the same way, in any (abstract) *metric space* and the proofs of most theorems are basically copies of the ones we gave for the special metric space $\mathbb{R}$, if one replaces $|x - y|$ by $d(x, y)$ throughout. Many proofs will therefore be brief or will be left as exercises for the reader.

5.1 Metrics and Metric Spaces

In this section we define the concepts of *metric*, *metric space*, and *subspace*. The reader should constantly compare the material to the corresponding one for the set of real numbers and its subsets, as most of the ideas developed here have their origin in the study of the real line and its *topological* properties.

© Springer Science+Business Media New York 2014
H.H. Sohrab, *Basic Real Analysis*, DOI 10.1007/978-1-4939-1841-6_5

Definition 5.1.1 (Metric Space, Subspace). A *metric space* (M, d) is a set M, whose elements will be called *points*, together with a map $d : M \times M \to \mathbb{R}$, called a *distance* (or *metric*), such that for any x, y, $z \in M$, the following properties are satisfied:

1. $d(x, y) \geq 0$;
2. $d(x, y) = 0 \iff x = y$;
3. $d(x, y) = d(y, x)$;   and
4. $d(x, z) \leq d(x, y) + d(y, z)$ (Triangle Inequality).

For each subset $S \subset M$, let $d_S := d | S \times S$ be the restriction of the metric d to the subset $S \times S \subset M \times M$. The metric space (S, d_S) is called a (metric) *subspace* of (M, d).

Remarks and Notation 5.1.2. We will usually abuse the notation and write "metric space M," instead of "metric space (M, d)," unless more than one such space is involved or the same set M is endowed with different distances. Also, for a subspace (S, d_S) of (M, d), we shall omit the subscript S in d_S and say that (S, d) is a (metric) subspace of (M, d) or even that S is a subspace of M.

Example 5.1.3.

1. For our purposes, it is obvious that the most important example is the space $\mathbb{R}$ of real numbers with its usual distance. There are, however, many *unusual* metric spaces and some of them will be introduced later in this chapter, possibly in exercises.
2. **(Extended Real Line)** Consider the function $f : \mathbb{R} \to (-1, 1)$ given by $f(x) := x/(1 + |x|) \; \forall x \in \mathbb{R}$. This map is a *homeomorphism* of $\mathbb{R}$ onto $(-1, 1)$ with inverse $f^{-1} : x \mapsto x/(1 - |x|)$ (Exercise 4.6.16). We extend f to $\overline{\mathbb{R}} := [-\infty, +\infty]$ by setting $f(-\infty) = -1$ and $f(+\infty) = 1$. Now on $\overline{\mathbb{R}}$ we define the metric d by $d(x, y) := |f(x) - f(y)| \; \forall x, y \in \overline{\mathbb{R}}$. It is easy to see that this is in fact a distance (why?); note that, for $x \geq 0$, we have $d(x, +\infty) = 1/(1 + |x|)$ and, for $x \leq 0$, $d(-\infty, x) = 1/(1 + |x|)$. With this metric, whose restriction to $\mathbb{R}$ is *not* the usual distance, $\overline{\mathbb{R}}$ becomes a metric space called the *extended real line*.
3. **(Discrete Metric)** Let $M \neq \emptyset$ be an arbitrary set and, for any x, $y \in M$, let $d(x, y) = 1$ if $x \neq y$ and $d(x, y) = 0$ if $x = y$. It is easy to check that all the above properties $(1) - (4)$ are satisfied, so that d is indeed a metric on M, called the *discrete metric*. With this metric, M is called a *discrete* metric space.
4. **(Uniform Metric)** Let S be any nonempty set, and let $M = B(S, \mathbb{R})$ be the set of all *bounded* functions from S to $\mathbb{R}$; i.e., the set of all $f : S \to \mathbb{R}$ such that $|f(s)| \leq A$ for all $s \in S$ and some $A > 0$. Note that, for any f, $g \in M$, we also have $f - g \in M$. (Why?) Now define

$$d_\infty(f, g) := \sup(\{|f(s) - g(s)| : s \in S\}).$$

The reader can check at once that this d_∞ is indeed a metric on M. We call it the *uniform metric* for reasons to be explained later.

Exercise 5.1.4 (Bounded Metrics). Let (M, d) be a metric space. Define the maps d_1, $d_2 : M \times M \to \mathbb{R}$ by $d_1 := d/(1 + d)$ and $d_2 := \min(1, d)$; i.e., for any x, $y \in M$, $d_1(x, y) = d(x, y)/(1 + d(x, y))$ and $d_2(x, y) = \min\{1, d(x, y)\}$. Show that d_1 and d_2 are both metrics on M and that we have $d_1 \leq d_2 \leq 2d_1$.

Exercise 5.1.5 (Product Metrics). Consider the set $\mathbb{R}^n := \{(x_1, x_2, \ldots, x_n) : x_k \in \mathbb{R} \ 1 \leq k \leq n\}$. Define the maps d_{euc}, d_{max}, $d_{\text{sum}} : \mathbb{R}^n \times \mathbb{R}^n \to \mathbb{R}$, as follows. For each $x = (x_1, x_2, \ldots, x_n)$ and $y = (y_1, y_2, \ldots, y_n)$, let

$$d_{\text{euc}}(x, y) := \sqrt{\sum_{k=1}^{n}(x_k - y_k)^2},$$

$$d_{\text{max}}(x, y) := \max\{|x_1 - y_1|, \ldots, |x_n - y_n|\}, \text{ and}$$

$$d_{\text{sum}}(x, y) := \sum_{k=1}^{n}|x_k - y_k|.$$

Show that d_{euc}, d_{max}, and d_{sum} are metrics on $\mathbb{R}^n$ and that we have the inequalities $d_{\text{max}} \leq d_{\text{euc}} \leq d_{\text{sum}} \leq n d_{\text{max}}$.

Definition 5.1.6 (Euclidean n-Space). The set $\mathbb{R}^n$ together with the metric d_{euc} defined in Exercise 5.1.5 is called the *Euclidean n-space* (or the *n-dimensional Euclidean space*).

Exercise 5.1.5 suggests a way of defining metrics on a (Cartesian) *product* of metric spaces:

Definition 5.1.7 (Product Spaces, Projections, Diagonal).

(a) Let M_1, M_2, $\ldots$, M_n be metric spaces with metrics d_1, d_2, $\ldots$, d_n, respectively, and let $M := M_1 \times M_2 \times \cdots \times M_n$. Define the maps d_{euc}, d_{max}, $d_{\text{sum}} : M \times M \to \mathbb{R}$ as follows:

$$d_{\text{euc}}(x, y) := \sqrt{\sum_{k=1}^{n} d_k^2(x_k, y_k)},$$

$$d_{\text{max}}(x, y) := \max\{d_1(x_1, y_1), \ldots, d_n(x_n, y_n)\}, \text{ and}$$

$$d_{\text{sum}}(x, y) := \sum_{k=1}^{n} d_k(x_k, y_k),$$

for every $x = (x_1, \ldots, x_n)$ and $y = (y_1, \ldots, y_n)$ in M. Then d_{euc}, d_{max}, and d_{sum} are metrics on M (Exercise 5.1.5), and the set M (with *any* one of these metrics) is called the (metric space) *product* of the metric spaces $M_1, \ldots, M_n$.

(b) With notation as in (a), we define the *k*th *projection* $\pi_k : M \to M_k$ by $\pi_k(x) :=$ x_k, $1 \le k \le n$, for each $x = (x_1, \ldots, x_n) \in M$.

(c) Given a metric space (M, d), the *diagonal* of $M \times M$ is defined to be

$$\Delta_M := \{(x, x) : x \in M\}.$$

Remark 5.1.8.

(a) The concepts of *open set, closed set,* and *topology* will be defined in the next section, and we shall see later that the three metrics d_{euc}, d_{max}, and d_{sum} produce the same topology (i.e., the same collection of open sets) and hence the same notions of limit and continuity on the product space $M = \prod_{k=1}^{n} M_k$. As we shall see on several occasions, depending on circumstances, one of the above metrics may be preferable and may simplify the analysis at hand in a significant way.

(b) It is easy to see (cf. Theorem 5.2.2(6) below) that the *diagonal* Δ_M is a *closed* subset of $M \times M$ for any metric space M.

Exercise 5.1.9. Let (M, d) be a metric space.

(a) Show that, for any $x_1, x_2, \ldots, x_n \in M$, $n \ge 2$, we have

$$d(x_1, x_n) \le d(x_1, x_2) + d(x_2, x_3) + \cdots + d(x_{n-1}, x_n).$$

(b) Show that, for any x, y, $z \in M$, we have

$$|d(x, z) - d(y, z)| \le d(x, y).$$

Exercise 5.1.10 (Ultrametric Space). A metric space (M, d) is called an *ultrametric space* if its metric d satisfies the *ultrametric inequality*:

$$d(x, z) \le \max\{d(x, y), d(y, z)\} \qquad (\forall \ x, y, z \in M).$$

Show that we then have $d(x, z) = \max\{(d(x, y), d(y, z)\}$ if $d(x, y) \ne d(y, z)$.

The distance between pairs of *points* in a metric space can be used to define distances between arbitrary pairs of *subsets* of the space as follows.

Definition 5.1.11 (Distance Between Subsets, Diameter, Bounded Set). Let (M, d) be a metric space.

1. For any *nonempty* subsets A, $B \subset M$, the *distance* between them is defined by

$$d(A, B) := \inf\{d(a, b) : a \in A, \ b \in B\}.$$

We write $d(a, B) := d(\{a\}, B)$.

2. The *diameter* of a *nonempty* subset $A \subset M$, denoted by $\delta(A)$, is defined to be the *extended* real number

$$\delta(A) := \sup\{d(x, y) : x, \ y \in A\} \in [0, \infty].$$

3. A *nonempty* subset $A \subset M$ is said to be *bounded* if its diameter is *finite*.

Exercise 5.1.12. Let (M, d) be a metric space. Show that, for any nonempty subsets $A, \ B \subset M$ and any points $x, \ y \in M$, the following are true.

1. If $A \cap B \neq \emptyset$, then $d(A, B) = 0$, but the converse *need not* be true. *Hint:* Use subsets of $\mathbb{R}$ to give a counterexample.
2. $|d(x, A) - d(y, A)| \leq d(x, y)$. Thus, $x \mapsto d(x, A)$ is *Lipschitz*.
3. If A and B are *bounded*, then so is $A \cup B$; i.e.,

$$\delta(A) < \infty \quad \text{and} \quad \delta(B) < \infty \quad \Longrightarrow \quad \delta(A \cup B) < \infty.$$

4. Define (on M) the metrics $d_1 := d/(1 + d)$ and $d_2 := \min(1, d)$ as in Exercise 5.1.4. Show that in the metric spaces (M, d_1) and (M, d_2) all subsets are *bounded*.

Remark 5.1.13.

(a) Although, for nonempty subsets $A, \ B \subset M, \ d(A, B)$ is called the *distance* between A and B, it does *not* define a metric on $\mathcal{P}(M) \setminus \{\emptyset\}$. For example, consider the subsets $A := [0, 1], \ B := [1, 2]$, and $C := [2, 3]$ of $\mathbb{R}$. Then we obviously have $d(A, B) = d(B, C) = 0$, but $d(A, C) = 1$.

(b) If (M, d) is a metric space and if d_1 and d_2 are the metrics defined in Exercise 5.1.4 above, then, as we shall see soon, these metrics give the set M the *same topology* (i.e., the same collection of *open* sets) as does the metric d, but they have the advantage of being both *bounded*.

Next, we define, for a general metric space, the concepts of open ball, closed ball, and sphere. Open balls are extensions to general metric spaces of the *ε-neighborhoods* we introduced for the real line $\mathbb{R}$.

Definition 5.1.14 (Open Ball, Closed Ball, Sphere). Let $x \in M$ (where M is a metric space) and let $\varepsilon > 0$ be arbitrary.

1. The *open ball* of radius ε centered at x is defined to be the set $B_\varepsilon(x) := \{y \in M : d(x, y) < \varepsilon\}$, and the corresponding *deleted open ball* is defined by $\mathring{B}_\varepsilon(x) := B_\varepsilon(x) \setminus \{x\}$.
2. The *closed ball* of radius ε centered at x is defined to be the set $B'_\varepsilon(x) := \{y \in M : d(x, y) \leq \varepsilon\}$.
3. The *sphere* of radius ε centered at x is defined to be the set $S_\varepsilon(x) := \{y \in M : d(x, y) = \varepsilon\}$.

Example 5.1.15.

(a) In the metric space $\mathbb{R}$ with its usual metric, the open ball $B_\varepsilon(x) := \{y \in \mathbb{R} : |y - x| < \varepsilon\} = (x - \varepsilon, x + \varepsilon)$ is simply the ε-neighborhood of x. The corresponding closed ball and sphere are $B'_\varepsilon(x) = [x - \varepsilon, x + \varepsilon]$ and $S_\varepsilon(x) = \{x - \varepsilon, x + \varepsilon\}$, respectively.

(b) Let M be a *discrete* metric space (cf. Example 5.1.3(3)). Then, for $\varepsilon < 1$, we have $B_\varepsilon(x) = B'_\varepsilon(x) = \{x\}$ and $S_\varepsilon(x) = \emptyset$. For $\varepsilon = 1$, $B_1(x) = \{x\}$, $B'_1(x) = M$, and $S_1(x) = M \setminus \{x\}$. Finally, for $\varepsilon > 1$, we have $B_\varepsilon(x) = B'_\varepsilon(x) = M$ and $S_\varepsilon(x) = \emptyset$.

(c) Consider the product $M := M_1 \times M_2 \times \cdots \times M_n$ of the metric spaces (M_k, d_k), $1 \le k \le n$, with metric $d_{\max}$ defined as in Definition 5.1.7. Then, as the reader may easily check, we have

$$B_\varepsilon(x) = B_{1,\varepsilon}(x_1) \times \cdots \times B_{n,\varepsilon}(x_n), \qquad (*)$$

where $B_{k,\varepsilon}(x_k)$ denotes the open ball of radius ε centered at $x_k \in M_k$, $1 \le k \le n$. Note that, by $(*)$, we have $\pi_k(B_\varepsilon(x)) = B_{k,\varepsilon}(x_k)$, $1 \le k \le n$. In particular, in $\mathbb{R}^n$ with the above metric, $B_\varepsilon(x)$ is a *cube* with sides parallel to the axes.

5.2   Topology of a Metric Space

We now look at the *topological structure* of a metric space, i.e., the structure deduced from the collection of its *open sets*. These open sets will then be used to define the limits of sequences and functions and the related concept of continuity.

Definition 5.2.1 (Open, Closed, Limit Point, Isolated Point, Perfect). Let (M, d) be a metric space.

1. A subset $U \subset M$ is called *open* if, given any $x \in U$, we have $B_\varepsilon(x) \subset U$ for some $\varepsilon = \varepsilon(x) > 0$.
2. A subset $F \subset M$ is called *closed* if its complement $F^c = M \setminus F$ is open.
3. A point $x \in M$ is called a *limit point* (or *accumulation point*) of a subset $E \subset M$ if, $\forall \varepsilon > 0$, $\dot{B}_\varepsilon(x) \cap E := (B_\varepsilon(x) \setminus \{x\}) \cap E \neq \emptyset$. (Note that x *need not* be in E.)
4. A point $x \in E \subset M$ is called an *isolated point* of E if it is *not* a limit point of E. Equivalently, $x \in E$ is an isolated point of E if $\exists \varepsilon > 0$ such that $B_\varepsilon(x) \cap E = \{x\}$.
5. A set $E \subset M$ is called *perfect* if it is *closed* and if *every* point of E is a limit point of E.

Theorem 5.2.2. *Let M be a metric space.*

1. *Ø and M are simultaneously open and closed.*
2. *For any collection $\{U_\lambda\}_{\lambda \in \Lambda}$ of open subsets of M, the union $\bigcup_{\lambda \in \Lambda} U_\lambda$ is open and for any finite collection $\{U_k\}_{k=1}^n$ of open subsets of M, the intersection $\bigcap_{k=1}^n U_k$ is open.*
3. *For any collection $\{F_\lambda\}_{\lambda \in \Lambda}$ of closed subsets of M, the intersection $\bigcap_{\lambda \in \Lambda} F_\lambda$ is closed and for any finite collection $\{F_k\}_{k=1}^n$ of closed subsets of M, the union $\bigcup_{k=1}^n F_k$ is closed.*
4. *Any singleton $\{x\} \subset M$ is closed and hence so is any finite subset of M.*
5. *Every open ball $B_\varepsilon(x)$ is open and every closed ball $B_\varepsilon'(x)$ is closed. Also, every sphere $S_\varepsilon(x)$ is closed.*
6. *The diagonal $\Delta_M := \{(x, x) : x \in M\}$ is closed in $M \times M$.*

Exercise 5.2.3. Prove Theorem 5.2.2. *Hint:* For (6), use the distance $d_{\max}$ and the comments in Example 5.1.15(c).

Definition 5.2.4 (Topology of (M, d), Equivalent Metrics).

(a) Given a metric space (M, d), the collection $\mathcal{T}(M, d) := \{U \subset M : U$ is open$\}$ of all *open* subsets of M is called the *topology* of (M, d). We sometimes write $\mathcal{T}(M)$ if there is no danger of confusion.
(b) Two metrics d and d' on a set M are said to be *equivalent* if they define the *same* topology on M, i.e., if $\mathcal{T}(M, d) = \mathcal{T}(M, d')$.

Theorem 5.2.5 (Open & Closed Relative to a Subspace). *Let (M, d) be a metric space and let $X \subset M$. A set $U \subset X$ is open (in X) if and only if $U = X \cap U'$ for an open set U' in M. A set $F \subset M$ is closed (in X) if and only if $F = X \cap F'$ for a closed set F' in M.*

Proof. Let $x \in X$. Then, in the (metric) *subspace* X, the open ball of radius $\varepsilon > 0$ centered at x is $B_{\varepsilon, X}(x) := \{y \in X : d(x, y) < \varepsilon\} = X \cap B_\varepsilon(x)$. Now, if U' is open (in M), then for any $x \in X \cap U'$ we have $B_\varepsilon(x) \subset U'$ for some $\varepsilon > 0$ and hence $B_{\varepsilon, X}(x) := X \cap B_\varepsilon(x) \subset X \cap U'$. Therefore, $X \cap U'$ is *open* (in X). Conversely, if $U \subset X$ is open (in X), then for each $x \in U$ there exists $\varepsilon_x > 0$ such that $B_{\varepsilon_x, X}(x) = X \cap B_{\varepsilon_x}(x) \subset U$. If we let $U' := \bigcup_{x \in U} B_{\varepsilon_x}(x)$, then U' is open (in M) and $U = X \cap U'$. Finally, a set $F \subset X$ is *closed* (in X) if and only if $X \setminus F$ is open (in X). By what we just proved, this is so if and only if $X \setminus F = X \cap U'$ for some open set $U' \subset M$. But then $F = X \cap F'$, where $F' := U'^c = M \setminus U'$ is indeed *closed* in M. $\qquad\Box$

Definition 5.2.6 (Relative Topology). Let X be a subset of a metric space (M, d). Then the topology of X induced by the metric d *restricted* to X is called the *relative topology* on X and, with this topology, X is then a (topological) *subspace* of M. It follows from Theorem 5.2.5 that

$$\mathcal{T}(X, d) = X \cap \mathcal{T}(M, d) := \{X \cap U : U \subset M \text{ is open}\}.$$

Remark 5.2.7. Note that, in a subspace X of a metric space M, a subset $S \subset X$ which is *open* (resp., *closed*) in X need *not* (in general) be open (resp., closed) in M. For example, $(1/2, 1]$ is open in the subspace $(0, 1]$ of $\mathbb{R}$, but is certainly not open in $\mathbb{R}$. Similarly, $(0, 1/2]$ is closed in $(0, 1]$ but not in $\mathbb{R}$. However, we have the following:

Proposition 5.2.8. *Let M be a metric space and let X be a subspace. In order that all open (resp., closed) subsets of X be open (resp., closed) in M, it is necessary and sufficient that X be open (resp., closed) in M.*

Exercise 5.2.9. Prove Proposition 5.2.8.

Definition 5.2.10 (Interior, Exterior, Closure, Boundary). For any metric space M and any subset $A \subset M$, we define the following:

1. A point $x \in M$ is called an *interior point* of A if, for some $\varepsilon > 0$, we have $B_\varepsilon(x) \subset A$. The set of all interior points of A is denoted by A° and is called the *interior* of A. Clearly, $A^\circ \subset A$.
2. A point $x \in M$ is called a *cluster point* of A if x is a limit point of A, or $x \in A$. The set of all cluster points of A is called the *closure* of A and is denoted by A^-. We obviously have $A \subset A^-$.
3. A point $x \in M$ is called an *exterior point* of A if it is an interior point of the complement $A^c = M \setminus A$. The set $(A^c)^\circ$ of all exterior points of A is called the *exterior* of A and is denoted by $\text{Ext}(A)$.
4. A point $x \in M$ is called a *boundary point* of A if x is a cluster point of *both* A and A^c. The set of all boundary points of A is denoted by $\text{Bd}(A)$ and is called the *boundary* of A. It is clear that $\text{Bd}(A) = A^- \cap (A^c)^- = \text{Bd}(A^c)$.

Example 5.2.11.

(a) In $\mathbb{R}$ with its usual metric, let $A := [a, b)$, $a < b$. Then $A^\circ = (a, b)$, $A^- = [a, b]$, $\text{Bd}(A) = \{a, b\}$, and $\text{Ext}(A) = (-\infty, a) \cup (b, \infty)$.
(b) Consider the set $\mathbb{N} \subset \mathbb{R}$. We have $\mathbb{N}^\circ = \emptyset$ and $\mathbb{N}^- = \mathbb{N} = \text{Bd}(\mathbb{N})$.
(c) Let M be a *discrete* metric space. Then, for $\varepsilon \in (0, 1]$, we have $B_\varepsilon(x) = \{x\} \ \forall x \in M$. Hence $B_1(x) = (B_1(x))^- = \{x\} \ \forall x \in M$. (Why?) On the other hand, the corresponding *closed ball* is $B_1'(x) = M$ and hence $\left(B_1'(x)\right)^\circ = M \ \forall x \in M$. This shows that the closure of an open ball (which is always a subset of the corresponding closed ball (why?)) is *not*, in general, equal to the closed ball. And the *interior* of the closed ball is *not* always the corresponding *open* ball.

Exercise 5.2.12. Let M be a metric space and let A, $B \subset M$.

1. Show that

$$A^\circ = \bigcup \{U \subset M : U \subset A \text{ and } U \text{ open}\}, \text{ and}$$

$$A^- = \bigcap \{F \subset M : A \subset F \text{ and } F \text{ closed}\}.$$

2. Show that A° is the *largest* open set *contained* in A and that A^- is the *smallest* closed set *containing* A. Here, "largest" and "smallest" are in the sense of *inclusion* "$\subset$." Conclude that A is open (resp., closed) if and only if $A = A^\circ$ (resp., $A = A^-$).

3. Show that, if M is *discrete*, then $A^\circ = A^- = A$; in other words *every* subset is simultaneously open and closed.

4. Let A' denote the set of all *limit points* of A. Show that $A^- = A \cup A'$. Deduce that A is closed if and only if $A' \subset A$.

5. Show that $x \in A^-$ if and only if $d(x, A) = 0$.

6. Show that $(A \cap B)^\circ = A^\circ \cap B^\circ$ and that $(A \cup B)^- = A^- \cup B^-$. Prove the inclusions $A^\circ \cup B^\circ \subset (A \cup B)^\circ$, $(A \cap B)^- \subset A^- \cap B^-$ and, using subsets of $\mathbb{R}$ with its usual metric, show that both inclusions may be *proper*.

7. Using induction, extend the facts in part (6) to any *finite* collection of subsets of M. Thus, given any sets $A_1, \ldots, A_n \subset M$,

$$\left(\bigcap_{j=1}^n A_j \right)^\circ = \bigcap_{j=1}^n A_j^\circ, \qquad \left(\bigcup_{j=1}^n A_j \right)^- = \bigcup_{j=1}^n A_j^-.$$

8. Show that $\text{Ext}(A) = (A^-)^c$. Also, show that $(A^c)^- = (A^\circ)^c$, and deduce that $\text{Bd}(A) = A^- \setminus A^\circ$.

9. Show that $\text{Bd}(A \cup B) \subset \text{Bd}(A) \cup \text{Bd}(B)$ and, using subsets of $\mathbb{R}$, show that the inclusion may be *proper*. Show, however, that if $A^- \cap B^- = \emptyset$, then $\text{Bd}(A \cup B) = \text{Bd}(A) \cup \text{Bd}(B)$.

10. Show that A is *open* if and only if the following holds for *every* set $B \subset M$:

$$A \cap B = \emptyset \implies A \cap B^- = \emptyset.$$

Remark 5.2.13. The *intersection* of a family of open sets is *not* open in general, as the example $\bigcap_{n=2}^\infty (-1/n, 1 + 1/n) = [0, 1]$ shows. Similarly, the *union* of a family of closed sets is *not* closed in general. For instance, $\bigcup_{n=2}^\infty [1/n, 1 - 1/n] = (0, 1)$. There are, however, exceptional cases as we shall see presently. First a definition:

Definition 5.2.14 (Locally Finite Family). A family $\mathcal{A}$ of subsets of a metric space M is said to be *locally finite* if, given any $x \in M$, there exists $\delta = \delta(x) > 0$ such that the open ball $B_\delta(x)$ has a *nonempty* intersection with (at most) a *finite* number of the sets $A \in \mathcal{A}$.

Proposition 5.2.15. *If $\mathcal{A} \subset \mathcal{P}(M)$ is a locally finite family of subsets of M, then*

$$\left(\bigcup_{A \in \mathcal{A}} A \right)^- = \bigcup_{A \in \mathcal{A}} A^-. \tag{†}$$

In particular, the union of a locally finite family of closed sets is closed.

Proof. The inclusion "⊃" in (†) is obvious. (Why?) To prove the reverse inclusion, suppose that $x \in (\bigcup_{A \in \mathcal{A}} A)^-$. We can pick $\delta > 0$ such that $B_\delta(x)$ has nonempty intersection with a *finite* number of the $A \in \mathcal{A}$, say $A_1, \ldots, A_m$. By Exercise 5.2.12(10), we then have $B_\delta(x) \cap A^- = \emptyset$ unless $A \in \{A_1, \ldots, A_m\}$. In particular, note that the family $\{A^- : A \in \mathcal{A}\}$ of all *closures* of the elements of $\mathcal{A}$ is also *locally finite*. (Why?) Using Exercise 5.2.12(7), it now follows that

$$x \in \left(\bigcup_{n=1}^m A_n \right)^- = \bigcup_{n=1}^m A_n^- \subset \bigcup_{A \in \mathcal{A}} A^-.$$

$\square$

Exercise 5.2.16 (Interior and Closure in a Subspace). Let X be a (metric) subspace of a metric space M and let $S \subset X$. Show that the *closure* of S *relative to* X is $X \cap S^-$ but that the *interior* of S *relative to* X need not be $X \cap S^\circ$. *Hint:* In $\mathbb{R}$, let $X := (0, 2]$ and $S := (1, 2]$.

Definition 5.2.17. Let (M, d) be a metric space.

(Dense).   A set $D \subset M$ is called *dense* (in M) if $D^- = M$.
(Nowhere Dense).   A set $A \subset M$ is called *nowhere dense* if $(A^-)^\circ = \emptyset$, i.e., if the closure of A contains no nonempty open balls.
(First Category, Second Category).   A set $E \subset M$ is said to be of *first category* (or *meager*) in M if it is a *countable union of nowhere dense sets*. A set which is *not* of first category is said to be of *second category*.
(Separable).   The metric space M is called *separable* if it contains a *countable dense* subset, i.e., if $D^- = M$ for a *countable* subset $D \subset M$.
(Second Countable).   The metric space M is called *second countable* (or is said to satisfy the *second axiom of countability*) if it has a *countable base*, in other words, if there is a countable collection $\{U_n\}_{n \in \mathbb{N}}$ of open sets in M such that each open set in M is the union of the U_j that it contains.

Exercise 5.2.18. Let (M, d) be a metric space and $A \subset M$. Show that

$$A \text{ is nowhere dense} \iff A^- \text{ is nowhere dense} \iff \text{Ext}(A) = (A^c)^\circ \text{ is dense.}$$

Example.

(a) Consider the subspace $[0, 1]$ of the metric space $\mathbb{R}$. The sets $\mathbb{Q} \cap [0, 1]$ and $\mathbb{Q}^c \cap [0, 1]$ are *dense*. Since the set $\mathbb{Q}$ of rationals is countable and is dense in $\mathbb{R}$, the metric space $\mathbb{R}$ is *separable* and so are its subspaces.
(b) As we saw in Chap. 4, the Cantor set $C \subset [0, 1]$ is *perfect* and *nowhere dense*. It is therefore an example of a perfect set of *first category*. It is obvious that any *singleton* $\{x\} \subset \mathbb{R}$ is *nowhere dense*. Thus any *countable* subset of $\mathbb{R}$ is of first category. In particular, the *dense* set $\mathbb{Q}$ of rationals is of first category. We shall see (Corollary 5.3.10 below) that $\mathbb{R}$ is of *second category* (in itself).

The next proposition shows that, for metric spaces, *separability* is equivalent to *second countability*:

Proposition 5.2.19. *A metric space is separable if and only if it is second countable.*

Proof. Suppose that (M, d) is separable and let $\{x_1, x_2, x_3, \ldots\} \subset M$ be a *dense* subset. For any j, $k \in \mathbb{N}$, set $U_{j,k} := B_{1/k}(x_j) = \{x \in M : d(x, x_j) < 1/k\}$. We claim that $\{U_{j,k} : j, k \in \mathbb{N}\}$ is a (countable) *base*. Indeed, given any open set $O \subset M$, let O' be the union of the $U_{j,k}$ contained in O. Then O' is obviously an open subset of M and we have $O' \subset O$. To prove the reverse inclusion, let $x \in O$ be arbitrary. Then there exists $k_0 \in \mathbb{N}$ such that $B_{1/k_0}(x) \subset O$. Let $j_0 \in \mathbb{N}$ be such that $d(x_{j_0}, x) < 1/(2k_0)$. Then $x \in U_{j_0, 2k_0} \subset O'$. Conversely, suppose that M is second countable; i.e., there is a sequence $(U_n)_{n \in \mathbb{N}}$ of open sets such that each open set in M is the union of some of the U_n. For each $n \in \mathbb{N}$, pick $x_n \in U_n$ and let $D = \{x_n : n \in \mathbb{N}\}$. Then, every (nonempty) open subset V of M contains at least one of the U_n and hence $V \cap D \neq \emptyset$. Therefore, D is dense and the proof is complete. $\qquad\square$

5.3 Limits, Cauchy Sequences, and Completeness

In Chaps. 2 and 3, we studied limits of sequences and functions in the metric space $\mathbb{R}$. Here, we shall do the same for general metric spaces. Since the basic definitions and properties are quite similar, the presentation will not be as extensive as before.

Definition 5.3.1 (Limit of a Sequence). Let (M, d) be a metric space. A sequence $(x_n) \in M^{\mathbb{N}}$ is said to *converge* to the *limit* $\xi \in M$ if

$$(\forall \varepsilon > 0)(\exists N \in \mathbb{N})(n \geq N \Rightarrow x_n \in B_\varepsilon(\xi)).$$

If (x_n) *converges* to ξ, we write $\lim_{n \to \infty} x_n = \lim(x_n) = \xi$. If (x_n) has no limit, we say that it is *divergent*.

The following proposition justifies the use of *the* limit (instead of limit) in the above definition.

Proposition 5.3.2 (Uniqueness of the Limit). *The limit of a convergent sequence in a metric space is unique.*

Proof. Suppose that a sequence (x_n) in a metric space (M, d) has two limits ξ, $\eta \in M$. Using the definition, for every $\varepsilon > 0$, we can find $N' \in \mathbb{N}$ (resp., $N'' \in \mathbb{N}$) such that $n \geq N'$ (resp., $n \geq N''$) implies $d(x_n, \xi) < \varepsilon/2$ (resp., $d(x_n, \eta) < \varepsilon/2$). If $N := \max\{N', N''\}$, then the Triangle Inequality gives

$$n \geq N \implies d(\xi, \eta) \leq d(\xi, x_n) + d(x_n, \eta) < \frac{\varepsilon}{2} + \frac{\varepsilon}{2} = \varepsilon.$$

Since this holds for *every* $\varepsilon > 0$, we get $d(\xi, \eta) = 0$ and hence $\xi = \eta$. ☐

Definition 5.3.3 (m-Tail, Ultimately True).

(a) Let (x_n) be a sequence in a metric space M. Given any integer $m \in \mathbb{N}$, the *m-tail* of (x_n) is the sequence $(x_m, x_{m+1}, x_{m+2}, \ldots)$.
(b) A property of sequences in a metric space M is said to be *ultimately true* for a sequence $(x_n) \in M^{\mathbb{N}}$ if, for some $m \in \mathbb{N}$, it holds for the m-tail $(x_m, x_{m+1}, x_{m+2}, \ldots)$.

Exercise 5.3.4. Let (M, d) be a metric space.

(a) Let $X \subset M$ and let x_0 be a *limit point* of X. Show that $x_0 = \lim(x_n)$ for a sequence $(x_n) \in X^{\mathbb{N}}$. Show that, if $x_0 \notin X$, then the *converse* is also true.
(b) Let $X \subset M$. Show that $x_0 \in X^-$ if and only if $x_0 = \lim(x_n)$ for some sequence $(x_n) \in X^{\mathbb{N}}$. Deduce that $X \subset M$ is *closed* if and only if $\lim(x_n) \in X$ for every *convergent* sequence $(x_n) \in X^{\mathbb{N}}$. *Hint:* Exercise 5.2.12(4).
(c) Assume that M is *discrete*. Show that a sequence $(x_n) \in M^{\mathbb{N}}$ is convergent if and only if it is *ultimately constant*, i.e., if and only if there exists $m \in \mathbb{N}$ such that $x_n = x_m \; \forall n \geq m$. What is the limit?

Definition 5.3.5 (Cauchy Sequence, Complete Metric Space). Let (M, d) be a metric space.

(a) A sequence $(x_n) \in M^{\mathbb{N}}$ is said to be a *Cauchy sequence* if

$$(\forall \varepsilon > 0)(\exists N \in \mathbb{N})(m, n \geq N \Rightarrow d(x_m, x_n) < \varepsilon).$$

(b) M is called a *complete* metric space if every Cauchy sequence in M converges to a point in M.

Example.

(a) The metric space $\mathbb{R}$ with its usual metric $d(x, y) = |x - y|$ is complete and separable. Indeed, the completeness is a consequence of Cauchy's Criterion, and the separability follows from the fact that the set $\mathbb{Q}$ of rational numbers, which we know is countable, is dense in $\mathbb{R}$. It follows that all *closed* subsets of $\mathbb{R}$ are also complete, separable metric spaces (Exercise 5.3.6 below).
(b) **(Uniform Approximation)** Consider the metric space $B[0, 1]$ of all *bounded* real-valued functions defined on $[0, 1]$ with the *uniform metric*

$$d_\infty(f, g) := \sup\{|f(x) - g(x)| : 0 \leq x \leq 1\}.$$

Let $C[0, 1]$, $Pol[0, 1]$, $Step[0, 1]$, and $PL[0, 1]$ denote the sets of *continuous, polynomial, step,* and *piecewise linear* functions on $[0, 1]$, respectively. It is obvious that these are all subspaces of the metric space $(B[0, 1], d_\infty)$ and that

$C[0, 1] \supset Pol[0, 1]$. (Why?) We shall see later that $B[0, 1]$ is *complete* and that $C[0, 1]$ is a *closed* subspace and hence (as we shall see below) is also complete. Now, as we saw in Chap. 4 (Theorems 4.7.2, 4.7.4, and 4.7.9), any continuous $f \in C[0, 1]$ is the limit (with respect to the uniform metric d_∞ above) of a sequence of step, piecewise linear, or polynomial functions. We express this by saying that f can be approximated *uniformly* by step, piecewise linear, or polynomial functions. Equivalently, the subspaces $Step[0, 1]$, $PL[0, 1]$, and $Pol[0, 1]$ are all *dense* in $C[0, 1]$. Since the continuous function $x \mapsto \sin x$, say, is *not* a step, piecewise linear, or polynomial function on $[0, 1]$, it follows that the subspaces $Step[0, 1]$, $PL[0, 1]$, and $Pol[0, 1]$ are *not* complete.

Exercise 5.3.6. Let M be a metric space. Prove the following statements:

1. If $(x_n) \in M^{\mathbb{N}}$ *converges*, then it is a Cauchy sequence.
2. For a sequence $(x_n) \in M^{\mathbb{N}}$, let $T_m = \{x_m, x_{m+1}, x_{m+2}, \ldots\}$ be the *set* of all terms in the m-tail of (x_n). Show that (x_n) is Cauchy if and only if $\lim_{m \to \infty} \delta(T_m) = 0$. Recall that $\delta(T_m)$ is the *diameter* of T_m.
3. Show that a Cauchy sequence $(x_n) \in M^{\mathbb{N}}$ is *bounded*; i.e., the diameter $\delta(\{x_1, x_2, x_3, \ldots\})$ is *finite*.
4. If $(x_n) \in M^{\mathbb{N}}$ is a *Cauchy* sequence, then $\lim(x_n) = \xi$ if and only if for some *subsequence* (x_{n_k}) of (x_n) we have $\lim(x_{n_k}) = \xi$.
5. A subset $A \subset M$ is *dense* in M if and only if for each $\xi \in M$ there is a sequence $(x_n) \in A^{\mathbb{N}}$ with $\lim(x_n) = \xi$.
6. If a (metric) subspace $A \subset M$ is *complete*, then it is *closed*. If the space M itself is *complete*, then (a subspace) $A \subset M$ is complete if and only if it is *closed*.

Our next goal is to prove the important *Baire Category Theorem*, which asserts that a complete metric space is of *second category* (in itself). We first prove a necessary and sufficient condition for the completeness of a metric space. This result, which is due to Cantor, will remind you of the *Nested Intervals Theorem* for the metric space $\mathbb{R}$.

Theorem 5.3.7 (Cantor's Theorem). *A metric space (M, d) is complete if and only if for any (decreasing) nested sequence $(F_n)_{n \in \mathbb{N}}$ of nonempty closed subsets of M (i.e., $M \supset F_1 \supset F_2 \supset \cdots$) satisfying $\lim(\delta(F_n)) = 0$, we have $\bigcap_{n=1}^{\infty} F_n = \{\xi\}$, for some $\xi \in M$.*

Proof. Suppose first that M is *complete* and that $(F_n)_{n \in \mathbb{N}}$ satisfies the conditions of the theorem. For each $n \in \mathbb{N}$, let $x_n \in F_n$. Then, given any $\varepsilon > 0$, we can pick $N \in \mathbb{N}$ such that $\delta(F_n) < \varepsilon \; \forall n \geq N$. Therefore, $m \geq n \geq N$ implies $d(x_m, x_n) \leq \delta(F_n) < \varepsilon$; i.e., (x_n) is a Cauchy sequence. By completeness, $\xi :=$ $\lim(x_n) \in M$ and hence each tail $(x_m, x_{m+1}, \ldots)$ also converges to ξ. Since F_m is *closed* for each $m \in \mathbb{N}$ and $\{x_m, x_{m+1}, \ldots\} \subset F_m$, we have $\xi \in F_m \; \forall m \in \mathbb{N}$; i.e., $\xi \in \bigcap_{n=1}^{\infty} F_n$. Furthermore, if $\xi' \in \bigcap_{n=1}^{\infty} F_n$, then $d(\xi, \xi') \leq \delta(F_n) \; \forall n \in \mathbb{N}$ implies $d(\xi, \xi') = 0$ and we get $\bigcap_{n=1}^{\infty} F_n = \{\xi\}$, as desired. Conversely, suppose M satisfies the *nested closed sets property* in the theorem. Let (x_n) be a *Cauchy* sequence and let $F_n := \{x_n, x_{n+1}, \ldots\}^{-}$. It is clear that the F_n are nonempty, closed, and nested. Also, since (x_n) is a Cauchy sequence, we have (Exercise 5.3.6(2)) $\lim(\delta(F_n)) = 0$. Therefore, $\bigcap_{n=1}^{\infty} F_n = \{\xi\}$ for some $\xi \in M$. Now, given any $\varepsilon > 0$ pick $N \in \mathbb{N}$ such that $\delta(F_N) < \varepsilon$. Since $\xi \in F_N$, we have $d(x_n, \xi) < \varepsilon \; \forall n \geq N$; thus $\lim(x_n) = \xi$. $\square$

We shall deduce the Baire Category Theorem from the following consequence of Cantor's Theorem which is of independent interest:

Theorem 5.3.8. *In a complete metric space the intersection of a countable collection of open, dense sets is itself dense.*

Proof. Let (M, d) be a *complete* metric space. Let $\{U_n\}_{n \in \mathbb{N}}$ be a countable collection of *open, dense* subsets of M and let $V := \bigcap_{n=1}^{\infty} U_n$. To show that V is *dense*, we must show that, for any *open* set $O \subset M$, we have $O \cap V \neq \emptyset$. Now, since U_1 is dense, $O \cap U_1$ is a nonempty open set. Let $x_1 \in U_1 \cap O$ and pick $\varepsilon_1 > 0$ so that $B_1^- \subset O \cap U_1$, where $B_1 := B_{\varepsilon_1}(x_1)$. Next, since U_2 is *dense*, $U_2 \cap B_1$ is a nonempty open set. Pick $x_2 \in U_2 \cap B_1$ and $0 < \varepsilon_2 < \min\{\varepsilon_1/2, \varepsilon_1 - d(x_1, x_2)\}$ such that $B_2^- \subset B_1 \cap U_2$, where $B_2 := B_{\varepsilon_2}(x_2)$. Continuing this construction inductively, we find a sequence (B_n) of open balls whose centers and radii form the sequences (x_n) and (ε_n), respectively. We have $B_{n+1}^- \subset B_n \cap U_{n+1} \; \forall n \in \mathbb{N}$, and $\lim(\varepsilon_n) = 0$. In particular, $\{B_n^-\}_{n \in \mathbb{N}}$ is a nested sequence of nonempty, closed subsets of the *complete* metric space M, and $\lim(\varepsilon_n) = 0$ implies $\lim(\delta(B_n^-)) = 0$. By Cantor's Theorem, we have $\bigcap_{n=1}^{\infty} B_n^- = \{\xi\}$, for some $\xi \in M$. From the inclusions $B_{n+1}^- \subset B_n \cap U_{n+1}$ it now follows that $\xi \in V \cap O$. $\square$

Corollary 5.3.9 (Baire Category Theorem). *A complete metric space is of second category in itself; in other words, it is not a countable union of nowhere dense sets.*

Proof. Let $\{E_n\}_{n \in \mathbb{N}}$ be a countable collection of *nowhere dense* subsets of a complete metric space (M, d). If we let $U_n := (E_n^-)^c \; \forall n \in \mathbb{N}$, then $\{U_n\}_{n \in \mathbb{N}}$ is a countable collection of *open dense* sets. (Why?) By Theorem 5.3.8, $V := \bigcap_{n=1}^{\infty} U_n$ is dense and, in particular, nonempty. If $\xi \in V$, then $\xi \notin \bigcup_{n=1}^{\infty} E_n$. $\square$

Corollary 5.3.10. *The metric space $\mathbb{R}$ with its usual metric is of second category in itself.*

Exercise 5.3.11. Show that the set $\mathbb{Q}^c$ of *irrational* numbers is of *second category* in $\mathbb{R}$. *Hint:* Show that the union of two sets of first category is of first category.

Next, let us prove a theorem concerning the *perfect sets* in *complete* metric spaces. Recall that a subset of a metric space is called perfect if it is *closed* and if *every* one of its points is a *limit point*.

Theorem 5.3.12 (Perfect $\Rightarrow$ Uncountable). *Let (M, d) be a complete metric space. If M is perfect, then it is uncountable.*

Proof. Let us first prove the following claim: *Given any $x \in M$ and any (nonempty) open set U (which may or may not contain x), there is an open set V such that $V \subset U$ and $x \notin V^-$.* Well, we first need a point $y \in U$ with $y \neq x$. If $x \notin U$, the existence of y is guaranteed by the assumption $U \neq \emptyset$. So, suppose that $x \in U$. Since x is a *limit point* and U is a neighborhood of x, the set U must contain at least one point $y \neq x$. Now pick $\varepsilon > 0$ such that $B_\varepsilon(x) \cap B_\varepsilon(y) = \emptyset$ and put $V := U \cap B_\varepsilon(y)$. This ends the proof of our claim. Now, to prove that M is *uncountable*, we will show that there are *no surjective sequences* $x : \mathbb{N} \to M$. Let (x_n) be a sequence in M. Applying the above claim to the open set $U := M$, we can find a nonempty open set V_1 such that $x_1 \notin V_1^-$ and $0 < \delta(V_1^-) < 1$. In general, having chosen a nonempty open set V_{n-1}, with $x_{n-1} \notin V_{n-1}^-$, we pick a nonempty open set V_n with $V_n \subset V_{n-1}$, $x_n \notin V_n^-$ and $\delta(V_n^-) < \delta(V_{n-1}^-)/2$. We therefore have a nested sequence of *nonempty closed* sets

$$V_1^- \supset V_2^- \supset \cdots \supset V_{n-1}^- \supset V_n^- \supset \cdots$$

with $\lim \left(\delta(V_n^-)\right) = 0$. By Cantor's Theorem, there is a point $\xi \in \bigcap_{n=1}^\infty V_n^-$. Since, for each $n \in \mathbb{N}$, we have $x_n \notin V_n^-$ while $\xi \in V_n^-$, we obviously have $\xi \neq x_n$ $\forall n \in \mathbb{N}$ and the proof is complete. □

For our next result, we recall that a metric space M is called *second countable* if it has a *countable base* of open sets, i.e., if there is countable collection $\mathcal{B} = \{U_n : n \in \mathbb{N}\}$ of open sets such that each open set in M is a union of some U_n's.

Theorem 5.3.13 (Cantor–Bendixon). *Let (M, d) be a second countable metric space and let $F \subset M$ be any closed subset. Then $F = P \cup C$, where $P \subset M$ is perfect and $C \subset M$ is countable.*

Proof. Let us call a point $x \in M$ a *condensation point of F* if $U \cap F$ is *uncountable* for each open set $U \ni x$. Let

$$P := \{x \in M : x \text{ is a condensation point of } F\},$$

and define $C := F \setminus P$. Since each condensation point is clearly a *limit point* and F is *closed*, we have $P \subset F$. It is then obvious that $F = P \cup C$ and $P \cap C = \emptyset$. Let $\mathcal{B} = \{U_n : n \in \mathbb{N}\}$ be a countable base of open sets in M. Since each $x \in C$ is *not* a condensation point of F, for each such x we can find an open set $U_{n(x)} \in \mathcal{B}$ containing x and such that $U_{n(x)} \cap F$ is *countable*. But then, $C \subset \bigcup_{x \in C} U_{n(x)} \cap F$ and hence C is also *countable*. Next, for each $x \in P$ and each open set $U \ni x$, the set $U \cap F$ is uncountable while the set $U \cap C$ is countable. It follows that $U \cap P = (U \cap F) \setminus (U \cap C)$ is *uncountable* and hence x is a *limit point* of P. Thus *all* points of P are limit points of P. Hence, to show that P is *perfect*, we need only show that it is *closed*. Well, let $x \notin P$. Then there is an open set $V \ni x$ such that $V \cap F$ is *countable*. We claim that $V \cap P = \emptyset$. Indeed, if $y \in V \cap P$, then V is a neighborhood of $y \in P$ and, since y is a condensation point of F, $V \cap F$ would be *uncountable*, which is absurd. Thus, no $x \in P^c$ is a limit point of P; i.e., P contains *all* its limit points. This proves that P is closed and hence *perfect*. The proof is now complete. □

Finally, before moving to the limits of functions, let us look at the completeness in a *product space*. First a lemma:

Lemma 5.3.14 (Convergence in Product Spaces). *Let* (M_1, d_1), (M_2, d_2), ..., (M_n, d_n) *be metric spaces and let* $M := M_1 \times M_2 \times \cdots \times M_n$. *A sequence* $(x_n) \in M^{\mathbb{N}}$ *converges to a point* $x_0 \in M$ *if and only if* $\lim(\pi_k(x_n)) = \pi_k(x_0)$ *for* $1 \leq k \leq n$.

Proof. We prove the lemma for $n = 2$, the arguments being the same in general. Let $((x_n, y_n))$ be a sequence in $M_1 \times M_2$. Then $\lim((x_n, y_n)) = (x_0, y_0)$ if and only if, given any $\varepsilon > 0$, there exists $N \in \mathbb{N}$ such that $n \geq N$ implies

$$d_{\max}((x_n, y_n), (x_0, y_0)) := \max\{d_1(x_n, x_0), d_2(y_n, y_0)\} < \varepsilon, \qquad (*)$$

where we have chosen the distance $d_{\max} = \max(d_1, d_2)$ on $M = M_1 \times M_2$. But $(*)$ is satisfied if and only if $d_1(x_n, x_0) < \varepsilon$ and $d_2(y_n, y_0)\} < \varepsilon$ are both satisfied for all $n \geq N$, which means precisely that $\lim(x_n) = x_0$ and $\lim(y_n) = y_0$. $\qquad \square$

Theorem 5.3.15 (Complete Product Spaces). *Let* (M_1, d_1), (M_2, d_2), ..., (M_n, d_n) *be metric spaces. Then the product* $M := M_1 \times M_2 \times \cdots \times M_n$ *is complete if and only if each* M_k, $1 \leq k \leq n$, *is complete.*

Proof. Once again, to simplify the notation, we give the proof for $n = 2$. The arguments in the general case are exactly the same. Using the distance $d_{\max}$ as in the lemma, it is easily seen that a sequence $((x_n, y_n))$ is Cauchy in $M = M_1 \times M_2$ if and only if the sequences (x_n) and (y_n) are Cauchy in M_1 and M_2, respectively. Now if M_1 and M_2 are complete metric spaces, then $\lim(x_n) = x_0$ and $\lim(y_n) = y_0$ for some $x_0 \in M_1$, $y_0 \in M_2$, and hence $\lim((x_n, y_n)) = (x_0, y_0)$. The converse follows from the above lemma. $\qquad \square$

We next look at limits of functions from one metric space to another. This was studied (in Chap. 3) for real-valued functions defined on subsets of the metric space $\mathbb{R}$.

Definition 5.3.16 (Limits of Functions). Let $f : X \to M'$ be a function from a subset X of a metric space (M, d) to a metric space (M', d'), and let x_0 be a *limit point* of X. We say that a point $y_0 \in M'$ is the *limit* of $f(x)$ *at* x_0 (or that $f(x)$ *converges* to y_0 as x approaches x_0), and we write $\lim_{x \to x_0} f(x) = y_0$, if the following is true:

$$(\forall \varepsilon > 0)(\exists \delta = \delta(\varepsilon) > 0)(\forall x \in X)(0 < d(x, x_0) < \delta \Rightarrow d'(f(x), y_0) < \varepsilon).$$
$$(*)$$

In terms of open balls, $(*)$ may be written as follows:

$$(\forall \varepsilon > 0)(\exists \delta = \delta(\varepsilon) > 0)(x \in X \cap \dot{B}_\delta(x_0) \Rightarrow f(x) \in B_\varepsilon(y_0)), \qquad (**)$$

where, we recall, $\dot{B}_\delta(x_0) := B_\delta(x_0) \setminus \{x_0\}$.

Remark 5.3.17. If $y_0 := \lim_{x \to x_0} f(x) \in M'$ (as defined above) exists, then it follows at once from $(**)$ that $y_0 \in (f(X))^-$.

Theorem 5.3.18 (Sequential Definition of Limit). *Let $f : X \to M'$ be a function from a subset X of a metric space (M, d) to a metric space (M', d'), and let x_0 be a limit point of X. Then $y_0 := \lim_{x \to x_0} f(x) \in M'$ if and only if $\lim(f(x_n)) = y_0$ for all sequences $(x_n) \in M^{\mathbb{N}}$ with $\lim(x_n) = x_0$.*

Proof. Suppose that $y_0 := \lim_{x \to x_0} f(x) \in M'$ and let $\varepsilon > 0$ be given. Then there is $\delta > 0$ such that $x \in X \cap \dot{B}_\delta(x_0)$ implies $f(x) \in B_\varepsilon(y_0)$. Now, if $(x_n) \in M^{\mathbb{N}}$ satisfies $\lim(x_n) = x_0$, then we can pick $N \in \mathbb{N}$ such that $x_n \in \dot{B}_\delta(x_0)$ $\forall n \geq N$ and hence $f(x_n) \in B_\varepsilon(y_0)$ $\forall n \geq N$. Thus (by $(**)$) we have $\lim(f(x_n)) = y_0$, as desired. Next, if $\lim_{x \to x_0} f(x) \neq y_0$, then there exists $\varepsilon_0 > 0$ such that, for each $n \in \mathbb{N}$, we can find $x_n \in X$ with $d(x_n, x_0) < 1/n$ and $d(f(x_n), y_0) \geq \varepsilon_0$. This, however, means that $\lim(x_n) = x_0$ but $\lim(f(x_n)) \neq y_0$. $\qquad\square$

The following theorem is similar to Theorem 3.3.3 and the proof is, with obvious modifications, exactly the same.

Theorem 5.3.19. *Let X be a subset of a metric space (M, d) and let x_0 be a limit point of X. Let $f, g : X \to \mathbb{R}$, and suppose that $\lim_{x \to x_0} f(x) = y_0$, $\lim_{x \to x_0} g(x) = z_0$. Then we have*

(a) $\lim_{x \to x_0} (f \pm g)(x) = y_0 \pm z_0$;
(b) $\lim_{x \to x_0} (fg)(x) = y_0 z_0$;
(c) $\lim_{x \to x_0} (cf)(x) = cy_0$ $\forall c \in \mathbb{R}$; and
(d) $\lim_{x \to x_0} (f/g)(x) = y_0/z_0$, if $z_0 \neq 0$.

Proof. Exercise! $\qquad\square$

We also have the analog of Theorem 3.3.14 for limits of composite functions:

Theorem 5.3.20. *Let (M, d), (M', d'), and (M'', d'') be metric spaces. Let $f : X \to Y$ and $g : Y \to M''$, where $X \subset M$ and $f(X) \subset Y \subset M'$. Suppose that $\lim_{x \to x_0} f(x) = y_0$ and $\lim_{y \to y_0} g(y) = z_0$, where x_0 (resp., y_0) is a limit point of X (resp., of $f(X)$). Assume in addition that there exists $\delta_0 > 0$ such that $f(x) \neq y_0$ for all $x \in \dot{B}_{\delta_0}(x_0) \cap X$. Then*

$$\lim_{x \to x_0} g(f(x)) = z_0.$$

Proof. Given $\varepsilon > 0$, pick $\delta_1 > 0$ such that $y \in Y \cap \dot{B}_{\delta_1}(y_0)$ implies $g(y) \in B_\varepsilon(z_0)$. Now, given this δ_1, pick $\delta > 0$ with $0 < \delta < \delta_0$ such that $x \in X$ and $0 < d(x, x_0) < \delta$ imply $f(x) \in Y \cap \dot{B}_{\delta_1}(y_0)$. We then have

$$x \in X \cap \dot{B}_\delta(x_0) \implies f(x) \in Y \cap \dot{B}_{\delta_1}(y_0) \implies g(f(x)) \in B_\varepsilon(z_0),$$

which completes the proof. $\qquad\square$

Finally, let us look at the limits of functions from a metric space into a *product* of metric spaces.

Proposition 5.3.21. *Let $f = (f_1, f_2)$ be a function from a subset X of a metric space (M, d) to the product $M_1 \times M_2$ of metric spaces (M_1, d_1) and (M_2, d_2); i.e., $f_j : X \to M_j$, $j = 1, 2$, and let x_0 be a limit point of X. Then $\lim_{x \to x_0} f(x) = (y_1, y_2) \in M_1 \times M_2$ if and only if $\lim_{x \to x_0} f_1(x) = y_1$ and $\lim_{x \to x_0} f_2(x) = y_2$.*

Proof. Using the *sequential* definition of limit (Theorem 5.3.18), the proposition is reduced to Lemma 5.3.14.   □

5.4   Continuity

The concept of *continuity*, which we defined in terms of *limits of functions*, was treated in detail (in Chap. 4) for real-valued functions defined on subsets of the metric space $\mathbb{R}$. It is possible, however, to define continuity *directly* on any abstract metric space as follows:

Definition 5.4.1 (Continuity at a Point, on a Set). Let M and M' be metric spaces with metrics d and d', respectively. Given a set $X \subset M$ and a point $x_0 \in X$, a function $f : X \to M'$ is said to be *continuous at x_0* if

$$(\forall \varepsilon > 0)(\exists \delta = \delta(\varepsilon, x_0) > 0)(\forall x \in X)(d(x, x_0) < \delta \Rightarrow d'(f(x), f(x_0)) < \varepsilon).$$
$$(*)$$

In terms of open balls, $(*)$ can be written as follows:

$$(\forall \varepsilon > 0)(\exists \delta = \delta(\varepsilon, x_0) > 0)(x \in X \cap B_\delta(x_0) \Rightarrow f(x) \in B_\varepsilon(f(x_0))). \qquad (**)$$

Given any *subset* $S \subset X$, the function f is said to be *continuous on S* if it is continuous at every point of S.

Remark 5.4.2.

(a) Note that the function f has to be *defined* at x_0 in order to be continuous at x_0.
(b) If $x_0 \in X$ is an *isolated point* of X, then $(*)$ implies that *any $f : X \to M'$* is automatically continuous at x_0. Indeed, given any $\varepsilon > 0$, we can choose $\delta > 0$ so small that $d(x, x_0) < \delta$ and $x \in X$ imply $x = x_0$, and then $d'(f(x), f(x_0)) = 0 < \varepsilon$ is trivially satisfied.
(c) Recall that, for a function $f : X \to M'$ from a *subset X of a metric space M* to a metric space M', we defined $\lim_{x \to x_0} f(x)$ for a *limit point x_0* of X. In particular, we may have $x_0 \in X^c$. As $(*)$ indicates, however, the definition of continuity of f *on the subset X* does *not* involve the complement X^c at all. Therefore, dropping this complement, we may as well talk about the continuity of functions from one metric space to another, rather than of functions defined on *subsets*. This will simplify matters in our *global* treatment of continuity (Theorem 5.4.7 below).

(d) The set $X \cap B_\delta(x_0)$ is the open ball of radius δ centered at x_0 in the subspace $X \subset M$, and $(**)$ implies that

$$X \cap B_\delta(x_0) \subset f^{-1}(B_\varepsilon(f(x_0))).$$

Recall that, for a subset $Y \subset M'$, its *inverse image under f* is the set $f^{-1}(Y) := \{x \in M : f(x) \in Y\}$.

The following theorem, which is an immediate consequence of the above definition and the definition of $\lim_{x \to x_0} f(x)$ given earlier, states that f is continuous at x_0 if and only if it *commutes* with "$\lim_{x \to x_0}$," i.e.,

$$\lim_{x \to x_0} f(x) = f(\lim_{x \to x_0} x).$$

Theorem 5.4.3. *Let (M, d) and (M', d') be metric spaces and $X \subset M$. If $x_0 \in X$ is a limit point of X, then a function $f : X \to M'$ is continuous at x_0 if and only if*

$$\lim_{x \to x_0} f(x) = f(x_0).$$

Proof. Exercise! $\qquad\square$

Corollary 5.4.4 (Sequential Definition of Continuity). *Let M and M' be metric spaces and $X \subset M$. A function $f : X \to M'$ is continuous at a point $x_0 \in X$ if and only if $\lim(f(x_n)) = f(x_0)$ for all sequences $(x_n) \in X^{\mathbb{N}}$ satisfying $\lim(x_n) = x_0$.*

Proof. If x_0 is a *limit point* of X, then we use Theorem 5.3.18. If, however, x_0 is an *isolated point* of X, then *any* function is automatically continuous at x_0. Note that, for an *isolated point* $x_0 \in X$, we have $\lim(x_n) = x_0$ if and only if (x_n) is *ultimately constant*, i.e., $x_n = x_0 \ \forall n \geq m$ for some $m \in \mathbb{N}$, which obviously implies that $(f(x_n))$ is ultimately constant and that $\lim(f(x_n)) = f(x_0)$. $\qquad\square$

The next two theorems are similar to Theorems 5.3.19 and 5.3.20 above.

Theorem 5.4.5. *Let $f, g : X \to \mathbb{R}$ be defined on a subset X of a metric space M, and let $x_0 \in X$. If f and g are continuous at x_0, then so are the functions $f \pm g$, fg, and cf for any constant $c \in \mathbb{R}$. If, in addition, $g(x_0) \neq 0$, then f/g is also continuous at x_0. In particular, if f and g are continuous on X, then so are the functions $f \pm g$, fg and cf, $\forall c \in \mathbb{R}$. If, in addition, $g(x) \neq 0 \ \forall x \in X$, then f/g is also continuous on X.*

Proof. Exercise! $\qquad\square$

Theorem 5.4.6. *Let (M, d), (M', d'), and (M'', d'') be metric spaces. Let $f : X \to Y$ and $g : Y \to M''$, where $X \subset M$ and $f(X) \subset Y \subset M'$. If f is continuous at $x_0 \in X$ and g is continuous at $y_0 := f(x_0) \in Y$, then $h := g \circ f : X \to M''$ is continuous at x_0.*

Proof. Let $\varepsilon > 0$ be given. Using the continuity of g at $y_0 = f(x_0)$, we have

$$(\exists \delta' > 0)(y \in Y \cap B_{\delta'}(y_0) \Rightarrow g(y) \in B_{\varepsilon}(h(x_0))), \qquad (*)$$

where, we recall, $h(x_0) = g(f(x_0)) = g(y_0)$. Since f is continuous at x_0, we have

$$(\exists \delta > 0)(x \in X \cap B_{\delta}(x_0) \Rightarrow f(x) \in B_{\delta'}(y_0)). \qquad (**)$$

It follows from $(*)$ and $(**)$ that

$$x \in X \cap B_{\delta}(x_0) \Rightarrow h(x) = g(f(x)) \in B_{\varepsilon}(h(x_0)).$$

$\square$

So far, we have looked at continuity *locally*, i.e., at a point. The definition of continuity in terms of open balls suggests a way of giving a *global* definition in terms of *open* sets. As was remarked after the definition of continuity, we may as well look at functions defined on a metric space rather than a subspace:

Theorem 5.4.7. *Let (M, d) and (M', d') be metric spaces. A function $f : M \to M'$ is continuous on M if and only if, for every open set $V \subset M'$, the inverse image $f^{-1}(V)$ is open in M. In particular, a function $f : X \to M'$ is continuous on a subset $X \subset M$ if and only if, for every open set $V \subset M'$, we have $f^{-1}(V) = X \cap U$ for an open subset $U \subset M$.*

Proof. Suppose that f is continuous on M and let $V \subset M'$ be open. For each $x \in f^{-1}(V)$ we have $f(x) \in V$ and V is *open*. Therefore, we can find $\varepsilon = \varepsilon(x) > 0$ such that $B_{\varepsilon}(f(x)) \subset V$. Now, using the continuity of f at x, we can pick $\delta > 0$ such that

$$B_{\delta}(x) \subset f^{-1}(B_{\varepsilon}(f(x))) \subset f^{-1}(V),$$

which proves that each point $x \in f^{-1}(V)$ is an *interior* point and hence $f^{-1}(V)$ is open. Conversely, suppose that $f^{-1}(V)$ is open in M for every open set $V \subset M'$. Given any fixed $x \in M$ and $\varepsilon > 0$, let $V := B_{\varepsilon}(f(x))$. Then V is open and hence $f^{-1}(V)$ is open. Since $x \in f^{-1}(V)$, we can find $\delta > 0$ such that

$$B_{\delta}(x) \subset f^{-1}(V) = f^{-1}(B_{\varepsilon}(f(x))),$$

which shows indeed that f is continuous at x. Since x was an arbitrary point of M, the proof is complete.

$\square$

Notation 5.4.8. Given any metric spaces (M, d) and (M', d') and any $X \subset M$, the set of all *continuous* functions $f : X \to M'$ will be denoted by $C(X, M')$. In particular, $C(M, M')$ will denote the set of all continuous functions $f : M \to M'$. If $M' = \mathbb{R}$, then we use the abbreviations $C(X)$ and $C(M)$ instead of $C(X, \mathbb{R})$ and $C(M, \mathbb{R})$.

Corollary 5.4.9. *Let (M, d) and (M', d') be metric spaces and let $X \subset M$. If $f \in C(M, M')$, then $f|X \in C(X, M')$. Here $f|X$ is the restriction of f to X.*

Proof. Since f is continuous on M, $f^{-1}(V)$ is *open* in M for every open set $V \subset M'$, and hence $(f|X)^{-1}(V) = X \cap f^{-1}(V)$ is open in X. $\qquad \square$

Remark 5.4.10. Note that the restriction of a function to a subspace may be continuous without the function itself being continuous at any point. For example, the Dirichlet function $\chi_{\mathbb{Q}}$ (Exercise 4.3.12) is nowhere continuous but its restriction to $\mathbb{Q}$ is identically 1, hence continuous.

Exercise 5.4.11. Let (M, d) and (M', d') be metric spaces and $f : M \to M'$. Show that the following statements are pairwise equivalent:

(a) $f \in C(M, M')$;
(b) for every *closed* set $Y \subset M'$, $f^{-1}(Y)$ is *closed* in M;
(c) for every set $Y \subset M'$, $f^{-1}(Y^\circ) \subset (f^{-1}(Y))^\circ$;
(d) for every set $Y \subset M'$, $(f^{-1}(Y))^- \subset f^{-1}(Y^-)$;
(e) for every set $X \subset M$, $f(X^-) \subset (f(X))^-$.

Exercise 5.4.12 (Continuity of Addition and Multiplication). Show that the maps $+ : \mathbb{R} \times \mathbb{R} \to \mathbb{R}$ and $\cdot : \mathbb{R} \times \mathbb{R} \to \mathbb{R}$ defined by $+(x, y) := x + y$ and $\cdot(x, y) := xy$ are continuous.

Remark 5.4.13. In fact, the function "$+$" is *uniformly continuous* and so is the *dilation* $x \mapsto ax$, where $a \in \mathbb{R}$ is a *constant*. (Why?)

Given a metric space M, we saw (Theorem 5.2.2(6)) that the *diagonal* $\Delta_M := \{(x, x) : x \in M\}$ of $M \times M$ is *closed* in $M \times M$. Now Δ_M is the *graph* of the *identity map* $\mathrm{id}_M : M \to M$ defined by $\mathrm{id}_M(x) := x \; \forall x \in M$. We shall presently see that this *closedness* of the graph is a consequence of *continuity*.

Notation 5.4.14 (Graph). The *graph* of a function f from a metric space M_1 to a metric space M_2 shall be denoted by

$$\Gamma_f := \{(x, f(x)) : x \in M_1\} \subset M_1 \times M_2.$$

We are now ready for our theorem. In fact, we shall give two proofs: the first uses the closedness of the *diagonal*, while the second is a direct one using a *sequential* argument which can be extended to other situations as well (Exercise 5.4.16 following the theorem).

Theorem 5.4.15 (Closedness of Graphs). *Let (M_1, d_1) and (M_2, d_2) be metric spaces and $f \in C(M_1, M_2)$. Then the graph of f (i.e., the set $\Gamma_f := \{(x_1, f(x_1)) : x_1 \in M_1\}$) is a closed subset of the product space $M := M_1 \times M_2$.*

First Proof. Let $\Phi : M_1 \times M_2 \to M_2 \times M_2$ be defined by

$$\Phi(x, y) := (f(x), y) \quad \forall (x, y) \in M_1 \times M_2.$$

From the continuity of f, it follows easily (e.g., using Theorem 5.4.3 and Proposition 5.3.21) that Φ is *continuous*. Since $\Gamma_f = \Phi^{-1}(\Delta_{M_2})$, the closedness of Γ_f follows from the fact that the diagonal Δ_{M_2} is closed (Exercise 5.4.11). $\qquad \square$

Second Proof. Let us show that Γ_f contains its *limit points*. Suppose $(x_0, y_0) \in M_1 \times M_2$ is a limit point of Γ_f. Then, $\forall n \in \mathbb{N}$, we have $\Gamma_f \cap B_{1/n}((x_0, y_0)) \neq \emptyset$; i.e., we can pick a point $(x_n, f(x_n)) \in B_{1/n}((x_0, y_0))$. Using the distance $d_{\max} := \max(d_1, d_2)$, this implies that

$$d_{\max}((x_n, f(x_n)), (x_0, y_0)) := \max\{d_1(x_n, x_0), d_2(f(x_n), y_0)\} < 1/n.$$

In particular, we have $d_1(x_n, x_0) < 1/n$ and $d_2(f(x_n), y_0) < 1/n$ for all $n \in \mathbb{N}$. Therefore, $\lim(x_n) = x_0$ and $\lim(f(x_n)) = y_0$. On the other hand, the *continuity* of f at x_0 implies that $y_0 = \lim(f(x_n)) = f(x_0)$ and we indeed have $(x_0, y_0) = (x_0, f(x_0)) \in \Gamma_f$. $\qquad \square$

Exercise 5.4.16 (Closedness of Level Curves). Let M_1, M_2, and M_3 be metric spaces and $f \in C(M_1 \times M_2, M_3)$. Show that, for each fixed point $z_0 \in M_3$, the set

$$K_{z_0} := \{(x, y) \in M_1 \times M_2 : f(x, y) = z_0\}$$

is *closed* in $M_1 \times M_2$. Deduce that, for a continuous function $f : \mathbb{R} \times \mathbb{R} \to \mathbb{R}$, the *level curve*

$$K_c := \{(x, y) \in \mathbb{R} \times \mathbb{R} : f(x, y) = c\}$$

is a closed subset of the plane $\mathbb{R}^2$ for each constant $c \in \mathbb{R}$. *Hint:* Use an argument similar to the second proof of Theorem 5.4.15.

Warning! The converse of Theorem 5.4.15 is *false* in general. For example, consider the function $f : \mathbb{R} \to \mathbb{R}$ defined by $f(x) = 1/x$ if $x \neq 0$, and $f(0) = 0$. Then f is *discontinuous* at $x = 0$ but its graph $\Gamma_f = \{(x, y) \in \mathbb{R} \times \mathbb{R} : xy = 1\} \cup \{(0, 0)\}$ is *closed*. Indeed, the singleton $\{(0, 0)\}$ is closed and (by Exercise 5.4.16) so is the *level curve* $\{(x, y) \in \mathbb{R} \times \mathbb{R} : xy = 1\}$ of the *multiplication* $(x, y) \mapsto xy$, which is continuous (Exercise 5.4.12). We shall see, however, that the converse is *true* if we assume that M_2 is *compact*.

As we have seen above, the *inverse images* of open (resp., closed) sets under a continuous function are open (resp., closed). On the other hand, the *direct images* of open (resp., closed) sets under a continuous function need *not* be open (resp., closed) in general. This motivates the following:

Definition 5.4.17 (Open Map, Closed Map). A map f from a metric space M to a metric space M' is called *open* (resp., *closed*) if $f(X)$ is open (resp., closed) in M' for each open (resp., closed) subset $X \subset M$.

Before giving examples of open and closed maps, let us mention an important class of open maps:

Proposition 5.4.18 (Openness of Projections). *Let* (M_1, d_1) *and* (M_2, d_2) *be metric spaces and consider their (metric space) product* $M := M_1 \times M_2$. *Then the projections* $\pi_1 : M \to M_1$ *and* $\pi_2 : M \to M_2$ *defined by* $\pi_1(x_1, x_2) = x_1$ *and* $\pi_2(x_1, x_2) = x_2$ *are both open maps.*

Proof. Since open sets are unions of open balls and since the (direct) image of a union of sets is the union of their images, it suffices to show that π_1 (resp., π_2) maps each *open ball* of M onto an open set in M_1 (resp., M_2). Now, by the Example 5.1.15(c), if we use the metric $d_{\max}((x_1, x_2), (y_1, y_2)) := \max\{d_1(x_1, y_1), d_2(x_2, y_2)\}$ on M, then, for any $(x_1, x_2) \in M$, we have

$$B_\varepsilon((x_1, x_2)) := \{(y_1, y_2) \in M : d_{\max}((x_1, x_2), (y_1, y_2)) < \varepsilon\} = B_{1,\varepsilon}(x_1) \times B_{2,\varepsilon}(x_2),$$

where $B_{j,\varepsilon}(x_j)$ is the open ball of radius ε centered at $x_j \in M_j$, $j = 1, 2$. It now follows that $\pi_1(B_\varepsilon((x_1, x_2))) = B_{1,\varepsilon}(x_1)$ and $\pi_2(B_\varepsilon((x_1, x_2))) = B_{2,\varepsilon}(x_2)$, which are open in M_1 and M_2, respectively. $\square$

Now we are ready for our examples which show that a function may be open without being closed or closed without being open. Also, a function may be simultaneously open and closed or neither open nor closed.

Example 5.4.19.

(a) If $f : M \to M'$ is a *continuous one-to-one correspondence* between two metric spaces, then f^{-1} is *both open and closed* (Exercise 5.4.11). If f is also open (or closed), then it is a *homeomorphism* (to be defined below).

(b) The function $f : [0, 11\pi/6] \to \mathbb{R}$ defined by $f(x) := \sin(x)$ is *neither open nor closed*. Indeed, $(\pi/3, 5\pi/3)$ is open in $[0, 11\pi/6]$ but $f((\pi/3, 5\pi/3)) = [-1, 1]$ is closed (in $\mathbb{R}$). Also, the interval $[3\pi/2, 11\pi/6]$ is closed in $[0, 11\pi/6]$ but $f([3\pi/2, 11\pi/6)) = [-1, -1/2)$ is not closed (in $\mathbb{R}$).

(c) Let $f : \mathbb{R} \to \mathbb{R}$ be a *constant* map, i.e., $f(x) = c \ \forall x \in \mathbb{R}$ and some $c \in \mathbb{R}$. Then f is obviously *closed*, but *not open*.

(d) Let $\pi_1 : \mathbb{R} \times \mathbb{R} \to \mathbb{R}$ be the projection $\pi_1(x, y) := x$, $\forall (x, y) \in \mathbb{R}^2$. Then π_1 is *open but not closed*. Indeed, by Proposition 5.4.18, π_1 is open. To show that it is not closed, consider the *hyperbola* $\Gamma := \{(x, y) \in \mathbb{R} \times \mathbb{R} : xy = 1\}$. As we pointed out in the "Warning" following Exercise 5.4.16, Γ is a level curve of the *multiplication* function and is therefore *closed* in $\mathbb{R}^2$. However, $\pi_1(\Gamma) = \mathbb{R} \setminus \{0\}$ is not closed in $\mathbb{R}$. The situation is better for *cross sections*, as we shall see later.

Exercise 5.4.20. Let (M_1, d_1), $(M_2, d_2), \ldots, (M_n, d_n)$ be metric spaces and let $M := M_1 \times M_2 \times \cdots \times M_n$ be their product with, say, the metric

$$d_{\max}(x, y) := \max\{d_1(x_1, y_1), \ldots, d_n(x_n, y_n)\}.$$

(a) Show that, if $\emptyset \neq X_k \subset M_k$, $1 \le k \le n$, then the product $X_1 \times \cdots \times X_n$ is *open* (resp., *closed*) in M if and only if each $X_k \subset M_k$ is open (resp., closed).

(b) With the X_k as in part (a), show that

$$(X_1 \times \cdots \times X_n)^- = X_1^- \times \cdots \times X_n^-, \text{ and}$$
$$(X_1 \times \cdots \times X_n)^\circ = X_1^\circ \times \cdots \times X_n^\circ.$$

In Chap. 4 we studied the behavior of one-to-one continuous functions between subsets of $\mathbb{R}$ and proved that such functions must necessarily be strictly monotone. We also looked at *homeomorphisms*, i.e., bijective functions that are continuous and have continuous inverse. We now define these concepts for a general metric space.

Definition 5.4.21 (Homeomorphism, Isometry). Let (M, d) and (M', d') be metric spaces and let $f : M \to M'$ be a *one-to-one correspondence*.

(a) We say that f is a *homeomorphism* if both f and f^{-1} are continuous. If such a function exists, then M and M' are said to be *homeomorphic*.
(b) We say that f is an *isometry* if

$$d'(f(x), f(y)) = d(x, y) \qquad \forall x, y \in M. \tag{$\ddagger$}$$

If such a function exists, the metric spaces M and M' are said to be *isometric*.

Exercise 5.4.22. Let $f : M \to M'$ and $g : M' \to M''$ be *one-to-one correspondences* between the metric spaces M, M', and M''.

(a) Show that if f, g are homeomorphisms (resp., isometries), then so are f^{-1} and $g \circ f$.
(b) Given a metric space M, let $Homeo(M)$ (resp., $Isom(M)$) denote the set of all homeomorphisms (resp., isometries) of M *onto itself.* Show that both these sets are *groups* under the operation of "composition." This means (for $Homeo(M)$) that

(G_1) $(f \circ g) \circ h = f \circ (g \circ h)$ $\forall f, g, h \in Homeo(M)$,
(G_2) $(\exists \iota \in Homeo(M))(\forall f \in Homeo(M))(\iota \circ f = f \circ \iota = f)$, and
(G_3) $(\forall f \in Homeo(M))(\exists f^{-1} \in Homeo(M))(f \circ f^{-1} = f^{-1} \circ f = \iota)$,

with similar properties for the set $Isom(M)$. (*Hint:* Let $\iota = \mathrm{id}_M$ be the *identity map:* $\mathrm{id}_M(x) = x, \forall x \in M$).

Remark 5.4.23.

(a) It is obvious that an isometry is a homeomorphism. (Why?) The converse is false, however, as the following trivial example shows. Consider the *dilation* $f : \mathbb{R} \to \mathbb{R}$ defined by $f(x) := 2x$. It is clear that both f and f^{-1}, which is given by $f^{-1}(x) = x/2, \forall x \in \mathbb{R}$, are continuous (even Lipschitz). Hence f is a *homeomorphism* of $\mathbb{R}$ onto itself. On the other hand, f is obviously *not* an *isometry*.

(b) **(Transported Distance)** Let (M, d) be a metric space and let M' be a *set*. Assume that there exists a *bijection* $f : M \to M'$. We can then *define* a distance d' on M' by the formula (‡) above. This distance is said to have been *transported* from M to M', and the metric spaces M and M' are then obviously *isometric*.

Example 5.4.24 (Extended Real Line). Recall that, in Example 5.1.3(2), we extended the homeomorphism $f : x \to x/(1 + |x|)$ of $\mathbb{R}$ onto $(-1, 1)$ to $\overline{\mathbb{R}} := [-\infty, +\infty]$ by setting $f(-\infty) = -1$ and $f(+\infty) = 1$. If, on $\overline{\mathbb{R}} \times \overline{\mathbb{R}}$, we define $d'(x, y) := |f(x) - f(y)|$, then d' is easily seen to be a metric and $f : \overline{\mathbb{R}} \to [-1, 1]$ is then an isometry of $(\overline{\mathbb{R}}, d')$ onto $([-1, 1], d)$, where d is the usual distance $(d(x, x') = |x - x'|)$ in $[-1, 1]$.

Theorem 5.4.25. *A bijection $f : M \to M'$ from a metric space M to a metric space M' is a homeomorphism if and only if, for any open sets $U \subset M$ and $V \subset M'$, $f(U)$ is open in M' and $f^{-1}(V)$ is open in M, i.e., if and only if f is continuous and open.*

Proof. This is an immediate consequence of Theorem 5.4.7. □

Exercise 5.4.26. Show that an *isometry* $f : \mathbb{R} \to \mathbb{R}$ is necessarily of the form $f(x) := \pm x + b$ for a constant $b \in \mathbb{R}$.

Recall that two metrics d and d' on a set M are said to be *equivalent* if they give M the same topology (i.e., the same collection of open sets). The following proposition is an immediate consequence of Theorem 5.4.25.

Proposition 5.4.27 (Equivalent Metrics). *Two metrics d and d' on a set M are equivalent if and only if the identity map $\mathrm{id}_M : (M, d) \to (M, d')$, defined by $\mathrm{id}_M(x) := x \; \forall x \in M$, is a homeomorphism.*

Corollary 5.4.28. *Let d and d' be two metrics on a set M and assume that for some constants $c > 0, c' > 0$ we have $d' \le cd \le c'd'$. Then d and d' are equivalent.*

Proof. Exercise! □

Exercise 5.4.29.

(a) Let (M, d) be a metric space and define the distances $d_1 := d/(1 + d)$ and $d_2 := \min(1, d)$. Show that d_1 and d_2 satisfy the conditions of the corollary and hence are equivalent.

(b) **(Product Spaces)** Let $M_1, M_2, \ldots, M_n$ be metric spaces with metrics $d_1, d_2, \ldots, d_n$, respectively, and let $M := M_1 \times M_2 \times \cdots \times M_n$. Recall that the metrics $d_{\mathrm{euc}}, d_{\max}, d_{\mathrm{sum}} : M \times M \to \mathbb{R}$ are defined as follows:

$$d_{\mathrm{euc}}(x, y) := \sqrt{\sum_{k=1}^{n} d_k^2(x_k, y_k)},$$

$$d_{max}(x, y) := \max\{d_1(x_1, y_1), \ldots, d_n(x_n, y_n)\}, \text{ and}$$

$$d_{sum}(x, y) := \sum_{k=1}^{n} d_k(x_k, y_k),$$

for every $x = (x_1, \ldots, x_n)$ and $y = (y_1, \ldots, y_n)$ in M. Show that d_{euc}, d_{max}, and d_{sum} are equivalent metrics on M. The space M together with any one of these metrics is called the *product* of the M_k $1 \le k \le n$. *Hint:* Show that $d_{max} \le d_{euc} \le d_{sum} \le n d_{max}$.

Example 5.4.30 (Joint vs. Separate Continuity). On $M := \mathbb{R} \times \mathbb{R} = \mathbb{R}^2$, consider the functions

$$f(x, y) := \begin{cases} \frac{2xy}{x^2+y^2} & \text{if } (x, y) \ne (0, 0) \\ 0 & \text{if } (x, y) = (0, 0) \end{cases}, \quad g(x, y) := \begin{cases} \frac{2xy}{\sqrt{x^2+y^2}} & \text{if } (x, y) \ne (0, 0) \\ 0 & \text{if } (x, y) = (0, 0) \end{cases}.$$

Now both f and g are *separately* continuous in each variable x, y if the other is kept fixed, as follows easily by using, e.g., the sequential definition of continuity and the fact that $f(0, y) = g(0, y) = 0$ for all y, while $f(x, 0) = g(x, 0) = 0$ for all x. Also, using sequences again, we see at once that both f and g are *jointly* continuous functions of (x, y) for all $(x, y) \ne (0, 0)$. On the other hand, f is *discontinuous* at $(0, 0)$ because if $y_n = x_n = 1/n$ for all $n \in \mathbb{N}$ and if $n \to \infty$, then $(x_n, x_n) \to (0, 0)$ and $(x_n, 0) \to (0, 0)$, but $f(x_n, x_n) \to 1$ while $f(x_n, 0) \to 0$. Finally, g is *(jointly) continuous* even at $(0, 0)$. Indeed, since $2xy \le x^2 + y^2$ for all $(x, y) \in \mathbb{R}^2$, we have

$$g(x, y) = \frac{2xy}{\sqrt{x^2 + y^2}} \le \frac{x^2 + y^2}{\sqrt{x^2 + y^2}} = \sqrt{x^2 + y^2} \quad \forall (x, y) \ne (0, 0),$$

and hence $\lim_{(x,y) \to (0,0)} g(x, y) = 0 = g(0, 0)$.

Here is an example of a function on $\mathbb{R}^2$ that is separately continuous, but has a *dense* discontinuity set:

Exercise 5.4.31. Let $\mathbb{Q}^2 = \{(r_n, s_n) : n = 1, 2, 3, \ldots\}$ be an enumeration of the dense subset $\mathbb{Q}^2 \subset \mathbb{R}^2$ and let f be the function in the above example. Consider the function

$$h(x, y) := \sum_{n=1}^{\infty} \frac{f(x - r_n, y - s_n)}{2^n}.$$

Show that h is coordinate-wise continuous, but the discontinuity set of h is precisely $\mathbb{Q}^2$. *Hint:* Use whatever you need from Chap. 8!

Definition 5.4.32 (Cross Sections, Horizontal and Vertical Fibers).   Let (M_1, d_1) and (M_2, d_2) be metric spaces, $M = M_1 \times M_2$ their product, and $S \subset M_1 \times M_2$.

1. The sets $\pi_2^{-1}(x_2) = M_1 \times \{x_2\}$ (resp., $\pi_1^{-1}(x_1) = \{x_1\} \times M_2$), where (x_1, x_2) runs through $M_1 \times M_2$, are called the *horizontal* (resp., *vertical) fibers* of $M_1 \times M_2$. Note that they are all *closed* subspaces of $M_1 \times M_2$.
2. For any (fixed) point $(a_1, a_2) \in M_1 \times M_2$ we define the a_1-*cross section* (resp., a_2-*cross section*) of S to be the set $S_{a_1} := \pi_2(S \cap \{a_1\} \times M_2)$ (resp., $S^{a_2} := \pi_1(S \cap M_1 \times \{a_2\})$). In other words, we have

$$S_{a_1} := \{x_2 \in M_2 : (a_1, x_2) \in S\} \quad \text{and} \quad S^{a_2} := \{x_1 \in M_1 : (x_1, a_2) \in S\}.$$

Exercise 5.4.33. Let M_1 and M_2 be metric spaces and $M = M_1 \times M_2$ their product. Show that for any (fixed) $a_2 \in M_2$ (resp., $a_1 \in M_1$) the function $x_1 \mapsto (x_1, a_2)$ (resp., $x_2 \mapsto (a_1, x_2)$) is an *isometry* of M_1 (resp., M_2) onto the horizontal fiber $M_1 \times \{a_2\}$ (resp., the vertical fiber $\{a_1\} \times M_2$) of M.

Recall that the projections π_1 and π_2 are *open* maps, but *not closed* ones (cf. Example 5.4.19(d)). However, we have the following

Proposition 5.4.34. *Let the notation be as in Definition 5.4.32. If S is open (resp., closed) then so are the cross sections S_{a_1} and S^{a_2} for any $(a_1, a_2) \in M_1 \times M_2$.*

Proof. For S_{a_1}, note that (in view of Exercise 5.4.33) we need only show that the set $S \cap (\{a_1\} \times M_2)$ is (relatively) open (resp., closed) in the vertical fiber $\{a_1\} \times M_2$ if S is open (resp., closed) in M. But this follows at once from Theorem 5.2.5. A similar proof can be given for the cross section S^{a_2}.   □

Let us end this section with a necessary and sufficient condition for the continuity of a map from a metric space to a product of metric spaces. We state the theorem for a product of *two* metric spaces, but the extension to any *finite* product is immediate.

Theorem 5.4.35. *Let M, M_1, and M_2 be metric spaces and $f = (f_1, f_2) : M \to M_1 \times M_2$. Then f is continuous if and only if $f_1 : M \to M_1$ and $f_2 : M \to M_2$ are both continuous, i.e., if and only if the composite functions $\pi_1 \circ f$ and $\pi_2 \circ f$ are both continuous. Here, π_j is, of course, the projection of $M_1 \times M_2$ onto M_j, $j = 1, 2$.*

Proof. This is an immediate consequence of Theorem 5.4.3.   □

5.5   Uniform Continuity and Continuous Extensions

In this section we define the concept of *uniform continuity* for functions from a general metric space to another. Recall that uniform continuity played an important role in the approximation of continuous real-valued functions on compact subsets of $\mathbb{R}$ by step, piecewise linear, and polynomial functions (Theorems 4.7.2, 4.7.4, and 4.7.9).

Definition 5.5.1 (Uniform Continuity). Let (M, d) and (M', d') be metric spaces. A function $f : M \to M'$ is said to be *uniformly continuous* on M if

$$(\forall \varepsilon > 0)(\exists \delta = \delta(\varepsilon) > 0)(\forall x, \ x' \in M)(d(x, x') < \delta \Rightarrow d'(f(x), f(x')) < \varepsilon).$$
$$(\dagger)$$

Remark 5.5.2.

(a) Unlike continuity, uniform continuity *cannot* be defined at a *point*; it is only defined on larger *subsets* of the domain.

(b) If we compare ($\dagger$) to ($*$) in Definition 5.4.1, we notice that in the *pointwise* definition of continuity at x, the number $\delta = \delta(\varepsilon, x)$ depends on *both ε and x*. In ($\dagger$), however, δ depends *only* on ε; i.e., for a given $\varepsilon > 0$, it is possible to find a $\delta > 0$ that works for *all* points $x \in M$.

(c) It is obvious that a uniformly continuous function is continuous. We shall see later that, as in the special case of $\mathbb{R}$, the converse (which is false in general) is true if M is *compact*.

The Lipschitz class, which was defined in Chap. 4 for real-valued functions of a real variable, can also be defined for functions from one metric space to another:

Definition 5.5.3 (Lipschitz Function, Contraction). For any metric spaces (M, d) and (M', d') and any $X \subset M$, we say that a function $f : X \to M'$ is *Lipschitz*, and we write $f \in Lip(X, M')$, if there exists a constant $c > 0$ such that

$$d'(f(x), f(x')) \le cd(x, x') \qquad \forall x, \ x' \in X.$$

The constant c is then called a *Lipschitz constant* for f. If $c < 1$, the function f is said to be a *contraction* (or a *contraction mapping*).

Example 5.5.4.

(a) Let (M, d) be a metric space and $S \subset M$ a *nonempty* subset. The function $x \mapsto d(x, S)$ is a *Lipschitz function* from M to $[0, \infty)$. Indeed, we have

$$|d(x, S) - d(x', S)| \le d(x, x') \quad \forall x, \ x' \in M \qquad (*)$$

To show ($*$), note that the Triangle Inequality implies $d(x, y) \le d(x', y) + d(x, x')$ for all $x, \ x' \in M, \ y \in S$, and hence (since $d(x, S) := \inf\{d(x, y) : y \in S\}$),

$$d(x, S) \le d(x', y) + d(x, x') \quad \forall x, \ x' \in M, \ \forall y \in S. \qquad (**)$$

Now, keeping x and x' fixed, ($**$) implies that

$$d(x, S) \le d(x', S) + d(x, x'),$$

and the same inequality also holds if x and x' are interchanged. The inequality (∗) now follows at once. Note that, since $d(x, S) = 0$ if and only if $x \in S^-$ (Exercise 5.2.12(5)), the set of *zeroes* of the function $x \mapsto d(x, S)$ is precisely the set S^-. In particular, if $M = \mathbb{R}$ and if $S = C$ is the *Cantor set* (which we know is *closed*), then the set of zeroes of the function $f(x) := d(x, C)$ (with the usual distance $d(x, y) = |x - y|$) is the Cantor set, which is *uncountable* and totally disconnected. The latter property implies that the function f is never *identically zero* on any open subinterval of $[0, 1]$.

(b) Let (M, d) be a metric space and consider the *product* space $M \times M$ with metric

$$d_{\mathrm{sum}}((x, y), (x', y')) := d(x, x') + d(y, y').$$

Then the *distance* function $d : M \times M \to \mathbb{R}$ is *Lipschitz*. Indeed, it follows from the Triangle Inequality that

$$|d(x, y) - d(x', y')| \leq d(x, x') + d(y, y') = d_{\mathrm{sum}}((x, y), x', y')).$$

(c) **(Projections)** Let $M := M_1 \times M_2 \times \cdots \times M_n$ be a product of metric spaces. Then for each $1 \leq k \leq n$, the *kth projection* $\pi_k : M \to M_k$, defined by $\pi_k(x_1, x_2, \ldots, x_n) := x_k$, is *Lipschitz*. Indeed, given any points $x = (x_1, \ldots x_n)$ and $y = (y_1, \ldots, y_n)$ in M, we obviously have

$$d_k(\pi_k(x), \pi_k(y)) = d_k(x_k, y_k) \leq \tilde{d}(x, y),$$

where $\tilde{d}$ is *any* one of the distances d_{euc}, d_{max}, d_{sum} defined in Exercise 5.1.5.

Exercise 5.5.5 (Urysohn's Lemma for Metric Spaces). Let (M, d) be a metric space and A, $B \subset M$ two *disjoint (nonempty) closed* sets. Define the function $f : M \to \mathbb{R}$ by

$$f(x) := \frac{d(x, B)}{d(x, A) + d(x, B)}.$$

Show that f is continuous on M and that $f(A) = \{1\}$, $f(B) = \{0\}$; i.e., f is identically 1 on A and identically 0 on B. Deduce that there are open sets U, $V \subset M$ such that $A \subset U$, $B \subset V$, and $U \cap V = \emptyset$.

Remark 5.5.6. It is obvious that a Lipschitz function is *uniformly continuous*, but, as we saw in Chap. 4, the converse is *false*. One can also define the classes $Lip^\alpha(X, M')$ of Lipschitz functions of *order* $\alpha \in (0, 1]$ from a subset X of a metric space (M, d) to a metric space (M', d') by requiring the inequalities

$$d'(f(x), f(x')) \leq c(d(x, x'))^\alpha \quad \forall x, x' \in X,$$

where $c > 0$ is again called a *Lipschitz constant*. All functions in these classes are also uniformly continuous on X and we obviously have $Lip^1(X, M') = Lip(X, M')$. Finally, one can *localize* the Lipschitz condition and define the classes $Lip^{\alpha}_{loc}(X, M')$ to be the classes of functions $f : X \to M'$ such that, for each $x \in X$, there exists $\varepsilon = \varepsilon(x) > 0$ such that the restriction $f | B_{\varepsilon}(x) \cap X$ is Lipschitz of order α. Note that the Lipschitz constant will (in general) vary with the neighborhood $B_{\varepsilon}(x)$, but *not* the order α.

The *Fixed Point Theorem* we proved in Chap. 4 (Theorem 4.6.17) can now be extended to *complete* metric spaces:

Theorem 5.5.7 (Banach's Fixed Point Theorem). *Let (M, d) be a complete metric space. If $f : M \to M$ is a contraction, then f has a unique fixed point; i.e., there exists a unique $\xi \in M$ such that $f(\xi) = \xi$.*

Proof. The proof is a copy of the one given for Theorem 4.6.17. We pick an arbitrary point $x_0 \in M$ and define the sequence $(x_0, x_1, x_2, \ldots)$ by the recursive formula

$$x_{n+1} = f(x_n) \qquad \forall n \in \mathbb{N}_0. \tag{$*$}$$

Using $(*)$ and the fact that (by assumption) $d(f(x), f(x')) \leq c d(x, x')$, $\forall x, x' \in M$ and a Lipschitz constant $c \in (0, 1)$, we inductively prove (as in the proof of Theorem 4.6.17) the inequalities

$$d(x_n, x_{n+1}) = d(f(x_{n-1}), f(x_n)) \leq c^n d(x_0, x_1) \qquad \forall n \in \mathbb{N}. \tag{$**$}$$

Repeated use of $(**)$ and the Triangle Inequality then imply that

$$d(x_m, x_n) \leq \frac{c^n}{1 - c} d(x_0, x_1) \qquad \forall m \geq n. \tag{$\dagger$}$$

Since $1 - c > 0$, we have $\lim(c^n) = 0$ and $(\dagger)$ implies that (x_n) is a Cauchy sequence. The *completeness* of M now implies that $\xi := \lim(x_n) \in M$. To show that ξ is a fixed point, we note that f is (*uniformly*) continuous and hence

$$f(\xi) = f(\lim(x_n)) = \lim(f(x_n)) = \lim(x_{n+1}) = \xi.$$

Finally, if we also have $f(\xi') = \xi'$ for another point $\xi' \in M$, then

$$d(\xi, \xi') = d(f(\xi), f(\xi')) \leq c d(\xi, \xi'),$$

from which it follows at once that $d(\xi, \xi') = 0$ and hence that $\xi = \xi'$.   □

In Chap. 4, we proved (Theorem 4.6.6) that a continuous function $f : (a, b) \to \mathbb{R}$ is uniformly continuous if and only if it has a *continuous extension* to the closed interval $[a, b]$. Note that (a, b) is *dense* in $[a, b]$ and that a function which is merely *continuous* on (a, b) *cannot* always be continuously extended to $[a, b]$. For example, the function $f(x) := 1/(1 - x^2)$ is continuous on $(-1, 1)$, but $f(-1 + 0) = f(1 - 0) = +\infty$ shows that it has no continuous extensions to $[-1, 1]$. We want to prove a result analogous to Theorem 4.6.6 in a general metric space. Before announcing this extension theorem, let us prove a couple of useful *extension* results:

Proposition 5.5.8 (Extension of Identities and Inequalities). *Let* (M, d) *and* (M', d') *be metric spaces.*

(a) *If* $f, g \in C(M, M')$, *then the set* $E := \{x \in M : f(x) = g(x)\}$ *is closed in* M. *In particular, if* E *is dense in* M, *then* $f = g$.
(b) *If* $f, g \in C(M, \mathbb{R})$, *then the set* $F := \{x \in M : f(x) \leq g(x)\}$ *is closed in* M. *In particular, if* F *is dense in* M, *then* $f \leq g$; *i.e.,* $f(x) \leq g(x) \; \forall x \in M$.

Proof. To prove (a), let us show that the set $E^c = \{x \in M : f(x) \neq g(x)\}$ is *open*. Let $\xi \in E^c$. Since $f(\xi) \neq g(\xi)$, we have $\varepsilon := d'(f(\xi), g(\xi)) > 0$. Now, by the continuity of f and g, we can pick $\delta > 0$ such that $x \in B_\delta(\xi)$ implies $d'(f(x), f(\xi)) < \varepsilon/2$ and $d'(g(x), g(\xi)) < \varepsilon/2$. But then we must have $f(x) \neq g(x) \; \forall x \in B_\delta(\xi)$, since otherwise we get $d'(f(\xi), g(\xi)) < \varepsilon$ by the Triangle Inequality. Thus, E^c is open and E is indeed *closed*. Moreover, if E is *dense*, then $E = E^- = M$, and we get $f(x) = g(x) \; \forall x \in M$. Next, we prove (b) by showing that the complement $F^c = \{x \in M : f(x) > g(x)\}$ is *open*. Again, let $\xi \in F^c$, so that $f(\xi) > g(\xi)$. Pick $\eta \in \mathbb{R}$ such that $f(\xi) > \eta > g(\xi)$. Since $(\eta, +\infty]$ and $[-\infty, \eta)$ are *open* in $\overline{\mathbb{R}}$, the continuity of f and g implies that the inverse images $U := f^{-1}((\eta, +\infty])$ and $V := g^{-1}([-\infty, \eta))$ are both open in M and hence so is $U \cap V$. But then we have $f(x) > \eta > g(x) \; \forall x \in U \cap V$; i.e., $\xi \in U \cap V \subset F^c$, which proves indeed that F^c is *open* and hence F is *closed*. If F is *dense*, then $F = F^- = M$ and we get $f \leq g$. $\square$

Remark 5.5.9. Note that, although *nonstrict* inequalities ($\leq$ and $\geq$) extend by continuity, this is *not* true (in general) for *strict* inequalities. For example, we have $x^2 < x$ on the dense subset $(0, 1)$ of $[0, 1]$, but $x^2 = x$ for $x = 0$ and $x = 1$.

The next theorem is an intuitively "obvious" necessary and sufficient condition for the existence of *continuous extensions* of continuous functions defined on *dense* subsets of a metric space.

Theorem 5.5.10. *Let* (M, d) *and* (M', d') *be metric spaces,* X *a dense subset of* M, *and* $f \in C(X, M')$. *Then* f *has a continuous extension* $\tilde{f} \in C(M, M')$ *(i.e.,* $\tilde{f}|X = f$) *if and only if, for any* $x_0 \in M \setminus X$, *the limit* $\lim_{x \to x_0} f(x)$ *exists in* M'. *The extension* $\tilde{f}$ *(if it exists) is then unique.*

Proof. If $\tilde{f} \in C(M, M')$ exists, then we obviously have $\tilde{f}(x_0) = f(x_0) \; \forall x_0 \in X$. On the other hand, the *density* of X implies that any $x_0 \in M \setminus X$ is a *limit point* of X. Hence, by Theorem 5.4.3, we must have $\tilde{f}(x_0) = \lim_{x \to x_0} f(x) \in M'$, which also proves the *uniqueness* of $\tilde{f}$ if it exists. Conversely, if the condition of the theorem is satisfied, we *define* the extension $\tilde{f}$ by $\tilde{f}(x_0) := \lim_{x \to x_0} f(x)$ for each $x_0 \in M \setminus X$ and $\tilde{f}(x_0) = f(x_0) \; \forall x_0 \in X$. To prove the continuity of $\tilde{f}$ at an arbitrary point $x_0 \in M$, let $\varepsilon > 0$ be given. We must show the existence of a

$\delta > 0$ such that $x \in B_\delta(x_0)$ implies $f(x) \in B_\varepsilon(y_0)$, where $y_0 := \tilde{f}(x_0)$. Since $\tilde{f}$ is automatically continuous at any *isolated point* $x_0 \in M$, we may as well assume that x_0 is a *limit point* of M. It is then true that $\tilde{f}(x_0) = \lim_{x \to x_0} f(x)$ (whether $x_0 \in X$ or not). Hence, we can pick $\delta > 0$ such that $d'(f(x), y_0) < \varepsilon/2$ for all $x \in X \cap B_\delta(x_0)$. For a point $\xi \in B_\delta(x_0) \setminus X$, since ξ is then a limit point of X, we have $\tilde{f}(\xi) := \lim_{x \to \xi} f(x)$, where the x's in "$x \to \xi$" are in X. We may, however, restrict these x's to be in $X \cap B_\delta(x_0)$. (Why?) It then follows that, for each such x, $f(x) \in B_{\varepsilon/2}(y_0)$ and hence, passing to the limit, $\tilde{f}(\xi) \in B'_{\varepsilon/2}(y_0) := \{y \in M' : d'(y, y_0) \le \varepsilon/2\}$, which is the corresponding *closed* ball (cf. Remark 5.3.17). It is now clear that $x \in B_\delta(x_0)$ implies $\tilde{f}(x) \in B_\varepsilon(y_0)$ and the continuity of the extension $\tilde{f}$ is established.   □

We are now going to prove the existence theorem for the extensions of *uniformly continuous* functions from a metric space to a *complete* metric space. The following proposition will be needed.

Proposition 5.5.11. *Let f be a uniformly continuous function from a metric space (M, d) to a metric space (M', d'). For any Cauchy sequence $(x_n) \in M^{\mathbb{N}}$ the sequence $(f(x_n))$ is Cauchy in M'.*

Proof. Let $\varepsilon > 0$ be given. Since f is *uniformly continuous*, there exists $\delta > 0$ such that

$$(\forall x, x' \in M)(d(x, x') < \delta \Rightarrow d'(f(x), f(x')) < \varepsilon). \qquad (*)$$

Now, if $(x_n) \in M^{\mathbb{N}}$ is a *Cauchy* sequence, then we can find $N \in \mathbb{N}$ such that $m, n \ge N$ implies $d(x_m, x_n) < \delta$ and hence, by $(*)$, $d'(f(x_m), f(x_n)) < \varepsilon$.   □

Remark 5.5.12. Note that if f is merely *continuous*, then the image under f of a Cauchy sequence need *not* be Cauchy. For example, the function $f(x) := 1/x$ is continuous on $(0, 1]$ and $(1/n)$ is a Cauchy sequence in $(0, 1]$, but $(f(1/n)) = (n)$ is *not* a Cauchy sequence.

Theorem 5.5.13 (Extensions of Uniformly Continuous Functions). *Let f be a uniformly continuous function from a dense subspace X of a metric space (M, d) to a complete metric space (M', d'). Then f has a unique extension $\tilde{f} \in C(M, M')$. Moreover, the function $\tilde{f}$ is also uniformly continuous.*

Proof. Let $x_0 \in M$. Since X is *dense*, we have $x_0 = \lim(x_n)$ for a sequence $(x_n) \in X^{\mathbb{N}}$. The sequence (x_n) converges to x_0, hence it is a Cauchy sequence. By Proposition 5.5.11, the sequence $(f(x_n))$ is also Cauchy in the *complete* metric space M'. Therefore, $\lim(f(x_n)) = y_0$ for some $y_0 \in M'$. We now *define* $\tilde{f}(x_0) := y_0$. To show that this is *well defined* (i.e., that y_0 depends only on x_0 and not on the particular sequence (x_n)), suppose we also have $\lim(x'_n) = x_0$ for another sequence $(x'_n) \in X^{\mathbb{N}}$. It then follows that $\lim(d(x_n, x'_n)) = 0$ and, since f is *uniformly continuous*, we have $\lim_{n \to \infty} d'(f(x_n), f(x'_n)) = 0$. (Why?) Thus $y_0 = \lim(f(x'_n))$, as desired. Finally, we show that $\tilde{f}$ is *uniformly continuous*. To this end, let $\varepsilon > 0$ be given. By the uniform continuity of f, we can find

$\delta > 0$ such that for any x, $x' \in X$ we have $d'(f(x), f(x')) < \varepsilon/2$ whenever $d(x, x') < 3\delta$. Now, given any ξ, $\xi' \in M$ with $d(\xi, \xi') < \delta$, pick sequences (x_n), $(x'_n) \in X^{\mathbb{N}}$ such that $\lim(x_n) = \xi$ and $\lim(x'_n) = \xi'$. It then follows from the inequalities

$$d(x_n, x'_n) \leq d(x_n, \xi) + d(\xi, \xi') + d(\xi', x'_n) \qquad \forall n \in \mathbb{N}$$

that $\exists N \in \mathbb{N}$ with $d(x_n, x'_n) < 3\delta \ \forall n \geq N$. (Why?) Thus $d'(f(x_n), f(x'_n)) < \varepsilon/2 \ \forall n \geq N$ and since $\tilde{f}(\xi) := \lim(f(x_n))$, $\tilde{f}(\xi') := \lim(f(x'_n))$, we get (e.g., from the continuity of the distance function and the *Extension of Inequalities*) that $d'(\tilde{f}(\xi), \tilde{f}(\xi')) \leq \varepsilon/2 < \varepsilon$. □

Example 5.5.14. Let $\{r_1, r_2, \dots\}$ be an *enumeration* of the set $\mathbb{Q} \cap [0, 1]$ and define the *jump function* $h(x) := \sum_{r_n < x} 2^{-n}$, $x \in [0, 1]$ (cf. Definition 4.4.10). Then the restriction f of h to the set $\mathbb{Q}^c \cap [0, 1]$ of irrational numbers in $[0, 1]$ is *continuous*, but has *no* continuous extension to $[0, 1]$. (Why?)

The following important consequence of the Baire Category Theorem is a version of the *Uniform Boundedness Principle* to be discussed later (cf. Theorem 9.2.29).

Theorem 5.5.15 (Osgood's Theorem). *Let (M, d) be a complete metric space and $\mathcal{F} \subset C(M)$. Suppose that, for each $x \in M$, there exists a constant $C_x > 0$ such that $|f(x)| \leq C_x \ \forall f \in \mathcal{F}$. Then there exists a nonempty open ball $B \subset M$ and a constant $C > 0$ such that*

$$|f(x)| \leq C \quad \forall f \in \mathcal{F}, \quad \forall x \in B.$$

Proof. For each $n \in \mathbb{N}$, let $F_{n,f} := \{x \in M : |f(x)| \leq n\}$ and set $F_n := \bigcap_{f \in \mathcal{F}} F_{n,f}$. Since f is continuous, each $F_{n,f}$ is *closed* and hence so is each F_n. Now, for each $x \in M$, the assumptions in the theorem imply that we can find some $n \in \mathbb{N}$ such that $|f(x)| \leq n \ \forall f \in \mathcal{F}$. This means that, for each $x \in M$, there is an integer $n \in \mathbb{N}$ with $x \in F_n$. Therefore, we have

$$M = \bigcup_{n=1}^{\infty} F_n.$$

Since M is *complete*, it follows from the Baire Category Theorem that, for at least one $n \in \mathbb{N}$, the (closed) set F_n is *not* nowhere dense. Therefore, there exists a nonempty open ball B with $B \subset F_n$. But then, for each $x \in B$, we have $|f(x)| \leq n$ $\forall f \in \mathcal{F}$. □

To end the section, we note that it is obviously desirable to work in *complete* metric spaces. Fortunately, *any* metric space can, in fact, be *completed* (in an essentially *unique* way) in the following sense:

Theorem 5.5.16 (Completion of a Metric Space). *Any metric space (M, d) is isometric to a dense subspace of a complete metric space (M^*, d^*); i.e., there is* an isometry $\iota : M \to M^*$ such that $\iota(M)$ is dense in M^*. The space M^*, which is known as a "completion" of M, is unique in the sense that, if M^{**} is another completion of M, then M^* and M^{**} are isometric.

Proof. Let $\mathcal{C}$ denote the set of all *Cauchy* sequences in M. Thus

$$\mathcal{C} := \{(x_n) \in M^{\mathbb{N}} : \lim_{m,n \to \infty} d(x_n, x_m) = 0\}.$$

We define the following relation between the elements of $\mathcal{C}$:

$$(x_n) \sim (y_n) \iff \lim(d(x_n, y_n)) = 0.$$

It is easily checked that $\sim$ is an *equivalence relation* on $\mathcal{C}$. (Why?) We denote the equivalence class of $(x_n) \in \mathcal{C}$ by $x^* := [(x_n)]$ and the set of all equivalence classes by M^*. For each pair of elements x^*, $y^* \in M^*$, we define

$$d^*(x^*, y^*) := \lim(d(x_n, y_n)) \qquad ((x_n) \in x^*, \ (y_n) \in y^*). \tag{$\dagger$}$$

Let us show that this limit *always exists* and is *independent* of the representatives (x_n) and (y_n) of x^* and y^*, respectively. First, we show that $(d(x_n, y_n))$ is a *Cauchy sequence* in $\mathbb{R}$. Indeed, as m, $n \to \infty$, the Triangle Inequality implies

$$|d(x_m, y_m) - d(x_n, y_n)| \leq |d(x_m, y_m) - d(x_n, y_m)| + |d(x_n, y_m) - d(x_n, y_n)|$$
$$\leq d(x_m, x_n) + d(y_m, y_n) \to 0.$$

Since $\mathbb{R}$ is *complete*, the limit in ($\dagger$) exists. Also, if $(x_n') \sim (x_n)$ and $(y_n') \sim (y_n)$, then (using the Triangle Inequality again) we have

$$|d(x_n, y_n) - d(x_n', y_n')| \leq d(x_n, x_n') + d(y_n, y_n') \to 0 \quad (n \to \infty),$$

which shows that the limit in ($\dagger$) is indeed independent of the choice of representatives. Next, let us show that d^* is a *metric* on M^*. First, it is obvious that $d^*(x^*, y^*) = d^*(y^*, x^*)$. Also, $d^*(x^*, y^*) = 0$ if and only if $x^* = y^*$, by the very definition of the relation $\sim$. To prove the Triangle Inequality, note that we have the Triangle Inequality

$$d(x_n, y_n) \leq d(x_n, z_n) + d(z_n, y_n) \qquad (\forall n \in \mathbb{N})$$

in M for all (x_n), (y_n), and (z_n) in $\mathcal{C}$ and hence, taking the limit as $n \to \infty$, we obtain

$$d^*(x^*, y^*) \leq d^*(x^*, z^*) + d^*(z^*, y^*).$$

Let us now define $\iota : M \to M^*$ by setting

$$\iota(x) := [(x, x, \ldots)] \in M^*.$$

Then ι is *injective* since no class can contain more than one *constant* sequence. (Why?) Also, it is obvious from the definition of d^* that

$$d^*(\iota(x), \iota(y)) = d(x, y)$$

and hence ι is indeed an *isometry*. To prove that $\iota(M)$ is *dense* in M^*, given any $x^* = [(x_n)] \in M^*$ and any $\varepsilon > 0$, pick $N \in \mathbb{N}$ such that $d(x_n, x_N) \leq \varepsilon/2$ for all $n \geq N$. Then it follows at once that $d^*(x^*, \iota(x_N)) = \lim_{n \to \infty} d(x_n, x_N) \leq \varepsilon/2 < \varepsilon$, which shows indeed that $x^* \in (\iota(M))^-$. Finally, let us show that M^* is *complete*. Let (x_n^*) be a *Cauchy* sequence in M^* and, using the density of $\iota(M)$ in M^*, for each $n \in \mathbb{N}$ pick $y_n \in M$ such that $d^*(x_n^*, \iota(y_n)) < 1/n$. Now observe that

$$d^*(\iota(y_m), \iota(y_n)) \leq d^*(\iota(y_m), x_m^*) + d^*(x_m^*, x_n^*) + d^*(x_n^*, \iota(y_n))$$

$$< \frac{1}{m} + \frac{1}{n} + d^*(x_m^*, x_n^*),$$

so that $(\iota(y_n))$ is a *Cauchy* sequence in $\iota(M)$ and hence $(y_n) \in \mathcal{C}$. Let $y^* := [(y_n)] \in M^*$. Then we have

$$d^*(x_n^*, y^*) \leq d^*(x_n^*, \iota(y_n)) + d^*(\iota(y_n), y^*) < \frac{1}{n} + d^*(\iota(y_n), y^*),$$

and since $d^*(\iota(y_n), y^*) = \lim_{m \to \infty} d(y_n, y_m) = 0$, it follows that $\lim(x_n^*) = y^*$. To complete the proof, we must show that any two completions of M are isometric. To show this, let us *identify* the space M with its image $\iota(M)$ and assume that M^* and M^{**} are complete and contain M as a *dense subspace*. For each $x^* \in M^*$, pick a Cauchy sequence $(x_n) \in M$ with $\lim(x_n) = x^*$. Since (x_n) is also Cauchy in the complete space M^{**}, we have $\lim(x_n) = x^{**} \in M^{**}$. We now define

$$\phi(x^*) := x^{**}.$$

It is easy to check that this construction is independent of the (x_n) that converges to x^* and that ϕ is a well-defined *bijection* of M^* onto M^{**}. (Why?) Now note that $\phi(x) = x$ for all $x \in M$. Therefore, if $\lim(x_n) = x^* \in M^*$ and $\lim(x_n) = x^{**} \in M^{**}$, while $\lim(y_n) = y^* \in M^*$ and $\lim(y_n) = y^{**} \in M^{**}$, then

$$d^*(x^*, y^*) = \lim(d(x_n, y_n)) = d^{**}(x^{**}, y^{**})$$

and hence $d^{**}(\phi(x^*), \phi(y^*)) = d^*(x^*, y^*)$, as desired. $\square$

5.6   Compact Metric Spaces

Our goal here will be to introduce, for subsets of an abstract metric space, the fundamental concept of *compactness* which replaces, for metric spaces, the concept of *finiteness* for general sets. We saw in Chap. 4 that, for the special metric space $\mathbb{R}$, a subset is compact precisely when it is *closed and bounded*. We showed that continuous functions map compact subsets of $\mathbb{R}$ onto compact subsets of $\mathbb{R}$ and that a continuous function on a compact subset of $\mathbb{R}$ is *uniformly continuous*. Most of these results have analogs in general metric spaces as we shall presently see.

Definition 5.6.1 (Open Cover, Subcover). Let (M, d) be a metric space and let $S \subset M$. A collection $\mathcal{U} = \{U_\lambda\}_{\lambda \in \Lambda}$ of *open* subsets of M is said to be an *open cover* of S if

$$S \subset \bigcup_{\lambda \in \Lambda} U_\lambda.$$

If we also have $S \subset \bigcup_{\lambda \in \Lambda'} U_\lambda$ for a *subset* $\Lambda' \subset \Lambda$, then the collection $\mathcal{U}' = \{U_\lambda\}_{\lambda \in \Lambda'}$ is called a *subcover* (of $\mathcal{U}$). If, in addition, Λ' is *finite*, then the subcover $\mathcal{U}'$ is called a *finite subcover*.

Definition 5.6.2 (Compact & Relatively Compact Sets). Let (M, d) be a metric space. A set $K \subset M$ is said to be *compact* if *every* open cover $\{U_\lambda\}_{\lambda \in \Lambda}$ of K contains a *finite* subcover; in other words, there are finitely many indices $\lambda_1, \ldots, \lambda_n \in \Lambda$ such that

$$K \subset \bigcup_{j=1}^{n} U_{\lambda_j}.$$

We say that $K \subset M$ is *relatively compact* if (the closure) K^- is *compact*.

Exercise 5.6.3. Let M be a metric space.

1. Show that any *finite* set $F \subset M$ is compact.
2. Show that, if $(x_n) \in M^{\mathbb{N}}$ is a *convergent* sequence with $\lim(x_n) = \xi \in M$, then the set $\{x_1, x_2, x_3, \ldots\} \cup \{\xi\}$ is compact.
3. Show that, if M is *discrete*, then every compact set $K \subset M$ is *finite*.
4. Show that the intersection of *any* collection of compact subsets of M is compact and that the union of any *finite* collection of compact subsets of M is compact.

Definition 5.6.4 (Finite Intersection Property). A family of sets is said to have the *finite intersection property* if each *finite* subfamily has *nonempty* intersection.

Proposition 5.6.5. *A metric space M is compact if and only if each family of closed subsets of M having the finite intersection property has a nonempty intersection. In particular, if (K_n) is a (decreasing) nested sequence of nonempty closed subsets of M (i.e., $K_{n+1} \subset K_n \ \forall n \in \mathbb{N}$), then $\bigcap_{n=1}^{\infty} K_n \neq \emptyset$.*

Proof. This is simply an application of *De Morgan's laws.* Indeed, $\mathcal{U}$ is an open cover of M if and only if $\mathcal{F} := \{U^c : U \in \mathcal{U}\}$ is a family of closed subsets of M with *empty* intersection. Therefore, every open cover has a *finite* subcover if and only if every family of closed sets with empty intersection has a *finite* subfamily with empty intersection.   □

Recall that, if M is a metric space and if $S \subset X \subset M$, then S may be *open* in the *subspace* X without being open in M, and the same can be said for *closed* subsets. This may be expressed by saying that "openness" and "closedness" are *relative* concepts. As the following proposition shows, however, *compactness* of a set is in fact an *absolute* topological property; i.e., it is independent of the space in which the set is embedded:

Proposition 5.6.6. *Let M be a metric space and $X \subset M$. A set $K \subset X$ is compact in M if and only if it is compact in the (metric) subspace X.*

Proof. Suppose that K is compact in M and that $K \subset \bigcup_{\lambda \in \Lambda} V_\lambda$, where each V_λ is *open in X.* Then, for each $\lambda \in \Lambda$, there is an open set $U_\lambda \subset M$ such that $V_\lambda = X \cap U_\lambda$ and we clearly have $K \subset \bigcup_{\lambda \in \Lambda} U_\lambda$. Using the compactness of K (in M), we get

$$K \subset \bigcup_{j=1}^{n} U_{\lambda_j} \qquad (*)$$

for a finite subset $\{\lambda_1, \ldots, \lambda_n\} \subset \Lambda$. But then we obviously have

$$K \subset \bigcup_{j=1}^{n} V_{\lambda_j} \qquad (**)$$

and K is indeed compact *in X.* Conversely, if K is compact in X and if $K \subset \bigcup_{\lambda \in \Lambda} U_\lambda$, where each U_λ is open in M, then $(**)$ is satisfied (with $V_\lambda := X \cap U_\lambda$) for a subset $\{\lambda_1, \ldots, \lambda_n\} \subset \Lambda$ and, since $K \subset X$, $(*)$ follows.   □

Theorem 5.6.7. *Let M be a metric space. The following statements are true:*

(a) Any compact subset $K \subset M$ is closed and bounded. Thus any relatively compact subset is bounded.

(b) Any closed subset F of a compact set $K \subset M$ is compact.

(c) If $K \subset M$ is compact and $F \subset M$ is closed, then $F \cap K$ is compact.

Proof. To prove (a) we show that, if $K \subset M$ is compact, then K^c is open. Now suppose $x \notin K$. For each $y \in K$, pick ε_y such that $0 < \varepsilon_y < d(x, y)/2$ and let $U_y := B_{\varepsilon_y}(x)$, $V_y := B_{\varepsilon_y}(y)$. Using the compactness of K, we can choose $y_1, y_2, \ldots, y_n \in K$ such that

$$K \subset V := V_{y_1} \cup V_{y_2} \cup \cdots \cup V_{y_n},$$

which already proves that K is *bounded*. (Why?) Now the set $U := U_{y_1} \cap U_{y_2} \cap \cdots \cap U_{y_n}$ is open and $U \cap V = \emptyset$. Therefore, we have $x \in U \subset K^c$ and K^c is indeed open. To prove (b), suppose that $F \subset K \subset M$, with F closed and K compact. Let $\{U_\lambda\}_{\lambda \in \Lambda}$ be an open cover of F. Then the open collection $\{U_\lambda\}_{\lambda \in \Lambda} \cup \{F^c\}$ covers K and hence F. Using the compactness of K, we can pick a *finite* subcover. If F^c is part of this subcover, we simply remove it to get a finite cover of F by the U_λ. Finally, (c) is an immediate consequence of (a) and (b).   □

Our next goal is to prove that a *compact* subset of a metric space is *complete*. First, let us introduce some older variants of the concept of compactness. It turns out that for *metric spaces* they are equivalent to the compactness defined above. This, however, is *not* true for more general *topological spaces*.

Definition 5.6.8 (Fréchet Compact, Bolzano–Weierstrass Property). A metric space M is called *Fréchet compact* (or is said to satisfy the *Bolzano–Weierstrass property*) if every *infinite* subset of M has a *limit point*.

Definition 5.6.9 (Sequentially Compact). A metric space (M, d) is called *sequentially compact* if every sequence in M has a *convergent subsequence*.

Definition 5.6.10 (Countably Compact). We say that a metric space (M, d) is *countably compact* if every *countable* open cover of M has a *finite* subcover.

Remark 5.6.11. It is obvious that every *compact* metric space is *countably compact*. The converse (which is not true for general topological spaces) turns out to be true for metric spaces (Theorem 5.6.25 below).

Let us begin by proving the following

Proposition 5.6.12. *Any countably compact metric space (M, d) is Fréchet compact; i.e., every infinite subset $S \subset M$ has a limit point.*

Proof. Since every infinite set contains a *countably infinite* subset, we may as well assume that S is countably infinite. So let $S = \{x_1, x_2, \ldots\}$. If no $x \in M$ is a limit point of S, then each x_n is an *isolated point* of S and S is *closed*. (Why?) For each $n \in \mathbb{N}$, let B_n be an open ball with $S \cap B_n = \{x_n\}$. The collection $\{B_n : n \in \mathbb{N}\} \cup S^c$ is then a countable open cover of M with no finite subcover, contradicting the countable compactness of M.   □

Remark 5.6.13.

1. The *converse* of Proposition 5.6.12 is also *true* and will be a consequence of Theorem 5.6.25 below.
2. Fréchet compact spaces have the following interesting property:

Theorem 5.6.14 (Lebesgue's Covering Lemma). *Let (M, d) be a Fréchet compact metric space. Then, given any open cover $\{U_\lambda\}_{\lambda \in \Lambda}$ of M, there exists $\varepsilon > 0$ such that each open ball $B_\varepsilon(x)$ is contained in some U_λ.*

Proof. Suppose, to get a contradiction, that the statement is *false* for an open cover $\mathcal{U} = \{U_\lambda\}_{\lambda \in \Lambda}$ of M. We can then find a sequence (x_n) in M such that $B_{1/n}(x_n) \not\subset$

U_λ for all $\lambda \in \Lambda$. If the set $T := \{x_1, x_2, \dots\}$ of terms is *finite*, then (at least) one term, say x_k, is repeated infinitely often. Since $\mathcal{U}$ covers M, we have $x_k \in U_\lambda$ for some $\lambda \in \Lambda$. Now U_λ is *open* so there exists $\delta > 0$ such that $B_\delta(x_k) \subset U_\lambda$. Pick $N \in \mathbb{N}$ such that $1/N < \delta$ and $x_N = x_k$ to get $B_{1/N}(x_N) \subset U_\lambda$, a contradiction! So let us assume that T is *infinite*. Since M is Fréchet compact, T has a limit point, say ξ, and we can pick $\lambda \in \Lambda$ such that $\xi \in U_\lambda$. There is a $\delta > 0$ such that $B_\delta(\xi) \subset U_\lambda$ and we may pick $N \in \mathbb{N}$ so large that $1/N < \delta/2$ and $x_N \in B_{\delta/2}(\xi)$. But then we get $B_{1/N}(x_N) \subset B_\delta(\xi) \subset U_\lambda$, a contradiction again!   □

Remark 5.6.15. The number $\varepsilon > 0$ in the above theorem *depends* on the open cover $\mathcal{U} = \{U_\lambda\}_{\lambda \in \Lambda}$. Note, however, that if every $B_\varepsilon(x)$ is contained in some U_λ, then the same is true for $B_{\varepsilon'}(x)$, where $0 < \varepsilon' < \varepsilon$. This suggests the following:

Definition 5.6.16 (Lebesgue Number). Let $\mathcal{U} = \{U_\lambda\}_{\lambda \in \Lambda}$ be an open cover of a metric space (M, d) and consider the set

$$E_\mathcal{U} := \{\varepsilon > 0 : (\forall \, x \in M) \, (\exists \lambda \in \Lambda) \text{ such that } B_\varepsilon(x) \subset U_\lambda\}.$$

If $E_\mathcal{U} \neq \emptyset$, then the number $\varepsilon_L = \varepsilon_L(\mathcal{U}) := \sup(E_\mathcal{U})$ is called the *Lebesgue number* of the covering $\mathcal{U}$.

The following corollary is now an immediate consequence of the Lebesgue's Covering Lemma:

Corollary 5.6.17. *Let (M, d) be a Fréchet compact metric space. Then any open cover $\{U_\lambda\}_{\lambda \in \Lambda}$ of M has a Lebesgue number ε_L.*

Here is another fundamental property of compact spaces:

Theorem 5.6.18 (Compact $\Longrightarrow$ Complete). *Let K be a compact subset of a metric space M. Then, as a (metric) subspace of M, K is complete.*

Proof. Let $(x_n) \in K^\mathbb{N}$ be a *Cauchy* sequence in K. If there is a subscript n_0 such that for each $k \geq n_0$ we can find $n_k \geq k$ with $x_{n_k} = x_{n_0}$, then $\lim(x_{n_k}) = x_{n_0}$ and, since (x_n) is Cauchy, we also have (Exercise 5.3.6(4)) $\lim(x_n) = x_{n_0} \in K$. If no such n_0 exists, then the set $\{x_1, x_2, x_3, \dots\}$ of all terms is *infinite* (why?) and hence, by Proposition 5.6.12, has a limit point ξ. Since the compact set K is *closed*, we have $\xi \in K$. Pick a subsequence (x_{n_k}) of (x_n) with $\lim(x_{n_k}) = \xi$. We then also have $\lim(x_n) = \xi$.   □

Recall that by Lindelöf's Theorem (Proposition 4.1.1) every open cover of a subset of $\mathbb{R}$ has a *countable* subcover. The proof used the fact that $\mathbb{R}$ is *separable*, i.e., has a countable dense subset (namely $\mathbb{Q}$). This property of $\mathbb{R}$, called the *Lindelöf property*, is in fact shared by all separable metric spaces. We shall see (Proposition 5.6.22 below) that countable compactness implies separability and hence the Lindelöf property. First, a couple of definitions:

Definition 5.6.19 (Lindelöf Space). A metric space (M, d) is said to be a *Lindelöf space* (or to have the *Lindelöf property*) if every open cover of M has a *countable* subcover.

Definition 5.6.20 (Totally Bounded, ε-Net). A subset X of a metric space (M, d) is said to be *totally bounded* if, given any $\varepsilon > 0$, there exists a *finite* subset $\{x_1, x_2, \ldots, x_n\} \subset M$, called an *$\varepsilon$-net*, such that $X \subset \bigcup_{k=1}^{n} B_\varepsilon(x_k)$; i.e., for each $\varepsilon > 0$, we can cover X by a *finite* number of open balls of radius ε.

Exercise 5.6.21.

1. Show that a set $S \subset \mathbb{R}^n$ is *totally bounded* if and only if it is *bounded*.
2. Let X be a totally bounded subset of a metric space (M, d).

   (a) Show that, if in the above definition we replace $\{x_1, x_2, \ldots, x_n\} \subset M$ by $\{x_1, x_2, \ldots, x_n\} \subset X$, we get an *equivalent* definition.
   (b) Show that X is bounded; i.e., $\delta(X) < \infty$.
   (c) Show that the closure X^- is totally bounded.
   (d) Show that any subset $S \subset X$ is totally bounded.

Proposition 5.6.22. *A separable metric space is a Lindelöf space.*

Proof. Let M be a *separable* metric space. By Proposition 5.2.19 M has a countable base $\mathcal{B}$. Let $\mathcal{U} = \{U_\lambda\}_{\lambda \in \Lambda}$ be any open cover of M. Since each U_λ is a union of members of $\mathcal{B}$, there is a subcollection $\mathcal{C}$ of $\mathcal{B}$ that covers M and each member of $\mathcal{C}$ is a subset of some U_λ. If for each $B \in \mathcal{C}$ we pick a U_λ such that $B \subset U_\lambda$, the resulting subcollection of $\mathcal{U}$ is the desired countable subcover. $\quad\square$

Lemma 5.6.23. *A countably compact metric space is totally bounded.*

Proof. Let (M, d) be a countably compact metric space. If M is *not* totally bounded, then we can find a number $\varepsilon_0 > 0$ and a *countably infinite* set $S := \{x_1, x_2, \ldots\} \subset M$ such that $d(x_m, x_n) \geq \varepsilon_0$. (Why?) Since each open ball of radius $\varepsilon_0/3$ can contain at most one point of S, the infinite set S has no limit points, contradicting Proposition 5.6.12. $\quad\square$

Proposition 5.6.24. *A countably compact metric space is separable and hence (by Proposition 5.6.22) a Lindelöf space.*

Proof. Let (M, d) be a countably compact metric space. Then, by the above lemma, M is totally bounded. For each $n \in \mathbb{N}$ let F_n be a $1/n$-net and hence $M \subset \bigcup_{x \in F_n} B_{1/n}(x)$. Then $D := \bigcup_{n=1}^{\infty} F_n$ is a countable dense subset of M. $\quad\square$

Theorem 5.6.25 (Equivalence of Compactness Notions). *Let (M, d) be a metric space. Then the following statements are pairwise equivalent:*

(a) M is compact;
(b) M is sequentially compact;
(c) M is Fréchet compact;
(d) M is countably compact.

Proof. Suppose M is compact and let $(x_n) \in M^{\mathbb{N}}$ be any sequence. Let $T_n :=$ $\{x_n, x_{n+1}, \ldots\}$ $\forall n \in \mathbb{N}$ and set $F_n := T_n^-$. Then $\{F_n\}_{n \in \mathbb{N}}$ is a family of closed subsets of M having the *finite intersection property* (in fact, it is even nested). Thus, by Proposition 5.6.5, $F := \bigcap_{n=1}^{\infty} \neq \emptyset$. Let $\xi \in F$. It is then easily seen that $\xi = \lim(x_{n_k})$ for a subsequence (x_{n_k}) of (x_n) and the implication (a) $\Rightarrow$ (b) follows. To prove (b) $\Rightarrow$ (c), suppose M is sequentially compact and let $X \subset M$ be an *infinite* set. Then X contains a *countably infinite* subset $S = \{x_1, x_2, \ldots\}$ with $x_j \neq x_k$ for $j \neq k$. If now $\xi = \lim(x_{n_k})$ for a subsequence (x_{n_k}) of (x_n), then ξ is clearly a *limit point* of X. (Why?) Before proving (c) $\Rightarrow$ (d), let us point out that (arguing as in Proposition 5.6.5) M is countably compact if and only if every *countable* family $\mathcal{F} = \{F_n\}_{n \in \mathbb{N}}$ of closed sets with finite intersection property has a nonempty intersection. Now suppose M is Fréchet compact and let $\mathcal{F} = \{F_n\}_{n \in \mathbb{N}}$ be a countable collection of closed sets with finite intersection property. Let $E_n := \bigcap_{k=1}^{n} F_k$ $\forall n \in \mathbb{N}$, and note that the E_n are *nonempty*, closed, and *nested*. For each $n \in \mathbb{N}$ pick $x_n \in E_n$ and note that, by the Bolzano–Weierstrass property, the sequence (x_n) has a convergent subsequence (x_{n_k}). If $\xi = \lim(x_{n_k})$, then, since $x_n \in F_k$ $\forall k \geq n$ and the F_n are *closed*, we have $\xi \in F_n$ $\forall n \in \mathbb{N}$. This establishes (c) $\Rightarrow$ (d). Finally, suppose M is countably compact. Then, by Proposition 5.6.24, it is a Lindelöf space. Thus, each open cover $\mathcal{U}$ of M has a *countable* subcover $\mathcal{U}'$ which, in turn, has a *finite* subcover in view of the countable compactness of M. This proves the implication (d) $\Rightarrow$ (a) and completes the proof. $\square$

We have seen that a compact (hence countably compact) space is complete and totally bounded. In fact the *converse* is also true:

Theorem 5.6.26 (Compact $\Longleftrightarrow$ Complete and Totally Bounded). *A metric space (M, d) is compact if and only if it is complete and totally bounded.*

Proof. If M is *compact*, then it is complete by Theorem 5.6.18 and totally bounded by Lemma 5.6.23. Conversely, suppose that M is complete *and* totally bounded. Let us show that it is *sequentially compact*. So let $(x_n) \in M^{\mathbb{N}}$ and let $T := \{x_1, x_2, \ldots\}$ be the set of its terms. If T is *finite*, then (at least) one of the terms, say x_k, is repeated an *infinite* number of times and the constant subsequence $(x_k, x_k, \ldots)$ is obviously convergent. Suppose then that T is *infinite*. Cover M with open balls of radius $\varepsilon = 1$ centered at the (*finite* set of) points of a 1-*net*. At least one of these balls, say B_1, contains an infinite number of the x_n; i.e., $B_1 \cap T$ is *infinite*. Pick $x_{n_1} \in B_1 \cap T$. Next, cover M with open balls of radius $\varepsilon = 1/2$ centered at the (finite set of) points of a $1/2$-*net* and pick one of the balls, say $B_{1/2}$, such that $B_1 \cap B_{1/2} \cap T$ is *infinite*. Now pick $x_{n_2} \in B_1 \cap B_{1/2} \cap T$ with $n_2 > n_1$. Continuing this process produces a subsequence (x_{n_k}) that is *Cauchy*. (Why?) Since M is assumed to be *complete*, (x_{n_k}) is convergent and the proof is complete. $\square$

We end this section with a quick look at the relationship between *continuity* and *compactness* in metric spaces.

Theorem 5.6.27 (Continuity and Compactness). *If f is a continuous function from a compact metric space (M, d) to a metric space (M', d'), then (the range) $f(M)$ is a compact subspace of M'.*

Proof. Let $\{V_\lambda\}_{\lambda \in \Lambda}$ be an open cover of $f(M)$. Then, by the continuity of f, each $U_\lambda := f^{-1}(V_\lambda)$ is *open* in M. Since M is *compact*, there is a finite set $\{\lambda_1, \lambda_2, \dots, \lambda_n\} \subset \Lambda$ such that

$$M \subset U_{\lambda_1} \cup \cdots \cup U_{\lambda_n}. \tag{$*$}$$

Since for each $Y \subset M'$ we have $f(f^{-1}(Y) \subset Y$, $(*)$ implies

$$f(M) \subset V_{\lambda_1} \cup \cdots \cup V_{\lambda_n},$$

which completes the proof.   □

Corollary 5.6.28 (Weierstrass's Extreme Value Theorem). *Let (M, d) be a compact metric space and $f \in C(M)$. Then f attains its maximum and minimum values. In other words, if $\alpha := \inf(f) = \inf\{f(x) : x \in M\}$ and $\beta := \sup(f) = \sup\{f(x) : x \in M\}$, then there exist a, $b \in M$ such that $f(a) = \alpha$ and $f(b) = \beta$.*

Proof. Indeed, by Theorem 5.6.27, $f(M)$ is a *compact* subset of $\mathbb{R}$ and hence, by the Heine–Borel Theorem (Theorem 4.1.10), is *closed* and *bounded*. In particular, $f(M)$ contains its cluster points $\alpha = \inf(f)$ and $\beta = \sup(f)$.   □

Exercise 5.6.29 (Completeness of $C(K)$). Let (K, d) be a *compact* metric space. Then $(C(K), d_\infty)$, where d_∞ is the *uniform metric*, is *complete*. *Hint:* If (f_n) is a *Cauchy sequence* in $C(K)$, then, for each $x \in K$, $(f_n(x))$ is *Cauchy* in $\mathbb{R}$ and hence converges to a number $f(x) \in \mathbb{R}$. Show that $f \in C(K)$, using an $\varepsilon/3$-argument and the inequalities

$$|f(x) - f(x_0)| \le |f(x) - f_n(x)| + |f_n(x) - f_n(x_0)| + |f_n(x_0) - f(x_0)|.$$

Exercise 5.6.30 (Metric Spaces $Lip^\alpha(K)$). Let (K, d) be a *compact* metric space and let $Lip^\alpha(K)$ denote the set of all real-valued *Lipschitz* functions of *order* $\alpha \in (0, 1]$ on K. For each $f \in Lip^\alpha(K)$, define

$$d_{\alpha,\infty}(f, 0) := d_\infty(f, 0) + \sup \left\{ \frac{|f(x) - f(y)|}{d(x, y)^\alpha} : x, y \in K, \ x \ne y \right\}$$

and, for each f, $g \in Lip^\alpha(K)$, let

$$d_{\alpha,\infty}(f, g) := d_{\alpha,\infty}(f - g, 0).$$

Show that $d_{\alpha,\infty}$ is a *metric* and $(Lip^\alpha(K), d_{\alpha,\infty})$ is *complete*.

Exercise 5.6.31. Let A and B be *nonempty* subsets of a metric space (M, d):

(a) Show that, if A is *compact*, then there is a point $a \in A$ such that $d(a, B) = d(A, B)$.

(b) Show that, if A and B are *both compact*, then there exist $a \in A$ and $b \in B$ such that $d(a, b) = d(A, B)$.

(c) Show that, if A is *compact* and B is *closed*, then $d(A, B) = 0$ if and only if $A \cap B \neq \emptyset$.

Corollary 5.6.32. *A continuous map f from a compact metric space M to a metric space M' is closed.*

Proof. Indeed, if $X \subset M$ is *closed*, then it is (by Theorem 5.6.7) a *compact* subspace of M and hence (by Theorem 5.6.27) $f(X)$ is *compact* in M'. Another application of Theorem 5.6.7 now shows that $f(X)$ is *closed*. □

Recall that two metric spaces M and M' are said to be *homeomorphic* if there is a *bijection* $f : M \to M'$ such that f and f^{-1} are both continuous. In general, the continuity of f does *not* imply the continuity of f^{-1}. For example, consider the metric space $\mathbb{R}$ with its usual metric and let $\tilde{\mathbb{R}}$ denote the set $\mathbb{R}$ with the *discrete* metric $\tilde{d}$; i.e., $\tilde{d}(x, y) = 1$ if $x \neq y$ and $\tilde{d}(x, x) = 0$. Then the identity map $\iota : \tilde{\mathbb{R}} \to \mathbb{R}$ (defined by $\iota(x) := x \ \forall x \in \mathbb{R}$) is clearly a *continuous bijection*, but the inverse (which is again the identity map) is *not* continuous. (Why?) The following theorem shows that the continuity of the inverse is automatic if the domain space M is *compact*:

Theorem 5.6.33. *A continuous bijection f of a compact metric space M onto a metric space M' is a homeomorphism.*

Proof. We must only prove that f^{-1} is continuous. Now, recall that a function is continuous if and only if the inverse image of every closed set is closed. Thus, we must show that for each closed set $X \subset M$, the inverse image of X under f^{-1} is *closed*. This, however, means that $(f^{-1})^{-1}(X) = f(X)$ is closed, which follows from Corollary 5.6.32. □

We proved (Theorem 5.4.15) that the graph of a continuous function is *closed*. We also gave an example to show that the converse is false in general. Now we prove what was promised, namely, that the converse is *true* if the *codomain* (i.e., the target space) is *compact*:

Theorem 5.6.34 (Closed Graph Theorem). *Let f be a map from a metric space M to a compact metric space M'. Then f is continuous if and only if its graph is closed.*

Proof. In view of Theorem 5.4.15, we need only show that, if the graph of f (i.e., the set $\Gamma_f := \{(x, f(x)) : x \in M\}$) is *closed* in $M \times M'$, then f is continuous. Since all functions are continuous at *isolated points* of M, it suffices to show that f is continuous at every *limit point* of M. So let x_0 be such a point and suppose, to get a contradiction, that f is *discontinuous* at x_0. We can then find $\epsilon_0 > 0$ and a sequence $(x_n) \in M^{\mathbb{N}}$ such that $x_0 = \lim(x_n)$ but $d'(f(x_n), f(x_0)) \geq \varepsilon_0 \; \forall n \in \mathbb{N}$. Since M' (being compact by assumption) is *sequentially compact*, the sequence $(f(x_n)) \in M'^{\mathbb{N}}$ has a convergent subsequence. Thus, there is an increasing sequence (n_k) of positive integers and a point $y_0 \in M'$ such that $\lim(f(x_{n_k})) = y_0$. But then the points $(x_{n_k}, f(x_{n_k}))$ form a sequence in the graph Γ_f with $\lim_{k \to \infty}(x_{n_k}, f(x_{n_k})) = (x_0, y_0)$. Since Γ_f is *closed*, we have $(x_0, y_0) \in \Gamma_f$; i.e., $y_0 = f(x_0)$, which contradicts $d'(f(x_{n_k}), f(x_0)) \geq \varepsilon_0 \; \forall k \in \mathbb{N}$. $\qquad\square$

We next prove the analog of Theorem 4.6.4 for abstract metric spaces. As before, we include two proofs that are essentially copies of the ones given for Theorem 4.6.4.

Theorem 5.6.35 (Uniform Continuity and Compactness). *Let f be a continuous function from a compact metric space (M, d) to a metric space (M', d'). Then f is uniformly continuous.*

First Proof. If f is *not* uniformly continuous on M, then $\exists \varepsilon_0 > 0$ and two sequences (x_n), $(x'_n) \in M^{\mathbb{N}}$ such that $\lim d(x_n, x'_n) = 0$ and $d'(f(x_n), f(x'_n)) \geq \varepsilon_0 \; \forall n \in \mathbb{N}$. (Why?) Since the compact space M is *sequentially compact*, there is a subsequence (x_{n_k}) such that $\lim(x_{n_k}) = x_0$ for some $x_0 \in M$. But then $\lim d(x_{n_k}, x'_{n_k}) = 0$ implies that we also have $\lim(x'_{n_k}) = x_0$. Therefore, by the *continuity* of f at x_0,

$$\lim_{k \to \infty} f(x_{n_k}) = \lim_{k \to \infty} f(x'_{n_k}) = f(x_0).$$

This, however, is impossible since

$$d'(f(x_{n_k}), f(x'_{n_k})) \geq \varepsilon_0 \quad \forall k \in \mathbb{N}.$$

$\qquad\square$

Second Proof. Let $\varepsilon > 0$ be given. For each $x \in M$, the continuity of f at x implies that

$$(\exists \delta_x > 0)(\forall x' \in M)(d(x, x') < \delta_x \Rightarrow d'(f(x), f(x')) < \varepsilon/2).$$

The open balls $B_{\delta_x/2}(x)$, $x \in M$ form an open cover of the *compact* space M and hence we can find finitely many points $x_1, x_2, \ldots, x_n \in M$ such that, with $B_k := B_{\delta_{x_k}/2}(x)$, we have

$$M \subset \bigcup_{k=1}^{n} B_k.$$

Let $\delta := \min\{\delta_{x_1}/2, \delta_{x_2}/2, \ldots, \delta_{x_n}/2\}$. Now note that, if x, $x' \in M$ satisfy $d(x, x') < \delta$, then x, $x' \in B_k$ for some k. Indeed, if $x \in B_k$, then we also have

$$d(x', x_k) \leq d(x', x) + d(x, x_k) < \delta + \delta_{x_k}/2 \leq \delta_{x_k}/2 + \delta_{x_k}/2 = \delta_{x_k}.$$

Thus, $d(x, x') < \delta$ implies

$$d'(f(x), f(x')) \leq d'(f(x), f(x_k)) + d'(f(x'), f(x_k)) < \varepsilon/2 + \varepsilon/2 = \varepsilon$$

and the proof is complete.   □

Let us end this section with a theorem that is a special case of the celebrated *Tychonoff Theorem*. This important theorem, which states that the product of an arbitrary collection of compact spaces is compact, requires the definition of the *product topology* for *infinite products* of topological spaces and is rather involved. The proof is much simpler for finite products of *metric* spaces where compactness and *sequential* compactness are identical. Before stating the theorem, we invite the reader to solve the following exercise!

Exercise 5.6.36. Let $(M_1, d_1), \ldots, (M_n, d_n)$ be metric spaces. Show that the spaces $M_1 \times \cdots \times M_n$ and $(M_1 \times \cdots \times M_{n-1}) \times M_n$ are *homeomorphic*. In fact, show that if we use the distance $d_{\max}$ throughout, then they are even *isometric*.

Theorem 5.6.37. *Let* $(M_1, d_1), \ldots, (M_n, d_n)$ *be metric spaces. Then the product* $M := M_1 \times \cdots \times M_n$ *is compact if and only if each* M_k, $1 \leq k \leq n$ *is compact.*

Proof. If M is compact, then, since the projections $\pi_k : M \to M_k$ $1 \leq k \leq n$ are *continuous* (even Lipschitz), it follows from Theorem 5.6.27 that $M_k = \pi_k(M)$ is compact for $1 \leq k \leq n$. For the converse, let us first consider the case $n = 2$; i.e., let us show that the product $M := M_1 \times M_2$ of compact spaces M_1 and M_2 is *sequentially compact*. Given any sequence $((x_n, y_n)) \in M^{\mathbb{N}}$, the compactness of M_1 implies that the sequence $(x_n) \in M_1^{\mathbb{N}}$ has a convergent subsequence (x_{n_k}). Let $x_0 := \lim(x_{n_k})$ and note that, by Exercise 5.4.33, the *vertical fiber* $\{x_0\} \times M_2$ is isometric to M_2 and hence is *compact*. Therefore, the sequence $((x_0, y_{n_k})) \in (\{x_0\} \times M_2)^{\mathbb{N}}$ has a convergent subsequence. Hence there is a sequence (k_j) in $\mathbb{N}$ with $k_1 < k_2 < k_3 < \cdots$, such that $(y_{n_{k_j}})$ converges to a point, say $y_0 \in M_2$. It is then clear that the subsequence $((x_{n_{k_j}}, y_{n_{k_j}}))$ of $((x_n, y_n))$ converges to $(x_0, y_0) \in M$. The general case now follows by induction (using Exercise 5.6.36) and the proof is complete.   □

Corollary 5.6.38. *A set* $K \subset \mathbb{R}^n$ *(with any one of the distances d_{euc}, $d_{\max}$, d_{sum}) is compact if and only if it is closed and bounded.*

Proof. If $K \subset \mathbb{R}^n$ is compact, then (Theorem 5.6.7(a)) it is *closed*. On the other hand, $\pi_k(K) \subset \mathbb{R}$ is a compact subset of $\mathbb{R}$ for each k and hence we have $\pi_k(K) \subset [a_k, b_k]$, $1 \leq k \leq n$, for some a_k, $b_k \in \mathbb{R}$, $a_k \leq b_k$. Therefore, $K \subset [a_1, b_1] \times \cdots \times [a_n, b_n]$ and hence is *bounded*. Conversely, if K is closed and bounded, then, as we just saw, K is a *closed* subset of a product $[a_1, b_1] \times \cdots \times [a_n, b_n]$ of compact subsets of $\mathbb{R}$, which is compact by Theorem 5.6.37. Theorem 5.6.7(b) now implies that K is compact.   □

Example 5.6.39 (Unit Sphere, Torus).   The *unit sphere*

$$S^{n-1} := \left\{ (x_1, \ldots, x_n) \in \mathbb{R}^n : \sum_{k=1}^{n} x_k^2 = 1 \right\},$$

which is the set of all points in $\mathbb{R}^n$ whose *Euclidean distance* from the origin $(0, 0, \ldots, 0)$ is 1, is closed and bounded hence compact in $\mathbb{R}^n$. In particular, the *unit circle* $S^1 = \{(x, y) \in \mathbb{R}^2 : x^2 + y^2 = 1\}$ is a compact subset of the plane $\mathbb{R}^2$. By Theorem 5.6.37, the *torus*

$$\mathbf{T}^n := (S^1)^n = S^1 \times \cdots \times S^1$$

is therefore a compact subset of $\mathbb{R}^{2n}$.

5.7   Connected Metric Spaces

The concept of *connectedness* was defined for subsets of $\mathbb{R}$ in Chap. 4, and it was proved that a set of real numbers is connected if and only if it is an *interval*. We also saw that connected sets are mapped onto connected sets by continuous functions. In this section we define connected metric spaces and prove some of their basic properties.

Definition 5.7.1 (Connected Space, Subspace).   A metric space (M, d) is said to be *connected* if there does *not* exist any *partition* of M into two (disjoint nonempty) open sets, i.e., if it is *not* possible to write $M = U \cup V$, where U, $V \subset M$ are *open*, $U \neq \emptyset \neq V$, and $U \cap V = \emptyset$. A set $X \subset M$ is said to be connected if (with the *relative topology*) the *subspace* X of M is connected.

Exercise 5.7.2.   Show that, for a metric space M, the following are pairwise equivalent.

(a) M is connected;
(b) M admits no partition into two (nonempty disjoint) *closed* sets;
(c) the only subsets of M that are *both* open and closed are $\emptyset$ and M.

Definition 5.7.3 (Separation).   Let (M, d) be a metric space. Two subsets X, $Y \subset M$ are said to form a *separation* of M if $\{X, Y\}$ is a *partition* of M (i.e., $M = X \cup Y$, $X \cap Y = \emptyset$, and $X \neq \emptyset \neq Y$), and $X \cap Y^- = X^- \cap Y = \emptyset$.

Remark 5.7.4 (Hausdorff–Lennes Separation Condition). We can combine the two conditions $X \cap Y^- = \emptyset$ and $X^- \cap Y = \emptyset$ and write them as

$$(X \cap Y^-) \cup (X^- \cap Y) = \emptyset. \tag{$*$}$$

We call $(*)$ the *Hausdorff–Lennes Separation Condition.*

Exercise 5.7.5. Let (M, d) be a metric space. Prove the following assertions.

(a) If $\{X, Y\}$ is a *separation* of M, then X and Y are both open and closed.
(b) M is connected if and only if it has *no* separation.

We defined the concept of connectedness for a metric space rather than a *subspace*. The reason is that, like compactness, connectedness is an *absolute* (topological) property of a set; i.e., it does *not* depend on the space in which the set is embedded:

Proposition 5.7.6. *Let M be a metric space and $S \subset X \subset M$. Then S is connected in X if and only if it is connected in M.*

Proof. By Exercise 5.7.5, we must prove that S has no separation *in X* if and only if it has no separation in M. Now recall (Exercise 5.2.16) that the *closure* of a subset $E \subset X$ *relative to X* (which we denote here by E_X^-) is given by $E_X^- = X \cap E^-$, where E^- is, of course, the closure of E in M. It follows from this fact that, if $U, V \subset S$, then

$$(U \cap V_X^-) \cup (U_X^- \cap V) = (U \cap X \cap V^-) \cup (U^- \cap X \cap V) = (U \cap V^-) \cup (U^- \cap V).$$

Thus, the *Hausdorff–Lennes Condition* is satisfied in the *relative topology* of X if and only if it is satisfied in M. □

Proposition 5.7.7. *Let X be a connected subset of a metric space M. Then any set Y satisfying $X \subset Y \subset X^-$ is connected.*

Proof. If Y is *not* connected, then there are nonempty sets U, V, *open in Y*, such that $Y = U \cup V$ and $U \cap V = \emptyset$. Since our assumption implies $X^- \cap Y = Y$ and $X^- \cap Y$ is the closure of X in Y, it follows that X is *dense* in Y. Thus, $X \cap U$ and $X \cap V$ are *nonempty* open subsets of X with $X = (X \cap U) \cup (X \cap V)$ and $(X \cap U) \cap (X \cap V) = \emptyset$, contradicting the assumption that X is *connected*. □

Corollary 5.7.8. *If X is a connected subset of a metric space M, then its closure X^- is also connected.*

It is intuitively clear that the union of a family of connected sets must be connected if these sets have a common point. That this is indeed the case will follow from

Theorem 5.7.9. *Let $\{X_\lambda\}_{\lambda \in \Lambda}$ be a family of connected subsets of a metric space M. If $\bigcap_{\lambda \in \Lambda} X_\lambda \neq \emptyset$, then $X := \bigcup_{\lambda \in \Lambda} X_\lambda$ is connected.*

Proof. Suppose that $X = U \cup V$ where U and V are nonempty open sets in X with $U \cap V = \emptyset$. Let $x \in \bigcap_{\lambda \in \Lambda} X_\lambda$, say $x \in U$. Since V is a nonempty subset of the union $\bigcup_{\lambda \in \Lambda} X_\lambda$, we must have $V \cap X_\lambda \neq \emptyset$ for at least one $\lambda \in \Lambda$. But then, $U \cap X_\lambda$ and $V \cap X_\lambda$ are nonempty open subsets of X_λ that are *disjoint* and

$$X_\lambda = (U \cap X_\lambda) \cup (V \cap X_\lambda),$$

contradicting the connectedness of X_λ.   □

Corollary 5.7.10. *Let $\{X_n\}_{n\in\mathbb{N}}$ be a countable family of connected subspaces of a metric space M such that $X_n \cap X_{n+1} \neq \emptyset$ $\forall n \in \mathbb{N}$. Then $\bigcup_{n=1}^{\infty} X_n$ is connected.*

Proof. Using induction and Theorem 5.7.9, one sees that $Y_n := \bigcup_{k=1}^{n} X_k$ is connected for each $n \in \mathbb{N}$. Also, $Y_1 \subset Y_2 \subset Y_3 \subset \cdots$, so that $\bigcap_{n=1}^{\infty} Y_n = Y_1 \neq \emptyset$. Another application of Theorem 5.7.9 now implies that $\bigcup_{n=1}^{\infty} X_n = \bigcup_{n=1}^{\infty} Y_n$ is connected.   □

The above properties may also be considered as corollaries of the following intuitively "obvious" fact:

Theorem 5.7.11. *Let $\{U, V\}$ be a separation of a metric space M; i.e., $U \neq \emptyset \neq V$, $(U \cap V^-) \cup (U^- \cap V) = \emptyset$, and $M = U \cup V$. If $X \subset M$ is connected, then either $X \subset U$ or $X \subset V$.*

Proof. First, we obviously have

$$X = X \cap M = X \cap (U \cup V) = (X \cap U) \cup (X \cap V).$$

Since $\{U, V\}$ is a *separation* of M,

$$((X \cap U) \cap (X \cap V)^-) \cup ((X \cap U)^- \cap (X \cap V)) \subset (U \cap V^-) \cup (U^- \cap V) = \emptyset.$$

Thus, the subsets $X \cap U$ and $X \cap V$ form a separation of the *connected* set X if they are both nonempty. Therefore, we must either have $X \cap U = \emptyset$ so that $X \subset V$, or $X \cap V = \emptyset$ so that $X \subset U$.   □

Corollary 5.7.12. *Let X be a subset of a metric space M. If every two points of X are contained in a connected subset of X, then X is connected.*

Proof. If X is not connected, let $\{U, V\}$ be a *separation* of X. Since U and V are both nonempty, we can pick $x \in U$ and $y \in V$. It follows from the hypothesis that $\{x, y\} \subset Y$ for a *connected* set $Y \subset X$. Theorem 5.7.11 now implies that either $Y \subset U$ or $Y \subset V$. Since U and V are *disjoint*, we have a contradiction.   □

The following theorem is also in agreement with our intuition: If a connected set intersects both a set and its complement, then it must also intersect its boundary.

Theorem 5.7.13. *Let X and Y be subsets of a metric space M and assume that X is connected. If $X \cap Y \neq \emptyset$ and $X \cap Y^c \neq \emptyset$, then $X \cap \mathrm{Bd}(Y) \neq \emptyset$. Here, $\mathrm{Bd}(Y) := Y^- \cap (Y^c)^-$ is the boundary of Y.*

Proof. To get a contradiction, let us assume that $X \cap \mathrm{Bd}(Y) = \emptyset$. First, we have

$$X = X \cap M = X \cap (Y \cup Y^c) = (X \cap Y) \cup (X \cap Y^c), \qquad (*)$$

and the two sets on the right side are both *nonempty* by hypothesis. Next, we note that our assumption implies

$$(X \cap Y) \cap (X \cap Y^c)^- \subset (X \cap Y^-) \cap (Y^c)^- = X \cap (Y^- \cap (Y^c)^-) \qquad (**)$$
$$= X \cap \mathrm{Bd}(Y) = \emptyset.$$

A similar argument shows that $(X \cap Y)^- \cap (X \cap Y^c) = \emptyset$, which together with $(*)$ and $(**)$ implies that $\{X \cap Y, X \cap Y^c\}$ is a separation of the *connected* set X. This contradiction completes the proof. $\qquad\square$

Exercise 5.7.14. Let X be a *nonempty* subset of a *connected* metric space M and assume that $X \neq M$. Show that $\mathrm{Bd}(X) \neq \emptyset$. *Hint:* Use Theorem 5.7.13.

If a metric space is *not* connected, it is natural to look for its connected *pieces*. An important role is played by the *maximal* connected pieces of the space, the so-called connected components:

Definition 5.7.15 (Connected Component). For every point x of a metric space M, the *union* $C(x)$ of all connected subsets of M that contain x is called the *connected component* of x.

Example.

(a) If M is connected, then there is only *one* connected component, namely M itself.
(b) Consider the space $\mathbb{Q}$ of rational numbers, which is a *dense* subspace of the metric space $\mathbb{R}$. Since the only connected subsets of $\mathbb{R}$ are *intervals*, for each $x \in \mathbb{Q}$ we have $C(x) = \{x\}$.
(c) The space $\mathbb{R} \setminus \{0\}$ has two connected components, namely $(-\infty, 0)$ and $(0, \infty)$.

Exercise 5.7.16. Show that, for each point x in a metric space M, the connected component $C(x)$ is *closed*, i.e., $C(x) = (C(x))^-$.

Exercise 5.7.17. On a metric space M, define a *binary relation* $\sim$ by "$x \sim y$ if and only if there exists a *connected* subset of M containing x and y." Show that $\sim$ is an *equivalence relation* and that, for each $x \in M$, the *equivalence class* of x is precisely $C(x)$. Deduce that the connected components form a *partition* of the space into *closed* connected subsets. In particular, if $y \in C(x)$, then $C(x) = C(y)$ and, if $y \notin C(x)$, then $C(x) \cap C(y) = \emptyset$.

Definition 5.7.18 (Locally Connected, Totally Disconnected). Let M be a metric space.

(a) We say that M is *locally connected* if, given any $x \in M$ and any open set V containing x, there exists a *connected* open set U with $x \in U \subset V$.
(b) We say that M is *totally disconnected* if $C(x) = \{x\} \; \forall x \in M$.

Example.

(a) As was pointed out above, the set $\mathbb{Q}$ is totally disconnected. Another example of a totally disconnected set is the Cantor set C (cf. Theorem 4.2.9).
(b) Every *interval* of $\mathbb{R}$ (and hence $\mathbb{R}$ itself) is locally connected. (Why?)
(c) The set $\mathbb{Z}$ is *both* totally disconnected and locally connected. (Why?) On the other hand, the set $\mathbb{Q}$ (which is totally disconnected) is *not* locally connected. (Why?)

Proposition 5.7.19. *A metric space M is locally connected if and only if for each open set $U \subset M$ the connected components of U are open.*

Proof. Assume first that M is *locally connected* and let $V \subset M$ be open. If C is a connected component of V, then for each $x \in C$ there exists a *connected open* set U with $x \in U \subset V$. By the very definition of connected components, we have $U \subset C$ and hence C is *open* as claimed. Conversely, if every connected component of every open set is open, then for any $x \in M$ and any open set V with $x \in V$, the connected component C of V containing x is a *connected, open* set with $x \in C \subset V$.   □

The following corollary of Proposition 5.7.19 is, of course, nothing but Theorem 4.1.2:

Corollary 5.7.20. *A set $O \subset \mathbb{R}$ is open if and only if it is a countable union of pairwise disjoint open intervals.*

Proof. The sufficiency of the condition is obvious. To prove its necessity, note that the connected components of O are open (since O is *locally connected*) and form a partition of O. Being *connected*, each component is therefore an open *interval*. Finally, since the (countable) set $O \cap \mathbb{Q}$ is dense in O, each component contains a *necessarily different* rational number.   □

Exercise 5.7.21. Let M be a *locally connected, separable* metric space. Show that the set $\{C(x) : x \in M\}$ of connected components of M is *countable*.

We now prove for general metric spaces what was proved for the special metric space $\mathbb{R}$, namely that *continuous* functions map connected sets onto connected sets:

Theorem 5.7.22 (Continuity and Connectedness). *For any continuous map f from a metric space M to a metric space M' and any connected subset $X \subset M$, the image $f(X)$ is a connected subset of M'.*

Proof. If $f(X) = U' \cup V'$, where U' and V' are nonempty open subsets of $f(X)$ with $U' \cap V' = \emptyset$, then $U := X \cap f^{-1}(U')$ and $V := X \cap f^{-1}(V')$ are *nonempty* sets *open* in X such that $X = U \cup V$ and $U \cap V = \emptyset$, contradicting the connectedness of X.   □

Exercise 5.7.23. Show that a metric space M is *connected* if and only if every *continuous* function from M to a *discrete* metric space M' having *at least two* elements is *constant*.

Perhaps the most intuitive notion of *connectedness* is that of *arcwise connectedness,* by which we mean that any pair of points can be *joined* by a continuous *arc.* The precise definitions follow.

Definition 5.7.24 (Arc, Arcwise Connected).

(a) Let M be a metric space and x, $y \in M$. A *continuous* function $\gamma : [a, b] \to M$, where a, $b \in \mathbb{R}$, $a \leq b$, is said to be an *arc joining* x *to* y if $\gamma(a) = x$ and $\gamma(b) = y$. Since $[a, b]$ is *connected* in $\mathbb{R}$, by Theorem 5.7.22, the image $\gamma([a, b])$ is a *connected* subset of M.

(b) A metric space M is said to be *arcwise connected* (or *path connected*) if for any pair of points x, $y \in M$ there exists an arc joining x to y.

Exercise 5.7.25 (Path Components). Let M be a metric space. Given any points x, $y \in M$, let us write $x \sim y$ if and only if there exists an arc joining x to y. Show that $\sim$ is an *equivalence relation* on M. For each $x \in M$, its *equivalence class*, $[x]$, is called the *path component* of x. Deduce that the path components of M form a partition of M into arcwise connected subsets.

Proposition 5.7.26. *An arcwise connected metric space is connected.*

Proof. Suppose M is an *arcwise connected* metric space. Then, for any pair of points x, $y \in M$, there is an arc $\gamma : [a, b] \to M$ with $\gamma(a) = x$ and $\gamma(b) = y$. Since γ is *continuous*, the image $\gamma([a, b])$ is a connected subset of M containing x and y. The proposition now follows from Corollary 5.7.12.   □

Remark 5.7.27 (Topologist's Sine Curve). The converse of Proposition 5.7.26 is *false.* Indeed, consider the set $X := \Gamma \cup Y \subset \mathbb{R}^2$, where $\Gamma := \{(x, \sin(1/x)) : 0 < x \leq 1\}$ is the graph of the function $f(x) = \sin(1/x)$ with domain $(0, 1]$, the so-called Topologist's Sine Curve and $Y := \{0\} \times [-1, 1]$. Since f is continuous, Γ is *connected* and hence so is its closure $\Gamma^- = X$. On the other hand, X is *not* arcwise connected:

Proposition 5.7.28. *Let X be as in the above remark. Then X is connected, but not arcwise connected.*

Proof. By the above remark, we need only show that X is *not* arcwise connected. Suppose there is a (continuous) path $\gamma : [0, 1] \to X$ with $\gamma(0) = (0, 0)$ and $\gamma(1) = (1/\pi, 0)$. We have $\gamma(t) = (x(t), y(t))$, where $x := \pi_1 \circ \gamma$ is *continuous* because both γ and the projection π_1 are. Therefore, the set $T := \{t \in [0, 1] : x(t) = 0\} = x^{-1}(0)$ is *closed.* Also, we have $0 \in T$. Thus $\tau := \sup(T) \in T$ and we have $0 \leq \tau < 1$ because $x(1) = 1/\pi > 0$. Hence $x(t) = 0$ for $t \in [0, \tau]$, while $x(t) > 0$ for $t \in (\tau, 1]$. Let $\varepsilon = 1$ and pick any $\delta > 0$ with $\tau + \delta \leq 1$. Let $N \in \mathbb{N}$ be so large that $x(\tau) = 0 < 1/(2N\pi + \pi/2) < 1/(2N\pi - \pi/2) < x(\tau + \delta)$. By the Intermediate Value Theorem, we can then pick t_1, $t_2 \in (\tau, \tau + \delta]$ with $x(t_1) = 1/(2N\pi + \pi/2)$ and $x(t_2) = 1/(2N\pi - \pi/2)$ and hence

$$y(t_1) = \sin\left(\frac{1}{x(t_1)}\right) = 1, \qquad y(t_2) = \sin\left(\frac{1}{x(t_2)}\right) = -1.$$

Thus, if $y(\tau) \geq 0$, then $|y(\tau) - y(t_2)| \geq 1 = \varepsilon$ and if $y(\tau) \leq 0$, then $|y(\tau) - y(t_1)| \geq 1 = \varepsilon$. We have reached the contradiction that y (and hence γ) is *discontinuous* at τ and the proof is complete.   □

The example in the above remark is a subset of the *product space* $\mathbb{R}^2 := \mathbb{R} \times \mathbb{R}$. The following theorem shows that a product of connected metric spaces is connected.

Theorem 5.7.29. *Let $M_1, \ldots, M_n$ be metric spaces. Then the product $M := M_1 \times \cdots \times M_n$ is connected if and only if each M_k, $1 \leq k \leq n$, is connected.*

Proof. If the product M is connected, then Theorem 5.7.22 and the continuity of the projections $\pi_k : M \to M_k$ imply that $M_k = \pi_k(M)$ is connected for $1 \leq k \leq n$. For the converse, let us first consider the case $n = 2$; i.e., let us show that, if M_1 and M_2 are *connected*, then so is $M := M_1 \times M_2$. Now choose a fixed *base point* $(a, b) \in M_1 \times M_2$. Then the *horizontal fiber* $M_1 \times \{b\}$ is isometric to M_1 and hence is connected. Similarly, for each $x \in M_1$, the *vertical fiber* $\{x\} \times M_2$ is isometric to M_2 and hence also connected. It follows that the set $T_x := (M_1 \times \{b\}) \cup (\{x\} \times M_2)$, which is the union of two connected sets with the point (x, b) in common, is connected for each $x \in M_1$. Finally, note that

$$M_1 \times M_2 = \bigcup_{x \in M_1} T_x$$

is connected, being the union of connected sets with the point (a, b) in common. The general case now follows from Exercise 5.6.36 and induction.   □

5.8   Problems

1. Define $d : \mathbb{R} \times \mathbb{R} \to \mathbb{R}$ by $d(x, y) := |x^2 - y^2|$. Is d a metric on $\mathbb{R}$? Is it a metric on $[0, \infty)$?

2. Let M be a *nonempty* set and suppose that $d : M \times M \to \mathbb{R}$ satisfies the following conditions:

(i) $d(x, y) = 0 \iff x = y$ ($\forall x, y \in M$).
(ii) $d(x, y) \leq d(x, z) + d(y, z)$ ($\forall x, y, z \in M$).

Show that (M, d) is a *metric space*.

3 (The Spaces ℓ^∞, ℓ^1, and ℓ^2).

(a) Let $\ell^\infty(\mathbb{N})$ denote the set of all *bounded* real sequences $x \in \mathbb{R}^\mathbb{N}$. For each x, $y \in \ell^\infty(\mathbb{N})$, define

$$d_\infty(x, y) := \sup\{|x_n - y_n| : n \in \mathbb{N}\}.$$

Show that $(\ell^\infty(\mathbb{N}), d_\infty)$ is a metric space.

(b) Let $\ell^1(\mathbb{N})$ denote the set of all real sequences $x \in \mathbb{R}^\mathbb{N}$ that are *summable* (i.e., $\sum_{n=1}^\infty |x_n| < \infty$). For each x, $y \in \ell^1(\mathbb{N})$, define

$$d_1(x, y) := \sum_{n=1}^\infty |x_n - y_n|.$$

Show that $(\ell^1(\mathbb{N}), d_1)$ is a metric space.

(c) Consider the space $\ell^2(\mathbb{N})$ of all real sequences $x \in \mathbb{R}^\mathbb{N}$ that are *square summable* (i.e., $\sum_{n=1}^\infty x_n^2 < \infty$). For each x, $y \in \ell^2(\mathbb{N})$, define

$$d_2(x, y) := \sqrt{\sum_{n=1}^\infty |x_n - y_n|^2}.$$

Show that $(\ell^2(\mathbb{N}), d_2)$ is a metric space.

4 (Washington D. C. Space). Let $\mathbb{D} := \{z \in \mathbb{C} : |z| \leq 1\} \subset \mathbb{C}$ be the *closed unit disk* and, for any z, $w \in \mathbb{D}$, define

$$d(z, w) := \begin{cases} |z - w| & \text{if } z/|z| = w/|w|, \\ |z| + |w| & \text{otherwise.} \end{cases}$$

Geometrically, if two points z, $w \in \mathbb{D} \setminus \{0\}$ are on a *radius* of the unit circle, then $d(z, w)$ is their Euclidean distance. Otherwise, $d(z, w)$ is the sum of the distances of z and w from the origin. Show that $(\mathbb{D}, d)$ is a metric space.

5 (Pseudometric). Given a set $M \neq \emptyset$, a map $d : M \times M \to \mathbb{R}$ is called a *pseudometric* if it satisfies the conditions (1), (3), and (4) of Definition 5.1.1 and the *weaker* condition (2)′ : $x = y \Rightarrow d(x, y) = 0$. The pair (M, d) is then called a *pseudometric space*. Given such a space, let us write $x \sim y$ if and only if $d(x, y) = 0$. Show that this defines an *equivalence relation* on M. On the set $M^* := \{[x] : x \in M\}$ of all equivalence classes, define $d^*([x], [y]) := d(x, y)$. Show that d^* is *well defined* (i.e., independent of the representatives of the classes) and that (M^*, d^*) is a *metric space*.

6. Let $f \in \mathbb{R}^\mathbb{R}$ be *bounded* and *continuous*. Define $d : \mathbb{R} \times \mathbb{R} \to \mathbb{R}$ by $d(x, y) := \sup\{|f(t - x) - f(t - y)| : t \in \mathbb{R}\}$. Show that d is a *pseudometric* on $\mathbb{R}$ and that it is a *metric* if and only if f is *not* periodic.

7 (Hausdorff Distance). Let $\mathfrak{C}$ denote the set of all (nonempty) *closed* subsets of a metric space (M, d), where d is assumed to be *bounded*. For each A, $B \in \mathfrak{C}$, let $d^*(A, B) := \sup\{d(x, B) : x \in A\}$ and define

$$d_H(A, B) := \max\{d^*(A, B), d^*(B, A)\}.$$

Show that $(\mathfrak{C}, d_H)$ is a metric space. Also, show that

$$d_H(A \cup B, C \cup D) \le \max\{d_H(A, C), d_H(B, D)\} \qquad (\forall A, \ B, \ C, \ D \subset \mathfrak{C}).$$

8. Show that, in any metric space (M, d), we have

$$|d(x, y) - d(x', y')| \le d(x, x') + d(y, y') \quad (\forall x, \ x', \ y, \ y' \in M).$$

9.

(a) Let M be a metric space. Show that

$$A \subset M \ \text{is open} \iff A \cap B^- \subset (A \cap B)^- \quad (\forall B \subset M).$$

(b) Find two *open* sets $A, \ B \subset \mathbb{R}$ such that the sets $A \cap B^-$, $B \cap A^-$, $(A \cap B)^-$, and $A^- \cap B^-$ are all *distinct*.

(c) Find two *intervals* $I, \ J \subset \mathbb{R}$ such that $I \cap J^- \not\subset (I \cap J)^-$.

10 (Derived Set). Let (M, d) be a metric space. For each $A \subset M$, the *derived set* of A, denoted by A', is the set of all *limit points* of A. For any $A, \ B \subset M$, prove the following.

(a) A' is *closed*.
(b) $(A \cup B)' = A' \cup B'$.
(c) $(A \cap B)' \subset A' \cap B'$.
(d) $A' \setminus B' \subset (A \setminus B)'$.
(e) If $A \subset B$ and $B \setminus A$ is *finite*, then $A' = B'$.

11. Let M be a metric space and $A \subset M$. Show that A is *closed* (resp., *open*) if and only if $\text{Bd}(A) \subset A$ (resp., $A \cap \text{Bd}(A) = \emptyset$).

12. Let A and B be subsets of a metric space M. Show that

$$\text{Bd}(A \cup B) \cup \text{Bd}(A \cap B) \cup [\text{Bd}(A) \cap \text{Bd}(B)] = \text{Bd}(A) \cup \text{Bd}(B).$$

13 ($\mathcal{F}_\sigma$ and $\mathcal{G}_\delta$). Let M be a metric space. A set $S \subset M$ is called an $\mathcal{F}_\sigma$ (resp., a $\mathcal{G}_\delta$) if there is a sequence (F_n) (resp., (G_n)) of *closed* (resp., *open*) subsets of M with $S = \bigcup_{n=1}^\infty F_n$ (resp., $S = \bigcap_{n=1}^\infty G_n$). Show that every *closed* set $F \subset M$ is a $\mathcal{G}_\delta$ and that every *open* set $G \subset M$ is an $\mathcal{F}_\sigma$.

14. Let (M, d) be a metric space and for every subset $A \subset M$ define $\alpha(A) := (A^-)^\circ$ and $\beta(A) := (A^\circ)^-$.

(a) Show that if A is *open*, then $A \subset \alpha(A)$ and that if A is *closed*, then $\beta(A) \subset A$.
(b) Using (a), show that we always have $\alpha(\alpha(A)) = \alpha(A)$ and $\beta(\beta(A)) = \beta(A)$.
(c) Give an example $A \subset \mathbb{R}$ such that $A, A^\circ, A^-, \alpha(A), \beta(A), \alpha(A^\circ)$, and $\beta(A^-)$ are all *distinct*.

15. Let M be a metric space and $A, \ B, \ G \subset M$.

(a) Show that $\text{Ext}(A^-) = \text{Ext}(A)$ and $\text{Ext}(A \cup B) = \text{Ext}(A) \cap \text{Ext}(B)$.
(b) Show that, if G is *open*, then $G \cup \text{Ext}(G)$ is dense (in M).

16. Let M be a metric space and $D \subset M$ a *dense* subset. Show that, for any *open* set $G \subset M$, we have $G \subset (D \cap G)^-$.

17. Show that the union of a *finite* number of *nowhere dense* subsets of a metric space is itself nowhere dense.

18.

(a) Show that, for any *closed* or *open* set S in a metric space M, its boundary Bd(S) is *nowhere dense*. Is this still true if S is *neither* closed *nor* open?

(b) Show that $\mathbb{Q}$ is *not* a $\mathcal{G}_\delta$. (See Problem 13).

19. For each $k \in \mathbb{N}$, let $\textbf{\textit{Pol}}_k[0, 1]$ denote the set of all polynomial functions on $[0, 1]$ of degree $\leq k$. Show that, as a subspace of the metric space $(C[0, 1], d_\infty)$, each $\textbf{\textit{Pol}}_k[0, 1]$ is *nowhere dense*. *Hints:* (i) $\textbf{\textit{Pol}}_k[0, 1]$ has *empty interior* since, for any $p \in \textbf{\textit{Pol}}_k[0, 1]$ and any $\varepsilon > 0$, we have $p(x) + (\varepsilon/2)x^{k+1} \in B_\varepsilon(p)$. (ii) $\textbf{\textit{Pol}}_k[0, 1]$ is *closed*. Indeed, let $p_n(x) = \sum_{j=0}^{k} a_{jn}x^j$, where $x^0 := 1$, and $\lim(p_n) = f$. Pick *distinct* points $t_j \in [0, 1]$, $0 \leq j \leq k$, and note that $\lim(p_n(t_j)) = f(t_j)$ for each j. Deduce that (a_{jn}) converges for each j. As we shall see later (cf. Theorem 9.2.14), the *closedness* of the $\textbf{\textit{Pol}}_k[0, 1]$ is actually a consequence of a general fact. Why doesn't $\textbf{\textit{Pol}}[0, 1] = \bigcup_{k=1}^{\infty} \textbf{\textit{Pol}}_k[0, 1]$ contradict the Baire Category Theorem?

20. Show that, in a metric space, any subset of a set of *first category* is itself a set of first category. Also, show that any *countable* union of sets of first category is of first category.

21. Let M be a *second countable* (e.g., *separable*) metric space and let $(U_\alpha)_{\alpha \in A}$ be a family of *nonempty, open* sets in M.

(a) Show that, if $\alpha \neq \beta$ implies $U_\alpha \cap U_\beta = \emptyset$, then the index set A is *countable*.

(b) Show that, if $(U_\alpha)_{\alpha \in A}$ is a *cover* of M, i.e., $M = \bigcup_{\alpha \in A} U_\alpha$, then there is a *countable* set $C \subset A$ such that $M = \bigcup_{\gamma \in C} U_\gamma$.

22 (Condensation Point). Let M be a metric space. A point $x \in M$ is said to be a *condensation point* of a set $A \subset M$ if $U \cap A$ is *uncountable* for each *open* set $U \ni x$. Assuming that M is *second countable*, prove each statement:

(a) If $A \subset M$ has *no* condensation points, then it is *countable*. *Hint:* Pick a *countable base* of open sets, $(U_n)_{n \in \mathbb{N}}$, and look at the sets $A \cap U_n$.

(b) The set C of all condensation points of a set $A \subset M$ is *closed*, every $x \in C$ is a condensation point of C, and $A \cap C^c$ is *countable*. *Hint:* Use part (a).

23.

(a) Show that the spaces (ℓ^1, d_1) and (ℓ^2, d_2) introduced in Problem 3 are *separable*. Show, however, that (ℓ^∞, d_∞) is *not* separable. *Hint:* For ℓ^1 and ℓ^2, look at the sequences $x \in \mathbb{Q}^{\mathbb{N}}$ with $x_n = 0$ except for a *finite* number of n's. For ℓ^∞, let $X := \{x \in \ell^\infty : x_n \in \{0, 1\}\}$ and note that X is *uncountable*. What is $d_\infty(x, y)$ for each $x, y \in X$?

(b) Show that $(C[a, b], d_\infty)$ is *separable*. *Hint:* Using the Weierstrass Approximation Theorem (Corollary 4.7.10), show that polynomials with *rational* coefficients are dense in $C[a, b]$. Show, however, that $(B[a, b], d_\infty)$ is *not* separable.

24. Let (x_n) be a sequence in a metric space (M, d). Show that, if the subsequences (x_{2n-1}), (x_{2n}), and (x_{3n}) are *convergent*, then so is (x_n).

25.

(a) Let (M, d) be an *ultrametric* space (see Exercise 5.1.10). Show that $(x_n) \in M^{\mathbb{N}}$ is a *Cauchy* sequence if and only if $\lim(d(x_n, x_{n+1})) = 0$.

(b) **(Baire Metric)** Let X be a nonempty set and $M := X^{\mathbb{N}}$. Given any sequences $x, y \in M$, let $k(x, y) := \min\{n \in \mathbb{N} : x_n \neq y_n\}$ and define

$$d(x, y) := \begin{cases} 1/k(x, y) & \text{if } x \neq y, \\ 0 & \text{if } x = y. \end{cases}$$

Show that (M, d) is a *complete, ultrametric* space.

26. Show that the metric space $(\mathbb{R}, d)$, where

$$d(x, y) := \left| \frac{x}{1 + |x|} - \frac{y}{1 + |y|} \right| \qquad (\forall x, y \in \mathbb{R}),$$

is *not* complete.

27. Let (M, d) be a metric space and $D \subset M$ a *dense* subset. Show that, if every *Cauchy* sequence in D converges to an element of M, then (M, d) is *complete*.

28. Let (M, d) be a *complete* metric space and let $x \in M^{\mathbb{N}}$. Show that if we have $\sum_{n=1}^{\infty} d(x_n, x_{n+1}) < \infty$, then (x_n) is *convergent*.

29. Let (M, d) be a metric space with the property that given any *closed* sets $A, B \subset M$ with $A \cap B = \emptyset$, we have $d(A, B) > 0$. Show that (M, d) is *complete*. *Hint:* Suppose that $(x_n) \in M^{\mathbb{N}}$ is a Cauchy sequence that is *not* convergent and assume that $x_m \neq x_n$ when $m \neq n$. Now look at the sets $\{x_{2n-1} : n \in \mathbb{N}\}$ and $\{x_{2n} : n \in \mathbb{N}\}$.

30. Let (M, d) be a metric space.

(a) Given a set $A \subset M$, at which points $a \in M$ is χ_A *continuous?*
(b) For which sets $A \subset M$ is χ_A *continuous?*

31. Consider the subspaces $\mathbb{N}$ and $\mathbb{Q}$ of $\mathbb{R}$ with its usual metric. Show that if $f : \mathbb{N} \to \mathbb{Q}$ is any *bijection*, then f is *everywhere continuous* while f^{-1} is *nowhere continuous*.

32. Let M and M' be metric spaces and $f, g \in C(M, M')$. Show that the set $E := \{x \in M : f(x) = g(x)\} \subset M$ is *closed* and so is the set $\{x \in M : f(x) = y\}$ for any *fixed* point $y \in M'$. In particular, if $f \in C(M) := C(M, \mathbb{R})$, then the set $Z(f) := \{z \in M : f(z) = 0\}$ of all *zeros* of f is closed in M.

33. Let M be a *separable* metric space and $f : M \to \mathbb{R}$. For each interval (p, q) with $p, q \in \mathbb{Q}$, let A_{pq} denote the set of all $a \in M$ such that $\lim_{x \to a} f(x)$ exists and $f(a) \leq p < q \leq \lim_{x \to a} f(x)$. Show that A_{pq} is *countable*. Deduce that the set of all points $a \in M$ such that $\lim_{x \to a} f(x)$ exists but does *not* equal $f(a)$ is *countable*.

34. Let M and M' be metric spaces. Show that a function $f : M \to M'$ is *continuous* if and only if the restriction $f|K$ is continuous for each *compact* set $K \subset M$.

35 (Oscillation on a Set, at a Point). Let M and M' be metric spaces, $f : M \to M'$, and $S \subset M$. We define the *oscillation of f on S* to be the nonnegative number $\omega_f(S) := \delta(f(S)) = \sup\{d'(f(s), f(t)) : s, t \in S\}$. Given a point $a \in M$, the *oscillation of f at a* is then the number $\omega_f(a) := \inf\{\omega_f(B_\varepsilon(a)) : \varepsilon > 0\}$. Prove the following:

(a) Show that f is *continuous* at a if and only if $\omega_f(a) = 0$. Deduce that the set of all $a \in M$ at which f is *continuous* is a $\mathcal{G}_\delta$.
(b) For each $c > 0$, the set $\{x \in M : \omega_f(x) < c\}$ is *open*.
(c) There is *no* function $f : \mathbb{R} \to \mathbb{R}$ that is *continuous* on $\mathbb{Q}$ and *discontinuous* on $\mathbb{Q}^c$.
(d) There is a function $f : \mathbb{R} \to \mathbb{R}$ that is *continuous* on $\mathbb{Q}^c$ and *discontinuous* on $\mathbb{Q}$.

36. Let M and M' be metric spaces and $f : M \to M'$. Suppose that $M = \bigcup_{\alpha \in A} F_\alpha$, where each F_α is *closed* and $f|F_\alpha$ is *continuous* for each $\alpha \in A$.

(a) Show that, if A is *finite*, then f is *continuous* on M.
(b) Give an example where A is *countable* and f is *not* continuous on M.
(c) Show that, if the collection $(F_\alpha)_{\alpha \in A}$ is *locally finite* (cf. Definition 5.2.14), then f is continuous on M.

37. Let M and M' be metric spaces and $f : M \to M'$. We say that f is *locally bounded* (resp., *locally open*, *locally closed*), if for each $x \in M$ there is an open $U \subset M$ containing x such that $f|U$ is *bounded* (resp., *open*, *closed*). We say that f is a *local homeomorphism* if, given any $x \in M$, there are open sets $U \subset M$ and $V \subset M'$ such that $x \in U$ and $f|U$ is a homeomorphism *onto* V. Prove the following statements:

(a) $f \in C(M, M') \Rightarrow f$ is *locally bounded*.
(b) f is *locally bounded* $\not\Rightarrow$ f is *bounded*.
(c) f is *locally open* $\Rightarrow$ f is *open*.
(d) f is *locally closed* $\not\Rightarrow$ f is *closed*.
(e) f is a *local homeomorphism* $\not\Rightarrow$ f is a *homeomorphism*. Can you give a condition for such a local homeomorphism to be a homeomorphism?

38. Show that the map $f(x) := x/(1 + x)$ is a *homeomorphism* of $[0, \infty)$ onto $[0, 1)$. Deduce that, given any metric space (M, d), the metric $d' := d/(1 + d)$ is *equivalent* to d, but has the advantage of being *bounded*: $d'(x, y) < 1$ for all $x, y \in M$.

39. In a metric space (M, d), let $A, B \subset M$ be *nonempty* subsets such that $A \cap B^- = B \cap A^- = \emptyset$. Show that there are *open* sets $U, V \subset M$ with $A \subset U$, $B \subset V$, and $U \cap V = \emptyset$. *Hint:* Consider the function $x \mapsto d(x, A) - d(x, B)$.

40. Show that given any metric space (M, d), the distance function $d : M \times M \to \mathbb{R}$ (given by $(x, y) \mapsto d(x, y)$) is *uniformly continuous*. Deduce that, if (x_n), $(y_n) \in M^{\mathbb{N}}$ are *Cauchy* sequences, then $(d(x_n, y_n))$ is a Cauchy sequence in $\mathbb{R}$.

41 (Topological, Metric, & Uniform Properties). Let M and M' be metric spaces. A *property* is said to be *topological* if it is *preserved* under *homeomorphisms* $f : M \to M'$ (i.e., if M and M' are homeomorphic, then M has the property if and only if M' does). It is called a *metric property* if it is preserved under all (bijective) *isometries* $f : M \to M'$. Finally, we call it a *uniform property* if it is preserved under *uniform homeomorphisms*, i.e., bijective maps $f : M \to M'$ such that both f and f^{-1} are *uniformly continuous*.

(a) Let $M = (0, 1] = M'$, $d(x, y) := |x - y|$ and $d'(x, y) := |1/x - 1/y|$. Show that d and d' are *equivalent* so that M and M' are *homeomorphic*. Show, however, that (M, d) is *not* complete while (M', d') *is*. It follows that *completeness* is *not* a topological property. By Proposition 5.5.11, however, it is a *uniform property*.
(b) **(Uniform Equivalence)** Let $M' = M$ and $d' = d/(1+d)$. Show that d and d' are *uniformly equivalent* in the sense that the *identity* map $x \mapsto x$ is a *uniform homeomorphism*. Since d' is *bounded* while d need not be (e.g., look at $(\mathbb{R}, d)$ with d the usual distance), deduce that *boundedness* is *not* a uniform property. However, it *is* obviously a *metric* one.
(c) Let $M = \mathbb{R} = M'$, $d(x, y) := |x - y|$, and $d'(x, y) := |x^3 - y^3|$. Show that d and d' are *equivalent* but *not uniformly* equivalent. Nevertheless, show that M and M' have the same *Cauchy* sequences. Thus completeness may be preserved *without* uniform equivalence.

42. Let M and M' be metric spaces. Show that a map $f : M \to M'$ is *uniformly continuous* if and only if, for any $A, B \subset M$, we have $d'(f(A), f(B)) = 0$ whenever $d(A, B) = 0$.

43. Let M be a metric space, $\emptyset \neq A \subset M$, and $f \in Lip(A, \mathbb{R})$ with Lipschitz constant c. For each $x \in M$ and each $a \in A$, define $f_a(x) := f(a) + cd(x, a)$ and $g(x) := \inf\{f_a(x) : a \in A\}$. Show that $g : M \to \mathbb{R}$, $g \in Lip(M, \mathbb{R})$ with Lipschitz constant c, and $g|A = f$.

44. Let K be a *compact* subset of a metric space (M, d). Show that there are two points $x, y \in K$ such that $\delta(K) = d(x, y)$.

45 (Local Compactness). A metric space (M, d) is called *locally compact* if for each $x \in M$ there is an *open* set $U \ni x$ such that U^- is *compact*.

(a) Show that, if M is locally compact, then every *open* subspace $G \subset M$ and every *closed* subspace $F \subset M$ is locally compact.

(b) In the metric space $\mathbb{R}$ with its usual metric, give an example of two locally compact subspaces $A, B \subset \mathbb{R}$ such that $A \cup B$ is *not* locally compact.

(c) Let M be a locally compact metric space. Show that M is *separable* if and only if $M = \bigcup_{n=1}^{\infty} K_n$, where each K_n is *compact*.

46. Let (M, d) be a metric space. Show that, if every *closed ball* in M is *compact,* then M is *locally compact, complete,* and *separable.*

47.

(a) Show that *total boundedness* is a *uniform property* but *not* a *topological* one. For definitions, cf. Problem 41.

(b) Show that a *totally bounded* metric space need not be *complete.*

48 (Pointwise Finite Cover). An open cover $(U_\alpha)_{\alpha \in A}$ of a metric space K is said to be *pointwise finite* if any $x \in K$ belongs to at most a *finite* number of the U_α. Show that K is *compact* if and only if each pointwise finite open cover of K has a *finite* subcover.

49 (Lebesgue's Covering Property). Let (K, d) be a *compact* metric space. Show that, given any open cover $(U_\alpha)_{\alpha \in A}$ of K, there is an $\varepsilon > 0$ such that for each $x \in K$ there is an $\alpha_x \in A$ with $B_\varepsilon(x) \subset U_{\alpha_x}$.

50. Let M be a metric space, $f : M \to M$, and let $f^{[n]} := f \circ f \circ \cdots \circ f$ (with n copies of f) denote the nth *iterate* of f. Show that, if $f^{[n]}$ has a *unique fixed point* x_0, then $f(x_0) = x_0$.

51. Let K be a *compact* metric space and $f : K \to \mathbb{R}$. Show that f is *continuous* if and only if its graph $\Gamma_f := \{(x, f(x)) : x \in K\}$ is a *compact* subset of $K \times \mathbb{R}$.

52. Let $f \in C(K, K)$, where K is a *compact* metric space. Show that, if f is *nilpotent,* i.e., $f^{[n]} = \mathrm{id}_K$ for some $n \in \mathbb{N}$ (where $f^{[n]}$ is the nth iterate of f), then f is a *homeomorphism.* Show that, if $K = [0, 1]$ (with the usual metric) and $f(0) = 0$, then $f = \mathrm{id}_{[0,1]}$. Give an example with $K = [0, 1]$ to show that, in general, $f \neq \mathrm{id}_K$.

53 (Contractive Map, Edelstein's Theorem). Let (K, d) be a *compact* metric space and let $f : K \to K$ be a *contractive map,* i.e., $d(f(x), f(y)) < d(x, y)$ for all $x, y \in K$. Show that f has a *unique* fixed point. *Hint:* Look at $\inf\{d(x, f(x)) : x \in K\}$.

54. Let M be a *complete* metric space and, for each $n \in \mathbb{N}$, let $f^{[n]}$ be the nth *iterate* of $f :$ $M \to M$. Show that, if $f^{[n]}$ is a *contraction* for some $n \in \mathbb{N}$, then f has a *unique* fixed point.

55. Let (K, d) be a *compact* metric space and let $f : K \to K$ be an *isometry,* i.e., $d(f(x), f(y)) = d(x, y)$ for all $x, y \in K$. Show that f is *onto.* *Hint:* If $M := f(K) \subsetneq K$, pick $x_0 \in K \setminus M$ and let $\delta := d(x_0, M)$. Now define the sequence $(x_n)_{n=0}^{\infty}$ inductively by $x_{n+1} := f(x_n)$ for all $n \geq 0$ and observe that $d(x_m, x_n) \geq \delta$ for all $m < n$. Give an example to show that, if K is *not* compact, then f need *not* be onto.

56. Let (K, d) and (K', d') be two *compact* metric spaces and let $f : K \to K'$ and $g : K' \to K$ be *isometries.* Show that $f(K) = K'$ and $g(K') = K$.

57 (Expansive Map). Let (K, d) be a *compact* metric space and let $f : K \to K$ be an *expansive map,* i.e., $d(f(x), f(y)) \geq d(x, y)$ for all $x, y \in K$. Show that f is an *isometry* of K onto itself. *Hint:* Pick any points $a_0, b_0 \in K$ and define the sequences $(a_n)_{n=0}^{\infty}$ and $(b_n)_{n=0}^{\infty}$ inductively by $a_{n+1} := f(a_n)$ and $b_{n+1} := f(b_n)$, respectively. Now, given any $\varepsilon > 0$, show that there is a sequence $(n_k) \in \mathbb{N}^{\mathbb{N}}$ with $d(a_0, a_{n_k}) < \varepsilon$ and $d(b_0, b_{n_k}) < \varepsilon$ for all $k \in \mathbb{N}$ and deduce that $d(a_1, b_1) = d(a_0, b_0)$.

58. Let (M, d) be a *connected* metric space. Show that, if the distance d is *not* bounded, then every *sphere* in M is *nonempty;* i.e., given any $x \in M$ and any $r > 0$, we have $S_r(x) := \{y \in M : d(x, y) = r\} \neq \emptyset$.

59. Let M be a metric space and $A, B \subset M$ two (nonempty) *connected* sets. Show that, if $A^- \cap B \neq \emptyset$, then $A \cup B$ is *connected*.

60. Let $E := \{(x, y) \in \mathbb{R}^2 : x \in \mathbb{Q}^c \text{ or } y \in \mathbb{Q}^c\}$. Show that E is *connected*.

61 (Convex Set). A subset $K \subset \mathbb{R}^n$ is said to be *convex* if, given any vectors $x, y \in K$ and any $t \in [0, 1]$, we have $tx + (1 - t)y \in K$. Show that a convex subset of $\mathbb{R}^n$ is *connected*.

62. Let A and B be two *closed* subsets of a metric space M such that $A \cup B$ and $A \cap B$ are both *connected*. Show that A and B are connected. Show (by an example in $\mathbb{R}$) that the *closedness* of A and B is necessary.

63 (Chain Connectedness). A metric space (M, d) is called *chain connected* if, given any $a, b \in M$ and any $\varepsilon > 0$, there are $x_0, x_1, \ldots, x_n \in M$ such that $x_0 = a$, $x_n = b$, and $d(x_j, x_{j+1}) < \varepsilon$ for $0 \leq j \leq n - 1$. Show that if M is *compact* and chain connected, then it is *connected*.

Chapter 6
The Derivative

The *derivative* is one of the two fundamental concepts introduced in calculus. The other one is, of course, the (Riemann) *integral*. For a real-valued function of a real variable, the derivative may be interpreted as an extension of the notion of *slope* defined for (nonvertical) straight lines. Recall that a (nonvertical) straight line is the graph of an *affine function* $x \mapsto ax+b$, where a, b are real constants and a is the slope of the line. Now, if $f(x) := ax+b \ \forall x \in \mathbb{R}$, then, for any x, $x_0 \in \mathbb{R}$, $x \neq x_0$, we have

$$\frac{f(x) - f(x_0)}{x - x_0} = \frac{ax + b - (ax_0 + b)}{x - x_0} = a. \qquad (*)$$

In other words, the slope a is the (average) rate of change of the *dependent variable* $y := f(x) = ax + b$ with respect to the *independent variable* x. For a general function $f : I \to \mathbb{R}$, where $I \subset \mathbb{R}$ is an interval, the quotient in $(*)$ is no longer a constant because the graph is a *curve*. Now, using a graphing calculator, which is a quite popular tool these days, if we zoom in repeatedly at a point $(x_0, f(x_0))$ where the graph is *smooth*, we observe that the graph becomes practically a straight line segment. In other words, at least *locally* [i.e., in a small neighborhood of a (smooth) point], the graph is *linear*. Therefore, in that small neighborhood, the graph of f and the *tangent line* to this graph at $(x_0, f(x_0))$ are practically identical. This suggests, once again, the "analytical" approach of *divide and conquer*. Our goal in this chapter will be to carry out this analysis by making the above intuitive approach mathematically rigorous. *Throughout the chapter, I, J will always denote intervals in $\mathbb{R}$ with nonempty interior.*

© Springer Science+Business Media New York 2014
H.H. Sohrab, *Basic Real Analysis*, DOI 10.1007/978-1-4939-1841-6_6

6.1   Differentiability

In this section we define the *derivative* of a real-valued function of a real variable and investigate its basic properties. We begin with the following definition.

Definition 6.1.1 (Differentiable, Derivative, Tangent Line). Let $f : I \to \mathbb{R}$ and let $x_0 \in I$.

1. We say that f is *differentiable at* x_0 if the limit

$$f'(x_0) := \lim_{x \to x_0} \frac{f(x) - f(x_0)}{x - x_0}$$

exists. The number $f'(x_0)$ is then called the *derivative* of f at x_0. This number is also called the *slope of the tangent line to the graph of f at the point* $(x_0, f(x_0))$. The equation of this tangent line is then

$$y - f(x_0) = f'(x_0)(x - x_0).$$

2. If $A \subset I$ and if $f'(x)$ exists for every $x \in A$, then we say that f is *differentiable on* A. We say that f is *differentiable* if it is differentiable on I. If f is differentiable on I, the *function* $x \mapsto f'(x)$ (defined on I) is called the *derivative* of f.

Remark 6.1.2.

1. The *difference quotient* $(f(x) - f(x_0))/(x - x_0)$ is the slope of the line segment joining the points $(x_0, f(x_0))$ and $(x, f(x))$ of the graph of f. It can also be interpreted as the *average rate of change* of $y := f(x)$ with respect to x on the interval with endpoints x_0 and x. The derivative $f'(x_0)$, if it exists, is then the *instantaneous rate of change* of $y = f(x)$ with respect to x at x_0.
2. If we set $h = x - x_0$ in the above definition, we may also write

$$f'(x_0) := \lim_{h \to 0} \frac{f(x_0 + h) - f(x_0)}{h},$$

if the limit exists.
3. It follows from the above definition that differentiability of a function is a *local property*. In other words, if $f : I \to \mathbb{R}$, $x_0 \in I$, and if g is a function such that $f(x) = g(x)$ $\forall x \in (x_0 - \delta, x_0 + \delta) \cap I$ is satisfied for some $\delta > 0$, then f is differentiable at x_0 if and only if g is differentiable at x_0, and $f'(x_0) = g'(x_0)$.

Notation 6.1.3. There are several commonly used forms to denote the derivative of a function. Depending on the situation, one may prefer one form to another. For example, there are cases in which Leibniz's df/dx is more convenient than Newton's (in fact, Lagrange's!) *prime notation* $f'(x)$. There are also situations where Arbogast's *operator* notation $Df(x)$ (or $D_x f(x)$) has definite advantages.

Most of these forms will be used in this book. Thus, if f is differentiable at x_0, we write

$$f'(x_0) = \frac{d}{dx} f(x_0) = \frac{df}{dx}(x_0) = \frac{df}{dx}\Big|_{x=x_0} = Df(x_0) = D_x f(x_0).$$

In fact, we shall even abuse the notation and write $(f(x))'$ instead of $f'(x)$, if this simplifies the exposition. For example, if $f(x) = x^n$, we may write $(x^n)'$ instead of $f'(x)$.

Example 6.1.4. The function $f(x) := x^3 \ \forall x \in \mathbb{R}$ is differentiable on $\mathbb{R}$ with derivative $f'(x) = 3x^2 \ \forall x \in \mathbb{R}$.

To see this, note that for each $x_0 \in \mathbb{R}$ we have

$$f'(x_0) = \lim_{x \to x_0} \frac{x^3 - x_0^3}{x - x_0} = \lim_{x \to x_0} \frac{(x - x_0)(x^2 + xx_0 + x_0^2)}{x - x_0}$$

$$= \lim_{x \to x_0} (x^2 + xx_0 + x_0^2) = 3x_0^2.$$

Definition 6.1.5.

(Left (Right) Derivative, Angular Point).   Let $f : I \to \mathbb{R}$ and let $x_0 \in I$. If x_0 is *not the right endpoint* of I, then we say that f is *right differentiable at* x_0 if the limit

$$f'_+(x_0) := \lim_{x \to x_0+} \frac{f(x) - f(x_0)}{x - x_0}$$

exists. The number $f'_+(x_0)$ is then called the *right derivative* of f at x_0. Similarly, if x_0 is *not the left endpoint* of I, then we say that f is *left differentiable at* x_0 if the limit

$$f'_-(x_0) := \lim_{x \to x_0-} \frac{f(x) - f(x_0)}{x - x_0}$$

exists. The number $f'_-(x_0)$ is then called the *left derivative* of f at x_0. If the left and right derivatives of f are both defined at $x_0 \in I^\circ$ but are *not* equal, then we say that $(x_0, f(x_0))$ is an *angular point* of the graph of f.

(Infinite Derivative, Vertical Tangent).   We say that f has an *infinite derivative* at x_0, and write $f'(x_0) = \pm\infty$, if

$$\lim_{h \to 0} \frac{f(x_0 + h) - f(x_0)}{h} = \pm\infty.$$

If f has an infinite derivative at x_0, we say that the graph of f has a *vertical tangent* at $(x_0, f(x_0))$. The equation of this line is, of course, $x = x_0$.

Remark 6.1.6. If $f : I \to \mathbb{R}$ and $x_0 \in I^\circ$, then it is obvious that f is differentiable at x_0 if and only if it is both right and left differentiable at x_0 and we have $f'_-(x_0) = f'_+(x_0)$. This common value is then $f'(x_0)$. Also, if $f'(x_0)$ exists and x_0 is the *left endpoint* (resp., *right endpoint*) of the interval I, then we automatically have $f'(x_0) = f'_+(x_0)$ (resp., $f'(x_0) = f'_-(x_0)$).

Example 6.1.7.

(a) The function $f(x) := |x| \; \forall x \in \mathbb{R}$ is differentiable on $\mathbb{R} \setminus \{0\}$ with derivative

$$f'(x) = \begin{cases} -1 & \text{if } x < 0, \\ 1 & \text{if } x > 0. \end{cases}$$

Indeed, this is an immediate consequence of the definition:

$$f'(x_0) = \lim_{x \to x_0} \frac{|x| - |x_0|}{x - x_0} = \begin{cases} \lim_{x \to x_0} \dfrac{x - x_0}{x - x_0} = 1 & \text{if } x_0 > 0, \\ \lim_{x \to x_0} \dfrac{-(x - x_0)}{x - x_0} = -1 & \text{if } x_0 < 0. \end{cases}$$

Note also that

$$f'(0) = \lim_{x \to 0} \frac{|x| - |0|}{x - 0} = \lim_{x \to 0} \frac{|x|}{x}$$

does *not* exist. In fact, $f'_-(0) = -1 \neq 1 = f'_+(0)$. The point $(0,0)$ is therefore an angular point of the graph.

(b) The function $f(x) := x^{1/3} \; \forall x \in \mathbb{R}$ is differentiable on $\mathbb{R} \setminus \{0\}$ with derivative

$$f'(x) = \frac{1}{3} x^{-2/3} \quad \forall x \neq 0.$$

Also, f has a *vertical tangent* at $x_0 = 0$. Indeed, for each $x_0 \in \mathbb{R} \setminus \{0\}$,

$$f'(x_0) = \lim_{x \to x_0} \frac{\sqrt[3]{x} - \sqrt[3]{x_0}}{x - x_0} = \lim_{x \to x_0} \frac{\sqrt[3]{x} - \sqrt[3]{x_0}}{(\sqrt[3]{x} - \sqrt[3]{x_0})(\sqrt[3]{x^2} + \sqrt[3]{x x_0} + \sqrt[3]{x_0^2})}$$

$$= \lim_{x \to x_0} \frac{1}{\sqrt[3]{x^2} + \sqrt[3]{x x_0} + \sqrt[3]{x_0^2}} = \frac{1}{3\sqrt[3]{x_0^2}},$$

which also implies that $\lim_{h \to 0} \frac{f(h) - f(0)}{h} = +\infty$. (Why?)

Exercise 6.1.8. Consider the function $f(x) := \sqrt{|x|}, \; \forall x \in \mathbb{R}$.

(a) Using the definition (and considering the cases $x_0 > 0$ and $x_0 < 0$ separately), find $f'(x_0)$ for all $x_0 \neq 0$.

(b) Show that $f'(0)$ does *not* exist (even as an infinite derivative). In fact, show that the left derivative at $x_0 = 0$ is $-\infty$ while the right derivative is $+\infty$. This shows that the graph of f does *not* have a vertical tangent at $(0, 0)$ in the sense of the above definition.

Exercise 6.1.9. Given a finite set $\{a_1, a_2, \ldots, a_n\} \subset \mathbb{R}$, use an appropriate (algebraic) combination of functions of the form $x \mapsto |x - c|$ to construct a *continuous* function $f : \mathbb{R} \to \mathbb{R}$ such that $f'(a_k)$ *does not exist* for $k = 1, 2, \ldots, n$.

Remark 6.1.10. In fact, it is even possible to construct functions that are continuous on $\mathbb{R}$ but are *nowhere differentiable*. We shall construct such a function later, when we study sequences and series of functions.

The following characterization of differentiability will be useful in many proofs. Before stating it, we briefly recall the definitions of *equivalent functions* and of Landau's *little "oh"* (see Sect. 3.5 for details). We say that two functions f and g (defined near a point x_0) are *equivalent* at x_0, and we write $f \sim g$ $(x \to x_0)$, if there is a function u (defined near x_0) such that $f(x) = g(x)u(x)$ and $\lim_{x \to x_0} u(x) = 1$. Also, we say that f is *negligible* compared to g as $x \to x_0$, and write $f = o(g)$ $(x \to x_0)$, if there is a function ζ (defined near x_0) such that $f(x) = g(x)\zeta(x)$ and $\lim_{x \to x_0} \zeta(x) = 0$.

Proposition 6.1.11 (Carathéodory). *Let $f : I \to \mathbb{R}$ and let $x_0 \in I$. Then f is differentiable at x_0 if and only if there exists a function $\phi : I \to \mathbb{R}$ such that ϕ is continuous at x_0 and we have*

$$f(x) = f(x_0) + (x - x_0)\phi(x) \qquad (\forall x \in I).$$

In this case, we have $f'(x_0) = \phi(x_0)$.

Proof. If ϕ exists, then $\phi(x) = (f(x) - f(x_0))/(x - x_0)$, $x \neq x_0$, and, since ϕ is *continuous* at x_0, we have $\lim_{x \to x_0}(f(x) - f(x_0))/(x - x_0) = \phi(x_0)$; i.e., $f'(x_0) = \phi(x_0)$. Conversely, if $f'(x_0)$ exists, then we define

$$\phi(x) := \begin{cases} \dfrac{f(x) - f(x_0)}{x - x_0} & \text{if } x \in I \setminus \{x_0\}, \\ f'(x_0) & \text{if } x = x_0. \end{cases}$$

It is then easily seen (why?) that ϕ satisfies the conditions of the proposition. □

Remark 6.1.12. By Remark 6.1.2(3), for $f'(x_0) = \phi(x_0)$ to exist, the continuous function ϕ in the above proposition need *not* be defined on *all* of I. We only need ϕ to be defined on a (nondegenerate) *subinterval* $J \subset I$ with $x_0 \in J$.

As we saw above, the function $f(x) := |x|$ $\forall x \in \mathbb{R}$ is *not differentiable* at $x_0 = 0$ even though it is obviously *continuous* there. The following corollary shows that differentiability is a *stronger* condition and, in general, *implies* continuity:

Corollary 6.1.13 (Differentiable ⟹ Continuous). *Let $f : I \rightarrow \mathbb{R}$ and let $x_0 \in I$. If f is differentiable at x_0, then it is continuous at x_0. In fact, if f is right (resp., left) differentiable at x_0, then it is right (resp., left) continuous at x_0. In particular, f is continuous at its angular points.*

Proof. Well, let ϕ be as in Proposition 6.1.11. Then

$$f(x) = f(x_0) + (x - x_0)\phi(x) \quad \forall x \in I,$$

so that $\lim_{x \to x_0} f(x) = f(x_0)$ as desired. Alternatively, we have

$$f(x) - f(x_0) = \frac{f(x) - f(x_0)}{x - x_0} \cdot (x - x_0).$$

So letting $x \rightarrow x_0$ or $x \rightarrow x_0+$ or $x \rightarrow x_0-$, we obtain the *continuity* or *right* (resp., *left*) *continuity* of f at x_0. The last statement is then obvious. □

The next consequence is in fact a rewording of Proposition 6.1.11 itself:

Corollary 6.1.14 (Local Linearity). *Let $f : I \rightarrow \mathbb{R}$ and let $x_0 \in I$. Then f is differentiable at x_0 if and only if there exists a number $m \in \mathbb{R}$ such that*

$$f(x) = f(x_0) + m(x - x_0) + (x - x_0)o(1) \qquad (x \to x_0) \qquad (*)$$

and we then have $m = f'(x_0)$. Thus, with the affine function $g(x) := mx + f(x_0) - mx_0$, whose graph is (by definition) the tangent line to the graph of f at $(x_0, f(x_0))$, we have $f(x) - g(x) = (x - x_0)o(1)$, as $x \to x_0$.

Proof. It is obvious (from $(*)$ and the definition of $f'(x_0)$) that, if m exists, then we indeed have $f'(x_0) = m$. Conversely, define $\zeta(x) := \phi(x) - \phi(x_0)$ for $x \in I$, where ϕ is as in Proposition 6.1.11. Then $\lim_{x \to x_0} \zeta(x) = 0$. Thus, $\zeta(x) = o(1)$ $(x \to x_0)$ and $(*)$ follows at once with $m = \phi(x_0) = f'(x_0)$. □

Remark 6.1.15. Note that, with the above notation, not only $f(x) - g(x) \to 0$ as $x \to x_0$, but even $[f(x) - g(x)]/(x - x_0) \to 0$ as $x \to x_0$.

Corollary 6.1.16. *Let $f : I \rightarrow \mathbb{R}$ be differentiable at $x_0 \in I$. If $f'(x_0) \neq 0$, then we have*

$$f(x_0 + h) - f(x_0) \sim hf'(x_0) \qquad (h \to 0).$$

Proof. Since $\lim_{h \to 0}(f'(x_0) + o(1)) = f'(x_0) \neq 0$, we have $f'(x_0) + o(1) \sim f'(x_0)$ $(h \to 0)$. Also, we obviously have $h \sim h$ $(h \to 0)$. Thus, by Corollary 6.1.14 and Theorem 3.5.11, we have

$$f(x_0 + h) - f(x_0) = h[f'(x_0) + o(1)] \sim hf'(x_0) \qquad (h \to 0).$$

□

6.2 Derivatives of Elementary Functions

We are now going to find the derivatives of some of the most commonly used functions. These include the *power functions*, the *trigonometric functions*, and the *exponential function*. As we have mentioned before, the rigorous definitions of trigonometric and exponential functions will be given later. In fact, the definition of the general power function $x \mapsto x^r$, where $x > 0$ and $r \in \mathbb{R}$, also depends on the exponential function. Thus, we are going to assume some of the properties of these functions whose proofs will not be given in this section. Once these properties are assumed, however, the rest of the arguments are quite straightforward.

Beginning with *constant functions*, we have the following trivial result:

Proposition 6.2.1. *If $f(x) := c \ \forall x \in I$ and some constant $c \in \mathbb{R}$, then $f'(x) = 0 \ \forall x \in I$. In other words, the derivative of a constant function (on an interval I) is the (identically) zero function (on I).*

Proof. Indeed, for every $x_0 \in I$, it follows from the definition that

$$\lim_{x \to x_0} \frac{f(x) - f(x_0)}{x - x_0} = \lim_{x \to x_0} \frac{c - c}{x - x_0} = 0.$$

$\square$

Next, we look at *power functions*. Recall that, if $r = m/n \in \mathbb{Q}$, where m, n are relatively prime integers (and, of course, $n \neq 0$), then we have $x^r := \sqrt[n]{x^m}$, where we assume $x \geq 0$ if n is *even* and $x \neq 0$ if $r \leq 0$. Recall also that, for $x \neq 0$, we have $x^0 := 1$ and that $x^{-r} := 1/x^r$, when x^r is well defined.

Proposition 6.2.2 (Power Rule). *Given any rational number $r \in \mathbb{Q}$, the function $x \mapsto x^r$ is differentiable, and we have*

$$(x^r)' = r x^{r-1},$$

for every x for which the two sides are defined. In fact, the rule remains valid for the function $x \mapsto x^r$ where $x > 0$ and $r \in \mathbb{R}$ is arbitrary.

Proof. (For $r \in \mathbb{Q}$) For $r = 0$ (resp., $r = 1$) we have $x^0 = 1 \ \forall x \neq 0$ (resp., $x^r = x \ \forall x \in \mathbb{R}$) and a direct application of the definition implies that $(x^0)' = 0$ (resp., $(x)' = 1$). Assume next that $0 < r = m/n \neq 1$, with relatively prime *positive* integers m and n. Also, assuming x^r and x_0^r are both defined, let $\xi := \sqrt[n]{x}$ and $\xi_0 := \sqrt[n]{x_0}$. Then we have

$$x^r - x_0^r = \xi^m - \xi_0^m = (\xi - \xi_0)(\xi^{m-1} + \xi^{m-2}\xi_0 + \cdots + \xi\xi_0^{m-2} + \xi_0^{m-1}),$$

as can be checked at once by expanding and simplifying the right-hand side. Similarly, we have

$$x - x_0 = \xi^n - \xi_0^n = (\xi - \xi_0)(\xi^{n-1} + \xi^{n-2}\xi_0 + \cdots + \xi\xi_0^{n-2} + \xi_0^{n-1}).$$

Therefore, assuming $x_0 \neq 0$ if $r \in (0, 1)$, we have

$$\lim_{x \to x_0} \frac{x^r - x_0^r}{x - x_0} = \lim_{\xi \to \xi_0} \frac{\xi^{m-1} + \xi^{m-2}\xi_0 + \cdots + \xi\xi_0^{m-2} + \xi_0^{m-1}}{\xi^{n-1} + \xi^{n-2}\xi_0 + \cdots + \xi\xi_0^{n-2} + \xi_0^{n-1}}$$

$$= \frac{m\xi_0^{m-1}}{n\xi_0^{n-1}} = \frac{m}{n}\xi_0^{m-n} = r(\xi_0^n)^{m/n-1}$$

$$= rx_0^{r-1}.$$

Finally, suppose that $r = -m/n$, with m, n as above. Then, using the previous case and assuming that all powers make sense, we have

$$\lim_{x \to x_0} \frac{x^r - x_0^r}{x - x_0} = - \lim_{x \to x_0} (x^r x_0^r) \left(\frac{x^{-r} - x_0^{-r}}{x - x_0} \right) = -x_0^{2r} \left(-rx_0^{-r-1} \right) = rx_0^{r-1},$$

and the proof is complete for the case $r \in \mathbb{Q}$. For $r \in \mathbb{R}$, the proof will be given later when the power function $x \mapsto x^r$ is defined rigorously.   □

Next, we consider the *(natural) exponential function* $\exp(x) = e^x \ \forall x \in \mathbb{R}$. As we pointed out above, this function will be rigorously defined later. One of the consequences of that definition will be the well-known property

$$\exp(x + y) = \exp(x)\exp(y) \qquad (\forall x, \ y \in \mathbb{R}).$$

Another important consequence is the following proposition. The proofs of these facts are postponed until the precise definition is given.

Proposition 6.2.3. *The (natural) exponential function $x \mapsto \exp(x) = e^x$ satisfies*

$$\lim_{h \to 0} \frac{e^h - 1}{h} = 1. \qquad (*)$$

In other words, since $e^0 := 1$, the function $x \mapsto \exp(x)$ is differentiable at $x = 0$ and we have $\exp'(0) = 1$.

An immediate consequence is then the following.

Proposition 6.2.4. *The exponential function $x \mapsto \exp(x)$ is differentiable on $\mathbb{R}$ and we have*

$$\exp'(x) = \exp(x) \qquad (\forall x \in \mathbb{R}).$$

Proof. Using the limit property $(*)$ in Proposition 6.2.3, we have that

$$\lim_{h \to 0} \frac{\exp(x+h) - \exp(x)}{h} = \lim_{h \to 0} \frac{\exp(x)\exp(h) - \exp(x)}{h}$$

$$= \exp(x) \lim_{h \to 0} \frac{e^h - 1}{h} = \exp(x).$$

□

Finally, we look at the derivatives of the trigonometric functions sin and cos. Once again, the rigorous definitions of these functions will be given later when we discuss *power series*. It will follow from those definitions that, for any real numbers x, $h \in \mathbb{R}$, we have

(i) $\sin(x + h) = \sin x \cos h + \cos x \sin h$.

Similarly, for all x, $h \in \mathbb{R}$, we have

(ii) $\cos(x + h) = \cos x \cos h - \sin x \sin h$.

We also have the following limit properties:

Proposition 6.2.5. *The functions $x \mapsto \sin x$ and $x \mapsto \cos x$ are continuous on $\mathbb{R}$ and we have*

$$(a) \qquad \lim_{h \to 0} \frac{\sin h}{h} = 1, \qquad\qquad (b) \qquad \lim_{h \to 0} \frac{\cos h - 1}{h} = 0.$$

In other words, since $\sin 0 = 0$ and $\cos 0 = 1$, the functions sin and cos are both differentiable at $x = 0$ with $(\sin)'(0) = 1$ and $(\cos)'(0) = 0$.

Proof. Postponed!

□

We can now prove that the functions sin and cos are differentiable on $\mathbb{R}$.

Proposition 6.2.6. *The functions $x \mapsto \sin x$ and $x \mapsto \cos x$ are differentiable on $\mathbb{R}$ and we have*

$$(a) \qquad (\sin)'(x) = \cos x, \qquad\qquad (b) \qquad (\cos)'(x) = -\sin x.$$

Proof. For (a), using the identity (i) above and Proposition 6.2.5, we have

$$(\sin)'(x) = \lim_{h \to 0} \frac{\sin(x+h) - \sin x}{h} = \lim_{h \to 0} \frac{\sin x \cos h + \cos x \sin h - \sin x}{h}$$

$$= \sin x \lim_{h \to 0} \frac{\cos h - 1}{h} + \cos x \lim_{h \to 0} \frac{\sin h}{h} = \cos x.$$

For (b), we use the identity (ii) and Proposition 6.2.5 to get

$$(\cos)'(x) = \lim_{h \to 0} \frac{\cos(x+h) - \cos x}{h} = \lim_{h \to 0} \frac{\cos x \cos h - \sin x \sin h - \cos x}{h}$$

$$= \cos x \lim_{h \to 0} \frac{\cos h - 1}{h} - \sin x \lim_{h \to 0} \frac{\sin h}{h} = -\sin x.$$

□

6.3 The Differential Calculus

In this section we shall derive the fundamental rules of differentiation. Some of these rules allow the differentiation of functions constructed from differentiable functions by means of simple algebraic operations. The *Chain Rule*, which is the most important and powerful of these rules, will handle the differentiation of *composite* functions.

Theorem 6.3.1. *Let f and g be real-valued functions defined on an interval I and assume that both functions are differentiable at a point $x_0 \in I$. Then the functions $f \pm g$, fg, cf (where $c \in \mathbb{R}$ is any constant), and f/g are differentiable at x_0 (for f/g we obviously assume $g(x_0) \neq 0$), and we have*

(a) $(f \pm g)'(x_0) = f'(x_0) \pm g'(x_0)$;
(b) $(fg)'(x_0) = f'(x_0)g(x_0) + f(x_0)g'(x_0)$ *(product rule)*;
(c) $(cf)'(x_0) = cf'(x_0)$;
(d) $(f/g)'(x_0) = \dfrac{f'(x_0)g(x_0) - f(x_0)g'(x_0)}{(g(x_0))^2}$ *(quotient rule)*.

Proof. These rules are immediate consequences of the definition of the derivative and the *limit properties* (cf. Theorem 3.3.3). Part (a) follows from the fact that

$$\frac{(f \pm g)(x) - (f \pm g)(x_0)}{x - x_0} = \frac{f(x) - f(x_0)}{x - x_0} \pm \frac{g(x) - g(x_0)}{x - x_0}.$$

Also, (c) follows from (b) and Proposition 6.2.1 or from the obvious observation

$$\frac{(cf)(x) - (cf)(x_0)}{x - x_0} = c\frac{f(x) - f(x_0)}{x - x_0}.$$

To prove (b), note that

$$\frac{(fg)(x) - (fg)(x_0)}{x - x_0} = \left(\frac{f(x) - f(x_0)}{x - x_0}\right) g(x_0) + f(x)\left(\frac{g(x) - g(x_0)}{x - x_0}\right). \quad (*)$$

Now, by Corollary 6.1.13, f is continuous at x_0 and we have $\lim_{x \to x_0} f(x) = f(x_0)$. Therefore, taking limits as $x \to x_0$ in $(*)$, we obtain (b). Finally, to prove (d), we first observe that, if $g(x_0) \neq 0$, then

$$\left(\frac{1}{g}\right)'(x_0) = -\frac{g'(x_0)}{(g(x_0))^2}. \quad (**)$$

Indeed,

$$\frac{\dfrac{1}{g(x)} - \dfrac{1}{g(x_0)}}{x - x_0} = -\frac{\dfrac{g(x) - g(x_0)}{x - x_0}}{g(x)g(x_0)}. \quad (\dagger)$$

But, by Corollary 6.1.13, g is continuous at x_0 and hence $\lim_{x \to x_0} g(x) = g(x_0)$. Taking limits in (†) as $x \to x_0$, we obtain (∗∗) as claimed. The property (d) is now an immediate consequence of (b) and (∗∗). $\qquad\Box$

Corollary 6.3.2. *Let the functions $f_j : I \to \mathbb{R}$ $j = 1, 2, \ldots,$ n be differentiable at $x_0 \in I$ and let $c_1, c_2, \ldots, c_n \in \mathbb{R}$ be arbitrary constants. Then the linear combination $\sum_{j=1}^{n} c_j f_j$ is differentiable at x_0 and we have*

$$(c_1 f_1 + c_2 f_2 + \cdots + c_n f_n)'(x_0) = c_1 f_1'(x_0) + c_2 f_2'(x_0) + \cdots + c_n f_n'(x_0).$$

Also, the product $f_1 f_2 \cdots f_n$ is differentiable at x_0, with derivative

$$(f_1 f_2 \cdots f_n)'(x_0) = f_1'(x_0) f_2(x_0) \cdots f_n(x_0) + f_1(x_0) f_2'(x_0) \cdots f_n(x_0)$$
$$+ \cdots + f_1(x_0) f_2(x_0) \cdots f_n'(x_0).$$

Exercise 6.3.3. (a) Prove the corollary. *Hint:* Use induction on n.
(b) Deduce the following extension of the *Power Rule* for integral exponents: Let $f : I \to \mathbb{R}$ and, for any integer $n \in \mathbb{Z}$, consider the function $g(x) := [f(x)]^n$ $\forall x \in I$ (where, for $n \le 0$, we have $\mathrm{dom}(g) = \{x \in I : f(x) \ne 0\}$). If f is differentiable at $x_0 \in I$, then so is g and we have

$$g'(x_0) = n[f(x_0)]^{n-1} f'(x_0),$$

where the formula is interpreted as $g'(x_0) = f'(x_0)$ if $n = 1$, and we assume $f(x_0) \ne 0$ if $n < 1$.

We are now going to state and prove the *Chain Rule,* which is the most important and powerful rule of differentiation. This rule, combined with the other rules and the well-known derivatives of the elementary functions, allows the differentiation of all functions one encounters in practice.

Theorem 6.3.4 (Chain Rule). *Let $f : I \to \mathbb{R}$, $f(I) \subset J$, and $g : J \to \mathbb{R}$. If f is differentiable at a point $x_0 \in I$ and g is differentiable at the point $y_0 := f(x_0) \in J$, then the composite function $h := g \circ f$ is differentiable at x_0 and we have*

$$(g \circ f)'(x_0) = g'(f(x_0)) f'(x_0).$$

Proof. By Proposition 6.1.11, there is a function $\phi : I \to \mathbb{R}$ such that ϕ is continuous at x_0 and $f(x) - f(x_0) = (x - x_0)\phi(x)$. Similarly, there exists a function $\psi : J \to \mathbb{R}$ such that ψ is continuous at $y_0 := f(x_0)$ and $g(y) - g(y_0) = (y - y_0)\psi(y)$. It follows that

$$h(x) - h(x_0) = g(y) - g(y_0) = (y - y_0)\psi(y) \qquad\qquad (†)$$
$$= (f(x) - f(x_0))\psi(f(x))$$
$$= (x - x_0)\phi(x)\psi(f(x)).$$

Since products and composites of continuous functions are continuous, the function $x \mapsto \phi(x)\psi(f(x))$ is continuous at x_0 and the theorem follows from (†) and Proposition 6.1.11.   □

Remark 6.3.5. Given that differentiability means *local linearity,* the Chain Rule should come as no surprise. Indeed, if $f(x) = ax + b$ and $g(x) = cx + d$ are both *affine* functions, then so is the composite $(g \circ f)(x) = cax + cb + d$, whose *slope* is precisely $ca = g'(f(x))f'(x)$, valid for *all x* in this case.

Exercise 6.3.6 (General Power Rule). Show that, if $f : I \to \mathbb{R}$ is differentiable at $x_0 \in I$, then so is the function $g : x \mapsto [f(x)]^r, r \in \mathbb{R}$, and we have

$$g'(x_0) = (f^r)'(x_0) = r[f(x_0)]^{r-1} f'(x_0).$$

Here, the domain of g depends on the exponent r. Thus, for arbitrary $r \in \mathbb{R}$, we have $\text{dom}(g) = \{x \in I : f(x) > 0\}$.

Example 6.3.7.

(a) The function

$$f(x) := \begin{cases} x \sin(1/x) & \text{if } x \neq 0, \\ 0 & \text{if } x = 0, \end{cases}$$

is continuous on $\mathbb{R}$ and differentiable on $\mathbb{R} \setminus \{0\}$. Indeed, the functions $x \mapsto x$, $x \mapsto \sin x$, and $x \mapsto 1/x$ are all continuous on $\mathbb{R} \setminus \{0\}$ and hence so is f. To prove the continuity at $x = 0$, we note that $|x \sin(1/x)| \leq |x| \ \forall x \neq 0$. Therefore, by the *Squeeze Theorem,* we have $\lim_{x \to 0} f(x) = \lim_{x \to 0} |x| = 0 = f(0)$. Next, for each $x \neq 0$, it follows from the Product Rule, the Quotient Rule, and the Chain Rule, that

$$f'(x) = \sin\left(\frac{1}{x}\right) - \frac{1}{x} \cos\left(\frac{1}{x}\right) \quad \forall x \neq 0.$$

Therefore, f is indeed differentiable on $\mathbb{R} \setminus \{0\}$ as stated and, in fact, f' is *continuous* there. At $x = 0$, we use the definition:

$$f'(0) = \lim_{x \to 0} \frac{f(x) - f(0)}{x - 0} = \lim_{x \to 0} \frac{x \sin(1/x)}{x} = \lim_{x \to 0} \sin(1/x).$$

Since this limit does *not* exist (why?), f is *not* differentiable at $x = 0$.

(b) The function

$$g(x) := \begin{cases} x^2 \sin(1/x) & \text{if } x \neq 0, \\ 0 & \text{if } x = 0, \end{cases}$$

is differentiable on $\mathbb{R}$ and $g'(0) = 0$. Moreover, g' is continuous at every $x \in \mathbb{R}$ except $x = 0$. To see this, note first that, applying the differential calculus, we have

$$g'(x) = 2x \sin(1/x) - \cos(1/x) \quad \forall x \neq 0,$$

so that g' is indeed continuous on $\mathbb{R} \setminus \{0\}$. At $x = 0$, we use the definition and obtain

$$g'(0) = \lim_{x \to 0} \frac{g(x) - g(0)}{x - 0} = \lim_{x \to 0} \frac{x^2 \sin(1/x)}{x} = \lim_{x \to 0} x \sin(1/x) = 0,$$

as was pointed out above. Finally, g' is *not* continuous at 0 because $\lim_{x \to 0} g'(x) = \lim_{x \to 0}(2x \sin(1/x) - \cos(1/x))$ does *not* exist. (Why?)

Our next goal will be to look at the derivative of an *inverse function*. Recall that a function $f : I \to \mathbb{R}$ is *invertible* if and only if it is *injective* (i.e., *one-to-one*). If this is the case, then the inverse function f^{-1} has domain $f(I)$ and is characterized by

$$y = f(x) \quad \Longleftrightarrow \quad x = f^{-1}(y).$$

When we are interested in differentiability, the natural question is whether or not injective, differentiable functions have differentiable inverses. The following theorem addresses this question.

Theorem 6.3.8 (Differentiability of Inverse Functions). *Let $I \neq \emptyset$ be an open interval and let $f : I \to \mathbb{R}$ be an injective, continuous function. If f is differentiable at $x_0 \in I$ and $f'(x_0) \neq 0$, then f^{-1} is differentiable at $y_0 := f(x_0)$, and we have*

$$(f^{-1})'(y_0) = \frac{1}{f'(x_0)} = \frac{1}{f'(f^{-1}(y_0))}.$$

In particular, if f is injective and differentiable on I and $f'(x) \neq 0 \; \forall x \in I$, then f^{-1} is differentiable on $J := f(I)$ and we have

$$(f^{-1})'(y) = \frac{1}{f'(f^{-1}(y))} \quad (\forall y \in J).$$

Proof. By Theorem 4.5.23, $J := f(I)$ is an *interval* and f is a *homeomorphism* of I onto J; in other words, $f^{-1} : J \to I$ is also continuous. In particular, f and f^{-1} are either both strictly increasing or both strictly decreasing. Using the *sequential definition of limit* (Theorem 3.3.1), we must show that, given any sequence (y_n) in $J \setminus \{y_0\}$ with $\lim(y_n) = y_0$, we have

$$\lim_{n \to \infty} \frac{f^{-1}(y_n) - f^{-1}(y_0)}{y_n - y_0} = \frac{1}{f'(x_0)}.$$

But, if $x_n := f^{-1}(y_n)$, the injectivity and continuity of f^{-1} imply that (x_n) is a sequence in $I \setminus \{x_0\}$ with $\lim(x_n) = x_0$. Since f is differentiable at x_0 and $f'(x_0) \neq 0$, we have

$$\lim_{n \to \infty} \frac{f^{-1}(y_n) - f^{-1}(y_0)}{y_n - y_0} = \lim_{n \to \infty} \frac{x_n - x_0}{f(x_n) - f(x_0)} = \frac{1}{f'(x_0)}.$$

The last statement now follows from the fact that, if f is differentiable on I, then (by Corollary 6.1.13) it is *continuous* on I.   □

Corollary 6.3.9 (Derivative of the Natural Logarithm).   *The natural logarithm $x \mapsto \log x$ is differentiable on $(0, \infty)$ and we have*

(i) $(\log x)' = \dfrac{1}{x}$ $(\forall x > 0)$.

In fact, the function $x \mapsto \log|x|$ is differentiable on $\mathbb{R} \setminus \{0\}$ and we have

(ii) $(\log|x|)' = \dfrac{1}{x}$ $(\forall x \neq 0)$.

More generally, if $u : I \to \mathbb{R}$ is differentiable on I and $u(x) \neq 0 \ \forall x \in I$, then the function $x \to \log|u(x)|$ is differentiable on I and we have

(iii) $(\log|u(x)|)' = \dfrac{u'(x)}{u(x)}$ $(\forall x \in I)$.

Proof. To prove (i), note that $x \mapsto \log x \ \forall x > 0$ is the inverse of the natural exponential $x \mapsto \exp(x)$. Since $(e^x)' = e^x \ \forall x \in \mathbb{R}$ and $e^x > 0 \ \forall x \in \mathbb{R}$, Theorem 6.3.8 implies that the inverse function $x \mapsto \log x$ is differentiable on its domain $(0, \infty)$ and we have

$$(\log x)' = \frac{1}{\exp'(\log x)} = \frac{1}{\exp(\log x)} = \frac{1}{x}.$$

Next, we have $\log|x| = \log x \ \forall x > 0$ so that we must only check (ii) for the case $x < 0$. But then, $|x| = -x$ and (i) together with the Chain Rule implies

$$(\log|x|)' = (\log(-x))' = \frac{1}{-x} \cdot (-x)' = \frac{-1}{-x} = \frac{1}{x}.$$

Finally, (iii) is an immediate consequence of (ii) and the Chain Rule.   □

Exercise 6.3.10.

(a) Consider the function $f(x) := x^n \ \forall x \in \mathbb{R}$, where n is an   *odd* integer (cf. Example 4.5.24). Using Theorem 6.3.8, prove the *Power Rule*

$$(x^{1/n})' = \frac{1}{n} x^{1/n - 1} \qquad (\forall x \neq 0).$$

(b) Prove the same rule for *even* integers n, using the function $x \mapsto x^n \ \forall x > 0$.
(c) Combining (a) and (b) (and the Chain Rule), give another proof of the Power Rule for rational exponents: $(x^r)' = rx^{r-1}$, $r \in \mathbb{Q}$.

Exercise 6.3.11. Let I be an open interval. Assume that $u, \ v : I \to \mathbb{R}$ are both differentiable on I and $u(x) > 0 \ \forall x \in I$. Using the Chain Rule, find the derivative of the function

$$u(x)^{v(x)} := e^{v(x)\log u(x)} \quad \forall x \in I.$$

6.4   Mean Value Theorems

Recall that the derivative of a function f at a point x_0 is defined to be the *instantaneous* rate of change of the values $f(x)$ with respect to x, as x approaches x_0. In other words, it is the limit (as $x \to x_0$) of the *average* rate of change $(f(x) - f(x_0))/(x - x_0)$ on the interval with endpoints x_0 and x. The main result of this section will be that, for a function that is continuous on a closed, bounded interval $[a, b]$ and differentiable inside, the average rate of change on the interval is in fact equal to the instantaneous rate of change at an interior point. As we shall see, this result turns out to play a fundamental role in the study of the behavior of real-valued functions of a real variable. We begin with a definition:

Definition 6.4.1 (Local Extrema). Let $f : I \to \mathbb{R}$ and let $x_0 \in I^\circ$. We say that f has a *local maximum* (resp., *local minimum*) at x_0 if there exists $\delta > 0$ such that $f(x) \le f(x_0)$ (resp., $f(x) \ge f(x_0)$) for all $x \in B_\delta(x_0) \cap I := (x_0 - \delta, x_0 + \delta) \cap I$. We say that f has a *local extremum* at x_0 if it has a local maximum or a local minimum at x_0.

Remark 6.4.2.

(a) Recall that $f : I \to \mathbb{R}$ is said to have a *(global* or *absolute) maximum* [resp., *(global* or *absolute) minimum*] at x_0 if $f(x) \le f(x_0)$ (resp., $f(x) \ge f(x_0)$) for all $x \in I$.
(b) The *plurals* for maximum, minimum, and extremum are *maxima, minima,* and *extrema*, respectively.

Proposition 6.4.3 (Fermat's Theorem). *Let $f : I \to \mathbb{R}$ and let $x_0 \in I^\circ$ be an interior point. If f has a local extremum at x_0 and is differentiable at x_0, then $f'(x_0) = 0$. In other words, the tangent line to the graph of f at the point $(x_0, f(x_0))$ is horizontal.*

Proof. Let us assume that f has a local *maximum* at x_0. For the local minimum, the proof is similar or one may use the function $-f$. Pick $\delta > 0$ so small that $B_\delta(x_0) \subset I$ and $f(x) \le f(x_0) \ \forall x \in B_\delta(x_0)$. Then we have

$$\frac{f(x) - f(x_0)}{x - x_0} \le 0 \qquad \forall x \in (x_0, x_0 + \delta). \tag{$*$}$$

Letting $x \to x_0$ in (∗), we get $f'(x_0) \le 0$. Similarly, we have

$$\frac{f(x) - f(x_0)}{x - x_0} \ge 0 \qquad \forall x \in (x_0 - \delta, x_0). \qquad (**)$$

Letting $x \to x_0$ in (∗∗), we get $f'(x_0) \ge 0$. Therefore, we must have $f'(x_0) = 0$ as claimed.   □

The following consequence of the proposition shows that derivatives have one fundamental property in common with continuous functions, namely, the *Intermediate Value Property*:

Theorem 6.4.4 (Darboux's Theorem).   *Let $f : I \to \mathbb{R}$ be differentiable on $I°$ and let $a < b$ in $I°$ be such that $f'(a) < \eta < f'(b)$. Then there exists $\xi \in (a, b)$ such that $f'(\xi) = \eta$. A similar result holds, of course, if $f'(a) > f'(b)$.*

Proof. The function $g(x) := f(x) - \eta x$ on I is differentiable on $I°$ and hence continuous there. In particular, Theorem 4.5.2 implies that g attains its minimum value on $[a, b]$. Now, we have $g'(a) < 0$ and $g'(b) > 0$. It follows from the definition of the derivative that, for $\delta > 0$ small enough, we have $g(x) < g(a) \; \forall x \in (a, a + \delta)$ and $g(x) < g(b) \; \forall x \in (b - \delta, b)$. Therefore, the minimum value of g on $[a, b]$ occurs at some point $\xi \in (a, b)$. (Why?) It now follows from Proposition 6.4.3 that we have $g'(\xi) = 0$.   □

Remark 6.4.5.   Recall that the function

$$g(x) := \begin{cases} x^2 \sin(1/x) & \text{if } x \ne 0, \\ 0 & \text{if } x = 0, \end{cases}$$

defined in Example 6.3.7(b), is differentiable on $\mathbb{R}$ and g' is in fact continuous at all x *except* $x = 0$. Darboux's theorem implies that, despite the discontinuity at $x = 0$, the function g' has the Intermediate Value Property. In particular, the discontinuity at $x = 0$ is *not* of the *first kind* (i.e., jump discontinuity). The reader may refer to Sect. 4.4 for the definitions of various discontinuities. In general, we can make the following statement: *If $f : I \to \mathbb{R}$ is differentiable on I, then all the discontinuities of f' are of the second kind.*

Our next result is a special form of the Mean Value Theorem but, in fact, is strong enough to be equivalent to it.

Theorem 6.4.6 (Rolle's Theorem).   *Let $f : [a, b] \to \mathbb{R}$ be continuous on $[a, b]$, be differentiable on (a, b), and satisfy $f(a) = f(b)$. Then there exists a point $c \in (a, b)$ such that $f'(c) = 0$.*

Proof. By Theorem 4.5.2, the continuous function f attains both its maximum and minimum values on $[a, b]$. If both of them are attained at the endpoints, then f is in fact *constant* (why?) and we have $f'(c) = 0 \ \forall c \in (a,b)$. If not, at least one of the extrema is an *interior* one, i.e., attained at a point $c \in (a,b)$. But then, by Proposition 6.4.3, we have $f'(c) = 0$. $\square$

Remark 6.4.7.

(a) The point $c \in (a,b)$ guaranteed by Rolle's Theorem is *not* necessarily *unique*, as the following example shows: Consider the function $f(x) = 3x^4 - 6x^2 + 1$. Then $f(-2) = f(2) = 25$ and $f'(x) = 12x^3 - 12x$, so that $f'(-1) = f'(0) = f'(1) = 0$.

(b) The assumption $f(a) = f(b)$ implies that the *chord* joining the points $(a, f(a))$ and $(b, f(b))$ on the graph of f is *horizontal*. The theorem then implies that, if this is the case, then the tangent line to the graph of f is horizontal (i.e., parallel to the above chord) at some point $(c, f(c))$ with (a not necessarily unique) $c \in (a,b)$.

If one rotates the graph of the function f in Rolle's Theorem, then the condition $f(a) = f(b)$ will no longer be satisfied and hence the *chord* joining the points $(a, f(a))$ and $(b, f(b))$ will not be horizontal. It is obvious, however, that the new graph will have the property that the tangent line will be parallel to the chord at least once between the endpoints $(a, f(a))$ and $(b, f(b))$. This suggests the following extension of Rolle's Theorem:

Theorem 6.4.8 (Mean Value Theorem). *Let $f : [a,b] \to \mathbb{R}$ be continuous on $[a,b]$ and differentiable on (a,b). Then there exists a point $c \in (a,b)$ such that*

$$f'(c) = \frac{f(b) - f(a)}{b - a}.$$

Proof. Consider the function

$$g(x) := f(x) - f(a) - \frac{f(b) - f(a)}{b - a}(x - a).$$

Note that g is simply the difference between the function f and the affine function

$$x \mapsto \frac{f(b) - f(a)}{b - a}(x - a) + f(a) \qquad \forall x \in [a,b],$$

whose graph is the line segment joining the points $(a, f(a))$ and $(b, f(b))$. The hypotheses of the theorem imply that g is continuous on $[a,b]$, is differentiable on (a,b), and satisfies $g(a) = g(b)$. It then follows from Rolle's Theorem that $g'(c) = 0$ holds for at least one $c \in (a,b)$. But this means precisely that

$$g'(c) = f'(c) - \frac{f(b) - f(a)}{b - a} = 0,$$

and the proof is complete. $\square$

Remark 6.4.9.

(a) As the above proof shows, the Mean Value Theorem (henceforth abbreviated MVT) is a consequence of Rolle's Theorem. Since the converse is obviously satisfied (why?), the two theorems are in fact *equivalent*.

(b) The Mean Value Theorem (MVT) can also be interpreted in terms of *motion* as follows: If $f(t)$ represents a car's position at time t (i.e., its (signed) distance from an initial point), then $f'(t)$ represents the *instantaneous velocity* at that time, and $(f(b) - f(a))/(b - a)$ represents the *average velocity* over the time interval $[a, b]$. Thus the MVT implies that, at some time $c \in (a, b)$, the instantaneous velocity is in fact equal to the average (i.e., mean) velocity.

The Mean Value Theorem is a fundamental tool in the study of the behavior of functions defined and differentiable on intervals. For instance, it is obvious (e.g., geometrically) that the derivative of a constant function on an interval is the identically zero function on that interval (cf. Proposition 6.2.1). That the *converse* is also true is an immediate consequence of the MVT, as we shall see below.

Corollary 6.4.10. *Let $h > 0$ and suppose that $f : [x, x + h] \to \mathbb{R}$ is continuous on $[x, x + h]$ and differentiable on $(x, x + h)$. Then there exists a number $\theta \in (0, 1)$ such that*

$$f(x + h) = f(x) + hf'(x + \theta h).$$

Proof. Simply note that any number in $(x, x + h)$ can be written as $x + \theta h$, for some $\theta \in (0, 1)$. □

Corollary 6.4.11. *Let $f : [a, b] \to \mathbb{R}$ be continuous on $[a, b]$ and differentiable on (a, b). If $f'(x) = 0 \ \forall x \in (a, b)$, then f is constant on $[a, b]$.*

Proof. We must show that, for any $x_1, \ x_2 \in [a, b]$, we have $f(x_1) = f(x_2)$. Assume (without loss of generality) that $x_1 < x_2$. Then, applying the MVT to the function f on the interval $[x_1, x_2]$, we can find a point $x_0 \in (x_1, x_2)$ such that

$$f(x_2) - f(x_1) = (x_2 - x_1) f'(x_0) = 0,$$

and the corollary follows. □

Exercise 6.4.12. Suppose that $f, \ g : \mathbb{R} \to \mathbb{R}$ are both differentiable and satisfy $f' = g$ and $g' = -f$. Show that $f^2 + g^2$ is a *constant function*. Give examples of f and g satisfying the given conditions.

Corollary 6.4.13. *Suppose that f and g are continuous on $[a, b]$, differentiable on (a, b), and $f'(x) = g'(x) \ \forall x \in (a, b)$. Then there exists a constant C such that $f = g + C$.*

Proof. Apply Corollary 6.4.11 to the function $f - g$. □

Corollary 6.4.14. *Let $f : I \rightarrow \mathbb{R}$ be continuous on I and differentiable on its interior I°. Then f is increasing (resp., strictly increasing) on I if $f'(x) \geq 0 \; \forall x \in I^{\circ}$ (resp., $f'(x) > 0 \; \forall x \in I^{\circ}$). Similarly, f is decreasing (resp., strictly decreasing) on I if $f'(x) \leq 0 \; \forall x \in I^{\circ}$ (resp., $f'(x) < 0 \; \forall x \in I^{\circ}$).*

Proof. We simply note that, for any $x_1 < x_2$ in I, we may apply the MVT on $[x_1, x_2]$ to find a point $x_0 \in (x_1, x_2)$ with

$$f(x_2) - f(x_1) = (x_2 - x_1) f'(x_0),$$

from which the corollary follows at once. □

Remark 6.4.15.

(a) Note that, although the converses of the statements in Corollary 6.4.14 are *true* for the *increasing* (resp., *decreasing*) cases (why?), they are *false* for the *strictly increasing* (resp., *strictly decreasing*) cases. The function $f(x) := x^3 \; \forall x \in \mathbb{R}$, e.g., is strictly increasing on $\mathbb{R}$, but $f'(x) = 3x^2$, so that $f'(0) = 0$.
(b) Using the above corollary, we can strengthen the last statement of Theorem 6.3.8 as follows.

Corollary 6.4.16 (Inverse Function Theorem). *If $I \subset \mathbb{R}$ is an open interval and if $f : I \rightarrow \mathbb{R}$ is a differentiable function such that $f'(x) \neq 0$ for all $x \in I$, then f is a homeomorphism onto the interval $J := f(I)$ and its inverse $f^{-1} : J \rightarrow I$ is differentiable at every $y = f(x) \in J$ with derivative*

$$(f^{-1})'(y) = \frac{1}{f'(x)}.$$

Proof. Since f' is *never* zero, Darboux's theorem (Theorem 6.4.4) implies that we either have $f'(x) > 0$ for all $x \in I$ or $f'(x) < 0$ for all $x \in I$. Thus f is either *strictly increasing* or *strictly decreasing* on I. The rest of the proof is identical to that of Theorem 6.3.8. □

Exercise 6.4.17.

(a) Using the MVT and the fact that $(e^x)' = e^x \; \forall x \in \mathbb{R}$, prove the inequality

$$e^x \geq 1 + x \qquad (\forall x \in \mathbb{R}).$$

(b) Using the MVT and the fact that $(\sin x)' = \cos x, (\cos x)' = -\sin x \; \forall x \in \mathbb{R}$, show that both sin and cos are *Lipschitz* functions with *Lipschitz constant* 1 (cf. Sect. 4.6, particularly Example 4.6.12(a)). In other words, show that we have

$$|\sin x - \sin y| \leq |x - y|, \quad |\cos x - \cos y| \leq |x - y| \qquad (\forall x, y \in \mathbb{R}).$$

Deduce, in particular, that $|\sin x| \leq x$ and $|\cos x - 1| \leq x \; \forall x \geq 0$.

(c) **(Bernoulli's Inequality)** Using the MVT and the *Power Rule*, prove the following extension of Bernoulli's inequality (cf. Proposition 2.1.23):

$$(1+x)^r \geq 1 + rx \quad \forall\, x > -1 \quad \text{if } r \leq 0 \text{ or } r \geq 1,$$
$$(1+x)^r \leq 1 + rx \quad \forall\, x > -1 \quad \text{if } 0 \leq r \leq 1.$$

Show that the above inequalities are *strict* if $x \neq 0$ and $r \neq 0, 1$. Also prove the following version:

$$(1-x)^r \geq 1 - rx \quad \forall\, x \in [0,1] \quad \text{if } r \geq 1.$$

Exercise 6.4.18. Let $f : I \to \mathbb{R}$. Recall that f is said to be *Lipschitz of order α* on I, $0 < \alpha \leq 1$, if there is constant $A > 0$ such that

$$|f(x_1) - f(x_2)| \leq A|x_1 - x_2|^\alpha \quad \forall x_1, x_2 \in I.$$

In this case, we write $f \in \mathbf{Lip}^\alpha(I)$. If $\alpha = 1$, then f is said to be *Lipschitz on I* and we write $f \in \mathbf{Lip}(I) = \mathbf{Lip}^1(I)$.

1. Let $a \in I^\circ$ and assume that $f'(a)$ exists. Show that there exists $\delta > 0$ such that $f \in \mathbf{Lip}(B_\delta(a))$. Show, by an example, that the converse is *false*.
2. Show that, if $f \in \mathbf{Lip}^\alpha(I)$ for some $\alpha > 1$, then f is *constant* on I.

Exercise 6.4.19 (A Version of Gronwall's Inequality). Let $f : [0, \infty) \to \mathbb{R}$ be continuous on $[0, \infty)$ and differentiable on $(0, \infty)$. If $f(0) = 0$ and $|f'(x)| \leq |f(x)| \; \forall x \in (0, \infty)$, show that $f(x) = 0 \; \forall x \geq 0$. *Hint:* Differentiate the function $g(x) := [f(x)]^2 e^{-2x}$.

Corollary 6.4.20. *A differentiable function $f : I \to \mathbb{R}$ is Lipschitz on I if and only if f' is bounded on I. In particular, if f' is continuous on I, then f is Lipschitz on every compact subset (e.g., closed, bounded subinterval) of I.*

Proof. Let $A := \sup\{|f'(x)| : x \in I\}$. Then, given any $x < x'$ in I, the MVT implies that $f(x') - f(x) = (x' - x)f'(\xi)$, for some $\xi \in (x, x')$. Therefore,

$$|f(x') - f(x)| = |f'(\xi)||x' - x| \leq A|x' - x|,$$

which shows that f is indeed Lipschitz and proves the last statement as well. (Why?) Conversely, if $|f(x) - f(x')| \leq A|x - x'|$ for all $x, x' \in I$, then for each $x_0 \in I$, $|(f(x) - f(x_0))/(x - x_0)| \leq A$ for all $x \neq x_0$. Since $f'(x_0) := \lim_{x \to x_0} (f(x) - f(x_0))/(x - x_0)$, we have $|f'(x_0)| \leq A$. $\qquad\square$

Our last version of the MVT extends all the previous ones but, as the proof shows, is in fact *equivalent* to them. This version, called *Cauchy's Mean Value Theorem* (henceforth abbreviated Cauchy's MVT), will be used in the proof of *l'Hôpital's Rule*.

Theorem 6.4.21 (Cauchy's Mean Value Theorem). *If two real-valued functions* f *and* g *are both continuous on* $[a, b]$ *and differentiable on* (a, b), *then there exists* $c \in (a, b)$ *such that*

$$[g(b) - g(a)]f'(c) = [f(b) - f(a)]g'(c).$$

Proof. Well, consider the function

$$h(x) := [g(b) - g(a)][f(x) - f(a)] - [f(b) - f(a)][g(x) - g(a)] \quad \forall x \in [a, b].$$

Then h is continuous on $[a, b]$ and differentiable on (a, b), and we have $h(a) = 0 = h(b)$. Therefore, by Rolle's Theorem, there exists $c \in (a, b)$ such that $h'(c) = 0$, and the theorem follows at once. □

Remark 6.4.22.

1. If, in Theorem 6.4.21, we assume that $g'(x) \neq 0 \; \forall x \in (a, b)$, then we must have $g(a) \neq g(b)$ (why?) and the conclusion of the theorem can also be written as

$$\frac{f(b) - f(a)}{g(b) - g(a)} = \frac{f'(c)}{g'(c)}.$$

2. Under the assumptions of Theorem 6.4.21, it follows from Theorem 6.4.8 that, for some $c_1, c_2 \in (a, b)$, we have $f(b) - f(a) = (b - a)f'(c_1)$, and $g(b) - g(a) = (b - a)g'(c_2)$. In particular, if $g'(c_2) \neq 0$, we have

$$\frac{f(b) - f(a)}{g(b) - g(a)} = \frac{f'(c_1)}{g'(c_2)}.$$

Note, however, that $c_1 \neq c_2$, in general. For example, consider the functions $f(x) := x^3 - 8x + 3$ and $g(x) := x^2 - 2x + 2$ on $[0, 4]$. Then, a computation shows that $c_1 = 4/\sqrt{3}$ and $c_2 = 2$. In this case, the number $c \in (0, 4)$ guaranteed by Theorem 6.4.21 is $c = 8/3$. The reader is invited to check these simple facts.

Exercise 6.4.23. Consider the functions $f(x) := x - x^2$ and $g(x) := 2x^3 - 3x^4$ on $[0, 1]$. Show that there is *no* number $c \in (0, 1)$ such that

$$\frac{f(1) - f(0)}{g(1) - g(0)} = \frac{f'(c)}{g'(c)}.$$

Does this contradict Cauchy's MVT?

Finally, as pointed out in Corollary 6.4.20, if a differentiable function $f : I \to \mathbb{R}$ has *bounded derivative*, say $m \leq f'(x) \leq M$ for all $x \in I$, then for any $a < b$ in I, the Mean Value Theorem gives $m(b - a) \leq f(b) - f(a) \leq M(b - a)$. It turns out that this can be obtained with much weaker assumptions on f:

Proposition 6.4.24. *Let D be a countable subset of an interval I and let $f : I \to \mathbb{R}$ be continuous. If f is right differentiable at every $x \in I \setminus D$ and $m \le f'_+(x) \le M$ for all $x \in I \setminus D$, then for any $a < b$ in I we have*

$$m(b - a) \le f(b) - f(a) \le M(b - a),$$

and the inequalities are strict when f is not an affine function on $[a, b]$.

Proof. Let us first show that if $f'_+(x) \ge 0$ for all $x \notin D$, then f is *increasing* on I. Indeed, given any $\varepsilon > 0$ and $x \notin D$, the assumption $f'_+(x) \ge 0$ implies that for every *small enough* number $h > 0$ we must have

$$f(x + h) - f(x) \ge -\varepsilon h.$$

It follows that the function $g(x) := f(x) + \varepsilon x$ satisfies the conditions of Proposition 4.4.13 (why?) and hence is *increasing*. Since $\varepsilon > 0$ was arbitrary, the function f itself is also increasing. Now suppose that $m \le f'_+(x) \le M$ for all $x \in I \setminus D$. Then the functions $h(x) := Mx - f(x)$ and $k(x) := f(x) - mx$ satisfy $h'_+(x) \ge 0$ and $k'_+(x) \ge 0$ for all $x \notin D$ and hence are increasing and the desired inequalities follow. Finally, if f is *not* an affine function with $f' = M$, then the function $h(x) = Mx - f(x)$ is *not* constant on $[a, b]$ and hence

$$Ma - f(a) < Mb - f(b).$$

A similar argument is used for $k(x) = f(x) - mx$. $\square$

6.5   L'Hôpital's Rule

Indeterminate forms were discussed in Sect. 3.5. Of particular importance were limits having the indeterminate forms $0/0$ and ∞/∞. In this section, we shall see how derivatives can be used to compute some such limits. The basic tool will be *Cauchy's MVT*.

Theorem 6.5.1 (L'Hôpital's Rule). *Let $-\infty \le a < b \le +\infty$, and let $f, g : (a, b) \to \mathbb{R}$ be differentiable functions on (a, b), with $g'(x) \ne 0 \; \forall x \in (a, b)$. Suppose that either*

(i) $\lim_{x \to a} f(x) = 0 = \lim_{x \to a} g(x)$

or

(ii) $\lim_{x \to a} g(x) = \pm \infty.$

If, for some $L \in [-\infty, +\infty]$, we have

(iii) $\lim_{x \to a} \dfrac{f'(x)}{g'(x)} = L,$

then we also have

(iv) $\displaystyle\lim_{x \to a} \frac{f(x)}{g(x)} = L.$

The same conclusion holds if $\lim_{x \to a}$ *is replaced by* $\lim_{x \to b}$ *throughout. Note that, for finite a, we obviously have* $\lim_{x \to a} = \lim_{x \to a+}$.

Proof. **(Case 1:** $a > -\infty$**).** If (i) holds and if we *define* $f(a) = g(a) := 0$, then both f and g become continuous on $[a, b]$. Applying Cauchy's MVT on $[a, x]$ where $x \in (a, b)$, we have

$$\frac{f(x)}{g(x)} = \frac{f(x) - f(a)}{g(x) - g(a)} = \frac{f'(\xi)}{g'(\xi)}, \tag{$*$}$$

for some $\xi \in (a, x)$. Since $x \to a$ implies $\xi \to a$, (iv) follows at once from (iii) and ($*$). Assume next that (ii) holds with $+\infty$ (for the case $\lim_{x \to a} g(x) = -\infty$, replace g by $-g$) and that L is *finite*. In view of (iii), for each $\varepsilon > 0$ we can find $t > a$ such that $g(u) > 0$ and $|f'(u)/g'(u) - L| < \varepsilon$ for all $u \in (a, t]$. Applying Cauchy's MVT on $[x, t] \subset (a, t]$, we can find $\eta \in (x, t)$ with

$$[f(t) - f(x)]g'(\eta) = [g(t) - g(x)]f'(\eta),$$

which can also be written as

$$\frac{f(x)}{g(x)} = \frac{f'(\eta)}{g'(\eta)} - \frac{g(t)}{g(x)} \cdot \frac{f'(\eta)}{g'(\eta)} + \frac{f(t)}{g(x)}. \tag{$**$}$$

Let $M = \sup\{|f'(u)/g'(u)| : u \in (a, t]\}$. (Why is M finite?) Then ($**$) implies that

$$\left| \frac{f(x)}{g(x)} - L \right| < \varepsilon + \left| \frac{g(t)}{g(x)} \right| M + \left| \frac{f(t)}{g(x)} \right|.$$

Letting $x \to a$, we obtain

$$\left| \frac{f(x)}{g(x)} - L \right| \le \varepsilon,$$

which implies (iv). If $L = +\infty$ and if $B > 0$ is arbitrary, we can pick $t > a$ such that $g(t) > 0$ and $f'(u)/g'(u) > B$ for all $u \in (a, t)$. Keeping t fixed, we can pick $t' \in (a, t)$ such that $0 < g(t) < g(x)$ for all $x \in (a, t')$. (Why?) It then follows from ($**$) that

$$\frac{f(x)}{g(x)} > B\left(1 - \frac{g(t)}{g(x)}\right) + \frac{f(t)}{g(x)} \qquad \forall x \in (a, t').$$

Since the right side converges to B as $x \to a$, (iv) follows. The case $L = -\infty$ is treated similarly. The proof of case (1) is now complete.

(Case 2: $a = -\infty$). Here we may assume that $b < 0$. Now observe that $x \mapsto -1/x$ is a homeomorphism of $(-\infty, b)$ onto $(0, -1/b)$ and that $x \to -\infty$ if and only if $-1/x \to 0+$. Therefore,

$$\lim_{x \to -\infty} \frac{f(x)}{g(x)} = \lim_{x \to 0+} \frac{f(-1/x)}{g(-1/x)},$$

and it suffices to show that the right side converges to L. This, however, follows from *Case 1*, because

$$\lim_{x \to 0+} \frac{[f(-1/x)]'}{[g(-1/x)]'} = \lim_{x \to 0+} \frac{[f'(-1/x)]/x^2}{[g'(-1/x)]/x^2} = \lim_{x \to -\infty} \frac{f'(x)}{g'(x)} = L.$$

$\square$

The above rule handles the cases in which $x \to a$ or $x \to b$, where a and b are the left and right endpoints of an interval, respectively. To have a rule which can also be applied to the cases $x \to c$, where c is an *interior* point, we have the following

Corollary 6.5.2. *Let f and g be differentiable on (a, c) and (c, b), with $g'(x) \neq 0$ $\forall x \in (a, c) \cup (c, b)$. Suppose that either*

(i) $\lim_{x \to c} f(x) = 0 = \lim_{x \to c} g(x)$

or

(ii) $\lim_{x \to c} g(x) = \pm\infty$.

If, for some $L \in [-\infty, +\infty]$, we have

(iii) $\lim_{x \to c} \dfrac{f'(x)}{g'(x)} = L,$

then we also have

(iv) $\lim_{x \to c} \dfrac{f(x)}{g(x)} = L.$

Proof. This follows at once from Theorem 6.5.1 by applying it to f and g on the intervals (a, c) and (c, b) separately and using the fact that $\lim_{x \to c} \frac{f'(x)}{g'(x)} = L$ if and only if $\lim_{x \to c-} \frac{f'(x)}{g'(x)} = L = \lim_{x \to c+} \frac{f'(x)}{g'(x)}$. $\square$

Remark 6.5.3.

1. As we saw in Sect. 3.5, one can always change an indeterminate form ∞/∞ to an indeterminate form $0/0$, by observing that $f(x)/g(x) = [1/g(x)]/[1/f(x)]$. If this is done, however, l'Hôpital's Rule becomes

$$\lim_{x \to a} f(x)/g(x) = \lim_{x \to a} [1/g(x)]'/[1/f(x)]',$$

which is not the same as the rule in Theorem 6.5.1. The case (ii) in l'Hôpital's Rule is therefore important in general. In fact, changing ∞/∞ to $0/0$ may actually complicate matters, as the following simple example shows. Let $f(x) := x$ and $g(x) := e^x$ on $(-\infty, \infty)$. Then, $\lim_{x \to +\infty} x = +\infty = \lim_{x \to +\infty} e^x$ and l'Hôpital's Rule implies that

$$\lim_{x \to +\infty} \frac{x}{e^x} = \lim_{x \to +\infty} \frac{(x)'}{(e^x)'} = \lim_{x \to +\infty} \frac{1}{e^x} = 0.$$

On the other hand, if we write $x/e^x = e^{-x}/(1/x)$, which has the indeterminate form $0/0$ as $x \to +\infty$, then the rule implies

$$\lim_{x \to +\infty} \frac{e^{-x}}{1/x} = \lim_{x \to +\infty} \frac{(e^{-x})'}{(1/x)'} = \lim_{x \to +\infty} \frac{e^{-x}}{1/x^2},$$

which is more complicated than $\lim_{x \to +\infty} e^{-x}/(1/x)$.

2. Although a powerful tool for computing limits of indeterminate forms, l'Hôpital's Rule is not necessarily the right one in all cases. The following simple example illustrates this point. Recall that $x \mapsto x/\sqrt{1+x^2}$ is a homeomorphism of $\mathbb{R}$ onto the open unit interval $(-1, 1)$. The inverse homeomorphism is $x \mapsto x/\sqrt{1-x^2}$. Now, $\lim_{x \to \infty} x = \lim_{x \to \infty} \sqrt{1+x^2} = +\infty$, so that $\lim_{x \to \infty} x/\sqrt{1+x^2}$ has the indeterminate form ∞/∞. If we use l'Hôpital's Rule, we get

$$\lim_{x \to \infty} \frac{x}{\sqrt{1+x^2}} = \lim_{x \to \infty} \frac{1}{x/\sqrt{1+x^2}} = \lim_{x \to \infty} \frac{\sqrt{1+x^2}}{x}.$$

Therefore, the rule does not help at all. In fact, a second application of it will send us back to the original limit. On the other hand, we can find the limit easily as follows:

$$\lim_{x \to \infty} \frac{x}{\sqrt{1+x^2}} = \lim_{x \to \infty} \frac{x}{|x|\sqrt{1+1/x^2}} = \lim_{x \to \infty} \frac{1}{\sqrt{1+1/x^2}} = 1.$$

3. L'Hôpital's Rule can be applied repeatedly *as long as the required conditions are all satisfied*. For example, we have

$$\lim_{x \to 0} \frac{1 - \cos x}{x^2} = \lim_{x \to 0} \frac{\sin x}{2x} = \lim_{x \to 0} \frac{\cos x}{2} = \frac{\cos(0)}{2} = \frac{1}{2}.$$

4. It should be noted that the converse of Theorem 6.5.1 (or its corollary) is *not* true. In other words, $\lim_{x \to a} \frac{f(x)}{g(x)}$ may very well exist even though $\lim_{x \to a} \frac{f'(x)}{g'(x)}$ does not. A simple example is the following. Consider the functions $f(x) := x - \sin x$ and $g(x) := x$ on $\mathbb{R}$. Then

$$\lim_{x \to +\infty} \frac{f(x)}{g(x)} = \lim_{x \to +\infty} \frac{x - \sin x}{x} = \lim_{x \to +\infty} \left(1 - \frac{\sin x}{x}\right) = 1,$$

because $|\sin x/x| \le 1/|x| \to 0$ as $x \to +\infty$. On the other hand,

$$\lim_{x \to +\infty} \frac{f'(x)}{g'(x)} = \lim_{x \to +\infty} (1 - \cos x)$$

does not exist.

Example 6.5.4.

1. Let α and β be arbitrary *positive* numbers. Then we have $x^\alpha = o(e^{\beta x})$ $(x \to \infty)$. We must show that $\lim_{x \to \infty} x^\alpha/e^{\beta x} = 0$. Now $x^\alpha/e^{\beta x} = (x/e^{\beta x/\alpha})^\alpha$, and $\lim_{t \to 0+} t^\alpha = 0$ for all $\alpha > 0$. The claim is therefore a consequence of l'Hôpital's Rule:

$$\lim_{x \to \infty} \frac{x}{e^{\beta x/\alpha}} = \lim_{x \to \infty} \frac{1}{\frac{\beta}{\alpha} e^{\beta x/\alpha}} = 0.$$

2. Given any $\alpha > 0$, we have $\lim_{x \to 0+} x^\alpha \log x = 0$. To see this, note that $\lim_{x \to 0+} x^\alpha = 0$ and $\lim_{x \to 0+} \log x = -\infty$. These facts will be proved later, when we define the logarithms and (general) power functions rigorously. Therefore, we are dealing with an indeterminate form $0 \cdot \infty$. To compute it, we use l'Hôpital's Rule as follows:

$$\lim_{x \to 0+} x^\alpha \log x = \lim_{x \to 0+} \frac{\log x}{x^{-\alpha}} = \lim_{x \to 0+} \frac{x^{-1}}{-\alpha x^{-\alpha-1}} = -\frac{1}{\alpha} \lim_{x \to 0+} x^\alpha = 0.$$

3. Let us show that $\lim_{x \to 0+} x^x = 1$. Note first that this limit has the indeterminate form 0^0. Now, by definition, $x^x := \exp(x \log x) \; \forall x > 0$. Also, by Example (2) above, $\lim_{x \to 0+} x \log x = 0$. Since $\exp$ is *continuous*, we obtain

$$\lim_{x \to 0+} x^x = \exp(\lim_{x \to 0+} x \log x) = \exp(0) = 1.$$

4. Show that $\lim_{x \to \infty} (1 + \alpha/x)^x = e^\alpha$. Here the limit has the indeterminate form 1^∞. By definition, we have $(1 + \alpha/x)^x := \exp[x \log(1 + \alpha/x)]$, so we must find $\lim_{x \to \infty} x \log(1 + \alpha/x)$, which has the indeterminate form $\infty \cdot 0$ (or $0 \cdot \infty$). Using l'Hôpital's Rule, we have

$$\lim_{x \to \infty} x \log(1 + \alpha/x) = \lim_{x \to \infty} \frac{\log(1 + \alpha/x)}{1/x} = \lim_{x \to \infty} \frac{\frac{-\alpha/x^2}{1 + \alpha/x}}{-1/x^2} = \lim_{x \to \infty} \frac{\alpha}{1 + \alpha/x} = \alpha,$$

and the claim follows from the continuity of $\exp$.

5. Let us show that $\log(1 + x) \sim x \; (x \to 0)$. Recall (cf. Sect. 3.5) that this is equivalent to $\lim_{x \to 0} \log(1 + x)/x = 1$, which follows at once from l'Hôpital's Rule:

$$\lim_{x \to 0} \frac{\log(1 + x)}{x} = \lim_{x \to 0} \frac{1/(1 + x)}{1} = 1.$$

Exercise 6.5.5. Find the following limits, where a, b, α, and β are arbitrary *positive* constants.

$$(1) \quad \lim_{x \to 0} \frac{a^x - 1}{b^x - 1}, \qquad\qquad (2) \quad \lim_{x \to \infty} \frac{(\log x)^\alpha}{x^\beta}.$$

Exercise 6.5.6. Find each limit.

$$(1) \quad \lim_{x \to 0+} \frac{e^{-1/x}}{x}, \qquad\qquad (2) \quad \lim_{x \to 0+} (\sin x)^x.$$

6.6 Higher Derivatives and Taylor's Formula

If $f : I \to \mathbb{R}$ is differentiable on I, then its derivative defines a new function f' : $I \to \mathbb{R}$ and it is quite legitimate to ask whether this new function f' is differentiable at a point $x_0 \in I$. Hence the following definition:

Definition 6.6.1 (Higher-Order Derivatives). Let $f : I \to \mathbb{R}$ and suppose that f is differentiable *near* $x_0 \in I$; i.e., that $f'(x)$ exists for all $x \in B_\delta(x_0) \cap I$ and some $\delta > 0$. If the derivative of f' exists at x_0, then we say that f is *twice differentiable* at x_0 and we write $f''(x_0) := (f')'(x_0)$. The number $f''(x_0)$ is called the *second derivative* of f at x_0. Inductively, we define $f^{(0)}(x_0) := f(x_0)$, $f^{(1)}(x_0) := f'(x_0)$, and, for each positive integer $n \in \mathbb{N}$, we define $f^{(n)}(x_0) := (f^{(n-1)})'(x_0)$. If $f^{(n)}(x_0)$ exists, we call it the *nth derivative* (or *nth-order derivative*) of f at x_0 and say that f is *n-times differentiable* at x_0. If $f^{(n)}(x)$ exists for all $x \in I$, we say that f is *n-times differentiable on I*.

Remark 6.6.2.

1. If $f : I \to \mathbb{R}$ and $f''(x_0)$ exists for some $x_0 \in I$, then f' must be defined *near* x_0. In other words, we can find $\delta > 0$ such that f is *differentiable on $B_\delta(x_0) \cap I$*. More generally, if $f^{(n)}(x_0)$ exists, then f is $(n-1)$-times differentiable on $B_\delta(x_0) \cap I$ for some $\delta > 0$.
2. If $f : I \to \mathbb{R}$ is n-times differentiable on I for some $n \in \mathbb{N}$, then the derivatives $f^{(k)}$, $0 \le k \le n - 1$ are all defined and *continuous* (why?) on I.

Notation 6.6.3. As in the case of the (first) derivative, there are several ways to denote higher derivatives of a function, each having its own merits. We shall use all these forms in this text. If $f : I \to \mathbb{R}$ and if $f^{(n)}(x_0)$ exists for some $x_0 \in I$ and $n \in \mathbb{N}$, then we write

$$f^{(n)}(x_0) = \frac{d^n f}{dx^n}(x_0) = \frac{d^n}{dx^n} f(x_0) = \left. \frac{d^n f}{dx^n} \right|_{x=x_0} = D^n f(x_0) = D_x^n f(x_0).$$

We even abuse the notation, occasionally, and write $(f(x))^{(n)}$ instead of $f^{(n)}(x)$.

Exercise 6.6.4. Let $f : I \to \mathbb{R}$ and assume that $f''(x_0)$ exists for some $x_0 \in I^\circ$. Show that

$$f''(x_0) = \lim_{h \to 0} \frac{f(x_0 + h) - 2f(x_0) + f(x_0 - h)}{h^2},$$

and give an example where this limit exists even though $f''(x_0)$ does not. *Hint:* Use l'Hôpital's Rule and, for example, consider an *odd* function.

Definition 6.6.5 (The Classes C^n). Let $f : I \to \mathbb{R}$ and $n \in \mathbb{N}$. We say that f is *of class C^n on I*, and write $f \in C^n(I)$, if $f^{(n)}$ is defined and *continuous* on all of I. We say that f is *of class C^∞ on I*, and write $f \in C^\infty(I)$, if $f \in C^n(I)$ $\forall n \in \mathbb{N}$. The class of *continuous* functions on I will be denoted by $C(I)$ instead of $C^0(I)$. We call $C^n(I)$ the class of *n-times continuously differentiable functions on I*. For $n = 1$, it is called the class of *continuously differentiable functions on I*. Finally, $C^\infty(I)$ is called the class of *infinitely differentiable functions on I*.

Remark 6.6.6. Note that, as was pointed out above, the existence of $f^{(n)}$ on I automatically guarantees the existence *and continuity* of f, f', ..., $f^{(n-1)}$ on I. Also, it is obvious that we have the inclusions

$$C^\infty(I) \subset \cdots \subset C^{n+1}(I) \subset C^n(I) \cdots \subset C^2(I) \subset C^1(I) \subset C(I) := C^0(I).$$

We should keep in mind that all the above inclusions are *proper*, as the following exercise demonstrates.

Exercise 6.6.7. For $n = 0, 1, 2, \ldots$, consider the functions $f_n : \mathbb{R} \to \mathbb{R}$ defined by $f_0(x) := |x|$, and $f_n(x) := x^n|x|$ $\forall n \geq 1$. Show that, for all $n \geq 0$, we have $f_n \in C^n(\mathbb{R})$ but $f_n \notin C^{n+1}(\mathbb{R})$. What are the successive derivatives of f_n? *Hint:* Note that $f_n = x f_{n-1}$ $\forall n \geq 1$, and use induction.

The following proposition is an extension of the *Product Rule* to higher-order derivatives.

Proposition 6.6.8 (Leibniz Rule). *Let f, $g : I \to \mathbb{R}$ be n-times differentiable functions on I for some $n \in \mathbb{N}$. Then the product fg is also n-times differentiable on I and we have*

$$D^n(fg) = \sum_{k=0}^{n} \binom{n}{k} D^{n-k}f \cdot D^k g, \qquad \text{(Leibniz Rule)} \qquad (\dagger)$$

where, for any k-times differentiable function h, $D^k h := h^{(k)}$ and $D^0 h = h^{(0)} := h$.

Proof. We use induction on n. For $n = 1$, the rule is reduced to the Product Rule: $D(fg) = Df \cdot g + f \cdot Dg$. Thus, we must only show that if ($\dagger$) is satisfied for any n and if f and g are $(n + 1)$-times differentiable, then so is fg and the rule holds for $n + 1$; i.e., we have

$$D^{n+1}(fg) = \sum_{j=0}^{n+1} \binom{n+1}{j} D^{n+1-j} f \cdot D^j g. \tag{‡}$$

Now, differentiating both sides of (†), we obtain

(i) $\quad D^{n+1}(fg) = \sum_{k=0}^{n} \binom{n}{k} D^{n+1-k} f \cdot D^k g + \sum_{k=0}^{n} \binom{n}{k} D^{n-k} f \cdot D^{k+1} g.$

If we set $k = j$ in the first sum on the right side of (i) and $k = j - 1$ in the second sum, we get

(ii) $\quad D^{n+1}(fg) = \sum_{j=0}^{n} \binom{n}{j} D^{n+1-j} f \cdot D^j g + \sum_{j=1}^{n+1} \binom{n}{j-1} D^{n+1-j} f \cdot D^j g.$

If we isolate the first term of the first sum and the last term of the second sum and combine the remaining sums, then the right side of (ii) is

$$= D^{n+1} f \cdot D^0 g + \sum_{j=1}^{n} \left[\binom{n}{j} + \binom{n}{j-1} \right] D^{n+1-j} f \cdot D^j g + D^0 f \cdot D^{n+1} g$$

$$= D^{n+1} f \cdot D^0 g + \sum_{j=1}^{n} \binom{n+1}{j} D^{n+1-j} f \cdot D^j g + D^0 f \cdot D^{n+1} g$$

$$= \sum_{j=0}^{n+1} \binom{n+1}{j} D^{n+1-j} f \cdot D^j g,$$

where we have used the identity $\binom{n}{j} + \binom{n}{j-1} = \binom{n+1}{j}$ (cf. Exercise 1.3.29). This establishes (‡) and completes the proof.   □

Corollary 6.6.9. *Suppose that $f : I \to \mathbb{R}$, $g : J \to \mathbb{R}$ and that $f(I) \subset J$. If f is n-times differentiable on I and g is n-times differentiable on J, then the composite function $g \circ f$ is n-times differentiable on I.*

Proof. We use induction on n, the case $n = 1$ being obviously true. (Why?) Now, it follows from the *Chain Rule* that $D(g \circ f) = (g' \circ f) \cdot f'$. Since f' is $(n-1)$-times differentiable on I, the corollary follows from the *Leibniz Rule* if we can show that $g' \circ f$ is also $(n - 1)$-times differentiable on I. This, however, follows from our inductive hypothesis.   □

Definition 6.6.10 (C^n-Diffeomorphism). Let I and J be open intervals. A function $f : I \to J$ is said to be a C^n-*diffeomorphism* if it is a bijection such that $f \in C^n(I)$ and $f^{-1} \in C^n(J)$.

The following extension of Corollary 6.4.16 is remarkable.

Corollary 6.6.11 (Smoothness of the Inverse Function). *Let $I \neq \emptyset$ be an open interval. If $f \in C^n(I)$ satisfies $f'(x) \neq 0 \; \forall x \in I$, then it is a C^n-diffeomorphism onto $f(I)$; i.e., the inverse function f^{-1} is n-times continuously differentiable on the interval $f(I)$.*

Proof. We proceed by induction again, the case $n = 1$ being Corollary 6.4.16 which also provides the formula $(f^{-1})' = 1/f' \circ f^{-1}$. Next, since f' is never zero on I, the same holds for $f' \circ f^{-1}$ on $f(I)$. In view of the *Quotient Rule*, it is therefore sufficient to show that $f' \circ f^{-1}$ is $(n-1)$-times differentiable on the interval $f(I)$. (Why?) By Corollary 6.6.9, this will follow if f^{-1} is $(n-1)$-times differentiable on the interval $f(I)$. But this is precisely the inductive step and the proof is complete.   □

Our last corollary will be an extension of the Leibniz Rule to products involving more than two functions:

Corollary 6.6.12. *If $f_j : I \to \mathbb{R}$, $j = 1, 2, \ldots, k$, are n-times differentiable on I, then so is their product, $f_1 f_2 \cdots f_k$, and we have*

$$D^n(f_1 f_2 \cdots f_k) = \sum_{n_1 + n_2 + \cdots + n_k = n} \frac{n!}{n_1! n_2! \cdots n_k!} D^{n_1} f_1 D^{n_2} f_2 \cdots D^{n_k} f_k. \qquad (*)$$

Proof. The case $k = 2$ is the Leibniz Rule. Inductively, we may assume that $f_2 f_3 \cdots f_k$ is n-times differentiable on I and apply the Leibniz Rule to obtain

$$D^n(f_1 f_2 \cdots f_k) = \sum_{n_1 + m = n} \frac{n!}{n_1! m!} D^{n_1} f_1 D^m(f_2 \cdots f_k). \qquad (**)$$

Applying $(*)$ (with n replaced by $m = n - n_1$) to the $k - 1$ functions $f_2, \; f_3, \ldots, \; f_k$, the right side of $(**)$ is then

$$= \sum_{n_1 + m = n} \frac{n!}{n_1! m!} D^{n_1} f_1 \cdot \left(\sum_{n_2 + \cdots + n_k = m} \frac{m!}{n_2! \cdots n_k!} D^{n_2} f_2 \cdots D^{n_k} f_k \right)$$

$$= \sum_{n_1 + n_2 + \cdots + n_k = n} \frac{n!}{n_1! n_2! \cdots n_k!} D^{n_1} f_1 D^{n_2} f_2 \cdots D^{n_k} f_k.$$

□

Our next goal is to prove *Taylor's formula*, which is an extension of the MVT, and plays an important role in the study of *local approximation* of functions by polynomials. In Sect. 4.7, we proved the *Weierstrass Approximation Theorem*, which asserts that a continuous function on a closed bounded interval can be *uniformly* approximated by *polynomials* on that interval. In fact, in Theorem 4.7.9, we approximated any continuous function on $[0, 1]$ by its *Bernstein polynomials*.

Despite the importance of this uniform approximation, one drawback is that the nth Bernstein polynomial depends on the values of the function at $n + 1$ equally spaced points in the interval. By contrast, the *Taylor polynomials* (defined below) depend only on the values of the function and some of its derivatives at a *single* point in the interval. Therefore, they are more suitable for *local* approximation, i.e., approximation *in a neighborhood of a given point*. Let us begin with the following.

Exercise 6.6.13.

(a) Using the *Power Rule*, show that, for each $k \in \mathbb{N}$ and each $j = 0, 1, \cdots, k$, we have

$$[(x - c)^k]^{(j)} = k(k - 1) \cdots (k - j + 1)(x - c)^{k-j} = \frac{k!}{(k - j)!}(x - c)^{k-j}.$$

In particular, $[(x - c)^k]^{(k)} = k!$ and $[(x - c)^k]^{(\ell)} \equiv 0 \; \forall \ell > k$. Deduce that the polynomial function

$$p(x) := \sum_{k=0}^{n} a_k(x - c)^k, \qquad (*)$$

where the a_k and c are real constants, has the property that

$$a_k = \frac{p^{(k)}(c)}{k!} \quad (0 \le k \le n), \qquad (**)$$

and $p^{(k)}(x) = 0$ for all $k > n$ and all $x \in \mathbb{R}$.

(b) Consider the function

$$f(x) := \begin{cases} e^{-1/x} & \text{if } x > 0, \\ 0 & \text{if } x \le 0. \end{cases}$$

Show that $f \in C^\infty(\mathbb{R})$ and that $f^{(n)}(0) = 0$ for all $n \in \mathbb{N}$. *Hint:* Concentrating at $x = 0$, use induction and l'Hôpital's Rule.

Remark 6.6.14. If, in the above exercise, we replace x by $\xi + \eta$ and c by ξ in $(*)$ and use $(**)$, then we obtain the identity

$$p(\xi + \eta) = \sum_{k=0}^{n} \frac{1}{k!} p^{(k)}(\xi)\eta^k \quad (\forall \xi, \eta \in \mathbb{R}).$$

Definition 6.6.15 (Taylor Polynomials). Let $f : I \to \mathbb{R}$ be n-times differentiable at $x_0 \in I$; i.e., suppose that $f^{(n)}(x_0)$ exists. The *nth Taylor polynomial of f at x_0* is then defined to be

$$P_{n,x_0}(x) := f(x_0) + \frac{f'(x_0)}{1!}(x - x_0) + \frac{f''(x_0)}{2!}(x - x_0)^2 + \cdots + \frac{f^{(n)}(x_0)}{n!}(x - x_0)^n.$$

The coefficients $f^{(j)}(x_0)/j!$, $0 \le j \le n$, are called the *Taylor coefficients of f at* x_0. It follows at once from Exercise 6.6.13 that we have

$$P_{n,x_0}^{(j)}(x_0) = f^{(j)}(x_0) \quad (0 \le j \le n).$$

Exercise 6.6.16.

(a) Let $f(x) := e^x \; \forall x \in \mathbb{R}$. Show that the nth Taylor polynomial of f at $x_0 = 0$ is

$$P_{n,0}(x) = 1 + \frac{x}{1!} + \frac{x^2}{2!} + \frac{x^3}{3!} + \cdots + \frac{x^n}{n!}.$$

(b) Let $g(x) := \sin x \; \forall x \in \mathbb{R}$. Show that the $(2n+1)$th Taylor polynomial of g at $x_0 = 0$ is

$$P_{2n+1,0}(x) = x - \frac{x^3}{3!} + \frac{x^5}{5!} - \frac{x^7}{7!} + \cdots + (-1)^n \frac{x^{2n+1}}{(2n+1)!}.$$

(c) Let $h(x) := \log x \; \forall x > 0$. Show that

$$\log^{(j)}(x) = \frac{(-1)^{j-1}(j-1)!}{x^j} \quad (j = 1, 2, \ldots).$$

Deduce that the nth Taylor polynomial of h at $x_0 = 1$ is

$$P_{n,1}(x) = (x-1) - \frac{(x-1)^2}{2} + \frac{(x-1)^3}{3} - \frac{(x-1)^4}{4} + \cdots + \frac{(-1)^{n-1}(x-1)^n}{n}.$$

If, instead of h, we consider the function $k(x) := \log(x+1) \; \forall x > -1$, show that the nth Taylor polynomial of k at $x_0 = 0$ is

$$P_{n,0}(x) = x - \frac{x^2}{2} + \frac{x^3}{3} - \frac{x^4}{4} + \cdots + \frac{(-1)^{n-1}x^n}{n}.$$

The following proposition shows how the Taylor polynomials of a function at a given point approximate the function in a neighborhood of that point.

Proposition 6.6.17. *Let* $f : I \to \mathbb{R}$ *be n-times differentiable at a point* $x_0 \in I$ *and let* P_{n,x_0} *be its nth Taylor polynomial at* x_0. *Then we have*

$$f(x) = P_{n,x_0}(x) + (x - x_0)^n o(1) \quad (x \to x_0).$$

More precisely, there exists a function $\zeta : I \to \mathbb{R}$, *with* $\lim_{x \to x_0} \zeta(x) = 0$, *such that*

$$f(x) = P_{n,x_0}(x) + (x - x_0)^n \zeta(x) \quad (\forall x \in I).$$

If, in addition, $f^{(n+1)}(x_0)$ exists, then we have

$$\lim_{x \to x_0} \frac{\zeta(x)}{(x - x_0)} = \frac{f^{(n+1)}(x_0)}{(n + 1)!}. \tag{†}$$

Proof. Consider the function

$$\zeta(x) := \frac{f(x) - P_{n,x_0}(x)}{(x - x_0)^n} \qquad (x \neq x_0). \tag{*}$$

As was pointed out above, the existence of $f^{(n)}(x_0)$ implies that there exists an interval J, with $x_0 \in J \subset I$, such that f is $(n - 1)$-times (continuously) differentiable on J. It is easily seen that all the derivatives of order $\leq n - 1$ of the numerator and denominator of $(*)$ are zero at x_0. Since $f^{(n-1)}(x)$ is defined for all $x \in J$, we can apply l'Hôpital's Rule $n - 1$ times to $(*)$ to obtain

$$\lim_{x \to x_0} \zeta(x) = \lim_{x \to x_0} \frac{f^{(n-1)}(x) - f^{(n-1)}(x_0) - (x - x_0) f^{(n)}(x_0)}{n!(x - x_0)}, \tag{**}$$

if the limit on the right side exists. But, by hypothesis, $f^{(n)}(x_0)$ exists; i.e., we have

$$\lim_{x \to x_0} \frac{f^{(n-1)}(x) - f^{(n-1)}(x_0)}{x - x_0} = f^{(n)}(x_0).$$

Therefore, the limit in $(**)$ is indeed zero, as desired. If we *define* $\zeta(x_0) := 0$, then the first part of the proposition is proved. To prove (†), note that the existence of $f^{(n+1)}(x_0)$ implies that f is n-times differentiable on an interval J, with $x_0 \in J \subset I$. We can therefore apply l'Hôpital's Rule n times to the function $\zeta(x)/(x - x_0)$ to obtain

$$\lim_{x \to x_0} \frac{\zeta(x)}{(x - x_0)} = \lim_{x \to x_0} \frac{f^{(n)}(x) - f^{(n)}(x_0)}{(n + 1)!(x - x_0)} = \frac{f^{(n+1)}(x_0)}{(n + 1)!}.$$

$\square$

Note that the above proposition only gives the behavior of the *remainder* $R_{n,x_0}(x) := f(x) - P_{n,x_0}(x)$ as $x \to x_0$. In particular, the remainder will be small in a sufficiently small neighborhood of x_0. If we impose more restrictions on the function f, we can find a more precise form of the remainder and give it an upper bound over the entire interval I.

Theorem 6.6.18 (Taylor's Formula with Lagrange's Remainder). *Let $f : I \to \mathbb{R}$ be $(n + 1)$-times differentiable on I and let $x_0 \in I$ be fixed. Then for each $x \in I, x \neq x_0$, there exists a point ξ between x_0 and x such that we have*

$$f(x) = P_{n,x_0}(x) + \frac{f^{(n+1)}(\xi)}{(n + 1)!}(x - x_0)^{n+1}. \tag{†}$$

The term $R_{n,x_0}(x) := f^{(n+1)}(\xi)(x-x_0)^{n+1}/(n+1)!$ is called Lagrange's remainder (or Lagrange's form of the remainder). In particular, if

$$M := \sup\{|f^{(n+1)}(x)| : x \in I\} < \infty,$$

then we have

$$|R_{n,x_0}(x)| \le M \frac{|x - x_0|^{n+1}}{(n+1)!} \qquad (\forall x \in I).$$

Proof. Assume $x_0 < x$; the other case is similar. On the interval $[x_0, x]$, consider the function

(i) $F(t) := f(x) - f(t) - \dfrac{f'(t)}{1!}(x-t) - \dfrac{f''(t)}{2!}(x-t)^2 - \cdots - \dfrac{f^{(n)}(t)}{n!}(x-t)^n.$

Computing $F'(t)$, all but one of the terms cancel out and we obtain

$$F'(t) = -\frac{f^{(n+1)}(t)}{n!}(x-t)^n.$$

Next, introduce the function

(ii) $G(t) := \dfrac{(x-t)^{n+1}}{(n+1)!} \qquad (\forall t \in [x_0, x]),$

so that $G'(t) = -(x-t)^n/n!$. Note, in particular, that we have $F(x) = G(x) = 0$, and

(iii) $\dfrac{F'(t)}{G'(t)} = f^{(n+1)}(t).$

Applying Cauchy's MVT to F and G on $[x_0, x]$, and using (iii), we can find a point ξ between x_0 and x such that

$$\frac{F(x_0)}{G(x_0)} = \frac{F(x) - F(x_0)}{G(x) - G(x_0)} = \frac{F'(\xi)}{G'(\xi)} = f^{(n+1)}(\xi).$$

In other words, we have $F(x_0) = G(x_0)f^{(n+1)}(\xi)$, which, in view of the definitions (i) and (ii), completes the proof of (†). The last statement is an obvious consequence of (†).   □

Remark 6.6.19 (Cauchy's Form of the Remainder).   If, in the above proof, we use the function $G(t) := x - t$ instead of (ii), then the remainder takes the form

$$R_{n,x_0}(x) = \frac{f^{(n+1)}(\xi)}{n!}(x-\xi)^n(x-x_0),$$

which is called *Cauchy's form of the remainder*. The reader is invited to supply the details. There is another important form of the remainder which requires *integration* and will be given in the next chapter.

The following corollaries are immediate consequences of Taylor's Formula.

Corollary 6.6.20. *Let* $f : I \to \mathbb{R}$ *be* $(n + 1)$-*times differentiable on* I. *Then for each* $x \in I$ *and each* $h \in \mathbb{R}$, *with* $x + h \in I$, *there exists a* $\theta \in (0, 1)$ *such that*

$$f(x+h)=f(x)+\frac{h}{1!}f'(x)+\frac{h^2}{2!}f''(x)+\cdots+\frac{h^n}{n!}f^{(n)}(x)+\frac{h^{n+1}}{(n+1)!}f^{(n+1)}(x+\theta h).$$

Corollary 6.6.21. *Let* $f : I \to \mathbb{R}$ *be* $(n + 1)$-*times differentiable on* I. *If* $f^{(n+1)}(x) = 0 \ \forall x \in I$, *then, on the interval* I, f *is a polynomial of degree at most* n.

Let us also include the following *uniqueness* property of Taylor's Formula:

Proposition 6.6.22. *Let* $f : I \to \mathbb{R}$ *be* n-*times differentiable at a point* $x_0 \in I$. *Suppose that for each* $x \in I$ *we have*

$$f(x) = a_0 + a_1(x - x_0) + a_2(x - x_0)^2 + \cdots + a_n(x - x_0)^n + (x - x_0)^n \zeta(x), \quad (*)$$

where $a_0, a_1, \ldots, a_n$ *are real constants and* $\zeta : I \to \mathbb{R}$ *satisfies* $\lim_{x \to x_0} \zeta(x) = 0$. *Then we have*

$$a_k = \frac{f^{(k)}(x_0)}{k!} \quad (0 \le k \le n). \quad (**)$$

Proof. Substituting $x = x_0$ in $(*)$, we get $a_0 = f(x_0)$. This implies that $[f(x) - f(x_0)]/(x - x_0) = a_1 + o(1)$, as $x \to x_0$, and hence $f'(x_0) = a_1$. Similarly, $[f(x) - f(x_0) - f'(x_0)(x - x_0)]/(x - x_0)^2 = a_2 + o(1)$, as $x \to x_0$, which (applying l'Hôpital's Rule twice) gives $a_2 = f''(x_0)/2$. Continuing, we deduce that the a_k are given by $(**)$. $\qquad\square$

As was pointed out before, Taylor's Formula can be used for local approximation of differentiable functions by polynomials. Here is an example:

Example 6.6.23. Let us approximate the function $f(x) := e^x$ by a polynomial on the interval $[-1, 1]$, with error less than 10^{-10}. Since $f'(x) = f(x)$, we have $f^{(n)}(x) = e^x \ \forall n \in \mathbb{N}$. In particular, $f^{(n)}(0) = 1 \ \forall n \in \mathbb{N}$. By Taylor's Formula, for each $x \ne 0$ in $[-1, 1]$, we can find a number ξ between 0 and x such that

$$e^x = 1 + x + \frac{x^2}{2!} + \frac{x^3}{3!} + \cdots + \frac{x^n}{n!} + e^\xi \frac{x^{n+1}}{(n+1)!}.$$

Now $|\xi| < 1$ implies $e^\xi < e < 3$. Therefore,

$$|R_{n,0}(x)| = e^\xi \frac{|x|^{n+1}}{(n+1)!} \leq \frac{e}{(n+1)!} < \frac{3}{(n+1)!}.$$

A calculation shows that $13! = 0.62270208 \times 10^{10} < 3 \times 10^{10} < 14! = 8.71782912 \times 10^{10}$. Therefore, if $n = 13$, the error is indeed less than 10^{-10}. In particular, for $x = 1$, we have

$$e \approx P_{13,0}(1) = 1 + 1 + \frac{1}{2!} + \cdots + \frac{1}{13!} \approx 2.718281828446759,$$

which is correct to 10 decimal places. In fact, $e \approx 2.718281828459045$.

Exercise 6.6.24. Using Taylor's Formula, find an approximate value of $\sin 1$ with error less than 10^{-5}.

Taylor's Formula can be used to prove a generalization of the Leibniz Rule. Before giving it, let us introduce some convenient terminology:

Definition 6.6.25.

(a) **(Differential Operator, Symbol)** Given any polynomial with real coefficients

$$p(\xi) = \sum_{k=0}^{n} a_k \xi^k = a_0 + a_1 \xi + a_2 \xi^2 + \cdots + a_n \xi^n, \qquad (*)$$

we can associate with it the corresponding *differential polynomial*

$$p(D) = \sum_{k=0}^{n} a_k D^k = a_0 + a_1 D + a_2 D^2 + \cdots + a_n D^n,$$

where $D = d/dx$. Given any n-times differentiable function $u : I \to \mathbb{R}$, the differential polynomial $p(D)$ can be applied to it in a natural way:

$$p(D)u = \sum_{k=0}^{n} a_k D^k u = \sum_{k=0}^{n} a_k \frac{d^k u}{dx^k}.$$

In this case, $p(D)$ is said to *operate* on u. When $p(D)$ operates on functions, we call it a *differential operator*. The polynomial $(*)$ is then called the *symbol* of $p(D)$. If $a_n \neq 0$, then $p(D)$ is said to be an *nth-order (ordinary) differential operator with (constant) coefficients* $a_0, a_1, \ldots, a_n$.

(b) **(Differential Equation, Solution)** An equation of the form

$$p(D)u = f,$$

where $f : I \to \mathbb{R}$ is a *given* function (with certain *differentiability conditions*) and $u : I \to \mathbb{R}$ is an *unknown* (n-times differentiable function) to be determined, is called a *differential equation*. A *solution* to this equation is any function u that satisfies it.

Remark 6.6.26. It is obvious that an nth-order differential operator $p(D)$ defines a map $p(D) : C^m(I) \to C^{m-n}(I)$ for each integer $m \geq n$. It is also clear that a differential operator $p(D)$ of *any* order defines a map $p(D) : C^\infty(I) \to C^\infty(I)$. The most natural setting for the study of differential operators is the theory of *distributions* (also known as *generalized functions*), because the differentiation is then *always* possible. Distribution theory is treated in more advanced courses on analysis and plays a fundamental role in the study of partial differential equations.

We are now ready to prove the extension of the Leibniz Rule mentioned above.

Theorem 6.6.27 (Hörmander's Generalized Leibniz Rule). *Let u, $v : I \to \mathbb{R}$ be n-times differentiable functions on I. Given any nth-order differential operator $p(D) = \sum_{k=0}^{n} a_k D^k$, we have*

$$p(D)(uv) = \sum_{k=0}^{n} \frac{1}{k!} p^{(k)}(D)u \cdot D^k v.$$

Proof. Well, first note that the Leibniz Rule gives

$$p(D)(uv) = \sum_{j=0}^{n} a_j D^j(uv) = \sum_{j=0}^{n} a_j \sum_{i=0}^{j} \binom{j}{i} D^{j-i}u \cdot D^i v. \qquad (*)$$

If, for each $k = 0, 1, \ldots, n$, we group all the terms on the right side of $(*)$ containing $D^k v$, then $(*)$ can be written as

$$p(D)(uv) = \sum_{k=0}^{n} q_k(D)u \cdot D^k v, \qquad (**)$$

where the q_k are polynomials to be determined. Next, note that, for each fixed ξ, we have $D_x e^{\xi x} = \xi e^{\xi x}$. Repeated use of this fact implies that

$$q(D)e^{\xi x} = q(\xi)e^{\xi x}$$

for every polynomial q. Therefore, if we apply $(*)$ with $u(x) := e^{\xi x}$ and $v(x) := e^{\eta x}$ for fixed ξ, $\eta \in \mathbb{R}$, we have

$$p(\xi + \eta)e^{(\xi + \eta)x} = \sum_{k=0}^{n} q_k(\xi) \cdot \eta^k e^{(\xi + \eta)x},$$

which implies the identity

$$p(\xi + \eta) = \sum_{k=0}^{n} q_k(\xi) \cdot \eta^k \quad \forall \xi, \, \eta \in \mathbb{R}. \tag{$\dagger$}$$

On the other hand, by Taylor's Formula (cf. Remark 6.6.14), we have

$$p(\xi + \eta) = \sum_{k=0}^{n} \frac{1}{k!} p^{(k)}(\xi) \cdot \eta^k \quad \forall \xi, \, \eta \in \mathbb{R}. \tag{$\ddagger$}$$

Comparing the identities ($\dagger$) and ($\ddagger$), we conclude that $q_k = p^{(k)}/k!$ and the proof is complete. □

Exercise 6.6.28. Show that the Leibniz Rule is an immediate consequence of the above generalized version.

6.7   Convex Functions

The reader is certainly familiar with the notion of *convexity* introduced in calculus, where it is usually referred to as *concavity*, and where one also encounters the terms *concave up* and *concave down*. The definitions given in calculus textbooks are often geometric and assume the differentiability of the function. The goal is then to explain the connection to the sign of the second derivative and to the extrema (via the *second derivative test*). The definition of convexity given below is more general and we shall see that convexity on an interval implies differentiability at *all but a countable set of points* in that interval.

Definition 6.7.1 (Convex Function, Concave Function). Let $f : I \to \mathbb{R}$. We say that f is *convex* on I if for every $s, \, t \in I$ and every $\lambda \in [0, 1]$, we have

$$f(\lambda s + (1 - \lambda)t) \leq \lambda f(s) + (1 - \lambda) f(t). \tag{$\dagger$}$$

We say that f is *concave* on I if $-f$ is convex on I. Since $\{(\lambda s + (1-\lambda)t, \lambda f(s) + (1 - \lambda) f(t)) : \lambda \in [0, 1]\}$ is simply the *chord* joining the points $(s, f(s))$ and $(t, f(t))$ on the graph of f, the inequality ($\dagger$) means, *geometrically*, that this chord is *above* the graph for every $s, \, t \in I$.

Proposition 6.7.2 (Jensen's Inequality). *If $f : I \to \mathbb{R}$ is convex on I, then it satisfies* Jensen's inequality:

$$f\left(\sum_{k=1}^{n} \lambda_k x_k\right) \leq \sum_{k=1}^{n} \lambda_k f(x_k),$$

for any $x_1, \ldots, x_n \in I$ and any $\lambda_1, \ldots, \lambda_n \in [0, 1]$, with $\sum_{k=1}^{n} \lambda_k = 1$.

Proof. We use (†) and induction, assuming the inequality for any $x_i \in I$ and $\lambda_i \in [0, 1]$, with $\sum_{i=1}^{m} \lambda_i = 1$ and $m < n$. We then have, for $\lambda < 1$,

$$f\left(\sum_{k=1}^{n} \lambda_k x_k\right) = f\left((1 - \lambda_n)\sum_{j=1}^{n-1} \frac{\lambda_j}{1 - \lambda_n} x_j + \lambda_n x_n\right)$$

$$\leq (1 - \lambda_n)f\left(\sum_{j=1}^{n-1} \frac{\lambda_j}{1 - \lambda_n} x_j\right) + \lambda_n f(x_n)$$

$$\leq \sum_{k=1}^{n} \lambda_k f(x_k).$$

$\square$

Proposition 6.7.3 (Three Chords Lemma). *Let* $f : I \to \mathbb{R}$. *Then,* f *is convex on* I *if and only if, for any points* a, b, $c \in I$ *with* $a < b < c$, *we have*

$$\frac{f(b) - f(a)}{b - a} \leq \frac{f(c) - f(a)}{c - a} \leq \frac{f(c) - f(b)}{c - b}, \tag{‡}$$

which is equivalent to saying that for any fixed $x_0 \in I$ *the function*

$$\phi(x) := \frac{f(x) - f(x_0)}{x - x_0} \qquad \forall x \in I \setminus \{x_0\}$$

[i.e., the slope of the chord joining $\left(x_0, f(x_0)\right)$ *and* $\left(x, f(x)\right)$*] is increasing.*

Proof. Note that, with

$$\lambda_1 = \frac{c - b}{c - a} \quad \text{and} \quad \lambda_2 = \frac{b - a}{c - a},$$

we have $b = \lambda_1 a + \lambda_2 c$. Applying Jensen's inequality, we have

$$f(b) \leq \frac{c - b}{c - a} f(a) + \frac{b - a}{c - a} f(c). \tag{$*$}$$

Now we subtract $f(a)$ from both sides of ($*$) to get the first inequality in (‡), and subtract $f(c)$ from both sides of ($*$) to get the second one. For the converse, we must show that (‡) implies (†) of the above definition. But a simple computation transforms the first inequality in (‡) into ($*$), which is precisely (†) with $a = s$, $c = t$, and $b = \lambda a + (1 - \lambda)c$. The last statement is an immediate consequence. $\square$

Exercise 6.7.4.

1. Show that an *affine* function $f(x) = ax + b$, a, $b \in \mathbb{R}$ is convex on $\mathbb{R}$. In fact, $f \in \mathbb{R}^{\mathbb{R}}$ is affine if and only it is *both* convex and concave.
2. Show that the quadratic function $f(x) = x^2$ is convex on $\mathbb{R}$.

Remark 6.7.5. It should be noted that a convex function *need not be* differentiable. For example, the function $f(x) := |x| \ \forall x \in \mathbb{R}$ is obviously convex (why?) but is not differentiable at $x = 0$. In fact, a convex function on a *closed* interval need not even be continuous. Indeed the function $f : [0, 1] \to \mathbb{R}$ defined by $f(x) := 0 \ \forall x \in (0, 1]$ and $f(0) := 1$ is convex on $[0, 1]$ but discontinuous at 0. It turns out, however, that a convex function on an *open* interval is automatically continuous there. In fact, such a function has finite left and right derivatives at every point of the interval and is differentiable except at a *countable* number of points:

Lemma 6.7.6. *Let* $f : (a, b) \to \mathbb{R}$ *be a convex function. Then the left and right derivatives* $f'_-(x)$ *and* $f'_+(x)$ *are finite at every* $x \in (a, b)$—*hence* f *is continuous on* (a, b)—*and* $f'_-(x) \le f'_+(x)$. *Moreover, the inequalities*

$$f'_-(s) \le f'_+(s) \le \frac{f(t) - f(s)}{t - s} \le f'_-(t) \le f'_+(t). \tag{$**$}$$

are satisfied for any $s < t$ *in* (a, b). *In particular,* f'_- *and* f'_+ *are both increasing on* (a, b) *and the set of* $x \in (a, b)$ *at which* f *is not differentiable is countable.*

Proof. Let $x_0 \in (a, b)$ be arbitrary and let ϕ be the *(slope)* function defined in Proposition 6.7.3. If $A := \{\phi(x) : x \in (a, x_0)\}$ and $B := \{\phi(x) : x \in (x_0, b)\}$, then $\alpha \le \beta$ for all $\alpha \in A$ and $\beta \in B$ because ϕ is *increasing* on the intervals (a, x_0) and (x_0, b). It follows (why?) that

$$f'_-(x_0) = \phi(x_0 - 0) = \sup(A) \le \inf(B) = \phi(x_0 + 0) = f'_+(x_0).$$

Thus f has finite one-sided derivatives at x_0 with $f'_-(x_0) \le f'_+(x_0)$. Since x_0 was arbitrary, the same is then true for every $x \in (a, b)$. In particular (by Corollary 6.1.13), f is *both* right and left continuous at each $x \in (a, b)$ and hence is *continuous* there. Next, let $s < t$ in (a, b) and let $x \in (s, t)$. Then, by what we just proved, we have

$$f'_-(s) \le f'_+(s) \le \frac{f(x) - f(s)}{x - s} \le \frac{f(x) - f(t)}{x - t} \le f'_-(t) \le f'_+(t),$$

from which $(**)$ follows. Moreover, $(**)$ implies that if f is *not* differentiable at s and t, then the open intervals $\big(f'_-(s), f'_+(s)\big)$ and $\big(f'_-(t), f'_+(t)\big)$ are *disjoint* and hence we *cannot* have more than a countable number of them. (Why?)   □

Corollary 6.7.7. *Let* I *be an open interval and let* $f : I \to \mathbb{R}$ *be convex on* I. *Given any* $x_0 \in I$ *and any number* m *with* $f'_-(x_0) \le m \le f'_+(x_0)$, *we have* $f(x) \ge f(x_0) + m(x - x_0) \ \forall x \in I$.

Proof. Since f is convex, the slope $\phi(x) := (f(x) - f(x_0))/(x - x_0)$ is *increasing*. Now, if $x > x_0$, then the definition of the right derivative implies that we have

$f(x) - f(x_0) \geq (x - x_0) f'_+(x_0) \geq m(x - x_0)$. If $x < x_0$, a similar argument shows that $f(x_0) - f(x) \leq (x_0 - x) f'_-(x_0) \leq m(x_0 - x)$, which is the desired inequality if we multiply the two sides by -1. □

We can now characterize convex functions on open intervals completely:

Theorem 6.7.8. *Let* $I \subset \mathbb{R}$ *be an open interval and* $f : I \to \mathbb{R}$*. Then,* f *is convex on* I *if and only if there is a countable subset* $D \subset I$ *such that* f *is continuous on* I*, has a finite right derivative* $f'_+(x)$ *at every* $x \in I \setminus D$*, and* f'_+ *is increasing on* $I \setminus D$*.*

Proof. If f is convex, then (by Lemma 6.7.6) the conditions of the theorem are all satisfied. To prove the converse, we show that the *slope* function ϕ in Proposition 6.7.3 is *increasing*. So let a, b, c be points of I with $a < b < c$ and define

$$m := \sup\{f'_+(x) : x \in (a, b) \setminus D\} \quad \text{and} \quad M := \inf\{f'_+(x) : x \in (b, c) \setminus D\}.$$

It then follows from Proposition 6.4.24 that

$$f(b) - f(a) \leq m(b - a) \quad \text{and} \quad M(c - b) \leq f(c) - f(b).$$

Since $m \leq M$ (why?), we get

$$\frac{f(b) - f(a)}{b - a} \leq \frac{f(c) - f(b)}{c - b},$$

and the proof is complete. □

Corollary 6.7.9. *Let* I *be an open interval and* $f : I \to \mathbb{R}$*.*

1. *Suppose* f *is differentiable on* I*. Then* f *is convex on* I *if and only if* f' *is increasing on* I*.*
2. *Suppose* f *is 2-times differentiable on* I*. Then* f *is convex on* I *if and only if* $f''(x) \geq 0 \ \forall x \in I$*.*

Proof. This is an immediate consequence of Theorem 6.7.8 and Corollary 6.4.14. □

Exercise 6.7.10. Prove Corollary 6.7.9 *directly* by using the Mean Value Theorem to show that the *(slope)* function ϕ in Proposition 6.7.3 is *increasing*.

Remark 6.7.11.

1. Note that, as was pointed out in Remark 6.4.15(a), the *increasing* in the above corollary *cannot* be replaced by *strictly increasing*.
2. (**Support Line**) Corollary 6.7.7 can be interpreted, geometrically, as follows: Given a convex function f on an open interval I, through each point $P_0 := (x_0, f(x_0))$ of the graph of f, we can draw a straight line lying entirely *below* the graph of f. Such a line is called a *support line* of f.

Exercise 6.7.12. Let I be an open interval, $x_0 \in I$, and let $f : I \to \mathbb{R}$ be convex on I. Show that, if $f'(x_0)$ exists, then there is a *unique* support line through $P_0 = (x_0, f(x_0))$, namely, the line *tangent* to the graph of f at P_0 whose equation is obviously $y = f(x_0) + f'(x_0)(x - x_0)$.

Example 6.7.13. Let $p > 1$ and let q be the positive number (necessarily > 1) such that $1/p + 1/q = 1$. The following inequalities are then satisfied for any $a \geq 0$, $b \geq 0$:

(i) $\quad a^{1/p} b^{1/q} \leq \dfrac{a}{p} + \dfrac{b}{q}$, $\qquad$ (ii) $\quad \left(\dfrac{a+b}{2}\right)^p \leq \dfrac{1}{2}(a^p + b^p)$.

First note that the inequalities are obvious if $ab = 0$. Now, to prove (i), note that $(-\log x)'' = (-1/x)' = 1/x^2 > 0$ for all $x > 0$. Thus, $-\log$ is *convex* on $(0, \infty)$. By Jensen's inequality, we have

$$-\log\left(\frac{1}{p}a + \frac{1}{q}b\right) \leq \frac{1}{p}(-\log a) + \frac{1}{q}(-\log b). \qquad (*)$$

Since

$$\frac{1}{p}(\log a) + \frac{1}{q}(\log b) = \log(a^{1/p}) + \log(b^{1/q}) = \log(a^{1/p} b^{1/q}),$$

the inequality (i) follows from $(*)$ and the fact that exp is *increasing*. The inequality (ii) is an immediate consequence of Jensen's inequality (with $\lambda_1 = \lambda_2 = 1/2$) applied to the function $f(x) := x^p \; \forall x \in (0, \infty)$, which is convex in view of the fact that $f''(x) = p(p-1)x^{p-2} > 0 \; \forall x > 0$.

Exercise 6.7.14 (Hölder and Minkowski Inequalities).   Given any finite sequences $(a_k)_{k=1}^n$ and $(b_k)_{k=1}^n$ in $\mathbb{R}$, prove the following inequalities:

(a) For any $p > 1$, $q > 1$ with $1/p + 1/q = 1$, we have

$$\left|\sum_{k=1}^n a_k b_k\right| \leq \left(\sum_{k=1}^n |a_k|^p\right)^{1/p} \left(\sum_{k=1}^n |b_k|^q\right)^{1/q}. \qquad \text{(Hölder)}$$

Hint: Show that we may assume $a_k \geq 0$, $b_k \geq 0$, for all k, and $\sum_{k=1}^n |a_k|^p = \sum_{k=1}^n |b_k|^q = 1$. Now use (i) of the above example with $a = a_k^p$ and $b = b_k^q$.

(b) For any $p \geq 1$, we have

$$\left(\sum_{k=1}^n |a_k + b_k|^p\right)^{1/p} \leq \left(\sum_{k=1}^n |a_k|^p\right)^{1/p} + \left(\sum_{k=1}^n |b_k|^p\right)^{1/p}. \qquad \text{(Minkowski)}$$

Hint: Assume $a_k \geq 0$, $b_k \geq 0$ for all k. Now, for $p > 1$, let $q := p/(p-1)$ and apply *Hölder's inequality* to the two sums on the right side of the identity

$$\sum_{k=1}^{n}(a_k + b_k)^p = \sum_{k=1}^{n} a_k(a_k + b_k)^{p-1} + \sum_{k=1}^{n} b_k(a_k + b_k)^{p-1}.$$

(c) Extend both inequalities in (a) and (b) to the case of infinite sequences $(a_n)_{n=1}^{\infty}$ and $(b_n)_{n=1}^{\infty}$. *Hint:* Look at *partial sums*.

Here is a more general definition of convexity that does *not* imply continuity:

Exercise 6.7.15. Suppose that $f : I \to \mathbb{R}$ satisfies the condition

$$f\left(\frac{s+t}{2}\right) \leq \frac{1}{2}f(s) + \frac{1}{2}f(t) \quad \forall s, t \in I. \tag{\dagger}$$

(a) Show that $f(\lambda s + (1 - \lambda)t) \leq \lambda f(s) + (1 - \lambda)f(t)$ holds for all s, $t \in I$ and all $\lambda \in [0, 1]$ of the form $\lambda = m/2^n$, with *integers* $m \geq 0$ and $n \geq 1$. *Hint:* Use induction and the identity

$$\frac{m}{2^n}s + \left(1 - \frac{m}{2^n}\right)t = \frac{1}{2}\left[\frac{m}{2^{n-1}}s + \left(1 - \frac{m}{2^{n-1}}\right)t + t\right].$$

(b) Show that if f satisfies (†) and is *continuous*, then f is *convex*. *Hint:* Show that $\{m/2^n : m/2^n \leq 1, m \in \mathbb{N}_0, n \in \mathbb{N}\}$ is *dense* in $[0, 1]$ and use part (a).

6.8 Problems

1. Let $f(x) := |x|^3$. Find $f'(x)$ and $f''(x)$. Show that $f'''(0)$ does *not* exist.

2. Give an example of a function $f : \mathbb{R} \to \mathbb{R}$ such that $f'''(x)$ exists for all $x \in \mathbb{R}$ but is *discontinuous* at $x = 0$.

3. Show that the function

$$f(x) := \begin{cases} x & \text{if } x \in \mathbb{Q}, \\ -x & \text{if } x \in \mathbb{Q}^c \end{cases}$$

is *nowhere differentiable*. Show, however, that $(f \circ f)(x) = x$ for all $x \in \mathbb{R}$.

4. Suppose that $f(x) = xg(x)$ where g is *continuous* at $x = 0$. Show that f is differentiable at $x = 0$ and find $f'(0)$.

5. Consider the function

$$f(x) = \begin{cases} x^2 & \text{if } x \in \mathbb{Q}, \\ 0 & \text{if } x \notin \mathbb{Q}. \end{cases}$$

Show that f is differentiable at $x = 0$ and find $f'(0)$.

6. Let $\alpha \in (0, 1)$ and $\delta > 0$ be constants and assume that $f(0) = 0$ and $|f(x)| \geq |x|^\alpha$ for $x \in (-\delta, \delta)$. Show that $f'(0)$ does *not* exist.

7 (Differentiable Periodic Function). Let $f : \mathbb{R} \to \mathbb{R}$ be a *differentiable, periodic* function with period a, i.e., $f(x + a) = f(x)$ for all $x \in \mathbb{R}$. Show that f' is also periodic. What is its period?

8. Let $f : \mathbb{R} \to \mathbb{R}$ be differentiable. Show *directly* (i.e., without using the Chain Rule) that $[f(cx)]' = cf'(cx)$ for all $c \in \mathbb{R}$.

9 (Euler's Theorem). A function $f \in \mathbb{R}^\mathbb{R}$ is said to be *homogeneous of order* $n \in \mathbb{R}$, if $f(tx) = t^n f(x)$ for all $t > 0$. If such a function is *differentiable*, show that $xf'(x) = nf(x)$, for all $x \in \mathbb{R}$.

10. Given a polynomial function $p(x) = a_0 + a_1 x + \cdots + a_n x^n$, find a polynomial $q(x)$ with $q'(x) = p(x)$ for all $x \in \mathbb{R}$.

11 (Diffeomorphism). Let I and J be open intervals. A map $f : I \to J$ is called a *diffeomorphism* if it is *bijective* and if f and f^{-1} are *both differentiable*. Show that $f(x) := x^3 + x$ is a diffeomorphism of $\mathbb{R}$ (onto $\mathbb{R}$) and find $(f^{-1})'(2)$.

12. Let $\arcsin x$ and $\arctan x$ denote the inverses of $\sin x$ (restricted to $[-\pi/2, \pi/2]$) and $\tan x$ (restricted to $(-\pi/2, \pi/2)$), respectively. Find the derivatives $(\arcsin)'(x)$ (for $x \in (-1, 1)$) and $(\arctan)'(x)$ (for $x \in \mathbb{R}$).

13.

(a) Let $f, g : (a, b) \to \mathbb{R}$ be differentiable. Show that, between any pair of *consecutive zeros* of f, there is always a zero of $f' + fg'$. *Hint:* Look at the function fe^g.
(b) Show that, between any pair of *consecutive* zeros of $f(x) := 1 - e^x \sin x$, there is at least one zero of $g(x) := 1 + e^x \cos x$.

14.

(a) Show that a polynomial of *even* degree attains its *absolute minimum*.
(b) Show that the polynomial

$$p(x) := 1 + x + \frac{x^2}{2!} + \frac{x^3}{3!} + \cdots + \frac{x^n}{n!}$$

has a *unique* real root if n is *odd* and *no* real roots if n is *even*. *Hint:* Note that, when $p'(x) = 0$, we have $p(x) = x^n/n!$.

15. Let $p(x) = a_0 + a_1 x + \cdots + a_n x^n$ and assume that $a_0 + a_1/2 + \cdots + a_n/(n + 1) = 0$. Show that $p(\zeta) = 0$ for some $\zeta \in (0, 1)$. *Hint:* Find a polynomial $q(x)$ with $q' = p$.

16. Show that, if a polynomial $p(x)$ with real coefficients has m *distinct* real roots, then $p'(x)$ has $m - 1$ distinct real roots.

17. Prove the inequalities

$$\frac{x}{1 + x} \leq \log(1 + x) \leq x \qquad (\forall x > -1).$$

18. Prove the following inequalities.

$$\frac{m(x-1)}{x^{1-m}} < x^m - 1 < m(x-1) \qquad (0 < m < 1,\ x > 1).$$

19. Show that, if $f : (a, b) \to \mathbb{R}$ is differentiable and f' is *bounded* on (a, b), then $f(a + 0)$ and $f(b - 0)$ exist.

20. Show that, if f is differentiable on I and $f' = kf$, then $f(x) = Ce^{kx}$ for some constant C and all $x \in I$.

21. Let $f : \mathbb{R} \to \mathbb{R}$ satisfy the functional equation

$$f(x + y) = f(x)f(y) \qquad (\forall x,\ y \in \mathbb{R}).$$

(a) Show that f is differentiable (on $\mathbb{R}$) if and only if $f'(0)$ exists.
(b) Show that, if f is *differentiable* and is *not* identically zero, then $f(x) = e^{cx}$ for a constant $c \in \mathbb{R}$.
(c) Show that the statement in (b) is true if f is merely *continuous* (instead of differentiable).

22. Find the following limit.

$$\lim_{x \to 0} \frac{\sin x - \tan x}{\tan^{-1} x - \sin^{-1} x}.$$

23 ("Sublinear" Function). Let us define a function $f : \mathbb{R} \to \mathbb{R}$ to be *sublinear* if $f(x) = o(x)$ as $|x| \to \infty$. Let $f \in \mathbb{R}^{\mathbb{R}}$ be a differentiable function. Show that if $\lim_{|x| \to \infty} f'(x) = 0$, then f is sublinear and we have $\lim_{|x| \to \infty}[f(x + y) - f(x)] = 0$ for each $y \in \mathbb{R}$.

24. Suppose that f is continuous on $[0, \infty)$, differentiable on $(0, \infty)$, $f(0) = 0$, and f' is *increasing*. Show that the function $g(x) := f(x)/x$ is increasing on $(0, \infty)$.

25. Let $f \in \mathbb{R}^{\mathbb{R}}$ be differentiable.

(a) Show that, if $|f'(x)| < 1$ for all $x \in \mathbb{R}$, then f has *at most one* fixed point.
(b) Show that the function $f(x) := x + 1/(1 + e^x)$ satisfies $|f'(x)| < 1$ for all $x \in \mathbb{R}$, but has *no* fixed point.

26. Let $f : (0, 1) \to \mathbb{R}$ be differentiable and $|f'(x)| \le 1$ for all $x \in (0, 1)$. Show that the sequence $(f(1/n))_{n \in \mathbb{N}}$ is *convergent*.

27. Show that, if f, $g : [0, \infty) \to \mathbb{R}$ are differentiable, $f(0) = g(0)$, and $f'(x) \le g'(x)$ for all $x > 0$, then $f(x) \le g(x)$ for all $x \ge 0$.

28. Let $f : [0, 1] \to \mathbb{R}$ be a differentiable function such that there is *no* point $x \in [0, 1]$ with $f(x) = f'(x) = 0$. Show that the set $Z := \{x \in [0, 1] : f(x) = 0\}$ of *zeros* of f is *finite*.

29. Let $f : [1, 3] \to \mathbb{R}$ be continuous on $[1, 3]$ and differentiable on $(1, 3)$, and assume that $f'(x) = [f(x)]^2 + 4$. Explain whether $f(3) - f(1) = 5$ is possible.

30. Can the Dirichlet function $\chi_{\mathbb{Q}}$ be the derivative of any function?

31 (Symmetric Derivative). Let $f \in \mathbb{R}^{\mathbb{R}}$. For each $x \in \mathbb{R}$, define the *symmetric derivative* of f at x by

$$f^s(x) := \lim_{h \to 0+} \frac{f(x + h) - f(x - h)}{2h},$$

if the limit exists. Show that, if $f'(x)$ exists, then $f^s(x) = f'(x)$. Let $g(x) := 2|x| + x$. Show that $g^s(x)$ exists for all $x \in \mathbb{R}$ even though $g'(0)$ does *not* exist. Also show that g attains its *absolute minimum* (i.e., $\min\{g(x) : x \in \mathbb{R}\}$) at $x = 0$, but $g^s(0) \neq 0$.

32 (Uniform Differentiability). Let f be differentiable on $[a, b]$. Show that f' is *continuous* on $[a, b]$ if and only if f is *uniformly differentiable* on $[a, b]$; i.e., given any $\varepsilon > 0$, there is a $\delta = \delta(\varepsilon) > 0$ such that for any $x_0 \in [a, b]$, we have

$$0 < |x - x_0| < \delta \Longrightarrow \left| \frac{f(x) - f(x_0)}{x - x_0} - f'(x_0) \right| < \varepsilon.$$

33. Let $f \in \mathbb{R}^{\mathbb{R}}$ be differentiable with *bounded* derivative, i.e., $|f'(x)| \leq M$ for all $x \in \mathbb{R}$ and some $M > 0$. Show that the function $g(x) := x + \varepsilon f(x)$ is *injective* for small enough $\varepsilon > 0$.

34. Suppose that $f : [a, \infty) \to \mathbb{R}$ satisfies $\lim_{x \to \infty}[f'(x) + \alpha f(x)] = 0$ for some $\alpha > 0$. Show that $\lim_{x \to \infty} f(x) = 0$. *Hint:* Apply Cauchy's MVT to $f(x)e^{\alpha x}$.

35. Show that if $f : \mathbb{R} \to [0, \infty)$ is twice differentiable and $f'' \leq 0$ on $\mathbb{R}$, then f is *constant*.

36. Let $f : (0, 1) \to \mathbb{R}$ be a differentiable function such that $\lim_{x \to 0+} f(x)$ and $\lim_{x \to 0+} x f'(x)$ both exist. Show that $\lim_{x \to 0+} x f'(x) = 0$

37 (Subexponential Function). Let us define a function $f : \mathbb{R} \to \mathbb{R}$ to be *subexponential* if

$$f(x) = o(e^{\varepsilon|x|}) \quad \forall \varepsilon > 0, \quad \text{as } |x| \to \infty.$$

(a) Show that, if $f : \mathbb{R} \to \mathbb{R}$ satisfies $|f(x)| > 0$ and $f'(x) = o(f(x))$ (as $x \to \infty$), then f is subexponential. *Hint:* Show that (assuming $f > 0$) $f(x)e^{-\varepsilon x}$ is *decreasing* (hence *bounded*) for all *large* $x > 0$ and use l'Hôpital's Rule.
(b) Let $\langle x \rangle := \sqrt{1 + x^2}$. Show that $\exp(\langle x \rangle^\alpha)$ is subexponential for $\alpha < 1$.
(c) Give an example of a (nontrivial) *bounded* function $f \in C^\infty(\mathbb{R})$ that satisfies $f'(x) = o(f(x))$, as $|x| \to \infty$.

38 (Schwarzian Derivative). Let $f : I \to \mathbb{R}$ and assume that $f'''(x)$ exists and $f'(x) \neq 0$ for all $x \in I$. Define the *Schwarzian derivative* of f at x by

$$\mathcal{D}f(x) := \frac{f'''(x)}{f'(x)} - \frac{3}{2} \left[\frac{f''(x)}{f'(x)} \right]^2 = \left[\frac{f''(x)}{f'(x)} \right]' - \frac{1}{2} \left[\frac{f''(x)}{f'(x)} \right]^2.$$

(a) Show that $\mathcal{D}(f \circ g) = (\mathcal{D}f \circ g) \cdot (g')^2 + \mathcal{D}g$.
(b) Show that, if $f(x) = (ax + b)/(cx + d)$, then $\mathcal{D}f = 0$.
(c) Show that $\mathcal{D}g = \mathcal{D}h$ if and only if $h = (ag + b)/(cg + d)$, where $ad - bc \neq 0$.
(d) Show that, if $fg = 1$, then $\mathcal{D}f = \mathcal{D}g$.
(e) Deduce the "if" part of (c) from (d). *Hint:* Note that, if $c \neq 0$, then $(ag + b)/(cg + d) = a/c + (bc - ad)/[c(cg + d)]$.

39. Let f be *continuous* on $[a, b]$ and *differentiable* on (a, b) except possibly at a point $x_0 \in (a, b)$. Show that, if $\lim_{x \to x_0} f'(x) = \ell \in \mathbb{R}$, then f is differentiable at x_0 and $f'(x_0) = \ell$ so that f' is actually *continuous* at x_0. *Hint:* Apply the MVT on $[x_0, x_0 + h]$ (resp., $[x_0 + h, x_0]$) for small $h > 0$ (resp., $h < 0$) or use l'Hôpital's Rule.

40. Consider the function

$$f(x) := \begin{cases} e^{-1/x^2} & \text{if } x \neq 0, \\ 0 & \text{if } x = 0. \end{cases}$$

Show that $f \in C^\infty(\mathbb{R})$ and $f^{(n)}(0) = 0$ for all $n \in \mathbb{N}$.

41 (Legendre's Polynomials). Define the polynomials

$$P_n(x) := \frac{1}{2^n(n!)} \frac{d^n}{dx^n}(x^2 - 1)^n \qquad (\forall n \in \mathbb{N}).$$

(a) Show that $P_n(x)$ has degree n and has n *distinct* (hence *simple*) real roots all of which are in $[-1, 1]$. *Hint:* Let $u := (x^2 - 1)^n$. Note that $u^{(k)}$ is *even* (resp., *odd*) for k even (resp., odd). Also, for $k \leq n - 1$, we have $u^{(k)}(\pm 1) = 0$ if k is even, and $u^{(k)}(\pm 1) = u^{(k)}(0) = 0$ if k is odd. Now use Rolle's Theorem repeatedly.

(b) Let $u := (x^2 - 1)^n$ as above. Show that

$$(x^2 - 1)\frac{du}{dx} = 2nxu$$

and, taking the $(n+1)$th derivatives of both sides, that $y := P_n = u^{(n)}/2^n(n!)$ satisfies *Legendre's differential equation:*

$$(x^2 - 1)\frac{d^2y}{dx^2} + 2x\frac{dy}{dx} - n(n+1)y = 0 \qquad (\forall x \in \mathbb{R}).$$

42. Show that if $f \in \mathbb{R}^{\mathbb{R}}$ is $(n + 1)$-times differentiable and $f^{(n+1)}(x) = 0$ for all $x \in \mathbb{R}$, then $f(x)$ is a *polynomial* of degree $\leq n$.

43. Let $(x_n) \in [a, b]^{\mathbb{N}}$, $x_n \neq x_m$, for $n \neq m$, and $\lim(x_n) = \xi$. Also, let $f : [a, b] \to \mathbb{R}$ be such that $f(x_n) = 0$ for all $n \in \mathbb{N}$.

(a) Show that, if f is *twice* differentiable, then $f(\xi) = f'(\xi) = f''(\xi) = 0$.

(b) Show that, if $f \in C^\infty([a, b])$, then $f^{(k)}(\xi) = 0$ for all $k \in \mathbb{N} \cup \{0\}$.

44. Let $f : I \to \mathbb{R}$ and assume that $f^{(n)}(x) = 0$ for all $x \in I$ and $f^{(k)}(x_0) = 0$ for $1 \leq k \leq n - 1$ (recall that $f^{(0)} := f$) and some $x_0 \in I$. Show that f is *constant* on I.

45 (The Newton–Raphson Process). Let $f \in \mathbb{R}^{\mathbb{R}}$ be a *strictly increasing, convex* function that is *differentiable* and $f(\zeta) = 0$. Given a fixed $x_1 > \zeta$, define $x_{n+1} := x_n - f(x_n)/f'(x_n)$ for all $n \in \mathbb{N}$. Show that $\lim(x_n) = \zeta$. *Hint:* Use Corollary 6.7.7 and Exercise 6.7.12.

46. Let $f \in C^n(I)$ and $x_0 \in I$. Suppose that, for some polynomial $p(x)$ of degree n, we have $|f(x) - p(x)| \leq c|x - x_0|^{n+1}$, for all $x \in I$ and some constant c. Show that $p(x) = P_{n,x_0}(x)$; i.e., p is the nth Taylor polynomial of f at x_0.

47. Let $\alpha \in \mathbb{R}$ and consider the function $f(x) := (1 + x)^\alpha$ on $I := (-1, \infty)$. Find the nth Taylor polynomial of f at $x_0 \in I$.

48 (Landau's Inequality). Let $f : (0, \infty) \to \mathbb{R}$ be *twice differentiable* and define $M_j := \sup\{f^{(j)}(x) : x > 0\}$, for $j = 0, 1, 2$. Show that $M_1^2 \leq 4M_0M_2$. *Hint:* Note that, by Taylor's Formula, $f'(x) = [f(x+2h) - f(x)]/(2h) - hf''(\xi)$, for some ξ between x and $x + 2h$. Deduce that $|f'(x)| \leq hM_2 + M_0/h$ for all $h > 0$ and *minimize* the right side.

49. Let f be *twice differentiable* on $(0, \infty)$ and assume that $f''(x) = O(1)$ and $f(x) = o(1)$ as $x \to \infty$. Show that $f'(x) = o(1)$ as $x \to \infty$. Show that the statement need *not* be true if f'' is *not* bounded on $(0, \infty)$.

50 (Difference Operators). Given any $f : \mathbb{R} \to \mathbb{R}$ and any $h \in \mathbb{R}$, define the *difference operators*: $\Delta_h f(x) := f(x + h) - f(x)$, and $\Delta_h^{n+1} f(x) := \Delta_h(\Delta_h^n f(x))$ for all $n \in \mathbb{N}$.

(a) Using the *binomial coefficients*, find an explicit formula for $\Delta_h^n f(x)$.
(b) Show that, if $f \in C^n(\mathbb{R})$, then

$$\Delta_h^n f(x) = h^n f^{(n)}(x + n\theta h),$$

for some $\theta \in [0, 1]$. Use this to give a definition of $f^{(n)}(x)$ that is independent of the preceding derivatives $f', f'', \ldots, f^{(n-1)}$.
(c) Let $f \in C(\mathbb{R})$. Show that f is a *polynomial* of degree $\leq n$ if and only if $\Delta_h^{n+1} f(x) = 0$ for all $x, h \in \mathbb{R}$.

51 (Littlewood). Let $f : \mathbb{R}^+ \to \mathbb{R}$ be $(n + 1)$-times differentiable, $\lim_{x \to \infty} f(x) = L \in \mathbb{R}$, and $f^{(n+1)} = O(1)$, as $x \to \infty$. Show that $f^{(n)}(x) = o(1)$ as $x \to \infty$. *Hint:* Use Problem 50.

52 (Local Extrema). Let $f : I \to \mathbb{R}$ and let $x_0 \in I$ be an *interior* point. Suppose that $f \in C^n(J)$ for some *open* interval J with $x_0 \in J \subset I$, and $f'(x_0) = f''(x_0) = \cdots = f^{(n-1)}(x_0) = 0$, but $f^{(n)}(x_0) \neq 0$.

(a) If n is *even* and $f^{(n)}(x_0) > 0$, then f has a *local minimum* at x_0.
(b) If n is *even* and $f^{(n)}(x_0) < 0$, then f has a *local maximum* at x_0.
(c) If n is *odd*, then f has *neither* a local maximum *nor* a local minimum at x_0. *Hint:* Use Taylor's Formula.

53 (The Maximum Principle). Let $f : [a, b] \to \mathbb{R}$ be continuous on $[a, b]$ and twice differentiable on (a, b). Show that, if for some constant $\alpha > 0$ we have $f''(x) = \alpha f(x)$ for all $x \in (a, b)$, then

$$|f(x)| \leq \max\{|f(a)|, |f(b)|\} \quad \forall \, x \in [a, b].$$

54 (Convex $\Rightarrow$ Locally Lipschitz). Show that any convex function $f : (a, b) \to \mathbb{R}$ is *locally Lipschitz*.

55.

(a) Let $\emptyset \neq I \subset \mathbb{R}$ be an interval and f and g be convex functions on I. Show that if g is *increasing*, then $g \circ f$ is convex.
(b) Show that if $f : I \to (0, \infty)$ is a positive function on an interval $I \neq \emptyset$ and if $\log(f)$ is convex, the so is f. Show by an example that the converse is *false*.

56. Prove the following inequality.

$$(\sin x)^{\sin x} < (\cos x)^{\cos x} \quad \forall \, x \in (0, \pi/4).$$

57. Show that, if $f \in \mathbb{R}^{\mathbb{R}}$ is differentiable, convex, and bounded, then it must be *constant*. *Hint:* Use Corollary 6.7.7.

58. Show that, if $f : I \to \mathbb{R}$ is *continuous* and satisfies $f((x + y)/2) \leq [f(x) + f(y)]/2$ for all $x, y \in I$, then for every $x_1, \ldots, x_n \in I$, we have

$$f\left(\frac{x_1 + x_2 + \cdots + x_n}{n}\right) \leq \frac{1}{n}[f(x_1) + f(x_2) + \cdots + f(x_n)]. \tag{$\dagger$}$$

Deduce the *Arithmetic–Geometric Means Inequality:*

$$\sqrt[n]{x_1 x_2 \cdots x_n} \le \frac{x_1 + \cdots + x_n}{n} \qquad (\forall x_1 \ge 0, \ldots, x_n \ge 0).$$

59. (Corollaries of Jensen's Inequality).   For $k = 1, 2, \ldots, n$, let $x_k > 0$, $0 y_k > 0$ and $\lambda_k > 0$ with $\sum_{k=1}^{n} \lambda_k = 1$. Define the following *means:*

$$M_{-\infty} = M_{-\infty}(x_1, x_2, \ldots, x_n) := \min\{x_1, x_2, \ldots, x_n\},$$

$$M_{\infty} = M_{\infty}(x_1, x_2, \ldots, x_n) := \max\{x_1, x_2, \ldots, x_n\},$$

$$M_0 = M_0(x_1, x_2, \ldots, x_n) := x_1^{\lambda_1} x_2^{\lambda_2} \cdots x_n^{\lambda_n},$$

$$M_t = M_t(x_1, x_2, \ldots, x_n) := (\lambda_1 x_1^t + \lambda_2 x_2^t + \cdots + \lambda_n x_n^t)^{1/t},$$

where $t \ne 0$. Using *Jensen's inequality* (Proposition 6.7.2), prove the following inequalities.
(Power Mean Inequality).   If $s \le t$, then we have

$$M_{-\infty} \le M_s \le M_t \le M_{\infty}.$$

(Weighted Arithmetic–Geometric Means Inequality).   We have $M_0 \le M_1$, i.e.,

$$x_1^{\lambda_1} x_2^{\lambda_2} \cdots x_n^{\lambda_n} \le \lambda_1 x_1 + \lambda_2 x_2 + \cdots + \lambda_n x_n.$$

In particular, with $\lambda_k = 1/n$ for $k = 1, \ldots, n$, we obtain the *Arithmetic–Geometric Means Inequality:*

$$\frac{1}{n} \sum_{k=1}^{n} x_k \ge (x_1 x_2 \cdots x_n)^{1/n}.$$

(Weighted Arithmetic–Harmonic Means Inequality).   We have $M_{-1} \le M_1$, i.e.,

$$\sum_{k=1}^{n} \lambda_k x_k \ge \frac{1}{\sum_{k=1}^{n} \lambda_k / x_k}.$$

In particular, with $\lambda_k = 1/n$ for $k = 1, \ldots, n$, we obtain the *Arithmetic–Harmonic Means Inequality:*

$$\frac{1}{n} \sum_{k=1}^{n} x_k \ge \frac{1}{\frac{1}{n} \sum_{k=1}^{n} 1/x_k}.$$

(Hölder's Inequality).   For any $p > 1$ and $q > 1$ with $1/p + 1/q = 1$, we have

$$\sum_{k=1}^{n} x_k y_k \le \Big(\sum_{k=1}^{n} x_k^p\Big)^{1/p} \Big(\sum_{k=1}^{n} x_k^q\Big)^{1/q}.$$

(Minkowski's Inequality).   For any $p \ge 1$, we have

$$\Big(\sum_{k=1}^{n} (x_k + y_k)^p\Big)^{1/p} \le \Big(\sum_{k=1}^{n} x_k^p\Big)^{1/p} + \Big(\sum_{k=1}^{n} y_k^p\Big)^{1/p}.$$

60. Find two *smooth, convex* functions f, $g : \mathbb{R} \to \mathbb{R}$ such that $f(x) = g(x)$ if and only if $x \in \mathbb{Z}$.

Chapter 7
The Riemann Integral

As was pointed out in the previous chapter, the second fundamental topic covered in calculus is the *Riemann integral*, the first being the *derivative*. For a (nonnegative) real-valued function of a real variable, this integral extends the notion of *area*, defined initially for *rectangles*: For a *nonnegative* constant function $f(x) := c \; \forall x \in [a, b]$, the area of the rectangle bounded by the graph of f, the x-axis, and the vertical lines $x = a$ and $x = b$ is defined to be the nonnegative number $A := (b - a)c$. This is then trivially extended to the case of *step functions* which are *piecewise constant*: Simply add the areas of the finite number of rectangles involved. This suggests the following *analytic* approach to the general case: Try to approximate the given function by step functions, find the areas corresponding to the latter functions as above, and *pass to the limit*. Our objective in this chapter is to provide a mathematically rigorous foundation for this intuitive approach. We begin with some basic definitions.

7.1 Tagged Partitions and Riemann Sums

In this section we shall state all the basic definitions and notation needed for the rest of the chapter. *Throughout the section, $[a, b]$ with $-\infty < a < b < \infty$ will be a fixed interval.*

Definition 7.1.1 (Partition, Tagged Partition, Refinement). By a *partition* of an interval $[a, b]$ we mean a *finite* sequence $\mathcal{P} := (x_k)_{k=0}^n$ of points such that

$$a = x_0 < x_1 < x_2 < \cdots < x_{n-1} < x_n = b.$$

Given a partition $\mathcal{P} := (x_k)_{k=0}^n$ of $[a, b]$ and a *finite* sequence $\tau := (t_j)_{j=1}^n$ such that $t_j \in [x_{j-1}, x_j]$, $j = 1, 2, \ldots, n$, the pair $\dot{\mathcal{P}} := (\mathcal{P}, \tau)$ is said to be a *tagged partition* of $[a, b]$. The number t_j is the *tag* of the jth subinterval $I_j := [x_{j-1}, x_j]$.

© Springer Science+Business Media New York 2014
H.H. Sohrab, *Basic Real Analysis*, DOI 10.1007/978-1-4939-1841-6_7

It is obvious that $\mathcal{P} \leftrightarrow \{x_1, x_2, \ldots, x_{n-1}\}$ is a one-to-one correspondence between the set $\mathcal{P} = \mathcal{P}([a, b])$ of all partitions of $[a, b]$ and the set $\mathcal{F} = \mathcal{F}([a, b])$ of all *finite* subsets $\{x_1, x_2, \ldots, x_{n-1}\}$ of (a, b). Using this bijection, we can (partially) order the partitions of $[a, b]$ as follows: Given two partitions $\mathcal{P} := (x_k)_{k=0}^n$ and $\mathcal{P}' := (x_i')_{i=0}^m$ of $[a, b]$, we write $\mathcal{P} \subset \mathcal{P}'$ if and only if $\{x_1, x_2, \ldots, x_{n-1}\} \subset \{x_1', x_2', \ldots, x_{m-1}'\}$. In this case, $\mathcal{P}'$ is said to be a *refinement* of $\mathcal{P}$.

Notation 7.1.2. The set of all *tagged partitions* of the interval $[a, b]$ will be denoted by $\dot{\mathcal{P}} = \dot{\mathcal{P}}([a, b])$. Also, given any partition $\mathcal{P} = (x_k)_{k=0}^n \in \mathcal{P}([a, b])$, the set of all possible sequences of *tags* associated with $\mathcal{P}$ is denoted by $\mathcal{T}(\mathcal{P}) := \{\tau = (t_j)_{j=1}^n : t_j \in [x_{j-1}, x_j], \ 1 \leq j \leq n\}$.

Remark 7.1.3. In view of the bijection $\mathcal{P} \leftrightarrow \mathcal{F}$, we shall denote by $\mathcal{P} \cup \mathcal{P}'$ the partition corresponding to the finite subset

$$\{x_1, x_2, \ldots, x_{n-1}\} \cup \{x_1', x_2', \ldots, x_{m-1}'\} \subset [a, b].$$

Note that, according to the above definition, $\mathcal{P} \cup \mathcal{P}'$ is a *common* refinement of the partitions $\mathcal{P}$ and $\mathcal{P}'$.

Definition 7.1.4 (Riemann Sum, Darboux Sum). Let $f : [a, b] \to \mathbb{R}$. Given any tagged partition $\dot{\mathcal{P}} = (\mathcal{P}, \tau)$ of $[a, b]$, with $\mathcal{P} = (x_k)_{k=0}^n$ and $\tau = (t_j)_{j=1}^n$, we define the *Riemann sum of f* corresponding to this tagged partition to be

$$S(f, \dot{\mathcal{P}}) := \sum_{j=1}^n f(t_j) \Delta x_j \qquad (\Delta x_j := x_j - x_{j-1}, \ 1 \leq j \leq n). \qquad (*)$$

If f is bounded, then the *lower* and *upper Darboux sums* of f corresponding to the partition $\mathcal{P}$ are defined to be the sums

$$L(f, \mathcal{P}) := \sum_{j=1}^n m_j \Delta x_j \quad \text{and} \quad U(f, \mathcal{P}) := \sum_{j=1}^n M_j \Delta x_j,$$

where the Δx_j are as in $(*)$ and where we have defined $m_j := \inf\{f(x) : x_{j-1} \leq x \leq x_j\}$, $M_j := \sup\{f(x) : x_{j-1} \leq x \leq x_j\}$. If we also define $m := \inf\{f(x) : a \leq x \leq b\}$ and $M := \sup\{f(x) : a \leq x \leq b\}$, then it is obvious that $m \leq m_j \leq f(t_j) \leq M_j \leq M$, $1 \leq j \leq n$. Therefore, when f is bounded, then $\forall \dot{\mathcal{P}} \in \dot{\mathcal{P}}$, we have the inequalities

$$m(b - a) \leq L(f, \mathcal{P}) \leq S(f, \dot{\mathcal{P}}) \leq U(f, \mathcal{P}) \leq M(b - a). \qquad (**)$$

The following lemma describes the effect of *refinements* on the lower and upper Darboux sums and provides an extension of the inequality $L(f, \mathcal{P}) \leq U(f, \mathcal{P})$ contained in $(**)$ above.

Lemma 7.1.5. *Let* $f : [a, b] \to \mathbb{R}$ *be bounded and let* $\mathcal{P}$, $\mathcal{P}' \in \mathcal{P}$ *be two partitions of* $[a, b]$. *If* $\mathcal{P}'$ *is a refinement of* $\mathcal{P}$, *then we have*

$$L(f, \mathcal{P}) \leq L(f, \mathcal{P}') \leq U(f, \mathcal{P}') \leq U(f, \mathcal{P}).$$

Proof. Let $\mathcal{P} = (x_k)_{k=0}^{n}$, and assume first that $\mathcal{P}' = (x_i')_{i=0}^{n+1}$ is obtained from $\mathcal{P}$ by adjoining a *single* point, say $\xi \in (a, b)$. Thus, $\{x_1', \ldots, x_n'\} = \{x_1, \ldots, x_{n-1}\} \cup \{\xi\}$, and we may assume that $\xi \in (x_{j-1}, x_j)$. Let $\mu' := \inf\{f(x) : x_{j-1} \leq x \leq \xi\}$ and $\mu'' := \inf\{f(x) : \xi \leq x \leq x_j\}$. Then, with $m_j := \inf\{f(x) : x_{j-1} \leq x \leq x_j\}$, we have $m_j \leq \min\{\mu', \mu''\}$ and hence

$$L(f, \mathcal{P}') - L(f, \mathcal{P}) = \mu'(\xi - x_{j-1}) + \mu''(x_j - \xi) - m_j(x_j - x_{j-1})$$
$$= (\mu' - m_j)(\xi - x_{j-1}) + (\mu'' - m_j)(x_j - \xi) \geq 0.$$

The proof of $U(f, \mathcal{P}') - U(f, \mathcal{P}) \leq 0$ is similar. Finally, if $\mathcal{P}'$ is obtained from $\mathcal{P}$ by adjoining p points, the above argument is repeated p times. $\square$

Corollary 7.1.6. *Let* $f : [a, b] \to \mathbb{R}$ *be bounded. Given any partitions* $\mathcal{P}$, $\mathcal{P}' \in \mathcal{P}$, *we have*

$$L(f, \mathcal{P}) \leq U(f, \mathcal{P}').$$

Proof. Let $\mathcal{P}''$ be a *common* refinement of $\mathcal{P}$ and $\mathcal{P}'$; for example, we can pick $\mathcal{P}'' := \mathcal{P} \cup \mathcal{P}'$. Then, (∗∗) and Lemma 7.1.5 imply that

$$L(f, \mathcal{P}) \leq L(f, \mathcal{P}'') \leq U(f, \mathcal{P}'') \leq U(f, \mathcal{P}').$$ $\square$

Definition 7.1.7 (Lower and Upper (Darboux) Integrals). Given a bounded function $f : [a, b] \to \mathbb{R}$, the *lower* and *upper Darboux integrals* of f, denoted by $\underline{\int} f$ and $\overline{\int} f$, respectively, are defined to be the real numbers

$$\underline{\int} f := \sup\{L(f, \mathcal{P}) : \mathcal{P} \in \mathcal{P}\}, \qquad \overline{\int} f := \inf\{U(f, \mathcal{P}) : \mathcal{P} \in \mathcal{P}\}.$$

Example 7.1.8. Define $f : [a, b] \to \mathbb{R}$ by $f(x) = 1$ if $x \in \mathbb{Q}$ and $f(x) = -1$ if $x \in \mathbb{Q}^c$. Then, for any $\mathcal{P} \in \mathcal{P}$, we have $m_j = -1$ and $M_j = 1$ for all j. (Why?) Therefore, $L(f, \mathcal{P}) = -(b - a)$ and $U(f, \mathcal{P}) = b - a$ and hence $\underline{\int} f = -(b - a)$ and $\overline{\int} f = b - a$.

Exercise 7.1.9. Let $f : [a, b] \to \mathbb{R}$ be bounded and let $\mathcal{P}_0 \in \mathcal{P}$. Show that

$$\underline{\int} f := \sup\{L(f, \mathcal{P}) : \mathcal{P} \in \mathcal{P}, \mathcal{P}_0 \subset \mathcal{P}\},$$

$$\overline{\int} f := \inf\{U(f,\mathcal{P}) : \mathcal{P} \in \boldsymbol{\mathcal{P}}, \mathcal{P}_0 \subset \mathcal{P}\}.$$

Hint: Use Lemma 7.1.5 which shows that, as we refine $\mathcal{P}$, the lower sum $L(f,\mathcal{P})$ *increases* while the upper sum $U(f,\mathcal{P})$ *decreases*.

Corollary 7.1.10. *Given any bounded function* $f : [a,b] \to \mathbb{R}$*, we have*

$$\underline{\int} f \le \overline{\int} f.$$

Proof. By Corollary 7.1.6, for any partitions $\mathcal{P}$, $\mathcal{P}' \in \boldsymbol{\mathcal{P}}$, we have

$$L(f,\mathcal{P}) \le U(f,\mathcal{P}').$$

Keeping $\mathcal{P}'$ fixed and taking the "sup" over all partitions $\mathcal{P} \in \boldsymbol{\mathcal{P}}$, this inequality implies

$$\underline{\int} f \le U(f,\mathcal{P}'),$$

from which the corollary follows if we take the "inf" over all $\mathcal{P}' \in \boldsymbol{\mathcal{P}}$. $\square$

We now define the *Riemann integral* of a function f over the interval $[a,b]$ as the "limit" of the above Riemann sums in the following way:

Definition 7.1.11 ((Riemann) Integrable, Integral). Let $f : [a,b] \to \mathbb{R}$. We say that f is *integrable* (or *Riemann integrable*) *on* $[a,b]$ if there exists a number $I(f) \in \mathbb{R}$ such that

$$(\forall \varepsilon > 0)(\exists \mathcal{P}_\varepsilon \in \boldsymbol{\mathcal{P}})(\forall \dot{\mathcal{P}} \in \dot{\boldsymbol{\mathcal{P}}})(\mathcal{P}_\varepsilon \subset \mathcal{P} \Rightarrow |S(f,\dot{\mathcal{P}}) - I(f)| < \varepsilon). \tag{†}$$

The number $I(f)$ is then called the *integral* (or *Riemann integral*) *of* f *over* $[a,b]$ and will be denoted by $\int_a^b f(x)\,dx$, or simply $\int_a^b f$. The set of *all* Riemann integrable functions on $[a,b]$ is denoted by $\mathcal{R}([a,b])$.

Definition 7.1.12 (Area Under the Graph). If f is a *nonnegative* and *integrable* function on $[a,b]$, then its (Riemann) integral $\int_a^b f$ is called the *area under the graph of* f *from* $x = a$ *to* $x = b$.

Exercise 7.1.13 (Integral of a Constant Function). Show that, if $f(x) := c \ \forall x \in [a,b]$, then $f \in \mathcal{R}([a,b])$ and $\int_a^b f = c(b-a)$. Note that, if $c > 0$, then the area under the graph is precisely the area of the rectangle with sides $b-a$ and c. *Hint:* What are the numbers $L(f,\mathcal{P})$, $U(\mathcal{P})$, $\underline{\int} f$, and $\overline{\int} f$?

Exercise 7.1.14 (Uniqueness of the Integral). Show that, if the Riemann integral of f exists, then it is *unique*.

Exercise 7.1.15. Show that, if $f \in \mathcal{R}([a, b])$ is bounded, then we have

$$\underline{\int} f \leq \int_a^b f \leq \overline{\int} f.$$

Remark 7.1.16.

1. The condition $\mathcal{P}_\varepsilon \subset \mathcal{P}$ in (†) imposes *no restriction* on the sequence τ of tags in the corresponding tagged partition $\dot{\mathcal{P}} := (\mathcal{P}, \tau)$.
2. Since the integral $I(f)$ depends only on f and $[a, b]$, the simplified notation $\int_a^b f$ is more natural than $\int_a^b f(x)\, dx$. Indeed the variable x in the latter notation is a *dummy variable,* in the sense that it can be given any other name whatsoever:

$$\int_a^b f(x)\, dx = \int_a^b f(s)\, ds = \int_a^b f(t)\, dt = \int_a^b f(u)\, du.$$

In practice, however, the presence of a dummy variable has its advantages. For instance, to write the integral $\int_a^b \sin(tx)\, dx$ without the variable x, we must first *define* the function $f_t : x \mapsto \sin(tx)$ and then write $\int_a^b f_t$.
3. The "limit" used in the above definition is of a sort we have not seen so far. Indeed, we are not dealing here with a sequence, and we have only seen the limits of functions whose domains are subsets of *metric spaces* where the notion of *distance* is available. To define this new limit, one needs the concept of *net* which is defined in most advanced courses on topology and will not be introduced here. The interested reader may, e.g., consult the book by Kelley (cf. [Kel55]).

Definition 7.1.17 (Absolutely Integrable). A function $f : [a, b] \to \mathbb{R}$ is said to be *absolutely integrable* on $[a, b]$ if $|f| \in \mathcal{R}([a, b])$.

We shall see later that *integrable* implies *absolutely integrable*. The converse is easily seen to be false:

Exercise 7.1.18 (Absolutely Integrable $\nRightarrow$ Integrable). Give an example of an absolutely integrable function that is *not* integrable.

The following proposition shows that *integrability* implies *boundedness*.

Proposition 7.1.19 (Integrable $\Rightarrow$ Bounded). *If $f : [a, b] \to \mathbb{R}$ is (Riemann) integrable over $[a, b]$, then it is bounded on $[a, b]$.*

Proof. Suppose that $f \in \mathcal{R}([a, b])$ and pick a (fixed) partition $\mathcal{P} = (x_j)_{j=0}^n \in \mathcal{P}$ such that $|S(f, \dot{\mathcal{P}}) - \int_a^b f| < 1$ for *any* sequence $\tau = (t_i)_{i=1}^n$ of tags (with $t_i \in [x_{i-1}, x_i]$). If f is *unbounded above* on $[a, b]$, then it must be unbounded above on at least one subinterval, say $I_k := [x_{k-1}, x_k]$, of the partition $\mathcal{P}$. This means that $M_k := \sup\{f(x) : x_{k-1} \leq x \leq x_k\} = +\infty$. But then, we can pick the kth tag t_k so that $f(t_k)$ is as large as we please. Keeping the remaining tags fixed, we can therefore make $|S(f, \dot{\mathcal{P}})|$ arbitrarily large, contradicting the inequality $|S(f, \dot{\mathcal{P}})| <$ $|\int_a^b f| + 1$. Therefore, f must be bounded above on $[a, b]$. A similar argument shows that f must also be bounded below on $[a, b]$. $\square$

It should be noted, however, that boundedness *does not*, in general, imply integrability:

Example 7.1.20. Consider the *Dirichlet function*

$$f(x) := \begin{cases} 1 & \text{if } x \in \mathbb{Q}, \\ 0 & \text{if } x \notin \mathbb{Q}, \end{cases}$$

which is *discontinuous* at every point of $[0, 1]$ (cf. Exercise 4.3.12). It is obvious that f is bounded on $[0, 1]$. On the other hand, since $[0, 1] \cap \mathbb{Q}$ and $[0, 1] \cap \mathbb{Q}^c$ are both *dense* in $[0, 1]$, for any partition $\mathcal{P}$ of $[0, 1]$, we can pick two sequences of tags, $\tau = (t_i)$ and $\tau' = (t_i')$, such that the t_i are all rational while the t_i' are all irrational. If $\dot{\mathcal{P}} := (\mathcal{P}, \tau)$ and $\dot{\mathcal{P}}' := (\mathcal{P}, \tau')$, then $S(f, \dot{\mathcal{P}}) = 1$ and $S(f, \dot{\mathcal{P}}') = 0$. Therefore, f is *not* integrable on $[0, 1]$.

Theorem 7.1.21 (Riemann–Darboux).   *For a bounded function $f : [a, b] \to \mathbb{R}$, the following are equivalent:*

(i) $f \in \mathcal{R}([a, b])$.

(ii) *Given any $\varepsilon > 0$ there exists a partition $\mathcal{P}_\varepsilon \in \mathcal{P}$ such that*

$$U(f, \mathcal{P}_\varepsilon) - L(f, \mathcal{P}_\varepsilon) < \varepsilon. \tag{$\ddagger$}$$

(iii)

$$\underline{\int} f = \overline{\int} f \in \mathbb{R}.$$

The common value in (iii) is, of course, the Riemann integral of f.

Proof. To prove (i) $\Rightarrow$ (ii), suppose that $f \in \mathcal{R}([a, b])$ and let $I(f) := \int_a^b f(x)\, dx$. Given $\varepsilon > 0$, we can pick $\mathcal{P}_\varepsilon = (x_k)_{k=0}^n \in \mathcal{P}$ such that $|S(f, \dot{\mathcal{P}}_\varepsilon) - I(f)| < \varepsilon/2$. By Proposition 7.1.19, f is *bounded* on $[a, b]$. In particular, $m_j := \inf\{f(x) : x_{j-1} \le x \le x_j\}$ and $M_j := \sup\{f(x) : x_{j-1} \le x \le x_j\}$ are all *finite*. Pick tags $\sigma = (s_j)_{j=1}^n$, $\tau = (t_j)_{j=1}^n \in \mathcal{T}(\mathcal{P})$ such that $f(s_j) - m_j < \varepsilon/2(b - a)$ and $M_j - f(t_j) < \varepsilon/2(b - a)$. Then we have the inequalities

$$U(f, \mathcal{P}_\varepsilon) - S(f, (\mathcal{P}_\varepsilon, \tau)) = \sum_{j=1}^n [M_j - f(t_j)](x_j - x_{j-1}) < \varepsilon/2,$$

$$S(f, (\mathcal{P}_\varepsilon, \sigma)) - L(f, \mathcal{P}_\varepsilon) = \sum_{j=1}^n [f(s_j) - m_j](x_j - x_{j-1}) < \varepsilon/2,$$

which, together with $|S(f,(\mathcal{P}_\varepsilon,\tau))-I(f)| < \varepsilon/2$ and $|S(f,(\mathcal{P}_\varepsilon,\sigma))-I(f)| < \varepsilon/2$, imply ‡. The implication (ii) $\Rightarrow$ (iii) is obvious, since $\varepsilon > 0$ is arbitrary and we have

$$\overline{\int} f - \underline{\int} f \le U(f,\mathcal{P}_\varepsilon) - L(f,\mathcal{P}_\varepsilon) < \varepsilon.$$

Finally, to prove (iii) $\Rightarrow$ (i), suppose (iii) holds and let $I(f)$ denote the common value in (iii). Then, given any $\varepsilon > 0$, we can pick partitions $\mathcal{P}'_\varepsilon$, $\mathcal{P}''_\varepsilon \in \mathcal{P}$ such that

$$I(f) - L(f,\mathcal{P}'_\varepsilon) < \varepsilon \quad \text{and} \quad U(f,\mathcal{P}''_\varepsilon) - I(f) < \varepsilon.$$

(Why?) Using the refinement $\mathcal{P}_\varepsilon := \mathcal{P}'_\varepsilon \cup \mathcal{P}''_\varepsilon$, we then obtain the inequalities

$$I(f) - L(f,\mathcal{P}_\varepsilon) < \varepsilon \quad \text{and} \quad U(f,\mathcal{P}_\varepsilon) - I(f) < \varepsilon. \qquad (*)$$

On the other hand, for any refinement $\mathcal{P} \supset \mathcal{P}_\varepsilon$, we have

$$L(f,\mathcal{P}_\varepsilon) \le L(f,\mathcal{P}) \le S(f,\dot{\mathcal{P}}) \le U(f,\mathcal{P}) \le U(f,\mathcal{P}_\varepsilon). \qquad (**)$$

Combining $(*)$ and $(**)$, we finally have

$$-\varepsilon < L(f,\mathcal{P}_\varepsilon) - I(f) \le L(f,\mathcal{P}) - I(f) \le S(f,\dot{\mathcal{P}}) - I(f)$$
$$\le U(f,\mathcal{P}) - I(f) \le U(f,\mathcal{P}_\varepsilon) - I(f) < \varepsilon,$$

i.e., $|S(f,\dot{\mathcal{P}}) - I(f)| < \varepsilon$ for any choice of tags in $\dot{\mathcal{P}}$. Thus $f \in \mathcal{R}([a,b])$ and the proof is complete.   □

We can deduce the following *sequential version* from the theorem.

Corollary 7.1.22. *A bounded function* $f : [a,b] \to \mathbb{R}$ *is Riemann integrable if and only if there exists a sequence* $\{\mathcal{P}_n : n \in \mathbb{N}\}$ *of partitions of* $[a,b]$ *such that* $\lim_{n\to\infty}(U(f,\mathcal{P}_n)-L(f,\mathcal{P}_n)) = 0$. *In this case, we have* $\lim_{n\to\infty} S(f,\dot{\mathcal{P}}_n) = \int_a^b f$, *regardless of the choices of tags in the* $\dot{\mathcal{P}}_n$.

Remark 7.1.23.

1. The equivalence (i) $\Leftrightarrow$ (ii) in the above theorem is known as **Riemann's Lemma** and is basically *Cauchy's Criterion* for integrability. The equivalence (i) $\Leftrightarrow$ (iii) (or Theorem 7.1.26 below) is usually referred to as **Darboux's Theorem**.
2. We may assume (by refining, if necessary) that the partition $\mathcal{P}_\varepsilon$ contains any *prescribed* point (or any *finite* set of such points) in $[a,b]$.

Exercise 7.1.24 (Translations & Reflections). Let $f \in \mathcal{R}([a,b])$ and let $c \in \mathbb{R}$ be arbitrary. Define $f_c(x) := f(x-c)$ and $\check{f}(x) := f(-x)$, for all $x \in [a,b]$. Show that $f_c \in \mathcal{R}([a+c,b+c])$, $\check{f} \in \mathcal{R}([-b,-a])$ and that we have

$$\int_a^b f(x)\, dx = \int_{a+c}^{b+c} f(x-c)\, dx, \quad \text{and}$$

$$\int_a^b f(x)\, dx = \int_{-b}^{-a} f(-x)\, dx.$$

Deduce that $x \mapsto f(c-x)$ is integrable on $[c-b, c-a]$ for all $c \in \mathbb{R}$, and

$$\int_a^b f(x)\, dx = \int_{c-b}^{c-a} f(c-x)\, dx.$$

Hint: Note that $(x_k)_{k=0}^n$ is a partition of $[a,b]$ if and only if $(x_k + c)_{k=0}^n$ (resp., $(-x_k)_{k=0}^n$) is a partition of $[a+c, b+c]$ (resp., $[-b,-a]$).

There is another definition of the Riemann integral which is equivalent to Definition 7.1.11. This version is the one usually used in calculus courses and is convenient in practice for computing and approximating the integral. To introduce it, we first need a definition.

Definition 7.1.25 (Mesh (or Norm) of a Partition). For each partition $\mathcal{P} = (x_k)_{k=0}^n \in \mathcal{P}([a,b])$, the number

$$\|\mathcal{P}\| := \max\{x_j - x_{j-1} : 1 \le j \le n\} = \max\{\Delta x_j : 1 \le j \le n\}$$

is called the *mesh* (or *norm*) of the partition $\mathcal{P}$.

Theorem 7.1.26. *Let* $f : [a,b] \to \mathbb{R}$. *Then* $f \in \mathcal{R}([a,b])$ *if and only if there exists a number* $I(f) \in \mathbb{R}$ *such that*

$$(\forall \varepsilon > 0)(\exists \delta > 0)(\forall \dot{\mathcal{P}} \in \dot{\mathcal{P}})(\|\mathcal{P}\| < \delta \Rightarrow |S(f, \dot{\mathcal{P}}) - I(f)| < \varepsilon). \qquad (\dagger)$$

Proof. Suppose f satisfies the condition of the theorem, $\varepsilon > 0$ is given, and δ is as in ($\dagger$) above. Pick a partition $\mathcal{P}_\varepsilon$ such that $\|\mathcal{P}_\varepsilon\| < \delta$. Then we have $|S(f, \dot{\mathcal{P}}_\varepsilon) - I(f)| < \varepsilon$. Now, given any refinement $\mathcal{P} \supset \mathcal{P}_\varepsilon$, we obviously have $\|\mathcal{P}\| \le \|\mathcal{P}_\varepsilon\| < \delta$. (Why?) Therefore, by ($\dagger$), we also have $|S(f, \dot{\mathcal{P}}) - I(f)| < \varepsilon$ and hence $f \in \mathcal{R}([a,b])$. To prove the converse, suppose that $f \in \mathcal{R}([a,b])$ and let $\varepsilon > 0$ be given. Pick a partition $\mathcal{P}'_\varepsilon = (x'_j)_{j=0}^n$ such that

(i) $U(f, \mathcal{P}'_\varepsilon) - L(f, \mathcal{P}'_\varepsilon) = \sum_{j=1}^n (M'_j - m'_j)\Delta x'_j < \dfrac{\varepsilon}{2}$,

where, as in Definition 7.1.4, m'_j and M'_j are the "inf" and "sup" of f on $[x'_{j-1}, x'_j]$, respectively, and $\Delta x'_j := x'_j - x'_{j-1}$. Now, given *any* partition $\mathcal{P} = (x_i)_{i=0}^k$, we have a sum similar to (i) given by

(ii) $U(f, \mathcal{P}) - L(f, \mathcal{P}) = \sum_{i=1}^{k} (M_i - m_i) \Delta x_i$,

where m_i, M_i, and Δx_i are, once again, as in Definition 7.1.4. For each subinterval $[x_{i-1}, x_i]$ of $\mathcal{P}$, let us call it *type 1* if $x_{i-1} < x'_j < x_i$, for some point x'_j $(1 \le j \le n - 1)$ of $\mathcal{P}'_\varepsilon$, and *type 2* if $[x_{i-1}, x_i] \subset [x'_{j-1}, x'_j]$, for some $1 \le j \le n$. Note that these are the only types possible and, since a and b belong to all partitions, we have at most $n - 1$ type 1 subintervals. The sum on the right side of (ii) can then be written as $\sum_1 + \sum_2$, where $\sum_1$ (resp., $\sum_2$) is the contribution of type 1 (resp., type 2) subintervals. Since $f \in \mathcal{R}([a, b])$, Proposition 7.1.19 implies that f is *bounded*. Thus, $m := \inf\{f(x) : a \le x \le b\}$ and $M := \sup\{f(x) : a \le x \le b\}$ are both *finite*. By the above remarks, we have

(iii) $\sum_1 \le (n - 1)(M - m)\|\mathcal{P}\|$.

On the other hand, if $[x_{i-1}, x_i] \subset [x'_{j-1}, x'_j]$, then we obviously have $(M_i - m_i)\Delta x_i \le (M'_j - m'_j)\Delta x'_j$. Therefore, by (i),

(iv) $\sum_2 \le \sum_{j=1}^{n} (M'_j - m'_j)\Delta x'_j = U(f, \mathcal{P}'_\varepsilon) - L(f, \mathcal{P}'_\varepsilon) < \dfrac{\varepsilon}{2}$.

Combining (ii), (iii), and (iv), we have

(v) $U(f, \mathcal{P}) - L(f, \mathcal{P}) = \sum_{i=1}^{k} (M_i - m_i) \Delta x_i < \dfrac{\varepsilon}{2} + (n - 1)(M - m)\|\mathcal{P}\|$.

Let $\delta := \varepsilon/[2n(M - m)]$, where we assume $M > m$; the case $M = m$ is trivial. (Why?) If $\|\mathcal{P}\| < \delta$, then (v) implies that $U(f, \mathcal{P}) - L(f, \mathcal{P}) < \varepsilon$. Since $L(f, \mathcal{P}) \le \int_a^b f \le U(f, \mathcal{P})$ and $L(f, \mathcal{P}) \le S(f, \dot{\mathcal{P}}) \le U(f, \mathcal{P})$ for any sequence of tags in $\dot{\mathcal{P}}$, the proof is complete. $\square$

The following *sequential criterion* for integrability can be deduced from the theorem and Corollary 7.1.22:

Corollary 7.1.27. *A bounded function* $f : [a, b] \to \mathbb{R}$ *is integrable if and only if for every sequence* $(\dot{\mathcal{P}}_n)$ *of tagged partitions of* $[a, b]$ *such that* $\lim_{n \to \infty} \|\mathcal{P}_n\| = 0$, *the sequence* $(S(f, \dot{\mathcal{P}}_n))$ *is convergent; or, equivalently,*

$$\lim_{n \to \infty} (U(f, \mathcal{P}_n) - L(f, \mathcal{P}_n)) = 0.$$

It is then clear that we have

$$\int_a^b f = \lim_{n \to \infty} L(f, \mathcal{P}_n) = \lim_{n \to \infty} S(f, \dot{\mathcal{P}}_n) = \lim_{n \to \infty} U(f, \mathcal{P}_n),$$

7.2   Some Classes of Integrable Functions

Our goal in this section is to introduce the most commonly encountered classes of integrable functions. In particular, we shall see that, for any interval $[a, b] \subset \mathbb{R}$, the classes of *continuous* and *monotone* functions on $[a, b]$ are both subclasses of $\mathcal{R}([a, b])$. In fact, the most *natural* class of functions to study here is the class of *regulated functions* to be defined below. Although this class does contain the monotone and continuous ones, we prefer to treat the latter separately, due to their particular importance. Let us begin with the following.

Definition 7.2.1 (Content Zero). We say that a set $S \subset \mathbb{R}$ has *content zero* (or *is of content zero*) if for every $\varepsilon > 0$ there is a *finite* sequence of *pairwise disjoint* open intervals $\{(a_i, b_i) : 1 \le i \le p\}$ such that $S \subset \bigcup_{i=1}^{p}(a_i, b_i)$ and $\sum_{i=1}^{p}(b_i - a_i) < \varepsilon$.

Remark 7.2.2. In the above definition, the requirement that the intervals be *pairwise disjoint* may be dropped. Indeed, if the intervals overlap, then a pair of overlapping open intervals may be replaced by a single interval (namely, their union). In fact, the intervals need *not* even be open, as one may replace them by slightly larger open ones.

Exercise 7.2.3. For a set $S \subset \mathbb{R}$, show the following:

1. If S is *finite*, then it has content zero.
2. An *infinite* set may have content zero. *Hint:* Look at a set with a limit point.
3. If S has content zero and $R \subset S$, then R has content zero.
4. If $\{S_j : 1 \le j \le k\}$ is a *finite* collection of sets of content zero, then $S := \bigcup_{j=1}^{k} S_j$ has content zero. Give an example of an (even countably) *infinite* union of sets of content zero which is *not* of content zero.
5. If S has content zero, then it has empty interior, i.e., $S^\circ = \emptyset$.
6. For any $-\infty \le \alpha < \beta \le +\infty$, the set $[\alpha, \beta] \cap \mathbb{Q}$ is *not* of content zero.

Theorem 7.2.4 (Zero-Content Discontinuity $\Rightarrow$ Integrable). *Let f be a bounded function on $[a, b]$ and let $D := \{x \in [a, b] : f \text{ is discontinuous at } x\}$. If D has content zero, then $f \in \mathcal{R}([a, b])$.*

Proof. Put $K := b - a + M - m$, with $m := \inf\{f(x) : x \in [a, b]\}$ and $M := \sup\{f(x) : x \in [a, b]\}$. Let $\varepsilon > 0$ be given and suppose that $D \subset \bigcup_{i=1}^{p}(a_i, b_i)$, where the (a_i, b_i) are pairwise disjoint and $\sum_{i=1}^{p}(b_i - a_i) < \varepsilon/K$. The set $C := [a, b] \setminus \bigcup_{i=1}^{p}(a_i, b_i)$ is a closed and hence *compact* subset of $[a, b]$. In fact, C is a finite union of *closed intervals*. (Why?) Since f is continuous on C, by Theorem 4.6.4, it is therefore *uniformly continuous* on C. In particular, we can find $\delta > 0$ such that $|f(x) - f(x')| < \varepsilon/K$ if x and x' belong to a closed subinterval of C and $|x - x'| < \delta$. Using suitable partition points in each of the closed subintervals of C, we can construct a partition $\mathcal{P} = (x_k)_{k=0}^{n}$ of $[a, b]$ such that for each $1 \le j \le n$, either $[x_{j-1}, x_j] \subset C$ and $x_j - x_{j-1} < \delta$, or $[x_{j-1}, x_j] \subset [a_i, b_i]$ for some i, $1 \le i \le p$. Let G_1 (resp., G_2) denote the set of all j for which the first (resp., second) alternative holds. Now, with $m_j := \inf\{f(x) : x \in [x_{j-1}, x_j]\}$

and $M_j := \sup\{f(x) : x \in [x_{j-1}, x_j]\}$, we have $M_j - m_j < \varepsilon/K \ \forall j \in G_1$ and $\sum_{j \in G_2}(x_j - x_{j-1}) \le \sum_{i=1}^{p}(b_i - a_i) < \varepsilon/K$. Thus, with $\Delta x_j := x_j - x_{j-1}$, we have

$$U(f, \mathcal{P}) - L(f, \mathcal{P}) = \sum_{j \in G_1}(M_j - m_j)\Delta x_j + \sum_{j \in G_2}(M_j - m_j)\Delta x_j$$

$$\le \sum_{j \in G_1} \frac{\varepsilon}{K}\Delta x_j + \sum_{j \in G_2}(M - m)\Delta x_j$$

$$< (b - a)\frac{\varepsilon}{K} + (M - m)\frac{\varepsilon}{K} = \varepsilon.$$

The theorem now follows from Riemann's Lemma (cf. Theorem 7.1.21).   □

Before stating the immediate consequences of this theorem, let us recall a few definitions.

Definition 7.2.5. Let $f : [a, b] \to \mathbb{R}$.

(a) **(Step Function)** We say that f is a *step function*, and we write $f \in Step([a, b])$, if there is a partition $\mathcal{P} = (x_k)_{k=0}^{n}$ of $[a, b]$ and a finite sequence $(c_j)_{j=1}^{n}$ such that $f(x) = c_j \ \forall x \in (x_{j-1}, x_j)$, $1 \le j \le n$.

(b) **(Piecewise Linear)** We say that f is a *piecewise linear function*, and we write $f \in PL([a, b])$, if there is a partition $\mathcal{P} = (x_k)_{k=0}^{n}$ of $[a, b]$ and two finite sequences $(\alpha_j)_{j=1}^{n}$ and $(\beta_j)_{j=1}^{n}$ such that $f(x) = \alpha_j x + \beta_j \ \forall x \in (x_{j-1}, x_j)$, $1 \le j \le n$.

(c) **(Piecewise Continuous)** We say that f is *piecewise continuous*, and we write $f \in PC([a, b])$, if there is a partition $\mathcal{P} = (x_k)_{k=0}^{n}$ of $[a, b]$ such that f is continuous on (x_{j-1}, x_j), for $1 \le j \le n$ and that $f(x_{j-1} + 0)$ and $f(x_j - 0)$ are both *finite* for $1 \le j \le n$. In particular, a *continuous* function on $[a, b]$ is clearly piecewise continuous; i.e., we have $C([a, b]) \subset PC([a, b])$.

It is obvious that we have the *proper* inclusions

$$Step([a, b]) \subset PL([a, b]) \subset PC([a, b]).$$

Remark 7.2.6. The importance of *step functions* in the study of the Riemann integral comes from the fact that such functions arise naturally from the definitions of Darboux and Riemann sums. Indeed, given any bounded $f : [a, b] \to \mathbb{R}$ and any tagged partition $\dot{\mathcal{P}} = ((x_k)_{k=0}^{n}, (t_j)_{j=1}^{n})$ of $[a, b]$, it is quite natural to introduce the following step functions: $\underline{\phi}(x) := m_j$, $\phi(x) := f(t_j)$, and $\overline{\phi}(x) := M_j$, $\forall x \in (x_{j-1}, x_j)$, where m_j and M_j have their usual meaning.

Corollary 7.2.7. *Let $f : [a, b] \to \mathbb{R}$. If the set D of discontinuities of f is finite, then $f \in \mathcal{R}([a, b])$. In particular, we have*

$$Step([a, b]) \subset PL([a, b]) \subset PC([a, b]) \subset \mathcal{R}([a, b]),$$

and $C([a, b]) \subset PC([a, b]) \subset \mathcal{R}([a, b])$.

Proof. This follows at once from Theorem 7.2.4 and Exercise 7.2.3.   □

Due to the importance of the inclusion $C([a,b]) \subset \mathcal{R}([a,b])$, we give an independent proof of it:

Theorem 7.2.8 (Continuous $\Rightarrow$ Integrable). *If $f : [a,b] \to \mathbb{R}$ is continuous on $[a,b]$, then it is integrable on $[a,b]$. Furthermore, for any sequence $(\mathcal{P}_n) \in \mathcal{P}^{\mathbb{N}}$ of partitions of $[a,b]$ such that $\lim_{n\to\infty} \|\mathcal{P}_n\| = 0$, we have $\int_a^b f(x)\, dx = \lim_{n\to\infty} S(f, \dot{\mathcal{P}}_n)$, regardless of the choice of tags in the $\dot{\mathcal{P}}_n$.*

Proof. Let $\varepsilon > 0$ be given. Since f is continuous on the *compact* set $[a,b]$, it is *uniformly continuous* (cf. Theorem 4.6.4). We can therefore pick $\delta > 0$ such that $|x - x'| < \delta \Rightarrow |f(x) - f(x')| < \varepsilon/(b-a)$. In particular, if $\mathcal{P}$ is any partition of $[a,b]$ with $\|\mathcal{P}\| < \delta$ and if m_j and M_j are as in Definition 7.1.4, then, for some $\alpha_j,\ \beta_j \in [x_{j-1}, x_j]$, we have $M_j - m_j = f(\beta_j) - f(\alpha_j) < \varepsilon/(b-a)$. Therefore,

$$U(f, \mathcal{P}) - L(f, \mathcal{P}) = \sum_{j=1}^{n} (M_j - m_j)\Delta x_j$$

$$< \frac{\varepsilon}{b-a} \sum_{j=1}^{n} \Delta x_j = \frac{\varepsilon}{b-a}(b-a) = \varepsilon.$$

This proves that $f \in \mathcal{R}([a,b])$. The last statement follows from Corollary 7.1.27 and the proof is complete.   □

Finding the *exact* value of the integral as a limit of Riemann (or Darboux) sums can only be achieved in a handful of cases and, even then, may require considerable ingenuity. The following example gives one of these rare cases and is due to the famous French mathematician *Pierre de Fermat*.

Example 7.2.9 (Power Rule). The power function $f(x) := x^p$, with any $p \neq -1$, is integrable on $[a,b]$ for any $0 < a < b$ and we have

$$\int_a^b x^p\, dx = \frac{b^{p+1} - a^{p+1}}{p+1}.$$

Indeed, as we shall see later, $f(x) = e^{p \log x}$ is *continuous* (even C^∞) on $(0, \infty)$. Therefore, by the above theorem, $\int_a^b f(x)\, dx$ exists for any $0 < a < b$. To compute it, let $n \in \mathbb{N}$ be fixed and let $x_k := a\delta^k$, with $\delta = \delta(n) := (b/a)^{1/n}$ and $0 \leq k \leq n$. The sequence $\mathcal{P} := (x_k)_{k=0}^n$ is then a partition of $[a,b]$. Note that $\Delta x_j := x_j - x_{j-1} = a\delta^j - a\delta^{j-1} = a(\delta - 1)\delta^{j-1}$, so that the corresponding subintervals *do not* have equal length. For $p > 0$, the upper sum can be computed as follows:

$$U(f, \mathcal{P}) = \sum_{j=1}^{n} f(x_j)\Delta x_j$$

$$= \sum_{j=1}^{n} a^p \delta^{jp} a(\delta - 1)\delta^{j-1}$$

$$= a^{p+1} \left(\frac{\delta - 1}{\delta} \right) \sum_{j=1}^{n} \delta^{j(p+1)}$$

$$= a^{p+1} (\delta - 1) \frac{\delta^{n(p+1)} - 1}{\delta^{p+1} - 1} \delta^p \quad \text{(note that } p \neq -1)$$

$$= a^{p+1} \frac{(b/a)^{p+1} - 1}{1 + \delta + \cdots + \delta^p} \delta^p.$$

Since $\lim_{n \to \infty} \delta = 1$, we have

$$\int_a^b f(x)\,dx = \lim_{n \to \infty} U(f, \mathcal{P}) = a^{p+1} \frac{(b/a)^{p+1} - 1}{p + 1} = \frac{b^{p+1} - a^{p+1}}{p + 1}.$$

All the functions in Corollary 7.2.7 have *finite* discontinuity sets. The next theorem shows that all *monotone* functions are (Riemann) integrable. Recall that (cf. Theorem 4.4.7) the discontinuity set of a monotone function is *countable*, and hence, possibly infinite.

Theorem 7.2.10 (Monotone $\Rightarrow$ Integrable). *If $f : [a,b] \to \mathbb{R}$ is a monotone (i.e., increasing or decreasing) function, then $f \in \mathcal{R}([a,b])$.*

Proof. Assume f is *increasing*; the other case is similar (or we can use $-f$). Given any $\varepsilon > 0$, pick $n \in \mathbb{N}$ such that $[f(b) - f(a)](b-a)/n < \varepsilon$. Consider the partition $\mathcal{P}_n = (x_k)_{k=0}^n$ with $x_k := a + k(b - a)/n$. In particular, $\|\mathcal{P}_n\| = (b - a)/n = \Delta x_j \ \forall j$. Let m_j and M_j be the "inf" and "sup" of f on $[x_{j-1}, x_j]$, respectively. Then we have $m_j = f(x_{j-1})$ and $M_j = f(x_j)$ for $1 \leq j \leq n$. Therefore,

$$U(f, \mathcal{P}_n) - L(f, \mathcal{P}_n) = \frac{b - a}{n} \sum_{j=1}^{n} [f(x_j) - f(x_{j-1})]$$

$$= \frac{b - a}{n} [f(b) - f(a)] < \varepsilon.$$

Since $\varepsilon > 0$ was arbitrary, the theorem follows at once from Corollary 7.1.27. $\qquad \square$

As the reader has certainly observed, we are trying to find the *largest* class of integrable functions. Our next goal will be to introduce a class of functions containing, simultaneously, all the classes introduced so far. Here it is:

Definition 7.2.11 (Regulated Function). Let $I \subset \mathbb{R}$ be a nonempty interval and let $f : I \to \mathbb{R}$. We say that f is *regulated* if the one-sided limits

$$f(x - 0) := \lim_{h \to 0-} f(x + h), \qquad f(x + 0) := \lim_{h \to 0+} f(x + h)$$

exist for *every* $x \in I°$. If x is the *left* (resp., *right*) endpoint of I, then we only require the existence of the *right* (resp. *left*) limit. The set of all regulated functions on I will be denoted by $Reg(I)$.

Example 7.2.12. Let $f : I \rightarrow \mathbb{R}$. If f is a *monotone* function, then (cf. Theorem 3.4.36) f has one-sided limits at every point of I and is therefore regulated. It is also obvious that *step functions, piecewise linear functions* and *piecewise continuous functions* have one-sided limits at every point of their domains. Therefore, all these functions are regulated. In particular, *continuous functions* are regulated.

Exercise 7.2.13. Let $f : I \rightarrow \mathbb{R}$ be a *regulated* function. Show that the set $D :=$ $\{x \in I : f$ is discontinuous at $x\}$ is *countable*.

The following theorem gives a complete characterization of regulated functions on bounded closed intervals in terms of step functions, showing once again the crucial role played by step functions.

Theorem 7.2.14. *Let* $f : [a, b] \rightarrow \mathbb{R}$. *Then* f *is regulated if and only if it can be uniformly approximated by step functions; i.e., given any* $\varepsilon > 0$ *there is a step function* $g_\varepsilon \in Step([a, b])$ *such that* $|f(x) - g_\varepsilon(x)| < \varepsilon \ \forall x \in [a, b]$.

Proof. If the condition of the theorem is satisfied and if $\varepsilon > 0$ is given, then we can pick $g_\varepsilon \in Step([a, b])$ such that $|f(x) - g_\varepsilon(x)| < \varepsilon/3 \ \forall x \in [a, b]$. Now, for each $x_0 \in [a, b]$, we can pick $\delta > 0$ such that $|g_\varepsilon(s) - g_\varepsilon(t)| < \varepsilon/3$ if $s, t \in (x_0 - \delta, x_0)$ or $s, t \in (x_0, x_0 + \delta)$. Thus, for such points s and t, we have

$$|f(s) - f(t)| \le |f(s) - g_\varepsilon(s)| + |g_\varepsilon(s) - g_\varepsilon(t)| + |g_\varepsilon(t) - f(t)| < 3\frac{\varepsilon}{3} = \varepsilon.$$

This shows (by Cauchy's Criterion) that $f(x_0 - 0)$ and $f(x_0 + 0)$ both exist. Conversely, suppose that f is regulated. Then, for each $x \in [a, b]$, we can find $\delta_x > 0$ such that whenever $s, t \in (x - \delta_x, x) \cap [a, b]$ (or $s, t \in (x, x + \delta_x) \cap [a, b]$), we have $|f(s) - f(t)| < \varepsilon$. (Why?) Since $[a, b]$ is *compact*, we can find $x_1, x_2, \ldots, x_n \in [a, b]$ such that, with $\delta_j := \delta_{x_j}$ and $B_j := B_{\delta_j}(x_j) :=$ $(x_j - \delta_j, x_j + \delta_j)$, we have $[a, b] \subset \bigcup_{j=1}^n B_j$. Let $z_0 < z_1 < \cdots < z_m$ denote all the points $a, b, x_j, x_j - \delta_j, x_j + \delta_j$ $(1 \le j \le n)$ that belong to $[a, b]$ in *increasing order*. Now, for each $1 \le j \le n$, we have $z_{j-1} \in B_i$ for some i. It follows that either $z_j \in B_i$ or $z_j = x_i + \delta_i$. Therefore, if $s, t \in (z_{j-1}, z_j)$, then $|f(s) - f(t)| < \varepsilon$. Let $g_\varepsilon \in Step([a, b])$ be defined by $g_\varepsilon(z_j) := f(z_j)$ and $g_\varepsilon(x) := f((z_{j-1} + z_j)/2)$ $\forall x \in (z_{j-1}, z_j)$. Then we obviously have $|f(x) - g_\varepsilon(x)| < \varepsilon \ \forall x \in [a, b]$. $\quad\square$

Remark 7.2.15. As we saw in Chap. 5, the set $B([a, b]) := B([a, b], \mathbb{R})$ of all *bounded* real-valued functions on $[a, b]$ is a *metric space* with the *uniform metric:* $d_\infty(f, g) := \sup\{|f(x) - g(x)| : x \in [a, b]\}$. We shall see later that this space is *complete*, i.e., every *Cauchy sequence* in it is convergent. It is easy to see that the space $Reg([a, b])$ of all regulated functions on $[a, b]$ is a *closed* (hence complete) subspace of $B([a, b])$. The above theorem can therefore be stated as follows: *the subspace $Step([a, b])$ of all step functions on $[a, b]$ is dense in $Reg([a, b])$.*

Before stating the last theorem of the section, let us prove a general lemma which is of independent interest and from which the theorem will follow at once. The lemma states that $\mathcal{R}([a, b])$ is a *closed* subspace of $B([a, b], \mathbb{R})$ with the metric defined in the above remark.

Lemma 7.2.16 (Closure Under Uniform Limits). *Suppose that $f : [a, b] \to \mathbb{R}$ can be uniformly approximated by integrable functions, i.e., that we have*

$$(\forall \delta > 0)(\exists g_\delta \in \mathcal{R}([a, b]))(\forall x \in [a, b])(|f(x) - g_\delta(x)| < \delta).$$

Then $f \in \mathcal{R}([a, b])$ and we have

$$\int_a^b f(x)\, dx = \lim_{\delta \to 0+} \int_a^b g_\delta(x)\, dx.$$

Proof. The proof is a standard "$\varepsilon/3$-argument." Given $\varepsilon > 0$, let $\delta := \varepsilon/3(b - a)$. We can pick $g_\delta \in \mathcal{R}([a, b])$ such that $|f(x) - g_\delta(x)| < \delta$ $(\forall x \in [a, b])$. Keeping this δ fixed, we pick $\mathcal{P} \in \mathcal{P}([a, b])$ such that

$$U(g_\delta, \mathcal{P}) - L(g_\delta, \mathcal{P}) < \varepsilon/3. \tag{$*$}$$

Now $f(x) < g_\delta(x) + \delta$ $\forall x \in [a, b]$ implies

$$U(f, \mathcal{P}) \leq U(g_\delta, \mathcal{P}) + (b - a)\delta = U(g_\delta, \mathcal{P}) + \varepsilon/3. \tag{$**$}$$

Similarly, $g_\delta(x) - \delta < f(x)$ $\forall x \in [a, b]$ implies

$$L(f, \mathcal{P}) \geq L(g_\delta, \mathcal{P}) - (b - a)\delta = L(g_\delta, \mathcal{P}) - \varepsilon/3. \tag{$***$}$$

Combining $(*)$, $(**)$, and $(***)$, we obtain

$$U(f, \mathcal{P}) - L(f, \mathcal{P}) = [U(f, \mathcal{P}) - U(g_\delta, \mathcal{P})] + [U(g_\delta, \mathcal{P}) - L(g_\delta, \mathcal{P})]$$

$$+ [L(g_\delta, \mathcal{P}) - L(f, \mathcal{P})] < 3\frac{\varepsilon}{3} = \varepsilon,$$

which proves that $f \in \mathcal{R}([a, b])$. But then, as we shall see shortly (cf. Theorem 7.4.9 and Corollary 7.4.14 below), we have $f - g_\delta \in \mathcal{R}([a, b])$ and

$$\left| \int_a^b f(x)\, dx - \int_a^b g_\delta(x)\, dx \right| \leq \int_a^b |f(x) - g_\delta(x)|\, dx \leq (b - a)\delta,$$

from which the last assertion follows. $\square$

We are now ready to show that the class of Riemann integrable functions contains the *regulated functions:*

Theorem 7.2.17 (Regulated $\Rightarrow$ Integrable). *We have*

$$Reg\,([a,b]) \subset \mathcal{R}([a,b]).$$

In other words, regulated functions on $[a,b]$ are integrable on $[a,b]$.

Proof. Since step functions are integrable and regulated functions can be uniformly approximated by step functions (cf. Theorem 7.2.14), the theorem follows at once from the above lemma. $\square$

Remark 7.2.18. There is something unsatisfactory about all the above existence theorems, namely, they all give *sufficient* conditions for the existence of the Riemann integral. None of them gives a *necessary and sufficient* condition. Also, all the functions we have considered so far have *countable* discontinuity sets. We may be tempted to conjecture that the latter condition is also *necessary.* In fact, it is not! It turns out that the ideal theorem we seek involves deeper ideas. The next section will be devoted to introducing these ideas and proving the celebrated *Lebesgue's Integrability Criterion.*

7.3   Sets of Measure Zero and Lebesgue's Integrability Criterion

In the preceding section, we defined what is meant by a set of *content zero.* Recall that a set has content zero if it can be covered by a *finite* collection of intervals of *total length* less than any prescribed positive number. In this section, we will introduce, for subsets of $\mathbb{R}$, the concept of *measure zero.* What we shall do is to relax the restriction that the collection of intervals covering the set be *finite.* It turns out that this new concept is much more useful in analysis and provides the right tool for our ideal existence theorem. Let us begin by recalling a definition:

Definition 7.3.1 (Length of an Interval). Let $I \subset \mathbb{R}$ be an interval. If I is *bounded* with endpoints $a \leq b$, then the *length* of I is defined to be the nonnegative number $\lambda(I) := b - a$. Thus,

$$\lambda((a,b)) = \lambda([a,b)) = \lambda((a,b]) = \lambda([a,b]) = b - a.$$

In particular, $\lambda([a,a]) = \lambda(\{a\}) = 0$. Also, since $(a,a) = [a,a) = (a,a] = \emptyset$, $\forall a \in \mathbb{R}$, we define $\lambda(\emptyset) := 0$. If I is *unbounded,* then we define $\lambda(I) := \infty$.

Definition 7.3.2 (Measure Zero). A set $S \subset \mathbb{R}$ is said to have *measure zero* (or to *be of measure zero*) if for every $\varepsilon > 0$ there is a *sequence* of *bounded open* intervals $I_1, I_2, I_3, \ldots$ such that

(i) $S \subset \bigcup_{n=1}^{\infty} I_n$, and
(ii) $\sum_{n=1}^{\infty} \lambda(I_n) < \varepsilon$.

Remark 7.3.3. Note that the requirement that the intervals $\{I_n : n \in \mathbb{N}\}$ be *open* is not necessary. In fact, the I_n may be closed or half open. Indeed, if each I_n is replaced by an open interval $J_n \supset I_n$ such that $\lambda(J_n) = \lambda(I_n) + \varepsilon/2^n$, then $\sum_{n=1}^{\infty} \lambda(J_n) < 2\varepsilon$.

Definition 7.3.4 (Almost Everywhere, Almost All). Let $S \subset \mathbb{R}$. Suppose that $P(x)$ is a *proposition* (or *property*), for each $x \in S$. We say that $P(x)$ holds *almost everywhere* (abbreviated *a.e.*) or for *almost all $x \in S$* (abbreviated *a.a. $x \in S$*), if the set $\{x : x \in S$ and P(x) does *not* hold$\}$ has *measure zero*.

Proposition 7.3.5. *The following statements are true.*

(a) *A subset of a set of measure zero has measure zero.*
(b) *If S_n has measure zero for all $n \in \mathbb{N}$ and if $S := \bigcup_{n=1}^{\infty} S_n$, then S has measure zero.*
(c) *A countable set (finite or infinite) has measure zero.*
(d) *A set of content zero has measure zero. The converse is not true.*

Proof. Part (a) is obvious since any cover of a set is also a cover of each of its subsets. To prove (b), let $\varepsilon > 0$. For each n, we can find intervals I_{nk} such that $S_n \subset \bigcup_{k=1}^{\infty} I_{nk}$ and $\sum_{k=1}^{\infty} \lambda(I_{nk}) < \varepsilon/2^n$. Now, the collection $\{I_{nk} : n, \ k \in \mathbb{N}\}$ obviously covers S and we have

$$\sum_{n=1}^{\infty} \sum_{k=1}^{\infty} \lambda(I_{nk}) < \sum_{n=1}^{\infty} \frac{\varepsilon}{2^n} = \varepsilon.$$

Next, since a set with a *single* element has measure zero (why?), part (c) follows from (b). Finally, the first statement in (d) follows directly from the definitions. For the second one, note that $\mathbb{Q}$ has measure zero (by (c)) but *not* content zero (cf. Exercise 7.2.3). The proof is now complete. $\square$

Remark 7.3.6. As was pointed out above, the set $\mathbb{Q}$ of rational numbers has measure zero. The same is of course true of $[0, 1] \cap \mathbb{Q}$. On the other hand, the set $[0, 1] \cap \mathbb{Q}^c$ of the irrationals in $[0, 1]$ *does not* have measure zero; otherwise, $[0, 1]$ would have measure zero. This, however, is absurd, as the following proposition shows.

Proposition 7.3.7. *Let I be a* bounded *interval with endpoints $a < b$ and let $\{I_n : n \in \mathbb{N}\}$ be a sequence of open intervals covering I; i.e., $I \subset \bigcup_{n=1}^{\infty} I_n$. Then*

$$\lambda(I) \le \sum_{n=1}^{\infty} \lambda(I_n). \tag{$*$}$$

Proof. We may (and do) assume that the I_n are all *bounded*, since $(*)$ is obvious otherwise. Let $0 < \varepsilon < (b - a)/2$ and let $J := [a + \varepsilon/2, b - \varepsilon/2]$. Since J is *compact*, we can cover it with a *finite* number of I_n's. One of them, say (a_1, b_1), must contain $a + \varepsilon/2$. If $b_1 \leq b - \varepsilon/2$, then $b_1 \notin (a_1, b_1)$ implies that for another one of the I_n, say (a_2, b_2), we have $a_2 < b_1 < b_2$. We continue this process and note that it must terminate because our cover of J contains a *finite* number of the I_n. But then the last interval in our process, say (a_m, b_m), satisfies $a_m < b - \varepsilon/2 < b_m$ and we have $J \subset \bigcup_{j=1}^{m}(a_j, b_j)$. Now note that

$$\sum_{n=1}^{\infty} \lambda(I_n) \geq \sum_{j=1}^{m} \lambda\big((a_j, b_j)\big) = \sum_{j=1}^{m}(b_j - a_j)$$

$$= b_m - a_1 + \sum_{j=2}^{m}(b_{j-1} - a_j)$$

$$> \lambda(J) = b - a - \varepsilon.$$

Since this holds for all $\varepsilon \in (0, (b - a)/2)$, the inequality $(*)$ follows.   □

Exercise 7.3.8 (Measure Zero $\Rightarrow$ Empty Interior). Show that a set of measure zero must have *empty interior*.

This exercise and Proposition 7.3.5 may lead us to make the following conjectures:

Conjecture 1. All sets of measure zero must be *countable*.

Conjecture 2. All sets with *empty interior* must have measure zero.

In fact, both conjectures are *false*! To *disprove* Conjecture 1, we look at our old *uncountable* friend, the *Cantor set*:

Example 7.3.9 (The Cantor Set Has Measure Zero).   Recall (cf. Sect. 4.2) that the Cantor set C is obtained from $C_0 := [0, 1]$ by *successive deletion of middle thirds* and can be written as

$$C = \bigcap_{k=0}^{\infty} C_k, \tag{$*$}$$

where, $C_1 := [0, 1/3] \cup [2/3, 1]$, $C_2 := [0, 1/9] \cup [2/9, 1/3] \cup [2/3, 7/9] \cup [8/9, 1]$, etc. In general, C_n is the union of 2^n disjoint closed intervals of length $1/3^n$. Now, given any $\varepsilon > 0$, pick n such that $(2/3)^n < \varepsilon$. Then, $C \subset C_n$ and the total length of all the subintervals in C_n is $2^n(1/3^n) = (2/3)^n < \varepsilon$. Therefore, C has measure zero as claimed.

To disprove Conjecture 2, we need a *generalized* version of the Cantor set:

Definition 7.3.10 (The Generalized Cantor Sets $C(\alpha)$). Let $\alpha \in (0, 1)$ be fixed, but *arbitrary* and, for each $n \in \mathbb{N}$, let $a_n := 2^{n-1}\alpha/3^n$. From the center of $C_0(\alpha) := [0, 1]$ we remove an open interval of length a_1. The resulting set, $C_1(\alpha)$, is the union of two *disjoint closed* intervals of equal length. From the center of each of these two intervals we remove an open interval of length $a_2/2$. The resulting set, $C_2(\alpha)$, is a union of 2^2 disjoint closed intervals of equal length. Repeating this process, at the nth stage $(n > 1)$, we obtain the set $C_n(\alpha)$ by removing from the center of each one of the 2^{n-1} disjoint closed intervals in $C_{n-1}(\alpha)$ an open interval of length $a_n/2^{n-1}$. We now define:

$$C(\alpha) := \bigcap_{n=0}^{\infty} C_n(\alpha);$$

in other words, $C(\alpha)$ is what is left of $[0, 1]$ after removing all the open intervals in the above process. The set $C(\alpha)$ will be called a *generalized Cantor set*.

As the following proposition shows, the generalized Cantor sets have most of the interesting properties of Cantor's ternary set which, by the way, is $C(\alpha)$ with $\alpha = 1$.

Proposition 7.3.11. *For each* $\alpha \in (0, 1)$, *the generalized Cantor set* $C(\alpha)$ *is compact, nowhere dense, and perfect.*

Proof. By its very definition, $C(\alpha)$ is a *closed* subset of the *compact* set $[0, 1]$ and hence is itself compact. To prove that it is nowhere dense amounts to showing that it contains *no open intervals*. But, if $I \subset C(\alpha)$ is an open interval, then we have $I \subset C_n(\alpha)$ for each $n \in \mathbb{N}$. Since each of the 2^n (disjoint) intervals in $C_n(\alpha)$ has length $< 1/2^n$, we must have $\lambda(I) < 1/2^n$ for all $n \in \mathbb{N}$, which is absurd. Finally, we must show that $C(\alpha)$ is *perfect*, meaning that it is closed (which we know to be true) and that every point in it is a *limit point*. Let $x \in C(\alpha)$ and let $\varepsilon > 0$ be given. Then $x \in C_n(\alpha)$ for all $n \in \mathbb{N}$. Pick n such that $1/2^n < \varepsilon$. Then $B_\varepsilon(x) := (x - \varepsilon, x + \varepsilon)$ contains one of the 2^n disjoint intervals in $C_n(\alpha)$. Thus, denoting this interval by $I_k := [a_k, b_k]$, we have $[a_k, b_k] \subset B_\varepsilon(x)$. Since $a_k, b_k \in C(\alpha)$, we see that x is indeed a limit point of $C(\alpha)$. □

Exercise 7.3.12 ($C(\alpha)$ Is Uncountable). Show that, for any $\alpha \in (0, 1)$, the generalized Cantor set $C(\alpha)$ is *uncountable. Hint:* Theorem 5.3.12.

Let us now show that Conjecture 2 is also false.

Proposition 7.3.13 ($C(\alpha)$ Is Not of Measure Zero). *The generalized Cantor set, $C(\alpha)$, is an uncountable set with empty interior that is not of measure zero.*

Proof. In view of Proposition 7.3.11 and Exercise 7.3.12, we need only show the last statement. Now the total length of the open subintervals removed in the construction of $C(\alpha)$ is

$$\sum_{n=1}^{\infty} \left(\frac{2^{n-1}}{3^n} \right) \alpha = \alpha.$$

Therefore, if $C(\alpha)$ could be covered by a countable set of open intervals of total length $< 1 - \alpha$, then, combining these intervals with the removed subintervals, we would be able to cover the unit interval $[0, 1]$ by a collection of open intervals of total length *less than* $1 = (1 - \alpha) + \alpha$. Since this is impossible (by Proposition 7.3.7), the proof is complete. $\square$

Before we state and prove Lebesgue's theorem on the existence of the Riemann integral, we need a couple of definitions:

Definition 7.3.14 (Oscillation: on a Set, at a Point). Let $I \subset \mathbb{R}$ be an interval and let $f : I \to \mathbb{R}$ be a *bounded* function.

(a) For each *set* $\emptyset \neq S \subset I$, the *oscillation of f on S* is defined to be the *nonnegative* real number

$$\omega_f(S) := \sup\{|f(s) - f(t)| : s, t \in S\};$$

i.e., $\omega_f(S)$ is simply the *diameter* of the image $f(S)$.

(b) Given any *point* (i.e., *element*) $x_0 \in I$, the *oscillation of f at x_0* is defined to be the *nonnegative* number

$$\omega_f(x_0) := \inf\{\omega_f(B_\delta(x_0) \cap I) : \delta > 0\} = \lim_{\delta \to 0+} \omega_f(B_\delta(x_0) \cap I), \qquad (\dagger)$$

where, as usual, $B_\delta(x_0) := \{x \in I : |x - x_0| < \delta\}$.

Remark 7.3.15.

(a) Note that for a *point* $x_0 \in I$, the numbers $\omega_f(\{x_0\})$ and $\omega_f(x_0)$ need not be the same. Indeed, the former is obviously *always* zero, while the latter is zero *if and only if f is continuous at x_0* (cf. Exercise 4.3.8(d)).

(b) The relevance of the concept of *oscillation* to the existence of the Riemann integral becomes clear if we observe that, for any partition $\mathcal{P} := (x_k)_{k=0}^n$ of $[a, b]$, the difference between the corresponding upper and lower sums is

$$U(f, \mathcal{P}) - L(f, \mathcal{P}) = \sum_{j=1}^{n}(M_j - m_j)\Delta x_j = \sum_{j=1}^{n}\omega_f(I_j)\Delta x_j,$$

where, M_j, m_j and Δx_j have their usual meaning. Indeed, for each subinterval $I_j := [x_{j-1}, x_j]$, we have $\omega_f(I_j) = M_j - m_j$.

Exercise 7.3.16. Let the notation be as in Definition 7.3.14.

(a) Show that

$$\omega_f(S) = \sup\{f(s) - f(t) : s, t \in S\}$$
$$= \sup\{f(x) : x \in S\} - \inf\{f(x) : x \in S\}.$$

(b) Show that

$$S \subset T \subset I \Rightarrow \omega_f(S) \le \omega_f(T) \le 2\sup\{|f(x)| : x \in I\}.$$

(c) Prove the equality of the "inf" and the "lim" in (†).
(d) Show that the above definition of $\omega_f(x_0)$ is equivalent to the one given in Chap. 4 (cf. Definition 4.3.7). In other words, show that

$$\lim_{\delta \to 0+} \omega_f(B_\delta(x_0) \cap I) = \overline{f}(x_0) - \underline{f}(x_0),$$

where, we recall,

$$\overline{f}(x_0) := \inf\{\sup\{f(x) : x \in B_\delta(x_0) \cap I\} : \delta > 0\} \quad \text{and}$$

$$\underline{f}(x_0) := \sup\{\inf\{f(x) : x \in B_\delta(x_0) \cap I\} : \delta > 0\}.$$

Next, we prove a couple of lemmas that will be used in our main theorem but are of independent interest as well.

Lemma 7.3.17. *Let* $f : [a, b] \to \mathbb{R}$. *Then, for each* $v > 0$, *the set*

$$\Omega_v := \{x \in [a, b] : \omega_f(x) < v\}$$

is (relatively) open (in $[a, b]$).

Proof. Consider an arbitrary point $x_0 \in \Omega_v$. We must show that, for some $\delta > 0$, we have $B_\delta(x_0) \cap [a, b] \subset \Omega_v$. Put $\omega_0 := \omega_f(x_0)$ and $\omega_x := \omega_f(x)$. We then have $\omega_0 < v$ and must find $\delta > 0$ such that $x \in [a, b]$ and $|x - x_0| < \delta$ imply $\omega_x < v$. Now, by the very definition of ω_0, we can pick $\delta > 0$ such that

$$|\omega_f(B_\delta(x_0) \cap [a, b]) - \omega_0| < v - \omega_0.$$

It then follows at once (cf. Exercise 7.3.16) that, for any $x \in B_\delta(x_0)$, we have

$$\omega_x \le \omega_f(B_\delta(x_0) \cap [a, b]) < v,$$

and the lemma is proved.   □

Our second lemma is an extension of Theorem 4.6.4, which said that a continuous function on a compact interval $[a, b]$ is *uniformly continuous* on $[a, b]$. In fact, the proof given below is almost identical to the second proof we gave for Theorem 4.6.4.

Lemma 7.3.18. *Let* $f : [a, b] \to \mathbb{R}$ *and* $\varepsilon > 0$. *If* $\omega_f(x) < \varepsilon$, *for all* $x \in [a, b]$, *then*

$$(\exists \delta > 0)(\forall s, t \in [a, b])(|s - t| < \delta \Rightarrow |f(s) - f(t)| < \varepsilon).$$

Proof. For each $x \in [a, b]$, using $\omega_f(x) < \varepsilon$ and Lemma 7.3.17, we can pick $\delta_x > 0$ such that $\omega_f(B_{\delta_x}(x) \cap [a, b]) < \varepsilon$. Since $[a, b]$ is *compact*, we can find a *finite* number of points $x_1, x_2, \ldots, x_n$ in $[a, b]$ such that, with $\delta_j := \delta_{x_j}$ and $B_j := (x_j - \delta_j/2, x_j + \delta_j/2)$, we have

$$[a, b] \subset \bigcup_{j=1}^{n} B_j.$$

Now let $\delta := \min\{\delta_1/2, \ldots, \delta_n/2\}$. If s, $t \in [a, b]$ satisfy $|s - t| < \delta$ and $s \in B_j$, then we have $t \in (x_j - \delta_j, x_j + \delta_j)$. Indeed,

$$|t - x_j| \leq |t - s| + |s - x_j| < \delta + \delta_j/2 \leq 2(\delta_j/2) = \delta_j.$$

Since $\omega_f(B_{\delta_j}(x_j) \cap [a, b]) < \varepsilon$, for all $1 \leq j \leq n$, we see that $|s - t| < \delta$ implies $|f(s) - f(t)| < \varepsilon$, as desired. $\square$

We are now ready to prove the main theorem of this section. The reader may find the (ingenious) proof of the theorem rather complicated. It should be noted, however, that the important quantity to be estimated is the difference between the upper and lower (Darboux) sums:

$$U(f, \mathcal{P}) - L(f, \mathcal{P}) = \sum_{j=1}^{n} (M_j - m_j)(x_j - x_{j-1}).$$

The idea is now to divide this sum into two sums: the sum $\sum_1$ corresponding to the subintervals *meeting the discontinuity set of* f, and the sum $\sum_2$ over the subintervals on which f *is continuous*. Now, if the discontinuity set has *measure zero*, then the total length $\sum_1 (x_j - x_{j-1})$ can be made arbitrarily small while, in the second sum, the *oscillations* $M_j - m_j$ can be made (uniformly) small due to the *continuity* of f.

Theorem 7.3.19 (Lebesgue's Integrability Criterion). *Let* $f : [a, b] \to \mathbb{R}$ *be a bounded function. Then* f *is Riemann integrable if and only if it is continuous almost everywhere.*

Proof. For each $N \in \mathbb{N}$, let $D_N := \{x \in [a, b] : \omega_f(x) \geq 1/N\}$ and put $D := \bigcup_{N=1}^{\infty} D_N$. Then D is the set of all *discontinuity points* of f in $[a, b]$. (Why?) Suppose that $f \in \mathcal{R}([a, b])$. We want to prove that each D_N has *measure zero*. By Riemann's Lemma, given any $\varepsilon > 0$ we can find a partition $\mathcal{P} = (x_k)_{k=0}^{n}$ of $[a, b]$ such that $U(f, \mathcal{P}) - L(f, \mathcal{P}) < \varepsilon/N$. Let us divide $\{1, 2, \ldots, n\}$ into two parts, $G_i = G_i(D_N)$, $i = 1, 2$:

$$G_1 := \{j : (x_{j-1}, x_j) \cap D_N \neq \emptyset\}, \quad G_2 := \{j : (x_{j-1}, x_j) \cap D_N = \emptyset\}.$$

Now, with $M_j := \sup\{f(x) : x \in [x_{j-1}, x_j]\}$, $m_j := \inf\{f(x) : x \in [x_{j-1}, x_j]\}$, and $\Delta x_j := x_j - x_{j-1}$, we have

$$U(f,\mathcal{P}) - L(f,\mathcal{P}) = \sum_{j \in G_1} (M_j - m_j)\Delta x_j + \sum_{j \in G_2} (M_j - m_j)\Delta x_j < \varepsilon/N.$$

Since $(x_{j-1}, x_j) \cap D_N \neq \emptyset$ implies $M_j - m_j \geq 1/N$, we have

$$\sum_{j \in G_1} \Delta x_j \leq N \sum_{j \in G_1} (M_j - m_j)\Delta x_j < N\varepsilon/N = \varepsilon.$$

But the intervals (x_{j-1}, x_j) with $j \in G_1$ cover D_N. Therefore, D_N has measure zero for each N, and hence, by Proposition 7.3.5(b), D has measure zero. To prove the converse, let us assume that D has measure zero and let $\varepsilon > 0$ be given. By Lemma 7.3.17, each $[a,b] \setminus D_N$ is (relatively) *open*. Therefore each D_N is a *closed* (hence *compact*) subset of $[a,b]$ and has measure zero. Let N be such that $(b - a)/N < \varepsilon/2$ and pick a partition $\mathcal{P} = (x_k)_{k=0}^n$ of $[a,b]$ such that $\sum_{j \in G_1}(x_j - x_{j-1}) < \varepsilon/(4M)$, where G_1 is as above and $M := \sup\{|f(x)| : x \in [a,b]\}$. Next, note that if $K := \bigcup_{j \in G_2}[x_{j-1}, x_j]$, with G_2 defined as above, then K is a compact subset of $[a,b]$ such that $x \in K$ implies $\omega_f(x) < 1/N$. By Lemma 7.3.18, we can pick a $\delta > 0$ such that $s, t \in K$ and $|s - t| < \delta$ imply $|f(s) - f(t)| < 1/N$. Let $\mathcal{P}' = (x'_{k'})_{k'=0}^{n'}$ be a *refinement* of $\mathcal{P}$ with mesh $\|\mathcal{P}'\| < \delta$. Then, with $M_{j'}$, $m_{j'}$, and $\Delta x'_{j'}$ defined as usual and the subsets G'_1, $G'_2 \subset \{1, 2, \ldots, n'\}$ defined as in the first part of the proof, we have

$$U(f, \mathcal{P}') - L(f, \mathcal{P}') = \sum_{j' \in G'_1} (M_{j'} - m_{j'})\Delta x'_{j'} + \sum_{j' \in G'_2} (M_{j'} - m_{j'})\Delta x'_{j'}$$

$$< 2M \sum_{j \in G_1} (x_j - x_{j-1}) + \frac{b - a}{N}$$

$$< 2M \frac{\varepsilon}{4M} + \frac{\varepsilon}{2} = \varepsilon,$$

which shows indeed that $f \in \mathcal{R}([a,b])$ and completes the proof. □

Corollary 7.3.20. *If f, $g : [a,b] \to \mathbb{R}$ are (Riemann) integrable on $[a,b]$, then so are the functions $\alpha f + \beta g$ (with arbitrary constants α and β), $\max\{f, g\}$, $\min\{f, g\}$, $|f|$, f^2, and fg. If, in addition, $\inf\{|g(x)| : x \in [a,b]\} > 0$, then $1/g$ and f/g are also integrable.*

Proof. Exercise! *Hint:* For each function h, let $D(h)$ denote the *discontinuity set* of h. How are $D(\alpha f + \beta g)$, $D(\max\{f, g\})$, etc. related to $D(f)$ and $D(g)$? Also, note that, with $f^+ := \max\{0, f\}$ and $f^- := \max\{0, -f\}$, we have $f = f^+ - f^-$ and $|f| = f^+ + f^-$. Finally, the integrability of fg may also be deduced from the other cases and the identity

$$fg = \frac{1}{4}[(f + g)^2 - (f - g)^2].$$

□

Exercise 7.3.21 ((Interval) Additivity). Let $a < b$ and $f \in \mathcal{R}([a,b])$. Show that, for any $c \in (a,b)$, we have $f\chi_{[a,c]} \in \mathcal{R}([a,c])$, $f\chi_{[c,b]} \in \mathcal{R}([c,b])$, and

$$\int_a^b f(x)\,dx = \int_a^c f(x)\,dx + \int_c^b f(x)\,dx.$$

We end this section by two more corollaries of Lebesgue's Integrability Theorem that are of independent interest.

Corollary 7.3.22. *Let $f : [a,b] \to \mathbb{R}$ be Riemann integrable and let $g : [c,d] \to [a,b]$ be a bijection such that g^{-1} is Lipschitz. Then $f \circ g$ is Riemann integrable on $[c,d]$.*

Proof. We prove that the set $D(f \circ g) \subset [c,d]$ of discontinuity points of $f \circ g$ has measure zero. Note that g^{-1} (and hence g) is actually a *homeomorphism*. Indeed not only g^{-1} is (uniformly) continuous, but its domain $[c,d]$ is *compact* and hence g^{-1} sends *closed* (hence compact) subsets of $[c,d]$ to compact (hence closed) subsets of $[a,b]$, so that $g = (g^{-1})^{-1}$ is continuous as well. It follows that $D(f \circ g) = g^{-1}(D(f))$. (Why?) So to complete the proof, we must show that $g^{-1}(D(f))$ has measure zero. This, however, follows from the general fact that *Lipschitz functions map sets of measure zero onto sets of measure zero.* Indeed, suppose that $A > 0$ is a Lipschitz constant for g^{-1}; i.e.,

$$|g^{-1}(\eta) - g^{-1}(\xi)| \le A|\eta - \xi|, \quad \forall\, \xi,\, \eta \in [c,d].$$

By assumption, the set $D(f) \subset [a,b]$ of discontinuity points of f has measure zero. So let $\varepsilon > 0$ be given and suppose that $D(f) \subset \bigcup_n (a_n, b_n)$ with $\sum_n (b_n - a_n) < \varepsilon/A$. We may assume that g is *strictly increasing* and note that the intervals $(a'_n, b'_n) := g^{-1}((a_n, b_n))$, with $g(a'_n) = a_n$ and $g(b'_n) = b_n$, cover $D(f \circ g)$. But then,

$$\sum_n (b'_n - a'_n) \le A \sum_n (b_n - a_n) < \varepsilon.$$

This shows indeed that $D(f \circ g)$ has measure zero and completes the proof. □

For our next corollary, recall (Definition 6.6.10) that a C^1-*diffeomorphism* is a bijection that is continuously differentiable and so is its inverse function.

Corollary 7.3.23. *Let f be Riemann integrable on $[a,b]$ and let $g : [c,d] \to [a,b]$ be a C^1-diffeomorphism. Then $f \circ g$ is Riemann integrable on $[c,d]$.*

Proof. By our assumption, g^{-1} has continuous (hence *bounded*) derivative and hence is Lipschitz by Corollary 6.4.20. Thus the assertion follows from Corollary 7.3.22. □

7.4   Properties of the Riemann Integral

In Corollary 7.3.20, we used the powerful Lebesgue Integrability Criterion to deduce some of the basic properties of the Riemann integral. We can, however, avoid that theorem entirely and give *direct* proofs of those (and other) properties using only the definition or Riemann's Lemma. This is what we shall do in this section. Let us begin with the *additivity* property *with respect to intervals* (cf. Exercise 7.3.21 above):

Theorem 7.4.1 ((Interval) Additivity Theorem).   *Let* $f : [a,b] \to \mathbb{R}$ *and let* $c \in (a,b)$. *Then* f *is (Riemann) integrable on* $[a,b]$ *if and only if its restrictions to* $[a,c]$ *and* $[c,b]$ *are both integrable. In this case, we have*

$$\int_a^b f(x)\,dx = \int_a^c f(x)\,dx + \int_c^b f(x)\,dx. \tag{$\dagger$}$$

Proof. Suppose $f \in \mathcal{R}([a,b])$ and let $\varepsilon > 0$ be given. We can pick a partition $\mathcal{P} = (x_k)_{k=0}^n$ of $[a,b]$ such that

$$U(f,\mathcal{P}) - L(f,\mathcal{P}) < \varepsilon.$$

Adjoining the point c to our partition, if necessary, we may assume that $c = x_j$ for some $1 \le j \le n$. Now $\mathcal{P}' := (x_k)_{k=0}^j$ and $\mathcal{P}'' := (x_k)_{k=j}^n$ are partitions of $[a,c]$ and $[c,b]$, respectively. Also, we have

(i) $L(f,\mathcal{P}) = L(f,\mathcal{P}') + L(f,\mathcal{P}'')$ and $U(f,\mathcal{P}) = U(f,\mathcal{P}') + U(f,\mathcal{P}'')$,

from which it follows at once that

(ii) $[U(f,\mathcal{P}') - L(f,\mathcal{P}')] + [U(f,\mathcal{P}'') - L(f,\mathcal{P}'')] < \varepsilon.$

Since each of the differences inside the brackets is *nonnegative*, each is less than ε and the restrictions of f to $[a,c]$ and $[c,b]$ are indeed integrable. Moreover, from the inequalities

$$L(f,\mathcal{P}') \le \int_a^c f \le U(f,\mathcal{P}') \quad \text{and} \quad L(f,\mathcal{P}'') \le \int_c^b f \le U(f,\mathcal{P}''),$$

we deduce that

$$L(f,\mathcal{P}) \le \int_a^c f + \int_c^b f \le U(f,\mathcal{P}).$$

Since this holds for every partition (with c adjoined), ($\dagger$) follows. Conversely, suppose that the restrictions of f to $[a,c]$ and $[c,b]$ are both integrable and let $\varepsilon > 0$ be given. Pick partitions $\mathcal{P}'$ and $\mathcal{P}''$ of $[a,c]$ and $[c,b]$, respectively, such that

(iii) $U(f,\mathcal{P}') - L(f,\mathcal{P}') < \dfrac{\varepsilon}{2}$ and $U(f,\mathcal{P}'') - L(f,\mathcal{P}'') < \dfrac{\varepsilon}{2}.$

If now $\mathcal{P}$ is the partition of $[a, b]$ containing all the points of $\mathcal{P}'$ and $\mathcal{P}''$, then (i) is satisfied again and hence (by (ii) and (iii)),

$$U(f, \mathcal{P}) - L(f, \mathcal{P}) < \varepsilon.$$

This shows indeed that $f \in \mathcal{R}([a, b])$ and completes the proof.   □

Terminology. If $f \in \mathcal{R}([a, b])$ and if $[c, d] \subset [a, b]$, then the above theorem implies that the *restriction* of f to $[c, d]$ is integrable on $[c, d]$. Henceforth, to simplify the exposition, we abuse the language and say that f *is integrable on* $[c, d]$.

Exercise 7.4.2. Show that, if $f \in \mathcal{R}([a, b])$, then

$$\lim_{c \to a+} \int_c^b f(x) \, dx = \int_a^b f(x) \, dx = \lim_{d \to b-} \int_a^d f(x) \, dx.$$

Definition 7.4.3. Let $f \in \mathcal{R}([a, b])$, where $a < b$. Then we define

$$\int_b^a f := -\int_a^b f \quad \text{and} \quad \int_a^a f := 0.$$

Corollary 7.4.4. *For any a, b, $c \in \mathbb{R}$, if any two of the integrals $\int_a^b f$, $\int_a^c f$, and $\int_c^b f$ exist, then so does the third one and we have*

$$\int_a^b f = \int_a^c f + \int_c^b f, \qquad (*)$$

which can also be written in the more symmetric form

$$\int_a^b f + \int_b^c f + \int_c^a f = 0.$$

Proof. This follows at once from Theorem 7.4.1 and Definition 7.4.3. For example, if $c < a < b$, then

$$\int_c^b f = \int_c^a f + \int_a^b f = -\int_a^c f + \int_a^b f,$$

which gives $(*)$. Other cases can be treated the same way.   □

Corollary 7.4.5. *Let $f \in \mathcal{R}([a, b])$ and let $\mathcal{P} = (x_k)_{k=0}^n$ be a partition of $[a, b]$. Then f is integrable on each $I_j := [x_{j-1}, x_j]$, $1 \le j \le n$, and we have*

$$\int_a^b f(x) \, dx = \sum_{j=1}^n \int_{x_{j-1}}^{x_j} f(t) \, dt.$$

Proof. Induction!   □

Exercise 7.4.6. Let $f : \mathbb{R} \to \mathbb{R}$ be a *periodic* function and let $p > 0$ be its *period*. Show that, if $f \in \mathcal{R}([0, p])$, then $f \in \mathcal{R}([a, a + p])$ for any $a \in \mathbb{R}$ and

$$\int_0^p f(x)\, dx = \int_a^{a+p} f(x)\, dx.$$

Hint: Use the *Interval Additivity* property and Exercise 7.1.24.

Corollary 7.4.7 (Integral of a Step Function). *Suppose that $f : [a, b] \to \mathbb{R}$ is a step function. In other words, there is a partition $\mathcal{P} = (x_k)_{k=0}^n$ of $[a, b]$ and a finite sequence $(c_j)_{j=1}^n$ such that, for each $1 \le j \le n$,*

$$f(x) = c_j \quad \forall x \in (x_{j-1}, x_j).$$

Then we have

$$\int_a^b f(x)\, dx = \sum_{j=1}^n c_j (x_j - x_{j-1}).$$

Proof. This follows at once from Corollary 7.4.5 and Exercise 7.1.13. $\square$

Exercise 7.4.8. Let $f \in \mathcal{R}([a, b])$.

1. Show that for any $\varepsilon > 0$ there exists a *step function* g on $[a, b]$ such that

$$\int_a^b |f(x) - g(x)|\, dx < \varepsilon.$$

2. Show that

$$\int_a^b f = \sup \left\{ \int_a^b g : g \in Step([a, b]),\ g \le f \right\} = \int_a^b f.$$

We next prove that, just like the derivative, the (Riemann) integral is *linear*; i.e., it is *additive* and *homogeneous*:

Theorem 7.4.9 (Linearity of the Integral). *Let f and g be integrable on $[a, b]$. Then, for any real constants α and β, the linear combination $\alpha f + \beta g$ is also integrable on $[a, b]$ and we have*

$$\int_a^b (\alpha f + \beta g) = \alpha \int_a^b f + \beta \int_a^b g.$$

Proof. Simply observe that, for any tagged partition $\dot{P} \in \dot{\mathcal{P}}([a, b])$, the corresponding Riemann sums of the functions $\alpha f + \beta g$, f, and g (cf. Definition 7.1.4) satisfy the equation

$$S(\alpha f + \beta g, \dot{P}) = \alpha S(f, \dot{P}) + \beta S(g, \dot{P})$$

and use Theorem 7.1.26. $\square$

The next theorem shows that integrability is *stable* under composition with *continuous* functions.

Theorem 7.4.10 (Stability Under Composition). *Let f be integrable on $[a, b]$ with $m \leq f(x) \leq M$ for all $x \in [a, b]$. If ϕ is continuous on $[m, M]$, then $g := \phi \circ f \in \mathcal{R}([a, b])$.*

Proof. Let $\varepsilon > 0$ be given. Since ϕ is continuous, it is *bounded* and *uniformly continuous* on $[m, M]$. Therefore we have $|\phi(s)| \leq K$ for some $K > 0$ and all $s \in [m, M]$, and we can find $\delta > 0$ such that $|\phi(s) - \phi(t)| < \varepsilon/2(b - a)$ for all $s, t \in [m, M]$ satisfying $|s - t| < \delta$. Also, since $f \in \mathcal{R}([a, b])$, we can pick a partition $\mathcal{P} = (x_k)_{k=0}^{n}$ of $[a, b]$ such that

$$U(f, \mathcal{P}) - L(f, \mathcal{P}) < \frac{\varepsilon \delta}{4K}. \tag{$*$}$$

Let m_j (resp., M_j) be the infimum (resp., supremum) of f on $[x_{j-1}, x_j]$ and let m'_j and M'_j be the corresponding numbers for g. Divide the set $\{1, 2, \ldots, n\}$ into two subsets:

$$G_1 := \{j : M_j - m_j < \delta\}, \qquad G_2 := \{j : M_j - m_j \geq \delta\},$$

and note that, for $j \in G_1$, the choice of δ implies $M'_j - m'_j < \varepsilon/2(b - a)$ while, for $j \in G_2$, we have $M'_j - m'_j \leq 2K$. Now, in view of $(*)$, we have

$$\delta \sum_{j \in G_2} \Delta x_j \leq \sum_{j \in G_2} (M_j - m_j) \Delta x_j < \frac{\varepsilon \delta}{4K},$$

where, as always, $\Delta x_j := x_j - x_{j-1}$. Thus $\sum_{j \in G_2} \Delta x_j < \varepsilon/(4K)$. It now follows that

$$U(g, \mathcal{P}) - L(g, \mathcal{P}) = \sum_{j \in G_1} (M'_j - m'_j) \Delta x_j + \sum_{j \in G_2} (M'_j - m'_j) \Delta x_j$$

$$\leq \frac{\varepsilon}{2(b - a)} (b - a) + 2K \frac{\varepsilon}{4K} = \varepsilon.$$

Since $\varepsilon > 0$ was arbitrary, we have $g \in \mathcal{R}([a, b])$. $\square$

Remark 7.4.11. Note that the *continuity* of ϕ in the above theorem is crucial. Indeed, if ϕ is simply assumed to be *integrable*, i.e., if $f, \phi \in \mathcal{R}([a, b])$, then the composite function $\phi \circ f$ need *not* be integrable. (Cf. Problem #4 at the end of the chapter.)

Using Theorems 7.4.9 and 7.4.10, we can now give another proof of Corollary 7.3.20, avoiding Lebesgue's Integrability Criterion.

Corollary 7.4.12. *If f, $g : [a, b] \to \mathbb{R}$ are (Riemann) integrable on $[a, b]$, then so are the functions $\alpha f + \beta g$ (with arbitrary constants α and β), $\max\{f, g\}$, $\min\{f, g\}$, $|f|$, f^2, and fg. If, in addition, $\inf\{|g(x)| : x \in [a, b]\} > 0$, then $1/g$ and f/g are also integrable.*

Proof. The integrability of $\alpha f + \beta g$ is guaranteed by Theorem 7.4.9. Next, using Theorem 7.4.10 with $\phi(t) := t^+ := \max\{0, t\}$ (resp., $\phi(t) := t^- := \max\{0, -t\}$), we see that f^+ (resp., f^-) is integrable on $[a, b]$. Theorem 7.4.9 now implies that the same holds for $|f| = f^+ + f^-$. Taking $\phi(t) := t^2$ in Theorem 7.4.10, we obtain the integrability of f^2 which, together with Theorem 7.4.9 and the identity

$$fg = \frac{1}{4}[(f + g)^2 - (f - g)^2],$$

imply $fg \in \mathcal{R}([a, b])$. For $\max\{f, g\}$ and $\min\{f, g\}$, we use the identities

$$\max\{f, g\} = \frac{1}{2}(f + g + |f - g|) \quad \text{and} \quad \min\{f, g\} = \frac{1}{2}(f + g - |f - g|).$$

Finally, if $\inf\{|g(x)| : x \in [a, b]\} > 0$, then we can use Theorem 7.4.10 with $\phi(t) := 1/t$ to deduce the integrability of $1/g$. The case of f/g now follows from the fact that the product of two integrable functions is itself integrable. The proof is complete. $\qquad\square$

Next, we look at the behavior of the integral with respect to inequalities:

Proposition 7.4.13. *Let f be Riemann integrable on $[a, b]$ and $m \leq f(x) \leq M$, for all $x \in [a, b]$. Then $m(b - a) \leq \int_a^b f \leq M(b - a)$. In particular, if $|f(x)| \leq K$, for all $x \in [a, b]$, then $|\int_a^b f(x)\, dx)| \leq K(b - a)$.*

Proof. For any tagged partition $\dot{\mathcal{P}}$ of $[a, b]$, we obviously have

$$m(b - a) \leq S(f, \dot{\mathcal{P}}) \leq M(b - a),$$

from which the first statement follows at once. The second statement then follows from the first one and the inequalities $-K \leq f(x) \leq K$, for all $x \in [a, b]$. $\qquad\square$

Corollary 7.4.14 (Stability of Inequalities). *Let f and g be integrable functions on $[a, b]$ such that $f(x) \leq g(x)$, for all $x \in [a, b]$. Then $\int_a^b f(x)\, dx \leq \int_a^b g(x)\, dx$. If f is Riemann integrable on $[a, b]$, then so is $|f|$ and we have*

$$\left| \int_a^b f(x)\, dx \right| \leq \int_a^b |f(x)|\, dx.$$

Proof. Since $g(x) - f(x) \geq 0$, for all $x \in [a, b]$, Proposition 7.4.13 implies

$$0 \leq \int_a^b (g(x) - f(x))\, dx = \int_a^b g(x)\, dx - \int_a^b f(x)\, dx,$$

proving the first statement. In view of Corollary 7.4.12, the second statement follows from the first one and the inequalities $-|f(x)| \leq f(x) \leq |f(x)|$, for all $x \in [a, b]$. □

Corollary 7.4.15. *Let* $f : [a, b] \to \mathbb{R}$ *be integrable on* $[a, b]$ *and consider the function*

$$F(x) := \int_a^x f(t) \, dt \quad (\forall x \in [a, b]).$$

Then F *is Lipschitz (and hence uniformly continuous) on* $[a, b]$.

Proof. Since f is integrable on $[a, b]$, it is *bounded*. Let $K := \sup\{|f(x)| : x \in [a, b]\}$. If x, $x' \in [a, b]$, then (by Corollary 7.4.4 and Proposition 7.4.13) we have

$$|F(x') - F(x)| = \left| \int_a^{x'} f(t) \, dt - \int_a^x f(t) \, dt \right|$$

$$= \left| \int_x^{x'} f(t) \, dt \right| \leq K |x' - x|. \qquad \Box$$

It is obvious that, if f is *identically* zero on $[a, b]$, then its integral is also zero. Our next theorem answers the following natural question: When is the converse true?

Theorem 7.4.16 ($\int |f| = 0 \Leftrightarrow f = 0$ **a.e.**)**.** *Let* f *be integrable on* $[a, b]$. *Then* $\int_a^b |f(x)| \, dx = 0$ *if and only if* $f(x) = 0$ *almost everywhere.*

Proof. Suppose that $\int_a^b |f(x)| \, dx = 0$. Since f is integrable, by Lebesgue's Integrability Criterion, it is *continuous a.e.* Pick $x_0 \in (a, b)$ at which f is continuous. If $|f(x_0)| > 0$, then we can find $\delta > 0$ such that $[x_0 - \delta, x_0 + \delta] \subset (a, b)$ and $|f(x) - f(x_0)| < |f(x_0)|/2$, for all $x \in [x_0 - \delta, x_0 + \delta]$. Therefore, $|f(x)| > |f(x_0)|/2$ on $[x_0 - \delta, x_0 + \delta]$. But then, by Theorem 7.4.1 and Corollary 7.4.14, we have

$$\int_a^b |f(x)| \, dx \geq \int_{x_0 - \delta}^{x_0 + \delta} |f(x)| \, dx \geq \frac{1}{2} \int_{x_0 - \delta}^{x_0 + \delta} |f(x_0)| \, dx = \delta |f(x_0)|,$$

contradicting $\int_a^b |f(x)| \, dx = 0$. Thus $f(x_0) = 0$, for each $x_0 \in (a, b)$ at which f is continuous, and we have $f(x) = 0$ for a.a. $x \in [a, b]$. Conversely, suppose that $f(x) = 0$ a.e. and let $\mathcal{P} = (x_k)_{k=0}^n$ be *any* partition of $[a, b]$. Then, for each $1 \leq j \leq n$, there exists $t_j \in [x_{j-1}, x_j]$ such that $f(t_j) = 0$. (Why?) In particular, $m_j := \inf\{|f(x)| : x \in [x_{j-1}, x_j]\} = 0$, for all $1 \leq j \leq n$. But then $L(|f|, \mathcal{P}) = 0$, for all $\mathcal{P} \in \mathcal{P}([a, b])$, and we have

$$\int_a^b |f(x)| \, dx = \sup\{L(|f|, \mathcal{P}) : \mathcal{P} \in \mathcal{P}([a, b])\} = 0. \qquad \Box$$

Corollary 7.4.17. *Let f and g be integrable on $[a,b]$. If $f(x) = g(x)$ for almost all $x \in [a,b]$, then $\int_a^b f = \int_a^b g$.*

Proof. In view of Corollary 7.4.14, this follows from Theorem 7.4.16 applied to $f - g$.   □

The following important inequality is usually referred to as the *Cauchy–Schwarz* (or the *Cauchy–Bunyakovsky–Schwarz*) inequality:

Theorem 7.4.18 (Cauchy–Schwarz Inequality).   *Let f and g be integrable on $[a,b]$. Then we have*

$$\left(\int_a^b fg \right)^2 \le \left(\int_a^b f^2 \right)\left(\int_a^b g^2 \right). \tag{†}$$

Proof. Since (by Corollary 7.4.12) f^2, g^2, and fg are integrable, so is the (nonnegative) function $(f + tg)^2$, for every $t \in \mathbb{R}$. Now, by Corollary 7.4.14, we have

$$\left(\int_a^b g^2 \right)t^2 + 2\left(\int_a^b fg \right)t + \left(\int_a^b f^2 \right) = \int_a^b (f + tg)^2 \ge 0, \tag{‡}$$

for all $t \in \mathbb{R}$. But the quadratic function of t on the left side of (‡) cannot be ≥ 0 for *all real t* unless its discriminant is *nonpositive*. Writing this in detail gives (†).□

We end this section with the *First* and *Second Mean Value Theorems for Integrals*, henceforth abbreviated "First MVT for Integrals" and "Second MVT for Integrals," respectively:

Theorem 7.4.19 (First Mean Value Theorem for Integrals).   *If g is a nonnegative integrable function on $[a,b]$, then for any continuous function f on $[a,b]$ there is a point $\xi \in [a,b]$ such that*

(i) $\displaystyle \int_a^b f(x)g(x)\,dx = f(\xi) \int_a^b g(x)\,dx.$

In particular, we have

(ii) $\displaystyle \int_a^b f(x)\,dx = f(\xi)(b - a).$

Proof. Since f is continuous on the compact interval $[a,b]$, it is *bounded*. Let $m := \min\{f(x) : x \in [a,b]\}$ and $M := \max\{f(x) : x \in [a,b]\}$. Now, $g(x) \ge 0$ for all $x \in [a,b]$ implies $mg(x) \le f(x)g(x) \le Mg(x)$, for all $x \in [a,b]$. Therefore, by Corollary 7.4.14, we have

$$m \int_a^b g(x)\,dx \le \int_a^b f(x)g(x)\,dx \le M \int_a^b g(x)\,dx. \tag{*}$$

In particular, if $\int_a^b g(x)\,dx = 0$, then $\int_a^b f(x)g(x)\,dx = 0$ and (i) holds with any ξ. If $\int_a^b g(x)\,dx > 0$, dividing the inequalities (∗) by it, we get

$$m \le \frac{\int_a^b f(x)g(x)\,dx}{\int_a^b g(x)\,dx} \le M. \qquad (**)$$

But then, using (∗∗) and applying the Intermediate Value Theorem (cf. Theorem 4.5.6) to the continuous function f, we can find $\xi \in [a,b]$ such that

$$\frac{\int_a^b f(x)g(x)\,dx}{\int_a^b g(x)\,dx} = f(\xi),$$

which gives (i). Finally, for (ii), we simply use the *constant* function $g = 1$. $\qquad \square$

Theorem 7.4.20 (Second Mean Value Theorem for Integrals).   *For every function* $g \in \mathcal{R}([a,b])$, *the following are true:*

1. *If* f *is nonnegative and decreasing on* $[a,b]$, *then there is a point* $\xi \in [a,b]$ *such that*

(iii) $\displaystyle \int_a^b f(x)g(x)\,dx = f(a) \int_a^\xi g(x)\,dx.$

2. *If* f *is nonnegative and increasing on* $[a,b]$, *then there is a point* $\eta \in [a,b]$ *such that*

(iv) $\displaystyle \int_a^b f(x)g(x)\,dx = f(b) \int_\eta^b g(x)\,dx.$

Proof.

1. To prove (iii), given a partition $\mathcal{P} := (x_j)_{j=0}^n$ of $[a,b]$, let

$$S_n(f) = \sum_{k=1}^n f(x_{k-1})\Delta x_k, \qquad (\Delta x_k := x_k - x_{k-1}).$$

Also, put $\Delta f_k := f(x_k) - f(x_{k-1})$ and $\Delta G_k := G(x_k) - G(x_{k-1})$, where $G(x) := \int_a^x g(t)\,dt$. Finally, put $K := \sup\{|g(x)|\}$ on $[a,b]$ and let μ and M denote the infimum and supremum of G on $[a,b]$, respectively. Note that $g(x) + K \ge 0$ on $[a,b]$ and we have

$$\int_a^b f(g+K) = \sum_{k=1}^n \int_{x_{k-1}}^{x_k} f(g+K) \le \sum_{k=1}^n f(x_{k-1}) \int_{x_{k-1}}^{x_k} (g+K) \qquad (*)$$

$$= \sum_{k=1}^n f(x_{k-1})\Delta G_k + K S_n(f).$$

Now $G(a) = 0$, f is *decreasing*, and $f(b) \geq 0$. Therefore, by Abel's partial summation formula (Proposition 2.3.33), we have

$$\sum_{k=1}^{n} f(x_{k-1})\Delta G_k = -\sum_{k=1}^{n} G(x_k)\Delta f_k + f(b)G(b) \qquad (**)$$

$$\leq -M\sum_{k=1}^{n} \Delta f_k + f(b)G(b)$$

$$\leq M[f(a) - f(b)] + f(b)G(b) \leq f(a)M.$$

If we let $\|\mathcal{P}\| \to 0$, then $S_n(f) \to \int_a^b f$ so that by (*) and (**),

$$\int_a^b f(g + K) \leq f(a)M + K\int_a^b f,$$

which implies

$$\int_a^b fg \leq f(a)M = f(a)\sup\{G(x) : a \leq x \leq b\}. \qquad (\dagger)$$

Applying ($\dagger$) with g replaced by $-g$, we also get

$$\int_a^b fg \geq f(a)\mu = f(a)\inf\{G(x) : a \leq x \leq b\}. \qquad (\ddagger)$$

Since G is *continuous*, ($\dagger$) and ($\ddagger$) imply that, for some $\xi \in [a, b]$, we indeed have (iii):

$$\int_a^b fg = f(a)G(\xi) = f(a)\int_a^\xi g.$$

(2) To prove (iv), we apply (iii) to $f(b - x)$ and $g(b - x)$ to find a number $\xi \in [0, b - a]$ with

$$\int_0^{b-a} f(b - x)g(b - x)\, dx = f(b)\int_0^\xi g(b - x)\, dx.$$

Putting $\eta := b - \xi$ and using Exercise 7.1.24, we obtain (iv):

$$\int_a^b f(x)g(x)\, dx = f(b)\int_\eta^b g(x)\, dx.$$

$\square$

Corollary 7.4.21. *Let* $g \in \mathcal{R}([a,b])$. *If* f *is monotone on* $[a,b]$, *then there is a point* $\xi \in [a,b]$ *such that*

$$\int_a^b fg = f(a) \int_a^\xi g + f(b) \int_\xi^b g.$$

Proof. Suppose that f is *decreasing*. Then $f - f(b)$ is *nonnegative* and decreasing. By Theorem 7.4.20, we can therefore find a $\xi \in [a,b]$ such that

$$\int_a^b [f - f(b)]g = [f(a) - f(b)] \int_a^\xi g,$$

which means

$$\int_a^b fg = f(a) \int_a^\xi g + f(b) \left[\int_a^b g - \int_a^\xi g \right] = f(a) \int_a^\xi g + f(b) \int_\xi^b g,$$

and the proof is complete.   □

7.5   Fundamental Theorem of Calculus

In this section we prove the two fundamental theorems known as *the Fundamental Theorem of Calculus*. The reason why they are fundamental is that they relate the two basic concepts of *differentiation* and *integration* and provide a natural way of evaluating the integral for most "reasonable" functions obtained from "elementary functions" by simple operations. Before stating the first of these theorems, we need a definition:

Definition 7.5.1 (Primitive, Antiderivative). Let I be an interval and $f : I \to \mathbb{R}$. We say that $F : I \to \mathbb{R}$ is a *primitive* (or *antiderivative*) of f if $F'(x) = f(x)$ for all $x \in I$. A primitive of f is also called an *indefinite integral* of f.

The following proposition is in fact Corollary 6.4.13 of the Mean Value Theorem (Theorem 6.4.8).

Proposition 7.5.2. *Let* $f : [a,b] \to \mathbb{R}$. *If* F *is any primitive of* f *(on* $[a,b]$*), then any other primitive of* f *has the form* $F + C$ *for some constant* C.

Theorem 7.5.3 (First Fundamental Theorem). *Let* f *be integrable on* $[a,b]$ *and let* $C \subset [a,b]$ *be a finite set. If* $F : [a,b] \to \mathbb{R}$ *is a continuous function such that* $F'(x) = f(x)$ *for all* $x \in [a,b] \setminus C$, *then we have*

$$\int_a^b f(x)\,dx = F(b) - F(a).$$

Proof. Assume first that $C = \{a, b\}$. Given any partition $\mathcal{P} = (x_k)_{k=0}^n$ of $[a, b]$, the MVT and $F' = f$ on (a, b) imply that $F(x_j) - F(x_{j-1}) = f(t_j)(x_j - x_{j-1})$, for some $t_j \in (x_{j-1}, x_j)$, $1 \le j \le n$. Therefore, with $\dot{\mathcal{P}} := ((x_k)_{k=0}^n, (t_j)_{j=1}^n)$, we have

$$S(f, \dot{\mathcal{P}}) = \sum_{j=1}^n f(t_j)(x_j - x_{j-1}) \tag{$\dagger$}$$

$$= \sum_{j=1}^n [F(x_j) - F(x_{j-1})] = F(b) - F(a).$$

Since the right side of ($\dagger$) is independent of the partition $\mathcal{P} \in \mathcal{P}([a, b])$ and f is *integrable*, we have $\int_a^b f(x)\, dx = F(b) - F(a)$, as desired. In general, let $C := \{c_1, c_2, \ldots, c_m\}$ with $c_1 < c_2 < \cdots < c_m$. Applying what we just proved on the subintervals $[a, c_1], [c_1, c_2], \ldots, [c_m, b]$, we have

$$\int_a^b f = \int_a^{c_1} f + \int_{c_1}^{c_2} f + \cdots + \int_{c_m}^b f$$

$$= [F(c_1) - F(a)] + \sum_{k=2}^m [f(c_k) - F(c_{k-1})] + [F(b) - F(c_m)]$$

$$= F(b) - F(a).$$

$\square$

Notation 7.5.4. In what follows we shall often use the abbreviation

$$[F(x)]_a^b := F(b) - F(a).$$

Example 7.5.5. Since $\arcsin x$ is a primitive of $1/\sqrt{1 - x^2}$ on $(-1, 1)$, we have

$$\int_0^{1/2} \frac{1}{\sqrt{1 - x^2}}\, dx = [\arcsin x]_0^{1/2} = \arcsin(1/2) - \arcsin(0) = \frac{\pi}{6}.$$

Remark 7.5.6. The function f in the First Fundamental Theorem must be *integrable* on $[a, b]$. In fact, even if F is *differentiable* on $[a, b]$, its derivative $f = F'$ need *not* be integrable, as the following exercise shows.

Exercise 7.5.7. Consider the function

$$F(x) := \begin{cases} x^2 \sin(1/x^2) & \text{if } x \ne 0 \\ 0 & \text{if } x = 0. \end{cases}$$

Show that $f(x) := F'(x)$ exists for all $x \in \mathbb{R}$ and find it. Show, however, that f is *not* integrable on $[0, 1]$ so that the First Fundamental Theorem *cannot* be applied to F and f on $[0, 1]$.

Theorem 7.5.8 (Second Fundamental Theorem). *Let I be an interval and f : $I \to \mathbb{R}$. Suppose that f is integrable on any closed, bounded subinterval of I. If a is any point in I, then the function*

$$F(x) := \int_a^x f(t) \, dt \qquad (\forall x \in I)$$

is continuous on I. Moreover, if f is continuous at $x_0 \in I$, then $F'(x_0) = f(x_0)$.

Proof. Let $x_0 \in I$. If x_0 is an *interior* point, we pick $\delta > 0$ so that $J := [x_0 - \delta, x_0 + \delta] \subset I$. If x_0 is the left (resp., right) endpoint of I, then we pick $\delta > 0$ such that $J := [x_0, x_0 + \delta] \subset I$ (resp., $J := [x_0 - \delta, x_0] \subset I$). The continuity of F at x_0 now follows at once from Corollary 7.4.15 applied on the interval J. Next, assume that f is *continuous* at x_0 and let $\varepsilon > 0$ be given. Pick $\delta > 0$ so that $t \in I$ and $|t - x_0| < \delta$ imply $|f(t) - f(x_0)| < \varepsilon$. If h is such that $|h| < \delta$ and $x_0 + h \in I$, then

$$\left| \frac{F(x_0 + h) - F(x_0)}{h} - f(x_0) \right| = \left| \frac{1}{h} \int_{x_0}^{x_0 + h} f(t) \, dt - f(x_0) \right|$$

$$= \left| \frac{1}{h} \int_{x_0}^{x_0 + h} [f(t) - f(x_0)] \, dt \right|$$

$$\leq \frac{1}{|h|} |h| \varepsilon = \varepsilon.$$

Therefore, $F'(x_0) = f(x_0)$, as desired.   □

In view of Theorem 7.3.19, the following corollary is an immediate consequence of the theorem.

Corollary 7.5.9. *Let $f : I \to \mathbb{R}$ and $F(x) := \int_a^x f(t) \, dt$ be as in Theorem 7.5.8. Then $F'(x) = f(x)$ for almost all $x \in I$. In particular, if f is continuous on I, then the function $F(x)$ is the primitive of f on I with $F(a) = 0$.*

Remark 7.5.10. Note that the First Fundamental Theorem (Theorem 7.5.3) is a consequence of the second one (Theorem 7.5.8) if we assume that f is *continuous* on $[a, b]$. Indeed, under this assumption, Corollary 7.5.9 implies that $G(x) := \int_a^x f(t) \, dt$ is *the* primitive of f on $[a, b]$ satisfying $G(a) = 0$. Now, given any other primitive F of f on $[a, b]$, Proposition 7.5.2 implies that $G - F$ is *constant*, and hence, $G(x) - F(x) = G(a) - F(a) = -F(a)$, for all $x \in [a, b]$. With $x = b$, this gives

$$G(b) = \int_a^b f(t) \, dt = F(b) - F(a).$$

Corollary 7.5.11. *Let a and b be two differentiable functions on an open interval I. Let J be an interval such that $a(I) \cup b(I) \subset J$ and let f be a continuous function on J. Then the function $G(x) := \int_{a(x)}^{b(x)} f(t)\, dt$ is differentiable on I and we have*

$$G'(x) = \left(\int_{a(x)}^{b(x)} f(t)\, dt \right)' = f(b(x))b'(x) - f(a(x))a'(x), \qquad (\dagger)$$

for all $x \in I$.

Proof. Pick a fixed $c \in J$ and observe that

$$\int_{a(x)}^{b(x)} f(t)\, dt = \int_{c}^{b(x)} f(t)\, dt - \int_{c}^{a(x)} f(t)\, dt = F(b(x)) - F(a(x)), \qquad (\ddagger)$$

where $F(y) := \int_{c}^{y} f(t)\, dt$, for all $y \in J$. Since (by the Second Fundamental Theorem) $F'(y) = f(y)$, we obtain ($\dagger$) by differentiating ($\ddagger$) and using the *Chain Rule*. $\qquad \square$

Our next theorem, sometimes referred to as *integration by substitution*, is a useful tool for evaluating many integrals.

Theorem 7.5.12 (Change of Variables). *Let ϕ be a C^1 function on $[\alpha, \beta]$, and let $a := \phi(\alpha)$, $b := \phi(\beta)$. If f is continuous on the interval $\phi([\alpha, \beta])$, then*

$$\int_{a}^{b} f(x)\, dx = \int_{\alpha}^{\beta} f(\phi(t))\phi'(t)\, dt. \qquad (*)$$

Proof. Note that $f \circ \phi$ is continuous on $[\alpha, \beta]$. Now, if F is a primitive of f, then (using the Chain Rule) we have

$$\frac{d}{dt} F(\phi(t)) = F'(\phi(t))\phi'(t) = f(\phi(t))\phi'(t).$$

Therefore, by the First Fundamental Theorem, the right side of ($*$) is

$$[F(\phi(t))]_{\alpha}^{\beta} = F(\phi(\beta)) - F(\phi(\alpha)) = F(b) - F(a) = [F(x)]_{a}^{b},$$

which equals the left side $\int_{a}^{b} f(x)\, dx = [F(x)]_{a}^{b}$. $\qquad \square$

Example 7.5.13. Let $1 < \alpha < \beta$. Evaluate the integral

$$\int_{\alpha}^{\beta} \frac{1}{t \log t}\, dt.$$

Simply note that the integral has the form $\int_{\alpha}^{\beta} f(\phi(t))\phi'(t)\, dt$, where $\phi(t) := \log t$ and $f(x) := 1/x$. Therefore,

$$\int_{\alpha}^{\beta} \frac{1}{t \log t} \, dt = \int_{\log \alpha}^{\log \beta} \frac{1}{x} \, dx = \log(\log \beta) - \log(\log \alpha).$$

The *continuity* of f in the above Change of Variables Theorem may be replaced by the weaker and more natural assumption that f be Riemann integrable on $[a, b]$. So here is another version of the theorem:

Theorem 7.5.14 (Integration by Substitution). *Let f be Riemann integrable on $[a, b]$ and let $g : [c, d] \to [a, b]$ be a C^1-diffeomorphism with $g'(t) > 0$ for all $t \in [c, d]$. Then we have*

(i) $\displaystyle \int_a^b f(x) \, dx = \int_c^d f(g(t)) g'(t) \, dt.$

Proof. First note that $f(g(t)) g'(t)$ is Riemann integrable because it is the product of the continuous (hence integrable) function g' and the composite $f \circ g$ which is integrable by Corollary 7.3.23. Now let $\mathcal{P} = (t_k)_{k=0}^n$ be a partition of $[c, d]$ and, using the MVT, pick the tags $\tau_j \in [t_{j-1}, t_j]$ such that

$$g(t_j) - g(t_{j-1}) = g'(\tau_j)(t_j - t_{j-1}), \quad 1 \le j \le n.$$

Since g is a (strictly increasing) diffeomorphism, we have the corresponding partition $\mathcal{Q} = (x_k)_{k=0}^n$ of $[a, b]$ with $x_k = g(t_k)$ for all k and $\|\mathcal{P}\| \to 0$ if and only if $\|\mathcal{Q}\| \to 0$. If we set $\xi_j := g(\tau_j)$ for $1 \le j \le n$, then we have

(ii) $\displaystyle \sum_{j=1}^n f(\xi_j)(x_j - x_{j-1}) = \sum_{j=1}^n f((g(\tau_j)) g'(\tau_j)(t_j - t_{j-1}).$

Taking the limit in (ii) as $\|\mathcal{P}\| \to 0$, we obtain (i). $\qquad\qquad \square$

The following theorem is another valuable tool for evaluating integrals:

Theorem 7.5.15 (Integration by Parts). *Let f and g be integrable functions on $[a, b]$. Then, for any primitives F and G of f and g, respectively, we have*

$$\int_a^b F(x) g(x) \, dx = F(b) G(b) - F(a) G(a) - \int_a^b f(x) G(x) \, dx. \qquad (**)$$

Proof. Well, since $(FG)' = F'G + G'F = fG + gF$, the First Fundamental Theorem implies

$$\int_a^b [f(x) G(x) + g(x) F(x)] \, dx = F(b) G(b) - F(a) G(a),$$

from which $(**)$ follows at once. $\qquad\qquad \square$

Example 7.5.16. Let $0 < a < b$. Evaluate the integral $\int_a^b \log x \, dx$.

Note that the integral has the form $\int_a^b F(x)g(x)\,dx$, where $F(x) := \log x$ and $g(x) := 1$. Thus, $f(x) := F'(x) = 1/x$ and $G(x) := x$. Using integration by parts, we obtain

$$\int_a^b \log x\,dx = [x \log x]_a^b - \int_a^b \left(\frac{1}{x}\right)x\,dx = b \log b - a \log a - (b - a).$$

We end this section with another version of Taylor's Theorem. This time, as we promised in Chap. 6, we shall give another form of the remainder, called the *integral remainder:*

Theorem 7.5.17 (Taylor's Formula with Integral Remainder). *If $f : I \to \mathbb{R}$ is of class C^{n+1} [i.e., $(n + 1)$-times continuously differentiable] on I, then for any x_0, $x \in I$ we have*

$$f(x) = P_{n,x_0}(x) + \frac{1}{n!} \int_{x_0}^x (x - t)^n f^{(n+1)}(t)\,dt, \tag{†}$$

with the nth Taylor polynomial

$$P_{n,x_0}(x) := f(x_0) + \frac{f'(x_0)}{1!}(x - x_0) + \cdots + \frac{f^{(n)}(x_0)}{n!}(x - x_0)^n,$$

and the integral remainder

$$R_{n,x_0}(x) := \frac{1}{n!} \int_{x_0}^x (x - t)^n f^{(n+1)}(t)\,dt.$$

Proof. We use induction and integration by parts. For $n = 1$, the result is obvious:

$$f(x) = f(x_0) + (1/1!) \int_{x_0}^x f'(t)\,dt.$$

Assume that (†) holds with $n = k$; i.e., that we have

$$f(x) = P_{k,x_0}(x) + \frac{1}{k!} \int_{x_0}^x (x - t)^k f^{(k+1)}(t)\,dt. \tag{‡}$$

Then we must show that (‡) also holds for $n = k + 1$. Now, if f is $(k + 2)$-times continuously differentiable on I, then $f^{(k+1)}$ is *continuously differentiable.* The integral on the right side of (‡) has the form $\int_{x_0}^x u(t)v'(t)\,dt$, where $u(t) := f^{(k+1)}(t)$ and $v'(t) := (x - t)^k$. Therefore, we have $u'(t) = f^{(k+2)}(t)$ and $v(t) = -(x - t)^{k+1}/(k + 1)$. Thus, integrating by parts, we get

$$\frac{1}{k!}\int_{x_0}^{x}(x-t)^k f^{(k+1)}(t)\, dt = -\frac{1}{(k+1)!}[(x-t)^{k+1} f^{(k+1)}(t)]_{t=x_0}^{t=x}$$

$$+\frac{1}{(k+1)!}\int_{x_0}^{x}(x-t)^{k+1} f^{(k+2)}(t)\, dt$$

$$=\frac{f^{(k+1)}(x_0)}{(k+1)!}(x-x_0)^{k+1}$$

$$+\frac{1}{(k+1)!}\int_{x_0}^{x}(x-t)^{k+1} f^{(k+2)}(t)\, dt,$$

which proves the case $n = k + 1$ and completes the proof.   □

7.6 Functions of Bounded Variation

In this section we study an interesting class of functions that plays an important role in differentiation, rectifiability of curves, Fourier series, and many more situations. This class was introduced by the French mathematician *Camille Jordan* in his work on the convergence of Fourier series. We saw that *monotone* functions enjoy many nice properties. It will be seen that functions of *bounded variation* are closely related to monotone functions and hence share some of these properties as well. For example, the set of discontinuity points of a function of bounded variation is countable. We already know this to be true for monotone functions.

Definition 7.6.1 (Total Variation, Bounded Variation).   Given a function $f :$ $[a,b] \to \mathbb{R}$ and any partition $\mathcal{P} = (x_k)_{k=0}^{n}$ of $[a,b]$, let us put

$$V(f,\mathcal{P}):=\sum_{j=1}^{n}|f(x_j)-f(x_{j-1})| = \sum_{j=1}^{n}|\Delta f_j|,$$

where $\Delta f_j := f(x_j) - f(x_{j-1})$. The *total variation of f on $[a,b]$* is then defined to be the *extended* real number

$$V_a^b(f):= \sup\{V(f,\mathcal{P}) : \mathcal{P} \in \mathcal{P}([a,b])\}.$$

The function f is said to be of *bounded variation on $[a,b]$* if $V_a^b(f)$ is *finite*. The set of all functions of bounded variation on $[a,b]$ will be denoted by $BV([a,b])$.

Remark 7.6.2.

1. It is easy to see (cf. Proposition 7.6.10 below) that, if $f \in BV([a,b])$, then its *restriction* to any subinterval $[c,d] \subset [a,b]$ is of bounded variation on $[c,d]$. We shall abuse the language (and notation), however, by saying that f is of bounded variation on $[c,d]$ and write $f \in BV([c,d])$.

2. A function may be *continuous* (even *differentiable*) without being of bounded variation, as the following example shows.

Example 7.6.3 (Differentiable $\not\Rightarrow$ Bounded Variation). Consider the function

$$f(x) = \begin{cases} x^2 \sin \frac{\pi}{2x^2} & \text{if } x \in (0, 1], \\ 0 & \text{if } x = 0. \end{cases}$$

Then f is differentiable but *not* of bounded variation on $[0, 1]$.

Since $x \mapsto x^2 \sin(\pi/2x^2)$ is obviously differentiable on $\mathbb{R} \setminus \{0\}$, we need only check the differentiability of f at $x = 0$. Now, by the *Squeeze Theorem*,

$$\lim_{x \to 0} \frac{f(x) - f(0)}{x - 0} = \lim_{x \to 0} x \sin(\pi/2x^2) = 0;$$

i.e., $f'(0) = 0$. To show that f is *not* of bounded variation on $[0, 1]$, consider the partition

$$\mathcal{P} = \left(0, \frac{1}{\sqrt{2n-1}}, \frac{1}{\sqrt{2n-3}}, \ldots, \frac{1}{\sqrt{5}}, \frac{1}{\sqrt{3}}, 1\right).$$

A simple computation shows that $|f(x_1) - f(x_0)| = 1/(2n-1)$ and

$$|f(x_j) - f(x_{j-1})| = \frac{1}{2n - (2j-1)} + \frac{1}{2n - (2j-3)} \quad (2 \le j \le n).$$

Therefore,

$$\sum_{j=1}^n |\Delta f_j| = \frac{1}{2n-1} + \sum_{j=2}^n \left(\frac{1}{2n - (2j-1)} + \frac{1}{2n - (2j-3)}\right)$$

$$= 1 + \frac{2}{3} + \frac{2}{5} + \cdots + \frac{2}{2n-1}$$

$$> 1 + \frac{1}{2} + \frac{1}{3} + \cdots + \frac{1}{n}.$$

Since $\sum_{n=1}^\infty (1/n) = +\infty$, we see indeed that f is *not* of bounded variation on $[0, 1]$. The reader can check that the function in this example has an *unbounded* derivative. In fact, this *must* be the case because, if f' is bounded, then f is *Lipschitz* and hence necessarily of bounded variation (cf. Proposition 7.6.5 below).

Proposition 7.6.4 (Bounded Variation $\Rightarrow$ Bounded). *If $f : [a, b] \to \mathbb{R}$ is of bounded variation, then it is bounded on $[a, b]$.*

Proof. For each $x \in (a, b)$, consider the partition $\mathcal{P} := (a, x, b)$. Then we obviously have $|f(x) - f(a)| \le V_a^b(f)$ and the boundedness of f follows. $\qquad \square$

Proposition 7.6.5 (Monotone or Lipschitz $\Rightarrow$ Bounded Variation). *Every monotone function on $[a,b]$ is of bounded variation and so is every Lipschitz function. In particular, if $f : [a,b] \to \mathbb{R}$ is integrable, then the function*

$$F(x) := \int_a^x f(t)\,dt \quad (\forall x \in [a,b])$$

is of bounded variation on $[a,b]$.

Proof. If f is monotone (i.e., increasing or decreasing), then, for any partition $\mathcal{P} = (x_k)_{k=0}^n$ of $[a,b]$, we obviously have

$$\sum_{j=1}^n |f(x_j) - f(x_{j-1})| = |f(b) - f(a)|,$$

so that $V_a^b(f) = |f(b) - f(a)| < \infty$. If f is Lipschitz, then there is a constant $A > 0$ such that $|f(s) - f(t)| \le A|s - t|$, for all $s,\ t \in [a,b]$. It follows that, for any partition $\mathcal{P}$ as above,

$$\sum_{j=1}^n |f(x_j) - f(x_{j-1})| \le A \sum_{j=1}^n (x_j - x_{j-1}) = A(b - a),$$

which implies $V_a^b(f) \le A(b-a) < \infty$. This also proves the last statement because, by Corollary 7.4.15, F is Lipschitz on $[a,b]$. $\qquad\qquad\square$

Exercise 7.6.6. Show that, if f is *continuous* on $[a,b]$ and has a *bounded derivative* on (a,b), then it is *of bounded variation*. *Hint:* Use the MVT.

Remark 7.6.7. The sum and product of two *monotone* functions *need not* be monotone. For example, both $x \mapsto x$ and $x \mapsto x^2$ are (strictly) increasing on $[0,1]$, but $x - x^2$ is not monotone there. Also, $x \mapsto x$ is (strictly) increasing on $[-1,1]$, but $x \mapsto x^2$ is not monotone there. As the next proposition shows, however, the class of functions of bounded variation is *stable* under most algebraic operations including the operations of addition and multiplication:

Proposition 7.6.8. *If f and g are functions of bounded variation on $[a,b]$, then so are $|f|$, $\alpha f + \beta g$ (for any constants $\alpha,\ \beta \in \mathbb{R}$), $\min\{f,g\}$, $\max\{f,g\}$, and fg. If, in addition, $\inf\{|g(x)| : x \in [a,b]\} > 0$, then $1/g$ and f/g are also of bounded variation on $[a,b]$.*

Proof. Given any partition $\mathcal{P} = (x_k)_{k=0}^n$ of $[a,b]$ and any function $\phi : [a,b] \to \mathbb{R}$, let us write $\Delta\phi_j := \phi(x_j) - \phi(x_{j-1})$. Since $|\Delta|f|_j| \le |\Delta f_j|$, we have $V_a^b(|f|) \le V_a^b(f)$, which shows that $|f|$ is of bounded variation. Next, we have $\Delta(\alpha f + \beta g)_j = \alpha \Delta f_j + \beta \Delta g_j$, so that

$$\sum_{j=1}^n |\alpha \Delta f_j + \beta \Delta g_j| \le |\alpha| \sum_{j=1}^n |\Delta f_j| + |\beta| \sum_{j=1}^n |\Delta g_j|.$$

It follows that $V_a^b(\alpha f + \beta g) \leq |\alpha| V_a^b(f) + |\beta| V_a^b(g)$, and $\alpha f + \beta g$ is of bounded variation. For fg, let $A := \sup\{|f(x)| : x \in [a,b]\}$ and $B := \sup\{|g(x)| : x \in [a,b]\}$. Note that, by Proposition 7.6.4, A and B are *finite*. Now, if $h := fg$, then $\Delta h_j = f(x_j)\Delta g_j + g(x_{j-1})\Delta f_j$, from which we get

$$\sum_{j=1}^{n} |\Delta h_j| \leq A \sum_{j=1}^{n} |\Delta g_j| + B \sum_{j=1}^{n} |\Delta f_j|.$$

Therefore, we have $V_a^b(fg) \leq A V_a^b(g) + B V_a^b(f)$. For $\min\{f, g\}$ and $\max\{f, g\}$, the assertion follows from the identities used in Corollary 7.4.12. Note, in particular, that both $f^+ := \max\{f, 0\}$ and $f^- := \max\{-f, 0\}$ are of bounded variation. Finally, suppose that $m := \inf\{|g(x)| : x \in [a,b]\} > 0$. Then

$$\left| \frac{1}{g(x_j)} - \frac{1}{g(x_{j-1})} \right| = \frac{|g(x_j) - g(x_{j-1})|}{|g(x_j)g(x_{j-1})|} \leq \frac{1}{m^2}|\Delta g_j|,$$

and hence $V_a^b(1/g) \leq \frac{1}{m^2} V_a^b(g)$. Since $f/g = f \cdot (1/g)$, it also follows that $V_a^b(f/g)$ is finite and the proof is complete. $\qquad \square$

Corollary 7.6.9. *If f and g are two increasing functions on $[a,b]$, then $f - g$ is of bounded variation on $[a,b]$.*

Proof. This follows at once from Propositions 7.6.5 and 7.6.8. $\qquad \square$

Next we show that, like the Riemann integral, the total variation is *additive with respect to intervals*:

Proposition 7.6.10 (Interval Additivity). *Let $f : [a,b] \to \mathbb{R}$, with $a < b$, and let $c \in (a,b)$. Then $f \in BV([a,b])$ if and only if $f \in BV([a,c]) \cap BV([c,b])$, and we have*

$$V_a^b(f) = V_a^c(f) + V_c^b(f). \tag{†}$$

Proof. Suppose that $f \in BV([a,b])$ and let $\mathcal{P}_1$ and $\mathcal{P}_2$ be partitions of $[a,c]$ and $[c,b]$, respectively. If $\mathcal{P}$ is the partition of $[a,b]$ containing all the points of $\mathcal{P}_1$ and $\mathcal{P}_2$, then

$$V(f, \mathcal{P}_1) + V(f, \mathcal{P}_2) = V(f, \mathcal{P}) \leq V_a^b(f).$$

Therefore, f is of bounded variation on $[a,c]$ and on $[c,b]$, and we have

$$V_a^c(f) + V_c^b(f) \leq V_a^b(f). \tag{*}$$

Conversely, let $\mathcal{P}$ be any partition of $[a,b]$ and let $\mathcal{P}' := \mathcal{P} \cup \{c\}$. If $\mathcal{P}_1$ and $\mathcal{P}_2$ are the partitions of $[a,c]$ and $[c,b]$, respectively, induced by $\mathcal{P}'$, then we have

$$V(f, \mathcal{P}) \leq V(f, \mathcal{P}') = V(f, \mathcal{P}_1) + V(f, \mathcal{P}_2) \leq V_a^c(f) + V_c^b(f).$$

Hence $f \in BV([a, b])$ and we also have

$$V_a^b(f) \le V_a^c(f) + V_c^b(f).$$ $\qquad (**)$

Combining $(*)$ and $(**)$, we obtain $(\dagger)$ and the proof is complete. $\qquad\square$

Remark 7.6.11. If $f : [a, b] \to \mathbb{R}$ and if $\mathcal{P} = (x_k)_{k=0}^n$ is any partition of $[a, b]$, then Proposition 7.6.10 implies that

$$V_a^b(f) = \sum_{j=1}^n V_{x_{j-1}}^{x_j}(f).$$

In particular, if f is *monotone* on each subinterval $[x_{j-1}, x_j]$, then (in view of Proposition 7.6.5) $V_a^b(f)$ can be easily computed:

$$V_a^b(f) = \sum_{j=1}^n |f(x_j) - f(x_{j-1})|.$$

For example, we have

$$V_{-\pi/2}^{3\pi/2}(\cos) = [\cos(0) - \cos(-\pi/2)] + [\cos(0) - \cos(\pi)] + [\cos(3\pi/2) - \cos(\pi)]$$

$$= 1 + 2 + 1 = 4.$$

Definition 7.6.12 (Total Variation Function). Let f be of bounded variation on $[a, b]$. The function $v_f(x) := V_a^x(f)$, for all $x \in (a, b]$, and $v_f(a) := 0$ is called the *total variation function* of f on $[a, b]$.

Proposition 7.6.13. *Let $f : [a, b] \to \mathbb{R}$ be of bounded variation on $[a, b]$. Then the total variation function v_f is increasing on $[a, b]$ and, for each $c \in (a, b]$ (resp., $c \in [a, b)$), v_f is left (resp., right) continuous at c if and only if f is left (resp., right) continuous at c.*

Proof. That v_f is *increasing* is obvious. (Why?) Suppose that v_f is *left continuous* at $c \in (a, b]$. Since

$$|f(c) - f(x)| \le V_x^c(f) = v_f(c) - v_f(x),$$

for all $a \le x < c$, we see that $f(x) \to f(c)$ as $x \to c-$, and f is indeed left continuous at c. A similar argument shows that, if v_f is *right continuous* at $c \in [a, b)$, then so is f. Now suppose that f is *left continuous* at $c \in (a, b]$. Then, given any $\varepsilon > 0$, we can pick a partition $\mathcal{P} = (x_k)_{k=0}^n$ of $[a, c]$ such that

$$\sum_{j=1}^n |f(x_j) - f(x_{j-1})| > v_f(c) - \frac{\varepsilon}{2},$$

and $|f(c) - f(x_{n-1})| < \varepsilon/2$. (Why?) It then follows that $v_f(x) > v_f(c) - \varepsilon$, for all $x \in (x_{n-1}, c)$. (Why?) Therefore, v_f is indeed left continuous at c. A similar argument shows that, if f is right continuous at $c \in [a, b)$, then so is v_f and the proof is complete.   □

We are now ready for our main theorem:

Theorem 7.6.14 (Jordan Decomposition Theorem). *A function $f : [a, b] \to \mathbb{R}$ is of bounded variation on $[a, b]$ if and only if it is the difference of two increasing functions.*

Proof. If f is the difference of two increasing functions, then it is of bounded variation (cf. Corollary 7.6.9). Conversely, suppose $f \in BV([a, b])$ and let $w_f := v_f - f$, where v_f is the total variation function of f defined above. Then $f = v_f - w_f$. Also, we already know that v_f is increasing. Thus, we need only show that w_f is increasing as well. Now, if $a \leq x < x' \leq b$, then, by Proposition 7.6.10,

$$w_f(x') - w_f(x) = [v_f(x') - v_f(x)] - [f(x') - f(x)]$$
$$= V_x^{x'}(f) - [f(x') - f(x)] \geq 0,$$

and the proof is complete.   □

Corollary 7.6.15. *A continuous function $f : [a, b] \to \mathbb{R}$ is of bounded variation on $[a, b]$ if and only if it is the difference of two increasing, continuous functions on $[a, b]$.*

Proof. This follows at once from Theorem 7.6.14 and Proposition 7.6.13.   □

Corollary 7.6.16. *If f is increasing (resp., decreasing) on $[a, b]$ and if g is of bounded variation on $[f(a), f(b)]$ (resp., $[f(b), f(a)]$), then $g \circ f$ is of bounded variation on $[a, b]$.*

Proof. Suppose f is *increasing*. By Theorem 7.6.14, we can pick two increasing functions ϕ and ψ on $[f(a), f(b)]$ such that $g = \psi - \phi$. It then follows that $g \circ f = \psi \circ f - \phi \circ f$ is the difference of two increasing functions and hence is of bounded variation as claimed.   □

Corollary 7.6.17 (Bounded Variation $\Rightarrow$ Regulated). *If f is of bounded variation on $[a, b]$, then it is regulated on $[a, b]$. In other words, $BV([a, b]) \subset Reg([a, b])$. In particular, the set of discontinuity points of f is countable.*

Proof. This follows from Theorem 7.6.14, because *monotone* functions are regulated and have countable discontinuity sets.   □

7.7 Problems

1. Let $f, h \in \mathcal{R}([a, b])$ and let $g : [a, b] \to \mathbb{R}$. Show that, if $f(x) \le g(x) \le h(x)$ for all $x \in [a, b]$ and $\int_a^b f = A = \int_a^b h$, then $g \in \mathcal{R}([a, b])$ and $\int_a^b g = A$.

2. Let $f \in \mathcal{R}([a, b])$. Show that

$$\int_a^b f(x)\, dx = \lim_{\delta \to 0+} \int_{a+\delta}^b f(x)\, dx = \lim_{\delta \to 0+} \int_a^{b-\delta} f(x)\, dx = \lim_{\delta \to 0+} \int_{a+\delta}^{b-\delta} f(x)\, dx.$$

3. Give an example of a *positive* function f on $[0, 1]$ such that $f \in \mathcal{R}([0, 1])$ but $1/f \notin \mathcal{R}([0, 1])$.

4 ($\mathcal{R}([a, b])$ Is Not Closed Under Composition).

(a) Consider the function

$$f(x) = \begin{cases} \dfrac{1}{q} & \text{if } x = \dfrac{p}{q} \in (0, 1] \quad (p,\, q \in \mathbb{N},\ \gcd(p, q) = 1), \\ 0 & \text{if } x \in \mathbb{Q}^c \cap (0, 1], \end{cases}$$

and $f(0) := 1$. Show that $f \in \mathcal{R}([0, 1])$ and that $\int_0^1 f(x)\, dx = 0$.

(b) Let f be the function in part (a) and let $g := \chi_{(0,1]}$. Show that $g \circ f \notin \mathcal{R}([0, 1])$ even though $f, g \in \mathcal{R}([0, 1])$.

5. Show that a *bounded, infinite* set with a *finite* set of limit points has content zero.

6. Let $f, g : [a, b] \to \mathbb{R}$. Assume that f is *bounded*, $g \in \mathcal{R}([a, b])$, and the set $\{x \in [a, b] : f(x) \ne g(x)\}$ has *content zero*. Show that $f \in \mathcal{R}([a, b])$ and $\int_a^b f = \int_a^b g$.

7. Show that the functions

$$f(x) := \begin{cases} x \sin(1/x) & \text{if } 0 < |x| \le 1, \\ 0 & \text{if } x = 0 \end{cases} \quad \text{and} \quad g(x) := \begin{cases} x/|x| & \text{if } 0 < |x| \le 1, \\ 0 & \text{if } x = 0 \end{cases}$$

are both *regulated* on $[-1, 1]$ but the composite function $g \circ f$ is *not* regulated.

8. Show that, if $f \in \mathcal{R}([a, b])$, then there exists an $x \in [a, b]$ such that $\int_a^x f(t)\, dt = \int_x^b f(t)\, dt$. Give an example where the *only* such x is an *endpoint*. *Hint:* Look at the function $g(x) := \int_a^x f(t)\, dt - \int_x^b f(t)\, dt$.

9 (Cauchy's Integral Test).

(a) Let $k \in \mathbb{N}$ and let f be a *nonnegative, decreasing* function such that $[k, \infty) \subset \operatorname{dom}(f)$. Show that $\sum_{n=k}^{\infty} f(n)$ is convergent if and only if $\lim_{b \to \infty} \int_k^b f(x)\, dx$ exists (in $[0, \infty)$). *Hint:* Compare the *left* and *right* Riemann sums of f on $[k, n]$ to $\int_k^n f(x)\, dx$.

(b) Show that the series $\sum_{n=2}^{\infty} \dfrac{1}{n (\log n)^p}$ converges if and only if $p > 1$.

10. Let $a > 0$ and $f \in \mathcal{R}([-a, a])$. Show that, if f is *even* (i.e., $f(-x) = f(x)$ for all $x \in [-a, a]$), then $\int_{-a}^a f(x)\, dx = 2 \int_0^a f(x)\, dx$. Similarly, show that if f is *odd* (i.e., $f(-x) = -f(x)$ for all $x \in [-a, a]$), then $\int_{-a}^a f(x)\, dx = 0$.

11 (Average Value). Let $a < b$ and $f \in \mathcal{R}([a, b])$. Define the *average value* of f on $[a, b]$ to be the number

$$\mathrm{Av}(f) := \frac{1}{b - a} \int_a^b f(x)\, dx.$$

(a) Show that

$$\mathrm{Av}(f) = \lim_{n \to \infty} \frac{1}{n} \sum_{k=1}^n f\left(a + k \frac{b - a}{n}\right). \tag{$*$}$$

(b) Let L denote the limit on the right side of $(*)$ in (a). Give an example of a function $f \notin \mathcal{R}([0, 1])$ for which L exists. This will show that $(b - a)L$ *cannot* be used as the *definition* of $\int_a^b f(x)\, dx$.

12. Find each limit.

$$(a) \quad \lim_{n \to \infty} \sum_{k=1}^n \frac{1}{n + k}, \qquad\qquad (b) \quad \lim_{n \to \infty} \sum_{k=1}^n \frac{n}{n^2 + k^2}.$$

Hint: For (a), note that $1/(n + k) = (1/n)/(1 + k/n)$ and interpret the sum as a Riemann sum for $\int_0^1 dx/(1 + x)$.

13. Find each limit.

$$(a) \quad \lim_{n \to \infty} \frac{1}{n} \sum_{k=1}^n \sin \frac{k\pi}{n}, \qquad\qquad (b) \quad \lim_{n \to \infty} \frac{1}{n} \sum_{k=1}^n \frac{k}{\sqrt{n^2 + k^2}}.$$

14. Let $f \in C[0, \infty)$ and let $\mathrm{Av}(f, b) := (\int_0^b f(x)\, dx)/b$ be the *average value* of f on $[0, b]$. Show that if $\lim_{x \to \infty} f(x) = \ell$, then $\lim_{b \to \infty} \mathrm{Av}(f, b) = \ell$.

15. Let f be *strictly increasing* on $[a, b]$. Consider a partition $\mathcal{Q} := (y_j)_{j=0}^n$ of $[f(a), f(b)]$ and the corresponding partition $\mathcal{P} := (x_j)_{j=0}^n$ of $[a, b]$, where $f(x_j) = y_j$ for $0 \le j \le n$. Show that

$$U(f, \mathcal{P}) + L(f^{-1}, \mathcal{Q}) = bf(b) - af(a).$$

Deduce that

$$\int_{f(a)}^{f(b)} f^{-1}(x)\, dx = bf(b) - af(a) - \int_a^b f(x)\, dx.$$

Hint: Draw a picture! Show a similar result for a *strictly decreasing* function f on $[a, b]$. Namely, given the partitions $\mathcal{P} := (x_j)_{j=0}^n$ and $\mathcal{Q} := (y_j)_{j=0}^n$ as above, show that

$$L(f, \mathcal{P}) + U(f^{-1}, \mathcal{Q}) = bf(b) - af(a).$$

Deduce that we have

$$\int_{f(b)}^{f(a)} f^{-1}(x)\, dx = af(a) - bf(b) + \int_a^b f(x)\, dx.$$

16. Given any *positive* numbers p and q, show that

$$\int_0^1 (1 - x^p)^{1/q} \, dx = \int_0^1 (1 - x^q)^{1/p} \, dx.$$

17 (Young's Inequality).   Let $f \in C([0, \infty))$ be *strictly increasing*, $f(0) = 0$, and $\lim_{x \to \infty} f(x) = \infty$. Show that

$$ab \le \int_0^a f(x) \, dx + \int_0^b f^{-1}(x) \, dx \qquad (\forall a, \, b \in [0, \infty)),$$

with equality holding if and only if $b = f(a)$. *Hint:* Let $g := f^{-1}$ and consider the corresponding primitives $F(x) = \int_0^x f(t) \, dt$ and $G(x) := \int_0^x g(t) \, dt$, vanishing at $x = 0$. For fixed $b \ge 0$, define $H(a) := F(a) + G(b) - ab$ and note that $H(a) = 0$ if and only if $f(a) = b$. (Why?)

18. Let $p > 1$ and let $q := p/(p-1)$ so that $1/p + 1/q = 1$. Show that, for any $a, \, b \in [0, \infty)$,

$$ab \le \frac{a^p}{p} + \frac{b^q}{q},$$

with equality holding if and only if $a^p = b^q$. *Hint:* Let $f(x) := x^{p-1}$ and use *Young's Inequality*.

19 (Jensen's Inequality).

(a) Show that, if $\phi \in \mathbb{R}^{\mathbb{R}}$ is a *convex* function and if $f \in \mathcal{R}([0, 1])$, then we have

$$\phi\left(\int_0^1 f(x) \, dx\right) \le \int_0^1 \phi(f(x)) \, dx.$$

(b) Show that, for any $f \in \mathcal{R}([0, 1])$, we have

$$\exp\left(\int_0^1 f(x) \, dx\right) \le \int_0^1 \exp(f(x)) \, dx.$$

   Hint: Note that ϕ is *continuous*.

20. Let $f : [a, b] \to [0, \infty)$ be *continuous*. Show that

$$\lim_{n \to \infty} \left(\int_a^b (f(x))^n\right)^{1/n} = \sup\{f(x) : x \in [a, b]\}.$$

Hint: Let $M := \sup\{f(x) : x \in [a, b]\}$ and note that, given any $\varepsilon > 0$, we have $f(x) > M - \varepsilon$ in some interval $[c, d] \subset [a, b]$.

21. Let $f : [a, b] \to \mathbb{R}$ be a *continuous* function such that, $\int_\alpha^\beta f(x) \, dx = 0$ for all $[\alpha, \beta] \subset [a, b]$. Show that $f(x) = 0$ for all $x \in [a, b]$.

22. Show that, given any $f \in \mathcal{R}([a, b])$ and any $\varepsilon > 0$, there is a *polynomial* function $p_\varepsilon(x)$ such that

$$\int_a^b |f(x) - p_\varepsilon(x)| \, dx < \varepsilon.$$

23 (Lerch's Theorem).   Let $f \in C([0, 1])$. Show that, if $\int_0^1 x^n f(x) \, dx = 0$ for all $n \in \mathbb{N} \cup \{0\}$, then $f(x) = 0$ for all $x \in [0, 1]$.

24. Let $f \in C([0, 1])$. Show that, if $\int_0^1 x^k f(x) \, dx = 0$ for $k = 0, 1, \ldots, n - 1$ and $\int_0^1 x^n f(x) \, dx = 1$, then $|f(x_0)| \ge 2^n (n + 1)$ for some $x_0 \in [0, 1]$. *Hint:* $\int_0^1 (x - 1/2)^n f(x) \, dx = 1$.

25. Let $f \in C([a, b])$ and suppose that $\int_a^b f(x)g(x)\, dx = 0$ for all $g \in \textbf{\textit{Step}}([a, b])$. Show that $f(x) = 0$ for all $x \in [a, b]$.

26. Show that, if $f \in \mathcal{R}([0, 1])$, then

$$\left(\int_0^1 f(x)\, dx \right)^2 \le \int_0^1 [f(x)]^2\, dx.$$

27 (Poincaré–Wirtinger Inequality). Let $a > 0$ and $f : [-a, a] \to \mathbb{R}$ a continuously differentiable function. Show that we have

$$\int_{-a}^a \left(f(x) \right)^2 dx \le 4a^2 \int_{-a}^a \left(f'(x) \right)^2 dx. \tag{†}$$

28 (Lyapunov's Inequality). Let $a < b$ and let $p : [a, b] \to \mathbb{R}$ be a nonzero, continuous function. Suppose that $f \ne 0$ is a twice continuously differentiable function satisfying $f''(x) + p(x)f(x) = 0$ on $[a, b]$ and $f(a) = f(b) = 0$. Show that we then have

$$\int_a^b |p(x)|\, dx = \int_a^b \left| \frac{f''(x)}{f(x)} \right| dx > \frac{4}{b - a}.$$

29. Show that, if $0 < a < b$, then

$$\left| \int_a^b \frac{\sin x}{x}\, dx \right| \le 2 \left(\frac{1}{a} + \frac{1}{b} \right).$$

Deduce that $\lim_{b \to \infty} \int_a^b x^{-1} \sin x\, dx$ exists. *Hint:* Use Corollary 7.4.21.

30. Using the *integral remainder* in Theorem 7.5.17 and the First MVT for Integrals (Theorem 7.4.19) deduce the *Cauchy* form of the remainder for Taylor's Formula:

$$R_{n,x_0}(x) = \frac{f^{(n+1)}(\xi)}{n!} (x - \xi)^n (x - x_0).$$

31. Show that

$$F(x) := \int \frac{dx}{(1 + x^2)^{3/2}} = \sin(\arctan x) + C.$$

32. Show that

$$\int_0^1 \frac{x^4(1 - x)^4}{1 + x^2}\, dx = \frac{22}{7} - \pi.$$

Hint: Use "long division."

33. Let $f \in C([0, \infty))$ and $f(x) \ne 0$ for all $x > 0$. Show that, if

$$[f(x)]^2 = 2 \int_0^x f(t)\, dt \qquad (\forall x > 0),$$

then $f(x) = x$ for all $x \in [0, \infty)$.

34. Show that, if $\alpha < 1$, then

$$\lim_{x \to \infty} x^\alpha \int_x^{x+1} \sin(t^2)\, dt = 0.$$

Hint: Using *integration by parts*, estimate $|\int_x^{x+1} \sin(t^2)\, dt|$.

35. Let $f \in C^1([a, b])$ and $f(a) = f(b) = 0$.

(a) Show that

$$\int_a^b x f(x) f'(x)\, dx = -\frac{1}{2} \int_a^b [f(x)]^2\, dx.$$

(b) Show that, if we also have $\int_a^b [f(x)]^2\, dx = 1$, then

$$\left(\int_a^b [x f(x)]^2\, dx \right) \left(\int_a^b [f'(x)]^2\, dx \right) \geq \frac{1}{4}.$$

36. Let $b > 0$ and suppose that $f \in C([0, b])$ satisfies $f(x) + f(b - x) \neq 0$ for all $x \in [0, b]$. Evaluate the integral

$$\int_0^b \frac{f(x)}{f(x) + f(b - x)}\, dx.$$

Hint: Let $g(x) := f(x) + f(b - x)$. Look at the integrals $\int_0^b [f(x)/g(x)]\, dx$ and $\int_0^b [f(b - x)/g(x)]\, dx$.

37. Let $b > 0$ and let $f : [0, b] \to \mathbb{R}$ be a *differentiable* function such that $f'(b - x) = f'(x)$ for all $x \in [0, b]$. Evaluate $\int_0^b f(x)\, dx$. Can you give an example of such a function?

38. Let $b > 0$ and consider the set of all functions $f : [0, b] \to \mathbb{R}$ that satisfy the functional equation

$$f(x) f(b - x) = 1 \qquad (\forall x \in [0, b]). \tag{$*$}$$

(a) Show that there are infinitely many functions $f \in C([0, b])$ satisfying $(*)$. *Hint:* Use the exponential function.

(b) If $f \in \mathcal{R}([0, b])$ is any *positive* function satisfying $(*)$, for all $x \in [0, b]$, calculate the integral

$$\int_0^b \frac{dx}{1 + [f(x)]^{\sqrt{2}}}.$$

Hint: Use the substitution $u := b - x$.

39. Find all *differentiable* functions $f \in \mathbb{R}^{\mathbb{R}}$ which satisfy $[f(x)]^3 = \int_1^x [f(t)]^2\, dt$.

40. Find the derivative of each function.

(a) $\displaystyle \int_0^{x^2} e^{-\sqrt{t}}\, dt$, (b) $\displaystyle \int_{\sqrt{1+x^2}}^1 \sin(t^2)\, dt$, (c) $\displaystyle \int_x^{x^2} \sec(e^t)\, dt$.

41 (Euler's Beta Function). Show that

$$B(m,n) := \int_0^1 x^{m-1}(1-x)^{n-1}\, dx = \frac{(m-1)!(n-1)!}{(m+n-1)!} \qquad (\forall m, n \in \mathbb{N}).$$

Hint: Use repeated integration by parts, first getting $B(m,n) = \frac{n-1}{m} B(m+1, n-1)$.

42. Show that the function

$$f(x) := \begin{cases} x \sin(\log x) & \text{if } x \in (0, 1], \\ 0 & \text{if } x = 0 \end{cases}$$

is *differentiable* on $(0, 1]$ (but *not* at $x = 0$), that (defining $f'(0)$ arbitrarily) $f' \in \mathcal{R}([0, 1])$, and that $\int_0^1 f'(x)\, dx = 0$.

43. Given any $n \in \mathbb{N} \cup \{0\}$, show that the function

$$f_n(x) := \begin{cases} \dfrac{\sin(2n+1)x}{\sin x} & \text{if } x \in (0, \pi/2], \\ 2n+1 & \text{if } x = 0 \end{cases}$$

is *integrable* on $[0, \pi/2]$ and find $J_n := \int_0^{\pi/2} f_n(x)\, dx$ by considering $J_{n+1} - J_n$. Deduce the value of the integral

$$I_n := \int_0^{\pi/2} \frac{\sin^2 nx}{\sin^2 x}\, dx,$$

using the differences $I_{n+1} - I_n$. Here, $\sin^2 nx / \sin^2 x := n^2$ at $x = 0$.

44 (Wallis' Formula).

(a) Let $I_0 := \pi/2$ and $I_n := \int_0^{\pi/2} \sin^n x\, dx$ for all $n \in \mathbb{N}$. Show that $I_n = (n-1)(I_{n-2} - I_n)$ for all $n \geq 2$, and deduce that

$$I_{2n} = \frac{1 \cdot 3 \cdot 5 \cdots (2n-1)}{2 \cdot 4 \cdot 6 \cdots (2n)} \frac{\pi}{2} \qquad (\forall n \geq 1).$$

(b) Similarly, show that $I_1 = 1$ and hence

$$I_{2n+1} = \frac{2 \cdot 4 \cdot 6 \cdots (2n)}{1 \cdot 3 \cdot 5 \cdots (2n+1)} \qquad (\forall n \geq 1).$$

(c) Show that

$$\frac{I_{2n}}{I_{2n+1}} = \left(\frac{1 \cdot 3 \cdot 5 \cdots (2n-1)}{2 \cdot 4 \cdot 6 \cdots (2n)} \right)^2 \frac{(2n+1)\pi}{2}.$$

(d) Prove the inequalities $0 \leq I_{2n+1} \leq I_{2n} \leq I_{2n-1}$ and show that

$$1 \leq \frac{I_{2n}}{I_{2n+1}} \leq \frac{I_{2n-1}}{I_{2n+1}} = 1 + \frac{1}{2n}.$$

(e) Show that

$$\lim_{n \to \infty} \left(\frac{2 \cdot 4 \cdot 6 \cdots (2n)}{1 \cdot 3 \cdot 5 \cdots (2n-1)\sqrt{2n+1}} \right) = \sqrt{\frac{\pi}{2}}.$$

(f) Show that

$$\lim_{n\to\infty} \sqrt{n} \cdot \frac{2\cdot 4\cdot 6\cdots(2n)}{1\cdot 3\cdot 5\cdots(2n+1)} = \frac{\sqrt{\pi}}{2}.$$

45.

(a) Show that

$$\int_0^1 (1-x^2)^n \, dx = \frac{2\cdot 4\cdot 6\cdots(2n)}{1\cdot 3\cdot 5\cdots(2n+1)} \qquad (\forall n \geq 1).$$

Hint: Use the substitution $x = \sin t$.

(b) Similarly, show that

$$\int_0^\infty \frac{1}{(1+x^2)^n} \, dx := \lim_{b\to\infty} \int_0^b \frac{1}{(1+x^2)^n} \, dx = \frac{1\cdot 3\cdot 5\cdots(2n-3)}{2\cdot 4\cdot 6\cdots(2n-2)}\frac{\pi}{2} \qquad (\forall n \geq 2).$$

Hint: Make the substitution $x = \cot u$.

(c) Using the derivative, prove the inequalities $1 - x^2 \leq e^{-x^2}$ for all $x \in [0,1]$ and $e^{-x^2} \leq 1/(1+x^2)$ for all $x \geq 0$. Deduce that, for all $n \in \mathbb{N}$,

$$(1-x^2)^n \leq e^{-nx^2} \quad (\forall x \in [0,1]), \qquad \text{and} \qquad e^{-nx^2} \leq 1/(1+x^2)^n \quad (\forall x \geq 0).$$

(d) Integrating the inequalities in (c) and using the substitution $\xi := x\sqrt{n}$, deduce that

$$\sqrt{n} \cdot \frac{2\cdot 4\cdot 6\cdots(2n)}{1\cdot 3\cdot 5\cdots(2n+1)} \leq \int_0^{\sqrt{n}} e^{-\xi^2} \, d\xi \leq \sqrt{n} \cdot \frac{1\cdot 3\cdot 5\cdots(2n-3)}{2\cdot 4\cdot 6\cdots(2n-2)}\frac{\pi}{2}.$$

(e) Finally, conclude that

$$\int_0^\infty e^{-\xi^2} \, d\xi := \lim_{b\to\infty} \int_0^b e^{-\xi^2} \, d\xi = \frac{\sqrt{\pi}}{2}.$$

Hint: Use Problem 44.

46. Let $\alpha > 0$ and $\beta > 0$ be given. Show that the function

$$f(x):= \begin{cases} x^\alpha \sin\left(\dfrac{\pi}{2x^\beta}\right) & \text{if } x \in (0,1], \\ 0 & \text{if } x = 0 \end{cases}$$

is *continuous* but *not* of *bounded variation* if $\alpha \leq \beta$. *Hint:* First prove the case $\alpha = \beta = 1$ as in Example 7.6.3, then use Corollary 7.6.16.

47.

(a) Show that the function

$$f(x) := \begin{cases} x \cos\left(\dfrac{\pi}{2x}\right) & \text{if } x \in (0,1], \\ 0 & \text{if } x = 0 \end{cases}$$

is *continuous* but *not* of *bounded variation*. *Hint:* Consider the partition $\mathcal{P}_n :=$ $(0, 1/(2n), 1/(2n-1), \ldots, 1/2, 1)$.

(b) Show that the function

$$g(x) := \begin{cases} x^2 \cos\left(\dfrac{1}{x}\right) & \text{if } x \in (0, 1], \\ 0 & \text{if } x = 0 \end{cases}$$

is of bounded variation. *Hint:* Look at $g'(x)$.

48. Let $\mathcal{P} = (x_k)_{k=0}^n$ be a partition of $[a, b]$ and let $\phi \in Step([a, b])$ be *constant* on each $I_j := (x_{j-1}, x_j)$, $1 \leq j < n$. Show that $\phi \in BV([a, b])$ and we have

$$V_a^b(f) = |d_0^+| + |d_n^-| + \sum_{k=1}^{n-1}(|d_k^-| + |d_k^+|),$$

where $d_k^- := f(x_k) - f(x_k-)$ and $d_k^+ := f(x_k+) - f(x_k)$ are the *left* and *right jumps* of f at x_k, for $1 \leq k \leq n-1$, $d_0^+ := f(a+) - f(a)$, $d_n^- := f(b) - f(b-)$.

49. Show that, if $f \in BV([a, b])$ has the *Intermediate Value Property*, then $f \in C([a, b])$. *Hint:* Note that $f(x + 0)$ and $f(x - 0)$ exist for all $x \in [a, b]$.

50. Show that if $f \in C([a, b])$ has a *finite* number of (local) *maxima* and *minima*, then $f \in BV([a, b])$.

51. Show that, if $f \in C^1([a, b])$ (i.e., both f and f' are continuous on $[a, b]$), then $f \in BV([a, b])$ and we have

$$V_a^b(f) = \int_a^b |f'(x)| \, dx.$$

Hint: Use the MVT.

52 (Length of a Rectifiable Curve). Let $f : I \to \mathbb{R}$, where $I := [a, b]$. For each partition $\mathcal{P} := (x_k)_{k=0}^n$ of I, consider the sum

$$\ell(f, \mathcal{P}) := \sum_{j=1}^n \sqrt{(x_j - x_{j-1})^2 + [f(x_j) - f(x_{j-1})]^2}$$

which represents the length of the corresponding *polygonal arc* inscribed in the graph of f [with successive vertices $(x_0, f(x_0)), \ldots, (x_n, f(x_n))$]. Now define the *length of f on I* to be

$$L_a^b(f) := \sup\{\ell(f, \mathcal{P}) : \mathcal{P} \in \mathcal{P}(I)\}.$$

(a) Show that we have the inequalities

$$V_a^b(f) + (b - a) \geq L_a^b(f) \geq \sqrt{[V_a^b(f)]^2 + (b - a)^2},$$

so that $f \in BV([a, b])$ if and only if it is *rectifiable*, i.e., has *finite length*.
(b) Show that, if $a < c < b$, then we have

$$L_a^b(f) = L_a^c(f) + L_c^b(f).$$

(c) Show that, if $f \in C^1([a,b])$, then f is rectifiable and we have

$$L_a^b(f) = \int_a^b \sqrt{1 + [f'(x)]^2}\, dx.$$

Hint: Rewrite $\ell(f, \mathcal{P})$ using the MVT.
(d) Find the length of $f(x) := \log(\cos x)$ on $[0, \pi/4]$.

53. Let f be a real-valued function with *continuous*, *nonnegative* derivative such that $f(0) = 0$ and $f(1) = 1$. Show that if ℓ denotes the length of f on $[0, 1]$, then we have

$$\sqrt{2} \le \ell < 2.$$

54 (Positive & Negative Variations). Let $f : [a, b] \to \mathbb{R}$. Then for any partition $\mathcal{P} = (x_k)_{k=0}^n$ of $[a, b]$ we define, with $\Delta f_j := f(x_j) - f(x_{j-1})$,

$$P(f, \mathcal{P}) := \sum_{j=1}^n [f(x_j) - f(x_{j-1})]^+ = \sum_{j=1}^n (\Delta f_j)^+,$$

$$N(f, \mathcal{P}) := \sum_{j=1}^n [f(x_j) - f(x_{j-1})]^- = \sum_{j=1}^n (\Delta f_j)^-,$$

$$P_a^b(f) := \sup\{V^+(f, \mathcal{P}) : \mathcal{P} \in \mathcal{P}([a, b])\},$$

$$N_a^b(f) := \sup\{V^-(f, \mathcal{P}) : \mathcal{P} \in \mathcal{P}([a, b])\},$$

$$p_f(x) = P_a^x(f), \quad \forall\, x \in [a, b], \quad \text{and}$$

$$n_f(x) = N_a^x(f), \quad \forall\, x \in [a, b],$$

where $\forall\, t \in \mathbb{R}$, we have $t^+ := \max(t, 0)$ and $t^- := \max(-t, 0)$ so that $t = t^+ - t^-$ and $|t| = t^+ + t^-$. We call $P_a^b(f)$ and $N_a^b(f)$ the *positive* and *negative* variations of f on $[a, b]$, respectively. The functions p_f and n_f are called the *positive* and *negative variation functions* of f. Show the following for any $f \in BV([a, b])$.

(a) $\max\left(P_a^b(f), N_a^b(f)\right) \le V_a^b(f) \le P_a^b(f) + N_a^b(f).$
(b) $f(b) - f(a) = P_a^b(f) - N_a^b(f).$
(c) $V_a^b(f) = P_a^b(f) + N_a^b(f.$
(d) $f(x) - f(a) = p_f(x) - n_f(x)$ and $v_f(x) = p_f(x) + n_f(x).$

55. Recall (Problem 4.8.#53) that a function $F : [a, b] \to \mathbb{R}$ is said to be *absolutely continuous* (on $[a, b]$) if for each $\varepsilon > 0$ there is a $\delta > 0$ such that, given any collection $\{(a_k, b_k) : 1 \le k \le n\}$ of *pairwise disjoint* open subintervals of $[a, b]$, we have

$$\sum_{k=1}^n (b_k - a_k) < \delta \implies \sum_{k=1}^n |F(b_k) - F(a_k)| < \varepsilon.$$

(a) Show that, if $F : [a, b] \to \mathbb{R}$ is *absolutely continuous*, then $F \in BV([a, b])$.
(b) Show that every $F \in Lip([a, b])$ is *absolutely continuous*. In particular, $F(x) := \int_a^x f(t)\, dt$ is absolutely continuous for every $f \in \mathcal{R}([a, b])$.
(c) Show that $F(x) := \sqrt{x}$ is absolutely continuous on $[0, 1]$. *Hint:* Look at the *improper integral* $\int_0^x t^{-1/2}\, dt := \lim_{a \to 0+} \int_a^x t^{-1/2}\, dt$.

Chapter 8
Sequences and Series of Functions

In Chap. 2, we studied sequences and series of (*constant*) *real numbers*. In most problems, however, it is desirable to approximate *functions* by more elementary ones that are easier to investigate. We have already done this on a few occasions. For example, in Chap. 4, we looked at the *uniform* approximation of continuous functions by step, piecewise linear, and polynomial functions. Also, in Chap. 7, we proved that each bounded continuous function on a closed bounded interval is a uniform limit of regulated functions. Now all these approximations involve estimates on the *distance* between the given continuous function and the elementary functions that approximate it. This in turn suggests the introduction of *sequences* (and hence also *series*) whose terms are *functions* defined, in most cases, on the same interval. *Throughout this chapter, we shall assume that I, possibly with subscript, is an interval of* $\mathbb{R}$. Although we are studying *real analysis* here, we should at least introduce the field $\mathbb{C}$ of complex numbers and even use it in some definitions if this clarifies the concepts. Our presentation will be brief and most of the proofs are left as simple exercises for the reader.

8.1 Complex Numbers

On the Cartesian plane $\mathbb{R}^2 := \{(x, y) : x, \ y \in \mathbb{R}\}$, let us introduce the following operations of *addition* and *multiplication*:

$(+)$ $(x, y) + (x', y') := (x + x', y + y')$,
$(\cdot)$ $(x, y) \cdot (x', y') := (xx' - yy', yx' + xy')$.

Definition 8.1.1 (Complex Numbers). The set of *complex numbers*, denoted by $\mathbb{C}$, is defined to be the set $\mathbb{R}^2$ together with the binary operations $(+)$ and $(\cdot)$ defined above. Given two complex numbers $z = (x, y)$ and $z' = (x', y')$, their *sum* and *product* will be denoted by $z + z'$ and zz' (instead of $z \cdot z'$), respectively. Also, z and z' are said to be *equal* if and only if $x = x'$ and $y = y'$.

© Springer Science+Business Media New York 2014
H.H. Sohrab, *Basic Real Analysis*, DOI 10.1007/978-1-4939-1841-6_8

Definition 8.1.2 (Imaginary Unit, Identity, Zero). The complex numbers $i :=$ $(0, 1)$, $1 := (1, 0)$, and $0 := (0, 0)$ are called the *imaginary unit*, the *identity*, and the *zero* of $\mathbb{C}$, respectively. The definition of multiplication in $\mathbb{C}$ implies at once that $1^2 = (1, 0)^2 = (1, 0)$ and $i^2 = (0, 1)(0, 1) = (-1, 0)$. In view of the definition of *equality* given above, for a complex number $z = (x, y)$, we have $z = 0$ (i.e., $(x, y) = (0, 0)$) if and only if $x = y = 0$.

Proposition 8.1.3. *The operations of addition and multiplication in $\mathbb{C}$ satisfy the commutative, associative, and distributive laws. In other words, given any complex numbers $z = (x, y)$, $z' = (x', y')$, and $z'' = (x'', y'')$, we have*

(a) $z + z' = z' + z$ and $zz' = z'z$,
(b) $(z + z') + z'' = z + (z' + z'')$ and $(zz')z'' = z(z'z'')$,
(c) $z(z' + z'') = zz' + zz''$.

Proof. Exercise! $\square$

Proposition 8.1.4.

(a) For each $z \in \mathbb{C}$, we have $z + 0 = z$, $z \cdot 0 = 0$, and $z \cdot 1 = z$, where $0 := (0, 0)$ and $1 := (1, 0)$.
(b) If z, z', $z'' \in \mathbb{C}$ and $z + z' = z + z''$, then $z' = z''$.
(c) Given any $z \in \mathbb{C}$, there is a unique $z' \in \mathbb{C}$ such that $z + z' = 0$. This unique z' is denoted by $-z$.

Proof. For (c), simply define $-z = (-x, -y)$ if $z = (x, y)$. Parts (a) and (b) are obvious consequences of the definitions and the properties of real numbers. $\square$

Definition 8.1.5 (Subtraction). For any complex numbers z, $z' \in \mathbb{C}$, we write $z - z' := z + (-z')$, where if $z' := (x', y')$, then $-z' := (-x', -y')$ is the unique complex number satisfying $z' + (-z') = 0$.

Proposition 8.1.6. *For any complex numbers z, $z' \in \mathbb{C}$, we have $z - z = 0$ and*

$$(-z)z' = z(-z') = -(zz') = (-1)(zz'),$$

where $-1 = (-1, 0)$.

Proof. Exercise! $\square$

Definition 8.1.7 (Absolute Value). Given any $z = (x, y) \in \mathbb{C}$, the *absolute value* of z is defined to be the nonnegative number

$$|z| := \sqrt{x^2 + y^2}.$$

In other words, $|z|$ is simply the *Euclidean distance* between the points (x, y) and $(0, 0)$ in the plane $\mathbb{R}^2$. In particular, $|(x, 0)| = |x|$, where the right side is the usual absolute value of the *real* number x.

Using the absolute value, we can define the notions of *distance* and *convergence* for complex numbers:

Definition 8.1.8 (Distance, Limit). For any complex numbers $z = (x, y)$ and $w = (u, v)$, the nonnegative number $|z - w| = \sqrt{(x - u)^2 + (y - v)^2}$ is defined to be the *distance* between them. A sequence $(z_n) \in \mathbb{C}^{\mathbb{N}}$ is said to *converge* to a complex number $\zeta = (\xi, \eta) \in \mathbb{C}$ if

$$(\forall \varepsilon > 0)(\exists N = N(\varepsilon) \in \mathbb{N})(n \geq N \Rightarrow |z_n - \zeta| < \varepsilon).$$

The number ζ is then called the *limit* of (z_n) and we write

$$\lim_{n \to \infty} z_n = \lim(z_n) = \zeta.$$

In this case, the sequence (z_n) is called *convergent*; if (z_n) has no limit, we call it *divergent*. If x_0 is a *limit point* of a set $D \subset \mathbb{R}$ and if $f : D \to \mathbb{C}$, we say that $f(x)$ converges to $w_0 \in \mathbb{C}$ as $x \to x_0$, and we write $\lim_{x \to x_0} f(x) = w_0$, if

$$(\forall \varepsilon > 0)(\exists \delta = \delta(\varepsilon, x_0) > 0)(\forall x \in D)(0 < |x - x_0| < \delta \Rightarrow |f(x) - w_0| < \varepsilon).$$

The following proposition lists some of the properties of the absolute value:

Proposition 8.1.9. *Given any complex numbers $z = (x, y)$ and $w = (u, v)$, we have*

(a) $|z| > 0$ *if and only if $z \neq 0$,*
(b) $|zw| = |z||w|$,
(c) $|z + w| \leq |z| + |w|$ *(Triangle Inequality).*

Proof. (a) is obvious! (b) follows at once from *Lagrange's Identity*:

$$(xu - yv)^2 + (xv + yu)^2 = (x^2 + y^2)(u^2 + v^2),$$

which can be checked easily. Finally, the *Triangle Inequality* (c) follows from Proposition 2.1.23. □

Corollary 8.1.10. (a) *If z, $w \in \mathbb{C}$ and $zw = 0$, then $z = 0$ or $w = 0$.*
(b) *If z, u, $v \in \mathbb{C}$ satisfy $zu = zv$ and if $z \neq 0$, then $u = v$.*

Proof. To prove (a), note that $zw = 0$ if and only if $|zw| = |z||w| = 0$. For (b), observe that $zu = zv$ if and only if $z(u - v) = 0$ and use part (a). □

Corollary 8.1.11. *If $z = (x, y) \in \mathbb{C}$ and $z \neq 0$, then there is a unique complex number w (which is denoted $1/z$) such that*

$$zw = wz = 1.$$

Proof. Since $z \neq 0$, we have $|z|^2 = x^2 + y^2 > 0$. It is then easily checked that

$$w = \frac{1}{z} := \left(\frac{x}{x^2 + y^2}, \frac{-y}{x^2 + y^2} \right)$$

satisfies the desired property. □

Corollary 8.1.12. *Given any complex numbers u and v $\neq$ 0, there is a unique complex number z (which is denoted u/v) such that vz = u.*

Proof. Simply note that $z := u \cdot (1/v)$ satisfies the property.   □

We can now summarize all the algebraic properties highlighted above in the following.

Proposition 8.1.13. *The set $\mathbb{C}$ of complex numbers is a field.*

Proposition 8.1.14. *(1) A sequence $(z_n) = ((x_n, y_n)) \in \mathbb{C}^{\mathbb{N}}$ converges to $\zeta = (\xi, \eta) \in \mathbb{C}$ if and only if $\lim(x_n) = \xi$ and $\lim(y_n) = \eta$.*
(2) Let x_0 be a limit point of a set $D \subset \mathbb{R}$ and let $f = (u, v) : D \to \mathbb{C}$, i.e., $f(x) = (u(x), v(x))$, for all $x \in D$ and two real-valued functions u and v defined on D. Then $\lim_{x \to x_0} f(x) = w_0 = (u_0, v_0) \in \mathbb{C}$ if and only if $\lim_{x \to x_0} u(x) = u_0$ and $\lim_{x \to x_0} v(x) = v_0$.

Proof. Obvious!   □

Corollary 8.1.15 (Cauchy's Criterion). *A sequence $(z_n) = ((x_n, y_n)) \in \mathbb{C}^{\mathbb{N}}$ is convergent if and only if it is a Cauchy sequence, i.e.,*

$$(\forall \varepsilon > 0)(\exists N = N(\varepsilon) \in \mathbb{N})(m, n \geq N \Rightarrow |z_m - z_n| < \varepsilon).$$

Proof. Indeed, it is easily seen that (z_n) is a Cauchy sequence if and only if (x_n) and (y_n) are.   □

Proposition 8.1.16. *The set $\mathbf{R} := \{(x, 0) : x \in \mathbb{R}\} \subset \mathbb{C}$ is a subfield of $\mathbb{C}$ and the map $\phi : \mathbb{R} \to \mathbf{R}$ given by $\phi(x) := (x, 0)$ is a field isomorphism, i.e., a one-to-one correspondence satisfying the properties*

$$\phi(x + y) = \phi(x) + \phi(y), \qquad \phi(xy) = \phi(x)\phi(y) \quad \forall x, y \in \mathbb{R}. \tag{†}$$

Proof. That $\mathbf{R}$ is a *subfield* (i.e., a subset that is itself a field) is easily checked. Indeed, for any $x, y \in \mathbb{R}$, we have $(x + y, 0) = (x, 0) + (y, 0)$ and $(xy, 0) = (x, 0)(y, 0)$. Also, $y \neq 0$ implies $(1/y, 0) = 1/(y, 0)$ and hence $(x/y, 0) = (x, 0)/(y, 0)$. These relations show that (†) holds. Finally, ϕ is obviously a bijection.
□

Notation 8.1.17. Henceforth, we shall identify the sets $\mathbf{R}$ and $\mathbb{R}$ and write $\mathbf{R} = \mathbb{R}$. In fact, we write $(x, 0) = x$ for each $x \in \mathbb{R}$. With this agreement, we can then write $\mathbb{R} \subset \mathbb{C}$. Now recall (Definition 8.1.2) that $i := (0, 1)$ and hence, according to our identification,

$$i^2 = (0, 1)^2 = (-1, 0) = -1.$$

Also, it follows that each complex number $(x, y) \in \mathbb{C}$ can be written as

$$(x, y) = (x, 0) + (0, y) = (x, 0) + (y, 0)(0, 1) = x + yi.$$

Definition 8.1.18 (Real and Imaginary Parts). For each $z = (x, y) = x + yi \in \mathbb{C}$, the real numbers x and y are called the *real* and *imaginary parts* of z, respectively, and we write $\mathrm{Re}(z) = x$ and $\mathrm{Im}(z) = y$.

Definition 8.1.19 (Complex Conjugate). Given any $z = x + yi \in \mathbb{C}$, the complex number $\bar{z} := x - yi$ is called the *complex conjugate* (or simply *conjugate*) of z.

Proposition 8.1.20. *If $z = x + iy$, $w = u + iv \in \mathbb{C}$, then:*

(a) $\overline{(\bar{z})} = z$,
(b) $\overline{z + w} = \bar{z} + \bar{w}$,
(c) $\overline{zw} = \bar{z}\bar{w}$,
(d) $z\bar{z} = |z|^2$,
(e) $z + \bar{z} = 2\mathrm{Re}(z)$ *and* $z - \bar{z} = 2i\,\mathrm{Im}(z)$, *and*
(f) $z = \bar{z} \Leftrightarrow z \in \mathbb{R}$.

Proof. Exercise!   □

Corollary 8.1.21 ((Complex) Cauchy's Inequality). *Given any complex numbers $z_1, z_2, \ldots, z_n$ and $w_1, w_2, \ldots, w_n$, we have*

$$\left| \sum_{j=1}^{n} z_j \bar{w}_j \right|^2 \leq \sum_{j=1}^{n} |z_j|^2 \sum_{j=1}^{n} |w_j|^2.$$

Exercise 8.1.22. Prove Corollary 8.1.21. *Hint:* Note that the inequality can be written as $|V|^2 \leq ZW$, where $Z := \sum_{j=1}^{n} |z_j|^2$, $W := \sum_{j=1}^{n} |w_j|^2$, and $V := \sum_{j=1}^{n} z_j \bar{w}_j$. Now, using Proposition 8.1.20, show that

$$\sum_{j=1}^{n} |W z_j - V w_j|^2 = W(ZW - |V|^2).$$

8.2   Pointwise and Uniform Convergence

For a set $E \subset \mathbb{R}$, let us denote by $\mathcal{F}(E, \mathbb{R})$ the set of all functions from E to $\mathbb{R}$. We are interested in *sequences* and *series* in the sets $\mathcal{F}(E, \mathbb{R})$. For each sequence $(f_n) \in \mathcal{F}(E, \mathbb{R})^{\mathbb{N}}$ and each $x \in E$, the *numerical* sequence $(f_n(x)) \in \mathbb{R}^{\mathbb{N}}$ may or may not converge.

Definition 8.2.1 (Pointwise Convergence). For each $(f_n) \in \mathcal{F}(E, \mathbb{R})^{\mathbb{N}}$, let $E_0 \subset E$ be the set of all points $x \in E$ such that the *numerical* sequence $(f_n(x)) \in \mathbb{R}^{\mathbb{N}}$ converges and let

$$f(x) := \lim_{n \to \infty} f_n(x) = \lim(f_n(x)) \qquad \forall x \in E_0, \qquad (*)$$

which, in detail, means that

$$(\forall \varepsilon > 0)(\exists N = N(x, \varepsilon) \in \mathbb{N})(n \geq N \Rightarrow |f_n(x) - f(x)| < \varepsilon). \qquad (\dagger)$$

The sequence (f_n) is then said to be *pointwise convergent* (or simply *convergent*) on E_0 and the function $f \in \mathcal{F}(E_0, \mathbb{R})$, defined by $(*)$, is called the *pointwise limit* (or simply *limit*) of (f_n) on E_0.

Remark 8.2.2.

(1) It is obvious that the same definition can be given for *complex-valued* functions of a *complex* variable. Simply replace $\mathbb{R}$ by $\mathbb{C}$ in the above definition! For the most part, however, we shall be concerned with *real functions defined on subsets of the real line* and the complex case will only be used in connection with *Fourier series*.

(2) It is important to note that, in general, the integer $N = N(x, \varepsilon)$ depends on *both x and ε*, as indicated in $(\dagger)$.

Example 8.2.3. (a) Let $f_n(x) := x^n$ for all $x \in [0, 1]$. For $x \in [0, 1)$, we then have $\lim(f_n(x)) = 0$. On the other hand, $\lim(f_n(1)) = 1$. The sequence is therefore pointwise convergent on $[0, 1]$. We note, however, that the limit function

$$f(x) = \begin{cases} 0 & \text{if } 0 \leq x < 1, \\ 1 & \text{if } x = 1 \end{cases}$$

is *discontinuous* at $x = 1$, even though *all* functions $f_n(x) := x^n$ are *continuous* on $[0, 1]$.

(b) Let $g_n(x) := \sqrt{x^2 + 1/n}$ for all $x \in \mathbb{R}$. Here we clearly have $\lim(g_n(x)) = |x|$ for all $x \in \mathbb{R}$. Thus, the sequence is pointwise convergent on $\mathbb{R}$. We also observe that g_n is *differentiable* on $\mathbb{R}$ for all $n \in \mathbb{N}$ with $g_n'(x) = x/\sqrt{x^2 + 1/n}$. On the other hand, the limit function $g(x) := |x|$ is *not* differentiable at $x = 0$.

(c) Consider the sequence $h_n(x) := \sin(n^2 x)/n$ for all $x \in \mathbb{R}$ and $n \in \mathbb{N}$. Here the *limit* function, h, is the *(identically) zero function*. Indeed, $|\sin(n^2 x)/n| \leq 1/n$ holds for all $x \in \mathbb{R}$ and $n \in \mathbb{N}$ and $\lim(1/n) = 0$. Therefore, $h(x) = h'(x) = 0$ for all $x \in \mathbb{R}$. On the other hand,

$$h_n'(x) = \frac{n^2 \cos(n^2 x)}{n} = n \cos(n^2 x)$$

does *not* converge to 0. In fact, $\lim(h_n'(0)) = \lim(n) = +\infty$.

(d) Let $u_n(x) := [\cos^2(n!\pi x)] \ \forall x \in [0, 1]$, where, for each $t \in \mathbb{R}$, $[t]$ denotes the *greatest integer* $\leq t$. If $x = p/q$ with (relatively prime) positive integers p and q, then $n!x$ is an integer for all $n \geq q$ and hence $\cos^2(n!\pi x) = 1$. On the other hand, if $x \notin \mathbb{Q}$, then $\cos^2(n!\pi x) \in (0, 1)$. It follows that the (pointwise) limit function, u, is given by

$$u(x) = \begin{cases} 1 & \text{if } x \in \mathbb{Q} \cap [0, 1], \\ 0 & \text{if } x \in [0, 1] \setminus \mathbb{Q}. \end{cases}$$

In other words, u is the *Dirichlet function* which is *nowhere continuous* on $[0, 1]$. In particular, u is *not* Riemann integrable. On the other hand, each u_n has only a *finite* (in fact $n! + 1$) number of discontinuity points (why?) and hence is Riemann integrable.

(e) Consider the functions $v_n(x) := nx(1 - x^2)^n$, for all $x \in [0, 1]$. The (pointwise) limit, v, is the identically zero function: $v(x) = 0 \; \forall x \in [0, 1]$. This is obvious for $x = 0$ and $x = 1$, and for $x \in (0, 1)$ it follows from the fact that $\lim_{n \to \infty} n\alpha^n = 0$, for all $\alpha \in (0, 1)$. (Why?) Now a simple computation gives

$$\int_0^1 v_n(x) \, dx = -\frac{n}{2} \left[\frac{(1 - x^2)^{n+1}}{n+1} \right]_0^1 = \frac{n}{2(n+1)}.$$

It follows that $\lim_{n \to \infty} \int_0^1 v_n(x) \, dx = 1/2$, and yet $\int_0^1 v(x) \, dx = 0$.

Remark 8.2.4. The above examples show that, if f is the *pointwise* limit of a sequence of functions f_n, then the following may happen:

(1) Even if all the functions f_n are continuous at a point x_0, the limit f may be discontinuous there.
(2) Even if all the f_n are differentiable at x_0, f need not be differentiable at x_0. And, even if $f'(x_0)$ exists, the sequence of derivatives $f_n'(x_0)$ need not converge to $f'(x_0)$.
(3) Even if all the f_n are Riemann integrable on some interval $[a, b]$, the limit f need not be integrable on $[a, b]$. And, even if $\int_a^b f(x) \, dx$ exists, it need not be the limit of the sequence of integrals $\int_a^b f_n(x) \, dx$.

It turns out that, in order for the limit function f to inherit some of the nice properties shared by all the f_n, one must replace the *pointwise* convergence with a stricter one, called *uniform* convergence:

Definition 8.2.5 (Uniform Convergence). Let $E \subset \mathbb{R}$. We say that a sequence $(f_n) \in \mathcal{F}(E, \mathbb{R})^{\mathbb{N}}$ converges *uniformly on* $E_0 \subset E$ to a function $f : E_0 \to \mathbb{R}$ if

$$(\forall \varepsilon > 0)(\exists N = N(\varepsilon) \in \mathbb{N})(\forall x \in E_0)(n \geq N \Rightarrow |f_n(x) - f(x)| < \varepsilon). \qquad (\ddagger)$$

Remark 8.2.6.

(a) It is obvious that uniform convergence implies pointwise convergence. The converse is false, as some of the examples below will show.
(b) Comparing the above ($\ddagger$) to ($\dagger$) in Definition 8.2.1, we note that in *uniform convergence*, the integer $N = N(\varepsilon)$ depends *only* on ε and *not* on $x \in E_0$. In other words, the *same* N works for *all* $x \in E_0$. This turns out to have a major impact on the behavior of the limit function.

Exercise 8.2.7. Show that a sequence $(f_n) \in \mathcal{F}(E, \mathbb{R})^{\mathbb{N}}$, where $E \subset \mathbb{R}$, converges uniformly on $E_0 \subset E$ if and only if $\lim_{n \to \infty} \sup\{|f_n(x) - f(x)| : x \in E_0\} = 0$. In other words,

$$(\forall \varepsilon > 0)(\exists N = N(\varepsilon) \in \mathbb{N})(n \geq N \Rightarrow \sup\{|f_n(x) - f(x)| : x \in E_0\} < \varepsilon).$$

Example 8.2.8.

(1) Consider, once again, the sequence $(x^n)_{n \in \mathbb{N}}$, where $x \in [0, 1]$. As we saw above, the (pointwise) limit function, f, is

$$f(x) = \begin{cases} 0 & \text{if } 0 \leq x < 1, \\ 1 & \text{if } x = 1. \end{cases}$$

Here the convergence is *not uniform*. Indeed, if for some $\varepsilon \in (0, 1)$ there exists an integer $N = N(\varepsilon) \in \mathbb{N}$ such that $n \geq N$ implies $|x^n - f(x)| < \varepsilon$ for all $x \in [0, 1]$, then, for each $x \in [0, 1)$, we have $x^N < \varepsilon$. But this would imply that $1 = \lim_{x \to 1^-} x^N \leq \varepsilon$, which is absurd. Equivalently, using Exercise 8.2.7,

$$\lim_{n \to \infty} \sup\{|x^n - f(x)| : x \in [0, 1]\} = \lim_{n \to \infty} 1 = 1 \neq 0.$$

(2) As we saw in the above examples, the sequence $(g_n(x)) = (\sqrt{x^2 + 1/n})$ for all $x \in \mathbb{R}$ has pointwise limit $g(x) = |x|$ for all $x \in \mathbb{R}$. Let us show that this limit *is uniform*. Indeed, a computation shows that

$$\sup\left\{|\sqrt{x^2 + 1/n} - |x|| : x \in \mathbb{R}\right\} = \sup\left\{\frac{1/n}{\sqrt{x^2 + 1/n} + |x|} : x \in \mathbb{R}\right\}$$

$$= \frac{1/n}{\sqrt{1/n}} = \frac{1}{\sqrt{n}},$$

and the uniform convergence follows at once from Exercise 8.2.7.

(3) Consider the sequence $(h_n(x)) = (\sin(n^2 x)/n)$, where $x \in \mathbb{R}$. Here, as pointed out above, the pointwise limit is $h(x) = 0 \ \forall x \in \mathbb{R}$. As in the previous example, the limit *is uniform*. This follows from the simple inequality $|\sin(n^2 x)/n| \leq 1/n$, for all $x \in \mathbb{R}$ and all $n \in \mathbb{N}$.

Theorem 8.2.9 (Cauchy's Criterion). *Let $E \subset \mathbb{R}$ and let $(f_n) \in \mathcal{F}(E, \mathbb{R})^{\mathbb{N}}$. Then (f_n) converges uniformly on $E_0 \subset E$ (to some function f) if and only if*

$$(\forall \varepsilon > 0)(\exists N = N(\varepsilon) \in \mathbb{N})(\forall x \in E_0)(m, n \geq N \Rightarrow |f_m(x) - f_n(x)| < \varepsilon).$$

Proof. If (f_n) converges uniformly to f and if $\varepsilon > 0$ is given, then we can find $N = N(\varepsilon) \in \mathbb{N}$ such that $n \geq N$ implies $|f_n(x) - f(x)| < \varepsilon/2$ for all $x \in E_0$. If we also have $m \geq N$, then $|f_m(x) - f(x)| < \varepsilon/2$ as well. Thus, for any $m, n \geq N$ and any $x \in E_0$,

$$|f_m(x) - f_n(x)| \leq |f_m(x) - f(x)| + |f_n(x) - f(x)| < \varepsilon/2 + \varepsilon/2 = \varepsilon.$$

Conversely, if the condition of the theorem is satisfied, then, for each $x \in E_0$, the numerical sequence $(f_n(x))$ is a *Cauchy sequence* in $\mathbb{R}$ and hence converges. Let

$$f(x) := \lim_{n \to \infty} f_n(x) \qquad (\forall x \in E_0).$$

Now, given any $\varepsilon > 0$, we can find $N = N(\varepsilon)$ such that we have

$$|f_m(x) - f_n(x)| < \varepsilon \qquad (\forall x \in E_0, \ \forall m, \ n \geq N). \tag{$*$}$$

Keeping m fixed in $(*)$ and letting $n \to \infty$, we find that

$$|f_m(x) - f(x)| \leq \varepsilon \qquad (\forall x \in E_0, \ \forall m \geq N).$$

Since $\varepsilon > 0$ was arbitrary, it follows that (f_n) converges to f *uniformly* on E_0, as desired. $\square$

Exercise 8.2.10. Show that a sequence $(f_n) \in \mathcal{F}(E, \mathbb{R})^{\mathbb{N}}$, where $E \subset \mathbb{R}$, converges uniformly on $E_0 \subset E$ if and only if $\sup\{|f_m(x) - f_n(x)| : x \in E_0\} \to 0$, as $m, n \to \infty$. In other words,

$$(\forall \varepsilon > 0)(\exists N = N(\varepsilon) \in \mathbb{N})(m, \ n \geq N \Rightarrow \sup\{|f_m(x) - f_n(x)| : x \in E_0\} < \varepsilon).$$

Example 8.2.8(1) shows that pointwise convergence does *not* imply uniform convergence. There are exceptional situations, however, as the following theorem shows.

Theorem 8.2.11 (Dini's Theorem). *Let $I \subset \mathbb{R}$ be a compact (i.e., closed and bounded) interval and suppose that $(f_n) \in \mathcal{F}(I, \mathbb{R})^{\mathbb{N}}$ is a sequence of continuous functions converging pointwise to a continuous function $f : I \to \mathbb{R}$. If (f_n) is increasing (i.e., $f_n(x) \leq f_{n+1}(x)$, for all $x \in I$ and $n \in \mathbb{N}$) or decreasing (i.e., $f_n(x) \geq f_{n+1}(x)$, for all $x \in I$ and $n \in \mathbb{N}$), then (f_n) converges to f uniformly on I.*

Proof. The uniform convergence of (f_n) to f is equivalent to the uniform convergence of $(f - f_n)$ (or $(f_n - f)$) to 0. (Why?) Let $g_n := f - f_n$ (resp., $g_n := f_n - f$) if (f_n) is increasing (resp., decreasing). Then (g_n) is a *decreasing* sequence of *continuous nonnegative* functions converging *pointwise* to 0 on I. The theorem is proved if we show that this convergence is in fact *uniform* on I. Let $\varepsilon > 0$ be given. For each $x \in I$, $\lim(g_n(x)) = 0$ implies that we can pick $N(x) \in \mathbb{N}$ with $g_{N(x)}(x) < \varepsilon/2$. Since $g_{N(x)}$ is *continuous* at x, there is a $\delta(x) > 0$ such that

$$g_{N(x)}(t) < \varepsilon \qquad \forall t \in (x - \delta(x), x + \delta(x)). \tag{$*$}$$

Since I is *compact*, Proposition 4.1.7 implies that we can cover I by a finite number of intervals $I_j := (x_j - \delta(x_j), x_j + \delta(x_j))$, $1 \leq j \leq k$. Let $N := \max\{N(x_1), \ldots, N(x_k)\}$. Now, for any $t \in I$, we have $t \in I_j$ for some j and, by $(*)$, $g_{N(x_j)}(t) < \varepsilon$. But since $N \geq N(x_j)$ and (g_n) is *decreasing*, we have

$$0 \leq g_N(t) \leq g_{N(x_j)}(t) < \varepsilon \qquad \forall t \in I.$$

Therefore, we indeed have

$$(\forall x \in I)(n \geq N \Rightarrow g_n(x) \leq g_N(x) < \varepsilon),$$

and the proof is complete.   □

Exercise 8.2.12. Show that, in Theorem 8.2.11, the *compact interval* I can be replaced by any *compact set* $K \subset \mathbb{R}$.

So far, we have only looked at *sequences* of functions. Since the study of (*infinite*) *series* is equivalent to the study of the corresponding sequences of *partial sums*, all the above results have analogs for *infinite series of functions*.

To begin, let $E \subset \mathbb{R}$ and let $(f_n) \in \mathcal{F}(E, \mathbb{R})^{\mathbb{N}}$. Then the formal sum

$$f_1 + f_2 + \cdots + f_n + \cdots = \sum_{n=1}^{\infty} f_n$$

is called an *infinite series of functions* with *general term f_n*. For each $x \in E$, we have a *numerical series*

$$\sum_{n=1}^{\infty} f_n(x).$$

For each $n \in \mathbb{N}$, we can then define the *partial sum*

$$s_n(x) := \sum_{k=1}^{n} f_k(x).$$

This defines a sequence

$$(s_n) \in \mathcal{F}(E, \mathbb{R})^{\mathbb{N}}. \tag{$**$}$$

Definition 8.2.13 (Pointwise Convergent Series of Functions). With notation as above, the series $\sum_{n=1}^{\infty} f_n$ is said to be *pointwise convergent* (or simply *convergent*) on $E_0 \subset E$ with *sum $s \in \mathcal{F}(E_0, \mathbb{R})^{\mathbb{N}}$* if the sequence $(**)$ of partial sums converges (pointwise) to the function s on E_0. In other words, if

$$s(x) = \lim_{n \to \infty} s_n(x) \qquad (\forall x \in E_0).$$

Example 8.2.14. For each $n \in \mathbb{N}$, let $f_n(x) := x^n$, where $-1 < x < 1$ and $f_0 := 1$. Then the series $\sum_{n=0}^{\infty} f_n$ is (pointwise) convergent on $(-1, 1)$ with sum

$$s(x) = \sum_{n=0}^{\infty} f_n(x) = \sum_{n=0}^{\infty} x^n = \frac{1}{1-x}.$$

Definition 8.2.15 (Uniformly Convergent Series of Functions). With notation as in Definition 8.2.13, the series $\sum_{n=1}^{\infty} f_n$ is said to be *uniformly convergent* on E_0 if the sequence (s_n) of *partial sums* is *uniformly convergent* on E_0.

Example 8.2.16. In the above example, we saw that $\sum_{n=0}^{\infty} x^n = 1/(1-x)$, for each $x \in (-1, 1)$. To show that this convergence is *not uniform*, note that, with $s_n(x) := 1 + x + \cdots + x^{n-1} = (1 - x^n)/(1 - x)$, we have

$$|s(x) - s_n(x)| = \frac{|x|^n}{1-x} \qquad (-1 < x < 1),$$

and $\sup\{|x|^n/(1-x) : -1 < x < 1\} = +\infty$ for all $n \geq 0$. (Why?)

The following version of *Dini's Theorem* for series follows at once from Theorem 8.2.11.

Theorem 8.2.17 (Dini's Theorem for Series). *Let $I \subset \mathbb{R}$ be a compact (i.e., closed and bounded) interval and suppose that $(f_n) \in \mathcal{F}(I, \mathbb{R})^{\mathbb{N}}$ is a sequence of continuous functions such that the series $\sum_{n=1}^{\infty} f_n$ converges pointwise to a continuous sum $s : I \to \mathbb{R}$. If the f_n are all nonnegative (resp., nonpositive) on I; i.e., $f_n(x) \geq 0$ (resp., $f_n(x) \leq 0$) for all $x \in I$ and $n \in \mathbb{N}$, then $\sum_{n=1}^{\infty} f_n$ converges to s uniformly on I.*

Proof. Simply apply Theorem 8.2.11 to the sequence (s_n) of *partial sums*, where $s_n = f_1 + f_2 + \cdots + f_n$. □

For infinite series of functions, we also define the following (stricter) notion of convergence:

Definition 8.2.18 (Normally Convergent Series of Functions). With notation as in Definition 8.2.13, the series $\sum_{n=1}^{\infty} f_n$ is said to be *normally convergent* on E_0 if the series $\sum_{n=1}^{\infty} |f_n|$ of *absolute values* is *uniformly convergent* on E_0.

The following proposition is an immediate consequence of the above definitions.

Proposition 8.2.19. *With notation as above, normal convergence of the series $\sum_{n=1}^{\infty} f_n$ on E_0 implies its uniform convergence on E_0 which, in turn, implies its pointwise convergence on E_0.*

Proof. Exercise. □

Theorem 8.2.20 (Weierstrass M-Test). *Let $E \subset \mathbb{R}$ and $(f_n) \in \mathcal{F}(E, \mathbb{R})^{\mathbb{N}}$. Suppose that for each $n \in \mathbb{N}$ there exists a constant $M_n \geq 0$ such that $|f_n(x)| \leq M_n$, for all $x \in E$ and that $\sum_{n=1}^{\infty} M_n$ converges. Then (f_n) is normally (hence uniformly) convergent on E.*

Proof. It follows from our assumptions and Theorem 2.3.9 (*First Comparison Test*), that the series $\sum_{n=1}^{\infty} |f_n(x)|$ is convergent for each $x \in E$. In other words, $\sum_{n=1}^{\infty} f_n(x)$ is *absolutely convergent*; we must show that this convergence is *uniform* on E. Let us introduce the partial sums $s_n(x) := \sum_{k=1}^{n} |f_k(x)|$ and $\sigma_n := \sum_{k=1}^{n} M_k$. Since (σ_n) is convergent, it satisfies *Cauchy's Criterion:* Given any $\varepsilon > 0$, there exists $N \in \mathbb{N}$ such that

$$m > n \geq N \Rightarrow |\sigma_m - \sigma_n| = \sum_{k=n+1}^{m} \sigma_k < \varepsilon. \qquad (*)$$

Since $s_n(x) \leq \sigma_n$, for all $x \in E$ and all $n \in \mathbb{N}$, $(*)$ implies that

$$m > n \geq N \Rightarrow |s_m(x) - s_n(x)| = \sum_{k=n+1}^{m} s_k(x) \leq \sum_{k=n+1}^{m} \sigma_k < \varepsilon,$$

for all $x \in E$ and the proof is complete.   □

8.3 Uniform Convergence and Limit Theorems

As was pointed out in the previous section, even if all functions in a sequence have a nice property (such as continuity, differentiability, etc.), the (pointwise) limit function, if it exists, need not (in general) share this property. Our goal now is to show that, if the convergence is *uniform*, then many nice properties satisfied by all the functions in the sequence will also be satisfied by their (uniform) limit.

Theorem 8.3.1 (Uniform Convergence & Continuity). *Let $E_0 \subset E \subset \mathbb{R}$ and let $(f_n) \in \mathcal{F}(E, \mathbb{R})^{\mathbb{N}}$. If each f_n is continuous at some $x_0 \in E_0$ and (f_n) converges uniformly on E_0 to a function $f \in \mathcal{F}(E_0, \mathbb{R})$, then f is also continuous at x_0. Thus, if each f_n is continuous on E_0, then so is the limit function f.*

Proof. What we need here is a standard $\varepsilon/3$-*proof*. Let $\varepsilon > 0$ be given. Since f is the *uniform* limit of (f_n) on E_0, we can find $N \in \mathbb{N}$ such that

(i) $|f_n(x) - f(x)| < \varepsilon/3 \qquad (\forall x \in E_0, \ \forall n \geq N)$.

With N as in (i), the *continuity* of f_N at x_0 implies that we can find $\delta > 0$ with

(ii) $|f_N(x) - f_N(x_0)| < \varepsilon/3 \qquad \forall x \in E_0 \cap (x_0 - \delta, x_0 + \delta)$.

Also, (i) implies that

(iii) $|f(x) - f_N(x)| < \varepsilon/3 \qquad \forall x \in E_0 \cap (x_0 - \delta, x_0 + \delta)$.

Now (i), (ii), and (iii) imply that, for each $x \in E_0 \cap (x_0 - \delta, x_0 + \delta)$, we have

$$|f(x) - f(x_0)| \le |f(x) - f_N(x)| + |f_N(x) - f_N(x_0)| + |f_N(x_0) - f(x_0)| < \varepsilon,$$

and hence f is continuous at x_0. The last statement is now obvious. $\qquad \square$

Corollary 8.3.2. *Let $E_0 \subset E \subset \mathbb{R}$ and let $(f_n) \in \mathcal{F}(E, \mathbb{R})^{\mathbb{N}}$. If each f_n is continuous at a point $x_0 \in E_0$ and the series $\sum_{n=1}^{\infty} f_n$ converges uniformly on E_0 to a sum $s \in \mathcal{F}(E_0, \mathbb{R})$, then s is also continuous at x_0. In particular, if each f_n is continuous on E_0, then so is the sum s.*

Proof. Apply Theorem 8.3.1 to the sequence $(s_n) := (\sum_{k=1}^{n} f_k)$ of *partial sums.* $\qquad \square$

Next, we show that *Riemann integrability* is preserved when we pass to *uniform* limits:

Theorem 8.3.3 (Uniform Convergence & Integrability). *Let (f_n) be a sequence of Riemann integrable functions on $[a, b] \subset \mathbb{R}$. If $\lim(f_n) = f$, uniformly on $[a, b]$, then f is also Riemann integrable on $[a, b]$ and we have*

$$\int_a^x f(t)\, dt = \lim_{n \to \infty} \int_a^x f_n(t)\, dt \quad \forall \, x \in [a, b].$$

Proof. For $\varepsilon = 1$, the uniform convergence of (f_n) to f implies that, for some $N \in \mathbb{N}$, we have

$$|f(x) - f_n(x)| < 1 \qquad (\forall x \in [a, b],\ \forall n \ge N).$$

In particular, $|f(x) - f_N(x)| < 1$, for all $x \in [a, b]$. Therefore,

$$|f(x)| \le |f_N(x)| + 1 \qquad (\forall x \in [a, b]). \qquad (*)$$

Now, by Proposition 7.1.19, $f_N \in \mathcal{R}([a, b])$ implies that f_N is *bounded* on $[a, b]$. Hence, by $(*)$, so is f. In view of *Lebesgue's Integrability Criterion* (Theorem 7.3.19), to prove $f \in \mathcal{R}([a, b])$, we need only show that f is *continuous almost everywhere* on $[a, b]$. Now, for each $n \in \mathbb{N}$, f_n is Riemann integrable and hence continuous on $[a, b]$ except on a set D_n of measure zero. Let $D := \bigcup_{n=1}^{\infty} D_n$. Then D has measure zero. For each $x \in [a, b] \setminus D$, *all* the f_n are continuous at x. Since f_n converges to f *uniformly*, Theorem 8.3.1 implies that f is also continuous at x. Thus, f is indeed continuous on $[a, b] \setminus D$ and hence Riemann integrable.

Next, given any $\varepsilon > 0$, by uniform convergence, we can find $N \in \mathbb{N}$ such that $|f_N(t) - f(t)| < \varepsilon/(b - a)$, for all $t \in [a, b]$. It then follows from Proposition 7.4.13 that

$$\left| \int_a^x f_N(t)\, dt - \int_a^x f(t)\, dt \right| \le \int_a^b |f_N(t) - f(t)|\, dt < \varepsilon,$$

and the proof is complete. $\square$

Corollary 8.3.4 (Term-by-Term Integration). *Let (f_n) be a sequence of Riemann integrable functions on $[a, b] \subset \mathbb{R}$. If the series $\sum_{n=1}^{\infty} f_n$ converges uniformly on $[a, b]$ to a sum s, then $s \in \mathcal{R}([a, b])$ and we have*

$$\int_a^x s(t)\, dt = \sum_{n=1}^{\infty} \int_a^x f_n(t)\, dt \quad \forall\, x \in [a, b].$$

In other words, we can integrate the series term by term.

Finally, we look at the *differentiability* properties of the limit of a *uniformly convergent* sequence of differentiable functions. Here, the situation is more complicated. In fact, even the *uniform limit* of a sequence of differentiable functions *need not* be differentiable. Actually, we have already given an example above. Indeed, as we have seen, the sequence $(f_n(x)) = (\sqrt{x^2 + 1/n})$ converges *uniformly* to $f(x) := |x|$ on $\mathbb{R}$. It is also obvious that each f_n is continuously differentiable on $\mathbb{R}$ and yet f is not differentiable at 0. Therefore, we need stronger conditions. Before treating the general case, let us use Theorem 8.3.3 to handle a special case that is quite useful in many situations.

Theorem 8.3.5. *Let (f_n) be a sequence of continuously differentiable functions on $[a, b]$. Suppose that $(f_n(x_0))$ converges for some $x_0 \in [a, b]$ and that the sequence (f_n') of derivatives converges uniformly on $[a, b]$ to a function g. Then (f_n) converges uniformly to a continuously differentiable function f on $[a, b]$ and we have $f'(x) = g(x)$, for all $x \in [a, b]$.*

Proof. Since each f_n' is continuous and $\lim(f_n') = g$, uniformly on $[a, b]$, it follows from Theorem 8.3.1 that g is *continuous* on $[a, b]$. Also, by the *First Fundamental Theorem* (Theorem 7.5.3), we have

$$f_n(x) = f_n(x_0) + \int_{x_0}^x f_n'(t)\, dt \qquad (\forall x \in [a, b]).$$

Now Theorem 8.3.3 and the fact that $\lim(f_n(x_0)) = f(x_0)$ imply that

$$f(x) = f(x_0) + \int_{x_0}^x g(t)\, dt \qquad (\forall x \in [a, b]).$$

Finally, from the *Second Fundamental Theorem* (Theorem 7.5.8), we deduce that f is differentiable on $[a, b]$ and $f' = g$ as claimed. $\square$

We now give the more general result where the derivatives f_n' are *no longer* assumed to be *continuous* on $[a, b]$.

Theorem 8.3.6 (Uniform Convergence & Differentiability). *Let (f_n) be a sequence of differentiable functions on $[a, b]$ such that $(f_n(x_0))$ converges for some $x_0 \in [a, b]$. If the sequence (f_n') of derivatives converges to a function g uniformly on $[a, b]$, then the sequence (f_n) converges uniformly on $[a, b]$ to a differentiable function f, and we have*

$$f'(x) = \lim_{n \to \infty} f_n'(x) = g(x) \qquad (\forall x \in [a, b]).$$

Proof. For each *bounded* real-valued function h on $[a, b]$, let

$$\|h\|_\infty := \sup\{|h(x)| : x \in [a, b]\}. \tag{$\dagger$}$$

Since (f_n') converges *uniformly* on $[a, b]$, Exercise 8.2.10 implies that

(i) $m, n \to \infty \Rightarrow \|f_m' - f_n'\|_\infty \to 0$.

Now, for each $x \neq x_0$ in $[a, b]$ and any integers $m, n \in \mathbb{N}$, we can apply the MVT (Theorem 6.4.8) to the function $f_m - f_n$ on the interval with endpoints x_0 and x to obtain

(ii) $f_m(x) - f_n(x) = f_m(x_0) - f_n(x_0) + (x - x_0)[f_m'(\xi) - f_n'(\xi)]$,

for some ξ between x_0 and x. Using ($\dagger$), we deduce from (ii) that

(iii) $\|f_m - f_n\|_\infty \leq |f_m(x_0) - f_n(x_0)| + (b - a)\|f_m' - f_n'\|_\infty$.

Thus, in view of (i) and (iii), we have

$$m, n \to \infty \implies \sup\{|f_m(x) - f_n(x)| : x \in [a, b]\} \to 0,$$

which implies that (f_n) is uniformly convergent on $[a, b]$. Let f denote its limit. Fix any $t \in [a, b]$ and let $x \in [a, b] \setminus \{t\}$. Applying the MVT to $f_m - f_n$ on the interval with endpoints x and t, we have

$$[f_m(x) - f_n(x)] - [f_m(t) - f_n(t)] = (x - t)[f_m'(\tau) - f_n'(\tau)],$$

for some τ between x and t. Dividing both sides by $x - t$, we get

(iv) $\left| \dfrac{f_m(x) - f_m(t)}{x - t} - \dfrac{f_n(x) - f_n(t)}{x - t} \right| \leq \|f_m' - f_n'\|_\infty$.

Now, given $\varepsilon > 0$, pick $N \in \mathbb{N}$ such that $m, n \geq N$ implies $\|f_m' - f_n'\|_\infty < \varepsilon/3$. It then follows from (iv) that

(v) $\left| \dfrac{f_m(x) - f_m(t)}{x - t} - \dfrac{f_n(x) - f_n(t)}{x - t} \right| \le \varepsilon/3,$

for all m, $n \ge N$. Fixing $n = N$ and letting $m \to \infty$ in (v), we deduce from $\lim(f_m) = f$ that

(vi) $\left| \dfrac{f(x) - f(t)}{x - t} - \dfrac{f_N(x) - f_N(t)}{x - t} \right| \le \varepsilon/3.$

On the other hand, $\lim(f_n') = g$ uniformly on $[a, b]$ implies that the integer N in (vi) can be selected so large that we also have

(vii) $\left| f_N'(t) - g(t) \right| \le \varepsilon/3.$

Finally, since $\lim_{x \to t}(f_N(x) - f_N(t))/(x - t) = f_N'(t)$, we can pick $\delta > 0$ such that, for any $x \in [a, b]$,

(viii) $0 < |x - t| < \delta \implies \left| \dfrac{f_N(x) - f_N(t)}{x - t} - f_N'(t) \right| < \varepsilon/3.$

Combining (vi), (vii), and (viii), we see that, if $x \in [a, b]$ satisfies $0 < |x - t| < \delta$, then we have

$$\left| \frac{f(x) - f(t)}{x - t} - g(t) \right| < \varepsilon$$

and the proof is complete.   □

As before we can immediately deduce a corresponding result for *series* of differentiable functions:

Corollary 8.3.7 (Term-by-Term Differentiation).   *Let (f_n) be a sequence of differentiable functions on $[a, b]$ such that the series $\sum_{n=1}^{\infty} f_n(x_0)$ converges for some $x_0 \in [a, b]$. If the series $\sum_{n=1}^{\infty} f_n'$ of derivatives converges uniformly on $[a, b]$, then the series $\sum_{n=1}^{\infty} f_n$ converges uniformly on $[a, b]$ to a differentiable sum s and we have*

$$s'(x) = \left(\sum_{n=1}^{\infty} f_n(x) \right)' = \sum_{n=1}^{\infty} f_n'(x) \qquad (\forall x \in [a, b]).$$

We end this section by giving an example of a continuous function on $\mathbb{R}$ that is *nowhere differentiable*. Consider the *sawtooth function*:

$$f_0(x) := \begin{cases} x - [x] & \text{if } x \le [x] + 1/2, \\ [x] + 1 - x & \text{if } x > [x] + 1/2, \end{cases}$$

where $[x]$ denotes the *greatest integer* $\le x$. Then $f_0(x)$ is the *distance from x to the nearest integer*, i.e., $f_0(x) = d(x, \mathbb{Z})$, and is a *continuous, periodic* function on $\mathbb{R}$ with period 1. Now define $f_n(x) = 4^{-n} f_0(4^n x)$, for all $x \in \mathbb{R}$ and $n = 0, 1, 2, \ldots$.

Then f_n is also a continuous *sawtooth function* (with period $1/4^n$), whose graph consists of line segments of slope ± 1. Since $0 \le f_0 \le 1/2$, we have $0 \le f_n(x) \le 1/(2 \cdot 4^n)$, for all $x \in \mathbb{R}$ and $n \in \mathbb{N}_0$.

Theorem 8.3.8 (Van der Waerden). *The function*

$$f(x) := \sum_{n=0}^{\infty} f_n(x) \qquad (\forall x \in \mathbb{R}),$$

where the f_n are the sawtooth functions defined above, is continuous and nowhere differentiable on $\mathbb{R}$.

Proof. Since $0 \le f_n(x) \le 1/(2 \cdot 4^n)$ and $\sum_{n=0}^{\infty} 4^{-n} < \infty$, it follows from the Weierstrass M-Test (Theorem 8.2.20) that $\sum f_n$ converges uniformly on $\mathbb{R}$. Thus, if $f := \sum f_n$, then f is continuous on $\mathbb{R}$ because the f_n are. Now fix any $x \in \mathbb{R}$ and for each $n \in \mathbb{N}$ let $h_n = \pm(1/4^{n+1})$, where the sign is chosen so that $4^n x$ and $4^n(x + h_n) = 4^n x \pm 1/4$ both belong to the same interval $[k/2, (k+1)/2]$. Since on this interval f_0 has slope ± 1, we have

$$\varepsilon_n := \frac{f_n(x + h_n) - f_n(x)}{h_n} = \frac{f_0(4^n x + 4^n h_n) - f_0(4^n x)}{4^n h_n} = \pm 1.$$

If $m < n$, then a tooth of f_n is entirely below a *rise* or *fall* of a tooth of f_m. So the graph of f_m also has slope ± 1 on the interval with endpoints x and $x + h_n$:

$$\varepsilon_m := \frac{f_m(x + h_n) - f_m(x)}{h_n} = \pm 1 \qquad (\forall m < n).$$

For $m \ge n + 1$, however, $4^m(x + h_n) - 4^m x = \pm 4^{m-n-1}$ is an integer and f_0 has period 1; hence $f_m(x + h_n) - f_m(x) = 0$ for all $m > n$. Therefore,

$$\frac{f(x + h_n) - f(x)}{h_n} = \sum_{m=0}^{n} \frac{f_m(x + h_n) - f_m(x)}{h_n} = \sum_{m=0}^{n} \varepsilon_m, \qquad (*)$$

which is an even (resp., odd) integer if n is odd (resp., even). Since $h_n \to 0$ as $n \to \infty$ and $\lim_{n \to \infty} \sum_{m=0}^{n} \varepsilon_m$ does *not* exist, $(*)$ implies that f is *not* differentiable at x. $\qquad \square$

8.4 Power Series

Power series are probably the most frequently used series of functions and have many applications. The main reason for their importance is that their partial sums are *polynomials*. As we have already seen, *nice* functions can often be approximated by polynomials. We shall presently see that power series may be thought of as a *generalization* of polynomials and define a class of functions that includes most elementary functions we constantly use.

Definition 8.4.1 (Power Series). Let $(c_n)_{n=0}^{\infty}$ be a real sequence and $x_0 \in \mathbb{R}$. A series of the form

$$\sum_{n=0}^{\infty} c_n(x - x_0)^n = c_0 + c_1(x - x_0) + c_2(x - x_0)^2 + \cdots \qquad (\dagger)$$

is said to be a *power series about* x_0.

Remark 8.4.2.

(1) To simplify the exposition, we usually assume that $x_0 = 0$ (which can be achieved by the translation $x' := x - x_0$). The power series $(\dagger)$ is then reduced to

$$\sum_{n=0}^{\infty} c_n x^n = c_0 + c_1 x + c_2 x^2 + \cdots . \qquad (\ddagger)$$

(2) It is obvious that the power series $(\dagger)$ (resp., $(\ddagger)$) converges at $x = x_0$ (resp., at $x = 0$) and has sum c_0.

Proposition 8.4.3. *If the power series $\sum_{n=0}^{\infty} c_n x^n$ converges for $x = x_0 \neq 0$, then it is normally (i.e., uniformly and absolutely) convergent on any compact interval $[a, b] \subset (-|x_0|, |x_0|)$.*

Proof. Since $\sum_{n=0}^{\infty} c_n x_0^n$ is convergent, we can find $M > 0$ such that $|c_n x_0^n| \leq M$ for all $n \in \mathbb{N}$. (Why?) It follows that

$$|c_n x^n| = |c_n x_0^n||x/x_0|^n \leq M|x/x_0|^n. \qquad (*)$$

Now, for $|x| < |x_0|$ (i.e., $|x/x_0| < 1$), the geometric series $\sum_{n=0}^{\infty} M|x/x_0|^n$ is convergent (Proposition 2.3.8). Thus the proposition follows at once from $(*)$ and the Weierstrass M-Test (Theorem 8.2.20). $\qquad \square$

Theorem 8.4.4 (Radius of Convergence). *For any power series $\sum_{n=0}^{\infty} c_n x^n$, there exists a unique (extended) number $R \in [0, \infty]$, called the "radius of convergence," such that the power series converges absolutely for $|x| < R$ and diverges for $|x| > R$. In fact, the convergence is normal on any compact interval $[a, b] \subset (-R, R)$.*

Proof. Let E denote the set of all x for which the series $\sum_{n=0}^{\infty} c_n x^n$ converges. Since $0 \in E$, we have $E \neq \emptyset$. Let $R := \sup\{|x| : x \in E\}$. If $(R > 0$ and$)$ $|x| < R$, the definition of R implies that $|x| < |x_0|$ for some $x_0 \in E$ and hence the power series is *absolutely convergent* at x by Proposition 8.4.3. If $(R < \infty$ and$)$ $|x| > R$, then the definition of R implies that $x \notin E$ and hence $\sum_{n=0}^{\infty} c_n x^n$ is *divergent.* To prove the uniqueness of R, suppose that R' also satisfies the conditions of the theorem. If, say, $R < R'$, then the series $\sum_{n=0}^{\infty} c_n x^n$ is simultaneously convergent *and* divergent for each $x \in (R, R')$, which is absurd. Finally, to prove the last statement, pick $R_0 \in (0, R)$ such that $[a, b] \subset (-R_0, R_0)$ and use Proposition 8.4.3. □

Corollary 8.4.5 (Interval of Convergence). *Let $R \in [0, \infty]$ be the radius of convergence of a power series $\sum_{n=0}^{\infty} c_n x^n$. Then the set of all x for which the series converges is a possibly degenerate interval (centered at $x = 0$) with endpoints $-R$ and R. It is called the "interval of convergence" of the power series.*

Proof. This follows at once from the above theorem because the power series converges in $(-R, R)$ and diverges outside $[-R, R]$. □

Remark 8.4.6. If $R = 0$, then the series converges only at $x = 0$, and the interval of convergence is then the *degenerate* interval $[0, 0] := \{0\}$. If $R = \infty$, then the interval of convergence is obviously $\mathbb{R} = (-\infty, \infty)$. For $R \in (0, \infty)$, the convergence at the endpoints $\pm R$ must be checked *separately* because these are precisely the values of x for which the Root Test is inconclusive.

Our next theorem shows how to obtain the radius of convergence R. Before stating it, recall that, given a real sequence $(u_n)_{n \in \mathbb{N}}$, its *upper* and *lower* limits are the *extended real numbers*

$$\lim \sup(u_n) = \lim(\bar{u}_n) \quad \text{and} \quad \lim \inf(u_n) = \lim(\underline{u}_n),$$

respectively, where $\bar{u}_n := \sup\{u_k : k \geq n\}$ and $\underline{u}_n := \inf\{u_k : k \geq n\}$. Note that $(\bar{u}_n)$ is *decreasing* and $(\underline{u}_n)$ is *increasing*, as easily seen.

Theorem 8.4.7 (Cauchy–Hadamard). *Let R be the radius of convergence of the power series $\sum_{n=0}^{\infty} c_n x^n$ and let $\rho := \lim \sup_{n \to \infty} |c_n|^{1/n} \in [0, \infty]$. Then we have $R = 1/\rho$; i.e.,*

$$R = \begin{cases} 0 & \text{if } \rho = \infty, \\ 1/\rho & \text{if } 0 < \rho < \infty, \\ \infty & \text{if } \rho = 0. \end{cases}$$

Proof. Simply note that

$$\lim \sup(|c_n x^n|^{1/n}) = |x| \lim \sup(|c_n|^{1/n}) = \rho |x|.$$

The theorem is therefore an immediate consequence of the *Root Test* (Theorem 2.3.24) and the uniqueness of R. □

In many cases it is more convenient to use the *Ratio Test* to find the radius of convergence:

Theorem 8.4.8. *Suppose that, for some $N \in \mathbb{N}$, we have $c_n \neq 0$ for all $n \geq N$ and $\lim_{n \to \infty} |c_n/c_{n+1}|$ exists (as an extended nonnegative number). If R is the radius of convergence of the power series $\sum_{n=0}^{\infty} c_n x^n$, then we have $R = \lim_{n \to \infty} |c_n/c_{n+1}|$.*

Proof. We note that, for $x \neq 0$ and $n \geq N$,

$$\lim_{n \to \infty} \left| \frac{c_{n+1} x^{n+1}}{c_n x^n} \right| = |x| \lim_{n \to \infty} \left| \frac{c_{n+1}}{c_n} \right|.$$

Therefore, by the *Ratio Test* (Theorem 2.3.25), the power series converges if we have $|x| \lim_{n \to \infty} |c_{n+1}/c_n| < 1$ and diverges if $|x| \lim_{n \to \infty} |c_{n+1}/c_n| > 1$. The theorem now follows from the uniqueness of R.   □

Example 8.4.9.

(1) Since $\lim_{n \to \infty} n!/(n+1)! = 0$, we have $R = 0$ and the power series $\sum_{n=0}^{\infty} n! x^n$ diverges for all x except, obviously, $x = 0$.

(2) For the power series $\sum_{n=1}^{\infty} x^n/n^n$, we have $\limsup \sqrt[n]{1/n^n} = 0$. It follows that $R = +\infty$ and the interval of convergence is $(-\infty, \infty)$.

(3) Consider the *geometric series* $\sum_{n=0}^{\infty} x^n$. Here the interval of convergence is $(-1, 1)$ because the series diverges at both endpoints $x = \pm 1$.

(4) For the power series $\sum_{n=1}^{\infty} x^n/n$, we have $\lim_{n \to \infty} (1/n)/(1/(n+1)) = 1$. Therefore, the radius of convergence is $R = 1$ and the interval of convergence is $[-1, 1)$. Indeed, the series converges at the left endpoint $x = -1$ but diverges at the right endpoint $x = 1$.

(5) For $\sum_{n=1}^{\infty} x^n/n^2$, we have $\lim_{n \to \infty} (1/n^2)/(1/(n+1)^2) = 1$ and the interval of convergence is $[-1, 1]$ since the series converges at both endpoints $x = \pm 1$.

Remark 8.4.10. As was pointed out in Remarks 8.4.6 and Examples (3), (4), and (5) show, nothing can be said in advance about the convergence or divergence of a power series at the (finite) endpoints $\pm R$. Also, although (by Theorem 8.4.4) a power series converges uniformly on any *compact subinterval* of $(-R, R)$, it *does not* follow in general that the convergence is uniform *throughout* $(-R, R)$. Indeed, in Example (3) above, the geometric series $\sum_{n=0}^{\infty} x^n$ is *not* uniformly convergent on $(-1, 1)$ (cf. Example 8.2.16).

The following theorems are concerned with term-by-term differentiation and integration of power series.

Theorem 8.4.11. *Let $\sum_{n=0}^{\infty} c_n (x - x_0)^n$ be a power series with radius of convergence $R > 0$ and define $f(x) = \sum_{n=0}^{\infty} c_n (x - x_0)^n$, for all $x \in I := (x_0 - R, x_0 + R)$. Then f is differentiable on I and, for each $x \in I$, we have*

$$f'(x) = \sum_{n=1}^{\infty} n c_n (x - x_0)^{n-1}, \tag{$*$}$$

where the series on the right has the same radius of convergence R.

Proof. It suffices to consider the case $x_0 = 0$ so that $I = (-R, R)$. The last statement about the radius of convergence follows from the fact that $\lim(\sqrt[n]{n}) = 1$, which implies

$$\limsup \sqrt[n]{|nc_n|} = \lim(\sqrt[n]{n}) \limsup \sqrt[n]{|c_n|} = 1/R.$$

Therefore, $\sum_{n=1}^{\infty} nc_n x^{n-1}$ converges *normally* on any compact interval $[-R_0, R_0]$ with $R_0 < R$. Corollary 8.3.7 now implies that f is differentiable and $(*)$ holds for all $x \in (-R, R)$. $\square$

Corollary 8.4.12. *Let* $f(x) := \sum_{n=0}^{\infty} c_n (x - x_0)^n$, *where the series has radius of convergence* $R > 0$. *Then* f *is infinitely differentiable on* $I = (x_0 - R, x_0 + R)$ *(i.e.,* $f \in C^{\infty}(I)$*) and* $c_n = f^{(n)}(x_0)/n!$ *for all* $n \geq 0$.

Proof. Applying Theorem 8.4.11 repeatedly, we deduce that f', f'', ... are obtained from f by repeated term-by-term differentiation and that all the resulting power series have the same radius of convergence R. In fact, by induction, we have

$$f^{(k)}(x) = \sum_{n=k}^{\infty} n(n-1) \cdots (n-k+1)c_n(x-x_0)^{n-k}, \qquad (**)$$

for every $k = 0, 1, 2, \ldots$ and every $x \in (x_0 - R, x_0 + R)$. Putting $x = x_0$ in $(**)$, we have $f^{(k)}(x_0) = k!c_k$. $\square$

Theorem 8.4.13. *Let* $f(x) := \sum_{n=0}^{\infty} c_n (x - x_0)^n$, *where the series has radius of convergence* $R > 0$. *Then* f *is (Riemann) integrable over any compact subinterval,* K, *of the interval* $I := (x_0 - R, x_0 + R)$ *and its integral (over* K*) is obtained by integrating the series term by term. In particular, for each* $x \in I$, *we have*

$$\int_{x_0}^{x} f(t) \, dt = \sum_{n=0}^{\infty} \frac{c_n}{n+1}(x - x_0)^{n+1}, \qquad (\dagger)$$

where the series on the right has the same radius of convergence R.

Proof. Simply note that $\sum_{n=0}^{\infty} c_n (x - x_0)^n$ is obtained from term-by-term differentiation of the series in $(\dagger)$. The latter series is easily seen to have radius of convergence R and, if $F(x)$ denotes its sum, we have $F' = f$. Since $F(x_0) = 0$, the equation $(\dagger)$ follows. $\square$

Example 8.4.14.

(1) Consider the geometric series $\sum_{n=0}^{\infty} x^n$. Its interval of convergence is $(-1, 1)$ and we have

$$\frac{1}{1-x} = \sum_{n=0}^{\infty} x^n \qquad (\forall x \in (-1, 1)).$$

Term-by-term differentiation gives

$$\frac{1}{(1-x)^2} = 1 + 2x + 3x^2 + \cdots + nx^{n-1} + \cdots \qquad (\forall x \in (-1, 1)).$$

(2) We have

$$\frac{1}{1+x} = 1 - x + x^2 - x^3 + \cdots \qquad (\forall x \in (-1, 1)).$$

Integrating term by term, we obtain

$$\log(1 + x) = x - \frac{x^2}{2} + \frac{x^3}{3} - \cdots \qquad (\forall x \in (-1, 1)).$$

Next, we investigate the behavior of power series under *algebraic operations* and *composition*.

Theorem 8.4.15. *Let $f(x) := \sum_{n=0}^{\infty} a_n (x - x_0)^n$ and $g(x) := \sum_{n=0}^{\infty} b_n (x - x_0)^n$ have radii of convergence R' and R'', respectively, and let c be a real constant. Then (1) $cf(x) = \sum_{n=0}^{\infty} ca_n(x - x_0)^n$ for $|x - x_0| < R'$, (2) $f(x) + g(x) = \sum_{n=0}^{\infty} (a_n + b_n)(x - x_0)^n$ for $|x - x_0| < R := \min\{R', R''\}$, and (3) $f(x)g(x) = \sum_{n=0}^{\infty} c_n (x - x_0)^n$ for $|x - x_0| < R := \min\{R', R''\}$, where $c_n = \sum_{k=0}^{n} a_k b_{n-k}$, for all $n = 0, 1, 2, \ldots$.*

Proof. (1) and (2) follow from Theorem 2.3.39 and (3) is a consequence of Theorem 2.3.43 (*Mertens' Theorem*), since both $\sum_{n=0}^{\infty} a_n (x-x_0)^n$ and $\sum_{n=0}^{\infty} b_n (x-x_0)^n$ converge absolutely for $|x - x_0| < R$. $\qquad\qquad\qquad\qquad\qquad\qquad\qquad\qquad\qquad\quad \square$

Remark 8.4.16.

(1) It should be noted that, in Theorem 8.4.15, the interval of convergence of $f(x) + g(x)$ and $f(x)g(x)$ may be *larger*, as follows at once from the trivial case $g(x) = -f(x)$.

(2) Using induction, we can extend Theorem 8.4.15 to sums and products of several power series:

Corollary 8.4.17. *Suppose that $f_k(x) := \sum_{n=0}^{\infty} a_{kn}(x - x_0)^n$ has radius of convergence $R_k > 0$ for $k = 1, 2, \ldots, m$ and let $R := \min\{R_1, \ldots, R_m\}$. Then $\sum_{k=1}^{m} f_k(x) = \sum_{n=0}^{\infty} (\sum_{k=1}^{m} a_{kn})(x - x_0)^n$ and $f_1(x) \cdots f_m(x) = \sum_{n=0}^{\infty} c_n (x - x_0)^n$, for $|x - x_0| < R$ and*

$$c_n = \sum_{n_1 + \cdots + n_m = n} a_{1n_1} \cdots a_{mn_m} \qquad (\forall n \in \mathbb{N}_0).$$

Theorem 8.4.18 (Substitution Theorem). *Let R and R_1 be extended positive numbers, $f(x) := \sum_{n=0}^{\infty} a_n (x - x_0)^n$ for $|x - x_0| < R$, $g(x) - x_0 := \sum_{n=0}^{\infty} b_n (x - x_1)^n$, and suppose that $\sum_{n=0}^{\infty} |b_n| |x - x_1|^n < R$ if $|x - x_1| < R_1$. Then $f(g(x)) = \sum_{m=0}^{\infty} c_m (x - x_1)^m$ for $|x - x_1| < R_1$. Here $c_m = \sum_{n=0}^{\infty} a_n b_{nm}$, where, for each $n \in \mathbb{N}_0$, the b_{nm} are given by $[g(x) - x_0]^n = \sum_{m=0}^{\infty} b_{nm} (x - x_1)^m$.*

Proof. We have $f(y) = \sum_{n=0}^{\infty} a_n (y - x_0)^n$ for $|y - x_0| < R$. Let $y = g(x)$ with $|x - x_1| < R_1$ and note that, by assumption,

$$|y - x_0| = |g(x) - x_0| \leq \sum_{n=0}^{\infty} |b_n||x - x_1|^n < R.$$

It follows that

$$f(g(x)) = \sum_{n=0}^{\infty} a_n [g(x) - x_0]^n = \sum_{n=0}^{\infty} a_n \left(\sum_{m=0}^{\infty} b_{nm}(x - x_1)^m \right). \qquad (*)$$

In view of the definition of the c_m, the theorem follows if we can interchange the order of summation in $(*)$. To justify the interchange, we show that

$$\sum_{n=0}^{\infty} \sum_{m=0}^{\infty} |a_n b_{nm}(x - x_1)^m| < \infty \qquad (|x - x_1| < R_1), \qquad (**)$$

and use Theorem 2.4.26. Now $b_{nm} = \sum_{m_1 + \cdots + m_n = m} b_{m_1} \cdots b_{m_n}$, by Corollary 8.4.17. Thus, with $B_{nm} := \sum_{m_1 + \cdots + m_n = m} |b_{m_1}| \cdots |b_{m_n}|$, we have $|b_{nm}| \leq B_{nm}$. Let $h(x) := \sum_{m=0}^{\infty} |b_m||x - x_1|^m$. Then $\sum_{m=0}^{\infty} B_{nm}|x - x_1|^m = [h(x)]^n$ and we have $\sum_m |a_n b_{nm}(x - x_1)^m| \leq |a_n|[h(x)]^n$. Since $h(x) < R$ for $|x - x_1| < R_1$, we have $\sum_{n=0}^{\infty} |a_n|[h(x)]^n < \infty$ for $|x - x_1| < R_1$ and $(**)$ follows from the *First Comparison Test* (Theorem 2.3.9). $\qquad \square$

The following theorem on the ratio of two power series is our first application of Theorem 8.4.18.

Theorem 8.4.19. *Let $0 < R \leq \infty$ and let $f(x) := \sum_{n=0}^{\infty} a_n (x - x_0)^n$, $g(x) := \sum_{n=0}^{\infty} b_n (x - x_0)^n$, for $|x - x_0| < R$. If $g(x_0) \neq 0$, then there exists $R_1 \in (0, R]$ and a sequence $(c_n)_{n=0}^{\infty}$ such that $f(x)/g(x) = \sum_{n=0}^{\infty} c_n (x - x_0)^n$, for $|x - x_0| < R_1$.*

Proof. Let us assume that $x_0 = 0$. Since $f(x)/g(x) = f(x) \cdot (1/g(x))$, Theorem 8.4.15 implies that we need only consider the special case $f(x) \equiv 1$. Also, replacing $g(x)$ by $g(x)/g(0)$, we may assume that $g(0) = 1$. We then have $g(x) = 1 + h(x)$ with $h(x) = \sum_{n=1}^{\infty} b_n x^n$. Pick $R_1 \in (0, R]$ such that $\sum_{n=1}^{\infty} |b_n||x|^n < 1$ for $|x| < R_1$. Now the expansion $1/(1+x) = \sum_{n=0}^{\infty} (-1)^n x^n$ is valid for $|x| < 1$. Since $1/g(x) = 1/(1 + h(x))$, the theorem follows at once from the *Substitution Theorem*. $\qquad \square$

The class of all functions that can be represented (locally) by convergent power series plays a very important role in analysis:

Definition 8.4.20 ((Real) Analytic Functions). Let $I \neq \emptyset$ be an open interval. A function $f : I \to \mathbb{R}$ is said to be *(real) analytic (in I)* if for each $x_0 \in I$ there exists a real sequence $(c_n)_{n=0}^{\infty}$ and a number $\delta > 0$ such that $(x_0 - \delta, x_0 + \delta) \subset I$ and $f(x) = \sum_{n=0}^{\infty} c_n (x - x_0)^n$ for all $x \in (x_0 - \delta, x_0 + \delta)$.

As another application of the Substitution Theorem, let us show that the sum of a power series with positive radius of convergence is real analytic in the interior of its interval of convergence. This is an extension of *Taylor's Formula*.

Theorem 8.4.21 (Taylor's Theorem). *Let $f(x) := \sum_{n=0}^{\infty} a_n x^n$, where the series has radius of convergence $R > 0$. Then f is (real) analytic on $(-R, R)$. In fact, for each $x_0 \in (-R, R)$, we have*

$$f(x) = \sum_{n=0}^{\infty} \frac{f^{(n)}(x_0)}{n!}(x - x_0)^n \qquad (|x - x_0| < R_1), \tag{†}$$

where $R_1 := R - |x_0|$.

Proof. We use Theorem 8.4.18 (Substitution Theorem) with $g(x) := x$. Note that $|g(x)| = |x| = |x_0 + (x - x_0)| < R$ whenever $|x - x_0| < R_1 = R - |x_0|$. Now, using the *binomial formula* (Proposition 1.3.30), where $\binom{n}{k} = n!/k!(n - k)!$ for $0 \le k \le n$ and $\binom{n}{k} := 0$ for all $k > n$, we have

$$x^n = [g(x)]^n = [x_0 + (x - x_0)]^n \tag{*}$$

$$= \sum_{k=0}^{n} \binom{n}{k} x_0^{n-k}(x - x_0)^k = \sum_{k=0}^{\infty} \binom{n}{k} x_0^{n-k}(x - x_0)^k.$$

Substituting (∗) in $f(x)$ and interchanging the order of summation as in Theorem 8.4.18, we deduce (for $|x - x_0| < R_1$) that

$$f(x) = \sum_{n=0}^{\infty} a_n \sum_{k=0}^{\infty} \binom{n}{k} x_0^{n-k}(x - x_0)^k$$

$$= \sum_{k=0}^{\infty} \left(\sum_{n=k}^{\infty} \binom{n}{k} a_n x_0^{n-k} \right) (x - x_0)^k = \sum_{k=0}^{\infty} \frac{f^{(k)}(x_0)}{k!}(x - x_0)^k,$$

where the last equation follows from $\sum_{n=k}^{\infty} \binom{n}{k} a_n x_0^{n-k} = f^{(k)}(x_0)/k!$, a consequence of Corollary 8.4.12 (cf. the equation (∗∗) in the proof of that corollary). The proof is now complete.   □

Definition 8.4.22 (Taylor & Maclaurin Series (Expansions)). Supposing that f has derivatives of *all orders* at x_0, the series (†) in Theorem 8.4.21 is said to be the *Taylor series* (or *Taylor expansion*) of f at x_0 (or *about* x_0). If $x_0 = 0$, the Taylor series is called the *Maclaurin series* (or *Maclaurin expansion*).

Remark 8.4.23 (C^∞ vs. Analytic). Let $x_0 \in (a, b)$. If $f(x) = \sum_{n=0}^{\infty} c_n (x - x_0)^n$ for all $x \in (a, b)$, then $f \in C^\infty(a, b)$ and the series is necessarily the Taylor series of f at x_0 (cf. Corollary 8.4.12). Now, for *any* function $f \in C^\infty(a, b)$, its Taylor expansion at $x_0 \in (a, b)$ is obviously well defined. However, it is *not* true in general that $f(x) = \sum (f^{(n)}(x_0)/n!)(x - x_0)^n$ on (a, b) or even near x_0. In other words, a C^∞ function on (a, b) is *not* necessarily *analytic* on (a, b), although the converse is always true. As an example, the function

$$f(x) := \begin{cases} e^{-1/x} & \text{if } x > 0, \\ 0 & \text{if } x \le 0, \end{cases}$$

defined in Exercise 6.6.13, is in $C^\infty(\mathbb{R})$. It is *not*, however, equal to the sum of its Maclaurin series. Indeed, we have $f^{(n)}(0) = 0$ for all $n \in \mathbb{N}$ and hence the Maclaurin expansion is *identically* 0 whereas f is certainly not (identically) zero in *any* open interval about $x = 0$. A more interesting example can be constructed using the same type of functions:

Exercise 8.4.24.

(1) Given $a < b$, define the function

$$f(x, a, b) := \begin{cases} \exp\left[\frac{-1}{(x-a)(b-x)}\right] & \text{if } x \in (a, b), \\ 0 & \text{if } x \notin (a, b). \end{cases}$$

Note that this function has *compact support* $[a, b]$, where the *support* of a function f, denoted $\mathrm{supp}(f)$, is the closure of the set of all x with $f(x) \ne 0$:

$$\mathrm{supp}(f) := \{x : f(x) \ne 0\}^-.$$

Show that, as a function of x with fixed $a < b$, we have $f \in C^\infty(\mathbb{R})$. *Hint:* Show that, with $u := (x - a)^{-1}$, $v := (b - x)^{-1}$, and $n \in \mathbb{N}$, we have

$$D^n(e^{-uv}) = P_n(u, v)e^{-uv}, \tag{$*$}$$

where $D = d/dx$ and P_n is a polynomial (in two variables), and deduce that the derivatives $(*)$ converge to 0 as $x \to a+$ or $x \to b-$.

(2) Given any fixed $\delta \in (0, 1)$, let $g(x) := f(x, -1, -1 + \delta)$, with f as in part (1), and define

$$h(x) := \frac{1}{A} \int_{-1}^x g(t)\, dt,$$

where $A := \int_{-1}^{-1+\delta} g(t)\, dt$. Show that $h \in C^\infty(\mathbb{R})$ and $h(x) \equiv 1$ for all $x \ge -1 + \delta$, while $h(x) \equiv 0$ for all $x \le -1$.

(3) Let $\phi(x) := h(-|x|)$ for all $x \in \mathbb{R}$ and h as is part (2). Show that $\phi \in C^\infty(\mathbb{R})$, $\phi(x) \equiv 1$ for all $x \in [-1 + \delta, 1 - \delta]$, and $\phi(x) \equiv 0$ for all $|x| \ge 1$.

Theorem 8.4.25. *Let* $x_0 \in (a, b) \subset \mathbb{R}$ *and* $f \in C^\infty(a, b)$. *For each* $n \in \mathbb{N}_0$, *define* $M_n := \sup\{|f^{(n)}(x)| : x \in (a, b)\}$, *and suppose that* $\lim_{n \to \infty} M_n(b - a)^n/n! = 0$. *Then,*

$$f(x) = \sum_{n=0}^{\infty} \frac{f^{(n)}(x_0)}{n!}(x - x_0)^n \qquad (\forall x \in (a, b)). \tag{†}$$

In particular, if $M_n \leq CM^n$ for all $n \in \mathbb{N}_0$ and some constants $C > 0$ and $M > 0$ (i.e., $M_n = O(M^n)$ $(n \to \infty)$), then (†) holds.

Proof. For each $x \in (a, b) \setminus \{x_0\}$, Theorem 6.6.18 (*Taylor's Formula*) gives

$$f(x) = \sum_{k=0}^{n} \frac{f^{(k)}(x_0)}{k!}(x - x_0)^k + \frac{f^{(n+1)}(\xi)}{(n+1)!}(x - x_0)^{n+1},$$

for some ξ between x_0 and x. Since, by assumption,

$$\left| \frac{f^{(n+1)}(\xi)}{(n+1)!}(x - x_0)^{n+1} \right| \leq \frac{M_{n+1}}{(n+1)!}(b - a)^{n+1} \to 0,$$

as $n \to \infty$, the first statement in the theorem follows. To prove the last statement, we note that $\sum_{n=0}^{\infty} M^n (b - a)^n / n! < \infty$ (why?) and hence $\lim_{n \to \infty} M^n (b - a)^n / n! = 0$. $\square$

Example 8.4.26.

(1) Let $f(x) = e^x$. Then $f^{(n)}(x) = e^x$ for all $n \in \mathbb{N}$. In particular, $f^{(n)}(0) = 1$ for all $n \in \mathbb{N}$. Thus, for any $R > 0$, we have $\sup\{|f^{(n)}(x)| : x \in (-R, R)\} = e^R = O(1)$, as $n \to \infty$. Applying Theorem 8.4.25, we have

$$e^x = \sum_{n=0}^{\infty} \frac{x^n}{n!} \qquad \forall x \in \mathbb{R}.$$

(2) Let $f(x) = \sin x$. Here $|f^{(n)}(x)| = |\sin x|$ if n is *even* and $|f^{(n)}(x)| = |\cos x|$ if n is *odd*. It follows that $|f^{(n)}(x)| \leq 1$ for all $n \in \mathbb{N}$ and all $x \in \mathbb{R}$. Since $\sin 0 = 0$ and $\cos 0 = 1$, Theorem 8.4.25 (with $x_0 = 0$) gives

$$\sin x = \sum_{k=0}^{\infty} \frac{(-1)^k}{(2k+1)!} x^{2k+1} = x - \frac{x^3}{3!} + \frac{x^5}{5!} - \cdots \qquad \forall x \in \mathbb{R}.$$

Similarly, we find the Maclaurin series of $x \mapsto \cos x$:

$$\cos x = \sum_{k=0}^{\infty} \frac{(-1)^k}{(2k)!} x^{2k} = 1 - \frac{x^2}{2!} + \frac{x^4}{4!} - \cdots \qquad \forall x \in \mathbb{R}.$$

In the next section, we shall use the above Maclaurin expansions for e^x, $\sin x$, and $\cos x$ to *define* these functions. We shall also define the general exponential b^x for arbitrary $b > 0$ and $x \in \mathbb{R}$.

Definition 8.4.27 (Binomial Coefficients). For each $\alpha \in \mathbb{R}$ and $n \in \mathbb{N}_0$, the *binomial coefficient* $\binom{\alpha}{n}$ is defined as follows:

$$\binom{\alpha}{n} := \frac{\alpha(\alpha - 1) \cdots (\alpha - n + 1)}{n!}, \qquad \binom{\alpha}{0} := 1.$$

Exercise 8.4.28. Show that

$$(n + 1)\binom{\alpha}{n + 1} + n\binom{\alpha}{n} = \alpha\binom{\alpha}{n} \qquad (\alpha \in \mathbb{R}, \, n \in \mathbb{N}_0).$$

Theorem 8.4.29 (Newton's Binomial Theorem). *Suppose that $x \in (-1, 1)$ and $\alpha \in \mathbb{R}$. Then*

$$(1 + x)^\alpha = \sum_{n=0}^\infty \binom{\alpha}{n} x^n. \tag{$\dagger$}$$

If $\alpha \in \mathbb{N}_0$, then $(\dagger)$ holds for all $x \in \mathbb{R}$.

Proof. First note that, for $\alpha \in \mathbb{N}$, $(\dagger)$ is the usual *binomial formula* (Proposition 1.3.30). For $\alpha \notin \mathbb{N}$, it follows easily from the *Ratio Test* that the power series in $(\dagger)$ converges absolutely on $(-1, 1)$. Let $s(x)$ denote its sum. Then the function $x \mapsto s(x)$ is differentiable on $(-1, 1)$, and $s'(x)$ is obtained by term-by-term differentiation:

$$s'(x) = \sum_{n=1}^\infty n\binom{\alpha}{n} x^{n-1}.$$

Thus, using Exercise 8.4.28, we have

$$(1 + x)s'(x) = \sum_{n=1}^\infty n\binom{\alpha}{n} x^{n-1} + \sum_{n=1}^\infty n\binom{\alpha}{n} x^n$$

$$= \alpha + \sum_{n=2}^\infty n\binom{\alpha}{n} x^{n-1} + \sum_{n=1}^\infty n\binom{\alpha}{n} x^n$$

$$= \alpha + \sum_{n=1}^\infty \left[(n + 1)\binom{\alpha}{n + 1} + n\binom{\alpha}{n} \right] x^n$$

$$= \alpha \sum_{n=0}^\infty \binom{\alpha}{n} x^n$$

$$= \alpha s(x).$$

Put $g(x) := (1+x)^{-\alpha} s(x)$ for all $x \in (-1, 1)$. A simple calculation gives $g'(x) \equiv 0$ on $(-1, 1)$. It follows that g is *constant* on $(-1, 1)$. Since $g(0) = 1$, we indeed have $s(x) = (1 + x)^{\alpha}$ for $|x| < 1$. $\square$

We end this section with an important theorem on uniform convergence of power series. As was pointed out before, the *uniform* convergence of a power series about x_0 with radius of convergence R is (in general) only guaranteed on *compact* subsets of $(x_0 - R, x_0 + R)$. There is, however, an important result (due to Abel) that guarantees the uniform convergence *throughout* the interval of convergence.

Theorem 8.4.30 (Abel). *Let* $f(x) := \sum_{n=0}^{\infty} c_n (x - x_0)^n$, *and assume that the series converges at* $x = x_0 + R$, *for some* $R > 0$. *Then the series is in fact uniformly convergent on* $[x_0, x_0 + R]$ *and we have*

$$\lim_{x \to (x_0 + R)-} f(x) = f(x_0 + R) = \sum_{n=0}^{\infty} c_n R^n. \tag{$\dagger$}$$

Proof. We may (and do) assume that R is the *radius of convergence*. Indeed, R is *at most* equal to that radius and, if it is *strictly smaller*, then the result is obvious. (Why?) Also, using the substitution $x' = (x - x_0)/R$, we may (and do) assume that $x_0 = 0$ and $R = 1$. Thus, $f(x) = \sum_{n=0}^{\infty} c_n x^n$ and $\sum_{n=0}^{\infty} c_n$ is convergent. For $n > m$, Abel's *partial summation formula* (Proposition 2.3.33) implies

$$\sum_{k=m}^{n} c_k x^k = C_n x^n + \sum_{k=m}^{n-1} C_k (x^k - x^{k+1}), \tag{$*$}$$

where $C_k := \sum_{j=m}^{k} c_j$ for each $k \geq m$. Since $\sum c_n$ converges, given any $\varepsilon > 0$, we can pick $N = N(\varepsilon) \in \mathbb{N}$ such that $k \geq m \geq N$ implies $|C_k| < \varepsilon$. Now, for $x \in [0, 1]$, the sequence (x^n) is *decreasing* and hence, for each $x \in [0, 1]$ and $n > m \geq N$, $(*)$ gives

$$\left| \sum_{k=m}^{n} c_k x^k \right| \leq \varepsilon x^n + \sum_{k=m}^{n-1} \varepsilon (x^k - x^{k+1})$$

$$\leq \varepsilon x^n + \varepsilon (x^m - x^n) = \varepsilon x^m \leq \varepsilon,$$

which shows (by Cauchy's Criterion for uniform convergence) that $\sum c_n x^n$ converges *uniformly* on $[0, 1]$. In particular (Corollary 8.3.2), f is *continuous* on $[0, 1]$ and hence $\lim_{x \to 1-} f(x) = f(1)$, which proves $(\dagger)$ and completes the proof. $\square$

Exercise 8.4.31. If $f(x) := \sum_{n=0}^{\infty} c_n x^n$ has radius of convergence $R = 1$ and if the series converges at $x = \pm 1$, show that it is *uniformly* convergent (and hence *continuous*) on $[-1, 1]$.

As a corollary of Abel's theorem, let us prove Theorem 2.3.45 (also due to Abel, of course) as was promised:

Corollary 8.4.32 (Abel). *If $\sum_{n=0}^{\infty} a_n$ and $\sum_{n=0}^{\infty} b_n$ are convergent real series and if their Cauchy product $\sum_{n=0}^{\infty} c_n$ (with $c_n := \sum_{k=0}^{n} a_k b_{n-k}$) is also convergent, then we have*

$$\left(\sum_{n=0}^{\infty} a_n \right) \left(\sum_{n=0}^{\infty} b_n \right) = \sum_{n=0}^{\infty} c_n.$$

Proof. The series $\sum_{n=0}^{\infty} a_n x^n$ and $\sum_{n=0}^{\infty} b_n x^n$ are both absolutely convergent for $|x| < 1$ and hence (Theorem 8.4.15) so is $\sum_{n=0}^{\infty} c_n x^n$ and we have

$$\left(\sum_{n=0}^{\infty} a_n x^n \right) \left(\sum_{n=0}^{\infty} b_n x^n \right) = \sum_{n=0}^{\infty} c_n x^n. \tag{$*$}$$

In view of Theorem 8.4.30, $(*)$ implies

$$\sum_{n=0}^{\infty} c_n = \lim_{x \to 1-} \sum_{n=0}^{\infty} a_n x^n \cdot \lim_{x \to 1-} \sum_{n=0}^{\infty} b_n x^n = \left(\sum_{n=0}^{\infty} a_n \right) \left(\sum_{n=0}^{\infty} b_n \right)$$

and the proof is complete. $\qquad \square$

8.5 Elementary Transcendental Functions

Our goal in this section is to define the elementary transcendental functions (i.e., the *exponential, logarithmic,* and *trigonometric* functions) rigorously, using power series. We have already used these functions in many examples, assuming their basic properties. Here, we shall prove these properties and justify what was used without proof. We begin with the exponential function, which we define for *complex* variables first.

Definition 8.5.1 (Complex Exponential Function). For each $z \in \mathbb{C}$, we define

$$E(z) := \sum_{n=0}^{\infty} \frac{z^n}{n!}. \tag{$\dagger$}$$

It follows at once from the *Ratio Test* that the series is absolutely convergent (i.e., $\sum_{n=0}^{\infty} |z|^n / n! < \infty$ for all $z \in \mathbb{C}$). Therefore, ($\dagger$) converges for all $z \in \mathbb{C}$ (e.g., by Cauchy's Criterion). In fact, given any $R > 0$, the series converges *normally* on $[-R, R]$. This follows from the Weierstrass M-Test (Theorem 8.2.20), whose proof can be followed *verbatim* for the complex case, and the fact that $|z| \leq R$ implies $|z|^n/n! \leq R^n/n!$ and $\sum R^n/n!$ converges.

Theorem 8.5.2. *Consider the function* $E : \mathbb{C} \to \mathbb{C}$, *where* $E(z)$ *is given by* (†) *above. Then we have* $E(0) = 1$, $E(\bar{z}) = \overline{E(z)}$, $E(z) \neq 0$ *and* $E(-z) = 1/E(z)$, *for every* $z \in \mathbb{C}$. *Also, for any* z, $w \in \mathbb{C}$,

$$E(z + w) = E(z)E(w). \qquad (*)$$

More generally, for any $z_1, z_2, \ldots, z_n \in \mathbb{C}$,

$$E(z_1 + z_2 + \cdots + z_n) = E(z_1)E(z_2) \cdots E(z_n). \qquad (**)$$

Finally, for each $q \in \mathbb{Q}$, *we have* $E(q) = e^q$, *where* $e := E(1) = \sum_{n=0}^{\infty} 1/n!$.

Proof. That $E(0) = 1$ is obvious from (†) and so is $E(\bar{z}) = \overline{E(z)}$. To prove (*), we note that the proof of Mertens' Theorem on Cauchy products (Theorem 2.3.43) can be repeated for complex series. Since (†) is *absolutely convergent*, we therefore have

$$E(z)E(w) = \sum_{n=0}^{\infty} \frac{z^n}{n!} \sum_{m=0}^{\infty} \frac{w^m}{m!} = \sum_{n=0}^{\infty} \sum_{k=0}^{n} \frac{z^k w^{n-k}}{k!(n-k)!}$$

$$= \sum_{n=0}^{\infty} \frac{1}{n!} \sum_{k=0}^{n} \binom{n}{k} z^k w^{n-k} = \sum_{n=0}^{\infty} \frac{(z+w)^n}{n!}$$

$$= E(z + w).$$

The equation (**) now follows by induction. Next, (*) implies

$$E(z)E(-z) = E(z - z) = E(0) = 1 \qquad (\forall z \in \mathbb{C}), \qquad (***)$$

which shows that $E(z) \neq 0$ for all $z \in \mathbb{C}$ and $E(-z) = 1/E(z)$ as claimed. To prove the last statement, note that, taking $z_k = 1$ for $1 \leq k \leq n$ in (**), we have $E(n) = (E(1))^n = e^n$, for all $n \in \mathbb{N}$. Now, for each $q = n/m$ with $m, n \in \mathbb{N}$, we have

$$[E(q)]^m = E(mq) = E(n) = e^n,$$

which implies $E(q) = e^q$ for all $q > 0$. For $q < 0$, we have $-q > 0$ and (***) implies $E(q) = 1/E(-q) = 1/e^{-q} = e^q$. Since $e^0 := 1 = E(0)$, the proof is complete. □

Notation 8.5.3. In view of the equation $E(q) = e^q$ for all $q \in \mathbb{Q}$, we henceforth define $e^z := E(z)$ for all $z \in \mathbb{C}$.

Definition 8.5.4 (Real Exponential Function). For each $x \in \mathbb{R}$, we define

$$e^x = \exp(x) := E(x) = \sum_{n=0}^{\infty} \frac{x^n}{n!}. \tag{$\ddagger$}$$

Theorem 8.5.5. *The map $x \mapsto e^x$ defined by $(\ddagger)$ is a strictly positive and strictly increasing function from $\mathbb{R}$ onto $(0, \infty)$ satisfying the following conditions:*

(1) $\exp \in C^{\infty}(\mathbb{R})$ *and* $d^n(e^x)/dx^n = e^x$ *$(\forall n \in \mathbb{N})$; in particular,*

$$\lim_{x \to 0} \frac{e^x - 1}{x} = \exp'(0) = e^0 = 1;$$

(2) $e^{x+y} = e^x e^y$ *and* $e^{x-y} = e^x/e^y$ *$(\forall x, y \in \mathbb{R})$;*
(3) $e^x \geq 1 + x$ *$\forall x \in \mathbb{R}$, with equality only at $x = 0$;*
(4) $\lim_{x \to +\infty} e^x = +\infty$, $\lim_{x \to -\infty} e^x = 0$;*
(5) $\lim_{x \to +\infty} x^n e^{-x} = 0$ *$(\forall n \in \mathbb{N}_0)$.*

Proof. The series in $(\ddagger)$ converges uniformly on compact subsets of $\mathbb{R}$. Therefore, e^x is continuously differentiable on $\mathbb{R}$ and we can differentiate the series term by term to obtain $(e^x)' = e^x$. This proves (1). Also, (2) follows from Theorem 8.5.2. From $(\ddagger)$, it is obvious that $e^x > 0$ for all $x \geq 0$ which, using $e^{-x} = 1/e^x$, implies $e^x > 0$ for all $x < 0$ as well. Since $(e^x)' = e^x > 0$ for all $x \in \mathbb{R}$, the exponential function is *strictly positive* and *strictly increasing* as claimed. Next, for each $x \in \mathbb{R}$, the Mean Value Theorem implies $e^x - 1 = x e^{\xi}$ for some ξ *between 0 and x*. If $x > 0$, then $0 < \xi < x$ implies $e^{\xi} > e^0 = 1$ and hence $e^x - 1 = x e^{\xi} > x$. If $x < 0$, we have $\xi \in (x, 0)$ and hence $e^{\xi} < 1$, which implies $e^x - 1 = x e^{\xi} > x$. Since (3) is true for $x = 0$, it is therefore proved for all $x \in \mathbb{R}$. The first limit in (4) follows from (3) and the second one follows from the first and $e^{-x} = 1/e^x$. These limits and the continuity of e^x show that the range of $\exp$ is indeed $(0, \infty)$. Finally, by $(\ddagger)$, we have $e^x > x^{n+1}/(n+1)!$ and hence $x^n e^{-x} < (n+1)!/x$, for all $x > 0$, from which (5) follows at once. $\qquad \square$

Since e^x is *strictly increasing* and differentiable on $\mathbb{R}$, it has an *inverse* function which is also strictly increasing and differentiable on the *range* of e^x, i.e., on $(0, \infty)$.

Definition 8.5.6 (Natural Logarithm). The inverse function of $\exp : \mathbb{R} \to \mathbb{R}^+ := (0, \infty)$ is called the *(natural) logarithm function* and is denoted by $\log x$. Thus

$$y = \log x \iff x = e^y \qquad (\forall x \in \mathbb{R}^+, \ \forall y \in \mathbb{R}).$$

Equivalently,

$$e^{\log x} = x \quad (\forall x > 0), \qquad \log(e^x) = x \quad (\forall x \in \mathbb{R}).$$

The properties of the natural logarithm are immediate consequences of the corresponding properties of the (natural) exponential function.

Proposition 8.5.7. *The natural logarithm is a strictly increasing, infinitely differentiable function from $\mathbb{R}^+$ onto $\mathbb{R}$ satisfying, for all x, u, $v \in \mathbb{R}^+$, the following properties:*

(1) $(\log x)' = 1/x$ *and* $\log x = \int_1^x dt/t$;
(2) $\log x \leq x - 1$ *with equality precisely when $x = 1$;*
(3) $\log(uv) = \log u + \log v$ *and* $\log(u/v) = \log u - \log v$;
(4) $\lim_{x \to +\infty} \log x = +\infty$, $\lim_{x \to 0+} \log x = -\infty$.

Proof. Differentiating the relation $\exp(\log x) = x$ and using the fact that $(e^x)' = e^x$, we obtain

$$\exp(\log x)(\log x)' = x(\log x)' = 1 \quad (\forall x > 0),$$

which implies $(\log x)' = 1/x$. Since $e^0 = 1$, we have $\log 1 = 0$ and hence the second equation in (1) follows from the Fundamental Theorem of Calculus. For (2), note that by part (3) of Theorem 8.5.5, we have

$$x = e^{\log x} \geq 1 + \log x \quad (\forall x > 0).$$

Next, since e^x is the inverse of $\log x$, the first equation in (3) is equivalent to

$$e^{\log u + \log v} = e^{\log u} e^{\log v} = uv = e^{\log(uv)}.$$

The second equation is proved similarly or deduced from the first one applied to the product $v \cdot (u/v)$. Finally, the limits in (4) are a consequence of the ones in part (4) of Theorem 8.5.5. $\qquad \square$

Exercise 8.5.8. Let $\phi \in C^1(0, \infty)$ satisfy the condition

$$\phi(st) = \phi(s) + \phi(t) \quad (\forall s, t \in \mathbb{R}^+). \tag{$**$}$$

Show that there is a constant C such that $\phi(x) = C \log x$ for all $x \in (0, \infty)$. Deduce that log is the *unique* continuously differentiable function on $(0, \infty)$ that satisfies $(**)$ and whose derivative at 1 equals 1. *Hint:* Differentiate $(**)$ with respect to s and, fixing s, set $t = 1/s$. Alternatively, define $\psi := \phi \circ \exp$ and observe that ψ satisfies *Cauchy's functional equation*, i.e., $\psi(x + y) = \psi(x) + \psi(y)$ for all x, $y \in \mathbb{R}$ (cf. Theorem 4.3.11).

Using the natural logarithm, we can now define *general exponentials*:

Definition 8.5.9 (General Exponential & Power). Given any (fixed) $b \in \mathbb{R}^+ :=$ $(0, \infty)$ with $b \neq 1$, we define the *general exponential function*:

$$x \mapsto b^x := e^{x \log b} \quad (\forall x \in \mathbb{R}). \tag{$*$}$$

Using (∗), we define, for any fixed $\alpha \in \mathbb{R}$, the *general power function*:

$$x \mapsto x^\alpha := e^{\alpha \log x} \quad (\forall x \in \mathbb{R}^+). \tag{∗∗}$$

Proposition 8.5.10. *Fix any* $0 < b \neq 1$ *and* $\alpha \in \mathbb{R}$ *and define the general exponential and power functions by* (∗) *and* (∗∗)*, respectively. Then we have*

(1) $b^0 = 1$ *and* $\log b^x = x \log b$ $(\forall x \in \mathbb{R})$;
(2) $(b^x)' = b^x \cdot \log b$ *and* $\int b^x\, dx = b^x / \log b + C$;
(3) $(x^\alpha)' = \alpha x^{\alpha-1}$ *and* $\int x^\alpha\, dx = x^{\alpha+1}/(\alpha + 1) + C$ $(\forall \alpha \neq -1)$;
(4) b^x *(resp.,* x^α*) is* C^∞ *on* $\mathbb{R}$ *(resp.,* $\mathbb{R}^+$*)*;
(5) if $0 < b < 1$*, then* b^x *is strictly decreasing and* $b^x \to 0$ *as* $x \to +\infty$*, while* $b^x \to +\infty$ *as* $x \to -\infty$;
(6) if $b > 1$*, then* b^x *is strictly increasing and* $b^x \to +\infty$ *as* $x \to +\infty$*, while* $b^x \to 0$ *as* $x \to -\infty$;
(7) $\lim_{x \to +\infty} x^{-\alpha} \log x = 0$ $(\forall \alpha > 0)$.

Note that (3) *is an extension of the "Power Rule" to general exponents.*

Proof. Exercise! *Hint:* For the derivatives (and integrals) use the *Chain Rule* and the properties of e^x and $\log x$. For (7), note that $x^\alpha \to \infty$ as $x \to \infty$, for $\alpha > 0$ (why?), and use *L'Hôpital's Rule*. □

Exercise 8.5.11. Show that the general power function (given by Definition 8.5.9) satisfies the properties (1) $x^\alpha \cdot x^\beta = x^{\alpha+\beta}$, (2) $x^\alpha/x^\beta = x^{\alpha-\beta}$, (3) $(x^\alpha)^\beta = x^{\alpha\beta}$, (4) $x^{-\alpha} = 1/x^\alpha$, (5) $(xy)^\alpha = x^\alpha y^\alpha$, and (6) $(x/y)^\alpha = x^\alpha/y^\alpha$, for all x, $y \in \mathbb{R}^+$ and all α, $\beta \in \mathbb{R}$.

Next, we want to define the trigonometric functions *sine* and *cosine* without introducing the notion of *angle*. To do so, let us look at the complex exponential function $z \mapsto e^z$ applied to a *purely imaginary* number, i.e., a number of the form $z = ix$, where $x \in \mathbb{R} \setminus \{0\}$ and $i := \sqrt{-1}$. We then have

$$e^{ix} = \sum_{n=0}^\infty \frac{(ix)^n}{n!} = \sum_{k=0}^\infty (-1)^k \frac{x^{2k}}{(2k)!} + i \sum_{k=0}^\infty (-1)^k \frac{x^{2k+1}}{(2k+1)!} \tag{∗∗∗}$$

$$= \left(1 - \frac{x^2}{2!} + \frac{x^4}{4!} - \cdots\right) + i\left(x - \frac{x^3}{3!} + \frac{x^5}{5!} - \cdots\right).$$

It then follows from the Ratio Test that both series on the right side of (∗∗∗) have radius of convergence $R = +\infty$ and hence define infinitely differentiable functions on $\mathbb{R}$.

Definition 8.5.12 (Sine & Cosine). For each $x \in \mathbb{R}$, we define

$$\sin x := \sum_{k=0}^\infty (-1)^k \frac{x^{2k+1}}{(2k+1)!}, \qquad \cos x := \sum_{k=0}^\infty (-1)^k \frac{x^{2k}}{(2k)!};$$

i.e., $\cos x := \text{Re}(e^{ix})$ and $\sin x := \text{Im}(e^{ix})$. In particular, we have *Euler's formula*:

$$e^{ix} = \cos x + i \sin x \qquad (\forall x \in \mathbb{R}). \qquad (\dagger)$$

The following identities (also due to Euler) follow from ($\dagger$) and the fact that $\overline{e^{ix}} = e^{-ix}$.

$$\cos x = \frac{e^{ix} + e^{-ix}}{2}, \qquad \sin x = \frac{e^{ix} - e^{-ix}}{2i}. \qquad (\ddagger)$$

Notation 8.5.13. It is customary to write $\cos^k x := (\cos x)^k$ and $\sin^k x := (\sin x)^k$, for all $k \in \mathbb{N}$.

Theorem 8.5.14. $\sin x$ *and* $\cos x$ *are infinitely differentiable (in fact, real analytic) functions on* $\mathbb{R}$ *having the following properties for all* $x, y \in \mathbb{R}$:

(1) $\sin 0 = 0, \quad \cos 0 = 1$;
(2) $\sin(-x) = -\sin x, \quad \cos(-x) = \cos x$;
(3) $(\sin x)' = \cos x, \quad (\cos x)' = -\sin x$;
(4) $\cos^2 x + \sin^2 x = 1$ *(i.e.,* $|e^{ix}| = 1$);
(5) $\sin(x + y) = \sin x \cos y + \cos x \sin y$;
(6) $\cos(x + y) = \cos x \cos y - \sin x \sin y$;
(7) $\cos(2x) = \cos^2 x - \sin^2 x, \quad \sin(2x) = 2 \sin x \cos x$.

Proof. (1) and (2) follow at once from the definition and (3) follows from term-by-term differentiation of the power series defining sine and cosine. The identities (4), (5), and (6) follow easily from ($\ddagger$) above. We may also prove (4) as follows: let $f(x) := \cos^2 x + \sin^2 x$. Then (3) implies that $f'(x) = 0$ for all $x \in \mathbb{R}$ and hence f is *constant*. But then (1) implies that $f(x) \equiv 1$. Finally, (7) follows from (5) and (6) with $x = y$. $\qquad \square$

Proposition 8.5.15 (De Moivre's Formula). *We have*

$$(\cos x + i \sin x)^n = \cos(nx) + i \sin(nx) \quad \forall n \in \mathbb{Z} \quad \forall x \in \mathbb{R}.$$

First Proof. Using Euler's formula, we have (for all $n \in \mathbb{Z}$ and all $x \in \mathbb{R}$)

$$(\cos x + i \sin x)^n = (e^{ix})^n = e^{inx} = \cos(nx) + i \sin(nx).$$

Second Proof. Given any $x, y \in \mathbb{R}$, let $u = \cos x + i \sin x$ and $v = \cos y + i \sin y$. Then, using Theorem 8.5.14, we have

$$uv = (\cos x \cos y - \sin x \sin y) + i (\sin x \cos y + \cos x \sin y)$$

$$= \cos(x + y) + i \sin(x + y).$$

Using this repeatedly with $y = x$, the result follows for all $n > 0$. The case $n = 0$ is obvious from the definition $z^0 := 1$ for all $z \neq 0$. For $n < 0$, simply note that $(\cos x + i \sin x)^{-1} = \cos x - i \sin x = \cos(-x) + i \sin(-x)$. $\square$

Remark 8.5.16. Combining the *binomial formula*, the (Euler's) identities ($\ddagger$) and De Moivre's formula, one can express (positive) powers $\sin^m x$ and $\cos^m x$ as linear combinations of $\sin(kx)$ and/or $\cos(kx)$ for suitable integers k:

Theorem 8.5.17. *The following identities hold for all $x \in \mathbb{R}$ and all $n \in \mathbb{N}$.*

$$\cos^{2n} x = \frac{1}{4^n}\left[\binom{2n}{n} + 2\sum_{j=1}^{n}\binom{2n}{n-j}\cos(2jx)\right],$$

$$\sin^{2n} x = \frac{1}{4^n}\left[\binom{2n}{n} + 2\sum_{j=1}^{n}\binom{2n}{n-j}(-1)^j\cos(2jx)\right],$$

$$\cos^{2n+1} x = \frac{1}{4^n}\sum_{j=0}^{n}\binom{2n+1}{n-j}\cos(2j+1)x,$$

$$\sin^{2n+1} x = \frac{1}{4^n}\sum_{j=0}^{n}\binom{2n+1}{n-j}(-1)^j\sin(2j+1)x.$$

Proof. With $z = \cos x + i \sin x$ we have $1/z = \cos x - i \sin x$ and Euler's identities become $\cos x = (z + 1/z)/2$ and $\sin x = (z - 1/z)/2i$. Using the *binomial formula*, for any integer $m \in \mathbb{N}$, we get

$$\cos^m x = \frac{1}{2^m}\sum_{k=0}^{m}\binom{m}{k}z^{m-k}(z^{-1})^k = \frac{1}{2^m}\sum_{k=0}^{m}\binom{m}{k}z^{m-2k}.$$

Now, by De Moivre's formula, we have $z^{m-2k} = \cos(m-2k)x + i\sin(m-2k)x$ and hence

$$\cos^m x = \frac{1}{2^m}\sum_{k=0}^{m}\binom{m}{k}\cos(m-2k)x + \frac{i}{2^m}\sum_{k=0}^{m}\binom{m}{k}\sin(m-2k)x.$$

Since the left side is *real*, we get

$$\cos^m x = \frac{1}{2^m}\sum_{k=0}^{m}\binom{m}{k}\cos(m-2k)x. \tag{$*$}$$

If $m = 2n$ is even, then the term with $k = n$ is $\binom{2n}{n}\cos(2n-2n)x = \binom{2n}{n}$. Also, pairing the k-th term with the $(2n-k)$-th for $0 \leq k \leq n-1$ and noting that

$$\binom{2n}{k}\cos(2n-2k)x + \binom{2n}{2n-k}\cos(2k-2n)x = 2\binom{2n}{k}\cos(2n-2k)x,$$

we get

$$\cos^{2n} x = \frac{1}{4^n}\left[\binom{2n}{n} + 2\sum_{k=0}^{n-1}\binom{2n}{k}\cos(2n-2k)x\right],$$

which, substituting $j = n - k$, establishes the first identity in the proposition. The identity for $\cos^m x$ when m is *odd* and the identities for the even and odd cases of $\sin^m x$ are proved similarly.   □

Exercise 8.5.18. Prove the following formulas of Wallis for all $n \in \mathbb{N}_0$:

$$\int_0^{\pi/2} \cos^{2n} x \, dx = \int_0^{\pi/2} \sin^{2n} x \, dx = \frac{(2n)!}{4^n(n!)^2} \cdot \frac{\pi}{2}.$$

Proposition 8.5.19. *The set of all positive numbers x with $\cos x = 0$ is nonempty. In fact, there is a smallest $\xi > 0$ with $\cos \xi = 0$.*

Proof. It follows from the definition of cosine that

$$1 - \cos 2 = \sum_{k=0}^{\infty} \frac{2^{4k+2}}{(4k+2)!}\left(1 - \frac{4}{(4k+3)(4k+4)}\right),$$

and the convergent series on the right side has nonnegative terms. Therefore,

$$1 - \cos 2 \geq \left(1 - \frac{1}{3}\right)\frac{2^2}{2!} = \frac{4}{3} > 1,$$

and hence $\cos 2 < 0$. Since $\cos 0 = 1 > 0$ and cosine is *continuous*, the Intermediate Value Theorem (Theorem 4.5.6) implies that we must have $\cos x = 0$ for some $x \in (0, 2)$. Let $Z(\cos) := \{x \in \mathbb{R} : \cos x = 0\}$. Since $\cos x$ is continuous, $Z(\cos)$ is *closed* (Exercise 4.3.8), hence so is $Z^+ := Z(\cos) \cap [0, \infty)$ and note that $0 \notin Z^+$. Therefore, Z^+ has a *smallest* element, say $\xi > 0$, which is given by

$$\xi := \inf\{x > 0 : \cos x = 0\}. \qquad\qquad (*)$$

□

Definition 8.5.20 (The Number π). We define

$$\pi := 2\xi,$$

where ξ is given by $(*)$ above. In particular, $\cos(\pi/2) = 0$.

Theorem 8.5.21 (Periodicity of Sine & Cosine). $\sin x$ and $\cos x$ are periodic with period 2π; i.e., 2π is the smallest positive number such that

$$\sin(x + 2\pi) = \sin x, \quad \cos(x + 2\pi) = \cos x \quad (\forall x \in \mathbb{R}). \tag{\dagger}$$

Proof. First note that $\cos x > 0$ on $[0, \pi/2)$ so that $\sin x$ is *(strictly) increasing* on $[0, \pi/2]$. Since $\cos(\pi/2) = 0$, we have $\sin^2(\pi/2) = 1$ (Theorem 8.5.14) and hence $\sin(\pi/2) = 1$. In particular,

$$e^{i\pi/2} = \cos\frac{\pi}{2} + i\sin\frac{\pi}{2} = i,$$

which implies

$$e^{2\pi i} = \left(e^{i\pi/2}\right)^4 = i^4 = 1.$$

Therefore,

$$\cos(x + 2\pi) + i\sin(x + 2\pi) = e^{i(x+2\pi)} = e^{ix} = \cos x + i\sin x,$$

and ($\dagger$) follows. To complete the proof, we must show that *no* number in $(0, 2\pi)$ is a period for sine or cosine. Now Theorem 8.5.14 implies that $\cos\pi = -1$ and $\sin\pi = 0$. It also implies that $\cos(x + \pi/2) = -\sin x$ which shows, on the one hand, that a number α is a period of $\cos x$ if and only if it is a period of $\sin x$ (and hence, if and only if it is a period of $\exp(ix)$, which happens if and only if $\exp(i\alpha) = 1$). On the other hand, it implies that $\cos x \leq 0$ on $[\pi/2; \pi]$ and, since $\cos(x+\pi) = -\cos x$, we have $\cos x < 1 = \cos 0$ on $(0, 2\pi)$. Therefore, $\exp(ix) \neq 1$ for all $x \in (0, 2\pi)$ and the proof is complete. $\qquad\square$

Once $\sin x$ and $\cos x$ are defined, the remaining trigonometric functions are defined as usual: $\tan x := \sin x / \cos x$, $\cot x := 1/\tan x$, $\sec x := 1/\cos x$, and $\csc x := 1/\sin x$. The properties of these functions may be deduced from the corresponding properties of $\sin x$ and $\cos x$ described above.

8.6 Fourier Series

In the preceding section, we saw that a function that can be represented by a *power series* is *analytic*, hence *infinitely differentiable*, on the interior of the interval of convergence of that series. In practice, however, most important functions we encounter are hardly even *continuous*. Yet, these functions can in many cases be represented by *trigonometric series*. This type of representation was introduced by the French mathematician Fourier in his study of heat conduction. In view of Euler's formula (Definition 8.5.12), it will be convenient to work with *complex-valued* functions of a *real variable*. The notions of derivative and integral can immediately be extended to such functions. Thus, if $I \subset \mathbb{R}$ is an open interval and $f : I \to \mathbb{C}$, then the derivative of f at $x_0 \in I$ is defined to be

$$f'(x_0) := \lim_{x \to x_0} \frac{f(x) - f(x_0)}{x - x_0}, \tag{$*$}$$

if the limit exists. Now $f = u + iv$, where $u := \text{Re}(f)$ and $v := \text{Im}(f)$ are *real-valued* functions on I, so that $(*)$ and Proposition 8.1.14 give

$$
\begin{aligned}
f'(x_0) &= \lim_{x \to x_0} \left(\frac{u(x) - u(x_0)}{x - x_0} + i \frac{v(x) - v(x_0)}{x - x_0} \right) \\
&= \lim_{x \to x_0} \frac{u(x) - u(x_0)}{x - x_0} + i \lim_{x \to x_0} \frac{v(x) - v(x_0)}{x - x_0} \\
&= u'(x_0) + i v'(x_0).
\end{aligned}
$$

In particular, $f : I \to \mathbb{C}$ is differentiable (on I) if and only if both u and v are and we then have $f' = u' + iv'$ on I. Next, for any $f : [a, b] \to \mathbb{C}$ and any tagged partition $\dot{\mathcal{P}}$ of $[a, b]$, the corresponding Riemann sum $S(f, \dot{\mathcal{P}})$ of $f = u + iv$ is defined exactly as in Definition 7.1.4, and one sees at once that

$$S(f, \dot{\mathcal{P}}) = S(u, \dot{\mathcal{P}}) + i S(v, \dot{\mathcal{P}}). \tag{$**$}$$

f is then said to be Riemann integrable on $[a, b]$ (and one then writes $f \in \mathcal{R}([a, b])$) if there is a number $I(f) \in \mathbb{C}$ such that

$$(\forall \varepsilon > 0)(\exists \mathcal{P}_\varepsilon \in \mathcal{P})(\forall \dot{\mathcal{P}} \in \dot{\mathcal{P}})(\mathcal{P}_\varepsilon \subset \mathcal{P} \Rightarrow |S(f, \dot{\mathcal{P}}) - I(f)| < \varepsilon). \tag{$***$}$$

It follows from $(**)$ and $(***)$ that $f \in \mathcal{R}([a, b])$ if and only if $u, v \in \mathcal{R}([a, b])$ and, in this case, we have

$$\int_a^b f(x)\, dx = \int_a^b u(x)\, dx + i \int_a^b v(x)\, dx \tag{$\dagger$}$$

or, equivalently,

$$\text{Re}\left(\int_a^b f(x)\, dx \right) = \int_a^b \text{Re}(f(x))\, dx, \quad \text{Im}\left(\int_a^b f(x)\, dx \right) = \int_a^b \text{Im}(f(x))\, dx.$$

In particular, we note the effect of *complex conjugation*:

$$\overline{\int_a^b f(x)\, dx} = \int_a^b \overline{f(x)}\, dx.$$

One may also use (†) as the *definition* of $\int_a^b f(x)\,dx$. It follows at once that, for any $f,\ g \in \mathcal{R}([a,b])$ and $\alpha,\ \beta \in \mathbb{C}$, we have

$$\int_a^b [\alpha f(x) + \beta g(x)]\,dx = \alpha \int_a^b f(x)\,dx + \beta \int_a^b g(x)\,dx.$$

In fact, most of the properties of the integral can be deduced from (†) and the ones already proved for real-valued functions. For instance, if $F = U + iV : [a,b] \to \mathbb{C}$ with $U' = u$ and $V' = v$ on $[a,b]$, then F is a *primitive* of $f = u + iv$ (i.e., $F' = f$), and we have

$$\int_a^b f(x)\,dx = F(b) - F(a).$$

Also, if $f : [a,b] \to \mathbb{C}$ is *continuous* on $[a,b]$, then $f \in \mathcal{R}([a,b])$ and $\int_a^x f(t)\,dt$ is a primitive of f:

$$\frac{d}{dx} \int_a^x f(t)\,dt = f(x).$$

If $f = u + iv \in \mathcal{R}([a,b])$, then $|f| \in \mathcal{R}([a,b])$ (why?) and we have

$$\left| \int_a^b f(x)\,dx \right| \le \int_a^b |f(x)|\,dx.$$

To prove this, let $\alpha = \int_a^b u(x)\,dx$ and $\beta = \int_a^b v(x)\,dx$. Then

$$\left| \int_a^b f(x)\,dx \right|^2 = \int_a^b (\alpha u(x) + \beta v(x))\,dx$$

$$\le \int_a^b \sqrt{\alpha^2 + \beta^2}\sqrt{u(x)^2 + v(x)^2}\,dx$$

$$= \left| \int_a^b f(x)\,dx \right| \int_a^b |f(x)|\,dx,$$

where we have used Corollary 7.4.14 and Cauchy's inequality (Corollary 8.1.21):

$$|\alpha u(x) + \beta v(x)| \le \sqrt{\alpha^2 + \beta^2}\sqrt{u(x)^2 + v(x)^2}.$$

Exercise 8.6.1 (Cauchy–Schwarz Inequality). Given any f, $g \in \mathcal{R}([a,b])$, show that

$$\left| \int_a^b f(x)\overline{g(x)}\, dx \right|^2 \leq \left(\int_a^b |f(x)|^2\, dx \right) \left(\int_a^b |g(x)|^2\, dx \right).$$

Definition 8.6.2 (Trigonometric Polynomial). A finite sum of the form

(i) $f(x) = a_0 + \displaystyle\sum_{n=1}^{N}(a_n \cos nx + b_n \sin nx)$ $(\forall x \in \mathbb{R})$,

where the *coefficients* $a_0, \ldots, a_N$ and $b_1, \ldots, b_N$ are, in general, *complex* numbers, is said to be a *trigonometric polynomial*. Using the identities $\cos \theta = (e^{i\theta} + e^{-i\theta})/2$ and $\sin \theta = (e^{i\theta} - e^{-i\theta})/(2i)$, we may write (i) in the more convenient form

(ii) $f(x) = \displaystyle\sum_{n=-N}^{N} c_n e^{inx}$ $(\forall x \in \mathbb{R})$,

where $c_n \in \mathbb{C}$. It is obvious that $f(x)$ is periodic with period 2π.

Since, for $n \neq 0$, we have $[e^{inx}/(in)]' = e^{inx}$ and $e^{\pm i\pi} = -1$,

(iii) $\dfrac{1}{2\pi} \displaystyle\int_{-\pi}^{\pi} e^{inx}\, dx = \begin{cases} 1 & \text{if } n = 0, \\ 0 & \text{if } n = \pm 1,\ \pm 2, \ldots. \end{cases}$

If we multiply the trigonometric polynomial (ii) by e^{-imx}, where $m \in \mathbb{Z}$, and integrate this product on $[-\pi, \pi]$, Then (iii) implies that

(iv) $c_m = \dfrac{1}{2\pi} \displaystyle\int_{-\pi}^{\pi} f(x)e^{-imx}\, dx$

for $|m| \leq N$; for $|m| > N$, the integral in (iv) is 0.

In particular, it follows from (iv) that the trigonometric polynomial $f(x)$ in (ii) is *real* if and only if $c_{-n} = \overline{c_n}$, for $n = 0, \ldots, N$. The following definition is now motivated by (ii):

Definition 8.6.3 (Fourier Coefficient, Fourier Series). A series of the form

(v) $\displaystyle\sum_{n=-\infty}^{\infty} c_n e^{inx}$ $(x \in \mathbb{R})$

is said to be a *trigonometric series* and its *nth partial sum* is defined to be the right side of (ii). If $f \in \mathcal{R}([-\pi, \pi])$, then the numbers c_n defined by (iv) are called the *Fourier coefficients* of f and are also denoted by $\hat{f}(n)$:

$$\hat{f}(n) := \frac{1}{2\pi} \int_{-\pi}^{\pi} f(x)e^{-inx}\, dx.$$

The series (v) with *these* coefficients is called the *Fourier series* of f. In this case we write

$$(vi) \quad f(x) \sim \sum_{n=-\infty}^{\infty} \hat{f}(n)e^{inx} \qquad (x \in \mathbb{R}).$$

Remark 8.6.4. (1) Note that the symbol $\sim$ used in (vi) simply means that the coefficients c_n are given by (iv). It *does not* imply anything about the convergence or divergence of the series. The convergence of Fourier series is an extremely tricky business and has led many prominent mathematicians (including Dirichlet, Riemann, Cantor, and Lebesgue) to discover numerous fundamental results that play a crucial role in analysis. In this section we only look at some basic results that require the *Riemann integral*. For more advanced results, the *Lebesgue integral* is needed.

(2) As in Definition 8.6.2, we may write (vi) in the form

$$f(x) \sim a_0 + \sum_{m=1}^{\infty} a_m \cos mx + b_m \sin mx.$$

It is then easy to see (why?) that, for each $m \in \mathbb{N}$,

$$a_0 = \frac{1}{2\pi} \int_{-\pi}^{\pi} f(x) \, dx,$$

$$a_m = \frac{1}{\pi} \int_{-\pi}^{\pi} f(x) \cos mx \, dx,$$

$$b_m = \frac{1}{\pi} \int_{-\pi}^{\pi} f(x) \sin mx \, dx.$$

In particular, if f is *even*, then $b_m = 0$ for all $m \in \mathbb{N}$, $a_0 = \frac{1}{\pi} \int_0^{\pi} f(x) \, dx$, and $a_m = (2/\pi) \int_0^{\pi} f(x) \cos mx \, dx$, for all $m \in \mathbb{N}$, so that f has a *cosine expansion*: $f(x) \sim a_0 + \sum_{m=1}^{\infty} a_m \cos mx$. Similarly, if f is *odd*, then $a_0 = a_m = 0$ for all $m \in \mathbb{N}$ and $b_m = (2/\pi) \int_0^{\pi} f(x) \sin mx \, dx$ for all $m \in \mathbb{N}$ so that, in this case, f has a *sine expansion*: $f(x) \sim \sum_{m=1}^{\infty} b_m \sin mx$.

(3) The sequence $(e^{inx})_{n=-\infty}^{\infty}$ used above may be replaced by more general *systems of functions* that satisfy relations similar to (iii); such systems are called *orthogonal*. We shall look at this later when we discuss *Hilbert spaces*.

Notation 8.6.5. To simplify the exposition, we shall often use the notation

$$e_n(x) := e^{inx} \qquad (x \in \mathbb{R}).$$

The following theorem shows that, among all trigonometric polynomials, the partial sums of the Fourier series of a function f provide the *best mean square approximation* to f:

Theorem 8.6.6 (Best Approximation). *Let* $f \in \mathcal{R}([-\pi, \pi])$ *and let* $s_n(x) = \sum_{k=-n}^{n} c_k e^{ikx}$ *denote the n-th partial sum of its Fourier series. Then, given any trigonometric polynomials* $t_n(x) := \sum_{k=-n}^{n} c'_k e^{ikx}$, *we have*

$$\frac{1}{2\pi} \int_{-\pi}^{\pi} |f(x) - s_n(x)|^2 \, dx \leq \frac{1}{2\pi} \int_{-\pi}^{\pi} |f(x) - t_n(x)|^2 \, dx, \qquad (*)$$

and equality holds if and only if $c'_k = c_k$, *for* $k = -n, \ldots, n$. *In addition, we have*

$$\sum_{k=-n}^{n} |c_k|^2 \leq \frac{1}{2\pi} \int_{-\pi}^{\pi} |f(x)|^2 \, dx. \qquad (**)$$

Proof. For simplicity, let us write $\sum := \sum_{-n}^{n}$, $\int := \int_{-\pi}^{\pi}$, and $d'x := dx/2\pi$. Now, using the definition of the c_n, we have

$$\int f(x) \overline{t_n(x)} \, d'x = \int f(x) \sum \overline{c'_k} \, \overline{e_k(x)} \, d'x = \sum c_k \overline{c'_k}.$$

Also, (iii) implies that

$$\int |t_n(x)|^2 \, d'x = \int t_n(x) \overline{t_n(x)} \, d'x = \int \sum c'_j e_j(x) \sum \overline{c'_k} \, \overline{e_k(x)} \, d'x = \sum |c'_k|^2.$$

Therefore,

$$\int |f - t_n|^2 \, d'x = \int |f|^2 \, d'x - \int f \overline{t_n} \, d'x - \int \overline{f} t_n \, d'x + \int |t_n|^2 \, d'x$$

$$= \int |f|^2 \, d'x - \sum c_k \overline{c'_k} - \sum \overline{c_k} c'_k + \sum |c'_k|^2$$

$$= \int |f|^2 \, d'x - \sum |c_k|^2 + \sum |c'_k - c_k|^2,$$

which is obviously *minimized* if and only if $c'_k = c_k$ for $|k| \leq n$, proving $(*)$. Putting $c'_k = c_k$ and noting that $\int |f - t_n|^2 \, d'x \geq 0$, we obtain $(**)$ as well. $\square$

Theorem 8.6.7 (Bessel's Inequality). *If* $f \in \mathcal{R}([-\pi, \pi])$ *and if*

$$f(x) \sim \sum_{n=-\infty}^{\infty} c_n e^{inx},$$

then we have

$$\sum_{n=-\infty}^{\infty} |c_n|^2 = \lim_{n \to \infty} \sum_{k=-n}^{n} |c_k|^2 \leq \frac{1}{2\pi} \int_{-\pi}^{\pi} |f(x)|^2 \, dx. \qquad (\dagger)$$

In particular,

$$\lim_{n \to \pm\infty} c_n = \lim_{n \to \pm\infty} \frac{1}{2\pi} \int_{-\pi}^{\pi} f(x)e^{-inx} \, dx = 0. \tag{$\ddagger$}$$

Proof. Indeed (†) follows at once from the inequality (∗∗) in Theorem 8.6.6 if we let $n \to \infty$ and (‡) is then an immediate consequence. □

Remark 8.6.8. (1) As we shall see later, Bessel's inequality (†) is actually an *equality*, known as *Parseval's Relation*.

(2) That (‡) holds (i.e., that the limit of the n-th Fourier coefficient $\hat{f}(n)$ is *zero* as $n \to \infty$) is known as *Riemann's Lemma*. The following theorem is an extension of this lemma.

Theorem 8.6.9 (Riemann–Lebesgue Lemma). *If $f \in \mathcal{R}([a, b])$, then we have*

$$\lim_{\alpha \to \pm\infty} \int_a^b f(x)e^{i\alpha x} \, dx = 0.$$

In particular,

$$\lim_{\alpha \to \pm\infty} \int_a^b f(x) \sin(\alpha x) \, dx = \lim_{\alpha \to \pm\infty} \int_a^b f(x) \cos(\alpha x) \, dx = 0.$$

Proof. Suppose first that f is a *step function.* Thus, there is a partition $(x_k)_{k=0}^n$ of $[a, b]$ and constants $c_k \in \mathbb{R}$, $1 \le k \le n$, with $f(x) \equiv c_k$ on (x_{k-1}, x_k). Since, for $k = 1, 2, \ldots, n$ and $\alpha \in \mathbb{R}$, we have

$$\lim_{|\alpha| \to \infty} \int_{x_{k-1}}^{x_k} f(x)e^{i\alpha x} \, dx = \lim_{|\alpha| \to \infty} \frac{c_k}{i\alpha} \left(e^{i\alpha x_k} - e^{i\alpha x_{k-1}}\right) = 0,$$

the theorem follows in this case. (Why?) In general, let $f \in \mathcal{R}([a, b])$ and let $\varepsilon > 0$. Then (cf. Exercise 7.4.8) we can pick a step function g such that $\int_a^b |f(x) - g(x)| \, dx < \varepsilon/2$. Since the theorem is true for g, we can pick $A > 0$ such that $|\alpha| \ge A$ implies $|\int_a^b g(x)e^{i\alpha x} \, dx| < \varepsilon/2$. Therefore, if $|\alpha| \ge A$, we have

$$\left| \int_a^b f(x)e^{i\alpha x} \, dx \right| \le \left| \int_a^b [f(x) - g(x)]e^{i\alpha x} \, dx \right| + \left| \int_a^b g(x)e^{i\alpha x} \, dx \right|$$

$$\le \int_a^b |f(x) - g(x)| \, dx + \left| \int_a^b g(x)e^{i\alpha x} \, dx \right|$$

$$< \frac{\varepsilon}{2} + \frac{\varepsilon}{2} = \varepsilon,$$

and the proof is complete. □

There are two special sequences of trigonometric polynomials that play an important role in answering the question of convergence of Fourier series:

Definition 8.6.10 (Dirichlet's Kernel, Fejér's Kernel). For each integer $n \geq 0$, the trigonometric polynomials

$$D_n(x) := \sum_{k=-n}^{n} e^{ikx} \quad \text{and} \quad K_n(x) := \frac{1}{n+1} \sum_{j=0}^{n} D_j(x) \qquad (*)$$

are called *Dirichlet's kernel* and *Fejér's kernel*, respectively. Note that $K_n(x)$ is the *arithmetic mean* of the Dirichlet kernels $D_0(x), \ldots, D_n(x)$.

Theorem 8.6.11. *For each integer $n \geq 0$, we have*

(1) $D_n(x) = \dfrac{\sin(n + \frac{1}{2})x}{\sin(x/2)}$,

(2) $K_n(x) = \dfrac{1}{n+1} \cdot \dfrac{1 - \cos(n+1)x}{1 - \cos x} = \dfrac{1}{n+1} \cdot \dfrac{\sin^2[(n+1)x/2]}{\sin^2(x/2)}$,

(3) $D_n(2k\pi) = 2n + 1, \quad K_n(2k\pi) = n + 1 \quad (\forall k \in \mathbb{Z})$,

(4) $\dfrac{1}{2\pi} \displaystyle\int_{-\pi}^{\pi} D_n(x)\, dx = \dfrac{1}{2\pi} \int_{-\pi}^{\pi} K_n(x)\, dx = 1.$

Moreover, $K_n(x) \geq 0$ for all x and

(5) $\qquad K_n(x) \leq \dfrac{2}{(n+1)(1 - \cos \delta)} \qquad (0 < \delta \leq |x| \leq \pi).$

Proof. First note that $(*)$ implies

$$(e^{ix} - 1)D_n(x) = e^{i(n+1)x} - e^{-inx}. \qquad (**)$$

Multiplying both sides of $(**)$ by $e^{-ix/2}$ and using the identities ($\ddagger$) in Definition 8.5.12, we obtain (1). Next, we substitute $(**)$ in the definition of $K_n(x)$ and use the identity

$$(e^{-ix} - 1)\left(e^{i(k+1)x} - e^{-ikx}\right) = 2\cos kx - 2\cos(k+1)x$$

to obtain

$$(n+1)K_n(x)(2 - 2\cos x) = (n+1)K_n(x)(e^{ix} - 1)(e^{-ix} - 1)$$

$$= \sum_{k=0}^{n}(e^{-ix} - 1)\left(e^{i(k+1)x} - e^{-ikx}\right)$$

$$= 2 - 2\cos(n+1)x,$$

which (with the *double-angle* identity $1 - \cos 2\theta = 2 \sin^2 \theta$) implies (2). $K_n(x) \geq 0$ is then obvious and so is (5) because cos is *decreasing* on $[0, \pi]$. Finally, (3) and (4) follow directly from $(*)$.   □

Exercise 8.6.12.

(1) Using the identities $2 \sin u \sin v = \cos(v - u) - \cos(v + u)$ and $2 \sin u \cos v = \sin(v + u) - \sin(v - u)$, show that, for all $x \neq 2k\pi$ $(k \in \mathbb{Z})$, we have

$$\sum_{k=1}^{n} \sin kx = \frac{\cos(x/2) - \cos(n + \frac{1}{2})x}{2 \sin(x/2)},$$

$$\sum_{k=1}^{n} \cos kx = \frac{\sin(n + \frac{1}{2})x - \sin(x/2)}{2 \sin(x/2)}.$$

Use these identities to prove part (1) of Theorem 8.6.11.

(2) Show that, if (c_n) is a monotone real sequence with $\lim(c_n) = 0$, then the trigonometric series $\sum_{n=1}^{\infty} c_n \sin nx$ converges for all x and $\sum_{n=1}^{\infty} c_n \cos nx$ converges for all x except (possibly) $x = 2k\pi$, $k \in \mathbb{Z}$. *Hint:* Use the identities in part (1) and Dirichlet's Test (Theorem 2.3.35).

Remark 8.6.13. Henceforth, as we study the convergence of Fourier series, we shall consider functions $f : \mathbb{R} \to \mathbb{R}$ that are periodic with period 2π. We note that any function $f \in \mathcal{R}([-\pi, \pi])$ can be extended to $\mathbb{R}$ as a 2π-periodic function. Indeed, we may (if necessary) redefine $f(\pi)$ to be $f(-\pi)$. This will not affect the Riemann integrals involving f. Now any $x \in \mathbb{R}$ can be written as $x = x_0 + 2k\pi$ for some $x_0 \in [-\pi, \pi]$ and some $k \in \mathbb{Z}$. For this x, we define $f(x) := f(x_0)$.

Example 8.6.14. (1) Consider the function $f(x) := x$, $-\pi < x \leq \pi$, extended to $\mathbb{R}$ as a 2π-periodic function. Then f is *odd* so, as remarked before, $f(x) \sim \sum_{m=1}^{\infty} b_m \sin mx$, with $b_m = \frac{2}{\pi} \int_0^{\pi} x \sin mx \, dx$. Integrating by parts, we obtain

$$b_m = -\left[\frac{2x \cos mx}{m\pi}\right]_0^{\pi} + \frac{2}{m\pi} \int_0^{\pi} \cos mx \, dx = -\frac{2 \cos m\pi}{m} = \frac{2(-1)^{m-1}}{m}.$$

Thus,

$$f(x) \sim 2 \sum_{m=1}^{\infty} \frac{(-1)^{m-1} \sin mx}{m}.$$

(2) Let $f(x) := |x|$, $-\pi \leq x \leq \pi$, extended to $\mathbb{R}$ by 2π-periodicity. Then f is a continuous, *even* function. Therefore, $b_m = 0$ for all m, $a_0 = (1/\pi) \int_0^{\pi} x \, dx = \pi/2$ and (for $m > 0$) $a_m = \frac{2}{\pi} \int_0^{\pi} x \cos mx \, dx$. Thus,

$$a_m = \left[\frac{2x \sin mx}{m\pi}\right]_0^\pi - \frac{2}{\pi} \int_0^\pi \frac{\sin mx}{m} \, dx = \frac{2[(-1)^m - 1]}{m^2\pi},$$

which gives $a_m = 0$ if m is *even* and $a_m = -4/(m^2\pi)$ if m is *odd*. We therefore have

$$f(x) \sim \frac{\pi}{2} - \frac{4}{\pi} \sum_{k=1}^\infty \frac{\cos(2k-1)x}{(2k-1)^2}.$$

(3) Let us extend the function $f(x) := (\pi - |x|)^2$ on $[-\pi, \pi]$ to $\mathbb{R}$ by 2π-periodicity. Then f is a continuous, *even* function. In particular, $b_m = 0$ for all m and $a_0 = (1/\pi) \int_0^\pi (\pi - x)^2 \, dx = \pi^2/3$. For $m > 0$, integrating by parts, we have

$$\begin{aligned}
a_m &= \frac{2}{\pi} \int_0^\pi (\pi - x)^2 \cos mx \, dx \\
&= \left[\frac{2(\pi - x)^2 \sin mx}{\pi m}\right]_0^\pi + \frac{4}{\pi m} \int_0^\pi (\pi - x) \sin mx \, dx \\
&= -\left[\frac{4(\pi - x)\cos mx}{\pi m^2}\right]_0^\pi \\
&= \frac{4}{m^2}.
\end{aligned}$$

Therefore,

$$f(x) \sim \frac{\pi^2}{3} + \sum_{m=1}^\infty \frac{4}{m^2} \cos mx.$$

We now start our study of convergence with the following

Proposition 8.6.15 (Dirichlet's Integral). *Let $f : \mathbb{R} \to \mathbb{R}$ be a 2π-periodic function such that $f \in \mathcal{R}([-\pi, \pi])$ and let $s_n(x)$ be the n-th partial sum of its Fourier series: $s_n(x) = \sum_{-n}^n \hat{f}(k)e^{ikx}$. Then*

$$s_n(x) = \frac{1}{2\pi} \int_{-\pi}^\pi f(x-t)D_n(t) \, dt = \frac{1}{2\pi} \int_0^\pi [f(x+t)+f(x-t)]D_n(t) \, dt, \qquad (*)$$

with Dirichlet's kernel D_n as in Definition 8.6.10.

Proof. Using the definitions of $\hat{f}(n)$ and D_n, we have

$$s_n(x) = \sum_{k=-n}^n \hat{f}(k)e^{ikx} = \sum_{k=-n}^n \left(\int_{-\pi}^\pi f(t)e^{-ikt} \, d't\right) e^{ikx}$$

$$= \int_{-\pi}^{\pi} f(t) \left(\sum_{k=-n}^{n} e^{ik(x-t)} \right) d't = \int_{-\pi}^{\pi} f(t) D_n(x-t) \, d't,$$

where $d't := dt/(2\pi)$. Hence, the first equation in $(*)$ follows if we can prove

$$\int_{-\pi}^{\pi} f(t) D_n(x-t) \, d't = \int_{-\pi}^{\pi} f(x-t) D_n(t) \, d't.$$

For this, we take $x - t$ as a new variable and use Exercise 7.4.6, noting that D_n and f are both 2π-periodic functions. To show the second equation in $(*)$, note that D_n is an *even* function so that a change of variable from t to $-t$ gives

$$\int_{-\pi}^{0} f(x-t) D_n(t) \, d't = \int_{0}^{\pi} f(x+t) D_n(t) \, d't.$$

The second equation in $(*)$ now follows from

$$\int_{-\pi}^{\pi} f(x-t) D_n(t) \, d't = \int_{-\pi}^{0} f(x-t) D_n(t) \, d't + \int_{0}^{\pi} f(x-t) D_n(t) \, d't$$

and the proof is complete. $\qquad\square$

Corollary 8.6.16 (Fejér's Integral). *Let f and s_n be as in Proposition 8.6.15 and, for $n \geq 0$, consider the arithmetic means*

$$\sigma_n(x) := \frac{s_0(x) + s_1(x) + \cdots + s_n(x)}{n+1}.$$

Then, with Fejér's kernel K_n as in Definition 8.6.10, we have

$$\sigma_n(x) = \frac{1}{2\pi} \int_{-\pi}^{\pi} f(x-t) K_n(t) \, dt = \frac{1}{2\pi} \int_{0}^{\pi} [f(x+t) + f(x-t)] K_n(t) \, dt. \qquad (**)$$

Proof. Exercise! $\qquad\square$

The next theorem, due to Riemann, shows that, for a 2π-periodic function $f \in \mathcal{R}([-\pi, \pi])$, the behavior of the sequence $(s_n(x))$ of partial sums of the Fourier series of f at x depends only on the values of f in an *arbitrarily small* neighborhood of x.

Theorem 8.6.17 (Riemann's Localization Theorem). *Let f be a 2π-periodic function with $f \in \mathcal{R}([-\pi, \pi])$. Then, for any $\delta \in (0, \pi)$, we have*

$$\lim_{n \to \infty} \left(\int_{-\pi}^{-\delta} + \int_{\delta}^{\pi} \right) f(x-t) D_n(t) \, dt = 0. \qquad (\dagger)$$

Proof. As in the proof of Dirichlet's integral, we have

$$(i) \quad \left(\int_{-\pi}^{-\delta} + \int_{\delta}^{\pi} \right) f(x-t) D_n(t)\, dt = \int_{\delta}^{\pi} [f(x+t) + f(x-t)] D_n(t)\, dt.$$

By Theorem 8.6.11, the right side of (i) equals

$$(ii) \quad \int_{\delta}^{\pi} g_x(t) \sin[(n+1/2)t]\, dt, \qquad g_x(t) := \frac{f(x+t) + f(x-t)}{\sin(t/2)}.$$

Since $\sin(t/2) \geq \sin(\delta/2)$ on $[\delta, \pi]$, we have $g_x \in \mathcal{R}([\delta, \pi])$ and the Riemann–Lebesgue lemma (Theorem 8.6.9) implies that the integral in (ii) converges to 0 as $n \to \infty$ and (†) follows.   □

Corollary 8.6.18. *Under the conditions of Theorem 8.6.17, the Fourier series of f converges at x if and only if, for some (and hence any) $\delta \in (0, \pi]$,*

$$\lim_{n \to \infty} \frac{1}{2\pi} \int_0^{\delta} [f(x+t) + f(x-t)] D_n(t)\, dt$$

(or $\lim_{n\to\infty}(2\pi)^{-1} \int_{-\delta}^{\delta} f(x-t) D_n(t)\, dt$) exists (as a finite number); this limit is then the sum of the Fourier series of f at x.

Proof. By Dirichlet's integral,

$$s_n(x) - \int_{-\delta}^{\delta} f(x-t) D_n(t)\, d't = \left(\int_{-\pi}^{-\delta} + \int_{\delta}^{\pi} \right) f(x-t) D_n(t)\, d't,$$

where $d't = dt/(2\pi)$. Hence the corollary follows from Riemann's Localization Theorem which also implies that the limit is *independent* of δ. (Why?)   □

Theorem 8.6.19 (Dini's Criterion). *Let f be a 2π-periodic function on $\mathbb{R}$ with $f \in \mathcal{R}([-\pi, \pi])$. If*

$$\lim_{\delta \to 0+} \int_{\delta}^{\pi} \frac{|f(x+t) + f(x-t) - 2s|}{t}\, dt$$

exists (as a finite number), then the Fourier series of f converges to s at x.

Proof. Let $d't := dt/(2\pi)$ and note that, by Theorem 8.6.11,

$$\int_{-\pi}^{\pi} D_n(t)\, d't = 2 \int_0^{\pi} D_n(t)\, d't = 1.$$

Therefore, Dirichlet's integral implies

$$(1) \quad s_n(x) - s = \int_0^{\pi} [f(x+t) + f(x-t) - 2s] D_n(t)\, d't$$

$$= \int_0^{\pi} \phi_x(t) t D_n(t)\, d't,$$

where $\phi_x(t) := [f(x+t) + f(x-t) - 2s]/t$. Now the last integral in (1) can be written as

$$(2) \quad \lim_{\delta \to 0+} \int_\delta^\gamma \phi_x(t) t D_n(t) \, d't + \int_\gamma^\pi \phi_x(t) t D_n(t) \, d't,$$

where $\gamma \in (0, \pi]$ is arbitrary. Fix $\varepsilon > 0$ and note that $|t/\sin(t/2)| \leq \pi$ for all $t \in (0, \pi]$, as we see, e.g., by comparing the graph of $y = \sin x$ and the line segment joining $(0,0)$ and $(\pi/2, 1)$. Our assumption implies that the number $\gamma \in (0, \pi]$ can be chosen such that $\lim_{\delta \to 0+} \int_\delta^\gamma |\phi_x(t)| \, d't < \varepsilon/(2\pi)$. Since

$$|t D_n(t)| = |t/\sin(t/2)||\sin(n+1/2)t| \leq \pi \quad (\forall t \in (0, \pi]),$$

the first term in (2) satisfies

$$(3) \quad \left| \lim_{\delta \to 0+} \int_\delta^\gamma \phi_x(t) t D_n(t) \, d't \right| \leq \pi \lim_{\delta \to 0+} \int_\delta^\gamma |\phi_x(t)| \, d't < \varepsilon/2.$$

Next, $\phi_x(t)(t/\sin(t/2)) \sin(n+1/2)t \in \mathcal{R}([\gamma, \pi])$ and hence, by the Riemann–Lebesgue lemma, we can find $N \in \mathbb{N}$ such that the second term in (2) satisfies

$$(4) \quad n \geq N \Longrightarrow \left| \int_\gamma^\pi \phi_x(t) t D_n(t) \, d't \right| < \varepsilon/2.$$

In view of (1), (2), (3), and (4), we finally have

$$|s_n(x) - s| < \varepsilon \quad (\forall n \geq N)$$

and the proof is complete. $\qquad\qquad\square$

Before stating an important corollary of the above criterion, we invite the reader to solve the following.

Exercise 8.6.20. Let $f : (a, b] \to \mathbb{R}$. If $f \in \mathcal{R}([c, b])$ for all $c \in (a, b]$ and if f is *bounded* on $(a, b]$ (e.g., if $\lim_{c \to a+} f(x)$ exists), show that

$$\lim_{c \to a+} \int_c^b |f(x)| \, dx \qquad (*)$$

exists (as a finite number). *Hint:* Let $I(c)$ denote the integral in $(*)$. Show that $I(c)$ is *bounded* on $(a, b]$ and *increases* as c *decreases* to a.

Corollary 8.6.21. *Let f be as in Theorem 8.6.19. If $f(x+0)$ and $f(x-0)$ are both finite and*

$$\lim_{t \to 0+} \frac{f(x+t) - f(x+0)}{t} \quad and \quad \lim_{t \to 0+} \frac{f(x-t) - f(x-0)}{t}$$

both exist (and are finite), then the Fourier series of f converges to

$$\frac{f(x+0) + f(x-0)}{2}$$

at the point x. In particular, if $f'(x)$ exists, then the Fourier series of f converges to $f(x)$ at the point x.

Proof. If we take $s := [f(x+0) + f(x-0)]/2$ in Theorem 8.6.19, then a calculation shows that, with the notation of that theorem, we have

$$\phi_x(t) = \frac{f(x+t) - f(x+0)}{t} + \frac{f(x-t) - f(x-0)}{t}$$

and $\lim_{t\to 0+} \phi_x(t)$ is finite. Exercise 8.6.20 now gives $\lim_{\delta\to 0+} \int_\delta^\pi |\phi_x(t)| \, dt < \infty$ and the corollary follows from Dini's Criterion.   □

The following exercise refers to Example 8.6.14.

Exercise 8.6.22.

(1) Show that the function $f(x) = x$ on $(-\pi, \pi]$ extended to $\mathbb{R}$ by 2π-periodicity satisfies the conditions of Corollary 8.6.21. Deduce that

$$x = 2 \sum_{m=1}^\infty \frac{(-1)^{m-1} \sin mx}{m}, \qquad (\forall x \in (-\pi, \pi)).$$

Note that, at $x = \pi$, $[f(\pi+0) + f(\pi-0)]/2 = (-\pi + \pi)/2 = 0$. Also show that $\pi/4 = \sum_{k=1}^\infty (-1)^{k-1}/(2k-1)$.

(2) Show that the function $f(x) = |x|$ on $[-\pi, \pi]$ extended to $\mathbb{R}$ by 2π-periodicity satisfies the conditions of Corollary 8.6.21. Deduce that

$$|x| = \frac{\pi}{2} - \frac{\pi}{4} \sum_{k=1}^\infty \frac{\cos(2k-1)x}{(2k-1)^2} \qquad (\forall x \in [-\pi, \pi]).$$

Show, in particular, that $\pi^2/8 = \sum_{k=1}^\infty 1/(2k-1)^2$.

(3) Show that the function $f(x) = (\pi - |x|)^2$ on $[-\pi, \pi]$ extended to $\mathbb{R}$ by 2π-periodicity satisfies the conditions of Corollary 8.6.21. Deduce that

$$(\pi - |x|)^2 = \frac{\pi^2}{3} + \sum_{m=1}^\infty \frac{4}{m^2} \cos mx \quad (\forall x \in [-\pi, \pi]).$$

Show, in particular, that $\pi^2/6 = \sum_{m=1}^\infty 1/m^2$.

In fact, the differentiability condition in the above corollary may be replaced by weaker conditions:

Corollary 8.6.23. *Let* f *be as in Theorem 8.6.19. If* f *satisfies a Lipschitz condition of order* $\alpha \in (0, 1]$ *in a neighborhood of* x *(cf. Definition 4.6.9), i.e.,*

$$|f(x + h) - f(x)| \leq A|h|^{\alpha},$$

for all $|h|$ *sufficiently small, then the Fourier series of* f *at* x *converges to* $f(x)$.

Proof. Let $\phi_x(t) := [f(x + t) + f(x - t) - 2f(x)]/t$. Then our assumption implies that $|\phi_x(t)| = O(|t|^{\alpha - 1})$, as $|t| \to 0$, so that

$$\lim_{\delta \to 0+} \int_{\delta}^{\pi} \phi_x(t)\, dt$$

exists (why?) and hence Dini's Criterion may be used.   □

So far we have used *Dirichlet's kernel* (and integral) in our convergence theorems. We now look at *Fejér's kernel* (and integral):

Theorem 8.6.24 (Fejér's Theorem). *Let* f *and* $\sigma_n(x)$ *be as in Corollary 8.6.16 (Fejér's Integral).*

(1). If $f(x + 0)$ *and* $f(x - 0)$ *both exist (in* $\mathbb{R}$*), then*

$$\lim_{n \to \infty} \sigma_n(x) = \frac{f(x + 0) + f(x - 0)}{2}.$$

In particular, $\lim_{n \to \infty} \sigma_n(x) = f(x)$ *at any continuity point* x *of* f.
(2). If f *is continuous, then* $\lim_{n \to \infty} \sigma_n(x) = f(x)$ *uniformly for all* x.

Proof. Let $d't := dt/(2\pi)$. By Theorem 8.6.11, we have

$$\int_{-\pi}^{\pi} K_n(t)\, d't = 2 \int_{0}^{\pi} K_n(t)\, d't = 1.$$

Hence, using Fejér's Integral (cf. $(**)$ in Corollary 8.6.16),

(i) $\sigma_n(x) - \dfrac{f(x + 0) + f(x - 0)}{2} = \displaystyle\int_{0}^{\pi} \psi_x(t) K_n(t)\, d't,$

where

$$\psi_x(t) := [f(x + t) - f(x + 0)] + [f(x - t) - f(x - 0)].$$

Since by assumption f is 2π-periodic and $f \in \mathcal{R}([-\pi, \pi])$, we can find $M > 0$ with $|f(x)| \leq M$ for all $x \in \mathbb{R}$. (Why?) Therefore, $|\psi_x(t)| \leq 4M$ for all x and t. Also, $\lim_{t \to 0+} \psi_x(t) = 0$. Hence, for any $\varepsilon > 0$ we can find $\delta = \delta(\varepsilon) \in (0, \pi]$ such that $t \in (0, \delta)$ implies $|\psi_x(t)| < \varepsilon$. Thus, with $s := [f(x + 0) + f(x - 0)]/2$, (i) gives

(ii) $|\sigma_n(x) - s| \leq \left| \int_0^\delta \psi_x(t) K_n(t) \, d't \right| + \left| \int_\delta^\pi \psi_x(t) K_n(t) \, d't \right|$

$$\leq \varepsilon \int_0^\delta K_n(t) \, d't + 4M \int_\delta^\pi K_n(t) \, d't.$$

But $\int_0^\delta K_n(t) \, d't \leq \int_0^\pi K_n(t) \, d't = 1/2$ and $\lim_{n \to \infty} K_n(t) = 0$ *uniformly* on $[\delta, \pi]$ (by Theorem 8.6.11), so that there is $N_\varepsilon \in \mathbb{N}$ such that $n \geq N_\varepsilon$ implies $\int_\delta^\pi K_n(t) \, d't < \varepsilon/(8M)$. Using these estimates in (ii), we get

$$n \geq N_\varepsilon \implies \left| \sigma_n(x) - \frac{f(x+0) + f(x-0)}{2} \right| < \frac{\varepsilon}{2} + \frac{\varepsilon}{2} = \varepsilon.$$

Since $f(x + 0) = f(x) = f(x - 0)$ at any *continuity point* x of f, part (1) of the theorem follows. The proof of part (2) is almost identical. Here, we use the *uniform continuity* of f on $[-\pi, \pi]$ to pick $\delta = \delta(\varepsilon) > 0$ such that $t \in (0, \delta]$ implies $|\psi_x(t)| < \varepsilon$ for all $x \in [-\pi, \pi]$. Repeating the above argument, we get

$$|\psi_x(t) - f(x)| < \frac{\varepsilon}{2} + 4M \int_\delta^\pi K_n(t) \, d't, \quad (x \in [-\pi, \pi]).$$

Finally, we can again pick $N_\varepsilon \in \mathbb{N}$ as above to conclude that

$$n \geq N_\varepsilon \implies |\psi_x(t) - f(x)| < \varepsilon$$

and the proof is complete.   □

We now deduce some of the important consequences of Fejér's theorem. To begin, recall that, by the Weierstrass Approximation Theorem (Corollary 4.7.10), continuous functions can be *uniformly* approximated (on compact intervals) by *polynomials*. The following corollary (also due to Weierstrass) shows that *polynomials* can be replaced by *trigonometric polynomials*:

Corollary 8.6.25 (Weierstrass). *A continuous 2π-periodic function can be uniformly approximated (on $\mathbb{R}$) by trigonometric polynomials.*

Proof. This follows at once from Fejér's theorem if we use the trigonometric polynomials $\sigma_n(x)$.   □

Corollary 8.6.26. *If f and g are both 2π-periodic and continuous on $\mathbb{R}$ and have the same Fourier series, then $f(x) = g(x)$ for all $x \in \mathbb{R}$.*

Proof. Indeed, if $\sigma_n(x)$ is the n-th arithmetic mean of the partial sums of this Fourier series, then (by Fejér's theorem) $f(x) = \lim(\sigma_n(x)) = g(x)$ for all x.   □

Corollary 8.6.27. *If f is a 2π-periodic continuous function on $\mathbb{R}$ such that*

$$\hat{f}(n) = \frac{1}{2\pi} \int_{-\pi}^\pi f(x) e^{-inx} \, dx = 0 \quad (\forall n \in \mathbb{Z}),$$

then $f(x) = 0$ for all $x \in \mathbb{R}$.

Proof. Simply take $g = 0$ in Corollary 8.6.26. $\qquad\square$

Corollary 8.6.28. *If f is a 2π-periodic continuous function on $\mathbb{R}$ and if the Fourier series of f converges uniformly, then it converges to f.*

Proof. Let $s_n(x)$ denote the n-th partial sum of the Fourier series of f and let us write $d'x := dx/(2\pi)$. Now $s_n(x) \to g(x)$ uniformly implies $s_n(x)e^{-ikx} \to g(x)e^{-ikx}$ uniformly ($\forall\, k \in \mathbb{Z}$) and hence

$$\widehat{s_n}(k) = \int_{-\pi}^{\pi} s_n(x)e^{-ikx}\, d'x \to \int_{-\pi}^{\pi} g(x)e^{-ikx}\, d'x = \hat{g}(k).$$

But the *orthogonality* of the e^{inx} implies at once that $\widehat{s_n}(k) = \hat{f}(k)$ for all $n \geq k$. Therefore, $\hat{f}(k) = \hat{g}(k)$ for all k and, by Corollary 8.6.26, we have $g = f$. $\quad\square$

Before deducing a uniform convergence criterion from this corollary, we need the following.

Definition 8.6.29 (Piecewise Differentiable Function). A function $f : [a, b] \to \mathbb{C}$ is said to be *piecewise differentiable* if there is a partition $(x_j)_{j=0}^{n}$ of $[a, b]$ such that

(1) $f'(t)$ exists for all $t \in (x_{k-1}, x_k)$ and every k, $1 \leq k \leq n$;
(2) for each k, the restriction of f' to (x_{k-1}, x_k) has a *continuous extension* to $[x_{k-1}, x_k]$.

Example 8.6.30. It is obvious that any $f \in C^1([a, b])$ (i.e., any continuously differentiable function on $[a, b])$ is piecewise differentiable. Also, any *piecewise linear* function on $[a, b]$ is piecewise differentiable.

Theorem 8.6.31. *Let f be a continuous 2π-periodic function on $\mathbb{R}$ whose restriction to $[-\pi, \pi]$ is piecewise differentiable. Then the Fourier series of f converges uniformly on $\mathbb{R}$ to f.*

Proof. The assumptions imply that f' is *bounded* on $[-\pi, \pi]$ and *continuous* there, except possibly at finitely many points. In particular, $f' \in \mathcal{R}([-\pi, \pi])$. Therefore, an integration by parts (over the subintervals determined by the above points) shows that, for all $n \neq 0$, we have

$$\widehat{f'}(n) := \frac{1}{2\pi} \int_{-\pi}^{\pi} f'(x)e^{-inx}\, dx = \frac{in}{2\pi} \int_{-\pi}^{\pi} f(x)e^{-inx}\, dx = in\hat{f}(n).$$

Thus, $\hat{f}(n) = \widehat{f'}(n)/(in)$ for all $n \neq 0$ and hence, by Cauchy's inequality,

$$\sum_{k=-n}^{n} |\hat{f}(k)| = |\hat{f}(0)| + \sum_{0<|k|\leq n} \frac{|\widehat{f'}(k)|}{|k|}$$

$$\leq |\hat{f}(0)| + \left(\sum_{0<|k|\leq n} \frac{1}{k^2} \right)^{1/2} \left(\sum_{0<|k|\leq n} |\widehat{f'}(k)|^2 \right)^{1/2}.$$

Now Bessel's inequality (Theorem 8.6.7) implies that $\sum_{-\infty}^{\infty} |\widehat{f'}(k)|^2 < \infty$, and we have $\sum k^{-2} < \infty$ so that $\sum_{k=-\infty}^{\infty} |\hat{f}(k)| < \infty$. It now follows (from the Weierstrass M-Test) that the Fourier series $\sum_{k=-\infty}^{\infty} \hat{f}(k)e^{ikx}$ converges *uniformly* and, in view of Corollary 8.6.28, it must converge to f.   □

For our next theorem, we shall need the following lemma on the *mean square approximation* of a Riemann integrable function by continuous functions.

Lemma 8.6.32. *Let $f \in \mathcal{R}([a,b])$. Then, given any $\varepsilon > 0$, there is a continuous function $\phi_\varepsilon \in C([a,b])$ such that*

$$\int_a^b |f(x) - \phi_\varepsilon(x)|^2 \, dx < \varepsilon.$$

Proof. Pick a partition $(x_j)_{j=0}^n$ of $[a,b]$ such that, with M_k and m_k denoting the "sup" and "inf" of f on $[x_{k-1}, x_k]$, respectively, we have $\sum (M_k - m_k)\Delta x_k < \varepsilon/(2M)$, where $\Delta x_k := x_k - x_{k-1}$ and $M := \sup\{|f(x)| : x \in [a,b]\}$. Now let ϕ_ε be the *continuous, piecewise linear* function such that $\phi_\varepsilon(x_j) = f(x_j)$ for $0 \leq j \leq n$ and ϕ_ε is *affine* on each $[x_{k-1}, x_k]$, i.e., with $\Delta f_k := f(x_k) - f(x_{k-1})$, we have $\phi_\varepsilon(t) = f(x_{k-1}) + (\Delta f_k/\Delta x_k)(t - x_{k-1})$ for all $t \in [x_{k-1}, x_k]$. Now note that $M \geq \sup\{|\phi_\varepsilon(x)| : x \in [a,b]\}$ (why?), and hence

$$\int_a^b |f(x) - \phi_\varepsilon(x)|^2 \, dx \leq 2M \sum_{k=1}^n \int_{x_{k-1}}^{x_k} |f(x) - \phi_\varepsilon(x)| \, dx < 2M(\varepsilon/2M) = \varepsilon.$$

□

Theorem 8.6.33 (Parseval's Theorem). *Let f and g be 2π-periodic and f, $g \in \mathcal{R}([-\pi, \pi])$. If $s_n(x)$ denotes the n-th partial sum of the Fourier series of f, then we have*

(i) $\displaystyle \lim_{n\to\infty} \int_{-\pi}^{\pi} |f(x) - s_n(x)|^2 \, dx = 0,$

(ii) $\displaystyle \sum_{n=-\infty}^{\infty} \hat{f}(n)\overline{\hat{g}(n)} = \int_{-\pi}^{\pi} f(x)\overline{g(x)} \, dx,$

$$\sum_{n=-\infty}^{\infty} |\hat{f}(n)|^2 = \frac{1}{2\pi} \int_{-\pi}^{\pi} |f(x)|^2 \, dx. \qquad \text{(Parseval's Relation)}$$

Proof. Let us write $\int$ for $\frac{1}{2\pi} \int_{-\pi}^{\pi}$. Assume first that f is *continuous* and let $\varepsilon > 0$ be given. By Fejér's theorem, we can then pick $N \in \mathbb{N}$ such that $n \geq N$ implies

$|f(x) - \sigma_n(x)| < \varepsilon$ for all x. By Theorem 8.6.6, we have

$$\int |f - s_n|^2 \le \int |f - \sigma_n|^2 < \varepsilon^2 \quad (\forall n \ge N),$$

and (i) follows. Next,

(iii) $\displaystyle\int s_n(x)\overline{g(x)} = \sum_{-n}^{n} \hat{f}(k) \int \overline{g(x)} e^{ikx} = \sum_{-n}^{n} \hat{f}(k)\overline{\hat{g}(k)},$

and Cauchy–Schwarz inequality (Exercise 8.6.1) gives

(iv) $\displaystyle\left| \int f\bar{g} - \int s_n\bar{g} \right| \le \int |f - s_n||g| \le \left\{ \int |f - s_n|^2 \int |g|^2 \right\}^{1/2},$

which $\to 0$, as $n \to \infty$, by (i). Using (iii) and (iv), we obtain (ii), from which *Parseval's Relation* follows if we take $g = f$. Finally, if f is *not continuous*, then (using the above lemma) we pick a continuous function ϕ on $[-\pi, \pi]$, which we can extend by 2π-periodicity to a *continuous* function on $\mathbb{R}$ (why?), such that $\int |f - \phi|^2 < \varepsilon/4$. Let $\tau_n(x)$ be the n-th arithmetic mean of the partial sums of the Fourier series of ϕ and pick $N \in \mathbb{N}$ so that $n \ge N$ implies $\int |\phi - \tau_n|^2 < \varepsilon/4$. Then,

$$\int |f - \tau_n|^2 \le 2 \int |f - \phi|^2 + 2 \int |\phi - \tau_n|^2 < \varepsilon,$$

for all $n \ge N$. Thus (i) follows from Theorem 8.6.6 (*Best Approximation*) and the remaining assertions follow as above. The reader can provide the details. $\qquad\square$

Corollary 8.6.34 (Term-by-Term Integration). *If $f \in \mathcal{R}([-\pi, \pi])$ is 2π-periodic, then we have*

$$F(x) := \int_{-\pi}^{x} f(t)\, dt = \sum_{-\infty}^{\infty} \hat{f}(n) \int_{-\pi}^{x} e^{-int}\, dt,$$

where the convergence is uniform on $[-\pi, \pi]$.

Proof. Let $s_n(x)$ denote the n-th partial sum of the Fourier series of f. Then, by the *Cauchy–Schwarz* inequality, we have

$$\int_{-\pi}^{\pi} |f(x) - s_n(x)|\, dx \le \sqrt{2\pi} \left(\int_{-\pi}^{\pi} |f(x) - s_n(x)|^2\, dx \right)^{1/2}.$$

This, together with the inequality

$$\left| \int_{-\pi}^{x} f(t)\, dt - \int_{-\pi}^{x} s_n(t)\, dt \right| \le \int_{-\pi}^{\pi} |f(t) - s_n(t)|\, dt \quad (\forall x \in [-\pi, \pi])$$

and part (i) of *Parseval's Theorem*, implies that $\lim_{n\to\infty} \int_{-\pi}^{x} s_n(t)\, dt =$
$\int_{-\pi}^{x} f(t)\, dt$, with the convergence being *uniform* on $[-\pi, \pi]$.   □

Exercise 8.6.35.

(1) Applying Parseval's Relation to the function f in Exercise 8.6.22 (1), deduce
    that

$$\sum_{n=1}^{\infty} \frac{1}{n^2} = \frac{\pi^2}{6}.$$

(2) Applying Parseval's Relation to the function f in Exercise 8.6.22 (3), show that

$$\sum_{n=1}^{\infty} \frac{1}{n^4} = \frac{\pi^4}{90}.$$

Hint: Note that

$$\hat{f}(m) = \begin{cases} \frac{1}{2}(a_m - ib_m) & \text{if } m > 0, \\ a_0 & \text{if } m = 0, \\ \frac{1}{2}(a_{-m} + ib_{-m}) & \text{if } m < 0. \end{cases}$$

8.7 Problems

1 (Roots of Unity). Let $r_k := e^{2(k-1)\pi i/n}$, $k = 1, 2, \ldots, n$ be the nth roots of unity (i.e., the
solutions of $1 - z^n = 0$).

(a) Show that for all $z \in \mathbb{C}$ we have

$$1 - z^n = \prod_{k=1}^{n} (1 - r_k z) \qquad \text{and} \qquad (*)$$

$$1 = \frac{1}{n} \sum_{j=1}^{n} \prod_{k \neq j} (1 - r_k z). \qquad (**)$$

Hint: For ($**$) differentiate ($*$).
(b) Evaluate the following sums.

$$s_1 := \sum_{i=1}^{n} r_i,$$

$$s_2 := \sum_{i<j} r_i r_j,$$

$$s_3 := \sum_{i<j<k} r_i r_j r_k,$$

$$\vdots$$

$$s_{n-1} := \sum_j \prod_{k \ne j} r_k,$$

$$s_n := \prod_{k=1}^{n} r_k.$$

How are these sums affected if each r_j is replaced by $1/r_j$?

(c) Prove (**) using your answers to part (b).

2. (a) Suppose that $(f_n) \in \mathcal{F}(E, \mathbb{R})^{\mathbb{N}}$ converges to f *uniformly* on E and that each f_n is *bounded* on E. Show that (f_n) is *uniformly bounded*; i.e., there is an $M > 0$ such that $|f_n(x)| \le M$ for all $n \in \mathbb{N}$ and all $x \in E$.

(b) Let $f_n(x) := 1/x + x/n$ for all $x \in (0, 1]$. Show that (f_n) converges *uniformly* to $f(x) := 1/x$ on $(0, 1]$ and yet the f_n and f are all *unbounded* on $(0, 1]$.

3. Let (f_n), $(g_n) \in \mathcal{F}(E, \mathbb{R})^{\mathbb{N}}$ converge *uniformly* on E to f and g, respectively.

(a) Show that $(f_n \pm g_n)$ converges to $f \pm g$ uniformly on E.

(b) Show, by an example, that $(f_n g_n)$ need *not* converge to fg uniformly on E. If, however, the f_n and g_n are all *bounded* on E, show that $(f_n g_n)$ does converge to fg uniformly on E.

(c) Show (by an example) that, if $f_n(x) \ne 0$ for all $x \in E$ and all $n \in \mathbb{N}$, then $(1/f_n)$ need *not* converge to $1/f$ uniformly on E. Can you add an assumption that would guarantee an affirmative answer?

4. Show that $(\sin nx/(1 + nx))$ converges *uniformly* on $[a, \infty)$ for any $a > 0$ but *not* for $a = 0$.

5. Let $f_n(x) := (1 + x/n)^n$ for all $x \in \mathbb{R}$ and $n \in \mathbb{N}$. Show that $(f_n(x))$ converges *uniformly* to e^x on any compact interval $[a, b] \subset \mathbb{R}$. *Hint:* Use *Dini's* Theorem.

6 (Uniform Ratio Test). Let $(f_n) \in \mathcal{F}(E, \mathbb{R})^{\mathbb{N}}$ be a sequence of *bounded* functions with $f_n(x) \ne 0$ for all $x \in E$ and all $n \in \mathbb{N}$. Show that, if there is an $N \in \mathbb{N}$ and a constant $\rho \in (0, 1)$ such that $|f_{n+1}(x)/f_n(x)| \le \rho$ for all $n \ge N$, then $\sum f_n(x)$ is *uniformly* convergent on E.

7 (Uniform Dirichlet's Test).

(a) Let $(f_n) \in \mathcal{F}(E, \mathbb{R})^{\mathbb{N}}$, where $E \subset \mathbb{R}$, and assume that the partial sums $\sum_{k=1}^{n} f_k(x)$ are *uniformly bounded* on E. Show that, if $(g_n(x))$ is a *decreasing* sequence of *nonnegative* functions on E converging *uniformly* to *zero* on E, then $\sum_{n=1}^{\infty} f_n(x) g_n(x)$ converges *uniformly* on E.

(b) Show that, for any $\alpha > 0$, the series $\sum_{n=1}^{\infty} (\sin nx)/n^\alpha$ and $\sum_{n=1}^{\infty} (\cos nx)/n^\alpha$ are both *uniformly* convergent on any *compact* interval that does *not* contain $2k\pi$ for any $k \in \mathbb{Z}$. *Hint:* Use the identities in Problem 2.5.#63.

8 (Uniform Abel's Test). Let $(f_n) \in \mathcal{F}(E, \mathbb{R})^{\mathbb{N}}$ and assume that $\sum_{n=1}^{\infty} f_n(x)$ is *uniformly* convergent on $E \subset \mathbb{R}$. Then, for any *uniformly bounded* sequence $(g_n) \in \mathcal{F}(E, \mathbb{R})^{\mathbb{N}}$ such that $(g_n(x))$ is *monotone* for each $x \in E$, the series $\sum_{n=1}^{\infty} f_n(x) g_n(x)$ is *uniformly* convergent on E. *Hint:* Consider (separately) the subsets E_1 and E_2 of E on which $(g_n(x))$ is *increasing* and *decreasing*, respectively. Also, use Abel's partial summation formula (Proposition 2.3.33).

9. (a) Show that the sequence $(1/(nx + 1))$ converges *pointwise* but *not* uniformly on $(0, 1)$. What is the limit function?

(b) Show that the sequence $(x/(nx+1))$ converges *uniformly* on $(0, 1)$. What is the limit function?

10 (du Bois–Raymond Test). Let (f_n), $(g_n) \in \mathcal{F}(E, \mathbb{R})^{\mathbb{N}}$ be such that $\sum_{n=1}^{\infty} f_n$ and $\sum_{n=1}^{\infty} |g_n - g_{n+1}|$ are both *uniformly* convergent on $E \subset \mathbb{R}$ and (g_n) is *uniformly bounded* on E. Then $\sum_{n=1}^{\infty} f_n g_n$ is uniformly convergent on E. *Hint:* Use Abel's *partial summation formula* (Proposition 2.3.33).

11 (Uniform Leibniz's Test). Show that, if (f_n) is a *decreasing* sequence of functions converging *uniformly* to *zero* on $E \subset \mathbb{R}$, then $\sum_{n=1}^{\infty} (-1)^n f_n$ converges *uniformly* on E.

12. Let $f \in \mathbb{R}^{\mathbb{R}}$ be *uniformly continuous* (on $\mathbb{R}$) and define $f_n(x) := f(x + 1/n)$ for all $n \in \mathbb{N}$ and $x \in \mathbb{R}$. Show that (f_n) converges *uniformly* to f on $\mathbb{R}$.

13. For each $n \in \mathbb{N}$, consider the function

$$f_n(x) := \frac{x^n}{1 + x^{2n}} \qquad (\forall x \in \mathbb{R}).$$

(a) Show that the sequence (f_n) converges *uniformly* on $[a, b]$ if and only if $|x| \neq 1$ for all $x \in [a, b]$; (i.e., $[a, b]$ does *not* contain the points $+1$ and -1).
(b) Find all $x \in \mathbb{R}$ for which the series $\sum f_n(x)$ is convergent. Also, find the intervals on which the convergence is *uniform*.

14. Find the sum of the series

$$\sum_{n=0}^{\infty} \frac{x^{2^n}}{1 - x^{2^{n+1}}} \qquad (\forall x \in (0, 1)).$$

15. Let $(f_n) \in \mathcal{F}(E, \mathbb{R})^{\mathbb{N}}$ be *bounded* and suppose that (f_n) converges *uniformly* to f on E. Show that the sequence $(\sum_{k=1}^{n} f_k/n)$ of *arithmetic means* converges uniformly to f on E.

16. Show that the series $\sum_{n=1}^{\infty} (-1)^n/(n + x^2)$ is *uniformly* convergent on $[0, \infty)$ but is *not absolutely* convergent for any $x \geq 0$.

17. Show that, if $f_n \in C(\mathbb{R})$ for all $n \in \mathbb{N}$ and if (f_n) converges *uniformly* to f (on $\mathbb{R}$), then we have

$$\lim_{n \to \infty} f_n(x + 1/n) = f(x)$$

uniformly on any *bounded* interval.

18. (a) Show that, if $\sum_{n=1}^{\infty} |a_n| < \infty$, then $\sum_{n=1}^{\infty} a_n e^{-nx}$ is *uniformly* convergent on $[0, \infty)$.
(b) If we only assume that (a_n) is *bounded,* show that $\sum_{n=1}^{\infty} a_n e^{-nx}$ is *uniformly* convergent on $[\delta, \infty)$ for all $\delta > 0$.

19. Show that $\sum_{n=1}^{\infty} x^n(1 - x)$ converges *pointwise* (but *not* uniformly) on $[0, 1]$ and that $\sum (-1)^n x^n(1 - x)$ is *uniformly* convergent on $[0, 1]$. This shows that, if $\sum f_n(x)$ is uniformly convergent and $\sum |f_n(x)|$ is pointwise convergent, it does *not* follow (in general) that $\sum |f_n(x)|$ is uniformly convergent.

20. Show that the convergence is *not* uniform in the following limits.

(a) $\lim_{n \to \infty} \sin^n x$, $x \in [0, \pi]$;   (b) $\lim_{n \to \infty} e^{-nx^2}$, $x \in [-1, 1]$

Hint: Are the limits *continuous*?

21. Let f_n be *continuous* on $E \subset \mathbb{R}$ for all $n \in \mathbb{N}$, and assume that (f_n) converges *uniformly* to f on E. Show that, if $(x_n) \in E^{\mathbb{N}}$ and $\lim(x_n) = x \in E$, then $\lim(f_n(x_n)) = f(x)$.

22.

(a) Let $f_n \in C([a, b])$ for all $n \in \mathbb{N}$. Show that, if (f_n) converges *uniformly* on (a, b), then the convergence is actually uniform on $[a, b]$.

(b) More generally, let (f_n) be a sequence of *bounded, continuous* functions on $E \subset \mathbb{R}$ and let $D \subset E$ be *dense* in E (i.e., each $x \in E$ is the limit of a sequence (x_n) in D). Show that, if (f_n) converges *uniformly* on D, then the convergence is in fact uniform on E.

23. Prove each equation.

(a) $\displaystyle \int_1^2 \left(\sum_{n=1}^{\infty} n e^{-nx} \right) dx = \frac{e}{e^2 - 1}$;   (b) $\displaystyle \int_0^{\pi} \left(\sum_{n=1}^{\infty} \frac{n \sin nx}{e^n} \right) dx = \frac{2e}{e^2 - 1}$.

24. Prove each series can be differentiated term by term on the indicated domain.

(a) $\displaystyle \sum_{n=1}^{\infty} \frac{\sin nx}{n^3}$ $(\forall x \in \mathbb{R})$;   (b) $\displaystyle \sum_{n=1}^{\infty} \frac{1}{n^x}$ $(\forall x \in (1, \infty))$.

25. Let $f_n(x) := x/(1 + n^2 x^2)$ for all $x \in [-1, 1]$. Show that (f_n) converges *uniformly* to a *differentiable* function f, but that $(f_n'(x))$ does *not* converge to $f'(x)$ for all $x \in [-1, 1]$.

26. For each $n \in \mathbb{N}_0$, define the nth-order *Bessel function* by

$$J_n(x) := \sum_{k=0}^{\infty} \frac{(-1)^k}{k!(k + n)!} \left(\frac{x}{2} \right)^{2k+n} \qquad (\forall x \in \mathbb{R}).$$

(a) Show that $(J_n(x))$ converges *pointwise* on $\mathbb{R}$ and *uniformly* on any $[a, b] \subset \mathbb{R}$.

(b) Show that $(x^n J_n(x))' = x^n J_{n-1}(x)$ for all $x \in \mathbb{R}$ and all $n \in \mathbb{N}$.

(c) Show that $y := J_n(x)$ satisfies *Bessel's differential equation*

$$x^2 y'' + xy' + (x^2 - n^2)y = 0.$$

27. For each power series, find the *radius* of convergence and determine whether the series converges at the endpoints of the *interval* of convergence.

(a) $\displaystyle \sum_{n=1}^{\infty} (-1)^n \frac{x^n}{n^2}$;   (b) $\displaystyle \sum_{n=1}^{\infty} (-1)^n \frac{x^n}{\log(1 + n)}$;   (c) $\displaystyle \sum_{n=0}^{\infty} 2^n n^2 x^n$;

(d) $\displaystyle \sum_{n=1}^{\infty} (n!/n^n) x^n$;   (e) $\displaystyle \sum_{n=1}^{\infty} (1 - 1/n)^n x^n$;   (f) $\displaystyle \sum_{n=1}^{\infty} (-1/3)^n x^n$.

28. Show that the power series $\sum_{n=1}^{\infty} a_n x^n$, where

$$a_n := \begin{cases} 1 & \text{if } n \text{ is odd,} \\ n & \text{if } n \text{ is even,} \end{cases}$$

has radius of convergence $R = 1$, and that $\lim_{n \to \infty} |a_{n+1}/a_n|$ does *not* exist.

29. Consider the power series $s(x) := \sum_{n=1}^{\infty} x^{n+1}/(n(n + 1))$.

(a) Show that $s(x)$ has radius of convergence $R = 1$ and converges on $[-1, 1]$.

(b) Show that $s'(x) = \sum_{n=1}^{\infty} x^n/n$ has radius of convergence $R = 1$ and converges on $[-1, 1)$.

(c) Show that $s''(x) = \sum_{n=0}^{\infty} x^n$ has radius of convergence $R = 1$ and converges only on $(-1, 1)$.

30. If $a_n \in [m, M] \subset (0, \infty)$ for all $n \in \mathbb{N}$, show that the power series $\sum_{n=1}^{\infty} a_n x^n$ has radius of convergence $R = 1$.

31. If $\sum_{n=1}^{\infty} a_n x^n$ has radius of convergence R, find the radii of convergence of the series $\sum_{n=1}^{\infty} (a_n)^2 x^n$ and $\sum_{n=1}^{\infty} a_n x^{2n}$.

32. Using the fact that $(1 - x)^{-1} = \sum_{n=0}^{\infty} x^n$ on $(-1, 1)$, find the sum of the following series:

(a) $\sum_{n=1}^{\infty} n x^n;$

(b) $\sum_{n=1}^{\infty} n^2 x^n;$

(c) $\sum_{n=0}^{\infty} n^3 x^n;$

(d) $\sum_{n=1}^{\infty} \frac{n x^n}{n + 1};$

(e) $\sum_{n=1}^{\infty} \frac{x^n}{n(n + 1)};$

(f) $\sum_{n=0}^{\infty} \frac{n x^n}{(n + 1)(n + 2)}.$

33. Find the sum of each series.

(a) $\sum_{n=1}^{\infty} \frac{n}{(n + 1)!}$

(b) $\sum_{n=1}^{\infty} \frac{n^2}{2^n}$

(c) $\sum_{n=1}^{\infty} \frac{1}{3^n (n + 1)}$

Hint: For (a), use $[(e^x - 1)/x]'$.

34. Show that

$$\log\left(\frac{1 + x}{1 - x}\right) = 2\left(x + \frac{x^3}{3} + \frac{x^5}{5} + \cdots + \frac{x^{2n+1}}{2n + 1} + \cdots\right) \qquad (|x| < 1).$$

35. Show that the function $f(x) := 1/(1 + x^2)$ is (real) *analytic* on $\mathbb{R}$, even though its Maclaurin expansion converges only in $(-1, 1)$.

36 (Bernstein's Theorem). Recall (Remark 8.4.23) that C^∞ functions are *not* analytic in general. Prove, however, the following positive result. Let $I \neq \emptyset$ be an open interval and $f : I \to \mathbb{R}$. Suppose that $f(x) \geq 0$ and $f^{(n)}(x) \geq 0$ for all $x \in I$ and all $n \in \mathbb{N}$. Then f is *analytic* in I. *Hint:* Use Taylor's Formula (Theorem 7.5.17) in an interval $[a, b] \subset I$ with the *integral remainder* written as

$$R_{n,a}(x) = \frac{1}{n!} \int_a^x (b - t)^n \left(\frac{x - t}{b - t}\right)^n f^{(n+1)}(t) \, dt.$$

Note that $t \mapsto (x - t)/(b - t)$ is *decreasing* (hence attaining its maximum at $t = a$) and that $R_{n,a}(b) \leq f(b)$.

37. Let $I \neq \emptyset$ be an open interval and $f \in C^\infty(I, \mathbb{R})$. Suppose that for each $a \in I$, the Taylor series of f about a has a *positive* radius of convergence. Show that there is an open subinterval $J \subset I$ such that f is *analytic* in J.

38 (Principle of Isolated Zeroes).

(a) Let $f(x) := \sum_{n=0}^{\infty} a_n x^n$ on $(-R, R)$, $R > 0$, and assume that there is a sequence $(x_k)_{k \in \mathbb{N}}$ of *distinct* points in $(-R, R)$ with $\lim(x_k) = 0$ such that $f(x_k) = 0, \forall \, k \in \mathbb{N}$. Show that $a_n = 0$ for all $n \geq 0$.

(b) Show that the conclusion in (a) remains valid if $\lim(x_k) = x_0 \in (-R, R)$, even if $x_0 \neq 0$.

(c) Let (x_k) and x_0 be as in (b). If $g(x) := \sum_{n=0}^{\infty} b_n x^n$ on $(-R, R)$ and $f(x_k) = g(x_k)$ for all $k \in \mathbb{N}$, show that $a_n = b_n$ for all $n \geq 0$.

(d) Let f and g be as in (c). If there is a *compact* set $K \subset (-R, R)$ such that the set $\{x \in K : f(x) = g(x)\}$ is *infinite*, then $f(x) = g(x)$ for all $x \in (-R, R)$. *Hints:* For (a), let m be the *smallest* integer with $a_m \neq 0$. Note that $f(x) = a_m x^m + o(|x|^m)$, as $x \to 0$ (so that $f(x)/x^m \approx a_m$ for $x \approx 0$), and deduce that $f(x) \neq 0$ if $0 < |x| < \varepsilon$ for $\varepsilon > 0$ sufficiently small. For (b), use (a) and Taylor's Theorem (Theorem 8.4.21).

39 (Principle of Analytic Continuation). Let $I \subset \mathbb{R}$ be an open interval and let $f, g \in \mathbb{R}^I$ be *analytic* functions on I.

(a) If there is a point $x_0 \in I$ such that $f^{(k)}(x_0) = 0$ for all $k \in \mathbb{N}_0$ (note that $f^{(0)} := f$), then $f(x) = 0$ for all $x \in I$. *Hint:* Let $Z := \{x \in I : f^{(k)}(x) = 0 \ \forall k \geq 0\}$. Using the Taylor series of f, show that Z is both *open* and *closed* (in I).

(b) If there is an open set $U \subset I$ such that $f(x) = g(x)$ for all $x \in U$, then $f(x) = g(x)$ for all $x \in I$.

40. (a) Show that we have

$$\arctan x = \sum_{n=0}^{\infty} (-1)^n \frac{x^{2n+1}}{2n + 1} \qquad (\forall x \in (-1, 1)).$$

(b) Using Abel's theorem (Theorem 8.4.30), show that the above series is actually *uniformly* convergent on $[-1, 1]$ and deduce that

$$\frac{\pi}{4} = 1 - \frac{1}{3} + \frac{1}{5} - \frac{1}{7} + - \cdots .$$

41. Recall that, for any $\alpha \in \mathbb{R}$, we have (Newton's) *binomial series*:

$$(1 + x)^\alpha = \sum_{n=0}^{\infty} \binom{\alpha}{n} x^n \qquad (|x| < 1).$$

For $\alpha \in \mathbb{N}_0$, the series *terminates* and we get the *binomial formula*:

(a) Show that $\sum_{n=0}^{\infty} \binom{\alpha}{n}$ converges for $\alpha > -1$ and diverges for $\alpha \leq -1$.

(b) Show that $\sum_{n=0}^{\infty} (-1)^n \binom{\alpha}{n}$ converges for $\alpha \geq 0$ and diverges for $\alpha < 0$.

(c) Show (using Abel's theorem) that, if $\alpha \geq 0$, then the binomial series converges *uniformly* to $(1 + x)^\alpha$ on $[-1, 1]$ and, when $\alpha > -1$, it converges uniformly to $(1 + x)^\alpha$ on $[-1 + \delta, 1]$ for $\delta > 0$. In particular, we have $\sum_{n=0}^{\infty} \binom{\alpha}{n} = 2^\alpha$ for all $\alpha > -1$ and $\sum_{n=0}^{\infty} (-1)^n \binom{\alpha}{n} = 0$ for all $\alpha > 0$.

(d) Show that $f(x) := |x|$ can be *uniformly* approximated by *polynomials* on $[-1, 1]$.

42. Show that the *converse* of Abel's theorem is *false* by producing a convergent power series $s(x) := \sum_{n=0}^{\infty} a_n x^n$ in $(-1, 1)$ such that $\lim_{x \to 1-} s(x)$ exists but $\sum_{n=0}^{\infty} a_n$ is *divergent*. The next problem will show that, *under certain conditions*, the converse of Abel's theorem is *true*.

43 (Tauber's Theorem). Let $s(x) = \sum_{n=0}^{\infty} a_n x^n$ for all $x \in (-1, 1)$ and assume that $\lim(n a_n) = 0$. Show that, if $\lim_{x \to 1-} s(x) = S \in \mathbb{R}$, then $\sum_{n=0}^{\infty} a_n = S$. *Hint:* Given $\varepsilon > 0$, pick $m \in \mathbb{N}$ such that $n|a_n| < \varepsilon/3$ for all $n > m$. Note that $1 - x^n \leq n(1 - x)$ for all $x \in [0, 1]$ and hence

$$\left| \sum_{n=0}^{m} a_n - S \right| \leq |s(x) - S| + (1 - x) \sum_{n=0}^{m} n|a_n| + \frac{\varepsilon}{3(m + 1)} \sum_{n=m+1}^{\infty} x^n.$$

Now, the last sum on the right is at most $1/(1 - x)$. Let $x = 1 - (1/m)$ and show (using Proposition 2.2.32) that each term on the right is $< \varepsilon/3$ for sufficiently large m.

44 (Bernoulli Numbers). Since $e^x = \sum_{n=0}^{\infty} x^n/n!$, we have $(e^x - 1)/x = 1 + x/2! + x^2/3! + \cdots$ and hence $x/(e^x - 1)$ has a power series expansion. The coefficients B_n in the expansion

$$\frac{x}{e^x - 1} = \sum_{n=0}^{\infty} \frac{B_n}{n!} x^n$$

are called the *Bernoulli numbers*. Prove the following statements:

(a) We have $B_{2n+1} = 0$ for all $n \in \mathbb{N}$. *Hint:* Note that $x/(e^x - 1) + x/2$ is an *even* function.
(b) We have $B_0 = 1$ and the identity

$$\binom{n}{0} B_0 + \binom{n}{1} B_1 + \cdots + \binom{n}{n-1} B_{n-1} = 0 \qquad (\forall n \geq 2).$$

Hint: Note that $x = (\sum_{m=1}^{\infty} x^m/m!)(\sum_{n=0}^{\infty} B_n x^n/n!)$.
(c) Show that $B_1 = -1/2$, $B_2 = 1/6$, $B_4 = -1/30$, $B_6 = 1/42$, and $B_8 = -1/30$.

45 (Bernoulli Polynomials). The *Bernoulli polynomials* $B_n(x)$, $n \in \mathbb{N}_0$ are defined by the expansion

$$e^{xt} \cdot \frac{t}{e^t - 1} = \sum_{n=0}^{\infty} \frac{(xt)^n}{n!} \cdot \sum_{n=0}^{\infty} \frac{B_n}{n!} t^n = \sum_{n=0}^{\infty} \frac{B_n(x)}{n!} t^n,$$

where the B_n are the Bernoulli numbers defined in the preceding problem. Prove the following statements.

(a) $B_n(x)$ is a polynomial of degree n given by

$$B_n(x) := \binom{n}{0} B_0 x^n + \binom{n}{1} B_1 x^{n-1} + \cdots + \binom{n}{n-1} B_{n-1} x + \binom{n}{n} B_n,$$

$B_n(0) = B_n$ for all $n \geq 0$, and $B_n(1) = B_n$ for all $n \geq 2$.
(b) $B'_{n+1}(x) = (n + 1) B_n(x)$.
(c) $B_{n+1}(x) = B_{n+1} + (n + 1) \int_0^x B_n(t)\, dt$.
(d) $\int_0^1 B_n(x)\, dx = 0$.
(e) $B_0(x) = 1$, $B_1(x) = x - 1/2$, $B_2(x) = x^2 - x + 1/6$, $B_3(x) = x^3 - 3x^2/2 + x/2$, and $B_4(x) = x^4 - 2x^3 + x^2 - 1/30$.

46. Show that the following expansions are valid in the indicated intervals:

(a) $x = \pi - 2 \sum_{n=1}^{\infty} \frac{\sin nx}{n} \quad (0 < x < 2\pi)$,

(b) $x^2 = \frac{\pi^2}{3} + 4 \sum_{n=1}^{\infty} \frac{(-1)^n \cos nx}{n^2} \quad (-\pi \leq x \leq \pi)$.

47. Obtain the following Fourier expansions for the *Bernoulli polynomials* on the interval $[0, 1]$:

(a) $B_{2n}(x) = (-1)^{n+1} \frac{2(2n)!}{(2\pi)^{2n}} \sum_{k=1}^{\infty} \frac{\cos 2k\pi x}{k^{2n}} \quad (\forall n \in \mathbb{N})$,

(b) $B_{2n+1}(x) = (-1)^{n+1} \dfrac{2(2n+1)!}{(2\pi)^{2n+1}} \displaystyle\sum_{k=1}^{\infty} \dfrac{\sin 2k\pi x}{k^{2n+1}}$ $(\forall n \in \mathbb{N})$.

Hint: For $B_1(x)$ and $B_2(x)$, use Problem 46. For the general case, proceed inductively, using Problem 45 and term-by-term integration.

48 (Riemann Zeta Function). Recall that the series $\sum_{n=1}^{\infty} n^{-s}$ is convergent for $s > 1$ and divergent for $s \leq 1$. The *Riemann zeta function* is defined by

$$\zeta(s) := \sum_{n=1}^{\infty} \frac{1}{n^s} \qquad (\forall s > 1).$$

(a) Show that we have

$$\zeta(2n) := \sum_{k=1}^{\infty} \frac{1}{k^{2n}} = (-1)^{n+1} \frac{(2\pi)^{2n} B_{2n}}{2(2n)!} \qquad (\forall n \in \mathbb{N}).$$

(b) Show that $\zeta(6) = \pi^6/945$ and $\zeta(8) = \pi^8/9450$.

49. (a) Show that, if $\alpha \notin \mathbb{Z}$, then

$$\cos \alpha x = \frac{\sin \alpha \pi}{\pi} \left[\frac{1}{\alpha} - \frac{2\alpha}{\alpha^2 - 1^2} \cos x + \frac{2\alpha}{\alpha^2 - 2^2} \cos 2x - + \cdots \right]$$

and deduce that

$$\frac{\pi}{\sin \alpha \pi} = \frac{1}{\alpha} - \frac{2\alpha}{\alpha^2 - 1^2} + \frac{2\alpha}{\alpha^2 - 2^2} - + \cdots.$$

(b) Plugging $x = 0$ and $x = \pi$ in the above series for $\cos \alpha x$ and relabeling, prove the following *partial fractions expansions*:

(i) $\csc \pi x = \dfrac{1}{\pi x} + \dfrac{2x}{\pi} \displaystyle\sum_{n=1}^{\infty} \dfrac{(-1)^n}{x^2 - n^2}$,

(ii) $\cot \pi x = \dfrac{1}{\pi x} + \dfrac{2x}{\pi} \displaystyle\sum_{n=1}^{\infty} \dfrac{1}{x^2 - n^2}$.

50 (Term-by-Term Differentiation).

(a) Show that, if $f \in C(\mathbb{R})$ is 2π-periodic and *piecewise smooth* on $[-\pi, \pi]$, then the Fourier series of f' is obtained by differentiating the Fourier series of f *term by term*.

(b) Show that, if f is as in part (a), then $\hat{f}(n) = o(n^{-1})$, as $n \to \infty$. *Hint:* Apply the Riemann–Lebesgue lemma to f'.

(c) Let $f \in C^k(\mathbb{R})$ be a 2π-periodic function. Show that $\hat{f}(n) = o(n^{-k})$, as $n \to \infty$. Deduce that, for $k = 2$ (i.e., $f \in C^2(\mathbb{R})$), the Fourier series of f is *absolutely* convergent.

51. (a) **(Poincaré's Inequality)** Let $f \in C^1([-\pi, \pi])$ and let α denote its *average value*, i.e., $\alpha := \hat{f}(0) = (1/2\pi) \int_{-\pi}^{\pi} f(x)\, dx$. Show that we have

$$\int_{-\pi}^{\pi} |f(x) - \alpha|^2\, dx \leq \int_{-\pi}^{\pi} |f'(x)|^2\, dx,$$

and that equality holds if and only if $f(x) = \alpha + \beta \cos x + \gamma \sin x$ for some (possibly complex) constants α, β, and γ. *Hint:* Use Parseval's Relation.

(b) **(A Sobolev Inequality)** Show that, if $f \in C^2(\mathbb{R})$ is 2π-periodic, then

$$\int_{-\pi}^{\pi} |f'(x)|^2 \, dx \le \left(\int_{-\pi}^{\pi} |f(x)|^2 \, dx \right)^{1/2} \left(\int_{-\pi}^{\pi} |f''(x)|^2 \, dx \right)^{1/2} .$$

52. Show that the following trigonometric series are convergent on $\mathbb{R}$ but are *not* the Fourier series of any 2π-periodic functions that are (Riemann) integrable on $[-\pi, \pi]$.

(a) $\displaystyle\sum_{n=2}^{\infty} \frac{\sin nx}{\log n}$,   (b) $\displaystyle\sum_{n=1}^{\infty} \frac{\cos nx}{\sqrt{n}}$.

Hint: Use Exercise 8.6.12 and *Bessel's* inequality.

53 (Approximate Identity). A sequence (K_n) in $C(\mathbb{R}, \mathbb{R})$ is said to be an *approximate identity* or a *Dirac sequence* if it satisfies these conditions:

AI 1. $K_n(x) \ge 0$ for all $x \in \mathbb{R}$.
AI 2. We have

$$\int_{-\infty}^{\infty} K_n(x) \, dx = 1.$$

AI 3. Given $\varepsilon > 0$ and $\delta > 0$, there is an $N \in \mathbb{N}$ such that $n \ge N$ implies

$$\int_{-\infty}^{-\delta} K_n(x) \, dx + \int_{\delta}^{\infty} K_n(x) \, dx < \varepsilon.$$

Note that as n increases the K_n have increasingly higher peaks at the origin and the area under the curve in a small neighborhood $(-\delta, \delta)$ of 0 is almost 1. The *improper* integral $\int_{-\infty}^{\infty} K_n$ is *defined* to be

$$\int_{-\infty}^{\infty} K_n(x) \, dx := \lim_{s \to -\infty} \int_{s}^{0} K_n(x) \, dx + \lim_{t \to \infty} \int_{0}^{t} K_n(x) \, dx,$$

but we shall define $\int_{-\infty}^{\infty} K_n := \int_{-a}^{a} K_n$ if K_n is identically zero outside $[-a, a]$.
Show that each of the following sequences is an approximate identity.

(a) $K_n(x) := nf(nx)$ for all $n \in \mathbb{N}$, where (as in Exercise 8.4.24)

$$f(x) := \begin{cases} \frac{1}{A} e^{1/(x^2-1)} & \text{if } |x| < 1, \\ 0 & \text{if } |x| \ge 1, \end{cases}$$

with $A = \int_{-1}^{1} e^{1/(x^2-1)} \, dx$ and hence $\int_{-1}^{1} f(x) \, dx = 1$.

(b) **(Landau's Kernel).**

$$L_n(x) := \begin{cases} (1 - x^2)^n / c_n & \text{if } |x| < 1, \\ 0 & \text{if } |x| \ge 1, \end{cases}$$

where, as in Problem 7.7.#45 (a),

$$c_n = \int_{-1}^{1} (1 - x^2)^n \, dx = \frac{2[2 \cdot 4 \cdot 6 \cdot (2n)]}{1 \cdot 3 \cdot 5 \cdot (2n + 1)}.$$

(c) **(Fejér's Kernel).** Define

$$F_n(x) := \begin{cases} \sin^2[(n + 1)x/2]/[2\pi(n + 1)\sin^2(x/2)] & \text{if } x \in [-\pi, \pi], \\ 0 & \text{if } |x| > \pi. \end{cases}$$

Thus, $F_n(x) = K_n(x)/(2\pi)$, with $K_n(x)$ as in Definition 8.6.10.

54 (Uniform Approximation, Convolution). Let (K_n) be an *approximate identity* as in the previous problem and let $f : \mathbb{R} \to \mathbb{R}$ be a *bounded*, (piecewise) continuous function. For each $n \in \mathbb{N}$, define the function f_n by the *convolution*

$$f_n(x) = (K_n * f)(x) := \int_{-\infty}^{\infty} f(t)K_n(x - t) \, dt,$$

where the *improper integral* is defined as in the previous problem.

(a) Show that (f_n) converges *uniformly* to f on any *compact* set E on which f is *continuous*.
(b) Let $L_n, n \in \mathbb{N}$, be the *Landau's Kernel* defined above. Show that $f_n := L_n * f$ is a *polynomial* for each n. Deduce the *Weierstrass Approximation Theorem* (Corollary 4.7.10), which says that any continuous function $f : [a, b] \to \mathbb{R}$ can be uniformly approximated by polynomials. *Hint:* Reduce to the case where $[a, b] = [0, 1]$ and, replacing f by $g(x) := f(x) - f(0) - x[f(1) - f(0)]$, if necessary, assume that $f(0) = f(1) = 0$.
(c) Let $f : \mathbb{R} \to \mathbb{R}$ be a continuous, 2π-periodic function and let f_n be the function $f_n := F_n * f$, where F_n is the *Fejér's Kernel* as defined above. Show that f_n is a continuous, 2π-periodic function for each n and prove *Fejér's theorem*: The sequence of functions $\sigma_n(x) := [s_0(x) + s_1(x) + \cdots + s_n(x)]/(n + 1)$ (cf. Corollary 8.6.16) converges *uniformly* to f on $[-\pi, \pi]$.

55 (Poisson Kernel).

(a) Show that

$$\frac{1}{2} + \sum_{n=1}^{\infty} r^n \cos n\theta = \frac{1 - r^2}{2(1 - 2r \cos \theta + r^2)} \qquad (0 \le r < 1),$$

and that the series converges *uniformly* in θ for each fixed $r \in [0, 1)$. *Hint:* Let $z := re^{i\theta}$ and take *real parts* in the identity $(1/2) + \sum_{n=1}^{\infty} z^n = (1 - z)^{-1} - 1/2$.
(b) Show that $\lim_{r \to 1-}[(1/2) + \sum_{n=1}^{\infty} r^n \cos n\theta] = 0$ uniformly on $[\delta, 2\pi - \delta]$, $\forall \delta \in (0, \pi)$. Note, however, that $(1/2) + \sum_{n=1}^{\infty} \cos n\theta$ diverges for all θ.

The function

$$P_r(\theta) := \frac{1 - r^2}{2\pi(1 - 2r \cos \theta + r^2)} \qquad (0 \le r < 1)$$

is called the *Poisson kernel*. It satisfies the three conditions of an *approximate identity* given in Problem 53 if we replace n by r and $n \to \infty$ by $r \to 1$.

Chapter 9
Normed and Function Spaces

Metric spaces were defined and studied in Chap. 5. Despite their importance, *general* metric spaces do not share the *algebraic* properties of the fields $\mathbb{R}$ and $\mathbb{C}$, because a metric space need not be an *algebra* or even a *vector space*. Any (nonempty) subset of a metric space is again a metric space with the metric it borrows from the ambient space. Thus, curves and surfaces in the Euclidean space $\mathbb{R}^3$ are metric spaces but are almost never *vector subspaces* of $\mathbb{R}^3$. Even straight lines and planes are not vector subspaces unless they pass through the origin. There is an important class of metric spaces, however, that is a natural framework for the extension of topological as well as algebraic properties of $\mathbb{R}$ and $\mathbb{C}$: It is the class of *normed spaces*, which we now define. *Throughout this chapter, $\mathbb{F}$ will stand for either $\mathbb{R}$ or $\mathbb{C}$ and X, Y, Z, etc. will denote vector spaces over $\mathbb{F}$.*

9.1 Norms and Normed Spaces

In this first section we begin our study of *normed spaces* with the following

Definition 9.1.1 (Norm, Seminorm). Let X be a *vector space* over $\mathbb{F}$. A *norm* on X is a map $\| \cdot \| : x \mapsto \|x\|$ from X to $\mathbb{R}$ such that for all x, $y \in X$ and all $\alpha \in \mathbb{F}$, we have:

N1 $\|x\| \geq 0$,
N2 $\|x\| = 0 \Leftrightarrow x = 0$,
N3 $\|\alpha x\| = |\alpha| \|x\|$, and
N4 $\|x + y\| \leq \|x\| + \|y\|$ (Triangle Inequality).

The map $\| \cdot \| : x \mapsto \|x\|$ is said to be a *seminorm* on X if it only satisfies the conditions $N1$, $N3$, and $N4$.

© Springer Science+Business Media New York 2014
H.H. Sohrab, *Basic Real Analysis*, DOI 10.1007/978-1-4939-1841-6_9

Exercise 9.1.2.

(a) Show that $N1$ follows from $N3$ and $N4$.
(b) Show that, if $\| \cdot \|$ is a *seminorm* on X, then the set $Z := \{x \in X : \|x\| = 0\}$ is a (vector) *subspace* of X.

Definition 9.1.3 (Normed Space, Seminormed Space). Given any vector space X, and any *norm* (resp., *seminorm*) $\| \cdot \|$ on X, the pair $(X, \| \cdot \|)$ is said to be a *normed space* (resp., *seminormed space*). We say that a normed (or seminormed) space $(X, \| \cdot \|)$ is *real* (resp., *complex*) if $\mathbb{F} = \mathbb{R}$ (resp., $\mathbb{F} = \mathbb{C}$). By abuse of language, we shall refer to X itself as a normed (resp., seminormed) space if the norm (resp., seminorm) used on X is obvious from the context.

Definition 9.1.4 (Normed Algebra). A normed space $(X, \| \cdot \|)$ is said to be a *normed algebra* if X is an *algebra* over $\mathbb{F}$ (i.e., a vector space together with a binary operation $(x, y) \mapsto xy$ as in Definition 1.3.21) such that $\|xy\| \leq \|x\| \|y\|$, for all $x, y \in X$ and $\|e\| = 1$ if $e \in X$ is a *unit* element.

Exercise 9.1.5. Let $(X, \| \cdot \|)$ be a normed space.

(a) Show that the map $d : (x, y) \mapsto \|x - y\|$ from $X \times X$ to $\mathbb{R}$ is a *metric* on X such that, for all $x, y, z \in X$ and all $\alpha \in \mathbb{F}$, we have

$$d(x + z, y + z) = d(x, y) \quad \text{and} \quad d(\alpha x, \alpha y) = |\alpha| d(x, y).$$

(b) Show that the map $x \mapsto \|x\|$ is *Lipschitz* and hence *uniformly* continuous (with respect to the above metric d) on X.

The above exercise shows that norms behave nicely under *translations* and *dilations*:

Definition 9.1.6 (Translation, Dilation, Translated Dilation). Let X be a vector space over $\mathbb{F}$. For each $b \in X$ the map $\tau_b : X \to X$ such that $\tau_b(x) := x + b$ is said to be a *translation*. For each $\alpha \in \mathbb{F} \setminus \{0\}$, the map $\lambda_\alpha : X \to X$ such that $\lambda_\alpha(x) := \alpha x$ is called a *dilation*. Finally, composing the previous maps, we obtain the map $\tau_b \circ \lambda_\alpha : X \to X$ given by $\tau_b \circ \lambda_\alpha(x) = \alpha x + b$, which we call a *translated dilation*.

Remark 9.1.7. To simplify the exposition, we shall frequently use the same notation $\| \cdot \|$ to denote norms on different normed spaces as long as this introduces no confusion. The metric defined in Exercise 9.1.5 is said to be *associated* with the norm $\| \cdot \|$ on X. When we study X as a metric space, we shall always refer to the metric associated with the norm. Since in the above exercise we have $\|x\| = d(x, 0)$, it is tempting to claim that, in any vector space with a metric d, the map $x \mapsto d(x, 0)$ is a norm. In fact, this is *not the case* as the following exercise shows:

Exercise 9.1.8. Let d denote the *discrete metric* on the vector space $\mathbb{R}$; i.e., $d(x, y) = 0$ if $x = y$ and $d(x, y) = 1$ otherwise. Show that d is *not* associated with *any* norm on $\mathbb{R}$.

Definition 9.1.9 (Equivalent Norms). Two norms $\|\cdot\|_1$ and $\|\cdot\|_2$ on a vector space X are said to be *equivalent* if their associated metrics are equivalent in the sense of Definition 5.2.4 (i.e., define the *same topology* on X).

Remark 9.1.10. In view of Corollary 5.4.28, if $\|\cdot\|_1 \le c\|\cdot\|_2 \le c'\|\cdot\|_1$, for some *positive* constants c, c', then $\|\cdot\|_1$ and $\|\cdot\|_2$ are equivalent. We shall see that this condition is also *necessary*.

Example 9.1.11.

(1) The vector space $\mathbb{F}$ is a normed space (in fact a normed *algebra*) with $\|x\| = |x|$ for all $x \in \mathbb{F}$. More generally, the *Euclidean* space $\mathbb{F}^n$ is a normed space. One can take, e.g., the *Euclidean norm:* $\|x\|_{\text{euc}} = (\sum_{k=1}^{n} |x_k|^2)^{1/2}$ for each vector $x = (x_1, x_2, \ldots, x_n) \in \mathbb{F}^n$.

(2) Extending the example (1), we can look at sequences: Consider the vector space $\mathbb{F}^{\mathbb{N}}$ of all sequences in $\mathbb{F}$. There are many interesting subspaces of $\mathbb{F}^{\mathbb{N}}$ that are normed spaces. For example, consider the subspace of *bounded* sequences:

$$\ell^{\infty}(\mathbb{N}, \mathbb{F}) := \left\{ x = (x_n) \in \mathbb{F}^{\mathbb{N}} : \sup\{|x_n| : n \in \mathbb{N}\} < \infty \right\}.$$

Then $\ell^{\infty}(\mathbb{N}, \mathbb{F})$ is a subspace of $\mathbb{F}^{\mathbb{N}}$ and, as one can easily check, $\|x\|_{\infty} := \sup\{|x_n| : n \in \mathbb{N}\}$ defines a norm on it; we call it the *sup-norm*. Next, we may look at the *(absolutely) summable* sequences:

$$\ell^1(\mathbb{N}, \mathbb{F}) := \left\{ x = (x_n) \in \mathbb{F}^{\mathbb{N}} : \sum_{n=1}^{\infty} |x_n| < \infty \right\}.$$

Then $\ell^1(\mathbb{N}, \mathbb{F})$ is a subspace of $\mathbb{F}^{\mathbb{N}}$ and one can check that $\|x\|_1 := \sum |x_n|$ defines a norm on it; we call it the ℓ^1-*norm*. We may also look at *square-summable* sequences:

$$\ell^2(\mathbb{N}, \mathbb{F}) := \left\{ x = (x_n) \in \mathbb{F}^{\mathbb{N}} : \sum_{n=1}^{\infty} |x_n|^2 < \infty \right\},$$

which is again a subspace of $\mathbb{F}^{\mathbb{N}}$. Here, $\|x\|_2 := \sqrt{\sum |x_n|^2}$ defines a norm, the ℓ^2-*norm*, as is easily checked (cf. Example 2.3.52).

(3) As in example (2), one may consider the spaces $\ell^{\infty}(\mathbb{Z}, \mathbb{F})$, $\ell^1(\mathbb{Z}, \mathbb{F})$, and $\ell^2(\mathbb{Z}, \mathbb{F})$ of all *bounded*, *(absolutely) summable*, and *square-summable* sequences in $\mathbb{F}^{\mathbb{Z}}$, respectively. As we know (Theorem 8.6.33), for any 2π-periodic function f such that $f \in \mathcal{R}([-\pi, \pi])$, we have $(\hat{f}(n))_{n \in \mathbb{Z}} \in \ell^2(\mathbb{Z}, \mathbb{C})$ and

$$\|(\hat{f}(n))_{n \in \mathbb{Z}}\|_2^2 = \frac{1}{2\pi} \int_{-\pi}^{\pi} |f(x)|^2 dx.$$

(4) Let $X := \mathcal{R}([a, b])$ and, for each $f \in X$, define $\|f\|_1 := \int_a^b |f|$. Then $\|\cdot\|_1$ is
a *seminorm* on X. Indeed the properties N_1, N_3, and N_4 are obvious. (Why?)
On the other hand, $\|f\|_1 := \int_a^b |f| = 0$ only implies that $f(x) = 0$ for *almost
all* $x \in [a, b]$ (cf. Theorem 7.4.16). If we look at the subspace $Y := C([a, b])$
of all *continuous* functions on $[a, b]$, then $\|\cdot\|_1$ is a *norm* on Y. (Why?)

(5) Given *any* set $S \neq \emptyset$, let $B(S, \mathbb{F})$ denote the vector space of all *bounded*,
$\mathbb{F}$-valued functions on S. For each $f \in B(S, \mathbb{F})$, put $\|f\|_\infty := \sup\{|f(x)| : x \in S\}$.
Then $f \mapsto \|f\|_\infty$ is easily seen to be a *norm* on $B(S, \mathbb{F})$. We call it the *sup-
norm*. In fact, $(B(S, \mathbb{F}), \|\cdot\|_\infty)$ is a *normed algebra*.

(6) Given the normed spaces $X_1, X_2, \ldots, X_n$ (over $\mathbb{F}$) with respective norms
$\|\cdot\|_1, \ldots, \|\cdot\|_n$, consider their Cartesian product $X := \prod_{k=1}^n X_k$. This is a
vector space over $\mathbb{F}$ (with componentwise addition and scalar multiplication).
For each vector $x = (x_1, \ldots, x_n) \in X$, $\|x\|_{\text{euc}} := \sqrt{\sum_{k=1}^n \|x_k\|_k^2}$, $\|x\|_{\text{max}} :=$
$\max\{\|x_1\|_1, \ldots, \|x_n\|_n\}$, and $\|x\|_{\text{sum}} := \sum_{k=1}^n \|x_k\|_k$ are easily seen to be
norms on X. The associated metrics were considered in Exercise 5.1.5.
It follows from the inequalities

$$\|x\|_{\text{max}} \leq \|x\|_{\text{euc}} \leq \|x\|_{\text{sum}} \leq n\|x\|_{\text{max}}$$

that these norms are *equivalent* (cf. Exercise 5.4.29).

Proposition 9.1.12. *Let X be a normed space. The maps $(x, y) \mapsto x + y$, $x \mapsto \alpha x$
(for fixed $\alpha \in \mathbb{F}$), and $\alpha \mapsto \alpha x$ (for fixed $x \in X$) are Lipschitz (hence uniformly
continuous) on $X \times X$, X, and $\mathbb{F}$, respectively. Also, the map $(\alpha, x) \mapsto \alpha x$ [or,
more generally, $(\alpha, x) \mapsto \alpha x + b$, for any fixed $b \in X$] is continuous on $\mathbb{F} \times X$.*

Proof. For the map $(x, y) \mapsto x + y$, we simply note that, with $\|\cdot\|_{\text{sum}}$ as in
Example (6) above,

$$\|(x + y) - (x' + y')\| \leq \|x - x'\| + \|y - y'\| = \|(x, y) - (x', y')\|_{\text{sum}}.$$

For the maps $x \mapsto \alpha x$ and $\alpha \mapsto \alpha x$, the result follows from $\|\alpha x - \alpha x'\| = |\alpha| \|x - x'\|$ and $\|\alpha x - \alpha' x\| = \|x\| |\alpha - \alpha'|$, respectively. Finally, for $(\alpha, x) \mapsto \alpha x + b$, note
that

$$\|\alpha x - \alpha_0 x_0\| \leq |\alpha - \alpha_0| \|x - x_0\| + |\alpha - \alpha_0| \|x_0\| + |\alpha_0| \|x - x_0\|. \qquad \square$$

Corollary 9.1.13. *Let X be a normed space and let $b \in X$ and $\alpha \in \mathbb{F} \setminus \{0\}$ be
fixed. Then the translation $\tau_b(x) := b + x$ is an isometry of X onto itself (with
inverse τ_{-b}). Also, the translated dilation $x \mapsto \alpha x + b$ (and hence the dilation
$x \mapsto \alpha x$) is a (Lipschitz) homeomorphism of X onto itself with (Lipschitz) inverse
$x \mapsto \alpha^{-1}(x - b)$ (resp., $x \mapsto \alpha^{-1} x$).*

Proof. This follows at once from Exercise 9.1.5 and Proposition 9.1.12. $\qquad \square$

Corollary 9.1.14. *Let X be a normed space and $Y \subset X$ a subspace.*

(a) The closure Y^- is a subspace of X.

(b) Let $S \subset X$ be closed (i.e., $S^- = S$). Then, for any $b \in X$ and any $\alpha \in \mathbb{F}$, the set $b + \alpha S := \{b + \alpha s : s \in S\}$ is also closed and we have $(b + \alpha S)^- = b + \alpha \cdot S^-$. In particular, $b + S$ and αS are closed and so is the set $b + Y$ for any closed subspace Y.

Proof. Since (by Proposition 9.1.12) the map $\sigma(x, y) := x + y$ is continuous on $X \times X$, and since $(Y \times Y)^- = Y^- \times Y^-$, it follows from Exercise 5.4.11(e) that

$$\sigma(Y^- \times Y^-) = \sigma[(Y \times Y)^-] \subset [\sigma(Y \times Y)]^- = Y^-.$$

Similarly, since $h_\alpha : x \mapsto \alpha x$ is a homeomorphism for $\alpha \neq 0$, we have

$$h_\alpha(Y^-) = [h_\alpha(Y)]^- \subset Y^-$$

and *(a)* follows. For *(b)*, note that $b + \alpha S = h(S)$, where $h(x) := \alpha x + b$ is a *homeomorphism*, hence a *closed* map. $\qquad\square$

Exercise 9.1.15. Let X be a normed space and let $R, S \subset X$.

(1) Show that, if R is *open*, then so is $R + S := \{r + s : r \in R, s \in S\}$.

(2) Show that, if R is *compact* and S is *closed*, then $R + S$ is *closed*. Give an example of two *closed* sets $R, S \subset \mathbb{R}^2$ such that $R + S$ is *not closed*.

Definition 9.1.16 (Linear Map, Operator, Functional). Given any normed spaces X and Y, a map (or transformation) $T : X \to Y$ is said to be *linear* if

$$T(\alpha x + y) = \alpha T x + T y \qquad (\forall x, y \in X, \ \forall \alpha \in \mathbb{F}),$$

where $Tx := T(x)$. The set of all linear maps $T : X \to Y$ will be denoted by $\mathcal{L}(X, Y)$. A *linear map* $T : X \to X$ is called a *linear operator* on X. The set of all linear operators on X will be denoted by $\mathcal{L}(X)$. A *linear* map $\phi : X \to \mathbb{F}$ is called a *linear functional* (or linear *form*). The set of all linear functionals on X will be denoted by $\mathcal{L}(X, \mathbb{F})$ and is simply the *algebraic dual* of X.

Exercise 9.1.17 (Real vs. Complex Functional). Let X be a vector space over $\mathbb{C}$. A map $\phi : X \to \mathbb{C}$ is said to be a *real* (resp., *complex*) linear functional if $\phi(x + y) = \phi(x) + \phi(y)$ for all $x, y \in X$ and $\phi(\alpha x) = \alpha \phi(x)$ for all $x \in X$ and all $\alpha \in \mathbb{R}$ (resp., $\alpha \in \mathbb{C}$). Show that, if $\phi : X \to \mathbb{C}$ is a *complex* linear functional, then its real part $\psi := \mathrm{Re}(\phi)$ is a *real* linear functional and we have

$$\phi(x) = \psi(x) - i\psi(ix) \qquad (\forall x \in X). \qquad\qquad (*)$$

Conversely, show that, if $\psi : X \to \mathbb{R}$ is a *real* linear functional, then ϕ defined by $(*)$ is a *complex* linear functional on X.

Remark 9.1.18 (Balls in Normed Spaces). Recall (Definition 5.1.11) that, if (M, d) is a metric space, a subset $S \subset M$ is said to be *bounded* if it has *finite diameter;* i.e., if $\delta(S) := \sup\{d(x, y) : x, y \in M\} < \infty$. This is equivalent to requiring that S be contained in a ball of "sufficiently large" radius. Also, recall (Example 5.2.11) that the *closed ball* $B'_r(x) := \{y \in M : d(x, y) \le r\}$ (of *radius r centered at* $x \in M$) is *not*, in general, the *closure* $B^-_r(x)$ of the *open* ball $B_r(x) := \{y \in M : d(x, y) < r\}$, which in turn is *not* always the interior $(B'_r(x))°$ of the closed ball. This inconvenience disappears in normed spaces, where balls enjoy many nice properties; e.g., they are all *convex:*

Definition 9.1.19 (Convex Set). Let X be a vector space. A subset C of X is said to be *convex* if, for all $x, y \in C$ and $t \in [0, 1]$, we have $tx + (1 - t)y \in C$.

Proposition 9.1.20. *Let X be a normed space and let $B_r(x)$, $B^-_r(x)$, and $B'_r(x)$ be as above with $x \in X$ and $r > 0$. Then the following are true:*

(1) *We have $B^-_r(x) = B'_r(x)$ and $(B'_r(x))° = (B^-_r(x))° = B_r(x)$. Therefore, $\mathrm{Bd}(B'_r(x)) = \mathrm{Bd}(B^-_r(x)) = S_r(x) := \{x \in X : \|x\| = r\}$; i.e., the boundary of a closed ball is the corresponding sphere.*

(2) *For any map $h(x) := \alpha x + b$ with fixed $b \in X$ and $\alpha \in \mathbb{F} \setminus \{0\}$,*

$$\alpha B_r(x) + b = h(B_r(x)) = B_{|\alpha|r}(h(x)) = B_{|\alpha|r}(\alpha x + b). \tag{$*$}$$

Moreover, $(*)$ remains valid if B_r and $B_{|\alpha|r}$ are replaced by B^-_r and $B^-_{|\alpha|r}$.

(3) *$B_r(x) = x + B_r(0) = x + rB_1(0)$ and $B^-_r(x) = x + B^-_r(0) = x + rB^-_1(0)$.*

(4) *For any $x, y \in X$ and $r, s \in (0, \infty)$, we have $B_r(x) + B_s(y) = B_{r+s}(x + y)$ and $B^-_r(x) + B^-_s(y) = B^-_{r+s}(x + y)$. In particular, if $y = -x$, then*

$$B_r(x) - B_s(x) = B_r(x) + B_s(-x) = B_{r+s}(0) = (r + s)B_1(0)$$

and a similar statement for closed balls.

(5) *Every ball in X (open or closed) is convex. Also, every $B_r(0)$ or $B^-_r(0)$ is invariant under the dilations $x \mapsto \alpha x$ with $|\alpha| = 1$.*

(6) *We have $X = \bigcup_{k=1}^{\infty} B_k(0) = \bigcup_{k=1}^{\infty} kB_1(0)$.*

Proof.

(1) Since $B'_r(x)$ is closed and contains $B_r(x)$, we obviously have $B^-_r(x) \subset B'_r(x)$. For the reverse inclusion we need only show that $S_r(x) \subset B^-_r(x)$. (Why?) But if $\|y - x\| = r$ and if we set $y_n := x + \alpha_n(y - x)$, where $0 < \alpha_n < 1$ for all n and $\alpha_n \to 1$ as $n \to \infty$, then $y_n \in B_r(x)$ for all n and hence $y = \lim(y_n) \in B^-_r(x)$. Next, since $B_r(x)$ is an *open* subset of $B'_r(x)$, we certainly have $B_r(x) \subset (B'_r(x))°$. For the reverse inclusion we need only show that if $\|y - x\| = r$, then $y \notin (B'_r(x))°$. (Why?) But, given any $\varepsilon > 0$, let $z := y + \varepsilon(y - x)/(2r)$. Then $\|z - y\| = \varepsilon/2$ and hence $z \in B_\varepsilon(y)$. On the other hand, $z \notin B'_r(x)$ because

$$\|z - x\| = \left\| (y - x) + \frac{\varepsilon}{2r}(y - x) \right\| = \left(1 + \frac{\varepsilon}{2r}\right)r > r.$$

(2) $\forall\, z \in B_{|\alpha|r}(h(x))$, we have $z = h(y)$ with $y = \alpha^{-1}(z - b) \in B_r(x)$:

$$\|h(y) - h(x)\| = \|\alpha(y - x)\| = |\alpha|\,\|y - x\| < |\alpha|r \iff \|y - x\| < r,$$

and a similar equivalence with "$< r$" replaced by "$\le r$" for the closed balls.

(3) The first part follows from (2) if we set $x = 0$, $b = x$, and $\alpha = 1$. The second part then follows if we take the closure of both sides and use Corollary 9.1.13.

(4) This follows at once from (3) because

$$B_r(0) + B_s(0) = (r + s)B_1(0) = B_{r+s}(0).$$

(5) Indeed, if $y,\, z \in B_r(x)$, then for each $t \in [0, 1]$ we have

$$\|ty + (1 - t)z - x\| = \|t(y - x) + (1 - t)(z - x)\| \le t\|y - x\| + (1 - t)\|z - x\|$$
$$< tr + (1 - t)r = r,$$

and a similar statement with "$< r$" replaced by "$\le r$" if $y,\, z \in B_r^-(x)$. The statement about *invariance under dilations* is obvious.

(6) We simply note that for each $x \in X$ we have $x \in B_k(0)$ if $\|x\| < k$. $\qquad\square$

Proposition 9.1.21 (Bounded Set). *Let X be a normed space. A set $S \subset X$ is bounded if and only if it is contained in a closed ball centered at $0 \in X$.*

Proof. Exercise! $\qquad\square$

Theorem 9.1.22. *Let X, Y be normed spaces. For a linear map $T \in \mathcal{L}(X, Y)$, the following are pairwise equivalent:*

(1) T is bounded on each bounded subset of X.
(2) T is bounded on the (closed unit) ball $B_1^-(0)$ of X.
(3) T is bounded on the unit sphere $S_1(0)$ of X.
(4) There exists a constant $c \ge 0$ (called a bound for T) such that $\|Tx\| \le c\|x\| \quad \forall x \in X$.
(5) T is Lipschitz (hence uniformly continuous) on X.
(6) T is continuous on X.
(7) T is continuous at 0.

Proof. The implications (1) $\Rightarrow$ (2) $\Rightarrow$ (3) are obvious. If $\|Tx\| \le c$ for all $x \in S_1(0)$ and some $c > 0$, then $\|T(x/\|x\|)\| \le c$ for all $x \ne 0$ and hence $\|Tx\| \le c\|x\|$, which also holds for $x = 0$. This establishes (3) $\Rightarrow$ (4). Next, (4) $\Rightarrow$ (5) follows from $Tx - Ty = T(x - y)$ and (5) $\Rightarrow$ (6) $\Rightarrow$ (7) is obvious. To prove (7) $\Rightarrow$ (1), suppose that T is continuous at 0 and note that $T0 = 0$. Then, with $\varepsilon = 1$, we can pick $\delta > 0$ such that $\|x\| \le \delta$ implies $\|Tx\| < 1$. Now, if $S \subset X$ is *bounded*, then (Proposition 9.1.21) there is $r > 0$ such that $S \subset B_r^-(0)$. But $\|x\| \le r$ implies $\|\delta x/r\| \le \delta$ and hence $\|T(\delta x/r)\| < 1$, which gives $\|Tx\| < r/\delta$ and completes the proof. $\qquad\square$

Definition 9.1.23 (Bounded Linear Map, Bounded Operator, Dual). A map $T \in \mathcal{L}(X, Y)$ is called a *bounded linear map* if it satisfies any (and hence all) of the conditions in Theorem 9.1.22. A *bounded* linear map $T : X \to X$ is called a *bounded operator* on X. The set of all bounded linear maps from X to Y will be denoted by $\mathcal{B}(X, Y)$, and the set of all *bounded operators* on X will be denoted by $\mathcal{B}(X)$. Finally, the set of all *bounded linear functionals* on X will be denoted by X^* and will be called the (topological) *dual* of X; i.e., $X^* := \mathcal{B}(X, \mathbb{F})$.

Exercise 9.1.24 (Bounded Multilinear Map). Let Y and X_k, $1 \le k \le n$, be normed spaces and let $X := \prod_{k=1}^{n} X_k$. A map $T \in \mathcal{L}(X, Y)$ is said to be *multilinear* if $T(x_1, \ldots, x_n)$ is *linear* in each variable. Show that such a map T is *continuous* if and only if there is a constant $c > 0$ such that

$$\|T(x_1, \ldots, x_n)\| \le c \|x_1\| \cdots \|x_n\| \qquad (\forall (x_1, \ldots, x_n) \in X).$$

Definition 9.1.25 ((Topological) Isomorphism, Isomorphic). Let X and Y be normed spaces. A *bijective* linear map $T : X \to Y$ is said to be a *topological isomorphism* (or, simply, an *isomorphism*) if both T and T^{-1} are *bounded*. If such a map exists, we say that X and Y are (topologically) *isomorphic*.

Definition 9.1.26 (Isometry, Isometric Isomorphism). We say that a linear map $T : X \to Y$ is an *isometry* if $\|Tx\| = \|x\|$ for all $x \in X$. If T is *onto*, then it is an isomorphism (why?) and is called an *isometric isomorphism*. In this case, X and Y are said to be *isometrically isomorphic*.

Example 9.1.27. Note that any *dilation* $x \mapsto \alpha x$ with $|\alpha| = 1$ is an isometry of X onto itself. In fact, these maps are isometries of any open [resp., closed] ball $B_r(0)$ [resp., $B_r^-(0)$] onto itself because $\|\alpha x\| = |\alpha| \|x\| = \|x\|$.

Corollary 9.1.28 (Equivalent Norms). *Two norms $\|\cdot\|$ and $\|\cdot\|'$ on a vector space X are equivalent if and only if there are two positive numbers a and b such that*

$$a\|x\| \le \|x\|' \le b\|x\| \qquad (\forall x \in X).$$

Proof. The condition is *sufficient* in view of Remark 9.1.10. To show that it is *necessary*, note that the equivalence of the two norms means that the identity map $\iota : (X, \|\cdot\|) \to (X, \|\cdot\|')$ is an *isomorphism*; i.e., ι and its inverse ι^{-1} are bounded (i.e., continuous) *linear* maps. Theorem 9.1.22 now implies that there are positive constants c, c' such that for each $x \in X$,

$$\|x\|' = \|\iota(x)\|' \le c'\|x\| \quad \text{and} \quad \|x\| = \|\iota^{-1}(x)\| \le c\|x\|',$$

and the proof is complete. $\qquad\qquad\square$

Definition 9.1.29 (Kernel, Range, Nonzero, Injective, Surjective). The *kernel* (also called *null space*) and the *range* (also called *image*) of a linear map $T : X \to Y$ are defined to be the sets

$$\mathrm{Ker}(T) := T^{-1}(\{0\}) := \{x \in X : Tx = 0\},$$

$$\mathrm{Ran}(T) := T(X) := \{Tx : x \in X\}.$$

It is easily checked that $\mathrm{Ker}(T)$ is a (vector) *subspace* of X and that $\mathrm{Ran}(T)$ is a subspace of Y. Also, T is *nonzero* (i.e., $Tx \neq 0$ for at least one $x \in X$) if and only if $\mathrm{Ker}(T) \neq X$. Note that T is *injective* if and only if $\mathrm{Ker}(T) = \{0\}$ (why?) and it is *surjective* if and only if $\mathrm{Ran}(T) = Y$.

Proposition 9.1.30. *Given any linear map $T \in \mathcal{B}(X, Y)$, its kernel, $\mathrm{Ker}(T)$, is a closed subspace of X. The range $\mathrm{Ran}(T)$ is a (not necessarily closed) subspace of Y. A linear functional $\phi \in \mathcal{L}(X, \mathbb{F})$ is bounded if and only if $\mathrm{Ker}(\phi)$ is closed in X.*

Proof. That $\mathrm{Ker}(T)$ and $\mathrm{Ran}(T)$ are subspaces of X and Y, respectively, is obvious. Also, since T is *continuous* and $\{0\}$ is *closed* in $\mathbb{F}$, the subspace $\mathrm{Ker}(T) = T^{-1}(\{0\})$ is *closed* in X. In particular, $\mathrm{Ker}(\phi)$ is closed for each $\phi \in \mathcal{B}(X, \mathbb{F})$. Conversely, suppose that ϕ is *nonzero* and $H := \phi^{-1}(\{0\})$ is closed in X. Pick $x_0 \in X$ with $\phi(x_0) = 1$; hence $x_0 \notin H$. Since H is closed, so is $x_0 + H$, by Corollary 9.1.14, and we have $0 \notin x_0 + H$. Therefore, we can find $r > 0$ such that $B_r^-(0) \cap (x_0 + H) = \emptyset$. In particular, $\phi(x) \neq 1$ for all $x \in B_r^-(0)$. If $\phi(x) = \alpha$ for some $x \in B_r^-(0)$ and $|\alpha| > 1$, then $\|x/\alpha\| = \|x\|/|\alpha| < r$ (i.e., $x/\alpha \in B_r^-(0)$) and yet $\phi(x/\alpha) = 1$. This contradiction shows that $\|\phi(x)\| \leq 1$ for all $x \in B_r^-(0)$ and hence ϕ is continuous. $\square$

Definition 9.1.31 (Norm of a Bounded Linear Map). If X, Y are normed spaces and $T \in \mathcal{B}(X, Y)$, then the *norm* of T is defined to be the nonnegative number

$$\|T\| := \sup\{\|Tx\| : x \in X, \ \|x\| \leq 1\}. \tag{$*$}$$

Proposition 9.1.32. *For a linear map $T : X \to Y$, where $X \neq \{0\}$ and Y are normed spaces, we have*

$$\|T\| = \sup\{\|Tx\| : x \in X, \ \|x\| = 1\} \tag{$**$}$$

$$= \sup\{\|Tx\|/\|x\| : x \in X \setminus \{0\}\}$$

$$= \inf\{c > 0 : \|Tx\| \leq c\|x\|, \forall x \in X\}.$$

In particular, $\|Tx\| \leq \|T\|\|x\|, \forall x \in X$.

Proof. Let us write $\|T\|' := \sup\{\|Tx\| : x \in S_1(0)\}$, $\|T\|'' := \sup\{\|Tx\|/\|x\| : x \in X \setminus \{0\}\}$, and $\|T\|''' := \inf\{c > 0 : \|Tx\| \leq c\|x\|, \ \forall x \in X\}$. Then $\|T\|' = \|T\|''$ follows from $\|Tx\|/\|x\| = \|T(x/\|x\|)\|$, valid for $x \neq 0$. Next, $(*)$ implies that $\|T\| \geq \|T\|'$. Also, if $0 < \|x\| \leq 1$, then $\|T(x/\|x\|)\| \leq \|T\|'$ and hence $\|Tx\| \leq \|T\|'\|x\| \leq \|T\|'$, which also holds for $x = 0$. Thus, $\|T\|' \geq \|T\|$ and the first equality in $(**)$ follows. To prove $\|T\| = \|T\|'''$, note first that $(*)$ implies $\|T(x/\|x\|)\| \leq \|T\|$ for all $x \neq 0$ and hence $\|Tx\| \leq \|T\|\|x\|$ for all $x \in X$. This proves the inequality $\|T\| \geq \|T\|'''$. On the other hand, if $\|Tx\| \leq c\|x\|$ for all

$x \in X$, then $\|Tx\| \leq c$ for all x with $\|x\| \leq 1$ and hence (by $(*)$) $\|T\| \leq c$. This shows that we indeed have $\|T\| \leq \|T\|'''$ and completes the proof. $\qquad\qquad\square$

Exercise 9.1.33. Show that the "norm" $(*)$ in Definition 9.1.31 is indeed a norm on $\mathcal{B}(X, Y)$.

9.2 Banach Spaces

Unlike general metric spaces, normed spaces are vector spaces and hence the notions of convergent and absolutely convergent *series* make sense in any normed space. When we study sequences and series, however, it is natural to work with *complete* spaces, i.e., spaces in which every *Cauchy sequence* converges to a vector in the space. This is what we intend to do in this section.

Definition 9.2.1 (Banach Space, Banach Algebra). A normed space is said to be a *Banach space* if it is *complete*. A *complete* normed algebra is called a *Banach algebra*.

Example 9.2.2.

(1) The Euclidean spaces $\mathbb{F}^n$ for $n \in \mathbb{N}$ are all Banach spaces and $\mathbb{F}$ is in fact a Banach algebra.

(2) Given a set $S \neq \emptyset$, the normed algebra $X := \mathcal{B}(S, \mathbb{F})$ of all *bounded* $\mathbb{F}$-valued functions on S with the sup-norm $\|f\|_\infty := \sup\{|f(x)| : x \in S\}$ is a Banach algebra. Indeed, if $(f_n) \in X^{\mathbb{N}}$ is a Cauchy sequence, then so is the sequence $(f_n(x))$ in $\mathbb{F}$, for each $x \in S$. Let $f(x) := \lim(f_n(x))$. Then $\sup\{|f_n(x) - f_m(x)| : x \in S\} \to 0$, as $n, m \to \infty$ implies $\sup\{|f_n(x) - f(x)| : x \in S\} \to 0$, as $n \to \infty$ and hence f is bounded. (Why?)

(3) For each *compact* set $K \subset \mathbb{F}$, the space $C(K, \mathbb{F})$ of all continuous $\mathbb{F}$-valued functions on K with the sup-norm is a Banach algebra. Indeed, as in the previous example, any Cauchy sequence of continuous functions on K converges *uniformly* and hence (by Theorem 8.3.1, which also holds for complex-valued functions) has a *continuous* limit. In fact, the same is true if K is a *compact metric space*. Also, by Theorem 8.3.3, the space $\mathcal{R}([a, b], \mathbb{F})$ of all Riemann integrable (hence bounded) $\mathbb{F}$-valued functions on $[a, b]$ is a Banach space with the *sup-norm*.

(4) Let (K, d) be a *compact* metric space and let $Lip^\alpha(K)$ denote the set of all real-valued *Lipschitz* functions of *order* $\alpha \in (0, 1]$ on K. It follows from Exercise 5.6.30, that $Lip^\alpha(K)$ is a Banach space with norm

$$\|f\|_{\alpha,\infty} := \|f\|_\infty + \sup\left\{\frac{|f(x) - f(y)|}{d(x, y)^\alpha} : x, y \in K, x \neq y\right\}.$$

Exercise 9.2.3 (The Space c_0 of Banach). Consider the space

$$c_0 := \{x = (x_n) \in \mathbb{F}^{\mathbb{N}} : \lim(x_n) = 0\}.$$

Show that, with the sup-norm $\|x\|_\infty = \sup\{|x_n| : n \in \mathbb{N}\}$, the space c_0 is a *Banach space*. *Hint:* Let (x_k) be a *Cauchy sequence* in c_0, where $x_k = (x_{k,n})_{n=1}^\infty$ for each $k \in \mathbb{N}$. Show that, for each fixed $n \in \mathbb{N}$, the sequence $(x_{k,n})_{k=1}^\infty$ is *Cauchy* in $\mathbb{F}$ and let ξ_n denote its limit. Show that $\xi := (\xi_n) \in c_0$ and $\xi = \lim(x_k)$.

As we shall see presently, the *completeness* of a normed space is a very desirable property. As a nice example, the space $\boldsymbol{Pol}([a,b], \mathbb{R})$ of all polynomial functions on a (nontrivial) interval $[a,b]$ is a normed space with the *sup-norm*. It is *not* complete, but by the Weierstrass Approximation Theorem (Corollary 4.7.10), it is a *dense* subspace of the Banach space $C([a,b], \mathbb{R})$ of all *continuous* functions on $[a,b]$ with the "same" *sup-norm*. The following theorem shows that *any* normed space can be *completed* in an essentially unique way as follows:

Theorem 9.2.4 (Completion of a Normed Space). *For any normed space X, there is a Banach space $\hat{X}$ and an isometry (i.e., a norm-preserving linear map) $\iota : X \to \hat{X}$ such that $\iota(X)$ is dense in $\hat{X}$. The Banach space $\hat{X}$, which is known as the "completion" of X, is unique in the sense that, if $\hat{\hat{X}}$ is another completion of X, then $\hat{X}$ and $\hat{\hat{X}}$ are isometrically isomorphic.*

Proof. Since X is a metric space with metric $d(x,y) := \|x - y\|$, the theorem follows at once from Theorem 5.5.16. $\qquad\square$

Here is one way of constructing new Banach spaces:

Theorem 9.2.5. *Given any normed space X and any Banach space Y, the space $\mathcal{B}(X,Y)$ is a Banach space. In particular, the dual $X^* := \mathcal{B}(X,\mathbb{F})$ is a Banach space.*

Proof. Let (T_n) be a *Cauchy* sequence in $\mathcal{B}(X,Y)$ and let $\varepsilon > 0$. We can pick $N \in \mathbb{N}$ such that $m, n \geq N$ implies $\|T_m - T_n\| < \varepsilon$. It follows from Definition 9.1.31 that $\|T_m x - T_n x\| < \varepsilon$ if $m, n \geq N$ and $\|x\| \leq 1$. Thus $(T_n x)$ is Cauchy in the *complete* space Y if $\|x\| \leq 1$ and hence converges to an element $Tx \in Y$. This is then also true if $\|x\| > 1$, because $T_n x = \|x\| T_n(x/\|x\|)$. From the equations $T_n(\alpha x_1 + x_2) = \alpha T_n x_1 + T_n x_2$ and Proposition 5.5.8, it follows that $T(\alpha x_1 + x_2) = \alpha T x_1 + T x_2$; i.e., T is *linear*. On the other hand, $\|T_m - T_n\| < \varepsilon$ for all $m, n \geq N$ and $\|x\| \leq 1$ imply (as $m \to \infty$) that $\|(T - T_n)(x)\| = \|Tx - T_n x\| \leq \varepsilon$ and hence $\|Tx\| \leq \|T_n x\| + \varepsilon$ for all $n \geq N$ and $\|x\| \leq 1$. This proves that T is *bounded*. It also follows from Definition 9.1.31 that $\|T - T_n\| \leq \varepsilon$ so that (T_n) converges to T in $\mathcal{B}(X,Y)$. $\qquad\square$

Proposition 9.2.6. *Let X, Y, and Z be normed spaces, $T \in \mathcal{B}(X,Y)$, and $S \in \mathcal{B}(Y,Z)$. Then $ST := S \circ T \in \mathcal{B}(X,Z)$ and $\|ST\| \leq \|S\|\|T\|$.*

Proof. This follows at once from the inequalities

$$\|S(Tx)\| \leq \|S\|\|Tx\| \leq \|S\|\|T\|\|x\| \qquad (\forall x \in X). \qquad\square$$

Corollary 9.2.7. *Given a Banach space* X, *the space* $\mathcal{B}(X)$ *of all bounded operators on* X *is a Banach algebra.*

Proof. This follows from Theorem 9.2.5 and Proposition 9.2.6.   □

Now may be a good time to state Banach's Fixed Point Theorem (cf. Theorem 5.5.7) for Banach spaces.

Definition 9.2.8 (Contraction). Let X and Y be Banach spaces. A map $f : X \rightarrow Y$ is said to be a *contraction* if there is a constant $c \in (0, 1)$ such that

$$\|f(x) - f(y)\| \le c\|x - y\|.$$

Since a Banach space is a complete metric space, the following theorem is a special case of Theorem 5.5.7:

Theorem 9.2.9 (Banach's Fixed Point Theorem). *Let* X *be a Banach space and* $f : X \rightarrow X$ *a contraction. Then* f *has a unique fixed point; i.e., there exists a unique* $x \in X$ *such that* $f(x) = x$.

Example 9.2.10 (Fredholm Integral Equations).   Consider the Banach space $X :=$ $C([a, b])$ of continuous real-valued functions on a nondegenerate interval $[a, b]$ with the *sup-norm* $\|f\|_\infty := \sup\{|f(x)| : x \in [a, b]\}$. Let $k : [a, b] \times [a, b] \rightarrow \mathbb{R}$ be a *continuous* function. Now, given a function $g \in X$, the goal is to find a function $f \in X$ such that

$$f(x) - \int_a^b k(x, y)f(y)\, dy = g(x) \qquad (\forall\, x \in [a, b]). \qquad (*)$$

For each $f \in X$, define the function

$$(Kf)(x) := \int_a^b k(x, y)f(y)\, dy.$$

We claim that $K \in \mathcal{B}(X)$. Indeed, k is *uniformly continuous* on the (*compact*) square $R := [a, b] \times [a, b]$ so for each $\varepsilon > 0$ there is a $\delta > 0$ such that $|h| < \delta$ implies $|k(x + h, y) - k(x, y)| < \varepsilon$ for all $(x, y) \in R$ such that $x + h \in [a, b]$, and hence (by the MVT for integrals)

$$|(Kf)(x+h)-(Kf)(x)| = \left| \int_a^b [k(x+h, y)-k(x, y)]f(y)\, dy \right| < \varepsilon(b-a)\|f\|_\infty.$$

This shows that Kf is (*uniformly*) *continuous* on $[a, b]$. As for the *linearity* of K, it follows at once from the linearity of the integral. Now note that we can write $(*)$ as

$$f(x) = g(x) + \int_a^b k(x, y)f(y)\, dy = g(x) + (Kf)(x) \qquad (\forall\, x \in [a, b]).$$

$$(**)$$

Let $F : X \to X$ be the map $F(f) := g + Kf$. Then $(**)$ becomes $F(f) = f$; i.e., f is a *fixed point* of the map F. If $M := \sup\{|k(x,y)| : (x,y) \in R\}$, then for any $f_1, f_2 \in X$ and any $x \in [a,b]$, we have

$$|F(f_2)(x) - F(f_1)(x)| = \left| \int_a^b k(x,y)[f_2(y) - f_1(y)] \, dy \right| \le M(b-a)\|f_2 - f_1\|_\infty,$$

which gives

$$\|F(f_2) - F(f_1)\|_\infty \le M(b-a)\|f_2 - f_1\|_\infty \quad \forall \ f_1, \ f_2 \in X.$$

Thus, if we assume that $M < 1/(b-a)$, then the map F is a *contraction* and hence has a unique fixed point by Banach's Fixed Point Theorem. In other words, if $M(b-a) < 1$, then the integral equation $(*)$ has a *unique* solution $f \in X$ for each given $g \in X$.

Remark 9.2.11. As we shall demonstrate later (cf. Proposition 9.4.18), the linear map K in the above example is in fact *compact* in the following sense:

Definition 9.2.12 (Compact Map, Compact Operator). Let X and Y be normed spaces. A linear map $K \in \mathcal{L}(X,Y)$ is said to be *compact* if $K(B)$ is *relatively compact* (i.e., the closure $\big(K(B)\big)^-$ is *compact*) in Y for each *bounded* set $B \subset X$. A compact (linear) map $K : X \to X$ is called a *compact operator*. The set of all compact maps $K : X \to Y$ will be denoted by $\mathcal{K}(X,Y)$ and the set of all compact operators $K : X \to X$ by $\mathcal{K}(X)$.

Proposition 9.2.13. *With the above notation, we have* $\mathcal{K}(X,Y) \subset \mathcal{B}(X,Y)$ *and* $\mathcal{K}(X) \subset \mathcal{B}(X)$.

Proof. Indeed, if $K \in \mathcal{K}(X,Y)$, and if $B_1^- := B_1^-(0)$ denotes the closed unit ball in X, then the image $K(B_1^-)$ is relatively compact in Y and hence [by Theorem 5.6.7 (a)] *bounded*. $\qquad\square$

In the following theorem we shall use the fact that a subset of $\mathbb{R}^n$ is *compact* if and only if it is *closed and bounded* (cf. Corollary 5.6.38). Note that the same holds for $\mathbb{C}^n$, which is topologically $\mathbb{R}^{2n}$.

Theorem 9.2.14 (Finite-Dimensional Normed Space). *Every finite-dimensional normed space is a Banach space. In particular, every finite-dimensional subspace of a normed space is closed. In fact, if X is an n-dimensional normed space with basis $\{e_1, e_2, \ldots, e_n\}$, then the map $L\alpha := \sum_{k=1}^n \alpha_k e_k$, where $\alpha = (\alpha_1, \ldots, \alpha_n) \in \mathbb{F}^n$, is an isomorphism of the Euclidean space $\mathbb{F}^n$ onto X.*

Proof. It is a simple exercise to check that L is linear, one-to-one, and onto and that L^{-1} is also linear. That L is *bounded* follows from the inequalities

$$\|Lx\| = \left\| \sum_{k=1}^n \alpha_k e_k \right\| \le \sum_{k=1}^n |\alpha_k| \|e_k\| \le c\|\alpha\|, \qquad (*)$$

where $\|\alpha\| = \sqrt{\sum |\alpha_k|^2}$ and $c := \sqrt{\sum \|e_k\|^2}$. Now let $S_1 := \{\alpha \in \mathbb{F}^n : \|\alpha\| = 1\}$ be the unit sphere in $\mathbb{F}^n$ and recall that S_1 is closed and bounded hence *compact*. The continuous map $\alpha \mapsto \|L\alpha\|$ is *strictly positive* on S_1. (Why?) Thus,

$$\inf\{\|L\alpha\| : \|\alpha\| = 1\} = r > 0.$$

It follows that, if $\alpha \in \mathbb{F} \setminus \{0\}$, then $\|L(\alpha/\|\alpha\|)\| \geq r$ and hence $\|\alpha\| \leq (1/r)\|L\alpha\|$, which is also valid for $\alpha = 0$. This shows that L^{-1} is bounded as well. To show that X is *complete*, note that a Cauchy sequence (x_n) in X is mapped by the *Lipschitz* map L^{-1} onto the Cauchy (hence *convergent*) sequence $(L^{-1}x_n)$ in the *complete* Euclidean space $\mathbb{F}^n$. It follows that (x_n) converges to $L\alpha$, where $\alpha = \lim(L^{-1}x_n)$.□

Remark 9.2.15. Note in particular that, as pointed out in Chap. 5 (cf. Problem 5.8.#19), since the spaces $Pol_k[0, 1]$ are finite dimensional, they are *closed* subspaces of the Banach space (in fact *Banach algebra*) $X := C([0, 1])$ with the *sup-norm* $\|f\|_\infty := \sup\{|f(x)| : x \in [0, 1]\}$.

Corollary 9.2.16. *Any linear map T from a finite-dimensional normed space X to a normed space Y is compact, hence continuous.*

Proof. Since $\dim(X) < \infty$, the range $\mathrm{Ran}(T)$ is a *finite-dimensional*, hence *closed*, subspace of Y by the above theorem. In particular, T is *bounded* if and only if it is *compact*. Let $\{e_1, \ldots, e_n\}$ be a basis for X. For each $x \in X$ we have $x := \sum_{k=1}^n \alpha_k e_k$ with a *unique* vector $\alpha = (\alpha_1, \ldots, \alpha_n) \in \mathbb{F}^n$ and hence $Tx = \sum_{k=1}^n \alpha_k T e_k$. Thus $T = UL^{-1} := U \circ L^{-1}$, where $U \in \mathcal{L}(\mathbb{F}^n, Y)$ is given by $U\alpha := \sum \alpha_k T e_k$, while $L^{-1} \in \mathcal{L}(X, \mathbb{F}^n)$ is given by $L^{-1}(\sum \alpha_k e_k) := \alpha$ and is continuous by Theorem 9.2.14. Also, as in (∗), the boundedness of U follows from the inequalities

$$\|U\alpha\| = \left\| \sum_{k=1}^n \alpha_k T e_k \right\| \leq \sum_{k=1}^n |\alpha_k| \|T e_k\| \leq d \|\alpha\|, \quad d := \sqrt{\sum \|T e_k\|^2}.$$

□

Corollary 9.2.17. *Any two norms on a finite-dimensional space are equivalent.*

Proof. Indeed, if $\| \cdot \|_1$ and $\| \cdot \|_2$ are two norms on a finite-dimensional space X, then Corollary 9.2.16 implies that the identity map $I : (X, \| \cdot \|_1) \to (X, \| \cdot \|_2)$ is a topological isomorphism.   □

Definition 9.2.18 (Convergent and Absolutely Convergent Series). Let X be a normed space and $(x_n) \in X^{\mathbb{N}}$. We say that the infinite series $\sum_{n=1}^\infty x_n$ is *convergent* and has *sum* $s \in X$ if for each $\varepsilon > 0$ there exists $N \in \mathbb{N}$ such that $n \geq N$ implies $\|s - \sum_{k=1}^n x_k\| < \varepsilon$. The series $\sum_{n=1}^\infty x_n$ is said to be *absolutely convergent* if the numerical series $\sum_{n=1}^\infty \|x_n\|$ is convergent.

Proposition 9.2.19. *Let* X, Y *be normed spaces and* $T \in \mathcal{B}(X, Y)$. *If* $\sum x_n$ *is a convergent (resp., absolutely convergent) series in* X, *then* $\sum T x_n$ *is convergent (resp., absolutely convergent) in* Y *and we have* $\sum T x_n = T(\sum x_n)$.

Proof. Let $s = \sum x_n \in X$. Then

$$\left\| T s - T\left(\sum_{k=1}^{n} x_k \right) \right\| = \left\| T\left(s - \sum_{k=1}^{n} x_k \right) \right\|$$

$$\leq \|T\| \left\| s - \sum_{k=1}^{n} x_k \right\| \to 0,$$

as $n \to \infty$. The statement about absolute convergence is obvious. (Why?)   □

We have seen that an absolutely convergent series in $\mathbb{R}$ (or $\mathbb{C}$) is necessarily convergent. This is *not* true in general for series in normed spaces. In fact we have the following.

Theorem 9.2.20. *A normed space* X *is a Banach space if and only if every absolutely convergent series in* X *is convergent.*

Proof. Let X be a Banach space, $(x_n) \in X^{\mathbb{N}}$, and $\sum_{n=1}^{\infty} \|x_n\| = M < \infty$. Given any $\varepsilon > 0$ we can then find $N \in \mathbb{N}$ such that $\sum_{n=N}^{\infty} \|x_n\| < \varepsilon$. If $s_n := \sum_{k=1}^{n} x_n$ is the n-th partial sum of $\sum x_n$, then $n > m \geq N$ implies

$$\|s_n - s_m\| = \left\| \sum_{k=m}^{n} x_k \right\| \leq \sum_{k=m}^{n} \|x_k\| \leq \sum_{k=N}^{\infty} \|x_k\| < \varepsilon.$$

Therefore (s_n) is a *Cauchy sequence* in the *complete* space X and hence must converge. Conversely, suppose that every absolutely convergent series is convergent in X and let $(x_n) \in X^{\mathbb{N}}$ be a *Cauchy* sequence. We can then pick positive integers $n_1 < n_2 < \cdots$ such that $\|x_n - x_m\| < 2^{-k}$ if $\max\{m, n\} \geq n_k$. Put $y_1 := x_{n_1}$ and $y_k := x_{n_k} - x_{n_{k-1}}$ for $k > 1$. It then follows that x_{n_k} is the kth partial sum of the series $\sum y_k$. But then,

$$\sum \|y_k\| \leq \|y_1\| + \sum 2^{-k+1} = \|y_1\| + 1$$

and $\sum y_k$ is *absolutely convergent*. Our assumption now implies that $\sum y_k$ is *convergent*. This means that the subsequence (x_{n_k}) is convergent. (Why?) Since (x_n) is a Cauchy sequence, Exercise 5.3.6 (4) implies that it converges to the same limit.   □

Exercise 9.2.21 (Quotient Space). Let V be a vector space.

(1) Show that, if W is a (vector) subspace of V, then the relation $x \sim y \Leftrightarrow x - y \in W$ is an *equivalence relation* on V. The equivalence class of a vector $x \in V$

is $[x] := x + W$. The *quotient space* V/W is the space of all equivalence classes with vector addition and scalar multiplication defined by $[x] + [y] := [x + y]$ and $\alpha[x] := [\alpha x]$, respectively. Show that these operations are *well defined* in the sense that, for any $x' \in [x]$ and $y' \in [y]$, we have $[x] + [y] = [x'] + [y']$ and $\alpha[x'] = \alpha[x]$ and that V/W is indeed a vector space.

(2) Let $\| \cdot \|$ be a seminorm on V and, as in Exercise 9.1.2, consider the subspace

$$W := \{x \in V : \|x\| = 0\}.$$

For each $x \in V$, let $[x] = x + W$ be its equivalence class and define $\|[x]\| := \|x\|$. Show that this defines a *norm* on the quotient space V/W.

Here is another way of constructing new Banach spaces:

Theorem 9.2.22. *Let Y be a closed subspace of a normed space X and, for each $[x] \in X/Y$, define*

$$\|[x]\| := \inf\{\|x - y\| : y \in Y\} = d(x, Y). \tag{$*$}$$

Then $(*)$ defines a norm on X/Y. If X is a Banach space, then the quotient space X/Y, with the above norm, is also a Banach space.

Proof. First, $\|[x]\| \geq 0$ is obvious and $\|[x]\| = d(x, Y) = 0$ means $x \in Y^- = Y$ and hence $[x] = [0]$. Next, for any $\alpha \in \mathbb{F} \setminus \{0\}$,

$$\|[\alpha x]\| = \inf\{\|\alpha x - y\| : y \in Y\} = |\alpha| \inf\{\|x - y/\alpha\| : y \in Y\} = |\alpha| \|[x]\|.$$

Finally, $\|[x_1] + [x_2]\| \leq \|[x_1]\| + \|[x_2]\|$ follows from the fact that

$$\|x_1 + x_2 - (y_1 + y_2)\| \leq \|x_1 - y_1\| + \|x_2 - y_2\| \qquad (\forall y_1, y_2 \in Y).$$

Next, suppose that X is *complete*, hence a *Banach* space. In view of Theorem 9.2.20, to show that X/Y is *complete*, it suffices to show that, if $(x_n) \in X^{\mathbb{N}}$ satisfies $\sum \|[x_n]\| < \infty$, then $\sum [x_n]$ converges (in X/Y). To show this, note that we can pick a sequence $(y_n) \in Y^{\mathbb{N}}$ such that $\|x_n - y_n\| \leq \|[x_n]\| + 2^{-n}$. (Why?) Then $\sum \|x_n - y_n\| < \infty$ and hence (since X is a Banach space) $x := \sum (x_n - y_n) \in X$. But then

$$\left\| [x] - \sum_{k=1}^{n} [x_k] \right\| = \left\| [x] - \sum_{k=1}^{n} [x_k - y_k] \right\| = \left\| \left[x - \sum_{k=1}^{n} (x_k - y_k) \right] \right\|$$

$$\leq \left\| x - \sum_{k=1}^{n} (x_k - y_k) \right\| \to 0 \quad \text{as } n \to \infty,$$

and we have $[x] = \sum [x_n]$. $\square$

Remark 9.2.23 (Canonical Projection). With notation as above, the *surjective* map $\pi : X \to X/Y$ defined by $\pi(x) := [x]$ is called the *canonical projection*. It is a (linear) *Lipschitz* map, as $\|[x]\| = d(x, Y) \leq \|x\|$ for all $x \in X$.

Lemma 9.2.24. *Let $Y \neq X$ be a closed subspace of a normed space X. Then there is a sequence $(x_n) \in X^{\mathbb{N}}$ such that, $\|x_n\| = 1$ for all $n \in \mathbb{N}$, $\|[x_n]\|$ increases with n, and $\lim(\|[x_n]\|) = 1$. Here, $[x_n] = x_n + Y \in X/Y$.*

Proof. Let $x \in X \setminus Y$. It follows from the definition of $\|[x]\|$ that we can find a sequence $(y_n) \in Y^{\mathbb{N}}$ such that $(\|x - y_n\|)$ is decreasing and $\lim(\|x - y_n\|) = \|[x]\|$. Let $z_n := x - y_n$, $x_n := z_n/\|z_n\|$, and note that Y is *closed* and $x_n \notin Y$, so that $\|[x_n]\| > 0$ for all n. It is then easily checked that (x_n) is the desired sequence. $\quad\square$

As a corollary, we prove the following fundamental result which characterizes *finite dimensional* normed spaces (cf. Theorem 9.2.14):

Theorem 9.2.25 (F. Riesz's Lemma). *A normed space X is finite dimensional if and only if the closed unit ball $B_1^-(0) := \{x \in X : \|x\| \leq 1\}$ is compact.*

Proof. If $\dim(X) < \infty$, then the compactness of $B_1^-(0)$ follows from Theorem 9.2.14 and the fact that the closed unit ball in $\mathbb{F}^n$ is compact for any $n \in \mathbb{N}$. If $\dim(X) = \infty$, let us construct, inductively, a sequence (x_n) of (independent) vectors such that $\|x_n\| = 1$ for all n and $\|x_j - x_k\| \geq 1/2$ for $j \neq k$. Assuming that the x_i, $1 \leq i \leq n$, have been chosen, let $X_n := \mathrm{Span}(\{x_1, x_2, \ldots, x_n\})$ and note that X_n is a finite-dimensional (hence *closed*) subspace and $X_n \neq X$. By the above lemma, we can therefore pick $x_{n+1} \in X \setminus X_n$ such that $\|x_{n+1}\| = 1$ and $d(x_{n+1}, X_n) \geq 1/2$. Now note that (x_n) is a sequence in $B_1^-(0)$ with *no* convergent subsequence. $\quad\square$

Definition 9.2.26 (Total Set, Total Family). Let X be a normed space. A set $S \subset X$ is said to be *total* if its span is *dense* in X; i.e., if $\mathrm{Span}(S)^- = X$. A family $(x_j)_{j \in J}$ of vectors in X (i.e., a function $x \in X^J$) is called a *total family* if its *range* $\{x_j : j \in J\}$ is total. In particular, a sequence $(x_n) \in X^{\mathbb{N}}$ is total if the set $\{x_n : n \in \mathbb{N}\}$ is total.

Recall that a metric space is called *separable* if it contains a *countable dense* subset.

Theorem 9.2.27 (Separable Normed Space). *If a normed space X contains a total sequence, then it is separable. Conversely, if X is separable, then it contains a total sequence consisting of linearly independent vectors.*

Proof. Suppose that $(b_n) \in X^{\mathbb{N}}$ is a *total sequence* and let D denote the set of all *finite* linear combinations $\sum_{k=1}^n \rho_k b_k$ with *rational* coefficients; i.e., the ρ_k are in $\mathbb{Q}$ if $\mathbb{F} = \mathbb{R}$ and in $\mathbb{Q} + i\mathbb{Q}$ if $\mathbb{F} = \mathbb{C}$. It is easily seen that D is a countable union of countable sets and hence is itself *countable*. Since $S := \mathrm{Span}(\{b_n : n \in \mathbb{N}\})$ is dense in X, we need only show that D is dense in S. But this follows from

$$\left\|\sum_{k=1}^{n} \alpha_k b_k - \sum_{k=1}^{n} \rho_k b_k\right\| \leq \sum_{k=1}^{n} |\alpha_k - \rho_k| \|b_k\|$$

and the fact that $\mathbb{Q}$ (resp., $\mathbb{Q} + i\mathbb{Q}$) is dense in $\mathbb{R}$ (resp., $\mathbb{C}$). Conversely, suppose that X is *separable*. We may assume that X is *infinite dimensional* because otherwise every basis is already a *finite* total subset of independent vectors. Let (b_n) be an *infinite, dense* sequence of vectors in X and let $Y_1 := \text{Span}(\{b_{k_1}\})$, where k_1 is the *smallest* n with $b_n \neq 0$. Note that no *proper closed* subspace of X (hence no *finite-dimensional* subspace) can contain all the b_n. (Why?) So, let $k_2 > k_1$ be the smallest index with $b_{k_2} \notin Y_1$. Assuming we have chosen $k_1 < \cdots < k_n$, we put $Y_n := \text{Span}(\{b_{k_1}, \ldots, b_{k_n}\})$ and let $k_{n+1} > k_n$ be the smallest index with $b_{k_{n+1}} \notin Y_n$. It is then obvious that the sequence $(b_{k_n})_{n=1}^{\infty}$ is a total sequence of *linearly independent* vectors in X. (Why?)   □

Theorem 9.2.28 (Extension of Bounded Linear Maps). *Let X be a dense subspace of a normed space Y (i.e., $X^- = Y$) and let Z be a Banach space. Then any $T \in \mathcal{B}(X, Z)$ has a unique continuous extension $\tilde{T} \in \mathcal{B}(Y, Z)$ and we have $\|T\| = \|\tilde{T}\|$.*

Proof. Since T is *Lipschitz* (hence *uniformly continuous*), the existence of $\tilde{T}$ follows from Theorem 5.5.13. The linearity of $\tilde{T}$ follows from the continuity of $\tilde{T}$, Propositions 9.1.12, and 5.5.8. Finally, the equality of norms follows easily from Definition 9.1.31 and Proposition 5.5.8. (Why?)   □

We now prove a number of fundamental results that are consequences of the *Baire Category Theorem* (cf. Corollary 5.3.9). The first one, known as the *Principle of Uniform Boundedness* is essentially a consequence of *Osgood's Theorem* (Theorem 5.5.15) and is also known as the *Banach–Steinhaus Theorem*.

Theorem 9.2.29 (Uniform Boundedness Principle). *Let X be an arbitrary Banach space and let $\{T_j\}_{j \in J}$ be a family of bounded linear maps from X to a normed space Y. If for each $x \in X$ there is a constant $M_x > 0$ such that $\|T_j x\| \leq M_x$ for all $j \in J$, then there is a constant $M > 0$ such that $\|T_j\| \leq M$ for all $j \in J$.*

Proof. Let $f_j(x) := \|T_j x\|$. Then $\{f_j : j \in J\}$ is a set of *continuous real-valued* functions on X and, since X is *complete*, it follows from Theorem 5.5.15 that there is an open ball $B_\delta(x_0) \subset X$, $\delta > 0$, on which all the f_j are *uniformly bounded*; i.e., there is a constant $M' > 0$ such that $\|T_j x\| \leq M'$ for all $j \in J$ and all $x \in B_\delta(x_0)$. Now, if $\|x\| < \delta$, then $x + x_0 \in B_\delta(x_0)$ and hence $\|T_j x\| \leq \|T_j(x + x_0)\| + \|T_j x_0\| \leq M' + M_{x_0}$, for all $j \in J$. This implies that $\|T_j\| \leq M$ for all $j \in J$ with $M := (M' + M_{x_0})/\delta$.   □

Corollary 9.2.30. *Let X be a Banach space and let $(T_n)_{n \in \mathbb{N}}$ be a sequence of bounded linear maps from X to a normed space Y. If for each $x \in X$ the sequence $(T_n x)$ is convergent and if we let $Tx := \lim(T_n x)$, then $T \in \mathcal{B}(X, Y)$ and $\|T\| \leq \liminf_{n \to \infty} \|T_n\|$.*

Proof. The *linearity* of T follows at once from the linearity of the T_n and the *continuity* of the vector addition and scalar multiplication in Y. Also, $M_x := \sup\{\|T_n x\| : n \in \mathbb{N}\} < \infty$ for each $x \in X$, so (by Theorem 9.2.29) there is an $M > 0$ such that $\|T_n x\| \leq M\|x\|$ for all $x \in X$ and all $n \in \mathbb{N}$. Therefore, by the continuity of the norm,

$$\|Tx\| = \lim_{n \to \infty} \|T_n x\| \leq \liminf_{n \to \infty} \|T_n\|\|x\| \qquad (\forall x \in X).$$

$\square$

The next important theorem uses the Baire Category Theorem to show that surjective (bounded linear) maps between Banach spaces are *open*. Let us first prove a lemma.

Lemma 9.2.31. *Let* X, Y *be Banach spaces and let* $T \in \mathcal{B}(X, Y)$ *be surjective. If* $B_1 := B_1(0)$ *denotes the open unit ball in* X *(centered at 0), then the closure of its image under* T *(i.e.,* $(T(B_1))^-$*) contains an open ball* $B_\delta(0)$ *in* Y.

Proof. Since T is *onto* and $X = \bigcup_{k=1}^{\infty} k B_1$, it follows that $Y = \bigcup_{k=1}^{\infty} k T(B_1)$. But Y is *complete* and hence the Baire Category Theorem implies that $(NT(B_1))^-$ has *nonempty interior* for some $N \in \mathbb{N}$. Thus, we can find $y_0 \in Y$ and $r > 0$ such that $B_r(y_0) = \{y \in Y : \|y - y_0\| < r\} \subset (NT(B_1))^-$. But then, since $-B_1 = B_1$, we also have $B_r(-y_0) \subset (NT(B_1))^-$. This implies (by Proposition 9.1.20) that

$$B_{2r}(0) := \{y \in Y : \|y\| < 2r\} \subset (2NT(B_1))^- = 2N(T(B_1))^-, \qquad (*)$$

where the last equality follows from the *bicontinuity* of $y \mapsto \alpha y$ for all $\alpha \neq 0$. From $(*)$ we deduce that $B_\delta(0) \subset (T(B_1))^-$ with $\delta = r/N$. (Why?) $\square$

Theorem 9.2.32 (Open Mapping Theorem). *Let* X, Y *be Banach spaces and let* $T \in \mathcal{B}(X, Y)$ *be surjective. Then* T *is an open mapping, i.e., for any open set* $O \subset X$, $T(O)$ *is open in* Y.

Proof. Let $B_r := B_r(0)$ denote the open ball of radius r centered at 0 in X or Y. Using translations, it suffices to show that, if U is an open neighborhood of 0 in X, then $T(U)$ is an open neighborhood of 0 in Y. (Why?) This in turn follows (using dilations) if we show that $T(B_1)$ contains an open ball B_ε in Y. Thus we need only show that $(T(B_1))^- \subset T(B_2) = 2T(B_1)$ because, by Lemma 9.2.31, we know that $B_\delta \subset (T(B_1))^-$ for some $\delta > 0$. Now, given any $y \in (T(B_1))^-$, we can pick $x_1 \in B_1$ such that $y - Tx_1 \in B_{\delta/2} \subset (T(B_{1/2}))^-$. (Why?) Similarly, we can pick $x_2 \in B_{1/2}$ such that $(y - Tx_1) - Tx_2 \in B_{\delta/4} \subset (T(B_{1/4}))^-$. Continuing this process, we can pick $x_n \in B_{2^{1-n}}$ such that

$$y - \sum_{k=1}^{n} Tx_k \in B_{\delta/2^n} \subset (T(B_{1/2^n}))^-.$$

But then $x := \sum x_n \in B_2 = 2B_1$ and $y = Tx$. (Why?) This proves our claim and completes the proof. $\square$

Corollary 9.2.33. *If X, Y are Banach spaces and if $T \in \mathcal{B}(X,Y)$ is bijective, then it is in fact a topological isomorphism.*

Proof. By Theorem 9.2.32, T is *open* and hence T^{-1} is *continuous*.   □

Corollary 9.2.34 (Canonical Projection). *Let Y be a closed subspace of a Banach space X and let X/Y be the corresponding quotient space. Then the canonical projection $\pi : X \rightarrow X/Y$ defined by $\pi(x) := [x]$ is an open map.*

Proof. Indeed, X/Y is also a Banach space (Theorem 9.2.22) and π is *onto*.   □

Corollary 9.2.35. *Let $\| \cdot \|_1$ and $\| \cdot \|_2$ be two norms on a Banach space X. If there is a constant $c > 0$ such that $\|x\|_2 \le c\|x\|_1$, then the norms $\| \cdot \|_1$ and $\| \cdot \|_2$ are equivalent.*

Proof. Our assumption means that the identity map $I : (X, \| \cdot \|_1) \rightarrow (X, \| \cdot \|_2)$ is *continuous*. By Corollary 9.2.33, it is therefore a topological *isomorphism*.   □

Next, let X, Y be normed spaces and $T \in \mathcal{L}(X,Y)$. The *graph* of T is the set

$$\Gamma_T := \{(x, Tx) : x \in X\} \subset X \times Y.$$

Note that Γ_T is in fact a (vector) *subspace* of $X \times Y$. (Why?) By Theorem 5.4.15, if $T \in \mathcal{B}(X,Y)$, then Γ_T is *closed* in $X \times Y$. The following is a converse.

Theorem 9.2.36 (Closed Graph Theorem). *Let X, Y be Banach spaces and $T \in \mathcal{L}(X,Y)$. If Γ_T is closed, then $T \in \mathcal{B}(X,Y)$.*

Proof. It is easily seen that $(x, Tx) \mapsto \|x\| + \|Tx\|$ is a *norm* on the subspace Γ_T of $X \times Y$ and our hypothesis implies that Γ_T is in fact a *Banach space* with this norm. (Why?) Consider the projections $P_1 : \Gamma_T \rightarrow X$ and $P_2 : \Gamma_T \rightarrow T(X)$ defined by $P_1(x, Tx) := x$ and $P_2(x, Tx) := Tx$. They are both *continuous* linear maps. (Why?) Also, P_1 is *bijective* and hence a topological isomorphism of Γ_T onto X. But then, the composite map $T = P_2 P_1^{-1} := P_2 \circ P_1^{-1}$ is continuous.   □

We end the section with the celebrated *Hahn–Banach Theorem*, which guarantees the existence of (norm-preserving) extensions of linear functionals defined on subspaces and is one of the most fundamental results in functional analysis.

Theorem 9.2.37 (Hahn–Banach Theorem). *Let Y be a subspace of a real vector space X and let $p : X \rightarrow \mathbb{R}$ satisfy*

$$p(x + y) \le p(x) + p(y) \quad and \quad p(\alpha x) = \alpha p(x) \quad (\forall x, y \in X, \forall \alpha \ge 0).$$

If $\phi : Y \rightarrow \mathbb{R}$ is linear and $\phi(y) \le p(y)$ for all $y \in Y$, then there exists a linear functional $\Phi : X \rightarrow \mathbb{R}$ such that $\Phi(y) = \phi(y)$ for all $y \in Y$ (i.e., Φ is an extension of ϕ to X) and

$$-p(-x) \le \Phi(x) \le p(x) \quad (\forall x \in X).$$

Proof. We assume that $Y \neq X$, pick $x_0 \in X \setminus Y$, and let

$$Y_1 := Y \oplus \mathbb{R}x_0 = \{y + \alpha x_0 : y \in Y, \alpha \in \mathbb{R}\}$$

be the subspace spanned by Y and x_0. From the inequalities

$$\phi(y) + \phi(z) = \phi(y + z) \leq p(y + z) \leq p(y - x_0) + p(x_0 + z),$$

valid for all $y, z \in Y$, it follows that

$(i) \quad \phi(y) - p(y - x_0) \leq p(z + x_0) - \phi(z) \qquad (\forall y, z \in Y).$

If $\xi_0 := \sup\{\phi(y) - p(y - x_0) : y \in Y\}$, then (i) implies

$(ii) \quad \phi(y) - \xi_0 \leq p(y - x_0) \qquad (\forall y \in Y)$

and

$(iii) \quad \phi(z) + \xi_0 \leq p(z + x_0) \qquad (\forall z \in Y).$

Now define $\phi_1 : Y_1 \to \mathbb{R}$ by

$(iv) \quad \phi_1(y + \alpha x_0) := \phi(y) + \alpha \xi_0 \qquad (\forall y \in Y, \forall \alpha \in \mathbb{R}).$

Then ϕ_1 is *linear* and $\phi_1|Y = \phi$. If we replace y by $y/(-\alpha)$ in (ii) for $\alpha < 0$, z by z/α in (iii) for $\alpha > 0$, and multiply the resulting inequalities by $-\alpha$ and α, respectively, then (iv) implies that $\phi_1 \leq p$ on Y_1.

The second half of the proof requires the use of *Zorn's Lemma*. Let $\mathcal{E}$ denote the set of all pairs (Y', ϕ') such that Y' is a subspace of X with $Y' \supset Y$ and $\phi' : Y' \to \mathbb{R}$ is linear with $\phi'|Y = \phi$ and $\phi' \leq p$ on Y'. We partially order $\mathcal{E}$ by defining $(Y', \phi') \preceq (Y'', \phi'')$ to mean that $Y' \subset Y''$ and $\phi''|Y' = \phi'$. If $\mathcal{C}$ is a *chain* in $\mathcal{E}$, then an upper bound is the pair $(\tilde{Y}, \tilde{\phi})$, where $\tilde{Y} := \bigcup_{(Y',\phi') \in \mathcal{C}} Y'$ and $\tilde{\phi} : \tilde{Y} \to \mathbb{R}$ is defined by $\tilde{\phi}|Y' := \phi'$ for each $(Y', \phi') \in \mathcal{C}$. It is easily checked that $\tilde{Y}$ is a subspace of X and that $\tilde{\phi}$ is linear on $\tilde{Y}$. (Why?) By Zorn's Lemma, $\mathcal{E}$ has a *maximal element*, say (Z, ψ). We claim that $Z = X$ and that ψ is the desired extension Φ. Indeed, if $Z \neq X$, using the construction in the first half of our proof, we can extend ψ to the larger subspace $Z \oplus \mathbb{R}x_1$ where $x_1 \in X \setminus Z$, contradicting the maximality of (Z, ψ). Finally, $\psi \leq p$ implies that $-p(-x) \leq -\psi(-x) = \psi(x)$ for all $x \in X$. $\square$

The following consequence is an extension of the result to *complex* vector spaces and complex functionals.

Theorem 9.2.38. *Let $\| \cdot \|$ be a seminorm on a vector space X over $\mathbb{F}$. If Y is a subspace of X and $\phi : Y \to \mathbb{F}$ is a linear functional such that*

$$|\phi(y)| \leq \|y\| \qquad (\forall y \in Y),$$

then ϕ extends to a linear functional $\Phi : X \to \mathbb{F}$ such that

$$|\Phi(x)| \leq \|x\| \qquad (\forall x \in X).$$

Proof. If $\mathbb{F} = \mathbb{R}$, then the result is contained in the Hahn–Banach Theorem, where we define $p(x) := \|x\|$ and note that $\|-x\| = \|x\|$ for all $x \in X$. We therefore assume that $\mathbb{F} = \mathbb{C}$ and let $\psi := \operatorname{Re}(\phi)$. Then (Exercise 9.1.17) ψ is a *real* functional on Y and $\phi(y) = \psi(y) - i\psi(iy)$ for all $y \in Y$. By Theorem 9.2.37, there exists a *real* linear functional $\Psi : X \to \mathbb{R}$ such that $\Psi|Y = \psi$ and $|\Psi(x)| \leq \|x\|$ for all $x \in X$. Consider the corresponding *complex* linear functional: $\Phi(x) = \Psi(x) - i\Psi(ix)$ for all $x \in X$. We then have $\Phi|Y = \phi$ (by Exercise 9.1.17). Also, for each $x \in X$, we can pick $\alpha \in \mathbb{C}$ such that $|\alpha| = 1$ and $\alpha\Phi(x) = |\Phi(x)|$. It then follows that

$$|\Phi(x)| = \Phi(\alpha x) = \Psi(\alpha x) \leq \|\alpha x\| = |\alpha| \|x\| = \|x\|$$

and the proof is complete.   □

Corollary 9.2.39. *Let X be a normed space. Then, for each $x_0 \in X$, there exists $\Phi \in X^*$ such that*

$$\Phi(x_0) = \|x_0\| \quad and \quad |\Phi(x)| \leq \|x\| \qquad (\forall x \in X).$$

Proof. If $x_0 = 0$, we take $\Phi = 0$. Otherwise, applying Theorem 9.2.38 with $Y := \mathbb{F}x_0$, we can extend the linear functional $\phi(\alpha x_0) := \alpha\|x_0\|$ to the desired functional $\Phi \in X^*$ with the required property.   □

9.3 Hilbert Spaces

Recall that an *n-dimensional* normed space is *essentially* a copy of $\mathbb{F}^n$ and that all norms on such a space are *equivalent*. Since the *Euclidean norm* on $\mathbb{F}^n$ comes from the standard inner product $\langle \alpha, \beta \rangle := \sum_{k=1}^{n} \alpha_k \overline{\beta_k}$, it is natural to study normed spaces whose norms are derived from inner products. Such spaces are called *pre-Hilbert spaces* and their complete versions are called *Hilbert spaces*. In this section we explore some of the elementary properties of Hilbert spaces. These spaces form a particularly important class of Banach spaces and play a fundamental role in *functional analysis*. We begin with the notion of *inner product* on which most other properties will be based.

Definition 9.3.1 (Inner Product). Let X be a vector space over $\mathbb{F}$. By an *inner product* on X we mean a map $\langle \cdot, \cdot \rangle : X \times X \to \mathbb{F}$ such that, for any vectors x, y, $z \in X$ and any scalars α, $\beta \in \mathbb{F}$, the following conditions are satisfied:

IP1 $\langle x, y \rangle = \overline{\langle y, x \rangle}$,
IP2 $\langle \alpha x + \beta y, z \rangle = \alpha\langle x, x \rangle + \beta\langle y, z \rangle$,
IP3 $\langle x, x \rangle \geq 0$, and $\langle x, x \rangle = 0 \Leftrightarrow x = 0$.

Inner products are examples of *sesquilinear forms*:

Definition 9.3.2 (Sesquilinear Form). Given a normed space X, we say that a map $\psi : X \times X \to \mathbb{F}$ is a *sesquilinear form* if $\psi(x, y)$ is *linear* in x and *conjugate linear* in y, meaning that for all x, y, $y' \in X$ and $\alpha \in \mathbb{F}$, we have

$$\psi(x, \alpha y + y') = \overline{\alpha}\psi(x, y) + \psi(x, y').$$

Remark 9.3.3. Note that a sesquilinear form is always $\mathbb{R}$-bilinear (i.e., it is bilinear if we only use *real scalars*). Also, ψ is *continuous* if and only if it is *bounded*, i.e., $M := \sup\{|\psi(x, y)| : \|x\| = 1 = \|y\|\} < \infty$.

Definition 9.3.4 (Pre-Hilbert Space, Hilbert Space). A *pre-Hilbert space* is defined to be a pair $(X, \langle \cdot, \cdot \rangle)$, where X is a vector space over $\mathbb{F}$ and $\langle \cdot, \cdot \rangle$ is an inner product on X. A *complete* pre-Hilbert space is called a *Hilbert space*.

Example 9.3.5.

(1) The Euclidean space $\mathbb{F}^n$ is a pre-Hilbert (in fact a Hilbert) space with the standard inner product

$$\langle \alpha, \beta \rangle := \sum_{k=1}^{n} \alpha_k \overline{\beta_k}.$$

(2) The space $\ell^2(\mathbb{N}, \mathbb{C})$ of all sequences $x = (x_n) \in \mathbb{C}^{\mathbb{N}}$ that are *square summable*, i.e., $\sum |x_n|^2 < \infty$, is a pre-Hilbert (in fact a Hilbert) space with inner product

$$\langle x, y \rangle := \sum_{k=1}^{\infty} x_n \overline{y_n}. \tag{$*$}$$

Exercise 9.3.6. Show that the product $(*)$ in example (2) above is indeed an inner product.

Proposition 9.3.7. *Let* $(X, \langle \cdot, \cdot \rangle)$ *be a pre-Hilbert space. Then*

$$\|x\| := \langle x, x \rangle^{1/2}$$

defines a norm on X. *Moreover, we have the Cauchy–Schwarz inequality*

$$\langle x, y \rangle \le \|x\|\|y\| \qquad (\forall x, y \in X), \tag{$\dagger$}$$

and Minkowski's inequality

$$\|x + y\| \le \|x\| + \|y\| \qquad (\forall x, y \in X). \tag{$\ddagger$}$$

Proof. Let us first prove the Cauchy–Schwarz inequality. Using the properties $IP1 - IP3$ of inner products, for any x, $y \in X$ and any $\alpha \in \mathbb{F}$, we have

$$0 \le \langle x + \alpha y, x + \alpha y \rangle = \langle x, x \rangle + \langle x, \alpha y \rangle + \langle \alpha y, x \rangle + \langle \alpha x, \alpha y \rangle,$$

and hence

$$\|x\|^2 + \overline{\alpha}\langle x, y\rangle + \alpha\overline{\langle x, y\rangle} + |\alpha|^2\|y\|^2 \geq 0, \qquad (**)$$

where equality holds if and only if $x + \alpha y = 0$. If $y \neq 0$, then (†) follows by taking $\alpha = -\langle x, y\rangle/\|y\|^2$, and if $x \neq 0$ we can take $\alpha = -\langle x, y\rangle/\|x\|^2$. Since the inequality is trivially satisfied if $x = 0 = y$, (†) follows. For (‡), we take $\alpha = 1$ in $(**)$ and use (†) to get

$$\begin{aligned}
\langle x + y, x + y\rangle &= \|x\|^2 + 2\mathrm{Re}\langle x, y\rangle + \|y\|^2 \\
&\leq \|x\|^2 + 2|\langle x, y\rangle| + \|y\|^2 \\
&\leq \|x\|^2 + 2\|x\|\|y\| + \|y\|^2 \\
&= \left(\langle x, x\rangle^{1/2} + \langle y, y\rangle^{1/2}\right)^2.
\end{aligned}$$

Now (‡) is the Triangle Inequality for the norm $x \mapsto \langle x, x\rangle^{1/2}$ and the other properties are trivial.   □

Corollary 9.3.8. *Let X be a pre-Hilbert space. Then the inner product $(x, y) \to \langle x, y\rangle$ is a bounded sesquilinear form on $X \times X$. Moreover, for each $y \in X$, the linear map $x \to \langle x, y\rangle$ is a conjugate linear isometry of X into its dual X^*.*

Proof. The boundedness of the inner product follows from the Cauchy–Schwarz inequality. For a fixed $y \in X$, let $\psi^y(x) := \langle x, y\rangle$ and note that $y \to \psi^y$ is *conjugate linear*. The Cauchy–Schwarz inequality implies that ψ^y is a *bounded* linear functional (i.e., $\psi^y \in X^*$) with $\|\psi^y\| \leq \|y\|$. Since $\psi^y(y) = \|y\|^2$, we have $\|\psi^y\| = \|y\|$.   □

Remark 9.3.9. We shall see that the isometry $y \to \psi^y$ is in fact *onto* the dual X^* if X is *complete* (i.e., a *Hilbert space*).

Exercise 9.3.10 (Parallelogram Law). Let $(X, \langle \cdot, \cdot\rangle)$ be a pre-Hilbert space. Show that we have

$$\|x + y\|^2 + \|x - y\|^2 = 2\|x\|^2 + 2\|y\|^2 \qquad (\forall x, y \in X).$$

Proposition 9.3.11. *Let H be a Hilbert space and let $K \subset H$ be a (nonempty) closed, convex subset. Then, for each $x \in H$, there is a unique $z \in K$ such that*

$$\|x - z\| = d := d(x, K) = \inf\{\|x - y\| : y \in K\}.$$

In particular, the statement is true for any closed (hence Hilbert) subspace K of H.

Proof. Since $K - \{x\} := \{y - x : y \in K\}$ is also closed and convex (why?), we may assume that $x = 0$. Thus we have to show that K contains a *unique* element z of *minimal norm*, i.e.,

$$\|z\| = d := \inf\{\|y\| : y \in K\}.$$

Pick a sequence $(y_n) \in K^{\mathbb{N}}$ such that $\lim(\|y_n\|) = d$. Since K is *convex*, we have $(y_m + y_n)/2 \in K$ and hence $\|y_m + y_n\|^2 \geq 4d^2$ for all $m, n \in \mathbb{N}$. But then, using the *parallelogram law* (Exercise 9.3.10), we have

$$\|y_m - y_n\|^2 = 2\|y_m\|^2 + 2\|y_n\|^2 - \|y_m + y_n\|^2$$
$$\leq 2\|y_m\|^2 + 2\|y_n\|^2 - 4d^2 \to 0,$$

as $m, n \to \infty$. In other words, (y_n) is a *Cauchy* sequence and hence (H being *complete*) converges to a vector $z \in H$. Since K is *closed*, we have $z \in K$ and the proof is complete. $\qquad\square$

Definition 9.3.12 (Orthogonal Vectors). Let X be a pre-Hilbert space. We say that two vectors $x, y \in X$ are *orthogonal*, and we write $x \perp y$, if $\langle x, y \rangle = 0$. Given a nonempty subset $S \subset X$, we say that a vector $x \in X$ is *orthogonal to S*, and we write $x \perp S$, if $\langle x, y \rangle = 0$ for all $y \in S$. The set of all vectors orthogonal to S is denoted by $S^\perp$ (read "S perp"); thus

$$S^\perp := \{x \in X : x \perp S\} = \{x \in X : \langle x, y \rangle = 0 \; \forall y \in S\}.$$

We say that two subsets $S, T \subset X$ are *orthogonal*, and we write $S \perp T$, if $x \perp T$ for all $x \in S$.

Exercise 9.3.13. Let S be a nonempty subset of a pre-Hilbert space X. Show that $S^\perp = (S^-)^\perp = \mathrm{Span}(S)^\perp$ is a *closed subspace* of X. Deduce that $S^\perp = \{0\}$ if and only if S is *total*; i.e., $\mathrm{Span}(S)$ is *dense* in X. *Hint*: Note that $A \subset B \subset X$ implies $B^\perp \subset A^\perp$. Also, for each $x \in S$, $\{x\}^\perp$ is the *kernel* of the functional $y \mapsto \langle y, x \rangle$ and $S^\perp = \bigcap_{x \in S} \{x\}^\perp$.

The following characterization of orthogonality is a useful consequence of the arguments in Proposition 9.3.7 and is geometrically obvious (cf. Proposition 9.3.11, with K a *one-dimensional* subspace):

Proposition 9.3.14. *Let X be a pre-Hilbert space and $x, y \in X$. Then $x \perp y$ if and only if $\|y\| \leq \|\alpha x + y\|$ for all $\alpha \in \mathbb{F}$.*

Proof. We may assume that $x \neq 0$. Let $\beta := \langle x, y \rangle$. As in the proof of Proposition 9.3.7, we have

$$0 \leq \|\alpha x + y\|^2 = |\alpha|^2 \|x\|^2 + 2\mathrm{Re}(\alpha\beta) + \|y\|^2.$$

Taking $\alpha = -\overline{\beta}/\|x\|^2$, we get

$$0 \leq \|\alpha x + y\|^2 = \|y\|^2 - \frac{|\beta|^2}{\|x\|^2},$$

which shows that $\|y\| \leq \|\alpha x + y\|$ is *false* when $\beta \neq 0$. $\qquad\square$

Theorem 9.3.15 (Orthogonal Complement). *If K is a closed subspace of a Hilbert space H, then $K^\perp$ is also a closed subspace with $K \cap K^\perp = \{0\}$ and $H = K + K^\perp$. In other words, H is the orthogonal direct sum of K and $K^\perp$:*

$$H = K \oplus K^\perp.$$

The subspace $K^\perp$ is called the orthogonal complement of K.

Proof. That $K^\perp$ is a closed subspace follows from Exercise 9.3.13. If $x \in K \cap K^\perp$, then $\|x\|^2 = \langle x, x \rangle = 0$ so that $x = 0$. Next, given any $x \in H$, Proposition 9.3.11 implies that there is a *unique* $x_1 \in K$ that *minimizes* the distance $\|x - x_1\|$. Put $x_2 := x - x_1$, so that $x = x_1 + x_2$. Then $\|x_2\| \le \|x_2 + y\|$ for all $y \in K$ and hence, by Proposition 9.3.14, $x_2 \in K^\perp$ and the proof is complete.   □

Definition 9.3.16 (Orthogonal Projection). Let the notation be as in Theorem 9.3.15 and its proof. Then the map $P_K : H \to H$ defined by $P_K x = x_1$ (i.e., $P_K x$ is the point in K *closest* to x) is called the *orthogonal projection* onto the closed subspace K.

Corollary 9.3.17. *If K is a closed subspace of a Hilbert space H, then the orthogonal projection P_K is a linear operator from H onto K, satisfying $P_K^2 = P_K$. Also, $\|P_K x\| \le \|x\|$ for all $x \in H$ and, if $K \ne \{0\}$, then $\|P_K\| = 1$. Finally, we have*

$$\langle P_K x, P_K y \rangle = \langle P_K x, y \rangle = \langle x, P_K y \rangle \qquad (\forall x, y \in H). \qquad (*)$$

Proof. For any $x, y \in H$ and $\alpha \in \mathbb{F}$, note that $x_1 = P_K x$ and $y_1 = P_K y$ are the *unique* vectors in K with $x - x_1 \in K^\perp$ and $y - y_1 \in K^\perp$. It follows that $P_K x_1 = x_1$ and hence $P_K^2 = P_K$. Also, since $(\alpha x + y) - (\alpha x_1 + y_1) \in K^\perp$, the linearity of P_K follows. (Why?) Next, note that, for each $x \in H$,

$$\|P_K x\|^2 = \|x_1\|^2 \le \|x_1\|^2 + \|x_2\|^2 = \|x\|^2,$$

where $x_2 = x - x_1 \in K^\perp$. Therefore, $\|P_K\| \le 1$. If $K \ne \{0\}$ and $x_1 \in K \setminus \{0\}$, then $P_K x_1 = x_1$ and hence $\|P_K\| = 1$. Finally, to prove $(*)$, note that

$$\langle P_K x, y \rangle = \langle P_K x, P_K y + (y - P_K y) \rangle = \langle P_K x, P_K y \rangle,$$

as $y - P_K y \in K^\perp$. The other equality is similar.   □

Corollary 9.3.18. *If K is a closed subspace of a Hilbert space H, then $(K^\perp)^\perp = K$.*

Proof. By Theorem 9.3.15, we have

$$K \oplus K^\perp = H = K^\perp \oplus (K^\perp)^\perp.$$

Since for each $x \in H$ we have $x = x_1 + x_2$ with *unique* vectors $x_1 \in K$ and $x_2 \in K^\perp$, the corollary follows.   □

Exercise 9.3.19. Show that, for *any* subspace K of a Hilbert space H, we have $(K^\perp)^\perp = K^-$.

We now prove the characterization of the *dual of a Hilbert space* mentioned after the proof of Corollary 9.3.8.

Theorem 9.3.20 (Riesz Representation Theorem). *Given any Hilbert space H, the map $y \to \psi^y$, defined by $\psi^y(x) := \langle x, y \rangle$, is a conjugate linear isometry of H onto its dual H^*. In particular, for each $\psi \in H^*$, there is a unique $y \in H$ such that $\psi(x) = \langle x, y \rangle$ for all $x \in H$.*

Proof. In view of Corollary 9.3.8, we need only show that the given map is *onto*. So let $\psi \in H^*$. We must show that $\psi = \psi^y$ for some $y \in H$. If $\psi = 0$, then we take $y = 0$. If not, let $K := \mathrm{Ker}(\psi)$. It is a *closed, proper* subspace of H. By Theorem 9.3.15, there is a vector $z \in K^\perp \setminus \{0\}$. Now note that

$$\psi(x)z - \psi(z)x \in K \qquad (\forall x \in H)$$

gives $\langle \psi(x)z - \psi(z)x, z \rangle = 0$ for all $x \in H$. Thus $\psi(x) = (\psi(z)/\|z\|^2)\langle x, z \rangle$, so that $\psi = \psi^y$ with $y := \overline{\psi(z)}z/\|z\|^2$. Finally, the *uniqueness* of y follows from the fact that $\langle x, y \rangle = \langle x, y' \rangle$ for all $x \in X$ implies that $y - y' \in X^\perp = \{0\}$. $\qquad\square$

Remark 9.3.21. The *uniqueness* of y mentioned above implies that, for any $z, z' \in K^\perp \setminus \{0\}$, we have

$$\overline{\psi(z)}z/\|z\|^2 = \overline{\psi(z')}z'/\|z'\|^2.$$

This comes from the deeper fact that the kernel of a linear functional is a *hyperplane*:

Definition 9.3.22 (Hyperplane). Let $X \neq \{0\}$ be a vector space over $\mathbb{F}$. A proper subspace $Y \subset X$ is said to be a *hyperplane* if X is the (algebraic) direct sum of Y and $\mathbb{F}x_0 := \{\alpha x_0 : \alpha \in \mathbb{F}\}$ for some (and hence any) $x_0 \notin Y$; i.e., $X = Y \oplus \mathbb{F}x_0$.

Proposition 9.3.23. *Let $X \neq \{0\}$ be a normed space over $\mathbb{F}$ and let $\phi \in \mathcal{L}(X, \mathbb{F}) \setminus \{0\}$. Then $\mathrm{Ker}(\phi)$ is a hyperplane in X. Conversely, if $Y \subset X$ is a hyperplane and $x_0 \notin Y$, then there is a unique functional $\phi \in \mathcal{L}(X, \mathbb{F})$ such that $\mathrm{Ker}(\phi) = Y$ and $\phi(x_0) = 1$.*

Proof. Let $\phi \neq 0$ be a linear functional on X. If $x \in X$ and $x_0 \notin \mathrm{Ker}(\phi)$, then it is easily checked that $x = y + \alpha x_0$ with $\alpha \in \mathbb{F}$ and $y \in \mathrm{Ker}(\phi)$ if and only if $\alpha = \phi(x)/\phi(x_0)$. Therefore, $\mathrm{Ker}(\phi)$ is indeed a hyperplane. Conversely, if Y is a hyperplane and $x_0 \notin Y$, then each $x \in X$ can be written as $x = y + \alpha x_0$ with *unique* $y \in Y$ and $\alpha \in \mathbb{F}$. Define ϕ by setting $\phi(x) := \alpha$ and note that $\phi(x_0) = 1$. If $x, x' \in X$ and $\beta \in \mathbb{F}$, then, with unique $y, y' \in Y$ and $\alpha, \alpha' \in \mathbb{F}$, we have $x = y + \alpha x_0$ and $x' = y' + \alpha' x_0$. Therefore, $\beta x = \beta y + \beta \alpha x_0$ is the unique decomposition of βx and the corresponding one for $\beta x + x'$ is

$$\beta x + x' = (\beta y + y') + (\beta \alpha + \alpha')x_0.$$

Thus (by uniqueness), $\phi(\beta x + x') = \beta\phi(x) + \phi(x')$ and ϕ is linear. It is also clear that ϕ is unique and $Y = \mathrm{Ker}(\phi)$. (Why?) $\qquad\square$

Exercise 9.3.24. Deduce the Riesz Representation Theorem (Theorem 9.3.20) from Theorem 9.3.15 and Proposition 9.3.23.

Proposition 9.3.25 (Pythagorean Theorem). *If* x *and* y *are two orthogonal vectors in a pre-Hilbert space* X, *then we have*

$$\|x + y\|^2 = \|x\|^2 + \|y\|^2.$$

More generally, if $\{x_1, \ldots, x_n\} \subset X$, $x_i \perp x_j$ *for* $i \neq j$, *and* $\alpha_1, \ldots, \alpha_n \in \mathbb{F}$, *then*

$$\left\| \sum_{k=1}^{n} \alpha_k x_k \right\|^2 = \sum_{k=1}^{n} |\alpha_k|^2 \|x_k\|^2.$$

Proof. This follows at once from

$$\left\langle \sum \alpha_j x_j, \sum \alpha_k x_k \right\rangle = \sum_{j=1}^{n} \sum_{k=1}^{n} \alpha_j \overline{\alpha_k} \langle x_j, x_k \rangle. \qquad \square$$

Corollary 9.3.26. *Let* $\{e_1, \ldots, e_n\}$ *be a finite orthonormal set in a pre-Hilbert space* X; *i.e.,* $\langle e_i, e_j \rangle = \delta_{ij}$, *where* $\delta_{ij} = 0$ *if* $i \neq j$ *and* $= 1$ *if* $i = j$. *Then*

$$\|x\|^2 = \sum_{k=1}^{n} |\langle x, e_k \rangle|^2 + \left\| x - \sum_{k=1}^{n} \langle x, e_k \rangle e_k \right\|^2 \qquad (\forall x \in X). \qquad (*)$$

In particular, we have

$$\sum_{k=1}^{n} |\langle x, e_k \rangle|^2 \leq \|x\|^2 \qquad (Bessel's \ Inequality). \qquad (**)$$

Proof. A simple computation shows that $x - \sum_{k=1}^{n} \langle x, e_k \rangle e_k$ is orthogonal to each e_k and hence to $\sum_{k=1}^{n} \langle x, e_k \rangle e_k$. Thus $(*)$ follows from the Pythagorean theorem and $(**)$ is then an immediate consequence. $\qquad \square$

Definition 9.3.27 (Orthogonal & Orthonormal Systems). Let X be a pre-Hilbert space. We say that a set $S = \{x_j : j \in J\} \subset X$ is an *orthogonal system* if $x_j \neq 0$ for all $j \in J$ and $\langle x_j, x_k \rangle = 0$ for $j \neq k$. We say that S is an *orthonormal system* if it is an orthogonal system and $\|x_j\| = 1$ for all $j \in J$. It is clear that, if $\{x_j : j \in J\}$ is an orthogonal system, then $\{x_j / \|x_j\| : j \in J\}$ is an orthonormal system. A *sequence* $(x_n) \in X^{\mathbb{N}}$ is said to be orthogonal (resp., orthonormal) if its range $\{x_n : n \in \mathbb{N}\}$ is an orthogonal (resp., orthonormal) system.

Exercise 9.3.28.

(a) Let $S = \{x_j : j \in J\}$ be an *orthogonal system* in a pre-Hilbert space X. Show that the vectors in S are *linearly independent*.

(b) Show that if a pre-Hilbert space is *separable*, then every orthonormal system in X must be *countable*. *Hint:* If S is an orthonormal system in X, then $\|x-y\| = \sqrt{2}$ for all x, $y \in S$ with $x \neq y$.

Definition 9.3.29 (Complete System, Basis). Let S be an orthogonal (resp., orthonormal) system in a pre-Hilbert space X. We say that S is *complete* (or *total*) if it is *total* in the sense of Definition 9.2.26, i.e., if $X = \text{Span}(S)^{-}$. A complete orthogonal (resp., orthonormal) system is also called an *orthogonal (resp., orthonormal) basis*.

Example 9.3.30.

(1) In the Euclidean space $\mathbb{F}^n$, the *canonical* (orthonormal) basis is $\{e_1, \dots, e_n\}$, with $e_1 = (1, 0, 0, \dots, 0)$, $e_2 = (0, 1, 0 \dots, 0)$, $\dots$, $e_n = (0, 0, \dots, 0, 1)$.
(2) The pre-Hilbert space $\ell^2(\mathbb{N}, \mathbb{F})$ extends the above example. Here, the sequence (e_n), where $e_1 = (1, 0, 0, \dots)$, $e_2 = (0, 1, 0 \dots)$, $e_3 = (0, 0, 1, \dots)$, $\dots$ is obviously an *orthonormal basis*. Indeed, for any $x = (x_n)_{n=1}^{\infty} \in \ell^2(\mathbb{N}, \mathbb{F})$, let us define

$$x^k := (x_1, x_2, \dots, x_k, 0, 0, \dots) \qquad (\forall k \in \mathbb{N}).$$

Then $x^k = \sum_{j=1}^{k} x_j e_j$ and $\|x^k - x\| \to 0$ as $k \to \infty$.
(3) Let $C_2[-\pi, \pi]$ denote the space of all (complex-valued) *continuous* functions on $[-\pi, \pi]$ with the inner product

$$\langle f, g \rangle := \frac{1}{2\pi} \int_{-\pi}^{\pi} f(x)\overline{g(x)}dx$$

and the corresponding norm $\|f\|_2 := \sqrt{\langle f, f \rangle}$. Let us prove that the sequence $(e_n)_{n=-\infty}^{\infty}$, where $e_n(x) := e^{inx}$, is an *orthonormal basis*. That it is orthonormal, we already know (cf. Definition 8.6.2). Given any $f \in C_2[-\pi, \pi]$ and $N \in \mathbb{N}$, let $f_N \in C_2[-\pi, \pi]$ be such that $f_N = f$ on $[-\pi, \pi - 1/N]$, $f_N(\pi) = f(-\pi)$, and f_N is *affine* on $[\pi - 1/N, \pi]$. It is easily seen that, taking N large enough, we have $\|f - f_N\|_2 < \varepsilon$ for any prescribed $\varepsilon > 0$. Now, by *Parseval's Theorem* (Theorem 8.6.33), f_N can be approximated arbitrarily closely (in the above norm) by linear combinations of the $e_n(x)$ and hence so can f.

Proposition 9.3.31. *Any pre-Hilbert space $X \neq \{0\}$ has an orthonormal basis. In fact, any orthonormal system $S \subset X$ is contained in a complete orthonormal set.*

Proof. It suffices to prove the last statement. Let $S \subset X$ be orthonormal (e.g., $S := \{x/\|x\|\}$ for some $x \neq 0$) and let $\mathcal{S}$ denote the collection of all orthonormal sets containing S. Note that $\mathcal{S}$ is partially ordered by *inclusion*. Given any *chain* (i.e., totally ordered set) $\{S_j : j \in J\} \subset \mathcal{S}$, we let $T := \bigcup_{j \in J} S_j$. It is clear that $S \subset S_j \subset T$ for all $j \in J$ and T is orthonormal. Therefore T is an upper bound for $\{S_j : j \in J\}$. By Zorn's Lemma (Lemma 1.2.18), there exists a *maximal element* $S_0 \in \mathcal{S}$. If $Y := \text{Span}(S_0)^{-} \neq X$, then there is a vector $0 \neq y \in Y^{\perp}$ and we can obtain a bigger orthonormal system than S_0. $\square$

The above proposition is *not constructive*. For *separable* Hilbert spaces, there is a standard way of constructing an orthonormal system from any given (possibly finite) *total sequence* of *linearly independent* vectors. The existence of such a sequence is guaranteed, e.g., by Theorem 9.2.27.

Theorem 9.3.32 (Gram–Schmidt Orthogonalization). *Let (b_n) be a (possibly finite) total sequence of linearly independent vectors in a separable Hilbert space H. Then we can construct an orthonormal basis (e_n) for H having the same cardinality as the b_n.*

Proof. We note that the b_n are *nonzero* and define the sequences (c_n) and (e_n) as follows: $c_1 := b_1$, $c_2 := b_2 - \langle b_2, e_1 \rangle e_1$, and, in general, $c_n := b_n - \sum_{k=1}^{n-1} \langle b_n, e_k \rangle e_k$, where $e_k := c_k / \| c_k \|$ for all $k \in \mathbb{N}$. Observe that the c_n are all *nonzero* because the b_n are *independent*. Now the sequence (e_n) is easily seen to be orthonormal. Also, for each m, the finite sequences $(b_j)_{j=1}^m$ and $(e_j)_{j=1}^m$ span the same subspace. Therefore the linear span of the e_n is the same as the linear span of the b_n. $\square$

Definition 9.3.33 (Fourier Coefficient, Fourier Series). Let X be a pre-Hilbert space and $(e_j)_{j \in J}$ an orthonormal system. Then, given any $x \in X$ and $j \in J$, the number $\langle x, e_j \rangle$ is called the jth *Fourier coefficient* (or *coordinate*) of x (with respect to $(e_j)_{j \in J}$). The unordered series $\sum_{j \in J} \langle x, e_j \rangle e_j$ is called the *Fourier series of x*.

Exercise 9.3.34 (Best Approximation). Let $(e_j)_{j=1}^n$ be a finite orthonormal system in a pre-Hilbert space X (over $\mathbb{F}$) and let $x \in X$. Show that the *minimum* value of the number

$$\left\| x - \sum_{j=1}^n c_j e_j \right\|,$$

where $(c_j)_{j=1}^n \in \mathbb{F}^n$, is obtained when $c_j = \langle x, e_j \rangle$ for $1 \le j \le n$, and the minimum is then given by

$$\left\| x - \sum_{j=1}^n \langle x, e_j \rangle e_j \right\| = \sqrt{\| x \|^2 - \sum_{j=1}^n |\langle x, e_j \rangle|^2}.$$

Hint: Cf. Theorem 8.6.6.

Before proving our next result, let us briefly mention the notions of *summability* and *absolute summability* of *unordered series* in normed spaces. They were defined for the special case of $\mathbb{R}$ in Chap. 2 (cf. Sect. 2.4).

Definition 9.3.35 (Summable, Absolutely Summable). Let $(X, \| \cdot \|)$ be a normed space and J an arbitrary (nonempty) *index* set. We say that a function $(x_j)_{j \in J} \in X^J$ is *summable* with sum $x \in X$, and we write $x = \sum_{j \in J} x_j$, if given any $\varepsilon > 0$ there is a *finite* subset $J_\varepsilon \subset J$ such that $\| x - \sum_{j \in J'} x_j \| < \varepsilon$ for any (finite) $J' \subset J$ with $J' \supset J_\varepsilon$. We say that $(x_j)_{j \in J}$ is *absolutely summable* if the (numerical) unordered series $\sum_{j \in J} \| x_j \|$ is summable in the sense of Definition 2.4.1.

Definition 9.3.36 (Cauchy's Criterion). We say that $(x_j)_{j \in J} \in X^J$ satisfies *Cauchy's Criterion* if

$$(\forall \varepsilon > 0)(\exists J_\varepsilon \in \mathcal{F}_J)(\forall J' \in \mathcal{F}_J)(J' \cap J_\varepsilon = \emptyset \Rightarrow \|s_{J'}\| < \varepsilon),$$

where $\mathcal{F}_J$ denotes the set of all *finite* subsets of J and, for each $J' \in \mathcal{F}_J$, $s_{J'} := \sum_{j \in J'} x_j$ denotes the corresponding *partial sum*.

Exercise 9.3.37.

(1) Show that, if $(x_j)_{j \in J} \in X^J$ satisfies *Cauchy's Criterion*, then it has *bounded partial sums*; i.e., there is a constant $M > 0$ such that, with notation as in Definition 9.3.36, we have $\|s_{J'}\| \leq M$ for all $J' \in \mathcal{F}_J$. (*Hint:* Suppose not. For each $n \in \mathbb{N}$ pick $J_n \in \mathcal{F}_J$ such that $\|s_{J_n}\| > n$ and observe that, for each $J' \in \mathcal{F}_J$, $\|s_{J_n \setminus J'}\| > n - M_{J'}$, where $M_{J'} := \max\{\|s_{J''}\| : J'' \subset J'\}$.)

(2) Show that a *summable* function $(x_j)_{j \in J} \in X^J$ satisfies Cauchy's Criterion. (*Hint:* Imitate part (3) of Exercise 2.4.15.)

(3) Show that, if $(x_j)_{j \in J} \in X^J$ satisfies *Cauchy's Criterion*, then $\{j \in J : x_j \neq 0\}$ is *countable*. (*Hint:* Imitate Exercise 2.4.18.)

Theorem 9.3.38 (Cauchy's Criterion). *Let X be a Banach space and $J \neq \emptyset$ an arbitrary index set. Then a function $(x_j)_{j \in J} \in X^J$ is summable if and only if it satisfies Cauchy's Criterion.*

Proof. In view of Exercise 9.3.37, we need only show that the condition is *sufficient*. So suppose that $(x_j)_{j \in J} \in X^J$ satisfies Cauchy's Criterion. Using part (3) of the above exercise, let $\{j_1, j_2, \ldots\}$ be any *enumeration* of the set $\{j \in J : x_j \neq 0\}$ (which we may and do assume to be *infinite*) and consider the (*ordered*) series $\sum_{i=1}^{\infty} x_{j_i}$. It follows at once from Cauchy's Criterion that its partial sums $s_n := \sum_{i=1}^{n} x_{j_i}$ form a *Cauchy sequence*. (Why?) Since X is *complete*, (s_n) converges. Let $x := \sum_{i=1}^{\infty} x_{j_i} \in X$. We claim that x is the *sum* of the family $(x_j)_{j \in J}$. Let $\varepsilon > 0$ be given and let J_ε be a finite subset of J such that for any finite subset $J' \subset J$ with $J' \cap J_\varepsilon = \emptyset$, we have $\|s_{J'}\| < \varepsilon/2$. Also, pick $N \in \mathbb{N}$ such that $n \geq N$ implies $\|x - \sum_{i=1}^{n} x_{j_i}\| < \varepsilon/2$. We may (and do) assume that $\{x_{j_1}, \ldots, x_{j_N}\} \subset J_\varepsilon$ and that N is the *largest* i with $x_{j_i} \in J_\varepsilon$. If now $J'' \subset J$ is a finite subset with $J_\varepsilon \subset J''$, then

$$\left\| x - \sum_{j \in J''} x_j \right\| = \left\| x - \sum_{j_i \in J''} x_{j_i} \right\|$$

$$= \left\| \left(x - \sum_{i=1}^{N} x_{j_i} \right) - \sum_{j_i \in J'', \, i > N} x_{j_i} \right\|$$

$$\leq \frac{\varepsilon}{2} + \frac{\varepsilon}{2} = \varepsilon. \qquad \square$$

Theorem 9.3.39 (Parseval's Relation). *Let H be a Hilbert space. If $(e_j)_{j \in J}$ is an orthonormal basis, then for each $x \in H$ we have*

$$x = \sum_{j \in J} \langle x, e_j \rangle e_j, \qquad (Fourier\ series) \qquad (\dagger)$$

and

$$\|x\|^2 = \sum_{j \in J} |\langle x, e_j \rangle|^2. \qquad (Parseval's\ Relation) \qquad (\ddagger)$$

If the orthonormal system $(e_j)_{j \in J}$ *is not complete, then we still have the inequality*

$$\sum_{j \in J} |\langle x, e_j \rangle|^2 \leq \|x\|^2. \qquad (Bessel's\ Inequality)$$

Proof. Note that the series in (†) and (‡) are *unordered series* in H and $\mathbb{R}$, respectively. Bessel's inequality was proved in Corollary 9.3.26 for *finite* orthonormal sets. Thus, for any *finite* subset $J' \subset J$, we have $\sum_{j \in J'} |\langle x, e_j \rangle|^2 \leq \|x\|^2$. It follows from Proposition 2.4.8 that $\sum_{j \in J} |\langle x, e_j \rangle|^2$ is summable with sum bounded by $\|x\|^2$. This proves Bessel's inequality as stated, even if $(e_j)_{j \in J}$ is *not complete.* It also follows (cf. Corollary 2.4.17) that $\langle x, e_j \rangle \neq 0$ for at most a *countable* number of j's, say $j_1, j_2, \ldots$, and hence $\sum_{i=1}^{\infty} |\langle x, e_{j_i} \rangle|^2 < \infty$. Now let $x_n := \sum_{i=1}^{n} \langle x, e_{j_i} \rangle e_{j_i}$. Then, for $m < n$,

$$\|x_n - x_m\|^2 = \left\| \sum_{i=m+1}^{n} \langle x, e_{j_i} \rangle e_{j_i} \right\|^2 = \sum_{i=m+1}^{n} |\langle x, e_{j_i} \rangle|^2.$$

Thus, (x_n) is a *Cauchy* sequence and hence converges to a vector $x' \in H$. To prove (†), we show directly (i.e., without using Theorem 9.3.38) that $x = x'$. Since $(e_j)_{j \in J}$ is complete, it suffices to show that $(x - x') \perp e_j$, for all $j \in J$. But, for each j_m, we have

$$\langle x - x', e_{j_m} \rangle = \lim_{n \to \infty} \left\langle x - \sum_{i=1}^{n} \langle x, e_{j_i} \rangle e_{j_i}, e_{j_m} \right\rangle$$

$$= \langle x, e_{j_m} \rangle - \langle x, e_{j_m} \rangle = 0.$$

And, if $j \neq j_i$ for all i, then

$$\langle x - x', e_j \rangle = \lim_{n \to \infty} \left\langle x - \sum_{i=1}^{n} \langle x, e_{j_i} \rangle e_{j_i}, e_j \right\rangle = 0 - 0 = 0.$$

Therefore $x - x'$ is indeed orthogonal to each e_j and hence must be 0. In other words,

$$x = \lim_{n \to \infty} \sum_{i=1}^{n} \langle x, e_{j_i} \rangle e_{j_i} = \sum_{j \in J} \langle x, e_j \rangle e_j$$

and (†) holds. Moreover,

$$0 = \lim_{n\to\infty} \left\| x - \sum_{i=1}^{n} \langle x, e_{j_i} \rangle e_{j_i} \right\|^2$$

$$= \lim_{n\to\infty} \left(\|x\|^2 - \sum_{i=1}^{n} |\langle x, e_{j_i} \rangle|^2 \right)$$

$$= \|x\|^2 - \sum_{j\in J} |\langle x, e_j \rangle|^2$$

so that (‡) holds as well. □

Exercise 9.3.40 (Parseval's Identity). Let H be a Hilbert space and $(e_j)_{j\in J}$ an orthonormal basis. Show that, for every x, $y \in H$,

$$\langle x, y \rangle = \sum_{j\in J} \langle x, e_j \rangle \overline{\langle y, e_j \rangle}.$$

Hint: Use the continuity of the inner product.

Here is the converse to the above theorem:

Theorem 9.3.41 (Riesz–Fischer Theorem). *Let $(e_j)_{j\in J}$ be an orthonormal basis in a Hilbert space H. If $(\alpha_j) \in \mathbb{F}^J$ and $\sum_{j\in J} |\alpha_j|^2 < \infty$, then $\sum_{j\in J} \alpha_j e_j$ is summable; i.e., there is a vector $x \in H$ with $x = \sum_{j\in J} \alpha_j e_j$. Moreover, we have $\alpha_j = \langle x, e_j \rangle$ and $\sum_{j\in J} |\alpha_j|^2 = \|x\|^2$.*

Proof. The argument is basically the same as the one used in Theorem 9.3.39. Indeed, $\sum_{j\in J} |\alpha_j|^2 < \infty$ implies that the α_j are nonzero for at most a *countable* number of j's, say $j_1, j_2, \ldots$. Since

$$\left\| \sum_{i=m+1}^{n} \alpha_{j_i} e_{j_i} \right\|^2 = \sum_{i=m+1}^{n} |\alpha_{j_i}|^2 \to 0 \qquad \text{as} \quad m, n \to \infty,$$

the series $\sum_{i=1}^{\infty} \alpha_{j_i} e_{j_i}$ converges in H to a vector x and we have $x = \sum_{j\in J} \alpha_j e_j$. Also, using the continuity of the inner product, we have

$$\langle x, e_{j_m} \rangle = \lim_{n\to\infty} \left\langle \sum_{i=1}^{n} \alpha_{j_i} e_{j_i}, e_{j_m} \right\rangle = \alpha_{j_m}.$$

And, if $j \neq j_i$ for all i, then $\langle x, e_j \rangle = 0 = \alpha_j$. Finally, $\sum_{j\in J} |\alpha_j|^2 = \|x\|^2$ is nothing but Parseval's Relation. □

For the next result we need a couple of facts about cardinal numbers. The first one is the *Schröder–Bernstein* theorem (Theorem 1.4.21) which asserts that, if A

and B are sets with $|A| \le |B|$ and $|B| \le |A|$, then $|A| = |B|$. Here $|S|$ denotes the *cardinality* of the set S. The second fact is the assertion in Exercise 1.4.26: If A is an *infinite* set, then $|A \times \mathbb{N}| = |A|$.

Theorem 9.3.42 (Orthogonal Dimension). *Any two orthonormal bases in a Hilbert space H have the same cardinal number. This cardinal number is called the orthogonal dimension of H.*

Proof. Let $(e_i)_{i \in I}$ and $(f_j)_{j \in J}$ be two orthonormal bases in H. We must show that $|I| = |J|$. If H is *finite dimensional* with dimension n, then any orthonormal basis is just an algebraic basis and hence contains n elements. We therefore assume that I and J are *infinite*. For each $j \in J$, the set $D_j := \{i \in I : \langle e_i, f_j \rangle \ne 0\}$ is *countable*. Also, for each $i \in I$, $1 = \|e_i\|^2 = \sum_{j \in J} |\langle e_i, f_j \rangle|^2$ implies that $\langle e_i, f_j \rangle \ne 0$ for some $j \in J$. Therefore, $I = \bigcup_{j \in J} D_j$. But then,

$$|I| = \left| \bigcup_{j \in J} D_j \right| \le |J \times \mathbb{N}| = |J|.$$

By symmetry, we also have $|J| \le |I|$ and hence $|I| = |J|$.   □

Definition 9.3.43 (Isomorphic Hilbert Spaces). We say that two Hilbert spaces $(H, \langle \cdot, \cdot \rangle)$ and $(H', \langle \cdot, \cdot \rangle')$ are *isomorphic* if there is a linear bijection $T : H \to H'$ such that

$$\langle Tx, Ty \rangle' = \langle x, y \rangle \qquad (\forall x, y \in H).$$

Thus an isomorphism between Hilbert spaces is in fact an *isometric isomorphism*.

Example 9.3.44 (The Hilbert Space $\ell^2(J, \mathbb{F})$). Let J be a nonempty set and let

$$\ell^2(J, \mathbb{F}) := \left\{ \alpha := (\alpha_j)_{j \in J} \in \mathbb{F}^J : \sum_{j \in J} |\alpha_j|^2 < \infty \right\},$$

where we recall that, with $\mathcal{F}_J$ denoting the set of all *finite* subsets of J,

$$\sum_{j \in J} |\alpha_j|^2 := \sup \left\{ \sum_{j \in J'} |\alpha_j|^2 : J' \in \mathcal{F}_J \right\}.$$

Then,

$$\langle \alpha, \beta \rangle := \sum_{j \in J} \alpha_j \overline{\beta_j}, \qquad (\dagger)$$

with $\alpha = (\alpha_j)_{j \in J}$ and $\beta = (\beta_j)_{j \in J}$ in $\ell^2(J, \mathbb{F})$, defines an inner product on $\ell^2(J, \mathbb{F})$, making it a Hilbert space.

Indeed, the trivial inequality $|\alpha_j \overline{\beta_j}| \leq (|\alpha_j|^2 + |\beta_j|^2)/2$ shows that the unordered series in (†) is *absolutely summable* and hence summable. It is then easily checked that (†) satisfies the conditions of Definition 9.3.1. The corresponding norm will be denoted by $\|\cdot\|_2$. Thus, $\|\alpha\|_2 = (\sum_{j \in J} |\alpha_j|^2)^{1/2}$. Next, we define the *canonical orthonormal basis* in $\ell^2(J, \mathbb{F})$. For each $j \in J$, let $\chi^j := \chi_{\{j\}}$ denote the *characteristic function* of the singleton $\{j\}$; i.e., $\chi^j(j) = 1$ and $\chi^j(j') = 0$ for all $j' \neq j$. Then $(\chi^j)_{j \in J}$ is the desired orthonormal basis. That it is orthonormal is obvious. Also, its cardinality is clearly $|J|$. To prove its *completeness*, note first that, for each $\alpha \in \ell^2(J, \mathbb{F})$, we have $\langle \alpha, \chi^j \rangle = \alpha_j$ for all $j \in J$. Now, given any $\varepsilon > 0$, pick a finite $J_\varepsilon \subset J$ such that $\sum_{j \in J \setminus J_\varepsilon} |\alpha_j|^2 < \varepsilon$. Then,

$$\left\| \alpha - \sum_{j \in J_\varepsilon} \alpha_j \chi^j \right\|_2^2 = \left\| \alpha - \sum_{j \in J_\varepsilon} \langle \alpha, \chi^j \rangle \chi^j \right\|_2^2 = \sum_{j \in J \setminus J_\varepsilon} |\alpha_j|^2 < \varepsilon.$$

Theorem 9.3.45. *Every nonzero Hilbert space H is isomorphic to $\ell^2(J, \mathbb{F})$, for a set J such that $|J|$ is the orthogonal dimension of H. In particular, if two Hilbert spaces H and H' have the same orthogonal dimension, then they are isomorphic.*

Proof. Let $(e_j)_{j \in J}$ be an *orthonormal basis* in H. Then $|J|$ is the orthogonal dimension of H. Now let T be given by $Tx := (\langle x, e_j \rangle)_{j \in J}$. Then $Tx \in \ell^2(J, \mathbb{F})$ for each $x \in H$ (by Bessel's inequality) and it is easily checked that T is *linear*. That T is *onto* follows from the Riesz–Fischer Theorem (Theorem 9.3.41). Finally, Parseval's Relation (Theorem 9.3.39) implies that T is an *isometry*.   □

9.4   Function Spaces

We end this chapter with a brief look at some of the most basic facts about spaces of continuous functions on metric spaces. Our goal will be to prove two important theorems regarding families of functions. One is the *Arzelà–Ascoli Theorem* on *equicontinuous* families of functions and the other is the celebrated *Stone–Weierstrass Theorem* which characterizes the *dense* subalgebras of the algebra of continuous real (or complex)-valued functions on *compact* metric spaces. We recall that a subset B of a metric space (M, d) is said to be *bounded* if it has finite diameter, i.e., if $\sup\{d(x, y) : x, y \in B\} < \infty$. Also, given *any* nonempty set S and any metric space (M, d), a function $f : S \to M$ is said to be *bounded* if its *range* (i.e., $f(S)$) is bounded in M. *Throughout the section, M, M', etc. will denote metric spaces.*

Notation 9.4.1 (Bounded & Continuous Functions). Let (M, d) and (M', d') be metric spaces and $S \neq \emptyset$ an arbitrary set. The set of all *bounded* functions $f : S \to M$ will be denoted by $B(S, M)$. The set of all *continuous* functions from M to M' will be denoted by $C(M, M')$. Finally, the set of all *bounded, continuous* functions $f : M \to M'$ will be denoted by $BC(M, M')$. Thus $BC(M, M') = B(M, M') \cap C(M, M')$.

Theorem 9.4.2. *Let* (M, d) *be a metric space and* $S \neq \emptyset$ *an arbitrary set. For each pair of functions* $f,\ g \in B(S, M)$, *define*

$$d_\infty(f, g) := \sup\{d(f(x), g(x)) : x \in S\}. \tag{\dagger}$$

Then d_∞ *is a metric (called the uniform metric) on* $B(S, M)$ *and the metric space* $(B(S, M), d_\infty)$ *is complete if* M *is complete.*

Proof. That d_∞ is a metric is quite obvious. Indeed, d_∞ is *nonnegative* and *symmetric*, and $\sup\{d(f(x), g(x)) : x \in S\} = 0$ implies $f = g$. Also, the Triangle Inequality follows from the fact that

$$d(f(x), h(x)) \leq d(f(x), g(x)) + d(g(x), h(x)) \leq d_\infty(f, g) + d_\infty(g, h),$$

for all $x \in S$. If M is *complete* and if (f_n) is a *Cauchy sequence* in $(B(S, M), d_\infty)$, then $d(f_n(x), f_m(x)) \leq d_\infty(f_n, f_m)$ implies that, for each $x \in S$, $(f_n(x))$ is a Cauchy sequence in M and hence converges to some point $f(x) \in M$. We must show that $f \in B(S, M)$ and $d_\infty(f_n, f) \to 0$ as $n \to \infty$. Given any $\varepsilon > 0$, pick $N \in \mathbb{N}$ such that $m,\ n \geq N$ implies $d_\infty(f_m, f_n) < \varepsilon$ and hence $d(f_m(x), f_n(x)) < \varepsilon$ for each $x \in S$. Letting $m \to \infty$, we get $d(f(x), f_n(x)) \leq \varepsilon$ for all $x \in S$. This implies that

$$d(f(x), f(y)) \leq d(f_n(x), f_n(y)) + 2\varepsilon \qquad (\forall x,\ y \in S, \forall n \geq N),$$

and hence $f \in B(S, M)$. It also follows that $d_\infty(f, f_n) \leq \varepsilon$ for all $n \geq N$ and hence $\lim(f_n) = f$. $\qquad\Box$

Remark 9.4.3. It is obvious that a sequence of functions $f_n : S \to M$ converging with respect to the uniform metric d_∞ is *uniformly convergent.*

Corollary 9.4.4. *If* (M, d) *and* (M', d') *are metric spaces and* M' *is complete, then the space* $(BC(M, M'), d_\infty)$ *is a closed, hence complete, subspace of* $(B(M, M'), d_\infty)$. *In particular, if* M *is compact, then* $(C(M, M'), d_\infty)$ *is complete. Here, as in Theorem 9.4.2,*

$$d_\infty(f, g) := \sup\{d'(f(x), g(x)) : x \in M\} \qquad (\forall f,\ g \in B(M, M')).$$

Proof. We need only show that $BC(M, M')$ is *closed* in $B(M, M')$. Now, if (f_n) is a sequence in $BC(M, M')$ and $d_\infty(f, f_n) \to 0$, then (by Theorem 9.4.2) we have $f \in B(M, M')$. To prove the continuity of f, let $x_0 \in M$ and $\varepsilon > 0$ be given. Pick $n \in \mathbb{N}$ and $\delta > 0$ such that $d_\infty(f, f_n) < \varepsilon/3$ and $d'(f_n(x), f_n(x_0)) < \varepsilon/3$ for all $x \in B_\delta(x_0)$. For each such x, we then have

$$d'(f(x), f(x_0)) \leq d'(f(x), f_n(x_0)) + d'(f_n(x), f_n(x_0)) + d'(f_n(x_0), f(x_0))$$

$$< d_\infty(f, f_n) + \frac{\varepsilon}{3} + d_\infty(f_n, f) < \varepsilon.$$

Since x_0 and ε are arbitrary, we have $f \in C(M, M')$. Finally, if M is *compact*, then (Theorem 5.6.27) each $f \in C(M, M')$ has *compact* and hence (Theorem 5.6.7) *closed and bounded* range. Thus $C(M, M') \subset B(M, M')$ and the proof is complete. $\qquad\square$

Remark 9.4.5. Note that, in the above corollary, we have proved that the *uniform limit* of a sequence of *continuous* functions from one metric space to another is *continuous*. This extends Theorem 8.3.1 and the proof is basically the same.

Definition 9.4.6 (Equicontinuity). Let (M, d) and (M', d') be metric spaces. A set $\mathcal{F} \subset B(M, M')$ is said to be *equicontinuous at* $x_0 \in M$ if, given any $\varepsilon > 0$, there is $\delta = \delta(\varepsilon, x_0) > 0$ such that $d'(f(x_0), f(x)) < \varepsilon$ for all $f \in \mathcal{F}$ and all $x \in B_\delta(x_0) := \{x \in M : d(x, x_0) < \delta\}$. The family $\mathcal{F}$ is said to be *equicontinuous on* M if it is equicontinuous at *every* point $x \in M$.

Remark 9.4.7. Note that, if $\mathcal{F}$ is equicontinuous at x_0, then all the functions $f \in \mathcal{F}$ are *simultaneously continuous* at x_0. The converse is, of course, *not* true because if the $f \in \mathcal{F}$ are only simultaneously continuous, the δ in the above definition will depend on ε, x_0, *and* f. For $\mathcal{F}$ equicontinuous at x_0, δ is *independent of* $f \in \mathcal{F}$ (but, in general, it depends on $x_0 \in M$ and ε). If, however, $\mathcal{F}$ is equicontinuous on M and the space M is *compact*, we expect δ to be *independent of* $x \in M$ as well:

Exercise 9.4.8 (Uniform Equicontinuity). Let K and M be metric spaces with respective metrics d_K and d_M, and assume that K is *compact*. If $\mathcal{F} \subset B(K, M)$ is equicontinuous on K, show that it is in fact *uniformly equicontinuous;* i.e., given any $\varepsilon > 0$, there exists $\delta = \delta(\varepsilon) > 0$ (*independent of* $x \in K$ *and* $f \in \mathcal{F}$) such that

$$d_K(x, y) < \delta \Rightarrow d_M(f(x), f(y)) < \varepsilon \quad (\forall x, y \in K, \forall f \in \mathcal{F}).$$

Hint: Read the second proof of Theorem 5.6.35.

Example 9.4.9. As above, let (M, d) and (M', d') be metric spaces and $C(M, M')$ the set of continuous functions from M to M'.

(1) If $\mathcal{F} \subset C(M, M')$ is *finite*, then it is equicontinuous on M. (Why?)
(2) Let $\alpha \in (0, 1]$ and consider the set $Lip^\alpha(M, M')$ of all *Lipschitz functions of order* α from M to M'. Recall that $f \in Lip^\alpha(M, M')$ means that there is a constant $C > 0$ with

$$d'(f(x), f(y)) \leq C(d(x, y))^\alpha \quad (\forall x, y \in M).$$

Then $Lip^\alpha(M, M') \subset C(M, M')$ and *any* subset of $Lip^\alpha(M, M')$ is *uniformly equicontinuous* on M. (Why?)

Before stating the main theorem, we prove some lemmas. In the first one, we use G. Cantor's *diagonal method:*

Lemma 9.4.10. *Let D be a countable subset of a metric space M and let (f_n) be a sequence of functions from D to a metric space M'. If for each $x \in D$ the set*

$\{f_n(x) : n \in \mathbb{N}\}$ *is relatively compact (i.e., has compact closure), then there is a subsequence* (f_{n_k}) *such that* $(f_{n_k}(x))$ *converges for each* $x \in D$.

Proof. Let $D = \{x_k : k \in \mathbb{N}\}$. Since $\{f_n(x_1) : n \in \mathbb{N}\}$ is relatively compact, we can pick a subsequence (f_{1n}) such that $(f_{1n}(x_1))$ converges. Using a similar argument, we can then pick a subsequence (f_{2n}) of the sequence (f_{1n}) such that $(f_{2n}(x_2))$ converges. Continuing this process, for each $j \in \mathbb{N}$, we have a subsequence (f_{jn}) such that $(f_{jn}(x_j))$ is convergent. We now consider the *"diagonal"* sequence $(f_{nn})_{n=1}^{\infty}$. Since $(f_{nn})_{n=j}^{\infty}$ is a subsequence of (f_{jn}) for each $j \in \mathbb{N}$, the sequence $(f_{nn}(x_k))$ is convergent for all $k \in \mathbb{N}$. $\square$

Lemma 9.4.11. *Let* (f_n) *be an equicontinuous sequence of functions from a metric space* (M, d) *to a complete metric space* (M', d') *and let* $D \subset M$ *be a dense subset. If* $(f_n(x))$ *converges for each* $x \in D$, *then it converges for each* $x \in M$ *and the limit function* $f(x) := \lim(f_n(x))$ *is continuous (on* M).

Proof. Given any $x \in M$ and $\varepsilon > 0$, pick $\delta > 0$ so that $y \in B_\delta(x)$ implies $d'(f_n(x), f_n(y)) < \varepsilon/3 \; \forall n \in \mathbb{N}$. Since D is *dense*, we can pick a point $y \in D \cap B_\delta(x)$ and since (by our assumption) $(f_n(y))$ converges, we can pick $N \in \mathbb{N}$ such that $d'(f_n(y), f_m(y)) < \varepsilon/3$ for all $m, n \geq N$. Therefore,

$$d'(f_n(x), f_m(x)) \leq d'(f_n(x), f_n(y)) + d'(f_n(y), f_m(y)) + d'(f_m(y), f_m(x))$$

$$< \varepsilon \quad (\forall m, n \geq N).$$

In other words, $(f_n(x))$ is a *Cauchy sequence* in the *complete* space M' and hence is convergent. Define $f(x) := \lim(f_n(x))$ for each $x \in M$. To show that f is *continuous* at x, let $\varepsilon > 0$ be given. Using the *equicontinuity*, we can find $\delta > 0$ with $d'(f_n(x), f_n(y)) < \varepsilon$ for all $n \in \mathbb{N}$ and all $y \in B_\delta(x)$ and hence

$$d'(f(x), f(y)) = \lim(d'(f_n(x), f_n(y)) \leq \varepsilon \quad (\forall y \in B_\delta(x)),$$

which proves the continuity of f at x. $\square$

Lemma 9.4.12. *Let* (K, d_K) *be a compact metric space and* (f_n) *an equicontinuous sequence of functions from* K *to a metric space* (M, d_M). *If* (f_n) *converges pointwise to* f *(i.e.,* $\lim(f_n(x)) = f(x)$ *for all* $x \in K$*), then the convergence is in fact uniform (on* K).

Proof. First note that, by Exercise 9.4.8, (f_n) is actually *uniformly equicontinuous* on K. In other words, given any $\varepsilon > 0$, there is a $\delta = \delta(\varepsilon) > 0$ such that

$$d_K(x, y) < \delta \implies d_M(f_n(x), f_n(y)) < \varepsilon/3 \quad (\forall n \in \mathbb{N}). \qquad (*)$$

Letting $n \to \infty$ in $(*)$, we also have $d_M(f(x), f(y)) \leq \varepsilon/3$ whenever $d_K(x, y) < \delta$. Now pick $\{x_1, \ldots, x_m\} \subset K$ such that $K = \bigcup_{j=1}^{m} B_\delta(x_j)$. Also, for each $j \in \{1, \ldots, m\}$, pick $N_j \in \mathbb{N}$ such that $n \geq N_j$ implies $d_M(f(x_j), f_n(x_j)) < \varepsilon/3$. If $N := \max\{N_j : 1 \leq j \leq m\}$, then $\forall \, x \in K$ we have $x \in B_\delta(x_j)$ for some j, so that

$$d_M(f_n(x), f(x)) \le d_M(f_n(x), f_n(x_j)) + d_M(f_n(x_j), f(x_j))$$
$$+ d_M(f(x_j), f(x)) < \varepsilon \qquad (\forall n \ge N)$$

and the proof is complete. □

We now put together the above lemmas to prove the main theorem:

Theorem 9.4.13 (Arzelà–Ascoli). *Let M and M' be metric spaces. Assume that M is separable and M' is complete. Let $\mathcal{F} \subset B(M, M')$ be equicontinuous. If (f_n) is a sequence in $\mathcal{F}$ such that, for each $x \in M$, the set $\{f_n(x) : n \in \mathbb{N}\}$ is relatively compact (i.e., has compact closure), then there is a subsequence (f_{n_k}) that converges pointwise to a continuous function f. Moreover, this convergence is uniform on compact subsets of M.*

Proof. Indeed, let D be a *countable dense* subset of M. Then (f_n) satisfies the conditions of Lemma 9.4.10 and hence we have a subsequence (f_{n_k}) converging at each $x \in D$. It then follows from Lemma 9.4.11 that (f_{n_k}) converges pointwise (on M) to a *continuous* function f. Finally, the uniform convergence on compact subsets follows from Lemma 9.4.12. □

Recall (Theorem 9.4.2) that, if M and M' are metric spaces and M' is *complete*, then $(B(M, M'), d_\infty)$ and its closed subspace $(C(M, M'), d_\infty)$, where d_∞ is the *uniform metric*, are also complete. Since (by Theorem 5.6.25) for metric spaces *sequential compactness* is equivalent to *compactness*, the Arzelà–Ascoli Theorem characterizes *relatively compact* subsets of $(C(M, M'), d_\infty)$ with separable M and complete M'. We now prove some important special cases (of independent interest) as corollaries.

Corollary 9.4.14. *Let K and M be metric spaces, where K is compact and M is complete. Then $\mathcal{F} \subset C(K, M)$ is relatively compact if and only if (1) it is equicontinuous and (2) for each $x \in K$ the set $\{f(x) : f \in \mathcal{F}\}$ is relatively compact in M.*

Proof. If $\mathcal{F}$ is equicontinuous and each $\{f(x) : f \in \mathcal{F}\}$ is relatively compact, then the relative compactness of $\mathcal{F}$ follows from the theorem. Conversely, suppose that $\mathcal{F}$ is relatively compact. Given $\varepsilon > 0$, we can then pick a finite subset $\{f_1, \ldots, f_m\} \subset \mathcal{F}$ such that for each $f \in \mathcal{F}$ there exists $j \in \{1, \ldots, m\}$ with $d_\infty(f, f_j) < \varepsilon/3$. In particular, with d denoting the metric in M, for each $x \in K$ we have $d(f(x), f_j(x)) < \varepsilon/3$, $1 \le j \le m$. Therefore, the closure of the set $\{f(x) : f \in \mathcal{F}\}$ is *totally bounded* and *complete* (because M is) and hence is *compact* by Theorem 5.6.26. Next, given any $x \in K$, we can pick $\delta_j > 0$ such that $d(f_j(x), f_j(y)) < \varepsilon/3$ for all $y \in B_{\delta_j}(x)$. If $\delta = \min\{\delta_1, \ldots, \delta_m\}$, then for any $f \in \mathcal{F}$ we have $d_\infty(f, f_j) < \varepsilon/3$ for some j and hence

$$d(f(x), f(y)) \le d(f(x), f_j(x)) + d(f_j(x), f_j(y)) + d(f_j(y), f(y)) < \varepsilon,$$

whenever $y \in B_\delta(x)$ and the equicontinuity of $\mathcal{F}$ is also established. □

Corollary 9.4.15. *Let K and M be compact metric spaces. Then $\mathcal{F} \subset C(K, M)$ is relatively compact if and only if it is equicontinuous. In this case, any sequence (f_n) in $\mathcal{F}$ has a uniformly convergent subsequence.*

Proof. The if and only if statement follows at once from the above corollary because, M being *compact,* the condition (2); i.e., the relative compactness of the set $\{f(x) : f \in \mathcal{F}\}$ for each $x \in K$ is automatically satisfied. The last statement is obvious. $\square$

We next look at the special case of the above theorem where $M = \mathbb{F}^n$ for some $n \in \mathbb{N}$. Note that, in this case, $C(K, \mathbb{F}^n)$ is a *Banach space* with the sup-norm $\|f\|_\infty := \sup\{|f(x)| : x \in K\}$. Also, recall (Corollary 5.6.38) that a subset of the Euclidean space $\mathbb{F}^n$ is *compact* if and only if it is *closed and bounded.* In particular, any bounded subset $B \subset \mathbb{F}^n$ is *relatively compact.* Before stating the corollary, we invite the reader to try the following.

Exercise 9.4.16 (Pointwise & Uniform Boundedness). Let K be a *compact* metric space and $\mathcal{F} \subset C(K, \mathbb{F}^n)$ an *equicontinuous* family of functions. Show that if $\mathcal{F}$ is *pointwise bounded;* i.e., for each $x \in K$ there exists $M_x > 0$ such that $|f(x)| \le M_x$ for all $f \in \mathcal{F}$, then it is *uniformly bounded;* i.e., there exists a constant $M > 0$ (independent of $x \in K$ and $f \in \mathcal{F}$) such that $|f(x)| \le M$ for all $x \in K$ and all $f \in \mathcal{F}$.

Corollary 9.4.17. *Let K be a compact metric space and $n \in \mathbb{N}$. A set $\mathcal{F} \subset C(K, \mathbb{F}^n)$ is relatively compact if and only if it is pointwise bounded and equicontinuous. In particular, if $\mathcal{F} \subset C(K, \mathbb{F})$ is pointwise bounded and equicontinuous, then every sequence (f_n) in $\mathcal{F}$ has a uniformly convergent subsequence.*

Proof. Indeed, if $\mathcal{F}$ is pointwise bounded and equicontinuous, then (Exercise 9.4.16) it is *uniformly bounded:* $\exists M > 0$ with $|f(x)| \le M$ for all $f \in \mathcal{F}$ and $x \in K$. But then, $\mathcal{F}$ can be regarded as a subset of $C(K, B_M^-(0))$, with the closed ball $B_M^-(0) = \{\alpha \in \mathbb{F}^n : |\alpha| \le M\}$ and Corollary 9.4.15 may be applied. Conversely, if $\mathcal{F}$ is relatively compact, then it is equicontinuous by Corollary 9.4.15. In addition, if the f_j are as in the proof of that corollary and if $M := \max\{\|f_j\|_\infty : 1 \le j \le m\}$, then it is obvious that $\mathcal{F}$ is *uniformly bounded* by $M + \varepsilon/3$. $\square$

To apply the above results, let us now show, as was promised after Example 9.2.10, that the operator $K : f \mapsto Kf$ defined in that example is *compact,* i.e., maps bounded sets onto relatively compact sets:

Proposition 9.4.18 (Compactness of Integral Operators). *Consider the square $R := [a, b] \times [a, b]$, where $[a, b]$ is a nondegenerate interval and let $k \in C(R, \mathbb{R})$. Then the integral operator*

$$(Kf)(x) := \int_a^b k(x, y) f(y)\, dy, \quad f \in C([a, b])$$

is a compact operator of $X := (C([a, b]), \|\cdot\|_\infty)$, i.e., $K \in \mathcal{K}(X)$, and we have $\|K\| \le M(b - a)$, where $M := \sup\{|k(x, y)| : (x, y) \in R\}$.

Proof. That $K \in \mathcal{B}(X)$ was already proved in Example 9.2.10 and, as in that example, we have (by the MVT for integrals)

$$|(Kf)(x)| = \left| \int_a^b k(x, y) f(y)\, dy \right| \le M(b - a) \|f\|_\infty \quad (\forall x \in [a, b], \forall\, f \in X),$$

from which $\|K\| \le M(b - a)$ follows at once. So we need only show that K is a *compact* operator; i.e., if $\mathcal{B} \subset X$ is *bounded,* then $K(\mathcal{B})$ is *relatively compact.* Let $C > 0$ be such that $\|f\|_\infty \le C$ for all $f \in \mathcal{B}$. As pointed out in Example 9.2.10, k is *uniformly continuous* on the (*compact*) square $R := [a, b] \times [a, b]$ so for each $\varepsilon > 0$ there is a $\delta > 0$ such that $|h| < \delta$ implies $|k(x + h, y) - k(x, y)| < \varepsilon$ for all $(x, y) \in R$ with $x + h \in [a, b]$, and hence (by the MVT for integrals)

$$|(Kf)(x + h) - (Kf)(x)| < \varepsilon(b - a) \|f\|_\infty \le \varepsilon(b - a)C \qquad \forall\, f \in \mathcal{B}.$$

In other words, $K(\mathcal{B})$ is *equicontinuous* at each $x \in [a, b]$. Moreover, for each $x \in [a, b]$, we have

$$|(Kf)(x)| = \left| \int_a^b k(x, y) f(y)\, dy \right| \le M(b - a)C \qquad \forall\, f \in \mathcal{B}$$

so that $K(\mathcal{B})$ is also *pointwise bounded.* Therefore, the relative compactness of $K(\mathcal{B})$ follows from Corollary 9.4.17. $\square$

Our final goal in this chapter is to prove a fundamental result in approximation theory known as the *Stone–Weierstrass Theorem.* We have already seen Weierstrass's seminal contributions to the theory: He proved (cf. Corollary 4.7.10) that continuous functions on compact intervals can be *uniformly* approximated by *polynomials.* He also proved (Corollary 8.6.25) that continuous 2π-periodic functions can be *uniformly* approximated by *trigonometric polynomials.* Our objective is to give Stone's far-reaching generalization of these theorems.

Recall that, given a compact metric space K, the set $C(K, \mathbb{F})$ of all $\mathbb{F}$-valued continuous functions on K with the *sup-norm* is a *Banach algebra.* Stone's generalization is achieved by a careful study and ultimate characterization of the *dense subalgebras* of $C(K, \mathbb{F})$. We shall first look at the case $\mathbb{F} = \mathbb{R}$. Henceforth, K *will always denote a compact metric space.*

Definition 9.4.19 (Lattice). We say that a subset $\mathcal{L} \subset C(K, \mathbb{R})$ is a *lattice* if, for each f, $g \in \mathcal{L}$, we also have $f \vee g \in \mathcal{L}$ and $f \wedge g \in \mathcal{L}$, where

$$(f \vee g)(x) := \max\{f(x), g(x)\} \quad \text{and} \quad (f \wedge g)(x) := \min\{f(x), g(x)\}.$$

Definition 9.4.20 (Separating Points). A subset $\mathcal{F} \subset C(K, \mathbb{R})$ is said to *separate points* if, given any *distinct* points x, $y \in K$, there is an $f \in \mathcal{F}$ such that $f(x) \ne f(y)$.

Lemma 9.4.21. *Let $\mathcal{A}$ be a closed subalgebra of $C(K, \mathbb{R})$ such that $1 \in \mathcal{A}$. Then, $\mathcal{A}$ is a lattice.*

Proof. If suffices to show that $|f| \in \mathcal{A}$ for each $f \in \mathcal{A}$ and use

$$f \vee g = \frac{|f - g|}{2} + \frac{f + g}{2} \quad \text{and} \quad f \wedge g = -[(-f) \vee (-g)].$$

We may and do assume that $\|f\|_\infty = 1$. By the Weierstrass Approximation Theorem (Corollary 4.7.10), for each $n \in \mathbb{N}$ we can pick a polynomial $p_n(x)$ such that $\big|p_n(x) - |x|\big| < 1/n$ for all $x \in [0, 1]$. Since $\mathcal{A}$ is an $\mathbb{R}$-*algebra*, $1 \in \mathcal{A}$ and $f \in \mathcal{A}$ imply that $P(f) \in \mathcal{A}$ for any *polynomial* $P(x)$ (with real coefficients). Also, $\|f\|_\infty \le 1$ implies that $\big\|p_n(f) - |f|\big\|_\infty < 1/n$; i.e., $|f| = \lim_{n \to \infty} p_n(f)$. Since $\mathcal{A}$ is *closed*, it follows that $|f| \in \mathcal{A}$ and the proof is complete.   □

Lemma 9.4.22. *If $\mathcal{A} \subset C(K, \mathbb{R})$ is a subalgebra that separates points and $1 \in \mathcal{A}$, then for each pair x, y of distinct points of K and each pair of numbers α, $\beta \in \mathbb{R}$, there exists a function $g \in \mathcal{A}$ such that $g(x) = \alpha$ and $g(y) = \beta$.*

Proof. Since $\mathcal{A}$ separates points, we can pick $h \in \mathcal{A}$ such that $h(x) \ne h(y)$. Now define

$$g(t) := \alpha + (\beta - \alpha)\frac{h(t) - h(x)}{h(y) - h(x)}.$$

It is then clear that $g \in \mathcal{A}$ and satisfies the requirements.   □

Theorem 9.4.23 (Stone–Weierstrass Theorem). *Let $\mathcal{A}$ be a subalgebra of the Banach algebra $C(K, \mathbb{R})$. If $\mathcal{A}$ separates points and $1 \in \mathcal{A}$, then $\mathcal{A}$ is dense in $C(K, \mathbb{R})$.*

Proof. Since the *closure* $\mathcal{A}^-$ also separates points and $1 \in \mathcal{A}^-$, we may (and do) assume that $\mathcal{A}$ is *closed* and must then show that $\mathcal{A} = C(K, \mathbb{R})$. Let $f \in C(K, \mathbb{R})$ and $\varepsilon > 0$ be given. We must prove that $\|f - h\|_\infty < \varepsilon$ for some $h \in \mathcal{A}$. For each $g \in \mathcal{A}$, define

$$U(g) =: \{x \in K : g(x) < f(x) + \varepsilon\},$$

$$L(g) =: \{x \in K : g(x) > f(x) - \varepsilon\}.$$

Since f and g are *continuous*, both $U(g)$ and $L(g)$ are *open*. Now, given any $t \in K$, Lemma 9.4.22 implies that the sets $U(g)$ with $g \in \mathcal{A}$ and $g(t) = f(t)$ form an *open cover* of the *compact* metric space K. (Why?) Therefore, we can pick $g_1, \ldots, g_n \in \mathcal{A}$ such that $\{U(g_1), \ldots, U(g_n)\}$ covers K. Define

$$h_t := g_1 \wedge g_2 \wedge \cdots \wedge g_n.$$

By Lemma 9.4.21, we have $h_t \in \mathcal{A}$. Also, $h_t(x) < f(x) + \varepsilon$ for each $x \in K$ and $h_t(t) = f(t)$, so that $t \in L(h_t)$. Therefore, $\{L(h_t) : t \in K\}$ is an open cover of K. Let $\{L(h_{t_1}), \ldots, L(h_{t_m})\}$ be a finite subcover. Then (by Lemma 9.4.21)

$$h = h_{t_1} \vee h_{t_2} \vee \cdots \vee h_{t_m} \in \mathcal{A}$$

and we have

$$f(x) - \varepsilon < h(x) < f(x) + \varepsilon \qquad (\forall x \in K);$$

in other words, $\|f - h\|_\infty < \varepsilon$ and the proof is complete. □

Remark 9.4.24.

(1) In the proof of Lemma 9.4.21, we used the Weierstrass Approximation Theorem (Corollary 4.7.10) to approximate $|x|$ *uniformly* by polynomials on $[-1, 1]$. This, however, can be proved directly as in Problem 8.7.#41 (d).

(2) Let $Pol([a,b], \mathbb{R})$ and $C([a,b], \mathbb{R})$ denote the algebras of polynomial and continuous functions from $[a, b]$ to $\mathbb{R}$, respectively. Clearly, $Pol([a,b], \mathbb{R})$ is a subalgebra of $C([a,b], \mathbb{R})$, contains the constant functions, and separates points. It follows that the classical Weierstrass Approximation Theorem (Corollary 4.7.10) is a special case of the Stone–Weierstrass Theorem. The same holds for the algebra of continuous 2π-periodic functions and the subalgebra of *trigonometric* polynomials (Corollary 8.6.25).

(3) Since polynomial functions (on $[a, b]$) with *rational* coefficients form a *countable dense* subset of $Pol([a,b], \mathbb{R})$, it follows that the Banach algebra $C([a, b], \mathbb{R})$ is *separable*. In fact, this is true in general for any $C(K, \mathbb{R})$:

Corollary 9.4.25. *For any compact metric space (K, d_K), the Banach algebra $C(K, \mathbb{R})$ is separable.*

Proof. Let $D = \{x_1, x_2, \ldots\}$ be a *countable dense* subset of K and consider the continuous functions

$$f_{n,k}(t) := d_K\left(t, K \setminus B_{1/k}(x_n)\right) \qquad (\forall k, n \in \mathbb{N}).$$

The set $\mathcal{D}$ of all functions of the form

$$f_{n_1,k_1}^{j_1} \cdots f_{n_i,k_i}^{j_i}, \tag{†}$$

where $j_1, j_2, \ldots, j_i$ are nonnegative integers, is *countable*. Let $\mathcal{A}$ denote the subspace of $C(K, \mathbb{R})$ generated by the functions (†) (i.e., the set of all *finite linear combinations* of these functions with real coefficients). Then (Theorem 9.2.27) $\mathcal{A}$ is a *separable* subalgebra of $C(K, \mathbb{R})$ and $1 \in \mathcal{A}$. We claim that $\mathcal{A}^- = C(K, \mathbb{R})$. In view of the Stone–Weierstrass Theorem, we need only show that $\mathcal{A}$ *separates points*. Now, given any distinct points $x, y \in K$, we can pick integers n, k such that $x \in B_{1/k}(x_n)$ and $y \in K \setminus B_{1/k}(x_n)$. Since $K \setminus B_{1/k}(x_n)$ is *closed*, we have $f_{n,k}(x) > 0$ (why?) while $f_{n,k}(y) = 0$. □

Finally, let us prove the *Complex Stone–Weierstrass Theorem*. Here, the situation is not quite the same and the assumptions of the real case will not be enough.

Theorem 9.4.26 ((Complex)Stone–WeierstrassTheorem). *If $\mathcal{A}$ is a subalgebra of $C(K, \mathbb{C})$ such that $\mathcal{A}$ separates points, $1 \in \mathcal{A}$, and $\bar{f} \in \mathcal{A}$ for each $f \in \mathcal{A}$, then $\mathcal{A}$ is dense in $C(K, \mathbb{C})$.*

Proof. We first note that, for any $f \in \mathcal{A}$, the real-valued functions $\mathrm{Re}(f) = (f + \bar{f})/2$ and $\mathrm{Im}(f) = (f - \bar{f})/2i$ are both in $\mathcal{A}$. Let $\mathcal{A}_0$ denote the (*real*) subalgebra of $\mathcal{A}$ consisting of all *real-valued* functions, then it follows at once that $\mathcal{A}_0$ separates points and contains constant functions. By the (Real) Stone–Weierstrass Theorem, $\mathcal{A}_0$ is *dense* in $C(K, \mathbb{R})$. Since $\mathcal{A} = \mathcal{A}_0 + i\mathcal{A}_0$, its density in $C(K, \mathbb{C}) = C(K, \mathbb{R}) + iC(K, \mathbb{R})$ follows. $\square$

Remark 9.4.27. It should be noted that the complex conjugation condition imposed above is crucial. In fact, the theorem is *false* without it.

9.5 Problems

1. Let $\| \cdot \|$ be a norm on a vector space $X \neq \{0\}$. Define $\tilde{d} : X \times X \to \mathbb{R}$ by $\tilde{d}(x, y) := \|x - y\| + 1$ if $x \neq y$ and $\tilde{d}(x, x) := 0$, for all $x, y \in X$. Show that $\tilde{d}$ is a metric *not* associated with any norm on X.

2 (ℓ^p-Spaces). For each $p \geq 1$, consider the following subset of $\mathbb{F}^{\mathbb{N}}$:

$$\ell^p := \ell^p(\mathbb{N}, \mathbb{F}) := \left\{ x = (x_n) \in \mathbb{F}^{\mathbb{N}} : \sum_{n=1}^{\infty} |x_n|^p < \infty \right\}.$$

For each $(x_n) \in \ell^p$, $p \geq 1$, define its ℓ^p-*norm* by

$$\|(x_n)\|_p := \left(\sum_{n=1}^{\infty} |x_n|^p \right)^{1/p}.$$

For $p > 1$, define its *conjugate* to be the number $q > 1$ with $1/p + 1/q = 1$ and for $p = 1$, define $q := \infty$.

(a) Prove the *Hölder* and *Minkowski* inequalities for any $p \in [1, \infty]$ and any (x_n), $(y_n) \in \mathbb{F}^{\mathbb{N}}$:

$$\|(x_n y_n)\|_1 \leq \|(x_n)\|_p \|(y_n)\|_q, \tag{Hölder}$$

$$\|(x_n) + (y_n)\|_p \leq \|(x_n)\|_p + \|(y_n)\|_p. \tag{Minkowski}$$

Hint: Follow Exercise 6.7.14, treating the case $p = 1$ separately.

(b) Deduce that, for each $p \geq 1$, $\ell^p(\mathbb{N}, \mathbb{F})$ is a *normed space* with the ℓ^p-norm $\| \cdot \|_p$.

3. Let (B_n) be a *decreasing* sequence of *balls* in a normed space X; i.e., $B_n \supset B_{n+1}$ for all $n \in \mathbb{N}$. Show that the *centers* of the B_n form a *Cauchy* sequence. Show by an example that this may fail if X is only a *metric* space.

4. (a) Let $(X, \| \cdot \|)$ be a *seminormed* space. Show that $\| \cdot \|$ is a *norm* if and only if the open unit ball $B_1(0)$ (and hence any open ball $B_r(0)$, $r > 0$) does *not* contain a one-dimensional subspace.

(b) Show that, in a *normed* space $(X, \| \cdot \|)$ over $\mathbb{F}$, we have

$$\|x\| = \inf\{1/|\alpha| : \alpha \in \mathbb{F} \setminus \{0\}, \ \|\alpha x\| \le 1\} \qquad (\forall x \in X).$$

5. Recall that a subset C of a normed space X is *convex* if $tx + (1-t)y \in C$ for all x, $y \in C$ and $t \in [0, 1]$. Prove the following statements.

(a) $C \subset X$ is *convex* if and only if

$$sC + tC = (s + t)C \qquad (\forall s, \ t > 0).$$

(b) If $C \subset X$ is *convex*, then so are its closure C^- and its interior C°.

(c) If $(C_j)_{j \in J}$ is a collection of *convex* subsets of X, then so is $\bigcap_{j \in J} C_j$.

(d) If A, $B \subset X$ are *convex*, then so is $\alpha A + \beta B$, for any α, $\beta \in \mathbb{F}$.

(e) Let C be a (proper) *closed, convex* subset of X such that $C \cap B_r(x_0) = \emptyset$ for some $x_0 \notin C$ and $r > 0$. Show that $C + B_r(0)$ is *open* and convex, and that $x_0 \notin C + B_r(0)$.

6 (Convex Hull, Closed Convex Hull). Let X be a normed space. Given any $A \subset X$, we define the *convex hull* of A to be the set

$$\mathrm{co}(A) := \Big\{ \sum_{j=1}^{n} t_j a_j : a_j \in A, t_j \ge 0, \sum_{j=1}^{n} t_j = 1, n \in \mathbb{N} \Big\}.$$

(a) Show that $\mathrm{co}(A)$ is the *intersection* of all *convex* subsets of X *containing* A. Deduce that, if A is *bounded*, then so is $\mathrm{co}(A)$.

(b) Define the *closed convex hull* of A, denoted by $\overline{\mathrm{co}}(A)$, to be the *intersection* of all *closed, convex* subsets of X that *contain* A. Show that

$$\overline{\mathrm{co}}(A) = (\mathrm{co}(A))^-.$$

7. Let $(\| \cdot \|_k)_{k=1}^{n}$ be a (finite) sequence of *seminorms* on a vector space X and $(\alpha_k)_{k=1}^{n}$ a (finite) sequence of *nonnegative* numbers. Show that $\sum_{k=1}^{n} \alpha_k \| \cdot \|_k$ is also a seminorm on X.

8.

(a) Let $X := C^1([0, 1])$ and, for each $f \in X$, define $\| f \|_\infty = \sup\{|f(x)| : x \in [0, 1]\}$ and $\| f \|'_\infty := \sup\{|f'(x)| : x \in [0, 1]\}$. Show that $\| \cdot \|'_\infty$ is a *seminorm* on X which is *not* a norm but that $\| \cdot \|_{(1)} := \| \cdot \|_\infty + \| \cdot \|'_\infty$ is a *norm* on X.

(b) Define the sequence (f_n) in $C^1([0, 1])$ by $f_n(x) := n^{-1/2} \sin(nx)$. Show that $\lim(f_n) = 0$ with respect to $\| \cdot \|_\infty$, but that $\lim(f_n) \ne 0$ with respect to $\| \cdot \|_{(1)}$. Deduce that these norms are *not* equivalent.

(c) Show that, if $\| \cdot \|_0$ is a *norm* and $\| \cdot \|_j$, for $1 \le j \le n$, are *seminorms* on a vector space X, then $\sum_{k=0}^{n} \| \cdot \|_k$ is a *norm* on X.

9. As in Problem 8, consider the space $X := C^1([a, b])$ with norms $\| f \|_\infty := \sup\{|f(x)| : x \in [a, b]\}$ and $\| f \|_{(1)} := \| f \|_\infty + \| f \|'_\infty$, where $\| f \|'_\infty := \| f' \|_\infty$. Show that X is *not* a Banach space with norm $\| \cdot \|_\infty$, but *is* a Banach space with norm $\| \cdot \|_{(1)}$.

10. For each $f \in C([0, 1])$, define

(i) $\|f\|_1 := \displaystyle\int_0^1 |f(x)| dx,$

(ii) $\|f\|_2 := \left(\int_0^1 |f(x)|^2 dx \right)^{1/2}$.

(a) Show that $\| \cdot \|_1$ and $\| \cdot \|_2$ are both norms on $C([0,1])$ and that they are *not* equivalent. Also show that *neither* of these two norms is equivalent to the *sup-norm* $\| \cdot \|_\infty$.
(b) Show that $\left(C([0,1]), \| \cdot \|_1 \right)$ and $\left(C([0,1]), \| \cdot \|_2 \right)$ are *not* Banach spaces.

11.

(a) Show that the set $BV([a,b])$ of all functions of *bounded variation* on $[a,b]$ is a *seminormed* space with *seminorm* $\|f\| := V_a^b(f)$ (cf. Definition 7.6.1). Is this a norm?
(b) Show that $BV_0([a,b]) := \{ f \in BV([a,b]) : f(a) = 0 \}$ is a *Banach* space with *norm* $\|f\| := V_a^b(f)$.

12. Show that $\ell^p(\mathbb{N}, \mathbb{F})$ [cf. Problem 2] is a Banach space for all $p \geq 1$. *Hint:* Let (x_k) be a *Cauchy* sequence in ℓ^p, where $x_k := (x_{kn})_{n=1}^\infty$. Fixing n, show that $(x_{kn})_{k=1}^\infty$ is Cauchy in $\mathbb{F}$ and let $\xi_n \in \mathbb{F}$ be its limit. Let $\xi := (\xi_n)_{n=1}^\infty$, and note that $x_k - \xi \in \ell^p$, for k large enough, and hence $\xi \in \ell^p$. Finally, show that $(x_k) \to \xi$ in ℓ^p.

13 (Schauder Basis). A sequence (e_n) in a normed space X (over $\mathbb{F}$) is said to be a *Schauder basis* for X if, given any $x \in X$, there is a *unique* sequence $(\xi_n) \in \mathbb{F}^{\mathbb{N}}$ such that $x = \sum_{n=1}^\infty \xi_n e_n$; i.e.,

$$\lim_{n \to \infty} \|x - (\xi_1 e_1 + \cdots + \xi_n e_n)\| = 0.$$

(a) Show that, if X is a normed space with a Schauder basis, then X is *separable*. It is a known (though highly nontrivial) fact that the converse of this statement is *false*.
(b) Find Schauder bases for ℓ^p, $p \geq 1$ and $c_0 := \{(x_n) \in \ell^\infty : \lim(x_n) = 0\}$. *Hint:* Consider the sequences $e_n := (\delta_{nk})_{k=1}^\infty$, where $\delta_{nk} := 0$ if $k \neq n$ and $\delta_{nn} := 1$.

14.

(a) Show that $\ell^\infty(\mathbb{N}, \mathbb{F})$ with norm $\| \cdot \|_\infty$ is a *Banach algebra*. Note that all operations are *componentwise* and the multiplicative *identity* is obviously $(1, 1, 1, \cdots)$.
(b) Show that ℓ^∞ does *not* have a Schauder basis. *Hint:* Cf. Problem 5.8.#23.
(c) Let c_{00} be the set of all sequences in $\ell^\infty(\mathbb{N}, \mathbb{F})$ with only *finitely many* nonzero terms. Show that c_{00} is a subalgebra of $\ell^\infty(\mathbb{N}, \mathbb{F})$, but *not* a closed one.
(d) Show that the space c_0 (cf. Exercise 9.2.3) is a *closed subalgebra* of ℓ^∞, hence itself a Banach algebra. Note, however, that c_0 is *not* unital, i.e., has *no* multiplicative identity.
(e) Let c denote the set of all *convergent* sequences in $\mathbb{F}$. Show that c is a *closed* (hence *Banach*) subalgebra of ℓ^∞.

15. Consider the following subspaces of c_0 :

$$Y := \{(x_n) \in c_0 : x_{2n} = 0 \ \forall n \in \mathbb{N}\},$$

$$Z := \{(x_n) \in c_0 : x_{2n} = x_{2n-1}/n \ \forall n \in \mathbb{N}\}.$$

Show that Y and Z are *closed* and $Y + Z$ is *dense* in c_0, but $Y + Z \neq c_0$. Deduce that the sum of two closed subspaces of a Banach space need *not* be a closed subspace.

16. Let Y and Z be subspaces of a Banach space X. Show that if Y is *finite-dimensional* and Z is *closed*, then $Y + Z$ is also closed.

17 (Direct Sum, Complement, Complemented). Let Y and Z be subspaces of a normed space X. We say that X is the *direct sum* of Y and Z and write $X = Y \oplus Z$, if $X = Y + Z$

and $Y \cap Z = \{0\}$. The subspaces Y and Z are then said to be *complements* of each other. A given subspace Y is said to be *complemented* if it has a complement Z. Show that if Y is *finite-dimensional*, then it is complemented. In fact show that $X = Y \oplus Z$ with a *closed* subspace Z. *Hint:* Let $\{y_1, \ldots, y_n\}$ be a *basis* for Y. Let $\psi_k \in Y^*$ be the linear functional satisfying $\psi_k(y_j) = 1$ if $j = k$ and $\psi_k(y_j) = 0$ if $j \neq k$. Using the Hahn–Banach theorem, extend each ψ_k to a functional $\phi_k \in X^*$ and let $Z := \bigcap_{k=1}^{n} \mathrm{Ker}(\phi_k)$.

18.

(a) Show that a *proper, closed* subspace of a Banach space is *nowhere dense*.

(b) Consider the Banach algebra $X := C([a, b], \mathbb{R})$ with norm $\| \cdot \|_\infty$. For each $n \in \mathbb{N}_0$, let $X_n := Pol_n([a, b], \mathbb{R})$ denote the set of all *polynomials* of degree $\leq n$. Show that each X_n is *nowhere dense* in X and yet $\bigcup_{n=0}^{\infty} X_n$ is *dense* in X (cf. Problem 5.8#19).

(c) Prove that an *infinite-dimensional* Banach space cannot be spanned, as a vector space, by a *countable* subset; i.e., it *cannot* have a *countable* Hamel (i.e., algebraic) basis.

19 (Volterra Operator). Let $X := C([0, 1])$ (with norm $\| \cdot \|_\infty$) and for each $f \in X$ define the function Af by $Af(x) := \int_0^x f(t)dt$. Show that $A \in \mathcal{B}(X)$ and is *injective* but *not* surjective. What is $\|A\|$?

20. Let $X := C^\infty([0, 1])$ (with norm $\| \cdot \|_\infty$) and consider the linear map $Df := f'$ for all $f \in X$.

(a) Show that D is *onto* but *not* one-to-one. Also, show that $D \notin \mathcal{B}(X)$.

(b) Consider the subspace $X_0 := \{f \in X : f^{(k)}(0) = 0 \ \forall \ k \geq 0\}$ and the restriction $D_0 := D|X_0$. Show that $D_0 : X_0 \to X_0$ is a *bijective* linear map. Is $D_0 \in \mathcal{B}(X_0)$? Is $D_0^{-1} \in \mathcal{B}(X_0)$?

21. Let $X := \ell^\infty(\mathbb{N}, \mathbb{F})$ and define $T : X \to X$ by $Tx := (x_n/n)$ for each $x = (x_n) \in X$. Show that $T \in \mathcal{B}(X)$.

22 (Closed Range, Bounded Inverse). Let X and Y be normed spaces and suppose that $T \in \mathcal{B}(X, Y)$ satisfies $\|Tx\| \geq c\|x\|$ for some $c > 0$ and all $x \in X$.

(a) Show that if X is a *Banach* space, then $\mathrm{Ran}(T)$ is *closed*.

(b) Show that if $\mathrm{Ran}(T) = Y$, (i.e., T is *onto*), then T has a *bounded* inverse $T^{-1} \in \mathcal{B}(Y, X)$ with $\|T^{-1}\| \leq 1/c$. Deduce that this conclusion follows if X is a Banach space and $\mathrm{Ran}(T)$ is *dense*.

23. Show that the *range* $\mathrm{Ran}(T) := T(X)$ of a linear operator T on a normed space X need *not* be closed. Note, however, that if X is *finite dimensional*, then so is $\mathrm{Ran}(T)$ which *is* closed.

24 (Finite Rank Linear Map). Let X and Y be normed spaces. A linear map $T \in \mathcal{B}(X, Y)$ is said to be *finite rank* if $\mathrm{Ran}(T)$ is a *finite-dimensional* subspace of Y. Show that such a linear map is *compact*.

25. Let X be a normed space, Y a *Banach* spaces and let the $T_n \in \mathcal{K}(X, Y)$ be compact linear maps such that $\lim \|T_n - T\| = 0$ for some linear map $T \in \mathcal{B}(X, Y)$. Show that we have $T \in \mathcal{K}(X, Y)$. *Hint:* Use Cantor's *diagonal method* (cf. Lemma 9.4.10) and an $\varepsilon/3$-argument.

26 (The Ideal $\mathcal{K}(X)$). Let X be a normed space, $A \in \mathcal{K}(X)$ and $B \in \mathcal{B}(X)$. Show that $AB, BA \in \mathcal{K}(X)$. In particular, $\mathcal{K}(X)$ is a (two-sided) *ideal* of the Banach algebra $\mathcal{B}(X)$.

27. Let X be an *infinite-dimensional* Banach space and $A \in \mathcal{K}(X)$. Show that A *cannot* be a *bijection* onto X.

28 (Fredholm Operator). Let X be a *Banach* space and $T \in \mathcal{K}(X)$.

(a) Show that $\mathrm{Ker}(1_X - T)$ is *finite-dimensional*.

(b) Show that $\mathrm{Ran}(1_X - T)$ is *closed*.

(c) Given any $0 \neq \lambda \in \mathbb{F}$ and any $k \in \mathbb{N}$, show that $\mathrm{Ker}[(\lambda 1_X - T)^k]$ is *finite-dimensional* and $\mathrm{Ran}[(\lambda 1_X - T)^k]$ is *closed*.

29. Produce an *isometry* of $C([0,1])$ (with norm $\|\cdot\|_\infty$) into the space $\ell^\infty(\mathbb{N}, \mathbb{R})$. *Hint:* For each $f \in C([0,1])$, look at $(f(r_n))$ for a suitable dense sequence $(r_n) \in [0,1]^{\mathbb{N}}$.

30. Show that the canonical projection $\pi : X \to X/Y$, where X is a normed space and $Y \neq X$ a *closed* subspace, is a *bounded, open* linear map with $\|\pi\| = 1$.

31. Let X and Y be normed spaces, $T \in \mathcal{B}(X, Y)$, $Z := \mathrm{Ker}(T)$, and $\pi : X \to X/Z$ the canonical projection. Show that there is a *unique* $\tilde{T} \in \mathcal{B}(X/Z, Y)$ such that $T = \tilde{T} \circ \pi$ and $\|\tilde{T}\| = \|T\|$.

32. Let $X := C([a,b])$ with norm $\|\cdot\|_\infty$.

(a) Let $\phi(f) := \int_a^b f(t)dt$ for all $f \in X$. Show that $\phi \in X^*$ (i.e., is a bounded linear functional, on X) and that $\|\phi\| = b - a$.
(b) Let $g \in X$ be *fixed* and define $\phi_g(f) := \int_a^b f(x)g(x)dx$ for all $f \in X$. Show that $\phi_g \in X^*$ and $\|\phi_g\| = \int_a^b |g(x)|dx$. *Hint:* Use the functions $f_n(x) := ng(x)/(1 + n|g(x)|)$ for $n \in \mathbb{N}$.
(c) Given a fixed point $x_0 \in [a,b]$, define the *evaluation* map $\psi(f) := f(x_0)$ for all $f \in X$. Show that $\psi \in X^*$ and $\|\psi\| = 1$.

33. Consider the space $X := C^1([a,b])$ with norm $\|f\|_{(1)}$ defined in Problem 9. Show that the linear functional $\phi(f) := f'((a+b)/2)$ for all $f \in X$ is bounded on $(X, \|\cdot\|_{(1)})$, but *not* bounded on $(X, \|\cdot\|_\infty)$.

34 ($(c_0)^* \cong \ell^1$). Show that, if $\phi \in (c_0)^*$, then $(\phi(e_n)) \in \ell^1$; in other words, $\sum_{n=1}^\infty |\phi(e_n)| < \infty$. Here, $e_n := (\delta_{nk})_{k=1}^\infty$, where $\delta_{nk} := 0$ if $k \neq n$ and $\delta_{nn} := 1$. Conversely, given any $a := (a_n) \in \ell^1$, show that there is a *unique* functional $\phi_a \in (c_0)^*$ with $\phi_a(e_n) = a_n$ for all $n \in \mathbb{N}$ and we have $\|\phi_a\| = \|a\|_1$. Deduce that the dual $(c_0)^*$ can be *identified* with ℓ^1. *Hint:* Note that (e_n) is a Schauder basis for both c_0 and ℓ^1.

35 ($(\ell^1)^* \cong \ell^\infty$). Show that $(\ell^1)^*$ can be identified with ℓ^∞. *Hint:* Given any $x := (x_n) \in \ell^1$, we have $x = \sum_{n=1}^\infty x_n e_n$, with the Schauder basis (e_n) as in Problem 34. So, for each $\phi \in (\ell^1)^*$, we have $\phi(x) = \sum_{n=1}^\infty x_n b_n$ where $(b_n) := (\phi(e_n)) \in \ell^\infty$ and $\|(b_n)\|_\infty \leq \|\phi\|$. Conversely, given any $b := (b_n) \in \ell^\infty$, define $\phi_b \in (\ell^1)^*$ by $\phi_b(x) := \sum x_n b_n$ and show that $\|\phi_b\| = \|b\|_\infty$.

36. Let $(a_k)_{k=1}^\infty \in \mathbb{F}^{\mathbb{N}}$. Show that, if $\sum_{k=1}^\infty a_k x_k$ converges for all $x := (x_k) \in \ell^1$, then $(a_k) \in \ell^\infty$. *Hint:* For each $n \in \mathbb{N}$ define the linear functional ϕ_n on ℓ^1 by $\phi_n(x) := \sum_{k=1}^n a_k x_k$ and let $\phi(x) := \lim_{n \to \infty} \phi_n(x)$. Using the *Uniform Boundedness Principle*, show that $\phi \in (\ell^1)^*$ and $|a_k| \leq \|\phi\|$ for all $k \in \mathbb{N}$.

37. Let X be a normed space.

(a) Show that $X \neq \{0\}$ implies $X^* \neq \{0\}$. In fact, show that, if X has n (linearly) *independent* vectors, then so does X^*.
(b) Show that, given any $x \in X$, we have

$$\|x\| = \sup\{|\phi(x)| : \phi \in X^*, \|\phi\| \leq 1\}.$$

Hint: Use Corollary 9.2.39.

38 (Second Dual, Reflexivity). Given a normed space X, the space $X^{**} := (X^*)^*$ is called the *second dual* of X.

(a) For each $x \in X$, define the *evaluation map* $\hat{x}$ on X^* by $\hat{x}(\phi) := \phi(x)$. Show that $\hat{x} \in X^{**}$ and that the (natural) map $x \mapsto \hat{x}$ is an *isometry* of X into X^{**}. The space X is said to be *reflexive* if $x \mapsto \hat{x}$ is *onto*, i.e., $X \cong X^{**}$.

(b) Show that a reflexive normed space is a *Banach* space.
(c) Show that, if X is reflexive, then so is X^*.
(d) Show that the Banach space c_0 is *not* reflexive.

39 (Weak Convergence).   Let (x_n) be a sequence in a normed space X. We say that (x_n) converges to x *weakly* and write $x_n \rightharpoonup x$ as $n \to \infty$, if $\phi(x_n) \to \phi(x)$ for all $\phi \in X^*$.

(a) **(Uniqueness).** Show that weak limits are *unique*.
(b) Show that if $\lim(x_n) = x$, then $x_n \rightharpoonup x$.
(c) Show that if $x_n \rightharpoonup x$, then (x_n) is *bounded*.

40. Let X and Y be normed spaces and $A \in \mathcal{K}(X,Y)$ any compact operator. Show that if $x_n \rightharpoonup x$ as $n \to \infty$, then $\lim_{n\to\infty} A x_n = A x$ (in norm).

41. Let X be a Banach space, $T \in \mathcal{B}(X)$, and let $1_X := \mathrm{id}_X \in \mathcal{B}(X)$ denote the *identity* operator.

(a) Show that, if $\|T\| < 1$, then the operator $1_X - T$ is *invertible*, and we have

$$(1_X - T)^{-1} = \sum_{n=0}^{\infty} T^n \in \mathcal{B}(X). \tag{$*$}$$

Here, $T^0 := 1_X$ and $T^n := T \circ T \circ \cdots \circ T$ (with n copies of T) is the n-th iterate of T. *Hint:* Show that the series on the right side of $(*)$ is *absolutely* convergent and hence convergent. (Why?) Note that $\|T^n\| \le \|T\|^n \to 0$, as $n \to \infty$.
(b) More generally, show that, if $\|T\| < |\lambda|$ (where $\lambda \in \mathbb{F} \setminus \{0\}$), then $(\lambda 1_X - T)^{-1} \in \mathcal{B}(X)$, and we have

$$R_\lambda(T) := (\lambda 1_X - T)^{-1} = \sum_{n=0}^{\infty} \lambda^{-(n+1)} T^n.$$

42 (Resolvent Set, Spectrum).   With notation as in Problem 41, the set

$$\rho(T) := \{\lambda \in \mathbb{F} : R_\lambda(T) = (\lambda 1_X - T)^{-1} \in \mathcal{B}(X)\}$$

is called the *resolvent set* of T and the function $\lambda \mapsto R_\lambda(T)$ is called the *resolvent* of T. Furthermore, the complement

$$\sigma(T) := \mathbb{F} \setminus \rho(T)$$

is called the *spectrum* of T.

(a) Show that, if $\lambda,\ \mu \in \rho(T)$, then

$$R_\lambda(T) - R_\mu(T) = (\mu - \lambda) R_\mu(T) R_\lambda(T). \qquad \textbf{(Resolvent Equation)}$$

Deduce that $R_\mu(T) R_\lambda(T) = R_\lambda(T) R_\mu(T)$.
(b) Show that $\sigma(T)$ is *bounded;* in fact,

$$\sigma(T) \subset \{\lambda \in \mathbb{F} : |\lambda| \le \|T\|\}.$$

(c) Let $\lambda \in \rho(T)$. Show that, if $|\mu - \lambda| < \|R_\lambda(T)\|^{-1}$, then $\mu \in \rho(T)$. Deduce that $\rho(T)$ is *open*.
(d) Show that the spectrum $\sigma(T)$ is *compact*.

43. Let X be a *real* pre-Hilbert space and $x, \ y \in X$. Prove the following:

$$x \perp y \iff \|x + y\|^2 = \|x\|^2 + \|y\|^2, \quad \text{and} \tag{a}$$

$$\|x\| = \|y\| \iff \langle x + y, x - y \rangle = 0. \tag{b}$$

44. Let X be a pre-Hilbert space and $x, \ y \in X$. Show that

$$x \perp y \iff \|x + \alpha y\| = \|x - \alpha y\| \quad \forall \alpha \in \mathbb{F}.$$

45 (Appolonius' Identity). Show that, for any vectors x, y, and z in a pre-Hilbert space X, we have

$$\|z - x\|^2 + \|z - y\|^2 = \frac{1}{2}\|x - y\|^2 + 2\|z - (x + y)/2\|^2.$$

46 (Polarization Identity). Show that the *inner product* of a pre-Hilbert space X can be *rediscovered* from its *norm* by the following identities:

(a) $\langle x, y \rangle = \dfrac{1}{4}(\|x + y\|^2 - \|x - y\|^2) \qquad (\mathbb{F} = \mathbb{R})$,

(b) $\langle x, y \rangle = \dfrac{1}{4}(\|x + y\|^2 - \|x - y\|^2) + \dfrac{i}{4}(\|x + iy\|^2 - \|x - iy\|^2) \qquad (\mathbb{F} = \mathbb{C})$.

47.

(a) Show that the *Banach* space $C([0, 1])$ with norm $\| \cdot \|_\infty$ is *not* a Hilbert space; i.e., the norm $\| \cdot \|_\infty$ *cannot* be obtained from an inner product. *Hint:* Show that the *parallelogram law* (Exercise 9.3.10) fails for the functions $f(x) := 1$ and $g(x) := x$ on $[0, 1]$.

(b) Using part (a) and an *affine* transformation, show that $C([a, b])$ (with norm $\| \cdot \|_\infty$) is not a Hilbert space.

48. Show that the Banach space $\ell^p(\mathbb{N}, \mathbb{F})$, where $p \geq 1$, is a Hilbert space if and only if $p = 2$. *Hint:* Show that the parallelogram law fails if $p \neq 2$.

49. Let X be a pre-Hilbert space and $(x_n) \in X^{\mathbb{N}}$. Show that, if $\lim(\|x_n\|) = \|x\|$ and $\lim(\langle x_n, x \rangle) = \langle x, x \rangle$, then $\|x_n - x\| \to 0$, as $n \to \infty$.

50 (Completion). Show that, given any pre-Hilbert space X, there is a Hilbert space H and an isometric isomorphism ι of X onto a *dense* subspace of H. *Hint:* Let H be the completion of X guaranteed by Theorem 9.2.4 and ι the corresponding (isometric) isomorphism. Now for each $x, \ y \in H$, let $x = \lim(x_n)$ and $y = \lim(y_n)$ with $x_n, \ y_n \in X \cong \iota(X)$, and define $\langle x, y \rangle := \lim(\langle x_n, y_n \rangle)$. Using the continuity of $\langle \cdot, \cdot \rangle$, show that this is a well-defined inner product on H and that (in view of $\langle x, x \rangle = \|x\|^2$) ι is indeed an isometry.

51 (The Hilbert Space $L^2([a, b])$). Consider the pre-Hilbert space $C([a, b])$ with the inner product $\langle f, g \rangle := \int_a^b f(x)g(x)dx$. Let the Hilbert space $L^2([a, b])$ be the *completion* of $C([a, b])$ with the corresponding L^2-*norm*:

$$\|f\|_2 := \left(\int_a^b [f(x)]^2 dx \right)^{1/2}.$$

(a) Show that the system of trigonometric functions

$$1, \quad \cos \frac{2\pi nx}{b - a}, \quad \sin \frac{2\pi nx}{b - a} \qquad (n \in \mathbb{N})$$

is an *orthogonal basis* for $L^2([a, b])$. *Hint:* Using an affine transformation, reduce to the case of $L^2([-\pi, \pi])$. The system then consists of the functions 1, $\cos(nx)$, and $\sin(nx)$, for $n \in \mathbb{N}$. Now look at Example 9.3.30 (3).

(b) Show that, if (e_n) is an orthonormal sequence in $L^2([a, b])$ such that, given any $f \in C([a, b])$ and any $\varepsilon > 0$, there is an $N \in \mathbb{N}$ and scalars $a_1, \ldots, a_N$ such that

$$\left\| f - \sum_{n=1}^{N} a_n e_n \right\|_2 < \varepsilon,$$

then (e_n) is an orthonormal *basis* for $L^2([a, b])$.

52. Let (e_n) be an orthonormal *basis* for a separable Hilbert space H and let (f_n) be an orthonormal sequence in H such that $\sum_{n=1}^{\infty} \|e_n - f_n\|^2 < 1$. Show that (f_n) is also an orthonormal *basis*.

53. Let X and Y be *closed* subspaces of a Hilbert space H and Let $P : H \to X$ and $Q : H \to Y$ be the corresponding orthogonal projections. Show that, if $X \perp Y$, then $P + Q$ is the orthogonal projection onto $X \oplus Y$.

54. Show that, if P and Q are orthogonal projections in a Hilbert space H, and if $PQ = QP$, then $P + Q - PQ$ is also an orthogonal projection. What is its range?

55. Let X be a *complex* pre-Hilbert space and A, $B \in \mathcal{B}(X)$.

(a) Show that, if $\langle Ax, x \rangle = 0$ for all $x \in X$, then $A = 0$. *Hint:* Expanding $\langle A(x + y), x + y \rangle$ and $\langle A(x + iy), x + iy \rangle$, show that $\langle Ax, y \rangle = 0$ for all $x, y \in X$.
(b) Show that the assertion in (a) is *false* if X is a *real* pre-Hilbert space. *Hint:* Pick a suitable rotation in $\mathbb{R}^2$.
(c) Show that, if $\langle Ax, x \rangle = \langle Bx, x \rangle$ for all $x \in X$, then $A = B$.

56 (The Adjoint Operator). Let H be a Hilbert space and $A \in \mathcal{B}(H)$. For each fixed $y \in H$, the map $x \mapsto \langle Ax, y \rangle$ is easily seen to be a *bounded* linear functional and hence, by the Riesz Representation Theorem (Theorem 9.3.20), there is a *unique* $y^* \in H$ such that $\langle Ax, y \rangle = \langle x, y^* \rangle$. Define the map $A^* : H \to H$ by $A^* y := y^*$. We then have

$$\langle Ax, y \rangle = \langle x, A^* y \rangle \qquad (\forall x, \ y \in H). \tag{†}$$

We call A^* the *adjoint* of A. Prove the following assertions for any bounded operators A, $B \in \mathcal{B}(H)$.

(a) $A^* \in \mathcal{B}(H)$, $A^{**} := (A^*)^* = A$, and we have $\|A^*\| = \|A\|$. *Hint:* Note that

$$\|A^* y\|^2 = \langle A^* y, A^* y \rangle = \langle A(A^* y), y \rangle \leq \|A\| \|A^* y\| \|y\|.$$

(b) $\|A^* A\| = \|A\|^2$.
(c) $(A + B)^* = A^* + B^*$, $(AB)^* = B^* A^*$, and $(\alpha A)^* = \bar{\alpha} A^*$ for all $\alpha \in \mathbb{F}$.
(d) If $A^{-1} \in \mathcal{B}(H)$, then $(A^{-1})^* = (A^*)^{-1}$.

57. Let H be a Hilbert space and $A \in \mathcal{B}(H)$. Recall that $\mathrm{Ran}(A) := A(H)$.

(a) Show that $\mathrm{Ker}(A^*) = \mathrm{Ran}(A)^{\perp}$ and $\mathrm{Ker}(A) = \mathrm{Ran}(A^*)^{\perp}$. Deduce that $\mathrm{Ran}(A)$ is *dense* if and only if A^* is *injective,* and that $\mathrm{Ran}(A^*)$ is dense if and only if A is *injective*.
(b) Show that $H = \mathrm{Ker}(A^*) \oplus (\mathrm{Ran}(A))^-$.

58 (Hellinger–Toeplitz). Let T be a linear operator on a Hilbert space H such that

$$\langle Tx, y \rangle = \langle x, Ty \rangle \qquad (\forall x, \ y \in H).$$

Show that $T \in \mathcal{B}(H)$. *Hint:* Show that T has *closed graph;* i.e., if $\lim(x_n) = x \in H$ and $\lim(Tx_n)) = y \in H$, then $y = Tx$.

59 (Self-Adjoint Operator, Real Spectrum).   Let H be a (complex) Hilbert space. An operator $A \in \mathcal{B}(H)$ is said to be *self-adjoint* if $A^* = A$.

(a) Show that, if A is self-adjoint, then it has *real* spectrum, i.e., $\sigma(A) \subset \mathbb{R}$.
(b) Show that, if A and B are self-adjoint, then AB is self-adjoint if and only if $AB = BA$.
(c) Show that, for any $A \in \mathcal{B}(H)$, the operators A^*A and $A + A^*$ are self-adjoint.
(d) Show that, if a *self-adjoint* operator $A \in \mathcal{B}(H)$ is *surjective,* then it is also *injective.* Deduce that $A^{-1} \in \mathcal{B}(H)$ and is also self-adjoint.
(e) Show that if A is self-adjoint, then we have

$$\|A\| = \sup\{|\langle Ax, x \rangle| : \|x\| = 1\}.$$

60 (Eigenvalue, Eigenvector, Eigenspace).   Let $A \in \mathcal{B}(H)$. A scalar $\lambda \in \mathbb{F}$ is said to be an *eigenvalue* of A if $Ax = \lambda x$ for some $x \neq 0$; i.e., if $\mathrm{Ker}(\lambda 1_H - A) \neq \{0\}$. Any (nonzero) vector $x \in \mathrm{Ker}(\lambda 1_H - A)$ is then said to be an *eigenvector* of A corresponding to the eigenvalue λ. The subspace $\mathrm{Ker}(\lambda 1_H - A)$ is called the *eigenspace* of A corresponding to λ. Let $0 \neq A \in \mathcal{K}(H)$ be a *self-adjoint, compact* operator.

(a) **(Point Spectrum).** Let $\sigma_p(A)$ denote the set of all eigenvalues of A (the so-called point spectrum of A). Show that $\sigma_p(A) \subset \sigma(A)$ and deduce that all eigenvalues are *real.*
(b) Show that eigenvectors corresponding to *distinct* eigenvalues are *orthogonal.*
(c) Show that each eigenspace $\mathrm{Ker}(\lambda 1_H - A)$ with $\lambda \neq 0$ is *finite-dimensional.*
(d) Show that either $\|A\| > 0$ or $-\|A\| < 0$ is an eigenvalue of A.

61 (Multiplication Operator).   Let $H := \ell^2(\mathbb{N}, \mathbb{C})$. For each given sequence $a = (a_1, a_2, \ldots) \in \ell^\infty(\mathbb{N}, \mathbb{C})$, we define the corresponding *multiplication* operator $M_a : H \to H$ by

$$M_a x := ax = (a_1 x_1, a_2 x_2, a_3 x_3, \ldots) \quad \forall\, x = (x_1, x_2, x_3, \ldots) \in H.$$

(a) Show that $M_a \in \mathcal{B}(H)$ and $\|M_a\| = \|a\|_\infty$.
(b) Show that $M_a M_b = M_{ab}$ for all $a, b \in \ell^\infty(\mathbb{N}, \mathbb{C})$ and hence $M_a M_b = M_b M_a$. Deduce that M_a^{-1} exists if and only if $1/a := (1/a_1, 1/a_2, \ldots) \in \ell^\infty(\mathbb{N}, \mathbb{C})$ and we then have $M_a^{-1} = M_{1/a}$.
(c) Show that $M_a^* = M_{\bar{a}}$ and hence M_a is self-adjoint if and only if $a \in \ell^\infty(\mathbb{N}, \mathbb{R})$.
(d) Show that $\sigma_p(M_a) = \{a_1, a_2, a_3, \ldots\}$.
(e) Show that $\sigma(M_a) = \{a_1, a_2, a_3, \ldots\}^-$ is the *closure* of $\sigma_p(M_a)$. Deduce that $M_a^{-1} = M_{1/a} \in \mathcal{B}(H)$ if and only if 0 is *not* a limit point of $\sigma_p(M_a)$.
(f) Show that M_a is *compact* if and only if $\lim(a_n) = 0$.

62 (Shift Operators).   Let $H := \ell^2(\mathbb{N}, \mathbb{C})$ and define the *right shift* operator S_r and the *left shift* operator S_ℓ by letting for each $x = (x_1, x_2, \ldots) \in \ell^2(\mathbb{N}, \mathbb{C})$,

$$S_r(x_1, x_2, x_3, \ldots) := (0, x_1, x_2, x_3, \ldots),$$

$$S_\ell(x_1, x_2, x_3, \ldots) := (x_2, x_3, x_4, \ldots).$$

(a) Show that $S_r^* = S_\ell$ and hence S_r is *not* self-adjoint.
(b) Show that S_r and S_ℓ are *not* compact operators.
(c) Show that S_r has *no* eigenvalue, i.e., $\sigma_p(S_r) = \emptyset$. What is $\sigma(S_r)$?
(d) Show that $\sigma_p(S_\ell) = \{\lambda \in \mathbb{C} : |\lambda| < 1\}$. What is $\sigma(S_\ell)$?
(e) Consider the multiplication operator M_h, where $h := (1, 1/2, 1/3, \ldots)$ is the *harmonic sequence* and note that $h \in \ell^2(\mathbb{N}, \mathbb{R})$. Show that the operator $T := M_h S_r$ is compact, has *no* eigenvalue, and $\sigma(T) = \{0\}$.

63. Let $\mathcal{B}$ be a *bounded* subset of $C([a, b])$. Show that the set

$$\left\{ \int_a^x f(t) dt : f \in \mathcal{B} \right\}^-$$

is *compact*.

64. Let $X = C([0, 1], \mathbb{C})$ and let $A \in \mathcal{B}(X)$ be the *Volterra operator*

$$Af(x) := \int_0^x f(t) \, dt \qquad \forall \, f \in X.$$

Show that A is a *compact* operator with *no* eigenvalue.

65. Prove that $\{\sin(nx) : n \in \mathbb{N}\}$ is *not* an equicontinuous subset of $C([0, \pi])$.

66. Let $f : \mathbb{R} \to \mathbb{R}$ be *uniformly* continuous and, for each $a \in \mathbb{R}$, define $f_a(x) := f(x - a)$. Show that the family $\{f_a : a \in \mathbb{R}\}$ is *equicontinuous* on $\mathbb{R}$.

67. Define the sequence (f_n) in $\boldsymbol{BC}([0, \infty), \mathbb{R})$ by $f_n(x) = \sin(\sqrt{x + 4n^2\pi^2})$. Show that (f_n) is *equicontinuous* on $[0, \infty)$ and that $\lim(f_n(x)) = 0$ for all $x \in [0, \infty)$. Show, however, that $\{f_n : n \in \mathbb{N}\}$ is *not* relatively compact in $\boldsymbol{BC}([0, \infty), \mathbb{R})$. *Hint:* Show that (f_n) does *not* converge to zero *uniformly*.

68. Given $\alpha \in (0, 1]$, consider the subalgebra $\boldsymbol{Lip}^\alpha([0, 1])$ of $C([0, 1])$ consisting of all *Lipschitz* functions of *order* α. Recall [Example 9.2.2 (4)] that $\boldsymbol{Lip}^\alpha([0, 1])$ is actually a *Banach space* with norm

$$\|f\|_{\alpha,\infty} := \|f\|_\infty + \sup\{|f(x) - f(y)|/|x - y|^\alpha : x, y \in [0, 1], \ x \neq y\}.$$

Show that the set $\{f \in \boldsymbol{Lip}^\alpha([0, 1]) : \|f\|_{\alpha,\infty} \leq 1\}$ is a *compact* subset of $C([0, 1])$.

69. Let $(f_n)_{n=1}^\infty$ be a sequence of *differentiable* functions on $[0, 1]$ such that the sequence $(f_n(x_0))$ is bounded for some $x_0 \in [0, 1]$ and that $|f_n'(x)| \leq M$ for all $n \in \mathbb{N}$ and all $x \in [0, 1]$. Show that (f_n) has a *uniformly convergent* subsequence.

70. Let M be a metric space, X a normed space, and (f_n) an *equicontinuous* sequence in $\boldsymbol{BC}(M, X)$. Show that the set of all $x \in M$ such that $(f_n(x))$ is a *Cauchy* sequence in X is a *closed* subset of M.

71. Let M be a metric space, X a normed space, and (f_n) a sequence in $\boldsymbol{BC}(M, X)$ that is *equicontinuous* at a point $x_0 \in M$. Show that, if $\lim(f_n(x_0)) = y_0 \in X$, then we have $\lim(f_n(x_n)) = y_0$ for *any* sequence (x_n) in M with $\lim(x_n) = x_0$.

72. Let M be a metric space and $\mathcal{F}$ an *equicontinuous* subset of $\boldsymbol{BC}(M, \mathbb{R})$. Show that the set $B := \{x \in M : \mathcal{F}(x) \text{ is bounded}\}$ is *both* open and closed. Here $\mathcal{F}(x) := \{f(x) : f \in \mathcal{F}\}$.

73. This problem provides another way of showing that $|x|$ can be *uniformly* approximated by *polynomials* on $[-1, 1]$ and can be used in the proof of Lemma 9.4.21 instead of the Weierstrass Approximation Theorem (Corollary 4.7.10).

(a) Show that there is a sequence $(p_n(t))$ of polynomials that is *increasing* in $[0, 1]$ and converges uniformly to $\sqrt{t}$. *Hint:* Use the Babylonian method (cf. Exercise 2.2.25).
(b) Show that $|x|$ can be approximated *uniformly* on $[-1, 1]$ by *polynomials*.

74. Let $f \in C([0, 1])$ be *strictly increasing*. Show that the subalgebra $\mathcal{A}$ of $C([0, 1])$ generated by $\{1, f\}$ is *dense* in $C([0, 1])$.

75. Let (K, d_K) be a *compact* metric space containing at least *two* points. For each $y \in K$, consider the function $d_K(\cdot, y) \in C(K, \mathbb{R})$ defined by $d_K(\cdot, y) : x \mapsto d_K(x, y)$. Show that the *unital* subalgebra of $C(K, \mathbb{R})$ generated by the family $\{d_K(\cdot, y) : y \in K\}$ is *dense* in $C(K, \mathbb{R})$.

76. Let (K, d_K) be a *compact* metric space and $\mathcal{A} \subset C(K, \mathbb{R})$ an algebra that separates the points of K. Show that either $\mathcal{A}^- = C(K)$ or there is a point $x_0 \in K$ such that $\mathcal{A}^- = \{f \in C(K, \mathbb{R}) : f(x_0) = 0\}$.

77. Let K and M be *compact* metric spaces. Show that the subalgebra of $C(K \times M, \mathbb{R})$ generated by the functions of the form $(x, y) \mapsto g(x)h(y)$, where $g \in C(K, \mathbb{R})$, $h \in C(M, \mathbb{R})$, is *dense* in $C(K \times M, \mathbb{R})$; i.e., given any $f \in C(K \times M, \mathbb{R})$ and any $\varepsilon > 0$, there are functions $g_k \in C(K, \mathbb{R})$ and $h_k \in C(M, \mathbb{R})$, $1 \leq k \leq n$, such that

$$\left| f(x, y) - \sum_{k=1}^{n} g_k(x) h_k(y) \right| < \varepsilon \qquad (\forall x, y \in K \times M).$$

Chapter 10
Lebesgue Measure and Integral in $\mathbb{R}$

In Chap. 7 we saw that the Riemann integral of a (bounded) function $f : [a,b] \to \mathbb{R}$ can be obtained as a "limit" of integrals of step functions that approximate f. In fact, we have (cf. Exercise 7.4.8)

$$\underline{\int} f = \sup \left\{ \int_a^b \phi : \phi \in \textbf{\textit{Step}}([a,b]), \ \phi \le f \right\}, \qquad (*)$$

and for $f \in \mathcal{R}([a,b])$ the left side of $(*)$ is indeed $\int_a^b f$. If $(x_k)_{k=0}^n$ is a partition of $[a,b]$ and if ϕ is a step function with $\phi(x) = c_j$ for all $x \in I_j := (x_j - 1, x_j)$, $1 \le j \le n$, then we have

$$\int_a^b \phi := \sum_{j=1}^n c_j \lambda(I_j),$$

where $\lambda(I_j) := x_j - x_{j-1}$ is the *length* of I_j. Thus, the length of intervals (Definition 7.3.1) is all we need for Riemann's theory of integration. Now we did extend length (cf. the "temporary" Definition 4.2.4) to more complicated sets and also defined sets of *measure zero* (cf. Definition 7.3.2). The theory of integration we want to develop here, due to the French mathematician Henri Lebesgue, requires a more sophisticated *measure* that extends length and still has the most desirable properties we need: Ideally, with $2^{\mathbb{R}} = \mathcal{P}(\mathbb{R})$ denoting the power set of $\mathbb{R}$, what we want is a *set* function $\mu : 2^{\mathbb{R}} \to [0, \infty]$ that is:

1 an extension of length: $\mu(I)$ is the length of I if $I \subset \mathbb{R}$ is an interval,
2 monotone: $A \subset B$ implies $\mu(A) \le \mu(B)$,
3 translation invariant: $\mu(A + c) := \mu(\{a + c : a \in A\}) = \mu(A) \ \forall c \in \mathbb{R}$, and
4 countably additive: $\mu(\bigcup_{n \in \mathbb{N}} A_n) = \sum_{n=1}^{\infty} \mu(A_n)$ if $A_i \cap A_j = \emptyset$ for $i \ne j$.

© Springer Science+Business Media New York 2014
H.H. Sohrab, *Basic Real Analysis*, DOI 10.1007/978-1-4939-1841-6_10

Unfortunately, such a (set-) function μ does *not* exist unless we are willing to *reduce the domain* of μ from the *whole* power set $2^{\mathbb{R}}$ to a *proper subset* of it; namely, the σ-algebra $\mathcal{M}_\lambda(\mathbb{R})$ of the so-called (Lebesgue) measurable sets, to be defined below. So, to begin, we should define a *measure* precisely:

Definition 10.0.1 (Measure, Measurable Space, Measure Space).   Let X be a set and let $\mathcal{A}$ be a σ-algebra of subsets of X. A set function $\mu : \mathcal{A} :\to [0, \infty]$ is said to be a *countably additive measure* (or simply a *measure*) on $\mathcal{A}$ if it satisfies the following conditions:

(i) $\mu(\emptyset) = 0$, and
(ii)   Given any sequence (A_n) of *pairwise disjoint* elements of $\mathcal{A}$, we have

$$\mu\left(\bigcup_{n=1}^{\infty} A_n \right) = \sum_{n=1}^{\infty} \mu(A_n) \qquad \text{("countable additivity" or "}\sigma-\text{additivity").}$$

The pair $(X, \mathcal{A})$ is then called a *measurable space* and the triple $(X, \mathcal{A}, \mu)$ is said to be a *measure space*. Elements of $\mathcal{A}$ are called *measurable sets* and for each $A \in \mathcal{A}$, its *measure* is the (extended nonnegative) number $\mu(A)$.

(Operations on Measures). Let μ and v be measures on a σ-algebra $\mathcal{A}$ and let $c \geq 0$. Then one defines the set functions $\mu + v : \mathcal{A} :\to [0, \infty]$ and $c\mu : \mathcal{A} :\to [0, \infty]$ by

$$(\mu + v)(A) := \mu(A) + v(A) \quad \text{and} \quad (c\mu)(A) := c\mu(A) \qquad \forall A \in \mathcal{A}.$$

It is easily checked that the *"sum"* $\mu + v$ and the *"scalar multiple"* $c\mu$ are also measures. Thus, the set $\mathfrak{M}(\mathcal{A})$ of all measures on $\mathcal{A}$ is a *cone*.

Remark 10.0.2. Throughout this chapter (and the next), we shall obviously work with the space $X = \mathbb{R}$ and the σ-algebra $\mathcal{M}_\lambda(\mathbb{R})$, which will be defined shortly. General measure spaces will be studied in Chap. 12.

The Riemann integral was defined using step functions, which take on (constant) values on subintervals of a partition of an interval $[a, b]$. It is possible (as was done in the first edition of this book) to follow F. Riesz and develop Lebesgue's theory using step functions as well, without introducing measure theory except for the concept of sets of measure zero, already introduced and used in Chap. 7. But the success of the new theory comes from the fact that, instead of partitioning the *domain* of the function into subintervals, it partitions its *range* and assigns a measure (*"generalized length"*) to the *inverse image* $f^{-1}(J)$ for each subinterval J of such a partition. Now, if f is a *step* function, then $f^{-1}(J)$ is a (disjoint) union of intervals and its measure is simply the sum of their lengths. In general, however, $f^{-1}(J)$ must be *nice*, i.e., *measurable*. For this new approach the class of step functions must be replaced by the larger class of (measurable) *simple functions*. These are functions that take on (distinct) constant values on a finite collection of (pairwise disjoint) *measurable* sets. Thus, although step functions are simple, the converse is not true. The Lebesgue integral of f, if it exists, is then defined by modifying $(*)$ as follows:

$$\int_a^b f \, d\lambda = \sup \left\{ \int_a^b \phi \, d\lambda : \phi \in \textbf{\textit{Simp}}([a,b]), \; \phi \le f \right\}, \qquad (**)$$

where $\textbf{\textit{Simp}}([a,b])$ is the space of all *simple* functions on $[a,b]$. Here, the Lebesgue integral of a simple function ϕ in $(**)$ that takes on the (distinct) constant values c_j on the (pairwise disjoint) measurable sets $A_j := \phi^{-1}(c_j) \subset [a,b]$, $1 \le j \le n$, is defined to be the sum

$$\int_a^b \phi \, d\lambda := \sum_{j=1}^{n} c_j \lambda(A_j),$$

where $\lambda(A_j)$ is the *Lebesgue measure* of the measurable set A_j, $1 \le j \le n$.

Of course, a legitimate question to ask is this: Why even extend Riemann's theory of integration? The most logical answer seems to be this: As we have pointed out repeatedly, the most fundamental notion in analysis is that of *limit*. Therefore, the most desirable theory of integration is one that behaves nicely (i.e., *"continuously"*) when we deal with limits of functions and/or sets. The Riemann theory requires *uniform convergence* if we want nice properties to be preserved when we pass to the limit and the *uniformity* condition is too restrictive and hence an unnecessary nuisance in practice.

Example. Let $\mathbb{Q}_1 := \{r_1, r_2, r_3, \dots\}$ be an enumeration of the rational numbers in $[0, 1]$. Define the sequence (f_n) of functions on $[0, 1]$ by

$$f_n(x) = \begin{cases} 1 & \text{if } x \in \{r_1, r_2, \dots, r_n\}, \\ 0 & \text{otherwise.} \end{cases}$$

Since f_n is continuous *except* on $\{r_1, \dots, r_n\}$, we have $f_n \in \mathcal{R}([0, 1])$ for all $n \in \mathbb{N}$ and in fact $\int_0^1 f_n = 0$. On the other hand, $\lim f_n(x)$ is 1 if $x \in \mathbb{Q}_1$ and 0 otherwise, i.e., $\lim(f_n) = \chi_{\mathbb{Q}_1}$, which is *not* Riemann integrable. It *is*, however, Lebesgue integrable as we shall prove in what follows.

10.1 Outer Measure

To define Lebesgue measure we first introduce the (Lebesgue) *outer measure* of a subset of $\mathbb{R}$:

Definition 10.1.1 (Lebesgue Outer Measure). Given a set $A \subset \mathbb{R}$, its (Lebesgue) *outer measure* is defined to be the extended real number

$$\lambda^*(A) = \inf\left\{\sum_{n=1}^{\infty} \lambda(I_n) : A \subset \bigcup_{n=1}^{\infty} I_n\right\}, \qquad (\dagger)$$

where $(I_n)_{n=1}^{\infty}$ is a sequence of *open* intervals that *covers* the set A, i.e., $A \subset \bigcup_{n=1}^{\infty} I_n$, and the "inf" is taken over all such covers of A.

Remark 10.1.2.

(a) In Definition 10.1.1 the intervals I_n that cover A need *not* be open. Indeed, if I_n is not open, we may replace it with an open interval J_n with $I_n \subset J_n$ and $\lambda(J_n) = \lambda(I_n) + \varepsilon/2^n$ for any prescribed $\varepsilon > 0$. We then have $\sum_{n=1}^{\infty} \lambda(J_n) \leq \sum_{n=1}^{\infty} \lambda(I_n) + \varepsilon$ and hence the "inf" will be the same.

(b) **(Jordan Outer Measure).** If in Definition 10.1.1 we use *finite* covers of A by open intervals instead of *countable* ones, then the resulting infimum is called *Jordan* outer measure of A and will be denoted by $\mu^{*(J)}(A)$. As in the previous remark, open intervals may be replaced by closed (or half-open) ones.

Exercise 10.1.3. For any set $A \subset \mathbb{R}$ and any $x \in \mathbb{R}$ we define its *translate* $A + x := \{a + x : a \in A\}$ and its *reflection* $-A := \{-a : a \in A\}$. Prove the following assertions, where A, $B \subset \mathbb{R}$ and $x \in \mathbb{R}$ are arbitrary:

(a) $(A \cap B) + x = (A + x) \cap (B + x)$, $(A \cup B) + x = (A + x) \cup (B + x)$, and $A^c + x = (A + x)^c$.

(b) $-(A \cap B) = (-A) \cap (-B)$, $-(A \cup B) = (-A) \cup (-B)$, and $-A^c = (-A)^c$.

Proposition 10.1.4. *The outer measure satisfies the following properties:*

(1) If $A \subset \mathbb{R}$ has measure zero, then $\lambda^*(A) = 0$. In particular, any countable set has outer measure zero.

*(2) (λ^* is **Monotone**) If $A \subset B$, then $\lambda^*(A) \leq \lambda^*(B)$.*

*(3) (λ^* is **Translation Invariant**) $\lambda^*(A + x) = \lambda^*(A)$ for all $x \in \mathbb{R}$.*

*(4) (λ^* is **Reflection Invariant**) $\lambda^*(-A) = \lambda^*(A)$.*

(5) For any interval $I \subset \mathbb{R}$ (bounded or not), $\lambda^*(I) = \lambda(I)$ is the length of I.

Proof. Statements (1) and (2) follow at once from the definition ($\dagger$) of λ^* and the same is also true for property (3). (Why?) For (4) we simply note that if $I = (a, b)$, then $-I = (-b, -a)$ and $\lambda(I) = \lambda(-I) = b - a$. Also, a sequence (I_n) of open intervals covers a set $A \subset \mathbb{R}$ if and only if the corresponding sequence $(-I_n)$ covers $-A$. To prove (5), let us first consider the case of a closed, bounded interval $I = [a, b]$ with $a < b$. Since $I \subset (a - \varepsilon, b + \varepsilon)$ for each $\varepsilon > 0$ and $\lambda((a - \varepsilon, b + \varepsilon)) = b - a + 2\varepsilon$, we have $\lambda^*(I) \leq b - a + 2\varepsilon$ for all $\varepsilon > 0$ and hence $\lambda^*(I) \leq b - a$. Therefore, we need only show that $\lambda^*(I) \geq b - a$, which follows at once from Proposition 7.3.7. If I has endpoints $a < b$, but is *not* closed, then $I_\varepsilon := [a + \varepsilon, b - \varepsilon] \subset I$ for all small $\varepsilon > 0$ and hence, by property (2), $\lambda^*(I_\varepsilon) = b - a - 2\varepsilon \leq \lambda^*(I) \leq \lambda^*([a, b]) = b - a$ and the result follows at once. Finally, if I is *unbounded*, then for any $\ell > 0$ we can pick a (bounded) closed interval

$J \subset I$ with $\lambda(J) = \lambda^*(J) = \ell$. But then $\lambda^*(I) \geq \ell$ for all $\ell > 0$ and we get $\lambda^*(I) = \infty = \lambda(I)$. □

Exercise 10.1.5.

a. Let A, $Z \subset \mathbb{R}$ and assume that Z has measure zero. Show that $\lambda^*(A \cup Z) = \lambda^*(A)$ and that $\lambda^*(A \setminus Z) = \lambda^*(A)$.

b. **(The "Distance" $\lambda^*(A \triangle B)$).** For any A, $B \subset \mathbb{R}$ define

$$d_\lambda(A, B) := \lambda^*(A \triangle B).$$

Prove the following properties for arbitrary subsets A, B, and C of $\mathbb{R}$.

$$d_\lambda(A, B) = d_\lambda(B, A), \quad d_\lambda(A, A) = 0, \quad d_\lambda(A, B) \leq d_\lambda(A, C) + d_\lambda(B, C).$$

Deduce that $A \sim B \iff d_\lambda(A, B) = 0$ defines an *equivalence relation* on $\mathcal{P}(\mathbb{R})$. Also prove that

$$d_\lambda(A, B) = 0 \implies \lambda^*(A) = \lambda^*(B)$$

and give a simple example (with $\lambda^*(A) = \lambda^*(B) < \infty$) to show that the converse is *false* and hence $d_\lambda(A, B)$ is *not* a distance on $\mathcal{M}_{\lambda^* < \infty} := \{E \subset \mathbb{R} : \lambda^*(E) < \infty\}$. Show, however, that the set of all equivalence classes $\{[E] : E \in \mathcal{M}_{\lambda^* < \infty}\}$ is a *metric space* with metric $d_\lambda([A], [B]) := d_\lambda(A, B)$.

Unfortunately, λ^* is *not* countably additive, but it is *countably subadditive* in the following sense.

Proposition 10.1.6 (Countable Subadditivity of λ^*). *Given any sequence (A_n) of subsets of $\mathbb{R}$, we have*

$$\lambda^*\left(\bigcup_{n=1}^\infty A_n\right) \leq \sum_{n=1}^\infty \lambda^*(A_n).$$

Proof. We may (and do) assume that the right side is *finite* and hence that $\lambda^*(A_n) < \infty$ for all $n \in \mathbb{N}$. Now for each n we can pick a sequence $(I_{n,m})_{m=1}^\infty$ of open intervals such that $A_n \subset \bigcup_{m=1}^\infty I_{n,m}$ and $\sum_{m=1}^\infty \lambda(I_{n,m}) < \lambda^*(A_n) + \varepsilon/2^n$. Since $\{I_{n,m} : n, m \in \mathbb{N}\}$ is a countable collection of open intervals satisfying $\bigcup_{n=1}^\infty A_n \subset \bigcup_{(n,m) \in \mathbb{N} \times \mathbb{N}} I_{n,m}$ and $\varepsilon > 0$ is arbitrary, our claim follows from

$$\lambda^*\left(\bigcup_{n=1}^\infty A_n\right) \leq \sum_{n=1}^\infty \sum_{m=1}^\infty \lambda(I_{n,m}) < \sum_{n=1}^\infty \left(\lambda^*(A_n) + \frac{\varepsilon}{2^n}\right)$$

$$= \sum_{n=1}^\infty \lambda^*(A_n) + \varepsilon. \qquad \square$$

Before proving an important consequence of the countable subadditivity, let us give a definition that will be useful.

Definition 10.1.7 ($\mathcal{F}_\sigma$ and $\mathcal{G}_\delta$). A set $A \subset \mathbb{R}$ is said to be an $\mathcal{F}_\sigma$ if it is a *countable union* of *closed* sets. A set $B \subset \mathbb{R}$ is said to be a $\mathcal{G}_\delta$ if it is a *countable intersection* of *open* sets.

Remark 10.1.8. It is clear that any *countable* set, e.g., the set $\mathbb{Q}$ of rational numbers, is an $\mathcal{F}_\sigma$. It follows that the set $\mathbb{Q}^c$ of *irrational* numbers, which is an *uncountable*, dense set that contains no nondegenerate intervals, is a $\mathcal{G}_\delta$. In fact, if $\{r_1, r_2, \dots\}$ is an enumeration of the rational numbers, then

$$\mathbb{Q}^c = \bigcap_{n=1}^{\infty} \{r_n\}^c = \bigcap_{n=1}^{\infty} \left((-\infty, r_n) \cup (r_n, \infty) \right).$$

This shows, in particular, that the complement of a countable union of intervals may be a huge, complicated set, a fact we also encountered when we defined the Cantor set and its generalized versions.

We can now prove the following *regularity* result for the outer measure.

Corollary 10.1.9 (Outer Regularity of λ^*). *Given any set $A \subset \mathbb{R}$ we have*

$$\lambda^*(A) = \inf\{\lambda^*(O) : A \subset O \text{ and } O \text{ is open}\}. \tag{$\ddagger$}$$

In particular, for any $\varepsilon > 0$, there is an open set O such that $A \subset O$ and $\lambda^*(O) < \lambda^*(A) + \varepsilon$. Also, there is a set $G \in \mathcal{G}_\delta$ such that $\lambda^*(A) = \lambda^*(G)$.

Proof. Let $\lambda_O^*(A)$ denotes the right side of ($\ddagger$). Since every open set is a countable union of (pairwise disjoint) open intervals (cf. Theorem 4.1.2), we obviously have $\lambda^*(A) \leq \lambda_O^*(A)$. For the reverse inequality we may (and do) assume that $\lambda^*(A) < \infty$. Then, given any $\varepsilon > 0$, the definition of $\lambda^*(A)$ guarantees the existence of a sequence $(I_n)_{n=1}^{\infty}$ of open intervals with $A \subset \bigcup_{n=1}^{\infty} I_n$ and such that

$$\sum_{n=1}^{\infty} \lambda(I_n) < \lambda^*(A) + \varepsilon. \tag{$*$}$$

If we define $O := \bigcup_{n=1}^{\infty} I_n$, then ($*$) and the countable subadditivity of λ^* give

$$\lambda^*(O) \leq \sum_{n=1}^{\infty} \lambda^*(I_n) = \sum_{n=1}^{\infty} \lambda(I_n) < \lambda^*(A) + \varepsilon,$$

and hence $\lambda_O^*(A) < \lambda^*(A) + \varepsilon$, from which $\lambda_O^*(A) \leq \lambda^*(A)$ follows because $\varepsilon > 0$ was arbitrary. Finally, for each $k \in \mathbb{N}$, we can (by the first assertion) pick an open set O_k such that $A \subset O_k$ and $\lambda^*(O_k) \le \lambda^*(A) + 1/k$. If we set $G := \bigcap_{k=1}^{\infty} O_k$, then $G \in \mathcal{G}_\delta$ and we have $A \subset G \subset O_k$ for all $k \in \mathbb{N}$. Therefore,

$$\lambda^*(A) \le \lambda^*(G) \le \lambda^*(O_k) \le \lambda^*(A) + \frac{1}{k} \quad \forall\, k \in \mathbb{N},$$

which implies $\lambda^*(A) = \lambda^*(G)$, as desired. $\qquad\square$

Remark 10.1.10. Given any $A \subset \mathbb{R}$, Corollary 10.1.9 implies that for any $\varepsilon > 0$, we can pick an open set O such that $A \subset O$ and

$$\lambda^*(O) - \lambda^*(A) < \varepsilon.$$

Since the subadditivity of λ^* implies that $\lambda^*(O) - \lambda^*(A) \le \lambda^*(O \setminus A)$, the stronger condition

$$\lambda^*(O \setminus A) < \varepsilon \qquad\qquad (*)$$

need not be true. The condition $(*)$ is one of the many logically equivalent ways to define the *(Lebesgue) measurability* of A. It is basically *Littlewood's first principle*, which says that a measurable set is *nearly open* or, as Tao puts it in [Tao11], that it can be *efficiently contained* in open sets. If one adopts this definition, which is one of the most *intuitive* ones, then open sets are obviously measurable. There is another definition (due to Carathéodory) that is less intuitive, but in it A and A^c play symmetric roles and hence are either both measurable or both nonmeasurable. It also has the advantage that it can be used in abstract spaces, as we shall see in Chap. 12.

Definition 10.1.11 (Lebesgue Measurable, Lebesgue Measure). A set $E \subset \mathbb{R}$ is said to be *(Lebesgue) measurable* if for any set $A \subset \mathbb{R}$ we have

$$\lambda^*(A) = \lambda^*(A \cap E) + \lambda^*(A \cap E^c).$$

We then define the *(Lebesgue) measure* of E to be

$$\lambda(E) := \lambda^*(E).$$

Notation 10.1.12 ($\mathcal{M}_\lambda(\mathbb{R})$, $\mathcal{M}_{\lambda<\infty}(\mathbb{R})$). The set of all Lebesgue measurable subsets of $\mathbb{R}$ is denoted $\mathcal{M}_\lambda(\mathbb{R})$ or simply $\mathcal{M}_\lambda$ if no confusion results. We use $\mathcal{M}_{\lambda<\infty}(\mathbb{R})$ (or simply $\mathcal{M}_{\lambda<\infty}$) to denote the set of all $E \in \mathcal{M}_\lambda$ such that $\lambda(E) < \infty$.

Remark 10.1.13. Since $\lambda^*(A) \le \lambda^*(A \cap E) + \lambda^*(A \cap E^c)$ is always true because λ^* is countably (hence finitely) subadditive, we have

$$E \in \mathcal{M}_\lambda \iff \lambda^*(A) \ge \lambda^*(A \cap E) + \lambda^*(A \cap E^c) \quad \forall A \subset \mathbb{R}.$$

As the example of Cantor set and its generalized versions show, we are naturally led to the measure (i.e., generalized length) of countable unions and intersections of sets. Therefore, the natural domain of a measure should be a σ-*algebra* of subsets of $\mathbb{R}$ and hence we expect this to be true for the set $\mathcal{M}_\lambda(\mathbb{R})$ of Lebesgue measurable sets we just defined. Let us recall (Definition 1.1.5) that, given a set X, a *nonempty* subset $\mathcal{A}$ of the set $\mathcal{P}(X)$ of *all* subsets of X is said to be an *algebra* if it contains X as well as the *complements* and *finite unions* of its members. If an algebra $\mathcal{A}$ contains all *countable unions* of its members, then it is said to be a σ-algebra. The most trivial σ-algebras are, of course, $\{\emptyset, X\}$ and $\mathcal{P}(X)$. Also recall (Proposition 1.1.9) that, given any $\mathcal{C} \subset \mathcal{P}(X)$, there is a *smallest* σ-algebra, $\mathcal{A}_\mathcal{C}$, with $\mathcal{C} \subset \mathcal{A}_\mathcal{C}$; we call it the σ-algebra *generated by* $\mathcal{C}$. It is, in fact, the intersection of all σ-algebras (of subsets of X) that contain $\mathcal{C}$. Here is an important example:

Definition 10.1.14 (Borel Algebra, Borel Set). Let $\mathcal{O}_\mathbb{R}$ denote the collection of all *open* subsets of $\mathbb{R}$. The *Borel algebra* of $\mathbb{R}$, denoted by $\mathcal{B}_\mathbb{R}$, is the σ-algebra generated by $\mathcal{O}_\mathbb{R}$. A set $B \in \mathcal{B}_\mathbb{R}$ is said to be a *Borel set* (of $\mathbb{R}$).

Exercise 10.1.15.

(a) Show that, in the above definition of $\mathcal{B}_\mathbb{R}$, we may replace $\mathcal{O}_\mathbb{R}$ by the collection $\mathcal{C}_\mathbb{R}$ of all *closed* subsets of $\mathbb{R}$.

(b) Let $\mathcal{A}_{\mathcal{C}_1}$ and $\mathcal{A}_{\mathcal{C}_2}$ be the σ-algebras generated by $\mathcal{C}_1 := \{(a,b] : a, b \in \mathbb{R}, a \le b\}$ and $\mathcal{C}_2 := \{(-\infty, b] : b \in \mathbb{R}\}$, respectively. Show that

$$\mathcal{A}_{\mathcal{C}_1} = \mathcal{B}_\mathbb{R} = \mathcal{A}_{\mathcal{C}_2}.$$

Theorem 10.1.16 ($\mathcal{M}_\lambda(\mathbb{R})$ is a σ-Algebra). *The collection $\mathcal{M}_\lambda(\mathbb{R})$ of all (Lebesgue) measurable subsets of $\mathbb{R}$ is a σ-algebra containing every set $Z \subset \mathbb{R}$ with $\mu^*(Z) = 0$. Also, Lebesgue measure $\lambda := \lambda^*|\mathcal{M}_\lambda$ is monotone, translation invariant, reflection invariant, and countably (hence finitely) additive.*

Proof. If $\lambda^*(Z) = 0$, then for each $A \subset \mathbb{R}$ we have $0 \le \lambda^*(A \cap Z) \le \lambda^*(Z) = 0$ and hence $\lambda^*(A \cap Z) = 0$. Since λ^* is monotone, we have

$$\lambda^*(A) \ge \lambda^*(A \cap Z^c) = \lambda^*(A \cap Z) + \lambda^*(A \cap Z^c)$$

and $Z \in \mathcal{M}_\lambda$ follows. In particular, $\emptyset \in \mathcal{M}_\lambda$ and hence $\mathbb{R} \in \mathcal{M}_\lambda$ because it is obvious from the definition that $E \in \mathcal{M}_\lambda$ if and only if $E^c \in \mathcal{M}_\lambda$. Next, if $E, F \in \mathcal{M}_\lambda$ and $A \subset \mathbb{R}$, then we have

$$\lambda^*(A) = \lambda^*(A \cap E) + \lambda^*(A \cap E^c)$$
$$= \lambda^*(A \cap E \cap F) + \lambda^*(A \cap E \cap F^c)$$
$$+ \lambda^*(A \cap E^c \cap F) + \lambda^*(A \cap E^c \cap F^c)$$
$$\ge \lambda^*(A \cap (E \cup F)) + \lambda^*(A \cap (E \cup F)^c),$$

which implies that $E \cup F \in \mathcal{M}_\lambda$ and hence $\mathcal{M}_\lambda$ is an *algebra*. It also shows that, if $E, F \in \mathcal{M}_\lambda$ and $E \cap F = \emptyset$, then

$$\lambda^*(E \cup F) = \lambda^*((E \cup F) \cap E) + \lambda^*((E \cup F) \cap E^c) = \lambda^*(E) + \lambda^*(F).$$

Thus, by induction, $\lambda^*|\mathcal{M}_\lambda$ is *finitely additive*. To prove that $\mathcal{M}_\lambda$ is a σ-algebra, let $(E_n)_{n=1}^\infty$ be a sequence of *pairwise disjoint* sets in $\mathcal{M}_\lambda$ and put $E := \bigcup_{k=1}^\infty E_k$ and $F_n := \bigcup_{k=1}^n E_k$. Then we have $F_n \in \mathcal{M}_\lambda$, $F_n \cap E_n = E_n$, and $F_n \cap E_n^c = F_{n-1}$ (with $F_0 := \emptyset$). Now, for each $A \subset \mathbb{R}$, a simple induction gives

$$\lambda^*(A \cap F_n) = \lambda^*(A \cap E_n) + \lambda^*(A \cap F_{n-1}) = \sum_{k=1}^n \lambda^*(A \cap E_k).$$

(Why?) Since $E^c \subset F_n^c$, we therefore obtain

$$\lambda^*(A) = \lambda^*(A \cap F_n) + \lambda^*(A \cap F_n^c) \geq \sum_{k=1}^n \lambda^*(A \cap E_k) + \lambda^*(A \cap E^c),$$

which (letting $n \to \infty$) implies

$$\lambda^*(A) \geq \lambda^*(A \cap E^c) + \sum_{n=1}^\infty \lambda^*(A \cap E_n) \tag{$\dagger$}$$

$$\geq \lambda^*(A \cap E^c) + \lambda^*(A \cap E),$$

and hence $E \in \mathcal{M}_\lambda$. If the $E_n \in \mathcal{M}_\lambda$ are *not* pairwise disjoint, we define $F_1 := E_1$ and $F_n := E_n \setminus \bigcup_{k=1}^{n-1} F_k$ for all $n \geq 2$ and note that $F_n \in \mathcal{M}_\lambda$ for all $n \in \mathbb{N}$, $F_i \cap F_j = \emptyset$ if $i \neq j$, and $\bigcup_{n=1}^\infty E_n = \bigcup_{n=1}^\infty F_n$. We have thus proved that $\mathcal{M}_\lambda$ is a σ-algebra. Next, we already know (cf. Proposition 10.1.4) that λ^* is monotone, translation invariant, and reflection invariant. That $E \in \mathcal{M}_\lambda$ implies $E + x \in \mathcal{M}_\lambda$ for any $x \in \mathbb{R}$ can be seen by noting that (cf. Exercise 10.1.3)

$$\lambda^*(A) = \lambda^*(A - x) = \lambda^*((A - x) \cap E) + \lambda^*((A - x) \cap E^c)$$

$$= \lambda^*\big(((A - x) \cap E) + x\big) + \lambda^*\big(((A - x) \cap E^c) + x\big)$$

$$= \lambda^*(A \cap (E + x)) + \lambda^*(A \cap (E + x)^c),$$

for every $A \subset \mathbb{R}$. A similar argument shows that $E \in \mathcal{M}_\lambda$ if and only if $-E \in \mathcal{M}_\lambda$ and we then have $\lambda(E) = \lambda(-E)$. Finally, the *countable additivity* of λ follows from ($\dagger$) if we replace A by $E := \bigcup_{n=1}^\infty E_n$. $\qquad\square$

Exercise 10.1.17 (Effect of Scaling). Given any $E \in \mathcal{M}_\lambda$ and any *affine* function $f(x) := mx + b$, show that $f(E) = mE + b := \{mx + b : x \in E\} \in \mathcal{M}_\lambda$ and $\lambda(f(E)) = |m|\lambda(E)$. *Hint:* Reduce to the case $m > 0$ and $b = 0$.

Remark 10.1.18.

1. **(Lebesgue Measure Is Complete).** The measure λ is *complete* in the sense that $\mathcal{M}_\lambda(\mathbb{R})$ contains *all* subsets of *any* set of measure zero. We also say that the σ-algebra $\mathcal{M}_\lambda(\mathbb{R})$ is complete. The Borel algebra $\mathcal{B}_\mathbb{R}$, on the other hand, is *not* complete. For example, Cantor's ternary set contains subsets that are *not* Borel sets (cf. Problem #18 at the end of this chapter or Halmos's *Measure Theory* [Hal50], p. 67).
2. **(Lebesgue Measure Generalizes Length).** We have already seen (cf. Proposition 10.1.4) that the outer measure of an interval is its length. Thus $\lambda^*(I) = \lambda(I) := b - a$ if I is an interval with endpoints $a \le b$ and $\lambda^*(I) = \lambda(I) := \infty$ if I is *unbounded*. To justify our use of the symbol λ to denote both the *length* (of intervals) and the *measure* (of Lebesgue measurable sets), we must show that each interval $I \subset \mathbb{R}$ is measurable (i.e., $I \in \mathcal{M}_\lambda$) and hence its length and (Lebesgue) measure are *identical*. In fact, we have the following stronger assertion.

Theorem 10.1.19 ($\mathcal{B}_\mathbb{R} \subset \mathcal{M}_\lambda(\mathbb{R})$). *Every Borel set in $\mathbb{R}$ is Lebesgue measurable. In particular, all open and closed sets are measurable.*

Proof. In view of Exercise 10.1.15, it suffices to show that $(-\infty, b] \in \mathcal{M}_\lambda$ for every $b \in \mathbb{R}$. Given any set $A \subset \mathbb{R}$, we must show that

$$\lambda^*(A) \ge \lambda^*(A') + \lambda^*(A''), \tag{$*$}$$

where $A' := A \cap (-\infty, b]$ and $A'' := A \cap (b, \infty)$. We may (and do) assume that $\lambda^*(A) < \infty$ and, given any $\varepsilon > 0$, pick a sequence (I_n) of open intervals such that $A \subset \bigcup_{n=1}^\infty I_n$ and

$$\sum_{n=1}^\infty \lambda(I_n) \le \lambda^*(A) + \varepsilon,$$

where $\lambda(I_n)$ is the (usual) *length* of the interval I_n. For each $n \in \mathbb{N}$ consider the intervals $I_n' := I_n \cap (-\infty, b]$ and $I_n'' := I_n \cap (b, \infty)$, which may be empty, and note that we have $\lambda(I_n) = \lambda(I_n') + \lambda(I_n'')$. Now $A' \subset \bigcup_{n=1}^\infty I_n'$ and $A'' \subset \bigcup_{n=1}^\infty I_n''$ and hence we have

$$\lambda^*(A') \le \lambda^* \left(\bigcup_{n=1}^\infty I_n' \right) \le \sum_{n=1}^\infty \lambda(I_n') \quad \text{and}$$

$$\lambda^*(A'') \le \lambda^* \left(\bigcup_{n=1}^\infty I_n'' \right) \le \sum_{n=1}^\infty \lambda(I_n''),$$

from which we deduce

$$\lambda^*(A') + \lambda^*(A'') \leq \sum_{n=1}^{\infty} \left(\lambda(I'_n) + \lambda(I''_n) \right)$$

$$\leq \sum_{n=1}^{\infty} \lambda(I_n) \leq \lambda^*(A) + \varepsilon.$$

Since $\varepsilon > 0$ was arbitrary, (∗) follows and the proof is complete. □

Remark 10.1.20. As it turns out, the inclusion $\mathcal{B}_{\mathbb{R}} \subset \mathcal{M}_\lambda(\mathbb{R})$ is a *proper* one; i.e., there are Lebesgue measurable sets that are *not* Borel sets. One way to see this is given in Problem #18 at the end of the chapter. But, as was pointed out before, we have Littlewood's first principle, which says that a Lebesgue measurable set is *nearly* a finite union of (disjoint) open intervals and hence *nearly Borel*. Before making this and related approximations precise, let us give an example of a *nonmeasurable set:*

Example 10.1.21 (A Nonmeasurable Set). On the real line $\mathbb{R}$, let us set $x \sim y$ if $x - y \in \mathbb{Q}$. This is easily seen to be an *equivalence relation.* Now let $E \subset (0, 1)$ be a set containing *exactly one* representative from each equivalence class (the Axiom of Choice is needed here). We claim that E is *not* Lebesgue measurable. Indeed, given any $x \in (0, 1)$, there is a $y \in E$ such that $x - y \in (-1, 1) \cap \mathbb{Q}$. Therefore, if $(-1, 1) \cap \mathbb{Q} = \{r_1, r_2, \ldots\}$, then

$$(0, 1) \subset \bigcup_{n=1}^{\infty} (E + r_n) \subset (-1, 2), \tag{†}$$

where $E + r_n := \{y + r_n : y \in E\}$. Now, if $j \neq k$, then $(E + r_j) \cap (E + r_k) = \emptyset$. Indeed, if $x \in (E + r_j) \cap (E + r_k)$, then we have $x = y + r_j = z + r_k$ with $y, z \in E$. It follows that $y - z = r_k - r_j \in \mathbb{Q}$ and hence $y \sim z$. But then, by the definition of E, we must have $y = z$, which is absurd. Now suppose that E is Lebesgue measurable. Then we have $\lambda(E) = \lambda(E + r_n)$ for all $n \in \mathbb{N}$ and, by the countable additivity, (†) implies

$$1 = \lambda((0, 1)) \leq \sum_{n=1}^{\infty} \lambda(E + r_n) \leq \lambda((-1, 2)) = 3. \tag{‡}$$

If $\lambda(E) = 0$, then the first inequality in (‡) gives $1 \leq 0$ and if $\lambda(E) > 0$, then the second inequality in (‡) implies $\infty \leq 3$. It follows from these contradictions that E is indeed nonmeasurable as claimed.

Theorem 10.1.22 (Littlewood's First Principle). *If $E \in \mathcal{M}_{\lambda < \infty}(\mathbb{R})$ (i.e., $\lambda(E) < \infty$), then for any $\varepsilon > 0$ there exists a finite sequence $(I_k)_{k=1}^n$ of (disjoint) open intervals such that, with $U := \bigcup_{k=1}^n I_k$, we have*

$$d_\lambda(U, E) := \lambda^*(U \triangle E) < \varepsilon.$$

Conversely, if for each $\varepsilon > 0$ there is a finite sequence of open intervals with union U such that $\lambda^*(U \Delta E) < \varepsilon$, then $E \in \mathcal{M}_\lambda(\mathbb{R})$, possibly with $\lambda(E) = \infty$.

Proof. Suppose $E \in \mathcal{M}_{\lambda < \infty}$ and let $\varepsilon > 0$ be given. Using the *outer regularity* of λ (cf. Corollary 10.1.9), we can pick an open set O with $E \subset O$ and $\lambda(O) < \lambda(E) + \varepsilon/2$. In particular, $\lambda(O) < \infty$. Let $(I_k)_{k=1}^\infty$ be a sequence of *disjoint* open intervals with $O = \bigcup_{k=1}^\infty I_k$ and note that the (finite) additivity of λ gives

$$\lambda(O \setminus E) = \lambda\left(\bigcup_{k=1}^\infty I_n \setminus E\right) = \lambda(O) - \lambda(E) < \frac{\varepsilon}{2}. \qquad (*)$$

By countable additivity, $\lambda(O) = \sum_{k=1}^\infty \lambda(I_k) < \infty$ so we can pick $n \in \mathbb{N}$ such that $\sum_{k=1}^n \lambda(I_k) > \lambda(O) - \varepsilon/2$; i.e., $\lambda(O \setminus U) = \lambda(O) - \lambda(U) < \varepsilon/2$, where we have defined $U := \bigcup_{k=1}^n I_k$. Since $E \setminus U \subset O \setminus U$ and λ is *monotone*, we have $\lambda(E \setminus U) \le \lambda(O \setminus U) < \varepsilon/2$. On the other hand, using $U \setminus E \subset O \setminus E$ and $(*)$, we also have $\lambda(U \setminus E) < \varepsilon/2$ and hence

$$\lambda(U \Delta E) = \lambda(U \setminus E) + \lambda(E \setminus U) < \frac{\varepsilon}{2} + \frac{\varepsilon}{2} = \varepsilon.$$

For the converse we shall prove that

$$\exists G \in \mathcal{G}_\delta \quad \text{with} \quad E \subset G \quad \text{and} \quad \lambda^*(G \setminus E) = 0, \qquad (**)$$

from which $E = G \setminus (G \setminus E) \in \mathcal{M}_\lambda$ follows because any $\mathcal{G}_\delta$ is measurable and so is any set with (outer) measure zero. Now note that $(**)$ follows from

$$\forall \varepsilon > 0 \text{ there is an open set } O \supset E \text{ such that } \lambda^*(O \setminus E) < \varepsilon. \qquad (***)$$

Indeed, applying $(***)$ with $\varepsilon = 1/k, k \in \mathbb{N}$, we can find an open set $O_k \supset E$ with $\lambda^*(O_k \setminus E) < 1/k$. If we set $G := \bigcap_{k=1}^\infty O_k$, then $E \subset G \in \mathcal{G}_\delta$ and $\lambda^*(G \setminus E) < 1/k$ for all $k \in \mathbb{N}$, implying $(**)$. So suppose that for every $\varepsilon > 0$ there is a finite union U of open intervals with $\lambda^*(U \Delta E) < \varepsilon$. Then we have $\lambda^*(E \setminus U) < \varepsilon$ (as well as $\lambda^*(U \setminus E) < \varepsilon$) and, using *outer regularity*, we can pick an open set $O' \supset E \setminus U$ such that $\lambda^*(O') \le \lambda^*(E \setminus U) + \varepsilon < 2\varepsilon$. But then, with $O := O' \cup U$, we have $O \supset E$ and $\lambda^*(O \setminus E) \le \lambda^*(U \setminus E) + \lambda^*(O') < 3\varepsilon$, which implies $(***)$ and completes the proof. $\qquad \square$

In our next theorem we list a few statements that are logically equivalent to the (Lebesgue) measurability of a set. Actually, some of them are already contained in (the proof of) Littlewood's first principle (Theorem 10.1.22).

Theorem 10.1.23 (Criteria for Measurability). *For a set $E \subset \mathbb{R}$ the following statements are pairwise equivalent:*

   (i) *E is measurable.*
   (ii) *For every $\varepsilon > 0$ there is an open set $O \supset E$ such that $\lambda^*(O \setminus E) < \varepsilon$.*

(iii) *For every $\varepsilon > 0$ there is a closed set $F \subset E$ such that $\lambda^*(E \setminus F) < \varepsilon$.*
(iv) *There is a set $G \in \mathcal{G}_\delta$ such that $E \subset G$ and $\lambda^*(G \setminus E) = 0$.*
(v) *There is a set $F \in \mathcal{F}_\sigma$ such that $E \supset F$ and $\lambda^*(E \setminus F) = 0$.*

If $\lambda^*(E) < \infty$, then the above statements are equivalent to
(vi) *For every $\varepsilon > 0$ there is a finite union U of open intervals such that $d_\lambda(E, U) := \lambda^*(U \Delta E) < \varepsilon$.*

Proof. The implications (i) $\Rightarrow$ (ii) $\Rightarrow$ (iv) $\Rightarrow$ (i) are already contained in the proof of Theorem 10.1.22. Also, $E \in \mathcal{M}_\lambda \Leftrightarrow E^c \in \mathcal{M}_\lambda$ and applying (ii) to E^c, for any $\varepsilon > 0$ we can find an open set $O \supset E^c$ with $\lambda(O \setminus E^c) = \lambda(O \cap E) = \lambda(E \cap F^c) = \lambda(E \setminus F) < \varepsilon$, where $F := O^c$ is a closed set with $F \subset E$. It follows that (ii) $\Leftrightarrow$ (iii). Similarly, applying (iv) to E^c, we can find $G \in \mathcal{G}_\delta$ such that $E^c \subset G$ and $\lambda^*(G \setminus E^c) = 0$. Thus, if we set $F := G^c$, then $F \in \mathcal{F}_\sigma, E \supset F$ and $\lambda^*(E \setminus F) = 0$ and we obtain (iv) $\Leftrightarrow$ (v). Since the last statement also follows from Theorem 10.1.22, the proof is complete. $\square$

Remark 10.1.24.

1 (**Measurable Means "Almost Borel"**). Since $\mathcal{G}_\delta \subset \mathcal{B}_\mathbb{R}$ and $\mathcal{F}_\sigma \subset \mathcal{B}_\mathbb{R}$, criteria (iv) and (v) show that a set $E \subset \mathbb{R}$ is (Lebesgue) measurable if and only if it can be approximated by a $\mathcal{G}_\delta$ or an $\mathcal{F}_\sigma$ (i.e., by a Borel set) to within a set of measure zero. Indeed, (iv) means that $E = G \setminus (G \setminus E)$ with $G \in \mathcal{G}_\delta$ and $\lambda^*(G \setminus E) = 0$, while (v) gives $E = F \cup (E \setminus F)$ with $F \in \mathcal{F}_\sigma$ and $\lambda^*(E \setminus F) = 0$.

2 ($\mathcal{M}_\lambda(\mathbb{R})$ **is the Completion of** $\mathcal{B}_\mathbb{R}$). Let $\mathcal{A}$ denote the completion of $\mathcal{B}_\mathbb{R}$, i.e., the *smallest* σ-algebra that contains $\mathcal{B}_\mathbb{R}$ as well as all subsets of any Borel set of measure zero. Now $\mathcal{M}_\lambda(\mathbb{R})$ is complete and (by Theorem 10.1.19) $\mathcal{B}_\mathbb{R} \subset \mathcal{M}_\lambda(\mathbb{R})$. Thus we need only show that $\mathcal{M}_\lambda(\mathbb{R}) \subset \mathcal{A}$. Given any $E \in \mathcal{M}_\lambda$, pick (using the above remark) $G \in \mathcal{B}_\mathbb{R}$ with $G = E \cup (G \setminus E)$. Similarly pick $Z \in \mathcal{B}_\mathbb{R}$ with $G \setminus E \subset Z$ and $\lambda(Z) = 0$. But then $G \setminus E \in \mathcal{A}$ and hence $E = G \setminus (G \setminus E) \in \mathcal{A}$.

Recall that a numerical function f is continuous at a point a if we have $\lim_{n \to \infty} f(x_n) = f(a) = f(\lim_{n \to \infty} x_n)$, for every sequence (x_n) with $\lim(x_n) = a$. Even though Lebesgue measure is a *set function* $\lambda : \mathcal{M}_\lambda \to [0, \infty]$, we shall now show that it also behaves nicely (i.e., *continuously*) with respect to *limits of sets* in the following sense:

Theorem 10.1.25 (Continuity of λ, Monotone Convergence).

(a) *Let (E_n) be a sequence in $\mathcal{M}_\lambda$ that is increasing, i.e., $E_1 \subset E_2 \subset E_3, \ldots$ and let $E = \lim E_n := \bigcup_{n=1}^\infty E_n$. Then we have*

$$\lambda(E) = \lambda\left(\lim_{n \to \infty} E_n\right) = \lim_{n \to \infty} \lambda(E_n).$$

(b) Let (E_n) be a sequence in $\mathcal{M}_\lambda$ that is decreasing, i.e., $E_1 \supset E_2 \supset E_3, \ldots$ and let $E = \lim E_n := \bigcap_{n=1}^{\infty} E_n$. If $\lambda(E_1) < \infty$, then we have

$$\lambda(E) = \lambda\left(\lim_{n \to \infty} E_n\right) = \lim_{n \to \infty} \lambda(E_n).$$

Proof.

(a) Let $F_1 := E_1$ and $F_n := E_n \setminus E_{n-1}$ for all $n \geq 2$. Then (F_n) is a sequence of *pairwise disjoint* measurable sets with $E_n = \bigcup_{k=1}^{n} E_k = \bigcup_{k=1}^{n} F_k$ for every $n \in \mathbb{N}$ and $E = \bigcup_{k=1}^{\infty} E_k = \bigcup_{k=1}^{\infty} F_k$. Thus, by countable (and finite) additivity, we have $\lambda(E) = \sum_{k=1}^{\infty} \lambda(F_k)$, $\lambda(E_n) = \sum_{k=1}^{n} \lambda(F_k)$ and hence

$$\lambda(E) = \sum_{k=1}^{\infty} \lambda(F_k) = \lim_{n \to \infty} \sum_{k=1}^{n} \lambda(F_k) = \lim_{n \to \infty} \lambda(E_n).$$

(b) Let $F_n := E_n \setminus E_{n+1}$ for all $n \in \mathbb{N}$. Then (F_n) is an *increasing* sequence of measurable sets with $F := \bigcup_{k=1}^{\infty} F_k = E_1 \setminus E$ (where $E := \bigcap_{k=1}^{\infty} E_k$) and $F_n = \bigcup_{k=1}^{n} F_k = E_1 \setminus E_{n+1}$. Therefore, applying part (a), we have

$$\lambda(E_1) - \lambda(E) = \lambda(E_1 \setminus E) = \lambda(F) = \lim_{n \to \infty} \lambda(F_n) = \lim_{n \to \infty} \left(\lambda(E_1) - \lambda(E_{n+1})\right),$$

which, in view of the assumption $\lambda(E_1) < \infty$, completes the proof. $\square$

Exercise 10.1.26. Show that the assumption $\lambda(E_1) < \infty$ in part (b) is *necessary* by giving an example of a decreasing sequence (E_n) in $\mathcal{M}_\lambda$ with $\lambda(E_n) = \infty$ for all $n \in \mathbb{N}$ and $E := \bigcap_{n=1}^{\infty} E_n = \emptyset$.

10.2   (Lebesgue) Measurable Functions

As we saw in Chap. 7, the Riemann integral was initially defined for step functions before it was extended to more general functions that could be approximated by step functions, e.g., the regulated functions. For Lebesgue integral we follow the same kind of construction, but the natural substitute for a step function is now a *simple function*, to be defined shortly. These functions will be the building blocks of more general *measurable functions* that include all functions one encounters in practice. Lebesgue integral, initially defined for simple functions, will be extended to various subspaces of the space of all measurable functions by a limiting process. These spaces are much larger than the space of Riemann integrable functions and Lebesgue integral is proved to be a true *extension* of the Riemann integral when the latter is defined.

Definition 10.2.1 (Simple Function). We say that a function $f : \mathbb{R} \to \mathbb{R}$ is a *simple function* if it can be written as a *finite* linear combination

$$\phi = \sum_{k=1}^{n} a_k \chi_{A_k}, \tag{$\dagger$}$$

with real constants $a_1, \ldots, a_n$ and *measurable* sets $A_1, \ldots, A_n$. The set of all simple functions on $\mathbb{R}$ will be denoted by *Simp*$(\mathbb{R})$.

Example 10.2.2 (Step$(\mathbb{R}) \subset$ Simp$(\mathbb{R})$). Since *intervals* are *measurable* (including those reduced to points), it is obvious that all step functions are simple functions. The converse is *false*. For example (the Dirichlet function) $\chi_{\mathbb{Q}}$ is certainly *not* a step function, but it *is* clearly a simple function.

Remark 10.2.3 (Canonical Representation). Note that, for a simple function ϕ, the representation ($\dagger$) in Definition 10.2.1 is *not unique*. But if ϕ is simple and if $\{a_1, \ldots, a_n\}$ is the set of *distinct* values of ϕ, then

$$\phi = \sum_{k=1}^{n} a_k \chi_{A_k}, \tag{$\ddagger$}$$

where the measurable sets $A_k := \{x : \phi(x) = a_k\}$ are *pairwise disjoint* and $\bigcup_{k=1}^{n} A_k = \mathbb{R}$. The representation ($\ddagger$), which *is* unique, is called the *canonical representation* of ϕ.

Exercise 10.2.4. Given any $\phi \in$ *Simp*$(\mathbb{R})$, show that we have $|\phi| \in$ *Simp*$(\mathbb{R})$. More generally, show that if $f(x)$ is any real-valued function whose domain contains the (finite) range of ϕ, then $f \circ \phi \in$ *Simp*$(\mathbb{R})$.

Proposition 10.2.5. *Given any* ϕ, $\psi \in$ *Simp*$(\mathbb{R})$, *and* $c \in \mathbb{R}$, *the functions* $c\phi + \psi$, $\phi\psi$, $\max\{\phi, \psi\}$, *and* $\min\{\phi, \psi\}$ *are simple functions. In particular,* $\phi^+ := \max\{\phi, 0\}$, $\phi^- := \max\{-\phi, 0\}$, *and* $|\phi| = \phi^+ + \phi^-$ *are simple functions.*

Proof. Let $\phi = \sum_{j=1}^{m} a_j \chi_{A_j}$ and $\psi = \sum_{k=1}^{n} b_k \chi_{B_k}$ be the canonical representations of ϕ and ψ, respectively, and let $C_{jk} := A_j \cap B_k$, for $1 \le j \le m$ and $1 \le k \le n$. Since (A_j) and (B_k) are both (finite) sequences of pairwise disjoint measurable sets, so is the (double) sequence (C_{jk}). Also, $\chi_{A_j} = \sum_{k=1}^{n} \chi_{C_{jk}}$ and $\chi_{B_k} = \sum_{j=1}^{m} \chi_{C_{jk}}$. Therefore,

$$c\phi + \psi = \sum_{k=1}^{n} \sum_{j=1}^{m} (ca_j + b_k)\chi_{C_{jk}} \quad \text{and} \quad \phi\psi = \sum_{k=1}^{n} \sum_{j=1}^{m} a_j b_k \chi_{C_{jk}} \tag{$*$}$$

are indeed simple. For $\max\{\phi, \psi\}$ and $\min\{\phi, \psi\}$ we can now use Exercise 10.2.4 and the fact that $\max\{a, b\} = (a+b+|a-b|)/2$ and $\min\{a, b\} = (a+b-|a-b|)/2$ for any a, $b \in \mathbb{R}$. $\qquad\square$

Remark 10.2.6.

(1) Note that the representation of $c\phi + \psi$ (resp., $\phi\psi$) given in (∗) is *not* necessarily canonical because the numbers $ca_j + b_k$ (resp., $a_j b_k$) need not be distinct.
(2) Using the above proposition repeatedly, it follows that if $\phi_1, \ldots, \phi_n$ are simple functions and if $P(x_1, \ldots, x_n)$ is a real-valued *polynomial* function of n variables, then $P(\phi_1, \ldots, \phi_n)$, $\max\{\phi_1, \ldots, \phi_n\}$, and $\min\{\phi_1, \ldots, \phi_n\}$ are also simple functions.

As pointed out above, simple functions are the building blocks of the class of *measurable functions*. Before defining the latter, we need the following:

Proposition 10.2.7. *Let $E \in \mathcal{M}_\lambda(\mathbb{R})$ and $f : E \to \mathbb{R}$. Then the following statements are pairwise equivalent:*

(i) $\{x \in E : f(x) > b\} \in \mathcal{M}_\lambda \quad (\forall b \in \mathbb{R})$,
(ii) $\{x \in E : f(x) \geq b\} \in \mathcal{M}_\lambda \quad (\forall b \in \mathbb{R})$,
(iii) $\{x \in E : f(x) < b\} \in \mathcal{M}_\lambda \quad (\forall b \in \mathbb{R})$,
(iv) $\{x \in E : f(x) \leq b\} \in \mathcal{M}_\lambda \quad (\forall b \in \mathbb{R})$.

Moreover, any of these conditions implies that

(v) $f^{-1}(\{b\}) := \{x \in E : f(x) = b\} \in \mathcal{M}_\lambda \qquad (\forall b \in \mathbb{R})$.

The equivalence of (i)–(iv) *also holds if $f : E \to \overline{\mathbb{R}} := [-\infty, \infty]$ is extended real-valued and* (v) *can then be replaced by*

(vi) $f^{-1}(\{b\}) := \{x \in E : f(x) = b\} \in \mathcal{M}_\lambda \qquad (\forall b \in \overline{\mathbb{R}})$.

Proof. Whether $f : E \to \mathbb{R}$ or $f : E \to \overline{\mathbb{R}}$, for each $b \in \mathbb{R}$ we have

(1) $\{x \in E : f(x) \geq b\} = \bigcap_{n=1}^{\infty} \{x \in E : f(x) > b - \frac{1}{n}\}$,
(2) $\{x \in E : f(x) < b\} = E \setminus \{x \in E : f(x) \geq b\}$,
(3) $\{x \in E : f(x) \leq b\} = \bigcap_{n=1}^{\infty} \{x \in E : f(x) < b + \frac{1}{n}\}$, and
(4) $\{x \in E : f(x) > b\} = E \setminus \{x \in E : f(x) \leq b\}$,

from which the equivalence of the four statements follows at once. For statement (v), note that (i) and (ii) imply

$$f^{-1}(\{b\}) = \{x \in E : f(x) \geq b\} \setminus \{x \in E : f(x) > b\} \in \mathcal{M}_\lambda \quad (\forall\, b \in \mathbb{R}).$$

Finally, if $f : E \to \overline{\mathbb{R}}$, then (v) implies that for (vi) we need only consider the cases $b = \pm\infty$. But then note that we have

$$\{x \in E : f(x) = +\infty\} = \bigcap_{n=1}^{\infty} \{x \in E : f(x) > n\} \in \mathcal{M}_\lambda, \quad \text{and}$$

$$\{x \in E : f(x) = -\infty\} = \bigcap_{n=1}^{\infty} \{x \in E : f(x) < -n\} \in \mathcal{M}_\lambda,$$

which complete the proof.   □

Definition 10.2.8 ((Lebesgue) Measurable Function). Let $E \in \mathcal{M}_\lambda(\mathbb{R})$. A function $f : E \to \mathbb{R}$ [resp., $f : E \to \overline{\mathbb{R}}$] is said to be *(Lebesgue) measurable* if any one of the four statements (i)–(iv) in Proposition 10.2.7 is satisfied.

Notation 10.2.9 ($\mathcal{L}^0(E, \mathbb{R})$, $\mathcal{L}^0(E, \overline{\mathbb{R}})$). The set of all measurable real-valued [resp., extended real-valued] functions with domain E will be denoted $\mathcal{L}^0(E, \mathbb{R})$ [resp., $\mathcal{L}^0(E, \overline{\mathbb{R}})$]. We also write $\mathcal{L}^0(E, \overline{\mathbb{R}}_+)$ for the space of all *nonnegative*, extended real-valued functions $f : E \to \overline{\mathbb{R}}_+$, where $\overline{\mathbb{R}}_+ := [0, \infty]$. Also, we may write $\mathcal{L}^0(\mathbb{R})$ instead of $\mathcal{L}^0(\mathbb{R}, \mathbb{R})$.

Remark 10.2.10.

1. **(Measurability vs. Continuity).** Note that the sets in statements (i)–(iv) of Proposition 10.2.7 are all *inverse images*. For instance, in statement (i), $\{x \in E : f(x) > b\}$ is (by definition) $f^{-1}((b, \infty))$ [resp., $f^{-1}((b, \infty])$] if $f : E \to \mathbb{R}$ [resp., $f : E \to \overline{\mathbb{R}}$]. In other words, measurability imposes conditions on the inverse images of a special class of sets. This reminds us of *continuity*: Recall that a function $f : \mathbb{R} \to \mathbb{R}$ is continuous if and only if $f^{-1}(I)$ is *open* for each open interval I.

2. **(Global vs. Local).** To study measurable functions and their integrals it is possible and convenient to work with functions defined *globally*, i.e., on the *entire* real line $\mathbb{R}$. One can then *localize* the results using *trivial extensions:*

Definition 10.2.11 (Trivial Extension). Given any set $S \subset \mathbb{R}$ and any function $f : S \to \mathbb{R}$ (or $f : S \to \overline{\mathbb{R}}$), the *trivial extension* of f is the function $\tilde{f} : \mathbb{R} \to \mathbb{R}$ defined as follows:

$$\tilde{f}(x) := \begin{cases} f(x) & \text{if } x \in S, \\ 0 & \text{if } x \notin S. \end{cases}$$

Proposition 10.2.12. *Let $E \in \mathcal{M}_\lambda$. A function $f : E \to \mathbb{R}$ (or $f : E \to \overline{\mathbb{R}}$) is measurable if and only if its trivial extension $\tilde{f}$ is.*

Proof. For a given $b \in \mathbb{R}$, note that we have

$$\{x \in \mathbb{R} : \tilde{f}(x) \le b\} = \{x \in E : f(x) \le b\} \cup F,$$

where $F = \emptyset$ if $b < 0$ and $F = E^c$ if $b \ge 0$. It is then easily seen (why?) that $\{x \in E : f(x) \le b\} \in \mathcal{M}_\lambda$ if and only if $\{x \in \mathbb{R} : \tilde{f}(x) \le b\} \in \mathcal{M}_\lambda$. $\qquad \square$

Example 10.2.13.

(1) Constant functions are measurable. Indeed, if $f(x) = c \in \mathbb{R}$ for all $x \in \mathbb{R}$ and if $b \in \mathbb{R}$ is any given number, then

$$\{x : f(x) < b\} = \begin{cases} \emptyset & \text{if } b \le c, \\ \mathbb{R} & \text{if } b > c. \end{cases}$$

If $f(x) = \infty$ for all $x \in \mathbb{R}$, then $\{x : f(x) < b\} = \emptyset$ for every $b \in \mathbb{R}$. Finally, if $f(x) = -\infty$ for all $x \in \mathbb{R}$, then $\{x : f(x) < b\} = \mathbb{R}$ for all $b \in \mathbb{R}$.

(2) We have $\boldsymbol{Simp}(\mathbb{R}) \subset \mathcal{L}^0(\mathbb{R})$. In fact $\phi \in \boldsymbol{Simp}(\mathbb{R})$ if and only if ϕ is *measurable* and takes on only a *finite* number of values. Indeed, if ϕ is simple with canonical representation $\phi = \sum_{j=1}^{m} a_j \chi_{A_j}$, then for any $b \in \mathbb{R}$ we have

$$\{x : \phi(x) \le b\} = \bigcup_{\{j : a_j \le b\}} A_j \in \mathcal{M}_\lambda.$$

Conversely, if $\phi \in \mathcal{L}^0(\mathbb{R})$ and if $a_1, a_2, \ldots, a_m$ are the distinct values of ϕ, then $A_j := \phi^{-1}(\{a_j\}) \in \mathcal{M}_\lambda$ for each j and we have $\phi = \sum_{j=1}^{m} a_j \chi_{A_j}$.

(3) Since *step functions* are simple, every step function is measurable.

(4) Every *continuous* function $f : E \to \mathbb{R}$ with measurable domain $E \in \mathcal{M}_\lambda$ is measurable. Indeed, the interval (b, ∞) is *open* for every $b \in \mathbb{R}$ and hence (by Theorem 4.3.4) we have $\{x \in E : f(x) > b\} = f^{-1}((b, \infty)) = E \cap O$, for some open set $O \subset \mathbb{R}$. Since $O \in \mathcal{M}_\lambda$, we have $E \cap O \in \mathcal{M}_\lambda$ and $f \in \mathcal{L}^0(E, \mathbb{R})$ follows.

(5) If $f \in \mathcal{L}^0(E, \mathbb{R})$ and if $F \subset E$ is measurable, then the restriction $g := f|F$ is a measurable function because $\{x \in F : g(x) > b\} = \{x \in E : f(x) > b\} \cap F \in \mathcal{M}_\lambda$ for all $b \in \mathbb{R}$.

Exercise 10.2.14. Let $E \in \mathcal{M}_\lambda$. Show that $f \in \mathcal{L}^0(E, \overline{\mathbb{R}})$ if and only if the sets

$$E_\infty := \{x \in E : f(x) = \infty\} \quad \text{and} \quad E_{-\infty} := \{x \in E : f(x) = -\infty\}$$

are both measurable and the *real-valued* function

$$g(x) = \begin{cases} f(x) & \text{if } x \notin E_\infty \cup E_{-\infty}, \\ 0 & \text{if } x \in E_\infty \cup E_{-\infty} \end{cases}$$

is measurable.

Exercise 10.2.15 (Monotone $\Rightarrow$ Measurable). Show that, if $f : [a, b] \to \mathbb{R}$ is *monotone* (or, more generally, of *bounded variation*), then it is measurable.

Exercise 10.2.16 (Composition with Continuous Functions). Show that, if $f \in \mathcal{L}^0(\mathbb{R})$ and $g \in C(\mathbb{R})$, then $g \circ f \in \mathcal{L}^0(\mathbb{R})$.

The next proposition shows that we can modify a measurable function on a set of measure zero without destroying its measurability.

Proposition 10.2.17. *Let $E \in \mathcal{M}_\lambda$, $f \in \mathcal{L}^0(E, \mathbb{R})$, and $g : E \to \mathbb{R}$. If the set $Z := \{x \in E : f(x) \ne g(x)\}$ has measure zero, then $g \in \mathcal{L}^0(E, \mathbb{R})$.*

Proof. Simply note that for any $b \in \mathbb{R}$ we have

$$\{x \in E : g(x) > b\} = \{x \in E \setminus Z : f(x) > b\} \cup \{x \in Z : g(x) > b\}.$$

Now the first set on the right side is measurable because $E \setminus Z \in \mathcal{M}_\lambda$ and, f being measurable, so is its restriction $f | E \setminus Z$. As for the second set on the right side, it has measure zero and hence is also in $\mathcal{M}_\lambda$. □

Remark 10.2.18. Given any $E \in \mathcal{M}_\lambda(\mathbb{R})$, we can check at once that "$f(x) = g(x)$ *for almost all* $x \in E$" is an equivalence relation on $\mathcal{L}^0(E, \mathbb{R})$ (as well as $\mathcal{L}^0(E, \overline{\mathbb{R}})$). Thus, in view of Proposition 10.2.17, it is natural to *identify* measurable functions that are *equal almost everywhere* or even allow functions to be *defined almost everywhere*. This can be done more formally:

Notation 10.2.19 (The Spaces $L^0(E, \mathbb{R})$ and $L^0(E, \overline{\mathbb{R}})$). Given any $E \in \mathcal{M}_\lambda(\mathbb{R})$, let $\mathcal{N}(E, \mathbb{R})$ [resp. $\mathcal{N}(E, \overline{\mathbb{R}})$] denote the set of all *null functions,* i.e., all $f : E \to \mathbb{R}$ [resp., $f : E \to \overline{\mathbb{R}}$] with $f(x) = 0$ for *almost all* $x \in E$. We then define $L^0(E, \mathbb{R}) := \mathcal{L}^0(E, \mathbb{R}) / \mathcal{N}(E, \mathbb{R})$ [resp., $L^0(E, \overline{\mathbb{R}}) := \mathcal{L}^0(E, \overline{\mathbb{R}}) / \mathcal{N}(E, \overline{\mathbb{R}})$]. Thus, for each $f \in \mathcal{L}^0(E, \mathbb{R})$, say, its class $[f] \in L^0(E, \mathbb{R})$ will consist of all functions in $\mathcal{L}^0(E, \mathbb{R})$ that agree with f for almost all $x \in E$. If $E = \mathbb{R}$, we may write $L^0(\mathbb{R})$ instead of $L^0(\mathbb{R}, \mathbb{R})$. Actually, we shall often *identify* any pair of measurable functions on E that are equal almost everywhere. Therefore, we may (occasionally) even write $L^0(E, \mathbb{R})$ instead of $\mathcal{L}^0(E, \mathbb{R})$, etc.

As in the case of continuous functions, measurable functions are stable under algebraic and lattice operations:

Proposition 10.2.20. *Given any* $E \in \mathcal{M}_\lambda$, *the set* $\mathcal{L}^0(E, \mathbb{R})$ *is an algebra with identity as well as a lattice. In other words, all constant functions are measurable and, for any measurable functions f and g on E and any $c \in \mathbb{R}$, the functions*

$$cf + g, \quad f^2, \quad |f|, \quad fg, \quad \max\{f, g\}, \quad \min\{f, g\}, \quad f^+ \quad and \quad f^- \quad (\dagger)$$

are also measurable. Finally, if $g(x) \neq 0$ for all $x \in E$, or if the (measurable) set $Z := \{x \in E : g(x) = 0\}$ has measure zero and $g(x)$ is reassigned an arbitrary nonzero value for each $x \in Z$, then f/g is also measurable on E.

Proof. We have already seen that constant functions on E are measurable. To prove the measurability of the functions listed in ($\dagger$), it suffices to show that $f + g$, f^2 and $|f|$ are measurable. Indeed, we can then use the identities $fg = [(f + g)^2 - f^2 - g^2]/2$, $\max\{f, g\} = (f + g + |f - g|)/2$, and $\min\{f, g\} = (f + g - |f - g|)/2$. Now, given any $b \in \mathbb{R}$, consider the set $\{x \in E : f(x) + g(x) < b\}$. Since $\mathbb{Q}$ is *dense* in $\mathbb{R}$, we can find a number $r \in \mathbb{Q}$ such that $f(x) < r < b - g(x)$ and hence

$$\{x : f(x) + g(x) < b\} = \bigcup_{r \in \mathbb{Q}} \left(\{x : f(x) < r\} \cap \{x : g(x) < b - r\} \right) \in \mathcal{M}_\lambda$$

because the right side is a *countable* union of measurable sets. Therefore, $f + g \in \mathcal{L}^0(E, \mathbb{R})$. Next note that

$$b \geq 0 \implies \{x : f^2(x) > b\} = \{x : f(x) > \sqrt{b}\} \cup \{x : f(x) < -\sqrt{b}\} \in \mathcal{M}_\lambda \text{ and}$$

$$b < 0 \Longrightarrow \{x : f^2(x) > b\} = E \in \mathcal{M}_\lambda$$

imply $f^2 \in \mathcal{L}^0(E, \mathbb{R})$. Also, $|f| \in \mathcal{L}^0(E, \mathbb{R})$ because for each $b \geq 0$ we have

$$\{x : |f(x)| > b\} = \{x : f(x) > b\} \cup \{x : f(x) < -b\} \in \mathcal{M}_\lambda.$$

Finally, for f/g, we may assume (by Proposition 10.2.17) that $g(x) \neq 0$ for all $x \in E$ and need only show that $1/g \in \mathcal{L}^0(E, \mathbb{R})$. However, we then have

$$\left\{x : \frac{1}{g(x)} > b\right\} = \begin{cases} \{x : 0 < g(x) < 1/b\} \in \mathcal{M}_\lambda & \text{if } b > 0, \\ \{x : g(x) > 0\} \in \mathcal{M}_\lambda & \text{if } b = 0, \\ \{x : g(x) < 1/b\} \cup \{x : g(x) > 0\} \in \mathcal{M}_\lambda & \text{if } b < 0, \end{cases}$$

which shows indeed that $1/g \in \mathcal{L}^0(E, \mathbb{R})$. $\square$

Remark 10.2.21.

(1) Proposition 10.2.20 and its proof remain mostly valid for $\mathcal{L}^0(E, \overline{\mathbb{R}})$ with a couple of exceptions. First, recall that we have adopted the convention that $0 \cdot \pm\infty = 0$ but have left the expression $\infty - \infty$ undefined. Therefore, $cf = 0$ if $c = 0$, even for $f : E \to \overline{\mathbb{R}}$. However, $f + g$ is *not* well defined on the *measurable* set

$$F := \{x \in E : f(x) = \infty, \; g(x) = -\infty\} \cup \{x \in E : f(x) = -\infty, \; g(x) = \infty\}.$$

Thus $\mathcal{L}^0(E, \overline{\mathbb{R}})$ is *not* a vector space. However, if we *define* $f + g$ to be *any* (*fixed*) number (e.g., zero) on F, or if F has *measure zero* and we define $f + g$ *arbitrarily* on F, then $f + g \in \mathcal{L}^0(E, \overline{\mathbb{R}})$ follows. (Why?) Next, our proof of the measurability of fg for *real-valued* measurable functions f and g relied on the identity $fg = [(f + g)^2 - f^2 - g^2]/2$. We shall use this and a limit argument to prove the measurability of fg when $f, g \in \mathcal{L}^0(E, \overline{\mathbb{R}})$ after the next proposition.

(2) Note that in view of the identities

$$f = f^+ - f^-, \; |f| = f^+ + f^-, \; f^+ = \frac{|f| + f}{2}, \quad \text{and} \quad f^- = \frac{|f| - f}{2},$$

we have $f \in \mathcal{L}^0(E, \mathbb{R})$ if and only if $f^+, f^- \in \mathcal{L}^0(E, \mathbb{R})$.

(3) If $E \in \mathcal{M}_\lambda$, then for any $f_1, f_2, \ldots, f_n \in \mathcal{L}^0(E, \mathbb{R})$ and any constants $c_1, c_2, \ldots, c_n \in \mathbb{R}$, repeated use of Proposition 10.2.20 implies that we have $\sum_{j=1}^n c_j f_j \in \mathcal{L}^0(E, \mathbb{R})$, $\max\{f_1, \ldots, f_n\} \in \mathcal{L}^0(E, \mathbb{R})$, and $\min\{f_1, \ldots, f_n\} \in \mathcal{L}^0(E, \mathbb{R})$. Also, for any polynomial $P(x_1, \ldots, x_n)$, we have $P(f_1, \ldots, f_n) \in \mathcal{L}^0(E, \mathbb{R})$.

Fortunately for integration, measurable functions turn out to behave a lot better than continuous functions. As we have seen, a *pointwise* limit of continuous functions need *not* be continuous. The situation is different for sequences of measurable functions:

Proposition 10.2.22. *Let* $(f_n)_{n \in \mathbb{N}}$ *be a sequence in* $\mathcal{L}^0(E, \overline{\mathbb{R}})$. *Then the functions* $\inf\{f_1, \ldots, f_n\}$, $\sup\{f_1, \ldots, f_n\}$, $\inf\{f_n : n \in \mathbb{N}\}$, $\sup\{f_n : n \in \mathbb{N}\}$, $\underline{\lim}(f_n)$, *and* $\overline{\lim}(f_n)$ *are all measurable. In particular, if* $f(x) := \lim(f_n(x))$ *exists (as an extended real number) for all* $x \in E$ *then* f *is measurable.*

Proof. Let $g_n(x) := \sup\{f_1(x), \ldots, f_n(x)\}$ for all $x \in E$. Then $\{x : g_n(x) > b\} = \bigcup_{j=1}^{n}\{x : f_j(x) > b\} \in \mathcal{M}_\lambda$, which shows g_n is measurable. Similarly, if we define $g(x) := \sup\{f_k(x) : k \in \mathbb{N}\}$ for all $x \in E$, then $\{x : g(x) > b\} = \bigcup_{k=1}^{\infty}\{x : f_k(x) > b\} \in \mathcal{M}_\lambda$ and hence g is measurable. For the measurability of $\inf\{f_1, \ldots, f_n\}$ and $\inf\{f_n : n \in \mathbb{N}\}$ we can use a similar argument. Next, by what we just established, $h_n := \sup\{f_k : k \geq n\}$ is measurable for each $n \in \mathbb{N}$ and hence so is $\overline{\lim}(f_n) = \inf\{h_n : n \in \mathbb{N}\}$. Again, a similar argument establishes the measurability of $\underline{\lim}(f_n)$. Finally, if $f := \lim(f_n)$ exists, then $\underline{\lim}(f_n) = \overline{\lim}(f_n) = \lim(f_n)$ shows that it is measurable and the proof is complete. □

Proposition 10.2.23. $E \in \mathcal{M}_\lambda(\mathbb{R})$ *and* $f, g \in \mathcal{L}^0(E, \overline{\mathbb{R}})$ *imply* $fg \in \mathcal{L}^0(E, \overline{\mathbb{R}})$.

Proof. For each $m, n \in \mathbb{N}$, let us introduce the *truncated* functions

$$f_n(x) := \begin{cases} f(x) & \text{if } |f(x)| \leq n, \\ n & \text{if } f(x) > n, \\ -n & \text{if } f(x) < -n, \end{cases} \quad \text{and} \quad g_m(x) := \begin{cases} g(x) & \text{if } |g(x)| \leq m, \\ m & \text{if } g(x) > m, \\ -m & \text{if } g(x) < -m. \end{cases}$$

Then f_n and g_m are (real-valued) *measurable* functions on E (why?) and hence (by Proposition 10.2.20) so are the products $f_n g_m$ for all $m, n \in \mathbb{N}$. Since

$$f(x)g_m(x) = \lim_{n \to \infty} f_n(x)g_m(x) \quad \forall x \in E,$$

the measurability of fg_m follows from Proposition 10.2.22. But then another application of the same proposition and the fact that

$$f(x)g(x) = \lim_{m \to \infty} f(x)g_m(x) \quad \forall x \in E$$

show that $fg \in \mathcal{L}^0(E, \overline{\mathbb{R}})$. □

We now show that, as pointed out before, simple functions are indeed the *building blocks* of measurable functions:

Theorem 10.2.24. *Let* $E \in \mathcal{M}_\lambda(\mathbb{R})$. *Then* $f \in \mathcal{L}^0(E, \overline{\mathbb{R}})$ *if and only if there is a sequence* (ϕ_n) *of simple functions on* E *such that* $|\phi_n| \leq |f|$ *for all* n *and* $f(x) = \lim(\phi_n(x))$ *for all* $x \in E$. *In fact, if* $f : E \to [0, \infty]$, *then there exists* *an increasing sequence of simple functions that converges to f at every $x \in E$ and the convergence is uniform if f is bounded.*

Proof. If $f(x) = \lim(\phi_n(x))$ for all $x \in E$, then by Proposition 10.2.22 we have $f \in \mathcal{L}^0(E, \overline{\mathbb{R}})$. For the converse, let us first prove the last statement. So suppose that $f : E \to [0, \infty]$ is measurable and let $n \in \mathbb{N}$. Then, for $1 \le k \le n2^n$, the sets

$$E_{n,k} := \left\{ x \in E : \frac{k-1}{2^n} \le f(x) < \frac{k}{2^n} \right\}, \qquad F_n := \{ x \in E : f(x) \ge n \}$$

are all measurable. Define the sequence (ϕ_n) as follows:

$$\phi_n := \sum_{k=1}^{n2^n} \frac{k-1}{2^n} \chi_{E_{n,k}} + n \chi_{F_n}.$$

Then (ϕ_n) is a sequence of simple functions such that

$$0 \le \phi_1 \le \phi_2 \le \cdots \le f, \quad 0 \le \phi_n \le n, \quad \text{and} \quad 0 \le f(x) - \phi_n(x) < \frac{1}{2^n}$$

for all $x \notin F_n$. Thus $(\phi_n) \uparrow f = \sup\{\phi_n : n \in \mathbb{N}\}$. Also, if $f(x) \le N$ for all $x \in E$ and some $N \in \mathbb{N}$, then $0 \le f(x) - \phi_n(x) < 1/2^n$ for all $n \ge N$ and hence $\lim(\phi_n) = f$ uniformly. Finally, for a measurable function $f : E \to [-\infty, \infty]$, we apply the above construction to f^+ and f^- and use $f = f^+ - f^-$. $\square$

Exercise 10.2.25. Let $E \in \mathcal{M}_\lambda(\mathbb{R})$ and $f, g \in \mathcal{L}^0(E, \overline{\mathbb{R}})$. Use Theorem 10.2.24 to deduce the measurability of fg. Also deduce the measurability of $f + g$ provided it is well defined, e.g., if $f, g \in \mathcal{L}^0(E, \mathbb{R})$. Note that $f + g$ (which is not well defined in general) can be suitably redefined to become measurable, as we pointed out in Remarks 10.2.21.

Finally, in some applications we may have to work with complex-valued functions and need the following:

Definition 10.2.26. Let $E \in \mathcal{M}_\lambda(\mathbb{R})$. A function $f : E \to \mathbb{C}$ is said to be measurable if its real and imaginary parts are. Thus with $f = \text{Re}(f) + i\text{Im}(f)$, where $\text{Re}(f)$ and $\text{Im}(f)$ are real-valued, we have $\text{Re}(f)$, $\text{Im}(f) \in \mathcal{L}^0(E, \mathbb{R})$.

Notation 10.2.27 ($\mathcal{L}^0(E, \mathbb{C})$, $L^0(E, \mathbb{C})$). Given any $E \in \mathcal{M}_\lambda(\mathbb{R})$, we write $\mathcal{L}^0(E, \mathbb{C})$ for the *algebra* of all Lebesgue measurable functions $f : E \to \mathbb{C}$. We also write $L^0(E, \mathbb{C})$ for the corresponding algebra of equivalence classes of functions in $\mathcal{L}^0(E, \mathbb{C})$ that are equal almost everywhere on E.

10.3   The Lebesgue Integral

As promised above, we shall first define the Lebesgue integral for *simple* functions and then extend it to the largest possible subspace of Lebesgue measurable functions defined on a given measurable set. As was pointed out before, if $\mathbb{Q}_1 := \mathbb{Q} \cap [0, 1]$, then the simple functions $\chi_{\mathbb{Q}_1}$ and $\chi_{[0,1]\backslash\mathbb{Q}_1}$ are *not* Riemann integrable. And yet, it is natural to expect the "integral" of the characteristic function χ_E of a measurable set $E \in \mathcal{M}_\lambda$ to be the measure $\lambda(E)$ of that set:

$$\int \chi_E \, d\lambda = \lambda(E).$$

This will basically serve as the starting point in defining our new integral. Thus the integral of a (finite) linear combination of characteristic functions of measurable sets shall be the corresponding linear combinations of their measures. The following definition will be needed.

Definition 10.3.1 (*Simp*$_0(\mathbb{R})$, *Simp*$_0^+(\mathbb{R})$). We say that a function $f : \mathbb{R} \to \mathbb{R}$ *vanishes outside a set of finite measure* if there is a set $E \in \mathcal{M}_\lambda(\mathbb{R})$ with $\lambda(E) < \infty$ such that $f(x) = 0$ for all $x \notin E$. The set of all simple functions that vanish outside a set of finite measure will be denoted *Simp*$_0(\mathbb{R})$ and the set all simple functions $\phi : \mathbb{R} \to [0, \infty)$ that vanish outside a set of finite measure will be denoted *Simp*$_0^+(\mathbb{R})$. Note that, if ϕ, $\psi \in$ *Simp*$_0(\mathbb{R})$ and $c \in \mathbb{R}$, then we clearly have $c\phi + \psi \in$ *Simp*$_0(\mathbb{R})$ and $|\phi| \in$ *Simp*$_0^+(\mathbb{R})$. It is a simple exercise to check that *Simp*$_0(\mathbb{R})$ is a *real vector space*.

Definition 10.3.2 (Integral of a Simple Function). Let $\phi \in$ *Simp*$_0(\mathbb{R})$ have canonical representation $\phi = \sum_{j=1}^n a_j \chi_{A_j}$. Then we define its Lebesgue integral to be the number

$$\int \phi \, d\lambda := \sum_{j=1}^n a_j \lambda(A_j). \tag{†}$$

If E is any measurable set, then we also define

$$\int_E \phi \, d\lambda := \int \phi \cdot \chi_E \, d\lambda = \sum_{j=1}^n a_j \lambda(A_j \cap E). \tag{‡}$$

In particular, for any A, $E \in \mathcal{M}_\lambda(\mathbb{R})$, we have $A \cap E \in \mathcal{M}_\lambda$, $\chi_A \cdot \chi_E = \chi_{A \cap E}$ and hence

$$\int \chi_A \, d\lambda = \lambda(A) \quad \text{and} \quad \int_E \chi_A \, d\lambda = \lambda(A \cap E).$$

Remarks and Notation 10.3.3.

(1) Note that we may have to use $0 \cdot \infty = 0$ in (†) because it may happen that for some j we have $a_j = 0$ and $\lambda(A_j) = \infty$. Also, it follows from (‡) that

$$\lambda(E) = 0 \implies \int_E \phi = 0.$$

(2) If it is necessary to display the ("dummy") integration variable, then we write, e.g., $\int \phi(x) \, d\lambda(x)$. On the other hand, we may occasionally omit $d\lambda$ and simply write $\int \phi$ (or $\int_E \phi$) if no confusion results.

(3) Since simple functions have many possible representations, the following lemma will be useful.

Lemma 10.3.4. *Let* $\phi \in \mathbf{Simp}_0(\mathbb{R})$ *have canonical representation* $\phi = \sum_{j=1}^m a_j \chi_{A_j}$ *and suppose that* $\phi = \sum_{k=1}^n b_k \chi_{B_k}$ *is another representation of* ϕ, *where the* $B_k \in \mathcal{M}_\lambda$ *are also pairwise disjoint and* $\bigcup_{k=1}^n B_k = \mathbb{R}$. *Then*

$$\int \phi = \sum_{k=1}^n b_k \lambda(B_k).$$

Proof. Simply note that for each $j = 1, \ldots, m$, we have $A_j = \{x : \phi(x) = a_j\} = \bigcup_{\{k : b_k = a_j\}} B_k$ so (by the additivity of λ) we have $a_j \lambda(A_j) = \sum_{\{k : b_k = a_j\}} a_j \lambda(B_k) = \sum_{\{k : b_k = a_j\}} b_k \lambda(B_k)$. But then we have

$$\sum_{j=1}^m a_j \lambda(A_j) = \sum_{j=1}^m \sum_{\{k : b_k = a_j\}} b_k \lambda(B_k) = \sum_{k=1}^n b_k \lambda(B_k). \qquad \square$$

Theorem 10.3.5 (Properties of $\int \phi$). *Let* $\phi, \psi \in \mathbf{Simp}_0(\mathbb{R})$, $c \in \mathbb{R}$, *and* $E \in \mathcal{M}_\lambda(\mathbb{R})$ *be arbitrary. Then we have*

(1) $\int_E (c\phi + \psi) = c \int_E \phi + \int_E \psi$ *(Linearity)*,
(2) $\phi \leq \psi$ *a.e. on* $E \implies \int_E \phi \leq \int_E \psi$ *(Monotonicity)*,
(3) $\left| \int_E \phi \right| \leq \int_E |\phi|$ *(Triangle Inequality)*,
(4) $\phi = \psi$ *a.e. on* $E \implies \int_E \phi = \int_E \psi$, *and*
(5) $\phi \geq 0$ *a.e. and* $\int_E \phi = 0 \implies \phi = 0$ *a.e. on* E.

Proof. Let $\phi = \sum_{j=1}^m a_j \chi_{A_j}$ and $\psi = \sum_{k=1}^n a_k \chi_{B_k}$ be the canonical representations of ϕ and ψ. Then, the $A_j \cap B_k$ are disjoint measurable sets and, as in Proposition 10.2.5, we have

$$\phi = \sum_{j=1}^m \sum_{k=1}^n a_j \chi_{A_j \cap B_k}, \quad \psi = \sum_{j=1}^m \sum_{k=1}^n b_k \chi_{A_j \cap B_k}, \quad \text{and}$$

$$c\phi + \psi = \sum_{j=1}^{m}\sum_{k=1}^{n}(ca_j + b_k)\chi_{A_j \cap B_k}.$$

Applying Lemma 10.3.4, we therefore have

$$\int_E \phi = \sum_{j=1}^{m}\sum_{k=1}^{n}a_j\lambda(A_j \cap B_k \cap E), \quad \int_E \psi = \sum_{j=1}^{m}\sum_{k=1}^{n}b_k\lambda(A_j \cap B_k \cap E), \quad \text{and}$$

$$\int_E (c\phi + \psi) = \sum_{j=1}^{m}\sum_{k=1}^{n}(ca_j + b_k)\lambda(A_j \cap B_k \cap E),$$

from which (1) follows. To prove (2), note that by linearity we have $\int_E \phi - \int_E \psi = \int_E(\phi - \psi)$, so it suffices to show that $\phi \geq 0$ a.e. on E implies $\int_E \phi \geq 0$. But then, $\lambda(A_j \cap E) = 0$ if $a_j < 0$ and hence

$$\int_E \phi = \sum_{\{j : a_j < 0\}} a_j\lambda(A_j \cap E) + \sum_{\{j : a_j \geq 0\}} a_j\lambda(A_j \cap E) = \sum_{\{j : a_j \geq 0\}} a_j\lambda(A_j \cap E) \geq 0,$$

as desired. Next, (3) follows from (2) and $-|\phi| \leq \phi \leq |\phi|$ and (4) follows from (2) as well. Finally, to prove (5), note that we have $\phi(x) \geq 0$ for a.e. $x \in E$, so it suffices to show that $\lambda(E \cap \{x : \phi(x) > 0\}) = 0$. But if for some $a > 0$ we had $\lambda(E \cap \{x : \phi(x) = a\}) > 0$, then we would get $\int_E \phi \geq a\lambda(E \cap \{x : \phi(x) = a\}) > 0$, a contradiction. $\square$

Corollary 10.3.6. *Let $\phi \in Simp_0(\mathbb{R})$ be represented by $\phi = \sum_{k=1}^{n}b_k\chi_{B_k}$, where the B_k are not necessarily disjoint. Then for any $E \in \mathcal{M}_\lambda(\mathbb{R})$ we have*

$$\int_E \phi = \sum_{k=1}^{n}b_k\lambda(B_k \cap E).$$

In other words, in Lemma 10.3.4, the assumption that the B_k be pairwise disjoint is unnecessary.

Exercise 10.3.7 (Set Additivity). Let $\phi \in Simp_0(\mathbb{R})$ and $E, F \in \mathcal{M}_\lambda(\mathbb{R})$. Show that if $E \cap F$ has measure zero, then

$$\int_{E \cup F} \phi = \int_E \phi + \int_F \phi.$$

Hint: First show this for the case $E \cap F = \emptyset$.

Exercise 10.3.8 (Set Monotonicity). Let $\phi \in Simp_0^+(\mathbb{R})$ and E, $F \in \mathcal{M}_\lambda(\mathbb{R})$. Show that

$$E \subset F \implies \int_E \phi \le \int_F \phi.$$

Exercise 10.3.9. Let $\phi \in Simp_0(\mathbb{R})$ and E, $F \in \mathcal{M}_\lambda(\mathbb{R})$. Show that

$$\phi(x) = 0 \quad \forall x \notin F \implies \int_E \phi = \int_{E \cap F} \phi.$$

Before we go any further, let us include a useful lemma:

Lemma 10.3.10. *Let $\phi \in Simp^+(\mathbb{R})$. Then the set function $v : \mathcal{M}_\lambda(\mathbb{R}) \to [0, \infty]$ defined by*

$$v(E) = \int_E \phi, \qquad E \in \mathcal{M}_\lambda(\mathbb{R})$$

is a (countably additive) measure.

Proof. Since we obviously have $v(E) \ge 0$ for all $E \in \mathcal{M}_\lambda(\mathbb{R})$ and $v(\emptyset) = 0$, we need only show that v is countably additive. Let $\phi = \sum_{j=1}^m a_j \chi_{A_j}$ be the canonical representation of ϕ with $A_j \in \mathcal{M}_\lambda(\mathbb{R})$ for all j. If $B = \bigcup_{n=1}^\infty B_n$ is a disjoint union of measurable sets, then

$$v(B) := \int_B \phi = \sum_{j=1}^m a_j \lambda(A_j \cap B) = \sum_{j=1}^m \sum_{n=1}^\infty a_j \lambda(A_j \cap B_n)$$

$$= \sum_{n=1}^\infty \left(\sum_{j=1}^m a_j \lambda(A_j \cap B_n) \right) = \sum_{n=1}^\infty \int_{B_n} \phi = \sum_{n=1}^\infty v(B_n). \qquad \square$$

Remark 10.3.11. The above lemma is an extension of Theorem 10.1.25, which asserts the *continuity* of Lebesgue measure λ and is in fact the *Monotone Convergence Theorem for Measurable Sets*. It will be extended (in Sect. 10.3) to one of the fundamental convergence theorems for the Lebesgue integral.

We now define the upper and lower (Lebesgue) integrals of *bounded* functions:

Definition 10.3.12 (Upper and Lower (Lebesgue) Integrals). Let $E \in \mathcal{M}_{\lambda<\infty}(\mathbb{R})$ and let $f : E \to \mathbb{R}$ be a *bounded* function. We define the *upper (Lebesgue) integral* and *lower (Lebesgue) integral* of f on E by

$$\overline{\int}_E f \, d\lambda := \inf \left\{ \int_E \psi : f \le \psi \text{ on } E, \ \psi \in Simp_0(\mathbb{R}) \right\} \quad \text{and}$$

$$\underline{\int}_E f \, d\lambda := \sup\left\{ \int_E \phi : \phi \le f \text{ on } E, \ \phi \in \textit{Simp}_0(\mathbb{R})\right\}.$$

Proposition 10.3.13. *Let* $E \in \mathcal{M}_{\lambda < \infty}$ *and let* $f : E \to \mathbb{R}$ *be bounded. Then*

$$\underline{\int}_E f \, d\lambda = \overline{\int}_E f \, d\lambda$$

if and only if f *is measurable.*

Proof. Suppose first that f is measurable and let $M > 0$ be such that $|f(x)| \le M$ for all $x \in E$. Then for each $n \in \mathbb{N}$ the sets

$$E_k := \left\{ x \in E : \frac{M(k-1)}{n} < f(x) \le \frac{Mk}{n}\right\}, \quad -n \le k \le n$$

are measurable, pairwise disjoint and $E = \bigcup_{k=-n}^n E_k$. In particular, $\lambda(E) = \sum_{k=-n}^n \lambda(E_k)$. Now consider the simple functions

$$\phi_n = \frac{M}{n} \sum_{k=-n}^n k \chi_{E_k} \quad \text{and} \quad \psi_n = \frac{M}{n} \sum_{k=-n}^n (k-1)\chi_{E_k}.$$

Then $\phi_k, \ \psi_k \in \textit{Simp}_0(\mathbb{R})$ and we have

$$\int_E (\psi_n - \phi_n) = \frac{M}{n}\left(\sum_{-n}^n k\lambda(E_k) - \sum_{-n}^n (k-1)\lambda(E_k)\right) = \frac{M}{n}\lambda(E).$$

Since $\phi_n(x) \le f(x) \le \psi_n(x)$ for all $x \in E$, we deduce the inequalities

$$0 \le \overline{\int}_E f \, d\lambda - \underline{\int}_E f \, d\lambda \le \int_E (\psi_n - \phi_n) \le \frac{M}{n}\lambda(E) \quad \forall n \in \mathbb{N},$$

from which $\underline{\int}_E f \, d\lambda = \overline{\int}_E f \, d\lambda$ follows. Conversely, if $\underline{\int}_E f \, d\lambda = \overline{\int}_E f \, d\lambda$, then for each $n \in \mathbb{N}$ we can find simple functions $\phi_n, \ \psi_n \in \textit{Simp}_0(\mathbb{R})$ such that

$$\phi_n(x) \le f(x) \le \psi_n(x) \quad (\forall x \in E) \qquad \text{and} \qquad \int_E (\psi_n - \phi_n) \le \frac{1}{n}. \qquad (*)$$

Define the functions

$$g = \sup\{\phi_n : n \in \mathbb{N}\} \quad \text{and} \quad h = \inf\{\psi_n : n \in \mathbb{N}\}.$$

By Proposition 10.2.22 we have g, $h \in \mathcal{L}^0(E, \mathbb{R})$ and $g \leq f \leq h$. Now consider the measurable sets

$$Z_m := \{x \in E : h(x) - g(x) > 1/m\} \quad \text{and} \quad Z = \{x \in E : h(x) - g(x) > 0\}.$$

We claim that $\lambda(Z) = 0$. Indeed, $Z = \bigcup_{m=1}^{\infty} Z_m$ and it suffices to show that $\lambda(Z_m) = 0$ for all $m \in \mathbb{N}$. But, with $F_{m,n} := \{x \in E : \psi_n(x) - \phi_n(x) > 1/m\}$, it follows from $(*)$ that

$$\frac{1}{m}\lambda(F_{m,n}) = \frac{1}{m}\int_E \chi_{F_{m,n}} \leq \int_E (\psi_n - \phi_n) \leq \frac{1}{n},$$

which gives $\lambda(F_{m,n}) < m/n$. Since $Z_m \subset F_{m,n}$ for all $n \in \mathbb{N}$, we have $\lambda(Z_m) \leq m/n$ for all n and hence $\lambda(Z_m) = 0$. Thus we have proved that $g(x) = h(x)$ for *almost all* $x \in E$ and hence $f = g = h$ except on a set of measure zero. But then Proposition 10.2.17 implies that $f \in \mathcal{L}^0(E, \mathbb{R})$ and the proof is complete. □

Notation 10.3.14 ($\mathcal{BL}^0(E)$, $\mathcal{BL}_0^0(E)$). For any $E \in \mathcal{M}_\lambda(\mathbb{R})$ we let $\mathcal{BL}^0(E)$ denote the set of all *bounded, measurable* functions $f : E \to \mathbb{R}$. We write $\mathcal{BL}_0^0(E)$ for all functions $f \in \mathcal{BL}^0(E)$ that vanish outside a measurable subset $F \subset E$ with $\lambda(F) < \infty$. As pointed out before (cf. Remark 10.2.10 and Proposition 10.2.12), we may assume that all functions we study have domain $\mathbb{R}$ by using their *trivial extensions* if necessary. For a bounded, measurable function $f : E \to \mathbb{R}$ with $E \in \mathcal{M}_{\lambda < \infty}(\mathbb{R})$ we then have $\tilde{f} \in \mathcal{BL}_0^0(\mathbb{R})$.

We are now ready to define the Lebesgue integral of bounded measurable functions on a set of finite measure:

Definition 10.3.15 (Lebesgue Integral of Bounded Functions). Given any $E \in \mathcal{M}_{\lambda < \infty}$ and any *bounded, measurable* function $f \in \mathcal{BL}^0(E)$, we define its Lebesgue integral over E by

$$\int_E f \, d\lambda := \underline{\int}_E f \, d\lambda = \sup\left\{ \int_E \phi : \phi \leq f \text{ on } E, \, \phi \in \mathbf{Simp}_0(\mathbb{R}) \right\},$$

which, of course, coincides with the upper (Lebesgue) integral of f.

Notation 10.3.16. We sometimes use the abbreviated $\int_E f$ instead of $\int_E f \, d\lambda$. However, if the dummy variable of integration has to be displayed, then we write $\int_E f(x) \, d\lambda(x)$. If $E = [a, b]$, we write $\int_a^b f \, d\lambda$ instead of $\int_E f \, d\lambda$. Also, if $f \in \mathcal{BL}^0(\mathbb{R})$ and $f \equiv 0$ outside $E \in \mathcal{M}_{\lambda < \infty}$, then we may even write $\int f$ instead of $\int_E f$.

Exercise 10.3.17.

(1) Show that for a simple function $f \in \mathbf{Simp}_0(\mathbb{R})$ the above definition of the integral agrees with Definition 10.3.2.

(2) Show that if in Definition 10.3.15 $\lambda(E) = 0$, then $\int_E f \, d\lambda = 0$.

(3) Let E, $F \in \mathcal{M}_{\lambda < \infty}$ with $F \subset E$ and let $f \in \mathcal{BL}^0(E)$. Show that

$$\int_F f \, d\lambda = \int_E f \chi_F \, d\lambda.$$

Notation 10.3.18 ($\mathcal{BL}^1(E) = \mathcal{BL}^0(E)$, $BL^1(E) = BL^0(E)$). The set of all bounded Lebesgue integrable functions $f : E \to \mathbb{R}$ is denoted by $\mathcal{BL}^1(E)$. The above definition implies that we have $\mathcal{BL}^1(E) = \mathcal{BL}^0(E)$. Also, as in 10.2.19, we define $BL^1(E) := \mathcal{BL}^1(E)/\mathcal{N}(E)$, where $\mathcal{N}(E)$ denotes the set of real-valued, null functions on E.

Recall that for any closed, bounded interval $[a, b] \subset \mathbb{R}$, a Riemann integrable function $f \in \mathcal{R}([a, b])$ is necessarily *bounded*. As the following proposition shows, f is actually (Lebesgue) measurable (hence Lebesgue integrable) and its Lebesgue and Riemann integrals coincide:

Proposition 10.3.19 (Riemann Integrable $\Rightarrow$ Lebesgue Integrable). *If f is Riemann integrable on $[a, b]$, then it is measurable and we have*

$$\int_a^b f(x) \, dx = \int_a^b f(x) \, d\lambda(x),$$

with the Riemann integral on the left and the Lebesgue integral on the right.

Proof. Let $\underline{\int_a^b} f(x) \, dx$ and $\overline{\int_a^b} f(x) \, dx$ denote the lower and upper Darboux integrals of f, respectively. Since $Step([a, b]) \subset Simp([a, b])$, we have

$$\underline{\int_a^b} f(x) \, dx \leq \underline{\int_{[a,b]}} f \, d\lambda \leq \overline{\int_{[a,b]}} f \, d\lambda \leq \overline{\int_a^b} f(x) \, dx.$$

But $f \in \mathcal{R}([a, b])$ implies that its lower and upper Darboux integrals coincide and hence all the above inequalities are equalities. The (Lebesgue) measurability of f now follows from Proposition 10.3.13. $\square$

Theorem 10.3.20 (Properties of $\int f \, d\lambda$ for Bounded f). *If $E \in \mathcal{M}_{\lambda < \infty}$, then $\mathcal{BL}^1(E) = \mathcal{BL}^0(E)$ is a real vector space. Also, for any f, $g \in \mathcal{BL}^1(E)$ and any $c \in \mathbb{R}$, we have*

(1) $\int_E (cf + g) \, d\lambda = c \int_E f \, d\lambda + \int_E g \, d\lambda.$ *(Linearity)*
  In particular, we have

$$\int_E f \, d\lambda = \int_E f^+ \, d\lambda - \int_E f^- \, d\lambda \quad and$$

$$\int_E |f| \, d\lambda = \int_E f^+ \, d\lambda + \int_E f^- \, d\lambda.$$

(2) If A and B are measurable subsets of E with $\lambda(A \cap B) = 0$ (e.g., if $A \cap B = \emptyset$),
then

$$\int_{A \cup B} f \, d\lambda = \int_A f \, d\lambda + \int_B f \, d\lambda.$$

(3) $f \leq g$ a.e. on $E \Longrightarrow \int_E f \, d\lambda \leq \int_E g \, d\lambda$ (Monotonicity),
(4) $f = g$ a.e. on $E \Longrightarrow \int_E f \, d\lambda = \int_E g \, d\lambda$,
(5) $\left| \int_E f \, d\lambda \right| \leq \int_E |f| \, d\lambda$ (Triangle Inequality),
(6) If $m \leq f(x) \leq M$ for almost all $x \in E$, then we have

$$m\lambda(E) \leq \int_E f \, d\lambda \leq M\lambda(E), and$$

(7) $f \geq 0$ a.e. and $\int_E f \, d\lambda = 0 \Longrightarrow f = 0$ a.e. on E.

Proof. This is basically an extension of Theorem 10.3.5, where we saw that the
above statements are true if f and g are simple functions. In what follows all Greek
letters (ϕ, ψ, etc.) will always denote simple functions. Also, note that if $c \neq 0$, then
$\phi \in Simp_0(\mathbb{R})$ if and only if $c\phi \in Simp_0(\mathbb{R})$. For (1) we first show that $\int_E cf \, d\lambda = c \int_E f \, d\lambda$. Now for $c > 0$, we have

$$\int_E cf \, d\lambda = \sup \left\{ \int_E c\phi : c\phi \leq cf \right\} = c \sup \left\{ \int_E \phi : \phi \leq f \right\} = c \int_E f \, d\lambda.$$

If $c < 0$, then $c\psi \leq cf$ if and only if $\psi \geq f$ so (by Proposition 10.3.13) we get

$$\int_E cf \, d\lambda = \sup \left\{ c \int_E \psi : \psi \geq f \right\} = c \inf \left\{ \int_E \psi : \psi \geq f \right\} = c \int_E f \, d\lambda.$$

Next, note that $f + g \in \mathcal{BL}^0(E)$ and hence

$$\int_E (f + g) \, d\lambda = \sup \left\{ \int_E \phi : \phi \leq f + g \right\} = \inf \left\{ \int_E \psi : \psi \geq f + g \right\}. \qquad (*)$$

Now, if ϕ_1, $\phi_2 \in Simp_0(\mathbb{R})$ satisfy $\phi_1 \leq f$ and $\phi_2 \leq g$, then $\phi_1 + \phi_2 \leq f + g$
and $(*)$ gives

$$\int_E \phi_1 + \int_E \phi_2 = \int_E (\phi_1 + \phi_2) \leq \int_E (f + g) \, d\lambda.$$

Taking the sup of the left side over all $\phi_1 \leq f$ and $\phi_2 \leq g$, we get $\int_E f \, d\lambda + \int_E g \, d\lambda \leq \int_E (f + g) \, d\lambda$. For the reverse inequality, we note that if ψ_1, $\psi_2 \in Simp_0(\mathbb{R})$ satisfy $\psi_1 \geq f$ and $\psi_2 \geq g$, then $\psi_1 + \psi_2 \geq f + g$ and $(*)$ implies

$$\int_E \psi_1 + \int_E \psi_2 = \int_E (\psi_1 + \psi_2) \geq \int_E (f + g) \, d\lambda.$$

Taking the inf of the left side over all $\psi_1 \geq f$ and $\psi_2 \geq g$, we have our reversed inequality $\int_E f \, d\lambda + \int_E g \, d\lambda \geq \int_E (f + g) \, d\lambda$ and the proof of (1) is complete. Now, by Exercise 10.3.17, (2) follows from (1) and the fact that $\chi_{A \cup B} = \chi_A + \chi_B - \chi_{A \cap B}$. Also, since the *linearity* is established, to prove (3) we need only show that $f \geq 0$ a.e. on E implies $\int_E f \, d\lambda \geq 0$. But for any $\psi \in \boldsymbol{Simp}_0(\mathbb{R})$ with $\psi \geq f$ on E we have $\psi \geq 0$ a.e. on E and hence Theorem 10.3.5 gives $\int_E \psi \geq 0$. Thus $\int_E f \, d\lambda = \inf\{\int_E \psi : \psi \geq f\} \geq 0$, as desired. This also proves (4) and (5) because $f = g$ a.e. on E means $f \leq g \leq f$ a.e. on E and we have $-|f| \leq f \leq |f|$. Next, (6) follows from (1) and (3) because we have $m\chi_E \leq f \leq M\chi_E$ a.e. on E and $\int_E \chi_E \, d\lambda = \lambda(E)$. Finally, (7) follows if we can show that the set $\{x \in E : f(x) > 0\}$ has measure zero. Since

$$\{x \in E : f(x) > 0\} = \bigcup_{n=1}^{\infty} \{x \in E : f(x) > 1/n\},$$

it suffices to show that $\lambda(E_n) = 0$ for every $n \in \mathbb{N}$, where $E_n := \{x \in E : f(x) > 1/n\}$. But $(1/n)\chi_{E_n} \in \boldsymbol{Simp}_0(\mathbb{R})$ and $(1/n)\chi_{E_n} \leq f$, so if $\lambda(E_n) > 0$ for some $n \in \mathbb{N}$, then we have the contradiction

$$0 < \frac{1}{n}\lambda(E_n) = \int_E \frac{1}{n}\chi_{E_n} = 0$$

because the assumption $\int_E f \, d\lambda = \sup\{\int_E \phi : 0 \leq \phi \leq f\} = 0$ implies $\int_E \phi = 0$ for every simple function $0 \leq \phi \leq f$.   □

We first defined the Lebesgue integral for *simple functions* and then extended it to *bounded, measurable functions* vanishing outside a set of finite measure. Our final step will be an extension of the integral to the general case where the function is not necessarily bounded and need not vanish outside a set of finite measure. This we do in two steps: We first define the general integral for *nonnegative*, measurable functions and then use it to define the general integral.

Definition 10.3.21 (Lebesgue Integral of Nonnegative Functions). Let $E \in \mathcal{M}_\lambda(\mathbb{R})$ and let $f : E \to [0, \infty]$ be measurable. We define the Lebesgue integral of f on E to be the (extended, nonnegative) number

$$\int_E f \, d\lambda := \sup\left\{\int_E h \, d\lambda : h \leq f, \, h \in \boldsymbol{\mathcal{BL}}_0^0(E)\right\} \qquad (*)$$

$$= \sup\left\{\int_E h \, d\lambda : 0 \leq h \leq f, \, h \in \boldsymbol{\mathcal{BL}}_0^0(E)\right\},$$

where the second equality in $(*)$ follows from the fact that $\int_E h \, d\lambda \leq \int_E h^+ \, d\lambda$.

Since any nonnegative, measurable function is the (pointwise) limit of an increasing sequence of simple functions (cf. Theorem 10.2.24), the following equivalent definition of the integral, which is an extension of Definition 10.3.15, seems more natural and is, in fact, more convenient to use in most situations.

Proposition 10.3.22. *Let $E \in \mathcal{M}_\lambda(\mathbb{R})$ and let $f : E \to \overline{\mathbb{R}}_+ := [0, \infty]$ be measurable. Then we have*

$$\int_E f \, d\lambda = \sup \left\{ \int_E \phi \, d\lambda : 0 \le \phi \le f \text{ on } E, \ \phi \in Simp(\mathbb{R}) \right\}. \qquad (\dagger)$$

Proof. Let R denote the right side of $(\dagger)$ and assume first that $\int_E f \, d\lambda = \infty$. Then for any $N > 0$ there exists $h \in \mathcal{BL}_0^0(E)$ such that $0 \le h \le f$, $h \equiv 0$ outside a measurable set $F \subset E$ with $\lambda(F) < \infty$ and $\int_E h \, d\lambda \ge N$. Since $\int_E h \, d\lambda = \sup \left\{ \int_E \phi \, d\lambda : 0 \le \phi \le h \text{ on } E, \ \phi \in Simp(\mathbb{R}) \right\}$, we can pick $\phi_N \in Simp(\mathbb{R})$ such that $\int_E \phi_N > \int_E h \, d\lambda - 1 \ge N - 1$. This gives $R \ge N - 1$ and hence $R = \infty$ because $N > 0$ was arbitrary. Assume next that $\int_E f \, d\lambda < \infty$. Then the inclusion

$$\{\phi | E : 0 \le \phi \le f, \ \phi \in Simp(\mathbb{R})\} \subset \{h : 0 \le h \le f, \ h \in \mathcal{BL}_0^0(E)\}$$

implies that $R \le \int_E f \, d\lambda$. Therefore, we need only show the reverse inequality $\int_E f \, d\lambda \le R$. Now, given any $\varepsilon > 0$, we pick a function $h_\varepsilon \in \mathcal{BL}_0^0(E)$ such that

(i) $0 \le h_\varepsilon \le f$ and $\displaystyle\int_E f \, d\lambda < \int_E h_\varepsilon \, d\lambda + \frac{\varepsilon}{2}.$

Also, as above, we can pick $\phi_\varepsilon \in Simp(\mathbb{R})$ such that

(ii) $0 \le \phi_\varepsilon \le h_\varepsilon$ and $\displaystyle\int_E h_\varepsilon \, d\lambda < \int_E \phi_\varepsilon \, d\lambda + \frac{\varepsilon}{2}.$

Combining (i) and (ii), we have $\int_E f \, d\lambda < \int_E \phi_\varepsilon \, d\lambda + \varepsilon \le R + \varepsilon$. Since $\varepsilon > 0$ was arbitrary, the proof is complete. $\qquad\square$

The following theorem summarizes some of the properties of the integral we just defined.

Theorem 10.3.23 (Properties of $\int f \, d\lambda$ for $f \ge 0$). *Let $E \in \mathcal{M}_\lambda(\mathbb{R})$. Then for any $f, g \in \mathcal{L}^0(E, \overline{\mathbb{R}}_+)$ and $c > 0$ we have*

(1) $\int_E cf \, d\lambda = c \int_E f \, d\lambda$ *(Homogeneity).*
(2) $\int_E (f + g) \, d\lambda = \int_E f \, d\lambda + \int_E g \, d\lambda$ *(Additivity).*
(3) *If A and B are measurable subsets of E with $\lambda(A \cap B) = 0$ (e.g., if $A \cap B = \emptyset$), then*

$$\int_{A \cup B} f \, d\lambda = \int_A f \, d\lambda + \int_B f \, d\lambda.$$

(4) $f \le g$ *a.e. on $E \implies \int_E f \, d\lambda \le \int_E g \, d\lambda$ (Monotonicity).*

(5) If A and B are measurable subsets of E, then

$$A \subset B \implies \int_A f \, d\lambda \leq \int_B f \, d\lambda.$$

(6) $f = g$ a.e. on $E \implies \int_E f \, d\lambda = \int_E g \, d\lambda$.
(7) $\int_E f \, d\lambda = 0 \iff f = 0$ a.e. on E.

Proof. We prove only the *additivity:* $\int_E (f + g) \, d\lambda = \int_E f \, d\lambda + \int_E g \, d\lambda$. The remaining parts can be proved as in Theorem 10.3.20 and are left as exercises for the reader. Now given any h, $k \in \mathcal{BL}_0^0(E)$ with $0 \leq h \leq f$ and $0 \leq k \leq g$, we have $0 \leq h + k \leq f + g$ and hence

$$\int_E h \, d\lambda + \int_E k \, d\lambda = \int_E (h + k) \, d\lambda \leq \int_E (f + g) \, d\lambda.$$

Taking the sup of the left side over all such h and k, we get $\int_E f \, d\lambda + \int_E g \, d\lambda \leq \int_E (f + g) \, d\lambda$. For the reverse inequality, let $\ell \in \mathcal{BL}_0^0(E)$ satisfy $0 \leq \ell \leq f + g$. Define $h := \min(\ell, f)$ and $k := \ell - h$ and note that h, $k \in \mathcal{BL}_0^0(E)$ are both bounded by any upper bound for ℓ and vanish where ℓ vanishes. Also, $\ell = h + k$, $0 \leq h \leq f$ and $0 \leq k \leq g$. But then

$$\int_E \ell \, d\lambda = \int_E h \, d\lambda + \int_E k \, d\lambda \leq \int_E f \, d\lambda + \int_E g \, d\lambda,$$

and taking the sup of the left side gives $\int_E (f + g) \, d\lambda \leq \int_E f \, d\lambda + \int_E g \, d\lambda$. □

Exercise 10.3.24. Prove the remaining parts of Theorem 10.3.23. Also, show that for any $E \in \mathcal{M}_\lambda(\mathbb{R})$ and any $f \in \mathcal{L}^0(E, \overline{\mathbb{R}})$, we have

$$\int_E |f| \, d\lambda = \int_{\mathbb{R}} |f| \chi_E \, d\lambda \quad \text{and} \quad \int_E |f| \, d\lambda = \int_E f^+ \, d\lambda + \int_E f^- \, d\lambda.$$

Definition 10.3.25 (Nonnegative (Lebesgue) Integrable Functions). Given any $E \in \mathcal{M}_\lambda(\mathbb{R})$, a function $f : E \to \overline{\mathbb{R}}_+ := [0, \infty]$ is said to be *Lebesgue integrable* if $\int_E f \, d\lambda < \infty$.

Notation 10.3.26 ($\mathcal{L}^1(E, \overline{\mathbb{R}}_+)$, $L^1(E, \overline{\mathbb{R}}_+)$). Given any $E \in \mathcal{M}_\lambda(\mathbb{R})$, the set of all Lebesgue integrable functions $f : E \to \overline{\mathbb{R}}_+$ is denoted by $\mathcal{L}^1(E, \overline{\mathbb{R}}_+)$. Using the equivalence relation "$f = g$ a.e." on $\mathcal{L}^1(E, \overline{\mathbb{R}}_+)$, the corresponding space of all equivalence classes will be denoted $L^1(E, \overline{\mathbb{R}}_+)$.

Exercise 10.3.27 (Integrable $\Rightarrow$ Almost Real-Valued). Let $E \in \mathcal{M}_\lambda(\mathbb{R})$. Show that if $f \in \mathcal{L}^1(E, \overline{\mathbb{R}}_+)$, then the set $E_\infty := \{x \in E : f(x) = \infty\}$ has *measure zero* by integrating over the sets $E_n := \{x \in E : f(x) \geq n\}$.

We are now ready for the general case:

Definition 10.3.28 (The General Lebesgue Integral). Let $E \in \mathcal{M}_\lambda(\mathbb{R})$. Given any measurable function $f : E \to [-\infty, \infty]$, we define the Lebesgue integral of f by

$$\int_E f \, d\lambda := \int_E f^+ \, d\lambda - \int_E f^- \, d\lambda, \tag{$\dagger$}$$

provided at least one of the integrals on the right side is *finite*, where, as always, $f^+ = f \vee 0 := \max\{f, 0\}$ and $f^- = (-f) \vee 0 := \max\{-f, 0\}$. If both integrals on the right side of ($\dagger$) are finite, then we say that f is *Lebesgue integrable*.

Notation 10.3.29 ($\mathcal{L}^1(E)$, $\mathcal{L}^1(E, \overline{\mathbb{R}})$, $L^1(E)$, $L^1(E, \overline{\mathbb{R}})$). Given any $E \in \mathcal{M}_\lambda(\mathbb{R})$, the set of all Lebesgue integrable functions $f : E \to \mathbb{R}$ [resp., $f : E \to \overline{\mathbb{R}}$] will be denoted by $\mathcal{L}^1(E) := \mathcal{L}^1(E, \mathbb{R})$ [resp., $\mathcal{L}^1(E, \overline{\mathbb{R}})$]. Thus, using Exercise 10.3.24, we have

$$f \in \mathcal{L}^1(E, \overline{\mathbb{R}}) \iff \int_E |f| \, d\lambda = \int_E f^+ \, d\lambda + \int_E f^- \, d\lambda < \infty. \tag{$*$}$$

As in 10.2.19, 10.3.18, and 10.3.26, we write $L^1(E) := \mathcal{L}^1(E)/\mathcal{N}(E)$ [resp., $L^1(E, \overline{\mathbb{R}}) := \mathcal{L}^1(E, \overline{\mathbb{R}})/\mathcal{N}(E, \overline{\mathbb{R}})$] for the corresponding set of all equivalence classes modulo the null functions.

Proposition 10.3.30. *Let* $E \in \mathcal{M}_\lambda(\mathbb{R})$ *and any* f, $g \in \mathcal{L}^0(E, \overline{\mathbb{R}})$. *Then*

(1) $f \in \mathcal{L}^1(E, \overline{\mathbb{R}})$ *if and only if* $|f| \in \mathcal{L}^1(E, [0, \infty])$ *and we then have*

$$\left| \int_E f \, d\lambda \right| \le \int_E |f| \, d\lambda.$$

(2) $g \in \mathcal{L}^1(E, \overline{\mathbb{R}})$ *and* $|f| \le |g|$ *on* E *imply that* $f \in \mathcal{L}^1(E, \overline{\mathbb{R}})$ *and we have*

$$\int_E |f| \, d\lambda \le \int_E |g| \, d\lambda.$$

Proof. Note that (1) is a restatement of ($*$) and the Triangle Inequality. The latter can also be proved as follows:

$$\left| \int_E f \, d\lambda \right| = \left| \int_E f^+ \, d\lambda - \int_E f^- \, d\lambda \right| \le \int_E f^+ \, d\lambda + \int_E f^- \, d\lambda = \int_E |f| \, d\lambda.$$

For (2), $|f| \le |g|$ on E implies (by Theorem 10.3.23) that $\int_E |f| \, d\lambda \le \int_E |g| \, d\lambda < \infty$.   □

Theorem 10.3.31 (Properties of the General Lebesgue Integral). *For any $E \in \mathcal{M}_\lambda(\mathbb{R})$, f, $g \in \mathcal{L}^1(E, \mathbb{R})$, and $c \in \mathbb{R}$, we have cf, $f + g \in \mathcal{L}^1(E, \mathbb{R})$. In fact $\mathcal{L}^1(E, \mathbb{R})$ is a real vector space. Also, the integral is a (positive) linear functional on $\mathcal{L}^1(E, \mathbb{R})$, satisfying the following properties:*

(1) $\int_E cf \, d\lambda = c \int_E f \, d\lambda$. *(Homogeneity)*
(2) $\int_E (f + g) \, d\lambda = \int_E f \, d\lambda + \int_E g \, d\lambda$ *(Additivity),*
(3) *if A and B are measurable subsets of E with $\lambda(A \cap B) = 0$ (e.g., if $A \cap B = \emptyset$), then*

$$\int_{A \cup B} f \, d\lambda = \int_A f \, d\lambda + \int_B f \, d\lambda,$$

(4) $f \leq g$ *a.e. on* $E \implies \int_E f \, d\lambda \leq \int_E g \, d\lambda$ *(Monotonicity),*
(5) $\left| \int_E f \, d\lambda \right| \leq \int_E |f| \, d\lambda$ *(Triangle Inequality),*
(6) $f = g$ *a.e. on* $E \implies \int_E f \, d\lambda = \int_E g \, d\lambda$, *and*
(7) $\int_E |f| \, d\lambda = 0 \iff f = 0$ *a.e. on* E.

Proof. First, by the *additivity* and *monotonicity* in Theorem 10.3.23, we have $\int_E |f + g| \, d\lambda \leq \int_E |f| \, d\lambda + \int_E |g| \, d\lambda < \infty$ and hence $f + g \in \mathcal{L}^1(E, \mathbb{R})$. Checking the vector space axioms is now routine and left to the reader. Also, part (1) follows from the definition of the integral and Proposition 10.3.22. We prove (2) and leave the remaining parts as exercises for the reader. Since

$$(f + g)^+ - (f + g)^- = f + g = f^+ - f^- + g^+ - g^-,$$

we have

$$(f + g)^+ + f^- + g^- = (f + g)^- + f^+ + g^+$$

and all functions on both the left and right sides are nonnegative. Integrating the two sides and using Theorem 10.3.23, we therefore have

$$\int_E (f+g)^+ \, d\lambda + \int_E f^- \, d\lambda + \int_E g^- \, d\lambda = \int_E (f+g)^- \, d\lambda + \int_E f^+ \, d\lambda + \int_E g^+ \, d\lambda,$$

from which the additivity property follows at once. □

Exercise 10.3.32.

(a) Prove the remaining parts of Theorem 10.3.31.
(b) Let $E \in \mathcal{M}_\lambda(\mathbb{R})$ and $f \in \mathcal{L}^1(E, \overline{\mathbb{R}})$. Show that if $f = f_1 - f_2$ where f_1, $f_2 \in \mathcal{L}^1(E, \overline{\mathbb{R}}_+)$, then

$$\int_E f \, d\lambda = \int_E f_1 \, d\lambda - \int_E f_2 \, d\lambda$$

and use it to give another proof of part (2), i.e., the additivity of the integral.

(c) Let $E \in \mathcal{M}_\lambda(\mathbb{R})$ and $f \in \mathcal{L}^1(E, \overline{\mathbb{R}})$. Show that if $\int_A f \, d\lambda = 0$ for every measurable $A \subset E$, then $f = 0$ almost everywhere. *Hint:* Look at $A := \{x \in E : f(x) > 0\}$.

Remark 10.3.33 ($\mathcal{L}^1(E, \overline{\mathbb{R}})$ Is Almost a Vector Space). There is a reason why we didn't state (and prove) Theorem 10.3.31 for f, $g \in \mathcal{L}^1(E, \overline{\mathbb{R}})$, although it is *almost true* in this case as well. Indeed, as pointed out in Remark 10.2.21, $\mathcal{L}^0(E, \overline{\mathbb{R}})$ is *not* a vector space and hence neither is $\mathcal{L}^1(E, \overline{\mathbb{R}})$. In fact, for f, $g \in \mathcal{L}^1(E, \overline{\mathbb{R}})$, the function $f + g$ is *undefined* on the sets $\{x \in E : f(x) = \infty, \, g(x) = -\infty\}$ and $\{x \in E : f(x) = -\infty, \, g(x) = \infty\}$. However, these sets have *measure zero* (by Exercise 10.3.27) and hence $f + g$ is *almost well defined*. In other words, if we define it arbitrarily on the above sets, then our arbitrary choice will neither affect the integrability of $f + g$ nor the value of its integral.

Finally, we can define the Lebesgue integral of a *complex-valued*, measurable function as follows.

Definition 10.3.34 (Lebesgue Integral of Complex Functions). Let $E \in \mathcal{M}_\lambda(\mathbb{R})$ and $f = \mathrm{Re}(f) + i\mathrm{Im}(f) \in \mathcal{L}^0(E, \mathbb{C})$; i.e., both $\mathrm{Re}(f)$ and $\mathrm{Im}(f)$ are *real-valued*, measurable functions on E. We say that f is *Lebesgue integrable* if $\mathrm{Re}(f)$, $\mathrm{Im}(f) \in \mathcal{L}^1(E, \mathbb{R})$ and we then define

$$\int_E f \, d\lambda = \int_E \mathrm{Re}(f) \, d\lambda + i \int_E \mathrm{Im}(f) \, d\lambda;$$

i.e., $\mathrm{Re}\left(\int_E f \, d\lambda\right) = \int_E \mathrm{Re}(f) \, d\lambda$ and $\mathrm{Im}\left(\int_E f \, d\lambda\right) = \int_E \mathrm{Im}(f) \, d\lambda$.

Notation 10.3.35 (The Lebesgue Spaces $\mathcal{L}^1(E, \mathbb{C})$ and $L^1(E, \mathbb{C})$). Given $E \in \mathcal{M}_\lambda(\mathbb{R})$, the set of all complex valued, integrable functions on E is denoted by $\mathcal{L}^1(E, \mathbb{C})$. If we identify integrable functions that are equal almost everywhere on E, the corresponding space of equivalence classes modulo the null functions will be denoted by $L^1(E, \mathbb{C})$.

Here are some of the properties of the complex Lebesgue integral.

Theorem 10.3.36 (Properties of the Complex Lebesgue Integral). *Let $E \in \mathcal{M}_\lambda(\mathbb{R})$, f, $g \in \mathcal{L}^1(E, \mathbb{C})$, and $c \in \mathbb{C}$. Then we have cf, $f + g \in \mathcal{L}^1(E, \mathbb{C})$, $|f| \in \mathcal{L}^1(E, [0, \infty))$ (which in turn implies $f \in \mathcal{L}^1(E, \mathbb{C})$). In fact $\mathcal{L}^1(E, \mathbb{C})$ is a complex vector space. Also, the integral is a linear functional on $\mathcal{L}^1(E, \mathbb{C})$, satisfying the following properties, where $c \in \mathbb{C}$ and f, $g \in \mathcal{L}^1(E, \mathbb{C})$ are arbitrary:*

(1) $\int_E cf \, d\lambda = c \int_E f \, d\lambda$ *(Homogeneity),*
(2) $\int_E (f + g) \, d\lambda = \int_E f \, d\lambda + \int_E g \, d\lambda$ *(Additivity),*
(3) if A and B are measurable subsets of E with $\lambda(A \cap B) = 0$ (e.g., if $A \cap B = \emptyset$), then

$$\int_{A \cup B} f \, d\lambda = \int_A f \, d\lambda + \int_B f \, d\lambda,$$

(4) $\left| \int_E f \, d\lambda \right| \le \int_E |f| \, d\lambda$ *(Triangle Inequality)*,
(5) $\left| \int_E f \, d\lambda \right| = \int_E |f| \, d\lambda$ if and only if $|f| = \alpha f$ a.e. on E for some $\alpha \in \mathbb{C}$
     with $|\alpha| = 1$,
(6) $\int_E |f| \, d\lambda = 0 \Longleftrightarrow f = 0$ a.e. on E, and
(7) $f = g$ a.e. on $E \Longrightarrow \int_E f \, d\lambda = \int_E g \, d\lambda$.

Proof. First, $f \in \mathcal{L}^1(E, \mathbb{C}) \Leftrightarrow |f| \in \mathcal{L}^1(E, [0, \infty))$ follows from the inequalities

$$|f| = |\mathrm{Re}(f) + i\,\mathrm{Im}(f)| \le |\mathrm{Re}(f)| + |\mathrm{Im}(f)| \le 2|f|.$$

The integrability of cf and $f + g$ now follows from $|cf| = |c||f|$ and $|f + g| \le |f| + |g|$. In view of Theorem 10.3.31, we need only prove (1) for $c = i = \sqrt{-1}$. But then

$$\int_E (if) \, d\lambda = i \int_E \mathrm{Re}(f) \, d\lambda - \int_E \mathrm{Im}(f) \, d\lambda$$

$$= i \left(\int_E \mathrm{Re}(f) \, d\lambda + i \int_E \mathrm{Im}(f) \, d\lambda \right)$$

$$= i \int_E f \, d\lambda.$$

We prove (4) and (5) and leave the remaining parts for the reader as an exercise. For any complex number $z \in \mathbb{C} \setminus \{0\}$, we have $z = |z|e^{i\theta}$ for some $\theta \in \mathbb{R}$ and hence $|z| = e^{-i\theta}z > 0$. So we can pick $\alpha \in \mathbb{C}$ with $|\alpha| = 1$ such that $\left| \int_E f \, d\lambda \right| = \alpha \int_E f \, d\lambda$ and note that

$$\left| \int_E f \, d\lambda \right| = \alpha \int_E f \, d\lambda = \mathrm{Re}\left(\int_E \alpha f \, d\lambda \right) = \int_E \mathrm{Re}(\alpha f) \, d\lambda \le \int_E |f| \, d\lambda,$$

where we have used the fact that $\mathrm{Re}(\alpha f) \le |\alpha f| = |f|$ and the monotonicity of the real integral. If, as assumed in (5), the last inequality is actually an *equality*, then we have $\int_E [|f| - \mathrm{Re}(\alpha f)] \, d\lambda = 0$, which (the integrand being nonnegative) implies $|f| - \mathrm{Re}(\alpha f) = 0$ a.e. on E. Thus $|f| = \mathrm{Re}(\alpha f)$ a.e. on E and hence (why?) $|f| = |\alpha f| = \mathrm{Re}(\alpha f) = \alpha f$ a.e. on E. Since this condition is clearly sufficient, the proof is complete.   □

Exercise 10.3.37.

(a) Prove the remaining parts of Theorem 10.3.36.
(b) Let $E \in \mathcal{M}_\lambda(\mathbb{R})$ and $f \in \mathcal{L}^1(E, \mathbb{C})$. Show that if $\int_A f \, d\lambda = 0$ for every measurable $A \subset E$, then $f = 0$ almost everywhere.

We end the section with the following nice property of *average values* of an integrable function, which provides a converse to property (6) in Theorem 10.3.20.

Theorem 10.3.38 (Average Value Theorem). *Let $E \in \mathcal{M}_\lambda(\mathbb{R})$ with $\lambda(E) < \infty$ and let $f \in L^1(E, \mathbb{C})$. If $S \subset \mathbb{C}$ is a closed set such that*

$$\mathbf{av}_A(f) := \frac{1}{\lambda(A)} \int_A f \, d\lambda \in S$$

for every measurable $A \subset E$ with $\lambda(A) > 0$, then $f(x) \in S$ for almost all $x \in E$. In particular, if f is real-valued and $S = [a, b]$, then $a \leq f(x) \leq b$ for almost all $x \in E$.

Proof. Since S^c is *open*, for any $\zeta \in S^c$ we can pick $r > 0$ such that $B_r^-(\zeta) := \{z \in \mathbb{C} : |z - \zeta| \leq r\} \subset S^c$. In fact, S^c is a *countable* union of such closed disks. So it suffices to show that $\lambda(f^{-1}(D)) = 0$ if $D := B_r^-(\zeta)$. Let $F := f^{-1}(D)$ and note that if $\lambda(F) > 0$, then we have

$$|\mathbf{av}_F(f) - \zeta| = \frac{1}{\lambda(F)} \left| \int_F (f - \zeta) \, d\lambda \right| \leq \frac{1}{\lambda(F)} \int_F |f - \zeta| \, d\lambda \leq r,$$

which is impossible because $\mathbf{av}_F(f) \in S$. $\square$

10.4 Convergence Theorems

As already pointed out a number of times, the greatest advantage of using Lebesgue's theory of integration is that it behaves nicely under limit operations. In this section we state and prove the most important convergence theorems and some of their consequences. We begin with the first fundamental convergence theorem:

Theorem 10.4.1 (Monotone Convergence Theorem). *Let $E \in \mathcal{M}_\lambda(\mathbb{R})$ and let (f_n) be a sequence in $\mathcal{L}^0(E, \overline{\mathbb{R}}_+)$. If (f_n) is increasing (i.e., $f_1 \leq f_2 \leq f_3 \leq \cdots$) and converges pointwise to f [i.e., $\lim (f_n(x)) = f(x)$ for all $x \in E$], then $f \in \mathcal{L}^0(E, \overline{\mathbb{R}}_+)$ and we have*

$$\int_E f \, d\lambda = \lim_{n \to \infty} \int_E f_n \, d\lambda.$$

Proof. That f is measurable follows from Proposition 10.2.22. By *monotonicity*, the numerical sequence $\left(\int_E f_n \, d\lambda \right)$ is increasing and bounded above by $\int_E f \, d\lambda$ (because $f_n \leq f$ for all $n \in \mathbb{N}$). Hence $\lim \left(\int_E f_n \, d\lambda \right)$ exists as an extended nonnegative number and satisfies

$$\lim_{n \to \infty} \int_E f_n \, d\lambda \leq \int_E f \, d\lambda.$$

To prove the reverse inequality, it suffices to show that

$$\int_E \phi \, d\lambda \le \lim_{n \to \infty} \int_E f_n \, d\lambda \quad \forall \, \phi \in Simp^+(\mathbb{R}), \quad 0 \le \phi \le f, \tag{$*$}$$

because $\int_E f \, d\lambda := \sup\{\int_E \phi : 0 \le \phi \le f, \ \phi \in Simp(\mathbb{R})\}$. So let $\phi \in Simp^+(\mathbb{R})$ satisfy $0 \le \phi \le f$ and pick $\varepsilon \in (0, 1)$. Consider the measurable sets

$$E_n = \{x \in E : f_n(x) \ge (1 - \varepsilon)\phi(x)\}, \quad (n \in \mathbb{N}).$$

Since (f_n) is increasing, so is the sequence (E_n), i.e., $E_n \subset E_{n+1}$ for all $n \in \mathbb{N}$. We claim that $E = \bigcup_{n=1}^{\infty} E_n$. Indeed, if $f(x) = 0$, then $x \in E_n$ for all n. And if $f(x) > 0$, then we must have $(1 - \varepsilon)\phi(x) \le f_n(x)$ for large enough n because, otherwise, $f_n(x) < (1 - \varepsilon)\phi(x) \le (1 - \varepsilon)f(x)$ for all $n \in \mathbb{N}$ would give the contradiction $0 < f(x) \le (1 - \varepsilon)f(x)$. Next, integrating $f_n \ge (1 - \varepsilon)\phi$ over E_n, we have

$$(1 - \varepsilon) \int_{E_n} \phi \, d\lambda \le \int_{E_n} f_n \, d\lambda \le \int_E f_n \, d\lambda \quad (\forall n \in \mathbb{N}).$$

Taking limits as $n \to \infty$ and using Lemma 10.3.10, we have

$$(1 - \varepsilon) \int_E \phi \, d\lambda \le \lim_{n \to \infty} \int_E f_n \, d\lambda,$$

which, sending $\varepsilon \to 0$, gives $(*)$ and completes the proof. $\qquad\square$

Corollary 10.4.2. *Let $E \in \mathcal{M}_\lambda(\mathbb{R})$ and $f \in \mathcal{L}^0(E, \overline{\mathbb{R}}_+)$. Then, for any increasing sequence (ϕ_n) in $Simp^+(\mathbb{R})$ with $\lim (\phi_n(x)) = f(x) \ \forall x \in E$, we have*

$$\int_E f \, d\lambda = \lim_{n \to \infty} \int_E \phi_n \, d\lambda.$$

In particular, one can use the sequence (ϕ_n) constructed in Theorem 10.2.24.

Proof. Obvious!

Remark 10.4.3 ((f_n) Increasing Is Crucial).

(1) In the Monotone Convergence Theorem the assumption that (f_n) be *increasing* (at least a.e.) is essential. Indeed, on $E = \mathbb{R}$, the sequences $\left(\frac{1}{n}\chi_{(0,n)}\right)$, $\left(\chi_{(n,n+1)}\right)$, and $\left(n\chi_{(0,1/n)}\right)$ all converge to 0 everywhere and yet for every $n \in \mathbb{N}$ we have $\int \frac{1}{n}\chi_{(0,n)} \, d\lambda = \int \chi_{(n,n+1)} d\,\lambda = \int n\chi_{(0,1/n)}d\,\lambda = 1$.
(2) On $E = \mathbb{R}$, consider the *decreasing* sequence of functions $f_n := \chi_{[n,\infty)}$ for all $n \in \mathbb{N}$. Then $\lim(f_n) = 0$, but $\int_{\mathbb{R}} f_n \, d\lambda = \infty$ for all $n \in \mathbb{N}$.

Corollary 10.4.4 (Tonelli's Theorem). *Let* (f_n) *be a sequence in* $\mathcal{L}^0(E, \overline{\mathbb{R}}_+)$, *where* $E \in \mathcal{M}_\lambda(\mathbb{R})$, *and define* $f := \sum_{n=1}^\infty f_n$. *Then we have* $f \in \mathcal{L}^0(E, \overline{\mathbb{R}}_+)$ *and*

$$\int_E f \, d\lambda = \int_E \left(\sum_{n=1}^\infty f_n \right) d\lambda = \sum_{n=1}^\infty \int_E f_n \, d\lambda.$$

Proof. Let $g_n := \sum_{k=1}^n f_k$. Then (g_n) is an increasing sequence in $\mathcal{L}^0(E, \overline{\mathbb{R}}_+)$ and we have $\lim (g_n(x)) = f(x)$ for all $x \in E$. Thus the corollary follows from the Monotone Convergence Theorem and the (finite) additivity of the integral.   □

Corollary 10.4.5. *Let* (f_n) *be an increasing sequence in* $\mathcal{L}^0(E, \overline{\mathbb{R}}_+)$, *where* $E \in \mathcal{M}_\lambda(\mathbb{R})$, *and let* $f \in \mathcal{L}^0(\overline{\mathbb{R}}_+)$. *If* $\lim_{n \to \infty} f_n(x) = f(x)$ *for almost all* $x \in E$, *then*

$$\int_E f \, d\lambda = \lim_{n \to \infty} \int_E f_n \, d\lambda.$$

Proof. Let $F \subset E$ be the set of all $x \in E$ such that $f_n(x)$ increases to $f(x)$. Then $\lambda(E \setminus F) = 0$ and we have $f - f\chi_F = f_n - f_n\chi_F = 0$ a.e. on E. Thus $\int_E f \, d\lambda = \int_E f\chi_F \, d\lambda$ and $\int_E f_n \, d\lambda = \int_E f_n\chi_F \, d\lambda$ for all $n \in \mathbb{N}$. Since $(f_n\chi_F) \to f\chi_F$ *everywhere* (on E), the Monotone Convergence Theorem gives

$$\int_E f \, d\lambda = \int_E f\chi_F \, d\lambda = \lim_{n \to \infty} \int_E f_n\chi_F \, d\lambda = \lim_{n \to \infty} \int_E f_n \, d\lambda.$$

Corollary 10.4.6. *Let* $f \in \mathcal{L}^0(E, \overline{\mathbb{R}}_+)$, *where* $E \in \mathcal{M}_\lambda(\mathbb{R})$. *Then the set function* $\mu_f : \mathcal{M}_\lambda(\mathbb{R}) \to \overline{\mathbb{R}}_+$ *defined by*

$$\mu_f(A) := \int_{A \cap E} f \, d\lambda = \int_E f\chi_{A \cap E}, \quad \forall A \in \mathcal{M}_\lambda(\mathbb{R})$$

is a (countably additive) measure. Also, $\mu_f(A) = 0$ *whenever* $\lambda(A) = 0$.

Proof. Since we obviously have $\mu_f(\emptyset) = 0$, we need only check the countable additivity. So assume that $A = \bigcup_{n=1}^\infty A_n$ is a disjoint union of measurable sets and define $f_n := f\chi_{A_n \cap E} \in \mathcal{L}^0(E, \overline{\mathbb{R}}_+)$. Then $f\chi_{A \cap E} = \sum_{n=1}^\infty f_n$ and, using Tonelli's theorem, we have

$$\mu_f(A) := \int_{A \cap E} f \, d\lambda = \sum_{n=1}^\infty \int_E f\chi_{A_n \cap E} \, d\lambda = \sum_{n=1}^\infty \int_{A_n \cap E} f \, d\lambda = \sum_{n=1}^\infty \mu_f(A_n).$$

Since the last statement is obvious, the proof is complete.   □

The implication "$\lambda(A) = 0 \implies \mu_f(A) = 0$" is referred to as "the *absolute continuity* of μ_f with respect to λ" and is equivalent to the condition in the following:

Corollary 10.4.7 (Absolute Continuity). *Let $E \in \mathcal{M}_\lambda(\mathbb{R})$ and $f \in \mathcal{L}^1(E, \mathbb{C})$. Then, given any $\varepsilon > 0$, there is a $\delta > 0$ such that for any $A \in \mathcal{M}_\lambda(\mathbb{R})$ with $A \subset E$, we have*

$$\lambda(A) < \delta \implies \mu_{|f|}(A) := \int_A |f| \, d\lambda < \varepsilon.$$

Proof. If f is *bounded*, then the assertion is trivially satisfied. (Why?) Otherwise, define

$$f_n(x) := \begin{cases} |f(x)| & \text{if } |f(x)| \le n, \\ n & \text{if } |f(x)| > n. \end{cases}$$

Then (f_n) is an *increasing* sequence of *bounded,* measurable functions with $\lim(f_n) = |f|$ on E. By the Monotone Convergence Theorem, we can pick $N \in \mathbb{N}$ so large that $\int_E f_N \, d\lambda > \int_E |f| \, d\lambda - \varepsilon/2$ and hence $\int_E (|f| - f_N) \, d\lambda < \varepsilon/2$. If $0 < \delta < \varepsilon/(2N)$ and $\lambda(A) < \delta$, then

$$\int_A |f| \, d\lambda = \int_A (|f| - f_N) \, d\lambda + \int_A f_N \, d\lambda \le \int_E (|f| - f_N) \, d\lambda + N\lambda(A) < \frac{\varepsilon}{2} + \frac{\varepsilon}{2} = \varepsilon$$

and the proof is complete.   □

The Monotone Convergence Theorem can also be used to show that Lebesgue integral is *translation invariant.*

Proposition 10.4.8 (Translation Invariance of Lebesgue Integral). *Let $f \in \mathcal{L}^1(\mathbb{R})$ and, given any $b \in \mathbb{R}$, consider the translated function $f_b(x) := f(x + b)$. Then $f_b \in \mathcal{L}^1(\mathbb{R})$ and we have*

$$\int_\mathbb{R} f_b(x) \, d\lambda(x) = \int_\mathbb{R} f(x + b) \, d\lambda(x) = \int_\mathbb{R} f(x) \, d\lambda(x).$$

More generally, for any real numbers $m \ne 0$ and b, we have

$$|m| \int_\mathbb{R} f(mx + b) \, d\lambda(x) = \int_\mathbb{R} f(x) \, d\lambda(x). \tag{†}$$

In particular, taking $m = \pm 1$ shows that Lebesgue integral is invariant under the isometries of $\mathbb{R}$.

Proof. If $\phi = a\chi_A$ with $a > 0$ and $A \in \mathcal{M}_\lambda(\mathbb{R})$, then $\phi(mx+b) = a\chi_A(mx+b) = a\chi_{m^{-1}(A-b)}(x)$, and hence Exercise 10.1.17 gives

$$\int \phi(mx+b)\, d\lambda(x) = a\lambda\big(m^{-1}(A-b)\big) = \frac{a}{|m|}\lambda(A) = \frac{1}{|m|}\int \phi(x)\, d\lambda(x),$$

so (†) holds in this case. It then follows at once that (†) is satisfied for every $\phi \in \boldsymbol{Simp}^+(\mathbb{R})$. Next, assuming $f \geq 0$, we can pick an increasing sequence (ϕ_n) in $\boldsymbol{Simp}^+(\mathbb{R})$ such that $\lim(\phi_n) = f$. By the Monotone Convergence Theorem we then have

$$|m| \int f(mx+b)\, d\lambda(x) = |m| \lim_{n\to\infty} \int \phi_n(mx+b)\, d\lambda(x) = \lim_{n\to\infty} \int \phi_n(x)\, d\lambda(x)$$

$$= \int f(x)\, d\lambda(x),$$

proving (†) for $f \geq 0$. We now apply this to f^+ and f^- for a general $f = f^+ - f^- \in \mathcal{L}^1(\mathbb{R})$ to complete the proof. □

Despite its importance, the Monotone Convergence Theorem is not directly applicable in cases where one must deal with *nonmonotone* sequences. For such sequences, one should be particularly cautious when integrating term by term:

Example 10.4.9.

(1) For each $n \in \mathbb{N}$, consider the function $f_n(x) := ne^{-nx}$ on $[0,1]$. We have $\lim_{n\to\infty} f_n(x) = 0$ for all $x \in (0,1]$ and hence $\lim(f_n) = 0$ a.e. On the other hand, $\int_0^1 f_n = 1 - e^{-n} \to 1$, as $n \to \infty$. Therefore,

$$\lim_{n\to\infty} \int_0^1 f_n(x)\, dx \neq \int_0^1 \lim(f_n(x))\, dx.$$

(2) For each $n \in \mathbb{N}$, let $g_n(x) := n^2 e^{-nx}$ on $[0,1]$. Then $\lim_{n\to\infty} g_n(x) = 0$ for all $x \in (0,1]$ so that $\lim(g_n) = 0$ a.e. But a computation gives $\int_0^1 g_n = n - ne^{-n} \to +\infty$ and we obviously have

$$\lim_{n\to\infty} \int_0^1 g_n(x)\, dx \neq \int_0^1 \lim(g_n(x))\, dx.$$

The next consequence is another one of the fundamental convergence results, where the sequence of functions is *not* assumed to be increasing.

Lemma 10.4.10 (Fatou's Lemma). *If $E \in \mathcal{M}_\lambda(\mathbb{R})$ and (f_n) is a sequence in $\mathcal{L}^0(E, \overline{\mathbb{R}}_+)$, then we have*

$$\int_E \underline{\lim}(f_n)\, d\lambda \leq \underline{\lim}\left(\int_E f_n\, d\lambda\right).$$

Proof. For each $n \in \mathbb{N}$, let $g_n := \inf\{f_k : k \geq n\}$ and note that $g_n \in \mathcal{L}^0(E, \overline{\mathbb{R}}_+)$, (g_n) is *increasing,* and $\underline{\lim}(f_n) = \lim(g_n)$. Also, $g_n \leq f_k$ for all $k \geq n$ implies

$$\int_E g_n \, d\lambda \leq \int_E f_k \, d\lambda, \qquad \forall k \geq n,$$

which implies that the increasing sequence $\left(\int_E g_n \, d\lambda\right)$ satisfies

$$\int_E g_n \, d\lambda \leq \inf\left\{\int_E f_k \, d\lambda : k \geq n\right\} = \underline{\lim}\left(\int_E f_n \, d\lambda\right), \qquad \forall n \in \mathbb{N}.$$

By the Monotone Convergence Theorem, we therefore have

$$\int_E \underline{\lim}(f_n) = \int_E \lim(g_n) \, d\lambda = \lim\left(\int_E g_n \, d\lambda\right) \leq \underline{\lim}\left(\int_E f_n \, d\lambda\right).$$

$\square$

Example 10.4.11. On $E = \mathbb{R}$ consider the sequence of functions $f_n := n\chi_{(0,1/n)}$ for all $n \in \mathbb{N}$. Then

$$0 = \int_{\mathbb{R}} \lim(f_n) \, d\lambda < 1 = \underline{\lim}\left(\int_{\mathbb{R}} f_n \, d\lambda\right),$$

so that the inequality in Fatou's lemma is *strict* in this case.

Corollary 10.4.12. *Let $E \in \mathcal{M}_\lambda(\mathbb{R})$ and (f_n) a sequence in $\mathcal{L}^0(E, \overline{\mathbb{R}}_+)$ such that $\lim_{n \to \infty} f_n(x) = f(x)$ for almost all $x \in E$ and some $f \in \mathcal{L}^0(E, \overline{\mathbb{R}}_+)$. Then*

$$\int_E f \, d\lambda \leq \underline{\lim}\left(\int_E f_n \, d\lambda\right).$$

Proof. If $\lim\left(f_n(x)\right) = f(x)$ for *all* $x \in E$, then the assertion follows from Fatou's lemma. In general, we need only modify the f_n and f on a set of measure zero, which (by Theorem 10.3.23) has no effect on the integrals. $\square$

Exercise 10.4.13. Deduce the Monotone Convergence Theorem from Fatou's lemma.

So far we have only looked at sequences of *nonnegative* measurable functions. What if the f_n are not nonnegative? What conditions would guarantee the integrability of the limit of a sequence of (complex-valued) measurable functions? Also, will the integral of the limit function equal the limit of the integrals of the functions in the sequence? The most important theorem that answers these questions is the following.

Theorem 10.4.14 (Dominated Convergence Theorem). *Let $E \in \mathcal{M}_\lambda(\mathbb{R})$ and let (f_n) be a sequence in $\mathcal{L}^0(E, \mathbb{C})$ such that $\lim_{n \to \infty} f_n(x) = f(x)$ for almost all $x \in E$, where $f \in \mathcal{L}^0(E, \mathbb{C})$. If there exists a function $g \in \mathcal{L}^1(E, [0, \infty))$ such that $|f_n(x)| \le g(x)$ for all $n \in \mathbb{N}$ and almost all $x \in E$, then $f \in \mathcal{L}^1(E, \mathbb{C})$ and*

$$\int_E f \, d\lambda = \lim_{n \to \infty} \int_E f_n \, d\lambda.$$

In fact, we even have

$$\lim_{n \to \infty} \int_E |f - f_n| \, d\lambda = 0.$$

Proof. Modifying f and f_n on a set of measure zero, if necessary, we may (and do) assume that $|f_n(x)| \le g(x) \; \forall \, n \in \mathbb{N}$ and $\lim (f_n(x)) = f(x)$ hold for *all* $x \in E$. Also, considering the real and imaginary parts, we may (and do) assume that the f_n and f are *real-valued*. Since $|f| \le g$, Proposition 10.3.30 gives $f \in \mathcal{L}^1(E, \mathbb{R})$ and hence $\int_E (g \pm f) \, d\lambda = \int_E g \, d\lambda \pm \int_E f \, d\lambda$. Now the *nonnegative*, measurable functions $g \pm f_n$ satisfy $\lim (g \pm f_n) = g \pm f$ on E. By Fatou's lemma we therefore have

$$\int_E g \, d\lambda + \int_E f \, d\lambda \le \underline{\lim} \int_E (g + f_n) \, d\lambda = \int_E g \, d\lambda + \underline{\lim} \int_E f_n \, d\lambda,$$

$$\int_E g \, d\lambda - \int_E f \, d\lambda \le \underline{\lim} \int_E (g - f_n) \, d\lambda = \int_E g \, d\lambda - \overline{\lim} \int_E f_n \, d\lambda.$$

Thus $\overline{\lim} \int_E f_n \, d\lambda \le \int_E f \, d\lambda \le \underline{\lim} \int_E f_n \, d\lambda$, from which the assertion follows. For the last assertion, note that the assumption $|f_n(x)| \le g(x)$ for all n and almost all x implies that $|f(x)| \le g(x)$ for almost all x. Now we have $\lim |f(x) - f_n(x)| = 0$ and $|f(x) - f_n(x)| \le 2g(x)$ for almost all x (and for all n). Therefore, by the first part of the proof, we have $\int_E |f - f_n| \, d\lambda \to 0$ as $n \to \infty$. $\square$

Remark 10.4.15. The assumption that all the functions f_n be bounded above by the *same, fixed* integrable function $g \ge 0$ can be weakened as follows:

Corollary 10.4.16. *Let $E \in \mathcal{M}_\lambda(\mathbb{R})$ and let (g_n) be a sequence in $\mathcal{L}^1(E, [0, \infty))$ such that $\lim (g_n) = g \in \mathcal{L}^1(E, [0, \infty))$ and $\lim \left(\int_E g_n \, d\lambda \right) = \int_E g \, d\lambda$. If (f_n) is a sequence in $\mathcal{L}^0(E, \mathbb{C})$ such that $|f_n(x)| \le g_n(x)$ for all $n \in \mathbb{N}$ and almost all $x \in E$ and if for some $f \in \mathcal{L}^0(E, \mathbb{C})$ we have $\lim (f_n(x)) = f(x)$ for almost all $x \in E$, then $f \in \mathcal{L}^1(E, \mathbb{C})$ and*

$$\int_E f \, d\lambda = \lim_{n \to \infty} \int_E f_n \, d\lambda.$$

Once again, we even have the stronger condition

$$\lim_{n\to\infty} \int_E |f - f_n| \, d\lambda = 0.$$

Proof. The proof of the Dominated Convergence Theorem works in this case as well if we replace the functions $g \pm f_n$ by $g_n \pm f_n$. As for the last statement, simply note that $\lim(|f_n - f|) = 0$ and $|f_n - f| \le |f_n| + |f| \le g_n + g$ both hold almost everywhere and we have $\int_E (g_n + g) \, d\lambda \to \int_E 2g \, d\lambda$. Thus applying the first part implies $\int_E |f - f_n| \, d\lambda \to 0$. $\square$

The complex-valued case of the Dominated Convergence Theorem obviously includes the real-valued case. However, the theorem is true for *extended* real-valued functions as well:

Theorem 10.4.17 (Dominated Convergence Theorem in $\mathcal{L}^1(E,\overline{\mathbb{R}})$). *Let $E \in \mathcal{M}_\lambda(\mathbb{R})$ and (f_n) a sequence in $\mathcal{L}^0(E,\overline{\mathbb{R}})$ such that $\lim \big(f_n(x)\big) = f(x)$ for almost all $x \in E$ and a function $f \in \mathcal{L}^0(E,\overline{\mathbb{R}})$. If there exists a function $g \in \mathcal{L}^1(E,\overline{\mathbb{R}}_+)$ such that $|f_n(x)| \le g(x)$ for almost all $x \in E$, then $f \in \mathcal{L}^1(E,\overline{\mathbb{R}})$ and we have*

$$\int_E f \, d\lambda = \lim_{n\to\infty} \int_E f_n \, d\lambda.$$

In fact,

$$\lim_{n\to\infty} \int_E |f - f_n| \, d\lambda = 0.$$

Proof. It follows from the assumptions that the subsets of E on which the functions $|f|$, g, and $|f_n|$ take on the value $+\infty$ have measure zero and hence we may assume that all these functions are *real-valued*. The assertions then follow from Theorem 10.4.14. $\square$

Corollary 10.4.18 (Beppo-Levi). *Let $E \in \mathcal{M}_\lambda(\mathbb{R})$. Suppose that (f_n) is a sequence in $\mathcal{L}^1(E,\overline{\mathbb{R}})$ such that $\sum_{n=1}^\infty \int_E |f_n| \, d\lambda < \infty$. Then there is a function $f \in \mathcal{L}^1(E,\overline{\mathbb{R}})$ such that $\sum_{n=1}^\infty f_n(x) = f(x)$ for almost all $x \in E$ and we have*

$$\int_E \left(\sum_{n=1}^\infty f_n\right) d\lambda = \sum_{n=1}^\infty \int_E f_n \, d\lambda.$$

Proof. We have $\int_E \left(\sum_{n=1}^\infty |f_n|\right) d\lambda = \sum_{n=1}^\infty \int_E |f_n| \, d\lambda < \infty$, as follows at once from Tonelli's Theorem (Corollary 10.4.4). In particular, $g := \sum_{n=1}^\infty |f_n| \in \mathcal{L}^1(E,\overline{\mathbb{R}})$ and Exercise 10.3.27 implies that $\sum_{n=1}^\infty f_n(x)$ converges for almost all $x \in E$. Let $f(x)$ denote the sum for such x's and define $f(x)$ arbitrarily for any other x. Now note that $|\sum_{k=1}^{n} f_k| \le g$ for all n, so by the Dominated Convergence Theorem (applied to the partial sums of $\sum_n f_n$) we have $f \in \mathcal{L}^1(E, \overline{\mathbb{R}})$ and $\int_E f \, d\lambda = \sum_{n=1}^{\infty} \int_E f_n \, d\lambda$.   □

Here is another useful consequence (cf. Corollary 10.4.6):

Corollary 10.4.19. *Let $E \in \mathcal{M}_\lambda(\mathbb{R})$ and $f \in \mathcal{L}^1(E, \mathbb{C})$. Suppose that $E = \bigcup_{n=1}^{\infty} E_n$, where the E_n are pairwise disjoint measurable sets. Then f is integrable on each E_n and*

$$\int_E f \, d\lambda = \sum_{n=1}^{\infty} \int_{E_n} f \, d\lambda.$$

Moreover,

$$\sum_{n=1}^{\infty} \int_{E_n} |f| \, d\lambda < \infty. \tag{†}$$

Proof. Indeed, we have $f = \sum_{n=1}^{\infty} f\chi_{E_n}$ on E and hence for each partial sum $f_n := \sum_{k=1}^{n} f\chi_{E_k}$, we have $|f_n| \le |f|$. Since $f = \lim(f_n)$, and $\int f_n \, d\lambda = \sum_{k=1}^{n} \int_{E_k} f \, d\lambda$, the result follows from the Dominated Convergence Theorem. Also, (†) follows if we apply the same argument to the integrable function $|f|$.   □

Note that the converse of Corollary 10.4.19 is *not* true; i.e., the integrability of f on each E_n does *not* in general imply that f is integrable on E. However, it is true if the condition (†) is satisfied.

Proposition 10.4.20. *Let E and the E_n be as in Corollary 10.4.19 and suppose that $f : E \to \mathbb{C}$ is integrable on each E_n. If (†) is satisfied, then f is integrable on E and we have*

$$\int_E f \, d\lambda = \sum_{n=1}^{\infty} \int_{E_n} f \, d\lambda.$$

Proof. Since $\int_{E_n} f \, d\lambda = \int_E f\chi_{E_n} \, d\lambda$, the proposition follows from Corollary 10.4.18 with $f_n := f\chi_{E_n}$.   □

Finally, the following consequence of the Dominated Convergence Theorem can be applied in many situations and may be worth stating separately.

Theorem 10.4.21 (Bounded Convergence Theorem). *Let $E \in \mathcal{M}_{\lambda < \infty}(\mathbb{R})$ and (f_n) a sequence in $\mathcal{L}^0(E, \mathbb{C})$. Suppose that there is a constant $M > 0$ such that $|f_n(x)| \le M$ for all $n \in \mathbb{N}$ and almost all $x \in E$. If for some $f \in \mathcal{L}^0(E, \mathbb{C})$ we have $\lim(f_n(x)) = f(x)$ for almost all $x \in E$, then $f \in \mathcal{L}^1(E, \mathbb{C})$ and*

$$\int_E f \, d\lambda = \lim_{n \to \infty} \int_E f_n \, d\lambda.$$

In fact, we have

$$\lim_{n \to \infty} \int_E |f - f_n| \, d\lambda = 0.$$

Proof. This follows at once from the Dominated Convergence Theorem if we use $g := M \chi_E$.   □

Exercise 10.4.22.

(1) Deduce Fatou's lemma from the Bounded Convergence Theorem.
(2) Let $f_n = n^{-2} \chi_{[0,n^2]}$ for each $n \in \mathbb{N}$. Show that $\lim(f_n) = 0$ *uniformly* and that $\int f_n = 1$ for all $n \in \mathbb{N}$. Explain why this does *not* contradict Lebesgue's Dominated Convergence Theorem.

Remark 10.4.23 (Escape to Infinity, Moving Bumps). The basic reason why in our convergence theorems we need the sequence (f_n) to be *increasing* (in the Monotone Convergence Theorem) or *dominated by an integrable function* (in the Dominated Convergence Theorem) is to prevent the areas (masses) under the graphs of the f_n to *escape to infinity* in some sense. The standard examples that demonstrate this phenomenon are the so-called *moving bump* functions we saw above, namely, the functions $\chi_{(n,n+1)}$, $n\chi_{(0,1/n)}$, and $\frac{1}{n}\chi_{(0,n)}$. To use Tao's terminology (in [Tao11]), we say that for $\chi_{(n,n+1)}$ the mass escapes to *horizontal infinity*, for $n\chi_{(0,1/n)}$ it escapes to *vertical infinity*, and for $\frac{1}{n}\chi_{(0,n)}$ to *width infinity*. The reader is invited to sketch the graphs to see why the terminology fits the situation. The convergence theorems show that if we impose conditions that prevent such escapes to infinity, then the limit function has the desired properties.

10.5 Littlewood's Other Principles and Modes of Convergence

In Sect. 10.1 we established Littlewood's first principle (Theorem 10.1.22), which says that a measurable set is "nearly" a finite union of open intervals. There are two more principles that we want to look at in this section. For the record, here is the list of all three principles:

Remark 10.5.1 (Littlewood's Three Principles).

 (i) Every measurable set is *nearly* a finite union of intervals.
 (ii) Every measurable function is *nearly* continuous.
(iii) Every pointwise convergent sequence of measurable functions is *nearly* uniformly convergent.

The point is, of course, to give the precise meaning of the adverb *nearly* in each case. The first principle was basically a more precise version of "outer regularity" (Corollary 10.1.9). That a (Lebesgue) measurable set can be approximated by open sets from without and by closed sets from within [cf. Theorem 10.1.23, (ii) and (iii)] can be stated and proved in many ways and we begin this section by giving another version of this so-called *regularity* of Lebesgue measure which relates measurability to the topology of $\mathbb{R}$.

Theorem 10.5.2 (Regularity of Lebesgue Measure). *Let* $E \in \mathcal{M}_\lambda(\mathbb{R})$. *Then we have*

(a) *(Outer Regularity).* $\lambda(E) = \inf\{\lambda(U) : U \text{ is open and } E \subset U\}$.
(b) *(Inner Regularity).* $\lambda(E) = \sup\{\lambda(K) : K \text{ is compact and } K \subset E\}$.
(c) *(Lusin's Criterion).* $E \in \mathcal{M}_{\lambda<\infty}(\mathbb{R})$ *if and only if for each* $\varepsilon > 0$ *there is a compact set* K *and an open set* U *such that*

$$K \subset E \subset U \quad and \quad \lambda(U \setminus K) < \varepsilon.$$

Proof. Part (a) is simply Corollary 10.1.9 (the *outer regularity* of λ^*). To prove (b), note that if K is a compact subset of E, then $K \in \mathcal{B}_\mathbb{R} \subset \mathcal{M}_\lambda(\mathbb{R})$ and by monotonicity of λ we have $\lambda(K) \le \lambda(E)$. Taking the supremum we certainly have $\sup\{\lambda(K) : K \text{ is compact}, K \subset E\} \le \lambda(E)$. For the reverse inclusion, suppose first that E is *bounded* and let $\varepsilon > 0$ be given. Then by part (iii) of Theorem 10.1.23 we can pick a *closed* (hence *compact*) set $K \subset E$ such that $\lambda(E \setminus K) < \varepsilon$. Since $\varepsilon > 0$ was arbitrary, the equality in (b) follows for the *bounded* case. If E is *unbounded*, define $E_n := \{x \in E : n - 1 < |x| \le n\}$, $n \in \mathbb{N}$, and let $\varepsilon > 0$ be given. Using the preceding argument pick a compact set $K_n \subset E_n$ with $\lambda(K_n) > \lambda(E_n) - \varepsilon 2^{-n}$. Let $F_n := \bigcup_{k=1}^n K_k$ and note that F_n is compact and $F_n \subset E$ for each n. Also, $\lambda(F_n) > \lambda\left(\bigcup_{k=1}^n E_k\right) - \varepsilon$. Since the *continuity* of λ (Theorem 10.1.25) gives $\lambda(E) = \lim_{n \to \infty} \lambda(\bigcup_{k=1}^n E_k)$, the result holds for the unbounded case as well. Finally, (c) follows from (a) and (b).   □

Corollary 10.5.3. *If* $E \in \mathcal{M}_\lambda(\mathbb{R})$, *then there is an increasing sequence* $(F_n)_{n=1}^\infty$ *of closed sets and a decreasing sequence* $(U_n)_{n=1}^\infty$ *of open sets such that* $F_n \subset E \subset U_n$ *for all* $n \in \mathbb{N}$ *and*

$$\lim_{n \to \infty} \lambda(F_n) = \lambda(E) = \lim_{n \to \infty} \lambda(U_n).$$

If, in addition, $\lambda(E) < \infty$, *then the closed sets* F_n *may be assumed to be compact.*

Proof. For each $n \in \mathbb{N}$, the regularity of λ guarantees the existence of a *closed* set F_n' and an *open* set U_n' such that

$$F_n' \subset E \subset U_n' \quad and \quad \lambda(U_n' \setminus F_n') < \frac{1}{n}.$$

Now let $F_n := \bigcup_{k=1}^n F_k'$ and $U_n := \bigcap_{k=1}^n U_k'$. Note that $U_n \setminus F_n \subset U_n' \setminus F_n'$ and hence $\lim(\lambda(U_n \setminus F_n)) = 0$. It follows at once that the sequences (F_n) and (U_n) satisfy the required properties.   □

Here is another form of the measurability criterion:

Corollary 10.5.4. *Let* $E \subset \mathbb{R}$*. If* $E \in \mathcal{M}_\lambda(\mathbb{R})$*, then*

$$\sup\{\lambda(F) : F \subset E; \ F \ closed\} = \inf\{\lambda(U) : E \subset U; \ U \ open\}, \qquad (\dagger)$$

and $\lambda(E)$ *is then the (extended) number in* (†). *Conversely, if* (†) *holds and is finite, then* E *is measurable.*

Proof. Exercise!

Remark 10.5.5. Note that the *finiteness* of (†) for the converse in the above corollary is *necessary*. Indeed, let $A \subset [0, 1]$ be *nonmeasurable* and let $E := A \cup [2, \infty)$. Then both sides of (†) are $+\infty$ and yet E is *not* measurable.

Exercise 10.5.6 (Squeeze Theorem). Let $E \subset \mathbb{R}$ and $\lambda^*(E) < \infty$. Show that $E \in \mathcal{M}_{\lambda<\infty}(\mathbb{R})$ if and only if for any $\varepsilon > 0$ there are measurable sets $A_\varepsilon, B_\varepsilon \in \mathcal{M}_{\lambda<\infty}(\mathbb{R})$ such that

$$A_\varepsilon \subset E \subset B_\varepsilon \quad and \quad \lambda(B_\varepsilon \setminus A_\varepsilon) < \varepsilon.$$

Let us now look at a version of Littlewood's second principle.

Theorem 10.5.7 (Littlewood's Second Principle). *Let* $f : [a, b] \to \overline{\mathbb{R}}$ *be measurable and almost finite; i.e., the set* $\{x \in [a, b] : |f(x)| = \infty\}$ *has measure zero. Then for every* $\varepsilon > 0$ *there is a step function* ψ *and a continuous function* g *satisfying* $|g| \leq |\psi| \leq |f|$ *such that*

$$|f(x) - \psi(x)| < \varepsilon \quad and \quad |f(x) - g(x)| < \varepsilon$$

except on a set of measure less than ε*. Thus* $\lambda(\{x \in [a, b] : |f(x) - \psi(x)| \geq \varepsilon\}) < \varepsilon$ *and* $\lambda(\{x \in [a, b] : |f(x) - g(x)| \geq \varepsilon\}) < \varepsilon$*.*

Proof. Assume first that f is *bounded*. By Theorem 10.2.24, we can pick a sequence (ϕ_n) of *simple* functions such that $|\phi_n| \leq |f|$ for all n and $\lim(\phi_n) = f$ (uniformly) on $[a, b]$. Thus, there is a simple function $\phi : [a, b] \to \mathbb{R}$ with $|\phi| \leq |f|$ such that $|f(x) - \phi(x)| < \varepsilon/2$. Let $\phi = \sum_{j=1}^m a_j \chi_{A_j}$ be the canonical representation of ϕ. By Littlewood's first principle (Theorem 10.1.22), for each j there is a finite union $U_j = \bigcup_{i=1}^{m_j} I_{j,i}$ of (disjoint) open intervals such that $\lambda(U_j \triangle A_j) < \varepsilon/(2n)$. Consider the *step* function $\psi := \sum_{j=1}^m a_j \chi_{U_j \setminus \bigcup_{k=1}^{j-1} U_k}$. If $\psi(x) \neq \phi(x)$, then either for some i we have $\psi(x) = a_i \neq \phi(x)$, in which case $x \in U_i \setminus A_i$, or $\psi(x) = 0$ and $\phi(x) = a_i$, in which case $x \in A_i \setminus U_i$. Therefore,

$$\phi = \psi \quad \text{except possibly on} \quad A := \bigcup_{j=1}^n U_j \triangle A_j,$$

where $\lambda(A) < \varepsilon/2$. But then, the *Triangle Inequality* gives $|f(x) - \psi(x)| < \varepsilon/2 + \varepsilon/2 = \varepsilon$ except on the set A whose measure is less than $\varepsilon/2$. Also, since both ψ and ϕ take values in $\{a_1, \ldots, a_m\}$, we have $|\psi| \leq |f|$ as well. To construct the *continuous* function g, we first pick a *step* function ψ with $|\psi| \leq |f|$ and $|f(x) - \psi(x)| < \varepsilon/2$ except on a set of measure less than $\varepsilon/2$. We now define a *piecewise linear* function g as follows: Suppose $\psi(x) = c_1$ for $x \in [x_0, x_1]$ and $\psi(x) = c_j$ for $x \in (x_{j-1}, x_j]$, $2 \leq j \leq \ell$, where $a = x_0 < x_1 < \cdots < x_\ell = b$. Let $\mu = \max\{|x_j - x_{j-1}| : 1 \leq j \leq \ell\}$ denote the mesh of this partition and pick $\delta > 0$ with $\delta < \min\{\mu/2, \varepsilon/4(\ell - 1)\}$. Define $g := \psi$ outside the intervals $(x_j - \delta, x_j + \delta)$, $1 \leq j \leq \ell - 1$, and on each interval $(x_j - \delta, x_j + \delta)$ let the graph of g consist of the two line segments that join $(x_j, 0)$ to the points $(x_j - \delta, c_j)$ and $(x_j + \delta, c_{j+1})$. It then follows at once that $|f(x) - g(x)| < \varepsilon$ except on a set of measure less than ε and $|g| \leq |\psi| \leq |f|$. Finally, if f is *not* bounded, set $E_n := \{x \in [a, b] : |f(x)| > n\}$ and note that (E_n) is a decreasing sequence of measurable sets with $\lambda(\bigcap_{n=1}^\infty E_n) = 0$. We can therefore pick $N \in \mathbb{N}$ such that $\lambda(E_N) < \varepsilon/2$. The truncated function $f_N := (-N \vee f) \wedge N = \min\{\max\{-N, f\}, N\}$ is then *bounded* and $f = f_N$ except on a set of measure less than $\varepsilon/2$. We can now find a step function ψ and a continuous function g, satisfying $|g| \leq |\psi| \leq |f_N| \leq |f|$, such that $|f_N - \psi| < \varepsilon/2$ and $|f_N - g| < \varepsilon/2$ except on sets of measure less that $\varepsilon/2$. So, by the *Triangle Inequality*, $|f - \psi| < \varepsilon$ and $|f - g| < \varepsilon$ except on sets of measure less that ε. $\square$

Corollary 10.5.8. *Let $f : [a, b] \to \overline{\mathbb{R}}$ be measurable and almost finite. Then there is a sequence (ψ_n) of step functions and a sequence (g_n) of continuous functions on $[a, b]$ with $|g_n| \leq |\psi_n| \leq |f|$ for all $n \in \mathbb{N}$ such that $\lim_{n \to \infty} \psi_n(x) = f(x) = \lim_{n \to \infty} g_n(x)$ for almost all $x \in [a, b]$. In particular, this is true when f is integrable and we then also have*

$$\lim_{n \to \infty} \int_a^b |f(x) - \psi_n(x)| \, d\lambda(x) = \lim_{n \to \infty} \int_a^b |f(x) - g_n(x)| \, d\lambda(x) = 0. \qquad (\ddagger)$$

In fact, if E is measurable and $f \in \mathcal{L}^1(E, \overline{\mathbb{R}})$, then for each $\varepsilon > 0$ there is a step function ψ and a continuous function g such that

$$\int_E |f - \psi| \, d\lambda < \varepsilon \quad and \quad \int_E |f - g| \, d\lambda < \varepsilon.$$

Proof. We prove the existence of (g_n); the case of (ψ_n) is similar. Since $\lambda(\{x : |f(x)| = \infty\}) = 0$, we may (and do) assume that $f : [a, b] \to \mathbb{R}$. For each $n \in \mathbb{N}$ we use the above theorem with $\varepsilon = 1/2^n$ to pick a continuous function g_n on $[a, b]$ and a set $E_n \subset [a, b]$ such that $\lambda(E_n) < 1/2^n$ and $|f(x) - g_n(x)| < 1/2^n$ for all $x \in [a, b] \setminus E_n$. If $F_n := \bigcup_{k=n+1}^\infty E_k$, then (F_n) is a *decreasing* sequence of measurable sets with $\lambda(F_n) < \sum_{k=n+1}^\infty 1/2^k = 1/2^n$. Thus, with $Z := \bigcap_{n=1}^\infty F_n$, we have $\lambda(Z) = \lim_{n \to \infty} \lambda(F_n) = 0$. Moreover, for each $x \in [a, b] \setminus Z$, we have $x \notin F_n$ for some n; i.e., $x \notin E_k$ for all $k > n$ and hence $|f(x) - g_k(x)| < 1/2^k$ for all $k > n$. We therefore have $\lim_{n \to \infty} g_n(x) = f(x)$ for all $x \in [a, b] \setminus Z$. Finally, if f is *integrable*, then it is automatically *almost finite* by Exercise 10.3.27 and ($\ddagger$) follows from the inequalities $|f - \psi_n| \le 2|f|$, $|f - g_n| \le 2|f|$ and the Dominated Convergence Theorem. We now prove the existence of the step function ψ in the last assertion; the case of g is similar. For each $N \in \mathbb{N}$ define $f_N := f\chi_{[-N,N]}$. Then $\lim_{N \to \infty} |f - f_N| = 0$ and $|f - f_N| \le 2|f|$. It follows from the Dominated Convergence Theorem that $\lim_{N \to \infty} \int_E |f - f_N|\, d\lambda = 0$. So, given $\varepsilon > 0$ we can pick N so large that $\int_E |f - f_N|\, d\lambda < \varepsilon/2$. Now by the first part of the corollary, there is a step function ψ such that $\int_{-N}^{N} |f_N - \psi|\, d\lambda < \varepsilon/2$. If we define ψ to be zero outside $[-N, N]$, then we have

$$
\int_E |f - \psi|\, d\lambda \le \int_E |f_N - \psi|\, d\lambda + \int_E |f - f_N|\, d\lambda < \frac{\varepsilon}{2} + \frac{\varepsilon}{2} = \varepsilon. \qquad \square
$$

Having looked at Littlewood's first and second principles, we turn our attention to Littlewood's third principle. Before giving a version of the principle, we need the following.

Definition 10.5.9 (Uniform and Locally Uniform Convergence). Let $X \subset \mathbb{R}$ and consider a sequence of functions $f_n : X \to \mathbb{R}$ (resp., $f_n : X \to \mathbb{C}$). We say that (f_n) *converges to* $f : X \to \mathbb{R}$ *(resp.,* $f : X \to \mathbb{C}$) *uniformly (on X)* if for every $\varepsilon > 0$ there exists $N = N(\varepsilon) \in \mathbb{N}$ such that $|f_n(x) - f(x)| < \varepsilon$ for all $n > N$ and all $x \in X$.

We say that (f_n) *converges to* f *locally uniformly* (on X) if for every *bounded* set $B \subset X$, (f_n) converges uniformly to f on B.

Exercise 10.5.10. Let $X \subset \mathbb{R}$. Show that $f_n : X \to \mathbb{R}$ (or $f_n : X \to \mathbb{C}$) converges locally uniformly to f (on X) if and only if for every $x_0 \in X$ there is an open set $U \ni x_0$ such that (f_n) converges *uniformly* to f on $X \cap U$. *Hint:* Use the Heine–Borel theorem.

Remark 10.5.11. Note that uniform convergence is stronger than locally uniform convergence, which in turn is stronger that pointwise convergence.

Example 10.5.12.

(1) Let $f_n(x) := \sum_{k=0}^{n} x^k / k!$. Then $f_n(x) \to e^x$ locally uniformly (on $\mathbb{R}$), but *not* uniformly.

(2) Let $f_n(x) := \begin{cases} 1/(nx) & \text{if } x > 0 \\ 0 & \text{otherwise} \end{cases}$. Then $\lim_{n \to \infty} f_n(x) = 0$ for all $x \in \mathbb{R}$ (i.e., (f_n) converges to *zero* pointwise everywhere on $\mathbb{R}$), but (f_n) does *not* converge locally uniformly.

Here now is a version of Littlewood's third principle.

Proposition 10.5.13. *Let $E \in \mathcal{M}_{\lambda < \infty}(\mathbb{R})$ and let (f_n) be a sequence of real (or complex) measurable functions on E. Suppose that (f_n) converges (pointwise)*

almost everywhere (on E) to a (measurable) function f. Then, given any numbers
$\varepsilon > 0$ *and* $\delta > 0$, *there is a measurable set* $A \subset E$ *with* $\lambda(A) < \delta$ *and an integer* N
such that

$$x \in E \setminus A \quad and \quad n \geq N \Longrightarrow |f_n(x) - f(x)| < \varepsilon.$$

Proof. Let $B \subset E$ be such that $\lambda(B) = 0$ and $f_n(x) \to f(x)$ for all $x \in E \setminus B$.
Consider the measurable sets

$$E_k := \{x \in E \setminus B : |f_n(x) - f(x)| \geq \varepsilon \text{ for some } n \geq k\}.$$

Then we have $E_{k+1} \subset E_k$ for all k and, for each $x \in E \setminus B$, there is a $k_x \in \mathbb{N}$
such that $x \notin E_{k_x}$ because $f_n(x) \to f(x)$. We therefore have $\bigcap_{k=1}^{\infty} E_k = \emptyset$ and
the *continuity* of λ gives $\lim_{k \to \infty} \lambda(E_k) = 0$. Thus, given any $\delta > 0$, we can pick
$N \in \mathbb{N}$ so large that $\lambda(E_N) < \delta$. If we now set $A := B \cup E_N$, we then have
$\lambda(A) < \delta$ and

$$x \in E \setminus A \Longrightarrow |f_n(x) - f(x)| < \varepsilon \quad (\forall\, n \geq N). \qquad \square$$

Remark 10.5.14. Note that in the above proposition the convergence of (f_n) to f is
not uniform on $E \setminus A$ because the integer N depends on the prescribed $\delta > 0$. The
following important consequence is stronger and does indeed achieve the uniform
convergence.

Theorem 10.5.15 (Egorov's Theorem). *Let* (f_n) *be a sequence of real (or com-*
plex) measurable functions on a set $E \in \mathcal{M}_{\lambda < \infty}(\mathbb{R})$. *If* (f_n) *converges to a function*
f *almost everywhere (on E), then for each* $\varepsilon > 0$ *there is a measurable set* $A \subset E$
with $\lambda(A) < \varepsilon$ *and such that* (f_n) *converges to* f *uniformly on* $E \setminus A$.

Proof. Let $\varepsilon > 0$ be given. Using Proposition 10.5.13, for each $k \in \mathbb{N}$, we can pick
a measurable set $A_k \subset E$ with $\lambda(A_k) < \varepsilon/2^k$ and an integer N_k such that

$$x \in E \setminus A_k \quad and \quad n \geq N_k \Longrightarrow |f_n(x) - f(x)| < 1/k.$$

Now let $A := \bigcup_{k=1}^{\infty} A_k$ and note that $\lambda(A) < \sum_{k=1}^{\infty} \varepsilon/2^k = \varepsilon$. If we pick $k_0 \in \mathbb{N}$
such that $1/k_0 < \varepsilon$, then we have

$$x \in E \setminus A \quad and \quad n \geq N_{k_0} \Longrightarrow |f_n(x) - f(x)| < 1/k_0 < \varepsilon$$

and hence (f_n) converges to f *uniformly* on $E \setminus A$. $\qquad \square$

Remark 10.5.16. The convergence in Egorov's theorem *does not mean uniform*
convergence outside of a set of measure zero; so (in general) the set $A = A(\varepsilon)$
in Egorov's theorem *cannot* be chosen to have measure zero. For example, if $g_n :=$
$\chi_{(0,1/n)}$, then $\lim(g_n(x)) = 0$ for all $x \in [0, 1]$ so that Egorov's theorem guarantees
uniform convergence *outside a set of arbitrarily small measure.* But (g_n) does *not*

converge uniformly outside *any* set of measure zero. (Why?) As another example, consider the sequence of functions $f_n(x) = x^n$ on $[0, 1]$. We have $\lim(f_n) = f$, where $f(x) = 0$ for all $x \in [0, 1)$ and $f(1) = 1$. Since the f_n are all *continuous* on $[0, 1]$, the convergence is *not* uniform on $[0, 1]$. In fact, it is *not* uniform on $[0, 1] \setminus Z$, for *any* $Z \subset [0, 1]$ with $\lambda(Z) = 0$ (why?). On the other hand, the convergence is certainly uniform on $[0, 1 - \varepsilon]$ for any $\varepsilon \in (0, 1)$.

The following version of the theorem now follows from the one we just proved.

Theorem 10.5.17 (Egorov's Theorem, Locally Uniform Version). *Let $E \in \mathcal{M}_\lambda(\mathbb{R})$ and let (f_n) be a sequence of measurable real (or complex) functions converging almost everywhere (on E) to a function f. Then, for each $\varepsilon > 0$ there is a measurable set $A \subset E$ with $\lambda(A) < \varepsilon$ such that (f_n) converges to f locally uniformly on $E \setminus A$.*

Proof. For each $k \in \mathbb{N}$ let $E_k := E \cap [-k, k]$ and note that by Egorov's theorem applied on E_k we can find a set $A_k \subset E_k$ with $\lambda(A_k) < \varepsilon/2^k$ such that (f_n) converges to f uniformly on $E_k \setminus A_k$. Define $A := \bigcup_{k=1}^\infty A_k$ and note that $\lambda(A) < \sum_{k=1}^\infty \varepsilon/2^k = \varepsilon$. If now B is any *bounded* subset of E, then $B \subset E_k$ for some k and (f_n) does indeed converge to f uniformly on $B \setminus A \subset E_k \setminus A_k$. $\square$

Remark 10.5.18. The locally uniform convergence in Egorov's theorem cannot (in general) be upgraded to uniform convergence. Indeed, consider the *moving bump* sequence $f_n := \chi_{[n,n+1]}$ on $\mathbb{R}$. Then (f_n) converges pointwise and locally uniformly to $f = 0$ but *not* uniformly outside any set A with $\lambda(A) < \varepsilon \in (0, 1)$ because we then have $|f_n(x) - f(x)| = 1 > \varepsilon$ for all $x \in [n, n + 1]$. Thus, to get uniform convergence, the exceptional set A is forced to contain a set of measure 1. In fact, A must contain $[n, n + 1]$ for all sufficiently large n and hence we must have $\lambda(A) = \infty$. Note that this is caused by the fact that (f_n) escapes to *horizontal infinity*, which cannot happen under the condition $\lambda(E) < \infty$ imposed in Theorem 10.5.15.

In fact the *converse* of Egorov's theorem is also true:

Proposition 10.5.19 (Converse of Egorov's Theorem). *Let E be measurable with $\lambda(E) < \infty$ and let $f, f_n \in \mathcal{L}^0(E)$ $\forall n \in \mathbb{N}$. Suppose that for each $\varepsilon > 0$, there is a measurable subset $A(\varepsilon) \subset E$ with $\lambda(A(\varepsilon)) < \varepsilon$ such that $\lim(f_n) = f$ uniformly on $E \setminus A(\varepsilon)$. Then $\lim(f_n) = f$ almost everywhere on E.*

Proof. For each $k \in \mathbb{N}$, let A_k be a measurable subset of E with $\lambda(A_k) < 1/k$ such that $\lim(f_n) = f$ uniformly on $E \setminus A_k$. Define $Z = \bigcap_{k=1}^\infty A_k$. Then we have

$$\lambda(Z) \le \lambda(A_k) < \frac{1}{k} \qquad (\forall k \in \mathbb{N}).$$

It follows that $\lambda(Z) = 0$ and it is easily checked that $\lim(f_n(x)) = f(x)$ for all $x \in E \setminus Z$. $\square$

Let us now use Egorov's theorem to prove the following important version of Littlewood's second principle.

Theorem 10.5.20 (Lusin's Theorem). *Let $f : [a, b] \to \mathbb{R}$ (resp., $f : \mathbb{R} \to \mathbb{R}$) be a measurable function and let $\varepsilon > 0$ be given. Then there is a measurable set $A \subset [a, b]$ (resp. $A \subset \mathbb{R}$) with $\lambda(A) < \varepsilon$ such that the restriction of f to the complement $[a, b] \setminus A$ (resp., A^c) is continuous on that set.*

Proof. Consider first the case $f : [a, b] \to \mathbb{R}$. By Corollary 10.5.8, there is a sequence (g_n) of continuous (hence measurable) functions on $[a, b]$ such that (g_n) converges to f almost everywhere. Now for a given $\varepsilon > 0$ we use Egorov's theorem to find a set $A \subset [a, b]$ with $\lambda(A) < \varepsilon$ such that (g_n) converges to f *uniformly* on $[a, b] \setminus A$. Since uniform limits of continuous functions are continuous, it follows that f restricted to $[a, b] \setminus A$ is indeed continuous. If $f : \mathbb{R} \to \mathbb{R}$, then for each $n \in \mathbb{Z}$ we can (using the previous case) pick a measurable set $A_n \subset [n, n + 1]$ satisfying $\lambda(A_n) < 2^{-|n|}\varepsilon/3$ such that the restricted function $f |([n, n+1] \setminus A_n)$ is continuous. If $A := \bigcup_{n=1}^{\infty} A_n$, then $\lambda(A) < (\varepsilon/3) \sum_{n \in \mathbb{Z}} 2^{-|n|} = \varepsilon$ and the restriction $f | A^c$ is continuous. $\square$

Remark 10.5.21.

(1) Note that, as the trivial example $f := \chi_\mathbb{Q}$ shows, the *original* measurable function f in Lusin's theorem may in fact be *discontinuous everywhere*.

(2) Lusin's theorem remains valid if $f : [a, b] \to \mathbb{R}$ (resp., $f : \mathbb{R} \to \mathbb{R}$) is replaced by $f : [a, b] \to \overline{\mathbb{R}}$ (resp., $f : \mathbb{R} \to \overline{\mathbb{R}}$), provided we assume that f is *finite almost everywhere*, i.e., the set $\{x : |f(x)| = \infty\}$ has measure zero.

Exercise 10.5.22 (Converse of Lusin's Theorem). Show that the *converse* of Lusin's theorem is also true: *If $f : [a, b] \to \mathbb{R}$ and if for each $\varepsilon > 0$ there is a measurable set $A = A(\varepsilon) \subset [a, b]$ with $\lambda(A) < \varepsilon$ such that $f |([a, b] \setminus A)$ is continuous, then f is measurable.*

Before stating the last theorem of the section, let us recall (Definition 5.1.11) that, if (M, d) is a metric space and $S \subset M$, then, for each $x \in M$, the *distance between x and S* is the nonnegative number

$$d(x, S) := \inf\{d(x, s) : s \in S\}.$$

Note (cf. Exercise 5.1.12) that the function $x \mapsto d(x, S)$ is *Lipschitz* (hence *uniformly continuous*) on M. In particular, for any $\delta > 0$, the set $\{x \in M : d(x, S) < \delta\}$ is *open*. The theorem we are about to prove shows that, although a set $S \subset \mathbb{R}$ with *positive* measure may have *empty interior* (e.g., a *generalized Cantor set*), the set $S - S := \{s - t : s, t \in S\}$ has *nonempty* interior:

Theorem 10.5.23 (Steinhaus's Theorem). *Let $E \subset \mathbb{R}$ be a measurable set with $\lambda(E) > 0$. Then there exists $\varepsilon > 0$ such that*

$$(-\varepsilon, \varepsilon) \subset E - E := \{x - y : x, y \in E\}.$$

Proof. Let $E_n := E \cap [-n, n]$. Then $E_n \in \mathcal{M}_{\lambda < \infty}(\mathbb{R})$ for all n and $\lim(\lambda(E_n)) = \lambda(E) > 0$. Thus $\lambda(E_N) > 0$ for sufficiently large N and we may (and do) assume that $\lambda(E) < \infty$. Next, by Lusin's Criterion (cf. Theorem 10.5.2) for each $\varepsilon > 0$ there is a *compact* set K and an *open* set U such that $K \subset E \subset U$ and $\lambda(U \backslash K) < \varepsilon$. So we may pick $K \subset E$ with $\lambda(K) > 0$ as close to $\lambda(E)$ as we wish. Thus we may (and do) assume that $E = K$ is *compact*. For each $n \in \mathbb{N}$, consider the open set

$$U_n := \{x : d(x, K) < 1/n\}. \tag{$*$}$$

Note that the U_n are *bounded* and *decreasing*, i.e., $U_{n+1} \subset U_n$ for all $n \in \mathbb{N}$. (Why?) Also, we have

$$K = \bigcap_{n=1}^{\infty} U_n.$$

Indeed, $K \subset \bigcap_{n=1}^{\infty} U_n$ because $K \subset U_n$ for all n. On the other hand, if $x \notin K$, then, since K^c is *open*, we can pick m so large that $(x - 1/m, x + 1/m) \subset K^c$. Whence $x \notin U_m$. Now the *continuity* of λ (Theorem 10.1.25) implies that

$$\lim_{n \to \infty} \lambda(U_n) = \lambda(K).$$

Therefore, given any fixed $\delta \in (0, 1/2)$, we can pick N so large that

$$(1 - \delta)\lambda(U_N) < \lambda(K). \tag{$**$}$$

Let $\varepsilon := 1/N$. Then $|z| < \varepsilon$ implies that

$$K - z := \{x - z : x \in K\} \subset U_N.$$

Since $\lambda(K - z) = \lambda(K)$, writing $K_z := K - z$, $(**)$ implies that

$$\begin{aligned}
\lambda(U_N \backslash (K \cap K_z)) &= \lambda\big((U_N \backslash K) \cup (U_N \backslash K_z)\big) \\
&\leq \lambda(U_N \backslash K) + \lambda(U_N \backslash K_z) \\
&= 2\lambda(U_N) - 2\lambda(K) \\
&< 2\delta\lambda(U_N) < \lambda(U_N).
\end{aligned}$$

It follows that $K \cap (K - z)$ has *positive* measure and hence is *nonempty* for each $z \in (-\varepsilon, \varepsilon)$. Thus, for each such z, we can find $x, y \in K$ such that $y = x - z$ and hence $z = x - y \in K - K$ as desired. □

Corollary 10.5.24. *If $A \in \mathcal{M}_\lambda(\mathbb{R})$ and $\lambda(A) > 0$, then A contains a nonmeasurable set.*

Proof. For each $k \in \mathbb{Z}$, set $A_k := A \cap [k, k + 1]$ and note that we must have $\lambda(A_k) > 0$ for some k. Therefore, since λ is invariant under translations, $\lambda(A_k - k) = \lambda(\{x - k : x \in A_k\}) > 0$ and $A_k - k \subset [0, 1]$. So we may (without loss of generality) assume that $A \subset [0, 1]$. Now let $E \subset [0, 1]$ be the *nonmeasurable* set constructed in Example 10.1.21 and let the $E_n := E + r_n$, $n \in \mathbb{N}$, be the disjoint nonmeasurable sets in that example. Then we have the disjoint union $A = \bigcup_n A_n$, where $A_n := A \cap E_n$. If A_n is measurable for every n, then we have $\sum_n \lambda(A_n) = \lambda(A) > 0$ and hence can pick N with $\lambda(A_N) > 0$. Now given any distinct points $x, y \in A_N$, we have $x = \xi + r_N$ and $y = \eta + r_N$ with $\xi, \eta \in E$ and hence $x - y = \xi - \eta \notin \mathbb{Q}$. But this implies that $A_N - A_N$ contains *no* rational numbers, contradicting Steinhaus' theorem. So some A_n is nonmeasurable. $\square$

We now give an interesting application of Steinhaus's theorem. Recall (Theorem 4.3.11) that, if an *additive* function on the real line is continuous at a point, then it is continuous everywhere and is *linear*. In particular, if $f : \mathbb{R} \to \mathbb{R}$ is additive and *monotone*, then it is linear (cf. Exercise 4.4.8). The same also holds if the additive function f has *bounded variation*. The following proposition shows that, in fact, *measurability* (which holds in the above cases) is enough:

Proposition 10.5.25 (Cauchy's Functional Equation). *Let* $f : \mathbb{R} \to \mathbb{R}$ *be a measurable function satisfying Cauchy's functional equation:*

$$f(x + y) = f(x) + f(y) \qquad (\forall x, y \in \mathbb{R}).$$

Then f *is linear, i.e.,* $f(x) = ax$ *for all* $x \in \mathbb{R}$ *and* $a := f(1)$.

Proof. We have already seen (Theorem 4.3.11) that f satisfies $f(r) = ar$ for all $r \in \mathbb{Q}$ and $a := f(1)$. Let us first assume that f is *bounded* on $(-\varepsilon, \varepsilon)$ for some $\varepsilon > 0$; i.e., there is a constant $M > 0$ such that $|f(x)| \le M$ for all $x \in (-\varepsilon, \varepsilon)$. Given any fixed $x \in \mathbb{R}$ and any $n \in \mathbb{N}$, we can find a rational $r \in \mathbb{Q}$ such that $|x - r| < \varepsilon/n$. It then follows that

$$|f(x) - ax| = |f(x - r) - a(x - r)| \le \frac{M + a\varepsilon}{n}.$$

Since this holds for all $n \in \mathbb{N}$, we have $f(x) = ax$, as desired. In general, for each $n \in \mathbb{N}$, let $E_n := \{x \in \mathbb{R} : |f(x)| \le n\}$. Then (E_n) is an *increasing* sequence of measurable sets with $\bigcup_{n=1}^{\infty} E_n = \mathbb{R}$. Thus $\lambda(E_N) > 0$ for a large enough N. By Steinhaus's theorem, we have $(-\varepsilon, \varepsilon) \subset E_N - E_N$ for some $\varepsilon > 0$. Since $|f(z)| \le 2N$ for all $z \in E_N - E_N$, the linearity of f follows as before. $\square$

10.6 Problems

1. Show that if (E_n) is a sequence of pairwise disjoint measurable sets in $\mathbb{R}$, then for any set $A \subset \mathbb{R}$ we have

$$\lambda^* \left(A \cap \bigcup_{n=1}^{\infty} E_n \right) = \sum_{n=1}^{\infty} \lambda^*(A \cap E_n).$$

2 (Inclusion–Exclusion Principle). Show that if E and F are measurable subsets of $\mathbb{R}$ with $\lambda(E \cap F) < \infty$, then $\lambda(E \cup F) = \lambda(E) + \lambda(F) - \lambda(E \cap F)$.

3. Let $E_j \subset (0, 1)$, $1 \leq j \leq n$, be measurable sets with $\sum_{j=1}^{n} \lambda(E_j) > n - 1$. Show that $\lambda(\bigcap_{j=1}^{n} E_j) > 0$. *Hint:* Look at the complements.

4. Show that a set $Z \subset \mathbb{R}$ of measure *zero* may or may not be dense but that Z^c must be *dense*. Deduce that if f and g are *continuous* on $\mathbb{R}$ and if $f(x) = g(x)$ for all $x \in Z^c$, then $f = g$. Also deduce that any measurable set $E \subset [0, 1]$ with $\lambda(E) = 1$ is dense in $[0, 1]$.

5. Show that, if $E \subset \mathbb{R}$ is measurable and $\lambda(E) = 1$, then for any number $\alpha \in [0, 1]$, there is a measurable set $A_\alpha \subset E$ with $\lambda(A_\alpha) = \alpha$. *Hint:* Show that the function $F(x) := \lambda(E \cap (-\infty, x])$ is continuous on $\mathbb{R}$ with $\lim_{x \to -\infty} F(x) = 0$ and $\lim_{x \to \infty} F(x) = 1$.

6.

(a) Let $f \in \mathbf{Lip}(I)$. Show that, if $Z \subset I$ and $\lambda(Z) = 0$, then $\lambda(f(Z)) = 0$.
(b) Show that, if $\lambda(E) = 0$, then $\lambda(\{x^2 : x \in E\}) = 0$.

7. Show that $f^2 \in \mathcal{L}^0(I)$ implies $|f| \in \mathcal{L}^0(I)$. Find an interval I and a function $f \in \mathbb{R}^I$ such that $f^2 \in \mathcal{L}^0(I)$, but $f \notin \mathcal{L}^0(I)$.

8. Show that, if $f \in \mathbb{R}^I$ is *differentiable*, then $f' \in \mathcal{L}^0(I)$. *Hint:* Consider the functions $g_n(x) := n[f(x + 1/n) - f(x)]$, for all $x \in I^\circ$.

9. Let $E \in \mathcal{M}_\lambda([a, b])$. Show that, given any $\varepsilon > 0$, there are (relatively) *open* sets O_1, $O_2 \subset [a, b]$ such that $E \subset O_1$, $E^c := [a, b] \setminus E \subset O_2$, and $\lambda(O_1 \cap O_2) < \varepsilon$.

10. Let E_1, $E_2 \subset [a, b]$ and $\lambda(E_1) = 0$. Show that, if $E_1 \cup E_2 \in \mathcal{M}_\lambda([a, b])$, then $E_2 \in \mathcal{M}_\lambda([a, b])$.

11. Show that E_1, $E_2 \in \mathcal{M}_\lambda([0, 1])$ and $\lambda(E_1) = 1$ imply $\lambda(E_1 \cap E_2) = \lambda(E_2)$.

12. Show that there is a *nonmeasurable* set that is *dense* in $[0, 1]$.

13. Show that, if $E \in \mathcal{M}_\lambda([-1, 1])$ and $\lambda(E) > 1$, then there is a measurable set $F \subset E$ such that $F = -F$ and $\lambda(F) > 0$.

14.

(a) Let $E \subset [0, 1]$ be the nonmeasurable set in Example 10.1.21. Show that, if $F \subset E$ is *measurable*, then $\lambda(F) = 0$. *Hint:* Let $(r_j)_{j=0}^{\infty}$ be an enumeration of $\mathbb{Q} \cap [0, 1)$ with $r_0 = 0$, and let $F_j := F + r_j$. Show that $(F_j)_{j=0}^{\infty}$ is a *pairwise disjoint* sequence of measurable sets with $\lambda(F_j) = \lambda(F)$ for all j so that $\sum_{j=0}^{\infty} \lambda(F_j) = \lambda(\bigcup_{j=0}^{\infty} F_j) \leq 2$.
(b) Let $A \subset \mathbb{R}$ with $\lambda^*(A) > 0$. Show that there is a *nonmeasurable* set $B \subset A$. *Hint:* Suppose that $A \subset (0, 1)$ and set $A_j := A \cap (E + r_j)$ with E and (r_j) as in part (a). If $A_j \in \mathcal{M}_\lambda$, $\forall\ j$, then $\lambda(A_j) = 0$, $\forall\ j$, but $\sum \lambda(A_j) \geq \lambda^*(A) > 0$.

15. Show that, if $E \in \mathcal{M}_\lambda(\mathbb{R})$ is *bounded* and $\lambda(E) > 0$, then there exist x, $y \in E$ with $x - y \in \mathbb{Q}$. *Hint:* Assume $E \subset [0, 1]$. If $E \cap (E + r) = \emptyset$ for all $r \in \mathbb{Q} \cap [0, 1]$, then note that the sets $E_n := E + r_n$, $r_n \in \mathbb{Q} \cap [0, 1]$ are pairwise disjoint measurable subsets of $[0, 2]$.

16. Show that there *does not* exist a measurable set $E \subset \mathbb{R}$ such that for *any* interval I, we have $\lambda(E \cap I) = \lambda(I)/2$. *Hint:* If E exists, then $\lambda(E \cap [0, 1]) = 1/2$. Cover $E \cap [0, 1]$ by a sequence (I_n) of pairwise disjoint intervals such that $\sum_{n=1}^\infty \lambda(I_n) < 1$ and derive a contradiction from countable subadditivity.

17. Let $\kappa : [0, 1] \to [0, 1]$ be the *Cantor function* (cf. Example 4.3.13) and consider the function $f(x) := x + \kappa(x)$ for all $x \in [0, 1]$. Show that f is a *homeomorphism* of $[0, 1]$ onto $[0, 2]$. If $C \subset [0, 1]$ is the *Cantor set,* show that $\lambda(f(C)) = 1$ even though $\lambda(C) = 0$. *Hint:* Note that f^{-1} cannot have any *jump* discontinuities. Also, κ is *constant* on the subintervals that make up $C^c := [0, 1] \setminus C$ and f maps each of these intervals onto an interval of equal length. Thus $\lambda(f(C^c)) = 1$.

18 ($\mathcal{B}_\mathbb{R} \subsetneq \mathcal{M}_\lambda(\mathbb{R})$).

(a) Let $f : [0, 1] \to [0, 2]$ be the homeomorphism in the preceding problem. Show that f maps each *Borel* set onto a *Borel* set. Show, however, that f maps a *measurable* set E onto a *nonmeasurable* set B and hence that measurability is *not* a topological property. Also conclude that $f^{-1}(B) \in \mathcal{M}_\lambda(\mathbb{R}) \setminus \mathcal{B}_\mathbb{R}$ and hence that $\mathcal{B}_\mathbb{R} \subsetneq \mathcal{M}_\lambda(\mathbb{R})$. *Hint:* Pick a *nonmeasurable* subset B of $f(C)$.

(b) With f and the nonmeasurable set B as in part (a), let $g := \chi_B \circ f$. Show that $g(x) = 0$ a.e. Thus, χ_B is *nonmeasurable* and yet $\chi_B = g \circ f^{-1}$, with g *measurable* and f^{-1} *continuous* (cf. Exercise 10.2.16).

19. Let E be a *nonmeasurable* subset of $(0, 1)$ and define the function $f : (0, 1) \to (0, 2)$ by $f(x) := x + 1$ if $x \in E$ and $f(x) := x$ if $x \notin E$.

(a) Show that, if $Z \subset (0, 1)$ and $\lambda(Z) = 0$, then $\lambda(f(Z)) = 0$.
(b) Show that there is a *measurable* set $A \subset (0, 1)$ such that $f(A)$ is *nonmeasurable.*

20.

(a) Show that $f \in \mathcal{L}^0(I)$ if and only if $\{x \in I : f(x) > r\}$ is measurable for every $r \in \mathbb{Q}$.
(b) More generally, let $E \in \mathcal{M}_\lambda(\mathbb{R})$ and let D be a *dense* subset of $\mathbb{R}$. Show that $f \in \mathcal{L}^0(E)$ if and only if $\{x \in E : f(x) > d\} \in \mathcal{M}_\lambda$ for all $d \in D$.

21. Let $E \in \mathcal{M}_\lambda(\mathbb{R})$. Show that, if $f \in \mathcal{L}^0(E)$ and if $g \in \mathbb{R}^\mathbb{R}$ is *monotone,* then $g \circ f \in \mathcal{L}^0(E)$.

22. Let $E \in \mathcal{M}_\lambda(\mathbb{R})$ and f, $g \in \mathcal{L}^0(E, \mathbb{R})$. If $F : \mathbb{R}^2 \to \mathbb{R}$ is *continuous,* show that the composite function $h = F(f, g)$ is a measurable function on E. Deduce, in particular, that $f + g$ and fg are measurable.

23 (Borel vs. Lebesgue). Define a function $g \in \mathbb{R}^\mathbb{R}$ to be a *Borel function* (or *Borel measurable*) if $g^{-1}(B) \in \mathcal{B}_\mathbb{R}$ for any $B \in \mathcal{B}_\mathbb{R}$.

(a) Show that $g \in \mathbb{R}^\mathbb{R}$ is a Borel function if and only if $g^{-1}(O) \in \mathcal{B}_\mathbb{R}$ for each *open* set $O \subset \mathbb{R}$. In fact, show that *open set* can be replaced by *open interval.*
(b) Show that, if $f \in \mathcal{L}^0(\mathbb{R})$, then there is a *Borel function* $g \in \mathbb{R}^\mathbb{R}$ such that $f = g$ a.e. *Hint:* For each $r \in \mathbb{Q}$, let $E_r := \{x : f(x) < r\}$. Show that we can write $E_r = B_r \triangle Z_r$, where $B_r \in \mathcal{B}_\mathbb{R}$ and $\lambda(Z_r) = 0$. Now pick $Z \subset \mathcal{B}_\mathbb{R}$ with $\lambda(Z) = 0$ and $\bigcup_{r \in \mathbb{Q}} Z_r \subset Z$ (Theorem 10.1.23), and define

$$g(x) := \begin{cases} 0 & \text{if } x \in Z, \\ f(x) & \text{if } x \notin Z. \end{cases}$$

24. Let (f_n) be a sequence in $\mathcal{L}^0$. Show that the set $\{x : (f_n(x))$ converges$\}$ is *measurable*.

25. Let g, $h \in \mathcal{L}^1([a, b])$. Show that, if $f \in \mathcal{L}^0([a, b])$ and $g \le f \le h$ a.e. on $[a, b]$, then $f \in \mathcal{L}^1([a, b])$.

26. Let f, $g \in \mathcal{L}^1(\mathbb{R})$ have *compact support*; i.e., they vanish outside a pair of compact sets. Is it true in general that $g \circ f \in \mathcal{L}^1(\mathbb{R})$?

27. Show that the function

$$f(x) := \begin{cases} 1/\sqrt{x} & \text{if } x > 0, \\ 0 & \text{if } x = 0 \end{cases}$$

is (Lebesgue) *integrable* on $[0, 1]$ and find $\int_0^1 f(x) \, d\lambda(x)$. *Hint:* For each $n \in \mathbb{N}$, consider the function

$$f_n(x) := \begin{cases} 0 & \text{if } x \in [0, 1/n^2), \\ 1/\sqrt{x} & \text{if } x \in [1/n^2, 1]. \end{cases}$$

28. Show that, for each $b > 0$, the function $f(x) := xe^{-bx}$ is (Lebesgue) integrable on $[0, \infty)$ and find $\int_0^\infty f(x) \, d\lambda(x)$.

29. Show that the function $f(x) := x/(e^x - 1)$, defined to be 1 at $x = 0$, is Lebesgue integrable on $[0, \infty)$ and we have

$$\int_0^\infty \frac{x}{e^x - 1} \, d\lambda(x) = \frac{\pi^2}{6}.$$

Hint: Expand $(e^x - 1)^{-1}$, and then use the previous problem and Tonelli's Theorem (Corollary 10.4.4).

30. Let $f \in \mathcal{L}^1([0, 1])$. Show that $\lim_{n \to \infty} \int_0^1 x^n f(x) \, d\lambda(x) = 0$.

31 (Borel–Cantelli Lemma). Let (E_n) be a sequence of measurable sets in $\mathbb{R}$ such that

$$\sum_{n=1}^\infty \lambda(E_n) < \infty. \tag{$*$}$$

Show that almost every $x \in \mathbb{R}$ belongs to at most a *finite* number of the E_n. What if the condition $(*)$ is removed? *Hint:* Consider the function $g(x) := \sum_n \chi_{E_n}(x)$.

32. Let $E \subset \mathbb{R}$ be measurable. Show that if (f_n) is a sequence in $\mathcal{L}^1(E)$ with $f_n \to f$ almost everywhere and $f \in \mathcal{L}^1(E)$, then $\int |f_n - f| \, d\lambda \to 0$ if and only if $\int |f_n| \, d\lambda \to \int |f| \, d\lambda$.

33. Show that if (f_n) is a sequence in $\mathcal{L}^1(\mathbb{R})$ and if there is a function $f \in \mathcal{L}^1(\mathbb{R})$ such that

$$\int_\mathbb{R} |f_n - f| \, d\lambda \le \frac{1}{n^2} \quad \forall n \in \mathbb{N},$$

then $\lim_{n \to \infty} f_n(x) = f(x)$ for almost all $x \in \mathbb{R}$.

34. Evaluate the following limit using the Monotone Convergence Theorem.

$$\lim_{n \to \infty} \int_0^n (1 + x/n)^n e^{-2x} \, dx.$$

35. Let $f \in \mathcal{L}^0(\mathbb{R})$ and assume that $f(x) \geq 0$ for almost all $x \in \mathbb{R}$. Define the numbers

$$m_k = \lambda(\{x \in \mathbb{R} : 2^{k-1} < f(x) \leq 2^k\}), \quad \forall k \in \mathbb{Z}.$$

Show that $f \in \mathcal{L}^1(\mathbb{R})$ if and only if $\sum_{k=-\infty}^{\infty} 2^k m_k < \infty$.

36. Let $E \subset \mathbb{R}$ be measurable and let (f_n) be a sequence of nonnegative functions in $\mathcal{L}^0(E)$ such that $\lim(f_n) = f$ almost everywhere on E and $f_n(x) \leq f(x)$ for all n and almost all $x \in E$. Show that we have

$$\int_E f \, d\lambda = \lim_{n \to \infty} \int_E f_n \, d\lambda.$$

Show that the same holds if instead of assuming $f_n \geq 0$ for all n we assume that $f_n \geq g$ for a function $g \in \mathcal{L}^1(E)$.

37. Let (E_n) be an *increasing* sequence of measurable sets: $E_1 \subset E_2 \subset \cdots$. Show that, if $f \in \mathcal{L}^1(E_n)$ and $\lim_{n \to \infty} \int_{E_n} |f| < \infty$, then $f \in \mathcal{L}^1(E)$, where $E = \bigcup_{n=1}^{\infty} E_n$, and we have

$$\int_E f(x) \, d\lambda(x) = \lim_{n \to \infty} \int_{E_n} f(x) \, d\lambda(x).$$

Hint: Write $E = E_1 \cup (E_2 \setminus E_1) \cup (E_3 \setminus E_2) \cup \cdots$.

38 (Chebyshev's Inequality).   Let $E \in \mathcal{M}_\lambda(\mathbb{R})$ and $0 \leq f \in \mathcal{L}^0(E)$. Show that, for each $c > 0$,

$$\lambda(\{x \in E : f(x) \geq c\}) \leq \frac{1}{c} \int_E f(x) \, d\lambda(x).$$

Hint: Note that, with $E_c := \{x \in E : f(x) \geq c\}$, we have $\int_E f \geq \int_{E_c} f$.

39. Let $E \in \mathcal{M}_\lambda(\mathbb{R})$ and $f \in \mathcal{L}^1(E)$. Use *Chebyshev's inequality* to show that $\int_E |f(x)| \, d\lambda(x) = 0$ implies $f = 0$ a.e. *Hint:* Look at $\lambda(\{x \in E : |f(x)| > 1/n\})$.

40. Let $f \in \mathcal{L}^1([a, b])$ and assume that $\int_a^x f(t) \, d\lambda(t) = 0$ for all $x \in [a, b]$. Show that $f = 0$ almost everywhere.

41. Let $f \in \mathcal{L}^1([a, b])$ and assume that $\int_a^b x^n f(x) \, d\lambda(x) = 0$ for all $n \in \mathbb{N}_0$. Show that $f = 0$ almost everywhere. *Hint:* Use the previous problem and the Weierstrass Approximation Theorem (Corollary 4.7.10).

42. Let $f \in \mathcal{L}^1([a, b])$ and, for each $n \in \mathbb{N}$, let $E_n := \{x \in [a, b] : f(x) > n\}$. Show that $\lim_{n \to \infty} \int_{E_n} f = 0$.

43 (Jensen's Inequality).   Let $\phi : \mathbb{R} \to \mathbb{R}$ be a *convex* function and let $f \in \mathcal{L}^1([0, 1])$. Then we have

$$\int_0^1 \phi(f(x)) \, d\lambda(x) \geq \phi\left(\int_0^1 f(x) \, d\lambda(x)\right).$$

Deduce that

$$\int_0^1 \exp(f(x)) \, d\lambda(x) \geq \exp\left[\int_0^1 f(x) \, d\lambda(x)\right].$$

Hint: Look at a support line to the graph of ϕ at $(\xi, \phi(\xi))$, where $\xi := \int_0^1 f(x)\, d\lambda(x)$.

44. Show that, if $f \in \mathcal{L}^1([a, b])$, if f is *bounded* on $[a, b]$, and if

$$F(x) := \int_a^x f(t)\, d\lambda(t),$$

then $F \in Lip([a, b])$.

45. Show that, if in the preceding problem the assumption that f be *bounded* is removed, then we still have $F \in C([a, b])$. *Hint:* Approximate f by *continuous* functions.

46. Show that, if $f \in \mathcal{L}^1([a, b])$ and $F(x) := \int_a^x f(t)\, d\lambda(t)$, then $F = G - H$, where G and H are *continuous* and *increasing*. Deduce that $F \in BV([a, b])$. *Hint:* Note that $F(x) = \int_a^x f^+(t)\, d\lambda(t) - \int_a^x f^-(t)\, d\lambda(t)$.

47. Let $f \in \mathcal{L}^1([a, b])$ and $F(x) := \int_a^x f(t)\, d\lambda(t)$. Given any partition $\mathcal{P} := (x_i)_{i=0}^n$ with $a = x_0 < x_1 < \cdots < x_n = b$, show that

$$\sum_{j=1}^n |F(x_j) - F(x_{j-1})| \le \int_a^b |f|.$$

Deduce that $F \in BV([a, b])$ and that

$$V_a^b(F) \le \int_a^b |f|.$$

In fact, $V_a^b(F) = \int_a^b |f|$, as we shall see below.

48. Let $f \in \mathcal{L}^1([a, b])$ and $F(x) := \int_a^x f(t)\, d\lambda(t)$. Show that $F \in BV([a, b])$ and

$$V_a^b(F) = \int_a^b |f|.$$

Hint: Let $(\phi_n) \in Step([a, b])^{\mathbb{N}}$ with $\phi_n \to f$ a.e., and define the functions

$$\varepsilon_n(x) := \begin{cases} +1 & \text{if } \phi_n(x) > 0, \\ 0 & \text{if } \phi_n(x) = 0, \\ -1 & \text{if } \phi_n(x) < 0. \end{cases}$$

Now consider the integrals $\int_a^b \varepsilon_n(x) f(x)\, d\lambda(x)$.

49. Show that, if $F' = f$ is *bounded* on $[a, b]$, then $f \in \mathcal{L}^1([a, b])$ and we have

$$\int_a^b f(x)\, d\lambda(x) = F(b) - F(a).$$

Hint: Redefine F (if necessary) so that $F(x) = F(b)$ for all $x > b$ and define

$$f_n(x) := n[F(x + 1/n) - F(x)] \qquad (\forall x \in [a, b]).$$

Now use the *Bounded Convergence Theorem* to show that $\int_a^b f_n \to \int_a^b f$.

50. Let $f \in \mathcal{L}^1(\mathbb{R})$ and, for each $h \in \mathbb{R}$, define $f_h(x) := f(x + h)$. Show that

$$\lim_{h \to 0} \int_{\mathbb{R}} |f_h - f| \, d\lambda = 0.$$

51. Let (f_n) be a *decreasing* sequence of *nonnegative* functions with $f_1 \in \mathcal{L}^1(\mathbb{R})$ and $\lim(\int f_n) = 0$. Show that $\lim(f_n(x)) = 0$ for *almost all* $x \in \mathbb{R}$. *Hint:* Let $f := \lim(f_n)$.

52. Let $f \in \mathcal{L}^1([a,b])$ and suppose that $0 \le f(x) < 1$ for all $x \in [a,b]$. Show that $\lim_{n \to \infty} \int_a^b [f(x)]^n \, d\lambda(x) = 0$.

53. Show that if $f : \mathbb{R} \to \overline{\mathbb{R}}_+$ is integrable, then the function $F(x) := \int_{-\infty}^x f \, d\lambda$ is *continuous*.

54. Prove the Bounded Convergence Theorem (Theorem 10.4.21) using Egorov's theorem (Theorem 10.5.15).

55. Let $f \in \mathcal{L}^1(\mathbb{R})$. Show that for any $\varepsilon > 0$ there is a continuous function g with *compact support* such that

$$\int_{\mathbb{R}} |f - g| \, d\lambda < \varepsilon.$$

56. Let $f : \mathbb{R} \to \mathbb{R}$ be a measurable function such that

$$\lim_{|x| \to \infty} [f(x + y) - f(x)] = 0 \qquad (\forall y \in \mathbb{R}). \tag{$\dagger$}$$

Show that the convergence in ($\dagger$) is in fact *uniform* (in y) on *bounded* sets. *Hint:* Given $\varepsilon > 0$, consider the sets

$$E_n := \{y \in \mathbb{R} : |x| > n \Rightarrow |f(x + y) - f(x)| < \varepsilon\}.$$

Show that (E_n) is an *increasing* sequence of *measurable* sets with $\mathbb{R} = \bigcup_{n=1}^{\infty} E_n$. Deduce that $\lambda(E_N) > 0$ for sufficiently large N and hence $I := [-\delta, \delta] \subset E_N - E_N$ for some $\delta > 0$. Show that, $\forall y \in I$, we have $|f(x + y) - f(x)| < 2\varepsilon$ if $|x| > N + \delta$. Using this observation repeatedly, deduce the *uniform* convergence if y belongs to a *bounded* set.

57. Let $f : \mathbb{R} \to \mathbb{R}$ be a *nonzero* measurable function satisfying $f(x + y) = f(x)f(y)$ for all $x, y \in \mathbb{R}$. Show that $f(x) = a^x$, where $a = f(1)$.

Chapter 11
More on Lebesgue Integral and Measure

Our objective in this chapter is to add a few more topics to Lebesgue's theory of measure and integration introduced in Chap. 10. We begin by revisiting the connection to Riemann's theory and give a short discussion of *improper Riemann integrals*. Next, we look at integrals *depending on a parameter* and give sufficient conditions under which the order of limits and integrals may be interchanged as well as conditions that guarantee the possibility of differentiating *under the integral sign*. The third section includes a short introduction to L^p-spaces which are important examples of *classical Banach spaces*. The fourth section gives a brief treatment of additional modes of convergence including the notion of *convergence in measure*. Finally, the last section deals with the *differentiation problem* and includes Lebesgue's theorem on the differentiability of *monotone* functions as well as his versions of the *Fundamental Theorem(s) of Calculus*.

11.1 Lebesgue vs. Riemann

Now that we have the Lebesgue theory of integration with its powerful convergence theorems, it would be instructive to take another look at its relation to Riemann integral and use the convergence theorems to give shorter proofs. We begin by giving Lebesgue's own definition of his integral in terms of partitions and show that it is equivalent to the one given in Chap. 10. Next, we give another proof of Lebesgue's Criterion for a bounded function to be Riemann integrable. Finally, we obtain convergence criteria for *improper* Riemann integrals and explore their relation to the corresponding Lebesgue integrals.

© Springer Science+Business Media New York 2014
H.H. Sohrab, *Basic Real Analysis*, DOI 10.1007/978-1-4939-1841-6_11

Definition 11.1.1 (Lebesgue Sum). Let $E \in \mathcal{M}_{\lambda < \infty}(\mathbb{R})$ and $f : E \to \mathbb{R}$ a *bounded measurable* function with $m := \inf\{f(x) : x \in E\}$ and $M := \sup\{f(x) : x \in E\}$. Given any tagged partition $\dot{\mathcal{P}} = (\mathcal{P}, \eta)$ of $[m, M]$, where $\mathcal{P} = (y_k)_{k=0}^n$ and $\eta = (\eta_j)_{j=1}^n$ satisfy

$$y_0 := m < y_1 < \cdots < y_n := M \quad \text{and} \quad \eta_j \in [y_{j-1}, y_j], \quad 1 \le j \le n,$$

we define the corresponding *Lebesgue sum* of f to be the number

$$S_\lambda(f, \dot{\mathcal{P}}) := \sum_{j=1}^n \eta_j \lambda(E_j),$$

where $E_j := f^{-1}([y_{j-1}, y_j)), 1 \le j \le n-1$, and $E_n := f^{-1}([y_{n-1}, y_n])$.

Remark. Note that, since f is measurable, so are the sets E_j and the finite sequence $(E_j)_{j=1}^n$ is in fact a *measurable partition* of E into *pairwise disjoint* (measurable) subsets. Also note that, while for Riemann sums the *domain* of f was partitioned, for Lebesgue sums we partition the *range* of f. This was in fact already used in the proof of Theorem 10.2.24. Finally, let us recall that the *mesh* (or *norm*) of the partition $\mathcal{P}$ is the number $\|\mathcal{P}\| := \max\{y_j - y_{j-1} : 1 \le j \le n\}$.

Definition 11.1.2 (Lebesgue's Definition). Let $E \in \mathcal{M}_{\lambda < \infty}(\mathbb{R})$. Then, given a *bounded measurable* function $f : E \to \mathbb{R}$, the *Lebesgue integral of f on E*, denoted by $\int_E f \, d\lambda$, is defined to be

$$\int_E f \, d\lambda := \lim_{\|\mathcal{P}\| \to 0} S_\lambda(f, \dot{\mathcal{P}}), \tag{†}$$

where the limit is defined as in Theorem 7.1.26.

We now have the following.

Proposition 11.1.3. *Let $E \in \mathcal{M}_{\lambda < \infty}(\mathbb{R})$ and $f : E \to \mathbb{R}$ be a bounded, measurable function. Then the integral in (†) agrees with the Lebesgue integral of f, as defined in Chap. 10.*

Proof. Let the notation be as in Definition 11.1.2 and define the *simple function* $\phi_{\dot{\mathcal{P}}}$ by setting

$$\phi_{\dot{\mathcal{P}}} := \sum_{j=1}^n \eta_j \chi_{E_j}.$$

Then, integrating, we have

$$\int_E \phi_{\dot{\mathcal{P}}} \, d\lambda = S_\lambda(f, \dot{\mathcal{P}}).$$

Now observe that $|f(x) - \phi_{\dot{\mathcal{P}}}(x)| \leq \|\mathcal{P}\|$ for all $x \in E$. Thus, if $(\dot{\mathcal{P}}_n)$ is a sequence of tagged partitions of $[m, M]$ with $\|\mathcal{P}_n\| < 1/n$ and $\mathcal{P}_n \subset \mathcal{P}_{n+1}$ for all n, and if $\phi_n := \phi_{\dot{\mathcal{P}}_n}$, then $\lim(\phi_n) = f$ and the Bounded Convergence Theorem (Theorem 10.4.21) implies that

$$\lim_{n \to \infty} S_\lambda(f, \dot{\mathcal{P}}_n) = \lim_{n \to \infty} \int_E \phi_n(x) \, d\lambda(x) = \int_E f(x) \, d\lambda(x). \qquad \square$$

Our next goal is to give a shorter proof of Lebesgue's Integrability Criterion (Theorem 7.3.19). Recall that if $\mathcal{P} = (x_i)_{i=0}^n$ is a partition of $[a, b]$ and if $f : [a, b] \to \mathbb{R}$ is a bounded function, then with $m_j := \inf\{f(x) : x \in [x_{j-1}, x_j]\}$ and $M_j := \inf\{f(x) : x \in [x_{j-1}, x_j]\}$, the corresponding lower and upper Darboux sums of f are

$$L(f, \mathcal{P}) := \sum_{j=1}^n m_j(x_j - x_{j-1}) \quad \text{and} \quad U(f, \mathcal{P}) := \sum_{j=1}^n M_j(x_j - x_{j-1}).$$

Thus, with the step functions

$$\ell_{\mathcal{P}} := f(a)\chi_{\{a\}} + \sum_{j=1}^n m_j \chi_{(x_{j-1}, x_j]} \quad \text{and} \qquad (*)$$

$$u_{\mathcal{P}} := f(a)\chi_{\{a\}} + \sum_{j=1}^n M_j \chi_{(x_{j-1}, x_j]},$$

we have $\ell_{\mathcal{P}} \leq f \leq u_{\mathcal{P}}$ on $[a, b]$ and

$$\int_a^b \ell_{\mathcal{P}}(x) \, d\lambda(x) = L(f, \mathcal{P}) \quad \text{and} \quad \int_a^b u_{\mathcal{P}}(x) \, d\lambda(x) = U(f, \mathcal{P}).$$

Now, pick a sequence $(\mathcal{P}_k)$ of partitions of $[a, b]$ such that $\mathcal{P}_k \subset \mathcal{P}_{k+1}$ and $\|\mathcal{P}_k\| := \max\{x_j - x_{j-1} : 1 \leq j \leq n\} \to 0$ as $k \to \infty$. Let $\ell_k := \ell_{\mathcal{P}_k}$ and $u_k := u_{\mathcal{P}_k}$ be the corresponding step functions as in $(*)$. Then (ℓ_k) is increasing, (u_k) is decreasing, and $\ell_k \leq f \leq u_k$ on $[a, b]$ for all k. Define the functions $\ell := \lim(\ell_k)$ and $u = \lim(u_k)$. Then ℓ and u are bounded Borel functions and we have $l \leq f \leq u$ on $[a, b]$. Also, the Bounded Convergence Theorem gives

$$\int_{[a,b]} \ell \, d\lambda = \lim_{k \to \infty} \int_{[a,b]} \ell_k \, d\lambda = \lim_{k \to \infty} L(f, \mathcal{P}_k) = \underline{\int} f, \qquad (**)$$

$$\int_{[a,b]} u \, d\lambda = \lim_{k \to \infty} \int_{[a,b]} u_k \, d\lambda = \lim_{k \to \infty} U(f, \mathcal{P}_k) = \overline{\int} f.$$

Theorem 11.1.4 (Lebesgue's Integrability Criterion). *A bounded function f :* $[a, b] \to \mathbb{R}$ *is Riemann integrable if and only if it is continuous at almost every* $x \in [a, b]$.

Proof. Let $Q := \bigcup_{k=1}^{\infty} \mathcal{P}_k$, with the sequence of partitions $(\mathcal{P}_k)$ as above. Then Q has measure zero and if $x \notin Q$, then f is continuous at x, i.e., the *oscillation* of f at x is *zero* [cf. Exercise 4.3.8 (d)], if and only if $\ell(x) = u(x)$. Now, if f is Riemann integrable, then $(**)$ and the equality of the lower and upper Darboux integrals show that $\int_{[a,b]} (u - \ell) \, d\lambda = 0$. But then $u - \ell \geq 0$ implies that $\ell(x) = f(x) = u(x)$ for almost all $x \in Q^c$ and hence for almost all $x \in [a, b]$. Conversely, if f is continuous almost everywhere, then $\ell = f = u$ almost everywhere and hence $\int_{[a,b]} \ell \, d\lambda = \int_{[a,b]} u \, d\lambda$. Therefore, given $\varepsilon > 0$, we can pick k so large that

$$U(f, \mathcal{P}_k) - L(f, \mathcal{P}_k) = \int_{[a,b]} u_k \, d\lambda - \int_{[a,b]} \ell_k \, d\lambda < \varepsilon$$

and hence f is Riemann integrable. $\square$

Our objective for the rest of this section is to look briefly at *improper* Riemann integrals and the corresponding Lebesgue integrals.

Definition 11.1.5 (Improper Riemann Integral (1)). Let $-\infty < a < b < \infty$ and let $f : (a, b] \to \mathbb{R}$. Suppose that $f \in \mathcal{R}([c, b])$ for all $c \in (a, b)$. The *improper Riemann integral* of f on $[a, b]$ is then defined to be

(i) $\displaystyle \int_a^b f(x) \, dx := \lim_{c \to a+} \int_c^b f(x) \, dx,$

provided the limit exists, in which case we say that $\int_a^b f(x) \, dx$ is *convergent* (or *exists*). Otherwise, we say that the integral is *divergent*. Similarly, we define the improper Riemann integral

(ii) $\displaystyle \int_a^b f(x) \, dx := \lim_{c \to b-} \int_c^b f(x) \, dx,$

provided that $f \in \mathcal{R}([a, c])$ for all $c \in (a, b)$ and the limit in (ii) exists. Finally, let $c \in (a, b)$ be such that $f \in \mathcal{R}([a, d])$ and $f \in \mathcal{R}([e, b])$ for all $d \in (a, c)$ and all $e \in (c, b)$. Then we define the improper Riemann integral

(iii) $\displaystyle \int_a^b f(x) \, dx := \int_a^c f(x) dx + \int_c^b f(x) \, dx,$

if the improper Riemann integrals $\int_a^c f(x) \, dx$ and $\int_c^b f(x) \, dx$ (defined as in (ii) and (i), respectively) are *both convergent*.

Exercise 11.1.6. Show that, if $f \in \mathcal{R}([a, b])$, then the integrals defined by (i), (ii), and (iii) above coincide with the *proper* (i.e., usual) Riemann integral $\int_a^b f(x) \, dx$.

Example 11.1.7.

(1) For each $p \in (0, 1)$ the improper integral $\int_0^1 x^{-p} \, dx$ exists. Indeed, we have

$$\int_0^1 \frac{1}{x^p} \, dx = \lim_{c \to 0+} \int_c^1 x^{-p} \, dx = \lim_{c \to 0+} \left(\frac{1}{1-p} - \frac{c^{1-p}}{1-p} \right) = \frac{1}{1-p}.$$

(2) The improper Riemann integral $\int_0^1 x^{-p} \, dx$ is *divergent* for all $p \geq 1$. For $p > 1$ the computation is exactly as in the example (1) above, but, in this case, the limit is obviously *infinite*. For $p = 1$, we have

$$\int_0^1 \frac{1}{x} \, dx = \lim_{c \to 0+} \int_c^1 \frac{dx}{x} = \lim_{c \to 0+} (-\log c) = +\infty.$$

(3) Consider the improper integral $\int_0^1 \log x \, dx$. A simple integration by parts shows that $\int \log x \, dx = x \log x - x$. Therefore,

$$\int_0^1 \log x \, dx = \lim_{c \to 0+} \int_c^1 \log x \, dx = \lim_{c \to 0+} (\log 1 - 1 - c \log c + c) = -1,$$

where we have used the fact that $\lim_{c \to 0+} c \log c = 0$. (Why?)

Next, we consider improper Riemann integrals on *unbounded intervals*:

Definition 11.1.8 (Improper Riemann Integral (2)). Let $f : [a, \infty) \to \mathbb{R}$ and suppose that $f \in \mathcal{R}([a, b])$ for all $b > a$. The *improper Riemann integral* of f on $[a, \infty)$ is then defined to be

(iv) $$\int_a^\infty f(x) \, dx := \lim_{b \to +\infty} \int_a^b f(x) \, dx,$$

provided the limit exists, in which case we say that $\int_a^\infty f(x) \, dx$ is *convergent* (or *exists*). Otherwise, we say that the integral is *divergent*. Similarly, we define the improper Riemann integral

(v) $$\int_{-\infty}^b f(x) \, dx := \lim_{a \to -\infty} \int_a^b f(x) \, dx,$$

provided that $f \in \mathcal{R}([a, b])$ for all $a < b$ and the limit in (v) exists. Finally, if $f : \mathbb{R} \to \mathbb{R}$ and if $f \in \mathcal{R}([a, b])$ for all $-\infty < a < b < \infty$, then we define the improper Riemann integral

(vi) $$\int_{-\infty}^\infty f(x) \, dx := \int_{-\infty}^c f(x) \, dx + \int_c^\infty f(x) \, dx,$$

where c is *any fixed number*, if the improper Riemann integrals $\int_{-\infty}^c f(x) \, dx$ and $\int_c^\infty f(x) \, dx$ (defined as in (v) and (iv), respectively) are *both convergent*. $\int_{-\infty}^\infty f(x) \, dx$ is then *independent* of c. (Why?)

Example 11.1.9.

(1) For each $p > 1$, the improper integral $\int_1^\infty x^{-p}\,dx$ is convergent. Indeed, we have

$$\int_1^\infty \frac{1}{x^p}\,dx = \lim_{b\to\infty}\int_1^b x^{-p}\,dx = \lim_{b\to\infty}\left(\frac{b^{1-p}}{1-p} - \frac{1}{1-p}\right) = \frac{1}{p-1}.$$

(2) The improper Riemann integral $\int_1^\infty x^{-p}\,dx$ is *divergent* for all $p \in (0,1]$. For $p \in (0,1)$ we compute exactly as in example (4) above, but find that the limit is $+\infty$. For $p = 1$, we have

$$\int_1^\infty \frac{1}{x}\,dx = \lim_{b\to\infty}\int_1^b \frac{dx}{x} = \lim_{b\to\infty}\log b = +\infty.$$

(3) Consider the improper integral $\int_2^\infty 1/(x\log x)\,dx$. Here, we have

$$\int_2^\infty \frac{1}{x\log x}\,dx = \lim_{b\to\infty}[\log(\log b) - \log(\log 2)] = +\infty.$$

Remark 11.1.10.

1. Note that, if $f(x) \geq 0$ on $[a,\infty)$ and $f \in \mathcal{R}([a,b])$ for all $b > a$, then $\int_a^b f(x)\,dx$ is an *increasing function of* b. It follows that

$$\int_a^\infty f(x)\,dx = \sup\left\{\int_a^b f(x)\,dx : b \geq a\right\} \in [0,\infty].$$

Therefore, in this case, the improper integral $\int_a^\infty f(x)\,dx$ is either convergent or diverges to $+\infty$. The same comment can be made about the improper integrals $\int_{-\infty}^b f(x)\,dx$ and $\int_{-\infty}^\infty f(x)\,dx$ if f is a *nonnegative* function.

2. (Integral Test). If f is a *nonnegative, decreasing* function whose domain contains $[1,\infty)$, then (cf. Problem 7.7.#9)

$$\sum_1^\infty f(n) < \infty \iff \int_1^\infty f(x)\,dx < \infty.$$

Indeed, this follows at once from the inequalities

$$f(2) + f(3) + \cdots + f(n) \leq \int_1^n f(x)dx \leq f(1) + f(2) + \cdots + f(n-1).$$

Definition 11.1.11 (Absolute vs. Conditional Convergence). An improper integral $\int_a^\infty f(x)\,dx$ is said to be *absolutely convergent* if $\int_a^\infty |f(x)|\,dx$ is convergent. In this case, f is said to be *absolutely integrable on* $[a,\infty)$.

If $\int_a^\infty f(x)\, dx$ exists but $\int_a^\infty |f(x)|\, dx = +\infty$, then $\int_a^\infty f(x)\, dx$ is said to be *conditionally convergent*. For the improper integrals $\int_{-\infty}^b f(x)\, dx$ and $\int_{-\infty}^\infty f(x)\, dx$, the absolute and conditional convergence are defined similarly.

To simplify the exposition, we formulate the following results for functions defined on $[a, \infty)$. It is obvious, however, that similar results hold for the other types of improper integrals as well.

Theorem 11.1.12 (Cauchy's Criterion). *The improper integral $\int_a^\infty f(x)\, dx$ is convergent if and only if, given any $\varepsilon > 0$, there exists $B > 0$ such that*

$$x, \; y \geq B \Rightarrow \left| \int_a^x f(t)dt - \int_a^y f(t)dt \right| = \left| \int_x^y f(t)dt \right| < \varepsilon.$$

Proof. Let $F(x) := \int_a^x f(t)dt$. Then $S := \int_a^\infty f(x)\, dx$ exists if and only if $\lim_{x \to \infty} F(x) = S$. The theorem now follows from Exercise 3.4.18.   □

Corollary 11.1.13 (Absolute Convergence $\Rightarrow$ Convergence). *An absolutely integrable function f on $[a, \infty)$ is integrable on $[a, \infty)$.*

Proof. Simply note that if $x \leq y$, then $|\int_x^y f(t)\, dt| \leq \int_x^y |f(t)|\, dt$.   □

Remark 11.1.14. As this corollary shows, *improper* Riemann integrals behave like (ordered) series, where *absolute convergence* implies *convergence*. By contrast, only the converse holds for *proper* Riemann integrals (cf. Corollary 7.3.20 or Corollary 7.4.12).

Here is a lemma which provides positive integrable functions that can be used as *dominating* functions in Lebesgue's Dominated Convergence Theorem.

Lemma 11.1.15. *Let $I \subset \mathbb{R}$ be an interval, $f : I \to \mathbb{R}$, and let $(I_n)_{n=1}^\infty$ be an increasing sequence of intervals (i.e., $I_1 \subset I_2 \subset \cdots$) with $I = \bigcup_{n=1}^\infty I_n$. If $f|I_n \in \mathcal{L}^1(I_n)$ for all $n \geq 1$ and if the sequence $(\int_{I_n} |f|)$ is bounded, then $f \in \mathcal{L}^1(I)$ and*

$$\int_I f = \lim_{n \to \infty} \int_{I_n} f.$$

In fact, we have

$$\lim_{n \to \infty} \int |f - f\chi_{I_n}| = 0. \tag{$*$}$$

Proof. Consider the *truncated* functions $f_n := f\chi_{I_n}$ for all $n \in \mathbb{N}$. If $f \geq 0$, then (f_n) is an *increasing* sequence of nonnegative functions in $\mathcal{L}^1(I)$ that converges to f. The numerical sequence $(\int_I f_n\, d\lambda)$ is then increasing, nonnegative, and *bounded above* by assumption, hence convergent. The Monotone Convergence Theorem now implies that $\lim_{n \to \infty} \int_{I_n} f\, d\lambda = \int_I f\, d\lambda < \infty$. Moreover, $(f - f_n)$ is a *decreasing*, nonnegative sequence with $\lim_{n \to \infty}(f - f_n) = 0$, from which $(*)$ follows at once.

In general, we have $f = f^+ - f^-$ and may apply the previous case to the functions f^+ and f^- to deduce that $f \in \mathcal{L}^1(I)$. Also, since $(*)$ holds for f^+ and f^-, we have

$$\lim_{n\to\infty} \int_I |f - f_n| \leq \lim_{n\to\infty} \left(\int_I (f^+ - .f_n^+) + \int_I (f^- - f_n^-) \right) = 0. \qquad \Box$$

We now use the lemma to prove the following convergence result for improper Riemann integrals.

Proposition 11.1.16. *Let f be defined on $[a, \infty)$. If f is Riemann integrable on $[a, b]$ for each $b \geq a$ and $\int_a^b |f(x)|\, dx \leq M$ for some $M > 0$ and all $b \geq a$, then the improper Riemann integrals $\int_a^\infty f(x)\, dx$ and $\int_a^\infty |f(x)|\, dx$ are both convergent. Also, $f \in \mathcal{L}^1([a, \infty))$ and the Lebesgue integral of f is equal to its improper Riemann integral. In particular, the same conclusion holds if $|f| \leq g$, where g is Lebesgue (or improperly Riemann) integrable on $[a, \infty)$.*

Proof. This follows from the above lemma with $I := [a, \infty)$, $I_n := [a, a + n]$ and Proposition 10.3.19, which implies that for each $b \geq a$, the Riemann integral $\int_a^b f(x)\, dx$ and the Lebesgue integral $\int_a^b f(x)\, d\lambda(x)$ coincide. $\qquad \Box$

Example 11.1.17. The improper integral $\int_1^\infty \log x / x^p\, dx$ is convergent for all $p > 1$.

Indeed, we note that $\log x < x^\alpha / \alpha$ for all $x > 0$ and all $\alpha > 0$. (Why?) Thus, if $0 < \alpha < p - 1$, we have $\log x / x^p < 1/(\alpha x^{p-\alpha})$. Since, as we saw earlier, $\int_1^\infty 1/(\alpha x^{p-\alpha})\, dx$ exists, the assertion follows from the above proposition.

Exercise 11.1.18.

(a) Let $f(x) := e^{-|x|}$ for all $x \in \mathbb{R}$. Show that $f \in \mathcal{L}^1(\mathbb{R})$ and that its Lebesgue integral coincides with its improper Riemann integral. Conclude that

$$\int_{-\infty}^\infty e^{-|x|}\, dx = 2.$$

(b) State and prove Proposition 11.1.16 for the other types of improper Riemann integrals: $\int_{-\infty}^b f(x)\, dx = \lim_{a\to-\infty} \int_a^b f(x)\, dx$, $\int_a^b f(x)\, dx = \lim_{c\to a+} \int_c^b f(x)\, dx$, and $\int_a^b f(x)\, dx = \lim_{c\to b-} \int_a^c f(x)\, dx$.

Here is an exercise containing an example of an *improperly* Riemann integrable function that is *not* Lebesgue integrable.

Exercise 11.1.19. Let $f(x) := \sin x / x$ on $[1, \infty)$. Show that the improper Riemann integral

$$\int_1^\infty f(x)\, dx$$

is *convergent*. Show, however, that f is *not* Lebesgue integrable on $[1, \infty)$. *Hint:* Use integration by parts to obtain

$$\int_1^b \frac{\sin x}{x} \, dx = \left[\frac{-\cos x}{x} \right]_1^b - \int_1^b \frac{\cos x}{x^2} \, dx.$$

On the other hand, show that $\int_1^\infty |\sin x|/x \, dx$ is *divergent* by observing that

$$\int_{(k-1)\pi}^{k\pi} |\sin x|/x \, dx \geq \frac{1}{k\pi} \int_{(k-1)\pi}^{k\pi} |\sin x| \, dx = \frac{2}{k\pi}.$$

For each $x > 0$, consider the function $f(t) := t^{x-1}e^{-t}$ for all $t \in (0, \infty)$. If $t \in (0, 1]$, then $t^{x-1}e^{-t} < t^{x-1}$ and hence

$$\int_0^1 t^{x-1}e^{-t} dt = \sup\left\{ \int_\delta^1 t^{x-1}e^{-t} \, dt : 0 < \delta < 1 \right\} \leq \int_0^1 t^{x-1} \, dx = \frac{1}{x}.$$

If $t > 1$, then one easily checks that the function $t \mapsto t^{x+1}e^{-t}$ has a *maximum* at $t = x + 1$ and hence

$$\int_1^\infty t^{x-1}e^{-t} dt = \int_1^\infty (t^{x+1}e^{-t})t^{-2} dt$$

$$\leq (x+1)^{x+1}e^{-(x+1)} \int_1^\infty \frac{1}{t^2} dt = (x+1)^{x+1}e^{-(x+1)}.$$

In view of these estimates, it is legitimate to make the following.

Definition 11.1.20 (Euler's Gamma Function). For each $x > 0$ we define

$$\Gamma(x) := \int_0^\infty t^{x-1}e^{-t} dt.$$

Exercise 11.1.21 ($\Gamma(x+1) = x!$). Show that $\Gamma(1) = 1$ and that

$$\Gamma(x+1) = x\Gamma(x) \qquad (\forall x > 0).$$

Deduce that $\Gamma(n+1) = n!$ for all $n \in \mathbb{N}$. *Hint:* Use integration by parts:

$$\int_0^b t^x e^{-t} dt = \left[-t^x e^{-t} \right]_0^b + x \int_0^b t^{x-1}e^{-t} dt.$$

11.2 Dependence on a Parameter

As the example of the Gamma function shows, there are situations where we must consider integrals with integrands depending on a real parameter. In this section we show how Lebesgue's Dominated Convergence Theorem may help investigate the properties of such integrals. To simplify the exposition, we make the following

Assumption. Throughout the section, we assume that $E \in \mathcal{M}_\lambda(\mathbb{R})$, that $f : E \times [a, b] \to \mathbb{R}$, where $a < b$, and that the function $x \mapsto f(x, t)$ is (Lebesgue) measurable on E for each $t \in [a, b]$.

Proposition 11.2.1 (Interchanging Limit and Integral). *If for some $t_0 \in [a, b]$ we have*

$$f(x, t_0) = \lim_{t \to t_0} f(x, t) \qquad \forall\, x \in E,$$

and if $|f(x, t)| \le g(x)$ for all $(x, t) \in E \times [a, b]$ and a function $g \in \mathcal{L}^1_+(E)$, then

$$\int_E f(x, t_0)\, d\lambda(x) = \lim_{t \to t_0} \int_E f(x, t)\, d\lambda(x);$$

i.e., the order of $\lim_{t \to t_0}$ and $\int_E$ may be interchanged.

Proof. For any sequence (t_n) in $[a, b]$ with $\lim(t_n) = t_0$, define $f_n(x) := f(x, t_n)$ for each $n \in \mathbb{N}$ and note that our assumption implies that $\lim_{n \to \infty} f_n(x) = f(x, t_0)$ for each $x \in E$. Thus the proposition follows from Lebesgue's Dominated Convergence Theorem. $\square$

The following corollary is an immediate consequence.

Corollary 11.2.2. *If the function $t \mapsto f(x, t)$ is continuous on $[a, b]$ for each $x \in E$ and $|f(x, t)| \le g(x)$ for all $(x, t) \in E \times [a, b]$ and a function $g \in \mathcal{L}^1_+(E)$, then the function*

$$F(t) := \int_E f(x, t)\, d\lambda(x)$$

is continuous on $[a, b]$.

For the next proposition we shall need the following.

Definition 11.2.3 (Partial Derivative). Given a function $f : E \times [a, b]$, we define its *partial derivative* with respect to t, denoted $\partial f / \partial t$, to be the function

$$\frac{\partial f}{\partial t}(x, t) := \frac{d}{dt} f(x, t),$$

with domain the set of all $(x, t) \in E \times [a, b]$ for which the derivative exists.

Proposition 11.2.4 (Differentiating Under the Integral Sign). *Suppose that for some $t_0 \in [a,b]$ the function $x \mapsto f(x,t_0)$ is (Lebesgue) integrable on E, that $\partial f/\partial t$ exists on $E \times [a,b]$, and that there is a function $g \in \mathcal{L}_+^1(E)$ such that*

$$\left| \frac{\partial}{\partial t} f(x,t) \right| \le g(x) \qquad \forall \, (x,t) \in E \times [a,b].$$

Then the function $F(t) := \int_E f(x,t) \, d\lambda(x)$ is differentiable on $[a,b]$ and we have

$$\frac{dF}{dt}(t) = \frac{d}{dt} \int_E f(x,t) \, d\lambda(x) = \int_E \frac{\partial f}{\partial t}(x,t) \, d\lambda(x).$$

Proof. Let $t \in [a,b]$ and pick any sequence (t_n) in $[a,b]$ such that $t_n \ne t$ for all $n \in \mathbb{N}$ and $\lim(t_n) = t$. Then we have

$$\frac{\partial f}{\partial t}(x,t) = \lim_{n \to \infty} \frac{f(x,t_n) - f(x,t)}{t_n - t} \qquad \forall \, x \in E,$$

and hence the function $x \mapsto (\partial f)/(\partial t)(x,t)$ is measurable on E. Next, for any $x \in E$ and any $t \in [a,b]$ with $t \ne t_0$, we can apply the MVT (Theorem 6.4.8) to find a number s between t_0 and t such that

$$f(x,t) - f(x,t_0) = (t - t_0) \frac{\partial f}{\partial t}(x,s)$$

and hence

$$|f(x,t)| \le |f(x,t_0)| + |t - t_0| g(x),$$

which shows that $x \mapsto f(x,t)$ is integrable for every $t \in [a,b]$. Thus, if $t_n \ne t$, then we have

$$\frac{F(t_n) - F(t)}{t_n - t} = \int_E \frac{f(x,t_n) - f(x,t)}{t_n - t} \, d\lambda(x).$$

Since the integrand is dominated by $g \in \mathcal{L}_+^1(E)$, taking the limit as $n \to \infty$ and using Lebesgue's Dominated Convergence Theorem complete the proof. $\qquad \square$

The next proposition deals with the possibility of interchanging the order of the Riemann integral $\int_a^b$ and the Lebesgue integral $\int_E$ in the *iterated* integral $\int_a^b [\int_E f(x,t) \, d\lambda(x)] dt$.

Proposition 11.2.5 (Interchanging the Order of Integration). *Suppose, as in Corollary 11.2.2, that $t \mapsto f(x,t)$ is continuous on $[a,b]$ for each $x \in E$ and that $|f(x,t)| \le g(x)$ for all $(x,t) \in E \times [a,b]$ and a function $g \in \mathcal{L}_+^1(E)$. Then, with $F(t) := \int_E f(x,t) \, d\lambda(x)$, we have*

$$\int_a^b F(t)\, dt = \int_a^b \left[\int_E f(x,t)\, d\lambda(x) \right] dt$$

$$= \int_E \left[\int_a^b f(x,t)\, dt \right] d\lambda(x),$$

where the integrals with respect to t are Riemann integrals.

Proof. Consider the function

$$h(x,t) := \int_a^t f(x,s)\, ds \qquad \forall\, (x,t) \in E \times [a,b],$$

where the Riemann integral on the right side is a limit of Riemann sums and hence the function $x \mapsto h(x,t)$ is measurable on E. The (Second) Fundamental Theorem of Calculus (Theorem 7.5.8) implies that $(\partial h/\partial t)(x,t) = f(x,t)$. Also, since $|f(x,t)| \le g(x)$ for all $(x,t) \in E \times [a,b]$, it follows that $|h(x,t)| \le (b-a)g(x)$ and hence $x \mapsto h(x,t)$ is integrable on E for each $t \in [a,b]$. Therefore, if we define

$$H(t) := \int_E h(x,t)\, d\lambda(x),$$

then Proposition 11.2.4 implies that

$$\frac{dH}{dt}(t) = \int_E \frac{\partial h}{\partial t}(x,t)\, d\lambda(x) = \int_E f(x,t)\, d\lambda(x) = F(t).$$

But then, the (First) Fundamental Theorem of Calculus (Theorem 7.5.3) gives

$$\int_a^b F(t)\, dt = H(b) - H(a)$$

$$= \int_E [h(x,b) - h(x,a)]\, d\lambda(x)$$

$$= \int_E \left[\int_a^b f(x,t)\, dt \right] d\lambda(x).$$

$\square$

Example 11.2.6. We have

$$\int_0^\infty e^{-tx} \sin x\, dx = \frac{1}{1+t^2} \qquad \forall\, t > 0. \tag{$*$}$$

Here the integrand $f(x,t) := e^{-tx} \sin x$ is differentiable (in x and t), $|f(t,x)| \le e^{-tx}$ for all (x,t), and $\int_0^\infty e^{-tx}\, dx = 1/t < \infty$ for all $t > 0$. So the improper integral is convergent for all $t > 0$. To evaluate it, an integration by part gives

$$\int_0^b e^{-tx} \sin x \, dx = \left[-\frac{(t \sin x + \cos x)e^{-tx}}{1+t^2} \right]_0^b$$

$$= -\frac{(t \sin b + \cos b)e^{-tb}}{1+t^2} + \frac{1}{1+t^2},$$

and (∗) follows if we take the limit as $b \to \infty$.

Here is a classical example where the introduction of a parameter helps evaluate an improper Riemann integral.

Example 11.2.7 (Dirichlet). We have

$$\int_0^\infty \frac{\sin x}{x} \, dx = \frac{\pi}{2}.$$

First, by Exercise 11.1.19, the integral is *convergent*. Let us introduce the related function

$$F(t) := \int_0^\infty e^{-tx} \frac{\sin x}{x} \, dx, \qquad t \geq 0.$$

Note that $F(0)$ is the integral to be evaluated. Since $\lim_{x \to 0}(\sin x/x) = 1$, the integrand $f(x,t) := e^{-tx} \sin x/x$ may also be defined at $x = 0$ by setting $f(0,t) := 0$ and is then defined (and continuous) for all (x,t). Also, the inequality $|\sin x| \leq |x|$, which follows at once from the MVT [cf. Exercise 6.4.17(b)], implies that, with our convention $\sin 0/0 := 1$,

$$|f(x,t)| = e^{-tx} \frac{|\sin x|}{|x|} \leq e^{-tx} \qquad \forall (x,t). \qquad (**)$$

Now for all $t > 0$ we have $\int_0^\infty e^{-tx} \, dx = 1/t < \infty$, so that $F(t)$ is well defined and continuous on $(0, \infty)$. Also, (∗∗) implies that

$$|F(t)| \leq \int_0^\infty e^{-tx} \left| \frac{\sin x}{x} \right| dx \leq \int_0^\infty e^{-tx} \, dx = \frac{1}{t} \qquad \forall t > 0. \qquad (***)$$

In fact, $F(t)$ is differentiable for all $t > 0$. Indeed, $\partial f(x,t)/\partial t = -e^{-tx} \sin x$ and Example 11.2.6 shows that we have

$$F'(t) = -\int_0^\infty e^{-tx} \sin x \, dx = -\frac{1}{1+t^2} \qquad \forall t > 0.$$

We therefore have $F(t) = C - \arctan t$ for $t > 0$ and a constant C. To find the constant, note that (∗∗∗) implies $\lim_{t \to \infty} F(t) = 0$ and hence $C = \lim_{t \to \infty} \arctan t = \pi/2$. Thus

$$F(t) = \int_0^\infty e^{-tx} \frac{\sin x}{x} \, dx = \frac{\pi}{2} - \arctan t, \qquad \forall \, t > 0.$$

We already know that $F(t)$ is continuous (even differentiable) on $(0, \infty)$, but to be able to use $\lim_{t \to 0+} F(t) = F(0)$, i.e.,

$$\int_0^\infty \frac{\sin x}{x} \, dx = \lim_{t \to 0+} \int_0^\infty e^{-tx} \frac{\sin x}{x} \, dx = \lim_{t \to 0+} \left(\frac{\pi}{2} - \arctan t \right) = \frac{\pi}{2}, \qquad (\dagger)$$

we must show that F is *continuous* at $t = 0$. This will follow if we can show that $\int_0^\infty e^{-tx} (\sin x / x) \, dx$ converges *uniformly* to $\pi/2 - \arctan t$ on $[0, \infty)$. But the integral $\int_0^\infty (\sin x / x) \, dx$ being *convergent*, given any $\varepsilon > 0$, we can find $N > 0$ so large that $B > A \geq N$ implies $| \int_A^B (\sin x / x) \, dx | < \varepsilon$. Since e^{-tx} is a *decreasing*, positive function (of x), we can use Theorem 7.4.20 (the Second MVT for Integrals) to find $\xi \in [A, B]$ with

$$\left| \int_A^B e^{-tx} \frac{\sin x}{x} \, dx \right| = e^{-At} \left| \int_A^\xi \frac{\sin x}{x} \, dx \right| < \varepsilon \qquad \forall \, t \geq 0,$$

which proves the desired uniform convergence on $[0, \infty)$ and justifies $(\dagger)$.

Exercise 11.2.8. Show that for any $\alpha \in \mathbb{R}$ we have

$$\int_0^\infty \frac{\sin^2(\alpha x)}{x^2} \, dx = \frac{\pi}{2} |\alpha|.$$

Hint: Integrate by parts.

11.3 L^p-Spaces

The spaces we want to introduce in this section are often called *classical Banach spaces*, and their study will provide an opportunity to use some of the results we obtained for abstract Banach spaces in Chap. 9. In fact, we have already introduced one of these spaces before, namely the space $L^1(E, \mathbb{F})$, where $E \in \mathcal{M}_\lambda(\mathbb{R})$ and $\mathbb{F}$ is either $\mathbb{R}$ or $\mathbb{C}$. As was pointed out in Notation 10.3.29, we look at equivalence classes of measurable functions modulo the class of *null functions*.

Definition 11.3.1 (L^p-Spaces, $1 \leq p < \infty$). Let $\mathbb{F}$ denote either $\mathbb{R}$ or $\mathbb{C}$. Given any $E \in \mathcal{M}_\lambda(\mathbb{R})$ and any $p \in [1, \infty)$, we denote by $\mathcal{L}^p(E, \mathbb{F})$ the set of all measurable functions $f \in \mathcal{L}^0(E, \mathbb{F})$ such that $\int_E |f|^p \, d\lambda < \infty$. The L^p-*norm* of f is then defined to be

$$\|f\|_p := \left[\int_E |f(x)|^p \, d\lambda(x) \right]^{1/p}. \qquad (\| \cdot \|_p)$$

Thus $f \in \mathcal{L}^p(E,\mathbb{F})$ if and only if $|f|^p \in \mathcal{L}^1(E,\mathbb{F})$. The space $L^p(E,\mathbb{F}) := \mathcal{L}^p$ $(E,\mathbb{F})/\mathcal{N}(E,\mathbb{F})$ is then the set of all equivalence classes $[f] := f + \mathcal{N}(E,\mathbb{F})$, where $f \in \mathcal{L}^p(E,\mathbb{F})$ and $\mathcal{N}(E,\mathbb{F})$ is the set of all ($\mathbb{F}$-valued) *null functions* on E. Abusing the notation, we usually write $f \in L^p(E,\mathbb{F})$ instead of $f \in \mathcal{L}^p(E,\mathbb{F})$.

To prove that $\|\cdot\|_p$ is indeed a *norm* on $L^p(E)$, we need a couple of well-known inequalities which we now establish.

Proposition 11.3.2 (Hölder's Inequality). *Suppose that p, $q \in (1,\infty)$ are related by $1/p + 1/q = 1$. Then for any $f \in \mathcal{L}^p(E,\mathbb{F})$ and $g \in \mathcal{L}^q(E,\mathbb{F})$ we have $fg \in \mathcal{L}^1(E,\mathbb{F})$ and*

$$\|fg\|_1 \le \|f\|_p \|g\|_q.$$

Proof. We may (and do) assume that $\|f\|_p > 0$ and $\|g\|_q > 0$. Now recall (cf. Example 6.7.13) that we have

$$a^{1/p} b^{1/q} \le \frac{a}{p} + \frac{b}{q} \qquad \forall\, a \ge 0,\, b \ge 0. \tag{$*$}$$

Applying $(*)$ with $a = |f(x)|^p / \|f\|_p^p$ and $b = |g(x)|^q / \|g\|_q^q$, we have

$$\frac{|f(x)g(x)|}{\|f\|_p \|g\|_q} \le \frac{|f(x)|^p}{p\|f\|_p^p} + \frac{|g(x)|^q}{q\|g\|_q^q}.$$

Since both functions on the right side are integrable, $fg \in \mathcal{L}^1(E,\mathbb{F})$ follows. But then, integrating the two sides gives

$$\frac{\|fg\|_1}{\|f\|_p \|g\|_q} \le \frac{1}{p} + \frac{1}{q} = 1.$$

$\square$

Exercise 11.3.3. Show that if $E \in \mathcal{M}_{\lambda<\infty}(\mathbb{R})$, i.e., $\lambda(E) < \infty$, then $\mathcal{L}^p$ $(E,\mathbb{F}) \subset \mathcal{L}^1(E,\mathbb{F})$ for all $p \in [1,\infty)$.

For $p = q = 2$, Hölder's inequality becomes

Corollary 11.3.4 (Cauchy–Schwarz Inequality). *Let $E \in \mathcal{M}_\lambda(\mathbb{R})$. Then, for any f, $g \in \mathcal{L}^2(E,\mathbb{F})$, we have $fg \in \mathcal{L}^1(E,\mathbb{F})$ and*

$$\left| \int_E f(x)g(x)\, d\lambda(x) \right| \le \|fg\|_1 \le \|f\|_2 \|g\|_2.$$

We are now ready to prove the second important inequality that will show that $\|\cdot\|_p$ is indeed a *norm* on $L^p(E)$.

Proposition 11.3.5 (Minkowski's Inequality). *Let $E \in \mathcal{M}_\lambda(\mathbb{R})$ and $p \in [1, \infty)$. Given any functions f, $g \in \mathcal{L}^p(E, \mathbb{F})$, we have $f + g \in \mathcal{L}^p(E, \mathbb{F})$ and*

$$\|f + g\|_p \leq \|f\|_p + \|g\|_p.$$

Proof. For $p = 1$, we have already seen (cf. the proof of Theorem 10.3.31) that f, $g \in \mathcal{L}^1(E, \mathbb{F})$ implies $f + g \in \mathcal{L}^1(E, \mathbb{F})$ and we then have $\|f + g\|_1 \leq \|f\|_1 + \|g\|_1$. Also, we may (and do) assume that $\|f + g\|_p > 0$ because the inequality is trivially satisfied otherwise. Now, if $p > 1$, then it follows from Example 6.7.13 that

$$|f + g|^p \leq (|f| + |g|)^p \leq 2^{p-1}(|f|^p + |g|^p)$$

and hence $f + g \in \mathcal{L}^p(E, \mathbb{F})$. Also,

$$|f + g|^p = |f + g||f + g|^{p-1} \leq |f||f + g|^{p-1} + |g||f + g|^{p-1}. \qquad (**)$$

But $f + g \in \mathcal{L}^p(E)$ means $|f + g|^p \in \mathcal{L}^1(E, \mathbb{F})$ and, with $q = p/(p - 1)$, we have $|f + g|^{p-1} \in \mathcal{L}^q(E, \mathbb{F})$. Thus Hölder's inequality gives

$$\int_E |f||f + g|^{p-1} \, d\lambda \leq \|f\|_p \left[\int_E |f + g|^{(p-1)q} \right]^{1/q}$$

$$= \|f\|_p \|f + g\|_p^{p/q}.$$

Treating the second term on the right side of $(**)$ similarly, we deduce that

$$\|f + g\|_p^p \leq \|f\|_p \|f + g\|_p^{p/q} + \|g\|_p \|f + g\|_p^{p/q}$$

$$= (\|f\|_p + \|g\|_p)\|f + g\|_p^{p/q}.$$

Dividing the two sides by $\|f\|_p + \|g\|_p$ and noting that $p - p/q = 1$, Minkowski's inequality follows. $\qquad \square$

Corollary 11.3.6. *Let $E \in \mathcal{M}_\lambda(\mathbb{R})$ and $p \geq 1$. Then the Lebesgue space $L^p(E, \mathbb{F})$ is a normed vector space (over $\mathbb{F}$) with the operations*

$$[f] + [g] := [f + g] \quad and \quad c[f] := [cf] \qquad \forall \, f, \, g \in \mathcal{L}^p(E, \mathbb{F}), \quad \forall \, c \in \mathbb{F},$$

and the norm

$$\|[f]\|_p := \left[\int_E |f|^p \, d\lambda \right]^{1/p}.$$

Proof. It is easily checked that the operations are well defined. Also, the above proposition shows that for any f, $g \in \mathcal{L}^p(E, \mathbb{F})$ and any $c \in \mathbb{R}$, we have $f + g$, $cf \in \mathcal{L}^p(E, \mathbb{F})$. Checking the vector space axioms is also a simple exercise left to the reader. As for the norm, the *Triangle Inequality* is simply Minkowski's inequality and $\|[f]\|_p = 0$ if and only if $\int_E |f|^p \, d\lambda = 0$, which is the case if and only if $f(x) = 0$ for almost all $x \in E$, i.e., $[f] = [0]$. $\qquad\qquad\qquad\square$

In fact, as we shall now prove, $L^p(E, \mathbb{F})$ is *complete*, i.e., a *Banach* space.

Theorem 11.3.7 (L^p is a Banach Space). *Let $E \in \mathcal{M}_\lambda(\mathbb{R})$ and $p \in [1, \infty)$. Then $L^p(E, \mathbb{F})$ with norm $\| \cdot \|_p$ is a Banach space.*

Proof. We identify a class $[f] \in L^p(E, \mathbb{F})$ with the representative $f \in \mathcal{L}^p(E, \mathbb{F})$. By Theorem 9.2.20, it suffices to show that if (f_n) is a sequence in $L^p(E, \mathbb{F})$ such that $\sum_{n=1}^\infty \|f_n\|_p = S < \infty$, then $\sum_{n=1}^\infty f_n \in L^p(E, \mathbb{F})$. Set $G_n := \sum_{k=1}^n |f_k|$ and $G := \sum_{k=1}^\infty |f_k|$. Then $\|G_n\|_p \le \sum_{k=1}^n \|f_k\|_p \le S$ for all $n \in \mathbb{N}$. Since (G_n) is increasing, the Monotone Convergence Theorem implies that $\int_E G^p \, d\lambda = \lim_{n\to\infty} \int_E G_n^p \, d\lambda \le S^p$ and hence $G \in L^p(E, \mathbb{F})$. In particular, $G(x) < \infty$ for almost all $x \in E$, which implies that $F(x) := \sum_{n=1}^\infty f_n(x)$ converges for almost all $x \in E$. Since $|F| \le G$, we also have $F \in L^p(E, \mathbb{F})$. Furthermore, $|F - \sum_{k=1}^n f_k|^p \le (2G)^p \in L^1(E, \mathbb{F})$, so by the Dominated Convergence Theorem, we have

$$\left\| F - \sum_{k=1}^n f_k \right\|_p^p = \int_E \left| F - \sum_{k=1}^n f_k \right|^p \, d\lambda \to 0, \quad \text{as} \quad n \to \infty.$$

Thus the series $\sum_{n=1}^\infty f_n$ converges in $L^p(E, \mathbb{F})$ and the proof is complete. $\quad\square$

Remark 11.3.8. Given any $f = u + iv \in \mathcal{L}^0(E, \mathbb{C})$, we have $f \in \mathcal{L}^p(E, \mathbb{C})$ if and only if $u, v \in \mathcal{L}^p(E, \mathbb{R})$. Therefore, to simplify the exposition we may (and usually do) prove the results for *real-valued* functions. The complex-valued case then follows by looking at the real and imaginary parts of the functions involved.

Let us now show that simple functions that vanish outside sets of finite measure form a *dense* subspace of L^p for any $p \ge 1$.

Proposition 11.3.9 ($Simp_0(\mathbb{R})$ and $Step(\mathbb{R})$ are Dense in $L^p(\mathbb{R})$). *For any $p \ge 1$ the set $Simp_0(\mathbb{R}, \mathbb{F})$ of simple functions $f = \sum_{j=1}^m a_j \chi_{A_j}$, where the A_j are pairwise disjoint measurable sets with $\lambda(A_j) < \infty$ and $\mathbb{F} \ni a_j \ne 0$ for all j, is dense in $L^p(\mathbb{R}, \mathbb{F})$. The same conclusion holds if $Simp_0(\mathbb{R}, \mathbb{F})$ is replaced by $Step(\mathbb{R}, \mathbb{F})$.*

Proof. We may (and do) assume that $\mathbb{F} = \mathbb{R}$ and use the abbreviations $L^p(\mathbb{R}) := L^p(\mathbb{R}, \mathbb{R})$ and $\mathcal{L}^p(\mathbb{R}) := \mathcal{L}^p(\mathbb{R}, \mathbb{R})$. Clearly we have $Simp_0(\mathbb{R}) \subset L^p(\mathbb{R})$. Now, given any $f \in \mathcal{L}^p(\mathbb{R})$, there is (by Theorem 10.2.24) a sequence (f_n) in $Simp_0(\mathbb{R})$ such that f_n converges to f almost everywhere and we may assume $|f_n| \le |f|$ for all n. But then $f_n \in \mathcal{L}^p(\mathbb{R})$ and $|f - f_n|^p \le 2^p |f|^p \in \mathcal{L}^1(\mathbb{R})$, so we can use the Dominated Convergence Theorem to conclude that $\|f - f_n\|_p \to 0$, as $n \to \infty$. Also, if $f_n = \sum_j a_j \chi_{A_j}$ is the *canonical representation* of f_n, then $\int |f_n|^p \, d\lambda = \sum_j |a_j|^p \lambda(A_j) < \infty$ implies that we must have $\lambda(A_j) < \infty$ for all j. To prove the statement for $Step(\mathbb{R}) \subset L^p(\mathbb{R})$, for a given $f \in \mathcal{L}^p(\mathbb{R})$ we

first pick $N > 0$ so large that, with $f_N := f\chi_{[-N,N]}$, we have $\| f - f_N \|_p < \varepsilon/2$.
So (by Minkowski's inequality) we only need a step function ψ such that $\| f_N - \psi \|_p < \varepsilon/2$. But, by Corollary 10.5.8, there is a sequence (ψ_n) of step functions
with $|\psi_n| \leq |f_N|$ for all $n \in \mathbb{N}$ and such that $\lim(\psi_n) = f_N$ almost everywhere
on $[-N, N]$. Since $|f_N - \psi_n|^p \leq 2^p |f_N|^p$, the Dominated Convergence Theorem
implies that $\lim_{n\to\infty} \| f_N - \psi_n \|_p = 0$. Thus, for n large enough, we indeed have
$\| f_N - \psi_n \|_p < \varepsilon/2$ and the proof is complete. $\square$

We now use this proposition to extend the Riemann–Lebesgue lemma
(Theorem 8.6.9) to $\mathcal{L}^1(\mathbb{R})$. The proof is basically the same. First, we extend the
definition of Fourier coefficients:

Definition 11.3.10 (Fourier Transform). Given any $f \in \mathcal{L}^1(\mathbb{R})$, its *Fourier
transform* is defined to be the function

$$\hat{f}(\xi) := \int_{-\infty}^{\infty} f(x)e^{-ix\xi} \, d\lambda(x) \qquad \forall \, \xi \in \mathbb{R}.$$

The integral is convergent because $|f(x)e^{-ix\xi}| \leq |f(x)|$ for all x, $\xi \in \mathbb{R}$. In
particular, the (real) integrals

$$\int_{-\infty}^{\infty} f(x)\sin(\xi x) \, d\lambda(x) \quad \text{and} \quad \int_{-\infty}^{\infty} f(x)\cos(\xi x) \, d\lambda(x),$$

called the *Fourier sine transform of f* and *Fourier cosine transform of f*,
respectively, are both convergent for all $\xi \in \mathbb{R}$.

Theorem 11.3.11 (Riemann–Lebesgue Lemma). *Given any $f \in L^1(\mathbb{R})$, its
Fourier transform $\hat{f}$ is continuous (on $\mathbb{R}$) and we have*

$$\lim_{|\xi|\to\infty} \hat{f}(\xi) = \lim_{|\xi|\to\infty} \int_{-\infty}^{\infty} f(x)e^{-ix\xi} \, d\lambda(x) = 0. \qquad (\dagger)$$

In particular,

$$\lim_{|\xi|\to\infty} \int_{-\infty}^{\infty} f(x)\sin(\xi x) \, d\lambda(x) = 0 = \lim_{|\xi|\to\infty} \int_{-\infty}^{\infty} f(x)\cos(\xi x) \, d\lambda(x). \qquad (\ddagger)$$

Proof. Suppose that $\lim(\xi_n) = \xi$. Then the sequence of functions $g_n(x) := f(x)e^{-ix\xi_n}$ converges to $f(x)e^{-ix\xi}$ for all x and $|g_n| \leq |f|$ for all n. Therefore,
$\lim_{n\to\infty} \hat{f}(\xi_n) = \hat{f}(\xi)$ follows from the Dominated Convergence Theorem.
To prove $(\dagger)$, assume first that $f = \chi_{[a,b]}$ with $a < b$. Then we have

$$\int_{-\infty}^{\infty} f(x)e^{-i\xi x} \, d\lambda(x) = \int_{a}^{b} e^{-i\xi x} \, d\lambda(x) = \frac{1}{i\xi}(e^{-i\xi a} - e^{-i\xi b}) \to 0,$$

as $|\xi| \to \infty$. Since a step function is a *finite linear combination* of characteristic functions of bounded intervals, it follows that (†) holds for all *step functions*. In general, given $f \in L^1(\mathbb{R})$, pick (Proposition 11.3.9) a sequence (ψ_n) of step functions with $\lim(\psi_n) = f$ a.e. and $\|f - \psi_n\|_1 \to 0$ as $n \to \infty$. Then we also have $\lim_{n \to \infty} \psi_n(x)e^{-i\xi x} = f(x)e^{-i\xi x}$ for almost all x and $\left\| fe^{-ix\xi} - \psi_n e^{-ix\xi} \right\|_1 = \|f - \psi_n\|_1 \to 0$ as $n \to \infty$. Now, given $\varepsilon > 0$, pick $N \in \mathbb{N}$ so large that

$$\left\| fe^{-ix\xi} - \psi_N e^{-ix\xi} \right\|_1 = \|f - \psi_N\|_1 := \int_{-\infty}^{\infty} |f(x) - \psi_N(x)| \, d\lambda(x) < \frac{\varepsilon}{2}.$$

Next, keeping N fixed, pick $A > 0$ such that $|\xi| \geq A$ implies

$$\left| \int_{-\infty}^{\infty} \psi_N(x)e^{-i\xi x} \, d\lambda(x) \right| < \frac{\varepsilon}{2}.$$

It now follows that for $|\xi| \geq A$,

$$\left| \int_{-\infty}^{\infty} f(x)e^{-i\xi x} \, d\lambda(x) \right| \leq \int_{-\infty}^{\infty} |f(x) - \psi_N(x)| \, d\lambda(x) + \left| \int_{-\infty}^{\infty} \psi_N(x)e^{-i\xi x} \, d\lambda(x) \right|$$

$$< \frac{\varepsilon}{2} + \frac{\varepsilon}{2} = \varepsilon$$

and (†) is established in general. Since (‡) is an immediate consequence, the proof is complete. □

Remark 11.3.12. In view of Corollary 10.5.8, we can easily modify the proof of Proposition 11.3.9 to show that the set of all continuous functions on $\mathbb{R}$ that vanish outside a *compact* (hence bounded) set is also dense in $L^p(\mathbb{R}, \mathbb{F})$. However, as we shall see below, we can even show that (infinitely) *smooth* functions with *compact support* form a dense subspace of $L^p(\mathbb{R}, \mathbb{F})$ for all $p \in [1, \infty)$. Note, by the way, that if I is a *noncompact* interval, then $C(I, \mathbb{F}) \not\subset L^p(I, \mathbb{F})$. (Why?)

Definition 11.3.13 (Support of a Function). Let $I \subset \mathbb{R}$ be an open interval and $f : I \to \mathbb{F}$ a continuous function. The *support* of f, denoted by $\text{supp}(f)$, is the *closure* (relative to I) of the set $\{x \in I : f(x) \neq 0\}$. We say that f has *compact support* if $\text{supp}(f)$ is compact.

Notation 11.3.14 ($C_c(I)$, $C_c^k(I)$, $C_c^\infty(I)$). If I is an open interval of $\mathbb{R}$, then $C_c(I, \mathbb{F})$ denotes the set of all *continuous* ($\mathbb{F}$-valued) functions on I with *compact support*. For each $k \in \mathbb{N}$, $C_c^k(I, \mathbb{F})$ will denote the set of k-*times continuously differentiable* functions on I with *compact support*. Finally, $C_c^\infty(I, \mathbb{F})$ will denote the set of *infinitely differentiable* functions on I with *compact support*. It is obvious that $C_c^\infty(I, \mathbb{F}) \subset C_c^k(I, \mathbb{F}) \subset C_c(I, \mathbb{F})$ and that all these spaces are (vector) subspaces of $L^p(I, \mathbb{F})$ for each $p \in [1, \infty)$.

Theorem 11.3.15 ($C_c^\infty(\mathbb{R}, \mathbb{F})$ is Dense in $L^p(\mathbb{R}, \mathbb{F})$). *The space $C_c^\infty(\mathbb{R}, \mathbb{F})$ is dense in $L^p(\mathbb{R}, \mathbb{F})$ and hence so are $C_c(\mathbb{R}, \mathbb{F})$ and $C_c^k(\mathbb{R}, \mathbb{F})$ (for all $k \in \mathbb{N}$). The same is also true if $\mathbb{R}$ is replaced by any open interval I of $\mathbb{R}$.*

Proof. We may and do assume that $\mathbb{F} = \mathbb{R}$ and all functions are real-valued. Let $f \in L^p(\mathbb{R})$ and let $\varepsilon > 0$ be given. As in Proposition 11.3.9, pick N so large that, with $f_N := f \chi_{[-N,N]}$, we have $\| f - f_N \|_p < \varepsilon/3$. Then pick a *step* function ψ such that $\| f_N - \psi \|_p < \varepsilon/3$. If we can show the existence of a function $u \in C_c^\infty(\mathbb{R})$ such that $\| u - \psi \|_p < \varepsilon/3$, then the theorem follows from the Triangle Inequality in L^p. Since ψ is a *finite* sum of functions of the form $c\chi_{[a,b]}$ with a constant $c \in \mathbb{R}$, we need only look at the case $\psi = \chi_{[a,b]}$ and show the existence of a function $u \in C_c^\infty(\mathbb{R})$ such that $\| \chi_{[a,b]} - u \|_p < \varepsilon/3$. For this, let $h \in C^\infty(\mathbb{R})$ be as in Exercise 8.4.24 and recall that $h(x) \equiv 1$ for all $x > -1 + \delta$ and $h(x) \equiv 0$ for $x \leq -1$, where $\delta \in (0, 1)$. Now define

$$u(x) := h\left(-\left|\frac{2x - a - b}{b - a}\right|\right).$$

Then $u \in C_c^\infty(\mathbb{R})$, and we have

$$u(x) = \begin{cases} 1 & \text{if } a + \delta(b-a)/2 < x < b - \delta(b-a)/2; \\ 0 & \text{if } x \notin (a,b). \end{cases}$$

If $[\delta(b - a)/2]^{1/p} < \varepsilon/6$, then it follows that $\| \chi_{[a,b]} - u \|_p < \varepsilon/3$ as desired. □

Let us also include a nice inequality.

Proposition 11.3.16 (Chebyshev's Inequality). *Let $E \in \mathcal{M}_\lambda(\mathbb{R})$ and $p \in [1, \infty)$. If $f \in L^p(E, \mathbb{F})$, then for any $\alpha > 0$ we have*

$$\lambda(\{x \in E : |f(x)| > \alpha\}) \leq (\| f \|_p/\alpha)^p.$$

Proof. Let $E_\alpha := \{x \in E : |f(x)| > \alpha\}$. Then $E_\alpha \in \mathcal{M}_\lambda(\mathbb{R})$ and we have

$$\| f \|_p^p = \int_E |f|^p \, d\lambda \geq \int_{E_\alpha} |f|^p \, d\lambda \geq \alpha^p \int_{E_\alpha} 1 \, d\lambda = \alpha^p \lambda(E_\alpha). \qquad \square$$

In our study of L^p with $p \geq 1$, to each *index $p > 1$*, we associated a *conjugate index $q > 1$* such that $1/p + 1/q = 1$. To complete this picture, note that at least *formally* we have $1/1 + 1/\infty = 1$, so the conjugate index of $p = 1$ should be "$q = \infty$." This motivates the search for the space L^∞.

Definition 11.3.17 (Essentially Bounded, Essential Supremum). Let $E \in \mathcal{M}_\lambda(\mathbb{R})$ and let $f : E \to \mathbb{F}$ be measurable. We say that f is *essentially bounded* if it is *bounded almost everywhere*, i.e., there is a set $Z \subset E$ with $\lambda(Z) = 0$ and a constant $B \geq 0$ such that $|f(x)| \leq B$ for all $x \notin Z$. The *essential supremum* of f is then defined to be the number

$$\|f\|_\infty = \text{ess sup}_{x\in E}|f(x)| := \inf\{B \geq 0 : \lambda(\{x : |f(x)| > B\}) = 0\}, \qquad (\dagger)$$

with the convention $\inf\emptyset := \infty$.

Exercise 11.3.18. Show that if $f : [a,b] \to \mathbb{R}$ is *continuous*, then $\|f\|_\infty = \sup\{|f(x)| : x \in [a,b]\}$.

Notation 11.3.19 ($\mathcal{L}^\infty(E,\mathbb{F})$, $L^\infty(E,\mathbb{F})$). Given any $E \in \mathcal{M}_\lambda(\mathbb{R})$, the set of all essentially bounded measurable functions $f : E \to \mathbb{F}$ is denoted by $\mathcal{L}^\infty(E,\mathbb{F})$. Also, if $\mathcal{N}(E,\mathbb{F})$ denotes the set of all *null* functions on E, then we define the quotient space $L^\infty(E,\mathbb{F}) := \mathcal{L}^\infty(E,\mathbb{F})/\mathcal{N}(E,\mathbb{F})$.

Theorem 11.3.20. *For each $E \in \mathcal{M}_\lambda(E)$, the space $L^\infty(E,\mathbb{F})$ is a Banach space, i.e, a complete, normed vector space over $\mathbb{F}$ with the operations $[f]+[g] := [f+g]$ and $c[f] := [cf]$, and the norm $\|[f]\|_\infty := \|f\|_\infty$ as in $(\dagger)$ of Definition 11.3.17.*

Proof. Let $\mathbb{F} = \mathbb{R}$ and note that $(\dagger)$ is well defined on $L^\infty(E)$. Indeed, let $f_1, f_2 \in \mathcal{L}^0(E)$ satisfy $f_1 = f_2$ outside a set $Z_1 \subset E$ of measure zero. If $|f_2(x)| \leq B_2$ for all x outside a set $Z_2 \subset E$ of measure zero, then $|f_1(x)| = |f_2(x)| \leq B_2$ outside $Z_1 \cup Z_2$ gives $\|f_1\|_\infty \leq \|f_2\|_\infty$. Interchanging f_1 and f_2 gives the reverse inequality and hence $\|f_1\|_\infty = \|f_2\|_\infty$. Next, $(\dagger)$ implies that for each $n \in \mathbb{N}$ we can find $Z_n \subset E$ with $\lambda(Z_n) = 0$ and $|f(x)| \leq \|f\|_\infty + 1/n$ for all $x \notin Z_n$. If we set $Z := \bigcup_{n=1}^\infty Z_n$, then $\lambda(Z) = 0$ and we have $|f(x)| \leq \|f\|_\infty$ for all $x \notin Z$; in other words, the infimum $\|f\|_\infty$ in $(\dagger)$ is actually *attained*. Now, given $f, g \in \mathcal{L}^\infty(E)$, pick subsets Z_1, Z_2 of E with $\lambda(Z_1) = \lambda(Z_2) = 0$ such that $|f(x)| \leq \|f\|_\infty$ for all $x \notin Z_1$ and $|g(x)| \leq \|g\|_\infty$ for all $x \notin Z_2$. It then follows that $|f(x)| + |g(x)| \leq \|f\|_\infty + \|g\|_\infty$ for all $x \notin Z_1 \cup Z_2$ and hence the Triangle Inequality $\|f + g\|_\infty \leq \|f\|_\infty + \|g\|_\infty$ is established. Also, the properties $\|f\|_\infty \geq 0$, $\|0\|_\infty = 0$, and $\|cf\|_\infty = |c|\|f\|_\infty$ are trivially satisfied. Next, $\|f\|_\infty = 0$ implies that for each $n \in \mathbb{N}$ there is a set $Z_n \subset E$ with $\lambda(Z_n) = 0$ and such that $|f(x)| \leq 1/n$ for all $x \notin Z_n$. If we set $Z := \bigcup_{n=1}^\infty Z_n$, then $\lambda(Z) = 0$ and $|f(x)| \leq 1/n$ for all n and all $x \notin Z$, i.e., $f(x) = 0$ for all $x \notin Z$ and hence $[f] = [0]$. The vector space properties are also easily checked. Finally, to prove the *completeness*, let (f_n) be a Cauchy sequence in $\mathcal{L}^\infty(E)$. Then we can pick a set $Z \subset E$ with $\lambda(Z) = 0$ and such that $|f_n(x)| \leq \|f_n\|_\infty$ for all $n \in \mathbb{N}$ and all $x \notin Z$. In fact, we may even arrange for $|f_n(x) - f_m(x)| \leq \|f_n - f_m\|_\infty$ to be satisfied for all $x \notin Z$ and all $m, n \in \mathbb{N}$. But then the sequence (f_n) converges *uniformly* on $E \setminus Z$. If we define

$$f(x) := \begin{cases} \lim_{n\to\infty} f_n(x) & \text{for } x \notin Z \\ 0 & \text{for } x \in Z, \end{cases}$$

then f is measurable and it follows easily that $\lim_{n\to\infty} \|f - f_n\|_\infty = 0$. $\square$

The following exercise contains Hölder's inequality for the limiting case $p = 1$, $q = \infty$.

Exercise 11.3.21 (Hölder's Inequality, Again). Show that Hölder's inequality also holds when $p = 1$ and $q = \infty$; i.e., if $f \in L^1(E, \mathbb{F})$ and $g \in L^\infty(E, \mathbb{F})$, where $E \in \mathcal{M}_\lambda(\mathbb{R})$, then $fg \in L^1(E, \mathbb{F})$ and we have

$$\|fg\|_1 = \int_E |fg| \, d\lambda \le \|f\|_1 \|g\|_\infty.$$

11.4   More on Modes of Convergence

As we have seen (in Chap. 10), a sequence (f_n) of measurable functions on a measurable set $E \in \mathcal{M}_\lambda(\mathbb{R})$ may converge to a function f in a number of ways, namely, *pointwise, almost everywhere, uniformly,* and *almost uniformly.* And in this chapter we have also introduced convergence *in L^p* for $p \in [1, \infty]$. In this section we want to introduce another mode of convergence (introduced by F. Riesz) that is *weaker* than convergence almost everywhere (at least on sets of *finite measure*), but will be needed later when we discuss the *Weak Law of Large Numbers* in Chap. 12. This convergence is motivated by the following observation: If $\int_E |f_n| \, d\lambda \to 0$, then for each $\varepsilon > 0$, $\lambda(\{x \in E : |f_n(x)| \ge \varepsilon\})$ should go to zero.

Definition 11.4.1 (Convergence in Measure, Cauchy in Measure). Let $E \in \mathcal{M}_\lambda(\mathbb{R})$. A sequence (f_n) of real (or complex-)-valued measurable functions on E is said to converge *in measure* to a function f if, given any $\varepsilon > 0$, there is an $N \in \mathbb{N}$ such that

$$n \ge N \implies \lambda(\{x \in E : |f(x) - f_n(x)| \ge \varepsilon\}) < \varepsilon.$$

We say that (f_n) is *Cauchy in measure* if, given any $\varepsilon > 0$, there is an $N \in \mathbb{N}$ such that

$$n \ge m \ge N \implies \lambda(\{x \in E : |f_n(x) - f_m(x)| \ge \varepsilon\}) < \varepsilon.$$

Exercise 11.4.2. Let (f_n) be a sequence of real (or complex) measurable functions on $E \in \mathcal{M}_\lambda(\mathbb{R})$. Show that if (f_n) converges to f in measure (resp., is Cauchy in measure), then the same holds for any subsequence (f_{n_k}).

Unfortunately, convergence in measure (resp., almost everywhere or even everywhere) does *not* imply convergence almost everywhere (resp., in measure), as the following examples show.

Example 11.4.3.

1 **(Escape to Horizontal Infinity).** Let $f_n := \chi_{[n,n+1]}$ for all $n \in \mathbb{N}$. Then $f_n(x) \to 0$ for all $x \in [0, 1]$, but f_n does *not* converge in measure.
2 **(Typewriter Sequence).** Divide the interval $[0, 1]$ successively into $1, 2, 3, 4, \dots$ equal parts and enumerate the resulting subintervals in succession: $I_1 := [0, 1]$,

$I_2 := [0, 1/2], I_3 := (1/2, 1], I_4 := [0, 1/3], I_5 := (1/3, 2/3], I_6 := (2/3, 1],$
$I_7 := [0, 1/4]$, etc. Now let $f_n := \chi_{I_n}$, for all $n \in \mathbb{N}$. Then $f_n \to 0$ *in measure*.
Indeed, if $n \geq m(m + 1)/2 = 1 + 2 + \cdots + m$, then $f_n = \chi_{I_n}$ and $\lambda(I_n) \leq 1/m$.
In fact we even have

$$\| f_n - 0 \|_p^p = \int |f_n|^p \, d\lambda = \int f_n \, d\lambda \leq 1/m,$$

so that $f_n \to 0$ in L^p for all $p \in [1, \infty)$. However, (f_n) *diverges* everywhere. In
fact, for each $x \in [0, 1]$, the sets $\{n \in \mathbb{N} : f_n(x) = 1\}$ and $\{n \in \mathbb{N} : f_n(x) = 0\}$
are both infinite.

Despite the fact that convergence in measure does not imply convergence almost
everywhere, we have the following.

Theorem 11.4.4. *Let $E \in \mathcal{M}_\lambda(\mathbb{R})$ and let (f_n) be a sequence of real (or complex)
measurable functions on E. If $f_n \to f$ in measure, then (f_n) is Cauchy in measure.
Conversely, if (f_n) is Cauchy in measure, then there is a measurable function f
such that $f_n \to f$ in measure and there is a subsequence (f_{n_k}) that converges to
f almost everywhere on E. Moreover, if we also have $f_n \to g$ in measure, then
$f(x) = g(x)$ for almost all $x \in E$.*

Proof. Suppose that $f_n \to f$ in measure and, given $n \in \mathbb{N}$ and $\varepsilon > 0$, consider the
set $E_{n,\varepsilon} := \lambda(\{x \in E : |f_n(x) - f(x)| \geq \varepsilon/2\})$. Then

$$\lambda(\{x \in E : |f_n(x) - f_m(x)| \geq \varepsilon\}) \leq \lambda(E_{n,\varepsilon}) + \lambda(E_{m,\varepsilon}) \to 0, \quad \text{as } m, n \to \infty$$

and hence (f_n) is Cauchy in measure. Conversely, suppose (f_n) is Cauchy in
measure and pick positive integers n_k with $n_k < n_{k+1}$ such that

$$n \geq m \geq n_k \implies \lambda(\{x \in E : |f_n(x) - f_m(x)| \geq 2^{-k}\}) < 2^{-k}.$$

Define $g_k := f_{n_k}$ and $E_k := \{x \in E : |g_{k+1}(x) - g_k(x)| \geq 2^{-k}\}$. If $F_k :=
\bigcup_{j=k}^{\infty} E_j$, then (F_k) is a *decreasing* sequence of measurable sets with $\lambda(F_k) < 2^{1-k}$
and, for $j \geq i \geq k$, we have

$$|g_j(x) - g_i(x)| \leq \sum_{\ell=i}^{j-1} |g_{\ell+1}(x) - g_\ell(x)| \leq \sum_{\ell=i}^{j-1} 2^{-\ell} \leq 2^{1-k}. \qquad (*)$$

Thus (g_k) is (pointwise) Cauchy on $E \setminus F_k$. Now, with $Z := \bigcap_{k=1}^{\infty} F_k$, we have
$\lambda(Z) = 0$, so if we set $f(x) := \lim_{k \to \infty} g_k(x)$ for $x \in E \setminus Z$ and $f(x) = 0$ for
$x \in Z$, then f is measurable (cf. Proposition 10.2.12) and $f_{n_k} = g_k \to f$ almost
everywhere. Also, taking $j = k$ and letting $i \to \infty$ in $(*)$, we have $|f_{n_k}(x) -
f(x)| \leq 2^{1-k}$ for all $x \in E \setminus F_k$. Since $\lambda(F_k) \to 0$ as $k \to \infty$, we deduce that
$f_{n_k} \to f$ in measure; in fact, $(f_{n_j})_{j=k}^{\infty}$ converges to f *uniformly* on $E \setminus F_k$. But
then we actually have $f_n \to f$ in measure because for any $\varepsilon > 0$, we have

$$\{x : |f_n(x) - f(x)| \geq \varepsilon\} \subset \left\{x : |f_n(x) - f_{n_k}(x)| \geq \frac{\varepsilon}{2}\right\} \cup \left\{x : |f_{n_k}(x) - f(x)| \geq \frac{\varepsilon}{2}\right\},$$

and (f_n) being Cauchy in measure, both sets on the right have measures tending to zero, as $n, k \to \infty$. Finally, if $f_n \to g$ in measure, then the inclusions

$$\{x : |f(x) - g(x)| \geq \varepsilon\} \subset \{x : |f(x) - f_n(x)| \geq \varepsilon/2\} \cup \{x : |f_n(x) - g(x)| \geq \varepsilon/2\}$$

hold for all $n \in \mathbb{N}$ and every $\varepsilon > 0$ and hence $\lambda(\{x : |f(x) - g(x)| \geq \varepsilon\}) = 0$. Letting $\varepsilon \to 0$ (through a decreasing sequence of values), we get $f = g$ almost everywhere. □

The subsequence (f_{n_k}) in the above proof converges *uniformly* outside a set of *small measure*. We now give a name to the this type of convergence which was also seen in the conclusion of Egorov's theorem (Theorem 10.5.15):

Definition 11.4.5 (Almost Uniform Convergence). Let $E \in \mathcal{M}_\lambda(\mathbb{R})$ and let (f_n) be a sequence of real (or complex) measurable functions on E. We say that (f_n) converges to f *almost uniformly* if, given any $\varepsilon > 0$, there is a measurable set $F \subset E$ such that $\lambda(F) < \varepsilon$ and (f_n) converges to f *uniformly* on $E \setminus F$. We also say that (f_n) is *almost uniformly Cauchy* if for each $\varepsilon > 0$ there exists a set $F \subset E$ with $\lambda(F) < \varepsilon$ such that $f_n - f_m \to 0$ uniformly on $E \setminus F$ as $m, n \to \infty$.

Having this definition, we can state the following corollary of Theorem 11.4.4, whose proof is basically contained in the proof of the theorem.

Corollary 11.4.6. *Let $E \in \mathcal{M}_\lambda(\mathbb{R})$ and let (f_n) be a sequence of real (or complex) measurable functions on E. If $f_n \to f$ in measure, then there is a subsequence (f_{n_k}) such that $f_{n_k} \to f$ almost uniformly.*

Proof. Exercise!

Some of the easy implications involving the relation between different modes of convergence are summarized in the following theorem. To simplify the notation we use $\mathbb{R}$ as the domain of the functions, but one can obviously use any $E \in \mathcal{M}_\lambda(\mathbb{R})$.

Theorem 11.4.7. *Let (f_n) be a sequence of (complex-valued) measurable functions on $\mathbb{R}$ and let $p \in [1, \infty)$.*

(i) $f_n \to f$ *pointwise* $\implies$ $f_n \to f$ *almost everywhere.*
(ii) $f_n \to f$ *uniformly* $\implies$ $f_n \to f$ *pointwise.*
(iii) $f_n \to f$ *almost uniformly* $\implies$ $f_n \to f$ *almost everywhere.*
(iv) $f_n \to f$ *almost uniformly* $\implies$ $f_n \to f$ *in measure.*
(v) $f_n \to f$ *in L^p* $\implies$ $f_n \to f$ *in measure.*
(vi) (f_n) *is Cauchy in L^p* $\implies$ (f_n) *is Cauchy in measure.*
(vii) $f_n \to f$ *in L^∞* $\iff$ $f_n \to f$ *uniformly outside a set of measure zero.*
(viii) (f_n) *is Cauchy in L^∞* $\iff$ (f_n) *is uniformly Cauchy outside a set of measure zero.*

(ix) $f_n \to f$ in $L^\infty \Longrightarrow f_n \to f$ almost uniformly.
(x) $f_n \to f$ uniformly $\Longrightarrow f_n \to f$ in L^∞.

Proof. (i) and (ii) are obvious. For (iii), note that with $\varepsilon = 1/2^n$ we can pick a measurable set E_n with $\lambda(E_n) < 1/2^n$ such that $f_n(x) \to f(x)$ uniformly on E_n^c. If $F_n := \bigcup_{k=n+1}^\infty E_k$, then (F_n) is a *decreasing* sequence of measurable sets with $\lambda(F_n) < \sum_{k=n+1}^\infty 1/2^k = 1/2^n$. Thus, with $Z := \bigcap_{n=1}^\infty F_n$, we have $\lambda(Z) = \lim_{n\to\infty} \lambda(F_n) = 0$ and $f_n(x) \to f(x)$ uniformly (hence pointwise) on Z^c. To prove (iv), pick $Z \subset \mathbb{R}$ with $\lambda(Z) = 0$ such that $f_n \to f$ uniformly on Z^c. Then, given any $\varepsilon > 0$, we have $\{x \in Z^c : |f_n(x) - f(x)| \geq \varepsilon\} = \emptyset$ for all sufficiently large $n \in \mathbb{N}$. Thus, for all such n, we have $\lambda(\{x \in \mathbb{R} : |f_n(x) - f(x)| \geq \varepsilon\}) = 0$, i.e., $f_n \to f$ in measure. Next, suppose that $\|f_n - f\|_p \to 0$, as $n \to \infty$ and for a given $\varepsilon > 0$ let $E_n(\varepsilon) := \{x \in \mathbb{R} : |f_n(x) - f(x)| \geq \varepsilon\}$. Then we have

$$\varepsilon^p \lambda\big(E_n(\varepsilon)\big) \leq \int_{E_n(\varepsilon)} |f_n - f|^p \, d\lambda \leq \int_{\mathbb{R}} |f_n - f|^p \, d\lambda \to 0, \quad \text{as } n \to \infty,$$

which shows that $f_n \to f$ in measure and proves (v). In fact, the same argument also proves (vi). For (vii), note that if $\|f_n - f\|_\infty \to 0$, then for any integer $m \in \mathbb{N}$ we have $\|f_n - f\|_\infty < 1/m$ for all sufficiently large n. But then, for each such n, we can find a set Z_n with $\lambda(Z_n) = 0$ and such that $|f_n(x) - f(x)| < 1/m$ for all $x \in Z_n^c$. Thus, if $Z := \bigcup_{n=1}^\infty Z_n$, then $\lambda(Z) = 0$ and $f_n \to f$ *uniformly* on Z^c. Conversely, if there is a set Z with $\lambda(Z) = 0$ and such that $f_n \to f$ *uniformly* on Z^c, then for each $\varepsilon > 0$ we have $|f_n(x) - f(x)| < \varepsilon$ for all $x \in Z^c$ and all sufficiently large n. Therefore, $\|f_n - f\|_\infty < \varepsilon$ for all sufficiently large n and hence $f_n \to f$ in L^∞. An identical proof works for (viii) as well. Finally, (ix) and (x) follow from (vii) and the proof is complete.   □

Corollary 11.4.8. *Let $E \in \mathcal{M}_\lambda(\mathbb{R})$ and let (f_n) be a sequence in $L^p(E, \mathbb{F})$, where $p \in [1, \infty)$ and $\mathbb{F}$ is either $\mathbb{R}$ or $\mathbb{C}$. If $f_n \to f$ in L^p, then there is a subsequence (f_{n_k}) such that $f_{n_k} \to f$ almost everywhere.*

Proof. By Theorem 11.4.7 (v), $f_n \to f$ in measure and hence the corollary follows from Theorem 11.4.4.   □

The following exercise (where we use the terminology in [Tao11]) shows that the relation between different modes of convergence is far from straightforward.

Exercise 11.4.9. Prove the following statements, where $p \in [1, \infty)$.

[a] **(Escape to Horizontal Infinity).** Let $f_n := \chi_{[n,n+1]}$ for all $n \in \mathbb{N}$. Then the sequence (f_n) converges to zero pointwise (hence almost everywhere), but it does *not* converge uniformly, almost uniformly, in L^p, in L^∞, or in measure.

[b] **(Escape to Width Infinity).** Let $f_n := \frac{1}{n}\chi_{[0,n]}$ for all $n \in \mathbb{N}$. Then the sequence (f_n) converges to zero uniformly (hence also almost uniformly, in L^∞, pointwise, almost everywhere, and in measure), but *not* in L^p.

[c] **(Escape to Vertical Infinity).** Let $f_n := n\chi_{[\frac{1}{n},\frac{2}{n}]}$ for all $n \in \mathbb{N}$. Then the sequence (f_n) converges to zero pointwise (hence almost everywhere) and almost uniformly (and hence in measure), but *not* uniformly, in L^∞, or in L^p.

[d] **(Typewriter Sequence).** Let $f_n := \chi_{[j/2^k,(j+1)/2^k]}$, with $n = 2^k + j$ and $0 \le j < 2^k$. Then the sequence (f_n) converges to zero in measure and in L^p, but *not* almost everywhere (and hence not pointwise, not uniformly, not almost uniformly, and not in L^∞).

Remark 11.4.10. As we saw above, convergence almost everywhere does *not* in general imply convergence in measure. However, the situation is different if our functions are defined on a set of *finite* measure.

Proposition 11.4.11. *Let* $E \in \mathcal{M}_\lambda(\mathbb{R})$ *with* $\lambda(E) < \infty$ *and let* f_n, $n \in \mathbb{N}$, *and* f *be (real or complex) measurable functions on* E. *Then* $f_n \to f$ *almost everywhere if and only if* $f_n \to f$ *almost uniformly. Also, if* $f_n \to f$ *almost everywhere, then* $f_n \to f$ *in measure.*

Proof. Indeed, the "if and only if" assertion follows from Egorov's theorem (Theorem 10.5.15) and Theorem 11.4.7 (iii). The second assertion then follows from Theorem 11.4.7 (iv).   □

Remark 11.4.12. In fact, if $E \in \mathcal{M}_\lambda(\mathbb{R})$ and $\lambda(E) < \infty$, then we have the following characterization of *almost everywhere convergence:*

Proposition 11.4.13. *Let* $E \in \mathcal{M}_\lambda(\mathbb{R})$ *and* $\lambda(E) < \infty$ *and let* f_n, $n \in \mathbb{N}$, *and* f *be (real or complex) measurable functions on* E. *Then* $f_n \to f$ *almost everywhere if and only if*

$$\lim_{n\to\infty} \lambda\left(\bigcup_{k=n}^{\infty} \{x \in E : |f_k(x) - f(x)| \ge \varepsilon\}\right) = 0 \qquad \forall\, \varepsilon > 0.$$

Proof. Let $E_{n,\varepsilon} := \{x \in E : |f_n(x) - f(x)| \ge \varepsilon\}$ and $E_\varepsilon := \limsup_n E_{n,\varepsilon} := \bigcap_{n=1}^{\infty}\bigcup_{k=n}^{\infty} E_{k,\varepsilon}$. Then $\lim_{n\to\infty} \lambda\left(\bigcup_{k=n}^{\infty} E_{k,\varepsilon}\right) = \lambda(E_\varepsilon)$ because $\bigcup_{k=n}^{\infty} E_{k,\varepsilon}$ decreases to E_ε. Also, since $E_{\varepsilon_1} \subset E_{\varepsilon_2}$ for $\varepsilon_1 > \varepsilon_2$, we have

$$\{x \in E : \lim_n f_n(x) \ne f(x)\} = \bigcup_{\varepsilon>0} E_\varepsilon = \bigcup_{m=1}^{\infty} E_{1/m}.$$

Therefore,

$$f_n \xrightarrow{\text{ae.}} f \iff \lambda(E_\varepsilon) = 0 \quad \forall\, \varepsilon > 0 \iff \lim_{n\to\infty} \lambda\left(\bigcup_{k=n}^{\infty} E_{k,\varepsilon}\right) = 0 \quad \forall\, \varepsilon > 0.$$

    □

The next proposition shows that the Dominated Convergence Theorem (Theorem 10.4.14) remains valid if "convergence almost everywhere" is replaced by "convergence in measure."

Proposition 11.4.14. *Let* (f_n) *be a sequence in* $L^p(E, \mathbb{F})$, *where* $\mathbb{F}$ *is* $\mathbb{R}$ *or* $\mathbb{C}$ *and* $E \in \mathcal{M}_\lambda(\mathbb{R})$. *If* (f_n) *converges in measure to* f *and if* $|f_n(x)| \leq g(x)$ *for almost all* $x \in E$ *and some* $g \in L^p(E, [0, \infty))$, *then* (f_n) *converges to* f *in* $L^p(E, \mathbb{F})$.

Proof. If not, then there exists $\varepsilon_0 > 0$ and a subsequence $(g_k) := (f_{n_k})$ with

$$\|g_k - f\|_p \geq \varepsilon_0 \qquad \forall\, k \in \mathbb{N}. \tag{$*$}$$

Now (by Exercise 11.4.2) $g_k \to f$ in measure and hence (by Theorem 11.4.4) there is a subsequence (g_{k_j}) of (g_k) such that (g_{k_j}) converges almost everywhere and in measure to a function h. By the *uniqueness* part of Theorem 11.4.4, we then have $h = f$ almost everywhere. But $g_{k_j} \to f$ and $|g_{k_j}|^p \leq g^p$ almost everywhere imply (by the Dominated Convergence Theorem) that $\|g_{k_j} - f\|_p \to 0$, as $j \to \infty$, contradicting ($*$). $\qquad\qquad\square$

Exercise 11.4.15. Show that the Monotone Convergence Theorem (Theorem 10.4.1) and Fatou's lemma (Lemma 10.4.10) remain valid if "convergence almost everywhere" is replaced by "convergence in measure."

11.5 Differentiation

When we studied the Riemann integral in Chap. 7, we looked at the relation between differentiation and integration and proved the two fundamental theorems of calculus:

FTC 1 (First Fundamental Theorem). *Let* f *be Riemann integrable on* $[a, b]$ *and let* $C \subset [a, b]$ *be a finite set. If* $F : [a, b] \to \mathbb{R}$ *is a continuous function such that* $F'(x) = f(x)$ *for all* $x \in [a, b] \setminus C$, *then we have*

$$\int_a^b f(x)\, dx = F(b) - F(a).$$

FTC 2 (Second Fundamental Theorem). *Let* I *be an interval and* $f : I \to \mathbb{R}$. *Suppose that* f *is Riemann integrable on any closed, bounded subinterval of* I. *If* a *is any point in* I, *then the function*

$$F(x) := \int_a^x f(t)\, dt \qquad (\forall x \in I)$$

is continuous on I. *Also, if* f *is continuous at* $x_0 \in I$, *then* $F'(x_0) = f(x_0)$.

It is quite natural to look for similar results in the Lebesgue theory and this is our objective in this section. One of the fundamental results is *Lebesgue's Differentiation Theorem,* which says that *monotone* functions are differentiable almost everywhere. Let us begin by defining the *derivative(s)* of a function.

Definition 11.5.1 (Dini Derivatives). Let I be an open interval, $f : I :\to \mathbb{R}$, and $x \in I$. Then the four *Dini derivatives* of f at x are the following limits:

$$D^+ f(x) = \overline{\lim}_{h\to 0+}\frac{f(x+h) - f(x)}{h},$$

$$D^- f(x) = \overline{\lim}_{h\to 0-}\frac{f(x+h) - f(x)}{h},$$

$$D_+ f(x) = \underline{\lim}_{h\to 0+}\frac{f(x+h) - f(x)}{h},$$

$$D_- f(x) = \underline{\lim}_{h\to 0-}\frac{f(x+h) - f(x)}{h}.$$

We obviously have $D^+ f(x) \geq D_+ f(x)$ and $D^- f(x) \geq D_- f(x)$. If $D^+ f(x) = D_+ f(x) = D^- f(x) = D_- f(x) \neq \pm\infty$, then we say that f is *differentiable* at x and the common value of the four Dini derivatives is then denoted by $f'(x)$. Sometimes we even write $f'(x) = \infty$ (resp., $f'(x) = -\infty$) if all four Dini derivatives of f are ∞ (resp., $-\infty$).

Remark 11.5.2.

(1) If $D^+ f(x) = D_+ f(x) \neq \pm\infty$, then this common value is denoted $f'_+(x)$ and is called the *right derivative* of f at x. Similarly, if $D^- f(x) = D_- f(x) \neq \pm\infty$, then the common value is denoted $f'_-(x)$ and is called the *left derivative* of f at x. Thus $f'(x)$ exists $\Leftrightarrow f'_+(x) = f'_-(x) \neq \pm\infty$.
(2) If f is *increasing,* then all four Dini derivatives are nonnegative and hence "exist" as extended numbers in $[0, \infty]$.
(3) The four Dini derivatives are all equal if and only if

$$D_- f(x) \geq D^+ f(x) \quad \text{and} \quad D_+ f(x) \geq D^- f(x). \tag{$*$}$$

We first prove Lebesgue's Differentiation Theorem for *continuous,* monotone functions, using the *Rising Sun Lemma* of F. Riesz. We begin with the following.

Definition 11.5.3 (Shadow Point). Let $a < b$ and let $f : [a,b] \to \mathbb{R}$ be *continuous.* A point $x \in [a,b]$ is said to be a *shadow point* (of f) if there is a point y such that $x < y \leq b$ and $f(x) < f(y)$.

Remark 11.5.4. Think of the graph of a continuous function as a number of "hills" and "valleys"; draw a picture! If the (horizontal) rays of the sun [located at "$(+\infty, 0)$"] hit the graph, then the *shaded* parts of the graph consist of the points $(x, f(x))$ that are in the *shadow* of the rising sun.

Lemma 11.5.5 (Rising Sun Lemma). *Given $a < b$, let $f : [a,b] \to \mathbb{R}$ be a continuous function. Then the set O_f of all shadow points in (a,b) is open. In fact, either $O_f = \emptyset$ or $O_f = \bigcup_n I_n$ is a disjoint union of a (finite or denumerable) sequence of nonempty open intervals $I_n := (a_n, b_n) \subset (a,b)$ such that $f(a_n) \leq f(b_n)$ for all $n \in \mathbb{N}$.*

Proof. By definition, we have

$$O_f := \{x \in (a,b) : f(x) < f(\xi) \text{ for some } \xi \in (x,b)\}.$$

If $x_0 \in O_f$, then $f(x_0) < f(\xi)$ for some $\xi > x_0$. Since f is continuous, we then have $f(x) < f(\xi)$ if $|x - x_0| < \delta$ for $\delta > 0$ small enough. Therefore, O_f is *open* and hence (cf. Theorem 4.1.2) $O_f = \bigcup_n I_n$, where the $I_n = (a_n, b_n) \subset (a,b)$ are disjoint, nonempty intervals. To show that $f(a_n) \leq f(b_n)$ it suffices to show that $f(x) \leq f(b_n)$ for all $x \in (a_n, b_n)$ because the continuity of f at a_n will then give $f(a_n) = \lim_{x \to a_n+} f(x) \leq f(b_n)$. Now for a given $x \in (a_n, b_n)$, consider the set

$$F := \{y \in [x, b_n] : f(x) \leq f(y)\}.$$

Then F is a closed, bounded (i.e., *compact*) set containing x. Let $\eta := \sup(F) \in F$ and note that $f(x) \leq f(\eta)$. Thus $f(x) \leq f(b_n)$ follows if we show that $\eta = b_n$. But if $\eta < b_n$ then $\eta \in (a_n, b_n)$ is a shadow point. We can then pick $\xi > \eta$ with $f(\eta) < f(\xi)$. Since $\eta = \sup(F)$, the maximality of η forces $\xi > b_n$. Also $b_n > \eta$ gives $b_n \notin F$ and hence $f(\xi) > f(\eta) \geq f(x) > f(b_n)$. But then b_n is a shadow point, a contradiction. $\qquad\square$

We now use this lemma to prove Lebesgue's theorem for *continuous* functions. Before giving the proof, let us make a few remarks.

Remark 11.5.6. If f is an *increasing* function, then so is $-\check{f}$, where $\check{f}(x) := f(-x)$. Now we can easily check that $D_-\check{f}(-x) = -D_+f(x)$ and $D^+\check{f}(-x) = -D^-f(x)$. Using this observation in Remark 11.5.2 (3), we see that the first inequality in $(*)$ implies the second one. Therefore, an increasing function f is differentiable at x if and only if the two conditions

$$D^+f(x) \leq D_-f(x) \quad \text{and}| \tag{$\dagger$}$$

$$D^+f(x) < \infty \tag{$\ddagger$}$$

are satisfied because we then have

$$D^+f(x) \leq D_-f(x) \leq D^-f(x) \leq D_+f(x) \leq D^+f(x) < \infty.$$

Theorem 11.5.7 (Lebesgue's Differentiation Theorem 1). *Let I be an interval and $f : I \to \mathbb{R}$ a continuous, monotone function. Then f is differentiable at almost all $x \in I$.*

Proof. We may assume that f is *increasing* and, since I is a countable union of *compact* intervals, we may (and do) assume that $I = [a, b]$. Now we must show that (†) and (‡) hold for almost all $x \in [a, b]$. Let us first look at (‡) and show that, if $E^{\ddagger} := \{x \in (a, b) : D^+ f(x) = \infty\}$, then $\lambda(E^{\ddagger}) = 0$. We note that $E^{\ddagger} \subset E_n := \{x \in (a, b) : D^+ f(x) > n\}$ for every $n \in \mathbb{N}$. But $D^+ f(x) > n$ implies that $[f(y) - f(x)]/(y - x) > n$ for some $y > x$, which we can write as $g_n(y) > g_n(x)$ with the continuous function $g_n(x) := f(x) - nx$. In other words, $E^{\ddagger} \subset E_n \subset O_{g_n}$. Thus, using (Rising Sun) Lemma 11.5.5, we can cover $E^{\ddagger}$ by a sequence of disjoint open intervals (a_k, b_k) such that $g_n(a_k) \leq g_n(b_k)$ for all k. In other words, $n(b_k - a_k) \leq f(b_k) - f(a_k)$, and summing these inequalities over all k gives

$$n \sum_k (b_k - a_k) \leq \sum_n [f(b_k) - f(a_k)] \leq f(b) - f(a).$$

Therefore, $E^{\ddagger}$ can be covered by a sequence of intervals with total length $\leq [f(b) - f(a)]/n$. Since n was arbitrary, we indeed have $\lambda(E^{\ddagger}) = 0$. Next, we must show that (†) holds almost everywhere; i.e., if $E^{\dagger} := \{x \in (a, b) : D_- f(x) < D^+ f(x)\}$, then $\lambda(E^{\dagger}) = 0$. However, we note that the collection of sets

$$E^q_p := \{x \in (a, b) : D_- f(x) < p < q < D^+ f(x)\}, \quad 0 \leq p < q, \quad p, q \in \mathbb{Q}$$

is *countable* and $E^{\dagger} = \bigcup_{p,q} E^q_p$, so it suffices to show that $\lambda(E^q_p) = 0$ for each pair of rationals $0 \leq p < q$. Now note that

$$E^q_p = E_p \cap E^q, \quad \text{with} \quad E_p := \{x : D_- f(x) < p\}, \quad E^q := \{x : D^+ f(x) > q\}.$$

Assuming first that $D_- f(x) < p$, there exists $y \in (a, x)$ with $[f(y) - f(x)]/(y - x) < p$ and hence $f(x) - px < f(y) - py$. This can be written as $\check{g}_p(-x) < \check{g}_p(-y)$ with $-b < -x < -y < -a$, where $g_p(x) := f(x) - px$ and, as before, $\check{g}_p$ is the *flipped* function $\check{g}_p(z) = g_p(-z)$ for all $z \in (-b, -a)$. Therefore, $-E_p := \{-x : x \in E_p\} \subset O_{\check{g}_p}$ and, by Lemma 11.5.5, we can cover $-E_p$ by a sequence of disjoint intervals $(-b_i, -a_i)$ such that $\check{g}_p(-b_i) \leq \check{g}_p(-a_i)$ for all i. In other words, $f(b_i) - pb_i \leq f(a_i) - pa_i$ and hence E_p is covered by the sequence of disjoint intervals (a_i, b_i) with

$$f(b_i) - f(a_i) \leq p(b_i - a_i) \quad \forall i. \tag{11.1}$$

Next, if $x \in (a_i, b_i) \cap E^q$, i.e., $D^+ f(x) > q$, then for some $y \in (x, b_i)$, we have $[f(y) - f(x)]/(y - x) > q$ and hence $f(x) - qx < f(y) - qy$, which means $g_q(x) < g_q(y)$ with $g_q(x) := f(x) - qx$. Thus, $(a_i, b_i) \cap E^q \subset (a_i, b_i) \cap O_{g_q}$ and hence (by Lemma 11.5.5) can be covered by a sequence (a_{ij}, b_{ij}) (indexed by j) of disjoint subintervals of (a_i, b_i) such that $g_q(a_{ij}) \leq g_q(b_{ij})$, i.e.,

$$q(b_{ij} - a_{ij}) \leq f(b_{ij}) - f(a_{ij}) \quad \forall j. \tag{11.2}$$

Summing (11.2) over all j and using (11.1), we get

$$q \sum_j (b_{ij} - a_{ij}) \leq \sum_j [f(b_{ij}) - f(a_{ij})] \leq f(b_i) - f(a_i) \leq p(b_i - a_i). \qquad (11.3)$$

Since $E_p^q \subset \bigcup_{i,j} (a_{ij}, b_{ij})$, summing (11.3) over all i and dividing by q, we see that

$$\lambda(E_p^q) \leq \sum_{i,j} (b_{ij} - a_{ij}) \leq (p/q)(b - a).$$

If we now repeat the above argument with (a_{ij}, b_{ij}) instead of (a, b), then we obtain a sequence of intervals $(a_{ijk\ell}, b_{ijk\ell})$, whose union contains E_p^q, with inequalities

$$\lambda(E_p^q) \leq \sum_{i,j,k,\ell} (b_{ijk\ell} - a_{ijk\ell}) \leq (p/q) \sum_{i,j} (b_{ij} - a_{ij}) \leq (p/q)^2 (b - a).$$

Thus, iterating the process gives $\lambda(E_p^q) \leq (p/q)^n (b - a)$ for all $n \in \mathbb{N}$. Since $0 \leq p/q < 1$, we have $\lambda(E_p^q) = 0$ and hence $\lambda(E^\dagger) = 0$. $\qquad \square$

Lebesgue's theorem is true *without* the restriction that f be *continuous* on I. One way to prove it is to modify the Rising Sun Lemma and show that it is still valid without the continuity assumption. But there is another method, due to Rubel [Rub63], that we shall use instead. It requires the following lemma, which is intuitively plausible and was part of the above proof with the additional continuity assumption.

Lemma 11.5.8. *If $f : [a, b] \to \mathbb{R}$ is monotone, then the set*

$$E_\infty := \{x \in (a, b) : |f'(x)| = \infty\}$$

is of measure zero.

Proof. We assume that f is *increasing*. In fact, replacing $f(x)$ by $f(x) + x$, if necessary, we may (and do) assume that f is *strictly* increasing and satisfies

$$x < y \implies f(y) - f(x) > y - x.$$

If $x \in E_\infty$, then $D^+ f(x) = D^- f(x) = \infty$ and hence for any $c > 0$ we have $D^+ f(x) > c$ and $D^- f(x) > c$. Thus we can find $s, t \in (a, b)$ with $s < x < t$ such that $f(t) - f(x) > c(t - x)$ and $f(x) - f(s) > c(x - s)$, and hence

$$x \in E_\infty \implies f(t) - f(s) > c(t - s) \quad \text{for some} \quad a < s < x < t < b.$$

Now introduce the set

(i) $E_c := \{x \in (a,b) : f(t_x) - f(s_x) > c(t_x - s_x)$ for some $a < s_x < x < t_x < b\}$.

Then $E_\infty \subset E_c$ for all $c > 0$. Also, E_c is easily seen to be *open* (why?) and hence $E_c = \bigcup_n (a_n, b_n)$ is a countable union of disjoint open intervals. Let the intervals $[a'_n, b'_n]$ be chosen such that

(ii) $[a'_n, b'_n] \subset (a_n, b_n)$ and $2(b'_n - a'_n) = b_n - a_n$ $\forall n$,

and note that we have

$$[a'_n, b'_n] \subset \bigcup_{x \in [a'_n, b'_n]} (s_x, t_x) \subset (a_n, b_n).$$

Since $[a'_n, b'_n]$ is *compact*, there is a finite subcover, say, $[a'_n, b'_n] \subset \bigcup_{k=1}^N (s_k, t_k)$. Proceeding as in the proof of Proposition 7.3.7, we may (after a relabel, if necessary) assume that $s_1 < s_2 < t_1 < s_3 < t_2 < s_4 < \cdots$, so that both $\{(s_{2j-1}, t_{2j-1}) : j \in \mathbb{N}\}$ and $\{(s_{2j}, t_{2j}) : j \in \mathbb{N}\}$ consist of pairwise disjoint intervals. It then follows that

(iii) $\displaystyle\sum_{k=1}^N [f(t_k) - f(s_k)] \le 2[f(b_n) - f(a_n)]$.

Now using (i) and (iii), we obtain the inequalities

$$b'_n - a'_n \le \sum_{k=1}^N (t_k - s_k) < \frac{1}{c} \sum_{k=1}^N [f(t_k) - f(s_k)] \le (2/c)[f(b_n) - f(a_n)].$$

Summing over all n and using (ii), we finally have

$$\sum_n (b_n - a_n) = 2 \sum_n (b'_n - a'_n) < \frac{4}{c} \sum_n [f(b_n) - f(a_n)] \le (4/c)[f(b) - f(a)].$$

Since $E_\infty \subset E_c$ for every $c > 0$, we conclude that $\lambda(E_\infty) = 0$. $\square$

We are now ready to prove Lebesgue's Differentiation Theorem in its full generality.

Theorem 11.5.9 (Lebesgue's Differentiation Theorem 2). *Let I be an interval. If $f : I \to \mathbb{R}$ is a monotone function, then $f'(x)$ exists (as a finite number) for almost all $x \in I$.*

Proof. As before, we assume that $I = [a, b]$ and that f is *increasing*. In fact, since $f(x)$ is differentiable for almost all $x \in (a, b)$ if and only if $f(x) + x$ is, and the latter is *strictly increasing*, we may (and do) assume that f is strictly increasing on $[a, b]$. But then, by Proposition 4.5.19, f has a *continuous (left) inverse;* i.e., there is a *continuous* function $F : [f(a), f(b)] \to \mathbb{R}$ with $F(f(x)) = x$ for all $x \in [a, b]$.

By Theorem 11.5.7, $F'(y) < \infty$ exists for almost all $y \in [f(a), f(b)]$. Next, note that, as in Theorem 6.3.8, we have

$$\frac{f(y) - f(x)}{y - x} = \frac{f(y) - f(x)}{F(f(y)) - F(f(x))} = \left[\frac{F(f(y)) - F(f(x))}{f(y) - f(x)}\right]^{-1}.$$

Since the (strictly) *increasing* function f has at most *countably many* discontinuity points, we have $\lim_{y \to x} f(y) = f(x)$ for almost all $x \in [a, b]$ and hence

$$\lim_{y \to x} \frac{f(y) - f(x)}{y - x} = \lim_{f(y) \to f(x)} \left[\frac{F(f(y)) - F(f(x))}{f(y) - f(x)}\right]^{-1} = \frac{1}{F'(f(x))}$$

for almost all $x \in [a, b]$. In other words, we have $f'(x) \leq \infty$ for almost all $x \in [a, b]$. Since, by Lemma 11.5.8, the set of all such x for which $f'(x) = \infty$ has measure zero, the proof of the theorem is complete. □

Recall that, by Jordan Decomposition Theorem (cf. Theorem 7.6.14), if $f : [a, b] \to \mathbb{R}$ has *bounded variation* (cf. Definition 7.6.1), then it is the difference of two increasing functions. The following corollary is therefore an immediate consequence of Lebesgue's theorem.

Corollary 11.5.10 (Bounded Variation $\Rightarrow$ Differentiable a.e.). *If $f : [a, b] \to \mathbb{R}$ has bounded variation, then it is differentiable almost everywhere.*

Corollary 11.5.11 (Lipschitz $\Rightarrow$ Differentiable a.e.). *If $f : [a, b] \to \mathbb{R}$ is Lipschitz, then it is differentiable almost everywhere.*

Proof. This follows from Corollary 11.5.10 because a Lipschitz function is of bounded variation by Proposition 7.6.5. But we can also give a direct proof. Indeed, suppose that f is Lipschitz with Lipschitz constant A, i.e.,

$$|f(y) - f(x)| \leq A|y - x| \quad \forall \, x, \, y \in [a, b].$$

Then the function $g(x) := f(x) + Ax$ is *increasing*. Thus $g'(x)$ exists for almost all x and hence so does $f'(x) = g'(x) - A$. □

Let us also recall the definition of *absolutely continuous* functions (cf. Problem 4.8.# 53), which play a crucial role in Lebesgue's *Second* Fundamental Theorem of Calculus.

Definition 11.5.12 (Absolutely Continuous Function). A function $F : [a, b] \to \mathbb{R}$ is said to be *absolutely continuous* if for each $\varepsilon > 0$ there is a $\delta > 0$ such that given any *finite* sequence $(I_k)_{k=1}^n$ of *pairwise disjoint* open intervals $I_k := (a_k, b_k) \subset [a, b]$, we have

$$\sum_{k=1}^n (b_k - a_k) < \delta \implies \sum_{k=1}^n |F(b_k) - F(a_k)| < \varepsilon.$$

Since δ is independent of n, "*finite* sequence" may be replaced by "*countable* sequence." Hence if $(a_n, b_n) \subset [a, b]$, $n \in \mathbb{N}$, are pairwise disjoint, then

$$\sum_{n=1}^{\infty}(b_n - a_n) < \delta \Longrightarrow \sum_{n=1}^{\infty}|F(b_n) - F(a_n)| \leq \varepsilon.$$

The set of all absolutely continuous functions on $[a, b]$ will be denoted $AC([a, b])$.

Exercise 11.5.13. Show that $AC([a, b])$ is an *algebra* by showing that if f, $g \in AC([a, b])$ and $c \in \mathbb{R}$, then $cf + g \in AC([a, b])$ and $fg \in AC([a, b])$. Also, show that $|f| \in AC([a, b])$ and that if $g(x) \neq 0$ for all $x \in [a, b]$, then $1/g \in AC([a, b])$.

Lemma 11.5.14 ($Lip \Rightarrow AC \Rightarrow BV$). *We have* $F \in Lip([a, b]) \Longrightarrow F \in AC([a, b]) \Longrightarrow F \in BV([a, b])$.

Proof. If $F \in Lip([a, b])$ with Lipschitz constant $A > 0$, then with notation as in the above definition and $\delta < \varepsilon/A$, we have

$$\sum_{k=1}^{n}|F(b_k) - F(a_k)| \leq A\sum_{k=1}^{n}(b_k - a_k) < A\delta < \varepsilon$$

and hence $F \in AC([a, b])$. But then, assuming $x_j - x_{j-1} < \delta$ for all j, we have $V_{x_{j-1}}^{x_j}(F) < \varepsilon$ for all j and hence

$$V_a^b(F) = \sum_{j=1}^{n}V_{x_{j-1}}^{x_j}(F) < n\varepsilon < \infty.$$

$\square$

Example 11.5.15.

(1). The function $F(x) := x^2 \sin(\pi/x)$ for $x \in (0, 1]$ and $F(0) := 0$ is absolutely continuous. Indeed, it is easily checked F is differentiable on $[0, 1]$ and that F' is *bounded*. Therefore (cf. Corollary 6.4.20), F is Lipschitz and hence absolutely continuous.

(2). The function $G(x) := x^2 \sin(\pi/x^2)$ for $x \in (0, 1]$ and $G(0) := 0$ is *uniformly* continuous but is *not* absolutely continuous. The uniform continuity follows from the fact that G is continuous (even differentiable) on the compact set $[0, 1]$ (cf. Theorem 4.6.4). On the other hand, for each $n \in \mathbb{N}$, let $a_n = (2n + 1/2)^{-1/2}$ and $b_n = (2n)^{-1/2}$. Given $\delta > 0$, pick integers $N > M > 0$ such that $1/\sqrt{2M} < \delta$ and $\sum_{n=M}^{N} a_n^2 > 1$. Then the intervals (a_n, b_n), $M \leq n \leq N$, are pairwise disjoint and $\sum_{n=M}^{N}(b_n - a_n) < \delta$, but

$$\sum_{n=M}^{N} |G(b_n) - G(a_n)| = \sum_{n=M}^{N} a_n^2 > 1.$$

Thus G is *not* absolutely continuous on $[0, 1]$. In fact, as we have seen (cf. Example 7.6.3), F is *not* of bounded variation and hence cannot be absolutely continuous.

Corollary 11.5.16 (Absolutely Continuous $\Rightarrow$ Differentiable a.e.). *If $f : [a, b] \to \mathbb{R}$ is absolutely continuous, then it is differentiable almost everywhere.*

Proof. In view of the above lemma, this follows from Corollary 11.5.10. $\quad\square$

Next, we prove Lebesgue's *first* Fundamental Theorem of Calculus (1st FTC):

Theorem 11.5.17 (Lebesgue's 1st FTC). *Let $f : [a, b] \to \mathbb{R}$ be increasing. Then its derivative f' (defined almost everywhere) is measurable and we have*

$$\int_a^b f'(x) \, d\lambda(x) \le f(b) - f(a). \tag{$*$}$$

In particular, f' is integrable on $[a, b]$.

Proof. Extend f to $\mathbb{R}$ by setting $f(x) := f(a)$ for all $x < a$ and $f(x) := f(b)$ for all $x > b$. Let $g_n(x)$ be the slope of the secant line joining $(x, f(x))$ and $(x + 1/n, f(x + 1/n))$; i.e.

$$g_n(x) := \frac{f(x + 1/n) - f(x)}{1/n} = n[f(x + 1/n) - f(x)].$$

Now f is measurable (in fact, continuous almost everywhere) and hence so is g_n. Also, by Theorem 11.5.9, $\lim_{n \to \infty} g_n(x) = f'(x)$ for almost all x and for each such x we have $0 \le f'(x) < \infty$. Therefore, f' is measurable and Fatou's lemma gives

$$\int_a^b f'(x) \, d\lambda(x) = \int_a^b \liminf_{n \to \infty} g_n(x) \, d\lambda(x) \le \liminf_{n \to \infty} \int_a^b g_n(x) \, d\lambda(x).$$

So $(*)$ follows if we can show that $\int_a^b g_n(x) \, d\lambda(x) \le f(b) - f(a)$. However,

$$\int_a^b g_n(x) \, d\lambda(x) = n \int_b^{b+1/n} f(x) \, d\lambda(x) - n \int_a^{a+1/n} f(x) \, d\lambda(x) \le f(b) - f(a),$$

because $f(x) = b$ for all $x \ge b$ gives $n \int_b^{b+1/n} f(x) \, d\lambda(x) = f(b)$ while $f(x) \ge f(a)$ for all x implies $n \int_a^{a+1/n} f(x) \, d\lambda(x) \ge f(a)$. $\quad\square$

Remark 11.5.18.

(1) Following the proof with $[a, b]$ replaced by $[\alpha, \beta]$, where $a < \alpha < \beta < b$, and letting $\alpha \to a+$ and $\beta \to b-$, we even have the stronger inequality

$$\int_a^b f'(x) \, d\lambda(x) \le f(b-) - f(a+).$$

(2) The inequality in (∗) may be *strict*. For example, if $[a, b] := [0, 1]$ and $f := \kappa$ is Cantor's ternary function (cf. Example 4.3.13), then f is a continuous, increasing function that is constant on all the middle thirds removed from $[0, 1]$ to obtain the Cantor set C. In particular, $f' = 0$ on $[0, 1] \setminus C$. Since $\lambda(C) = 0$, we have $f' = 0$ almost everywhere. On the other hand $f(0) = 0$ and $f(1) = 1$ so that $0 = \int_0^1 f'(x) \, d\lambda(x) < f(1) - f(0) = 1$.

Let us now look at the Second Fundamental Theorem. Here, we will be dealing with functions of the form $F(x) := \int_a^x f(t) \, d\lambda(t)$, where $f : [a, b] \to \mathbb{R}$ is at least measurable. We begin with the following.

Proposition 11.5.19. *If f is integrable on $[a, b]$, then the function*

$$F(x) := \int_a^x f(t) \, d\lambda(t), \qquad x \in [a, b]$$

is absolutely continuous (hence uniformly continuous and of bounded variation) on $[a, b]$. Also, $F'(x)$ exists for almost all $x \in [a, b]$ and is integrable.

Proof. Since f is integrable on $[a, b]$, for each $\varepsilon > 0$, there exists (by Corollary 10.4.7) $\delta > 0$ such that $A \subset [a, b]$ and $\lambda(A) < \delta$ imply $\int_A |f| \, d\lambda < \varepsilon$. Now let $I_k := (a_k, b_k) \subset [a, b]$, $1 \le k \le n$, be pairwise disjoint intervals and set $A := \bigcup_{k=1}^n I_k$. If we have $\lambda(A) = \sum_{k=1}^n (b_k - a_k) < \delta$, then

$$\sum_{k=1}^n |F(b_k) - F(a_k)| = \sum_{j=1}^n \left| \int_{a_k}^{b_k} f \, d\lambda \right| \le \sum_{k=1}^n \int_{a_k}^{b_k} |f| \, d\lambda = \int_A |f| \, d\lambda < \varepsilon.$$

Next, F is differentiable almost everywhere by Corollary 11.5.16. Finally, F is of bounded variation and hence $F = F_1 - F_2$ with increasing functions F_1 and F_2 and we have $|F'(x)| = |F_1'(x) - F_2'(x)| \le F_1'(x) + F_2'(x)$. By Theorem 11.5.17, we therefore have

$$\int_a^b |F'(x)| \, d\lambda(x) \le F_1(b) + F_2(b) - F_1(a) - F_2(a). \qquad \square$$

Remark 11.5.20.

(1) That $F(x)$ is of *bounded variation* may also be seen by using Theorem 7.6.14. Indeed, $F(x) = \int_a^x f^+ \, d\lambda - \int_a^x f^- \, d\lambda$ and both functions on the right side are *increasing*.

(2) Of course, the important question now is whether or not $F'(x) = f(x)$. This is indeed the case for almost all x, as we shall see below.

Lemma 11.5.21. *If $f : [a, b] \to \mathbb{R}$ is integrable and if*

$$\int_a^x f(t)\, d\lambda(t) = 0 \quad \forall\, x \in [a, b],$$

then $f(t) = 0$ for almost all $t \in [a, b]$.

Proof. First note that for any $a \le c \le d \le b$, we have $\int_c^d f\, d\lambda = \int_a^d f\, d\lambda - \int_a^c f\, d\lambda = 0$. Now let $E^+ := \{t \in [a, b] : f(t) > 0\}$ and note that $E^+ = \bigcup_{n=1}^\infty E_n$, where $E_n := \{t \in [a, b] : f(t) > 1/n\}$ and we have $E_n \subset E_{n+1}$ for all $n \in \mathbb{N}$. Thus, if $\lambda(E^+) > 0$, then $\lambda(E_n) > 0$ for n large enough. Using the *regularity* of λ (Theorem 10.5.2), we can pick a compact set $K \subset E_n$ with $\lambda(K) > 0$. Now note that $(a, b) \setminus K = \bigcup_n (a_n, b_n)$, where the intervals $(a_n, b_n) \subset (a, b)$ are pairwise disjoint. By Corollary 10.4.19, we are then led to the contradiction

$$0 = \int_a^b f\, d\lambda = \sum_n \int_{a_n}^{b_n} f\, d\lambda + \int_K f\, d\lambda = \int_K f\, d\lambda > (1/n)\lambda(K),$$

where the last inequality follows from the L^1-version of Chebyshev's inequality (Proposition 11.3.16). Therefore, $\lambda(E^+) = 0$. Similarly, we can show that $\lambda(E^-) = 0$, where $E^- := \{t \in [a, b] : f(t) < 0\}$. $\qquad\square$

We can now prove Lebesgue's 2nd Fundamental Theorem. We first consider the case of *bounded* functions.

Lemma 11.5.22 (Lebesgue's 2nd FTC; Bounded Case). *Let $f : [a, b] \to \mathbb{R}$ be a bounded, measurable function and define*

$$F(x) := F(a) + \int_a^x f(t)\, d\lambda(t).$$

Then we have $F'(x) = f(x)$ for almost all $x \in [a, b]$.

Proof. By Proposition 11.5.19, F is absolutely continuous, $F'(x)$ exists for almost all $x \in [a, b]$ and is integrable. Suppose that $|f| \le M$ and define

$$f_n(x) := n[F(x + 1/n) - F(x)] = n \int_x^{x+1/n} f(t)\, d\lambda(t).$$

Then we have $|f_n| \le M$ for all n and $\lim_{n\to\infty} f_n(x) = F'(x)$ for almost all $x \in [a, b]$. Thus, the Bounded Convergence Theorem (Theorem 10.4.21) and the (uniform) continuity of F (which is actually *Riemann integrable*) imply

$$\int_a^\xi F'(x)\, d\lambda(x) = \lim \int_a^\xi f_n(x)\, d\lambda(x) = \lim \int_a^\xi n[F(x+1/n) - F(x)]\, d\lambda(x)$$

$$= \lim \left[n \int_\xi^{\xi+1/n} F(x)\, d\lambda(x) - n \int_a^{a+1/n} F(x)\, d\lambda(x) \right]$$

$$= F(\xi) - F(a) = \int_a^\xi f(x)\, d\lambda(x).$$

Therefore, we have

$$\int_a^\xi [F'(x) - f(x)]\, d\lambda(x) = 0 \qquad \forall\, \xi \in [a,b].$$

But then, Lemma 11.5.22 gives $F'(x) = f(x)$ for almost all $x \in [a,b]$.   □

Now we use the above special case to prove the general one:

Theorem 11.5.23 (Lebesgue's 2nd FTC; General Case). *Let $f : [a,b] \to \mathbb{R}$ be an integrable function. Then the function*

$$F(x) := F(a) + \int_a^x f(t)\, d\lambda(t)$$

is absolutely continuous and we have $F'(x) = f(x)$ for almost all $x \in [a,b]$.

Proof. The theorem is proved if we can prove it for f^+ and f^-. So we assume that $f \geq 0$. Define f_n by $f_n(x) := f(x)$ if $f(x) \leq n$ and $f_n(x) := n$ otherwise. If $F_n(x) := \int_a^x f_n(t)\, d\lambda(t)$, then $f - f_n \geq 0$ implies that $G_n(x) := F(x) - F_n(x) = \int_a^x [f(t) - f_n(t)]\, d\lambda(t)$ is *increasing* and hence $G_n'(x) \geq 0$ exists for almost all $x \in [a,b]$. Also, applying Lemma 11.5.22 to F_n, we have $F_n'(x) = f_n(x)$ for almost all $x \in [a,b]$ and hence

$$F'(x) = G_n'(x) + F_n'(x) \geq f_n(x) \quad \text{for almost all } x \in [a,b].$$

Since $n \in \mathbb{N}$ is arbitrary and $\lim(f_n) = f$, we obtain the inequality

$$F'(x) \geq f(x) \quad \text{for almost all } x \in [a,b]. \tag{$*$}$$

Integrating over $[a,b]$ and using Theorem 11.5.17, we have

$$F(b) - F(a) = \int_a^b f(x)\, d\lambda(x) \leq \int_a^b F'(x)\, d\lambda(x) \leq F(b) - F(a)$$

and hence

$$\int_a^b F'(x)\, d\lambda(x) = F(b) - F(a) = \int_a^b f(x)\, d\lambda(x),$$

which gives

$$\int_a^b [F'(x) - f(x)] \, d\lambda(x) = 0.$$

Since $F' - f \geq 0$ almost everywhere by $(*)$, we must have $F'(x) = f(x)$ for almost all $x \in [a, b]$. □

Remark 11.5.24. The above theorem can be strengthened to an "if and only if" statement: *A function is an indefinite integral if and only if it is absolutely continuous.* To prove it, we need a few lemmas of independent interest (cf. [Gor94]). First, it is a simple exercise to show that *Lipschitz* functions map sets of measure zero onto sets of measure zero. What is more interesting is that the same is also true for *absolutely continuous* functions.

Exercise 11.5.25. Show that if $f : [a, b] \to \mathbb{R}$ is Lipschitz, then $\lambda(f(Z)) = 0$ for any $Z \subset [a, b]$ with $\lambda(Z) = 0$.

Lemma 11.5.26. *An absolutely continuous function $f : [a, b] \to \mathbb{R}$ maps sets of measure zero onto sets of measure zero.*

Proof. Let $f : [a, b] \to \mathbb{R}$ be absolutely continuous and pick $Z \subset (a, b)$ with $\lambda(Z) = 0$. Then, given $\varepsilon > 0$, we can find $\delta > 0$ and a sequence of pairwise disjoint intervals $(a_k, b_k) \subset (a, b)$ such that

$$Z \subset \bigcup_n (a_k, b_k), \quad \sum_k (b_k - a_k) < \delta \quad \text{and} \quad \sum_k |F(b_k) - F(a_k)| < \varepsilon.$$

Since f is (uniformly) continuous, we have $m_k := \min\{f(x) : x \in [a_k, b_k]\} = f(\alpha_k)$ and $M_k := \max\{f(x) : x \in [a_k, b_k]\} = f(\beta_k)$ for some $\alpha_k, \beta_k \in [a_k, b_k]$. But then the absolute continuity gives

$$\sum_k |\beta_k - \alpha_k| \leq \sum_K (b_k - a_k) < \delta \implies \sum_k (M_k - m_k) < \varepsilon.$$

Since $f(Z) \subset \bigcup_k f[(a_k, b_k)] \subset \bigcup_k (m_k, M_k)$, we have

$$\lambda(f(Z)) \leq \sum_k (M_k - m_k) < \varepsilon.$$

□

Corollary 11.5.27. *An absolutely continuous function $f : [a, b] \to \mathbb{R}$ maps measurable sets onto measurable sets.*

Proof. If $E \subset [a, b]$ is measurable, then (by Corollary 10.5.3) we can pick an *increasing* sequence (F_n) of *closed* sets and a set Z of measure zero such that $E = (\bigcup_n F_n) \cup Z$. Since $f(E) = f(\bigcup_n F_n) \cup f(Z) = (\bigcup_n f(F_n)) \cup f(Z)$

and $\lambda\big(f(Z)\big) = 0$ by the above lemma, we need only show that each $f(F_n)$ is measurable. But f is continuous and F_n is *compact*, so $f(F_n)$ is also compact, hence measurable.   □

Lemma 11.5.28. *Let* $f : [a,b] \to \mathbb{R}$ *be continuous and set* $E := \{x \in [a,b] : D^+ f(x) \le 0\}$. *If* $f(E)$ *contains no intervals, then* f *is increasing on* $[a,b]$.

Proof. Suppose that $f(d) < f(c)$ for some $c < d$ in $[a,b]$. Since $f(E)$ contains no intervals, there is a point $y_0 \in \big(f(d), f(c)\big) \setminus E$. If $x_0 := \sup\{x \in [c,d] : f(x) \ge y_0\}$, then the *continuity* of f implies that $x_0 \in (c,d)$ and $y_0 = f(x_0)$. (Why?) Also, $x_0 \notin E$ gives $D^+ f(x_0) > 0$. But $f(x) < y_0$ for all $x \in (x_0, d]$ implies that $D^+ f(x_0) \le 0$. This contradiction completes the proof.   □

Lemma 11.5.29. *If* $f : [a,b] \to \mathbb{R}$ *is absolutely continuous and if* $D^+ f(x) \ge 0$ *for almost all* $x \in [a,b]$, *then* f *is increasing.*

Proof. Let $f_\varepsilon(x) := f(x) + \varepsilon x$, where $\varepsilon > 0$ is arbitrary. Then f_ε is *absolutely continuous* and the set $E := \{x \in [a,b] : D^+ f_\varepsilon(x) \le 0\}$ has measure zero. By Lemma 11.5.26, $f_\varepsilon(E)$ has measure zero and hence contains no intervals. Therefore f_ε is *increasing* by Lemma 11.5.28. Since $\varepsilon > 0$ was arbitrary, the function f is also increasing.   □

Corollary 11.5.30. *If* $f : [a,b] \to \mathbb{R}$ *is absolutely continuous and if* $f'(x) = 0$ *for almost all* $x \in [a,b]$, *then* f *is constant.*

Proof. Indeed, by Lemma 11.5.29, the functions f and $-f$ are both increasing. Thus f is both increasing and decreasing, hence constant.   □

We are now ready to prove the final version of Lebesgue's 2nd FTC.

Theorem 11.5.31 (Lebesgue's 2nd FTC; Final Version). *A function* $F : [a,b] \to \mathbb{R}$ *is an indefinite integral, i.e., has the form*

$$F(x) := F(a) + \int_a^x f(t) \, d\lambda(t) \qquad\qquad (\dagger)$$

for an integrable function f *on* $[a,b]$, *if and only if it is absolutely continuous.*

Proof. In view of Theorem 11.5.23, we need only show that if F is absolutely continuous, which we now assume, then it is an indefinite integral. Now, as we know (cf. Corollary 11.5.16 and Theorem 11.5.17), $F'(x)$ is defined for almost all x and is *integrable*. So if we define

$$G(x) := \int_a^x F'(t) \, d\lambda(t) \quad x \in [a,b],$$

then (by Theorem 11.5.23) G is absolutely continuous and $G'(x) = F'(x)$ for almost all $x \in [a,b]$. But then $F - G$ is absolutely continuous and we have $(F - G)'(x) = 0$ for almost all $x \in [a,b]$. Therefore, by Corollary 11.5.30, $F - G$ is *constant* and ($\dagger$) follows.   □

11.6 Problems

1. Evaluate each improper integral.

(a) $\int_0^{\pi/2} \sqrt{\sin x \tan x}\, dx$;

(b) $\int_{-1}^{1} \dfrac{dx}{\sqrt[3]{x}}$;

(c) $\int_0^{\pi/2} x \cot x\, dx$;

(d) $\int_0^1 \dfrac{\log x}{\sqrt{x}}\, dx$.

2.

(a) Using integration by parts, show that $\int_0^1 x^k \log x\, dx = -1/(k+1)^2$ for all $k \in \mathbb{N}_0 := \mathbb{N}\cup\{0\}$, where we define $x^0 := 1$.

(b) Using part (a), show that

$$\int_0^1 \frac{\log x}{1-x}\, dx = -\frac{\pi^2}{6}.$$

Hint: Note that $(\log x)/(1-x) = \sum_{k=0}^{\infty} x^k \log x$ for $x \in (0,1)$ and, considering partial sums, justify term-by-term integration of the series. Note that $x^k \log x \le 0$ for all $x \in (0,1]$.

3. Show that the integrals $\int_0^\infty \sin x\, dx$ and $\int_0^\infty \cos x\, dx$ are divergent.

4. Show that, for any $\alpha > 0$ and $\beta \in \mathbb{R}$, we have

(a) $\int_0^\infty e^{-\alpha x} \cos(\beta x)\, dx = \dfrac{\alpha}{\alpha^2 + \beta^2}$;

(b) $\int_0^\infty e^{-\alpha x} \sin(\beta x)\, dx = \dfrac{\beta}{\alpha^2 + \beta^2}$.

5. Evaluate each improper integral if it exists. If it doesn't, explain why.

(a) $\int_0^\infty \dfrac{dx}{x\sqrt{1+x^2}}$;

(b) $\int_{-1}^1 \dfrac{dx}{\sqrt{1-x^2}}$;

(c) $\int_0^\infty \dfrac{dx}{(x+\pi)\sqrt{x}}$;

(d) $\int_1^\infty \dfrac{\sqrt{x}}{(1+x)^2}\, dx$.

6. Show that the integral

$$\int_0^\infty \frac{x^{\alpha-1}}{1+x}\, dx$$

is *convergent* if and only if $0 < \alpha < 1$.

7 (Dirichlet's Test). Let f be *continuous* and ϕ *decreasing* on $[a, \infty)$. Show that, if $F(x) :=$ $\int_a^x f(t)\, dt$ is *bounded* on $[a, \infty)$ and $\lim_{x\to\infty} \phi(x) = 0$, then $\int_a^\infty f(x)\phi(x)\, dx$ is *convergent*.
Hint: Note that $|F(x)| \le M$ for some $M > 0$ and all $x \ge a$. Given $\varepsilon > 0$, pick $A \ge a$ such

that $\phi(A) < \varepsilon/(2M)$. Now let $B > A$. Use the fact that $\phi \geq 0$ (why?) and the *Second MVT for Integrals* (Theorem 7.4.20 (1)) to find a $\xi \in [A, B]$ with $\int_A^B f(x)\phi(x)\,dx = \phi(A)\int_A^\xi f(x)\,dx$. Deduce that $|\int_A^B f(x)\phi(x)\,dx| < \varepsilon$ and use *Cauchy's Criterion*.

8. Given any $p > 0$, show that the integrals $\int_1^\infty (\sin x/x^p)\,dx$ and $\int_1^\infty (\cos x/x^p)\,dx$ are both convergent. Show that, for $p > 1$, both integrals are *absolutely* convergent. That this is *false* if $p \in (0, 1]$ can be proved by an argument similar to the one used in Exercise 11.1.19.

9 (Fresnel Integrals). Show that the integrals $\int_0^\infty \sin(x^2)\,dx$ and $\int_0^\infty \cos(x^2)\,dx$ are both convergent. *Hint:* Make the substitution $t = x^2$ in $\int_1^\infty \sin(x^2)\,dx$.

10.

(a) Show that

$$\int_0^\pi \frac{\sin(n + \frac{1}{2})x}{\sin(x/2)}\,dx = \pi.$$

Hint: Use the identity $1 + 2\sum_{k=1}^n \cos(kx) = \sin(n + \frac{1}{2})x / \sin(x/2)$.

(b) Show that

$$\lim_{\alpha \to \infty} \int_0^\pi \left[\frac{2}{x} - \frac{1}{\sin(x/2)}\right] \sin\left(\alpha + \frac{1}{2}\right)x\,dx = 0$$

and hence, by (a),

$$\lim_{\alpha \to \infty} \int_0^\pi \frac{2\sin(\alpha + \frac{1}{2})x}{x}\,dx = \pi.$$

Hint: Note that the expression in brackets is *bounded* and use the *Riemann–Lebesgue lemma*.

(c) Using the substitution $t = (\alpha + \frac{1}{2})x$ and (b), show that

$$\int_0^\infty \frac{\sin x}{x}\,dx = \frac{\pi}{2}.$$

11. For each $\alpha \in \mathbb{R}$, prove the following:

(a) $\displaystyle\int_0^\infty \frac{\sin^2(\alpha x)}{x^2}\,dx = \frac{\pi}{2}|\alpha|$;

(b) $\displaystyle\int_0^\infty \frac{1 - \cos(\alpha x)}{x^2}\,dx = \frac{\pi}{2}|\alpha|$;

(c) $\displaystyle\int_0^\infty \frac{\sin^4 x}{x^2}\,dx = \frac{\pi}{4}$;

(d) $\displaystyle\int_0^\infty \frac{\sin^4 x}{x^4}\,dx = \frac{\pi}{3}$.

Hint: For (a), integrate by parts and use the preceding problem. For (c), note that $\cos^2 x + \sin^2 x = 1$, and for (d), use integration by parts and (c).

12.

(a) Using the substitution $u = t^x$ in the definition of the *Gamma function*, show that

$$\Gamma(x) = \frac{1}{x}\int_0^\infty e^{-u^{1/x}}\,du.$$

(b) Using (a) and the fact that $\int_0^\infty e^{-x^2}\,dx = \sqrt{\pi}/2$ (Problem 7.7.#45), show that $\Gamma(1/2) = \sqrt{\pi}$. Deduce that, in general,

$$\Gamma\left(n + \frac{1}{2}\right) = \left(n - \frac{1}{2}\right)\left(n - \frac{3}{2}\right)\cdots\frac{1}{2}\sqrt{\pi}.$$

13. Evaluate the following integrals:

(a) $\displaystyle\int_0^\infty x^2 e^{-x^2}\,dx$;

(b) $\displaystyle\int_0^\infty (\sqrt{x})e^{-2x}\,dx$.

14.

(a) Show that, for each $\alpha > 0$, we have

$$I_{n,\alpha} := \int_0^1 (1 - t)^n t^{\alpha - 1}\,dt = \frac{n!}{\alpha(\alpha + 1)\cdots(\alpha + n)} \qquad (\forall n \in \mathbb{N}_0).$$

   Hint: Show that $I_{n,\alpha} = \frac{n}{\alpha}I_{n-1,\alpha+1}$.

(b) Substituting $x = nt$, deduce that

$$\int_0^n \left(1 - \frac{x}{n}\right)^n x^{\alpha - 1}\,dx = \frac{n!\,n^\alpha}{\alpha(\alpha + 1)\cdots(\alpha + n)}.$$

15. Consider the function

$$F(t) := \int_0^\infty e^{-tx}\frac{\sin x}{x}\,dx \qquad (\forall t > 0),$$

where $\sin x/x := 1$ if $x = 0$.

(a) Differentiating under the integral sign, show that $F'(t) = -1/(1 + t^2)$ and deduce that $F(t) = C - \arctan t$, for all $t > 0$ and some constant $C \in \mathbb{R}$.

(b) Using the sequence $(F(n))_{n\in\mathbb{N}}$, show that $C = \pi/2$ and deduce that

$$\int_0^\infty \frac{\sin x}{x}\,dx = \frac{\pi}{2}.$$

16 (Fourier Transform).   Recall (Theorem 11.3.11) that, for any $f \in \mathcal{L}^1(\mathbb{R})$, its *Fourier transform*

$$\hat{f}(\xi) := \int_{\mathbb{R}} f(x)e^{-ix\xi}\,dx \qquad (\forall \xi \in \mathbb{R})$$

is *continuous* on $\mathbb{R}$ and that $\lim_{|\xi|\to\infty}\hat{f}(\xi) = 0$. Show that, if $\int_{\mathbb{R}} |xf(x)|\,dx < \infty$, then $\hat{f}$ is *continuously differentiable* on $\mathbb{R}$ and we have

$$(\hat{f})'(\xi) = \int_{\mathbb{R}} (-ix)f(x)e^{-ix\xi}\,dx \qquad (\forall \xi \in \mathbb{R}).$$

17. Show that

$$\Gamma(\alpha) = \lim_{n \to \infty} \frac{n! n^\alpha}{\alpha(\alpha+1)\cdots(\alpha+n)} \qquad (\forall \alpha > 0).$$

Hint: Note that, $(1 - x/n)^n \leq e^{-x}$ for all $x \in [0, n]$ and $\lim_{n \to \infty} (1 - x/n)^n = e^{-x}$. Now use Problem 14 and Lebesgue's Dominated Convergence Theorem.

18. Let $S_1 := \{ f \in \mathcal{L}^1([0,1]) : \|f\|_1 = 1 \}$. Find a sequence $(f_n) \in S_1^{\mathbb{N}}$ that has no convergent subsequence (with respect to the L^1-norm). This shows, by F. Riesz's Lemma (Theorem 9.2.25), that $L^1([0,1])$ is an *infinite dimensional* Banach space. *Hint:* Consider a sequence of functions with *disjoint supports*.

19. Let $E \in \mathcal{M}_{\lambda < \infty}(\mathbb{R})$. Show that, if $f_n \in \mathcal{L}^0(E, \mathbb{R})$ for all $n \in \mathbb{N}$ and if (f_n) converges *uniformly* to f, then $f_n \to f$ in $L^p(E, \mathbb{R})$, $p \in [1, \infty)$.

20. Let $E \in \mathcal{M}_{\lambda < \infty}(\mathbb{R})$ and $p \in [1, \infty)$. Suppose that $f_n \in \mathcal{L}^p(E, \mathbb{F})$ for all $n \in \mathbb{N}$ and $\lim(f_n) = f \in \mathcal{L}^p(E, \mathbb{R})$ *almost everywhere*. Show that $f_n \to f$ in $L^p(E, \mathbb{R})$ if and only if $\|f_n\|_p \to \|f\|_p$.

21. Show that in the previous problem $f_n \to f$ *almost everywhere* can be replaced by $f_n \to f$ *in measure*.

22. Let $E \in \mathcal{M}_\lambda(\mathbb{R})$ and $\lambda(E) < \infty$. If $0 < p < q \leq \infty$, show that $L^q \subset L^p$ and that we have

$$\|f\|_p \leq \|f\|_q \lambda(E)^{(1/p)-(1/q)}.$$

23 (L^p Interpolation). Let $0 < p < q < r \leq \infty$ and $E \in \mathcal{M}_\lambda(\mathbb{R})$. Then $L^p(E) \cap L^r(E) \subset L^q(E)$ and, with $\alpha := (1/q - 1/r)/(1/p - 1/r)$, we have

$$\|f\|_q \leq \|f\|_p^\alpha \|f\|_r^{1-\alpha}.$$

24. Let $E \in \mathcal{M}_{\lambda < \infty}(\mathbb{R})$ and $f \in L^p(E) \cap L^\infty(E)$ for some $p \in [1, \infty)$. Show that

$$\lim_{q \to \infty} \|f\|_q = \|f\|_\infty.$$

25 (Weak Convergence). Let $E \in \mathcal{M}_\lambda(\mathbb{R})$ and let (f_n) be a sequence in $L^p(E, \mathbb{R})$, where $p \in [1, \infty)$. We say that f_n converges to f *weakly* if $\int_E f_n g \to \int_E fg$ for all $g \in L^q(E, \mathbb{R})$, where $1/p + 1/q = 1$. Show that if $f_n \to f$ in $L^p(E, \mathbb{R})$ then $f_n \to f$ weakly.

26. Let $E \in \mathcal{M}_\lambda(\mathbb{R})$ and let (f_n) be a sequence in $\mathcal{L}^2(E, \mathbb{R})$. Show that if $f_n \to f \in \mathcal{L}^2(E, \mathbb{R})$ *weakly* and if $\lim_{n \to \infty} \|f_n\|_2 = \|f\|_2$, then $f_n \to f$ in $L^2(E, \mathbb{R})$.

27. Let $E \in \mathcal{M}_\lambda(\mathbb{R})$ and $p \in (1, \infty)$. Let (f_n) be a sequence in $\mathcal{L}^p(E, \mathbb{R})$ that converges *almost everywhere* to a function $f \in \mathcal{L}^p(E, \mathbb{R})$. If $\|f_n\|_p \leq M$ for all n and some constant M, show that $(f_n) \to f$ weakly. Is the assertion true for $p = 1$? *Hint:* Use a combination of *absolute continuity* and *Egorov's theorem*.

28 (A Converse to Hölder's Inequality). Let $E \in \mathcal{M}_{\lambda < \infty}(\mathbb{R})$, $g \in \mathcal{L}^0(E, \mathbb{R})$, and $p \in [1, \infty)$. Suppose that $fg \in \mathcal{L}^1(E, \mathbb{R})$ for all $f \in \mathcal{L}^p(E, \mathbb{R})$ and that for some constant M we have

$$\left| \int_E fg \, d\lambda \right| \leq M \|f\|_p \qquad \forall f \in \mathcal{L}^p(E, \mathbb{R}).$$

Then $g \in \mathcal{L}^q(E, \mathbb{R})$, where $1/p + 1/q = 1$, and $\|g\|_q \leq M$. *Hint:* First look at the case $p \in (1, \infty)$ and consider the sequence of functions $f_n := |g_n|^{q/p} \operatorname{sgn}(g_n)$, where $g_n(x) = g(x)\chi_{E_n}$

with $E_n := \{x \in E : |g(x)| \le n\}$. Here, for any $t \in \mathbb{R}$, its *signum* is defined to be $\operatorname{sgn}(t) := |t|/t$ if $t \ne 0$ and $\operatorname{sgn}(0) := 0$.

29 (Multiplication Operator). Consider the Hilbert space $H := L^2(E, \mathbb{C})$, where $E \in \mathcal{M}_\lambda(\mathbb{R})$ and $\lambda(E) > 0$. Given a function $\phi \in \mathcal{L}^0(E, \mathbb{C})$, we define the corresponding *multiplication* "operator" M_ϕ:

$$M_\phi f := \phi f \qquad \forall \, f \in \mathcal{L}^2(E, \mathbb{C}).$$

Show that $M_\phi : H \to H$, i.e., $\phi f \in \mathcal{L}^2(E, \mathbb{C})$ for every $f \in \mathcal{L}^2(E, \mathbb{C})$, if and only if $\phi \in \mathcal{L}^\infty(E, \mathbb{C})$ and that M_ϕ is then a *bounded* operator with $\|M_\phi\| = \|\phi\|_\infty$. *Hint:* Use the Closed Graph Theorem.

30. Let $E \in \mathcal{M}_\lambda(\mathbb{R})$, $\lambda(E) > 0$, and let $H := L^2(E, \mathbb{C})$. For each $\phi \in \mathcal{L}^\infty(E, \mathbb{C})$, let $M_\phi \in \mathcal{B}(H)$ be the corresponding multiplication operator as in the previous problem.

(a) Show that $M_\phi^* = M_{\bar{\phi}}$ and that M_ϕ is self-adjoint (i.e., $M_\phi^* = M_\phi$) if and only if ϕ is *almost real*; i.e., $\bar{\phi} = \phi$ almost everywhere.

(b) Show that $M_\phi M_\psi = M_\psi M_\phi$ for all $\phi, \psi \in \mathcal{L}^\infty(E, \mathbb{C})$. In particular, $M_\phi^* M_\phi = M_\phi M_\phi^* = M_{|\phi|^2}$ and hence M_ϕ is a *normal operator*, i.e., commutes with its adjoint.

(c) Show that M_ϕ is *unitary*, i.e., $M_\phi^* M_\phi = 1_H$, if and only if $|\phi| = 1$ almost everywhere.

(d) Show that M_ϕ is a *projection*, i.e., $M_\phi^2 = M_\phi$, if and only if $\phi = \chi_F$ for a measurable set $F \subset E$.

(e) Show that if ϕ is real-valued, then the *spectrum* $\sigma(M_\phi)$ is the *essential range* of ϕ defined as follows:

$$\sigma(M_\phi) = \{x \in \mathbb{R} : \lambda(\phi^{-1}(x - \varepsilon, x + \varepsilon)) > 0 \ \forall \, \varepsilon > 0\}.$$

31. Let $E \in \mathcal{M}_\lambda(\mathbb{R})$, $f_n \in \mathcal{L}^p(E, \mathbb{R})$, and $g_n \in L^q([E, \mathbb{R})$ for all $n \in \mathbb{N}$, where $p, q \in [1, \infty)$ and $1/p + 1/q = 1$. Show that if $f_n \to f$ in $L^p(E, \mathbb{R})$ and $g_n \to g$ in $L^q(E, \mathbb{R})$, then $\int_E f_n g_n \to \int_E fg$.

32. Let $f \in L^2([a, b])$ and let us denote its trivial extension (i.e., the one defined to be 0 outside $[a, b]$) by f as well. Show that

$$\lim_{h \to 0} \int_a^b [f(x + h) - f(x)]^2 dx = 0.$$

Hint: First show this for $f \in C([a, b])$.

33. Let f be as in the preceding problem and define $F(x) := \int_a^b f(x + t) f(t) dt$ for all $x \in \mathbb{R}$. Show that F is *continuous* at $x = 0$. *Hint:* Apply Cauchy–Schwarz to $F(x) - F(0)$ and use the preceding problem.

34 (Calderon's Proof of Steinhaus's Theorem). Let E be a *measurable* subset of $[a, b]$ with $\lambda(E) > 0$ and consider the function

$$F(x) := \int_a^b \chi_E(t) \chi_E(x + t) dt \qquad (\forall x \in \mathbb{R}).$$

(a) Show that F is *continuous* at $x = 0$ and deduce that there is a $\delta > 0$ such that $F(x) > 0$ for all $x \in (-\delta, \delta)$.

(b) Show that, given any $x \in (-\delta, \delta)$, there is a $t_0 = t_0(x)$ such that $\chi_E(t_0) \chi_E(x + t_0) = 1$. Conclude that $t_0 \in E$ and $x + t_0 \in E$ and hence that $(-\delta, \delta) \subset E - E$.

35 (M. Fréchet). Given any $f \in \mathcal{L}^0([a, b])$, define

$$\| f \| := \int_a^b \frac{|f|}{1 + |f|}.$$

(a) Show that $\| f \| = 0$ if and only if $f = 0$ a.e.
(b) Show that, if $f_n,\ f \in \mathcal{L}^0([a, b])$ for all $n \in \mathbb{N}$, then $\lim_{n \to \infty} \| f_n - f \| = 0$ if and only if $f_n \to f$ *in measure*.
(c) Recall that $L^0([a, b]) := \mathcal{L}^0([a, b])/\mathcal{N}$, where $\mathcal{N} := \{ f \in \mathcal{L}^0([a, b]) : f = 0\ a.e.\}$. Show that $d(f + \mathcal{N}, g + \mathcal{N}) := \| f - g \|$ defines a *metric* on $L^0([a, b])$, making it a *complete* metric space.

36. Let $E \in \mathcal{M}_{\lambda < \infty}(\mathbb{R})$ and let (f_n) be a sequence in $\mathcal{L}^0(E, \mathbb{R})$ such that *any* of its subsequences admits a (further) subsequence that converges *almost everywhere* to a function $f \in \mathcal{L}^0(E, \mathbb{R})$. Show that $f_n \to f$ *in measure*.

37. Let $F : \mathbb{R}^2 \to \mathbb{R}$ be a continuous function and let (f_n) and (g_n) be two sequences in $\mathcal{L}^0(E, \mathbb{R})$, where $E \in \mathcal{M}_{\lambda < \infty}(\mathbb{R})$. If $f_n \to f$ and $g_n \to g$ *in measure* for two functions $f,\ g \in \mathcal{L}^0(E, \mathbb{R})$, then $F(f_n, g_n) \to F(f, g)$ in measure. Deduce that $f_n + g_n \to f + g$ and $f_n g_n \to fg$ in measure.

38. Show that the condition $\lambda(E) < \infty$ in the previous problem *cannot* be removed in general. In fact, show that if $f_n \to f$ and $g_n \to g$ in measure, where f_n, g_n, f, and g are in $\mathcal{L}^0(E, \mathbb{R})$, then $f_n + g_n \to f + g$ in measure even if $\lambda(E) = \infty$, but that in this case $f_n g_n$ need *not* converge to fg in measure.

39. Show that the Monotone Convergence Theorem and Fatou's lemma remain valid if $f_n \to f$ *almost everywhere* is replaced by $f_n \to f$ *in measure*.

40. Find the Dini derivatives of the following functions at $x = 0$:

(a) $f(x) := \begin{cases} x \sin(1/x) & \text{if } x \neq 0, \\ 0 & \text{if } x = 0. \end{cases}$

(b) $g(x) := \begin{cases} |x| & \text{if } x \in \mathbb{Q}, \\ 2|x| & \text{if } x \notin \mathbb{Q}. \end{cases}$

41.

(a). Show that $D^+[-f(x)] = -D_+ f(x)$.
(b). Show that if $g := -\check{f}$, where $\check{f}(x) := f(-x)$, we have $D^+ g(-x) = D^- f(x)$ and $D_- g(-x) = D_+ f(x)$.

42. Show that if f is continuous on $[a, b]$ and has a *local maximum* at $c \in (a, b)$, then $D^+ f(c) \leq 0 \leq D_- f(c)$.

43. Let $f : \mathbb{R} \to \mathbb{R}$ be a continuous function and $a < b$. Suppose that (1) every point of (a, b) is a *shadow point* of f and (2) a and b are *not* shadow points. Show that

$$f(x) \leq f(b) \quad \forall x \in (a, b) \quad \text{and} \quad f(a) = f(b).$$

Deduce that in the Rising Sun Lemma (Lemma 11.5.5) we actually have $f(a_n) = f(b_n)$ except possibly when $a_n = a$ for some n.

44. Show that the Cantor's ternary function $\kappa : [0, 1] \to [0, 1]$ (Example 4.3.13) is *not* absolutely continuous.

45. Show that if $f \in AC([a, b])$, then its total variation function $v_f := V_a^x(f)$ is also absolutely continuous. In particular, we have

$$v_f(x) = \int_a^x v_f' \, d\lambda.$$

Hint: Note that v_f is *increasing* so that $V_x^y(f) = v_f(y) - v_f(x)$.

46. Show that if $f \in AC([a, b])$, then $f = g - h$, where g and h are both *increasing* and absolutely continuous.

47. Show that if $f \in AC([a, b])$, then $V_a^b(f) = \int_a^b |f'| \, d\lambda$.

48. Let $f \in BV([a, b])$. Show that $f \in AC([a, b])$ if and only if

$$V_a^b(f) = \int_a^b |f'| \, d\lambda. \tag{†}$$

49. Let f be absolutely continuous and *strictly increasing* on $[a, b]$. Show that if g is absolutely continuous on $[f(a), f(b)]$, then the composite function $g \circ f$ is absolutely continuous on $[a, b]$. Show by an example that the assertion is false if f in *not* strictly increasing.

50. Let $a > 0$ and $b > 0$ be given. Show that the function

$$f(x) := \begin{cases} x^a \sin\left(\dfrac{1}{x^b}\right) & \text{if } x \in (0, 1], \\ 0 & \text{if } x = 0 \end{cases}$$

is *absolutely continuous* if and only if $a > b$. Deduce that f has *bounded variation* if and only if $a > b$.

51. Suppose that $f : [0, 1] \to \mathbb{R}$ is *continuous* at $x = 0$ and *absolutely continuous* on $[\alpha, 1]$ for every $\alpha \in (0, 1)$. Does it follow that f is absolutely continuous on $[0, 1]$? What if we add the assumption $f \in BV([0, 1])$?

52 (Singular Function). Define a function $h \in BV([a, b])$ to be *singular* if $h'(x) = 0$ almost everywhere; e.g., Cantor's ternary function κ (cf. Example 4.3.13) is a *continuous*, singular function. Show that if $f : [a, b] \to \mathbb{R}$ is *increasing*, then $f = g + h$ with two increasing functions g and h such that g is *absolutely continuous* and h is *singular*. Can an absolutely continuous function be singular?

Chapter 12
General Measure and Probability

Our goal in this final chapter is to extend the notions of measure and integral to general sets. As an application, we shall include a brief discussion of some basic facts in probability theory. We saw in Chap. 10 that Lebesgue measure, λ, can be defined by first introducing the Lebesgue *outer measure*, λ^* (Definition 10.1.1), which is defined on $\mathcal{P}(\mathbb{R})$ and then restricting it by means of Carathéodory's definition (Definition 10.1.11). As was pointed out there, this construction has the advantage that it can be carried out in general sets and this is what we intend to do here. Most of the results on Lebesgue measure and integral will therefore be extended and, since the proofs are in many cases almost identical, we may omit such proofs and assign them as exercises for the reader. *Throughout this chapter, X will denote an arbitrary (nonempty) set.* Also recall that, in the set $[-\infty, \infty]$ of extended real numbers, we have $\pm\infty \cdot 0 := 0$.

12.1 Measures and Measure Spaces

Definition 12.1.1 (Measure). Let X be an arbitrary set and $\mathcal{A} \subset \mathcal{P}(X)$ a σ-algebra of subsets of X. A function $\mu : \mathcal{A} \to [0, \infty]$ is said to be a *measure* on $\mathcal{A}$ if it satisfies $\mu(\emptyset) = 0$ and is *countably additive*; i.e., for any sequence $(A_n)_{n=1}^{\infty}$ of *pairwise disjoint* sets in $\mathcal{A}$, we have

$$\mu\left(\bigcup_{n=1}^{\infty} A_n\right) = \sum_{n=1}^{\infty} \mu(A_n).$$

Definition 12.1.2 (Finite, σ-Finite). Let $\mathcal{A} \subset \mathcal{P}(X)$ be a σ-algebra. We say that a measure $\mu : \mathcal{A} \to [0, \infty]$ is *finite* if $\mu(X) < \infty$. We then call $\mu(X)$ the *total mass* of μ. If $X = \bigcup_{n=1}^{\infty} A_n$, where $A_n \in \mathcal{A}$ and $\mu(A_n) < \infty$ for all $n \in \mathbb{N}$, then we say that μ is σ-finite.

© Springer Science+Business Media New York 2014
H.H. Sohrab, *Basic Real Analysis*, DOI 10.1007/978-1-4939-1841-6_12

Definition 12.1.3 (Measurable and Measure Spaces). Given a set X and a σ-algebra $\mathcal{A} \subset \mathcal{P}(X)$, the pair $(X, \mathcal{A})$ is called a *measurable space* and the elements of $\mathcal{A}$ are called *measurable sets*. If $\mu : \mathcal{A} \to [0, \infty]$ is a measure, then the triple $(X, \mathcal{A}, \mu)$ is called a *measure space*.

Theorem 12.1.4. *Given a measure space* $(X, \mathcal{A}, \mu)$, *the following properties hold for any measurable sets* A, B, *and* A_n, $n \in \mathbb{N}$.

1. **(Monotonicity)**

$$A \subset B \Longrightarrow \mu(A) \leq \mu(B).$$

2. **(Finite Additivity)**

$$A_i \cap A_j = \emptyset \ \ (1 \leq i, j \leq n, \ i \neq j) \Longrightarrow \mu\left(\bigcup_{j=1}^{n} A_j \right) = \sum_{j=1}^{n} \mu(A_j).$$

3. **(Finite Subadditivity)**

$$\mu\left(\bigcup_{j=1}^{n} A_j \right) \leq \sum_{j=1}^{n} \mu(A_j).$$

4. **(Countable Subadditivity)**

$$\mu\left(\bigcup_{j=1}^{\infty} A_j \right) \leq \sum_{j=1}^{\infty} \mu(A_j).$$

5. **(Continuity)**

$$A_1 \subset A_2 \subset A_3 \subset \cdots \Longrightarrow \mu\left(\bigcup_{n=1}^{\infty} A_n \right) = \lim_{n \to \infty} \mu(A_n), \quad and$$

$$\mu(A_1) < \infty \ \ \& \ \ A_1 \supset A_2 \supset A_3 \supset \cdots \Longrightarrow \mu\left(\bigcap_{n=1}^{\infty} A_n \right) = \lim_{n \to \infty} \mu(A_n).$$

Proof. Since $\mu(\emptyset) = 0$, (2) follows from the *countable additivity* of μ if we take $A_j = \emptyset$ for all $j \geq n+1$. Now, if $A \subset B$, then $B = A \cup (B \setminus A)$ and $A \cap (B \setminus A) = \emptyset$. Therefore, by (2), we have $\mu(B) = \mu(A) + \mu(B \setminus A) \geq \mu(A)$ and (1) follows. To prove (3), let $A_1' := A_1$ and $A_k' := A_k \setminus (\bigcup_{j=1}^{k-1} A_j)$ for all $k > 1$. Then the A_k' are measurable and *pairwise disjoint*. Also, $\bigcup_{k=1}^{n} A_k = \bigcup_{k=1}^{n} A_k'$. Since $A_k' \subset A_k$, (1) and (2) imply

$$\mu\left(\bigcup_{k=1}^{n} A_k \right) = \mu\left(\bigcup_{k=1}^{n} A_k' \right) = \sum_{k=1}^{n} \mu(A_k') \leq \sum_{k=1}^{n} \mu(A_k).$$

To prove the first part of (5), let $B_1 := A_1$ and $B_n := A_n \setminus A_{n-1}$ for all $n \geq 2$. Then (B_n) is a sequence of pairwise disjoint measurable sets. Also, $A_n = \bigcup_{k=1}^{n} B_k$ and $\bigcup_{n=1}^{\infty} A_n = \bigcup_{n=1}^{\infty} B_n$. Therefore, $\mu(\bigcup_{n=1}^{\infty} A_n) = \sum_{n=1}^{\infty} \mu(B_n)$ and (by (2)) $\mu(A_n) = \sum_{k=1}^{n} \mu(B_k)$ so that

$$\mu\left(\bigcup_{n=1}^{\infty} A_n\right) = \sum_{n=1}^{\infty} \mu(B_n) = \lim_{n \to \infty} \sum_{k=1}^{n} \mu(B_k) = \lim_{n \to \infty} \mu(A_n).$$

The second part of (5) now follows if we apply the first part to the *increasing* sequence $(A_1 \setminus A_n)_{n=1}^{\infty}$. (Why?) Finally, to prove (4), consider the increasing sequence (B_n) of measurable sets $B_n := \bigcup_{k=1}^{n} A_k$. Applying (5), we have $\mu(\bigcup_{n=1}^{\infty} A_n) = \lim_{n \to \infty} \mu(B_n)$. But, using (3), we have

$$\mu(B_n) \leq \sum_{k=1}^{n} \mu(A_k) \leq \sum_{n=1}^{\infty} \mu(A_n) \qquad (\forall n \in \mathbb{N}),$$

from which (4) follows.   □

Example 12.1.5 (Counting Measure). Let $X \neq \emptyset$ and define $\nu : \mathcal{P}(X) \to [0, \infty]$ by $\nu(A) = |A|$ (i.e., the number of elements of A) if A is *finite*, and $\nu(A) = \infty$ if A is *infinite*. Then ν is a measure on $\mathcal{P}(X)$ (why?), called the *counting measure*. Clearly, ν is finite (resp., σ-finite) if X is finite (resp., countable).

Example 12.1.6 (Dirac Measure). Let $X \neq \emptyset$ and $x \in X$ a *fixed* element. Define $\delta_x : \mathcal{P}(X) \to [0, \infty]$ by $\delta_x(A) = 1$ if $x \in A$ and $\delta_x(A) = 0$ if $x \notin A$; i.e., $\delta_x(A) = \chi_A(x)$. Then δ_x is a (finite) measure on $\mathcal{P}(X)$. (Why?) We call it the *Dirac measure at* x.

Sets of *measure zero* play a special role in general measure spaces as they did in the case of Lebesgue measure:

Definition 12.1.7 (Null Set, Almost Everywhere). Let $(X, \mathcal{A}, \mu)$ be a measure space. A set $Z \in \mathcal{A}$ is called a *null set* (or set of *measure zero*) if $\mu(Z) = 0$. To be more precise, we sometimes say μ-null (or μ-measure zero). A statement about elements $x \in X$ is said to be true *almost everywhere* (abbreviated *a.e.*) or μ-almost everywhere (abbreviated μ-*a.e.*) if it is true *except* on a (μ-) null set. We also say that the statement is true for *almost all* x (abbreviated *a.a.* x).

It is obviously desirable that all subsets of a null set be measurable and hence also null sets. This motivates the following.

Definition 12.1.8 (Complete Measure). Let $(X, \mathcal{A})$ be a measurable space. A measure $\mu : \mathcal{A} \to [0, \infty]$ (or the corresponding measure space $(X, \mathcal{A}, \mu)$) is said to be *complete* if every subset of a null set is measurable; i.e., if $A \in \mathcal{A}$ and $\mu(A) = 0$, then $Z \in \mathcal{A}$ for all $Z \subset A$.

Theorem 12.1.9 (Completion). *Let $(X, \mathcal{A}, \mu)$ be a measure space. Let $\bar{\mathcal{A}}$ denote the set of all $E \subset X$ for which there are A, $B \in \mathcal{A}$ with $A \subset E \subset B$ and $\mu(B \setminus A) = 0$ and define $\bar{\mu}(E) := \mu(A)$ in this case. Then $\bar{\mu}$ is an extension of μ to $\bar{\mathcal{A}}$ and $(X, \bar{\mathcal{A}}, \bar{\mu})$ is a complete measure space, called the completion of $(X, \mathcal{A}, \mu)$.*

Proof. First, it is obvious that $\mathcal{A} \subset \bar{\mathcal{A}}$ and $\bar{\mu}(A) = \mu(A)$ for all $A \in \mathcal{A}$. In particular, $X \in \bar{\mathcal{A}}$. If $A \subset E \subset B$, then $B^c \subset E^c \subset A^c$ and $A^c \setminus B^c = B \setminus A$. Therefore, $E \in \bar{\mathcal{A}}$ implies $E^c \in \bar{\mathcal{A}}$. Next, if $A_n \subset E_n \subset B_n$, for all $n \in \mathbb{N}$, and if $A := \bigcup A_n$, $E := \bigcup E_n$, and $B := \bigcup B_n$, then $A \subset E \subset B$ and

$$B \setminus A \subset \bigcup_{n=1}^{\infty} (B_n \setminus A_n),$$

so that $\mu(B_n \setminus A_n) = 0$ for all $n \in \mathbb{N}$ implies $\mu(B \setminus A) = 0$. Thus $E_n \in \bar{\mathcal{A}}$ for all n implies $E \in \bar{\mathcal{A}}$. It follows that $\bar{\mathcal{A}}$ is indeed a σ-algebra. To check that $\bar{\mu}$ is well defined on $\bar{\mathcal{A}}$, suppose that $A \subset E \subset B$, $A' \subset E \subset B'$, and $\mu(B \setminus A) = 0 = \mu(B' \setminus A')$. Then $A \setminus A' \subset B' \setminus A'$ and hence $\mu(A \setminus A') = 0$. Similarly, $\mu(A' \setminus A) = 0$. Therefore,

$$\mu(A) = \mu(A \cap A') = \mu(A').$$

Finally, the countable additivity of $\bar{\mu}$ on $\bar{\mathcal{A}}$ is obvious. (Why?) $\qquad\qquad\square$

Definition 12.1.10 (Operations, Partial Order). Let $\mathfrak{M}_{\mathcal{A}}$ denote the set of all (positive) measures on a measurable space $(X, \mathcal{A})$. For any μ, $\nu \in \mathfrak{M}_{\mathcal{A}}$ and any $t \in [0, \infty)$, the *addition* and *scalar multiplication* are given by

$$(\mu + \nu)(A) := \mu(A) + \nu(A) \quad \text{and} \quad (t\mu)(A) := t\mu(A) \qquad (\forall\, A \in \mathcal{A}).$$

Also, we define a partial ordering on $\mathfrak{M}_{\mathcal{A}}$ by

$$\mu \leq \nu \iff \mu(A) \leq \nu(A) \qquad (\forall\, A \in \mathcal{A}).$$

Exercise 12.1.11.

(a) Show that the set functions $\mu + \nu$ and $t\mu$ defined above are also measures on $\mathcal{A}$. Deduce that $\mathfrak{M}_{\mathcal{A}}$ is a *cone*. Show, however, that this cone is *not* a lattice by giving an example of two measures μ, $\nu \in \mathfrak{M}_{\mathcal{A}}$ such that $\mu \vee \nu := \max\{\mu, \nu\}$ is *not additive;* i.e., there are two sets A, $B \in \mathcal{A}$ with $A \cap B = \emptyset$ but

$$(\mu \vee \nu)(A \cup B) \neq (\mu \vee \nu)(A) + (\mu \vee \nu)(B)$$

and hence $\mu \vee \nu \notin \mathfrak{M}_{\mathcal{A}}$.

(b) For a fixed $A \in \mathcal{A}$, define the set function $\mu_A : \mathcal{A} \to [0, \infty]$ by $\mu_A(E) := \mu(E \cap A)$. Show that μ_A is a measure on $\mathcal{A}$.

The following theorem shows, however, that *directed* sets of measures are nicely behaved:

Theorem 12.1.12. *Let $\mathcal{A} \subset \mathcal{P}(X)$ be a σ-algebra and let $\mathfrak{M}$ be a set of measures on $\mathcal{A}$. If $\mathfrak{M}$ is directed, i.e., given any μ_1, $\mu_2 \in \mathfrak{M}$, there exists a measure $\mu_3 \in \mathfrak{M}$ such that $\mu_3 \geq \mu_1 \vee \mu_2$, then*

$$\nu(E) := \sup\{\mu(E) : \mu \in \mathfrak{M}\} \qquad (\forall E \in \mathcal{A})$$

defines a measure on $\mathcal{A}$.

Proof. It is obvious that ν is a well-defined function from $\mathcal{A}$ to $[0, \infty]$. Let $(A_n)_{n \in \mathbb{N}}$ be a sequence of *pairwise disjoint* sets in $\mathcal{A}$. Then

$$\mu\left(\bigcup_{n=1}^{\infty} A_n\right) = \sum_{n=1}^{\infty} \mu(A_n) \leq \sum_{n=1}^{\infty} \nu(A_n) \qquad (\forall \mu \in \mathfrak{M}),$$

and hence

$$\nu\left(\bigcup_{n=1}^{\infty} A_n\right) \leq \sum_{n=1}^{\infty} \nu(A_n).$$

To prove the opposite inequality, we may (and do) assume that $\nu(A_n) < \infty$ for all $n \in \mathbb{N}$. (Why?) Now, for fixed $n \in \mathbb{N}$ and $\varepsilon > 0$, the definition of ν implies that, for each k, $1 \leq k \leq n$, we can pick a measure $\mu_k \in \mathfrak{M}$ such that $\nu(A_k) - \varepsilon/n < \mu_k(A_k)$. Using the hypothesis (and induction), we can now find a $\mu \in \mathfrak{M}$ such that $\mu_k \leq \mu$ for $k = 1, 2, \ldots, n$. Thus

$$\sum_{k=1}^{n} \nu(A_k) - \varepsilon < \sum_{k=1}^{n} \mu_k(A_k) \leq \sum_{k=1}^{n} \mu(A_k)$$

$$= \mu\left(\bigcup_{k=1}^{n} A_k\right) \leq \nu\left(\bigcup_{k=1}^{n} A_k\right)$$

$$\leq \nu\left(\bigcup_{k=1}^{\infty} A_k\right).$$

Since $\varepsilon > 0$ was arbitrary, we have

$$\sum_{k=1}^{n} \nu(A_k) \leq \nu\left(\bigcup_{k=1}^{\infty} A_k\right)$$

and, letting $n \to \infty$, we finally obtain

$$\sum_{k=1}^{\infty} \nu(A_k) \leq \nu\left(\bigcup_{k=1}^{\infty} A_k\right).$$

$\square$

The following corollary is a useful special case:

Corollary 12.1.13. *Let $\mathcal{A} \subset \mathcal{P}(X)$ be a σ-algebra and let (μ_n) be an increasing sequence of measures on $\mathcal{A}$. For each $A \in \mathcal{A}$, define $\mu(A) := \sup\{\mu_n(A) : n \in \mathbb{N}\}$. Then μ is a measure on $\mathcal{A}$.*

Example 12.1.14. Let X be a (nonempty) set. Given any function $p : X \to [0, \infty)$, let us define $\mu : \mathcal{P}(X) \to [0, \infty]$ by $\mu(\emptyset) := 0$ and $\mu(E) := \sum_{x \in E} p(x)$, where the sum is an *unordered series* in the sense of Definition 2.4.1. Then μ is a measure on $\mathcal{P}(X)$. Indeed, if for each *finite* set $F \subset X$ we put $\mu_F := \sum_{x \in F} p(x) \delta_x$, where δ_x is the *Dirac measure* at x (defined above), then each μ_F is a measure by Exercise 12.1.11. Since

$$\mu = \sup\{\mu_F : F \in \mathcal{F}_X\},$$

where $\mathcal{F}_X$ denotes the set of all *finite subsets* of X, Theorem 12.1.12 implies that μ is a measure. In particular, if $p(x) = 1$ for all $x \in X$, then μ is simply the *counting measure* defined above. Also, if $p = \chi_{\{x_0\}}$ for some $x_0 \in X$, then we obtain the Dirac measure δ_{x_0}.

The last example can be used to give new proofs of some of the results on unordered series discussed in Chap. 2:

Exercise 12.1.15. Using the above example, prove Corollary 2.4.25 for nonnegative functions and Theorem 2.4.26 for *nonnegative* double series.

To construct measures on a σ-algebra $\mathcal{A} \subset \mathcal{P}(X)$, it is often more convenient to start with *outer measures* (which are defined on *all of* $\mathcal{P}(X)$) and then use Carathéodory's definition (cf. Theorem 10.1.16).

Definition 12.1.16 (Outer Measure). Given a set X, we say that a map $\mu^* : \mathcal{P}(X) \to [0, \infty]$ is an *outer measure* on X if it satisfies the following conditions:

1. $\mu^*(\emptyset) = 0$.
2. $A \subset B \Longrightarrow \mu^*(A) \leq \mu^*(B)$ (monotonicity).
3. For any sequence (A_n) of subsets of X, we have

$$\mu^*\left(\bigcup_{n=1}^{\infty} A_n\right) \leq \sum_{n=1}^{\infty} \mu^*(A_n)   \text{(countable subadditivity)}.$$

As we saw in the case of Lebesgue measure, the construction of an outer measure requires a set function, μ, defined initially on a subset of $\mathcal{P}(X)$. For the Lebesgue outer measure, μ was the *length* defined on the set of *intervals*. Here is an extension of Proposition 10.1.6 to more general sets with essentially the same proof:

Theorem 12.1.17. *Let X be a set and suppose that there is a collection $\mathcal{C} \subset \mathcal{P}(X)$ with $\emptyset \in \mathcal{C}$ and a sequence (X_n) in $\mathcal{C}$ such that $X = \bigcup_{n=1}^{\infty} X_n$. Let $\mu : \mathcal{C} \to [0, \infty]$ with $\mu(\emptyset) = 0$ and define $\mu^* : \mathcal{P}(X) \to [0, \infty]$ by*

$$\mu^*(A) := \inf \left\{ \sum_{n=1}^{\infty} \mu(C_n) : C_n \in \mathcal{C}, \; A \subset \bigcup_{n=1}^{\infty} C_n \right\}. \tag{$\dagger$}$$

Then μ^* is an outer measure.

Proof. First note that the existence of the sequence (X_n) implies that the set on the right side of ($\dagger$) is *not* empty. Also, the *monotonicity* of μ^* and the fact that $\mu^*(\emptyset) = 0$ are easily checked. (Why?) Therefore, we need only show the *countable subadditivity*. So let (A_n) be a sequence in $\mathcal{P}(X)$ and let $\varepsilon > 0$ be given. We may (and do) assume that $\mu^*(A_n) < \infty$ for all $n \in \mathbb{N}$. Now, for each n, we pick a sequence $(C_{nk})_{k=1}^{\infty}$ in $\mathcal{C}$ such that $A_n \subset \bigcup_{k=1}^{\infty} C_{nk}$ and

$$\sum_{k=1}^{\infty} \mu(C_{nk}) \leq \mu^*(A_n) + \frac{\varepsilon}{2^n}.$$

Since $\{C_{nk} : n, \, k \in \mathbb{N}\}$ is a *countable* collection of sets in $\mathcal{C}$ covering $\bigcup_{n=1}^{\infty} A_n$, we have

$$\mu^*\left(\bigcup_{n=1}^{\infty} A_n\right) \leq \sum_{n, \, k \in \mathbb{N}} \mu(C_{nk}) = \sum_{n=1}^{\infty} \sum_{k=1}^{\infty} \mu(C_{nk})$$

$$\leq \sum_{n=1}^{\infty} (\mu^*(A_n) + 2^{-n}\varepsilon) = \sum_{n=1}^{\infty} \mu^*(A_n) + \varepsilon.$$

Since $\varepsilon > 0$ was arbitrary, the countable subadditivity follows. $\square$

Example 12.1.18.

1. **(Lebesgue Outer Measure on $\mathbb{R}^n$).** The most important example is, of course, the *Lebesgue outer measure on $\mathbb{R}^n$:* Here $X := \mathbb{R}^n$, $\mathcal{C}$ is the collection $\mathcal{I}$: $\{(\mathbf{a}, \mathbf{b}) : \mathbf{a}, \, \mathbf{b} \in \mathbb{R}^n\}$ of all bounded open intervals:

$$(\mathbf{a}, \mathbf{b}) := (a_1, b_1) \times (a_2, b_2) \times \cdots \times (a_n, b_n),$$

where $\mathbf{a} = (a_1, \ldots, a_n)$, $\mathbf{b} = (b_1, \ldots, b_n)$, $-\infty < a_j \leq b_j < \infty$, for $1 \leq j \leq n$, and $\mu = \lambda_n$, where $\lambda_n((\mathbf{a}, \mathbf{b})) := \prod_{j=1}^{n}(b_j - a_j)$ is the *volume* of the interval $(\mathbf{a}, \mathbf{b})$. The corresponding outer measure is then denoted by λ_n^*. If $n = 1$, we recover the Lebesgue outer measure, λ^*, on $\mathbb{R}$. (cf. Definition 10.1.1).

2. **(Lebesgue–Stieltjes Outer Measures).** Let $F : \mathbb{R} \to \mathbb{R}$ be an *increasing, right-continuous* function: $\forall\ x_1,\ x_2 \in \mathbb{R},\ x_1 < x_2$ implies $F(x_1) \le F(x_2)$ and $F(x+0) = F(x)\ \forall x \in \mathbb{R}$. Such a function is often called a *distribution function*. Now let $\mathcal{C}$ be the set of all half-open intervals $(a, b]$ of $\mathbb{R}$ and let $\mu = \lambda_F$, where $\lambda_F((a,b]) := F(b) - F(a)$. The corresponding outer measure λ_F^* is then called the *Lebesgue–Stieltjes Outer Measure* associated with F.

3. **(Hausdorff Outer Measures).** Let (M, d) be a *separable* metric space and $p \ge 0$. For each $\varepsilon > 0$ introduce the collection

$$C_\varepsilon := \{C \subset M : 0 < \delta(C) < \varepsilon\},$$

where $\delta(C) := \sup\{d(x, y) : x,\ y \in C\}$ is the *diameter* of C. Define $\mu_{p\varepsilon}(\emptyset) := 0$ and $\mu_{p\varepsilon}(C) := (\delta(C))^p$, for each $C \in C_\varepsilon$. Now let $\mu_{p\varepsilon}^*$ be the corresponding outer measure. It is easily checked that $\mu_{p\varepsilon}^* \le \mu_{p\varepsilon'}^*$ if $0 < \varepsilon' < \varepsilon$. Next, for each $E \subset M$, define

$$\mu_p^*(E) := \sup\{\mu_{p\varepsilon}^*(E) : \varepsilon > 0\}.$$

It follows (e.g., from Theorem 12.1.12) that μ_p^* is an outer measure. It is called the *p-dimensional Hausdorff outer measure* on M and is also denoted by $\mathcal{H}^p$.

Remark 12.1.19. Note that $\mu_p^*(E)$ will not change if we assume that the members of the C_ε are all *closed* or all *open*. Indeed, for any set $A \subset M$, we have $\delta(A) = \delta(A^-)$. Also, if $A_\varepsilon := \{x \in M : d(x, A) < \varepsilon\}$, then A_ε is open and $\delta(A_\varepsilon) \le \delta(A) + 2\varepsilon$.

Exercise 12.1.20.

1. Show that μ_0^* is the *counting measure* on M.
2. Let $M := \mathbb{R}$ with its usual metric. Show that $\mu_1^* = \lambda^*$, where λ^* is the Lebesgue outer measure (Definition 10.1.1).

Exercise 12.1.21. Let (M, d) be a *separable* metric space and $A \subset M$. Show that, if $\mu_p^*(A) < \infty$, then $\mu_q^* = 0$ for all $q > p \ge 0$. Deduce that, if $\mu_p^*(A) > 0$, then $\mu_q^*(A) = \infty$ for all $0 \le q < p$. *Hint:* Show that, if $q > p$, then $\mu_{q\varepsilon}^*(A) \le \varepsilon^{q-p}[\mu_p^*(A) + 1]$.

Definition 12.1.22 (Hausdorff Dimension). Let (M, d) be a *separable* metric space and $A \subset M$. The (unique) number

$$\dim_H(A) := \inf\{p \ge 0 : \mu_p^*(A) = 0\} = \sup\{p \ge 0 : \mu_p^*(A) = \infty\} \qquad (\dagger)$$

is called the *Hausdorff dimension* of A. Note that the equality in $(\dagger)$ follows from Exercise 12.1.21, which also implies

$$\dim_H(A) = \sup\{p \ge 0 : \mu_p^*(A) > 0\} = \inf\{p \ge 0 : \mu_p^*(A) < \infty\}.$$

In particular, if $0 < \mu_p^*(A) < \infty$, then $\dim_H(A) = p$.

Computing the Hausdorff measure (or dimension) of a set is a tricky business and, as a rule, lower bounds are much more difficult to find than upper bounds. We shall only consider the case of Cantor's ternary set and refer the reader to Falconer's *The Geometry of Fractal Sets* [Fal85] for other interesting examples. The following lemma (used in the same reference) will be needed.

Lemma 12.1.23. *Let $p := \log 2 / \log 3$. If a, b, and c are any nonnegative numbers with $c \geq (a + b)/2$, then we have*

$$(a + b + c)^p \geq a^p + b^p.$$

Proof. Since $0 < p < 1$, the function $x \mapsto x^p$ is *concave*. Therefore,

$$(a + b + c)^p \geq (3(a + b)/2)^p = 2\left(\frac{a + b}{2}\right)^p$$

$$\geq a^p + b^p,$$

where we have used the fact that $3^p = 2$.   □

Proposition 12.1.24 ($\dim_H(C) = \log 2 / \log 3$). *Let $C \subset [0, 1]$ be the Cantor set and let $p := \log 2 / \log 3$. Then $\dim_H(C) = p$ and $\mu_p^*(C) = 1$.*

Proof. Recall that $C = \bigcap_{n=1}^{\infty} C_n$, where C_n is the *disjoint* union of 2^n closed intervals of length 3^{-n}. Therefore, if $p := \log 2 / \log 3$, we have

$$\mu_{p(1/3)^n}^*(C) \leq 2^n (1/3)^{np} = 1,$$

which (letting $n \to \infty$) implies that $\dim_H(C) \leq \log 2 / \log 3$. To complete the proof, we need the opposite inequality $\dim_H(C) \geq \log 2 / \log 3$. This follows if, given any collection $\mathcal{I}$ of intervals covering C (i.e., $C \subset \bigcup_{I \in \mathcal{I}} I$), we can prove that

$$\sum_{I \in \mathcal{I}} (\lambda(I))^p \geq 1. \qquad (*)$$

Now, enlarging each I slightly and using the *compactness* of C, it suffices to prove $(*)$ when the cover $\mathcal{I}$ is a *finite* collection of *closed* intervals all contained in $[0, 1]$. Next, let (α, β) be one of the open middle thirds (henceforth called *holes*) removed to construct C. If $I = [a, b] \in \mathcal{I}$ and if $a \in [\alpha, \beta)$ (resp., $b \in (\alpha, \beta]$), then we replace I by $I \setminus [\alpha, \beta)$ (resp., $I \setminus (\alpha, \beta]$). This does not increase the sum in $(*)$ and gives a cover by a finite collection of *pairwise disjoint* closed intervals which we still denote by $\mathcal{I}$. In addition, the left (resp., right) endpoint of each $I \in \mathcal{I}$ is the left (resp., right) endpoint of an interval in some C_m. Now, given any $I \in \mathcal{I}$, let J be the *largest hole* contained in I. Then I is contained in the interval (used in the construction of C) from which J was removed, and hence we have the disjoint union $I = I' \cup J \cup I''$, with $\lambda(J) \geq (\lambda(I') + \lambda(I''))/2$. By Lemma 12.1.23, the sum in (∗) does *not* increase if the term $(\lambda(I))^p$ is replaced by $(\lambda(I'))^p + (\lambda(I''))^p$. We can therefore replace I by the pair of intervals I' and I''. Repeating this a finite number of times, we arrive at the cover C_n of C for some $n \in \mathbb{N}$ and (∗) follows. □

We shall now construct measures using outer measures:

Definition 12.1.25 (Carathéodory). Let X be a set and μ^* an outer measure on X. We say that a set $E \subset X$ is μ^*-*measurable* if, for every set $A \subset X$, we have

$$\mu^*(A) = \mu^*(A \cap E) + \mu^*(A \cap E^c). \tag{∗}$$

The set of all μ^*-measurable subsets of X will be denoted by $\mathcal{M}_\mu$.

Remark 12.1.26. Note that, because μ^* is (countably) subadditive, (∗) may be replaced by the inequality

$$\mu^*(A) \geq \mu^*(A \cap E) + \mu^*(A \cap E^c). \tag{∗∗}$$

Theorem 12.1.27. *Let μ^* be an outer measure on a set X. Then the collection $\mathcal{M}_\mu$ of all μ^*-measurable subsets of X is a σ-algebra containing every set $Z \subset X$ with $\mu^*(Z) = 0$, and the restriction $\mu := \mu^*|\mathcal{M}_\mu$ is a complete measure.*

Proof. If $Z \subset X$ and $\mu^*(Z) = 0$, then it is obvious that (∗∗) is satisfied for all $A \subset X$ and $E = Z$. Thus $Z \in \mathcal{M}_\mu$; in particular, $\emptyset \in \mathcal{M}_\mu$. It is also obvious that $E \in \mathcal{M}_\mu$ if and only if $E^c \in \mathcal{M}_\mu$. Next, if $E, F \in \mathcal{M}_\mu$ and $A \subset X$, then we have

$$\mu^*(A) = \mu^*(A \cap E) + \mu^*(A \cap E^c)$$
$$= \mu^*(A \cap E \cap F) + \mu^*(A \cap E^c \cap F^c)$$
$$+ \mu^*(A \cap E^c \cap F) + \mu^*(A \cap E \cap F^c)$$
$$\geq \mu^*(A \cap (E \cup F)) + \mu^*(A \cap (E \cup F)^c),$$

which implies that $E \cup F \in \mathcal{M}_\mu$ and hence $\mathcal{M}_\mu$ is an *algebra*. It also shows that, if $E, F \in \mathcal{M}_\mu$ and $E \cap F = \emptyset$, then

$$\mu^*(E \cup F) = \mu^*((E \cup F) \cap E) + \mu^*((E \cup F) \cap E^c) = \mu^*(E) + \mu^*(F).$$

Thus, by induction, μ^* is *finitely additive*. To prove that $\mathcal{M}_\mu$ is a σ-algebra, let $(E_n)_{n=1}^\infty$ be a sequence of *pairwise disjoint* sets in $\mathcal{M}_\mu$ and put $E := \bigcup_{k=1}^\infty E_k$ and $F_n := \bigcup_{k=1}^n E_k$. Then we have $F_n \in \mathcal{M}_\mu$, $F_n \cap E_n = E_n$, and $F_n \cap E_n^c = F_{n-1}$ (with $F_0 := \emptyset$). Now, for each $A \subset X$, a simple induction gives

$$\mu^*(A \cap F_n) = \mu^*(A \cap E_n) + \mu^*(A \cap F_{n-1}) = \sum_{k=1}^n \mu^*(A \cap E_k).$$

Since $E^c \subset F_n^c$, we therefore obtain

$$\mu^*(A) = \mu^*(A \cap F_n) + \mu^*(A \cap F_n^c) \geq \sum_{k=1}^{n} \mu^*(A \cap E_k) + \mu^*(A \cap E^c),$$

which (letting $n \to \infty$) implies

$$\mu^*(A) \geq \mu^*(A \cap E^c) + \sum_{n=1}^{\infty} \mu^*(A \cap E_n) \qquad (\dagger)$$

$$\geq \mu^*(A \cap E^c) + \mu^*(A \cap E),$$

and hence $E \in \mathcal{M}_\mu$. Since any countable union in the algebra $\mathcal{M}_\mu$ can be written as a countable *disjoint* union (in $\mathcal{M}_\mu$), we have thus proved that $\mathcal{M}_\mu$ is a σ-algebra. Finally, the *countable additivity* of μ^* on $\mathcal{M}_\mu$ follows from ($\dagger$) if we replace A by $E := \bigcup_{n=1}^{\infty} E_n$. $\qquad \square$

Definition 12.1.28 (Semialgebra). Let X be a set. A collecting $\mathcal{S} \subset \mathcal{P}(X)$ is said to be a *semialgebra* if (i) $S \cap T \in \mathcal{S}$ for any S, $T \in \mathcal{S}$, and (ii) if $S \in \mathcal{S}$, then S^c is a *finite, disjoint* union of sets in $\mathcal{S}$.

Example 12.1.29 (Semiclosed Intervals). Given any a, $b \in \overline{\mathbb{R}}$ with $a \leq b$, consider the set of all *left-open, right-closed* intervals with endpoints a and b : If $b < \infty$, then $(a, b] := \{x \in \mathbb{R} : a < x \leq b\}$ and if $b = \infty$, then we define $(a, b] := (a, \infty)$. We obviously have $(a, a] = \emptyset$ for all $a \in \overline{\mathbb{R}}$. The set of all these right-semiclosed intervals will be denoted by $\mathcal{I}_{\text{sem}}$. For each $I \in \mathcal{I}_{\text{sem}}$, we either have $I^c \in \mathcal{I}_{\text{sem}}$ or $I^c = I_1 \cup I_2$, with I_1, $I_2 \in \mathcal{I}_{\text{sem}}$ and $I_1 \cap I_2 = \emptyset$. Since $I^c = I^c \cup \emptyset$, in either case we may write I^c as a disjoint union of two members of $\mathcal{I}_{\text{sem}}$. Therefore, $\mathcal{I}_{\text{sem}}$ is a semialgebra.

Here is a useful fact:

Lemma 12.1.30. *If $\mathcal{S}$ is a semialgebra, then the set $\tilde{\mathcal{S}}$ of all finite, disjoint unions of sets in $\mathcal{S}$ is an algebra; i.e., for any A, $B \in \tilde{\mathcal{S}}$, we have $A^c \in \tilde{\mathcal{S}}$ and $A \cup B \in \tilde{\mathcal{S}}$ [or, equivalently, $A \cap B = (A^c \cup B^c)^c \in \tilde{\mathcal{S}}$].*

Proof. Given any disjoint unions $A = \bigcup_{i=1}^{m} S_i$ and $B = \bigcup_{j=1}^{m} T_j$ with the S_i and T_j in $\mathcal{S}$, we have $S_i \cap T_j \in \mathcal{S}$ for all j, k and hence the disjoint union $A \cap B = \bigcup_{i,j} (S_i \cap T_j)$ is also in $\tilde{\mathcal{S}}$. Thus (inductively) $\tilde{\mathcal{S}}$ is closed under finite intersections. But then $A^c = \bigcap_i S_i^c \in \tilde{\mathcal{S}}$ because $S_i^c \in \tilde{\mathcal{S}}$ for all i. $\qquad \square$

Notation 12.1.31. Let $\mathcal{I}_{\text{sem}}$ be as in Example 12.1.29. Then the corresponding algebra $\tilde{\mathcal{I}}_{\text{sem}}$ of all *finite, disjoint* unions of semiclosed intervals (provided by Lemma 12.1.30) will be denoted by $\mathcal{A}_{\text{sem}}$. Note that the σ-algebra generated by $\mathcal{A}_{\text{sem}}$ is in fact the Borel algebra $\mathcal{B}_{\mathbb{R}}$. (Why?)

Definition 12.1.32 (Premeasure). Let $\mathcal{G} \subset \mathcal{P}(X)$ be an *algebra*. A function μ : $\mathcal{G} \rightarrow [0, \infty]$ is said to be a *premeasure* (on $\mathcal{G}$) if $\mu(\emptyset) = 0$ and μ is *countably additive on* $\mathcal{G}$; i.e., if $G_n \in \mathcal{G}$ for all $n \in \mathbb{N}$, $G_i \cap G_j = \emptyset$ for $i \neq j$, and $G := \bigcup_{n=1}^{\infty} G_n \in \mathcal{G}$, then $\mu(G) = \sum_{n=1}^{\infty} \mu(G_n)$.

Proposition 12.1.33. *Suppose that the class $\mathcal{C}$ in Theorem 12.1.17 is actually an algebra and that $\mu : \mathcal{C} \rightarrow [0, \infty]$ is a premeasure. If μ^* is the outer measure constructed in that theorem, then each $E \in \mathcal{C}$ is μ^*-measurable and $\mu^*(E) = \mu(E)$.*

Proof. Let $E \in \mathcal{C}$ and $A \subset X$. Given any $\varepsilon > 0$, we can pick a sequence (C_n) in $\mathcal{C}$ such that

$$A \subset \bigcup_{n=1}^{\infty} C_n \quad \text{and} \quad \mu^*(A) \leq \sum_{n=1}^{\infty} \mu(C_n) \leq \mu^*(A) + \varepsilon.$$

Define the *disjoint* sequence (C_n') in $\mathcal{C}$ by $C_1' := C_1$ and $C_n' := C_n \setminus (\bigcup_{k=1}^{n-1} C_k)$ for all $n > 1$. Since $C_n' \subset C_n$ and μ is *countably additive* on $\mathcal{C}$, we have

$$\mu^*(A) + \varepsilon \geq \sum_{n=1}^{\infty} \mu(C_n') = \sum_{n=1}^{\infty} [\mu(C_n' \cap E) + \mu(C_n' \cap E^c)] \qquad (\dagger)$$

$$\geq \mu^*(A \cap E) + \mu^*(A \cap E^c),$$

which ($\varepsilon > 0$ being arbitrary) implies that E is μ^*-measurable. Next, note that $\mu^*(E) \leq \mu(E)$ is obvious. For the opposite inequality, we take $A = E$ in $(\dagger)$, let $\varepsilon \rightarrow 0$, and use the countable additivity of μ on $\mathcal{C}$ to deduce

$$\mu^*(E) \geq \sum_{n=1}^{\infty} \mu(E \cap C_n') = \mu(E).$$

$\square$

The above results provide the following important theorem:

Theorem 12.1.34 (Extension Theorem). *Let $\mathcal{G} \subset \mathcal{P}(X)$ be an algebra, μ a premeasure on $\mathcal{G}$, and let $\mathcal{A} := \mathcal{A}_{\mathcal{G}}$ be the σ-algebra generated by $\mathcal{G}$. Then there exists a measure $\tilde{\mu}$ on $\mathcal{A}$ whose restriction to $\mathcal{G}$ is μ. If ν is another such measure, then $\nu(E) \leq \tilde{\mu}(E)$ for all $E \in \mathcal{A}$, with equality if $\tilde{\mu}(E) < \infty$. If μ is σ-finite, then $\tilde{\mu}$ is the unique extension of μ to $\mathcal{A}$.*

Proof. Let μ^* be the outer measure constructed in Theorem 12.1.17 (with $\mathcal{C} := \mathcal{G}$) and let $\mathcal{M}_\mu$ denote the σ-algebra of all μ^*-measurable sets as in Theorem 12.1.27. Then $\mathcal{A} \subset \mathcal{M}_\mu$ and hence, by Proposition 12.1.33, $\tilde{\mu} := \mu^*|\mathcal{A}$ satisfies the first assertion. Next, let $E \in \mathcal{A}$ and assume that $E \subset A := \bigcup_{n=1}^{\infty} A_n$, with $A_n \in \mathcal{G}$ for all $n \in \mathbb{N}$. Then, with ν as in the second assertion, we have $\nu(E) \leq \sum_{n=1}^{\infty} \nu(A_n) = \sum_{n=1}^{\infty} \mu(A_n)$. Therefore, $\nu(E) \leq \mu^*(E) = \tilde{\mu}(E)$. Moreover,

$$v(A) = \lim_{n \to \infty} v\left(\bigcup_{k=1}^{n} A_k\right) = \lim_{n \to \infty} \mu\left(\bigcup_{k=1}^{n} A_k\right) = \tilde{\mu}(A).$$

If $\tilde{\mu}(E) < \infty$ and $\varepsilon > 0$, then (by the definition of μ^*) we can pick the sequence (A_n) in such a way that $\tilde{\mu}(A) < \tilde{\mu}(E) + \varepsilon$. Thus $\tilde{\mu}(A \setminus E) < \varepsilon$ and hence

$$\tilde{\mu}(E) \le \tilde{\mu}(A) = v(A) = v(E) + v(A \setminus E)$$
$$\le v(E) + \tilde{\mu}(A \setminus E) \le v(E) + \varepsilon.$$

Since $\varepsilon > 0$ is arbitrary, we have $\tilde{\mu}(E) = v(E)$. Finally, suppose that $X = \bigcup_{n=1}^{\infty} X_n$ with $X_n \in \mathcal{A}$ and $\mu(X_n) < \infty$ for all $n \in \mathbb{N}$. We may and (do) assume that the X_n are *pairwise disjoint*. (Why?) Given any $E \in \mathcal{A}$, we then have

$$\tilde{\mu}(E) = \sum_{n=1}^{\infty} \tilde{\mu}(E \cap X_n) = \sum_{n=1}^{\infty} v(E \cap X_n) = v(E).$$

$\square$

Remark 12.1.35 (Lebesgue–Stieltjes and Hausdorff Measures). Let us go back to the three outer measures λ_n^*, λ_F^*, and μ_p^* (cf. Example 12.1.18). Each gives rise to a measure according to Theorem 12.1.27. λ_n^* produces the *Lebesgue measure*, λ_n, on $\mathbb{R}^n$. The measure λ_F corresponding to λ_F^* is called the *Lebesgue–Stieltjes measure induced by F*. As for μ_p^*, the corresponding measure, μ_p, is called the *p-dimensional Hausdorff measure* (on $\mathbf{M}$).

Due to the important role played by the Lebesgue–Stieltjes measures in probability theory, we follow the recipe in the *extension theorem* to construct them in detail. First, we include another useful

Lemma 12.1.36. *Let $\mathcal{S} \subset \mathcal{P}(X)$ be a semialgebra and $\tilde{\mathcal{S}}$ the corresponding algebra as in Lemma 12.1.30. Suppose that μ is a set function defined on $\mathcal{S}$ such that $\mu(\emptyset) = 0$ and $\mu(S) = \sum_{i=1}^{m} \mu(S_i)$ if $S_i \in \mathcal{S}$ are pairwise disjoint and $S := \bigcup_{i=1}^{m} S_i \in \mathcal{S}$. Let us extend μ to a set function $\tilde{\mu}$ on $\tilde{\mathcal{S}}$ by setting*

$$\tilde{\mu}\left(\bigcup_{j=1}^{n} S_j\right) := \sum \mu(S_j),$$

if the $S_j \in \mathcal{S}$ are pairwise disjoint. Then for any sets A, $B_j \in \tilde{\mathcal{S}}$, $1 \le j \le n$, we have

(a) $A = \bigcup_{j=1}^{n} B_j$ *is a disjoint union* $\implies \tilde{\mu}(A) = \sum_{j=1}^{n} \tilde{\mu}(B_j)$.
(b) $A \subset \bigcup_{j=1}^{n} B_j \implies \tilde{\mu}(A) \le \sum_{j=1}^{n} \tilde{\mu}(B_j)$.

Proof. For (a), note that $B_j = \bigcup_\ell S_{j,\ell}$ is a finite, disjoint union of sets in $\mathcal{S}$ for each j, and hence

$$\tilde{\mu}(A) = \sum_{j,\ell} \mu(S_{j,\ell}) = \sum_j \tilde{\mu}(B_j).$$

In particular, if we also have $A = \bigcup_k C_k$ for pairwise disjoint sets $C_k \in \tilde{\mathcal{S}}$, then $\sum_j \tilde{\mu}(B_j) = \sum_k \tilde{\mu}(C_k)$ and hence $\tilde{\mu}$ is *well defined* on $\tilde{\mathcal{S}}$. To prove (b), assume first that $n = 1$ and set $B := B_1$. Then $A \subset B$ gives the disjoint union $B = A \cup (B \cap A^c)$ with $B \cap A^c \in \tilde{\mathcal{S}}$, so that

$$\tilde{\mu}(A) \leq \tilde{\mu}(A) + \tilde{\mu}(B \cap A^c) = \tilde{\mu}(B).$$

For $n > 1$, let $C_1 := B_1$ and $C_k := B_k \cap B_1^c \cap \cdots \cap B_{k-1}^c$. Then the C_k are pairwise disjoint sets in $\tilde{\mathcal{S}}$ with $C_k \subset B_k$ and $\bigcup_{j=1}^n B_j = \bigcup_{j=1}^n C_j$. Also, we have the disjoint union $A = A \cap (\bigcup_j B_j) = \bigcup_j (A \cap C_j)$. It then follows from the case $n = 1$ and part (a) that

$$\tilde{\mu}(A) = \sum_{j=1}^n \tilde{\mu}(A \cap C_j) \leq \sum_{j=1}^n \tilde{\mu}(C_j) \leq \sum_{j=1}^n \tilde{\mu}(B_j).$$

$\square$

We can now prove our main result, where we use the notation in 12.1.31.

Proposition 12.1.37. *Let $F : \mathbb{R} \to \mathbb{R}$ be an increasing, right-continuous function. Given any pairwise disjoint intervals $(a_j, b_j] \in \mathcal{I}_{\text{sem}}$, $1 \leq j \leq n$, define*

$$\lambda_F\left(\bigcup_{j=1}^n (a_j, b_j]\right) := \sum_{j=1}^n [F(b_j) - F(a_j)],$$

and hence $\lambda_F(\emptyset) = \lambda_F((a, a]) = 0$. Then λ_F is a premeasure on $\mathcal{A}_{\text{sem}}$.

Proof. To begin, note that $F(\infty) := \lim_{x \uparrow \infty} F(x)$ and $F(-\infty) := \lim_{x \downarrow -\infty} F(x)$ both exist, as F is *increasing*. Also $-\infty < F(\infty)$ and $F(-\infty) < \infty$, so that $\lambda_F((a, b]) := F(b) - F(a)$ makes sense for all $-\infty \leq a \leq b \leq \infty$. Now, if $a < b$ and $(a, b] = \bigcup_{k=1}^m (a_k, b_k]$, where the union is *disjoint,* then we may assume (after possibly relabeling the intervals) that $a = a_1 < b_1 = a_2 < b_2 = \cdots < b_n = b$ and hence

$$\lambda_F((a, b]) = F(b) - F(a) = \sum_{k=1}^m [F(b_k) - F(a_k)] = \sum_{k=1}^m \lambda_F((a_k, b_k]).$$

Therefore, the conditions of Lemma 12.1.36 are satisfied with $\mathcal{S} := \mathcal{I}_{\text{sem}}, \tilde{\mathcal{S}} = \mathcal{A}_{\text{sem}}$ and $\tilde{\mu} := \lambda_F$. In particular, λ_F is well defined on $\mathcal{A}_{\text{sem}}$ and is *finitely additive*. To prove the *countable additivity*, let $(I_n)_{n \in \mathbb{N}}$ be a *disjoint* sequence in $\mathcal{I}_{\text{sem}}$ with $A := \bigcup_{n=1}^{\infty} I_n \in \mathcal{A}_{\text{sem}}$. Then A is a *finite* disjoint union of intervals in $\mathcal{I}_{\text{sem}}$, each of which being the union of a *subsequence* of (I_n). So, using finite additivity, we may as well assume that $A = (a, b]$ is a *single* interval. But then

$$\lambda_F(A) = \lambda_F\left(\bigcup_{k=1}^{n} I_k\right) + \lambda_F\left(A \setminus \bigcup_{k=1}^{n} I_k\right) \geq \lambda_F\left(\bigcup_{k=1}^{n} I_k\right) = \sum_{k=1}^{n} \lambda_F(I_k),$$

and letting $n \to \infty$ gives $\lambda(A) \geq \sum_{n=1}^{\infty} \lambda_F(I_n)$. For the reverse inequality, assume first that $-\infty < a < b < \infty$ and let $\varepsilon > 0$ be given. Then the *right-continuity* of F may be used to pick $\delta > 0$ with $a + \delta < b$ such that $F(a + \delta) - F(a) < \varepsilon$. Also, if $I_n := (a_n, b_n]$ (where we may assume that $-\infty < a_n < b_n < \infty$ for all n), we can pick $\eta_n > 0$ with $F(b_n + \eta_n) - F(b_n) < \varepsilon 2^{-n}$ for each $n \in \mathbb{N}$. Now the open intervals $(a_n, b_n + \eta_n)$ cover the *compact* interval $[a + \delta, b]$. We pick a *finite* subcover and relabeling its intervals, if necessary, assume that $[a + \delta, b] \subset \bigcup_{j=1}^{N}(a_j, b_j + \eta_j)$. By part (b) of Lemma 12.1.36, we then have

$$F(b) - F(a + \delta) \leq \sum_{j=1}^{N}[F(b_j + \eta_j) - F(a_j)] \leq \sum_{j=1}^{\infty}[F(b_j + \eta_j) - F(a_j)].$$

Our choice of δ and η_j now gives

$$\lambda_F((a, b]) = F(b) - F(a) \leq 2\varepsilon + \sum_{j=1}^{\infty}[F(b_j) - F(a_j)] = 2\varepsilon + \sum_{j=1}^{\infty} \lambda_F((a_j, b_j]).$$

Since $\varepsilon > 0$ was arbitrary, the countable additivity of λ_F is established when $-\infty < a < b < \infty$. If $(a, b]$ is *unbounded*, then for any *bounded* interval $(\alpha, \beta] \subset (a, b]$, the above argument gives $F(\beta) - F(\alpha) \leq \sum_{n=1}^{\infty}[F(b_n) - F(a_n)]$. Since $(\alpha, \beta]$ was arbitrary, the proof is complete.   □

Theorem 12.1.38. *Given any increasing, right-continuous function $F : \mathbb{R} \to \mathbb{R}$, there is a unique measure λ_F on $\mathcal{B}_{\mathbb{R}}$ such that $\lambda_F((a, b]) = F(b) - F(a)$ for all $a, b \in \mathbb{R}$. If G is another such function, then $\lambda_F = \lambda_G$ if and only if $F - G$ is constant. Conversely, if μ is a measure on $\mathcal{B}_{\mathbb{R}}$ that is finite on all bounded Borel sets, then the function*

$$F(x) := \begin{cases} \mu((0, x]) & \text{if } x > 0, \\ 0 & \text{if } x = 0, \\ -\mu((x, 0]) & \text{if } x < 0 \end{cases}$$

is increasing, right-continuous, and $\mu = \lambda_F$.

Proof. By Proposition 12.1.37, each increasing, right-continuous F induces a *premeasure* λ_F on $\mathcal{A}_{\text{sem}}$ and it is obvious that $\lambda_F = \lambda_G$ if and only if $F - G$ is constant. (Why?) Also, since $\mathbb{R} = \bigcup_{n=-\infty}^{\infty}(n, n+1]$, the premeasure λ_F is σ-*finite*. Thus the first two assertions follow from the *extension theorem* (Theorem 12.1.34). To prove the last one, note that the *monotonicity* of μ implies that F is *increasing* and the *continuity* of μ implies that F is *right-continuous*. (Why?) Also, we clearly have $\mu = \lambda_F$ on $\mathcal{A}_{\text{sem}}$ and hence $\mu = \lambda_F$ by the uniqueness assertion in the extension theorem.   □

Remark 12.1.39.

1. Given any *finite* measure μ on $\mathcal{B}_{\mathbb{R}}$, we have $\mu = \lambda_F$, where $F(x) := \mu((-\infty, x])$. (Why?) This function F is called the *(cumulative) distribution function* of μ.
2. Note that, in Remark 12.1.35, we used λ_F to denote the Lebesgue–Stieltjes measure, i.e., the *complete* measure induced by the Lebesgue–Stieltjes outer measure λ_F^*. It turns out that the Lebesgue–Stieltjes measure λ_F is in fact the *completion* of the unique measure λ_F on $\mathcal{B}_{\mathbb{R}}$ given by Theorem 12.1.38.

Exercise 12.1.40. Let $F : \mathbb{R} \to \mathbb{R}$ be an increasing, right-continuous function. Show that the corresponding Lebesgue–Stieltjes measure λ_F is the *completion* of the unique measure λ_F provided by Theorem 12.1.38.

We end this section by giving a necessary and sufficient condition that an outer measure μ^* on a metric space (M, d) must satisfy in order for each *Borel set* of M to be μ^*-measurable. We shall need the following.

Definition 12.1.41 (Metric Outer Measure). We say that an outer measure μ^* on a metric space (M, d) is a *metric outer measure* if, for any sets A, $B \subset M$, we have

$$d(A, B) > 0 \implies \mu^*(A \cup B) = \mu^*(A) + \mu^*(B),$$

where $d(A, B) := \inf\{d(a, b) : a \in A, b \in B\}$.

Before stating the theorem, let us prove a lemma.

Lemma 12.1.42. *Let μ^* be a metric outer measure on a metric space (M, d) and let $O \subset M$ be a nonempty open set. If $\emptyset \neq E \subset O$ and if, for each $n \in \mathbb{N}$, we define $E_n := E \cap \{x : d(x, O^c) \geq 1/n\}$, then*

$$\lim_{n \to \infty} \mu^*(E_n) = \mu^*(E). \tag{†}$$

Proof. First, it is obvious that $(E_n)_{n=1}^{\infty}$ is an *increasing* sequence of subsets of E; i.e., $E_n \subset E_{n+1} \subset E$ for all $n \in \mathbb{N}$. Also, since O^c is *closed*, we have $E = \bigcup_{n=1}^{\infty} E_n$. (Why?) Now define $E_0 := \emptyset$ and $D_n := E_{n+1} \setminus E_n$ for all $n \in \mathbb{N}_0$. If D_{n+1} and E_n are both nonempty, then $d(D_{n+1}, E_n) > 0$. Indeed, if $d(D_{n+1}, E_n) = 0$, then we can find $x \in D_{n+1}$ and $y \in E_n$ such that $d(x, y) < 1/n(n + 1)$. But then

$$d(y, O^c) \le d(x, y) + d(x, O^c) < \frac{1}{n(n+1)} + \frac{1}{n+1} = \frac{1}{n},$$

which is absurd. It follows that, for each $n \in \mathbb{N}_0$, we have $\mu^*(D_{n+1} \cup E_n) = \mu^*(D_{n+1}) + \mu^*(E_n)$, and hence

$$\mu^*(D_{n+1}) = \mu^*(D_{n+1} \cup E_n) - \mu^*(E_n) \le \mu^*(E_{n+2}) - \mu^*(E_n). \tag{$*$}$$

Applying ($*$) repeatedly, we obtain the inequalities

$$\mu^*(E_{2n+1}) \ge \sum_{j=1}^{n} \mu^*(D_{2j}) \quad \text{and} \quad \mu^*(E_{2n}) \ge \sum_{j=1}^{n} \mu^*(D_{2j-1}). \tag{$**$}$$

Thus ($\dagger$) is trivially satisfied if either one of the two series

$$\sum_{n=1}^{\infty} \mu^*(D_{2n}) \quad \text{and} \quad \sum_{n=1}^{\infty} \mu^*(D_{2n-1})$$

diverges. So assume that they *both converge* and note that $E_n = \bigcup_{j=1}^{n} E_j = \bigcup_{j=1}^{n} D_j$ for all $n \in \mathbb{N}$ and $E = \bigcup_{n=1}^{\infty} E_n = \bigcup_{n=1}^{\infty} D_n$. Therefore, using ($**$) and the countable subadditivity of μ^*, we have

$$\mu^*(E) \le \mu^*(E_{2n-1}) + \mu^*\left(\bigcup_{j=2n}^{\infty} D_j \right)$$

$$\le \mu^*(E_{2n-1}) + \sum_{j=n}^{\infty} \mu^*(D_{2j}) + \sum_{j=n+1}^{\infty} \mu^*(D_{2j-1}),$$

which, in view of the fact that $(\mu^*(E_n))$ is an *increasing sequence*, shows that ($\dagger$) holds (why?) and completes the proof. $\qquad\square$

Theorem 12.1.43. *Let μ^* be an outer measure on a metric space (M, d). Then μ^* is a metric outer measure if and only if every Borel set of M is μ^*-measurable.*

Proof. Assume first that μ^* is a metric outer measure and let $O \subset M$ be an open set. Given any $A \subset M$, let $E := A \cap O$ and define the sets E_n as in Lemma 12.1.42. Now note that, for each $n \in \mathbb{N}$, we have $d(E_n, A \cap O^c) > 0$. Therefore,

$$\mu^*(A) \ge \mu^*(E_n \cup (A \cap O^c)) = \mu^*(E_n) + \mu^*(A \cap O^c).$$

Letting $n \to \infty$ and using Lemma 12.1.42, we have

$$\mu^*(A) \ge \mu^*(A \cap O) + \mu^*(A \cap O^c),$$

which proves that (the *arbitrary* open set) O is measurable and hence so is every Borel set. Conversely, assume that all Borel sets are measurable and pick arbitrary sets A, $B \subset M$ with $\delta := d(A, B) > 0$. Then, for each $x \in A$, the set $O_x := \{y \in M : d(x, y) < \delta/2\}$ is open and hence so is the set $O := \bigcup_{x \in A} O_x$. Also, we have $A \subset O$ (i.e., $A \cap O = A$) and $B \cap O = \emptyset$. Since O is *measurable,* we now have

$$\mu^*(A \cup B) = \mu^*((A \cup B) \cap O) + \mu^*((A \cup B) \cap O^c)$$
$$= \mu^*(A) + \mu^*(B),$$

and the proof is complete. $\square$

Exercise 12.1.44.

(a) Show that the Lebesgue outer measure λ^* is a metric outer measure on $\mathbb{R}$. Show that the same holds for the Lebesgue outer measure λ_n^* on $\mathbb{R}^n$ for $n > 1$.
(b) Let (M, d) be a separable metric space and $p \geq 0$. Show that the Hausdorff outer measure μ_p^* is a metric outer measure on M. *Hint:* Let A, $B \subset M$ with $d(A, B) > 0$. For any $\varepsilon \in (0, d(A, B))$, pick a sequence $(C_n)_{n=1}^\infty$ of subsets of M with $A \cup B \subset \bigcup_{n=1}^\infty C_n$ and $\delta(C_n) \leq \varepsilon$ for all $n \in \mathbb{N}$. Observe that no C_n intersects both A and B and split $\sum_{n=1}^\infty (\delta(C_n))^p$ into two parts according to whether $C_n \cap A = \emptyset$ or $C_n \cap B = \emptyset$. Deduce that

$$\sum_{n=1}^\infty (\delta(C_n))^p \geq \mu_p^*(A) + \mu_p^*(B).$$

12.2   Measurable Functions

We now want to take the natural step of extending the notions of (Lebesgue) measurability and integrability from the measure space $(\mathbb{R}, \mathcal{M}_\lambda, \lambda)$ to a general measure space $(X, \mathcal{A}, \mu)$. Let us recall that, if $\mathcal{C} \subset \mathcal{P}(X)$, then the σ-algebra *generated by* $\mathcal{C}$ is denoted by $\mathcal{A}_\mathcal{C}$. In particular, if X is a *metric space* and $\mathcal{O}_X$ is the set of all *open* sets in X, then $\mathcal{B}_X := \mathcal{A}_{\mathcal{O}_X}$ is the *Borel algebra* of X. In this case, the measurable space $(X, \mathcal{B}_X)$ will be called a *Borel space*. We shall also need the Borel algebra of $\overline{\mathbb{R}} := [-\infty, \infty]$, defined by

$$\mathcal{B}_{\overline{\mathbb{R}}} := \{E \subset \overline{\mathbb{R}} : E \cap \mathbb{R} \in \mathcal{B}_\mathbb{R}\}.$$

Note that any open set in $\overline{\mathbb{R}}$ is a union of intervals of the form $(a, b), [-\infty, b)$, and $(a, \infty]$, where a, $b \in \mathbb{R}$.

Exercise 12.2.1. Recall (cf. Example 5.1.3(2)) that $\overline{\mathbb{R}}$ is a *metric space* with metric $d(x, y) := |f(x) - f(y)|$, where $f(x) := x/(1 + |x|)$ for all $x \in \mathbb{R}$ and $f(\pm\infty) := \pm 1$. Show that $\mathcal{B}_{\overline{\mathbb{R}}}$ defined above is indeed the Borel algebra of the metric space $(\overline{\mathbb{R}}, d)$.

Here is a routine but useful exercise:

Exercise 12.2.2. For any sets X, X' and any function $f : X \to X'$, prove the following:

1. If $\mathcal{A}' \subset \mathcal{P}(X')$ is a σ-algebra, then the collection

$$f^{-1}(\mathcal{A}') := \{f^{-1}(A') : A' \in \mathcal{A}'\} \subset \mathcal{P}(X)$$

is a σ-algebra (on X).
2. If $\mathcal{C}' \subset \mathcal{P}(X')$ and $\mathcal{C} = f^{-1}(\mathcal{C}') := \{f^{-1}(C') : C' \in \mathcal{C}'\}$, then $\mathcal{A}_{\mathcal{C}} = f^{-1}(\mathcal{A}_{\mathcal{C}'})$.

Definition 12.2.3 (Measurable Function, Borel Function). Let $(X, \mathcal{A})$ be a measurable space and $(Y, \mathcal{B}_Y)$ a Borel space. A function $f : X \to Y$ is said to be *measurable* if (with notation as in Exercise 12.2.2)

$$f^{-1}(\mathcal{B}_Y) \subset \mathcal{A}.$$

The set of all measurable functions from X to Y (or from $(X, \mathcal{A})$ to $(Y, \mathcal{B}_Y)$) will be denoted by $\mathcal{L}^0(X, Y)$ (or, more accurately, $\mathcal{L}^0_{\mathcal{A}, \mathcal{B}_Y}(X, Y)$). If X is also a metric space and $\mathcal{A} = \mathcal{B}_X$, then a measurable function $f : X \to Y$ is said to be *Borel measurable* or simply a *Borel function*.

Notation 12.2.4 (Nonnegative Measurable Functions). Given a measurable space $(X, \mathcal{A})$, the set of all *measurable* functions $f : X \to [0, \infty]$ will be denoted by $\mathcal{L}^0_+(X)$.

Definition 12.2.5 (Restriction of a σ-Algebra). Let $\mathcal{A} \subset \mathcal{P}(X)$ be a σ-algebra and let $Y \subset X$. The *restriction* of $\mathcal{A}$ to Y is defined to be

$$\mathcal{A}|_Y := \{A \cap Y : A \in \mathcal{A}\}.$$

It is easy to see (cf. Problem 1.5.#8) that $\mathcal{A}|_Y$ is a σ-algebra on Y.

Here is a nice way of constructing a measurable function using a sequence of measurable functions defined on members of a disjoint cover.

Proposition 12.2.6. *Let $(X, \mathcal{A})$ be a measurable space and $(Y, \mathcal{B}_Y)$ a Borel space. Let (X_n) be a sequence in $\mathcal{A}$ of pairwise disjoint sets with $X = \bigcup_{n=1}^{\infty} X_n$ and for each $n \in \mathbb{N}$, let f_n be a measurable function from $(X_n, \mathcal{A}|_{X_n})$ to $(Y, \mathcal{B}_Y)$. Define f by $f(x) := f_n(x)$ if $x \in X_n$. Then $f : X \to Y$ is measurable.*

Proof. Well, by assumption, given any Borel set $B \subset Y$ and any $n \in \mathbb{N}$, we have $f_n^{-1}(B) = A_n \cap X_n$ for some $A_n \in \mathcal{A}$. But then

$$f^{-1}(B) = \bigcup_{n=1}^{\infty} f_n^{-1}(B) = \bigcup_{n=1}^{\infty} A_n \cap X_n \in \mathcal{A}.$$

$\square$

Proposition 12.2.7. *Let the notation be as in Definition 12.2.3. Then we have* $f \in \mathcal{L}^0(X, Y)$ *if and only if* $f^{-1}(O) \in \mathcal{A}$ *for each open set* $O \subset Y$. *In particular, a function* $f : X \to [-\infty, \infty]$ *is measurable if and only if* $f^{-1}((a, \infty]) \in \mathcal{A}$ *for all* $a \in \mathbb{R}$.

Proof. Exercise! Note that $\mathcal{B}_{\overline{\mathbb{R}}}$ is generated by the intervals $(a, \infty]$, $a \in \mathbb{R}$ and use Exercise 12.2.2. In fact, there are other equivalent conditions for the measurability of $f : X \to \overline{\mathbb{R}}$.

Exercise 12.2.8. Show that $f \in \mathcal{L}^0(X, \overline{\mathbb{R}})$ if and only if $f^{-1}([a, \infty]) \in \mathcal{A}$ if and only if $f^{-1}([-\infty, b)) \in \mathcal{A}$ if and only if $f^{-1}([-\infty, b]) \in \mathcal{A}$, for all a, $b \in \mathbb{R}$. Deduce that, for any real $c \neq 0$, f is measurable if and only if cf is measurable. For $c = 0$, we have $cf \equiv 0$, which is measurable since any *constant* function is measurable. (Why?)

Corollary 12.2.9. *Let* $(X, \mathcal{B}_X)$ *and* $(Y, \mathcal{B}_Y)$ *be Borel spaces. Then every continuous function* $f : X \to Y$ *is a Borel function.*

Proposition 12.2.10 (Stability Under Limits). *Let* $(X, \mathcal{A})$ *be a measurable space and let* $f_n \in \mathcal{L}^0(X, \overline{\mathbb{R}})$ *for all* $n \in \mathbb{N}$. *Then the functions* $x \mapsto \sup\{f_n(x) : n \in \mathbb{N}\}$, $x \mapsto \inf\{f_n(x) : n \in \mathbb{N}\}$, $x \mapsto \limsup_{n \to \infty}\{f_n(x) : n \in \mathbb{N}\}$, *and* $x \mapsto \liminf_{n \to \infty}\{f_n(x) : n \in \mathbb{N}\}$ *are also measurable.*

Proof. Simply note that, if $f(x) := \sup\{f_n(x) : n \in \mathbb{N}\}$, then $f^{-1}((a, \infty]) = \bigcup_{n=1}^{\infty} f_n^{-1}((a, \infty])$. Since $\inf(f_n) = -\sup(-f_n)$, the function $x \mapsto \inf\{f_n(x) : n \in \mathbb{N}\}$ is also measurable. Next, note that

$$\limsup_{n \to \infty} f_n(x) = \inf\big\{\sup\{f_k(x) : k \geq n\} : n \in \mathbb{N}\big\},$$

which implies the measurability of $\limsup(f_n)$. The statement for $\liminf$ is proved similarly or by considering $(-f_n)$.   □

Corollary 12.2.11. *If* $(X, \mathcal{A})$ *is a measurable space, if* $f_n \in \mathcal{L}^0(X, \mathbb{C})$ *for all* $n \in \mathbb{N}$, *and if* $f(x) = \lim_{n \to \infty} f_n(x)$ *for all* $x \in X$, *then* $f \in \mathcal{L}^0(X, \mathbb{C})$.

Proposition 12.2.12 (Composition with Continuous Functions). *Suppose that* $(Y, \mathcal{B}_Y)$ *and* $(Z, \mathcal{B}_Z)$ *are Borel spaces and* $(X, \mathcal{A})$ *is a measurable space. Then, for any* $f \in \mathcal{L}^0(X, Y)$ *and any continuous function* $g : Y \to Z$, *the composite function* $g \circ f : X \to Z$ *is measurable.*

Proof. Exercise!   □

Proposition 12.2.13. *Let* $(X, \mathcal{A})$ *be a measurable space, and let* u *and* v *be real-valued measurable functions on* X. *If* $(Y, \mathcal{B}_Y)$ *is a Borel space and* $F : \mathbb{R}^2 \to Y$ *is a continuous function, then the function* $h : X \to Y$ *defined by* $h(x) := F(u(x), v(x))$ *is measurable.*

Proof. Since $h = F \circ g$, where $g(x) := (u(x), v(x))$, it suffices (by Proposition 12.2.12) to prove that $g : X \to \mathbb{R}^2$ is measurable. But, given any open set $O \subset \mathbb{R}^2$, there are two sequences (I_m) and (J_n) of (pairwise disjoint) *open intervals* such that, with $R_{mn} := I_m \times J_n$, we have $O = \bigcup_{(m,n) \in \mathbb{N} \times \mathbb{N}} R_{mn}$. (Why?) Since

$$g^{-1}(O) = g^{-1}\left(\bigcup_{m,n} R_{mn}\right) = \bigcup_{m,n} g^{-1}(R_{mn}) = \bigcup_{m,n}\left(u^{-1}(I_m) \cap v^{-1}(J_n)\right),$$

and since u and v are measurable, the measurability of h follows. □

Proposition 12.2.14 (Stability Under Algebraic Operations). *Let $(X, \mathcal{A})$ be a measurable space.*

(a) *A function $f : X \to \mathbb{C}$ is measurable if and only if $\mathrm{Re}(f)$ and $\mathrm{Im}(f)$ are measurable. In this case, $|f|$ is also measurable.*
(b) *If $f, g : X \to \mathbb{C}$ are measurable, then so are $f + g$ and fg. Moreover, if $g(x) \neq 0$ for all $x \in X$, then f/g is also measurable.*
(c) *If $f, g : X \to \overline{\mathbb{R}}$ are measurable, then so are the functions $f \vee g := \max\{f, g\}$ and $f \wedge g := \min\{f, g\}$. In particular, $f^+ := f \vee 0$ and $f^- := (-f) \vee 0$ are measurable.*

Proof. For (a), we simply use Proposition 12.2.13 with the continuous function $F(x, y) := (x, y)$, and Proposition 12.2.12 with the continuous functions $z \mapsto \mathrm{Re}(z)$, $z \mapsto \mathrm{Im}(z)$, and $z \mapsto |z|$. Also, (b) follows from Proposition 12.2.13 with the continuous functions $F(x, y) := x + y$, $F(x, y) := xy$ and, for $y \neq 0$, $F(x, y) := x/y$. Finally, for (c), we use Proposition 12.2.13 with $F(x, y) := \max\{x, y\}$ and $F(x, y) := \min\{x, y\}$. □

Exercise 12.2.15. Let $(X, \mathcal{A}, \mu)$ be a *complete* measure space. Prove the following assertions.

(a) If $f \in \mathcal{L}^0(X, \mathbb{C})$, $g : X \to \mathbb{C}$, and $f(x) = g(x)$ for *almost all* $x \in X$, then $g \in \mathcal{L}^0(X, \mathbb{C})$.
(b) If f and f_n, $n \in \mathbb{N}$, are in $\mathcal{L}^0(X, \mathbb{C})$ and $\lim(f_n(x)) = f(x)$ for *almost all* $x \in X$, then $f \in \mathcal{L}^0(X, \mathbb{C})$.

The following definition is an extension of Definition 10.2.1 and Remark 10.2.3:

Definition 12.2.16 (Simple Function, Canonical Representation). Let $(X, \mathcal{A})$ be a measurable space and let $\mathbb{F}$ denote either $\mathbb{R}$ or $\mathbb{C}$. A function $\phi : X \to \mathbb{F}$ is said to be a *simple function* if there are measurable sets $E_j \in \mathcal{A}$ and constants $c_j \in \mathbb{F}$, $1 \leq j \leq n$, such that

$$\phi := \sum_{j=1}^{n} c_j \chi_{E_j}. \tag{†}$$

The set of all $\mathbb{F}$-valued [resp., nonnegative] simple functions on X will be denoted by $Simp(X, \mathbb{F})$ [resp., $Simp^+(X)$]. The representation (†) is called the *canonical representation* of ϕ if the c_j are all *distinct* ($c_i \neq c_j$ for $i \neq j$) and $E_j := \phi^{-1}(\{c_j\})$, for $1 \leq j \leq n$. Note that we then obviously have $E_i \cap E_j = \emptyset$ for $i \neq j$ and $X = \bigcup_{j=1}^{n} E_j$.

Remark 12.2.17. It is clear that a function $\phi : X \to \mathbb{F}$ is simple if and only if it is *measurable* and assumes only a *finite* number of values. (Why?)

Proposition 12.2.18. *Let $(X, \mathcal{A})$ be a measurable space. Given any measurable function $f : X \to [0, \infty]$, there is a sequence (ϕ_n) of nonnegative simple functions such that (ϕ_n) is increasing (i.e., $\phi_{n+1} \geq \phi_n$) and $\lim(\phi_n(x)) = f(x)$ for all $x \in X$. If f is bounded, then $(\phi_n) \to f$ uniformly on X. Furthermore, if $\mu : \mathcal{A} \to [0, \infty]$ is a σ-finite measure, then we may choose the ϕ_n so that each vanishes outside a set of finite measure.*

Proof. Exercise! *Hint:* Define the ϕ_n as in the proof of Theorem 10.2.24. $\square$

12.3   Integration

We now want to define the integral of functions defined on a general measure space $(X, \mathcal{A}, \mu)$. To this end, we first define the integral for nonnegative simple functions and then extend it to general measurable functions using Proposition 12.2.18. Throughout this section, $(X, \mathcal{A}, \mu)$ will be a measure space.

Definition 12.3.1 (Integral of a Simple Function). If $\phi \in Simp^+(X)$ has canonical representation $\phi := \sum_{j=1}^{n} c_j \chi_{E_j}$ and $E \in \mathcal{A}$, then the *(Lebesgue) integral of ϕ over E (with respect to μ)* is defined to be the *extended* nonnegative number

$$\int_E \phi \, d\mu := \sum_{j=1}^{n} c_j \mu(E_j \cap E). \tag{$*$}$$

Remark 12.3.2. Note that the sum in $(*)$ may contain a term of the form $0 \cdot \infty := 0$. Also, as we saw in Chap. 10 (cf. Lemma 10.3.4 and Corollary 10.3.6), the value of the integral $\int_E \phi \, d\mu$ is independent of the representation of ϕ as a (finite) linear combination of characteristic functions of measurable sets.

Definition 12.3.3 (Integral of a Nonnegative Function). If $f \in \mathcal{L}^0_+(X)$ and $E \in \mathcal{A}$, then the *(Lebesgue) integral of f over E (with respect to μ)* is defined to be the (extended) nonnegative number

$$\int_E f \, d\mu := \sup \left\{ \int_E \phi \, d\mu : \phi \in Simp^+(X), \ \phi \leq f \right\}.$$

If $E = X$, then we simply write

$$\int f \, d\mu := \int_X f \, d\mu.$$

Exercise 12.3.4. Show that if $f \in Simp^+(X)$, then the two definitions of $\int_E f \, d\mu$ given by Definitions 12.3.1 and 12.3.3 are the same.

Proposition 12.3.5. *Let* f, $g : X \rightarrow [0, \infty]$ *be measurable and let* A, B, *and* E *be measurable sets. Then the following are true:*

(a) $f \leq g$ *on* E *implies* $\int_E f \, d\mu \leq \int_E g \, d\mu$.
(b) $A \subset B$ *implies* $\int_A f \, d\mu \leq \int_B f \, d\mu$.
(c) $\int_E cf \, d\mu = c \int_E f \, d\mu$ *for all* $c \in [0, \infty)$.
(d) $f(x) = 0$ *for all* $x \in E$ *implies* $\int_E f \, d\mu = 0$, *even if* $\mu(E) = +\infty$.
(e) $\mu(E) = 0$ *implies* $\int_E f \, d\mu = 0$, *even if* $f(x) = +\infty$ *for all* $x \in E$.
(f) $\int_E f \, d\mu = \int_X f \chi_E \, d\mu$.

Proof. Exercise! □

Remark 12.3.6. Note that, by part (f), there is no loss of generality to restrict the definition of the integral to the case $E = X$.

Exercise 12.3.7. Show that, if $\phi \in Simp^+(X)$, then the map

$$\mu_\phi(E) := \int_E \phi \, d\mu \qquad (\forall E \in \mathcal{A})$$

is a *measure* on $\mathcal{A}$ such that $\mu(E) = 0$ implies $\mu_\phi(E) = 0$.

The general integral defined above is also *additive*, as we expect, but this is not easy to see from Definition 12.3.3. We prove the additivity for *simple functions* first and deduce the general case from the fundamental *convergence theorems*.

Proposition 12.3.8. *For any* ϕ, $\psi \in Simp^+(X)$ *and any* $E \in \mathcal{A}$, *we have*

$$\int_E (\phi + \psi) \, d\mu = \int_E \phi \, d\mu + \int_E \psi \, d\mu.$$

Proof. Let $\phi = \sum_{i=1}^m c_i \mu(E_i)$ and $\psi = \sum_{j=1}^n d_j \mu(F_j)$ be the canonical representations of ϕ and ψ, respectively. Then the collections (E_i), (F_j), and $(E_i \cap F_j)$ are all mutually disjoint covers of X and $\phi + \psi$ has the constant value $c_i + d_j$ on $E_i \cap F_j$. Thus

$$\int_E (\phi + \psi) \, d\mu = \sum_{i,j} (c_i + d_j)\mu(E_i \cap F_j \cap E)$$

$$= \sum_i c_i \sum_j \mu(E_i \cap F_j \cap E) + \sum_j d_j \sum_i \mu(E_i \cap F_j \cap E)$$

$$= \sum_i c_i \mu(E_i \cap E) + \sum_j d_j \mu(F_j \cap E)$$

$$= \int_E \phi \, d\mu + \int_E \psi \, d\mu.$$

□

Here is a useful inequality:

Proposition 12.3.9 (Chebyshev's Inequality). *Let* $f : X \to [0, \infty]$ *be a measurable function,* $E \in \mathcal{A}$, *and* $c > 0$. *If* $E_c := \{x \in E : f(x) \geq c\}$, *then we have*

$$\mu(E_c) \leq \frac{1}{c} \int_E f \, d\mu.$$

Proof. Since $f \geq c$ on E_c, parts (a) and (b) of Proposition 12.3.5 imply

$$c\mu(E_c) = \int_{E_c} c \, d\mu \leq \int_{E_c} f \, d\mu \leq \int_E f \, d\mu.$$

$\square$

Corollary 12.3.10. *If* $f \in \mathcal{L}_+^0(X)$ *and* $\int_E f \, d\mu < \infty$, *then*

$$\mu(\{x \in E : f(x) = +\infty\}) = 0;$$

i.e., f *is finite almost everywhere on* E.

Proof. Let $A := \{x \in E : f(x) = +\infty\}$ and $A_n := \{x \in E : f(x) \geq n\}$ for all $n \in \mathbb{N}$. Then we have $A = \bigcap_{n=1}^{\infty} A_n$. Now, by Chebyshev's inequality, we have

$$\mu(A) \leq \mu(A_n) \leq \frac{1}{n} \int_E f \, d\mu \qquad (\forall n \in \mathbb{N}).$$

Since $\int_E f \, d\mu < \infty$, the corollary follows. $\square$

Corollary 12.3.11. *Let* $f \in \mathcal{L}_+^0(X)$ *and let* $E \in \mathcal{A}$. *Then*

$$\int_E f \, d\mu = 0 \iff f = 0 \text{ a.e. on } E.$$

In particular, if $f(x) > 0$ *for all* $x \in E$, *then*

$$\int_E f \, d\mu = 0 \iff \mu(E) = 0.$$

Proof. Let $A := \{x \in E : f(x) > 0\}$ and let $A_n := \{x \in E : f(x) > 1/n\}$, for all $n \in \mathbb{N}$. Then $A = \bigcup_{n=1}^{\infty} A_n$ and it suffices to show that $\mu(A_n) = 0$ for all $n \in \mathbb{N}$. But, by Chebyshev's inequality, we have

$$\mu(A_n) \leq n \int_E f \, d\mu = 0 \qquad (\forall n \in \mathbb{N}).$$

If $f(x) > 0$ for all $x \in E$, then $\int_E f \, d\mu = \int_X f \chi_E \, d\mu = 0$ implies $f(x)\chi_E(x) = 0$, and hence $\chi_E(x) = 0$, for μ-almost all $x \in X$, which gives $\mu(E) = 0$. $\square$

Theorem 12.3.12 (Monotone Convergence Theorem). *Let* (f_n) *be an increasing sequence in* $\mathcal{L}_+^0(X)$; *i.e.,* $f_n(x) \le f_{n+1}(x)$ *for all* $n \in \mathbb{N}$ *and* $x \in X$. *If* $\lim(f_n(x)) = f(x)$ *for all* $x \in X$, *then* $f \in \mathcal{L}_+^0(X)$ *and we have*

$$\lim_{n \to \infty} \int_X f_n \, d\mu = \int_X f \, d\mu.$$

Proof. Since $(\int f_n \, d\mu)$ is an increasing sequence in $[0, \infty]$, it converges. Also, $\int f_n \, d\mu \le \int f \, d\mu$ for all $n \in \mathbb{N}$ implies that $\lim(\int f_n \, d\mu) \le \int f \, d\mu$. To prove the reverse inequality, fix $\delta \in (0, 1)$. Pick any $\phi \in \mathbf{Simp}^+(X)$ with $\phi \le f$ and let $E_n := \{x : f_n(x) \ge \delta\phi(x)\}$. Then the E_n are measurable, $E_n \subset E_{n+1}$ for all $n \in \mathbb{N}$, and $\bigcup_{n=1}^\infty E_n = X$. Also, we have $\int f_n \, d\mu \ge \int_{E_n} f_n \, d\mu \ge \delta \int_{E_n} \phi$. By Exercise 12.3.7 and Theorem 12.1.4(5), we have $\lim(\int_{E_n} \phi \, d\mu) = \int \phi \, d\mu$ and hence $\lim(\int f_n \, d\mu) \ge \delta \int \phi \, d\mu$. Since this is true for all $\delta < 1$, it remains true for $\delta = 1$ and, taking the supremum over all $\phi \in \mathbf{Simp}^+(X)$ with $\phi \le f$, we obtain $\lim(\int f_n \, d\mu) \ge \int f \, d\mu$. $\qquad\square$

Remark 12.3.13.

1. The assumption that the sequence (f_n) is *increasing* (at least almost everywhere) is necessary. Indeed, in $\mathbb{R}$ with Lebesgue measure λ, the sequence $(\chi_{(n,n+1)})$ converges to zero pointwise, but $\int \chi_{(n,n+1)} \, d\lambda = 1$ for all $n \in \mathbb{N}$. Note, however, that we will always have $\int f \, d\mu \le \lim_{n \to \infty} \int f_n \, d\mu$ (cf. Fatou's Lemma below).

2. The fact that sets of measure zero are *negligible* in integration suggests that we allow functions that are *defined almost everywhere*. Thus, if $(X, \mathcal{A}, \mu)$ is a measure space and $A \in \mathcal{A}$ with $\mu(A^c) = 0$, we say that $f : A \to \mathbb{R}$ (resp., $f : A \to \overline{\mathbb{R}}$) is *measurable on* X if $f^{-1}(B) \cap A \in \mathcal{A}$ for every $B \in \mathcal{B}_\mathbb{R}$ (resp., $B \in \mathcal{B}_{\overline{\mathbb{R}}}$). It then follows that $\tilde{f} \in \mathcal{L}^0(X, \mathbb{R})$ (resp., $\tilde{f} \in \mathcal{L}^0(X, \overline{\mathbb{R}})$), where the *trivial extension* $\tilde{f}$ is defined by $\tilde{f}(x) := f(x)$ if $x \in A$ and $\tilde{f}(x) := 0$ if $x \notin A$. (Why?) In fact, if μ is *complete*, then we may define f on A^c arbitrarily and still get measurable extension to X. (Why?)

Corollary 12.3.14. *If, in Theorem 12.3.12, we have* $\lim(f_n(x)) = f(x)$ *for almost all* $x \in X$, *then we still have* $\int f \, d\mu = \lim_{n \to \infty} \int f_n \, d\mu$.

Proof. Suppose that the $f_n(x)$ increase to $f(x)$ for all $x \in E \subset X$ with $\mu(E^c) = 0$. Then $f = f\chi_E$ and $f_n = f_n\chi_E$ almost everywhere, and hence

$$\int f \, d\mu = \int f\chi_E \, d\mu = \lim_{n \to \infty} \int f_n\chi_E \, d\mu = \lim_{n \to \infty} \int f_n \, d\mu.$$

$\qquad\square$

Corollary 12.3.15 (Additivity of the Integral). *If* $f, g \in \mathcal{L}_+^0(X)$ *and* $E \in \mathcal{A}$, *then we have*

$$\int_E (f + g) \, d\mu = \int_E f \, d\mu + \int_E g \, d\mu.$$

Proof. Without loss of generality, we may (and do) assume that $E = X$. Pick increasing sequences (ϕ_n) and (ψ_n) of simple functions such that $\lim(\phi_n) = f$ and $\lim(\psi_n) = g$. Then $(\phi_n + \psi_n)$ is *increasing* and $\lim(\phi_n + \psi_n) = f + g$. Using Proposition 12.3.8 and the Monotone Convergence Theorem, we have

$$\int (f + g)\, d\mu = \lim_{n \to \infty} \int \phi_n\, d\mu + \lim_{n \to \infty} \int \psi_n\, d\mu = \int f d\mu + \int g\, d\mu.$$

$\square$

Corollary 12.3.16. *Let* $(f_n) \in \mathcal{L}_+^0(X)^{\mathbb{N}}$ *and let* $f(x) := \sum_{n=1}^{\infty} f_n(x)$ *for all* $x \in X$. *Then*

$$\int f\, d\mu = \sum_{n=1}^{\infty} \int f_n\, d\mu.$$

Proof. Let $g_n := \sum_{k=1}^{n} f_k$. Then a simple induction using Corollary 12.3.15 shows that $\int g_n\, d\mu = \sum_{k=1}^{n} \int f_k\, d\mu$. Applying the Monotone Convergence Theorem, the corollary follows at once. $\square$

Corollary 12.3.17. *Given any* $g \in \mathcal{L}_+^0(X)$, *the set function*

$$\mu_g(E) := \int_E g\, d\mu \qquad (\forall E \in \mathcal{A}) \tag{$*$}$$

is a measure on $\mathcal{A}$ *and, for each* $f \in \mathcal{L}_+^0(X)$,

$$\int f\, d\mu_g = \int fg\, d\mu. \tag{$**$}$$

In particular, μ_g *is absolutely continuous with respect to* μ *in the sense that* $\mu(E) = 0$ *implies* $\mu_g(E) = 0$.

Proof. Let (E_n) be a sequence of *pairwise disjoint* elements of $\mathcal{A}$ with $E := \bigcup_{n=1}^{\infty} E_n$ and note that $g\chi_E = \sum_{n=1}^{\infty} g\chi_{E_n}$. Since $\mu_g(E) = \int g\chi_E\, d\mu$ and $\mu_g(E_n) = \int g\chi_{E_n}\, d\mu$, Corollary 12.3.16 gives $\mu_g(E) = \sum_{n=1}^{\infty} \mu_g(E_n)$. Also, $\mu_g(\emptyset) = 0$ and hence the first part of the corollary is proved. Next, note that $(*)$ implies $(**)$ for $g = \chi_E$ with any $E \in \mathcal{A}$. Therefore, $(**)$ holds for any (nonnegative) *simple* function. The general case now follows from the Monotone Convergence Theorem (cf. Proposition 12.2.18). $\square$

Remark 12.3.18. The *absolute continuity* referred to in this corollary (cf. also Corollary 10.4.7) will soon be extended to a general relation between a pair of measures and will provide a sufficient condition for a measure to have the form μ_g for a suitable $g \geq 0$. This will be the content of the Radon–Nikodym theorem (cf. Definition 12.3.31 and Theorem 12.3.35 below).

Theorem 12.3.19 (Fatou's Lemma). *For any sequence (f_n) in $\mathcal{L}^0_+(X)$, we have*

$$\int (\liminf_{n\to\infty} f_n)\, d\mu \leq \liminf_{n\to\infty} \int f_n\, d\mu. \qquad (\dagger)$$

Proof. Define $g_n(x) := \inf\{f_k(x) : x \in X, k \geq n\}$. Then $g_n \leq f_n$ for all $n \in \mathbb{N}$ and hence

$$\int g_n\, d\mu \leq \int f_n\, d\mu \quad (\forall n \in \mathbb{N}). \qquad (*)$$

Now, by Proposition 12.2.10, we have $g_n \in \mathcal{L}^0_+(X)$ for all $n \in \mathbb{N}$. Also, $g_1 \leq g_2 \leq \cdots$ and $\liminf(f_n) = \lim(g_n)$. Therefore, the Monotone Convergence Theorem implies that the left side of $(*)$ converges to the left side of $(\dagger)$ and hence $(\dagger)$ follows from $(*)$.   □

We now define the integral for general real (or complex-)-valued functions defined on a measure space $(X, \mathcal{A}, \mu)$:

Definition 12.3.20 (The Spaces $\mathcal{L}^1_\mu(X, \mathbb{R})$ and $\mathcal{L}^1_\mu(X, \mathbb{C})$). Let $(X, \mathcal{A}, \mu)$ be a measure space. For any function $f \in \mathcal{L}^0(X, \mathbb{R})$, we define its (Lebesgue) integral by

$$\int f\, d\mu := \int f^+\, d\mu - \int f^-\, d\mu$$

if at least one of the integrals on the right side is *finite*. If $\int f^+\, d\mu$ and $\int f^-\, d\mu$ are *both finite*, then we say that f is *integrable* (on X). The set of all real-valued integrable functions on X will be denoted by $\mathcal{L}^1_\mu(X, \mathbb{R})$. Next, if $f \in \mathcal{L}^0(X, \mathbb{C})$, we say that f is *integrable* if $\mathrm{Re}(f)$ and $\mathrm{Im}(f)$ are both integrable. In this case, we define

$$\int f\, d\mu := \int \mathrm{Re}(f)\, d\mu + i \int \mathrm{Im}(f)\, d\mu. \qquad (*)$$

The set of all complex-valued integrable functions will be denoted by $\mathcal{L}^1_\mu(X, \mathbb{C})$.

Exercise 12.3.21.

(a) Show that $f \in \mathcal{L}^1_\mu(X, \mathbb{C})$ if and only if $\int |f|\, d\mu < \infty$.

(b) Let $\mathcal{A} := \mathcal{P}(X)$ and let $\mu := \nu$ be the *counting measure;* i.e., $\nu(A) = |A|$ is the cardinality of A if A is *finite* and $\nu(A) = \infty$ otherwise. Show that $f \in \mathcal{L}^1_\nu(X, \mathbb{C})$ if and only if $|f|$ is *summable* in the sense of Definition 2.4.1 and, in this case,

$$\int_X f\, d\nu = \sum_{x \in X} f(x).$$

Theorem 12.3.22 (Linearity of the Integral). *Given any* f, $g \in \mathcal{L}_{\mu}^{1}(X, \mathbb{C})$ *and any* a, $b \in \mathbb{C}$, *we have* $af + bg \in \mathcal{L}_{\mu}^{1}(X, \mathbb{C})$ *and*

$$\int (af + bg) \, d\mu = a \int f \, d\mu + b \int g \, d\mu. \tag{$\dagger$}$$

Proof. That $af + bg$ is measurable follows from Proposition 12.2.14. Also,

$$\int |af + bg| \, d\mu \leq \int (|a||f| + |b||g|) \, d\mu$$

$$= |a| \int |f| \, d\mu + |b| \int |g| \, d\mu < \infty,$$

by Proposition 12.3.5, and hence $af + bg \in \mathcal{L}_{\mu}^{1}(X, \mathbb{C})$. We next prove (†) for f, $g \in \mathcal{L}_{\mu}^{1}(X, \mathbb{R})$ and $a = b = 1$. If we let $h := f + g$, then $h^{+} - h^{-} = f^{+} - f^{-} + g^{+} - g^{-}$ and hence $h^{+} + f^{-} + g^{-} = h^{-} + f^{+} + g^{+}$, so that Corollary 12.3.15 implies

$$\int h^{+} \, d\mu + \int f^{-} \, d\mu + \int g^{-} \, d\mu = \int h^{-} \, d\mu + \int f^{+} \, d\mu + \int g^{+} \, d\mu,$$

from which $\int h \, d\mu = \int f \, d\mu + \int g \, d\mu$ follows at once. Next, by Proposition 12.3.5, we have $\int af \, d\mu = a \int f \, d\mu$ if f is real-valued and $a \geq 0$. For $a < 0$, note that $(-f)^{+} = f^{-}$ and $(-f)^{-} = f^{+}$ imply $\int (-f) \, d\mu = -\int f \, d\mu$. Thus, (†) holds for real-valued functions and real constants. Finally, for complex-valued functions, given any u, $v \in \mathcal{L}_{\mu}^{1}(X, \mathbb{R})$,

$$i \int (u + iv) \, d\mu = i \int u \, d\mu - \int v \, d\mu = \int (iu - v) \, d\mu = \int i(u + iv) \, d\mu,$$

where we have used the real case and the *definition* (∗) above. □

Proposition 12.3.23. *For each* $f \in \mathcal{L}_{\mu}^{1}(X, \mathbb{C})$, *we have*

$$\left| \int f \, d\mu \right| \leq \int |f| \, d\mu \qquad \text{(Triangle Inequlity)}$$

with equality holding if and only if $\alpha f = |f|$ *for some* $\alpha \in \mathbb{C}$ *with* $|\alpha| = 1$.

Proof. See the proof of properties (4) and (5) in Theorem 10.3.36. □

The *average value theorem* (cf. Theorem 10.3.38) also holds in general if the measure μ is *finite:*

Theorem 12.3.24 (Average Value Theorem). *Let* μ *be a finite measure on a measurable space* $(X, \mathcal{A})$. *If* $f \in \mathcal{L}_{\mu}^{1}(X, \mathbb{C})$ *and if* $S \subset \mathbb{C}$ *is a closed set with*

$$\mathbf{av}_A(f) := \frac{1}{\mu(A)} \int_A f \, d\mu \in S$$

for every $A \in \mathcal{A}$ with $\mu(A) > 0$, then $f(x) \in S$ for almost all $x \in X$. In particular, if f is real-valued and $S = [a, b]$, then $a \leq f(x) \leq b$ for almost all $x \in X$.

Proof. See the proof of Theorem 10.3.38. □

Exercise 12.3.25. Let $f, g \in \mathcal{L}^1_\mu(X, \mathbb{C})$. Show that

$$\int_X |f - g| \, d\mu = 0 \iff f = g \ a.e. \iff \int_E f \, d\mu = \int_E g \, d\mu \quad \forall E \in \mathcal{A}.$$

Hint: Let $h := f - g = u + iv$ and $E := \{x \in X : u(x) \geq 0\}$. Now note that, if $\int_E h \, d\mu = 0$, then $\int_E u^+ \, d\mu = \mathrm{Re}(\int_E h \, d\mu) = 0$ and Corollary 12.3.11 may be applied.

Remark 12.3.26 (The Normed Spaces $L^1_\mu(X, \mathbb{F})$). Let $\mathbb{F}$ be either $\mathbb{R}$ or $\mathbb{C}$ and let $(X, \mathcal{A}, \mu)$ be any measure space. For any $f, g \in \mathcal{L}^0(X, \mathbb{F})$, we define $f \sim g$ if $f = g$ (μ-) a.e. (on X). This is easily seen to be an *equivalence relation* on $\mathcal{L}^0(X, \mathbb{F})$ and the equivalence class of each function f will still be denoted by f. If $f, g \in \mathcal{L}^1_\mu(X, \mathbb{F})$ and $f \sim g$, then (by the above exercise) $\int_X f \, d\mu = \int_X g \, d\mu$. The vector space of all equivalence classes of functions in $\mathcal{L}^1_\mu(X, \mathbb{F})$ will be denoted by $L^1_\mu(X, \mathbb{F})$. For each $f \in L^1_\mu(X, \mathbb{F})$, its L^1-*norm* is defined by

$$\|f\|_1 := \int_X |f| \, d\mu.$$

That this is indeed a norm follows from Exercise 12.3.25. We shall see below that (as in Chap. 10) the above spaces are *complete* and hence *Banach spaces*.

We now prove (Lebesgue's) *Dominated Convergence Theorem*:

Theorem 12.3.27 (Dominated Convergence Theorem). *Let (f_n) be a sequence in $\mathcal{L}^0(X, \mathbb{C})$ such that $\lim(f_n(x)) = f(x)$ for all $x \in X$. If there is a (nonnegative) function $g \in \mathcal{L}^1_\mu(X, [0, \infty))$ such that $|f_n| \leq g$ for all $n \in \mathbb{N}$, then $f \in \mathcal{L}^1_\mu(X, \mathbb{C})$,*

$$\lim_{n \to \infty} \int_X |f_n - f| \, d\mu = 0, \tag{*}$$

and

$$\lim_{n \to \infty} \int_X f_n \, d\mu = \int_X f \, d\mu. \tag{**}$$

Proof. Since f is measurable and $|f| \leq g$, it follows that $f \in \mathcal{L}^1_\mu(X, \mathbb{C})$. Since $|f_n - f| \leq 2g$, Fatou's lemma can be applied to the sequence $(2g - |f_n - f|)$ and implies

$$\int 2g \, d\mu \le \liminf_{n \to \infty} \int (2g - |f_n - f|) \, d\mu$$

$$= \int 2g \, d\mu + \liminf_{n \to \infty} \left(-\int |f_n - f| \, d\mu \right)$$

$$= \int 2g \, d\mu - \limsup_{n \to \infty} \int |f_n - f| \, d\mu.$$

Now $\int 2g \, d\mu < \infty$ and hence we can cancel it to obtain

$$\limsup_{n \to \infty} \int |f_n - f| \, d\mu \le 0,$$

from which (∗) follows at once. (Why?) Finally, applying Proposition 12.3.23 to $f_n - f$, we deduce (∗∗) from (∗).   □

As we saw in Chap. 11, for each $E \in \mathcal{M}_\lambda(\mathbb{R})$ and $p \in [1, \infty]$, the normed spaces $L^p(E, \mathbb{F})$ are complete and hence Banach spaces. The L^p-spaces can also be defined in general measure spaces $(X, \mathcal{A}, \mu)$ and the corresponding results in the general case can be proved exactly the same way. We therefore give the definitions and the summary of the results. The proofs are left as exercises and can be provided by the reader.

Definition 12.3.28 (L_μ^p Spaces, $1 \le p < \infty$). Let $\mathbb{F}$ denote either $\mathbb{R}$ or $\mathbb{C}$ and let $(X, \mathcal{A}, \mu)$ be a measure space. For each $p \in [1, \infty)$, we denote by $\mathcal{L}_\mu^p(X, \mathbb{F})$ the set of all measurable functions $f \in \mathcal{L}^0(X, \mathbb{F})$ such that $\int_X |f|^p \, d\mu < \infty$. The L^p-norm of f is then defined to be

$$\|f\|_p := \left[\int_X |f(x)|^p \, d\mu(x) \right]^{1/p}. \qquad (\|\cdot\|_p)$$

(The Space L_μ^∞). For any $f \in \mathcal{L}^0(X, \mathbb{F})$, its L^∞-norm is defined to be

$$\|f\|_\infty := \inf \left\{ a \ge 0 : \mu(\{x : |f(x)| > a\}) = 0 \right\}. \qquad (\|\cdot\|_\infty)$$

We call $\|f\|_\infty$ the *essential supremum* of f. The space $\mathcal{L}_\mu^\infty(X, \mathbb{F})$ consists of all $f \in \mathcal{L}^0(X, \mathbb{F})$ such that $\|f\|_\infty < \infty$. The space of all equivalence classes of functions in $\mathcal{L}_\mu^p(X, \mathbb{F})$ [resp., $\mathcal{L}_\mu^\infty(X, \mathbb{F})$] modulo the set of all μ-null functions is then denoted by $L_\mu^p(X, \mathbb{F})$ [resp., $L_\mu^\infty(X, \mathbb{F})$].

Theorem 12.3.29. *Given any measure space* $(X, \mathcal{A}, \mu)$, *the following are true:*

1. If p, $q \in (1, \infty)$ satisfy $1/p + 1/q = 1$ and f, $g \in \mathcal{L}^0(X, \mathbb{F})$, then

$$\|fg\|_1 \le \|f\|_p \|g\|_q \qquad \text{(Hölder's inequality).}$$

Thus, if $f \in L_\mu^p(X, \mathbb{F})$ and $g \in L_\mu^q(X, \mathbb{F})$, then $fg \in L_\mu^1(X, \mathbb{F})$.

2. If $1 \leq p \leq \infty$ and $f, g \in L_\mu^p(X, \mathbb{F})$, then we have

$$\|f + g\|_p \leq \|f\|_p + \|g\|_p \qquad \text{(Minkowski's inequality)}.$$

3. Given any $a > 0$ and any $f \in L_\mu^p(X, \mathbb{F})$, we have

$$\mu(\{x : |f(x)| > a\}) \leq (\|f\|_p/a)^p \qquad \text{(Chebyshev's Inequality)}.$$

4. $L_\mu^p(X, \mathbb{F})$ is a Banach space for every $p \in [1, \infty]$.
5. The space $L_\mu^2(X, \mathbb{F})$ is a Hilbert space with inner product

$$\langle f, g \rangle := \int_X f \bar{g} \, d\mu$$

and norm $\|f\|_2 = \langle f, f \rangle^{1/2}$, and we have

$$|\langle f, g \rangle| \leq \|f\|_2 \|g\|_2 \qquad \text{(Cauchy-Schwarz inequality)}.$$

6. The set of all $\mathbb{F}$-valued simple functions that vanish outside sets of finite measure is dense in $L_\mu^p(X, \mathbb{F})$ for every $p \in [1, \infty)$.
7. The set of all $\mathbb{F}$-valued simple functions is dense in $L_\mu^\infty(X, \mathbb{F})$.
8. Let $p \in [1, \infty]$. Then any Cauchy sequence (f_n) in $L_\mu^p(X, \mathbb{F})$ with limit f has a subsequence that converges μ-almost everywhere to f.

Proof. Exercise!   □

An important consequence of the fact that $L_\mu^2(X, \mathbb{F})$ is a Hilbert space is the Riesz Representation Theorem:

Theorem 12.3.30 (Riesz Representation Theorem). *Let $(X, \mathcal{A}, \mu)$ be a measure space. Given any bounded (i.e., continuous) linear functional $\phi : L_\mu^2(X, \mathbb{C}) \to \mathbb{C}$, there exists a unique $g \in L_\mu^2(X, \mathbb{C})$ such that*

$$\phi(f) = \int_X f \bar{g} \, d\mu \qquad \forall f \in L_\mu^2(X, \mathbb{C}).$$

Before we prove the Radon–Nikodym theorem, as promised in Remark 12.3.18, we need the (general) definition of *absolute continuity:*

Definition 12.3.31.

(Absolutely Continuous, Mutually Singular). Let $\mathfrak{M}_\mathcal{A}$ denote the set of all (positive) measures on a measurable space $(X, \mathcal{A})$ and let $\mu, \nu \in \mathfrak{M}_\mathcal{A}$.
(Absolutely Continuous). We say that ν is *absolutely continuous* with respect to μ and write $\nu \ll \mu$, if we have

$$A \in \mathcal{A} \quad \text{and} \quad \mu(A) = 0 \Longrightarrow \nu(A) = 0.$$

(Mutually Singular).   We say that μ and ν are *mutually singular* and write $\mu \perp \nu$
if we have

$$\mu(A) = 0 \quad \text{and} \quad \nu(A^c) = 0 \quad \text{for some} \quad A \in \mathcal{A}.$$

Exercise 12.3.32. Given any measures ν_1, ν_2, ν, and μ on a measurable space
$(X, \mathcal{A})$, prove the following:

(a) $\mu \perp \mu \Longrightarrow \mu = 0$.
(b) $\nu_1 \ll \mu$ and $\nu_2 \ll \mu$ imply $\nu_1 + \nu_2 \ll \mu$.
(c) $\nu_1 \perp \mu$ and $\nu_2 \perp \mu$ imply $\nu_1 + \nu_2 \perp \mu$.
(d) $\nu_1 \ll \mu$ and $\nu_2 \perp \mu$ imply $\nu_1 \perp \nu_2$.
(e) $\nu \ll \mu$ and $\nu \perp \mu$ imply $\nu = 0$.

Here is a characterization of absolute continuity (cf. Corollary 10.4.7):

Proposition 12.3.33. *Let μ, $\nu \in \mathfrak{M}_{\mathcal{A}}$ be positive measures on a measurable space
$(X, \mathcal{A})$ and assume that ν is finite, i.e., $\nu(X) < \infty$. Then $\nu \ll \mu$ if and only if for
any $\varepsilon > 0$ there is a $\delta > 0$ such that $\mu(A) < \delta$ implies $\nu(A) < \varepsilon$.*

Proof. If the (ε, δ)-condition is satisfied and if $\mu(A) = 0$, then $\mu(A) < \delta$ for any
$\delta > 0$ and hence $\nu(A) < \varepsilon$ for all $\varepsilon > 0$, giving $\nu(A) = 0$. If the (ε, δ)-condition
is *false,* pick $\varepsilon_0 > 0$ such that for any $n \in \mathbb{N}$ there is a set $A_n \in \mathcal{A}$ with
$\mu(A_n) < 1/2^n$, but $\nu(A_n) \geq \varepsilon_0$. If $B_n := \bigcup_{k=n}^{\infty} A_k$ and $B := \bigcap_{n=1}^{\infty} B_n$, then we
have $B_{n+1} \subset B_n$ and $\mu(B_n) < 1/2^{n-1}$ for all $n \in \mathbb{N}$. In particular, $\mu(B_1) < 1$ and
$\nu(B_1) \leq \nu(X) < \infty$. By the *continuity* of μ and ν [cf. part (5) of Theorem 12.1.4,
which is a special case of the Monotone Convergence Theorem] we therefore have
$\mu(B) = 0$ and yet

$$\nu(B) = \lim_{n \to \infty} \nu(B_n) \geq \varepsilon_0 > 0,$$

so that $\nu \ll \mu$ is false as well. $\square$

We are now ready for our main theorem. Since its proof involves the sum $\mu + \nu$
of two measures, we invite the reader to solve the following.

Exercise 12.3.34. Given a pair of measures μ, $\nu \in \mathfrak{M}_{\mathcal{A}}$ on a measurable space
$(X, \mathcal{A})$, show that

$$\int_X f \, d(\mu + \nu) = \int_X f \, d\mu + \int_X f \, d\nu \qquad \left(\forall \, f \in \mathcal{L}_+^0(X) \right). \qquad (\dagger)$$

Also, using the inequalities $\mu \leq \mu + \nu$ and $\nu \leq \mu + \nu$, show that

$$f \in \mathcal{L}_{\mu+\nu}^1(X, \mathbb{C}) \Longleftrightarrow f \in \mathcal{L}_{\mu}^1(X, \mathbb{C}) \cap \mathcal{L}_{\nu}^1(X, \mathbb{C}).$$

Hint: Prove $(\dagger)$ for $f = \chi_A$, where $A \in \mathcal{A}$ (cf. Definition 12.1.10), then for all
simple functions, and finally for all nonnegative measurable functions. For a general
function $f \in \mathcal{L}^0(X, \mathbb{C})$, look at $u^{\pm}$ and $v^{\pm}$, where $u := \text{Re}(f)$ and $v := \text{Im}(f)$.

Theorem 12.3.35 (Radon–Nikodym). *Let $(X, \mathcal{A})$ be a measurable space and let μ, $\nu \in \mathfrak{M}_{\mathcal{A}}$ be finite (positive) measures with $\nu \ll \mu$. Then there is a unique function $g \in L_{\mu}^{1}(X, [0, \infty))$ such that $\nu = \mu_{g}$; i.e.,*

$$\nu(A) = \int_{A} g \, d\mu \qquad (\forall \, A \in \mathcal{A}).$$

Proof (von Neumann). Introduce the finite measure $\omega := \mu + \nu$ and note that for any $f \in L_{\omega}^{2}(X, \mathbb{C})$, whose norm will be denoted $\| f \|_{2,\omega}$, we have (by Cauchy–Schwarz)

$$\left| \int_{X} f \, d\nu \right| \leq \int_{X} |f| \, d\nu \leq \int_{X} |f| \, d\omega \leq \sqrt{\omega(X)} \| f \|_{2,\omega}.$$

In particular, the linear functional $\phi(f) := \int_{X} f \, d\nu$ is *bounded* on $L_{\omega}^{2}(X, \mathbb{C})$ and hence the Riesz Representation Theorem implies that

$$\int_{X} f \, d\nu = \int_{X} f h \, d\omega = \int_{X} f h \, d\mu + \int_{X} f h \, d\nu \qquad (*)$$

for a *unique* "equivalence class" $h \in L_{\omega}^{2}(X, \mathbb{C})$. Clearly, as a *function* in $\mathcal{L}_{\omega}^{2}(X, \mathbb{C})$, h is determined only ω-almost everywhere. Using $f = \chi_{A}$ in $(*)$ for any $A \in \mathcal{A}$ with $\omega(A) > 0$, we get $\int_{A} h \, d\omega = \nu(A) \leq \omega(A)$ and hence

$$0 \leq \frac{1}{\omega(A)} \int_{A} h \, d\omega \leq 1.$$

By the *Average value theorem* (Theorem 12.3.24) we therefore have $0 \leq h(x) \leq 1$ for almost all $x \in X$ and $(*)$ can be written as

$$\int_{X} f(1 - h) \, d\nu = \int_{X} f h \, d\mu \qquad \forall \, f \in L_{\omega}^{2}(X, \mathbb{C}). \qquad (**)$$

Since our construction of g will involve division by $1 - h$, let us set $Y := \{x \in X : 0 \leq h(x) < 1\}$ and $Z := \{x \in X : h(x) = 1\}$. If in $(**)$ we put $f = \chi_{Z}$, it follows that $\mu(Z) = 0$ and hence $\nu(Z) = 0$ as well, in view of the assumption $\nu \ll \mu$. Thus $\omega(Z) = 0$ and, given that the set $W := \{x \in X : h(x) < 0 \text{ or } h(x) > 1\}$ also satisfies $\omega(W) = 0$, we may (and do) assume that $0 \leq h(x) < 1$ for *all* $x \in X$ and that $(**)$ holds for this h and every $f \in L_{\omega}^{2}(X, \mathbb{C})$. Next, since h is *bounded* and ω is a *finite* measure, $f := (1 + h + h^{2} + \cdots + h^{n-1}) \chi_{A} \in L_{\omega}^{2}(X, \mathbb{C})$ for every $n \in \mathbb{N}$ and every $A \in \mathcal{A}$. So using it in $(**)$ and simplifying, we have

$$\int_{A} (1 - h^{n}) \, d\nu = \int_{A} \frac{h}{1 - h} (1 - h^{n}) \, d\mu \qquad (\forall \, n \in \mathbb{N}, \forall \, A \in \mathcal{A}), \qquad (***)$$

where the denominator $1 - h(x)$ on the right side is *strictly positive* for all $x \in X$. But $(1 - h^n)$ is an *increasing* sequence with $\lim(1 - h^n) = 1$ and hence using the Monotone Convergence Theorem in $(* * *)$ we finally obtain

$$v(A) = \int_A \frac{h}{1 - h} \, d\mu \qquad (\forall \, A \in \mathcal{A}).$$

In particular, setting $A := X$, we have $\int_X h/(1 - h) \, d\mu = v(X) < \infty$ and hence $h/(1 - h) \in L^1_\mu(X, [0, \infty))$. In other words, our desired function g is indeed $g := h/(1 - h)$. $\square$

Remark 12.3.36. Examining the above proof, we see that the assumption $v \ll \mu$ was only used to show that $\mu(Z) = 0$ implies $v(Z) = 0$, where $Z := \{x : h(x) = 1\}$. Thus, even if $v \ll \mu$ is *not* true, the arguments leading to $(**)$ and $\mu(Z) = 0$ are still valid. If we introduce the measures

$$v_{ac}(A) := v(A \cap Z^c) \quad \text{and} \quad v_s(A) := v(A \cap Z) \qquad (\forall \, A \in \mathcal{A}),$$

we then have $v = v_{ac} + v_s$. Also, $v_{ac}(Z) = \mu(Z) = v_s(Z^c) = 0$ shows that $v_s \perp \mu$ and $v_{ac} \perp v_s$ (i.e., v_s and μ are *mutually singular* in the sense of Definition 12.3.31 and so are v_{ac} and v_s). Finally, if $A \subset Z^c$ and $\mu(A) = 0$, then putting $f = \chi_A$ in $(**)$ gives $\int_A (1 - h) \, dv = \int_A h \, d\mu = 0$. But then $1 - h > 0$ on A implies (cf. Corollary 12.3.11) that $v(A) = 0$. Thus $\mu(A) = 0$ for $A \in \mathcal{A}$ implies $v_{ac}(A) = 0$ and hence $v_{ac} \ll \mu$; i.e., v_{ac} is *absolutely continuous* with respect to μ. To summarize, we have proved that

$$v = v_{ac} + v_s, \quad \text{where} \quad v_{ac} \ll \mu, \quad v_s \perp \mu, \quad \text{and} \quad v_{ac} \perp v_s.$$

Corollary 12.3.37 (Lebesgue Decomposition).   *Let μ and v be finite (positive) measures on a measurable space $(X, \mathcal{A})$. There is a unique pair of measures v_{ac} and v_s in $\mathfrak{M}_\mathcal{A}$ satisfying $v_{ac} \perp v_s$ such that*

$$v = v_{ac} + v_s, \quad v_{ac} \ll \mu, \quad \text{and} \quad v_s \perp \mu. \tag{$\dagger$}$$

Proof. In view of the above remark, we need only prove the *uniqueness*. Now, if another pair (v'_{ac}, v'_s) of measures also satisfy $(\dagger)$, then we have

$$v'_{ac} - v_{ac} = v_s - v'_s. \tag{$\ddagger$}$$

But $v'_{ac} - v_{ac} \ll \mu$ and $v_s - v'_s \perp \mu$, so it follows from Exercise 12.3.32(e) that both sides of $(\ddagger)$ are zero. $\square$

Remark 12.3.38. The restriction (in Theorem 12.3.35 and Corollary 12.3.37) that *both* measures μ and v be *finite* is too strong and both results are in fact true if the measure μ is only assumed to be σ-finite (cf. Definition 12.1.2) rather than finite. Indeed, we can then write $X = \bigcup_{n=1}^\infty X_n$, where $X_n \in \mathcal{A}$ and $\mu(X_n) < \infty$ for all n.

We may even assume that the X_n are *pairwise disjoint* and, for otherwise, we can use the sequence (X'_n), where $X'_1 := X_1$ and $X'_n := X_n \setminus \bigcup_{k=1}^{n-1} X_k$. If we define the measures $\nu^n(A) := \nu(A \cap X_n)$ and $\mu^n(A) := \mu(A \cap X_n)$, then (ν^n, μ^n) is a pair of finite measures for each $n \in \mathbb{N}$ and we have $\mu = \sum_n \mu^n$, $\nu = \sum_n \nu^n$.

Theorem 12.3.39 (Lebesgue–Radon–Nikodym).   *Let $(X, \mathcal{A})$ be a measurable space and let μ, $\nu \in \mathfrak{M}_A$, where μ is σ-finite and ν is finite.*

(a) *There is a unique pair of measures ν_{ac} and ν_s such that*

$$\nu = \nu_{ac} + \nu_s, \quad \nu_{ac} \ll \mu, \quad \nu_s \perp \mu, \quad \text{and} \quad \nu_{ac} \perp \nu_s. \tag{L}$$

(b) *There is a unique function $g \in L^1_\mu(X, [0, \infty))$ such that*

$$\nu_{ac}(A) = \int_A g \, d\mu \qquad (\forall \, A \in \mathcal{A}). \tag{R-N}$$

In particular, if $\nu \ll \mu$, then (R-N) *holds with $\nu_{ac} = \nu$ and we have an extension of Theorem 12.3.35 to the case where μ is σ-finite.*

Proof. With notation as in the above remark, if $\nu \ll \mu$, i.e., $\nu_{ac} = \nu$, then we have $\nu^n \ll \mu^n$ for all n. By Theorem 12.3.35, there is a (unique) μ^n-integrable function $g_n : X_n \to [0, \infty)$ such that $\nu(A \cap X_n) = \int_{A \cap X_n} g_n \, d\mu$ for every $A \in \mathcal{A}$. Now define $g : X \to [0, \infty)$ by setting $g(x) := g_n(x)$ if $x \in X_n$. Then (by Proposition 12.2.6) g is measurable and for each $A \in \mathcal{A}$, we have

$$\int_A g \, d\mu = \sum_{n=1}^\infty \int_{A \cap X_n} g \, d\mu = \sum_{n=1}^\infty \int_{A \cap X_n} g_n \, d\mu = \sum_{n=1}^\infty \nu(A \cap X_n) = \nu(A).$$

In particular, with $A = X$, we get $\int_X g \, d\mu = \nu(X) < \infty$ and hence $g \in L^1_\mu(X, [0, \infty))$ and the proof of part (b) is complete. Next, to establish (a), for each n we can use Corollary 12.3.37 to find a pair (ν^n_{ac}, ν^n_s) of measures such that

$$\nu^n = \nu^n_{ac} + \nu^n_s, \quad \nu^n_{ac} \ll \mu^n, \quad \nu^n_s \perp \mu^n, \quad \text{and} \quad \nu^n_{ac} \perp \nu^n_s.$$

If we set $\nu_{ac} := \sum_{n=1}^\infty \nu^n_{ac}$ and $\nu_s := \sum_{n=1}^\infty \nu^n_s$, then we can check easily that (ν_{ac}, ν_s) provides the desired Lebesgue decomposition of ν. $\square$

Exercise 12.3.40.   Show that the pair (ν_{ac}, ν_s) of measures constructed in the above proof is *unique* and the relations (L) in Theorem 12.3.39 are indeed satisfied.

Definition 12.3.41 (Radon–Nikodym Derivative, Density).   The function g in the Radon–Nikodym theorem (or the Lebesgue–Radon–Nikodym theorem when $\nu \ll \mu$) is called the *Radon–Nikodym derivative* or *density* of ν with respect to μ and is written $g = d\nu/d\mu$. The justification for this is the fact that $\nu(A) = \int_A g \, d\mu$ gives $\int_A \phi \, d\mu = \int_A \phi g \, d\mu$ for any simple function $\phi \geq 0$ and hence, approximating any measurable function $f \geq 0$ by simple functions, we have

$$\int_A f \, dv = \int_A fg \, d\mu,$$

so that the relation between v, μ, and g can be *symbolically* abbreviated as

$$dv = g \, d\mu.$$

12.4 Product Measures

In many applications, we must consider functions defined on product spaces. To integrate such functions, we need *product measures*. We begin with a few definitions:

Definition 12.4.1 (Rectangle, Product σ-Algebra). Let $(X, \mathcal{A})$ and $(Y, \mathcal{B})$ be measurable spaces. By a *(measurable) rectangle* (in $X \times Y$) we mean a set of the form $A \times B$ where $A \in \mathcal{A}$ and $B \in \mathcal{B}$. The σ-algebra on $X \times Y$ generated by all measurable rectangles will be denoted by $\mathcal{A} \otimes \mathcal{B}$ and called the *product σ-algebra*. Thus, if $\mathcal{R}$ denotes the set of all rectangles in $X \times Y$, then $\mathcal{A} \otimes \mathcal{B} := \mathcal{A}_{\mathcal{R}}$.

Definition 12.4.2 (Cross sections). Let $(X, \mathcal{A})$, $(Y, \mathcal{B})$ be measurable spaces and let $E \subset X \times Y$. Given any $(x, y) \in X \times Y$, we define the *x-section E_x* and the *y-section E^y* (of E) by

$$E_x := \{y \in Y : (x, y) \in E\}, \qquad E^y := \{x \in X : (x, y) \in E\}.$$

Also, for any function f defined on $X \times Y$, the *x-section f_x* and the *y-section f^y* (of f) are the functions defined (on Y and X, respectively) by

$$f_x(y) := f(x, y), \qquad f^y(x) := f(x, y).$$

Proposition 12.4.3. *Let the notation be as in Definition 12.4.2. Given any $E \in \mathcal{A} \otimes \mathcal{B}$, we have $E_x \in \mathcal{B}$ and $E^y \in \mathcal{A}$, for every $x \in X$ and $y \in Y$.*

Proof. We prove the statement for E_x; the proof for E^y is similar. Define the collection

$$\mathcal{C} := \{E \in \mathcal{A} \otimes \mathcal{B} : E_x \in \mathcal{B} \quad \forall x \in X\}.$$

Now, given any rectangle $R := A \times B$, we have $R_x = B$ if $x \in A$ and $R_x = \emptyset$ if $x \notin A$. Therefore $\mathcal{C}$ contains the collection $\mathcal{R}$ of all rectangles. We now prove that $\mathcal{C}$ is a σ-algebra and hence $\mathcal{C} = \mathcal{A} \otimes \mathcal{B}$. First, $X \times Y \in \mathcal{C}$. Next, if $E \in \mathcal{C}$, then $(E^c)_x = (E_x)^c$ and hence $E^c \in \mathcal{C}$. Finally, if $E_n \in \mathcal{C}$ for all $n \in \mathbb{N}$ and $E := \bigcup_{n=1}^{\infty} E_n$, then $E_x = \bigcup_{n=1}^{\infty}(E_n)_x$ implies that $E \in \mathcal{C}$ and the proof is complete. $\square$

Proposition 12.4.4. *Let $(X, \mathcal{A})$ and $(Y, \mathcal{B})$ be measurable spaces and let $(Z, \mathcal{B}_Z)$ be a Borel space. If $f : X \times Y \to Z$ is $(\mathcal{A} \otimes \mathcal{B})$ measurable, then $f_x : Y \to Z$ (resp., $f^y : X \to Z$) is $(\mathcal{B})$ measurable (resp., $(\mathcal{A})$ measurable) for each $x \in X$ (resp., $y \in Y$).*

Proof. This follows at once from Proposition 12.4.3 because $(f_x)^{-1}(C) = (f^{-1}(C))_x$ and $(f^y)^{-1}(C) = (f^{-1}(C))^y$, for each $C \in \mathcal{B}_Z$.   □

Definition 12.4.5 (Elementary Set).   Let the notation be as in Definition 12.4.1. By an *elementary set* (in $X \times Y$) we mean a *finite* union of *pairwise disjoint* rectangles. The collection of all elementary sets will be denoted by $\mathcal{E}(X \times Y)$ or simply $\mathcal{E}$. Thus

$$\mathcal{E} := \{R_1 \cup \cdots \cup R_n : R_j \in \mathcal{R},\ 1 \le j \le n \in \mathbb{N},\ R_i \cap R_j = \emptyset,\ i \ne j\}.$$

Proposition 12.4.6. *The collection $\mathcal{E}$ of elementary sets in $X \times Y$ is an algebra.*

Proof. Let $R_j := A_j \times B_j \in \mathcal{R}$ for $j = 1, 2$. Then we have

$$R_1 \cap R_2 = (A_1 \cap A_2) \times (B_1 \cap B_2) \in \mathcal{E}$$

and

$$R_1 \setminus R_2 = [(A_1 \setminus A_2) \times B_1] \cup [(A_1 \cap A_2) \times (B_1 \setminus B_2)] \in \mathcal{E}.$$

It follows that, for any $P,\ Q \in \mathcal{E}$, we have $P \cap Q \in \mathcal{E}$ and $P \setminus Q \in \mathcal{E}$. Since $P \cup Q = (P \setminus Q) \cup Q$ and $(P \setminus Q) \cap Q = \emptyset$, we have $P \cup Q \in \mathcal{E}$.   □

Definition 12.4.7 (Monotone Sequences of Sets). We call a sequence (A_n) in $\mathcal{P}(X)$ *increasing* (resp., *decreasing*) if $A_n \subset A_{n+1}$ (resp., $A_n \supset A_{n+1}$) for all $n \in \mathbb{N}$; we then write $\lim_{n \to \infty} A_n := \bigcup_{n=1}^{\infty} A_n$ (resp., $\lim_{n \to \infty} A_n := \bigcap_{n=1}^{\infty} A_n$). We say that (A_n) is *monotone* if it is either increasing or decreasing.

Definition 12.4.8 (Monotone Class).   A family $\mathcal{M} \subset \mathcal{P}(X)$ is said to be a *monotone class* if, given any *monotone* sequence $(A_n)_{n=1}^{\infty}$ in $\mathcal{M}$, we have $\lim(A_n) \in \mathcal{M}$.

Exercise 12.4.9.

1. Show that any σ-algebra $\mathcal{A} \subset \mathcal{P}(X)$ is a monotone class.
2. Show that, given any family $(\mathcal{M}_\alpha)_{\alpha \in A}$ of monotone classes (of subsets of X), their intersection $\bigcap_{\alpha \in A} \mathcal{M}_\alpha$ is also a monotone class.
3. Show that, if $\mathcal{A} \subset \mathcal{P}(X)$ is an algebra *as well as* a monotone class, then it is a σ-algebra. *Hint:* If (A_n) is a sequence in $\mathcal{A}$, consider the sequence (A_n'), where $A_n' := \bigcup_{j=1}^{n} A_j$.

Part (2) of the above exercise suggests the following.

Definition 12.4.10. For any $C \subset \mathcal{P}(X)$, the intersection of all monotone classes containing C is the *unique, smallest* monotone class containing C and is called the monotone class *generated by* C. We denote it by $\mathcal{M}_C$.

Proposition 12.4.11. *Let* $\mathcal{G} \subset \mathcal{P}(X)$ *be an algebra. Then* $\mathcal{A}_{\mathcal{G}} = \mathcal{M}_{\mathcal{G}}$. *In other words, the σ-algebra generated by $\mathcal{G}$ coincides with the monotone class generated by $\mathcal{G}$.*

Proof. To simplify the notation, let us write $\mathcal{A}$ and $\mathcal{M}$ instead of $\mathcal{A}_{\mathcal{G}}$ and $\mathcal{M}_{\mathcal{G}}$. Since (by Exercise 12.4.9(1)) $\mathcal{A}$ is a monotone class, we have $\mathcal{M} \subset \mathcal{A}$. By Exercise 12.4.9(3), the reverse inclusion will follow if we show that $\mathcal{M}$ is an *algebra*. (Why?) Now, given any $E \in \mathcal{M}$, define

$$\mathcal{M}(E) := \{F \in \mathcal{M} : E \cup F,\ E \setminus F,\ F \setminus E \in \mathcal{M}\}.$$

Observe that $\emptyset,\ E \in \mathcal{M}(E)$ and that $F \in \mathcal{M}(E)$ if and only if $E \in \mathcal{M}(F)$. Also, if (F_n) is a *monotone* sequence in $\mathcal{M}(E)$, then we have

$$\left(\lim_{n\to\infty} F_n \right) \setminus E = \lim_{n\to\infty} (F_n \setminus E) \in \mathcal{M},$$

$$E \setminus \left(\lim_{n\to\infty} F_n \right) = \lim_{n\to\infty} (E \setminus F_n) \in \mathcal{M},$$

and

$$E \cup \left(\lim_{n\to\infty} F_n \right) = \lim_{n\to\infty} (E \cup F_n) \in \mathcal{M}.$$

Therefore, $\mathcal{M}(E)$ is a monotone class for each $E \in \mathcal{M}$. Since $\mathcal{G}$ is an *algebra*, $E \in \mathcal{G}$ implies $F \in \mathcal{M}(E)$ for all $F \in \mathcal{G}$. Thus $\mathcal{G} \subset \mathcal{M}(E)$ for all $E \in \mathcal{G}$ and hence (by the very definition of $\mathcal{M}$) we have $\mathcal{M} \subset \mathcal{M}(E)$ for all $E \in \mathcal{G}$. Therefore, if $F \in \mathcal{M}$, then $F \in \mathcal{M}(E)$ for all $E \in \mathcal{G}$. But then $E \in \mathcal{M}(F)$ for all $E \in \mathcal{G}$ and hence $\mathcal{G} \subset \mathcal{M}(F)$ for each $F \in \mathcal{M}$. Since each $\mathcal{M}(F)$ is a monotone class, we now have $\mathcal{M} \subset \mathcal{M}(F)$ for all $F \in \mathcal{M}$. Since $X \in \mathcal{G} \subset \mathcal{M}$, the definition of $\mathcal{M}(F)$ now implies that $\mathcal{M}$ is an algebra.   □

   The following proposition is now an immediate consequence of Propositions 12.4.6 and 12.4.11:

Proposition 12.4.12. *Let* $(X, \mathcal{A})$ *and* $(Y, \mathcal{B})$ *be measurable spaces and let* $\mathcal{E}$ *denote the algebra of all elementary sets in* $X \times Y$. *Then* $\mathcal{A} \otimes \mathcal{B} = \mathcal{M}_{\mathcal{E}}$; *i.e., the product σ-algebra is the monotone class generated by the elementary sets.*

   Our next goal is the construction of *product measures:* Given two measure spaces $(X, \mathcal{A}, \mu)$ and $(Y, \mathcal{B}, \nu)$, we want to construct a measure on $\mathcal{A} \otimes \mathcal{B}$ which is the "natural" *product* of μ and ν.

Lemma 12.4.13. *Let* $(A_n \times B_n)$ *be a countable sequence of pairwise disjoint rectangles and let* $A \times B$ *be a rectangle such that* $A \times B = \bigcup A_n \times B_n$. *Then we have*

$$\mu(A)v(B) = \sum_n \mu(A_n)v(B_n). \tag{$*$}$$

Proof. Given any $(x, y) \in X \times Y$, we have

$$\chi_A(x)\chi_B(y) = \chi_{A \times B}(x, y) = \sum_{n=1}^{\infty} \chi_{A_n \times B_n}(x, y) = \sum_{n=1}^{\infty} \chi_{A_n}(x)\chi_{B_n}(y).$$

Integrating *with respect to* x, and using Corollary 12.3.16, we have

$$\mu(A)\chi_B(y) = \sum_{n=1}^{\infty} \chi_{B_n}(y) \int \chi_{A_n}(x)\, d\mu(x) = \sum_{n=1}^{\infty} \mu(A_n)\chi_{B_n}(y).$$

If we now integrate with respect to y, another application of Corollary 12.3.16 implies $(*)$. $\qquad\square$

Exercise 12.4.14. With $\mathcal{E} \subset \mathcal{P}(X \times Y)$ denoting the algebra of *elementary sets*, define the map $\mu \times v : \mathcal{E} \to [0, \infty]$ as follows: For any $E = \bigcup_{j=1}^n R_k \in \mathcal{E}$, where the $R_j = A_j \times B_j$ are pairwise disjoint rectangles, let

$$(\mu \times v)(E) := \sum_{j=1}^n \mu(A_j)v(B_j).$$

Show that $\mu \times v$ is well defined; i.e., if we also have $E = \bigcup_{k=1}^m A'_k \times B'_k$, with pairwise disjoint rectangles $R'_k := A'_k \times B'_k$, then

$$\sum_{j=1}^n \mu(A_j)v(B_j) = \sum_{k=1}^m \mu(A'_k)v(B'_k).$$

Deduce that $\mu \times v$ is a *premeasure* on the algebra $\mathcal{E}$ of elementary sets. *Hint:* $(\mu \times v)(R_j) = \sum_{k=1}^m (\mu \times v)(R_j \cap R'_k)$.

Definition 12.4.15 (Product Measure). Let $(X, \mathcal{A}, \mu)$ and $(Y, \mathcal{B}, v)$ be two measure spaces and let $\mu \times v : \mathcal{E} \to [0, \infty]$ be the premeasure defined in Exercise 12.4.14. The *product measure* of μ and v, denoted by $\mu \otimes v$, is the extension of $\mu \times v$ to the product σ-algebra $\mathcal{A} \otimes \mathcal{B} = \mathcal{A}_{\mathcal{E}}$ provided by the extension theorem (Theorem 12.1.34). If, in addition, μ and v are σ-*finite* measures, then $\mu \otimes v$ is the *unique* measure on $\mathcal{A} \otimes \mathcal{B}$ such that $(\mu \otimes v)(A \times B) = \mu(A)v(B)$, for all rectangles $A \times B$. In this case, $\mu \otimes v$ is also σ-finite. (Why?)

Remark 12.4.16. Even if both μ and v are complete measures, $\mu \otimes v$ is *almost never complete*. Indeed, suppose that there is an $A \in \mathcal{A}$ with $A \neq \emptyset$ and $\mu(A) = 0$. Also, suppose that there is a $B \subset Y$ such that $B \notin \mathcal{B}$. Then $A \times B \subset A \times Y$ and $\mu \otimes v(A \times Y) = 0$, but $A \times B \notin \mathcal{A} \otimes \mathcal{B}$. In particular, if $X = Y = \mathbb{R}$ and

$\mu = \nu = \lambda$ (Lebesgue measure), then $\lambda \otimes \lambda$ is *not* complete and hence $\lambda \otimes \lambda \neq \lambda_2$. It is a fact, however, that λ_2 is the *completion* of $\lambda \otimes \lambda$ (cf., e.g., Rudin's *Real & Complex Analysis* [Rud74]).

The following proposition will be needed in the proof of the main result (the Fubini–Tonelli Theorem) on the integration of functions on $X \times Y$.

Proposition 12.4.17. *Let $(X, \mathcal{A}, \mu)$ and $(Y, \mathcal{B}, \nu)$ be σ-finite measure spaces. Then, for each $E \in \mathcal{A} \otimes \mathcal{B}$, the functions $x \mapsto \nu(E_x)$ and $y \mapsto \mu(E^y)$ are measurable on X and Y, respectively, and we have*

$$(\mu \otimes \nu)(E) = \int \nu(E_x)\, d\mu(x) = \int \mu(E^y)\, d\nu(y). \qquad (*)$$

Proof. Let us first assume that μ and ν are *finite*, i.e., $\mu(X) < \infty$ and $\nu(Y) < \infty$. Let $\mathcal{C}$ denote the collection of all $E \in \mathcal{A} \otimes \mathcal{B}$ satisfying the conclusions of the proposition. If $E = A \times B$ is a rectangle, then $(\mu \otimes \nu)(E) = \mu(A)\nu(B)$, $\nu(E_x) = \chi_A(x)\nu(B)$, and $\mu(E^y) = \mu(A)\chi_B(y)$, so that $(*)$ holds and $E \in \mathcal{C}$. Therefore, by additivity, we have $\mathcal{E} \subset \mathcal{C}$. By Proposition 12.4.11, the general case is proved if we show that $\mathcal{C}$ is a *monotone class*. So let (E_n) be an *increasing* sequence in $\mathcal{C}$ and let $E := \bigcup_n E_n$. Then the functions $f_n(y) := \mu((E_n)^y)$ are measurable and form an increasing sequence converging pointwise to $f(y) := \mu(E^y)$. By the Monotone Convergence Theorem, f is measurable and (by $(*)$) we have

$$\int \mu(E^y)\, d\nu(y) = \lim_{n \to \infty} \int \mu((E_n)^y)\, d\nu(y) = \lim_{n \to \infty} (\mu \otimes \nu)(E_n) = (\mu \otimes \nu)(E).$$

A similar argument shows that $(\mu \otimes \nu)(E) = \int \nu(E_x)\, d\mu(x)$ and hence $E \in \mathcal{C}$. Next, suppose that (E_n) is a *decreasing* sequence in $\mathcal{C}$ and let $E := \bigcap_{n=1}^{\infty} E_n$. Since $\mu((E_1)^y) \leq \mu(X) < \infty$, we have $g(y) := \mu((E_1)^y) \in \mathcal{L}^1(Y)$. Applying the Dominated Convergence Theorem, we easily deduce that $E \in \mathcal{C}$ and the proposition is proved for *finite* measure spaces. Finally, suppose that μ and ν are σ-finite. We then have $X \times Y = \bigcup_{n=1}^{\infty} X_n \times Y_n$, with an *increasing* sequence $(X_n \times Y_n)$ of rectangles such that $\mu(X_n) < \infty$ and $\nu(Y_n) < \infty$ for all $n \in \mathbb{N}$. Given $E \subset \mathcal{A} \otimes \mathcal{B}$, we can apply our previous arguments to the sets $E_n := E \cap (X_n \times Y_n)$ and obtain

$$(\mu \otimes \nu)(E_n) = \int \chi_{X_n}(x)\nu(E_x \cap Y_n)\, d\mu(x) = \int \chi_{Y_n}(y)\mu(E^y \cap X_n)\, d\nu(y).$$

Another application of the Monotone Convergence Theorem now proves the proposition. $\qquad\qquad\qquad\qquad\qquad\qquad\qquad\qquad\qquad\qquad\qquad\qquad\qquad$ □

We can now prove the main theorem:

Theorem 12.4.18 (Fubini–Tonelli Theorem). *Let $(X, \mathcal{A}, \mu)$ and $(Y, \mathcal{B}, \nu)$ be σ-finite measure spaces.*

(a) **(Tonelli)** If $f \in \mathcal{L}_+^0(X \times Y)$, then the functions $g(x) := \int f_x \, dv$ and $h(y) := \int f^y \, d\mu$ are in $\mathcal{L}_+^0(X)$ and $\mathcal{L}_+^0(Y)$, respectively, and we have

$$\int f \, d(\mu \otimes v) = \int \left[\int f(x, y) \, dv(y) \right] d\mu(x) \qquad (\dagger)$$

$$= \int \left[\int f(x, y) \, d\mu(x) \right] dv(y).$$

(b) **(Fubini)** If $f \in \mathcal{L}_{\mu \otimes v}^1(X \times Y, \mathbb{C})$, then $f_x \in \mathcal{L}_v^1(Y, \mathbb{C})$ for a.e. $x \in X$, $f^y \in \mathcal{L}_\mu^1(X, \mathbb{C})$ for a.e. $y \in Y$, the functions $g(x) := \int f_x \, dv$ and $h(y) := \int f^y \, d\mu$ (defined almost everywhere) are in $\mathcal{L}_\mu^1(X, \mathbb{C})$ and $\mathcal{L}_v^1(Y, \mathbb{C})$, respectively, and $(\dagger)$ is satisfied.

Proof. First, if $f = \chi_E$ for some $E \in \mathcal{A} \otimes \mathcal{B}$, then part (a) reduces to Proposition 12.4.17. By linearity, (a) is therefore true for all $f \in Simp^+(X \times Y)$. Now, given any $f \in \mathcal{L}_+^0(X \times Y)$, we can pick an *increasing* sequence (ϕ_n) in $Simp^+(X \times Y)$ such that $\lim_{n \to \infty} \phi_n(x, y) = f(x, y)$ for all $(x, y) \in X \times Y$. From the Monotone Convergence Theorem, we deduce that the corresponding sequences of functions $\psi_n(x) := \int \phi_{nx} \, dv$ and $\theta_n(y) := \int \phi_n^y \, d\mu$ are increasing and converge to g and h, respectively, that g and h are measurable, and that

$$\int g \, d\mu = \lim_{n \to \infty} \int \psi_n \, d\mu = \lim_{n \to \infty} \int \phi_n \, d(\mu \otimes v) = \int f \, d(\mu \otimes v),$$

$$\int h \, dv = \lim_{n \to \infty} \int \theta_n \, dv = \lim_{n \to \infty} \int \phi_n \, d(\mu \otimes v) = \int f \, d(\mu \otimes v),$$

which yields $(\dagger)$. This completes the proof of (a) and also shows (by Corollary 12.3.10) that, if $f \in \mathcal{L}_+^0(X \times Y)$ and $\int f \, d(\mu \otimes v) < \infty$, then g and h are *finite almost everywhere*, i.e., $f_x \in \mathcal{L}_v^1(Y, [0, \infty))$ for almost all x and $f^y \in \mathcal{L}_\mu^1(X, [0, \infty))$ for almost all y. To prove part (b), we simply apply the above results to the functions $\text{Re}(f)^\pm$ and $\text{Im}(f)^\pm$ for a given $f \in \mathcal{L}_{\mu \otimes v}^1(X \times Y, \mathbb{C})$. $\quad\square$

Remark 12.4.19.

1. In practice, given an $f \in \mathcal{L}^0(X \times Y, \mathbb{C})$, one usually tries to prove that one of the two *iterated integrals*

$$\int \left[\int |f(x, y)| \, dv(y) \right] d\mu(x), \quad \int \left[\int |f(x, y)| \, d\mu(x) \right] dv(y)$$

is finite. The other one and the *double integral* $\int |f| \, d(\mu \otimes v)$ are then also finite and the three integrals coincide. Therefore, *the order of integration may be reversed* if $f \geq 0$ or if one of the iterated integrals of $|f|$ is finite.

2. The σ-finiteness assumption in the theorem is *necessary*. Indeed, suppose that $X = Y = [0, 1]$, $\mu := \lambda$ is Lebesgue measure, and v is the *counting measure*.

Let $f := \chi_D$, where $D : \{(x, y) \in X \times Y : x = y\}$ is the *diagonal*. Then we have $\int f(x, y) \, d\nu(y) = 1$ for all $x \in X$ and hence $\int [\int f(x, y) \, d\nu(y)] \, d\mu(x) = 1$, but $\int f(x, y) \, d\mu(x) = 0$ for all $y \in Y$, so that $\int [\int f(x, y) \, d\mu(x)] \, d\nu(y) = 0$.

12.5   Probability

Our objective in this section is a brief discussion of some of the most basic concepts in probability theory which, in its modern axiomatic form (introduced by Kolmogorov in 1933), uses measure theory as its foundation. For a complete treatment, the interested reader should consult more advanced texts, some of which are listed in the bibliography. The study of probability was first undertaken by the French mathematicians *Fermat* and *Pascal*, who were primarily motivated by a desire to answer some challenging questions posed by a number of professional gamblers. A game of chance is an example of an *experiment* with a *finite* set of *(simple) outcomes*, $\omega_1, \ldots, \omega_n$. The set $\Omega := \{\omega_1, \ldots, \omega_n\}$ is then the *sample space* of the experiment and each subset $E \subset \Omega$ is an *event*. The event E is said to have *occurred* if the experiment results in an outcome that *belongs to E*. It should be pointed out that, although the entire sample space Ω is known in advance, the occurrence of each individual $\omega \in \Omega$ is *random* in the sense that it cannot be predicted in advance. The goal of the theory is to assign a number to each event E that would represent the *probability* (or *likelihood*) of its occurrence. The simplest example of an experiment is *coin tossing:* If a coin is flipped once, the sample space is $\{H, T\}$, where H and T symbolize the occurrences of *head* and *tail*, respectively. More generally, one may consider the (practically impossible) experiment of flipping a coin indefinitely:

Definition 12.5.1 (Bernoulli Trial, Sequence). A *Bernoulli trial* is defined to be an experiment with *two* outcomes called *success* and *failure*. A *Bernoulli sequence* is an outcome of a sequence of Bernoulli trials. Thus, a Bernoulli sequence may be represented by a string of s's and f's such as

$$ssfffsffssssfssfff \ldots,$$

where s and f stand for success and failure, respectively. The set of all Bernoulli sequences will be denoted by $\mathfrak{B}$.

The set $\mathfrak{B}$ is *almost* in one-to-one correspondence with the unit interval $[0, 1]$:

Proposition 12.5.2. *There is a countable set $\mathfrak{C} \subset \mathfrak{B}$ such that $\mathfrak{B} \setminus \mathfrak{C}$ is in one-to-one correspondence with $(0, 1]$.*

Proof. For each $\omega \in (0, 1]$ consider its *binary* expansion

$$\omega = (0.d_1 d_2 d_3 \ldots)_2 := \sum_{n=1}^{\infty} \frac{d_n}{2^n} \qquad (d_n \in \{0, 1\}),$$

and define $\beta(\omega)$ to be the Bernoulli sequence whose nth term is an s if $d_n = 1$ or an f if $d_n = 0$. Unfortunately, β is not a well-defined function from $(0, 1]$ to $\mathfrak{B}$ because some numbers $\omega \in (0, 1]$ have two binary expansions. For example, $1/2 = (0.1000\ldots)_2 = (0.0111\ldots)_2$. To fix this, let us adopt the following rule: If a number has a *terminating* and a *nonterminating* binary expansion, we always pick the nonterminating one. It is then obvious that β is a *one-to-one* map from $(0, 1]$ onto the set $\mathfrak{B} \setminus \mathfrak{C}$, where $\mathfrak{C}$ is the set of all Bernoulli sequences that end with a string of f's. Since $\mathfrak{C}$ is countable (why?), the proof is complete. $\square$

It turns out that, to study more sophisticated experiments and their properties, it is necessary to introduce a measure-theoretic model. We therefore begin with the following.

Definition 12.5.3 (Probability Space, Probability Measure). We define a *probability space* to be a measure space $(X, \mathcal{A}, \mu)$ such that μ is finite with *total mass* $1 : \mu(X) = 1$. *Following the standard practice, we shall henceforth denote X by Ω and μ by P.* Thus a probability space will be denoted by $(\Omega, \mathcal{A}, P)$ and the measure P will be called a *probability measure.*

Definition 12.5.4 (Sample Space, Event, Probability). In any probability space $(\Omega, \mathcal{A}, P)$, the set Ω is called the *sample space* and each $\omega \in \Omega$ is called a *sample point*. Also, each measurable set $A \in \mathcal{A}$ is called an *event*. The measure $P(A)$ of an event $A \in \mathcal{A}$ is called the *probability of A.* It is obvious that $0 \leq P(A) \leq 1$, for all $A \in \mathcal{A}$.

Definition 12.5.5 (Almost Surely). Let $(\Omega, \mathcal{A}, P)$ be a probability space. A statement about sample points $\omega \in \Omega$ is said to be true *almost surely* (abbreviated *a.s.*) if it is true P-almost everywhere, i.e., if the event $F \subset \Omega$ of all $\omega \in \Omega$ for which it *fails* to be true has *probability zero*: $P(F) = 0$.

Example 12.5.6. Let $\Omega = I := [0, 1]$, let $\mathcal{A}$ be the σ-algebra $\mathcal{M}_\lambda(I)$ of Lebesgue measurable sets in I, and let P be the restriction λ_I of Lebesgue measure to I. Note that, although $P(\{\omega\}) = 0$ for all $\omega \in \Omega$, we have $P(\Omega) = 1$ and hence one of the events $\{\omega\}$ is *certain to occur*. More generally, we can consider any Lebesgue measurable $\Omega \subset \mathbb{R}$ with $0 < \lambda(\Omega) < \infty$, let $\mathcal{A}$ be the σ-algebra of all measurable subsets of Ω, and define $P(A) := \lambda(A)/\lambda(\Omega)$ for each $A \in \mathcal{A}$.

Example 12.5.7 (Classical Probability). Let $\Omega = \{\omega_1, \ldots, \omega_n\}$ be any finite set, $\mathcal{A} = \mathcal{P}(\Omega)$, and $P := \nu/n$, where ν is the *counting measure*; i.e., $\nu(E) = |E|$ is the cardinality of E for each $E \subset \Omega$. In this case, we have $P(E) = |E|/n$. For example, consider a Bernoulli trial in which *success* and *failure* are *equally likely*. We then have the sample space $\Omega := \{s, f\}$ with $P(\{s\}) = P(\{f\}) = 1/2$.

Remark 12.5.8 (Random Selection). In the case of classical probability, we may think of Ω as the sample space of the *experiment* of selecting an element from $\Omega = \{\omega_1, \ldots, \omega_n\}$ *at random*; i.e., in such a way that each ω_j is *equally likely* to be selected. This implies that $P(\{\omega_i\}) = P(\{\omega_j\})$ for all i, j. Since $P(\Omega) = \sum_{j=1}^n P(\{\omega_j\}) = 1$, we have $P(\{\omega_j\}) = 1/n$ for all j and hence $P(E) = |E|/n$ for each $E \subset \Omega$.

Exercise 12.5.9. Let m and n be positive integers and suppose that m chips are placed *randomly* in n boxes.

(a) If $m \geq n$, what is the probability that no box is empty?
(b) If $m \leq n$, what is the probability that no box contains more than one chip? *Hint:* Look at the functions from the set $C := \{c_1, \ldots, c_m\}$ of chips to the set $B := \{b_1, \ldots, b_n\}$ of boxes and use Exercises 1.3.35 and 1.3.36.

Exercise 12.5.10. Let $n \geq 2$ be an integer and let a number be *randomly* selected from the set $\{1, 2, 3, \ldots, n\}$. What is the probability that the selected number is *relatively prime* to n? *Hint:* Let $n = p_1^{r_1} p_2^{r_2} \cdots p_m^{r_m}$ be the prime factorization of n and use Exercise 1.3.47.

We now give an example that contains the classical probability as a special case:

Example 12.5.11. Let Ω be a (possibly uncountable) set and let $p : \Omega \to [0, \infty)$ be a *summable* function with *positive sum*, i.e., $0 < s := \sum_{\omega \in \Omega} p(\omega) < \infty$, where the sum is an *unordered series*. Let $\mathcal{A} := \mathcal{P}(\Omega)$ and define $P(A) := \sum_{\omega \in A} p(\omega)/s$ for each $A \subset \Omega$. Then $(\Omega, \mathcal{A}, P)$ is a probability space. Note that, by Corollary 2.4.17, we have $P(\{\omega\}) = p(\omega)/s = 0$ except for a *countable* set of ω's. When Ω is finite and $p(\omega) = 1 \; \forall \omega \in \Omega$, we recover the classical probability.

If we know that an event $B \in \mathcal{A}$ has occurred, how does this knowledge affect the probability of another event $A \in \mathcal{A}$? To answer this question, we need the following.

Definition 12.5.12 (Conditional Probability, Independence). Given a probability space $(\Omega, \mathcal{A}, P)$ and any $A, B \in \mathcal{A}$ with $P(B) > 0$, the number

$$P(A|B) := \frac{P(A \cap B)}{P(B)}$$

is called the *conditional probability of A given B*. Two events $A, B \in \mathcal{A}$ are said to be *independent* if

$$P(A \cap B) = P(A)P(B).$$

It is then obvious that $P(A|B) = P(A)$ if $P(B) > 0$.

Exercise 12.5.13. Show that, if A and B are independent events, then so are A^c and B.

Proposition 12.5.14. *Let the notation be as in Definition 12.5.12. Then the set function $P_B(E) := P(E|B)$ is a probability measure on $\mathcal{A}$.*

Proof. See Exercise 12.1.11.   □

Remark 12.5.15 (Reduced Sample Space). Let $(\Omega, \mathcal{A}, P)$ be a probability space and let $B \in \mathcal{A}$ be any fixed event with $P(B) > 0$. Consider the *restricted σ-algebra* $\mathcal{A}|_B := \{A \cap B : A \in \mathcal{A}\}$ and the *restricted* probability measure $P|_B(E) := P(E|B) = P(E)/P(B)$ for each $E \in \mathcal{A}|_B$. Then $(B, \mathcal{A}|_B, P|_B)$ is a probability space with the *reduced sample space B*.

Exercise 12.5.16 (The Total Probability Law). If $\Omega = \bigcup_{n=1}^{\infty} E_n$, with a sequence (E_n) of *pairwise disjoint events* such that $P(E_n) > 0$ for all $n \in \mathbb{N}$, show that

$$P(A) = \sum_{n=1}^{\infty} P(A|E_n)P(E_n) \qquad (\forall A \in \mathcal{A}).$$

Exercise 12.5.17 (Bayes's Formula). Under the assumptions of Exercise 12.5.16, show that for each event A with $P(A) > 0$, we have *Bayes's formula:*

$$P(E_k|A) = \frac{P(A|E_k)P(E_k)}{P(A)} = \frac{P(A|E_k)P(E_k)}{\sum_{n=1}^{\infty} P(A|E_n)P(E_n)} \qquad (\forall k \in \mathbb{N}).$$

The concept of *independence* can be extended to more than two events. In fact, it can even be defined for collections whose elements are families of events:

Definition 12.5.18 (Independence of Families of Events). Let $(\Omega, \mathcal{A}, P)$ be a probability space. A subset $\mathcal{C} \subset \mathcal{A}$ is said to be *independent* if, given *any* (distinct) events $C_1, C_2, \ldots, C_n \in \mathcal{C}$, we have

$$P(C_1 \cap C_2 \cap \cdots \cap C_n) = P(C_1) \cdot P(C_2) \cdots P(C_n).$$

Let J be an index set. A collection $\mathfrak{C} := \{\mathcal{C}_j : j \in J\}$ of subsets $\mathcal{C}_j \subset \mathcal{A}$ is said to be *independent* if, given *any* (distinct) indices $j_1, j_2, \cdots, j_n \in J$ and *any* events $C_{j_1} \in \mathcal{C}_{j_1}, \ldots, C_{j_n} \in \mathcal{C}_{j_n}$, we have

$$P(C_{j_1} \cap C_{j_2} \cap \cdots \cap C_{j_n}) = P(C_{j_1}) \cdot P(C_{j_2}) \cdots P(C_{j_n}).$$

Exercise 12.5.19. Let $\mathcal{C} \subset \mathcal{A}$ be an *independent* collection of events. Show that $\mathcal{C} \cup \{C^c : C \in \mathcal{C}\}$ is independent. *Hint:* Use induction and $P(C^c) = 1 - P(C)$.

Exercise 12.5.20. Let $\{A_k : k = 1, 2, \ldots, n\}$ be a finite collection of events. Show that, in order for this collection to be *independent,* we must impose a total of $2^n - n - 1$ conditions of the form

$$P(A_{i_1} \cap \cdots \cap A_{i_j}) = P(A_{i_1}) \cdots P(A_{i_j}).$$

Remark 12.5.21 (Pairwise Independence $\not\Rightarrow$ Independence). Consider the experiment of throwing two *fair* dice. The sample space is $\Omega = \{(i, j) : i, j = 1, 2, \ldots, 6\}$. Let $A := \{(i, j) : i = 1, 2, 3\}$, $B := \{(i, j) : j = 4, 5, 6\}$, and $C := \{(i, j) : i + j = 7\}$. Then we have $P(A) = P(B) = 1/2$, $P(C) = 1/6$, $P(A \cap B) = 1/4$, $P(A \cap C) = P(B \cap C) = P(A \cap B \cap C) = 1/12$. (Why?) It follows that $P(A \cap B) = P(A)P(B)$, $P(A \cap C) = P(A)P(C)$, and $P(B \cap C) = P(B)P(C)$, so that the events A, B, and C are *pairwise independent.*

On the other hand, $P(A \cap B \cap C) \neq P(A)P(B)P(C)$ and hence the collection $\{A, B, C\}$ is *not* independent.

Suppose that a coin is flipped indefinitely, so that the sample space is $\Omega = \{H, T\}^{\mathbb{N}}$ (or $\{0, 1\}^{\mathbb{N}}$). Let A_n denote the event of a *head* on the nth flip. It is natural to consider the event E of *infinitely many heads* or the event F of *all but finitely many heads*. How can E and F be represented in terms of the A_n? The following definition answers this question.

Definition 12.5.22 (Limsup and Liminf). Let (A_n) be a sequence of events. Then the *limit superior* (or *upper limit*) of the A_n is defined to be the event

$$\limsup_{n \to \infty} A_n := \bigcap_{n=1}^{\infty} \bigcup_{k=n}^{\infty} A_k.$$

Similarly, the *limit inferior* (or *lower limit*) of the A_n is defined to be the event

$$\liminf_{n \to \infty} A_n := \bigcup_{n=1}^{\infty} \bigcap_{k=n}^{\infty} A_k.$$

Exercise 12.5.23 (Infinitely Often, Eventually). Let the notation be as in Definition 12.5.22. Prove the following:

$$\limsup_{n \to \infty} A_n = \{\omega \in \Omega : \omega \in A_n \text{ infinitely often, i.e., for infinitely many } n\},$$

$$\liminf_{n \to \infty} A_n = \{\omega \in \Omega : \omega \in A_n \text{ eventually, i.e., for all but finitely many } n\}.$$

Theorem 12.5.24 (First Borel–Cantelli Lemma). *Let $(A_n)_{n \in \mathbb{N}}$ be a sequence of events and let $A := \limsup_{n \to \infty} A_n$. Then $\sum_{n=1}^{\infty} P(A_n) < \infty$ implies that $P(A) = 0$.*

Proof. Let $E_n := \bigcup_{k=n}^{\infty} A_k$ so that $A = \bigcap_{n=1}^{\infty} E_n$. Now let $\varepsilon > 0$ be given. Since $\sum_{n=1}^{\infty} P(A_n) < \infty$, we can pick n so large that $\sum_{k=n}^{\infty} P(A_k) < \varepsilon$ and hence, by subadditivity,

$$P(E_n) \leq \sum_{k=n}^{\infty} P(A_k) < \varepsilon.$$

But $A \subset E_n$ for all $n \in \mathbb{N}$ so that $P(A) < \varepsilon$. Since $\varepsilon > 0$ was arbitrary, the proof is complete.   □

The following theorem is a partial converse to the First Borel–Cantelli Lemma:

Theorem 12.5.25 (Second Borel–Cantelli Lemma). *Let $(\Omega, \mathcal{A}, P)$ be a probability space. If (A_n) is an independent sequence of events such that we have $\sum_{n=1}^{\infty} P(A_n) = \infty$, then $P(\limsup_{n \to \infty} A_n) = 1$.*

Proof. Let $A := \limsup_{n \to \infty} A_n = \bigcap_{n=1}^{\infty} \bigcup_{k=n}^{\infty} A_k$. We must show that, if $\sum_{n=1}^{\infty} P(A_n) = \infty$, then $P(A^c) = 0$. Since $A^c = \bigcup_{n=1}^{\infty} \bigcap_{k=n}^{\infty} A_k^c$, it is sufficient, by subadditivity, to show that $P(\bigcap_{k=n}^{\infty} A_k^c) = 0$ for each n. Now, by Exercise 12.5.19, the A_k^c are independent and hence, for each $m > n$,

$$P\left(\bigcap_{k=n}^{m} A_k^c \right) = \prod_{k=n}^{m} P(A_k^c) = \prod_{k=n}^{m} [1 - P(A_k)]. \qquad (*)$$

Since $1 - x \le e^{-x}$ for all $x \in \mathbb{R}$ (why?), $(*)$ implies that

$$P\left(\bigcap_{k=n}^{m} A_k^c \right) \le e^{-\sum_{k=n}^{m} P(A_k)}. \qquad (**)$$

But $\sum_{n=1}^{\infty} P(A_n) = \infty$ so that $e^{-\sum_{k=n}^{m} P(A_k)} \to 0$ as $m \to \infty$. Therefore, by $(**)$ and the continuity of P,

$$P\left(\bigcap_{k=n}^{\infty} A_k^c \right) = \lim_{m \to \infty} \prod_{k=n}^{m} P(A_k^c) = 0.$$

$\square$

So far we have only looked at *events,* i.e., measurable *sets* in probability theory. We now look at what (in probability theory) corresponds to *measurable functions*:

Definition 12.5.26 (Random Variable). Let $(\Omega, \mathcal{A}, P)$ be a probability space. By a *random variable* we mean a *measurable function* $X : \Omega \to \mathbb{R}$ (i.e., $X \in \mathcal{L}^0(\Omega, \mathbb{R})$); thus $X^{-1}(B) \in \mathcal{A}$ for each Borel set $B \in \mathcal{B}_{\mathbb{R}}$. One may also define $\overline{\mathbb{R}}$-*valued* (resp., $\mathbb{C}$-valued) random variables $X : \Omega \to \overline{\mathbb{R}}$ (resp., $X : \Omega \to \mathbb{C}$) by requiring $X^{-1}(B) \in \mathcal{A}$ for each $B \in \mathcal{B}_{\overline{\mathbb{R}}}$ (resp., $B \in \mathcal{B}_{\mathbb{C}}$).

Notation 12.5.27. Given a random variable $X : \Omega \to \mathbb{R}$ and a Borel set $B \in \mathcal{B}_{\mathbb{R}}$, the event $X^{-1}(B)$ will often be denoted by $\{X \in B\}$. Thus, $\{X \le b\} := X^{-1}((-\infty, b])$, $\{a < X \le b\} := X^{-1}((a, b])$, etc.

To each random variable there corresponds, in a natural way, a Borel measure on $\mathbb{R}$:

Definition 12.5.28 (Probability Distribution). Let $X : \Omega \to \mathbb{R}$ be a random variable. Then

$$\mathcal{A}_X := \{ X^{-1}(B) : B \in \mathcal{B}_{\mathbb{R}} \}$$

is a sub-σ-algebra of $\mathcal{A}$. (Why?) We define a (Borel) measure, P_X, on $\mathbb{R}$ by setting

$$P_X(B) := P(X^{-1}(B)) \qquad (\forall B \in \mathcal{B}_{\mathbb{R}}).$$

The measure P_X is called the *probability distribution* of X.

Proposition 12.5.29. *With notation as in Definition 12.5.28, $(\mathbb{R}, \mathcal{B}_{\mathbb{R}}, P_X)$ is a probability space.*

Proof. It is obvious that $P_X(\mathbb{R}) = 1$. Now, given a sequence (B_n) of *pairwise disjoint* Borel sets in $\mathbb{R}$, the corresponding sequence of events $(X^{-1}(B_n))$ is pairwise disjoint and we have

$$P_X\left(\bigcup_{n=1}^{\infty} B_n\right) = P\left(X^{-1}\left(\bigcup_{n=1}^{\infty} B_n\right)\right) = \sum_{n=1}^{\infty} P(X^{-1}(B_n)) = \sum_{n=1}^{\infty} P_X(B_n).$$

$\square$

Example 12.5.30 (Constant Variable).   Suppose that X is *constant*, i.e., $X(\omega) = a$ for all $\omega \in \Omega$ and a fixed $a \in \mathbb{R}$. Then $P_X = \delta_a$ is the Dirac measure at a. Indeed, $P_X(B) = \chi_B(a)$ for each Borel set $B \in \mathcal{B}_{\mathbb{R}}$. (Why?)

Example 12.5.31 (Bernoulli Variable).   A random variable X on Ω whose range is $\{0, 1\}$ is called a *Bernoulli random variable*. The events $A := \{X = 1\}$ and $A^c = \{X = 0\}$ are then called *success* and *failure*, respectively, and we have $X = \chi_A$. Let $p := P(X = 1) := P(A)$. Then $P(X = 0) := P(A^c) = 1 - p$, and we have

$$P_X(B) = p\delta_1(B) + (1 - p)\delta_0(B) \qquad (\forall B \in \mathcal{B}_{\mathbb{R}}).$$

   A Bernoulli variable is a special *simple variable*:

Definition 12.5.32 (Simple Variable).   A random variable $X : \Omega \to \mathbb{R}$ is called a *simple variable* if it takes a *finite* number of values. If $\{a_1, \ldots, a_n\}$ is the range of X and if $A_j := \{X = a_j\}$, then we have

$$X = \sum_{j=1}^{n} a_j \chi_{A_j}.$$

Let $p_j := P(A_j)$, for $1 \leq j \leq n$. Then we have $\sum_{j=1}^{n} p_j = 1$ and

$$P_X(B) = \sum_{j=1}^{n} p_j \delta_{a_j}(B) \qquad (\forall B \in \mathcal{B}_{\mathbb{R}}).$$

   Extending the previous case, we can consider random variables with *countable* range:

Definition 12.5.33 (Discrete Variable).   A random variable $X : \Omega \to \mathbb{R}$ is called *discrete* if it takes at most *countably many* distinct values. If $\{a_1, a_2, a_3, \ldots\}$ is an enumeration of the (distinct) values of X and if $p_n := P(X = a_n)$, then $\sum_n p_n = 1$ and

$$P_X(B) = \sum_n p_n \delta_{a_n}(B) \qquad (\forall B \in \mathcal{B}_\mathbb{R}).$$

Example 12.5.34 (Binomial Variable). We say that a random variable X has *binomial distribution* with *parameters* $n \in \mathbb{N}$ *and* $p \in [0,1]$, and we write $X \sim B(n,p)$, if the range of X is $\{0,1,2,\ldots,n\}$ and if

$$p_k := P_X(\{k\}) = \binom{n}{k} p^k (1-p)^{n-k} \qquad (k = 0,1,\ldots,n).$$

Example 12.5.35 (Poisson Variable). We say that a random variable X has *Poisson distribution* with *parameter* $\lambda > 0$ if the range of X is the set $\mathbb{N}_0$ and if

$$p_k := P_X(\{k\}) = \frac{\lambda^k}{k!} e^{-\lambda} \qquad (k = 0,1,2,\ldots).$$

Definition 12.5.36 (Continuous Variable). A random variable $X \in \mathbb{R}^\Omega$ is said to be *continuous* if

$$P_X(\{\omega\}) := P(X = \omega) = 0 \qquad (\forall \omega \in \Omega).$$

Example 12.5.37. Consider the probability space $(I, \mathcal{M}_\lambda(I), \lambda_I)$, where $I = [0,1]$, λ_I is the restriction of Lebesgue measure to I, and $\mathcal{M}_\lambda(I)$ is the set of all Lebesgue measurable subsets of I. Then, any (Lebesgue) *measurable* function $X : I \to \mathbb{R}$ is a random variable. For instance, the *affine* function $X(\omega) := a\omega + b$, where $a > 0$ and $b \in \mathbb{R}$, is a *continuous* random variable with range $[b, a+b]$ and probability distribution $P_X = \frac{1}{a} \lambda_{[b,a+b]}$.

As was pointed out before, Lebesgue–Stieltjes measures play an important role in probability. Indeed, by Theorem 12.1.38, each *finite* measure μ (in particular each probability measure) on $\mathcal{B}_\mathbb{R}$ is completely characterized by its *(cumulative) distribution function*: $F(x) := \mu((-\infty, x])$. To construct probability distributions, it is therefore natural to look at distribution functions:

Definition 12.5.38 (Distribution Function). Given any random variable $X : \Omega \to \mathbb{R}$, its *(cumulative) distribution function* is defined by

$$F_X(x) := P(X \le x) = P_X((-\infty, x]) \qquad (\forall x \in \mathbb{R}).$$

Proposition 12.5.39. *Given a random variable $X : \Omega \to \mathbb{R}$, its distribution function F_X is increasing, right-continuous and satisfies the following asymptotic properties:*

$$\lim_{t \to -\infty} F_X(t) = 0, \quad \lim_{t \to \infty} F_X(t) = 1. \qquad (*)$$

Proof. That F is increasing follows from the *monotonicity* of P. Also, the right-continuity of F follows from the *continuity* of P and $(-\infty, x] = \bigcap_{n=1}^\infty (-\infty, x +$

$1/n]$. Finally, since $P(\emptyset) = 0$ and $P(\Omega) = 1$, the limit properties $(*)$ are also consequences of the continuity of P. (Why?)   □

Exercise 12.5.40.

1. Show that $P_X((a, b]) = F_X(b) - F_X(a)$ and $P_X(\{a\}) = F_X(a) - F_X(a - 0)$. Deduce from the latter that X is a *continuous* random variable if and only if F_X is *continuous* on $\mathbb{R}$.
2. Find the distribution function F_X for (i) a *Bernoulli variable* X with parameter $p := P(X = 1)$, (ii) a *binomial variable* $X \sim B(n, p)$, and (iii) a *Poisson variable* X with parameter $\lambda > 0$.

Example 12.5.41 (Uniform Distribution). Let $a < b$ be real numbers. A random variable X is said to be *uniformly distributed* over $[a, b]$ if its probability measure is given by $P_X(B) = \lambda(B)/(b - a)$, for each Borel set $B \subset [a, b]$. The distribution function of X is therefore

$$F_X(x) := \begin{cases} 0 & \text{if } x < a, \\ \dfrac{x - a}{b - a} & \text{if } a \leq x < b, \\ 1 & \text{if } x \geq b. \end{cases}$$

Most important random variables belong to the class of *absolutely continuous* variables:

Definition 12.5.42 (Absolute Continuity, Density). A random variable X is said to be *absolutely continuous* if its probability distribution P_X is absolutely continuous (with respect to Lebesgue measure). It then follows from the Lebesgue–Radon–Nikodym theorem (Theorem 12.3.39) that there is a "unique" *nonnegative* function $f_X \in \mathcal{L}_\lambda^1(\mathbb{R})$ (which we may assume to be defined *everywhere*) such that

$$P_X(B) = \int_B f_X(x) \, d\lambda(x) \qquad (\forall B \in \mathcal{B}_\mathbb{R}).$$

The function f_X is then called the *density function* (or simply the *density*) of X and we clearly have $\int_\mathbb{R} f_X(x) \, d\lambda(x) = 1$.

Remark 12.5.43.

1. If X is an absolutely continuous random variable, then its distribution function is obviously given by

$$F_X(x) = \int_{-\infty}^x f_X(t) \, d\lambda(t).$$

In particular, an absolutely continuous variable is *continuous*. (Why?) The converse, however, is *false* as the next example will show.
2. It follows from the *Fundamental Theorem of Calculus* that, if f_X is *continuous* (on $\mathbb{R}$), then F_X is *continuously differentiable* (on $\mathbb{R}$) and we have

$$F_X'(x) = f_X(x) \qquad (\forall x \in \mathbb{R}). \tag{*}$$

In fact, if f_X is only *piecewise continuous*, then (*) remains valid except at the *discontinuity* points of f_X.

Example 12.5.44 (Cantor–Lebesgue Variable). A random variable X is said to have the *Cantor–Lebesgue distribution* if its distribution function is given by

$$F_X := \begin{cases} 0 & \text{if } x < 0, \\ \kappa(x) & \text{if } 0 \le x \le 1, \\ 1 & \text{if } x > 1, \end{cases}$$

where $\kappa(x)$ is *Cantor's ternary function* (cf. Example 4.3.13). Note that $F_X(x) = 1/2$ for $x \in [1/3, 2/3)$, $F_X(x) = 1/2$ for $x \in [1/3, 2/3)$, $F_X(x) = 1/4$ for $x \in [1/9, 2/9)$, $F_X(x) = 3/4$ for $x \in [7/9, 8/9)$, etc. Thus F_X is *constant* outside $[0, 1]$ and on all the intervals removed in the construction of the Cantor set. It is also *increasing* and *continuous*. In particular, X is a *continuous* variable. Now suppose that X has a density f_X. If (a, b) is any one of the *removed middle thirds*, then $\int_a^b f_X(x)\, dx = F_X(b) - F_X(a) = 0$ and hence (since $f_X \ge 0$) we have $f_X(x) = 0$ for almost all $x \in (a, b)$. Similarly, $f_X = 0$ a.e. on $(-\infty, 0)$ and $(1, \infty)$. Thus, since $\lambda(C) = 0$, we have $f_X(x) = 0$ for a.a. $x \in \mathbb{R}$, which is absurd. Alternatively, since F_X is constant outside C, we have $f_X(x) = F_X'(x) = 0$ for all $x \notin C$. Therefore, X is *not* absolutely continuous. Note also that $P(X \notin C) = 0$ and hence $P(X \in C) = 1$ despite the fact that $P(X = x) = 0$ for all $x \in C$.

Here are some important examples of absolutely continuous variables:

Example 12.5.45 (Uniform Variables). Given a set $B \in \mathcal{B}_\mathbb{R}$ with $0 < \lambda(B) < \infty$, a random variable X is said to be *uniformly distributed* over B if P_X has (the uniform) density function

$$f_X(x) := \begin{cases} \dfrac{1}{\lambda(B)} & \text{if } x \in B, \\ 0 & \text{otherwise.} \end{cases}$$

Example 12.5.46 (Normal or Gaussian Variables). Let $m \in \mathbb{R}$ and $\sigma > 0$ be given. We say that a random variable X is *normal* (or *Gaussian*) with *mean* m and *variance* σ^2, and we write $X \sim N(m, \sigma^2)$, if X has density

$$f_X(x) := \frac{1}{\sigma\sqrt{2\pi}} \exp\left(-\frac{(x - m)^2}{2\sigma^2} \right).$$

If $X \sim N(0, 1)$, then we say that X is *standard normal*. Given the well-known fact that $\int_{-\infty}^\infty e^{-x^2/2}\, dx = \sqrt{2\pi}$, we indeed have $\int_\mathbb{R} f_X(x)\, d\lambda(x) = 1$.

Example 12.5.47 (Cauchy Variable). A random variable X is said to have the *Cauchy distribution* if its density is

$$f_X(x) := \frac{1}{\pi(1 + x^2)}.$$

Example 12.5.48 (Exponential Variables). We say that X is an *exponential* variable with *parameter* $\lambda > 0$ if its density is the function

$$f_X(x) := \begin{cases} \lambda e^{-\lambda x} & \text{if } x \geq 0, \\ 0 & \text{if } x < 0. \end{cases}$$

Before discussing the integration of random variables, let us give two more important definitions.

Definition 12.5.49 (Identically Distributed Variables). Two random variables $X, Y : \Omega \to \mathbb{R}$ are said to be *identically distributed* if they have the same probability distribution, i.e., if $P_X = P_Y$.

Definition 12.5.50 (Independent Variables). Two random variables $X, Y : \Omega \to \mathbb{R}$ are said to be *independent* if the corresponding σ-algebras $\mathcal{A}_X := X^{-1}(\mathcal{B}_\mathbb{R})$ and $\mathcal{A}_Y := Y^{-1}(\mathcal{B}_\mathbb{R})$ are independent; i.e., given any Borel sets $B, C \in \mathcal{B}_\mathbb{R}$,

$$P(X^{-1}(B) \cap Y^{-1}(C)) = P(X^{-1}(B))P(Y^{-1}(C)).$$

More generally, if J is an index set and $X_j : \Omega \to \mathbb{R}$ is a random variable for each $j \in J$, then the collection $\{X_j : j \in J\}$ is said to be *independent* if the collection of σ-algebras $\{\mathcal{A}_{X_j} : j \in J\}$ is independent.

Exercise 12.5.51. Let $(I, \mathcal{M}_\lambda(I), \lambda_I)$ be the probability space of the Example 12.5.37. Show that the random variables $X := \chi_{[0,1/2]}$ and $Y := \chi_{[1/4,3/4]}$ are independent. *Hint:* Consider the corresponding σ-algebras. Note, e.g., that $\mathcal{A}_X = \{\emptyset, [0, 1/2], (1/2, 1], [0, 1]\}$.

Exercise 12.5.52. Let $(X_k)_{k=1}^n$ be *independent* random variables on Ω, and let $g_k : \mathbb{R} \to \mathbb{R}$ be (Borel-) measurable functions for $1 \leq k \leq n$. Show that the composites $g_k(X_k)$ are also independent random variables on Ω.

Having defined random variables (i.e., measurable functions on the sample space Ω), it is natural to ask whether such variables are *integrable*. Now, given a random variable X, its probability distribution P_X is a probability measure on $\mathbb{R}$ and enables us to transform the integrals over Ω into integrals over $\mathbb{R}$:

Theorem 12.5.53 (Change of Variables). *Let $(\Omega, \mathcal{A}, P)$ be a probability space and $X : \Omega \to \mathbb{R}$ a random variable. Then, for each measurable function $g : \mathbb{R} \to \mathbb{R}$, the composite function $g \circ X$ is a random variable. Moreover, $g \circ X \in \mathcal{L}_P^1(\Omega)$ if and only if $g \in \mathcal{L}_{P_X}^1(\mathbb{R})$ and we have*

$$\int_\Omega g(X(\omega)) \, dP(\omega) = \int_\mathbb{R} g(x) \, dP_X(x). \tag{$\dagger$}$$

Proof. The first statement is obvious. To prove $(\dagger)$, we first note that if $g = \chi_B$ for some $B \in \mathcal{B}_\mathbb{R}$, then $\chi_B \circ X = \chi_{X^{-1}(B)}$ and both sides of $(\dagger)$ reduce to

$P(X^{-1}(B))$. By linearity, (†) is therefore satisfied for each *simple function* g. Next, if g is a *nonnegative* integrable function, then $g = \lim(g_n)$ where (g_n) is an *increasing* sequence of simple functions and hence (†) follows from the Monotone Convergence Theorem. Finally, for a general $g \in \mathcal{L}^1_{P_X}(\mathbb{R})$, we note that $g = g^+ - g^-$ and apply the previous case to g^+ and g^-.   □

Corollary 12.5.54. *Under the assumptions of Theorem 12.5.53, if X is absolutely continuous with density function f_X, then we have*

$$\int_\Omega g(X)\, dP = \int_\mathbb{R} g(x) f_X(x)\, d\lambda(x).$$

Example 12.5.55.

1. If X is *constant,* i.e., $X(\omega) = a$ for all $\omega \in \Omega$ and some $a \in \mathbb{R}$, then $g \circ X \equiv g(a)$ and hence the left side of (†) is $g(a) P(\Omega) = g(a)$. Also, $P_X = \delta_a$ so that (†) becomes

$$\int_\mathbb{R} g(x)\, d\delta_a(x) = g(a).$$

2. More generally, consider a *discrete* random variable X taking the values $\{a_1, a_2, a_3, \ldots\}$ with probabilities $p_n := P(X = a_n)$. Then we have $P_X = \sum_n p_n \delta_{a_n}$ and (†) gives

$$\int_\Omega g(X)\, dP = \sum_n g(a_n) p_n.$$

The most important special case is when g is the *identity function:* $g(x) = x$ for all $x \in \mathbb{R}$:

Definition 12.5.56 (Expectation).   Given an *integrable* random variable $X : \Omega \to \mathbb{R}$, the *expectation* (or *mean*) of X is its integral:

$$E[X] := \int_\Omega X(\omega)\, dP(\omega) = \int_\mathbb{R} x\, dP_X(x).$$

Of course, $E[X]$ exists (as a *finite* number) if and only if $\int_\mathbb{R} |x|\, dP_X(x) < \infty$ and we then have $|E[X]| \leq \int_\mathbb{R} |x|\, dP_X(x) = E[|X|]$. More generally, given any $g \in \mathcal{L}^1_{P_X}(\mathbb{R})$, we have

$$E[g(X)] = \int_\Omega g(X(\omega))\, dP(\omega) = \int_\mathbb{R} g(x)\, dP_X(x).$$

For a *complex* variable $Z : \Omega \to \mathbb{C}$, we define $E[Z] := E[\text{Re}(Z)] + i E[\text{Im}(Z)]$, if $E[\text{Re}(Z)]$ and $E[\text{Im}(Z)]$ exist.

Exercise 12.5.57.

1. Let X be a *Poisson* variable with parameter $\lambda > 0$. Thus X takes only the values $k = 0, 1, 2, \ldots$ with probabilities $P(X = k) = \lambda^k e^{-\lambda}/k!$. Show that $E[X] = \lambda$.
2. Find $E[X]$ for an *exponential* variable with parameter $\lambda > 0$.
3. If X has the *Cauchy density* $f_X := 1/\pi(1 + x^2)$, show that $E[X]$ does *not* exist.

Another special case of importance is obtained if we take $g(x) := (x - m)^2$, where $m = E[X]$. To introduce it, we need the following.

Definition 12.5.58 (Square-Integrable Variable).   A random variable X on Ω is said to be *square integrable* if $X^2 \in \mathcal{L}_P^1(\Omega)$; i.e.,

$$\int_\Omega |X(\omega)|^2 \, dP(\omega) < \infty.$$

The set of all square integrable random variables will be denoted by $\mathcal{L}_P^2(\Omega)$.

Proposition 12.5.59. $\mathcal{L}_P^2(\Omega)$ *is a (vector) subspace of* $\mathcal{L}_P^1(\Omega)$.

Proof. That $\mathcal{L}_P^2(\Omega)$ is a vector space follows at once from the elementary inequality $|X + Y|^2 \le 2(|X|^2 + |Y|^2)$. Next, since $P(\Omega) = 1$, we have $1 \in \mathcal{L}_P^2(\Omega)$. Thus, if $X^2 \in \mathcal{L}_P^1(\Omega)$, then the trivial inequality $|X| \le (1 + X^2)/2$ implies that $X \in \mathcal{L}_P^1(\Omega)$. □

Definition 12.5.60 (Variance, Standard Deviation).   Given any random variable $X \in \mathcal{L}_P^2(\Omega)$, we define its *variance* to be the integral

$$\mathrm{Var}(X) := \int_\Omega (X - E[X])^2 \, dP = E[(X - E[X])^2] = E[X^2] - (E[X])^2,$$

where the last equation follows from the fact that $E[X]$ is *linear* in X. The notation $\sigma^2(X) := \mathrm{Var}(X)$ will also be used. The square root of the variance, i.e., $\sigma(X) = \sqrt{\mathrm{Var}(X)}$, is called the *standard deviation* of X.

Exercise 12.5.61.   Show that, if $X \in \mathcal{L}_P^2(\Omega)$, then

$$\mathrm{Var}(aX + b) = a^2 \mathrm{Var}(X) \qquad (\forall a, b \in \mathbb{R}).$$

Exercise 12.5.62.

1. Find $\mathrm{Var}(X)$ if X is a (i) *Poisson* variable with parameter $\lambda > 0$ or (ii) an *exponential* variable with parameter $\lambda > 0$. *Hint:* Use Exercise 12.5.57.
2. Let X be *uniformly distributed* over $[a, b]$. Show that $E[X] = (a + b)/2$ and $\mathrm{Var}(X) = (b - a)^2/12$.
3. Show that, for a normal variable $X \sim N(m, \sigma^2)$, we indeed have $E[X] = m$ and $\mathrm{Var}(X) = \sigma^2$. *Hint:* Use the substitution $z := (x - m)/\sigma$ and integration by parts.

Remark 12.5.63. Let X and Y be *identically distributed* random variables on Ω. Then, for any $f \in \mathcal{L}^1_{P_X}(\mathbb{R}) = \mathcal{L}^1_{P_Y}(\mathbb{R})$, the *Change of Variables Theorem* implies that

$$\int_\Omega f(X) \, dP = \int_\Omega f(Y) \, dP.$$

In particular, if $X, \ Y \in \mathcal{L}^2_P(\Omega)$, then $E[X] = E[Y]$ and $\mathrm{Var}(X) = \mathrm{Var}(Y)$.

Definition 12.5.64 (Joint Distribution). Let $X, \ Y$ be random variables on Ω and consider the *random vector* $(X, Y) : \Omega \to \mathbb{R}^2$, i.e., the measurable map

$$(X, Y)(\omega) := (X(\omega), Y(\omega)) \in \mathbb{R}^2.$$

Its *probability distribution*, denoted by $P_{(X,Y)}$, is called the *joint distribution* of X and Y and is defined by

$$P_{(X,Y)}(B) := P((X, Y) \in B) \qquad (\forall B \in \mathcal{B}_{\mathbb{R}^2}).$$

Exercise 12.5.65 (Marginal Distributions). Let X and Y be random variables on Ω. Show that the probability distributions P_X and P_Y can be obtained from the joint distribution $P_{(X,Y)}$ as follows:

$$P_X(B) = P_{(X,Y)}(B \times \mathbb{R}), \quad P_Y(B) := P_{(X,Y)}(\mathbb{R} \times B) \qquad (\forall B \in \mathcal{B}_{\mathbb{R}}).$$

When expressed this way, P_X and P_Y are called *marginal distributions*.

Definition 12.5.66 (Jointly Continuous). Two random variables $X, \ Y$ on Ω are said to be *jointly continuous* if their joint distribution $P_{(X,Y)}$ is *absolutely continuous;* i.e., there is a *nonnegative* function $f_{(X,Y)} \in \mathcal{L}^1_{\lambda_2}(\mathbb{R}^2)$ such that

$$P_{(X,Y)}(B) = \int_B f_{(X,Y)}(x, y) \, d\lambda_2(x, y) \qquad (\forall B \in \mathcal{B}_{\mathbb{R}^2}).$$

The function $f_{(X,Y)}$ is then called the *joint density* of X and Y.

Exercise 12.5.67 (Marginal Densities). Let $f_{(X,Y)}$ be the joint density of random variables X and Y. Show that both X and Y are then *absolutely continuous* with respective densities

$$f_X(x) := \int_{\mathbb{R}} f_{(X,Y)}(x, y) \, d\lambda(y), \quad f_Y(y) := \int_{\mathbb{R}} f_{(X,Y)}(x, y) \, d\lambda(x).$$

When expressed this way, f_X and f_Y are called *marginal densities*.

The analog of *Change of Variables* (Theorem 12.5.53) holds for joint distributions as well and is proved in exactly the same way:

Theorem 12.5.68 (Change of Variables).   *Let $P_{(X,Y)}$ be the joint distribution of the random variables X and Y on Ω. Then, for each measurable function $g : \mathbb{R}^2 \to \mathbb{R}$, the composite $g(X, Y)$ is a random variable on Ω. Moreover, $g(X, Y) \in \mathcal{L}^1_P(\Omega)$ if and only if $g \in \mathcal{L}^1_{P_{(X,Y)}}(\mathbb{R}^2)$ and we have*

$$\int_\Omega g(X(\omega), Y(\omega))\, dP(\omega) = \int_{\mathbb{R}^2} g(x, y)\, dP_{(X,Y)}(x, y).$$

Proof. Exercise!   □

Using joint distributions, we can characterize *independent* random variables:

Proposition 12.5.69.   *Two random variables X, Y on Ω are independent if and only if*

$$P_{(X,Y)} = P_X \otimes P_Y. \tag{†}$$

Proof. First note that $\mathcal{B}_{\mathbb{R}^2}$ is generated by the *Borel rectangles* $B \times C$, where B, $C \in \mathcal{B}_{\mathbb{R}}$. (Why?) Therefore, by the definition of the *product measure* $P_X \otimes P_Y$, we need only show that (†) holds when both sides are evaluated at Borel rectangles. Now, assuming that X and Y are *independent*, for any Borel rectangle $B \times C$, we have

$$P_{(X,Y)}(B \times C) = P((X, Y) \in B \times C) = P((X \in B) \cap (Y \in C))$$
$$= P(X \in B)P(Y \in C) = P_X(B)P_Y(C)$$
$$= (P_X \otimes P_Y)(B \times C).$$
                                                                                                                                            □

Theorem 12.5.70.   *Let X and Y be independent random variables with finite expectations $E[X]$ and $E[Y]$, respectively. Then $E[XY]$ exists and we have*

$$E[XY] = E[X]E[Y].$$

Proof. Applying Theorem 12.5.68 with $g(x, y) := xy$ and Proposition 12.5.69, we have

$$E[XY] = \int_{\mathbb{R}^2} xy\, dP_{(X,Y)}(x, y) = \int_{\mathbb{R}^2} xy\, dP_X(x) \otimes dP_Y(y)$$
$$= \left(\int_{\mathbb{R}} x\, dP_X(x) \right) \left(\int_{\mathbb{R}} y\, dP_Y(y) \right) = E[X]E[Y],$$

where we have used Fubini's theorem.   □

Corollary 12.5.71.   *Let $(X_k)_{k=1}^n$ be independent, square integrable random variables on Ω. Then*

$$\mathrm{Var}(X_1 + \cdots + X_n) = \sum_{k=1}^n \mathrm{Var}(X_k).$$

Proof. Introduce the *centered* variables $Y_k := X_k - E[X_k]$, for $1 \leq k \leq n$. Then the Y_k are also *independent* (why?) and we have $E[Y_k] = 0$ for $1 \leq k \leq 0$. Thus, by Theorem 12.5.70, we have

$$E[Y_j Y_k] = E[Y_j]E[Y_k] = 0 \quad (j \neq k),$$

and hence

$$\mathrm{Var}(X_1 + \cdots + X_n) = E[(Y_1 + \cdots + Y_n)^2] = \sum_{j,k} E[Y_j Y_k]$$

$$= \sum_k E[Y_k^2] = \sum_k \mathrm{Var}(X_k).$$

$\square$

Corollary 12.5.72. *Let $(X_k)_{k=1}^n$ be independent, square integrable variables on Ω and let $S_n := X_1 + \cdots + X_n$. Then we have*

$$E[S_n/n] = \frac{1}{n} \sum_{k=1}^n E[X_k], \quad \mathrm{Var}(S_n/n) = \frac{1}{n^2} \sum_{k=1}^n \mathrm{Var}(X_k).$$

Proof. The first equation follows from the *linearity* of E and the second from Corollary 12.5.71 and Exercise 12.5.61. $\square$

For the main results of this section, we shall need a few well-known and useful inequalities. Let us start with

Proposition 12.5.73 (Markov's Inequality). *Let X be a random variable. Then for any constant $c > 0$ we have*

$$P(|X| \geq c) \leq \frac{E[|X|]}{c}. \tag{†}$$

More generally, for any nonnegative (Borel-) measurable function $f : \mathbb{R} \to \mathbb{R}$ and any $c > 0$, we have

$$P(f(X) \geq c) \leq \frac{E[f(X)]}{c}. \tag{‡}$$

Proof. To prove (‡), let $A := \{f(X) \geq c\}$. Then $f(X) \geq c\chi_A$ and hence

$$E[f(X)] = \int_\Omega f(X) \, dP \geq c \int_\Omega \chi_A \, dP = cP(A).$$

The inequality (†) now follows if we take $f(x) := |x|$. $\square$

Here is an immediate corollary:

Proposition 12.5.74 (Chebyshev's Inequality). *Let* $X \in \mathcal{L}_P^2(\Omega)$. *Then, for any constant* $c > 0$, *we have*

$$P(|X - E[X]| \geq c) \leq \frac{\mathrm{Var}(X)}{c^2}.$$

Proof. This follows from Proposition 12.5.73 if we use $f(x) := (x - m)^2$ in ($\ddagger$), where $m := E[X]$. Or, note that $Y := X - E[X]$ is square integrable and $E[Y^2] = \mathrm{Var}(X)$. Applying Markov's inequality—the inequality ($\dagger$)—we have

$$P(|X - E[X]| \geq c) = P(Y^2 \geq c^2) \leq \frac{E[Y^2]}{c^2} = \frac{\mathrm{Var}(X)}{c^2}.$$

$\square$

We shall also need the following extension of Chebyshev's inequality:

Proposition 12.5.75 (Kolmogorov's Inequality). *Let* $(X_k)_{k=1}^n$ *be independent, square-integrable variables on* Ω *with* $E[X_k] = 0$ *and* $\mathrm{Var}(X_k) = \sigma_k^2$, *and define* $S_k := X_1 + \cdots + X_k$, *for* $1 \leq k \leq n$. *Then, for any constant* $c > 0$, *we have*

$$P\left(\max_{1 \leq k \leq n} |S_k| \geq c \right) \leq \frac{1}{c^2} \sum_{k=1}^n \sigma_k^2.$$

Proof. Let $A_1 := \{|S_1| \geq c\}$ and, for $2 \leq k \leq n$, define

$$A_k := \{|S_1| < c, |S_2| < c, \ldots, |S_{k-1}| < c, |S_k| \geq c\}.$$

Note that the A_k are *pairwise disjoint* and we have

$$B := \left\{ \max_{1 \leq k \leq n} |S_k| \geq c \right\} = \bigcup_{k=1}^n A_k.$$

Thus, setting $\chi_k := \chi_{A_k}$, for $1 \leq k \leq n$, we have $\chi_B = \sum_{k=1}^n \chi_k$. Now, by Corollary 12.5.71, we have

$$\sum_{k=1}^n \sigma_k^2 = E[S_n^2] \geq E[S_n^2 \chi_B] = \sum_{k=1}^n E[S_n^2 \chi_k]. \qquad (*)$$

Next, note that $S_k \chi_k$ and $S_n - S_k$ are *independent* and $E[S_n - S_k] = 0$. (Why?) Thus, by Theorem 12.5.70,

$$E[S_k \chi_k (S_n - S_k)] = 0. \qquad (**)$$

Using $(\ast\ast)$ and the fact that $S_n^2 = [S_k + (S_n - S_k)]^2$, we have

$$E[S_n^2 \chi_k] = E[S_k^2 \chi_k] + E[(S_n - S_k)^2 \chi_k] \geq E[S_k^2 \chi_k]$$
$$\geq c^2 E[\chi_k] = c^2 P(A_k),$$

which, by $(\ast)$ and the fact that the A_k are pairwise disjoint, gives

$$\sum_{k=1}^n \sigma_k^2 \geq c^2 \sum_{k=1}^n P(A_k) = c^2 P(B).$$

$\square$

Let us use this inequality to prove the following result on the convergence of a *series* of random variables.

Theorem 12.5.76. *If* (X_n) *is a sequence of independent random variables with* $E[X_n] = 0$ *for all n and* $\sum_{n=1}^\infty \mathrm{Var}(X_n) < \infty$, *then* $\sum_{n=1}^\infty X_n$ *converges almost surely.*

Proof. Let $S_n := \sum_{k=1}^n X_k$ and for each $m,\ k \in \mathbb{N}$, let $T_{m,k} := S_{m+k} - S_m = \sum_{j=1}^k X_{m+j}$. Since the X_j are independent, for each $\varepsilon > 0$, we can apply Kolmogorov's inequality to the $T_{m,k},\ 1 \leq k \leq n$, to get

$$P\left(\max_{1 \leq k \leq n} |T_{m,k}| \geq \varepsilon \right) \leq \frac{\mathrm{Var}(T_{m,n})}{\varepsilon^2} = \frac{1}{\varepsilon^2} \sum_{j=m+1}^{m+n} \mathrm{Var}(X_j). \qquad (\ast)$$

Next, note that $P(\max_{1 \leq k \leq n} |T_{m,k}| \geq \varepsilon) = P(\bigcup_{k=1}^n \{|T_{m,k}| \geq \varepsilon\})$ and that our assumption implies $\lim_{m \to \infty} \sum_{j=m+1}^\infty \mathrm{Var}(X_j) = 0$. Therefore, letting $m \to \infty$ in $(\ast)$, we get

$$\lim_{m \to \infty} P\left(\bigcup_{k=1}^\infty \{|T_{m,k}| \geq \varepsilon\} \right) = \lim_{m \to \infty} \left[\lim_{n \to \infty} P\left(\bigcup_{k=1}^n \{|T_{m,k}| \geq \varepsilon\} \right) \right]$$

$$\leq \lim_{m \to \infty} \left[\lim_{n \to \infty} \frac{1}{\varepsilon^2} \sum_{j=m+1}^{m+n} \mathrm{Var}(X_j) \right] \qquad (\ast\ast)$$

$$= \frac{1}{\varepsilon^2} \lim_{m \to \infty} \sum_{j=m+1}^\infty \mathrm{Var}(X_j) = 0.$$

We now use $(\ast\ast)$ to prove that (S_n) converges almost surely. Note that $(S_n(\omega))$ converges if and only if it is Cauchy. Therefore, if $E := \{\omega : (S_n(\omega))$ diverges$\}$, then we have

$$E = \bigcup_{j=1}^\infty \bigcap_{n=1}^\infty \bigcup_{k=1}^\infty \left\{ |T_{n,k}| \geq 1/j \right\}. \qquad (\dagger)$$

Since for any $m \in \mathbb{N}$ we have the inclusions

$$\bigcap_{n=1}^{\infty} \bigcup_{k=1}^{\infty} \{|T_{n,k}| \geq 1/j\} \subset \bigcap_{n=1}^{m} \bigcup_{k=1}^{\infty} \{|T_{n,k}| \geq 1/j\} \subset \bigcup_{k=1}^{\infty} \{|T_{m,k}| \geq 1/j\},$$

letting $m \to \infty$ and using (∗∗), we get $P(\bigcap_{n=1}^{\infty} \bigcup_{k=1}^{\infty} \{|T_{n,k}| \geq 1/j\}) = 0$, which in view of (†) gives $P(E) = 0$.   □

We are now ready to prove our main result: *the law of large numbers*. There are two versions of this law. For the *weak* one, we need the following.

Definition 12.5.77 (Convergence in Probability).   We say that a sequence $(X_n)_{n=1}^{\infty}$ of random variables converges to a random variable X *in probability* if

$$\lim_{n \to \infty} P(|X_n - X| > \varepsilon) = 0 \qquad (\forall \varepsilon > 0).$$

Remark 12.5.78. Convergence in probability is in fact *weaker* than *convergence almost surely*. In the context of measure theory, the former is referred to as *convergence in measure* and the latter corresponds, of course, to *convergence almost everywhere*. (cf. Proposition 11.4.11).

Proposition 12.5.79. *If $X_n \to X$ almost surely, then $X_n \to X$ in probability.*

Proof. Suppose that $X_n \to X$ almost surely; i.e., if $Z := \{\omega \in \Omega : X_n(\omega) \not\to X(\omega)\}$, then $P(Z) = 0$. This is equivalent to the inclusions

$$\bigcap_{n=1}^{\infty} \bigcup_{k=n}^{\infty} \{\omega : |X_k(\omega) - X(\omega)| > \varepsilon\} \subset Z \qquad (\forall \varepsilon > 0).$$

In particular, using the *continuity* of P,

$$P(|X_n - X| > \varepsilon) \leq P\left(\bigcup_{k=n}^{\infty} \{\omega : |X_k(\omega) - X(\omega)| > \varepsilon\} \right) \to 0,$$

as $n \to \infty$.   □

Theorem 12.5.80 (Weak Law of Large Numbers).   *Let (X_n) be a sequence of independent, square-integrable variables with means $(m_n)_{n=1}^{\infty}$ and variances $(\sigma_n^2)_{n=1}^{\infty}$. If $\lim_{n \to \infty} \sum_{k=1}^{n} \sigma_k^2/n^2 = 0$, then $n^{-1} \sum_{k=1}^{n} (X_k - m_k) \to 0$ in probability, as $n \to \infty$.*

Proof. By Corollary 12.5.72, the average $n^{-1} \sum_{k=1}^{n} (X_k - m_k)$ has mean *zero* and variance $n^{-2} \sum_{k=1}^{n} \sigma_k^2$. Therefore, by Chebyshev's inequality, for any $\varepsilon > 0$ we have

$$\lim_{n\to\infty} P\left(\left|n^{-1}\sum_{k=1}^{n}(X_k - m_k)\right| > \varepsilon\right) \leq \lim_{n\to\infty} (n\varepsilon)^{-2}\sum_{k=1}^{n}\sigma_k^2 = 0.$$

□

Corollary 12.5.81. *If* $(X_n)_{n=1}^{\infty}$ *is a sequence of independent, identically distributed, and square-integrable variables with mean* m *and variance* σ^2, *then, with* $S_n :=$ $\sum_{k=1}^{n} X_k$, *we have*

$$\lim_{n\to\infty} P\left(\left|\frac{S_n}{n} - m\right| > \varepsilon\right) = 0 \qquad (\forall \varepsilon > 0);$$

i.e., (S_n/n) *converges in probability to the constant variable* m.

Proof. Since (by Corollary 12.5.72) $E[S_n/n] = m$ and $\mathrm{Var}(S_n/n) = \sigma^2/n$, the corollary follows from Theorem 12.5.80. □

We now look at a version of the *strong* law of large numbers due to *Kolmogorov:*

Theorem 12.5.82 (Strong Law of Large Numbers). *Let* (X_n) *be a sequence of independent, square-integrable variables with means* $(m_n)_{n=1}^{\infty}$ *and variances* $(\sigma_n^2)_{n=1}^{\infty}$. *If* $\sum_{n=1}^{\infty}\sigma_n^2/n^2 < \infty$, *then*

$$\lim_{n\to\infty} n^{-1}\sum_{k=1}^{n}(X_k - m_k) = 0 \quad \text{almost surely.}$$

Proof. Let $S_n := \sum_{k=1}^{n}(X_k - m_k)$ and let $\varepsilon > 0$ be given. For each $k \in \mathbb{N}$, define

$$A_k := \{\omega \in \Omega : |S_n(\omega)|/n \geq \varepsilon \text{ for some } n \text{ with } 2^{k-1} \leq n < 2^k\}.$$

Then, for each $\omega \in A_k$, we have $|S_n(\omega)| \geq \varepsilon 2^{k-1}$ for some $n < 2^k$. Thus, by Kolmogorov's inequality, we have

$$P(A_k) \leq \frac{1}{(\varepsilon 2^{k-1})^2}\sum_{n=1}^{2^k}\sigma_n^2,$$

which (in view of $\sum_{k>m} 2^{-2k} < \int_m^{\infty} 2^{-2x}\,dx$) implies that

$$\sum_{k=1}^{\infty} P(A_k) \leq \frac{4}{\varepsilon^2}\sum_{k=1}^{\infty}\sum_{n=1}^{2^k-1}\sigma_n^2/2^{2k} = \frac{4}{\varepsilon^2}\sum_{n=1}^{\infty}\left(\sum_{k>\log_2 n} 2^{-2k}\right)\sigma_n^2$$

$$\leq \frac{8}{\varepsilon^2}\sum_{n=1}^{\infty}\frac{\sigma_n^2}{n^2} < \infty.$$

Therefore, it follows from the *First Borel–Cantelli Lemma* (Theorem 12.5.24) that we have $P(\limsup_{k \to \infty} A_k) = 0$. Since

$$\limsup_{k \to \infty} A_k = \{\omega \in \Omega : |S_n(\omega)|/n \geq \varepsilon \text{ for infinitely many } n\},$$

it follows that

$$P(\limsup_{n \to \infty}\{|S_n|/n < \varepsilon\}) = 1.$$

Setting $\varepsilon := 1/j$, $j \in \mathbb{N}$, and letting $j \to \infty$, we deduce that $\lim_{n \to \infty} S_n/n = 0$ almost surely. $\qquad\square$

Corollary 12.5.83. *Let $(X_n)_{n=1}^{\infty}$ be a sequence of independent, identically distributed, and square-integrable variables with mean m and variance σ^2. Then, with $S_n := \sum_{k=1}^{n} X_k$, we have*

$$\lim_{n \to \infty} \frac{S_n}{n} = m \quad almost\ surely.$$

Proof. Since $\sum_{n=1}^{\infty} \sigma^2/n^2 = \pi^2\sigma^2/6$, the corollary follows at once from Theorem 12.5.82. $\qquad\square$

Remark 12.5.84. In fact, the assumption $X_n \in \mathcal{L}_P^2(\Omega)$ in Corollary 12.5.83 can be replaced by the weaker assumption $X_n \in \mathcal{L}_P^1(\Omega)$, but the proof is then more involved. The reader is referred to more advanced texts for this and other extensions of the strong law of large numbers.

We end the chapter by giving a probabilistic proof of *Bernstein Approximation Theorem* (Theorem 4.7.9), i.e., the fact that *Bernstein polynomials* are dense in the space of all continuous real-valued functions on $[0, 1]$ with *uniform metric*. The following exercise will be needed:

Exercise 12.5.85. Let $(X_k)_{k=1}^{n}$ be independent *Bernoulli* variables with $E[X_k] = P(X_k = 1) = p$, $1 \leq k \leq n$, and let $S := \sum_{k=1}^{n} X_k$. Show that S is a *binomial* random variable with parameters n and p; i.e., $S \sim B(n, p)$. Deduce that

$$E[S] = np, \quad \text{Var}(S) = np(1 - p).$$

Theorem 12.5.86 (Bernstein Approximation Theorem). *For any continuous function $f : [0, 1] \to \mathbb{R}$, the Bernstein polynomials*

$$B_n(x) := \sum_{k=0}^{n} f\left(\frac{k}{n}\right)\binom{n}{k} x^k (1 - x)^{n-k}$$

converge to f uniformly on $[0, 1]$.

Proof. Let $\varepsilon > 0$ be given. Since f is *uniformly continuous* on $[0, 1]$, we can pick a $\delta > 0$ such that $|x - y| < \delta$ implies $|f(x) - f(y)| < \varepsilon/2$. Let $M := \sup\{|f(x)| : 0 \le x \le 1\}$ and pick $N \in \mathbb{N}$ so large that $M/N < \varepsilon\delta^2$. We shall prove that $|B_n(x) - f(x)| < \varepsilon$ for all $n \ge N$ and $x \in [0, 1]$. Let $n \ge N$ and $x \in [0, 1]$ be *fixed*. Let $(X_k)_{k=1}^{n}$ be an independent sequence of *Bernoulli* variables with $P(X_k = 1) = x$ and $P(X_k = 0) = 1 - x$, $1 \le k \le n$. Then $E[X_k] = x$ and $\text{Var}[X_k] = x(1 - x)$ for $1 \le k \le n$. If $S := \sum_{k=1}^{n} X_k$, then (by Exercise 12.5.85) $S \sim B(n, x)$ and we have $E[f(S/n)] = B_n(x)$. (Why?) Now, by the (weak) *law of large numbers,* we expect S/n to be close to x with large probability. To make this precise, note that

$$\left| f\left(\frac{S}{n}\right) - f(x) \right| \le \begin{cases} \varepsilon/2 & \text{on } \{|S/n - x| < \delta\}, \\ 2M & \text{on } \{|S/n - x| \ge \delta\}. \end{cases} \qquad (*)$$

But Exercise 12.5.85 and the fact that $x(1 - x) \le 1/4$ on $[0, 1]$ give $\text{Var}(S/n - x) \le 1/(4n)$ (why?) so, by *Chebyshev's inequality,* we have $P(|S/n - x| \ge \delta) \le 1/(4n\delta^2)$. Therefore, integrating $(*)$ and noting that $n \ge N$, we get

$$|B_n(x) - f(x)| \le E\left[\left| f\left(\frac{S}{n}\right) - f(x) \right|\right] \le \frac{\varepsilon}{2} + \frac{2M}{4n\delta^2} < 2\left(\frac{\varepsilon}{2}\right) = \varepsilon.$$

$\square$

12.6 Problems

1. Let X be an *uncountable* set and let $\mathcal{S} := \{\{x\} : x \in X\}$ be the set of all *singletons* in X. Show that the σ-algebra $\mathcal{A}_{\mathcal{S}}$ generated by $\mathcal{S}$ is given by

$$\mathcal{A}_{\mathcal{S}} := \{E \subset X : |E| \le \aleph_0 \text{ or } |E^c| \le \aleph_0\},$$

and that the map $\mu : \mathcal{A}_{\mathcal{S}} \to [0, \infty)$ given by $\mu(E) = 0$ if $|E| \le \aleph_0$ and $\mu(E) = 1$ if $|E^c| \le \aleph_0$ is a measure on $\mathcal{A}_{\mathcal{S}}$.

2. Let $(X, \mathcal{A})$ be a measurable space and suppose that $\mu : \mathcal{A} \to [0, \infty]$ is *finitely additive*. Show that, if μ is σ-subadditive, i.e., if $\mu(\bigcup A_n) \le \sum \mu(A_n)$ for any sequence (A_n) in $\mathcal{A}$, then μ is a *measure*. Deduce that any finitely additive *outer measure* is actually a measure.

3. Let $\mathcal{C}$ denote the collection of all *countable* subsets of $\mathbb{R}$ and define $\mu : \mathcal{C} \to [0, \infty]$ by $\mu(A) = 0$ if $|A| < \infty$ and $\mu(A) = \infty$ if $|A| = \aleph_0$. Show that μ is *finitely additive* but *not countably additive.*

4. Let ν be the *counting* measure on an *infinite* set X. Show that there is a *decreasing* sequence $(A_n) \in \mathcal{P}(X)^{\mathbb{N}}$ with $\lim(A_n) = \emptyset$ and yet $\lim(\nu(A_n)) \ne 0$.

5. Let X be a (nonempty) set and $f : X \to [0, \infty)$. Define the measure $\mu_f : \mathcal{P}(X) \to [0, \infty]$ by $\mu_f(\emptyset) := 0$ and $\mu_f(E) := \sum_{x \in E} f(x)$, where the sum is an *unordered series*. Find necessary and sufficient conditions (on f) for μ_f to be *finite* or σ-*finite*.

6. Let μ^* be an outer measure on X. Show that, if $A \subset X$ and $\mu^*(A) = 0$, then $\mu^*(A \cup B) = \mu^*(B)$ for all $B \subset X$.

7. Let μ^* be an outer measure on X. Show that $A \subset X$ is μ^*-measurable if and only if, given any $\varepsilon > 0$, there is an $E \in \mathcal{M}_\mu$ such that $E \subset A$ and $\mu^*(A \setminus E) < \varepsilon$.

8. Let μ^* be an outer measure on X and let $A \subset X$. Show that, if for each $\varepsilon > 0$ there is an $E \in \mathcal{M}_\mu$ such that $\mu^*(A \triangle E) < \varepsilon$, then $A \in \mathcal{M}_\mu$.

9. Let μ^* be an outer measure on X and $A \subset E \subset X$. Show that, if $A \notin \mathcal{M}_\mu$ and $E \in \mathcal{M}_\mu$, then $\mu^*(E \setminus A) > 0$.

10. Let μ^* be an outer measure on X that is *regular*; i.e., for each $B \subset X$ there is a $C \in \mathcal{M}_\mu$ such that $B \subset C$ and $\mu^*(B) = \mu^*(C)$, and let $A \subset X$. Show that, if there is an $E \in \mathcal{M}_\mu$ such that $\mu^*(E) < \infty$ and $\mu^*(E) = \mu^*(A) + \mu^*(E \setminus A)$, then $A \in \mathcal{M}_\mu$.

11. Let $(X, \mathcal{A}, \mu)$ be a measure space. For each sequence $(A_n) \in \mathcal{A}^{\mathbb{N}}$, define

$$\liminf(A_n) := \bigcup_{k=1}^{\infty} \bigcap_{n=k}^{\infty} A_n, \qquad \limsup(A_n) := \bigcap_{k=1}^{\infty} \bigcup_{n=k}^{\infty} A_n.$$

(a) Show that $\mu(\liminf(A_n)) \le \liminf(\mu(A_n))$.
(b) Show that, if $\mu(\bigcup A_n) < \infty$, then $\mu(\limsup(A_n)) \ge \limsup(\mu(A_n))$.
(c) **(Borel–Cantelli)** Show that, if $\sum_{n=1}^{\infty} \mu(A_n) < \infty$, then $\mu(\limsup(A_n)) = 0$.
(d) Show that, if $\liminf(A_n) = \limsup(A_n)$ and we denote it by A, then $\mu(A) = \lim(\mu(A_n))$.

12. Let (M, d) be a *complete* metric space and, for each $E \subset M$, define $\mu^*(E) = 0$ if E is of *first category* and $\mu^*(E) = 1$ otherwise. Show that μ^* is an outer measure. What are the μ^*-measurable sets?

13. Let X be an *uncountable* set and, for each $E \subset X$, define $\mu^*(E) = 0$ if E is *countable* and $\mu^*(E) = 1$ otherwise. Show that μ^* is an outer measure. What is the corresponding $\mathcal{M}_\mu$?

14. Let $(X, \mathcal{A}, \mu)$ be a *complete* measure space and $A \in \mathcal{A}$. Show that, if $B \subset X$ and $\mu(A \triangle B) = 0$, then $B \in \mathcal{A}$.

15. (Atom, Nonatomic). Let $(X, \mathcal{A})$ be a measurable space. A nonempty set $A \in \mathcal{A}$ is called an *atom* (of $\mathcal{A}$) if

$$A' \subset A \quad \text{and} \quad A' \in \mathcal{A} \Longrightarrow A' = A \quad \text{or} \quad A' = \emptyset.$$

A measurable space $(X, \mathcal{A})$ with *no* atoms is called *nonatomic*.

(a) Show that, if A, $A' \in \mathcal{A}$ are *distinct* atoms, then $A \cap A' = \emptyset$.
(b) Show that, if $\mathcal{A}$ is *finite* (i.e., contains a finite number of sets), then every *nonempty* $A \in \mathcal{A}$ is the union of the atoms it contains. *Hint:* Show first that every *nonempty* set $A \in \mathcal{A}$ contains *at least one* atom.
(c) Continuing (b), show that, if $A_1, \dots, A_n$ are the atoms of $\mathcal{A}$, then the A_k are *pairwise disjoint* and $X = \bigcup_{k=1}^{n} A_k$.
(d) Show that, if $X = \bigcup_{j=1}^{m} B_j$ where the B_j are *pairwise disjoint*, and if $\mathcal{B}$ is the (finite) σ-algebra generated by the B_j, then the B_j are the *atoms* of $\mathcal{B}$.

16. Let μ be a Borel measure on a metric space X such that $\mu(X) = 1$ and $\mu(\{x\}) = 0$ for each $x \in X$. Show that, given any $\varepsilon > 0$ and any $x \in X$, there is an open set U with $x \in U$ and $\mu(U) < \varepsilon$. Show that, if X is *separable*, then there is a *dense, open* set $O \subset X$ with $\mu(O) < \varepsilon$.

17. Let $F : \mathbb{R} \to \mathbb{R}$ be an *increasing, right-continuous* function. Show that the corresponding *Lebesgue–Stieltjes* outer measure λ_F^* is a *metric outer measure*.

18.

(a) Show that $\mu_1([a,b]) = b - a$ and that $\dim_H([a,b]) = 1$, where $\dim_H$ denotes the *Hausdorff dimension*.
(b) Show that $\dim_H(\mathbb{R}) = 1$.
(c) Show that $\dim_H(\{x\}) = 0$ for each $x \in \mathbb{R}$ and deduce that $\dim_H(E) = 0$ for every *countable* set $E \subset \mathbb{R}$.

19. Show that, if $F(x) := x$ for all $x \in \mathbb{R}$, then $\lambda_F^* = \lambda^*$; i.e., the corresponding Lebesgue–Stieltjes outer measure is identical to Lebesgue outer measure.

20. What is the Lebesgue–Stieltjes measure corresponding to the following function?

$$F(x) := \begin{cases} 0 & \text{if } x < 0, \\ x & \text{if } 0 \le x < 1, \\ 1 & \text{if } x \ge 1. \end{cases}$$

21. Let μ_1 denote the *one-dimensional Hausdorff measure* on $\mathbb{R}^2$. Show that, for any open set $O \subset \mathbb{R}^2$, we have $\mu_1(O) = \infty$.

22. Let $(X, \mathcal{A})$ be a measurable space and $f, g \in \mathcal{L}^0(X)$. Show that the following sets are measurable:

(a) $\{x \in X : f(x) < g(x)\}$;
(b) $\{x \in X : f(x) \le g(x)\}$;
(c) $\{x \in X : f(x) = g(x)\}$.

23. Let $(X, \mathcal{A})$ be a measurable space and $E \in \mathcal{A}$. We say that a function $f : X \to \mathbb{R}$ is measurable *on* E if the restriction $f|E$ is measurable on the measurable space $(E, \mathcal{A} \cap E)$, where $\mathcal{A} \cap E := \{A \cap E : A \in \mathcal{A}\}$.

(a) Show that, if $f \in \mathcal{L}^0(X)$, then f is measurable on E for *every* $E \in \mathcal{A}$.
(b) Let $E \in \mathcal{A}$, $f : E \to \mathbb{R}$, and let f_E be the *trivial extension* of f defined by $f_E(x) = f(x)$ if $x \in E$ and $f_E(x) = 0$ if $x \in E^c$. Show that, if f is measurable on the measurable space $(E, \mathcal{A} \cap E)$, then $f_E \in \mathcal{L}^0$.

24. Let $(X, \mathcal{A})$ be a measurable space. If $X = A \cup B$ with $A, B \in \mathcal{A}$, show that $f : X \to \mathbb{R}$ is measurable (on X) if and only if it is measurable on A and on B.

25. Let $(X, \mathcal{A})$ be a measurable space, $f, g \in \mathcal{L}^0(X)$, and $A \in \mathcal{A}$. Show that the function $h : X \to \mathbb{R}$ defined by $h(x) := f(x)$ if $x \in A$ and $h(x) := g(x)$ if $x \in A^c$ is measurable.

26. Let $(X, \mathcal{A}, \mu)$ be a measure space and $f_n \in \mathcal{L}^0(X)$ for each $n \in \mathbb{N}$.

(a) Show that $\{x \in X : \lim(f_n(x)) \text{ exists}\} \in \mathcal{A}$.
(b) Define the function

$$f(x) := \begin{cases} \lim(f_n(x)) & \text{if } \lim(f_n(x)) \text{ exists}, \\ 0 & \text{otherwise}. \end{cases}$$

Show that $f \in \mathcal{L}^0(X)$.
(c) Suppose that $f_n \to g$ a.e., where g is *not* necessarily measurable (unless μ is complete). Show that there is a function $f \in \mathcal{L}^0(X)$ such that $f_n \to f$ a.e.

27. Let $(X, \mathcal{A})$ be a measurable space and $f : X \to \overline{\mathbb{R}}$. Show that, if $f^{-1}((r, \infty]) \in \mathcal{A}$ for each $r \in \mathbb{Q}$, then $f \in \mathcal{L}^0(X, \overline{\mathbb{R}})$. More generally, show that, if D is a *dense* subset of $\mathbb{R}$ and $f^{-1}((d, \infty]) \in \mathcal{A}$ for each $d \in D$, then $f \in \mathcal{L}^0(X, \overline{\mathbb{R}})$.

28 (Convergence in Measure, Cauchy in Measure). Let $(X, \mathcal{A}, \mu)$ be a measure space and (f_n) a sequence in $\mathcal{L}^0(X)$. We say that $f_n \to f$ *in measure* if given any $\varepsilon > 0$ there is an $N \in \mathbb{N}$ such that

$$n \geq N \implies \mu(\{x \in X : |f_n(x) - f(x)| \geq \varepsilon\}) < \varepsilon.$$

We say that (f_n) is *Cauchy in measure* if for each $\varepsilon > 0$ there is an $N \in \mathbb{N}$ such that

$$m, n \geq N \implies \mu(\{x \in X : |f_m(x) - f_n(x)| \geq \varepsilon\}) < \varepsilon.$$

(a) Show that f_n converges in measure if and only if it is Cauchy in measure.
(b) Show that, if $f_n \to f$ in measure, then every subsequence (f_{n_k}) converges to f in measure.
(c) Show that, if $f_n \to f$ in measure, then there is a subsequence (f_{n_k}) such that $f_{n_k} \to f$ a.e.
(d) Show that $f_n \to f$ in measure if and only if every subsequence of (f_n) has in turn a subsequence that converges to f in measure.
(e) Show that $f_n \to f$ in measure if and only if every subsequence of (f_n) has in turn a subsequence that converges to f *almost everywhere*.
(f) Show that, if $(X, \mathcal{A}, \mu)$ is *complete* and if $f_n \to f$ in measure, then $f \in \mathcal{L}^0(X)$.
(g) Let $(X, \mathcal{A}, \mu) = (\mathbb{R}, \mathcal{M}_\lambda, \lambda)$ and consider the functions $f_n := \chi_{[n, n+1]}$ for all $n \in \mathbb{N}$. Show that $f_n \to 0$ *everywhere*, but $f_n \not\to 0$ in measure. See, however, Problem 32 below.

29. Show that, if ν is the *counting measure* on $X := \mathbb{Z}$, then convergence in measure is equivalent to *uniform convergence*.

30. Let $(X, \mathcal{A}, \mu)$ be a measure space with $\mu(X) < \infty$. For any $f, g \in \mathcal{L}^0(X, \overline{\mathbb{R}})$, define

$$d_\mu(f, g) := \inf \{\varepsilon > 0 : \mu(\{x \in X : |f(x) - g(x)| > \varepsilon\}) \leq \varepsilon\}.$$

(a) Show that $d_\mu(f, g) = 0$ if and only if $f = g$ μ-a.e.
(b) Show that d_μ induces a *metric* (still denoted by d_μ) on the quotient space $L^0(X, \overline{\mathbb{R}}) := \mathcal{L}^0(X, \overline{\mathbb{R}})/\mathcal{N}$, where $\mathcal{N} := \{f \in \mathcal{L}^0(X, \overline{\mathbb{R}}) : f(x) = 0 \text{ a.e.}\}$.
(c) Show that $d_\mu(f_n, f) \to 0$ if and only if $f_n \to f$ *in measure*.
(d) Show that the metric space $(L^0(X, \overline{\mathbb{R}}), d_\mu)$ is *complete*.

31. Let $(X, \mathcal{A})$ be a measurable space. Show that a map $\phi : X \to \mathbb{R}$ is *simple* if and only if $\phi^{-1}(\mathcal{B}_\mathbb{R}) \subset \mathcal{A}$ is a *finite* σ-algebra.

32. Let $(X, \mathcal{A}, \mu)$ be a *finite* measure space and let f and $(f_n)_{n=1}^\infty$ be measurable functions on X such that $f_n \to f$ a.e.

(a) **(Egorov's Theorem)** Show that, given any $\varepsilon > 0$, there is an $A \in \mathcal{A}$ such that $\mu(A^c) < \varepsilon$ and $f_n \to f$ uniformly on A.
(b) Show that $f_n \to f$ *in measure*. *Hint:* Given $\varepsilon > 0$, let $E_n := \{x \in X : |f_n(x) - f(x)| \geq \varepsilon\}$ and show that $\lim(\mu(E_n)) = 0$.

33. Let $(X, \mathcal{A}, \mu)$ be a measure space and $f \in \mathcal{L}^1(X)$. Show that, if $\int_E f \, d\mu = 0$ for every $E \in \mathcal{A}$, then $f = 0$ a.e. *Hint:* Consider the sets $\{x : f(x) > 0\}$ and $\{x : f(x) < 0\}$.

34. Let $(X, \mathcal{A}, \mu)$ be a measure space, $f \in \mathcal{L}^1(X)$, $(A_n) \in \mathcal{A}^\mathbb{N}$, and $A := \bigcup_{n=1}^\infty A_n$.

(a) Show that, if (A_n) is *increasing*; i.e., $A_n \subset A_{n+1}$ for all $n \in \mathbb{N}$, then

$$\int_A f \, d\mu = \lim_{n \to \infty} \int_{A_n} f \, d\mu.$$

(b) Show that, if $A_j \cap A_k = \emptyset$ for all $j \neq k$, then

$$\int_A f \, d\mu = \sum_{n=1}^{\infty} \int_{A_n} f \, d\mu.$$

35. Let $(X, \mathcal{A}, \mu)$ be a measure space and $f \in \mathcal{L}^1(X)$. Show that, given any $\varepsilon > 0$, there is an $A \in \mathcal{A}$ such that $\mu(A) < \infty$ and $\int_{A^c} |f| \, d\mu < \varepsilon$.

36. Let $(X, \mathcal{A}, \mu)$ be a measure space and $f \in \mathcal{L}^1(X)$. Show that the measure $\mu_{|f|}(E) := \int_E |f| \, d\mu$ is absolutely *continuous* (with respect to μ) in the sense that, given any $\varepsilon > 0$, there is a $\delta > 0$ such that

$$E \in \mathcal{A} \quad \text{and} \quad \mu(E) < \delta \Longrightarrow \int_E |f| \, d\mu < \varepsilon.$$

37. Let $(X, \mathcal{A}, \mu)$ be a measure space and (f_n) a sequence in $\mathcal{L}^0(X)$ such that $\sum_{n=1}^{\infty} \int |f_n| < \infty$. Show that $\sum_{n=1}^{\infty} f_n(x)$ converges almost everywhere to an *integrable* sum and that

$$\int \left(\sum_{n=1}^{\infty} f_n \right) d\mu = \sum_{n=1}^{\infty} \int f_n \, d\mu.$$

Deduce that, if $f_n \geq 0$ for all n, then $\sum_n \int f_n \, d\mu < \infty$ implies $\sum f_n(x) < \infty$ for almost all $x \in X$.

38. Let $(X, \mathcal{A}, \mu)$ be a *finite* measure space and $f \in \mathcal{L}^0(X)$. Define the map $\Phi : (0, \infty) \to [0, \infty]$ by $\Phi(t) := \mu(\{x \in X : |f(x)| > t\})$.

(a) Show that (even in the case $\mu(X) = \infty$), if $\int |f|^p \, d\mu < \infty$ for some $p > 0$, then $\Phi(t) \leq Ct^{-p}$ for some constant $C > 0$ and all $t > 0$.
(b) Show that $f \in \mathcal{L}^1(X)$ if and only if $\sum_{n=1}^{\infty} \Phi(n) < \infty$.
(c) Show that, if there exist $C > 0$ and $p > 0$ such that $\Phi(t) \leq Ct^{-p}$ for all $t > 0$, then $\int |f|^q \, d\mu < \infty$ for every $q \in (0, p)$.
(d) Show that, if $f \in \mathcal{L}^1(X)$, then $\lim_{t \to \infty} t\Phi(t) = 0$.

39. Let $(X, \mathcal{A}, \mu)$ be a measure space and $f \in \mathcal{L}^1(X)$. Show that the set $\{x \in X : f(x) \neq 0\}$ has σ-*finite* measure; i.e., it is the countable union of a sequence of measurable sets of finite measure.

40. Let $(X, \mathcal{A}, \mu)$ be a measure space and $A, B \in \mathcal{A}$. Show that, if $\mu(A \triangle B) = 0$, then for every $0 \leq f \in \mathcal{L}^0(X)$, we have $\int_A f \, d\mu = \int_B f \, d\mu$.

41. Let $(X, \mathcal{A}, \mu)$ be a measure space and $0 \leq f \in \mathcal{L}^0(X)$.

(a) Find conditions under which the measure $\mu_f(E) := \int_E f \, d\mu$ is (i) *finite* and (ii) σ-*finite*.
(b) Show that, if f is *bounded*, then $\mathcal{L}_\mu^1(E) \subset \mathcal{L}_{\mu_f}^1(E)$ for each $E \in \mathcal{A}$. Show that the *boundedness* of f is necessary.

42. Let $(X, \mathcal{A}, \mu)$ be a measure space, $f \in \mathcal{L}^1(X)$, and (A_n) a *decreasing* sequence in $\mathcal{A}$. Show that, with $A := \bigcap_{n=1}^{\infty} A_n$, we have $\int_A f \, d\mu = \lim_{n \to \infty} \int_{A_n} f \, d\mu$.

43. Let $(X, \mathcal{A}, \mu)$ be a measure space and $f \in \mathcal{L}^1(X)$. For each $\alpha > 0$, define $A_\alpha := \{x \in X : |f(x)| > \alpha\}$. Show that $\lim_{\alpha \to \infty} \int_{A_\alpha} |f| \, d\mu = 0$.

44. Let $(X, \mathcal{A}, \mu)$ be a *finite* measure space, $f_n \in \mathcal{L}^1(X)$ for all $n \in \mathbb{N}$, and $f_n \to f$ *uniformly* on X. Show that $f \in \mathcal{L}^1(X)$ and $\int f_n \, d\mu \to \int f \, d\mu$.

45. Let $(X, \mathcal{A}, \mu)$ be a *finite* measure space and, for any $f, g \in \mathcal{L}^0(X)$, define

$$d_\mu(f, g) := \int \frac{|f - g|}{1 + |f - g|} \, d\mu.$$

(a) Show that $d_\mu(f, g) = 0$ if and only if $f = g$ μ-a.e.
(b) Show that d_μ induces a *metric* (still denoted by d_μ) on the quotient space $L^0(X) :=$ $\mathcal{L}^0(X)/\mathcal{N}$, where $\mathcal{N} := \{f \in \mathcal{L}^0(X) : f(x) = 0 \text{ a.e.}\}$.
(c) Show that $d_\mu(f_n, f) \to 0$ if and only if $f_n \to f$ *in measure*.
(d) Show that the metric space $(L^0(X), d_\mu)$ is *complete*.

46 (Convergence Theorems for Convergence in Measure). Let $(X, \mathcal{A}, \mu)$ be a measure space and $f, f_n \in \mathcal{L}^0(X)$ for all $n \in \mathbb{N}$. Prove the *Bounded Convergence Theorem* (BCT), *Fatou's Lemma* (FL), the *Monotone Convergence Theorem* (MCT), and the *Dominated Convergence Theorem* (DCT) for convergence *in measure*:

(a) **(BCT)** Suppose that $\mu(X) < \infty$, that (f_n) is *uniformly bounded* on X, and that $f_n \to f$ in measure. Show that $\lim_{n\to\infty} \int f_n \, d\mu = \int f \, d\mu$. *Hint:* Given any $\varepsilon > 0$, define $E_n := \{x : |f_n(x) - f(x)| > \varepsilon\}$ and let $n \to \infty$ in the inequality

$$\int |f_n - f| \, d\mu = \int_{E_n^c} |f_n - f| \, d\mu + \int_{E_n} |f_n - f| \, d\mu \le \varepsilon \mu(X) + 2M\mu(E_n),$$

where $M = \sup\{|f_n(x)| : n \in \mathbb{N}, \, x \in X\}$.
(b) **(FL)** Suppose that $f_n \ge 0$ for all $n \in \mathbb{N}$ and $f_n \to f$ in measure. Then we have $\int f \, d\mu \le \liminf_{n\to\infty} \int f_n \, d\mu$.
(c) **(MCT)** If $f_n \ge 0$ for all $n \in \mathbb{N}$, (f_n) is *increasing*, and $f_n \to f$ in measure, then $\lim_{n\to\infty} \int f_n \, d\mu = \int f \, d\mu$.
(d) **(DCT)** If $f_n \to f$ in measure and if $|f_n| \le g$ for all $n \in \mathbb{N}$ and some $0 \le g \in \mathcal{L}^1(X)$, then $f \in \mathcal{L}^1(X)$ and $\lim(\int f_n \, d\mu) = \int(\lim(f_n)) \, d\mu$. *Hint:* For (b), (c), and (d), use Problem 28 (parts (b) and (c)), and the fact that $\int f_n \, d\mu \to \int f \, d\mu$ if and only if *every subsequence* $(\int f_{n_k})$ has in turn a subsequence converging to $\int f \, d\mu$.

47. Let $(X, \mathcal{A}, \mu)$ be a *finite* measure space and $f, f_n \in \mathcal{L}^0(X)$ for all $n \in \mathbb{N}$. Show that $f_n \to f$ a.e. if and only if, given any $\varepsilon > 0$, we have

$$\lim_{n\to\infty} \mu\left(\bigcup_{k=n}^{\infty} \{x \in X : |f_k(x) - f(x)| \ge \varepsilon\} \right) = 0.$$

48. Let $(X, \mathcal{A}, \mu)$ be a *finite* measure space and $f \in \mathcal{L}^0(X)$.

(a) Show that $\lim_{n\to\infty} \int |f|^n \, d\mu$ exists (in $\mathbb{R}^+$) if and only if $\mu(\{x \in X : |f(x)| > 1\}) = 0$.
(b) Suppose that $f^n \in \mathcal{L}^1(X)$ for all $n \in \mathbb{N}$. Show that we have $\int f^n \, d\mu = C$ for all $n \in \mathbb{N}$ and some constant $C \in \mathbb{R}$ if and only if $f = \chi_A$ (except possibly on a set of measure zero) for some $A \in \mathcal{A}$.

49. Let $(X, \mathcal{A}, \mu)$ be a measure space and $g, f_n \in \mathcal{L}^0(X)$ for all $n \in \mathbb{N}$. Suppose that $|f_n| \le g$ for all n, that $g^p \in \mathcal{L}^1(X)$ for some $p > 0$, and that $f_n \to f$ a.e. Show that $|f|^p \in \mathcal{L}^1(X)$ and that $\lim_{n\to\infty} \int |f_n - f|^p \, d\mu = 0$. *Hint:* Use the Dominated Convergence Theorem.

50 (The Banach Spaces $L_\mu^p(X, \mathbb{F})$). Let $(X, \mathcal{A}, \mu)$ be a measure space. Given any $p \in [1, \infty)$, let $\mathcal{L}_\mu^p(X, \mathbb{F})$ (or $\mathcal{L}^p(X, \mathbb{F})$) denote the set of all $f \in \mathcal{L}^0(X, \mathbb{F})$, where $\mathbb{F}$ is either $\mathbb{R}$ or $\mathbb{C}$, such that $|f|^p \in \mathcal{L}_\mu^1(X, \mathbb{F})$. Now define $L_\mu^p(X, \mathbb{F}) := \mathcal{L}_\mu^p(X, \mathbb{F})/\mathcal{N}$, where $\mathcal{N}$ is the set of all $f \in \mathcal{L}^0(X, \mathbb{F})$ with $f(x) = 0$ for almost all $x \in X$. For each $f \in \mathcal{L}^p(X, \mathbb{F})$, define its L^p-norm by

$$\|f\|_p := \left(\int |f|^p \, d\mu \right)^{1/p}.$$

(a) Show that $\mathcal{L}^p(X, \mathbb{F}))$ is a vector space (over $\mathbb{F}$). *Hint:* $|f + g|^p \le 2^{p-1}(|f|^p + |g|^p)$. (Why?)

(b) **(Hölder's Inequality)** Let p, $q \in (1, \infty)$ with $1/p + 1/q = 1$. Show that, given any $f \in \mathcal{L}^p(X, \mathbb{F})$ and $g \in \mathcal{L}^q(X, \mathbb{F})$, we have $fg \in \mathcal{L}^1(X, \mathbb{F})$ and

$$\int |fg|\, d\mu \le \|f\|_p \|g\|_q.$$

Hint: Using that $ab \le a^p/p + b^q/q$ for any a, $b \ge 0$, show that $\int |fg| \le \|f\|_p^p/p + \|g\|_q^q/q$. In it, replace f by tf and g by g/t and *minimize* over $t \in (0, \infty)$.

(c) **(Minkowski's Inequality)** Show that, for any $p \ge 1$ and f, $g \in \mathcal{L}^p(X, \mathbb{F})$, we have $f + g \in \mathcal{L}^p(X, \mathbb{F})$ and

$$\|f + g\|_p \le \|f\|_p + \|g\|_p.$$

Hint: For $p > 1$, note that $|f + g|^p \le |f||f + g|^{p-1} + |g||f + g|^{p-1}$ and that we have $|f + g|^{p-1} \in \mathcal{L}^q$ (where $1/q = 1 - 1/p$). Now apply Hölder's inequality.

(d) Deduce that $L_\mu^p(X, \mathbb{F})$ is a normed space with the L^p-norm.

(e) Show that $L_\mu^p(X, \mathbb{F})$ is a *Banach space*. *Hint:* Follow the proof of Theorem 11.3.7.

(f) Show that, if $\mu(X) < \infty$, then $L_\mu^p(X) \subset L_\mu^1(X)$ for all $p \ge 1$.

51 (The Hilbert Space $L_\mu^2(X, \mathbb{F})$). With notation as in the preceding problem, show that $L_\mu^2(X, \mathbb{F})$ is a *Hilbert space* with inner product

$$\langle f, g \rangle := \int_X f\bar{g}\, d\mu.$$

In this case, Hölder's inequality is reduced to the *Cauchy–Schwarz* inequality:

$$|\langle f, g \rangle| \le \|f\|_2 \|g\|_2.$$

52. Show that, if $(X, \mathcal{A}, \mu) = (\mathbb{N}, \mathcal{P}(\mathbb{N}), \nu)$, where ν denotes the *counting measure*, then $\mathcal{L}^p(\mathbb{N}, \mathbb{F})$ is in fact the Banach space $\ell^p(\mathbb{N}, \mathbb{F})$ already introduced in Sect. 9.5 (Problem 9.5.#2). More generally, considering the measure space $(X, \mathcal{P}(X), \nu)$ where ν is the counting measure on the set X, introduce the Banach spaces $\ell^p(X, \mathbb{F})$.

53. Let $(X, \mathcal{A})$ be a measurable space and let $\mathfrak{M}_A$ denote the set of all *positive* measures on $\mathcal{A}$. Show that the *binary* relation $\nu \ll \mu$ is *reflexive* and *transitive* and that $\mathfrak{M}_A$ is *directed* with $\ll$ in the sense that for any pair of measures ν_1, $\nu_2 \in \mathfrak{M}_A$, there exists a measure μ such that $\nu_1 \ll \mu$ and $\nu_2 \ll \mu$. In fact, given any measures $\nu_1, \ldots, \nu_n$, there exists a measure μ with $\nu_j \ll \mu$ for $1 \le j \le n$.

54. Given any measures ν_1, ν_2, ν, and μ on a measurable space $(X, \mathcal{A})$ and any constants $c_1 \ge 0$ and $c_2 \ge 0$, prove the following: except possibly on a set of measure zero)

(a) $\mu \perp \mu \Longrightarrow \mu = 0$.

(b) $\nu_1 \ll \mu$ and $\nu_2 \ll \mu$ imply $c_1\nu_1 + c_2\nu_2 \ll \mu$.

(c) $\nu_1 \perp \mu$ and $\nu_2 \perp \mu$ imply $c_1\nu_1 + c_2\nu_2 \perp \mu$.

(d) $\nu_1 \ll \mu$ and $\nu_2 \perp \mu$ imply $\nu_1 \perp \nu_2$.

(e) $\nu \ll \mu$ and $\nu \perp \mu$ imply $\nu = 0$.

55. Let $(\nu_k)_{k \in \mathbb{N}}$ be a sequence of measures on a measurable space $(X, \mathcal{A})$ such that $\nu_k(X) \le 1$ for all $k \in \mathbb{N}$, and define

$$\mu(A) := \sum_{k=1}^{\infty} 2^{-k}\nu_k(A) \qquad \forall\, A \in \mathcal{A}.$$

Show that μ is a measure and that we have $\nu_k \ll \mu$ for all $k \in \mathbb{N}$.

56 (Radon–Nikodym Derivatives). Let ν_1, ν_2, ν, μ, and ρ be σ-finite (positive) measures on a measurable space $(X, \mathcal{A})$.

(a) Show that if $\nu \ll \mu$ and if $f \in L_\nu^1$, then $f(d\nu/d\mu) \in L_\mu^1$ and we have

$$\int f \, d\nu = \int f \frac{d\nu}{d\mu} \, d\mu. \tag{$*$}$$

(b) Show that if $\nu_1 \ll \mu$ and $\nu_2 \ll \mu$, then we have $\nu_1 + \nu_2 \ll \mu$ and

$$\frac{d(\nu_1 + \nu_2)}{d\mu} = \frac{d\nu_1}{d\mu} + \frac{d\nu_2}{d\mu}.$$

(c) **(Chain Rule)** Show that if $\nu \ll \mu$ and $\mu \ll \rho$, then we have $\nu \ll \rho$ and

$$\frac{d\nu}{d\rho} = \frac{d\nu}{d\mu} \frac{d\mu}{d\rho} \qquad \rho - \text{almost everywhere.}$$

(d) If $\nu \ll \mu$ and $\mu \ll \nu$, then

$$\left(\frac{d\nu}{d\mu}\right)\left(\frac{d\mu}{d\nu}\right) = 1 \quad \text{almost everywhere (with respect to either } \mu \text{ or } \nu\text{).}$$

57 (Lebesgue–Stieltjes Measures). Let $G : \mathbb{R} \to \mathbb{R}$ be *increasing* and *absolutely continuous* (i.e., absolutely continuous on any interval $[a, b] \subset \mathbb{R}$). Show directly that the *Lebesgue–Stieltjes* measure λ_G is absolutely continuous (with respect to Lebesgue measure λ) and find its Radon–Nikodym derivative $g := d\lambda_G/d\lambda$.

58. Show that, if M and M' are *separable* metric spaces, then we have

$$\mathcal{B}_{M \times M'} = \mathcal{B}_M \times \mathcal{B}_{M'}.$$

Hint: Note that $M \times M'$ is a separable metric space and hence a *Lindelöf* space (Definition 5.6.19).

59. Let $X = Y = [0, 1]$, $\mu := \lambda$ is Lebesgue measure, and ν is the *counting measure*. Let $f := \chi_D$, where $D : \{(x, y) \in X \times Y : x = y\}$ is the *diagonal*. Show that $\int f(x, y) \, d\nu(y) = 1$ for all $x \in X$ and that $\int [\int f(x, y) \, d\nu(y)] \, d\mu(x) = 1$. Next, show that $\int f(x, y) \, d\mu(x) = 0$ for all $y \in Y$ and hence that we have $\int [\int f(x, y) \, d\mu(x)] \, d\nu(y) = 0$. Conclude that χ_D is *not* $\mu \otimes \nu$-integrable.

60. If $X = Y = \mathbb{N}$, $\mathcal{A} = \mathcal{B} = \mathcal{P}(\mathbb{N})$, and $\mu = \nu$ is the *counting measure* on $\mathbb{N}$, interpret Fubini–Tonelli's theorem in terms of *double series* of real numbers.

61. Show that, if f is *continuous* on $R := [a, b] \times [c, d]$, then, with $dx := d\lambda(x)$, $dy := d\lambda(y)$, and $dx \, dy := d(\lambda \otimes \lambda)(x, y)$, we have

$$\int_a^b \left[\int_c^d f(x, y) \, dy \right] dx = \iint_R f(x, y) \, dx \, dy = \int_c^d \left[\int_a^b f(x, y) \, dx \right] dy.$$

62. Show that, if $f \in \mathcal{L}^1([0, 1] \times [0, 1])$, then we have

$$\int_0^1 \left[\int_0^x f(x, y) \, dy \right] dx = \int_0^1 \left[\int_y^1 f(x, y) \, dx \right] dy.$$

63. Consider the function

$$f(x, y) := \begin{cases} \dfrac{x^2 - y^2}{(x^2 + y^2)^2} & \text{if } (x, y) \neq (0, 0), \\ 0 & \text{if } (x, y) = (0, 0). \end{cases}$$

(a) Show that

$$\int_0^1 \left[\int_0^1 f(x, y)\, dy \right] dx = \frac{\pi}{4}, \qquad \int_0^1 \left[\int_0^1 f(x, y)\, dx \right] dy = -\frac{\pi}{4}.$$

Hint: $\dfrac{d}{dt}[t/(a^2 + t^2)] = (a^2 - t^2)/(a^2 + t^2)^2$.

(b) Deduce that f is *not* integrable over $R := [0, 1] \times [0, 1]$ and that $\iint_R |f(x, y)|\, dx\, dy$ does *not* exist.

64. Let $Q := [-1, 1] \times [-1, 1]$ and define f on Q by $f(x, y) := xy/(x^2 + y^2)^2$ if $(x, y) \neq (0, 0)$ and $f(0, 0) := 0$.

(a) Show that

$$\int_{-1}^1 \left[\int_{-1}^1 f(x, y)\, dy \right] dx = 0 = \int_{-1}^1 \left[\int_{-1}^1 f(x, y)\, dx \right] dy.$$

(b) Show that $\iint_Q f(x, y)\, dx\, dy$ does *not* exist. *Hint:* If it did, then $\int_0^1 [\int_0^1 f(x, y)\, dy]\, dx$ would also exist. Show, however, that

$$\int_0^1 f(x, y)\, dy = \frac{1}{2x} - \frac{x}{2(x^2 + 1)} \qquad (\forall x \in (0, 1]).$$

65. Use Fubini's theorem and the fact that $\int_0^\infty e^{-xt}\, dt = 1/x$, for all $x > 0$, to prove that

$$\int_0^\infty \frac{\sin x}{x}\, dx = \frac{\pi}{2}.$$

66. Show that, if $f(x, y) := ye^{-(1+x^2)y^2}$, then

$$\int_0^\infty \left[\int_0^\infty f(x, y)\, dx \right] dy = \int_0^\infty \left[\int_0^\infty f(x, y)\, dy \right] dx,$$

and use this to prove that

$$\int_0^\infty e^{-x^2}\, dx = \frac{\sqrt{\pi}}{2}.$$

67. Let $(X, \mathcal{A}, \mu)$ and $(Y, \mathcal{B}, \nu)$ be σ-finite measure spaces, $f \in \mathcal{L}_\mu^1(X)$ and $g \in \mathcal{L}_\nu^1(Y)$, and define $h : X \times Y \to \mathbb{R}$ by $h(x, y) := f(x)g(y)$. Show that $h \in \mathcal{L}_{\mu \otimes \nu}^1(X \times Y)$ and

$$\int h\, d(\mu \otimes \nu) = \left(\int f\, d\mu \right) \left(\int g\, d\nu \right).$$

68 (Convolution of Functions).

(a) Given $f, g \in \mathcal{L}^1(\mathbb{R})$, show that

$$\int_{\mathbb{R}} |f(x-y)g(y)| \, dy < \infty$$

for *almost all* $x \in \mathbb{R}$ and, for each such x, define the *convolution* of f and g to be the function

$$(f * g)(x) := \int_{\mathbb{R}} f(x-y)g(y) \, dy.$$

(b) Show that $f * g \in \mathcal{L}^1(\mathbb{R})$ and that we have

$$\|f * g\|_1 \le \|f\|_1 \|g\|_1.$$

Hint: By Problem 10.6#23, there are *Borel* functions f_0, g_0 with $f = f_0$ and $g = g_0$ almost everywhere. Since the above integrals are unchanged if (for each x) we replace f and g by f_0 and g_0, one may assume that f and g are *Borel* functions. With this assumption, and using the fact that $(x, y) \mapsto x - y$ is *continuous*, show that the function $(x, y) \mapsto f(x-y)g(y)$ is *Borel* on $\mathbb{R}^2$. Now use *Fubini's theorem* together with the *translation invariance* of Lebesgue measure.

69 (Convolution of Borel Measures).   Let μ and ν be two σ-finite Borel measures on $\mathbb{R}$.

(a) Show that, for any $B \in \mathcal{B}_{\mathbb{R}}$, we have

$$B_2 := \{(x, y) \in \mathbb{R}^2 : x + y \in B\} \in \mathcal{B}_{\mathbb{R}^2}.$$

(b) Show that the functions $g(x) := \nu(B - x)$ and $h(y) := \mu(B - y)$, where $B - t := \{b - t : b \in B\}$, are Borel functions.

(c) Show that, with B and B_2 as above,

$$(\mu \otimes \nu)(B_2) = \int_{\mathbb{R}} \nu(B - x) \, d\mu(x) = \int_{\mathbb{R}} \mu(B - y) \, d\nu(y).$$

(d) Define the *convolution* of μ and ν by

$$(\mu * \nu)(B) := (\mu \otimes \nu)(B_2) \qquad (\forall B \in \mathcal{B}_{\mathbb{R}}).$$

Show that $\mu * \nu$ is a Borel measure on $\mathbb{R}^2$ and that $\mu * \nu = \nu * \mu$. Also show that, if $\mu = \delta := \delta_0$ is the *Dirac measure* at $x = 0$, then we have

$$\delta * \nu = \nu.$$

(e) Show that, given any $f \in \mathcal{L}^1_{\mu*\nu}(\mathbb{R}, \mathcal{B}_{\mathbb{R}})$, we have

$$\int_{\mathbb{R}^2} f(x + y) \, d(\mu \otimes \nu)(x, y) = \int_{\mathbb{R}} f(t) \, d(\mu * \nu)(t).$$

Hint: Use approximation by simple functions.

(f) Given any $0 \le f \in \mathcal{L}^1(\mathbb{R})$, the set function $\mu_f(B) := \int_B f \, d\lambda$ defines a finite Borel measure on $\mathbb{R}$. (Why?) Show that, if $f, g \in \mathcal{L}^1(\mathbb{R})$ are *nonnegative*, then

$$\mu_f * \mu_g = \mu_{f*g},$$

where the convolution $f * g$ is defined as in the preceding problem.

70 (Fourier Transform of a Measure).   Given a *finite* Borel measure μ on $\mathbb{R}$, define its *Fourier transform* $\hat{\mu} : \mathbb{R} \to \mathbb{C}$ by

$$\hat{\mu}(\xi) := \int_{\mathbb{R}} e^{ix\xi} \, d\mu(x).$$

(a) Show that $\hat{\mu}$ is well defined; i.e., the above integral exists for each $\xi \in \mathbb{R}$.
(b) Show that, if v is another finite Borel measure on $\mathbb{R}$, then we have

$$\widehat{\mu * v}(\xi) = \hat{\mu}(\xi)\hat{v}(\xi),$$

where $\mu * v$ is the convolution of μ and v defined in the preceding problem. *Hint:* Use Fubini's theorem.

71 (Ordinate Set, Area Under the Graph).   Given a σ-finite measure space $(X, \mathcal{A}, \mu)$ and a function $f \in \mathcal{L}^1(X, [0, \infty))$, introduce the set $A(f) := \{(x, y) \in X \times \mathbb{R} : 0 \le y \le f(x)\}$, called the *(upper) ordinate set* of f. Show that $A(f) \subset \mathcal{A} \otimes \mathcal{B}_{\mathbb{R}}$ and that we have

$$(\mu \otimes \lambda)(A(f)) = \int f \, d\mu.$$

Hint: Note that the map $(x, y) \mapsto (f(x), y)$ is measurable from $(X \times \mathbb{R}, \mathcal{A} \otimes \mathcal{B}_{\mathbb{R}})$ to $(\mathbb{R}^2, \mathcal{B}_{\mathbb{R}^2})$ and that $(z, y) \mapsto z - y$ is *continuous*. Also, with the cross section $A_x := \{y : (x, y) \in A(f)\}$, we have

$$\lambda(A_x) = \begin{cases} f(x) & \text{if } x \in A(f), \\ 0 & \text{if } x \notin A(f). \end{cases}$$

72. Bill and Barbara are among ten people seated at a round table. What is the probability that they are seated next to each other?

73. Two fair dice are rolled and the sum of the outcomes is S. Find $P(S = 9)$; $P(S \ge 4)$; $P(S < 10)$; and $P(S$ is an *odd* number).

74. What is the probability that among 24 people, at least two have the same birthday? *Hint:* Look at the "complementary event."

75. In a shipment of 2,000 light bulbs, it is known that 4 % are defective. If a *random* sample of 25 bulbs is selected, what is the probability that it contains three defective ones?

76. Show that, if $\{A_1, \ldots, A_n\}$ is an independent set of events in Ω, then $\bigcup_{k=1}^{n-1} A_k$ and A_n are independent. *Hint:* Use induction.

77. Show that, if $P(A) = \alpha$ and $P(B) = \beta > 0$, then $P(A|B) \ge (\alpha + \beta - 1)/\beta$.

78 (Law of Multiplication).   Show that, if $(A_k)_{k=1}^{n}$ is a sequence of events in Ω such that $P(\bigcap_{k=1}^{n-1} A_k) > 0$, then we have

$$P\left(\bigcap_{k=1}^{n} A_k\right) = P(A_1)P(A_2|A_1)P(A_3|A_1 \cap A_2) \cdots P(A_n|A_1 \cap \cdots \cap A_{n-1}).$$

79. Let A, B, and C be events in Ω. Show that

$$P(A|C) \ge P(B|C) \quad \text{and} \quad P(A|C^c) \ge P(B|C^c) \Longrightarrow P(A) \ge P(B).$$

80 (Favorable).   Given any events A, $B \subset \Omega$, we say that A is *favorable to* B if $P(A \cap B) \geq P(A)P(B)$. Show that A is favorable to B if and only if A^c is favorable to B^c.

81.

(a) Let $A \subset \Omega$ be an event. Show that, if A is *independent of itself*, then $P(A) = 0$ or $P(A) = 1$ and that, in this case, A is independent of *every* event $B \subset \Omega$.

(b) Show that, if $A \subset B \subset \Omega$ are two *independent* events, then either $P(A) = 0$ or $P(B) = 1$.

82. Let $\Omega := [0, 1]$ with Lebesgue measure and let $\emptyset \neq [a, b] \subsetneq \Omega$. Find all intervals $I \subset \Omega$ such that $\{I, [a, b]\}$ is independent.

83. Box B_1 contains nine red and six blue balls and box B_2 contains five red and seven blue balls. A ball is *randomly* picked from B_1 and put in B_2 and then, after mixing the balls, one ball is randomly selected from B_2.

(a) What is the probability that the selected ball is *red?*

(b) If the selected ball is *red,* what is the probability that the *transferred* ball was *red? Hint:* Bayes's formula.

84.

(a) Show that, if $X \sim B(n, p)$, then

$$P_X(\{k\}) = \frac{p}{1-p} \frac{n-k+1}{k} P_X(\{k-1\}) \qquad (k = 1, 2, \ldots, n).$$

(b) Using the above formula and induction (starting with $P_X(\{0\}) = (1-p)^n$), derive the formula $P_X(\{k\}) = \binom{n}{k} p^k (1-p)^{n-k}$ for $0 \leq k \leq n$.

85. Let $X_n \sim B(n, \lambda/n)$ for all $n \in \mathbb{N}$ and some $\lambda > 0$. Show that

$$\lim_{n \to \infty} P(X_n = k) = \frac{\lambda^k}{k!} e^{-\lambda}.$$

86.

(a) Let X be a *Poisson* variable with parameter $\lambda > 0$. Show that

$$P_X(\{k + 1\}) = \frac{\lambda}{k+1} P_X(\{k\}) \qquad (k = 0, 1, 2, \ldots).$$

(b) Using the above formula and induction (starting with $P_X(\{0\}) = e^{-\lambda}$), find $P_X(\{k\})$.

87 (Density of a Function of a Random Variable). Let X be an absolutely continuous variable on Ω with (continuous) density f_X.

(a) Find the density function of X^2. *Hint:* Differentiate $F_{X^2}(x) = F_X(\sqrt{x}) - F_X(-\sqrt{x})$.

(b) If $X > 0$, find the density of $\sqrt{X}$.

(c) Find the density functions of X^3 and e^X.

(d) Let $g : \mathbb{R} \to \mathbb{R}$ be a differentiable function with $g'(x) \neq 0$ for all $x \in \mathbb{R}$. Show that $g(X)$ is absolutely continuous with density

$$f_{g(X)}(y) = f_X(g^{-1}(y))|(g^{-1})'(y)|.$$

Hint: Note that, if g is *strictly increasing,* then $F_{g(X)}(y) := P(g(X) \leq y) = F_X(g^{-1}(y))$.

88. If two points X and Y are randomly selected from $[0, 1]$, what is the probability that the distance between them is *at least* $1/3$? *Hint:* Find $P(|X - Y| \geq 1/3)$, where (X, Y) is a *random vector* from the unit square $R := [0, 1]^2$.

89. Let X be a *positive* random variable on Ω such that

$$P(X > s + t \mid X > s) = P(X > t) \qquad (\forall s, \ t \in [0, \infty)).$$

Show that X has an *exponential* distribution. *Hint:* Let $\Phi(t) := P(X > t)$ and use Problem 10.6.#57.

90. Let $X \geq 0$ be a random variable on Ω and $n \in \mathbb{N}$. Show that $E[X^n] = n \int_0^\infty x^{n-1} P(X > x) \, dx$. Deduce, in particular, that $E[X] = \int_0^\infty P(X > x) \, dx$. *Hint:* Note that $x^n = \int_0^x n t^{n-1} \, dt$ and use Fubini–Tonelli's theorem.

91. A discrete random variable X takes the values $1, 2, 3, \ldots$, with probabilities $1/3, 1/9, 1/27,$ $\ldots$. What is $E[X]$? What is $E[e^X]$?

92 (Geometric Random Variable).   A sequence of independent Bernoulli trials, each with probability of success p, is performed. Let X be the number of trials until the first success occurs.

(a) Find $P_X(\{k\})$ for each $k \in \mathbb{N}$.
(b) Find $E[X]$ and $\operatorname{Var}(X)$.

93 (Negative Binomial Variable).   A sequence of independent Bernoulli trials, each with probability of success p, is performed. Let X be the number of trials until k successes are obtained, where $k \in \mathbb{N}$ is fixed.

(a) Find $P_X(\{n\})$.
(b) Find $E[X]$.

94. Let X be a random variable on Ω with distribution function

$$F_X(x) := \begin{cases} 0 & \text{if } x \leq -1, \\ a + b \arcsin x & \text{if } -1 < x \leq 1, \\ 1 & \text{if } x \geq 1. \end{cases}$$

(a) Find the constants a and b.
(b) Find $E[X]$ and $\operatorname{Var}(X)$.

95. Let a and b be arbitrary positive numbers and let $Z := (X, Y)$ be a randomly selected point from the rectangle $R := [0, a] \times [0, b]$. Show that the variables X and Y are *independent*.

96. Two random points X and Y from $[0, 1]$ divide $[0, 1]$ into three segments. What is the probability that the three segments can be used to form a triangle? *Hint:* Consider the case $X < Y$ first and note that the *Triangle Inequality* must hold.

97. Let $\Omega := [0, 1]$ with $P := \lambda_{[0,1]}$ (the restriction of Lebesgue measure to $[0, 1]$). Show that the random vector $(X, Y)(\omega) := (\omega, \omega)$ is *not* absolutely continuous. Show, however, that the *marginal* distributions P_X and P_Y are both absolutely continuous. What are their densities?

98. Show that two random variables X and Y on Ω are *independent* if and only if

$$P(X < a, Y < b) = P(X < a)P(Y < b) \qquad (\forall a, \ b \in \mathbb{Q}).$$

99. Show that a sequence $(X_n)_{n=1}^{\infty}$ of *discrete* random variables on Ω is independent if and only if, given any $\{i_1, i_2, \ldots, i_k\} \subset \mathbb{N}$ and any $x_{i_1} \in X_{i_1}(\Omega), \ldots, x_{i_k} \in X_{i_k}(\Omega)$, we have

$$P\big(X_{i_1} = x_{i_1}, \ldots, X_{i_k} = x_{i_k}\big) = P\big(X_{i_1} = x_{i_1}\big) \cdots P\big(X_{i_k} = x_{i_k}\big).$$

100. Let X and Y be *jointly continuous* random variables on Ω. Show that the variables $U := X + Y$ and $V := X - Y$ are also jointly continuous and we have

$$f_{U,V}(u, v) = \frac{1}{2} f_{(X,Y)}\left(\frac{u + v}{2}, \frac{u - v}{2}\right).$$

101. Let $X \sim B(m, p)$ and $Y \sim B(n, p)$ be *independent* binomial variables. Find the distribution of $Z := X + Y$.

102. Let X and Y be *independent* Poisson variables with parameters α and β, respectively. Show that $Z := X + Y$ is also a Poisson variable. What is its parameter?

103. Let $(X_k)_{k=1}^{n}$ be *independent* random variables on Ω with distribution functions $(F_k)_{k=1}^{n}$. If $\bigvee X_k := \max(X_1, \ldots, X_n)$ and $\bigwedge X_k := \min(X_1, \ldots, X_n)$, find the distribution functions $F_{\bigvee X_k}$ and $F_{\bigwedge X_k}$.

104. Let $(\Omega, \mathcal{A}, P) = (\mathbb{D}, \mathbb{D} \cap \mathcal{M}_2, \pi^{-1}\lambda_2)$, where $\mathbb{D} := \{(x, y) \in \mathbb{R}^2 : x^2 + y^2 \le 1\}$ is the unit disk in $\mathbb{R}^2$, $\mathcal{M}_2$ is the σ-algebra of Lebesgue measurable subsets of $\mathbb{R}^2$, and λ_2 is the Lebesgue measure on $\mathbb{R}^2$. Let (X, Y) be a randomly selected point from $\mathbb{D}$.

(a)　Find $E[D]$, where $D(X, Y) := \sqrt{X^2 + Y^2}$ is the distance from the origin to (X, Y).

(b)　Are the variables X and Y independent?

105 (Covariance).　Let X and Y be two random variables on Ω with *finite* variances. The *covariance* of X and Y, denoted by $\text{Cov}(X, Y)$, is defined by

$$\text{Cov}(X, Y) := E\big[(X - E[X])(Y - E[Y])\big].$$

(a)　Show that, if X and Y have finite variances, then we have

$$\text{Cov}(X, Y) = E[XY] - E[X]E[Y].$$

(b)　Show that, if $(X_j)_{j=1}^{m}$ and $(Y_k)_{k=1}^{n}$ are random variables on Ω with finite variances and if $(a_j)_{j=1}^{m} \in \mathbb{R}^m$ and $(b_k)_{k=1}^{n} \in \mathbb{R}^n$ are arbitrary, then

$$\text{Cov}\left(\sum_{j=1}^{m} a_j X_j, \sum_{k=1}^{n} b_k Y_k\right) = \sum_{j=1}^{m} \sum_{k=1}^{n} a_j b_k \text{Cov}(X_j, Y_k).$$

(c)　Show that, if $(X_k)_{k=1}^{n}$ are square integrable random variables and $(a_k)_{k=1}^{n} \in \mathbb{R}^n$ is arbitrary, then

$$\text{Var}\left(\sum_{k=1}^{n} a_k X_k\right) = \sum_{k=1}^{n} a_k^2 \text{Var}(X_k) + 2 \sum_{j < k} \sum a_j a_k \text{Cov}(X_j, X_k).$$

106 (Uncorrelated Variables).　Let X and Y be two random variables on Ω with *finite* variances. We say that X and Y are *uncorrelated* if $\text{Cov}(X, Y) = 0$. More generally, given any index set J and any family $(X_j)_{j \in J}$ of random variables on Ω with finite variances, we say that the X_j are uncorrelated if $\text{Cov}(X_i, X_j) = 0$ for all *distinct* $i, j \in J$.

(a) Show that, if $(X_k)_{k=1}^n$ are uncorrelated and $(a_k)_{k=1}^n \in \mathbb{R}^n$, then

$$\mathrm{Var}\Big(\sum_{k=1}^n a_k X_k \Big) = \sum_{k=1}^n a_k^2 \mathrm{Var}(X_k).$$

(b) Show that, given any square integrable variables X and Y, the random variables $X + Y$ and $X - Y$ are uncorrelated if and only if $\mathrm{Var}(X) = \mathrm{Var}(Y)$.

107 (Correlation Coefficient). Let X and Y be two random variables on Ω with finite, *positive* variances $\sigma^2(X)$ and $\sigma^2(Y)$, respectively. The *correlation coefficient* of X and Y is then defined by

$$\rho(X, Y) := \frac{\mathrm{Cov}(X, Y)}{\sigma(X)\sigma(Y)}.$$

(a) Show that

$$\mathrm{Var}\Big(\frac{X}{\sigma(X)} \pm \frac{Y}{\sigma(Y)} \Big) = 2 \pm 2\rho(X, Y).$$

(b) Show that $-1 \le \rho(X, Y) \le 1$.
(c) Show that $\rho(X, Y) = \pm 1$ if and only $Y = aX + b$ for some constants a, $b \in \mathbb{R}$. *Hint:* Cauchy–Schwarz inequality!

108. Let X and Y be jointly continuous on Ω with joint density

$$f_{(X,Y)}(x, y) := \begin{cases} \frac{1}{2}\sin(x + y) & \text{if } x, \ y \in [0, \pi/2], \\ 0 & \text{otherwise.} \end{cases}$$

Find the correlation coefficient $\rho(X, Y)$.

109. Let X and Y be *independent* variables on Ω

(a) Show that $P_{X+Y} = P_X * P_Y$, with the convolution $*$ defined as in Problem 69.
(b) Show that, if X is absolutely continuous with density f_X, then $Z := X + Y$ is also absolutely continuous and has density

$$f_{X+Y}(z) = \int f_X(z - y)\, dP_Y(y).$$

(c) Show that, if X and Y are *both* absolutely continuous, then $f_{X+Y} = f_X * f_Y$ with the convolution $*$ as in Problem 68.

110.

(a) If X and Y are both uniformly distributed over $[0, 1]$, find the density function of $X + Y$.
(b) If X and Y are two *exponential* random variables with parameters $\alpha = 1/2$ and $\beta = 1/3$, respectively, find the density function of $X + Y$.

111 (Characteristic Function). Given a random variable X on Ω, its *characteristic function*, denoted by ϕ_X, is defined to be the *Fourier transform* of its probability distribution P_X :

$$\phi_X(t) := \widehat{P_X}(t) = E[e^{itX}] = \int e^{itx}\, dP_X(x).$$

In particular, if X is absolutely continuous with density f_X, then

$$\phi_X(t) = \int f_X(x) e^{itx} \, d\lambda(x) = \widehat{f_X}(t).$$

(a) Show that ϕ_X is a *continuous* function with $\phi_X(0) = 1$, $|\phi_X(t)| \le 1$, and $\phi_X(-t) = \overline{\phi_X(t)}$.

(b) Show that, for any a, $b \in \mathbb{R}$, we have $\phi_{aX+b}(t) = e^{itb}\phi_X(at)$.

(c) Show that, if X and Y are independent, we have

$$\phi_{X+Y} = \phi_X \phi_Y.$$

(d) Find ϕ_X if (i) X is *Bernoulli* with $P(X = 1) = p$; (ii) $X \sim B(n, p)$; (iii) X is Poisson with parameter $\lambda > 0$; (iv) X is *uniformly* distributed over $[a, b]$; and (v) X is *exponential* with parameter $\lambda > 0$.

112 (Moments).   Let X be a random variable on Ω. For each integer $k \in \mathbb{N}$, the kth *moment* of X, denoted by $m_k = m_k(X)$, is the number

$$m_k := E[X^k] = \int x^k \, dP_X(x),$$

provided $E[|X|^k] := \int |x|^k \, dP_X(x) < \infty$. In particular, $m_1(X) = E[X]$ and $m_2(X_0) = \text{Var}(X)$, where the *centered* variable $X_0 := X - E[X]$ is the deviation of X from its mean. One also defines $m_0 := 1$ even if $P[X = 0] > 0$.

(a) Show that, if $m_k = E[X^k]$ exists for some $k > 0$, then m_j exists for all j with $0 \le j \le k$.

(b) Show that, if m_k exists for some $k > 0$, then

$$m_k = \left(i^{-k} \frac{d^k}{dt^k} \phi_X \right)(0).$$

Hint: For $k = 1$, note that the Mean Value Theorem implies $|(e^{i(t+h)X} - e^{itX})/h| = |X|$ and use the Dominated Convergence Theorem. For $k > 1$, use induction.

(c) Assuming that $m_n = E[X^n]$ is finite, deduce from Taylor's formula (Theorem 7.5.17) that we have

$$\phi_X(t) = \sum_{k=0}^{n} m_k \frac{(it)^k}{k!} + o(t^n) \qquad (t \to 0).$$

Also show that, if m_n is finite for *all* $n \in \mathbb{N}$, then

$$\phi_X(t) = \sum_{k=0}^{\infty} m_k \frac{(it)^k}{k!}$$

for all t in the interval of convergence of the series.

(d) Suppose that $\phi_X^{(2n)}(0)$ exists and is finite for some $n \in \mathbb{N}$. Show that the moments $m_k(X)$ exist for $0 \le k \le 2n$. *Hint:* With $\phi := \phi_X$, consider the difference operators $\Delta_h \phi(0) := \phi(h) - \phi(0)$, and $\Delta_h^{k+1}\phi(0) := \Delta_h(\Delta_h^k \phi)(0)$ for all $k \in \mathbb{N}$ as in Problem 6.8.#50. Show that

$$\Delta_h^{2n} e^{itX}|_{t=0} = (-1)^n 2^{2n} e^{inhX} \sin^{2n}(hX/2),$$

and hence that

$$\phi^{(2n)}(0) = \lim_{h \to 0} \frac{\Delta_h^{2n} \phi(0)}{h^{2n}} = \lim_{h \to 0} \int e^{inhx} \left(\frac{\sin(hx/2)}{h/2} \right)^{2n} dP_X(x).$$

Now use the fact that $\lim_{h \to 0} \sin(hx/2)/(h/2) = x$ and Fatou's lemma to deduce that $E[X^{2n}] \le |\phi^{(2n)}(0)|$.

(e) Suppose that $X \sim N(0,1)$; i.e., that X is *standard normal*. Show that $m_{2k+1} = 0$ and $m_{2k} = (2k)!/k!2^k$ for every integer $k \ge 0$. Deduce (using (c) and (d)) that

$$\phi_X(t) = \sum_{k=0}^{\infty} (-1)^k \frac{(t^2/2)^k}{k!} = e^{-t^2/2} \qquad (\forall t \in \mathbb{R}).$$

113. Recalling that $\Gamma(m+1) = \int_0^\infty x^m e^{-x} \, dx = m!$, fix $m \in \mathbb{N}$ and let X be a random variable with density

$$f_X(x) := \begin{cases} \dfrac{x^m}{m!} e^{-x} & \text{if } x \ge 0, \\ 0 & \text{otherwise.} \end{cases}$$

Show that $P(0 \le X \le 2(m+1)) > m/(m+1)$. *Hint:* Use Chebyshev's inequality.

114 (Chebyshev's Inequality). Let $X \ge 0$ be a random variable on Ω.

(a) Show that, for any $p > 0$ and $\varepsilon > 0$, we have

$$P(X \ge \varepsilon) \le \frac{E[X^p]}{\varepsilon^p}.$$

(b) Deduce that, if $E[X^p] < \infty$, then

$$\lim_{\varepsilon \to \infty} \varepsilon^p P(X \ge \varepsilon) = 0.$$

115. Show that, if $X \ge 0$ is a random variable on Ω, then

$$\sum_{n=1}^{\infty} P(X \ge n) \le E[X] \le \sum_{n=0}^{\infty} P(X \ge n).$$

Deduce that $E[X]$ is finite if and only if $\sum_{n=1}^\infty P(X \ge n) < \infty$.

116. Let (X_n) be an independent sequence of Bernoulli variables with $P(X_n = 1) = p_n$ and $P(X_n = 0) = 1 - p_n$ for all $n \in \mathbb{N}$.

(a) Show that $X_n \to 0$ *in probability* if and only if $\lim(p_n) = 0$.
(b) Show that $X_n \to 0$ *almost surely* if and only if $\sum_{n=1}^\infty p_n < \infty$.

117 (Monte Carlo Method). Let (X_n) be a sequence of randomly selected points from $\Omega :=$ $[0,1]$. Let $f \in \mathcal{L}^1([0,1])$ and let $S_n(f) := \frac{1}{n} \sum_{k=1}^n f(X_k)$, $\forall \, n \in \mathbb{N}$. Show that $S_n(f) \to \int_0^1 f(x) \, d\lambda(x)$ in probability. *Hint:* weak law of large numbers.

118. Let X be uniformly distributed over $[0, 2\pi]$ and define $X_k := \sin(kX)$ for all $k \in \mathbb{N}$. Show that

$$\lim_{n \to \infty} \frac{X_1 + X_2 + \cdots + X_n}{n} = 0 \quad \text{almost surely.}$$

119. Let $\Omega := (0, 1]$ with $P = \lambda_\Omega$. For each $\omega \in \Omega$, consider its binary expansion $\omega = (0.d_1 d_2 d_3 \ldots)_2$, where (cf. Proposition 12.5.2) we always pick the *nonterminating* expansion if there is also a terminating one. For each $n \in \mathbb{N}$, define $X_n(\omega) := d_n$.

(a) Show that $(X_n)_{n=1}^\infty$ is a sequence of *independent, identically distributed* random variables with $P(X_n = 0) = P(X_n = 1) = 1/2$ for all $n \in \mathbb{N}$. *Hint:* Given any $(0.d_1 d_2 \ldots)_2 \in (0, 1]$, let $A_n := \{\omega : X_k(\omega) = d_k, 1 \le k \le n\}$. Then

$$P(A_n) = \lambda\left(\left(\frac{d_1}{2} + \frac{d_2}{2^2} + \cdots + \frac{d_n}{2^n}, \frac{d_1}{2} + \frac{d_2}{2^2} + \cdots + \frac{d_n}{2^n} + \frac{1}{2^n}\right]\right) = \frac{1}{2^n}.$$

Using this, deduce that, given any positive integers $i_1 < \cdots < i_k$ and any $d_{i_j} \in \{0, 1\}$, with $1 \le j \le k$, we have

$$P\left(X_{i_1} = d_{i_1}, \ldots, X_{i_k} = d_{i_k}\right) = \frac{1}{2^k}.$$

(b) If $S_n := X_1 + \cdots + X_n$, for each $n \in \mathbb{N}$, show that

$$\lim_{n \to \infty} \left(\frac{S_n}{n} - \frac{1}{2}\right) = 0 \quad \text{almost surely.}$$

(c) Show that, if $f : [0, 1] \to \mathbb{R}$ is *continuous*, then

$$\lim_{n \to \infty} E[f(S_n/n)] = f(1/2).$$

120 (Rademacher Functions).   Let $\Omega = (0, 1]$ and, for each $\omega = (0.d_1 d_2 \ldots)_2$ as in the preceding problem, define $R_n(\omega) = 2d_n - 1$. The R_n are called the *Rademacher functions*.

(a) Show that $(R_n)_{n=1}^\infty$ is an independent sequence of identically distributed variables with $P(R_n = 1) = P(R_n = -1) = 1/2$ for all $n \in \mathbb{N}$.

(b) Show that

$$\sum_{n=1}^\infty 2^{-n} R_n(\omega) = 2\omega - 1.$$

(c) Show that the *random harmonic series*

$$\sum_{n=1}^\infty \frac{R_n}{n}$$

converges *with probability 1*.

(d) Show that $\int_0^1 R_n(\omega)\, d\omega = 0$, for each $n \in \mathbb{N}$, and

$$\int_0^1 R_m(\omega) R_n(\omega)\, d\omega = \begin{cases} 0 & \text{if } m \ne n, \\ 1 & \text{if } m = n. \end{cases}$$

(e) **(Vieta's Formula)** Find the *characteristic* functions $\phi_{R_n}(t)$ and, using part (b), show that

$$\frac{\sin t}{t} = \prod_{n=1}^\infty \cos\left(t/2^n\right),$$

where the right side is defined by $\lim_{n \to \infty} \prod_{k=1}^n \cos(2^{-k} t)$.

Appendix A
Construction of Real Numbers

The purpose of this appendix is to give a construction of the field $\mathbb{R}$ of *real* numbers from the field $\mathbb{Q}$ of *rational* numbers which, we assume, is known to the reader. Let us point out that we did not give the axiomatic construction of the set $\mathbb{N}$ of *natural numbers* from which one can first construct the set $\mathbb{Z}$ of *integers* and, subsequently, the set $\mathbb{Q}$ of *rationals*. These constructions may be found in most textbooks on abstract algebra, e.g., *A Survey of Modern Algebra*, by Birkhoff and MacLane [BM77].

Most authors use the so-called Dedekind Cuts to construct the set of real numbers from that of rational numbers. Since, however, the reader is now familiar with sequences and series, it is more natural to use Georg Cantor's method of construction, which is based on Cauchy sequences of rational numbers, and can be extended to more abstract situations. This abstraction, which is referred to as the *completion* of a *metric space*, was discussed in Chap. 5. We begin our discussion by introducing some notation and definitions.

Notation. We recall that the set of *rational* numbers is denoted by $\mathbb{Q}$ and the set of *positive rationals* by $\mathbb{Q}^+ = \{r \in \mathbb{Q} : r > 0\}$. Also, the set of all sequences of rational numbers, i.e., the set of all functions from $\mathbb{N}$ to $\mathbb{Q}$, is denoted by $\mathbb{Q}^{\mathbb{N}}$.

Next, we define the *Cauchy* sequences of rational numbers. Although the definition of Cauchy sequences was given earlier, since we have not yet constructed the set of real numbers, we must insist that the $\varepsilon > 0$ in our definition take on *rational values only*.

Definition A.1 (Cauchy Sequences in $\mathbb{Q}$). A sequence $x \in \mathbb{Q}^{\mathbb{N}}$ is called a *Cauchy* sequence if the following holds:

$$(\forall \varepsilon \in \mathbb{Q}^+)(\exists N \in \mathbb{N})(m, n \geq N \Rightarrow |x_m - x_n| < \varepsilon).$$

The set of all Cauchy sequences in $\mathbb{Q}$ will be denoted by $\mathcal{C}$.

© Springer Science+Business Media New York 2014
H.H. Sohrab, *Basic Real Analysis*, DOI 10.1007/978-1-4939-1841-6

Next, we define *null* sequences in $\mathbb{Q}$. Again, this was defined earlier, but we must be careful to use *only rational* $\varepsilon > 0$.

Definition A.2 (Null Sequences in $\mathbb{Q}$).   A sequence $x \in \mathbb{Q}^{\mathbb{N}}$ is called a *null* sequence if the following holds:

$$(\forall \varepsilon \in \mathbb{Q}^+)(\exists N \in \mathbb{N})(n \geq N \Rightarrow |x_n| < \varepsilon).$$

The set of all null sequences in $\mathbb{Q}$ will be denoted by $\mathcal{N}$.

Remark.   Note that $\mathcal{N}$ is precisely the set of all rational sequences that converge to *zero* and that we obviously have $\mathcal{N} \subset \mathcal{C}$. (Why?)

As we have seen, the set $\mathbb{Q}$ of rationals is *dense* in the set $\mathbb{R}$ of real numbers, which we introduced axiomatically. It follows that each real number ξ is the limit of a (*not unique*) sequence (x_n) of rational numbers. It is tempting, therefore, to take such a sequence (x_n) as the *definition* of the real number ξ. The nonuniqueness of (x_n) poses a problem, however, for *two* such sequences in fact represent the *same* ξ. This motivates the following definition.

Definition A.3 (Equivalent Cauchy Sequences). We say that two Cauchy sequences $x, y \in \mathcal{C}$ are *equivalent* and write $x \sim y$, if and only if $x - y \in \mathcal{N}$.

Exercise A.1.   Show that the relation $\sim$ is indeed an *equivalence relation* on the set $\mathcal{C}$.

Notation.   For each sequence $x \in \mathcal{C}$, its *equivalence class* is denoted by $[x]$ and, we recall, is defined by $[x] = \{y \in \mathcal{C} : y \sim x\}$. The set of all equivalence classes of elements of $\mathcal{C}$ is denoted by $\mathcal{C}/\mathcal{N}$.

Definition A.4 (Real Number).   The set $\mathbb{R}$ of *real numbers* is defined to be $\mathbb{R} := \mathcal{C}/\mathcal{N}$. Thus ξ is a real number if $\xi = [x]$ for some $x \in \mathcal{C}$. The sequence $x \in \mathcal{C}$ is then called a *representative* of ξ. Clearly, if x and y both represent ξ, then $x - y \in \mathcal{N}$.

Exercise A.2.

1. Show that a Cauchy sequence in $\mathbb{Q}$ is bounded.
2. Show that $\mathcal{C}$ is *closed* under addition and multiplication; i.e., $\forall x, \ y \in \mathcal{C}$, we have $x + y, \ xy \in \mathcal{C}$.
3. Show that $\mathcal{N}$ is an *ideal* in $\mathcal{C}$; i.e., it is closed under addition and satisfies the stronger condition that $\forall x \in \mathcal{N}$ and $\forall y \in \mathcal{C}$ we have $xy \in \mathcal{N}$. *Hints:* For the addition, use an $\varepsilon/2$-argument. For the multiplication, use the inequalities $|x_m y_m - x_n y_n| \leq |y_m||x_m - x_n| + |x_n||y_m - y_n| \leq B|x_m - x_n| + A|y_m - y_n|$, for some constants $A, \ B \in \mathbb{Q}^+$, where the second inequality follows from part (1).

Definition A.5 (Addition, Subtraction, Multiplication).   Let $\xi = [x]$ and $\eta = [y]$ be any real numbers. We define $\xi + \eta$, $-\eta$, $\xi - \eta$, and $\xi\eta$ (or $\xi \cdot \eta$) as follows:

1. $\xi + \eta := [x + y]$,
2. $-\eta := [-y]$,
3. $\xi - \eta := \xi + (-\eta) = [x - y]$, and
4. $\xi\eta := [xy]$.

Exercise A.3. Show that the definitions of $\xi + \eta$ and $\xi\eta$ are *independent* of the representatives x and y of ξ and η, respectively. In other words, show that, if $x \sim x'$ and $y \sim y'$, then we have $x + y \sim x' + y'$ and $xy \sim x'y'$. *Hint:* You will need arguments similar to those needed in Exercise A.2.

Proposition A.1 (Ring Properties of $\mathbb{R}$). *The set $\mathbb{R}$ of real numbers is a commutative ring with identity. In other words, for all real numbers ξ, η, and ζ, we have*

1. $\xi + \eta = \eta + \xi$,
2. $(\xi + \eta) + \zeta = \xi + (\eta + \zeta)$,
3. $\exists\, 0 \in \mathbb{R}$ with $0 + \xi = \xi$,
4. $\exists\, -\xi \in \mathbb{R}$ with $\xi + (-\xi) = 0$,
5. $\xi\eta = \eta\xi$,
6. $(\xi\eta)\zeta = \xi(\eta\zeta)$,
7. $\exists\, 1 \in \mathbb{R}$, $1 \neq 0$, with $1 \cdot \xi = \xi$, and
8. $\xi(\eta + \zeta) = \xi\eta + \xi\zeta$.

Proof. The proofs of these properties are straightforward. For example, to prove (2), note that if $\xi = [x]$, $\eta = [y]$, and $\zeta = [z]$, then $(\xi + \eta) + \zeta = [(x + y) + z]$, while $\xi + (\eta + \zeta) = [x + (y + z)]$. Since we obviously have $(x+y)+z = x+(y+z)$ in $\mathcal{C}$, (2) follows. Note that the *additive identity* ("0" in (3)) is in fact $0 = [(0, 0, 0, \ldots)] \in \mathcal{N}$ and that the *multiplicative identity* ("1" in (7)) is $1 = [(1, 1, 1, \ldots)]$. Also, $1 \neq 0$ is obvious, because the sequences $(0, 0, 0, \ldots)$ and $(1, 1, 1, \ldots)$ are *not* equivalent. $\square$

Proposition A.2. *Let $\phi : \mathbb{Q} \to \mathbb{R}$ be defined by $\phi(r) = [(r, r, r, \ldots)]$. Then ϕ is an injective "ring homomorphism." In other words, ϕ is a one-to-one map satisfying $\phi(r + s) = \phi(r) + \phi(s)$, $\phi(rs) = \phi(r)\phi(s)$, $\phi(0) = 0$, and $\phi(1) = 1$, $\forall\, r, s \in \mathbb{Q}$.*

Exercise A.4. Prove Proposition A.2.

Remark. By Proposition A.2, the map ϕ is a *field isomorphism* of $\mathbb{Q}$ onto its image $\phi(\mathbb{Q}) \subset \mathbb{R}$; i.e., a one-to-one correspondence between $\mathbb{Q}$ and $\phi(\mathbb{Q})$ that preserves all the algebraic properties of $\mathbb{Q}$. Therefore, we henceforth *identify* the two sets and, by abuse of notation, will write $\mathbb{Q} = \phi(\mathbb{Q}) \subset \mathbb{R}$. Based on this identification, the *field* $\mathbb{Q}$ of rational numbers becomes a *subfield* of the field $\mathbb{R}$ of real numbers. Here, by a *field* we mean a set $\mathbf{F}$ together with two *operations* "+" of *addition* and "·" of *multiplication*, i.e., two maps $+ : (x, y) \mapsto x + y$ and $\cdot : (x, y) \mapsto x \cdot y$, from $\mathbf{F} \times \mathbf{F}$ to $\mathbf{F}$, satisfying the nine (algebraic) axioms ($A_1 - A_4$, $M_1 - M_4$, D) stated for real numbers in Sect. 2.1 of Chapter 2.

Proposition A.1 only shows that $\mathbb{R}$ is a *commutative ring with identity*. To prove that $\mathbb{R}$ is actually a *field*, the only property we need to check is the existence of *reciprocals* for *nonzero* real numbers (cf. Axiom (M_4) at the beginning of Chap. 2). To this end, we shall need the following.

Proposition A.3. *Let ξ be a nonzero element of $\mathbb{R}$. Then, there exists a rational number $r \in \mathbb{Q}^+$ and a representative $x \in C$ of ξ such that either $x_n \geq r$ $\forall n \in \mathbb{N}$ or $x_n \leq -r$ $\forall n \in \mathbb{N}$.*

Proof. Let $y \in C$ be a representative of ξ. Since $\xi \neq 0$, the sequence (y_n) is *not* equivalent to $(0, 0, 0, \ldots)$ and we have

$$(\exists \varepsilon \in \mathbb{Q}^+)(\forall N \in \mathbb{N})(\exists n \geq N)(|y_n - 0| \geq \varepsilon). \qquad (*)$$

On the other hand, $(y_n) \in C$ implies that

$$(\exists K \in \mathbb{N})(m, n \geq K \;\Rightarrow\; |y_m - y_n| < \varepsilon/2). \qquad (**)$$

Now, by $(*)$, we can find $k \geq K$ such that $|y_k| \geq \varepsilon$. Changing ξ to $-\xi$, if necessary, we may assume that $y_k \geq \varepsilon$. Therefore, using $(**)$,

$$m \geq K \;\Rightarrow\; |y_m - y_k| < \varepsilon/2 \;\Rightarrow\; y_m \geq y_k - |y_m - y_k| \geq \varepsilon - \varepsilon/2 = \varepsilon/2.$$

Let $x_n := \varepsilon/2$ for $n < K$, and $x_n = y_n$ for $n \geq K$. It is then clear that $\xi = [(x_n)]$ and that, with $r := \varepsilon/2$, we have $x_n \geq r$ for all $n \in \mathbb{N}$. $\qquad\qquad\square$

Definition A.6 (Positive and Negative Cauchy Sequences). We say that a sequence $x \in C$ is *positive* (resp., *negative*) if it satisfies the first (resp., second) alternative in Proposition A.3. The set of all positive (resp., negative) sequences in C is denoted by C^+ (resp., C^-).

Remark. It is obvious that the two alternatives in Proposition A.3 are mutually exclusive, i.e., that $C^+ \cap C^- = \emptyset$. Moreover, the condition in the first (and hence also second) alternative needs only be satisfied *ultimately*; i.e., it can be replaced by

$$(\exists r \in \mathbb{Q}^+)(\exists N \in \mathbb{N})(n \geq N \Rightarrow x_n \geq r).$$

Indeed, one can always replace x by the equivalent sequence x' defined by $x'_k := r$ $\forall k < N$ and $x'_k := x_k$ $\forall k \geq N$.

Proposition A.4. *We have $C = C^+ \cup N \cup C^-$, where the union is disjoint. In other words, $\{C^+, N, C^-\}$ is a partition of C.*

Proof. This is an obvious consequence of Proposition A.3. $\qquad\qquad\square$

We are now going to prove that $\mathbb{R}$ is indeed a field.

Theorem A.1. *The set $\mathbb{R}$ of real numbers is a field. In other words, in addition to the ring properties (1)–(8) of Proposition A.1, we also have the following:*

$$(\forall \xi \in \mathbb{R} \setminus \{0\})(\exists \, 1/\xi \in \mathbb{R} \setminus \{0\})(\xi \cdot (1/\xi) = 1).$$

Proof. Suppose that $\xi \in \mathbb{R} \setminus \{0\}$. By Proposition A.3, we can then find $r \in \mathbb{Q}^+$ and a representative (x_n) of ξ such that $|x_n| \geq r \ \forall n \in \mathbb{N}$. If we can show that $(1/x_n) \in \mathcal{C}$, then, setting $1/\xi := [(1/x_n)]$, we clearly get $\xi \cdot (1/\xi) = 1$. However, $(x_n) \in \mathcal{C}$ implies

$$(\forall \varepsilon \in \mathbb{Q}^+)(\exists N \in \mathbb{N})(m, \ n \geq N \Rightarrow |x_m - x_n| < \varepsilon r^2).$$

Therefore,

$$m, \ n \geq N \ \Rightarrow \ |1/x_m - 1/x_n| = \frac{|x_m - x_n|}{|x_m||x_n|} < \frac{\varepsilon r^2}{r^2} = \varepsilon,$$

which proves indeed that $(1/x_n) \in \mathcal{C}$ and completes the proof. $\square$

Having established the field properties of $\mathbb{R}$, we now turn our attention to its *order* properties. Recall that this was treated axiomatically (cf. Axioms $(O)_1 - (O)_3$ at the beginning of Chap. 2) by means of a subset $P \subset \mathbb{R}$ called the subset of *positive* real numbers. In what follows we will define this subset and will denote it by $\mathbb{R}^+$, rather than P.

Definition A.7 (Positive and Negative Real Numbers). We define a real number $\xi \in \mathbb{R}$ to be *positive* (resp., *negative*) and write $\xi > 0$ (resp., $\xi < 0$), if $\xi = [x]$ for some $x \in \mathcal{C}^+$ (resp., $x \in \mathcal{C}^-$). The set of all positive (resp., negative) real numbers will be denoted by $\mathbb{R}^+$ (resp., $\mathbb{R}^-$).

Proposition A.5. *We have* $\mathbb{R}^- = -\mathbb{R}^+ := \{\xi \in \mathbb{R} : -\xi \in \mathbb{R}^+\}$, *and the set* $\mathbb{R}^+$ *of positive real numbers satisfies the following properties:*

1. $\mathbb{R}^+ + \mathbb{R}^+ \subset \mathbb{R}^+$,
2. $\mathbb{R}^+ \cdot \mathbb{R}^+ \subset \mathbb{R}^+$, *and*
3. $\mathbb{R} = \mathbb{R}^+ \cup \{0\} \cup \mathbb{R}^-$, *where the union is disjoint*   *(Trichotomy).*

Exercise A.5. Prove Proposition A.5.

Now that the existence of the set $\mathbb{R}^+$ of positive real numbers has been established and that, in view of Proposition A.5, the order axioms (O_1), (O_2), and (O_3) are satisfied, all the order properties of the set $\mathbb{R}$ of real numbers can be proved as before. For instance, given $\xi, \ \eta \in \mathbb{R}$, we write $\xi \leq \eta$ to mean $\eta - \xi \in \mathbb{R}^+ \cup \{0\}$ and the set $\mathbb{R}$ is then *totally* ordered by the ordering $\leq$.

Remark.

1. We have defined the notion of *Cauchy sequence* once for (axiomatically defined) *real* numbers in Chap. 2 and again, in this appendix, for *rational* numbers (which are real numbers), using exclusively *rational* $\varepsilon > 0$. To show that, for rational sequences, the two definitions are identical, we need only show the following:

$$(\forall \varepsilon \in \mathbb{R}^+)(\exists \varepsilon' \in \mathbb{Q})(0 < \varepsilon' \leq \varepsilon).$$

This, however, follows at once from Proposition A.3.

2. Since the set $\mathbb{R}$ we have constructed satisfies all the *algebraic* and *order* properties treated axiomatically in Chap. 2, the notion of *convergent sequence* can be defined as before. In other words, a sequence $(\xi_n) \in \mathbb{R}^{\mathbb{N}}$ of real numbers *converges* to the *limit* $\lambda \in \mathbb{R}$ (in symbols $\lim(\xi_n) = \lambda$), if the following holds:

$$(\forall \varepsilon \in \mathbb{R}^+)(\exists N \in \mathbb{N})(n \geq N \Rightarrow |\xi_n - \lambda| < \varepsilon).$$

Our construction of real numbers was motivated by the intuitive idea that a real number should be the limit of a convergent sequence of rationals. The following proposition shows that this is indeed the case.

Proposition A.6. *Let ξ be a real number. For a sequence $x \in \mathcal{C}$ to be a representative of ξ, it is necessary and sufficient that $\lim(x_n) = \xi$.*

Proof. Suppose that $\xi = [x]$, and let $\varepsilon \in \mathbb{R}^+$ be given. Then, we can find $\varepsilon' \in \mathbb{Q}^+$ with $\varepsilon' \leq \varepsilon$. We can also find $N \in \mathbb{N}$ such that

$$m, n \geq N \quad \Rightarrow \quad -\varepsilon' < x_m - x_n < \varepsilon'. \tag{$*$}$$

Given $m \geq N$, the real number $x_m - \xi$ is the class of the sequence $(x_m - x_1, x_m - x_2, \ldots)$ which, using $(*)$, can be replaced by an equivalent one, $(y_n) \in \mathcal{C}$, such that $x_m - \xi = [(y_n)]$ and $-\varepsilon' < y_n < \varepsilon'$ $\forall n \in \mathbb{N}$. Therefore, $-\varepsilon' < x_m - \xi < \varepsilon'$, and hence $|x_m - \xi| < \varepsilon$. This shows that we have $\lim(x_n) = \xi$. Conversely, suppose that $\lim(x_n) = \xi$ and that $\xi = [(y_n)]$ for a sequence $(y_n) \in \mathcal{C}$. Then, as we just proved, $\lim(y_n) = \xi$. It then follows that $(x_n) \sim (y_n)$ (why?), and we have $\xi = [(x_n)]$. $\square$

All the algebraic and order properties we have proved for the set $\mathbb{R} := \mathcal{C}/\mathcal{N}$ are also shared by its subfield $\mathbb{Q}$ of rational numbers. We are finally ready to prove the *completeness* of $\mathbb{R}$ which, in the axiomatic treatment, was called the *Supremum Property* or *Completeness Axiom*. This property is *not* satisfied by the subfield $\mathbb{Q}$. Since the Supremum Property is *equivalent* to *Cauchy's Criterion* [as was pointed out in Remark 2.2.47 (2)], all we need is to prove this criterion for our set $\mathbb{R} := \mathcal{C}/\mathcal{N}$.

Theorem A.2 (Cauchy's Criterion). *A sequence $(\xi_n) \in \mathbb{R}^{\mathbb{N}}$ is convergent if and only if it is a Cauchy sequence.*

Proof. The necessity of the condition is obvious, as we saw in the proof of Theorem 2.2.46. To prove the sufficiency, note that, by Proposition A.6, for each $n \in \mathbb{N}$, we can find a *rational* number $x_n \in \mathbb{Q}$ (recall that $x_n = [(x_n, x_n, \ldots)]$) such that $|\xi_n - x_n| < 1/n$. Now

$$(\forall \varepsilon \in \mathbb{R}^+)(\exists N \in \mathbb{N})(m, n \geq N \quad \Rightarrow \quad |\xi_m - \xi_n| < \varepsilon/3).$$

Thus, if m, $n \geq \max\{N, 3/\varepsilon\}$, then

$$|x_m - x_n| \leq |x_m - \xi_m| + |\xi_m - \xi_n| + |\xi_n - x_n|$$

$$< \frac{1}{m} + \frac{\varepsilon}{3} + \frac{1}{n} \leq \frac{\varepsilon}{3} + \frac{\varepsilon}{3} + \frac{\varepsilon}{3} = \varepsilon,$$

and hence $(x_n) \in C$. Let $\xi = [(x_n)]$. We then have $\lim(x_n) = \xi$ and, since $\lim(\xi_n - x_n) = 0$, we get $\lim(\xi_n) = \xi$.   □

Bibliography

[AG96] Adams, M., Guillemin, V.: Measure Theory and Probability. Birkhäuser, Boston (1996)

[AB90] Aliprantis, C.D., Burkinshaw, O.: Principles of Real Analysis, 2nd edn. Academic, New York (1990)

[Apo74] Apostol, T.: Mathematical Analysis, 2nd edn. Addison-Wesley Publishing Co., Inc., Reading (1974)

[Ash72] Ash, R.B.: Real Analysis and Probability. Academic, New York (1972)

[ABu66] Asplund, E., Bungart, L.: A First Course in Integration. Holt, Rinehart and Winston, New York (1966)

[Bar01] Bartle, R.G.: A Modern Theory of Integration. American Mathematical Society, Providence (2001)

[BS00] Bartle, R.G., Sherbert, D.R.: Introduction to Real Analysis, 3rd edn. Wiley, New York (2000)

[Bea97] Beardon, A.F.: Limits. A New Approach to Real Analysis. Springer, New York (1997)

[Ber94] Berberian, S.K.: A First Course in Real Analysis. Undergraduate Texts in Mathematics, Springer, New York (1994)

[BW90] Biler, P., Witkowski, A.: Problems in Mathematical Analysis. Dekker Inc., New York (1990)

[Bil95] Billingsley, P.: Probability and Measure, (3rd edn). Wiley, New York (1995)

[Bin77] Binmore, K.G.: Mathematical Analysis, A Straightforward Approach. Cambridge University Press, Cambridge (1977)

[BM77] Birkhoff, G., MacLane, S.: A Survey of Modern Algebra, 4th edn. Macmillan Publishing Co., New York (1977)

[Boa60] Boas, R.P.: A Primer of Real Functions. Carus Mathematical Monographs, vol. 13. Wiley, New York (1960)

[Bou68] Bourbaki, N.: General Topology. Addison-Wesley Publishing Co., Reading (1968)

[Bou74] Bourbaki, N.: Elements of Mathematics; Theory of Sets. Addison-Wesley Publishing Co., Reading (1974)

[Bre08] Bressoud, D.M.: A Radical Approach to Lebesgue's Theory of Integration. Cambridge University Press, New York (2008)

[Bri98] Bridges, D.S.: Foundations of Real and Abstract Analysis, An Introduction. Springer, New York (1998)

[Bro96] Browder, A.: Mathematical Analysis, An Introduction. Springer, New York (1996)

[BP95] Brown, A., Pearcy, C.: An Introduction to Analysis. Springer, New York (1995)

[BB70] Burkill, J.C., Burkill, H.: A Second Course in Mathematical Analysis. Cambridge University Press, Cambridge (1970)

© Springer Science+Business Media New York 2014
H.H. Sohrab, *Basic Real Analysis*, DOI 10.1007/978-1-4939-1841-6

[BK69] Burrill, C.W., Knudsen, J.R.: Real Variables. Holt, Rinehart and Winston, New York (1969)

[CK99] Capiński, M., Kopp, E.: Measure, Integral and Probability. Springer, London (1999)

[CZ01] Capiński, M., Zastawniak, T.: Probability Through Problems. Springer, New York (2001)

[Car00] Carothers, N.L.: Real Analysis. Cambridge University Press, Cambridge (2000)

[Cha95] Chae, S.B: Lebesgue Integration, 2nd edn. Springer, New York (1995)

[Cho66] Choquet, G.: Topology. Academic, New York (1966) [Translated by Amiel Feinstein]

[Chu74] Chung, K.L: Elementary Probability Theory with Stochastic Processes. Springer, New York (1974)

[Coo66] Cooper, R.: Functions of Real Variables. D. Van Nostrand, London (1966)

[DM90] Debnath, L., Mikusiński, P.: Introduction to Hilbert Spaces with Applications. Academic, New York (1990)

[DS88] Depree, J.D., Swartz, C.W.: Introduction to Real Analysis. Wiley, New York (1988)

[Die69] Dieudonné, J.: Foundations of Modern Analysis. Academic, New York (1969)

[Dud02] Dudley, R.M.: Real Analysis and Probability. Cambridge University Press, Cambridge (2002)

[DSc58] Dunford, N., Schwartz, J.T.: Linear Operators. Interscience Publishers, New York (1958)

[Dur10] Durrett, R.: Probability, Theory and Examples. Cambridge University Press, New York (2010)

[Edg90] Edgar, G.A.: Measure, Topology, and Fractal Geometry. Springer, New York (1990)

[Edg98] Edgar, G.A.: Integral, Probability, and Fractal Measure. Springer, New York (1998)

[ER92] Evans, L.C., Gariepy, R.F.: Measure Theory and Fine Properties of Functions. CRC Press, Inc., Boca Raton (1992)

[Fal85] Falconer, K.J.: The Geometry of Fractal Sets. Cambridge University Press, Cambridge (1985)

[Fel68] Feller, W.: An Introduction to Probability Theory and Its Applications, 3rd edn. Wiley, New York (1968)

[Fis83] Fischer, E.: Intermediate Real Analysis. Undergraduate Texts in Mathematics. Springer, New York (1983)

[Fla76] Flatto, L.: Advanced Calculus. The Williams & Wilkins Co., Baltimore (1976)

[Fol84] Folland, G.B.: Real Analysis: Modern Techniques and Their Applications. Wiley, New York (1984)

[Fri71] Friedman, A.: Advanced Calculus. Holt, Rinehart and Winston, New York (1971)

[Gal88] Galambos, J.: Advanced Probability Theory. Dekker, New York (1988)

[Gar68] Garsoux, J.: Analyse Mathématique. Dunod, Paris (1968)

[Gel92] Gelbaum, B.R.: Problems in Real and Complex Analysis. Springer, New York (1992)

[Ger06] Gerhardt, C.: Analysis II. International Press, New York (2006)

[Gha96] Ghahramani, S.: Fundamentals of Probability. Prentice Hall, New Jersey (1996)

[Gof67] Goffman, C.: Real Functions, revised edn. Prindle, Weber, and Schmidt, Boston (1967)

[Gol64] Goldberg, R.R.: Methods of Real Analysis. Blaisdell Publishing Co., New York (1964)

[Gor94] Gordon, R.A.: The Integrals of Lebesgue, Denjoy, Perron, and Henstock. Graduate Studies in Mathematics, vol. 4. American Mathematical Society, Providence (1994)

[Gor01] Gordon, R.A.: Real Analysis, A First Course. Addison-Wesley, New York (2001)

[Hal50] Halmos, P.: Measure Theory. Van Nostrand, Princeton (1950) [reprinted as Graduate Texts in Mathematics, Springer-Verlag, NY 1975]

[Hal58] Halmos, P.: Finite Dimensional Vector Spaces. Van Nostrand, Princeton (1958) [reprinted as Undergraduate Texts in Mathematics, Springer-Verlag, NY, 1974]

[Hal60] Halmos, P.: Naive Set Theory. Van Nostrand, Princeton (1960) [reprinted as Undergraduate Texts in Mathematics, Springer-Verlag, NY, 1974]

[Her75] Herstein, I.N.: Topics in Algebra, 2nd edn. Wiley, New York (1975)

[HS69] Hewitt, E., Stromberg, K.: Real and Abstract Analysis. Springer, Berlin (1969)

[KN00] Kaczor, W.J., Nowak, M.T.: Problems in Mathematical Analysis I. Real Sequences and Series. Student Mathematical Library, vol. 4. American Mathematical Society, Providence (2000)

[Kan03] Kantorovitz, S.: Introduction to Modern Analysis . Oxford University Press, New York (2003)

[Kel55] Kelley, J.L.: General Topology. Van Nostrand, New York (1955)

[KG82] Kirillov, A.A., Gvishiani, A.D.: Theorems and Problems in Functional Analysis. Springer, New York (1982) [Translated from the 1979 Russian original by Harold H. McFadden]

[Kna05] Knapp, A.W.: Basic Real Analysis. Birkhäuser, New York (2005)

[Kno63] Knopp, K.: Theory and Application of Infinite Series. Blackie & Son Ltd., London (1963)

[K75] Kolmogorov, A.N.: Foundations of the Theory of Probability, second English edn. Chelsea Publishing Co., New York (1975) [Translated from Russian by Nathan Morrison]

[KF75] Kolmogorov, A.N., Fomin, S.: Introductory Real Analysis. Dover Publications, Inc., New York (1975) [Translated from Russian by Richard A. Silverman]

[Kos95] Kosmala, W.A.J.: Introductory Mathematical Analysis. Wm. C. Brown Publishers, Dubuque (1995)

[Kra91] Krantz, S.G.: Real Analysis and Foundations. CRC Press, Inc., Boca Raton (1991)

[Kre78] Kreyszig, E.: Introductory Functional Analysis with Applications. Wiley, New York (1978)

[Lan51] Landau, E.: Foundations of Analysis. Chelsea, New York (1951)

[Lang69] Lang, S.: Real Analysis. Addison-Wesley Publishing Co., Reading (1969)

[Mal95] Malliavin, P.: Integration and Probability. Springer, New York (1995)

[MW99] McDonald, J.N., Weiss, N.A.: A Course in Real Analysis. Academic, New York (1999)

[Meg98] Megginson, R.E.: An Introduction to Banach Space Theory. Springer, New York (1998)

[Mun75] Munkres, J.R.: Topology, a First Course. Prentice Hall, Inc., New Jersey (1975)

[Munr71] Munroe, M.E.: Measure and Integration, 2nd edn. Addison-Wesley, Reading (1971)

[Nat64] Natanson, I.P.: Theory of Functions of a Real Variable. Frederick Ungar Publishing Co., New York (1964)

[Nev65] Neveu, J.: Mathematical Foundations of the Calculus of Probability. Holden-Day Inc., San Francisco (1965) [Translated by Amiel Feinstein]

[Olm59] Olmstead, J.M.H.: Real Variables. Appleton-Century-Crofts, Inc., New York (1959)

[Ped00] Pedersen, M.: Functional Analysis in Applied Mathematics and Engineering. Chapman & Hall/CRC, Boca Raton (2000)

[Pedr94] Pedrick, G.: A First Course in Analysis. Undergraduate Texts in Mathematics. Springer, New York (1994)

[Per64] Pervin, W.J.: Foundations of General Topology. Academic, New York (1964)

[PSz72] Pólya, G., Szegö, G.: Problems and Theorems in Analysis, vols. 1 and 2. Springer, Berlin (1972)

[PR69] Prohorov, Yu.V., Rozanov, Yu.A.: Probability Theory. Springer, New York (1969)

[Pro98] Protter, M.H.: Basic Elements of Real Analysis. Springer, New York (1998)

[Pug01] Pugh, C.C.: Real Mathematical Analysis. Springer, New York (2001)

[Rad09] Radulescu, T-L.T., Radulescu, V.D., Andreescu, T.: Problems in Real Analysis: Advanced Calculus on the Real Axis. Springer, New York (2009)

[Ran68] Randolph, J.F.: Basic Real and Abstract Analysis. Academic, New York (1968)

[Ree98] Reed, M.C.: Fundamental Ideas of Analysis. Wiley, New York (1998)

[RS72] Reed, M., Simon, B.: Methods of Mathematical Physics, vol. 1. Academic, New York (1972)

[RN56] Riesz, F., Nagy, B.Sz.: Functional Analysis, English edn. Ungar, New York (1956)

[Ros76] Ross, S.: A First Course in Probability. Macmillan, New York (1976)

[Ros80] Ross, K.A.: Elementary Analysis: The Theory of Calculus. Springer, New York (1980)

[Roy88] Royden, H.L.: Real Analysis, 3rd edn. Macmillan, New York (1988)

[Rub63] Rubel, L.A.: Differentiability of monotone functions. Collquium Mathematicum 10, 276–279 (1963)

[Rud64] Rudin, W.: Principles of Mathematical Analysis, 2nd edn. McGraw-Hill Book Co.,
New York (1964)

[Rud73] Rudin, W.: Functional Analysis. McGraw-Hill Book Co., New York (1973)

[Rud74] Rudin, W.: Real and Complex Analysis, 2nd edn. McGraw-Hill Book Co., New York
(1974)

[Sch96] Schramm, M.J.: Introduction to Real Analysis. Prentice Hall, Inc., New Jersey (1996)

[Shi84] Shiryayev, A.N.: Probability. Springer, New York (1984) [Translated from Russian by
R. P. Boas]

[Spi94] Spivak, M.: Calculus, 3rd edn. Publish or Perish Inc., Houston (1994)

[Sto01] Stoll, M.: Introduction to Real Analysis, 2nd edn. Addison-Wesley Longman, Inc,
Boston (2001)

[Sup60] Suppes, P.C.: Axiomatic Set Theory. Van Nostrand, New York (1960)

[Tao11] Tao, T.: An introduction to measure theory.
http://terrytao.files.wordpress.com/2011/01/measure-book1.pdf

[Tayl65] Taylor, A.A.: General Theory of Functions and Integration. Blaisdell Publishing
Company, Waltham (1965)

[Tay96] Taylor, M.E.: Partial Differential Equations, Basic Theory. Springer, New York (1996)

[Tay06] Taylor, M.E.: Measure Theory and Integration. American Mathematical Society,
Providence (2006)

[Tit39] Titchmarsh, E.C.: The Theory of Functions, 2nd edn. Oxford University Press, Oxford
(1939)

[Tor88] Torchinsky, A.: Real Variables. Addison-Wesley Publishing Co., Inc., Reading (1988)

[VS82] Van Rooij, A.C.M., Schikhof, W.H.: A Second Course on Real Functions. Cambridge
University Press, Cambridge (1982)

[Var01] Varadhan, S.R.S.: Probability Theory. A.M.S. Courant Lecture Notes, vol. 7. American
Mathematical Society, Providence (2001)

[Wad00] Wade, W.R.: An Introduction to Analysis. Prentice-Hall, Inc., New Jersey (2000)

[Wei73] Weir, A.J.: Lebesgue Integration and Measure. Cambridge University Press, Cambridge
(1973)

[Yos74] Yosida, K.: Functional Analysis, 4th edn. Springer, Berlin (1974)

Index

A

Abel's partial summation formula, 71
Abel's Test, 72
Abel's Theorem, 74, 372
Abelian (commutative), 18
absolute continuity, 505
absolute value, 42, 346
absolutely continuous, 179, 344, 605
absolutely continuous function, 559
absolutely convergent series, 67
absolutely summable, 81
accumulation point, 52, 186
additive (function) , 142
adjoint operator, 461
aleph naught ($\aleph_0$), 30
algebra, 22
   σ-, 5
   Banach, 420
   Borel, 472
   commutative, 22
   division, 22
   normed, 412
   sub-, 22
algebra of sets, 5
almost all (a.a.), 307
almost everywhere (a.e.), 307, 577
almost surely (a.s.), 617
almost uniform convergence, 550
alternating series, 72
angular point, 243
antiderivative (primitive), 324
Appolonius' identity, 460
approximate identity, 408
approximation (uniform), 192
Archimedean Property, 46
arcwise connected, 231

area under the graph, 294, 647
Arithmetic-Geometric Means Inequality, 43, 289
Arzelà–Ascoli Theorem, 449
associativity, 17, 84
asymptote
   horizontal, 112
   vertical, 110
at random, 617
atom, 638
Average Value, 337
Average Value Theorem, 502
Axiom of Choice, 10, 30

B

$B(n, p)$, 623
$\mathcal{BL}^0(E)$, $\mathcal{BL}^0_0(E)$, 492
$\mathcal{BL}^1(E)$, $\mathcal{BL}^1(E)$, 493
Baire Category Theorem, 194
Baire metric, 235
ball
   closed, 185
   open, 185
Banach
   the space c_0 of, 420
Banach algebra, 420
Banach space, 420
Banach spaces
   classical, 540
Banach's Fixed Point Theorem, 210, 422
Banach–Steinhaus Theorem, 428
base
   countable, 190
Basic Counting Principle, 24

© Springer Science+Business Media New York 2014
H.H. Sohrab, *Basic Real Analysis*, DOI 10.1007/978-1-4939-1841-6

basis, 22
  orthogonal, orthonormal, 439
  Schauder, 456
Bayes's formula, 619
Bernoulli
  random variable, 622
  sequence, 616
  trial, 616
Bernoulli numbers, 406
Bernoulli polynomials, 406
Bernoulli's inequality, 43, 260
Bernstein Approximation Theorem, 172, 636
Bernstein polynomials, 172
Bernstein's Theorem, 404
Bessel's inequality, 386, 438, 442
Best Approximation, 386, 440
big O, 121
binary expansion, 49
binary operation, 17
binomial coefficients, 25, 371
Binomial Formula, 25
binomial random variable, 623
Birkhoff and MacLane, 17
Bisection Method, 156
Bolzano–Weierstrass Property, 218
Bolzano–Weierstrass Theorem, 59
Borel algebra, 472
Borel function, 522, 593
Borel set, 472
Borel–Cantelli Lemma
  First, 620
  Second, 620
bound
  least upper, greatest lower, 9
  upper, lower, 9
boundary, 188
  point, 188
bounded
  above, below, 10
  essentially, 546
  function, 17
  pointwise, 450
  uniformly, 80, 450
bounded away from zero as, 123
Bounded Convergence Theorem, 510, 642
bounded functions
  metric space of, 192
bounded inverse, 457
bounded set, 10, 185, 417
bounded variation, 330
bounded, unbounded (sequence), 53

C
C^n, C^∞, 268
Calderon's proof
  of Steinhaus's Theorem, 571
canonical projection, 14, 427, 430
canonical representation, 479
Cantor set, 134, 135
  generalized, 308
  Hausdorff dimension of, 583
  measure of, 308
Cantor's diagonal method, 447
Cantor's ternary function, 140, 143
Cantor's Theorem, 34, 193
Cantor–Bendixon Theorem, 195
Carathéodory's definition, 584
Carathéodory's Theorem, 245
cardinal number (or cardinality), 30
Cartesian product, 6, 29
Cauchy in measure, 548, 640
Cauchy product, 73, 86
  Abel's Theorem on, 373
Cauchy sequence, 60, 192
  negative, 658
  positive, 658
Cauchy's Condensation Theorem, 67
Cauchy's Criterion, 60, 62, 81, 82, 106, 111,
    348, 441, 533, 660
  uniform, 352
Cauchy's functional equation, 129, 142, 520
Cauchy's inequality, 43, 349
Cauchy–Hadamard Theorem, 363
Cauchy–Schwarz inequality, 77, 321, 384, 433,
    643
Cauchy-Schwarz inequality, 541
chain (totally ordered set), 9
chain connected, 239
Chain Rule, 251
Change of Variables, 327, 626, 630
characteristic function, 16
  of a random variable, 651
characterization of intervals, 47
Chebyshev's Inequality, 546
Chebyshev's inequality, 524, 598, 632, 653
choice function, 29
class, 8
  equivalence, 8
  representative of a, 8
class C^n (function of), 268
classical Banach spaces, 540
closed ball, 185
Closed Graph Theorem, 223, 430

closed range, 457
closed set, 51, 186
closure, 100
    relative, 190
closure (of a set), 188
cluster point, 100, 188
coin-tossing, 616
commutative ring, 19
compact, 131, 216
    countably, 218
    Fréchet, 218
    relatively, 216
    sequentially, 218
compact map, 423
compact operator, 423, 450
compact support, 369
compactness, 217
complement, 2, 456
complement (of a subspace), 21
complemented, 456
complete, 192
Completeness Axiom (Supremum Property),
    45
completion, 213, 460
    of a normed space, 421
complex conjugate, 349
complex number, 345
composite function, 14
composition, 7
    of relations, 7
concave function, 278
condensation point, 195, 235
conditional probability, 618
conditionally convergent series, 67
congruence modulo n, 7
conjugate linear, 433
connected (metric space), 226
    arcwise, 231
    locally, 230
connected component, 229
connected, disconnected, 133
content zero (set of), 300
continuity
    at a point, on a set, 198
    global definition of, 200
    sequential definition of, 142, 199
continuous, 140, 198
    jointly, 206
    separately, 206
Continuous Extension Theorem, 162
continuous extensions, 211
continuum (c), 30
contraction (mapping), 163, 208, 422
contractive map, 178, 238

contractive sequence, 60
convergence
    absolute, 67, 532
    almost surely, 634
    almost uniform, 550
    conditional, 67, 532
    in measure, 548
    in probability, 634
    interval of, 363
    locally uniform, 515
    normal, 355
    of a sequence, 51
    of Fourier series, 392
    of series, 61
    pointwise, 349
    radius of, 362
    uniform, 351, 515
    weak, 570
convergence in measure, 548, 640
convergent
    weakly, 459
convergent, divergent, 51
convergent, divergent (series), 61
convex function, 278
convex hull, 455
convex set, 239, 416
convolution, 409
convolution of Borel measures, 646
convolution of functions, 646
correlation coefficient, 651
cosine function, 377
countable base, 190
countable set, 30
countably compact, 218
countably infinite, 30
covariance, 650
cover
    open, 131, 216
    pointwise finite, 238
Criterion
    Cauchy's, 60
    Dini's, 392
    Lebesgue's Integrability, 312
    Lusin's, 512
Cross Sections, 207

D
Darboux integrals, 293
Darboux sum, 292
Darboux's Theorem, 256, 297
De Moivre's formula, 378
De Morgan's Laws, 3
decimal expansion, 49

decreasing, 98
degenerate interval, 45
dense, 47, 190
    nowhere, 134
density, 609
density function, 624
    joint, 629
density of $\mathbb{Q}$ (in $\mathbb{R}$), 47
denumerable, 30
derivative, 242
    left, right, 243
    partial, 536
    Radon-Nikodym, 609
    Schwarzian, 286
    symmetric, 285
derivatives
    Dini, 554
derived set, 234
diagonal, 7, 184
diameter, 185
diffeomorphism, 284
    C^n-, 269
difference operator, 288
difference set, 2
differentiability of inverse functions, 253
differentiable, 242
    n-times, 267
    n-times continuously, 268
    infinitely, 268
    uniformly, 286
Differential Calculus, 250
differential equation
    Legendre's, 287
differential operator, 276
    symbol of, 276
differentiating under the integral sign, 537
differentiation
    term-by-term, 360
dilation, 412
dimension, 22
    orthogonal, 444
Dini derivatives, 554
Dini's Criterion, 392
Dini's Theorem, 353, 355
Dirac measure, 577
Dirac sequence, 408
direct (or Cartesian) product, 6, 29
    infinite, 29
direct image, 13
direct sum, 21, 456
directed set, 12, 579
Dirichlet function, 143, 296
Dirichlet's integral, 390
Dirichlet's Kernel, 388

Dirichlet's Test, 72, 567
    Uniform, 401
Dirichlet's Theorem, 88
discontinuity
    infinite, 148
    jump, 148
    of the first kind, 148
    of the second kind, 148
    removable, 148
discontinuous, 140
discrete, 145
    random variable, 622
distance (metric), 182
    Hausdorff, 233
    in $\mathbb{R}$, 51
    transported, 205
distribution function, 582
    cumulative, 590
    of a random variable, 623
divergent
    sequence, 191
    series, 61
division algebra, 22
Division Algorithm, 27
division ring, 19
domain, 6
Dominated Convergence Theorem, 509, 603,
        642
domination (set-), 32
double, multiple (sequence), 79
double, multiple (series), 79
du Bois–Raymond Test, 402
dual
    (algebraic), 415
    (topological), 418

E
$E[X]$, 627
ε-neighborhood, 51
e (natural base), 65
    irrationality of, 66
Edelstein's Theorem, 238
Egorov's Theorem, 516, 640
eigenspace, 462
eigenvalue, 462
eigenvector, 462
element, 1
    maximal (minimal), 9
elementary functions, 247
    derivatives of, 247
elementary set, 611
elementwise method, 2
enumeration, 30

envelope
   upper, lower, 142
Epsilon-net, 220
equation
   Kepler's, 178
equicontinuous, 447
   uniformly, 447
equivalence class, 8
equivalent (or equipotent, equipollent) sets, 30
equivalent functions, 119
equivalent metrics, 187, 205
equivalent norms, 413, 418
essential range, 571
essential supremum, 546, 604
essentially bounded, 546
Euclidean n-space, 30, 183
Euler's ϕ-function, 28
Euler's Beta Function, 341
Euler's Constant, 89
Euler's formula, 378
Euler's Theorem, 284
event, 617
events
   independent, 618
   limsup, liminf of, 620
eventually, 620
expansion
   binary, 49
   decimal, 49
   ternary, 49
expansive map, 178, 238
expectation, 627
experiment, 616
exponential function
   complex, 373
   derivative of, 248
   general, 376
   real, 375
extended real line, 50, 100, 205
extension
   trivial, 481
Extension Theorem, 586
exterior, 188
   point, 188
extrema
   global, 255
   local, 255, 288
Extreme Value Theorem, 153, 222

F
F. Riesz, 554
F. Riesz's Lemma, 427
Falconer, 583

Fatou's Lemma, 506, 601, 642
Fatou's lemma, 572
favorable, 648
Fejér's integral, 391
Fejér's Kernel, 388, 409
Fejér's Theorem, 395
Fermat's Theorem, 255
Fermat, Pierre de, 302
fiber
   horizontal, vertical, 207
   horizontal, 232
   vertical, 232
field, subfield, 20
finite (real number), 50
Finite Intersection Property, 174, 216
finite rank linear map, 457
finite set, 16
finite-dimensional, 21
first category (meager), 190
First Comparison Test, 63
First Fundamental Theorem, 324
First Fundamental Theorem of Calculus
   Lebesgue's, 561
Fixed Point Theorem, 155, 166
Formula
   Binomial, 25
   Multinomial, 25
   Taylor's, 273
Fourier coefficient, 384, 440
Fourier series, 385, 440
Fourier Transform, 544
Fourier transform, 569
   of a measure, 647
Fréchet compact, 218
fractional powers (roots), 47
Fredholm integral equation, 422
Fredholm operator, 457
Fresnel integrals, 568
Fubini–Tonelli Theorem, 614
function, 13
   nth iterate of, 238
   absolutely continuous, 179, 344, 559
   absolutely summable, 81
   additive, 142
   Borel, 522, 593
   bounded above, below, 17, 97
   bounded, unbounded, 17, 97
   Cantor's ternary, 140
   characteristic, 16
   choice, 29
   complex exponential, 373
   composite, 14
   continuous, 140, 192, 198
   contractive, 178, 238

function (*cont.*)
  convex, concave, 278
  differentiable, 242
  Dirichlet, 143, 296
  discontinuous, 140
  distribution, 582
  domain, range of, 13
  Euler's Beta, 341
  expansive, 178
  extended real-valued, 50
  Gamma, 535
  general exponential, 376
  general power, 376
  graph of, 201
  greatest integer, 148
  homogeneous, 284
  identity, 14, 146
  increasing at a point, 125
  increasing, decreasing, 98
  indicator, 16
  integrable, 601
  inverse of, 14
  jump, 150
  Lebesgue measurable, 481
  left continuous, 147
  limit of, 100, 196
  linear, 142
  Lipschitz, 163, 208, 260
  maximum, minimum of, 98
  measurable, 593
  monotone, 98
  natural logarithm, 375
  nowhere differentiable, 361
  of bounded variation, 330
  one-to-one (injective), 14
  onto (surjective), 14
  oscillation of, 142, 236
  periodic, 144
  piecewise continuous, 156, 301
  piecewise differentiable, 397
  piecewise linear, 168, 192, 301
  piecewise monotone, 156
  polynomial, 192
  rational, 146
  real analytic, 367
  real exponential, 375
  regulated, 303
  Riemann Zeta, 407
  right continuous, 147
  right differentiable, 243
  sawtooth, 360
  simple, 479, 595
  sine, cosine, 377
  singular, 573
  step, 168, 192, 301
  subexponential, 286
  sublinear, 285
  summable, 79
  support of, 545
  supremum, infimum of, 98
  total variation, 334
  uniformly continuous, 159, 208
  unordered sum of, 79
  with compact support, 545
functions
  equivalent, 119
  trigonometric, 377

G
Gamma function, 535
Gauss's Test, 70
geometric series, 62
  ratio of, 62
Geometric-Harmonic Means Inequality, 87
global extrema, 255
Gram–Schmidt Orthogonalization, 440
graph, 201
greatest common divisor (gcd), 27
greatest integer function, 148
greatest lower bound (inf), 9
Gronwall's inequality, 260
group, 17
  Abelian (commutative), 18
  symmetric, 18

H
Hölder's inequality, 541
Hahn–Banach Theorem, 430
Halmos, 17, 29, 474
harmonic series, 62
  alternating, 68
Hausdorff dimension, 582
  of the Cantor set, 583
Hausdorff distance, 233
Hausdorff measure, 587
Hausdorff outer measure, 582
Hausdorff–Lennes separation condition, 227
Heine–Borel Theorem, 132
Hellinger–Toeplitz Theorem, 461
Herstein, 17
higher derivatives, 267
Hilbert space, 433, 460
Hilbert spaces
  $L_\mu^2(X, \mathbb{F})$, 643
  $\ell^2(J, \mathbb{F})$, 444
  isomorphic, 444

Hölder's inequality, 282, 454, 643
homeomorphic, 158, 204
homeomorphism, 158, 204, 223
Homeomorphism Theorem, 159
homogeneous, 284
homomorphism, 24
Hörmander's Generalized Leibniz Rule, 277
horizontal asymptote, 112
horizontal fiber, 232
hyperplane, 437

I
ideal, 20
    maximal, 36
identity element, 17
identity function, 14
image (direct, inverse), 13
image (range), 418
imaginary part, 349
imaginary unit, 346
improper Riemann integral, 530
Inclusion-Exclusion Principle, 26, 521
increasing, 98
increasing at a point, 125
increasing, decreasing (sequence), 53
indefinite integral, 324
independent events, 618
independent families of events, 619
indeterminate forms, 117
index set, 29
indicator function, 16
Induction
    Principle of Mathematical, 11
    Principle of Strong, 11
    Principle of Transfinite, 11
inequality
    Arithmetic-Geometric Means, 43, 289
    Arithmetic-Harmonic Means, 289
    Bernoulli's, 43, 260
    Bessel's, 386, 438, 442
    Cauchy's, 43
    Cauchy–Schwarz, 77, 321, 384, 433, 643
    Cauchy-Schwarz, 541
    Chebyshev's, 524, 546, 598, 632, 653
    Geometric-Harmonic Means, 87
    Gronwall's, 260
    Hölder's, 282, 289, 454, 541, 643
    Jensen's, 278, 338, 524
    Kolmogorov's, 632
    Landau's, 287
    Lyapunov's, 339
    Markov's, 631
    Minkowski's, 78, 282, 289, 433, 454, 542, 643

Poincaré, 407
Poincaré-Wirtinger, 339
Power Mean, 289
Sobolev, 408
Triangle, 42, 43, 411
ultrametric, 184
Weighted Arithmetic-Geometric Means, 289
Young's, 338
Infimum Property, 45
infinite limit, 57
infinite set, 16
infinite-dimensional, 21
infinitely often, 620
infinitesimal, 122
    order of, 124
    principal part of, 124
infinity ($\pm\infty$), 50
initial segment, 11
injective, 14
inner product, 432
inner regularity, 512
integers, 4
integrable function, 294, 601
    absolutely, 295
integral
    indefinite, 324
    Lebesgue, 492, 495, 498, 528
    linearity of , 602
    lower Darboux, 293
    Riemann, 294
    upper Darboux, 293
integral equation
    Fredholm, 422
integral operator, 450
Integral Test (Cauchy's), 336
integration
    by parts, 328
    by substitution, 327
    term-by-term, 358, 399
Integration by Substitution, 328
interchanging limit and integral, 536
interchanging the order of integration, 537
interior, 188
    point, 100, 188
    relative, 190
Intermediate Value Property, 154, 256
Intermediate Value Theorem, 154
Interpolation
    L^p, 570
intersection, 2
interval, 44
    bounded, unbounded, 44
    endpoint(s) of, 44

interval (*cont.*)
    length of, 306
    open, half-open, closed, degenerate, 44
Interval Additivity Theorem, 315
interval of convergence, 363
inverse element, 17
inverse function, 14
    derivative of, 253
Inverse Function Theorem, 259
inverse image, 13
irrationality of $\sqrt{2}$, 47
irrationality of e, 66
isolated point, 52, 186
isometric, 204
isometric isomorphism, 418, 444
isometry, 204, 418
isomorphic
    algebras, 24
    fields, 24
    groups, 24
    rings, 24
    vector spaces, 24
isomorphic (topologically), 418
isomorphism, 24
isomorphism (topological), 418
iterated sum, 84

J
Jensen's inequality, 278, 338, 524
joint density, 629
joint distribution, 629
jointly continuous, 206
Jordan Decomposition Theorem, 335
Jordan outer measure, 468
Jordan, Camille, 330
jump (of a function), 148
jump function, 150

K
Kelley, John, 295
Kepler's equation, 178
kernel
    Dirichlet's, 388
    Fejér's, 388, 409
    Landau's, 408
    Poisson, 409
kernel (null space), 418
Kolmogorov, 616
Kolmogorov's inequality, 632
Kronecker's delta, 16
Kronecker's lemma, 71
Kummer's Test, 69

L
$L^0(E, \mathbb{C})$, $L^0(E, \mathbb{C})$, 486
$L^0(E, \mathbb{R})$, $L^0(E, \overline{\mathbb{R}}$, 483
L^∞ space, 604
L^∞ spaces, 547
L^∞-norm, 604
L^p interpolation, 570
L^p norm, 540, 604
L^p spaces, 540, 604
$L^p_\mu(X, \mathbb{F})$, 642
$\mathcal{L}^0(E, \mathbb{R})$, $\mathcal{L}^0(E, \overline{\mathbb{R}}$, 481
$\mathcal{L}^1(E)$, $\mathcal{L}^1(E, \overline{\mathbb{R}})$, $L^1(E)$, $L^1(E, \overline{\mathbb{R}})$, 498
$\mathcal{L}^1(E, \mathbb{C})$, $L^1(E, \mathbb{C})$, 500
$\mathcal{L}^1(E, \overline{\mathbb{R}_+})$, $L^1(E, \overline{\mathbb{R}_+})$, 497
$\mathcal{L}^\infty_\mu(X, \mathbb{F})$, 604
$\mathcal{L}^p(E, \mathbb{F})$, 540
$\mathcal{L}^p_\mu(X, \mathbb{F})$, 604
ℓ^1, ℓ^2, ℓ^∞, 233
ℓ^2, 77
$\mathcal{L}^0(X, Y)$, 593
$\mathcal{L}^1_\mu(X, \mathbb{R})$, $\mathcal{L}^1_\mu(X, \mathbb{C})$, 601
Lagrange's identity, 347
Lagrange's remainder, 274
Landau's inequality, 287
Landau's Kernel, 408
Landau's o, O, 121
lattice, 12, 451
    distributive, 12
lattice identities, 12
Law of Multiplication, 647
least upper bound (sup), 9
Lebesgue covering property, 238
Lebesgue decomposition, 608
Lebesgue integrable (function), 601
Lebesgue integrable function, 498
Lebesgue integral, 492, 495, 498, 528, 601
    general, 498
    lower, 490
    of a nonnegative function, 596
    of bounded functions, 492
    of nonnegative functions, 495
    upper, 490
Lebesgue Measurable, 471
Lebesgue measurable function, 481
Lebesgue Measure, 471
Lebesgue measure
    completeness of, 474
Lebesgue measure
    Regularity of, 512
Lebesgue number, 219
Lebesgue outer measure, 467
Lebesgue sum, 528
Lebesgue's 1st FTC, 561

Lebesgue's 2nd Fundamental Theorem, 563, 566
Lebesgue's Covering Lemma, 218
Lebesgue's Differentiation Theorem, 555, 558
Lebesgue's Integrability Criterion, 312, 530
Lebesgue–Stieltjes
    measure, 587
    outer measure, 582
Lebesgue-Radon-Nikodym theorem, 609
left continuous, 147
left limit, right limit, 107
Legendre's differential equation, 287
Legendre's Polynomials, 287
Leibniz Rule, 268
    Hörmander's Generalized, 277
Leibniz's Test, 72
    Uniform, 402
Lemma
    Lebesgue's Covering, 218
lemma
    Fatou's, 506
    Kronecker's, 71
    Riemann's, 387
    Riemann–Lebesgue, 544
    Rising Sun, 555
length, 135
Lerch's Theorem, 338
L'Hôpital's Rule, 262
limit, 51, 191
    infinite, 57, 109
    left, right, 107
    one-sided, 107
    properties of, 103
    sequential definition of, 103, 197
    uniqueness of, 191
    upper, lower, 58, 141
limit (of a function), 100
Limit Comparison Test, 64
limit point, 52, 100, 186
Limit Theorems, 55
lim sup, lim inf, 58
Lindelöf, 130
Lindelöf property, 219
Lindelöf space, 220
linear, 142
linear combination, 21
linear functional, 415
linear map, 24, 415
    bounded, 418
    kernel of, 418
    range of, 418
linear operator, 415
linearly independent, 21
Lip, *Lip*$^{\alpha}$, 163, 208, 420

Lipschitz, 163, 208, 260, 420
    condition, 163
    constant, 163, 208
    locally, 163, 210
    of order α, 163
little o, 121
Littlewood's Theorem, 288
local extrema, 255, 288
local homeomorphism, 237
locally bounded, 237
locally closed, 237
locally compact, 237
locally connected, 230
locally finite, 189
locally Lipschitz, 163
    of order α, 163
locally open, 237
locally uniform convergence, 515
Location of zeros Theorem, 155
logarithm (natural), 104
lower (Lebesgue) integral, 490
Lusin's Criterion, 512
Lyapunov's Inequality, 339

M
m-tail, 52, 192
Maclaurin series, 368
map
    compact, 423
map (or mapping), 13
    (bounded) multilinear, 418
    contraction, 208, 422
    linear, 24, 415
    open, closed, 202
marginal distributions, 629
Markov's inequality, 631
maximal (minimal) element, 9
maximal ideal, 36
Maximum Principle, 288
maximum, minimum, 9, 98
    global, 255
    local, 176, 255
meager (of first category), 190
mean, 627
mean square approximation, 398
Mean Value Theorem, 257
    Cauchy's, 261
    for Integrals (First), 321
    for Integrals (Second), 322
Mean Value Theorem for Integrals, 321
measurable
    Lebesgue, 471
measurable function, 593
    Lebesgue, 481

measurable set, 466
measurable space, 466, 576
measure, 466, 575
  σ-finite, 575
  complete, 474, 577
  completion of, 578
  continuity of, 576
  countable subadditivity of, 576
  counting, 577
  Dirac, 577
  finite, 575
  finite additivity of, 576
  finite subadditivity of, 576
  Fourier transform of, 647
  Hausdorff, 587
  Lebesgue, 471
  Lebesgue–Stieltjes, 587
  metric outer, 590
  monotonicity of, 576
  outer, 580
  probability, 617
  product, 613
measure space, 466, 576
measure zero, 306, 577
measures
  absolutely continuous, 605
  mutually singular, 605
Mertens' Theorem, 73
mesh (or norm), 298
metric
  associated with a norm, 412
  Baire, 235
  discrete, 182
  product, 183
  uniform, 182, 192
metric (distance), 182
metric outer measure, 590
metric property, 237
metric space, 182
  chain connected, 239
  complete, 192
  completion of, 213
  connected, 226
  countably compact, 218
  Fréchet compact, 218
  locally compact, 237
  product, 184
  second countable, 190
  separable, 190
  sequentially compact, 218
metrics
  equivalent, 205
  uniformly equivalent, 237
middle third (open), 135

Minkowski's inequality, 78, 282, 433, 454, 542, 643
module, 21
moments (of a random variable), 652
monotone (function), 98
monotone class, 611
Monotone Convergence Theorem, 54, 502, 572, 599, 642
Monotone Limit Theorem, 107, 116
monotone sequence, 53
  of sets, 611
Monte Carlo method, 653
multilinear map, 418
Multinomial Formula, 25
multiplication operator, 462, 571
mutually singular, 605
MVT, 258

N
n-space
  Euclidean, 30
  Unitary, 30
natural logarithm, 375
  derivative of, 254
negative variation, 344
Nested Intervals Theorem, 48
Newton's Binomial Theorem, 371
Newton-Raphson process, 287
nonatomic, 638
nonmeasurable, 475
norm, 411
  L^2-, 460
  L^∞, 604
  L^p, 540, 604
  L^p-, 643
  ℓ^1-, 413
  ℓ^2-, 77, 413
  ℓ^p-, 454
  Euclidean, 413
  sup-, 413
norm (or mesh), 298
normal operator, 571
normed algebra, 412
normed space, 412
  finite dimensional, 423
  quotient, 425
  separable, 427
nowhere dense, 134, 190
nowhere differentiable, 361
null sequence, 56, 656
null set, 577
null space (kernel), 418

numbers
    complex, 5, 345
    extended, 50
    irrational, 47
    natural, 3
    prime, 28
    rational, 5
    real, 5, 39

O

one-sided limits, 107
    infinite, 109
one-to-one, 14
one-to-one correspondence (bijective), 14
onto, 14
open, 51, 130
    locally, 237
open ball, 185
open cover, 131, 216
open interval, 44
    in $\mathbb{R}^n$, 581
open map, 202
Open Mapping Theorem, 429
open set, 51, 186
operation (binary), 17
operator
    adjoint, 461
    bounded, 418
    compact, 423, 450
    difference, 288
    differential, 276
    Fredholm, 457
    integral, 450
    multiplication, 462, 571
    normal, 571
    self-adjoint, 462
    shift, 462
    unitary, 571
    Volterra, 457
ordered
    n-tuple, 6
    pair, 6
    linearly, 9
    partially, 9
    totally, 9
    well, 10
ordering
    lexicographic (or dictionary), 12
    partial, 9
    total, 9
    well, 10
ordinate set, 647
orthogonal

basis, 439
    complement, 436
    projection, 436
    system, 385
    vectors, 435
orthogonal dimension, 444
orthogonal system, 438
    complete, 439
orthogonalization
    Gram–Schmidt, 440
orthonormal
    basis, 439
orthonormal system, 438
    complete, 439
oscillation, 142
    at a point, 236, 310
    on a set, 236, 310
Osgood's Theorem, 213, 428
outer measure, 580
    Hausdorff, 582
    Jordan, 468
    Lebesgue, 467
    Lebesgue (on $\mathbb{R}^n$), 581
    Lebesgue–Stieltjes, 582
    metric, 590
outer regularity, 512

P

π, 380
p-series, 62
parallelogram law, 434
Parseval's
    Identity, 443
    Relation, 398, 441
    Theorem, 398
partial derivative, 536
partial ordering, 9
partial sum (of a function), 79
partial sum (of series), 61
partition, 8, 226, 291
    mesh (or norm) of, 298
    refinement of, 292
    tagged, 291
path component, 231
path connected, 231
peak, 54
perfect set, 53, 186, 194
period, 144
periodic function, 144
    continuous, 145
permutation, 14
permutation, combination, 24
piecewise continuous function, 156, 301

piecewise differentiable function, 397
Piecewise Linear Approximation, 169
piecewise linear function, 168, 301
piecewise monotone function, 156
Poincaré inequality, 407
Poincaré-Wirtinger inequality, 339
point
    accumulation, 52
    angular, 243
    condensation, 195, 235
    isolated, 52
    limit, 52
    shadow, 554
point spectrum, 462
pointwise
    convergence, 349
    limit, 350
pointwise finite, 238
Poisson Kernel, 409
polarization identity, 460
Pólya–Szegö, 24
Polynomials
    Legendre's, 287
    Taylor, 271
positive variation, 344
power function (general), 376
Power Mean Inequality, 289
Power Rule, 247, 302
    General, 252
power series, 362
power set, 2
pre-Hilbert space, 433
premeasure, 586
prime
    factorization, 28
    number, 28
primitive (antiderivative), 324
Principle of Analytic Continuation, 405
Principle of Isolated Zeroes, 404
Principle of Mathematical Induction, 11
Principle of Strong Induction, 11
Principle of Transfinite Induction, 11
probability, 617
    classical, 617
    conditional, 618
probability distribution, 621
probability measure, 617
probability space, 617
product
    Cauchy, 73
    direct (or Cartesian), 6, 29
product σ–algebra, 610

product (metric) space, 184
    complete, 196
    convergence in, 196
product measure, 613
product metric, 183
Product Rule, 250
projection
    canonical, 14, 427, 430
    orthogonal, 436
proper inclusion, 1
pseudometric, 233
pseudometric space, 233
Pythagorean Theorem, 438

Q
quantifier, 3
    existential, 3
    universal, 3
quaternions (real), 22
Quotient Rule, 250
quotient set, 8
quotient space, 425

R
Raabe's Test, 70
Rademacher functions, 654
radius of convergence, 362
Radon-Nikodym derivative, 609
Radon-Nikodym Theorem, 607
random, 616
random selection, 617
random variable, 621
    absolutely continuous, 624
    Bernoulli, 622
    binomial, 623
    Cantor–Lebesgue, 625
    Cauchy, 625
    characteristic function of, 651
    constant, 622
    continuous, 623
    density function of, 624
    discrete, 622
    distribution function of, 623
    expectation of, 627
    exponential, 626
    geometric, 649
    mean of, 627
    negative binomial, 649
    normal (or Gaussian), 625
    Poisson, 623

probability distribution of, 621
simple, 622
square-integrable, 628
standard deviation of, 628
standard normal, 625
uniform, 625
uniformly distributed, 624
variance of, 628
random variables
identically distributed, 626
independent, 626
jointly continuous, 629
uncorrelated, 650
range
essential, 571
range (image), 418
range, 6
rate of change
average, 255
instantaneous, 255
Ratio Test, 68
Uniform, 401
rational function, 146
real analytic function, 367
real numbers, 39
addition of, 656
construction of, 655
multiplication of, 656
subtraction of, 656
real part, 349
real spectrum, 462
rearrangement, 74
rectangle, 610
rectifiable curve, 343
reduced sample space, 618
reflexive space, 458
regularity of Lebesgue measure, 512
regulated function, 303
relation, 6
antisymmetric, 7, 9, 398
composite, 7
domain of, 6
equivalence, 7
extension of, 7
inverse of , 7
range of, 6
restriction of, 7
relative interior, closure, 190
relative topology, 187
relatively compact, 216
relatively open, closed, 187
relatively prime, 28
remainder
Cauchy's form of, 274

Lagrange's, 274
repeated sum, 84
representative, 8
resolvent, 459
equation, 459
set, 459
restriction of a $\sigma-algebra$, 593
Riemann integrable, 294
Riemann integral, 294
improper, 530
Riemann sum, 292
Riemann Zeta Function, 407
Riemann's Lemma, 297, 387
Riemann's Localization Theorem, 391
Riemann's Theorem (on rearrangements),
75
Riemann–Darboux Theorem, 296
Riemann–Lebesgue Lemma, 387, 544
Riesz Representation Theorem, 437, 605
Riesz–Fischer Theorem, 443
right continuous, 147
right differentiable, 243
ring, 19
σ-, 5
commutative, 19
division, 19
with unit element, 19
ring of sets, 5
Rising Sun Lemma, 554, 555
Rolle's Theorem, 256
Root Test, 68
Rudin, 614

S
$\sigma-algebra$
restriction of a, 593
sample point, 617
sample space, 617
reduced, 618
sawtooth function, 360
scalar multiplication, 20
Schauder basis, 456
Schröder–Bernstein Theorem, 32
Schwarzian derivative, 286
second category, 190
Second Comparison Test, 66
second countable, 190
second dual, 458
Second Fundamental Theorem, 326
Lebesgue's, 563, 566
self-adjoint operator, 462
semialgebra, 585
seminorm, 412

seminormed space, 412
separable, 190
separately continuous, 206
separating points, 451
separation, 226
sequence, 13
    m-tail of, 52
    bounded, unbounded, 53
    Cauchy, 60, 192
    contractive, 60
    convergence of, 51
    convergent, divergent, 51, 191
    Dirac, 408
    double, multiple, 79
    increasing, decreasing, 53
    limit of, 51, 191
    monotone, 53
    null, 56
    pointwise convergent, 350
    strictly increasing, decreasing, 53
    uniformly convergent, 351
sequential definition
    of continuity, 142, 199
    of limit, 197
sequential definition of limit, 103
sequentially compact, 218
series, 61
    Abel's Test, 72
    absolutely convergent, 67, 424
    alternating, 72
    alternating harmonic, 68
    Cauchy product, 73
    conditionally convergent, 67
    convergent, divergent, 61, 424
    Dirichlet's Test, 72
    double, multiple, 79
    First Comparison Test, 63
    Fourier, 385
    Gauss's Test, 70
    geometric, 62
    harmonic, 62
    Kummer's Test, 69
    Leibniz's Test, 72
    Limit Comparison Test, 64
    Maclaurin, 368
    normally convergent, 355
    p-, 62
    partial sum of, 61
    pointwise convergent, 354
    power, 362
    Raabe's Test, 70
    Ratio Test, 68
    rearrangement of, 74

Riemann's Theorem, 75
Root Test, 68
Second Comparison Test, 66
square summable, 77
Taylor, 368
trigonometric, 384
uniformly convergent, 355
unordered, 79
sesquilinear form, 433
set, 1
    $\mathcal{F}_\sigma$, 234
    $\mathcal{G}_\delta$, 234
    Borel, 472
    boundary of, 188
    bounded, 10, 185, 417
    Cantor, 134
    closure of, 100, 188
    compact, 131, 216
    connected, disconnected, 133, 226
    convex, 239, 416
    countable, 30
    countably infinite, 30
    dense, 190
    denumerable, 30
    derived, 234
    diameter of, 185
    directed, 12
    discrete, 145
    elementary, 611
    exterior of, 188
    finite, infinite, 16
    interior of, 188
    interior point, interior (of), 100
    Lebesgue measurable, 471
    linearly ordered, 9
    measurable, 466
    nonmeasurable, 475
    nowhere dense, 134, 190
    null (or of measure zero), 577
    of first category, 190
    of measure zero, 306
    of second category, 190
    open, closed, 51, 186
    partially ordered, 9
    partition of, 8
    perfect, 53, 186
    quotient, 8
    relatively compact, 216
    totally bounded, 220
    totally disconnected, 134
    totally ordered, 9
    uncountable, 30
    universal, 2

sets, 2
   algebra of, 5
   disjoint, 2
   equivalent, 30
   pairwise disjoint, 8
   ring of, 5
shadow point, 554
shift operator, 462
σ-algebra, 5
   product, 610
σ-finite measure, 575
σ-ring, 5
simple function, 479, 595
   canonical representation of, 479
   integral of, 596
sine function, 377
singular function, 573
Sobolev inequality, 408
space
   n-dimensional Euclidean, 183
   Euclidean, 183
   measurable, 466, 576
   measure, 466, 576
   metric, 182
   normed, 412
   probability, 617
   pseudometric, 233
   sample, 617
   seminormed, 412
   Washington D. C., 233
spaces
   Lip, Lip^{α}, 420
   $L_{\mu}^{1}(X, \mathbb{R})$, $L_{\mu}^{1}(X, \mathbb{C})$, 603
   L^{∞}, 547
   L^{p}, 540, 604
   $L_{\mu}^{p}(X, \mathbb{F})$, 642
   $\mathcal{B}(X)$, $\mathcal{B}(X, Y)$, 418
   $\mathcal{L}(X)$, $\mathcal{L}(X, Y)$, 415
   $\mathcal{L}(X, \mathbb{F})$, 415
   ℓ^{p}-, 454
   ℓ^{∞}, ℓ^{1}, ℓ^{2}, 413
   $\mathcal{L}^{0}(X, Y)$, 593
   $\mathcal{L}_{\mu}^{1}(X, \mathbb{R})$, $\mathcal{L}_{\mu}^{1}(X, \mathbb{C})$, 601
   c_{0}, 420
   Banach, 420
   Hilbert, 433, 460
   pre-Hilbert, 433
span, 21
spectrum, 459
   point, 462
   real, 462
sphere, 185

unit, 226
square root (existence of), 46
square summable series, 77
Squeeze Theorem, 104, 114, 513
standard deviation, 628
Steinhaus's Theorem, 518
   Calderon's proof of, 571
step function, 168, 301
   integral of, 317
Step Function Approximation, 168
Stone–Weierstrass Theorem, 452
   Complex, 454
strictly increasing, decreasing (sequence), 53
Strong Law of Large Numbers, 635
subcover (open), 131
subexponential function, 286
subgroup, 18
sublinear function, 285
subring, 20
subsequence, 54
   monotone, 54
subset, 1
   proper, 1
subspace
   metric, 182
Substitution Theorem, 366
sufficiently close, 101, 111
sufficiently large, 111
sum
   Darboux, 292
   iterated, 84
   Lebesgue, 528
   repeated, 84
   Riemann, 292
   unordered, 79
summable, 440
   absolutely, 81, 440
summable function, 79
Suppes, 12
support, 369, 545
   compact, 545
support line, 281
supremum
   essential, 546, 604
Supremum Property (Completeness Axiom), 45
supremum, infimum, 9, 98
surjective, 14
symbol, 276
symmetric derivative, 285
symmetric difference, 2
symmetric group, 18

T

tag, 291
Tauber's Theorem, 405
Taylor coefficients, 272
Taylor Polynomials, 271
Taylor series, 368
Taylor's Formula
  with integral remainder, 329
Taylor's Formula with Lagrange's Remainder,
  273
Taylor's Theorem, 368
term
  nth , 13
ternary expansion, 49, 138
ternary set
  Cantor's, 134
Test
  Abel's, 72
  Dirichlet's, 72
  First Comparison, 63
  Gauss's, 70
  Kummer's, 69
  Leibniz's, 72
  Limit Comparison, 64
  Raabe's, 70
  Ratio, 68
  Root, 68
  Second Comparison, 66
Theorem
  (Lebesgue's) Dominated Convergence, 509
  Average Value, 502
  Radon-Nikodym, 607
  Abel's, 74, 372
  Abel's (on Cauchy Product), 373
  Arzelà–Ascoli, 449
  Baire Category, 194
  Banach's Fixed Point, 210, 422
  Banach–Steinhaus, 428
  Bernstein Approximation, 172, 636
  Bernstein's, 404
  Bolzano–Weierstrass, 59
  Bounded Convergence, 510, 642
  Cantor's, 34, 193
  Cantor–Bendixon, 195
  Carathéodory's, 245
  Cauchy's Condensation, 67
  Cauchy–Hadamard, 363
  Closed Graph, 223, 430
  Complex Stone–Weierstrass, 454
  Continuous Extension, 162
  Darboux's, 256, 297
  Dini's, 353, 355
  Dirichlet's, 88
  Dominated Convergence, 603, 642

  Edelstein's, 238
  Egorov's, 516, 640
  Euler's, 284
  Extension, 586
  Extreme Value, 153, 222
  Fejér's, 395
  Fermat's, 255
  First Fundamental, 324
  Fixed Point, 155, 166
  Fubini–Tonelli, 614
  Hahn–Banach, 430
  Heine–Borel, 132
  Hellinger–Toeplitz, 461
  Homeomorphism, 159
  Intermediate Value, 154
  Interval Additivity, 315
  Inverse Function, 259
  Jordan Decomposition, 335
  Lebesgue's Differentiation, 555
  Lebesgue-Radon-Nikodym, 609
  Lerch's, 338
  Littlewood's, 288
  Location of zeros, 155
  Mean Value, 257
  Mean Value (for Integrals), 321
  Mertens', 73
  Monotone Convergence, 54, 502, 599, 642
  Monotone Limit, 107, 116
  Nested Intervals, 48
  Newton's Binomial, 371
  Open Mapping, 429
  Osgood's, 213
  Parseval's, 398
  Pythagorean, 438
  Riemann, 75
  Riemann's Localization, 391
  Riemann–Darboux, 296
  Riesz Representation, 437
  Riesz representation, 605
  Riesz–Fischer, 443
  Rolle's, 256
  Schröder–Bernstein, 32
  Second Fundamental, 326
  Squeeze, 104, 513
  Steinhaus's, 518
  Stone–Weierstrass, 452
  Substitution, 366
  Tauber's, 405
  Taylor's, 368
  Tonelli's, 504
  Volterra's, 175
  Weierstrass Approximation, 174, 396
  Zermelo's Well Ordering, 12
Three Chords Lemma, 279

Tonelli's Theorem, 504
topological property, 237
    absolute, 217, 227
Topologist's Sine Curve, 231
topology, 187
    relative, 187
torus, 226
total
    family, 427
    mass, 575
    ordering, 9
    set, 427
Total Probability Law, 619
totally bounded, 220
totally disconnected, 134, 230
translated dilation, 412
translation, 412
transported distance, 205
Triangle Inequality, 42, 43, 411
trichotomy, 40
trigonometric function, 377
trigonometric polynomial, 384
trigonometric series, 384
trivial extension, 481
true near, 101, 111

U
ultimately equal, 52
ultimately true, 52, 192
ultrametric inequality, 184
ultrametric space, 184
uncountable set, 30, 194
uniform approximation, 192, 409
Uniform Boundedness Principle, 213, 428
uniform convergence, 351, 515
uniform distribution, 624
uniform limit, 351
uniform metric, 182, 192
uniform property, 237
uniformly bounded, 80
uniformly continuous, 159, 208
uniformly differentiable, 286
uniformly equivalent metrics, 237
union, 2
uniqueness of weak limits, 459
unit, 5
unit element, 19, 22
unit sphere, 226
unital, 21
unitary operator, 571
unordered pair, 6
unordered series, 79

unordered sum, 79
    associativity of, 84
upper (Lebesgue) integral, 490
upper bound, lower bound, 9
upper envelope, lower envelope, 142
upper limit, lower limit, 58, 141
Urysohn's lemma, 209

V
Var(X), $\sigma^2(X)$, 628
Van der Waerden, 361
variance, 628
variation
    bounded, 330
    negative, 344
    positive, 344
    total, 330
variation function
    negative, 344
    positive, 344
vector, 20
vector addition, 20
vector space, 20
    basis of, 22
    dimension of, 22
vertical asymptote, 110
vertical fiber, 232
vertical tangent, 243
Vieta's formula, 654
Volterra operator, 457
Volterra's Theorem, 175

W
Wallis' Formula, 341
Washington D. C. space, 233
weak convergence, 459, 570
Weak Law of Large Numbers, 634
weak limit
    uniqueness, 459
weakly convergent, 459
Weierstrass Approximation Theorem, 174, 396
Weierstrass M-test, 356
well ordering, 10
Well Ordering Axiom, 11
Well Ordering Theorem, 12

Y
Young's inequality, 338

Z
Zorn's Lemma, 10

Printed in the United States
By Bookmasters